AF831346

Carotenoids: Structure and Function in the Human Body

Muhammad Zia-Ul-Haq
Saikat Dewanjee • Muhammad Riaz
Editors

Carotenoids: Structure and Function in the Human Body

Springer

Editors
Muhammad Zia-Ul-Haq
Lahore College for Women University
Lahore, Pakistan

Saikat Dewanjee
Jadavpur University
Kolkata, India

Muhammad Riaz
Shaheed Benazir Bhutto University
Sheringal Dir Upper, Pakistan

ISBN 978-3-030-46458-5 ISBN 978-3-030-46459-2 (eBook)
https://doi.org/10.1007/978-3-030-46459-2

© Springer Nature Switzerland AG 2021
This work is subject to copyright. All rights are reserved by the Publisher, whether the whole or part of the material is concerned, specifically the rights of translation, reprinting, reuse of illustrations, recitation, broadcasting, reproduction on microfilms or in any other physical way, and transmission or information storage and retrieval, electronic adaptation, computer software, or by similar or dissimilar methodology now known or hereafter developed.
The use of general descriptive names, registered names, trademarks, service marks, etc. in this publication does not imply, even in the absence of a specific statement, that such names are exempt from the relevant protective laws and regulations and therefore free for general use.
The publisher, the authors, and the editors are safe to assume that the advice and information in this book are believed to be true and accurate at the date of publication. Neither the publisher nor the authors or the editors give a warranty, expressed or implied, with respect to the material contained herein or for any errors or omissions that may have been made. The publisher remains neutral with regard to jurisdictional claims in published maps and institutional affiliations.

This Springer imprint is published by the registered company Springer Nature Switzerland AG
The registered company address is: Gewerbestrasse 11, 6330 Cham, Switzerland

My parents and mentors, Qutb al-Aqtab Shaykh al-Hadees Muhammad Zakariya Kandhlawi Saharanpuri Muhajir Madani Rahmatullah and Shaikul-ul-Mashaikh Imam-e-Rabbani Piran-e-Pir Abdal-e-Waqt Imam-ul-Auliya Mujaddid-ul-Asar Dr. Shahid Awais Naqashbandi Damat Barakatohum (AAS academy).
Dr. Muhammad Zia-Ul-Haq

Ms. Aparna Dewanjee, the most important person in my life.
Dr. Saikat Dewanjee

My parents, family, and friends.
Dr. Muhammad Riaz

Preface

Carotenoids are found abundantly in nature in different foods and play various roles in biological, chemical, and physiological processes of life. They exhibit health effects by prevention of various disorders and diseases. The scientific evidence of the effects of low molecular weight antioxidants including carotenoids has been established by hundreds of research teams over a century of work. This research was driven forward by the awarding of two Nobel Prizes to Paul Karrer and Richard Kühn in 1937 and 1938, respectively, for their contribution on the chemistry of carotenoids. This book has a double objective: (a) to provide a state-of-the-art and scientific review of carotenoids and their health-related properties; and (b) to encourage research into this emerging topic. The intention is not to merely produce another book, but to give a comprehensive outline of carotenoids.

The book consists of 28 chapters. The carotenoids story, from the origin of this term to our current knowledge, is briefly reported in Chap. 1, along with a short overview of the research highlights and perspectives. The following 27 chapters, written by highly qualified scientists in their fields, provide a detailed and thorough study of the various domains of carotenoids. Chaps. 2 and 3 address the chemistry and synthesis of carotenoids, respectively. Chapter 4 offers an overview of apocarotenoids, their history, importance, roles, and properties in various living organisms. Perceptible visual aesthetics in terms of color is one of the most important aspects of product marketability and acceptability as the color itself is ubiquitous. Their role in photosynthesis and as bio-coloring agents is detailed in Chaps. 4 and 5, respectively. Carotenoids and processing and stability are the subjects of Chaps. 7 and 8, which report on the effects of food processing on carotenoids and techniques being utilized to enhance carotenoids stability.

The influence of encapsulation materials on the inhibition of carotenoids degradation is discussed with an attempt to facilitate the design of proper delivery systems for specific applications. Efficacy studies should be augmented by rigorous analytical data and carotenoid constituents should be named unequivocally. Therefore, researchers must know basic chemistry and analytical techniques of carotenoids, the most challenging compounds. Liquid chromatography (LC) augmented by absorbance detectors (UV, Vis, PDA) and/or mass spectrometers (MS) are currently the most common instrumental methods for carotenoids analysis. Supercritical fluid chromatography (SFC) and comprehensive two-dimensional LC (LC × LC) are interesting alternatives to conventional LC separations as they show extra capability to resolve complex mixtures of lipidic nature, including carotenoid isomers. All this is discussed in Chap. 9, namely analysis of carotenoids. Chapter 10 details various techniques and mechanism being used to enhance carotenoids contents in foods.

Chapter 11, after summarizing the factors that influence the fate of carotenoids, highlights an important gap in carotenoid research: a major part of ingested carotenoids are not bioavailable and they are fermented by colonic microbiota to become unknown metabolites. It also discusses the factors affecting the bioaccessibility and bioefficacy of carotenoids. Due to lipophilic nature, and interference with dietary and physical factors, the absorption of carotenoids at gastrointestinal level is very low. The chapter highlights dietary approaches as well as their merits and demerits in targeting and enhancing bioaccessibility and bioefficacy of carotenoids in relation to nutrition-related health benefits.

Carotenoids are also important dietary antioxidants. Chapter 12 provides data on carotenoids in different diets, together with structure–activity relationships for antioxidant capacity and associated health-related benefits. Anticancer and antidiabetic physiological effects, properties, activities, and properties are the focuses of Chaps. 13 and 14. In Chap. 15, their potential use against Parkinson's disease is discussed. Chapter 16 indicates their role as CNS protecting agents. Antiobesity effects are detailed in Chap. 17. Chapters 18 and 19 then present the efficacy of carotenoids in lung and liver diseases and eyesight. Their role in CVD is discussed in Chap. 20, suggesting a strong relation between carotenoids and cardiovascular diseases, and also the chemical and physiological approaches adopted in order to study these effects. Their positive effects on bone and periodontal health are described in Chaps. 21 and 22, respectively. Similarly, their effects on human skin and their use as cosmaceuticals is described in Chaps. 23 and 24, respectively. Their efficacy for women and infant health is explained in Chap. 25.

Chapter 26 describes provitamin A carotenoids. The commercialization potential is given in Chap. 27, as it summarizes the most significant patents and considers evidence supporting the health claims made by different industry sectors. It also discusses the significance of carotenoids in various sectors and describes existing methods for commercial production. It is essential to summarize and critically evaluate the human epidemiological evidence linking carotenoids with human health. An outline of the most significant studies is given in Chap. 28, the final chapter. It is expected that chemists, nutritionists, doctors, pharmacists, food scientists, and technologists as well as health writers will be interested in this book. Until now, most researchers of carotenoids usually came from chemistry and food sciences background. Researchers from other backgrounds (formulation experts, clinical researchers, metabolism specialists) should be encouraged to advance existing knowledge. We hope that our book will also bridge this gap by making available existing literature in easy language to researchers of all fields, even to the layman.

We would like to thank each of the authors for their contributions and for their dedicated efforts in providing carefully prepared chapters. It is the aim of the editors that this book will be of benefit and a reference source to anyone researching the area on the role of carotenoids in human health. Although we tried to include existing pertinent literature in the book, we might have missed some significant papers due to huge literature on this topic. We apologize to the authors of papers we could not include. We are obligated to the staff at Springer for their support in assembling this work and their efforts in keeping this book on schedule. Finally, we have a message for every reader of this book. These collaborative book projects of hundreds of thousands of words may always contain some errors or gaps. Therefore, instructive comments or even criticism are always welcome. So, please do not hesitate to contact us in order to discuss any issues.

Lahore, Pakistan — Muhammad Zia-Ul-Haq
Kolkata, India — Saikat Dewanjee
Sheringal Dir Upper, Pakistan — Muhammad Riaz

Contents

About the Authors

Muhammad Zia-Ul-Haq has a PhD from University of Karachi, Pakistan. He also has a LLB from Punjab University Pakistan and LLM-IP from Turin University Italy. Currently he is serving Office of Research, Innovation and Commercialization, Lahore College for Women University as Senior Manager. Previously he served as patent examiner at The Patent Office, IPO Pakistan (Ministry of Commerce) for 8 years. He received patent trainings from Japan, Korea, Malaysia, the USA, and France. He has published 2 books with Springer, more than 120 research and review papers with total IF of 120 and total Google Scholar Citations of 4500. He is peer reviewer of many journals published by Elsevier and Springer. He won RPA from Pakistan Council for Science and Technology (PCST), Ministry of Science and Technology (MOST), during 2010–2015.

Saikat Dewanjee a pharmacist by qualification with specialization in pharmacognosy, holds a postgraduate and a doctoral degree in Pharmacy from Jadavpur University, India. Dr. Dewanjee is an associate professor at the Department of Pharmaceutical Technology, Jadavpur University. He pursued his postdoctoral research at the University of Lisbon, Portugal through Fundação para a Ciência e a Tecnologia, Postdoctoral Research Fellowship. Dr. Dewanjee has been awarded the Endeavour Research Award by the Government of Australia. His research is based on a trans-disciplinary perspective integrating the classical pharmacognosy and modern biotechnology to develop natural product-based drugs or supplements. He is a principal investigator and collaborator in the Government of India-funded research projects involving bioactive products from natural sources. Dr. Dewanjee supervised more than 12 PhD students. His publication portfolio comprises 90+ peer-reviewed journal articles with an h index of 21 (Scopus) and 12+ book chapters. He has delivered scientific presentations including invited talks on different national and international platforms. Dr. Dewanjee is a recipient of several awards from different national and international organizations. Dr. Dewanjee serves as an associate editor of *BMC Complementary Medicine and Therapies*, an academic editor of *PLoS One*, and a guest associate editor for the special issue of *Frontiers in Pharmacology* on "Inspired by Nature: Towards Novel Anti-infective Agents." Dr. Dewanjee's scholarly activities include service on scientific and review panels of federal agencies, professional boards, learned societies, and expert committees in India and abroad.

Muhammad Riaz is currently an assistant professor in Shaheed Benazir Bhutto University Sheringal, Pakistan. He completed PhD in pharmacognosy at the University of Karachi, Pakistan, followed by a postdoctoral fellowship in Prof. Dou Deqiang laboratory, Liaoning University of Traditional Chinese Medicine, China. During his PhD, he worked at Prof. Michael Heinrich's laboratory, UCL School of Pharmacy, London, on extraction from medicinal plant *Pouzolzia indica.* He has many peer-reviewed publications in renowned journals at national and international level and authored a book on anthocyanins in 2016 for Springer.

Chapter 1
Historical and Introductory Aspects of Carotenoids

Muhammad Zia-Ul-Haq

1.1 Introduction

"Carotenoids" is a generic term used to indicate the diversity of non-nitrogenous bio-chromes that are virtually universally dispersed in living entities. The International Union of Pure and Applied Chemistry (IUPAC) defines a carotenoid as a compound that essentially contains the central C(20) and C(20′) intact methyl groups. Right and left end groups can differ while hydrogenation and integration of oxygen-containing functional groups generates a large family of natural compounds. They are nonpolar compounds that gather intracellularly in microbes and plant tissues [1, 2]. The dietary sources of carotenoids are plant-derived foods mainly fruits > vegetables > cereals > legumes > herbs. The dietary sources of carotenoids from foods of animal origin are egg yolk > milk > fish. Similarly dietary sources of carotenoids from commercial products are fruit juices > dairy products > confectionery.

They are, perhaps the first natural pigments being used since ancient times. Life in oxygenic atmosphere would have been impossible without them. From the green grass to ruddy autumn shades, colors surround us and deprived of pigments, we are nothing. Nature gifts us a kaleidoscope of colors via carotenoids. Since they color and cologne the natural world, they are aptly named as the "sensual molecules" leveraging the beauty in the endurance of life. They mainly absorb in the blue wavelengths, permitting the longer wavelengths to be spread and causing the specific colors. Their distinctive colors range from red to yellow, through orange and several intermediate shades. They bestow brownness to brown algae, purple-ness to purple bacteria, pinkish to salmon flesh, yellow stain to mussels, black carapace to crayfish and range of colors to feathers of birds and skins of reptiles. The water of some lakes is packed with carotenoids producing microorganisms thus imparting it color.

M. Zia-Ul-Haq (✉)
Office of Research, Innovation and Commercialization, Lahore College for Women University, Lahore, Pakistan

© Springer Nature Switzerland AG 2021
M. Zia-Ul-Haq et al. (eds.), *Carotenoids: Structure and Function in the Human Body*, https://doi.org/10.1007/978-3-030-46459-2_1

Being lipophilic compounds, they can be stored biologically in a lipophilic environment. These are generally insoluble in water unless strongly polar groups are present, as in the case of the dicarboxylic acid norbixin and the carotenoid sulphates. The c.d.b. interacts with each other by conjugation, permitting electrons in the molecule to travel freely. Increasing number of double bonds, provides more space to move electrons connected with conjugated systems demanding less energy to change states. This decreases the range of light energy captured by the molecule. As light frequency captured from the short end of the visible spectrum increases, carotenoids obtain a gradually red look [3]. Key properties of carotenoids are summarized in Fig. 1.1.

They are a key symbol of a suitable nutritional condition in birds and fishes suggesting a sign of fitness and consequently increasing sexual pull. In algae and higher plants, they maintain the configuration and task of the photosynthetic complex, quench the chlorophyll triplet states, scavenge ROS, dissipate of excess energy and help in harvesting light. As vital floral pigments, they entice pollinators and seed dispersers due to their striking rich color. Carotenoids stimulate production of adaptive enzymes and proteins that offer resistance to cellular stress. They stimulate hormonal and immune responses and stimulate detoxification enzymes. Besides their safety, they increase life quality, improve cognitive and physical performance and intensify self-esteem.

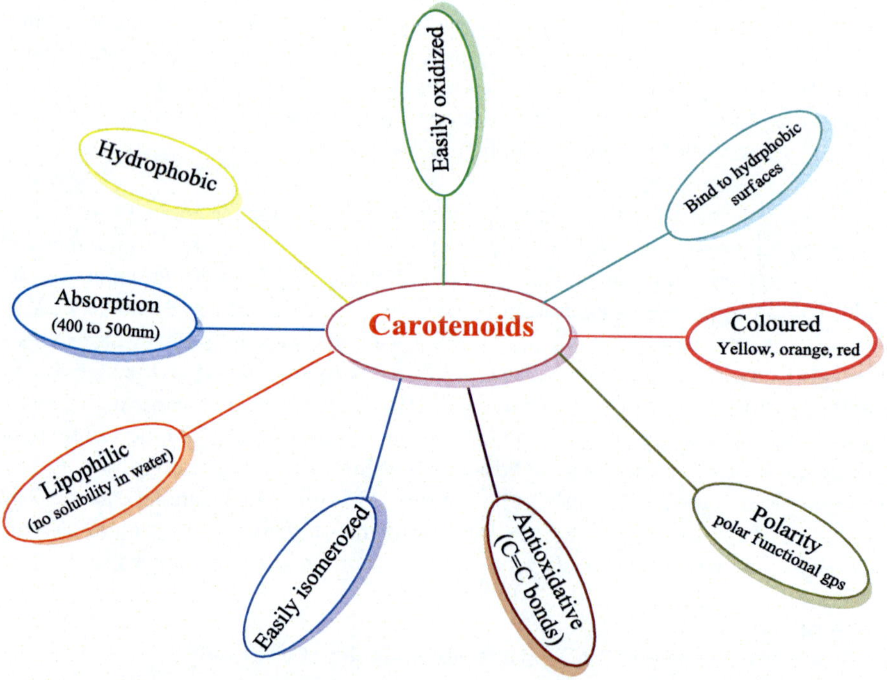

Fig. 1.1 Properties of carotenoids molecules

Carotenoid de novo synthesis takes place in specific plastids of plant organs. Maturation and ripening of carotenogenic plant parts is accompanied by increased carotenoids synthesis. Being accessory photosynthetic pigments, they gather obtainable light and forward the energy to the chlorophyll in low light. During more light, the reverse happens, they capture surplus energy from chlorophyll and disperse it therefore shielding chlorophyll from photodamage. Outside of photosynthesis machinery, they occur in association with biomembranes (antioxidant or membrane stabilizing agents), in vesicular domains (storage) like chromoplasts (flowers, fruits) or oil droplets (avian feathers, eyes). Unqualified for carotenogenesis, animals depend selectively or unselectively on food carotenoids, which are stored unchanged or slightly changed into specific carotenoids of animals.

They are the second most abundant pigment in nature after chlorophyll. Till today, 1183 carotenoids have been isolated from 702 source organisms (http://carotenoiddb.jp/). Although C30, C40, C45 and C50 carotenoids are identified, the more researched and most frequently found in nature are C40 carotenoids (>95%) with 1093 different molecules out total reported carotenoids.

Carotenoids are classified into three classes on the basis of their origin or source.

Class 1 This class mainly contains carotenoids in fresh forms or if they are processed there are very minor changes just to convert them in edible forms or make them safe from foodborne diseases. There are no food additives in this class of carotenoids. This category includes milk, coffee, infusions, herbal teas, fish, meat, eggs, honey, nuts, herbs, mushrooms, legumes, beans, stalks, rhizomes, tubers, roots, bulbs, rice, cereals, vegetables and fruits.

Class 2 This class includes processed foods which are extracted from raw materials or after subjection to technological processes that may have excess of food additives other than flavoring compounds. This class is comprises of meals, dairy products, processed fish, vitamins and juices, baked products, meat, sweeteners and sugars, animal produced fats i.e. cream, milk cream and butter, processed carotenoids of plants, flour, pasta, processed grains and vegetable oils.

Class 3 This class contains foods which are highly processed and may denature the nutritional ingredients in them. Other than coloration compounds, these foods may have some other additives that are not good for health. In turn, these additives have a huge impact on the sensory characters of carotenoids as estimated by WHO in 2013. These additives may impart aroma and flavor but they are not in the safety list and contain some chemicals that are not compatible for human gut. So these foods are named as highly processed as they contain a lot of flavoring and coloring agents. This excessive processing leads to the formation of the artificial or fantasy foods that are not similar to their raw forms in any aspect as they are subjected to various processing stages and additions due to which they are entirely in a different composition reducing the actual amount of carotenoids. This class contains plant compounds after processing, dairy products, pizzas, sandwiches, dehydrated soups, noodles and sauces, carbonated and non-carbonated drinks, chewing gums, chocolates and sweets, cookies, crackers and many products of processed white and red meat.

1.2 History of Carotenoids

They are one of the oldest known molecules, originating about 3 billion years ago. They first emerged in the archaea-bacteria as lipids reinforcing cell membranes. For this role, their long molecules possess very rigid backbone as their polyene skeleton is significantly rigid to bow or turn. They still possess this membrane-reinforcing task in certain fungi and animals. The evolution of carotenoids is believed to be from the ancient Egypt and Greece. Since then, it was discovered that night blindness could be cured by eating liver from mammals or fish, which is rich source of vitamin A. During evolution, they have been developed for different jobs and roles, possibly as photosynthetic and photo-protective pigments and later as sources of color, aroma and flavor [4]. Further details are provided in Chap. 28.

Carotenoids are one of the fastest paced saga in plant sciences. Research started on chemistry and then allied disciplines line biochemists, pharmacologist, physicians and nutritionists got involved. Currently the multidisciplinary research is led by –omics and nutraceutical potential, thanks to multidisciplinary approaches and technical advancements that have helped explore functions, roles, activities and properties of carotenoids. Carotenoids research can be divided broadly into five eras, the details event with chronological order are given in Table 1.1.

Discovered in nineteenth century carotenoids, the study of carotenoids surpasses 200 years. The French chemists, Edme Jean Baptiste Bouillon-Lagrange and Nicolas Louis Vauquelin started research for the very first time on carotenoids. Lagrange when failed to analyze the pigment of carrot-root, described it as 'yellow oily material'. Nicolas Louis Vauquelin's called it 'a resinous pigment'. In 1817, Was Braconnot performed the first investigation in paprika while in 1818, Aschoff isolated crocin now known as bixin from the saffron. In 1823, Goebel's work on crab (*Brachyura*) proposed their presence in animals for the first time. The term "carotene" was proposed by German pharmacist, Heinrich Wilhelm Ferdinand Wackenroder Wachen in 1831 for the hydrocarbon pigment crystals obtained from carrot roots while Berzelius in 1837 named the pigments isolated from autumn leaves as "xanthophylls"[5]. Kraus and Millardet in 1843 for the first time investigated carotenoids of cyanobacteria. Thudichum named un-saponifiable substance

Table 1.1 Research timescale of carotenoids

Era	Landmarks
1700s–1800s	Discovery, extraction, and light absorption measurements
1900–1927	Defining the structural formulae
1928–1949	Quick series of findings exploring the provitamin A function, β-carotene role and structure and function of about 80 carotenoids
1950s–2000	Increased interdisciplinary research led to understanding the genetics of the biosynthesis and isolation of many sequenced genes
2001–2020	Commercial production, clinical trials, intervention and cohort studies
Post COVID19	Unpredicted future

from egg yolk as lutein in 1869 [6]. Only 4 years later in 1873, lycopene for the first time was obtained from *Toafmus communis* by Hartsen. Willstaetter and Mieg in 1907 reported that carotenoids are made of isoprene units. Willstatter in 1907 chemically characterized the carotenoids for the first time and classified it into carotenes and xanthophyll. Russian Botanist Tswett, in 1906 and 1911 first time separated them by paper and column chromatography and termed the whole group "carotenoids". Harry Steenbock, proposed an association between yellow plant pigments (β-carotene) and vitamin A in 1919. In 1930s Kuhn demonstrated the relation between color and conjugation of the double bonds in the chain. Zechmeister in 1930s introduced the concept of polyene. Karrer documented the symmetrical feature of the several carotenoids (β –carotene, lycopene, zeaxanthin) and vitamin A is equal to 50% of the β -carotene in 1930 (Karrer, 1934; Karrer and Helfenstein, 1933). Kühn (1935) indicated that they absorb in the visible region (~480 nm) due to alternating single and double bonds [7]. Paul Karrer and Richard Kühn won Chemistry Nobel prizes successively in 1937 and 1938 for their contribution on the carotenoids chemistry. Strain (1938) used term "carotenes" for hydrocarbons and "xanthophylls" for their oxygenated derivatives. However, Bogert (1938) proposed that xanthophylls should be named carotenols due to their chemical structure and as they not limited only to leaves [8]. Rabinowitch (1945, 1951, and 1956) agreed the term carotenols; hence, violaxanthin was violaxanthol, lutein was luteol, and zeaxanthin was zeaxanthol. However, use of the -ol ending is obsolete now, and the terminology used by Harold Strain is in practice. Wagner and his co-workers suggested that the alteration of β-carotene into vitamin A occurs in the mucosa in 1939. In 1941, Palmer and Eckles discovered the carotene and xanthophylls in human blood plasma. By 1948, around 80 carotenoids were identified, and structures of about 40 of these were recognized. By 1950, Karrer and others performed total synthesis of β-carotene. Goodwin in 1952 investigated comparative biochemistry of the carotenoids.

Hoffmann-La Roche (Switzerland) started producing synthetic β-carotene and introduced it commercially in 1954 as a food colorant. The company also introduced synthetic ß-apo-8′- carotenal, Synthetic ß-apo-8′-carotenoic acid, and synthetic canthaxanthin in 1960, 1962 and 1964 respectively, as food and feed colorants. Synthetic citraxanthin was introduced by BASF SE (Germany) as a feed additive in 1968. By 1966, β-carotene was found suitable by FAO/WHO Expert Committee on Food Additives. In 1972, specifications for β-carotene used in food were established by the U.S. Food Chemical Codex. Carotenoids were recognized as "Generally Recognized as Safe (GRAS)" ingredient that can be employed as a dietary supplement or in food fortification in 1979. Bauernfeind in 1981 emphasized the practical aspects of the use of carotenoids. Between 1981 and 1982, review of the U.S. National Academy of Sciences showed that ingestion of carotenoid-rich foods can decrease the risk of some cancers. In 1988 the U.S. National Center of Cancer (NCI) advised Americans to ingest a range of vegetables and fruits via their daily diet. The EU approved canthaxanthin to be used as a zocro-technical feed additive in 2014.

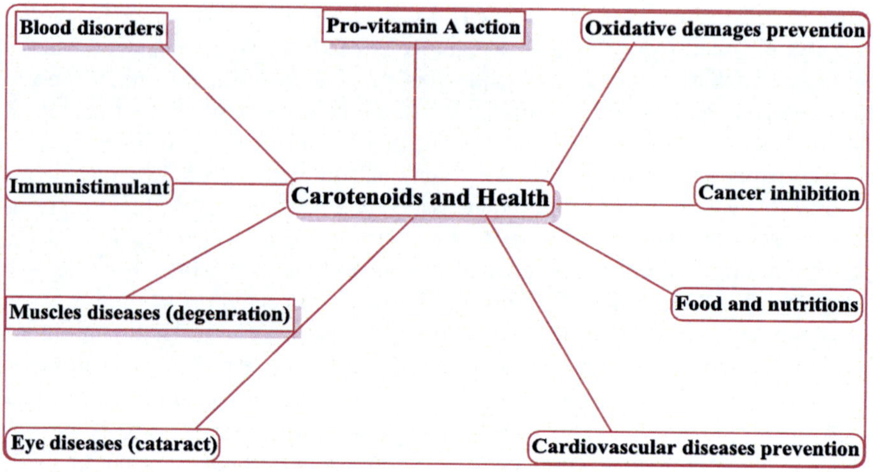

Fig. 1.2 Health benefits of carotenoids

1.3 Carotenoids Role in life

They play a significant part in the physiology of nearly all living organisms. They are responsible for the color of flowers (rapeseed, marigold, chrysanthemum, etc.) and fruits (tomatoes, oranges, peppers, apricot, peach, etc.) to favor the pollination and dispersion of seeds, or structures of animals like the feathers and beaks of some birds, the shellfish exoskeleton and the muscle or skin of some fish. Carotenoids of non-photosynthetic organisms, e.g. bacteria and fungi, protect them from photo-oxidative damage due to light and ample air. The c.d.b. imparts them light absorption ability and antioxidant potential. In plants and bacteria, they protect photosynthetic machinery from surplus light (photoprotection) by various mechanisms, including excitation energy quenching and as auxiliary structures to chlorophyll for light harvesting (photosynthesis). Key health benefits of carotenoids are summarized in Fig. 1.2.

1.4 Functions in Flora, Microbes and Other Organisms

There can hardly be a more interesting set of molecules which are inherently related to so many fundamental biological processes. Below are some key functions associated with carotenoids.

1.4.1 Light-Harvesting Accessory Pigments in Photosynthesis

In photosynthetic bacteria and plants, they capture light at wavelengths other than captured by the chlorophyll *a* and *b*, consequently enhancing the variety of sunlight absorbed and utilized in photosynthesis. Chloroplasts carotenes channelize the energy of absorbed light to the chlorophyll fragments of the reaction center. They protect prokaryotes from the harmful impact of light, and defend key cell functions in plants against the damaging properties of UV light, behaving like the "sunscreen".

Carotenoids makeup ~20–30% of total harvested light. In photosynthetic apparatus, they are integral part of Light Harvesting Chlorophyll Protein. Carotenoids of photosynthetic organisms are fixed with integral membrane proteins of chloroplasts and chromoplasts and are associated with light-harvesting complexes, where they capture light through a broad range of spectrum and transmit the energy to chlorophyll, instigating the photochemical proceedings of photosynthesis.

Photosynthetic carotenoids perform numerous tasks:

(a) Absorb incoming photons and forward this energy for utilization in photosynthesis (carotenoids contribute 20–30% of the absorbed light energy).
(b) Broaden the absorption spectrum of the photosystem because carotenoids have a wider spectrum than chlorophyll and the hydroxylated carotenoids (lutein and zeaxanthin) specifically have a bathochromic red shift in their absorbance characteristics.
(c) Absorb excess light energy (the intensity of sunlight obviously varies greatly) and remove it by dissipation as heat (*i.e.* by increased vibrational energy of the carotenoid chain).
(d) Quench the high energy of other excited molecules such as singlet state oxygen and triplet-state chlorophyll which protects the photosynthetic molecules from damage.
(e) Confer chemical protection by capture of excited singlet oxygen with chemical attachment to the carotenoid (the polyene structure is able to dissipate the free radical). This is a sacrificial action in which the carotenoid is chemically altered.
(f) Carotenoids are involved with membrane stabilization and may also conduct electrons (molecular wires) between other molecules such as cytochromes and chlorophylls. The central chain in most common carotenoids is unaltered. Any alteration causes a change of function or chain cleavage.

Carotenoids and chlorophylls are the main pigments involved in the photosynthesis. There are two well-known functions of the carotenoids performed in photosynthesis including their action as accessory pigments in harvesting light and being photoreceptors protecting the cells against damages caused by oxidative burst. It is observed that carotenoids which have function in light harvesting have their origin in anaerobic systems but have implementation in the aerobic systems. Carotenoids have their main role in absorption of light which helps the p-delocalized electrons to excite and have a light induced transformation after which they are of higher energy and ready to be transferred to chlorophyll. The structure of chloroplast is

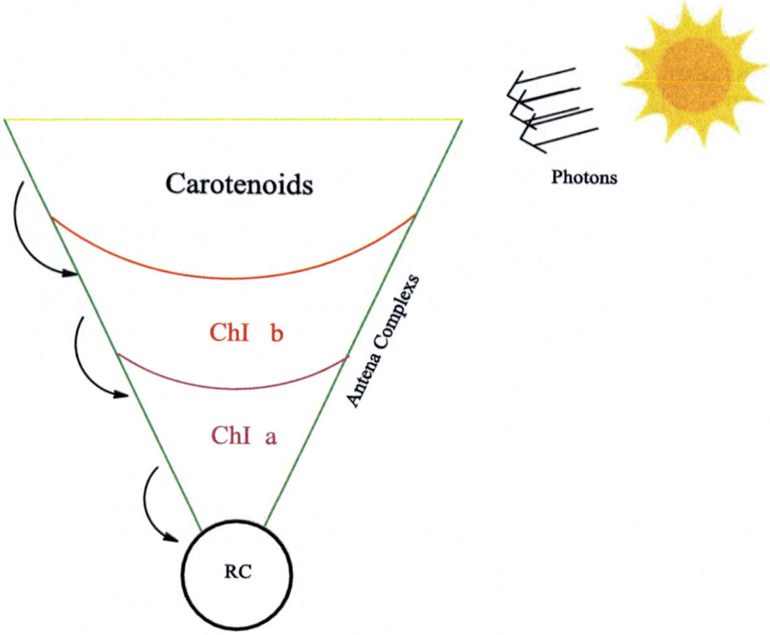

Fig. 1.3 Carotenoids role in photosynthesis (RC=reaction centre)

also in favor of this energy linked transformation. Moreover, photosystem I and II are formed in the thylakoid membranes without which the photosynthetic events are incomplete and they contain carotenoids and chlorophyll in adjuvant forms. Role of carotenoids in photosynthesis is summarized in Fig. 1.3.

1.4.2 Role in Reproduction

Carotenoids attract insects for pollination and lure animals for seed dissemination. They often work as aesthetic features e.g., the red color of lobsters' shells are because of carotenoids. They play a role in ornamental traits since, due to their physiological and chemical characteristics, they are considered as signs of individual health. Therefore, they are valuable markers during selection of tentative mates.

1.4.3 Blue Light Filtering

The alternating single and double bonds of carotenoids helps them in light absorption in the visible range of spectrum. This property is especially important for the eyes, where lutein, zeaxanthin, and meso-zeaxanthin effectively captivate blue light.

These molecules can absorb up to 90% of blue light depending on the carotenoid pigment density at the macula. Decreasing the extent of short-wavelength light that reaches the visual structures of eyes, helps shelter them from light-induced oxidative loss. Various observational and intervention studies have proved the potential of dietary and supplemental carotenoids in protection against age-related eye diseases. Blue light filtering property protects cellular components form environmental effects. The macular pigments (MP) shield the retina from light damage. They decrease the hostile effect of light scattering and chromatic aberration hence improving the retinal contrast sensitivity. This blue-light filtering potential also protects the skin from UV-induced erythema. Further details are provided in Chap. 19.

1.4.4 Intercellular Communication

Carotenoids ease communication between neighboring cells grown in culture by exciting the production of connexin proteins. Connexins make gap junctions (pores) in cell membranes, permitting cells to communicate via interchange of small molecules. This communication is central for sustaining cells in a differentiated state and is usually absent in cancer cells. Carotenoids enable intercellular communication by augmenting the expression of the genes encoding the connexin proteins.

1.4.5 Immune Function

Normal functioning of immune system depends upon vitamin A. It is challenging to define that the effect of provitamin A carotenoids are linked with their vitamin A activity or some other phenomenon. Moore discovered the formation of vitamin A from carotene in intestine of animals [9]. Although β-carotene supplementation increases various biomarkers of immune function as per several clinical trials, increased intakes of non-provitamin A carotenoids (like lycopene and lutein) did not result in analogous developments in biomarkers of immune function.

1.4.6 Cognitive Function

Observational studies indicate that dietary lutein may beneficial for cognitive health as it favorably accumulates in the brain. Lutein and zeaxanthin are the only carotenoids that cross the blood-retina barricade to form the macular pigments [5]. Macular lutein and zeaxanthin contents are connected with brain lutein and zeaxanthin profile and can be used as a biomarker to evaluate cognitive health. Further details are provided in Chap. 16.

1.4.7 Role in Membrane Structures

The first role in the oldest Earth organisms was reinforcing of bacterial cell membrane lipids. Their lengthy structures possess an exceptionally rigid backbone because of linear chain of generally 10–11 c.d.b, the length conforming to the width of the hydrophobic region of the membrane they infiltrate as "molecular rivets" [4]. Carotenoids are usually present deeply within the hydrophobic lipid core and are positioned equivalent to the membrane surface (carotenes) or span the membrane bilayer (xanthophylls). The oxygen-containing functional groups on the ionone-rings of xanthophylls interact the polar head of the membrane phospholipids, while the chromophore is positioned within the hydrophobic core of the membrane. Hence besides providing photoprotection against membrane damage, they also affect the skeleton and dynamics of membranes. In microbial membranes, they function like cholesterol for modulation of membrane characteristics. They act as additives to help membranes perform their function smoothly and can act as sterol alternatives by providing required firmness and integrity to the membrane. Xanthophylls limit the molecular movement of lipids and enhance the firmness of the membrane in its liquid crystalline state, modifying the membrane flexibility [10].

1.4.8 Chemotaxonomic and Intake Markers

Carotenoids producing flora and microbes display a distinct pattern of carotenoids during various growth phases which is quite helpful in their identification. Their huge structural diversity make them species-specific chemotaxonomic biomarkers. Their HPLC characterization is helpful in chemotaxonomy. It is provides information about physiological circumstances and the impact of climatic and anthropogenic activities. Humans obtain carotenoids chiefly (80–90%) from fruits and vegetables as human body cannot synthesize carotenoids. It is the reason of their use as biomarkers of vegetables and fruits intake, suggesting a direct correlation between the ingestion of vegetables and fruits and carotenoids quantity in blood.

1.4.9 Antioxidants/Pro-Oxidants

They also behave as antioxidants by engrossing energy from singlet oxygen generated during photosynthesis. When intersystem transmission of extra energy of oxygen to the carotenoid fails during confrontation with singlet oxygen, singlet oxygen can strike the carotenoid directly, known as 'sacrificial scavenging'. It changes the anti-oxidant nature of carotenoid to a pro-oxidant. Their role as tentative antioxidants or pro-oxidants *in vivo* is yet dubious. New techniques are being introduced that will distinguish the antioxidant capacity of both the lipophilic and the aqueous parts of tissues and fluids, thus resolving whether carotenoids possess in vivo antioxidant capability. This antioxidant/pro-oxidant potential can be linked to the effect of carotenoids on different signal transduction systems. Increased metabolic rate in early stages of life generates many ROS which need diet-derived antioxidants to cope the stress produced by ROS. Carotenoids exert their antiviral effect mostly by suppressing ROS stress. Further details are provided in Chap. 12.

1.4.10 Provitamin A Activity

Any discussion about vitamin A is incomplete without due regard to the carotenoids themselves. The main physiological role of carotenoids in humans is their provitamin A activity. In 1930, for the first time, Thomas Moore revealed that β-carotene can be changed into vitamin A, *in vivo*. Fruits and vegetables are cache of carotenoids and are vital constituent of diet owing to their activity as vitamin-A precursor. A carotenoid should be an intact un-substituted β-ionine ring and possess an unsaturated hydrocarbon chain to express provitamin A activity. Quantification of provitamin A carotenoids in foods helps understand fruit and vegetable sources of vitamin A. It is believed that 12μg of food β-carotene provides 1μg of retinol therefore a conversion factor of 12:1 is utilized to evaluate the quantity of retinol made from β-carotene. Similarly, for labeling purposes a conversion factor of 24:1 is utilized for asymmetrical provitamin A carotenoids. Crude palm oil, is the richest carotenoids source in nature in terms of retinol (provitamin A) equivalent. Provitamin A carotenoids are converted into vitamin A (retinol) in the liver or intestine [3, 11].

The xanthophylls don't possess provitamin A activity except β-cryptoxanthin. In developing countries provitamin A carotenoids from flora supply around 70% of daily vitamin A ingestion. Surprisingly, pre-formed vitamin A sources provide fewer than 30% of daily vitamin A ingestion. However, in west, the provitamin A carotenoids derived from plants offer less than 30% of daily vitamin A intake, while preformed vitamin A obtained from animals delivers higher than 70% daily vitamin A consumption [12]. Out of so many carotenoids, only about 50 molecules are vitamin A precursors. Pro-vitamin A is converted to vitamin A only when body needs it, hence escaping possible toxicity from an overdose of vitamin A. Daily recommendation of provitamin A carotenoids exists only if not consuming other sources of vitamin A [13]. It was earlier assumed that preventive effect of carotenoids against infections was due to provitmanin A (PVAC) potential. However, effects of nonprovitmanin A carotenoids (NPVAC) like lycopene, lutein and astaxanthin potential to increase cell-mediated and humoral immunity in animal models and humans. Animal models like cats which lack efficient conversion potential of PVAC, immunity was increased also. Since 1930s, it is obvious that vitamin A deficiency seriously impacts immune system. Similarly, it is established now that retinoic acid stimulates immune response rather than nonprovitamin A carotenoids especially astaxanthin, lutein and canthaxanthin. Further details are provided in Chap. 26.

1.5 Synthesis and Presence

They are naturally synthesized *de novo* by all photosynthetic organisms and some non-photosynthetic organisms. They are not made *de novo* by vertebrates and invertebrates. However, they are present in all animals and protozoans as they get them by ingestion. They are densely present in fatty tissues such as internal fat and egg yolk in terrestrial animals, while in fish they are linked to the flesh or present in fat [6, 14].

The only carotenoids-producing (mostly torulene) animals are red pea aphid (*Acyrthosiphon pisum*), spider mite (*Tetranychus urticae*) and gall midges (*Asteromyia carbonifera).* This production can be ascribed to transfer of genes through horizontal gene transfer (carotenoids synthesizing) from fungi. Similarly, their synthesis in human protist parasites (e.g. toxoplasma and plasmodium) is probably due to presence of a remnant plastid. Hence, these organisms can be viewed as "natural transgenic organisms".

Their biosynthesis volume is by plants> algae> yeasts>archaeabacteria> eubacteria. Only C40 or C40-derived carotenoids are synthesized by Eukaryotes. However, bacteria produce all C45 and majority of the C30 and C50 carotenoids discovered till date while the remaining are produced by archaea bacteria [13]. Some 307 carotenoids are reported from 170 bacterial species while 9 species of archaea contain 19 different carotenoids. More than 100 carotenoids have been obtained from marine organisms.

In plants, they can be present in roots, stems, leaves, flowers, fruits and seeds. They are strikingly noticeable in petals, pollens and fruits of the flowering plants. They are present in the membranes of plastids in a plant cell. Chloroplasts produce and store carotenoids. In photosynthetic organisms, they are necessary for photosynthesis and photoprotection, while in non-photosynthetic organisms they contribute in mitigating photo-oxidative damage [10].

1.6 Animal Carotenoids

Animals that cannot create carotenoids *de novo* can store them in specific tissues. Animals sources are egg yolk > poultry and allied products > milk and allied products > liver and adipose tissues of domesticated animals > fish (salmon& trout) > marine organism like oyster, clams, mussel, scallop. Mammals are usually separated into "yellow-fat" animals (cattle, horses etc) and "white-fat" animals (sheep, goat, carnivores, rodents) on the basis of their carotenoid's storage capacity in adipose tissues. White-fat animals rarely absorb carotenoids while "yellow-fat" animals chiefly captivate carotenes. Animals exhibit colors ingested as carotenoids. Ruminants accrue β-carotene in the milk which is consequently conveyed to dairy products. Egg yolks of hens contain macular carotenoids. Carnivorous animals, which do not consume plants, obtain carotenoids from the surplus stored in the fat stock of their target [13]. They also serve as precursors for animal-compounds such as cholesterol and various hormones essential for metabolism regulation. In marine organisms, carotenoids color can span to blue, green and purple as carotenoids are found as complex with proteins. For example, blue color of lobster converts to red when they are cooked as xanthophylls are liberated from xanthophyll-protein complex during heating.

It is believed that carotenoids present as ornamental features of animals can be an honest indicator of individual health, and therefore animals can use them when selecting prospective mates. The contribution of carotenoids from food products of animal origin, such as dairy products, eggs, some fishes and seafood, is less important. Several animals have established pathways to accumulate and exhibit these molecules. Flamingos, salmon, and lobsters have developed specific pigmentation due to storage of carotenoids. Naturally various animal parts are colored because of carotenoids e.g. egg yolk, pink flesh of salmonids, flamingo's pink plumage and red exoskeleton of crustaceans. Specific colors of many fish (salmon, trout), crustaceans (shrimp, lobster) and mollusks (mussel, clam) are also due to carotenoids. Marine organism like salmon fish which nourish on algae or on carotenoids rich products can display the color of these pigments. Birds store these molecules in their feathers as a social interface display. The colors of feathers of some birds like flamingos are also due to carotenoid rich diet [15].

Although they are not produced by animals or humans naturally, they are frequently found in animal and human body tissues in typical amounts as they are ingested, transferred, and dumped in the body through food consumption. Since human body lacks carotenoids synthesizing potential, they can be employed as biomarkers to reveal the ingestion of fruits and vegetables, suggesting a direct link between the ingestion of fruits and vegetables and the carotenoids quantity in blood. They are present in all tissues of human body, chiefly in fatty tissues>liver> plasma>serum [16]. The biggest body pool of carotenoids (~80–85% of carotenoids) is adipose tissue. Adipose tissues and liver are generally considered as the main storage sites for carotenoids, from where they are pooled with lipoprotein for blood circulation. High per gram concentration of carotenoids are also present in the adrenal gland (in the external layer), lungs, corpus luteum and testes (Leydig cells). Significant quantities of the colorless carotenoids, phytoene and phytofluene, are also found in human plasma and tissue [13]. Their serum contents remain usually constant and seldom change even during periods of decreased consumption. In serum, they are usually associated with specific tissues. For example, prostate contains mostly lycopene, corpus luteum contains chiefly β-carotene while neural retina and brain neocortex are the hub of lutein and zeaxanthin. Lutein, lycopene, zeaxanthin, β- cryptoxanthin, β-carotene, and α-carotene constitute 60%–70% of plasma carotenoids. β-carotene, lutein, lycopene, and canthaxanthin are present in the skin. About 90% of carotenoids are found in body tissues and 10% in plasma. Carotenoids have their highest concentration in colostrum, causing its deep yellow color. Their concentration in breast milk decreases by the third week, during the mature milk stage when the milk turns white. After this time, the level is dependent on the maternal diet [14].

They are most bioavailable when they are consumed simultaneously (or embedded in foods) containing a lipid base (as yolks of eggs) and when processing (e.g., blending, heating) disrupts the plant cell walls. The chemical extractability of carotenoids is usually enhanced by heating arrangements leading to increased carotenoid contents in processed food compared to raw products. The bioaccessibility and solublization of carotenoids in the gastrointestinal tract (GIT) affects the bioactivity of carotenoids. They do not mix well in aqueous medium of GIT due to their lipophilic nature [17].

Active, natural formulations marketed as "ready-to-eat carotenoid-rich products" are the outcome of the food-processing and carotenoid-extraction tools. Due to their lipid solubility, carotenoids should be taken with at least 3–5 g of fats so that small intestine can absorb them. Usually a typical plant food contains 1–5 main carotenoids with traces of some less-famous carotenoids. Extent of solubilization depends upon the polarity of carotenoids, more the polarity less will be its solubility within the lipid emulsion [16].

Humans get 90% of the carotenoids from fruits and vegetable. The human blood typically contains lycopene, lutein, zeaxanthin, α- and β-carotene and β-cryptoxanthin as major carotenoids although phytofluene and phytoene are also present substantially. Their profile in human blood is variable, depending chiefly on consumption styles. Their maximum contents in blood appears from 24 to 28 h after ingestion. The *liver* accumulates their highest quantity because of its huge size and higher amounts of carotenoid-binding proteins. The *adipose tissue also* stores their higher levels and is used to guess their long-term ingestion. In the *eyes*, they gather in the macula lutea of the central retina. Only lutein, zeaxanthin, and *meso*-zeaxanthin are present in the eyes. Lutein and zeaxanthin are shifted to the retina in the equal ration as that of plasma. Then they move to the macula, where lutein is favorably transformed to *meso*-zeaxanthin. That's why *meso*-zeaxanthin, a non-dietary carotenoid absents in serum; is only present in the retina. In *breast*, carotenoids are transferred from the blood to the breast tissues. Lactation can protect from breast cancer by increasing the delivery of chemo-preventive molecules like carotenoids. An inversely proportional relationship has been observed between tomato consumption and prostate cancer possibility. This protection is ascribed to the increased amounts of tomato lycopene [5].

1.7 Plant Carotenoids

Plants can produce carotenoids *de novo*, hence they are extensively found in foods of plants origin and their composition is inconstant and generally complex. In terrestrial plants, they are present quantitatively as leaves>other green parts> fruits> flowers> roots>seeds. Their quantity and type in green tissues is relatively uniform in plant species while there is considerably more variation in non-green parts like flowers, fruits and seeds.

They are found in both photosynthetic and non-photosynthetic tissues. In photosynthetic green tissues, they perform important roles in photosynthesis for photosystem assembly, light harvesting, and photoprotection. In non-photosynthetic tissues, they deliver colors and provide scents and flavors. Their production occurs in all kinds of differentiated plastids. However, they gather in high amounts in the chloroplasts of green tissues and the chromoplasts in non-green parts (roots, fruits, and flower petals). They usually don't occur as single compound in plants. Rather they are found as bound with chlorophyll. Their binding with chlorophylls can yield a range of beautiful colors in plants, fruits and vegetables. They produce the autumn colors of many leaves which are revealed during chlorophylls degradation [18]. In temperate regions during autumn, when deciduous trees halt making chlorophyll and preparing for winter, the carotenoids disclose colors as the green color disappears, exhibiting attractive fall foliage. Carotenoids help categorize plants, discover plant parts such as fruit, leaves, stems, roots, or tubers and define stages of growth such as ripeness or overall senescence. They impart beauty, attract insects for pollination, help photosynthesis and perhaps there would not have been life over the world without them. Carotenoids impart yellow color to corn, orange color to carrots and red color to tomatoes. Their concentration is quite small as compared to overall total pigments present in plants and absorption features of their spectra coincide with that of chlorophyll, therefore precise appraisal of carotenoid contents in plants from remotely sensed data is a challenging task. In most green plants and plant parts, usually the darker the green color, more will be the carotenoid contents. For instance, pale green cabbage contains less than 1% of carotenoid content than that of dark green cabbage [5].

Carotenoids present in terrestrial plants chiefly relate to yellow and red xanthophylls such as lutein, zeaxanthin, capsanthin, violaxanthin or neoxanthin which protect photosynthetic machinery. They are positioned in subcellular organelles (plastids), chiefly connected with proteins in the chloroplasts and dumped as oily droplets or crystals in chromoplasts. In chromoplasts, they are present either free or hydroxy groups of the ionone ring can be esterified. Higher plant typically contains analogous carotenoids yet their dissemination varies quantitatively. Leucoplast principally houses colorless carotenoids like phytoene and phytofluone which are the main precursors of all others carotenoids of the carotenoid kingdom. However, being colorless, limited scholars study them [15].

Fruits and vegetables can be categorized into three groups on the basis of carotenoids presence. Firstly, green vegetables containing high content of xanthophyll and carotenes such as broccoli, green beans and spinach. Second group contains carotenes in primary form mostly in red and yellow vegetables and fruits such as melon, tomato, plum and carrot. Third group chiefly contains xanthophyll esters largely present in colored fruits such as peach, orange and pumpkin.

Usually yellow-orange vegetables and fruit mostly provide α-and β-carotene, α-cryptoxanthin is present in orange fruits while dark green vegetables are a chief source of lutein. Carotenoids of fruits and vegetables can be found as crystals (e.g. in carrots or tomatoes), protein–carotenoid complexes (e.g. in green leafy vegetables), or in oil solution (e.g. in papaya and mango).

Table 1.2 Classification of fruit carotenoids

Group	Pigment attributes
I	Insignificant quantities
II	Minor amounts, usually of chloroplast carotenoids
III	Comparatively big quantities of lycopene and its hydroxyl derivatives
IV	Comparatively huge volumes of β,β-carotene and hydroxyl derivatives
V	Great quantities of epoxides, mainly furanoid epoxides
VI	Uncommon carotenoids e.g. capsanthin
VII	Poly-Z carotenoids e.g. prolycopene
VIII	Apocarotenoids e.g. β-citraurin

Table 1.2 indicates the classification of fruit and vegetable on the basis of carotenoids.

They chiefly occur in sub-cellular organelles (plastids), i.e. chloroplasts (photosynthetic tissues) and chromoplasts (flowers and fruits), but they are also found in amyloplasts of seeds and in etioplasts of plants grown in dark. They are deposited as oily droplets or in crystalline form in chromoplast while in chloroplasts, they are mostly linked with proteins and act as complementary pigments in photosynthesis [19]. Over millions of years, chloroplasts conserved pools of carotenoids to shelter the complex and delicate photosynthetic machinery from damage by photo-oxidation. In plants and bacteria, they protect photosynthetic machinery from surplus light (photoprotection) by various mechanisms, including excitation energy quenching and as auxiliary structures to chlorophyll for light harvesting (photosynthesis).

1.7.1 Vegetable Carotenoids

More than 60 diverse carotenoids are recognized in vegetable products and around 20 exist in measurable quantity in human tissues and serum. Leafy vegetables are outstanding source of carotenoids, particularly β-carotene and lutein (~80% of the total carotenoids), with very little quantity of α-carotene [5]. Green vegetables (leafy and non-leafy) display a distinct qualitative array whereby lutein, β−carotene, violaxanthin, and neoxanthin are the major carotenoids. The comparative fractions of these carotenoids are generally persistent; however, the absolute amounts vary significantly. Minor amounts of α−carotene, β−cryptoxanthin, zeaxanthin, antheraxanthin, and lutein-5,6-epoxide are also found in in green vegetables [13]. The xanthophylls of green vegetables are unesterified. Some carotenogenic roots (sweet potato & carrot) mostly contain carotenes while seeds (e.g. maize) mostly contain xanthophylls. Lactucaxanthin is a key xanthophyll in few species like lettuce. In leafy and non-leafy green vegetables, they are present in the chloroplasts in unesterified form. Carotenes lead in root crop vegetables, like carrots and sweet potatoes [11].

1.7.2 Fruit Carotenoids

Common carotenoids in fruits are α- and γ-carotene and lycopene. Ripening leads to fading or degradation of chlorophylls, degradation or transformation of chloroplasts into chromoplasts, and synthesis or presence of carotenoids. Unripe green fruits normally comprise chloroplast carotenoids, and during ripening, chromoplasts grow and carotenoids are created on a big scale, typically different from those of the chloroplast. If pigmentation is mainly due to carotenoids, color can be used as an indicator of their concentration. Due to the plenty of other powerful colorants e.g. anthocyanins or betalains, the carotenoids contents of red-colored fruits and vegetables cannot be assessed by color. Yellow-fleshed fruits like mango and apricot, have enormous contents in their chromoplasts, therefore the color may be used to classify fruit by carotenoids quantity [19].

Compared to leafy and green vegetables, fruits exhibit complex and inconstant carotenoids composition. In ripe fruit, they are found in the chromoplasts, chiefly esterified and their composition and proportion are highly variable and complex. Normally, peel contains more carotenoids contents than the pulp of fruits and vegetables with few exceptions. Palm fruits are particularly rich in carotenoids, predominantly provitamin A carotenes. β-carotene and β-cryptoxanthin are the key provitamin A carotenoids in fruits. Fruits such as pumpkins are outstanding source of carotenoids and vitamin A [11]. Tropical and subtropical fruits have higher carotenoids relative to temperate fruits which possess more anthocyanins. Mango and papaya are tropical fruits contain higher contents of β-cryptoxanthin [5]. Chlorophyll disappears during the development of many fruits thus specific fruit color appears in ripe fruits. These colors, along with the aromas produced by the breakage of some carotenoids, inform animals when the fruit is ready to eat it (and incidentally disperse the seeds inside). Key carotenoids pattern observed in fruits are given in Table 1.3.

Green fruits and vegetables, e.g. avocado and kiwi, maintain their chloroplasts until they are completely mature. Hence, their carotenoid pattern is comparable with that of leafy vegetables [19].

Table 1.3 Key carotenoids pattern in fruits

Pattern	Fruits
Trivial number of carotenoids	Pear
Minor contents generally of chloroplast carotenoids	Grapes
Considerable quantity of lycopene	Watermelon, tomato
Prevalence of β–carotene and/or β –cryptoxanthin	Peach, apricot
Bulk concentration of epoxides	mango, carambola
Majority of unusual or species-specific carotenoids e.g. Capsanthin and capsorubin	Capsicum genus (red pepper)
Extensive contents of poly-*cis*-carotenoids	Tangerine tomato
Major amount of apocarotenoids	Citrus species

1.7.3 Leaf Carotenoids

The intensity of green color in leaves may be a rough biomarker of carotenoids concentration. Usually dark-colored leaves have more contents. Since this is seemingly linked with light exposure and chloroplast amounts, they are higher in external leaves [19]. In leaves, they exist in free form while in other tissues they occur in esterified form. The stronger the color of a plant tissue or organ, the higher the carotenoids concentration it has. Carotenes usually exist in free form entrenched in chromoplast and chloroplast bodies. In green plant parts, because of leaf senescence and ripening (corresponding with the conversion of chloroplasts into chromoplasts), they get esterified with various fatty acids. This esterification does not change their chromatic properties. In green leaves, hydroxycarotenoids are unesterified. A strikingly constant carotenoid pattern is found in leaves, frequently mentioned as the chloroplast carotenoid pattern, the core carotenoids are lutein (~45%), β-carotene (~25–30%) and neoxanthin and violaxanthin (~15% for each). Minor contents of α-carotene, α-cryptoxanthin, β-cryptoxanthin, antheraxanthin, zeaxanthin, and lutein 5,6-epoxide are also observed.

1.8 Carotenoproteins Complexes

Due to their hydrophobic nature, they are linked with lipid parts of human cells, membranes and tissues. Due to their hydrophobic character they are normally found in lipophilic environments, such as in membranes, although their association with proteins or glycosylation reactions allow them also be present in aqueous media. The function of carotenoids in membranes is of a different nature, carotenes remain inside them while xanthophylls can be found in other locations in which they interact through their hydroxyl groups with phospholipid molecules. The association with proteins also allows carotenoids to remain in a correct position with respect to other molecules, e.g. the pigment-protein complexes that keep carotenoids and chlorophylls in proper positions for the energy transfer processes that occurring during photosynthesis. The carotenoproteins are very stable and soluble in water. The color of these complexes is stable for years at ambient temperature and in contact with the air, so they have a great interest as possible dyes. Carotenoids perform their functions mostly in carotenoids-protein form as they are soluble in aqueous cellular environment in this form. It also fine-tunes their electronic and vibrational attributes. This carotenoids-protein interaction yields a dark blue color and is responsible for a red shift of the carotenoid absorption maximum e.g. the shift in absorption of the astaxanthin from 480 to 630 nm in crustacyanin, the carotenoid protein complex of the lobster shell. The association of carotenoids with proteins stabilizes pigments in addition to extending the range of colors to green, blue and purple. Thus, the maximum absorption of astaxanthin in acetone is 478 nm, while that of crustacyanin is 630 nm, hence its bluish color. Fruit and vegetable carotenoids can occur as protein–carotenoid complexes [13].

1.9 Key Carotenoids

1.9.1 Carotenes (α-.β- and γ-Carotene)

Coloring by carotenes is exhibited by various animals and animal products, e.g. egg yolks, the shells of lobsters, and the yellow-colored milk of Guernsey cows. *β* -carotene is the most common carotene isomer and is present in a range of plants (Table). Being primary photosynthetic carotenoid, its presence in the reaction centre of photosystem II, helps shuttle excitation energy to chlorophyll and scavenging reactive oxygen species.

It exists as a violet crystalline powder ($C_{40}H_{56}$, mol. Wt. 536.9) with both ends of the molecule cyclized into β-rings. β-carotene is different from α-carotene as it contains β-ionone ring at both ends. It is present in several isomer forms, 2 of which (9-cis and all-trans) makeup nearly 80% of the total β-carotene. It possesses 11 c.d.b., 2 of which are present in the β-rings. The ring C=C are not coplanar with those of the polyene chain [20]. It is insoluble in H_2O and C_2H_5OH and not much soluble in vegetable fats. The maximum spectrometric absorption in $CHCL_3$, is between 466 and 496 nm. It is the first, the most extensively dispersed, most widely studied, the most famous nutraceutical carotenoid and is the most powerful provitamin A carotenoid.

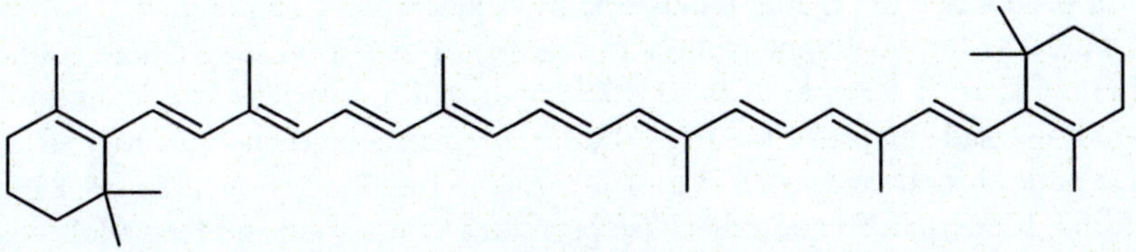

β-Carotene

It is the most frequently found carotenoid, as it is about 25–30% of the total carotenoid content of plants. Being the most active carotenoid, it has highest bioconvertibility in the human body accounting for 15–30% of all serum carotenoids. Although a single β-carotene molecule can extinguish 1000 singlet oxygen molecules prior to oxidation, the irreversible reaction occurs ending the quenching process [1, 10]. It is the first carotenoid whose diverse health-promoting effects especially anticancer roles were confirmed from 1980 to early 1990. The daily dose of 0–5 mg/kg body weight is safe even taken for long periods. The Recommended Daily Allowance (RDA) for β-carotene is rather small: 800 mg @ females, and 1000 mg @ males with marginally more quantities for lactating or pregnant women.

Orange carrot is the chief and the most popular β-carotene source while the best plant source for β-carotene, red-orange colored organic pigment, is red palm oil of the African palm (*Elaeis guineensi*). The outer mesocarp of the fruit casing the seeds yield the oil which is one of the richest sources of carotenoids (500–700 ppm).

It comprises almost 40 mg/g β-carotene and 20 mg/g α-carotene together with other minor carotenes [21]. This oil contains 15 times higher amounts of pro-vitamin A carotenes (500–800 mg) than those in carrot when compared on weight-by-weight basis. The carotenoids bioavailability from this oil is also more than other vegetable sources. Therefore, this oil is extensively studied as a dietary intervention tactic to handle vitamin A deficiency.

Commercial production of natural β-carotene is obtained from *Blakesleea trispora* and *Dunaliella salina* (or *D. bardawil*). The β-carotene yield obtained from *D. bardawil* is 1.65 pg/cell (3.0–5.0% dry weight basis) with an estimated value of 0.6 US$/g of β-carotene. Industrial production by *Dunaliella* sp. is performed in China, Australia, Japan, Israel, and the US while small scale production units are positioned in Cuba, Chile, Taiwan and Mexico. The β-carotene so obtained is in highly dilute form and its extraction requires organic solvents which are not a good choice for an entirely natural product. β-carotene obtained from *Dunaliella* is generally used in 3 forms: *Dunaliella* powder @ human use, dried *Dunaliella* @ feed use and β-carotene extracts. The price of natural β-carotene varies from US$300 to US$3000 kg^{-1}. Purified form is traded in vegetable oil form (1% to 20%) subject to food products. Commercial production is predominately based on chemical synthesis as trans-isomeric configuration. Currently much of the world's synthetic carotene is provided by a DSM manufacturing complex situated in Freeport, Texas. In Spain, β-carotene is obtained from *B. trispora* by Vitatene. In Australia, *D. salina* is used for production of organic β-carotene by Aquacarotene Limited.

It is used as food colorant (yellow to orange) or as a food supplement acting as provitamin A, @ 2–50 ppm. It is added to hydrophilic matrices (drinks and juices preparations) and lipophilic matrices (butter, margarine and cheese). It is an inexpensive natural food color as 3–5 g of β-carotene imparts yellow color to 1 ton of margarine. More than 50 countries have permitted it as a food and feed additive due to its double role as nutritive and coloring agent. In powder form, natural β-carotene is used for animal (cattle and poultry) and aquaculture (shrimp and fish) feed.

α-carotene is the second most common form of carotene and is present chiefly in carrots, tomatoes, squash, sweet potatoes, red peppers and dark green vegetables while γ-carotene is found in *Eugenia uniflora*. It is important to mention that α-carotene has 10 c.d.b., and 2 rings at ends but one β-ionone ring at one end and α-ionone ring at the other end containing four polymerized isoprene molecules between both. α-carotene (bicyclic) and γ-carotene (monocyclic) sometimes accompany β-carotene in foods, usually in much less amounts. α-carotene in human plasma distinctively specifies high carrot consumption. α-carotene can also provide vitamin A, but as only half of each molecule possesses the required unsubstituted β-ionone ring, it theoretically has 50 % the activity than β-carotene. α-carotene provide color and flavor to orange- and red-colored fruits and vegetables. Heating can change α-carotene to β-carotene. It is present in plants in lesser amounts than β-carotene [11].

Table 1.4 Major sources of key carotenoids [11]

Carotenoid	Major food source
β–carotene (β-rings at bothends)	Dark orange fruits and vegetables (e.g. apricot, cantaloupe, mango, carrots, red peppers, sweet potatoes, pumpkins) and all green vegetables (e.g. spinach, broccoli, chard, and kale)
α–carotene (always in combination with β-carotene)	Green leafy vegetables (e.g. carrots) and some varieties of squash and pumpkins
β –cryptoxanthin	Extensively found in ↓ amounts in many tropical orange-fleshed fruits (e.g. oranges, mangos, ripe red and orange peppers, mandarins, papayas, and persimmons)
Lutein (dihydroxy derivative of α –carotene)	Green vegetables (e.g. squash, broccoli, peas, brussels sprouts, string, beans)
Zeaxanthin (a dihydroxy derivative of β –carotene)	Egg yolks, certain yellow-orange fruits (e.g. squash, oranges) and dark leafy green vegetables
Lycopene (unsaturated acyclic hydrocarbon with an open hydrocarbon chain)	Tomatoes and tomato products, watermelon, guava and pink grapefruit
Typical chloroplast-associated carotenoids	Green parts of plants (leaves, stems, seeds, fruits)
Very diverse carotenoid profile	Non-green tissues of seeds, ripe fruits, and tubers subject to the species, the developmental stage, or the environmental factors
Free xanthophylls (non-esterified)	Green leaves
Esterified xanthophylls	Numerous ripe fruits and some leaves and tubers

γ–carotene (contain a β-end group) occurs as red crystals with blue reflexes chiefly found in carrots. It is less prevalent in plant kingdom. It contains pseudo-ionone ring at one end and β-ionone ring at the other end. Major sources of key carotenoids are given in Table 1.4.

1.9.2 Lutein and Zeaxanthin

Lutein and zeaxanthin are dicyclic, dihydroxy molecules derived from α- and β-carotene respectively. Both possess same molecular formula ($C_{40}H_{56}O_2$) and molecular weight (568.9). Both are geometric isomers of each other. The -OH of both are at 3 and 3′ carbons and they differ only in the position of one double bond in the end ring. Lutein has a single β-ring while zeaxanthin has two β-rings. Similarly, lutein has 10 c.d.b., 1 of which is in the β-ring while zeaxanthin has 11 c.d.b., 2 of which are in β-rings. Therefore, lutein is light yellow while zeaxanthin exhibits darker yellow shade [11, 20].

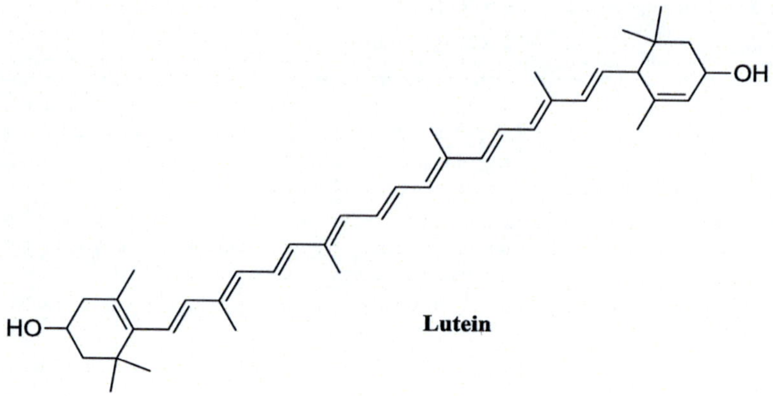

Lutein

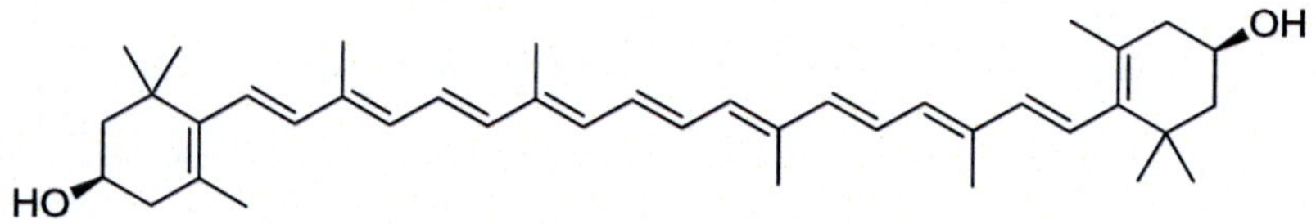

Zeaxanthin

The maximum spectrometric absorption of lutein is between 453 and 481 nm. It is a yellow crystalline element with purple gloss, found in all green plants along with β–carotene and chlorophyll. It's more soluble in ethanol than other carotenes. Lutein is for the most part insoluble in water. It is slightly less sensitive to heat degradation and oxidation than β-carotene. Color of chicken fat, egg yolk, and chicken feathers are due to lutein. It is the most abundant carotenoid in plants, present mainly in almost all fruits and vegetables. It is present as an organic color in leaves of green vegetables, e.g. spinach and black pepper.

Lutein is favored than its isomer zeaxanthin due to its greater percentage distribution in human serum. The neighborhood of the long chromophore of c.d.b. provides it specific light-retaining characteristics. Plant tissues usually possess much greater quantity of lutein than zeaxanthin. The main role of lutein in photosynthesis is due to its plenty than zeaxanthin in most green plant tissues.

Animal sources include egg yolk and animal tissues particularly the ovary. Mainly important dietary sources for humans are maize and egg yolk. The fat contents of eggs increase lutein bioavailability, although egg yolks contain less amount of lutein than kale and spinach. The best commercial source is marigold flowers as fresh petals, extracts, or dried powder which normally contains 0.6–2.5% by dry weight of xanthophylls, 92% of which is lutein. Petals contain lutein fatty acid esters as high as 3–6 mg/g. Oleoresin powder or extracts obtained from the marigold flowers are utilized as chicken feed to increase the egg yolk color intensity [21].

Lutein obtained from marigold is utilized as color for human food and as additive in poultry feed for the coloring of the bird's fat, skin, and egg yolk. Microalgae are an economical source as they show higher productivities with low land area and labor requirements when compared to the marigold cultivars. Microalgae *Muriellopsis* sp., *Scenedesmus almeriensis*, *Chlorella protothecoides*, *Chlorella zofingiensis*, *Chlorococcum citriforme*, and *Neospongiococcus gelatinosum* are usually used for its production. The lutein produced from *Scenedesmus almeriensis* is 4.77 mg/L/d, with an estimated value of 2.5 US$/g of lutein. As nutraceutical, free lutein or lutein esters purified from the oleoresin are commercially available. In USA, 2 products containing lutein, Aztec Marigold and Tagetes are being commercialized [21].

Zeaxanthin is concentrated at the central macula while lutein is spread all over the retina. Both can enhance macular pigmentation of ocular tissues, which inhibits harmful blue light from crossing the eye lens, consequently decreasing tissue harm in the eye. Both perform their role by two ways: (1) as blue light filters and (2) as antioxidants. Lutein is the chief carotenoid of human brain and impacts neural function in elder adults [16].

Zeaxanthin takes the form of orange crystals and is chiefly obtained from maize as its name proposes, though minor amounts are present in many foods. Zeaxanthin is normally present in egg yolk, corn, gul mohr, orange, berries, and marigold flowers. It is challenging to isolate chromatographically from lutein. Only few food sources especially orange peppers and Lycium chinense berries are the richest source of zeaxanthin. It is one of the most common carotenoid alcohols present in nature. Picrocrocin, is the break-down product of zeaxanthin, which gives saffron its specific aroma and taste. Main industrial applications of zeaxanthin are in cosmetics, pharmaceutical and food. Synthetic zeaxanthin, is favored than other carotenoids for increasing poultry and fish pigmentation as it accumulates evenly in the eggs and flesh. The usual production sources of zeaxanthin are microalgae *Nannochloropsis oculata* and *Scenedesmus almeriensis*. The zeaxanthin yield from *Scenedesmus almeriensis* is 0.34 mg/g with a market value of about 10 US$/g [22].

1.9.3 Lycopene

Lycopene ($C_{40}H_{56}$, mol. Wt. 536.9) is an uncyclized carotene with 11 c.d.b. and 2 nonconjugated double bonds. It is an acyclic isomer of β-carotene without provitamin A potential and it is the principal carotenoid in human plasma. Its singlet-oxygen-quenching potential is more than that of β-carotene. It has absorption maxima at 446, 472, and 505 nm for the *trans* form as is usually found as the all-trans form naturally. It is soluble in chloroform and benzene, and nearly insoluble in CH_3OH and C_2H_5OH. Lycopene has an aliphatic structure and is the most plentiful plant carotenoid.

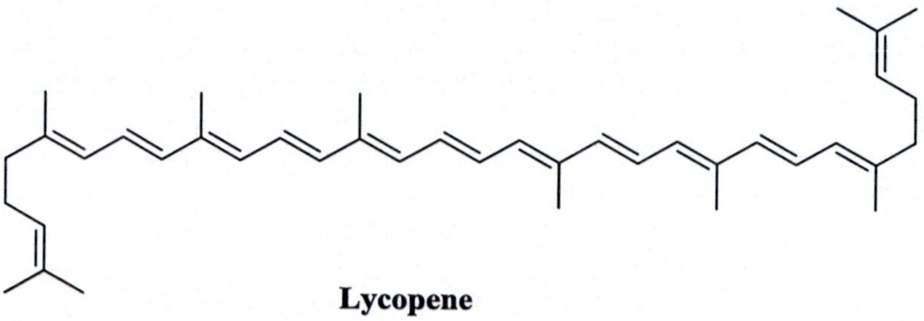

Lycopene

It is found in red-fleshed fruits of whom tomato and tomato products are the highly cited and most investigated food sources. The tomato fruit is one of the richest lycopene sources with contents higher than 100 µg/g fresh weight [21]. Vietnamese Gac and the Spanish sarsaparilla contains its highest known amounts. Lycopene gives the color to tomatoes, watermelon, blood oranges, papaya, and pink grapefruit. Often a synthetic carotenoid as a fine purple crystalline powder, insoluble in H2O, slightly soluble in C_2H_5OH and vegetable oils, and very soluble in $CHCL_3$ is used. Microbes like *Fusarium, Sporotrichioides*, and *B. trispora* are also used as a source. *Blakeslea trispora*, gives good yield through fermentation (156–578 mg/liter) and is used for commercial production.

It is included as a component in fruit/vegetable juices, sports drinks, foods proposed for energy-controlled diets for weight reduction, breakfast cereals, dressings, soups other than tomato soups and bread. It is also added in special medical diets to fulfill specific nutritional needs. Its maximum recommended intake in food supplements is 15 mg/day. It is used as meat colorant in USA, Australia and New Zealand [23].

It is listed as a permitted food colorant in Europe and Japan. It is the most powerful in vitro antioxidant of all carotenoids present in humans. It exhibits the highest singlet oxygen sequestration potentially, possibly due to the occurrence of 2 unconjugated C=C, which make it highly reactive [1, 11]. The health-promoting capacity of lycopene is ascribed mainly due to antioxidant potential, it however is believed to exert its effects via non-oxidative mechanisms. It is mainly stored in the fatty adipose tissue surrounding the heart. It inhibits the oxidation of low-density lipoprotein (LDL) in adipose tissue decreasing the chances of myocardial infraction. It is the highest investigated carotenoid in animal models of chronic liver diseases [16].

It is priced at over $6000 per kg. One reason for this higher price of natural lycopene is the decreased efficacy of traditional extraction methods with food-grade solvents e.g. $CH_3(CH_2)_4CH_3$, C_2H_5OH, and $C_4H_8O_2$, under the circumstances that usually preserve its activity. The second reason is the use of pesticide-free organic fresh tomatoes grown by companies extracting natural lycopene [24].

Its isolation from tomato pomace is very appropriate. The peel has five times more lycopene than the pulp by weight. More than 1,200,000 tons of tomato pomace are created yearly across the world, a number continuously increasing [24].

Schunck stated in 1903 reported that the red material obtained from tomato was apparently different from carotene and named it "lycopin". This term substituted the word "solanorubine" suggested in 1876 by Millardet. It absorbs majority of the visible light, except that of the lowest frequencies that's why its red color. Lycopene is also an efficient sunscreen agent. The yield of lycopene, achieved from chemical synthesis is poor as compared to β-carotene (36% and 45% respectively) [24].

1.9.4 Astaxanthin

Astaxanthin is 11 times more effective as a singlet oxygen scavenger than β-carotene. Its antioxidant activity is 10 times greater than zeaxanthin, lutein, and canthaxanthin. After β-carotene and lutein, it is the 3rd most significant molecule in international carotenoids market. The 3-hydroxyl and 4-keto functional groups of the terminal rings makes it more polar than other carotenoids [25].

It is a zeaxanthin metabolite, a red carotenoid produced by aquatic organisms like salmon fish and crustacean shells. Astaxanthin has important role in detoxifications, immunizing and have anticancer potentials, used in cosmetic, pharmaceutical and nutraceutical productions. *Xanthophyllomyces dendrorhous* and *Haematococcus pluvialis* are considered to be important microorganisms in the commercial productions of astaxanthin.

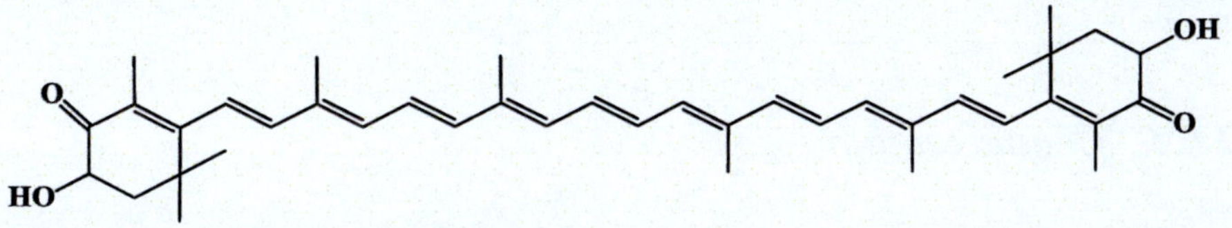

Astaxanthin

Being a secondary carotenoid, it has 2 extra oxygen atoms on every benzene ring as compared to β-carotene imparting it an extra red shade and highest antioxidant potential. As it possesses c.d.b. (keto and hydroxyl groups) it exhibits both hydrophilic and lipophilic features. It is more stable, has high antioxidant capacity, easily cross blood–brain barrier, and more tinctorial potential than other carotenoids. Natural astaxanthin is believed to be the "World's Strongest Natural Antioxidant". However, as it is not part of regular diet, therefore there is dearth of epidemiological and intervention studies about its health effects.

The red color of salmon, trout, Arctic charr, cooked shrimp, lobster, and crab is ascribed to astaxanthin. It is sold mainly for coloring the aquaculture, principally for salmonid fish and shrimp. Besides coloration and aesthetic features, it is central in aquaculture as a necessary nutritional ingredient for suitable growth and reproduction. Synthetic astaxanthin is not esterified, while algal analog is present in esterified form [20, 22].

Before its production by algae, it was naturally from krill oil and meal, crayfish oil, and *Phaffia rhodozyma* yeast. These sources possess small quantities fluctuating from 0.15% in oils to 0.40% in *Phaffia rhodozyma*, compared to 1.5–3.0% (dry weight) in *Haematococcus*. Japan, US, and India are the chief producers of *Haematococcus astaxanthin* [20, 22].

The algae *Haematococcus pluvialis* and the fungi *Xanthophyllomyces dendrorhous* are the most advantageous biological systems for astaxanthin generation. *Haematococcus* meal is permitted in Japan as a red food color and as fish feed colorant. The *H. pluvialis*, is the main natural source of astaxanthin as it produced 35 mg/g astaxanthin (4–5% of dry weight) with an estimated price of 1.8 US$ for 1,000 mg. It was allowed as a feed additive for aquaculture in 1987 and use as a nutraceutical by the FDA in 1999 [22, 25].

It is one of the most common colors of algae, fungi, and fish. It is used to impart the characteristic pink-red color to farmed salmon, trout and shrimp. It is equipped with 2 asymmetric carbon located at the 3 and 3′ position of the benzene rings on either end of the molecule. Its price depends on the %age contents in algae; the value of 5% astaxanthin is approximately US$1900 kg^{-1}. Astaxanthin, demand as feed additive is appraised to be 130 tons per annum for aquaculture and poultry breeding. Synthetic astaxanthin is not very common and is a mixture of three isomers, RR, RS and SS (1:2:1) and appears has low availability during digestion as compared to natural analogs.

1.9.5 *Canthaxanthin*

Canthaxanthin is a broad-based keto-carotene and is quite important in food and cosmetic businesses. Obtained from *Bradyrhizobium* Sepp, it is accepted as a food colorant and used in variety of foods, salmon and poultry feed. It is a used as colorant in farms (egg, yolk, fish and crustaceans) and cosmetic industry for the treatment of skin diseases. Astaxanthin and canthaxanthin are the least efficient reductive scavengers of radicals as per theoretical calculations and real-time kinetic investigations.

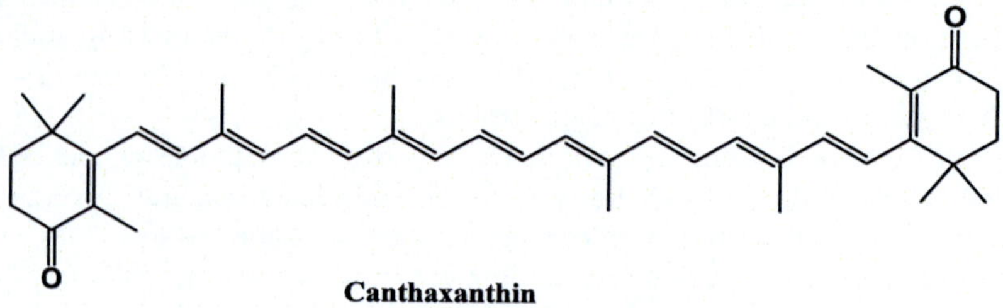

Canthaxanthin

1.9.6 *β*-Cryptoxanthin

β-cryptoxanthin ($C_{40}H_{56}O$) is a monohydroxy derivative of β-carotene with the OH group at the 3rd position of one of the β-rings. Like β-carotene, it has 11 c.d.b., 2 of which are present in the β-rings. Since the other ring is the β-ionone ring, it yields one molecule of vitamin A being a provitamin A molecule [20]. β-cryptoxanthin is the only xanthophyll possessing pro-vitamin A potential due to the presence of β-ionone ring structure in one half of its molecule. β-carotene is found in large concentrations in high number of fruits and vegetables, β-cryptoxanthin is found at high amounts in small number of food commodities.

It is the chief color of many orange-fleshed citrus fruits e.g. oranges, mangoes, peaches, nectarines, papayas, persimmons, tree tomatoes. Persimmon, squash/pumpkin, pepper (red, orange), and loquat are other sources of β-cryptoxanthin. Besides exhibiting provitamin A potential, β-cryptoxanthin exerts beneficial effects on human health. Being a chemopreventive agent, it is used in lung cancer. However, studies of this molecule have been limited to epidemiological, cell culture, and animal studies. Currently there is no commercial demand for the production of β-cryptoxanthin.

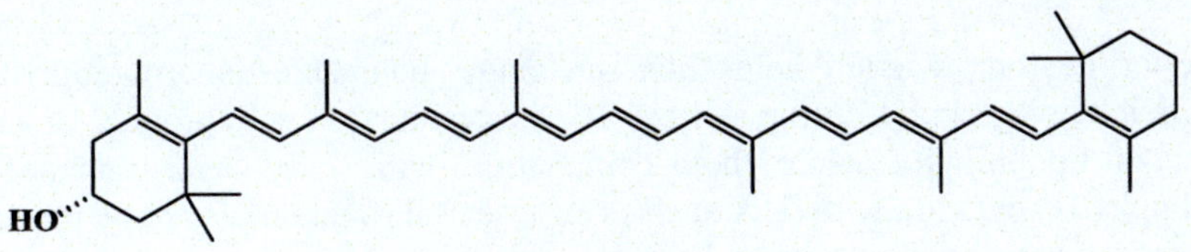

β-Cryptoxanthin

1.10 Properties, Activities and Functions

1.10.1 *Chirality*

A carotenoid characteristic associated with the large number of double bonds is the occurrence of multiple forms of isomerism, most frequently *cis-trans* isomerism. Carotenoids comprise of a conjugated backbone containing isoprene units, which are typically inverted at the center of the molecule, imparting symmetry. Variations in geometrical configuration around the double bonds produces many *cis* and *trans* isomers. In nature, carotenoids are present chiefly in the more stable *trans* (all-*E*) form, but minor contents of *cis* (*Z*) isomers are found in bread and durum wheats, which increase substantially during thermal processing and light exposure. Similarly bixin exists naturally in the *cis* form [20, 26].

They exist preferably in the *trans* stable configuration in their natural forms. However, the processing and storage conditions can rearrange its geometry to the unstable cis configuration resulting in oxidation, loss of coloring characteristics and antioxidant properties.

Their optical properties vary depending upon polarity of solvent, they usually have 3 peaked absorption spectra with well-defined maxima and minima named as "fine structure". The absorption spectrum of a *cis* isomer shows a subordinate peak known as *cis* peak in the near-ultraviolet, usually it is present 143 nm from the longest wavelength maximum. For example, *cis* peak will display at 330 nm if the longest wavelength maximum is 473 nm [7]. The spectrum of Z or cis isomers presents some peculiarities with respect to those of all-E or all-trans isomers. Thus, the maximum of absorption is located at between 2 and 6 nm lower in the case of mono-cis isomers, the fine structure decreases and a new absorption band appears in the ultraviolet region. About 50% of natural carotenoids are chiral, generally comprising one to six chiral centers [27]. Usually all-trans isomers have less energy and more stability than cis-trans and cis isomers. These variations intensely impact spectral and chromatographic characteristics of carotenoids and modify their optical activity if the molecule contains a chiral centre.

1.10.2 Esterification

The esterification procedure helps them in storage and enables incorporation in the lipid-rich plastoglobules during storing. Their esterification with proteins or sugars decreases lipophilicity while their conjugation with fatty acids enhances it. Xanthophylls acylation with fatty acids is an essential route employed by plants for storage in carotenogenic fruits and senescing vegetables, increasing and maintaining their outer color. Xanthophylls esterification also increases their stability against thermo-oxidative methods linked with food processing [13]. The esterification also increases carotenoids bioavailability via enhanced solubilization and bio-accessibility (extraction) during assimilation in presence of dietary fat. Esterified carotenoids exhibit more stability being resistant to oxidation. However, esterified forms are less bioavailable as they are not absorbed well by the intestinal epithelia except converted into free form in the intestines. The esterification mechanism helps the carotenoids in storage and assists assimilation within the lipid-rich plastoglobules during storage. Lutein, occurs free or predominantly esterified, in one (monoester) or both hydroxyl groups (diester) in nasturtium and marigold flowers. Esterification happens gradually during maturation, enhancing the lipophilic profile of xanthophylls and enabling their storage in the chromoplasts. Esterified carotenoids possess better stability, as experienced in red and hot chili peppers. The variation in esterification profile of the xanthophylls in red pepper fruits has been suggested as a ripening index.

1.10.3 Aggregation

Carotenoids always exist as separate molecules in organic solvent except when they are present in high amounts. If H_2O is added to this water miscible solution (e.g. acetone), color changes due to aggregation of carotenoids. In polar aqueous solutions, carotenoids form aggregates which are stabilized by various non-covalent interactions. The aggregates are of 2 type, J-type (head-to-tail with specific red-shifted absorption) and H-type (card-pack with specific blue-shifted absorption). The aggregates change color, spectra and functions of carotenoids by redox reactions and energy rakishness. These aggregates are very specific and are significant for functions of cells and organisms. For example they make carotenoids highly bioavailable. B-carotene of raw carrots and unprocessed tomato lycopene exist as crystals which are difficult to solubilize in lipids thus decreasing their absorption. This aggregation property also effects isomerization. The all-trans isomers aggregate more rapidly than cis isomers. There is discrepancy in structure-activity-relationships of carotenoids in organic solvents (*in vitro*) and in aqueous biological systems of living organism especially due to their aggregates-formation ability (*in vivo*).

1.10.4 Chemistry

Carotenoids are isoprenoids synthesized by tail to tail bond of 2 C20 geranyl diphosphate molecules. Phytofluene and phytoene, the acyclic carotenes which are the building blocks of other carotenoids. They may have open chain (aliphatic/acyclic) structures or closed (cyclic/alicyclic) structures. The cyclic carotenoids can be monocyclic or bicyclic. Carotenoids are highly lipophilic, with very limited exceptions, and they are typically present in hydrophobic milieus. Their lipophilicity increases by esterification while their association with proteins or sugars lowers it. Xanthophylls (zeaxanthin & lutein) contain oxygen as a functional group in the form of epoxy, keto & hydroxyl other than their basic hydrocarbon structure, whereas, carotenes (e.g. lycopene, α-carotene & β-carotene) only contain parent hydrocarbon structure. The chemical extractability of carotenoids is usually enhanced by heating arrangements leading to increased carotenoid contents in processed food compared to raw products. The conjugated chromophore is responsible for the light absorption and light-harvesting role and photoprotective characteristics against reactive oxygen species (ROS). Further details are provided in Chap. 2.

Carotenoids yield blue color when they react with concentrated H_2SO_4 and with a $CHCl_3$ solution of $SbCl_3$ (the Carr-Price reaction). Therefore, the Carr-Price reaction is utilized for quantitative estimation methods of carotenoids. The extended pi electron system of carotenoids helps them stabilize unpaired electrons after radical quenching. Carotenoids are synthesized and localized in cellular plastids. Acid, heat, or light can readily isomerize carotenoids. As they span from yellow to red color, detection wavelengths for detecting carotenoids usually range from about 430 to 480 nm. The higher wavelengths are mostly used for some xanthophylls to avoid intervention from chlorophylls. Theoretically, isomerization may lead to large numbers of geometrical configurations due to higher amounts of double bonds present in carotenoids. For example, β-carotene has tentatively 272 *cis* forms. The *cis* isomers creation yields trivial spectral shifts and consequently color of the product is generally same, although provitamin A activity decreases. The c.d.b. of carotenoids are responsible for their rapid oxidation leading to color loss of carotenoids in foods. In this c. d. b. system, π-electrons are efficiently delocalized over the complete distance of the polyene chain [5]. This c.d.b. system determines the distinct molecular shape, chemical activity and light absorbing characteristics. This c.d.b. produces the rigid and rod-like nature of the (all-*trans*)-isomers of carotenoids. Likewise, the size of ring substituent defines the activity of carotenoids. Since carotenoids can be easily oxidized, they possess antioxidant properties. Dietary xanthophylls contain radicals as canthaxanthin and capsanthin (carbonyl), violaxanthin and neoxanthin (epoxide) or lutein and zeaxanthin (hydroxyl) groups. Carotenoids can be sliced at one or both ends of the molecule, thus producing apocarotenoids. The colorless and less rigid characteristics of phytoene and phytofluene are due to a smaller number of c.d.b. Due to the usage of saponification as a regular step during analysis, carotenoid esters have been ignored. Therefore, little information is available about carotenoid ester amounts in vegetables, fruits, and other plant parts (seeds, flowers and tubers) [13]. C30 and C50 carotenoids containing 6 and 10 C5 isoprenoid units, correspondingly are synthesized by archaea and bacteria only. Similarly, only the bacteria can synthesize C45 carotenoids consisting of 9 isoprenoid units. Colorless carotenoids, such as phytoene and phytofluene, probably exert health benefits; however, these carotenoids are usually not considered in food databases and epidemiologic studies. Since they are present in foods with high contents of lycopene, like tomatoes and tomato products, it is possible that part of the beneficial effects attributed to this carotenoid could be due to the presence of these colorless carotenoids [6].

Although most abundant carotenoids in nature are C40 (with 1093 different structures), some shorter (C30) or longer (C45 or C50) carotenoids also exist. The C30 carotenoids merely have six C5 isoprenoid units, while C45 and C50 molecules have nine or ten isoprenoid units, respectively. The most common carotenoids in algae are β-carotene and zeaxanthin [13]. They are also recognized as major carotenoids in terrestrial plants. Among the carotenoids reported to date, more than 250 are of marine origin.

The 7 terminal groups of carotenoids are ψ, β, γ, ε, φ, χ, and κ, which establish the ends of the main polyene chain of the carotenoid structure. Usually, ψ ends constitute the terminal rings β, γ, and ε while β end groups form φ, χ, and κ rings [6].

Carotenoids were named for any distinct characteristic or for their source, e.g. carotene (from carrots), zeaxanthin (from *Zea mays*), and cryptoxanthin (hidden pigment). Revision and updating of carotenoid nomenclature rules seem necessary as currently existing rules were agreed since 40 years back and have hardly altered since then. It will resolve problems arising from newly discovered carotenoids with infrequent structures or those related to sugar, acyl or other moieties. The structure of carotenoids contributes a lot in their antioxidative functions. It is seen in accessibility analysis that higher the carotenoid content, lower will be its peroxide value being favored by longer chromophores in the reaction. Astaxanthin containing 13 conjugated double bonds is the best antioxidant in many living systems.

Only 8 carotenoids including lycopene, ß-carotene, astaxanthin, canthaxathin, zeaxanthin, and 3 apocarotenoids ß-apo-8′-carotenal, citranaxanthin and ethyl ß-apo-8′-carotenoate are synthesized at industrial level. Although Wittig reactions elaborated by Badische Anilin- & Soda-Fabrik (BASF) in 1960 and Grignard reactions developed by Hoffman-La Roche in 1954 are main reactions used for their commercial synthesis however Witting reaction is more frequently used than the Grignard reaction. They are effectively formed by double Wittig condensation of 2 equivalents of an appropriate C15-PH_4^+ salt with 1 symmetrical C10-dialdehyde as the central C10-building block. Besides these synthetic stages, these combinations of isomers are isomerized thermally, in C_2H_5OH or C_7H_{16}, for the full creation of all-trans/E configurations, as in the process, cis/Z stereoisomers are created to some extents. Further it is essential to combine 2 methanol molecules and 1 diketone molecule to use Grignard compounds and, resulting in C40 molecule. All synthetic C40 have symmetric structures as all have identical end groups [6].

1.10.5 Apocarotenoids

They contains a tetraterpene skeleton optionally skirted by terminating rings. The array of c.d.b. present in central chain of molecules makes them susceptible to enzymatic and oxidative breakdown and creation of more metabolites. The IUPAC defines apocarotenoids as “carotenoids in which the carbon skeleton has been shortened by the removal of fragments from one end or both ends”. The electron-rich polyene chain of carotenoids is vulnerable to enzymatic or non-enzymatic oxidative breakdown producing apocarotenoids. Information about structural transformations and stability of APOs help in using these chemical systems for applications in food, cosmetics, and other consumer products [2]. Apocarotenoids (APOs), have important functions in plants and in mammals, closely related to carotenoid roles. Lycopene, β-carotene, and zeaxanthin are the forerunners of the chief apocarotenoids APOs reported till today including bixin, crocin, picrocrocin, abscisic acid, strigolactone, and mycorradicin. In plants, apocarotenoids play role in growth, the flavor and aroma of fruits and flowers, and serve as antifungals agents [6, 10]. Despite wider distribution of carotenoids and apocarotenoids in nature, their cellular contents are very less. Three apocarotenoids are produced on an industrial scale:

Table 1.5 Major apocarotenoids and their functions

Apocarotenoid	Function
Bixin (C25)	Coloring component of annatto
Crocetin (C20)	Yellow coloring component of saffron
Crocin	The red pigmentation of saffron
Abscisic acid	Well-known phytohormone engaged in a variety of biological processes
Strigolactones	Necessary for a diverse range of biological mechanism
Vitamin A	Immunity-booster

citranaxanthin (animal feed additives), ethyl β-apo-8′-carotenoate and β-apo-8′-carotenal (food colorants). Table 1.5 shows major apocarotenoids and their functions [22]. Crocetin and bixin are present in highest level in food items. The first apocarotenoids discovered in plants is abscisic acid (ABA) which mediates stress responses.

Flower stigmas of saffron (*Crocus sativus*) and waxy seed arils of achiote (*Bixa orellana*) is the highest sources of apocarotenoids. More than 100 naturally occurring apocarotenoids are reported till now. Most of the apocarotenoids including bixin, crocetin, abscisic acid, strigolactone, and mycorradicin are the cleavage products of the lycopene, β-carotene, and zeaxanthin. Apocarotenoids can act as odorants (aroma of wine, tea and some flowers), retinoids (some of which can be vitamin A), the phytohormones (abscisic acid & strigolactones) and as insect repelelnts (e.g. grasshopper ketone). Carotenoid cleavage dioxygenases (CCDs) help carotenoids breaking at specific C=C bonds, mostly by integrating oxygen into adjacent carbon atoms along the conjugated carotenoid backbone. However, non-enzymatic apocarotenoid can be created through singlet oxygen attack, chiefly on β-carotene. Furthermore, peroxidases and lipoxygenases can also produce apocarotenoids [6]. Further details are provided in Chap. 4.

1.10.6 Stability

Although carotenoids are naturally stable in plant or animal matrices, these pigments can easily undergo degradation under heat, light, acid and enzymes exposure, among others factors, when tissue is damaged, with consequent modification of color and biological activity. Differences in degradation are related to the three-dimensional shape of the carotenoid structure, which determines hydrophobicity, crystalline state, ease of crystal formation, type of organization (multilayers or aggregates), as well as location of biosynthesis or storage in the cells.

Due to frequent proven and alleged health-promoting properties, understanding of chemical structures and related isomerization, and degradation mechanisms is very important for comprehension of carotenoid stability in food [10, 19]. They are very reactive molecules due to their polyene structure. The structure contains abundant electrons and vulnerable to electrophilic reagents attack leading to their instability against oxidation and providing them radical features. Carotenoids in the crystal form are liable to oxidation after their isolation quickly degrading even in the presence of traces of oxygen. Long-chain molecules are more prone to isomerization and oxidization. Classic storage and handling of maximum vegetables and fruits does not affect carotenoids stability. Mostly blanched plant products show a surge in carotenoid quantity comparative to raw tissues. Volatile fragments can be generated at high temperature. Oxidation (both enzymatic/non-enzymatic) is the key reason of carotenoid degradation during storage and processing of food. However, common household processing like microwave oven, steam and boiling usually does not affect carotenoid contents largely. But heating can extremely damage these bioactive molecules. Further details are provided in Chap. 8.

1.10.7 Analysis

Carotenoid molecules have strong light absorption and solutions may be visible to the human eye with only microgram quantities of pigment present. Absorption at the maximum wavelength and characteristic molar absorptivity's for individual carotenoids forms the basis for quantitative determination of carotenoid quantities. The intense light absorption and low detection limits are useful to monitor purification steps and chemical reactions. Loss or change of color provides a warning of decomposition or structural modifications. Previous 3 decades have increased understanding of chemistry and functions of carotenoids leading towards rapid developments in analytical techniques and instrumentation. Characterization helps to establish identity of species and screen carotenoid structures. Analysis of carotenoid faces serious challenges as they are present as a mixture, along with structurally related compounds and isomers and interfering compounds like diacylglycerides and chlorophylls. Currently, nearly 90 years after the first structural interpretation of β-carotene by Kuhn and Karrer in the 1930s, lot has been done in analytical field [15].

Despite their stability within living cells, carotenoids are unstable when extracted from the cells, as they are vulnerable to light, temperature, oxygen from the air, acids and bases (very few). Their laboratory handling needs special equipment and vigilant laboratory practices. Extraction techniques from tissues exploit organic solvents that essentially penetrates the hydrophilic matrix. For this purpose, hexane-acetone mixtures are regularly used, however occasionally special solvents and treatments are required for appropriate separation. Maximum wavelength up to which they are visible varies depending upon the functional groups. The ultraviolet and visible spectrum is the primary analytical apparatus for the recognition and quantification of carotenoids. The c.d.b. system of the chromophore contributes color and is source of visible absorption spectra. Chromatographic studies of carotenoids from leafy vegetables using reverse phase C-30 hydrophobic column have revealed that first eluted peak comprises of neoxanthin & violaxanthin. Afterwards, the eluted peaks are of lutein followed by zeaxanthin, β-cryptoxanthin, α-carotene and β-carotene. As a rule of thumb, *cis*-isomers are eluted firstly followed by *trans*-isomers mainly due to more polar nature of *cis* as compared to *trans* isomers. Nevertheless, this eluting pattern is reverse in case of C-18 columns. After separation by liquid chromatography procedures, carotenoids detection happens in a specific absorption zone between 400 nm and 500 nm wavelength while the detection in *cis*- or *Z*- isomers generally takes place between 330 and 340 nm. The absorption intensity is influenced by the solvent or the composition of the mobile phase [1].

Positive ion mode of atmospheric pressure chemical ionization-mass spectrometry (APCI-MS) is characterized as the most important methodology in identification of various carotenes and xanthophylls. HPLC techniques comprising of two-dimensional LC x LC (liquid chromatography) usually containing reverse or normal phases are used for separation of carotenoids. In this technique primary column is interconnected to 1 or even extra columns known as secondary columns. Part of effluent eluting from first column are generally allowed to pass through second column by using switched valve. According to outcomes of a study, 33 carotenoids that belonged to 10 various chemical classes were recognized through application of comprehensive NP-LC × RP-LC (normal phase-liquid chromatography x reversed phase-liquid chromatography) system. As the number of c.d.b. increases, the energy required to promote electron transitions of the valence layer reduces and the absorption wavelength escalates. Mobile phase composition and the solvent used affects the absorption intensity. Further details are provided in Chap. 9.

1.10.8 Coloring Agents

The history of carotenoids as colorants dates back to earlier nineteenth century. Carotenoid were found in paprika (1817), the crocin (bixin), annatto (1825), carrots (1831), and autumn leaves (1837). The key element of carotenoids is their long c.d.b. system. These C=C bonds by the conjugation process result in an overall decreased energy state of the carotenoid and as the number of c.d.b increases, color fluctuates from pale yellow, to orange, to red. Major physical and chemical properties including shape, color and reactivity depend on this polyene pillar. The acyclic carotenes phytoene and phytofluene, which are fewer as they have much less c.d.b. (5 &3 respectively) as compared to majority of carotenoids. As a result, both are colorless and possess specific actions and properties. Chlorophylls absorption spectra varies substantially from that of carotenoids although both are part of photosynthetic machinery. At least 7 c.d.b. are required for color however other factors like concentration and aggregation/association with other molecules like sugars and fatty acids.

Analytically, the color of the carotenoids is of great importance, since a color change during the analysis is indicative of degradation or structural modification of the pigments. Similarly, the color allows easy monitoring their separation during extraction, separation, identification or other laboratory work by column and layer chromatography. Recent studies propose the objective measurement of color as a powerful tool in the field of quality control for estimation rapid carotenoid content in various sources, like tomatoes, orange juice and apricots, mainly due to the advantages offered by such measures, such as speed, no destruction of samples, and versatility etc. Thus, the objective measurement of color has been proposed recently as an appropriate method for determination of vitamin A activity of orange juices from a more efficient, fast and realistic way in the field of quality control. Further details are provided in Chap. 6.

There is less variety of natural colors that can be used in foods as compared to an extensive range of synthetic analogs. Synthetic colors are mainly used due to color and price range, resilience to oxidative damage, and solubility. Although food and Drug administration (FDA) has not yet defined "natural colors" it is generally believed that natural color means those extracted and derived from natural sources. Carotenoids are used mainly as dyes to reinstate color lost during storage and subsequently to processing and to color pale food and to standardize the color of food products. The use of carotenoids colors is particularly pronounced in the UK, Scandinavia and the northern part of continental Europe. Generally, natural colors are 'exempt from certification' and can be made and advertised without certification from the FDA. Some carotenoids natural colorants used are annatto (bixin), turmeric (crocin), tomatoes (lycopene) and paprika (capsanthin). β-carotene imparts yellow color to food (e.g. butter) or liquids (e.g. soft drinks) after suitable formulation. For the yellow coloration of dairy products, water-soluble norbixin, a bixin derivative is employed [21]. Canthaxanthin or lycopene are suitable for red coloring of food and beverages. Table 1.6 indicates key carotenoids along with their specific colors.

Benefits of microbial production are the potential of microbes to use a range of inexpensive substrates, well-controlled cultivation, and the decreased production time. Research on natural colors should focus on procurement a range of shades, enhancing their shelf life, and depressing production expenses. There is need to invent methods that stabilize natural colors in various food matrices, and search economical organic substrates for the growth of the color producing microbes. Reduced stability or poor solubility can be managed by practices like micro-encapsulations and nano-formulations, facilitating diverse application of colors to different foods.

Table 1.6 Key carotenoids along with their specific colors

Carotenoid	Color
Acyclic carotene lycopene	Red
β-carotene (Bicyclic,2 β-rings) and/or its hydroxyl derivatives e.g. zeaxanthin (Bicyclic, 2 β-rings and 2 hydroxy groups) and β-cryptoxanthin (Bicyclic, 2 β- rings and 1 hydroxy group)	Orange color
α-carotene (Bicyclic, ε-ring and β-ring) and/or its hydroxyl derivatives e.g. lutein (Bicyclic, β-ring and ε ring and 2 hydroxy groups)	Yellow-orange
Carotenoid epoxides	Yellow
Violaxanthin (Bicyclic, 2 hydroxy groups and epoxy groups)	Yellow
Carotenoids that are unique to or specific for that species, e.g. capsorubin and capsanthin	Yellow, orange, or red

1.10.9 Cosmetics

Beauty comes from the inside. The effect of nutrition on skin aging is an exciting field of interest for humans throughout the ages. Skin aging consists of 2 biologically and clinically discrete processes. The first is intrinsic skin aging, denoting chronological aging and disturbs skin in the similar design it affects other internal organs. The second is extrinsic skin aging, seen as aged skin and is due to outward factors and environmental influences including poor nutrition. Since beauty and youthfulness are believed to be prerequisites for personal and professional success, hence the desire for a lasting youthful appearance is becoming even stronger. The skin and the eyes are the only two organs of the human body exposed continuously to damage caused by the environment and require protection. They shelter from UV-light harm and are useful outwardly via skin (as a topical application) and via nutritional sources besides their usage in tan lotions due to their color. Astaxanthin and phytoene are currently utilized as components of skin products advertised by AstaReal and IBR, correspondingly. Phytoene, being fairly light-stable molecule, is employed in dietary nutricosmetics for topical skin applications to shield against oxidative skin damage. Competition among cosmetic industries like Unilever, L'Oreal, Henkel, and Beiersdorf will increase the carotenoid market value in Europe [21]. Further details are provided in Chap. 24. Role of carotenoids in skin beauty is summarized in Fig. 1.4.

1.10.10 Nutraceuticals

The key carotenoids sources in the human food are fruits and vegetables > green leaves > dairy products > eggs > marine products [19]. Britton and Khachik [28], characterized carotenoid containing foods as shown in Table 1.7.

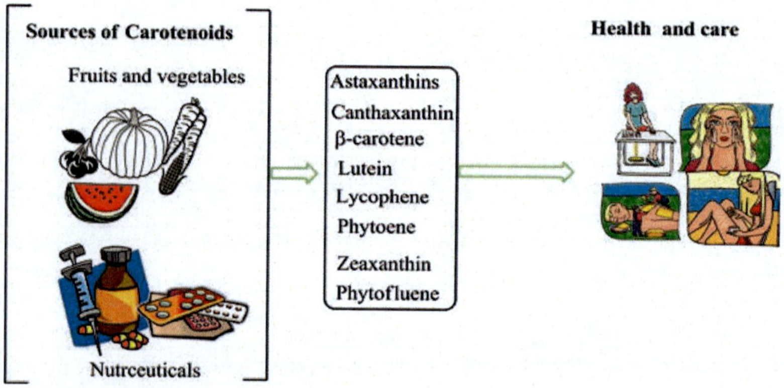

Fig. 1.4 Carotenoids role in skin beauty

Table 1.7 Characterization of carotenoids containing foods

Carotenoid contents (mg/100 g)	Remarks
0–0.1	Low
0.1–0.5	Moderate
0.5–2	High
>2	Very high

It is noteworthy that higher carotenoid contents in a given food are not always an indicator of a better dietary carotenoid supplier. The consumption rate is crucial to assess the chief dietary providers for carotenoid consumption. Thus, a food containing higher carotenoid contents can be rarely taken in the diet e.g. expensive fruits whose intake is much less as compared to a food with modest content consumed frequently (e.g. corn or bananas) being cheap and easily available. Few people would argue with the statement that carotenoids are good for us. The dazzling "rainbow diet" (@ colorful carotenoids) means the regular intake of colored fruits and vegetables.

1.10.11 Anticancer Effects

Anticancer effects of carotenoids first came in 1973 when growth of skin tumors induced in animal models was slowed when "unrestrained quantity of red carrots" diet was given to these animals. These animal model studies continued till early 1980 followed by human studies from 1981 onwards to observe the relationship between carotenoids intake and risk of cancer. The research team led by Peto and Doll in a study published in Nature in 1981 appraised the link between β-carotene consumption and reduced cancer risk in humans. This paper initiated second golden age of carotenoids research suggesting that carotenoids influence various cell and molecular processes, stimulate immunological system and protect from lipids peroxidation. Subsequently 6 huge placebo-controlled trials of carotene supplements however did not give expected results. In last decade of twentieth century, α-tocopherol β-carotene trial (ATBC) and β-carotene retinol-efficiency trial (CARET) indicated rather increase in coronary heart diseases in smokers or lung cancer. β-carotene supplementation in ATBC and CARET was 20mg/day and 30 mg/day respectively which is nearly 10–20 times more than normal daily consumption making results less applicable in bulk of population [29]. Further details are provided in Chap. 13.

1.10.12 Macular Pigments

Provitamin A potential of β-carotene (reported in 1930s) and its usefulness in vision was the sole recognized health benefit of carotenoids till 1970s.From there onward usefulness of other carotenoid molecules and nonprovitamin A carotenoids was realized. Waled in 1945, suggested that yellow color of macula lutea present in human eye was due to "xanthophyll", however Bone and Landrum in 1985 deciphered this xanthophyll to be made-up of lutein, zeaxanthin and meso- zeaxanthin. A clear link between decreased carotenoids contents of macula and age-related macular degeneration was established in 1980s [30]. Macular pigments (MP) consist of three isomeric xanthophyll molecules having common chemical formula of $C_{40}H_{56}O$, and are lutein, zeaxanthin, and *meso*zeaxanthin. The OH groups on 3 and 3′ location of terminal ionone rings of MP are joined by a rigid 22-carbon isoprenoid skeleton with nine c.d.b. These OH groups and the number of c.d.b. define their solubility, polarity, light absorption potential and antioxidant capacity. The λmax of lutein is ~445 nm, and for zeaxanthin is ~450. The peak λmax of MP around 460 nm parallels with "blue light hazard" wavelength of 450–500 nm. Hence these MP function as visible blue light filters by absorbing 40–90% of incoming high energy, short wavelength blue light, protecting retina from light-induced damage. Many C=C double bonds make various cis/trans (E/Z) confirmations possible although MP usually exist as all=trans. Lutein has one less C=C double bonds than zeaxanthin and meso-zeaxanthin. The presence of 3 stereo-centers at C-3, C-3′ and C-6′confer 8 various stereoisomers to lutein. Xanthophylls in human tissues and fluids are exclusively in free form as compared to their esterified form in plants. They are concentrated (1mM) in central retina. Lutein and β-cryptoxanthin are present in high concentration in serum as compared to low serum zeaxanthin concentration. Zeaxanthin (mostly in central region of macula) and lutein (mostly in peripheral macula) are the only carotenoids present in retina along with lutein transformational product, the meso-zeaxanthin. Their selection from a group of about 40 dietary and 15 circulatory carotenoids. Both zeaxanthin and meso-zeaxanthin dominate in fovea region. Human diet is zeaxanthin poor and lutein rich. Literature does not shows zeaxanthin and lutein concentration separately rather as one figure. Egg yolk and corn contain about same amount of both of them (molar ratio 1:1). The human serum zeaxanthin-to-lutein molar ratio varies from 1:7 to 1:14 due to lutein rich diet of fruits and vegetables. Only few foods like goji berries, orange pepper, and *Physalis alkekengi* fruit. Zeaxanthin is more easily bioavailable due to its presence in fruits chromoplast (in contrast to chloroplast of green vegetables) is more easily released from food matrix. Still we have limited understanding of deposition of only 3 MP in macula lutea. Further details are provided in Chap. 19.

1.11 RDA & Clinical Uses

Carotenoids do not participate directly in any key metabolic process and their relative deficit does not lead to any specific deficiency or chronic disease, therefore no FDA-approved "health claims" exits as such. However certain carotenoids (e.g. lutein and zeaxanthin) can be viewed as partially essential nutrients.

Carotenoids are seldom exploited as a particular remedy per se although several of them like lycopene is used against infertility in male while β-carotene decreases the harmful effects of light in erythropoietic protoporphyria patients. They are suggested as segment of a fit food or as supplements to decrease the possibility of both genetic and acquired ailments. Carotenoids are usually nontoxic, even when consumed in higher amounts as purified supplements with few exceptions. Some carotenoids are also suggested as palliatives, such as recommendation of carotenoids supplements by ophthalmologists for decreasing the risk of eye disease and improving visual function. There exits, however, a noticeable discrepancy: (a) an enormous data associating carotenoid consumption with well-being and the avoidance of a large number of acquired ailments; and (b) a medical community that cannot suggest dietary commendations due to contemporary evidentiary principles [17].

1.12 Natural vs Synthetic

Carotenoids are usually extracted from plants or algae or synthesized chemically however there is increasing trend of their production from biotechnological processes. This biotechnological production due to variety of microorganisms present in nature, diversity of available substrates and the prospects to regulate functional conditions.

Carotenoids, like lutein, are only available in natural forms. Commercial carotenoids are mostly obtained by extraction from plants and by chemical synthesis. However, challenges regarding seasonal and geographic variability are present in the case of production of colorants of plant origin. The chemical synthesis can create hazardous waste affecting the environment. Unlike these customary approaches, the microbial production of carotenoids is very prospective and safe. It uses low-cost substrates, decreasing the production prices [25].

Despite their nature of origin, both carotenoids either synthetic or natural are identical in their molecular structure. Synthetic carotenoids are more stable as compare to natural ones as they are designed to minimize oxidation. They are formulated and distributed in market as dispersions, colloids, colloidal suspensions, and emulsions so that their application in food products is much easier. In spite of these benefits, they possess carcinogenic & teratogenic properties and are reported to be toxic therefore increases hesitancy in consumer. Natural carotenoids being used are plant extracts or powder.

1.13 Commercial Production & Uses

Chemical synthesis of β-carotene, astaxanthin, canthaxanthin, lycopene, and a limited amount of zeaxanthin, is well-known on a bigger scale. β-carotene was first time synthesized in 1950 and Roche in 1954 commercialized it [21]. Now β-apo-8′-carotenal (E160e) (k), canthaxanthin (E161g), and astaxanthin (E161j) are synthesized chemically. Since 1960s, consumers have indicated a strong inclination for natural products, including colors, as nourishing and healthy features have been allied with them. The rapidly evolving metabolic engineering and synthetic biology techniques are alternate ways to obtain various carotenoids at comparatively high titers and yields using fast-growing microbes.

Carotenoids are used as functional foods ingredients due to their extraordinary melting points, creating crystals during food storage. Despite of that, endogenous carotenoids in foods are usually stable. Loss of double bond can degrade these molecules. Hence these molecules are handled in encapsulated forms rather than crystalline forms. Further details are provided in Chap. 27.

1.14 Conclusions

Molecules with less than 40 C atoms are categorized as C30 carotenoids or diapocarotenoids, as apocarotenoids that are properly in-chain oxidized compounds, or as nor-carotenoids where C atoms are properly detached from the structure. Elements other than carbon, hydrogen, and oxygen are not directly linked to the C skeleton of carotenoids found in nature. The unique structural property of carotene is a centrally situated, stretched system of alternating single and double bonds whereby the π-electrons are efficiently delocalized over the entire polyene chain. This c.d.b. system makes the light-absorbing chromophore (electrons in the double bonds are delocalized and possess less ground energy state). The c.d.b. is imparts the distinguishing shades, from colorless (phytoene), yellow (4,40-diaponeurosporene), orange (*β*-carotene), red (capsanthin), to pink (bacterioruberin) with increasing number of c.d. b. Basically carotenoids have backbone of symmetrical, tetraterpene with two terminal C20 moieties that shift the absorption wavelength to the visible region of the spectrum. The earliest role established for carotenoids in animals was as a vitamin A precursor. They exhibit limited presence in diets of animal origin. The plant carotenoids provide almost 82% vitamin A in developed countries and 68% of vitamin A diet globally. The key advantage of provitamin A carotenoids is that they are only adapted into vitamin A when the body needs, hence avoiding its addition.

References

1. De Carvalho LMJ, Ortiz GMD, de Carvalho JLV, de Oliveira ARG (2018) Carotenoids in raw plant materials. Progr Carotenoid Res 2018:107
2. Meléndez-Martínez AJ (2019) An overview of carotenoids, apocarotenoids, and vitamin A in agro-food, nutrition, health, and disease. Mol Nutr Food Res 63(15):1801045. https://doi.org/10.1002/mnfr.201801045
3. Speight JG (2011) Chapter 13 – pharmaceuticals. In: Speight JG (ed) Handbook of industrial hydrocarbon processes. Gulf Professional Publishing, Boston, pp 467–497
4. Vershinin A (1999) Biological functions of carotenoids – diversity and evolution. BioFactors 10(2–3):99–104. https://doi.org/10.1002/biof.5520100203
5. Yahia EM, JDJ O-P (2010) Chemistry, stability, and biological actions of carotenoids. In: Fruit and vegetable phytochemicals: chemistry, nutritional value and stability. Wiley Online Library, New York, pp 177–222
6. Fernandes AS, do Nascimento TC, Jacob-Lopes E, De Rosso VV, Zepka LQ (2018) Carotenoids: a brief overview on its structure, biosynthesis, synthesis, and applications. Progr Carotenoid Res 2018:1
7. Gross J (1991) Chlorophylls. Pigments in vegetables. Springer, New York, pp 3–74
8. Frank HA, Young A, Britton G, Cogdell RJ (2006) The photochemistry of carotenoids. Springer Science & Business Media, New York
9. Moore T (1930) Vitamin A and carotene: The absence of the liver oil vitamin A from carotene. VI. The conversion of carotene to vitamin A in vivo. Biochem J 24(3):692
10. Stange C (2016) Carotenoids in nature: biosynthesis, regulation and function. Springer, New York
11. Tanumihardjo SA (2012) Carotenoids and human health. Springer Science & Business Media, New York
12. Tang G (2010) Bioconversion of dietary provitamin A carotenoids to vitamin A in humans. Am J Clin Nutr 91(5):1468S–1473S. https://doi.org/10.3945/ajcn.2010.28674G
13. Llansola-Portoles MJ, Pascal AA, Robert B (2017) Electronic and vibrational properties of carotenoids: from in vitro to in vivo. J R Soc Interface 14(135):20170504. https://doi.org/10.1098/rsif.2017.0504
14. Langi P, Kiokias S, Varzakas T, Proestos C (2018) Carotenoids: from plants to food and feed industries. In: Microbial carotenoids. Springer, New York, pp 57–71
15. Barreiro C, Barredo J-L (2018) Microbial carotenoids. Springer, New York
16. Nagarajan J, Ramanan RN, Raghunandan ME, Galanakis CM, Krishnamurthy NP (2017) Chapter 8 – carotenoids. In: Galanakis CM (ed) Nutraceutical and functional food components. Academic Press, London, pp 259–296
17. Hammond BR Jr, Renzi LM (2013) Carotenoids. Adv Nutr 4(4):474–476. https://doi.org/10.3945/an.113.004028
18. Yuan H, Zhang J, Nageswaran D, Li L (2015) Carotenoid metabolism and regulation in horticultural crops. Horticult Res 2(1):15036. https://doi.org/10.1038/hortres.2015.36
19. Schieber A, Weber F (2016) 5 – carotenoids. In: Carle R, Schweiggert RM (eds) Handbook on natural pigments in food and beverages. Woodhead Publishing, Cambridge, pp 101–123
20. Rodriguez-Amaya DB (2015) Carotenes and xanthophylls as antioxidants. In: Handbook of antioxidants for food preservation. Elsevier, Amsterdam, pp 17–50
21. Sandmann G (2014) Carotenoids of biotechnological importance. In: Biotechnology of isoprenoids. Springer, Cham, pp 449–467
22. Rodriguez-Amaya DB (2015) Carotenoids as food colorants and precursors of aroma compounds. In: Rodriguez-Amaya DB (ed) Food carotenoids, pp 199–224
23. Berman J, Zorrilla-López U, Farré G, Zhu C, Sandmann G, Twyman RM et al (2015) Nutritionally important carotenoids as consumer products. Phytochem Rev 14(5):727–743. https://doi.org/10.1007/s11101-014-9373-1

24. Ciriminna R, Fidalgo A, Meneguzzo F, Ilharco LM, Pagliaro M (2016) Lycopene: emerging production methods and applications of a valued carotenoid. ACS Sustain Chem Eng 4(3):643–650. https://doi.org/10.1021/acssuschemeng.5b01516
25. Córdova P, Baeza M, Cifuentes V, Alcaíno J (2018) Microbiological synthesis of carotenoids: pathways and regulation. In: Progress in carotenoid research. IntechOpen, London
26. Rabasco Álvarez AM, González Rodríguez ML (2000) Lipids in pharmaceutical and cosmetic preparations. Grasas y Aceites 51(1–2):74–96
27. Brotosudarmo THP, Limantara L, Chandra RD (2018) Chloroplast pigments: structure, function, assembly and characterization. In: Plant growth and regulation-alterations to sustain unfavorable conditions. IntechOpen, London
28. Britton G, Khachik F (2009) Carotenoids in food. In: Carotenoids. Springer, Dordrecht, pp 45–66
29. Peto R, Doll R, Buckley JD, Sporn M (1981) Can dietary beta-carotene materially reduce human cancer rates? Nature 290(5803):201–208
30. Arunkumar R, Gorusupudi A, Bernstein PS (2020) The macular carotenoids: a biochemical overview. Biochim Biophys Acta (BBA)-Mol Cell Biol Lipids 2020:158617

Chapter 2
Chemistry of Carotenoids

Muhammad Riaz, Muhammad Zia-Ul-Haq, and Deqiang Dou

2.1 Introduction

Carotenoids are second most abundant natural compounds that occur in nature after chlorophyll, and the carotenoid database reported 1181 natural carotenoids of 700 source organisms [1]. The term carotenoid was given to this group after the discovery of carotene from carrot by Wackenroder in 1831 [2].

The polyene arrangement comprises up to 15 conjugated double bonds (CDB)s in carotenoids, which are responsible for their typical absorption spectra and particular photochemical properties [3–5].

Naturally, carotenoids are predominantly or entirely found in linear or all- configuration. Exposure of the *trans* to light or heat may facilitate *the cis* transformation of one or more double bonds [5]. The cis double bonds create more significant steric hindrance between nearby hydrogen atoms and methyl groups, and in general, are less stable thermodynamically than that of trans form [6]. The biological activities and physicochemical properties of carotenoids are closely related to their structures. The structure of known carotenoids can be categorically recognized by the combined use of chromatographic and spectroscopic techniques. UV-visible absorption spectra and specific chemical reactions may be used to confirm the type, location, and a number of functional groups in xanthophylls [7].

M. Riaz (✉)
Department of Pharmacy, Shaheed Benazir Bhutto University, Sheringal Dir (U), Pakistan

M. Zia-Ul-Haq
Office of Research, Innovation, and Commercialization (ORIC), Lahore College for Women University, Lahore, Pakistan

D. Dou
College of Pharmacy, Liaoning University of Traditional Chinese Medicine, Dalian, China

© Springer Nature Switzerland AG 2021

M. Zia-Ul-Haq et al. (eds.), *Carotenoids: Structure and Function in the Human Body*, https://doi.org/10.1007/978-3-030-46459-2_2

2.2 Structure of a Typical Carotenoid

Carotenoids, composed of the poly-isoprenoid skeleton, have a long conjugated bond system and bilateral symmetry around this conjugated system. β-carotene is a familiar carotenoid; it is crucial both in plant and animal kingdom both structurally and physiologically and acts as a precursor to several other carotenoids.

The central conjugated polyene structure has importance due to delocalized electrons, which are low in-ground energy state, thus a useful chromophore for visible light. These alternate single and double bonds are responsible for the colour of carotenoids. Carotenoids possess 3–13 CDBs, for example, the ζ-carotene has seven double bonds with pale yellow colour, b-carotene with nine double bonds bright orange colour, and deep red colour of canthaxanthin with 13 double bonds [8]. The increase in chromophore length increases resonance through electronic transitions, thus reducing the energy of molecular orbital and stabilize the system. The molecule will absorb light of a larger wavelength. Cyclization in the conjugated system causes the wavelength (λ_{max}) shift from longer to shorter wavelength called a hypsochromic shift [3]. The extensively conjugated system plays a vital role in colouring power of carotenoids; however, this unsaturation poses them to various degradation reactions and un-stability [9]. In the case of β-carotene, the end ring at both sides is identical, have one unsaturated bond and methyl group attached. This ring is usually termed as Ionones, which are of three types subject to the position of the double bond as α-ionone, β-ionone, and γ-ionone. As compared to the central chain, the ring commonly undergoes substitution reactions. The most common reactions are oxidation reaction, while reduction reactions are slightly less. The end ring is modified by hydroxy or epoxy or keto group. The presence of ionone ring in carotenoids represents that it is the precursor of vitamin A (Fig. 2.1).

2.3 Classification of Carotenoids [10]

Carotenoids can be classified based on structure, cyclization, functions, and modification in structures.

2.3.1 *Classification I: Chemical Structure*

Carotenoids are classified as follows:

(i) **Acyclic carotenes**: have open ends at both sides e.g. ζ-Carotene Phytoene (colourless), Lycopene (red), Neurosporene and Phytofluene.
(ii) **Cyclic carotenes**: containing one **(Monocyclic)** or two cyclic **(Bicyclic)** structures e.g. α-Carotene (orange), β-Carotene, (orange), E160a, γ-Carotene (orange), δ-Carotene, α-Zeacarotene and β-Zeacarotene

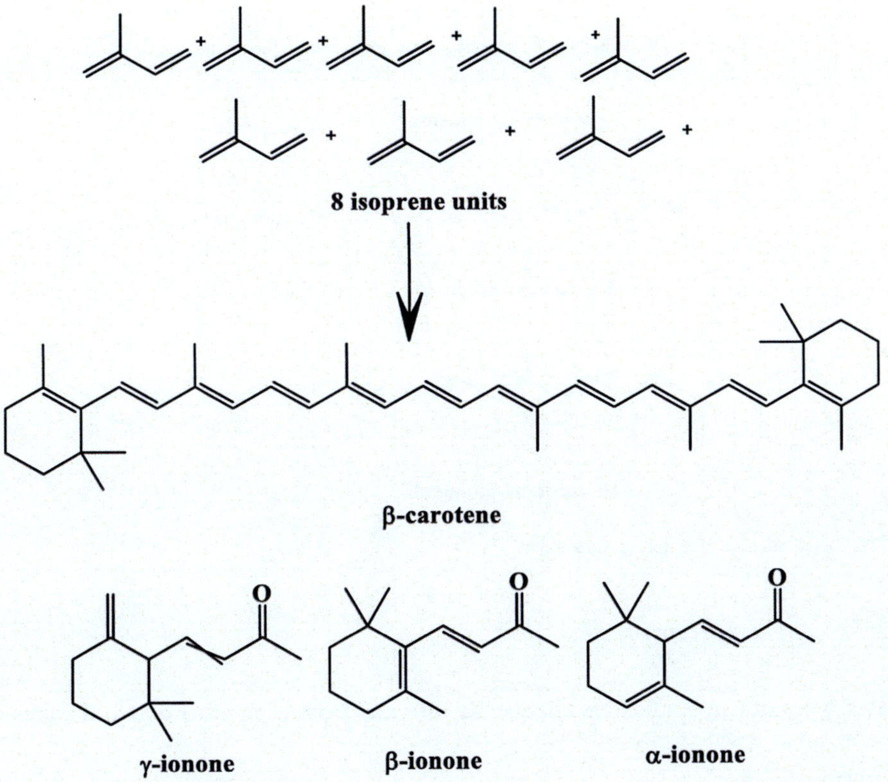

Fig. 2.1 Structure of β-carotene

(iii) **Carotenols:** They are also called hydroxy carotenoids, as they contain at least a hydroxyl group. The example of carotenols are α-Cryptoxanthin and Lutein, which are Yellow, β-Cryptoxanthin orange in colour, other examples include Lycofill, Lycoxanthin, Rubixanthin, Zeaxanthin (yellow-orange) and Zeinoxanthin.

(iv) **Epoxycarotenoids**: this subgroup of xanthophylls contains at least an epoxy group examples are Antheraxanthin, Auroxanthin, β-Carotene-5,6-epoxide, Lutein-5,6-epoxide, Luteoxanthin, Neoxanthin, and Violaxanthin (yellow). These compounds are also named as epoxide, and one oxygen is linked to two carbons, which may or may not connected, thus creating different kinds of epoxides in this group. Almost 180 epoxy carotenoids have been reported [1]. Violaxanthin and neoxanthin are present in nature sufficiently. However, other members are usually found in trace amounts this creates a sort of question on its presence in nature, examples of some are given below (Fig. 2.2).

(v) **Species-specific carotenoids** include Capsorubin, Capsanthin, Bixin, and Crocetin.

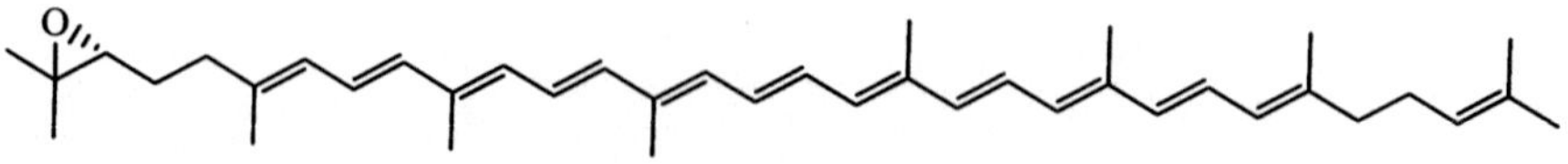

Lycopene 1,2 epoxide

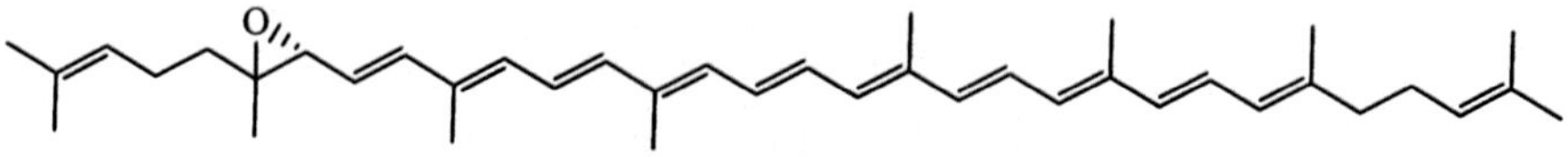

Lycopene 5,6 epoxide

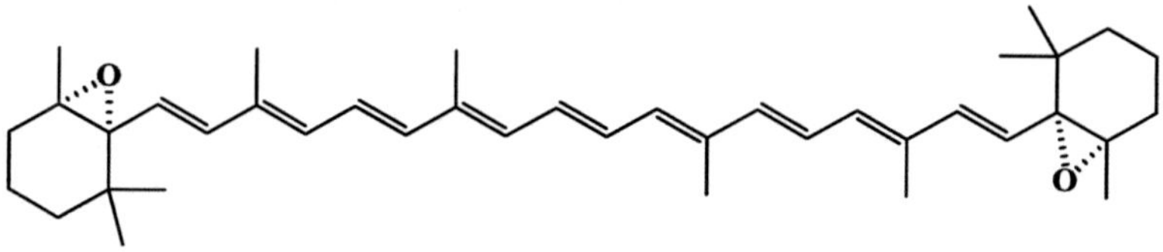

β-carotene-diepoxide

Fig. 2.2 Structure of epoxycarotenoids

2.3.2 Classification II

Two types of carotenoids are chemically different from each other due to the presence and absence of oxygen.

(i) **Hydrocarbon Carotenoids** or **Carotenes** composed of only hydrogen and carbon forming a polyunsaturated chain, either cyclic or linear with the chemical formula $C_{40}H_{56}$. These are usually red, yellow and orange e.g. α-Carotene, β-Carotene, β-cryptoxanthin. β-carotene and lycopene are among those carotenoids that are very closely related to animal nutrition. β-carotene shows referenced pro-vitamin A activity among carotenoids. The typical food sources of carotenes include peach, watermelon, cantaloupe, red palm oil, pumpkin, tomato, papaya, oranges, and Carrots. Carotenes reflect mostly red and orange light. Chlorophyll absorbed energy from sunlight and transmit some energy possessed by orange to red wavelength region, through carotenes. They also act as antioxidants in plants by reducing singlet oxygen that is generated during photosynthesis. Carotenes are highly soluble in hydrocarbon solvents, while xanthophylls are soluble in more hydrophilic solvents e.g., ethanol.

Carotenes usually exist as free and entrenched in chromoplast and chloroplast bodies. Carotenes exist in several isomers that have the same formula but different molecular structures.

Excitingly, both carotenes and xanthophylls are absorbed from food by animals, and their presence can easily be confirmed in human milk. However, other animals such as birds or amphibians absorb mainly xanthophylls.

(ii) **Xanthophylls** consist of carbon, hydrogen, and oxygen in the form of hydroxy, epoxy, or oxy groups are so also known as oxy-carotenoids. Xanthophylls are derived from carotenes by the addition of oxygen-containing functional groups such as COOH, -OH, =O at one or both ends and enhance their water-soluble property. The general chemical formula of xanthophyll is $C_{40}H_{56}O_{1-4}$. Xanthophylls are usually yellow and are present in high concentrations in leaves, especially during fall, and don not prepared directly during photosynthesis. Carotenoids show different absorptive wavelengths than chlorophyll and are mostly yellow or green e.g., Lutein, zeaxanthin, violaxanthin, fucoxanthin. On the other hand, xanthophyll molecules are excess in leaves and play its role indirectly during photosynthesis. They assist chlorophyll in photosynthetic tissues. Carotenes being non-polar are more solubilized and enters the lipid globule more efficiently as compared to xanthophylls (polar). Apart from polyene, the keto functional group integrates strong antioxidant capacity in xanthophylls then carotenes. Xanthophylls contain oxygen in the form of hydroxy, keto, epoxy, and methoxy groups. Xanthophylls are more comparative, more polar. The common food sources of xanthophylls are Yellow maize, mango, yellow sweet potato, zucchini, egg yolk, spinach, and broccoli. Xanthophylls can be divided further as

(a) **Plant origin** e.g., lutein
(b) **Marine/animal origin** e.g., astaxanthin, zeaxanthin, violaxanthin, cryptoxanthin, and capsanthin

2.3.3 *Classification III: Structural Modification*

1. **Allenic carotenoids** have continuous double bonds e.g., neoxanthin, crocoxanthin, diatoxanthin
2. **Acetylenic carotenoids** have triple bonds in their structure e.g., Dehydroapocarotenoids, peridinin, fucoxanthin
3. **Apocarotenoid** modified such that it has less than 40 carbons e.g., bixin. Carotenoid skeleton shortened by removal fragments at one end results in apocarotenoid generation while at both ends generate dia-apocarotenoids [11]. They are derivative of carotenoids usually form as a result of oxidative cleavage. The types of apocarotenoids generated during oxidative cleavage at specific positions. A group of enzyme name carotenoid specific cleavage oxygenase that targets the double bond of the carotenoid chain is responsible for the biosynthesis of apocarotenoids [12]. Two enzymes group are responsible for the oxidative cleavage of carotenoids e.g., 9-cis-epoxycarotenoid dioxygenases (NCEDs) and carotenoid cleavage dioxygenases (CCDs). Apart from enzymatic cleavage, the synthesis of apocarotenoids via nonenzymatic cleavage also occurs during incubation, extraction, and processing of α-carotene, β-cryptoxanthin, lutein, zeaxanthin, and lycopene [13].

Bixin from annatto and crocetin from saffron is a natural example of apocarotenoids. Crocetin, which has a molecular formula of $C_{20}H_{24}O_4$, is symmetrical

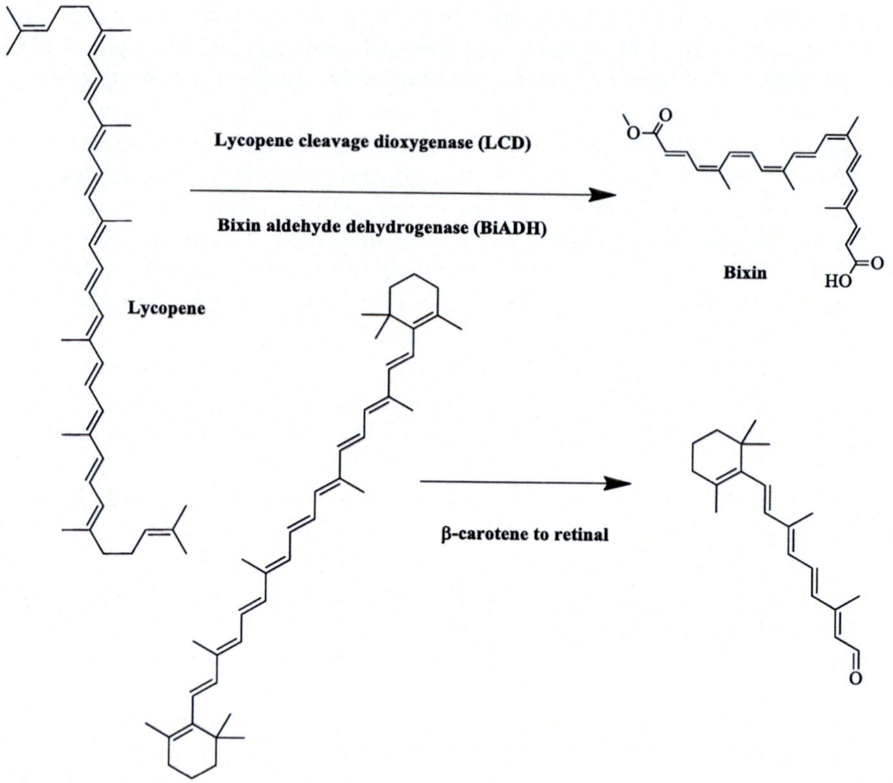

Fig. 2.3 Structure of apocarotenoids

diapocarotenoid with seven carbon-carbon double bonds and carboxylic groups at both ends Bixin ($C_{25}H_{30}O_4$) is the monomethyl ester of apocarotenoid, having a total of 11 CDBs, 9 carbon-carbon and two carbon-oxygen double bonds [14] (Fig. 2.3).

2.3.3.1 Functions of Apocarotenoids

Apocarotenoids have been reported for various functions e.g., as aromatics (rose scent) e.g. β, ionone and β-damascenone [15] and nor-isoprenoids (floral and fruity characteristics in wine) [16], as repellents e.g. β-ionone. They act as chemo-attractants, repellant freshwater nematodes [17], growth simulators, inhibitors [18, 19], and food-finding cues for freshwater herbivores [20]. They provide an architecture to plant body and play a role in growth, such as e.g. strigolactones [12]. Apocarotenoids affect glucose uptake, carotenoids biosynthesis [21], allelochemicals [22], and pheromones [23, 24].

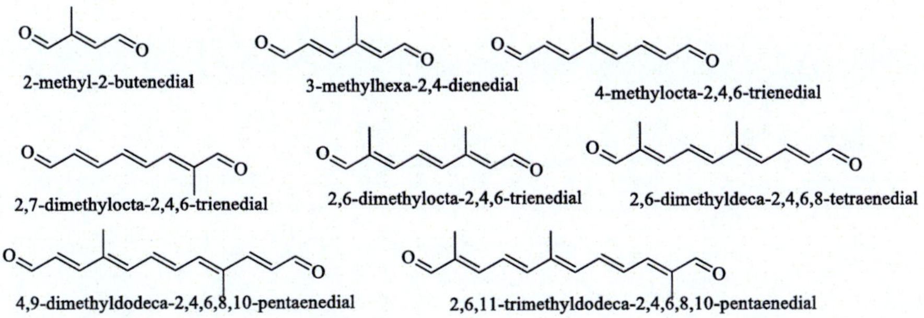

Fig. 2.4 Structure of dials

2.3.3.2 Dials

Dia-apocarotenoids derivatives or dials (dialdehydes) have attracted researchers as precursors of bixin and crocin, however, dials have regulatory functions both in humans and plants. Anchorene a C_{10} dial regulates root growth in rice and Arabidopsis and rosafluene (C_{14}) cause a reduction in cancer cell growth [25] (Fig. 2.4).

4. Higher carotenoid has more than 40 carbon, e.g., crocetin, bacterioruberin
5. **Nor-carotenoids** in which CH_3, CH_2 and CH group removed, e.g., pyrrhoxanthininol. These are derived from carotenoids with one or more carbon removed from the main skeleton of carotenoids, e.g., for example (Fig. 2.5)
6. Hydrogenated carotenoids or dehydrogenated carotenoids, e.g., lycopersene
7. **Seco-carotenoids**

These are compounds of carotenoids in which one bond between adjacent carbon atoms is broken except C1 and C6. e.g., triphasiaxanthin. They are reported to date to be 13 in numbers [1]. Few examples are (Fig. 2.6)

8. **Retro-carotenoids**: a shift of bond in chain occurs in one position. They are also reported to be 13 in numbers till date [1], e.g., eschscholtzxanthin and structure of some retro-carotenoids are given below (Fig. 2.7)

2.3.4 Classification IV: Function

(a) **Primary carotenoids**: These carotenoids are basic and useful components of photosynthetic apparatus, also known as photosynthetic pigments and usually present in leaves, e.g., β-Carotene, zeaxanthin, Lutein, violaxanthin, antheraxanthin, neoxanthin. Lutein as accessory pigment assist in the transfer of absorbed energy to chlorophylls, therefore increasing the light-absorbing spectrum of algae or plants.

(b) **Secondary Carotenoids**: are not necessary for plant survival; however, play an essential role in functions. They are produced after exposure to environmental stimuli through carotenogenesis, e.g., astaxanthin, capsanthin, bixin, lycopene,

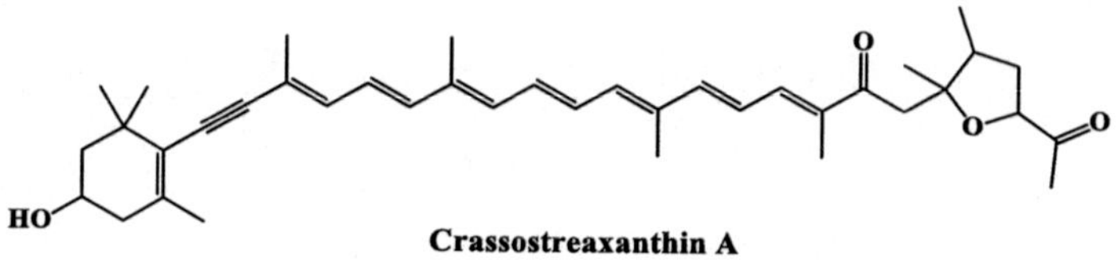

Crassostreaxanthin A

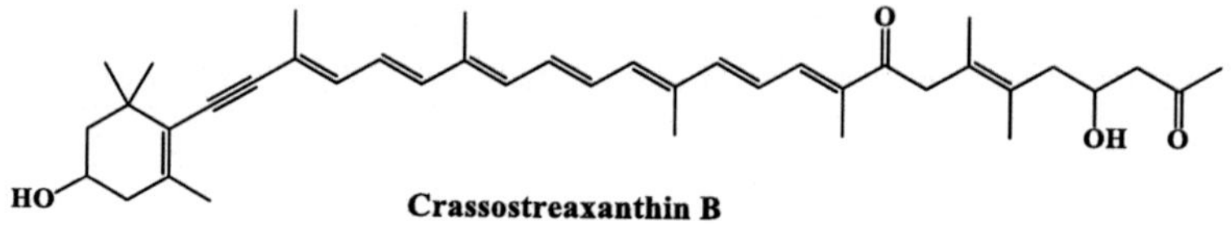

Crassostreaxanthin B

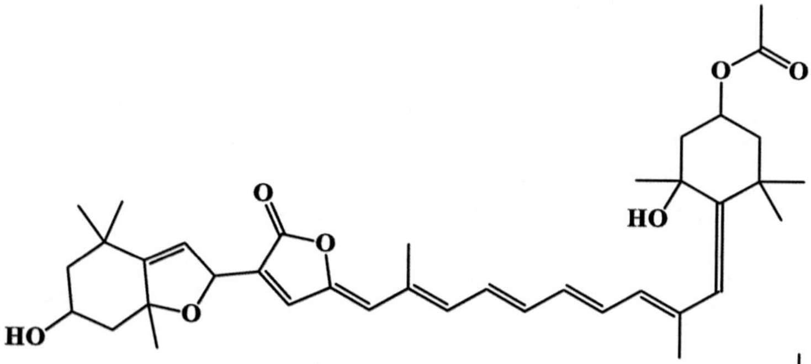

Peridinine 5,8-furanoside

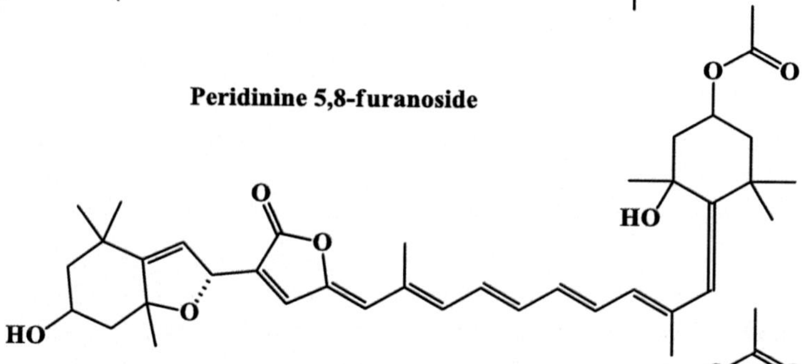

Nor-peridinin 1

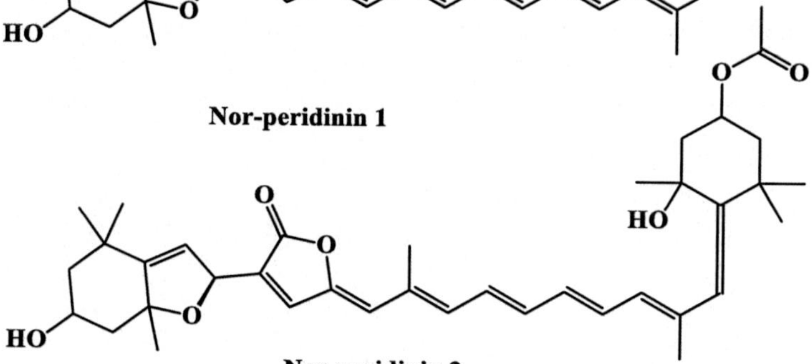

Nor-peridinin 2

Fig. 2.5 Structure of nor-carotenoids

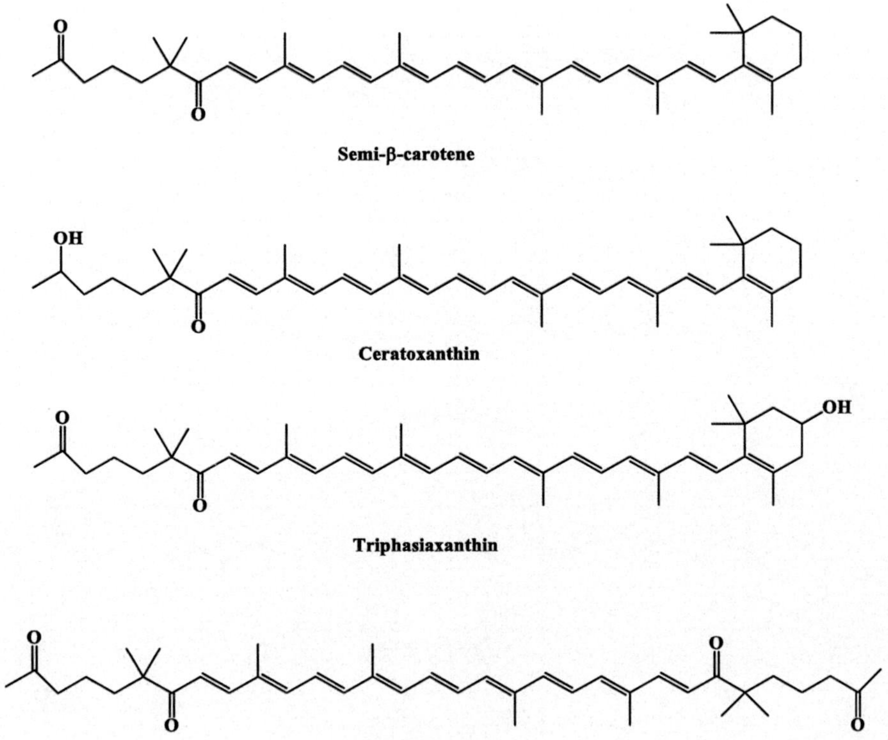

Semi-β-carotene

Ceratoxanthin

Triphasiaxanthin

β-carotenone

Fig. 2.6 Structure of Seco-carotenoids

and α-Carotene (carotenoids localized in fruits and flowers), lutein named after the source from macula lutea of the retina. Xanthophylls serve as accessory pigments or secondary carotenoids to plants. Xanthophylls are frequently found in a chloroplast (non-esterified). They are usually present as xanthophyll esters in chromoplasts. They are produced to high levels and are dispensed in oily droplets. Secondary carotenoids like canthaxanthin and astaxanthin play its part in cell-protection.

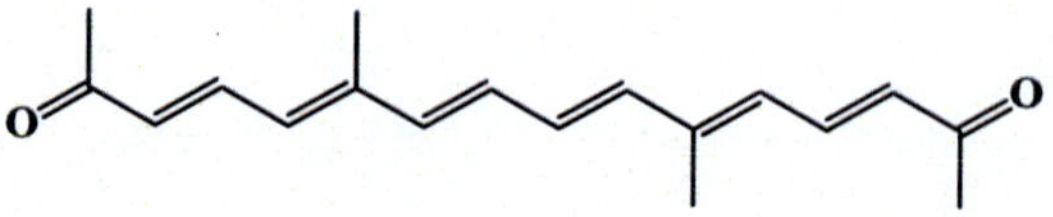

Retro-χ18-dione

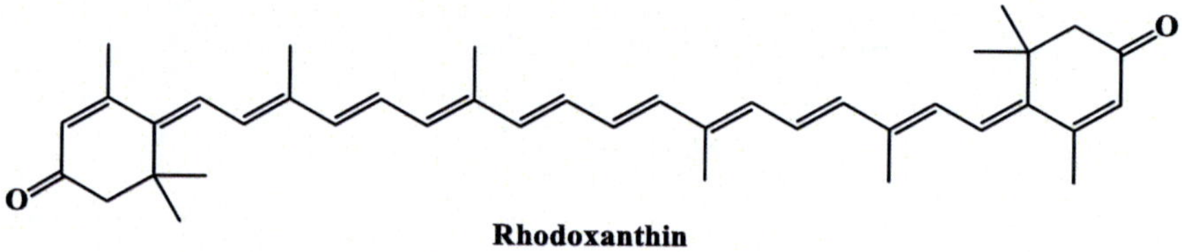

Rhodoxanthin

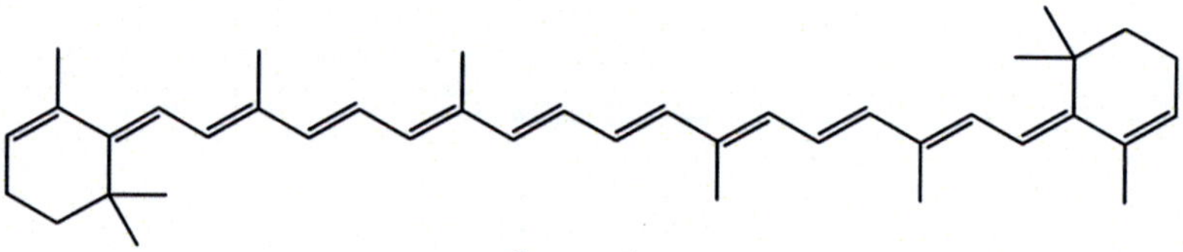

Isocarotene

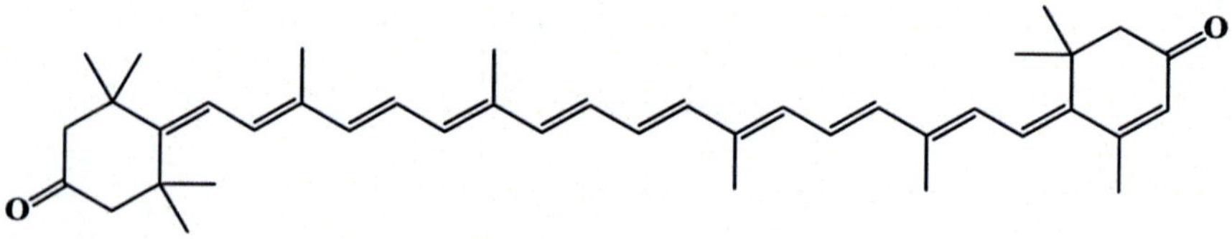

Loniceraxanthin

Fig. 2.7 Structure of retro-carotenoids

2.3.5 *Classification V: Carbon Number*

See Table 2.1.

2.3.6 *Classification VI: Domains of Life and Chemical Modification*

Carotenoids can be classified into three domains of life as given in Fig. 2.8.

Table 2.1 Classification of carotenoids based on Carbon numbers [1]

S.no	Type	Number of isoprene units	Source
1	C30	6	Archaea and bacteria
2	C40	8	Eukaryotes, archaea, and bacteria
3	C45	9	Bacteria
4	C50	10	Archaea and bacteria

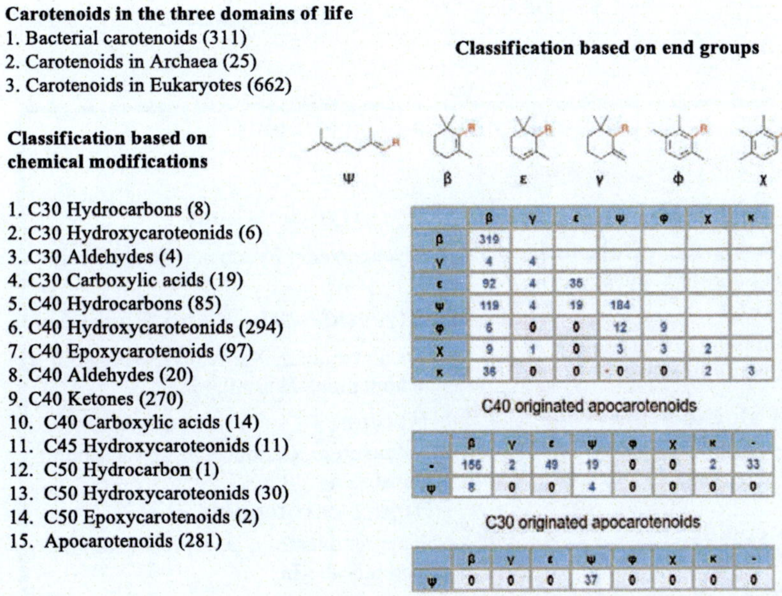

Fig. 2.8 Different types of carotenoids classifications [1]

2.3.7 *Classification VII: Partition Character*

Carotenoids can also be categorized based on their partition between the two immiscible solvents, i.e., 90% aqueous methanol and light petrol. See Table 2.2.

2.3.8 *Classification VIII: Natural Occurrence*

See Fig. 2.9.

Table 2.2 Classification based on the partition between two immiscible solvents

Carotenoid characteristics	Chemical nature	Phase
Hydrophasic carotenoids	Containing two or more hydroxyl groups	Lower (aqueous methanol) phase
Epiphasic carotenoids e.g., Hydrocarbons, xanthophyll esters	Containing an ether group or one oxo group	Upper (light petrol) phase
Biphasic carotenoids	Containing one hydroxyl group or two oxo groups or one carboxyl group	Distributed b/w 2 phases

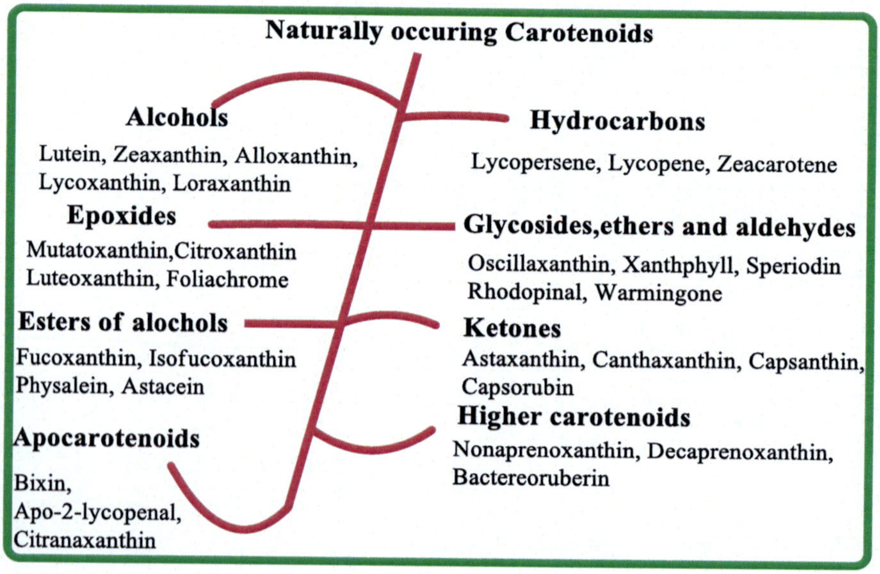

Fig. 2.9 Naturally occurring carotenoids based on chemical nature

2.3.9 *Classification IX: End Groups*

See Fig. 2.4 for details.

2.3.10 *Classification X: Based on Provitamin A*

Carotenoids are classified into two groups

1. **Provitamin A Carotenoids**: Carotenoids that are converted to retinol by the body, e.g., α-Carotene, β-carotene and β-cryptoxanthin
2. **Non-provitamin A carotenoids**: These carotenoids cannot be converted to retinol by the body, e.g. lycopene etc. (Fig. 2.10)

Examples of various groups of carotenoids

Hydrocarbons:
Hexahydrolycopene, Lycopersene, Phytofluene, Torulene, α-zeacarotene

Alcohols:
Alloxanthin, Crustaxanthin, Cryptomonaxanthin, Cynthiaxanthin, Gazaniaxanthin, Loroxanthin, Lutein, Lycoxanthin, OH-chlorobactene, Pectenoxanthin, Rhodopin, Rhodopinol or Warmingol, Saproxanthin and Zeaxanthin

Glycosides:
Oscillaxanthin, and Phleixanthophyll

Ethers:
Rhodovibrin, and Spheroidene

Epoxides:
Citroxanthin, Diadinoxanthin, Foliachrome, Luteoxanthin, Mutatoxanthin, Neochrome, Trollichrome, Vaucheriaxanthin, and Zeaxanthin

Aldehydes:
Rhodopinal, Torularhodinaldehyde, and Warmingone

Esters of alcohols:
Astacein, Fucoxanthin, Isofucoxanthin, Physalien, Siphonein, and Zeaxanthin

Acids and acid esters:
Torularhodin, and Torularhodin methyl ester

Ketones:
2,2'-diketospirilloxanthin, 3'-hydroxyechinenone, 3-OH- canthaxanthin or adonirubin or Phoenicoxanthin, Astacene, Astaxanthin, Canthaxanthin or Aphanicin, Chlorellaxanthin, Capsanthin, Capsorubin, Cryptocapsin, Echinenone, Flexixanthin, Hydroxyspheriodenone, Okenone, Pectenolone, Phoeniconone or Dehydroadonirubin, Phoenicopterone, Rubixanthone, and Siphonaxanthin

Apocarotenoids:
Apo-2-lycopenal, Apo-6'-lycopenal 6'-apo-y- caroten-6'-al, Azafrinaldehyde, Bixin, Citranaxanthin, Crocetin, Crocetinsemialdehyde, Crocin, Hopkinsiaxanthin, Methyl apo-6'- lycopenoate, Paracentrone, Sintaxanthin, and β-apo-2'-carotenal

Nor- and seco-carotenoids:
β-carotenone, Actinioerythrin, Peridinin, Triphasiaxanthin, Pyrrhoxanthininol, Semi-α-carotenone, and Semi-β-carotenone

Retro-carotenoids and retro-apo-carotenoids:
Eschscholtzxanthin, Eschscholtzxanthone, Rhodoxanthin, and Tangeraxanthin

Higher carotenoids:
Bacterioruberin, Decaprenoxanthin and nonaprenoxanthin

Fig. 2.10 Examples of various groups

2.4 Naming and Numbering of Carotenoids

2.4.1 Trivial Names

The carotenoids were named in the past as per their origin of discovery, like source, colour, or other related quality. Like carotene was named after isolation from carrot, and zeaxanthin from *Zea mays* (corn). These names are still in use more commonly, but these names provide no information about the structure and stereochemistry [26]. Despite less importance, these trivial names are used for up to 40 dietary carotenoids due to simplicity and ease of use.

2.4.2 *Structure of Carotenoids*

All carotenoids are considered to be derivatives of lycopene $C_{40}H_{56}$ that has a long central chain of 11 CDBs. The structure of carotenoids consists of the main carotenoid chain and end ring system. Mostly composed of eight (C5) isoprenoid units joined by head to tail arrangement except in the centre of molecules where the two units are connected tail to tail, thus positions the side chain methyl groups 1,6 relatively. In contrast, the remaining nonterminal groups are sited as 1,5 bonds.

The central main chain is a polyene structure of alternate single and double bonds. These alternate double bond electrons are of low energy capable of absorbing visible range light act as chromophore that gives them a characteristic colour. Modification in the chain modifies the chromophore [26].

The end rings groups are present that may be at one or both ends with saturation or unsaturation. These end rings are commonly substituted. The most common reaction at the end chain is an oxidation reaction [26].

The modified carotenoids can easily be recognized from the prefix used in systematic nomenclature like **apo** label the shortening of a carotenoid skeleton at both ends or one, **epoxy** for oxygen bridges, **dehydro** or **hydro** used for removal or addition of hydrogen.

The structure of carotenoids has high reducing potentials and electron distribution along with the entire structure. The transfer of electrons occurs via two ways: (1) **singlet-singlet transfer** is a lower energy state transfer that occurs during photosynthesis, from carotenoid to chlorophyll when polyene absorb lights, the energy caused the excitation and transfer of excited electrons to complete photosynthesis. The second way of electron transfer is termed as the **triplet-triplet transfer** is a higher energy transfer that occurs from chlorophyll to carotenoid and is essential in photoprotection.

2.5 IUPAC Naming of Carotenoids [27]

The carotenoids are named using IUPAC systematic naming as:

Rule 1
All carotenoids are derived from $C_{40}H_{56}$ acyclic structure having a long chain of CDBbs by hydrogenation, oxidation, cyclization or dehydrogenation or combination of these.

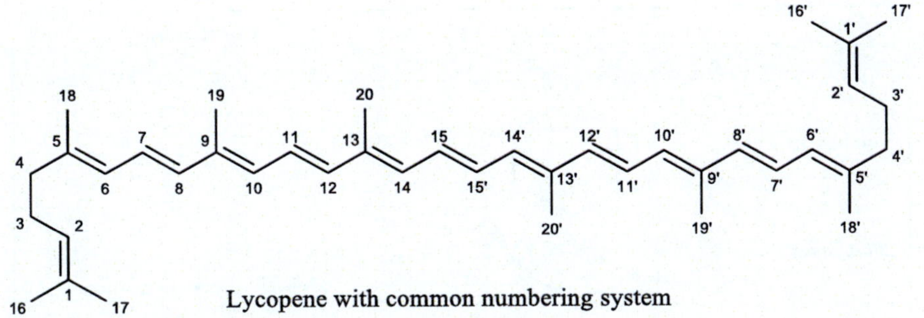

Fig. 2.11 Carotenoid with numbering

Rule 2
The parent chain/skeleton is numbered as given in Fig. 2.11; it can be seen that numbering starts at both ends with primed and unprimed, 1–15, 1′–15′ to the centre symmetrically. The attached methyl groups are number at both sides of the central position as 16 onward.

Rule 3
After numbering the main chain, the end groups are usually represented by a Greek letter as given in the figure.

Alphabetical order is followed in writing the Greek-letter prefixes separating the first one by a comma from the second one, while the second one is connected by a hyphen to the stem name. The Greek-letters are β (beta), γ (gamma), ε (epsilon), κ (kappa), π (phi), χ (chi), ψ (psi) (Figs. 2.12 and 2.13).

Rule 4
The carotenoid skeleton is numbered as shown in Fig. 2.11, the numbering of end groups shall be according to Rule 3, in case of different groups the order in naming should be alphabetical one.

Rule 5: Nor Carotenoids and Seco Carotenoids
In nor carotenoids the basic numbering of the skeleton and substituents remain the same however a prefix nor is added to the parent name

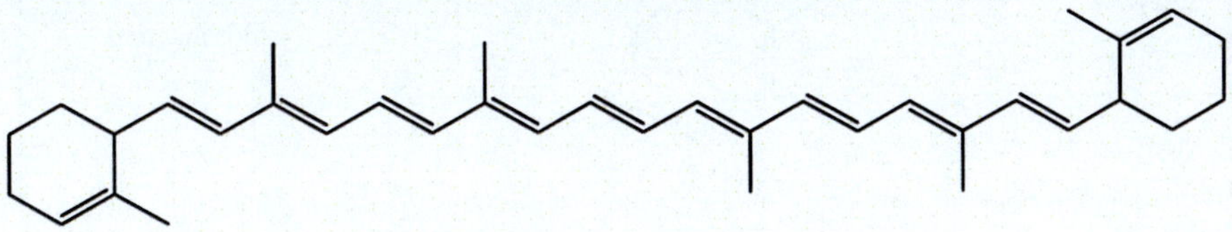

16, 17, 16′,17′-tetranor-ε,ε-carotene

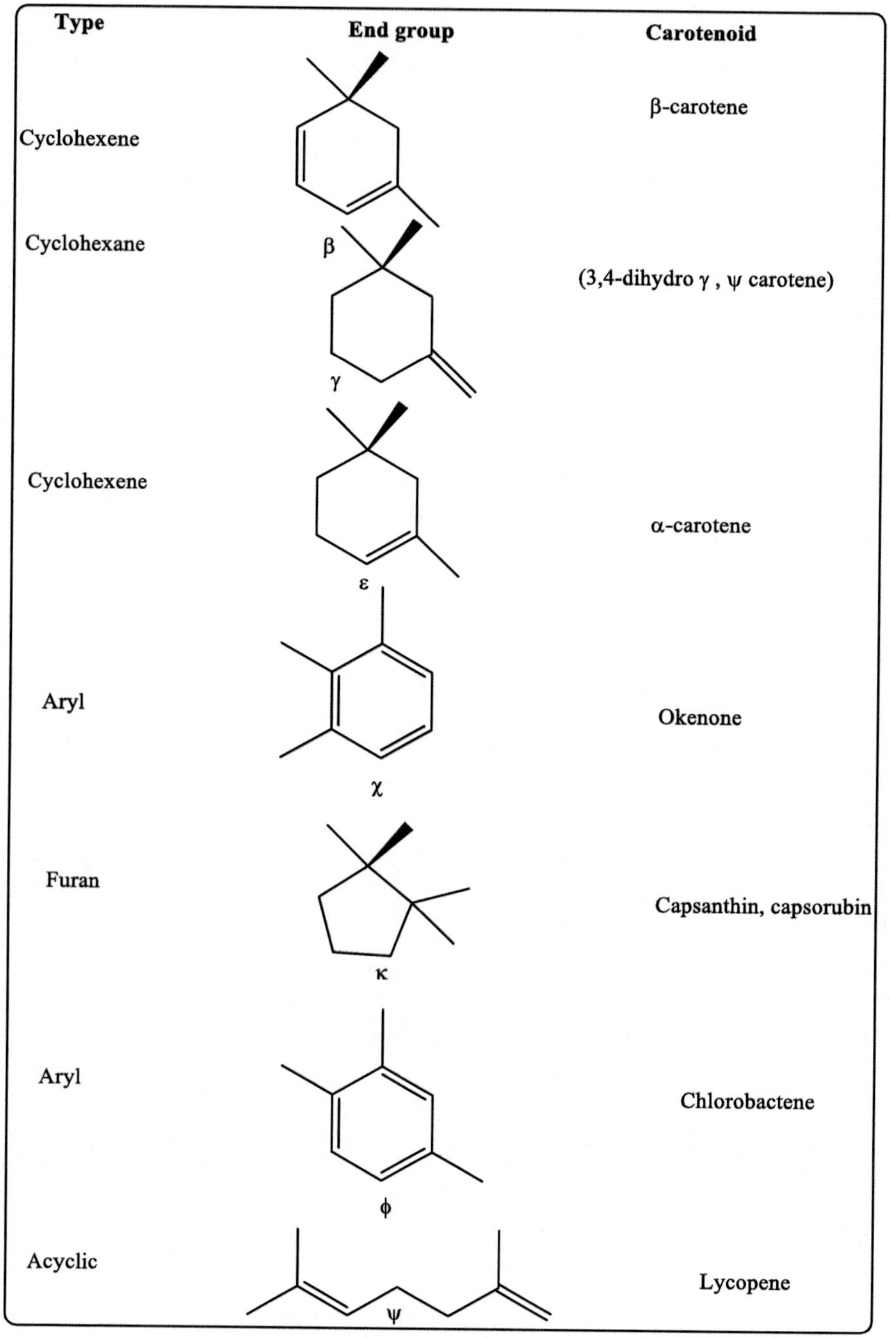

Fig. 2.12 End groups of carotenoids, Greek-letters: β (beta), γ (gamma), ε (epsilon), κ (kappa), π (phi), χ (chi), ψ (psi)

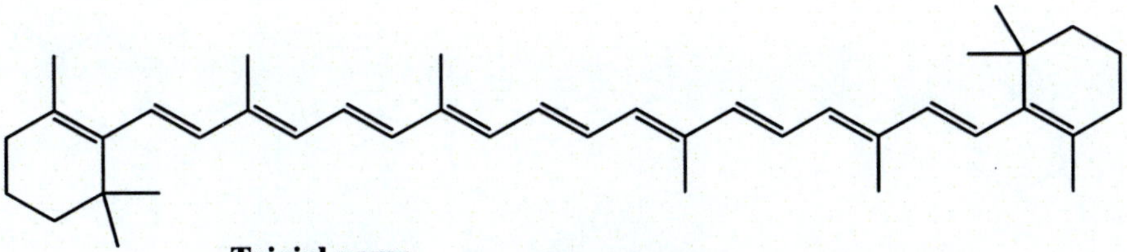

Trivial name **β-carotene**

IUPAC name **ε,ε-carotene**

Fig. 2.13 The common and IUPAC name example

In Seco-carotenoids the word "Seco" is added to the original carotenoid numbering

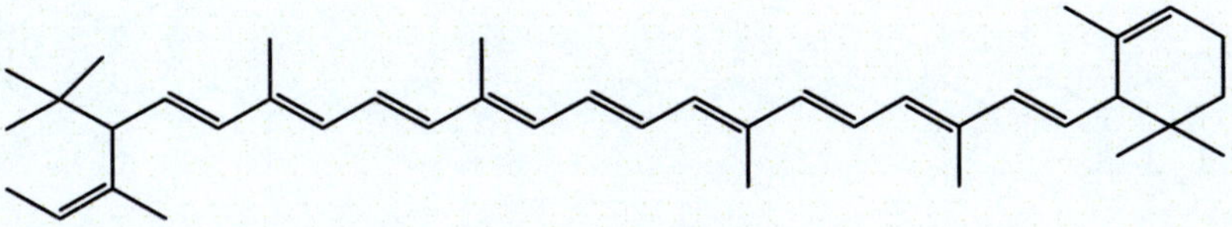

2,3-seco-ε,ε-carotene

Rule 6

After a change in hydrogenation level of carotenoids, the corresponding compounds are named as adding prefix "dehydro or hydro" as the case may be, preceded the Greek letters connected by a hyphen.

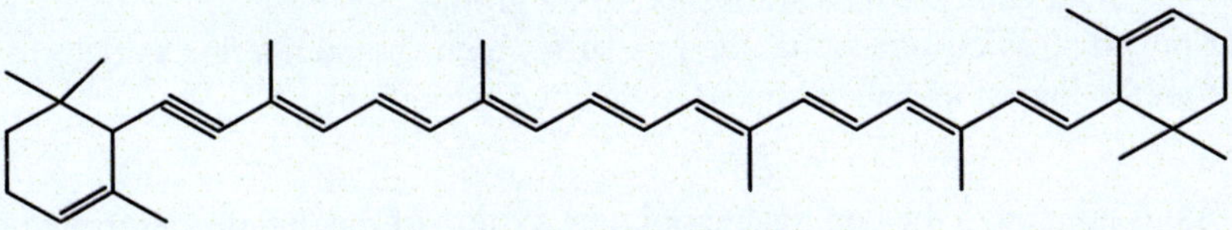

7, 8-didehydro-ε,ε-carotene

Rule 7

The oxygenated derivatives or other compounds are named by the use of prefix or suffix as per IUPAC rules for other organic compounds.

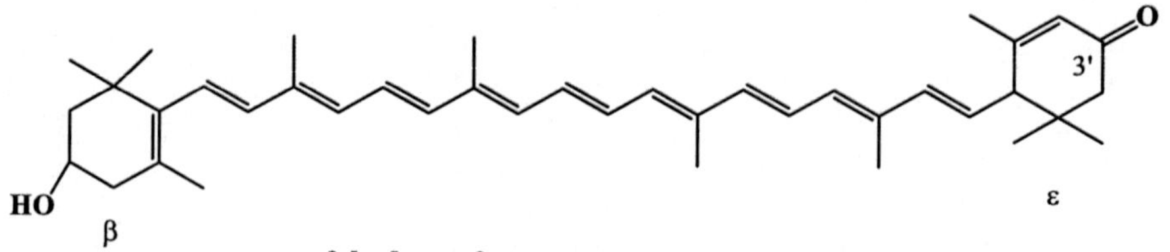

3-hydroxy-β,ε-carotene, 3'-one

Rule 8.1
If the end groups having 9 carbons are not similar their oxygenated derivatives are numbered as per the order of Greek letters the Greek letter first in the sequence will receive unprimed locant e.g.

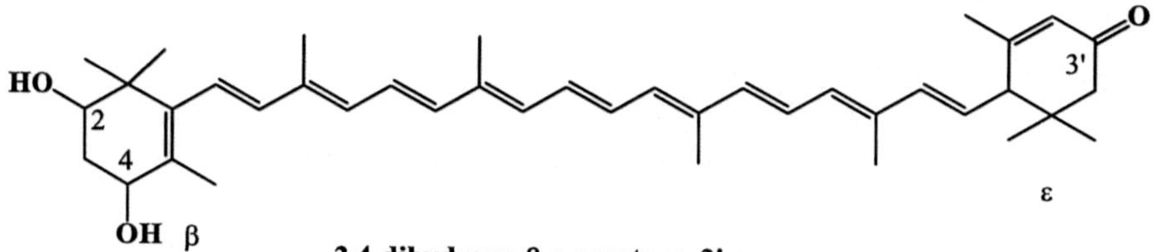

2,4-dihydroxy-β,ε-carotene, 3'-one

Rule 8.2

(A) If the end groups (C9) of the carotenoid are similar the lowest number will be assigned to the group which will be cited as a suffix
(B) If both or more end groups are chosen to be cited as suffix then numbering is determined from the lowest locants for all groups
(C) If no group is chosen to be cited as suffix then the numbering is determined from the lowest locants assigned to be cited as prefixes.

Rule 9
This rule is applied to the conjugated polyene system of the double and triple bond, the word retro (in an Italic form) is used to indicate the shift of proton between two positions.

The position from which transfer of proton occur is preceded, and followed by position number to which proton transfer, this pair positions is hyphenated to the Greek letter as given under Rule 3.

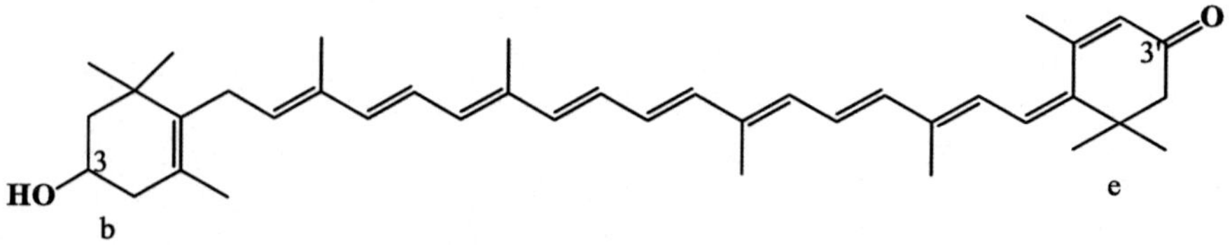

3-Hydroxy-6',7-retro-β,ε-caroten-3'-one

Rule 10

Rule 10 is related to naming apocarotenoid, one or both ends of carotenoids are usually shortened in apocarotenoids and so named accordingly.

The word apo (non-italicized) is used in naming apocarotenoid, however, the apo is preceded by position number at which the skeleton is modified to apocarotenoid. The rest of the skeleton is named as mention in rule 3. The cleaved end group is no more represented by Greek symbols as in Rule 3.

If the end group is shortened by 5 or less skeletal carbon atom, then the end group is termed as acyclic (ψ) end group.

The word diapo is used preceded by two locants in case if both ends of the carotenoid skeleton are cleaved. If both end groups are cleaved unequally, then the lower locant is written unprimed.

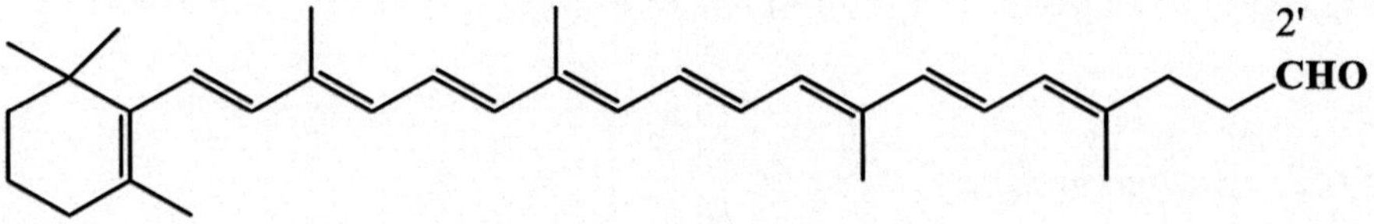

2'-Apo-β-ψ, carotene-2'-al

Rule 11

The higher carotenoids are named as mono or disubstituted while the numbering of normal carotenoid is retained.

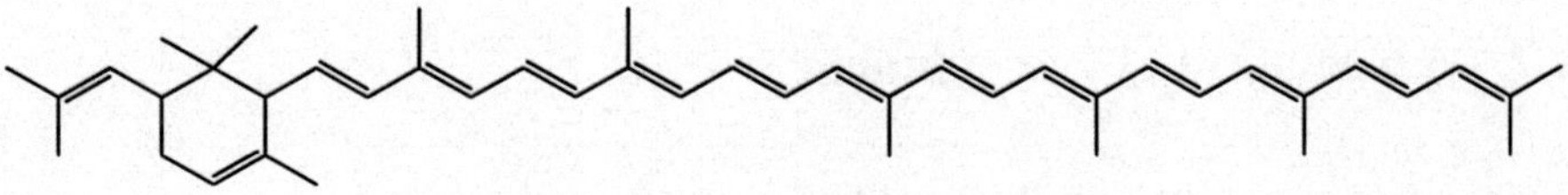

2-(3-Methyl-2-butenyl)-ε,ψ -carotene

Rule 12

The absolute stereo-configuration at the chiral centre is designated by using RS at corresponding locants before the carotenoid name.

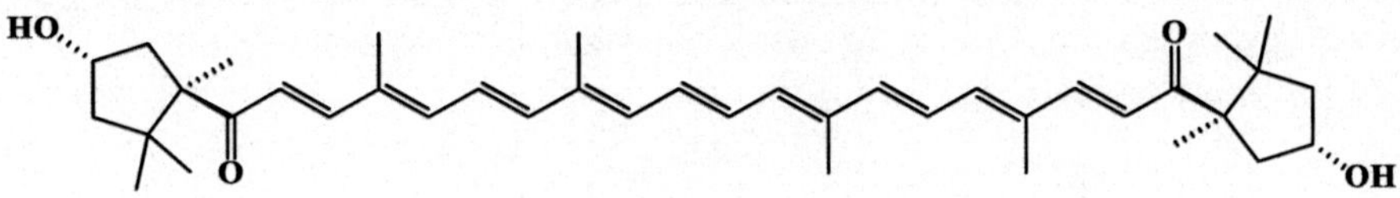

(3S,5R,3'S,5'R)-3,3'-Dihydroxy-κ,κ-carotene-6,6'-dione
(Capsorubin)

2.6 Carotenoids Aggregates

Carotenoids form aggregates with other compounds like through Van Dar Waals, hydrogen bonding, or other dipole interactions. These aggregates have different physicochemical properties than corresponding carotenoids [28].

There are two types of carotenoids aggregates

2.6.1 H-Aggregates

These are formed when carotenoids are in polar hydrated solvents so their unsaturated portion slanted to each other and form aggregate.

2.6.2 J-Aggregates

When the carotenoids molecules CDBs organized in head to tail arrangements and thus loosely bonded.

Both kinds of aggregates are formed in hydro-alcoholic mixtures with critical modifying factor-like pH, temperature, and concentration of carotenoids. These aggregates appear just like brickworks or ladder [28].

2.7 Association of Carotenoids

Carotenoids are found in associations with sugars, proteins, sulphates, and acyl esters.

2.7.1 Associations with Sugars

Carotenoids are associated with sugars like glucose or gentiobiose to form carotenoids glycosides e.g., crocins (crocetin and glucose), decaprenoxanthin diglucoside, astaxanthin glucoside [28–30].

2.7.2 Associations with Proteins

Carotenoids exist in association with proteins termed as carotenoproteins and reported in various animals like crustaceans where astaxanthin and canthaxanthin are part of the carotenoproteins named as crustacyanin in ovaries and eggs, while pink hue in salmon fish [31]. Asteriarubin and blue linckiacyanin have been isolated from starfish [32, 33]. In some plants like carrots, carotenoproteins have also been reported [34, 35].

The association is not usually covalently bonded; however, still, it affects the physicochemical properties of carotenoids. They are bound in precise stoichiometric ratios. Two types of crustacyanins are α-crustacyanin and β-crustacyanin. α-crustacyanin is a parent type of water-soluble carotenoprotein of 320 kDa, which on dissociation yields β-crustacyanin of 40 kDa [36]. Another carotenoprotein, asteriarubin 43 kDa, is isolated from the dorsal skin of starfish [32, 33]. Crustochrin is carotenoprotein contain 20 astaxanthin molecules attached to a protein, isolated from lobster [37, 38]. Dark green Ovoverdin is also present in lobster but ovaries and eggs and associated with lipoglycoprotein [39]. Ovorubin is glycoprotein carotenoprotein of astaxanthin isolated from eggs for river snail having a molecular weight of 330 kDa [40]. Velellacyanin is blue cartenoprotein of astaxanthin isolated from jellyfish similar absorption maximum to α-crustacyanin however the number of subunits and arrangement is different [41]. Phycobilisome is a carotenoprotein of 35 kDa isolated from Cyanobaceria that plays its role as a light-harvesting complex that has two forms one is orange coloured which is non-active and red coloured which is an active form [42, 43]. Peridinin is a carotenoid found in association or complex form with chlorophyll and proteins results in the formation of a light-harvesting complex that presents in dinoflagellates both water-soluble form and membrane-bounded form [44].

2.7.3 Associations with Sulphates

Carotenoids associations with sulphates have been reported mostly in bacteria like caloxanthin 3-sulphate, bastaxanthin D in sponge, and ophioxanthin in zebrafish [45–48].

2.7.4 Association with Fatty Acids

Carotenoids form an ester with fatty acids; the hydroxyl groups of carotenoids take part in bond formation. They are termed as a carotenoid ester or xanthophyll ester. They may be monoester, diester (homo-diester, hetero-diester). The type of esterification is based on the number of hydroxyl group acylated, and the type of fatty acids attached [49].

2.8 Physical Properties of Carotenoids

The rate and extent of carotenoids solubility are varied with solvents. Carotenoids are readily soluble in ethyl acetate, chloroform, and alcohol. Carotenes are readily soluble in petroleum ether, hexane, and toluene; xanthophylls dissolve better in alcoholic solvents like methanol and ethanol. Crystal-like carotenoids may be difficult to dissolve in the mentioned solvents but dissolve in benzene and dichloromethane. In the cell environment, carotenoids are found in the inner core of membranes, which is hydrophobic; however, carotenoid-protein complex permits them entree to aqueous environments. Carotenoids are solid at room temperatures and usually have characteristic colour. Melting points of carotenoids are higher and get higher with an increase in molecular weight and functional groups [50].

2.9 Physico-Chemical Properties of Carotenoids

The physicochemical characteristics of carotenoids in vivo are influenced by various elements, comprising contacts with other molecules in their microenvironment and can differ substantially from free carotenoids present in an organic solvent solution. Hence, extrapolations of in vitro results to in vivo circumstances should be done with care [51].

2.9.1 Melting Points

They are high, generally in the range 130–220 °C and the solubility of crystals generally small, being better in chlorinated organic solvents and benzene.

2.9.2 Size and Shape

The overall molecular geometry (size, shape, presence of functional groups) of carotenoids is central for their integration into cellular and subcellular structures in the right location and alignment for proper functioning. Cyclization shortens the length of long, linear acyclic carotenoids, enhancing the working bulk of the end groups and the space they occupy. The shape and size of Z- and E isomers differ significantly, influencing their properties and biological functions. The E isomers are linear and rigid molecules, while the Z-isomers are bent. The Z-isomers aggregate or crystallize to a lesser extent, thereby exhibiting lower melting points and increased solubility, absorption, and transportation than corresponding E isomers [51, 52].

2.9.3 Z-Isomerization

Z isomerization affects physicochemical properties (solubility, crystallinity, colour, melting point, stability) of carotenoids. Z-isomerization shifts the lambda max towards shorter wavelength, and thus, visibility and colour decreases. Similarly, all z-isomers are more soluble in organic solvents and oils comparative to all-cis-isomers. Z-isomerization also affects the crystallinity of carotenoids, for example, the crystallinity of lycopene, β-carotene, and astaxanthin decrease with z-isomerization. Most of the Z-isomers have lesser melting points than corresponding all-E-isomers. The stability of carotenoids is also affected by z-isomerization. It is reported that most of the all-E-isomers are more stable than z-isomers [53].

2.9.4 Absorption of Light and Colour

The absorption of light by carotenoids is a function of the alternate conjugated π bonds. π electrons of the polyene system are very delocalized and the excited state is of relatively less energy, the excitation energy required is comparatively low, conforming to blue and violet light absorption in the visible region (400–500 nm) and hence exhibit colours in the yellow to red range. Due to their CDB mechanism, carotenoids exhibit UV-visible absorption characteristics having maximum absorption at 3 wavelengths resulting in the tri-peak spectrum. Carotenoids exhibit different optical properties in different solvents, depending on the polarity of the solvent. This property of having a typical tri-peaked absorption spectrum with well-defined maxima and minima is called having high persistence and fine structure. Carotenoids exhibit 3 absorption maximum at 3 wavelengths, generating a 3 peaks spectrum (from 430 to 480 nm); therefore, carotenoids exhibit vibrant colours [54].

It is well-known that at least 7 CDBs are required by a carotenoid to produce colour. Thus, carotenoids with ≤5 CDB such as phytofluene and phytoene, absorb in the UV region and are colourless [55]. As the number of CDB increases, the energy required to promote electron transitions of the valence layer reduces, and the absorption wavelength enhances. An increasing number of CDB increases the maximum absorption wavelength (λ_{max}), changing the colour from light yellow (ζ-carotene, 7 CDB), orange (β-carotene, 11 CDB) to red (astaxanthin, 13 CDB) [54].

The expansion of this CDB leads towards colour intensification; hence, lycopene (11 CDB) yields red colour. Structures with different molecular weights, but with the same chromophore, show identical UV-visible spectra (*e.g.*, β-carotene and zeaxanthin). The absorption spectrum of β-carotene (a carotenoid pigment) includes violet and blue-green light, as is indicated by its peaks at around 450 and 475 nm [56].

The conjugation of carotenoids ended rings have little effect on the bathochromic shift, and it may be due to steric conflict between the ring and the polyene system. Conjugated carbonyl groups usually result in a considerable reduction of fine spectral structure, particularly in a solvent such as methanol [56].

2.9.5 *Effects of Substituents*

The position of the absorption maxima (λ_{max}) is affected by the length of the chromophore, the solvent used for the measurement, the number of C=C, the position of the end double bond in the chain or ring and the taking out of conjugation of one double bond in the ring or eliminating it through epoxidation. However, the maximum absorption of the carotenoids in vivo appears at wavelengths about 10 nm higher corresponding to the same in hexane or ethanol due to its presence in a protein environment or lipid [56, 57].

λ_{max} increases with the length of the chromophore, so that the unCDBs do not significantly affect the spectrum. However, when CDB exists in a ring, because this is not coplanar with the linear polynolic chain, λ_{max} appear at shorter wavelengths compared with non-cyclic carotenoids with the same number from CDB [56].

Carbonyl groups conjugated with the polyene chain also increase the length of the chromophore Thus, the presence of one of these groups in a ring causes the λmax to be located at approximately 10 nm higher while its presence in the chain polyenic causes them to travel at wavelengths in around 30 nm higher. The addition of conjugated C=C in a chromophore displaces the absorption maxima towards higher wavelengths, thus increasing the λ_{max} values. Therefore, the acyclic carotenoid lycopene (11 CDB) is red and absorbs at the longest wavelengths (λ_{max} at 444, 470, and 502 nm). Being also acyclic, β –carotene spectrum has 3 well-defined peaks, at wavelengths much lower than those of lycopene (λ_{max} at 378, 400, and 425 nm). Similarly, colourless phytoene (λ_{max} in hexane = 286 nm) is transformed by 4 dehydrogenation steps to red-coloured lycopene (λ_{max} in hexane = 473 nm). The spectrum shift of 186 nm is due to the addition of 8 C=C, resulting in a chromophore absorbing light at higher wavelengths [56] (Fig. 2.14).

Non-CDBs have little or no effect on the position of the λ_{max}, but a bathochromic shift of 2–3 nm results from the introduction of olefinic bonds some 3 positions away from, and on either side of the chromophore. The forwarded movement to longer wavelengths is illustrated by the absorption spectra of the acyclic carotenoid of increasing chromophore length. When there are carbonyl groups conjugated with the polyenic chain, a bathochromic shift occurs of the maximums, in addition to a loss of structure fine, so that the spectrum of these compounds, such as astaxanthin, cantaxanthin or capsorubin, among others, it is reduced to a symmetric curve or the main band with inflexions on either side [56].

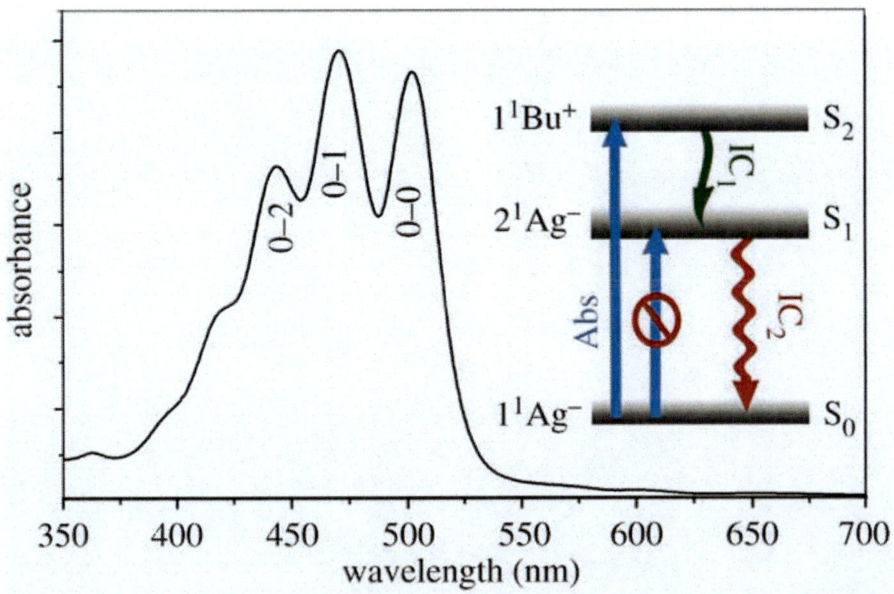

Fig. 2.14 The representative absorption band of carotenoids: lycopene at room temperature in *n*-hexane. Inset: a simplified energy diagram of carotenoids; the blue arrows denote absorption (Abs), which is forbidden for $S_0 \rightarrow S_1$, green and red arrows denote internal conversion by non-radiative decay for $S_2 \rightarrow S_1$ (IC_1) and $S_1 \rightarrow S_0$ (IC_2), respectively [58]

Although the number of CDBs remains constant, cyclization decreases the absorption maxima. The double bonds get enclosed in rings with cyclization and caused a change in the plane of the polyene chain, reducing their colouration. Thus, γ-carotene (1 CDB ring) is reddish-orange, while β-carotene (2 Cdb ring) is orange (carrot), although both have CDB as does lycopene [56].

The inclusion of carbonyl groups destroys high persistence and fine structure. This modification generates a rounded symmetrical absorption peak (for all-*trans* isomers). Other chemical modifications in the basic structures (except the introduction of epoxide groups) have a slight effect on absorption features (Fig. 2.15).

The introduction of epoxide groups decreases the absorption maxima by 10–20 nm (a process termed as hypsochromic shift). The spectrum of cyclic carotenoids, if that the chromophore does not extend to the rings, also presents persistent absorption bands, although when conjugation extends to rings there are steric impediments between the methyl group in carbon 5 of the ring and the carbon 8 of the polyenic chain, which makes the C=C of the rings not coplanar with those of the polyenic chain. As a consequence, a hypochromic displacement of the λmax, a hypochromic effect (decreased absorption) and a loss of fine structure occurs. So, the first band of absorption of two-ring carotenoids, such as â-carotene, α-Cryptoxanthin, and zeaxanthin, is reduced to a mere inflexion [56].

The influence of other substituents such as OH and methoxy groups is negligible; thus, both α-carotene and its dihydroxy derivative, lutein, are pale yellow. Similarly,

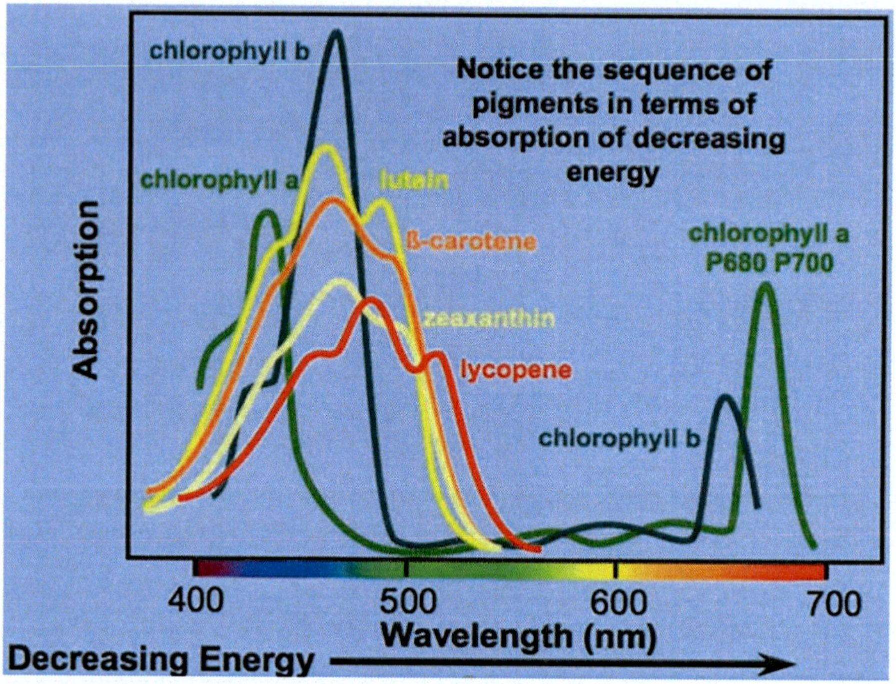

Fig. 2.15 Light absorption spectrum by different pigments [59]

the spectra of α-carotene and its hydroxy derivatives α-cryptoxanthin and zeaxanthin are practically identical. Likewise, the monohydroxy and dihydroxy derivatives of β-carotene, β-cryptoxanthin and zeaxanthin, have the same colour as β-carotene.

On the other hand, the shape of the spectrum and the persistence of the bands of absorption, commonly known as fine structure, reflects the degree of planarity of the chromophore. The CDB system of acyclic carotenoids can adopt an almost planar conformation; hence its spectra present perfectly defined maximums and minimums, although the persistence of the bands decreases when there are more than 9 C=C [56].

The absorption spectrum varies with the functional group's variation and rings. The increase in conjugated rings caused the absorption maximum towards longer wavelengths. Similarly, the hydroxyl groups shift the λmax towards longer wavelength. However, the influence on absorption is lesser with methylation, acetylation, and silylation. Lutein has its maximum absorption at 450 nm, cryptoxanthin at 453 nm, and zeaxanthin at 454 nm [52, 56] (Fig. 2.16).

2.10 Chemical Properties of Carotenoids

Carotenoids undergo various reactions like some properties are given in Table 2.3,

Oxidation
Halogenation
Cyclization

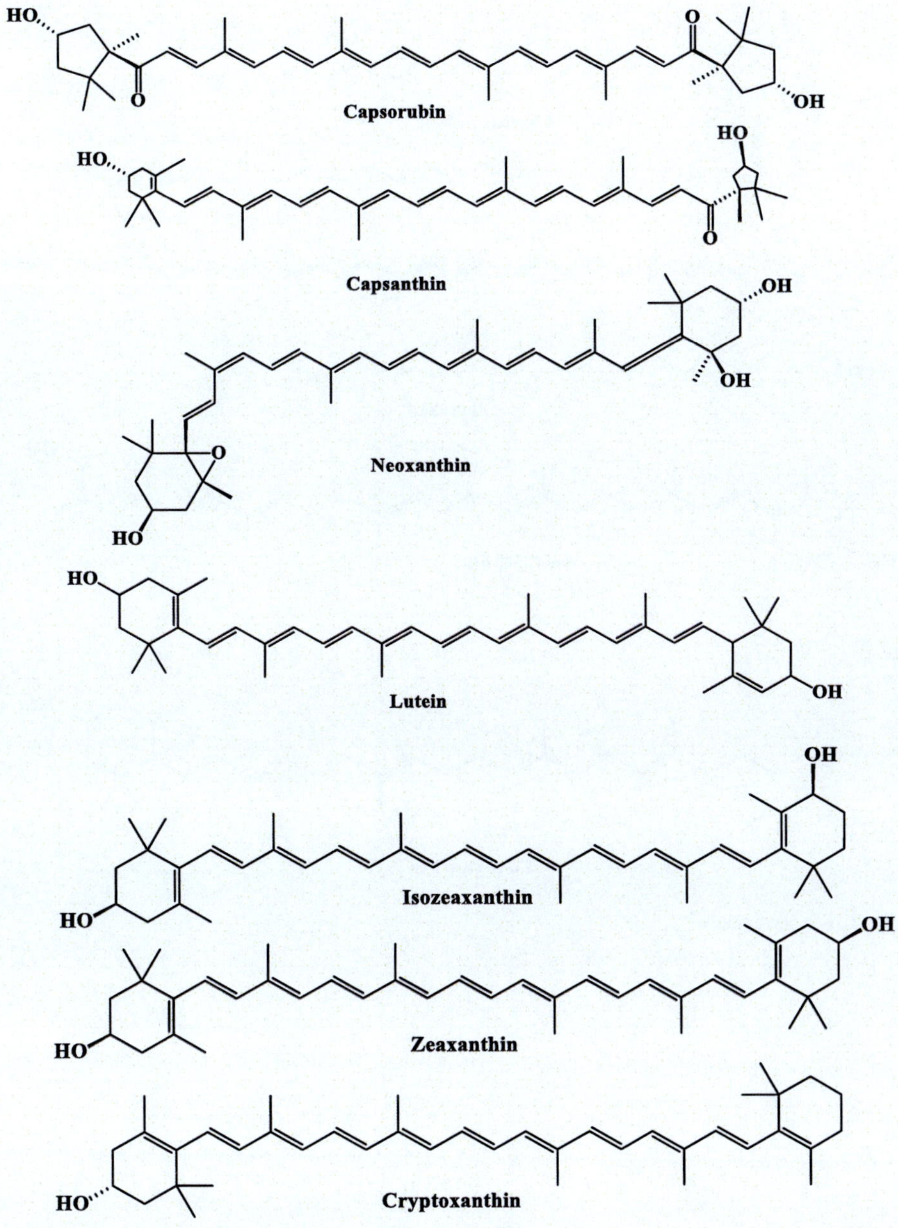

Fig. 2.16 Structures of common carotenoids

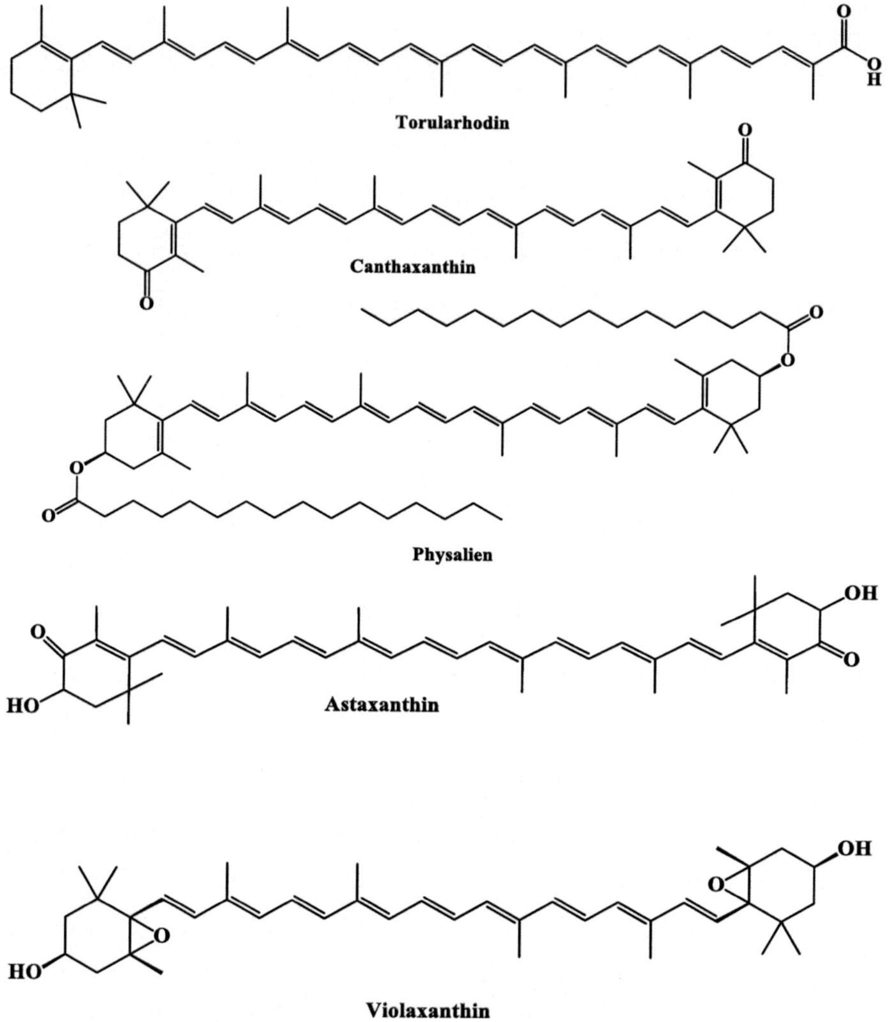

Fig. 2.16 (continued)

Table 2.3 Chemical characteristics of different carotenoids

Carotenoid	Solvent	$A^{1\%}_{1cm}$	Absorption maximum (nm)	Molecular weight	Molecular formulae
α–Carotene	Petroleum ether	2800	424,448,476	536.87	$C_{40}H_{56}$
ß–Carotene	Acetone	2620	430,453,477	536.87	$C_{40}H_{56}$
γ–Carotene	Hexane	2555	439,461,491	536.87	$C_{40}H_{56}$
ß– Cryptoxanthin	Petroleum ether	2386	425,449,476	552.87	$C_{40}H_{56}O$
Echinenone	Petroleum ether	2158	432,459,483	550.86	$C_{40}H_{56}O$
Antheraxanthin	Acetone	2349	424,448,476	584.89	$C_{40}H_{56}O_3$
Canthaxanthin	Petroleum ether	2200	472	564.84	$C_{40}H_{52}O_2$
Astaxanthin	Hexane	2100	468	596.84	$C_{40}H_{52}O_4$
Lycopene	Hexane	3450	446,474,504	536.87	$C_{40}H_{56}$
Lutein	Acetone	2340	427,448,472	568.87	$C_{40}H_6O_2$
Neoxanthin	Acetone	2050	416,438,466	600.87	$C_{40}H_{56}O_4$
Violaxantin	Acetone	2400	443,472	600.88	$C_{40}H_{56}O_4$
Zeaxanthin	Acetone	2340	427,453,477	568.87	$C_{40}H_6O_2$

2.10.1 *Oxidation*

Carotenoids undergo oxidation during storage, processing, or during another metabolic pathway. If carotenoids were used as food additives, then the stability of individual carotenoids varies like lycopene is more unstable compared to β-carotene. Canthaxanthin is more stable than β-carotene. Heat, Light, enzymes, and the presence of oxygen further facilitate the oxidation of carotenoids [50].

The products of oxidation vary as per the oxidizing agent used, e.g., oxygen, ozone, potassium permanganate, and chromic acid. Slow and stepwise degradation occurs in the presence of chromic acid and potassium permanganate [60].

Carotenoids —(Postassium Permanganate or Chromic acid)→ **Apocarotenals** + **Ketones**

β-carotene —(H_2O_2)→ **Retinal**

β-carotene —(MnO_2)→ **Retinal** (**10%**) + **β-apo-10'-carotenal** (2%)

and traces of **β-apo-8'-carotenal**

2.10.1.1 Autoxidation

Carotenoids usually undergo autooxidation during storage, processing, even in solution and crystallized form in the presence of oxygen (air). The presence of oxygen, high temperature, light, and moisture further aggravate the conditions [60].

$$\textbf{β-carotene} \xrightarrow{\text{autoxidation}} \textbf{Mono and diaepoxide of carotene}$$

2.10.2 Halogenation

Carotenoids react with halogens to form halogenated product as

$$\underset{\text{β-carotene}}{C_{40}H_{56}} \xrightarrow{Cl_2} \underset{\text{Halogenated carotene}}{C_{40}H_{56}Cl_{22}}$$

2.10.3 Cyclization

As the name indicates in this reaction, the normal chain molecules of carotenoids convert into a cyclic structure at one or both ends, during biosynthesis enzymes cyclases are responsible for cyclization of the normal carotenoids. In higher plants, these are two types of cyclases that convert lycopene into cyclic compounds having ionone rings. These are lycopene β-cyclase and lycopene ε-cyclase [61, 62]. These two enzymes have a common ancestor, so their nature or sequence of amino acids shows similarity in some points. Another two similar functions bearing enzyme, i.e., cyclases, have been reported, one in tomato and potato named neoxanthin synthase and another in the pepper called capsanthin-capsorubin synthase [63–65]. Capsanthin-capsorubin synthase of pepper catalyzes the formation k-ring [66].

Cyclized carotenoids are less coloured compared to the normal or straight-chain carotenoids. For example, lycopene with 11 carbon double bond absorbs @ 470–500 nm and is powerfully coloured red. But β-carotene, with the same number of double bonds, absorbs @ 450 nm and is orange-yellow due to cyclization.

2.11 Stereochemistry

Carotenoids exhibit stereoisomerism due to the presence of several CDBs and cyclic end groups. These stereoisomers are different in their chemical and physical properties. In stereoisomerism, the geometric (*E-*/*Z-*) type is very common to carotenoids. The position of double bond act as a center while the presence of groups on either side of the double bond led to the existence of isomers either of *E*-configuration with both parts on opposite sides, or of *Z*-configuration with both parts on the same side. These isomers are different in physical properties like melting points, solubility, stability, and colour intensity.

The stereoisomerism in carotenoids is involved in the double bond and absolute configuration of an asymmetric carbon atom. The type of isomerism is geometrical or cis-trans isomerism in carotenoids. The number of cis/trans isomers is directly related to the presence of double bonds; however, in carotenoids the side methyl group poses steric hindrance cause the numbers are practically lesser. For example, there 1056 possible stereoisomers of lycopene however only 72 sterically unhindered cis isomer exists in nature, similarly, the number of sterically unhindered cis isomers are 20 of 272 theoretical one for β-carotene [60].

The number of possible geometric isomer can be calculated as considering the carotenoids as two types symmetrical, having two halves of molecules identical in geometry while in un-symmetrical, they are different. Carotenoids generally occur in more thermodynamically stable form all-E form; however phytoene and phytofluene exist in 15-Z configuration in most of the natural sources. Bixin also exists in Z form. Theoretically, each double bond in polyene chain can undergo isomerization from E to Z however practically it is not the case some of the methyl groups produce hindrance to such isomerization some positions are C-7,8, C-11,12, C-7`,8` and C-11`,12` double bonds [67].

2.12 Conclusion

Carotenoids are isoprenoids, having typically 8 isoprenoid units joined head to tail fashion expect in the centre attachment, which is tail to tail. This second most abundant group of natural compounds isolated from higher plants, fungi, and cyanobacteria. Chemically carotenoids belong to almost all groups like aldehyde, epoxides, ketones, esters, carboxylic acids, hydrocarbons, glycosides, and associated with proteins. They are famous for their common or trivial names; however, the International Union of Pure and Applied Chemistry has set rules to name these compounds scientifically. They are mostly nonpolar and soluble easily in nonpolar solvents. They mostly exhibit stereoisomerism. They show characteristics spectrum under UV-visible spectrophotometer as they have several chromophores in their structures.

References

1. Yabuzaki J (2017) Caotenoids database; structures, chemical fingerprints and distribution among organisms. http://carotenoiddb.jp/. Accessed 28 Sept 2019
2. Namitha K, Negi P (2010) Chemistry and biotechnology of carotenoids. Crit Rev Food Sci Nutr 50(8):728–760
3. Britton G, Liaaen-Jensen S, Pfander H (1995) Carotenoids, Spectroscopy, vol 1. Birkhauser Verlag, Basel
4. Britton G, Liaaen-Jemsen S, Pfander H (1998) Carotenoids, biosynthesis and metabolism. Birkauser verlag, Basel
5. Krinsky NI (1989) Antioxidant functions of carotenoids. Free Radic Biol Med 7(6):617–635
6. Dutta D, Chaudhuri UR, Chakraborty R (2005) Structure, health benefits, antioxidant property and processing and storage of carotenoids. Afr J Biotechnol 4(13):1510–1520
7. Azevedo-Meleiro CH, Rodriguez-Amaya DB (2004) Confirmation of the identity of the carotenoids of tropical fruits by HPLC-DAD and HPLC-MS. J Food Compos Anal 17(3–4):385–396
8. Meléndez-Martínez AJ, Britton G, Vicario IM, Heredia FJ (2007) Relationship between the colour and the chemical structure of carotenoid pigments. Food Chem 101(3):1145–1150
9. Matioli G, Rodriguez-Amaya DB (2003) Microencapsulation of lycopene with cyclodextrins. Food Sci Technol 23:102–105
10. Vargas FD, Lopez OP (2003) Natural colorants for food and nutraceutical uses. Food Chem 4:93–109
11. Weedon B, Moss G (1995) Carotenoids, vol 1A. Birkh‰ user Verlag, Basel, pp 27–70
12. Beltran JCM, Stange C (2016) Apocarotenoids: a new carotenoid-derived pathway. In: Carotenoids in nature. Springer, New York, pp 239–272
13. Harrison EH, Quadro L (2018) Apocarotenoids: emerging roles in mammals. Annu Rev Nutr 38:153–172
14. Stange C (2016) Carotenoids in nature: biosynthesis, regulation and function. Springer, New York
15. Huang F-C, Horváth G, Molnár P, Turcsi E, Deli J, Schrader J et al (2009) Substrate promiscuity of RdCCD1, a carotenoid cleavage oxygenase from Rosa damascena. Phytochemistry 70(4):457–464
16. Mendes-Pinto MM (2009) Carotenoid breakdown products the—norisoprenoids—in wine aroma. Arch Biochem Biophys 483(2):236–245
17. Höckelmann C, Moens T, Jüttner F (2004) Odor compounds from cyanobacterial biofilms acting as attractants and repellents for free-living nematodes. Limnol Oceanogr 49(5):1809–1819
18. Bouvier F, Dogbo O, Camara B (2003) Biosynthesis of the food and cosmetic plant pigment bixin (annatto). Science 300(5628):2089–2091
19. Bouvier F, Isner J-C, Dogbo O, Camara B (2005) Oxidative tailoring of carotenoids: a prospect towards novel functions in plants. Trends Plant Sci 10(4):187–194
20. Fink P, Von Elert E, Jüttner F (2006) Volatile foraging kairomones in the littoral zone: attraction of an herbivorous freshwater gastropod to algal odors. J Chem Ecol 32(9):1867–1881
21. Watson SB (2003) Cyanobacterial and eukaryotic algal odour compounds: signals or by-products? a review of their biological activity. Phycologia 42(4):332–350
22. Kato-Noguchi H, Seki T, Shigemori H (2010) Allelopathy and allelopathic substance in the moss Rhynchostegium pallidifolium. J Plant Physiol 167(6):468–471
23. El-Sayed A (2014) The pherobase: database of pheromones and semiochemicals. See http://www.pherobase.com
24. Macías FA, López A, Varela RM, Torres A, Molinillo JM (2004) Bioactive apocarotenoids annuionones F and G: structural revision of annuionones A, B and E. Phytochemistry 65(22):3057–3063
25. Mi J, Jia K-P, Balakrishna A, Feng Q, Al-Babili S (2019) A highly sensitive SPE-derivatization-UHPLC-MS approach for quantitative profiling of carotenoid-derived dialdehydes from vegetables. J Agric Food Chem 67(20):5899–5907

26. Preedy VR (2012) Vitamin A and carotenoids: chemistry, analysis, function and effects, vol 1. Royal Society of Chemistry, London
27. Commission I (1972) The nomenclature of organic chemistry and IUPAC-IUB Commission on biochemical nomenclature: tentative rules for the nomenclature of carotenoids. J Biol Chem 247:2633–2643
28. Mercadante AZ (2019) Carotenoid esters in foods: physical, chemical and biological properties. Royal Society of Chemistry, London
29. Arpin N, Liaaen-Jensen S, Trouilloud M (1972) Bacterial carotenoids. XXXVIII. C 50-Carotenoids. 9. Isolation of decaprenoxanthin mono-and diglucoside from an Arthrobacter sp. Acta Chem Scand 26(6):2524–2526
30. Yokoyama A, Adachi K, Shizuri Y (1995) New carotenoid glucosides, astaxanthin glucoside and adonixanthin glucoside, isolated from the astaxanthin-producing marine bacterium, Agrobacterium aurantiacum. J Nat Prod 58(12):1929–1933
31. Shahidi F, Brown JA (1998) Carotenoid pigments in seafoods and aquaculture. Crit Rev Food Sci 38(1):1–67
32. Shone CC, Britton G, Goodwin TW (1979) The violet carotenoprotein of the starfish, Asterias rubens. Comparat Biochem Physiol Part B: Comparat Biochem 62(4):507–513
33. Zagalsky P, Haxo F, Hertzberg S, Liaaen-Jensen S (1989) Studies on a blue carotenoprotein, linckiacyanin, isolated from the starfish Linckia laevigata (Echinodermata: Asteroidea). Comparat Biochem Physiol Part B Comparat Biochem 93(2):339–353
34. Milicua J, Juarros J, De Las RJ, Ibarrondo J, Gomez R (1991) Isolation of a yellow carotenoprotein from carrot. Phytochemistry 30(5):1535–1537
35. Bryant JD, McCord JD, Unlu LK, Erdman JW (1992) Isolation and partial characterization of. alpha.-and. beta.-carotene-containing carotenoprotein from carrot (Daucus carota L.) root chromoplasts. J Agric Food Chem 40(4):545–549
36. Cheesman D, Zagalsky P, Ceccaldi H (1966) Purification and properties of crustacyanin. Proc R Soc Lond Ser B Biol Sci 164(994):130–151
37. Buchwald M, Jencks WP (1968) Properties of the crustacyanins and the yellow lobster shell pigment. Biochemistry 7(2):844–859
38. Salares V, Young N, Bernstein H, Carey P (1979) Mechanisms of spectral shifts in lobster carotenoproteins the resonance raman spectra of ovoverdin and the crustacyanins. Biochimica et Biophysica Acta (BBA)-Protein. Structure 576(1):176–191
39. Renstrøm B, Rønneberg H, Borch G, Liaaen-Jensen S (1982) Animal carotenoids—27. Further studies on the carotenoproteins crustacyanin and ovoverdin. Comparat Biochem Physiol Part B Comparat Biochem 71(2):249–252
40. Cheesman D (1958) Ovorubin, a chromoprotein from the eggs of the gastropod mollusc Pomacea canaliculata. Proc R Soc Lond Ser B-Biol Sci 149(937):571–587
41. Zagalsky P, Jones R (1982) Quaternary structures of the astaxanthin-proteins of Velella velella, and of α-crustacyanin of lobster carapace, as revealed in electron microscopy. Comparat Biochem Physiol Part B Comparat Biochem 71(2):237–242
42. Punginelli C, Wilson A, Routaboul J-M, Kirilovsky D (2009) Influence of zeaxanthin and echinenone binding on the activity of the orange carotenoid protein. Biochimica Biophysica Acta (BBA)-Bioenergetics 1787(4):280–288
43. Rakhimberdieva MG, Elanskaya IV, Vermaas WF, Karapetyan NV (2010) Carotenoid-triggered energy dissipation in phycobilisomes of Synechocystis sp. PCC 6803 diverts excitation away from reaction centers of both photosystems. Biochimica Biophysica Acta (BBA)-Bioenergetics 1797(2):241–249
44. Macpherson AN, Hiller RG (2003) Light-harvesting systems in chlorophyll c-containing algae. In: Light-harvesting antennas in photosynthesis. Springer, Dordrecht, pp 323–352
45. Takaichi S, Furihata K, Ishidsu J-I, Shimada K (1991) Carotenoid sulphates from the aerobic photosynthetic bacterium, Erythrobacter longus. Phytochemistry 30(10):3411–3415
46. Matsuno T (2001) Aquatic animal carotenoids. Fish Sci 67(5):771–783

47. Ramdahl T, Kazlauskas R, Bergquist P, Liaaen-Jensen S (1981) Carotenoids from the marine sponge lanthella basta. Biochem Syst Ecol 9(2–3):211–213
48. d'Auria M, Riccio R, Minale L (1985) Ophioxanthin, a new marine carototenoid sulphate from the ophiuroid ophioderma longicaudum. Tetrahedron Lett 26(15):1871–1872
49. Breithaupt DE, Bamedi A (2002) Carotenoids and carotenoid esters in potatoes (Solanum tuberosum L.): new insights into an ancient vegetable. J Agric Food Chem 50(24):7175–7181
50. Borenstein B, Bunnell R (1966) Carotenoids: properties, occurrence, and utilization in foods. In: Advances in food research. Elsevier, New York, pp 195–276
51. Britton G (1995) Structure and properties of carotenoids in relation to function. FASEB J 9(15):1551–1558
52. Minguez-Mosquera M, Gondul-Rojas B, Gallardo-Guerrero L, Chlorophylls MJ-G (2008) Methods of analysis for functional foods and nutraceuticals. CRC Press, Boca Raton
53. Honda M, Kageyama H, Hibino T, Zhang Y, Diono W, Kanda H et al (2019) Improved carotenoid processing with sustainable solvents utilizing Z-isomerization-induced alteration in physicochemical properties: a review and future directions. Molecules 24(11):2149
54. Britton G, Liaaen-Jensen S, Pfander H (1994) Carotenoids, volume 1B: spectroscopy. Birkhäuser, Basel
55. Meléndez-Martínez AJ, Mapelli-Brahm P, Stinco CM (2018) The colourless carotenoids phytoene and phytofluene: from dietary sources to their usefulness for the functional foods and nutricosmetics industries. J Food Compos Anal 67:91–103
56. Mariutti LR, Mercadante AZ (2019) Identification of carotenoids and carotenoid esters. In: Carotenoid esters in foods. MDPI, Basel, pp 322–350
57. Zang L-Y, Sommerburg O, Van Kuijk FJ (1997) Absorbance changes of carotenoids in different solvents. Free Radic Biol Med 23(7):1086–1089
58. Llansola-Portoles MJ, Pascal AA, Robert B (2017) Electronic and vibrational properties of carotenoids: from in vitro to in vivo. J R Soc Interface 14(135):20170504
59. Koning RE (1994) Light reactions. In: Plant physiology information. http://plantphys.info/plant_physiology/lightrxn.shtml. Accessed November 2019
60. Gross J (2012) Pigments in vegetables: chlorophylls and carotenoids. Springer Science & Business Media, New York
61. Pecker I, Gabbay R, Cunningham FX, Hirschberg J (1996) Cloning and characterization of the cDNA for lycopene β-cyclase from tomato reveals decrease in its expression during fruit ripening. Plant Mol Biol 30(4):807–819
62. Cunningham FX, Gantt E (2001) One ring or two? determination of ring number in carotenoids by lycopene ε-cyclases. Proc Natl Acad Sci 98(5):2905–2910
63. Cunningham FX, Pogson B, Sun Z, McDonald KA, DellaPenna D, Gantt E (1996) Functional analysis of the beta and epsilon lycopene cyclase enzymes of Arabidopsis reveals a mechanism for control of cyclic carotenoid formation. Plant Cell 8(9):1613–1626
64. Bouvier F, D'Harlingue A, Backhaus RA, Kumagai MH, Camara B (2000) Identification of neoxanthin synthase as a carotenoid cyclase paralog. Eur J Biochem 267(21):6346–6352
65. Al-Babili S, Hugueney P, Schledz M, Welsch R, Frohnmeyer H, Laule O et al (2000) Identification of a novel gene coding for neoxanthin synthase from Solanum tuberosum. FEBS Lett 485(2–3):168–172
66. Bouvier F, Camara B (1997) Molecular analysis of carotenoid cyclase inhibition. Arch Biochem Biophys 346(1):53–64
67. Zechmeister L, LeRosen A, Went F, Pauling L (1941) Prolycopene, a naturally occuring stereoisomer of lycopene. Proc Natl Acad Sci U S A 27(10):468–474

Chapter 3
Carotenoids Synthesis and Isomerism

Saima Zulfiqar, Shahzad Sharif, Muhammad Zia-Ul-Haq, and Nasir Rasool

3.1 Introduction

Carotenoids, also called tetra-terpenoids, are lipid soluble pigments, hydrophobic in nature, conjugated polyenes and exhibit color range from yellow to red in plants, light exposed tissues of the feeding animals as well as micro-organisms. These are synthesized in plants, while animals get through food. In plants, these are present in plastids and chloroplasts, where these help in stabilizing the structure and functioning of the complex involved in photosynthetic process. Bright color of plants due to presence of carotenoids contributes in pollination, dispersion of seeds and as sexual trait [1, 2].

These are very important class of molecules as they play crucial role in life. These inactivate free radicals in the body and lessen risks of oxidative stress, cancer, atherosclerosis and such other serious problems via reducing lipid peroxidation. These are responsible to determine various lipooxygenase activities, increase immune systems due to break down of their conjugated double bonds [3–7]. As a result, apocarotenoids and diapocarotenoids, produced as by-products during their oxidative cleavage, are being used as additives, vitamin supplements, food products, and cosmetics [8]. As apocarotenoids exist in the form of vitamin A and its derivatives, these play an important role as cell growth regulators throughout life as well as embryo differentiation during embryo development [9].

S. Zulfiqar · S. Sharif (✉)
Materials Chemistry Laboratory, Department of Chemistry, GC University Lahore, Lahore, Pakistan
e-mail: mssharif@gcu.edu.pk

M. Zia-Ul-Haq
Office of Research, Innovation and Commercialization, Lahore College for Women University, Lahore, Pakistan

N. Rasool
Department of Chemistry, GC University Faisalabad, Faisalabad, Pakistan

© Springer Nature Switzerland AG 2021

M. Zia-Ul-Haq et al. (eds.), *Carotenoids: Structure and Function in the Human Body*, https://doi.org/10.1007/978-3-030-46459-2_3

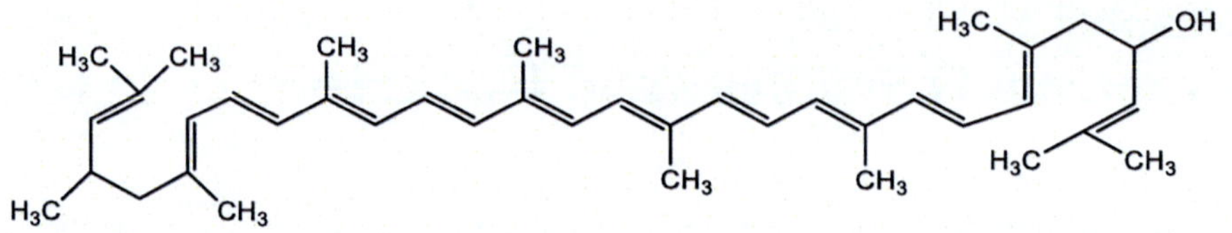

Fig. 3.1 Basic unit of carotenoid

Its basic unit consists of eight isoprene units. Each isoprene unit has 40 carbon atoms, where, single and double bonds in alternate fashion interact with other molecules and absorb excess energy. However, unsaturation increases, absorption maxima also increases [10] (Fig. 3.1).

These are mostly thermodynamically stable in *trans* forms but in some plants, *cis* forms are also present.

3.2 Structure, and Physico-Chemical Properties

Carotenoids, three dimensional molecules, are classified into two main groups: one having hydrogen and carbon in linear as well as cyclic form (β-carotene) having structures in Fig. 3.2)

The other class has hydrogen, carbon and oxygen in molecule (Lutein, and xanthophylls) having structures in Fig. 3.3.

As methyl groups around double bonds in isoprenoid unit cause steric hindrance, so *trans* conformation of carotenoids is more abundant than *cis*. Melting point of the *trans* forms is higher than that of *cis* form as it has high crystallinity and low solubility. Mostly, these are chiral molecules and show optical activity in the presence of plane polarized light, but some are achiral which also show optical rotation [11–13].

Structural rearrangements occur in the molecule of carotenoids due to heat. These show stability upto 50 °C after that these are decomposed. During decomposition, two phenomena occur simultaneously: isomerization and oxidation. As a result, two types of the products are obtained: low molecular weight fraction that is volatile in nature and other larger molecular weight fraction that is non volatile in nature [14]. These show stability in pH range 2–7 [15].

Several processes are responsible for their degradation, such as oxidation, transformation, irreversible rearrangements, etc. As these have three dimensional structures, these different processes halp in the study of their hydrophobic nature, crystallinity, how easily crystal is formed and type of association [16, 17] (Fig. 3.4).

Primary oxidation products include aldehydes and ketones, while secondary products volatile and polymeric compounds.

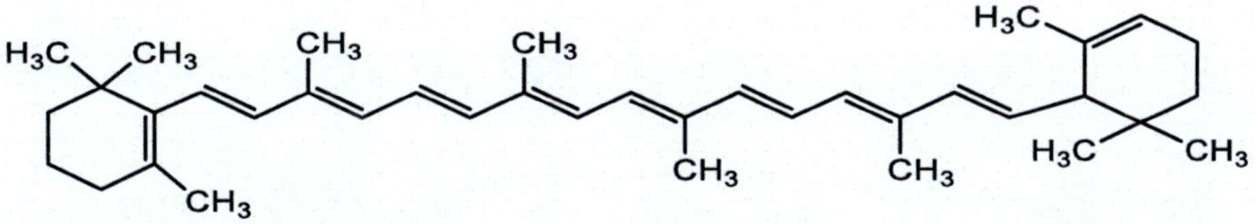

Fig. 3.2 Structures of β-carotene

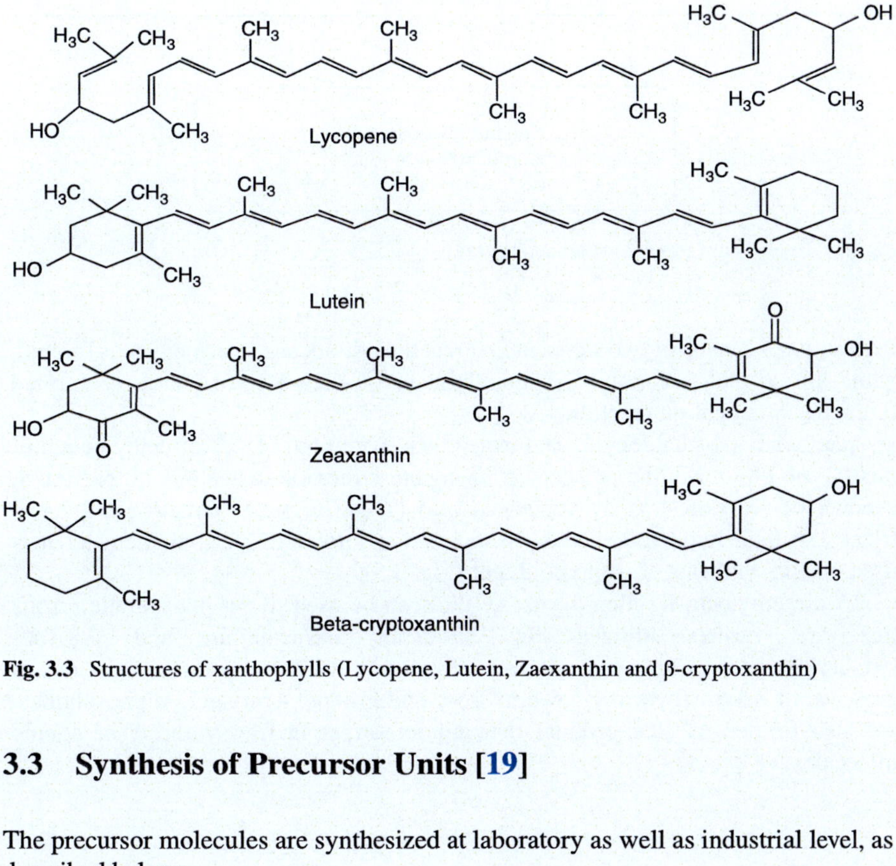

Fig. 3.3 Structures of xanthophylls (Lycopene, Lutein, Zaexanthin and β-cryptoxanthin)

3.3 Synthesis of Precursor Units [19]

The precursor molecules are synthesized at laboratory as well as industrial level, as described below:

3.3.1 Synthesis of Precursor in Laboratory

To synthesize of symmetrical building units, Grignard reactions occurs between acetylene dimagnesium dibromide and 2-methylprop-2-enal, in reaction 1, to get mixture of meso and recemate forms of 2,7-dimethylocta-I,7-dien-4-yne-3,6-diol, that is rapidly changed into dibromide form through reaction with PBr_3 in reaction

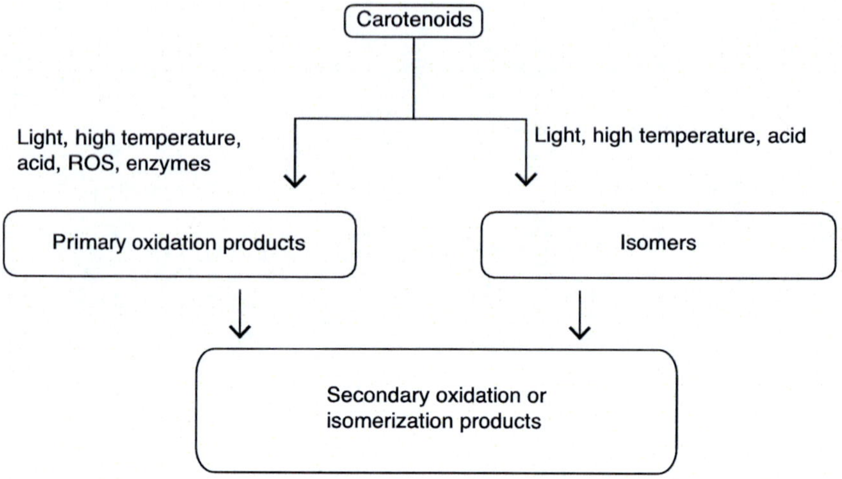

Fig. 3.4 Degradation scheme of carotenoids [18]

2. Reaction 3 consist of two steps: in first reaction, di-acetate intermediate is formed, while in second reaction, 2,7-dimethylocta-2,6-dien-4-yne-l,8-diol is formed through saponification reaction.

Acetylenic C_{10}-dialdehyde is formed via oxidation of 2,7-dimethylocta-2,6-dien-4-yne-l,8-diol in the presence of manganese dioxide in acetone in reaction 4. Product of reaction 4 is hydrogenised into (2E,4Z,6E)-2,7 -dimethylocta-2,4,6-triene-1,8-dialdehyde followed by isomerzation into (all-E)-C_{10}-diaidehyde. It is used for the synthesis of isoprenoid unit (Fig. 3.5).

2,7-dimethylocta-I,7-dien-4-yne-3,6-diol can be used to get phosphonium salts that are further utilized for 7,8-didehydroastaxanthin and 7,8,T,8′-tetradehydroastaxanthin synthesis and carotenoids that have optical activity due to presence of 3,5,6-trihydroxy-5,6-dihydro-γ-end groups. Acetylenic diphosphonate are also formed as side product that are important in the synthesis of allenic molecules.

3.3.2 *Synthesis of Precursor at Industrial Level*

Industrially, C_{10} components are being prepared through three different ways, as following:

3.3.2.1 By Butadiene

For this reaction gaseous phase is used. Here, the dialdehyde is formed though addition of bromine to butadiene followed by formation of isomers of dibromobutane. In gaseous phase, produced (E)-butene-I,4-diphosphonate undergoes condensation

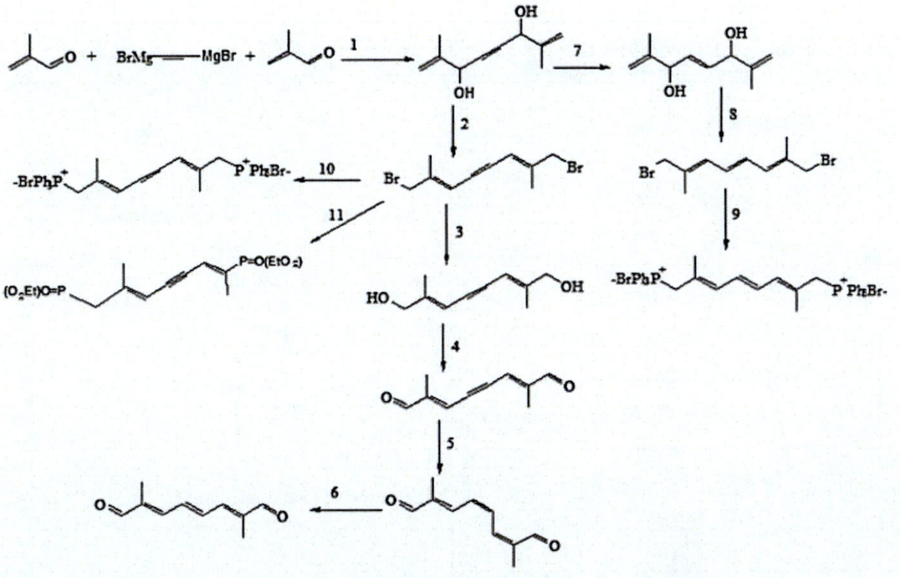

Fig. 3.5 Laboratory Synthesis of symmetrical C_{10} component

reaction with methylglyoxal dimethyl acetal to from C_{10} dialdehyde, as described in figure (Fig. 3.6).

3.3.2.2 By Furan

Bromination of furan occurs in the presence of methanol. Bromonated furan undergoes transformation into but-2-ene-1,4-dial bisdimethylacetal, which undergoes enol ether condensation with I-propenyl methyl ether. Afterwards, hydrolysis and elimination reaction takes place. This process is beneficial due to giving more than 50% yield (Fig. 3.7).

3.3.2.3 By Acetals

Acetals can also be used for this purpose through hydrolysis and condensation reaction (Fig. 3.8).

C_{10} dialdehyde is converted into C_{10} monoacetal which is further utilized as an intermediate for the synthesis of C_{25}-apoaldehyde through hydrolysis (Fig. 3.9).

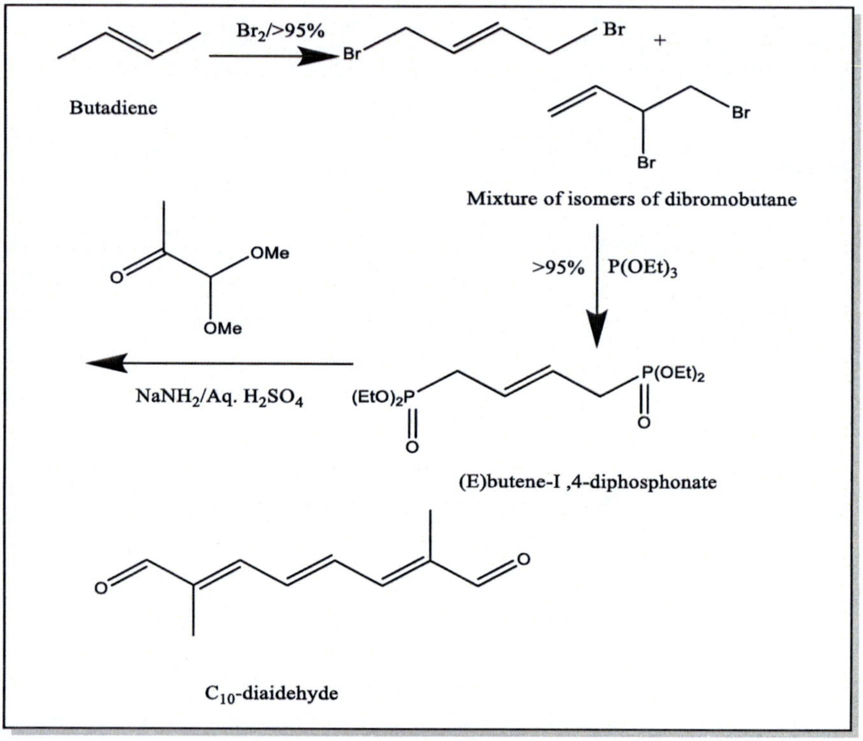

Fig. 3.6 Industrial synthesis of building component from Butadiene

3.3.3 Synthesis of C_{10} Monophosphonium Salt

C_{10}-dialdehyde is reduced into hydroxyaldehyde, via sodium borohydride, being used to preparemethyl ester of natural bixin. It is converted into (all-E)-8-bromo-2,7-dimethylocta-2,4,6-trienal in the presence of NBS/dimethyl sulphide in CH_2Cl_2 at low temperature. Phosphonium salt of (all-E)-8-bromo-2,7-dimethylocta-2,4,6-trienal is obtained by using triphenylphosphine in ethyl acetate. High yield is obtained at the end of reaction (Fig. 3.10).

C_{10}-phosphonium acetals are formed from phosphonium salts of (all-E)-8-bromo-2,7-dimethylocta-2,4,6-trienal as well as important in synthesis of 7,8-didehydroastaxanthin (Fig. 3.11).

3.3.4 Synthesis of Bifunctional Isoprenoid Precursor

For the synthesis of bifunctional building units, three different ways are commonly used;

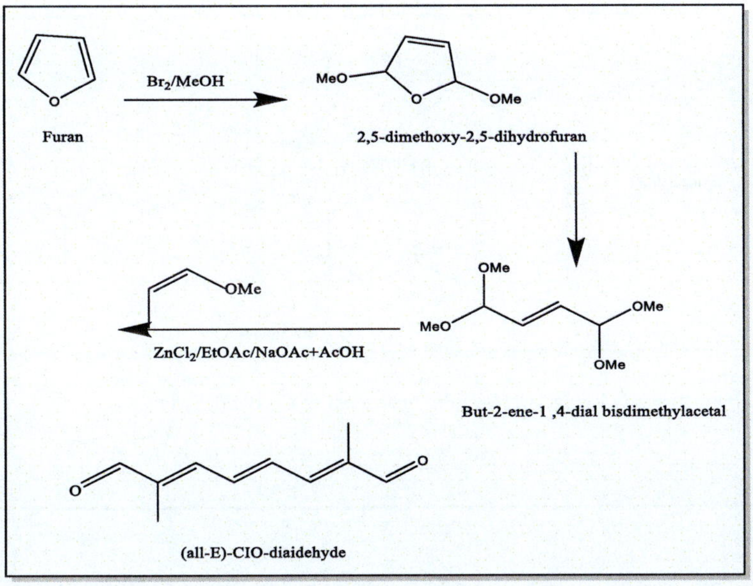

Fig. 3.7 Industrial synthesis of building component from Furan

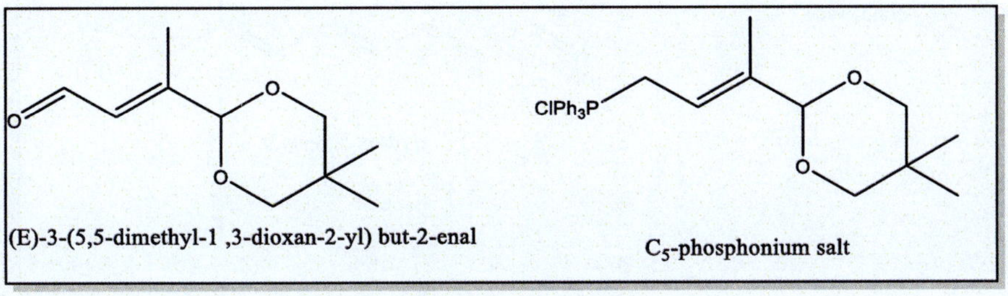

Fig. 3.8 Precursor molecules for the synthesis of C_{10} dialdehyde

1. One of them is (E)-4-Acetyloxy-2-methylbut-2-enal that is further used for the synthesis of Vitamin A. It is synthesized through acetylation in the presence of Copper. In case of butadiene, this molecules is formed as a by-product (Fig. 3.12).

   Industrially, it is also prepared from Acetoxyacetaldehyde and propanal, when acetic acid and aqueous dimethylamine (in equimolar ratio) are used as solvent system (Fig. 3.13).

2. (E)-4-Acetyloxy-2-methylbut-2-enal is used for the synthesis of bifunctional molecule that can be used as building block unit of carotenoids. Methanol as solvent increase the possibility of transesterification, where sodium methoxide is being used as a catalyst. As a result, hydroxyacetal is obtained that under oxidation conditions gives (E)-3-(5,5-dimethyl-1,3-dioxan-2-yl) but-2-enal. This six membered ring molecule does not decompose through oxidation, while acetals

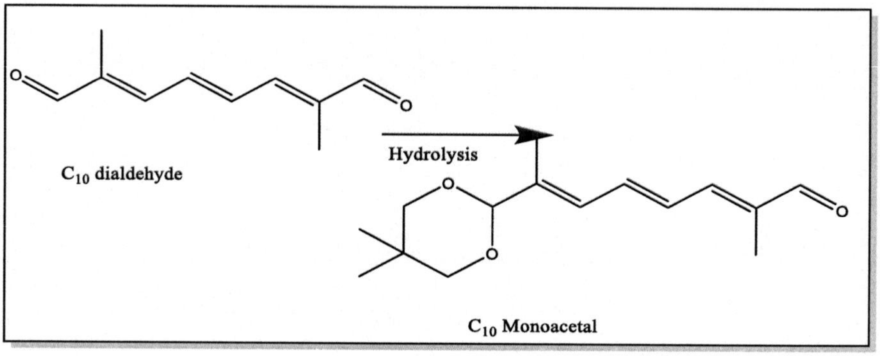

Fig. 3.9 Formation of C_{10} monoacetal from hydrolysis of C_{10} dialdehyde

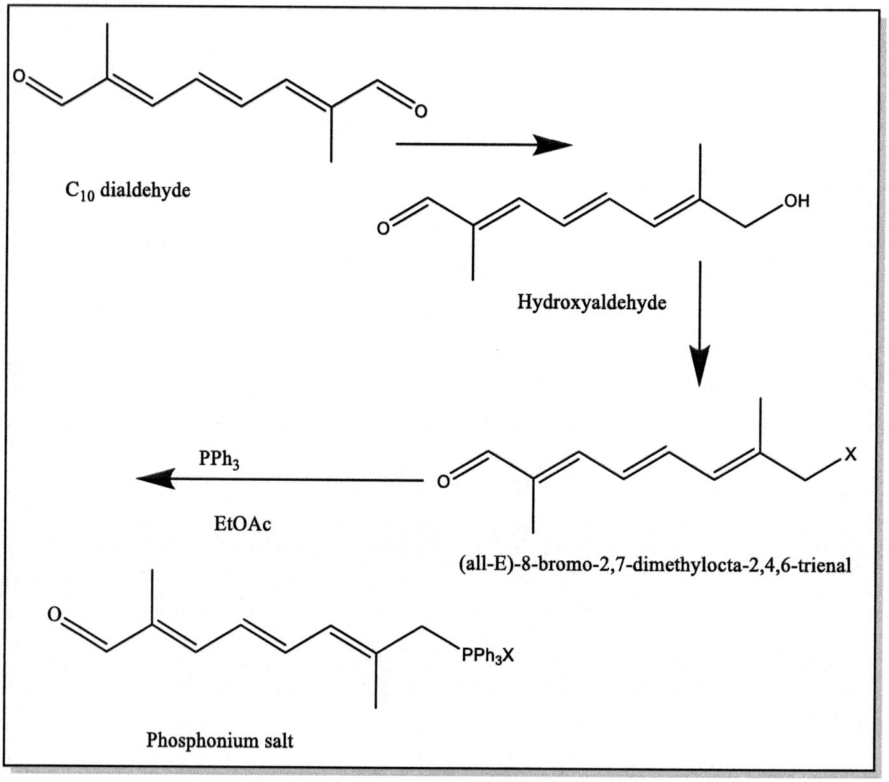

Fig. 3.10 Schematic pathway for Synthesis of C_{10} Monophosphonium salt from C_{10}-dialdehyde

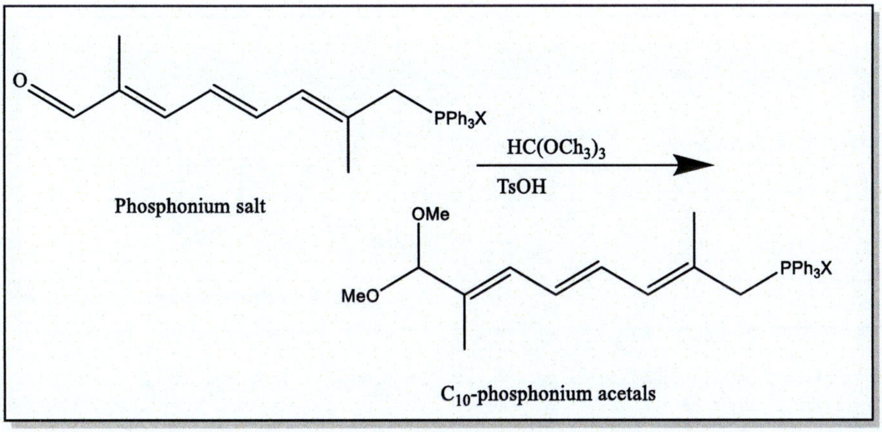

Fig. 3.11 Transformation of phosphonium salts into C_{10}-phosphonium acetals

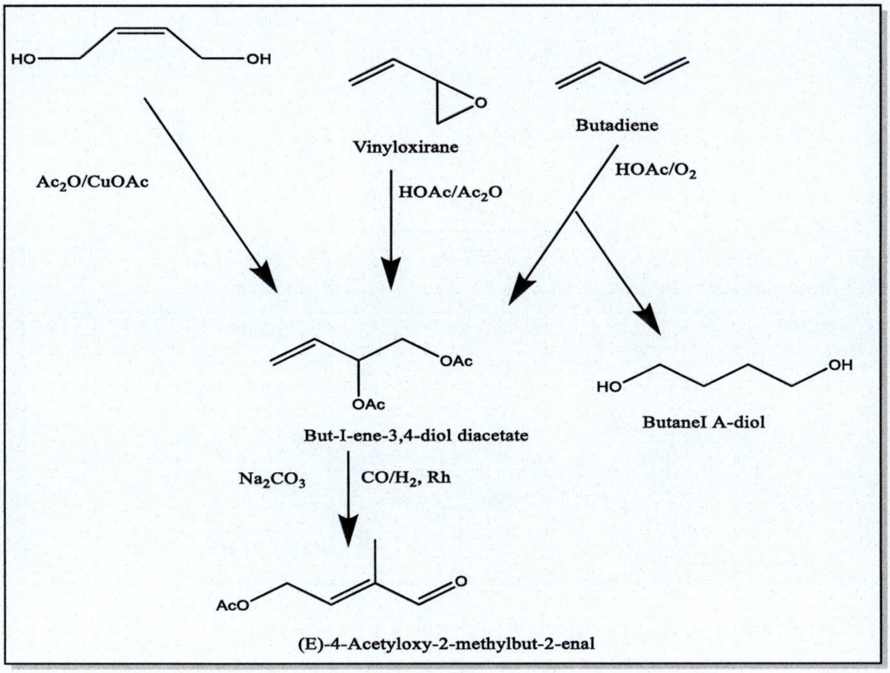

Fig. 3.12 Laboratory synthesis of(E)-4-Acetyloxy-2-methylbut-2-enal

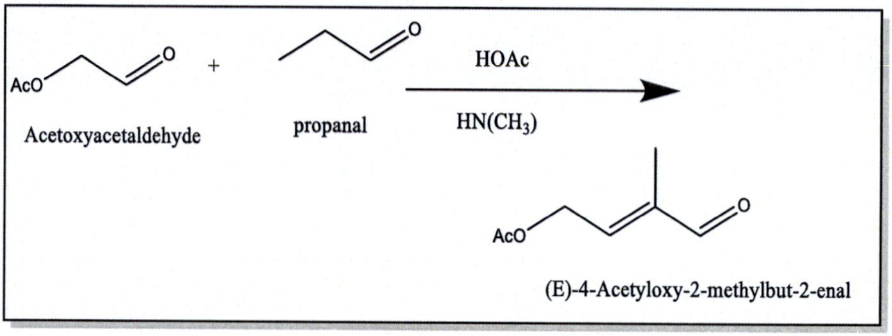

Fig. 3.13 Industrial synthesis of stereoisomers of 4-Acetyloxy-2-methylbut-2-enal

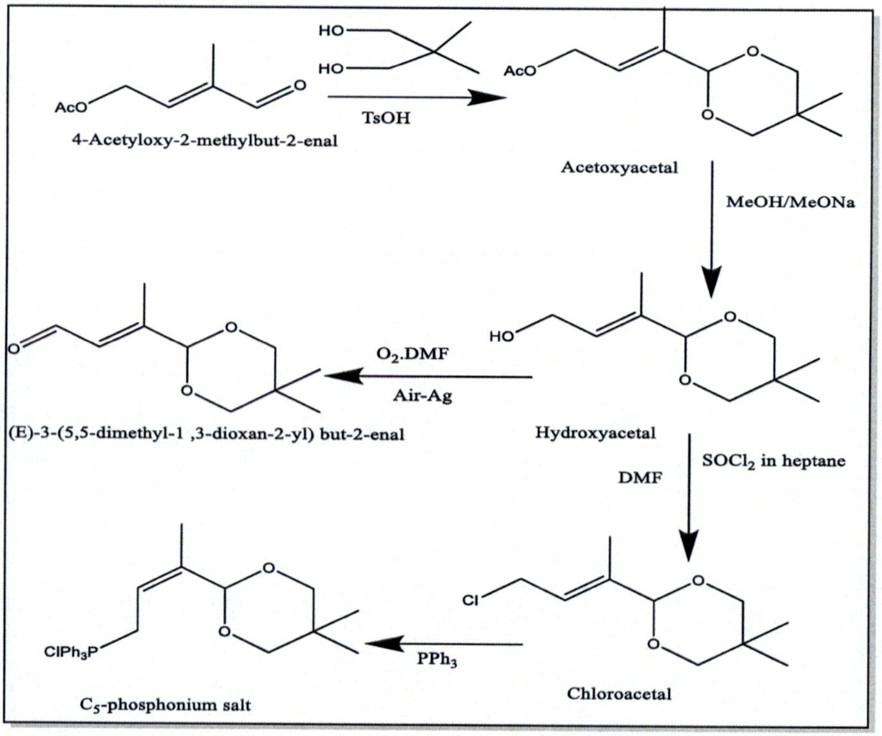

Fig. 3.14 Synthesis of (E)-4-Acetyloxy-2-methylbut-2-enal via oxidation

which have five- and seven-membered rings and all dialkylacetals decompose. Phosphonium salt formed as side product can be used, as a result, 12′-apo-p-caroten-12′-al is obtained (Fig. 3.14)

3. To synthesize carotenoids from polyenecarboxylic acid esters, key building blocks are C_5-phosphonate ester, methyl esters obtained from C_5-phosphonate ester and C_5-phosphonium ester salts

4-bromo-2-methylbut-2-enoic acid ethyl ester is an important building block unit. It is synthesized from bromination of 2-methylbut-3-enoic acid ethyl ester at −20 °C but in the absence of solvent. The product is dehydrogenated in the presence of ethanolic sodium hydroxide. The detailed reactions of synthesis of 4-bromo-2-methylbut2-enoic acid ethyl ester are following (Fig. 3.15):

2-hydroxy-2-methylbut-3-enoic acid ethyl ester is used as primary building block unit that undergoes a series of reactions in synthesis of 4-bromo-2-methylbut-2-enoic acid ethyl ester. This intermediate is synthesized from cyanohydrins under acidic conditions as well as ethanolic hydrolysis of methylvinyl ketone. Phosphonium chloride is prepared fr om 4-chloro-2-methyl-but-2-enoic acid ethyl ester through reaction of this intermediate with thionyl chloride (Fig. 3.16).

4-bromo-2-methylbut-2-enoic acid ethyl ester can be used as precursor for the synthesis of Horner reagent as well as phosphonium bromide salt can be obtained, when 4-bromo-2-methylbut-2-enoic acid ethyl ester reacts with different species (Fig. 3.17).

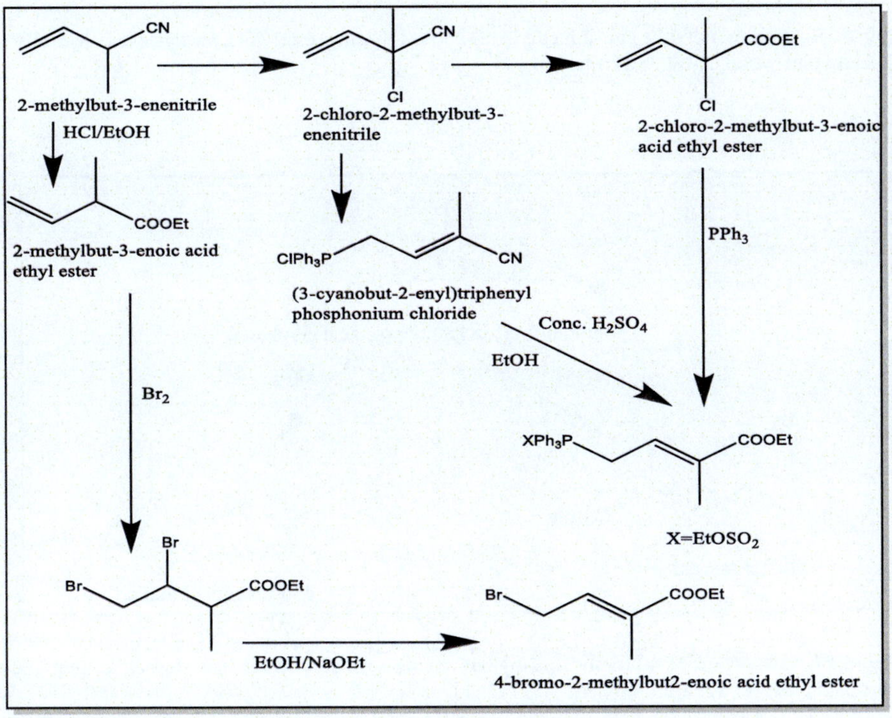

Fig. 3.15 Synthesis of 4-bromo-2-methylbut-2-enoic acid ethyl ester

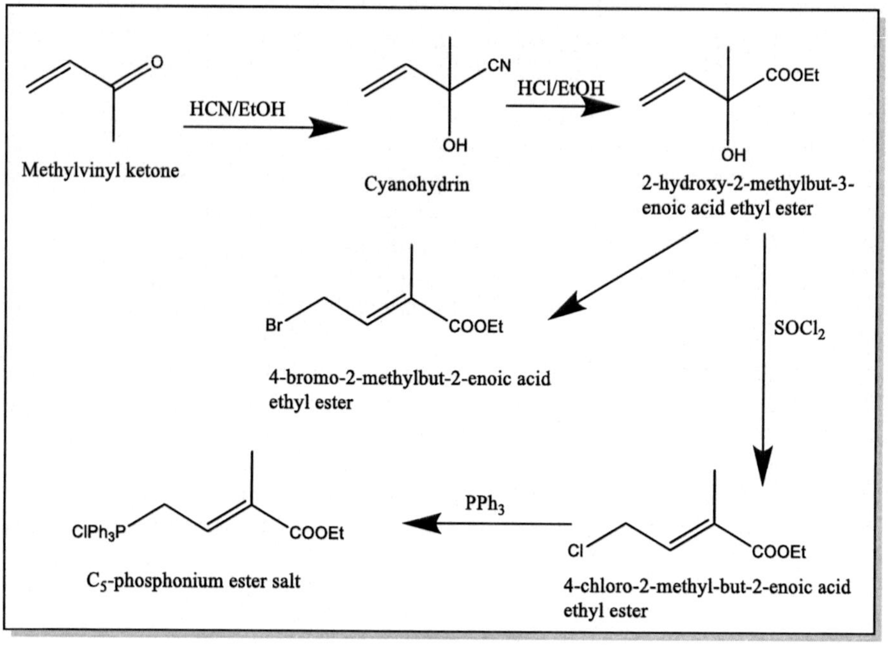

Fig. 3.16 Synthesis of 4-bromo-2-methylbut2-enoic acid ethyl ester via hydrolysis of 2-hydroxy-2-methylbut-3-enoic acid ethyl ester

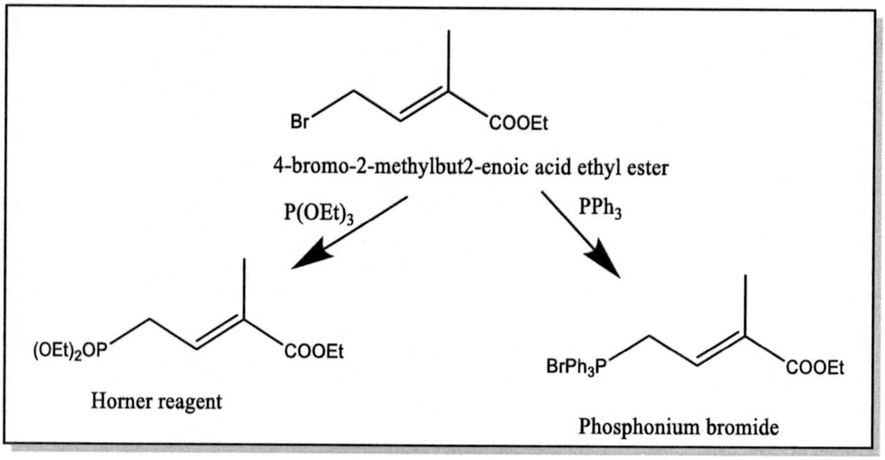

Fig. 3.17 Conversion of 4-bromo-2-methylbut2-enoic acid ethyl ester into Horner reagent and phosphonium bromide salt

3.4 Synthesis of Acyclic Carotenoids

Acyclic carotenoids consist of different end groups, some are below:

3.4.1 3,4-Didehydro-¥-End Group Carotenoids

3,4,3′,4’-Tetradehydrolycopene is acyclic molecule. It is prepared through condensation reaction of 2-methylbut-3-yn-2-ol with diketene. The resultant product undergoes pyrolysis followed by reaction with lithium acetylide forming an alcohol. Triene formed by hydrogenation reacts with triphenylphosphine hydrobromide, as a result, C_{10}-phosphonium salt is produced. C_{10}-phosphonium salt, sodium methoxide and crocetindialdehyde form 3,4,3′,4′-tetradehydrolycopene under Witting reaction [20] (Fig. 3.18).

3.4.2 1,2-Dihydro-¥-End Group Carotenoids

Carotenoids as 1,2-dihydrolycopene and 1, 2, I′,2′-tetrahydrolycopene bear 1,2-dihydro-¥-end group. For their synthesis, the precursor 6-methylhept-5-en-2-one is used. It is hydrogenated by palladium catalyst and condensed with ethyl diethylphosphonoacetate forming α,β-unsaturated ester. $LiAlH_4$ is used to reduce

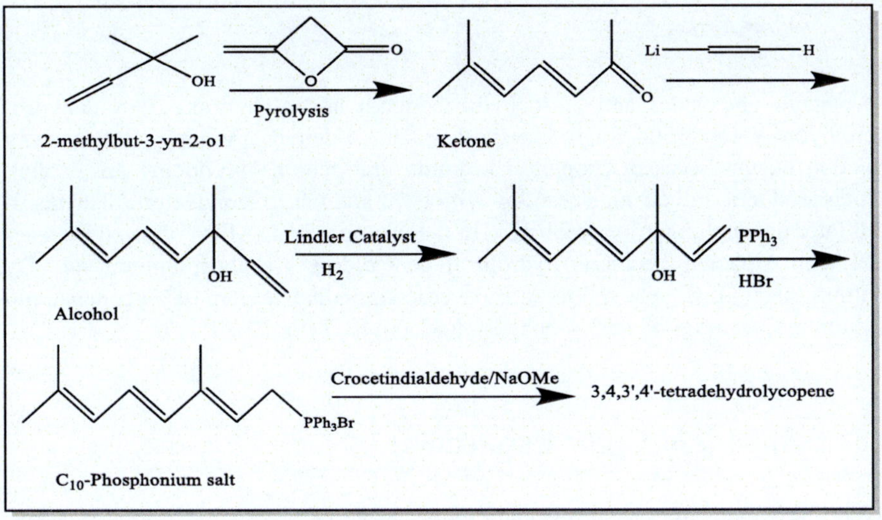

Fig. 3.18 Possible route for the synthesis of 3,4,3′,4′-tetrahydrolycopene

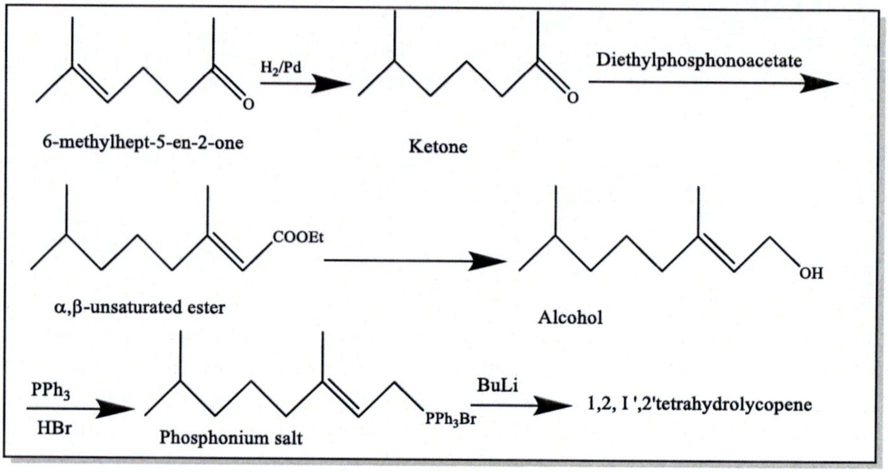

Fig. 3.19 Possible route for 1, 2, I′,2′-tetrahydrolycopene

ester into alcohol, which forms phosphonium salt on reaction with triphenylphosphine hydrobromide. Crocetindialdehyde with phosphonium salt and BuLi yields in 1,2,I′,2′-tetrahydrolycopene via Witting reaction [21] (Fig. 3.19).

For the synthesis of the unsymmetrical 1,2-dihydrolycopene, geranyltriphenylphosphonium bromide is used in place of Crocetindialdehyde.

3.4.3 1-Hydroxy-3,4-Didehydro-L,2-Dihydro-¥-End Group Carotenoids

Carotenoid glycosides and 3,4-dehydrorhodopin have 1-hydroxy-3,4-didehydro-1,2-dihydro-¥-end group. So, for the synthesis of 3,4-dehydrorhodopin, Reformatsky reaction occurs between propargyl bromide and acetone producing an alcohol, which produces diol through reaction with PhLi and methylvinylketone. The resultant intermediate ungerdoes reduction in the presence of $LiAIH_4$ followed by reaction with triphenylphosphine hydrobromide yielding C_{10}-phosphonium salt. By Witting reaction, C_{10}-phosphonium salt, crocetindialdehyde, apo-8′-lycopenal and sodium methoxide produce3,4-dehydrorhodopin [22] (Fig. 3.20).

3.5 Synthesis of Cyclic Carotenoids

Cyclic carotenoids have multiple end groups, as below:

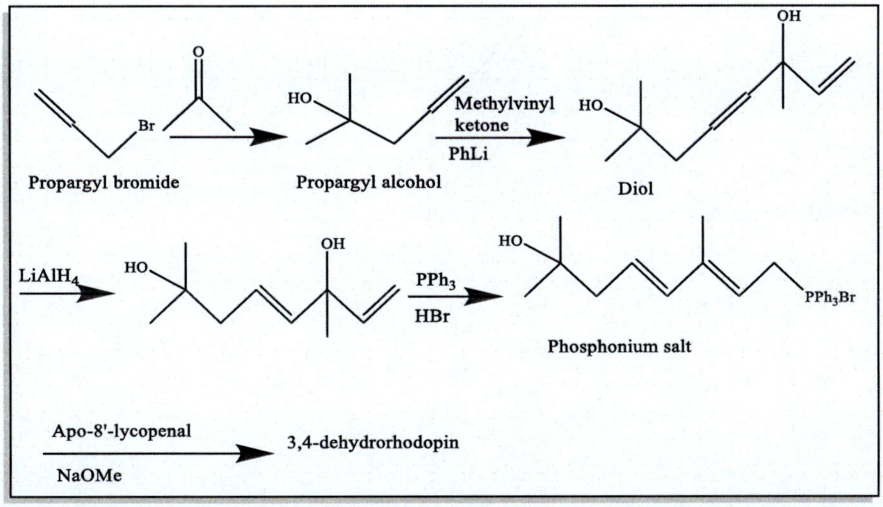

Fig. 3.20 3,4-dehydrorhodopin synthesis

3.5.1 Carotenoids with β-End Group [23]

There are different routes for the synthesis of such carotenoids, some of them are following:

3.5.1.1 Carotenoids with the 4-Hydroxy-β-End Group

Isozeaxanthin has 4-hydroxy-β-end group. It has chirality at two positions C(4) and C(4′). It is synthesized by following reactions steps:1) epoxidation of optically active (−)-(S)-α-ionone withmonoperoxyphthalic acid into cis epoxide; 2) chain is elongated bycamphenyl ester; 3) Horner-Emmons reaction and hydrolysis of camphenyl ester group; 4) use of TMS ether to protect hydroxyl groups, followed by redox reaction; 5) Wittig reaction of the resultant molecule, after which deprotection of the hydroxy groups takes place (Fig. 3.21).

3.5.1.2 3,4-Dihydroxy-β-End Group Carotenoids

Crustaxanthin has 3,4-dihydroxy-β-end group in its molecule. It is synthesized by using synthon (that can be used to synthesize violaxanthin) whose hydroxyl groups are oxidized and followed by opening of the epoxide ring along with Horner-Emmons reaction giving C_{15}-molecule. By reduction of this molecule via 9-borabicyclononane, diastereoisomeric allylic alcohols are obtained. These alcohols are hydrolyzed in the presence of alkaline solution into diols. (Allylic alcohol

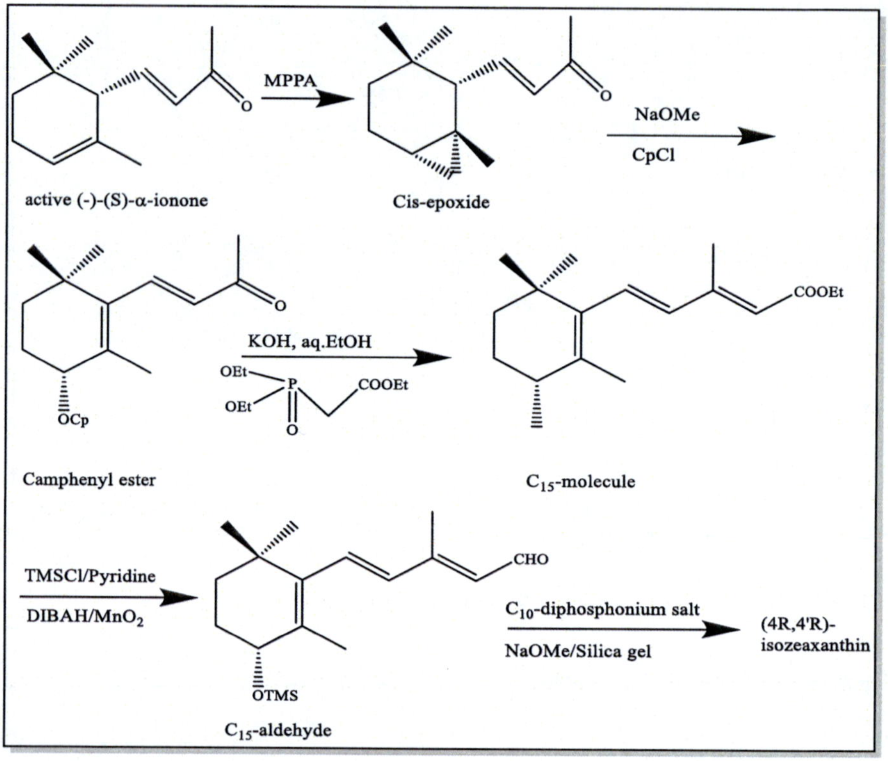

Fig. 3.21 Synthesis of (4R,4R′)-isozeaxanthin

and diols are sepersted via various chromatographic techniques.) TMS ethers is used to protect the hydroxyl groups of the resultant molecule, nitrile group is reduced and an aldehyde is formed.(3S,4R,3'S,4'R)-crustaxanthin isomers are formed after reaction of aldehyde with C_{10}-diphosphonate, partial hydrogenation, isomerization and protecting groups removal (Fig. 3.22).

3.5.1.3 5,6-Dihydroxy-5,6-Dihydro-β-End Group Carotenoids

C_{27}-apocarotenoic acid molecule, commonly called an 'azafrin', bears 5,6-dihydroxy-5,6-dihydro-β-end group. For the synthesis of C_{27}-apocarotenoic acid, (5S,6S)5,6-dihydroxy-5,6-dihydro-β-ionone is used as a precursor molecule that is synthesized from (R)-4-hydroxy-β-ionone through various reactions: its epoxidation into cis-hydroxyepoxy-β-ionone that is further oxidized into hydroxy ketone. During these reacrtions, chirality shifts from C(4) to C(6).Hydroxy ketone is again epoxidized with MCPBA and mixture of diastereoisomers is formed which are reduced with DIBAH and oxidized at C(9), finally building block element (5S,6S)5,6-dihydroxy-5,6-dihydro-β-ionone is formed. At carbon number 2, 5 and

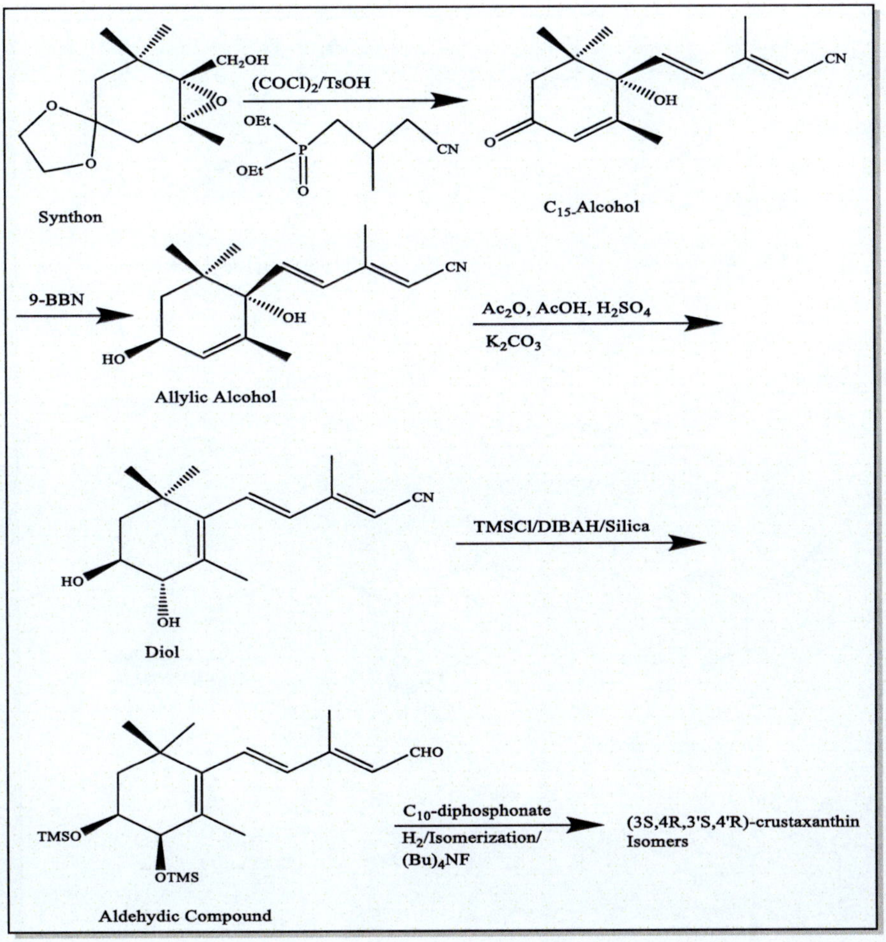

Fig. 3.22 Synthesis of (3S, 4R,3S′, 4R′)-crustaxanthin stereoisomers

7 of precursor, multipleHorner-Emmons reactions lead to formation of (5S,6S)-methyl ester (Fig. 3.23).

3.5.1.4 3,5,6-Trihydroxy-5,6-Dihydro-β-End Group Carotenoids

3,5,6-trihydroxy-5,6-dihydro-β-end group is not commonly present in carotenoids naturally, but rarely in heteroxanthin and mactraxanthin.

For their synthesis, (R)-3-acetoxy-β-ionone is epoxidized, in the presence of MCPBA, intocis- and trans formsin 4:1 respectively. These are separated bychromatography and fractional crystallization. Cis-epoxide undergoes Horner-Emmons reaction and forms C_{15}-ester. Sulphuric acid opens up the ring of epoxide resulting

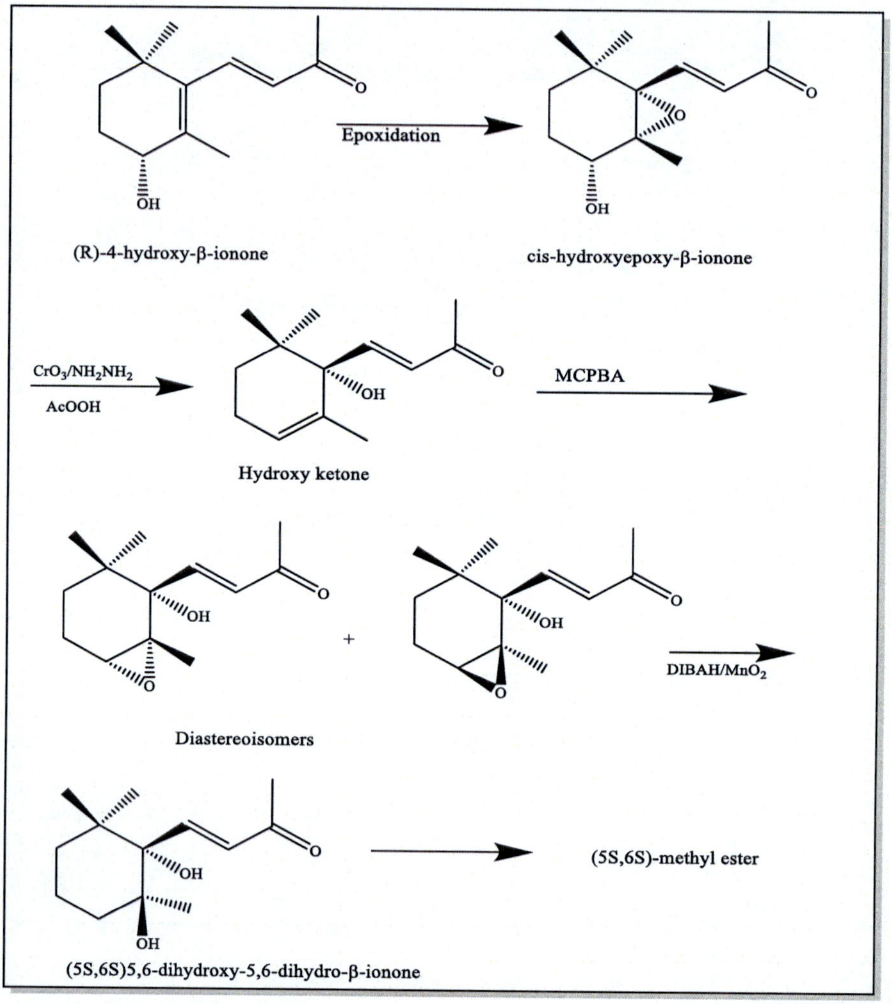

Fig. 3.23 Pathway for the synthesis of (5S,6S)-methyl ester

in formation of dihydroxyester which has inverted configuration at position 5. This dihydroxyester forms an aldehyde, through redox reaction, which reacts with C_{10}-diphosphonium salt as well as undergoes saponification reaction to form mactraxanthin (Fig. 3.24).

3.5.1.5 5,6-Epoxy-5,6-Dihydro-β-End Group Carotenoids

Naturally, two epoxy and dihydro end groups are present in β,β′-Carotene-5,6:5′,6′-diepoxide. Its isomeric forms can be partially synthesized through epoxidation of β,β′-carotene with MCPBA.

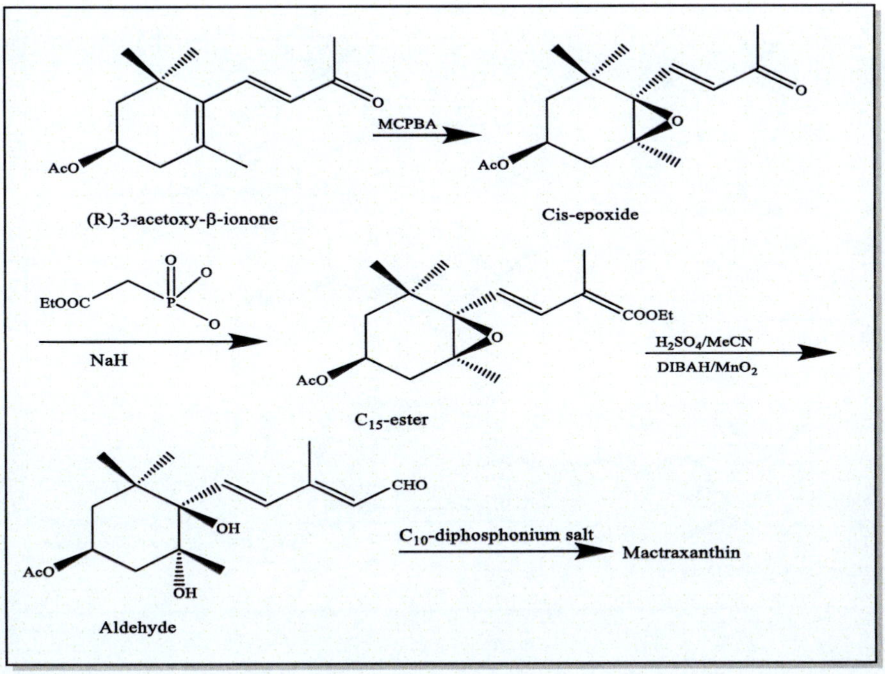

Fig. 3.24 Possible route for the synthesis of mectraxanthin

For the synthesis of mixture of isomers of (5R,6S,5'R,6'S)-5,6:5′,6′-diepoxy-5,6,5′,6′-tetrahydro-β,β′-carotene, C_{15}-aldehyde and C_{10}-diphosphonium salt are used as precursor molecule.C_{15}-aldehyde is synthesized through three step reaction: in first step, cis-hydroxyepoxy-β-ionone is undergone through tosylation, substitution in the presence of iodine and reduction by sodium cyanoborohydrate to form 5,6-epoxy compound; in second step, C_{15}-ethyl ester is formed through Horner Emmons reaction: and in the last step, C_{15}-ethyl ester is reduced to C_{15}-aldehyde, which undergoes Wittig reaction withC_{10}-diphosphonium salt forming the desired molecules isomers which can further be separated through various chromatographic techniques (Fig. 3.25).

3.5.1.6 4-Oxo-β-End Group Carotenoids

A carotenoid, canthaxanthin, contains 4-oxo-β-end group; it can be synthesized via various routes on industrial scale but the best one is sulphone method. In this method, racemic α-ionone forms ketone via three steps: formation of4,5-epoxide via reaction with peroxyacetic acid followed formation of 4-hydroxy-β compound by ring opening and oxidation into ketone. This ketone molecule forms sulphone molecule on reaction with vinyl magnesium chloride as well as sodium sulphinate.

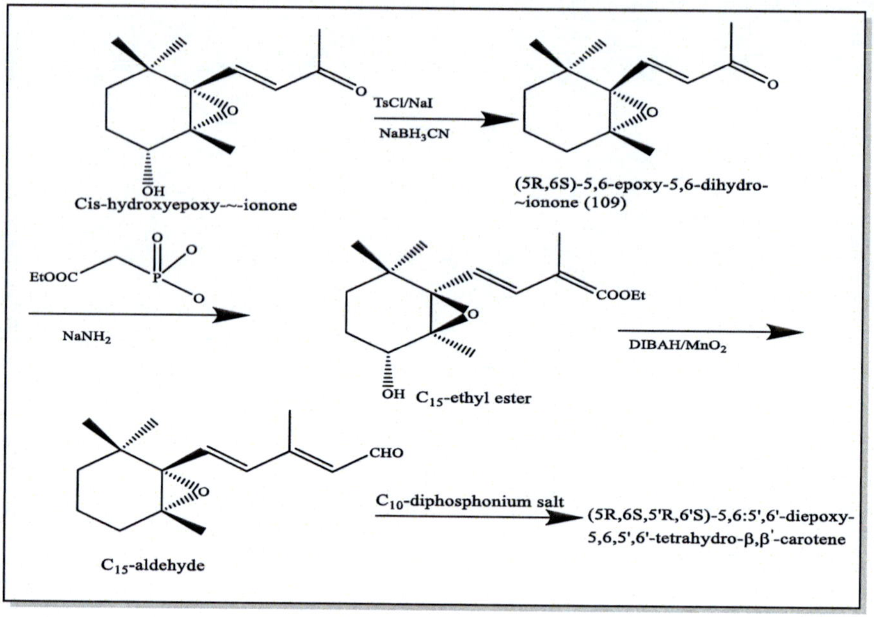

Fig. 3.25 Serial reaction for formation of isomers of (5R,6S,5’R,6’S)-5,6:5′,6′-diepoxy-5,6,5′,6′-tetrahydro-β,β′-carotene

Sulphone is converted into ketal which reacts with C_{10}-dialdehyde and sodium dithionite leading to formation of (all-E)canthaxanthin via isomerization (Fig. 3.26).

3.5.2 Carotenoids with Ɛ*-End Group [24]*

Lutin is an important carotenoid that has Ɛ-end group. For its synthesis, corresponding acetate compound is formed by acetylation of optically active hydroxyketone, a starting molecule, with tetrabutylammonium acetate. This acetate molecule has inverted stereochemistry at C(3). This is hydrolyzed in the presence of isopropenyl ether and the product is converted into epoxide with dimethylsulphonium methylide. In the reaction of epoxidation, axial C-C bond is formed through attack of ylide on carbon-oxygen double bond preferably from one side as protecting groups are in an equatorial position. Epoxide upon Grignard reaction, forms aldehyde, which after chain via HornerEmmons reaction, reacts with MeLi as well as undergo hydrolysis. As a result,3-hydroxy-ɑ-ionone is formed.3-hydroxy-ɑ-ionone on reaction with vinylmagnesium chloride forming C_{15}-compound. Phosphonium salt is formed by acetylation as well as transformation of C_{15}-compound (Fig. 3.27).

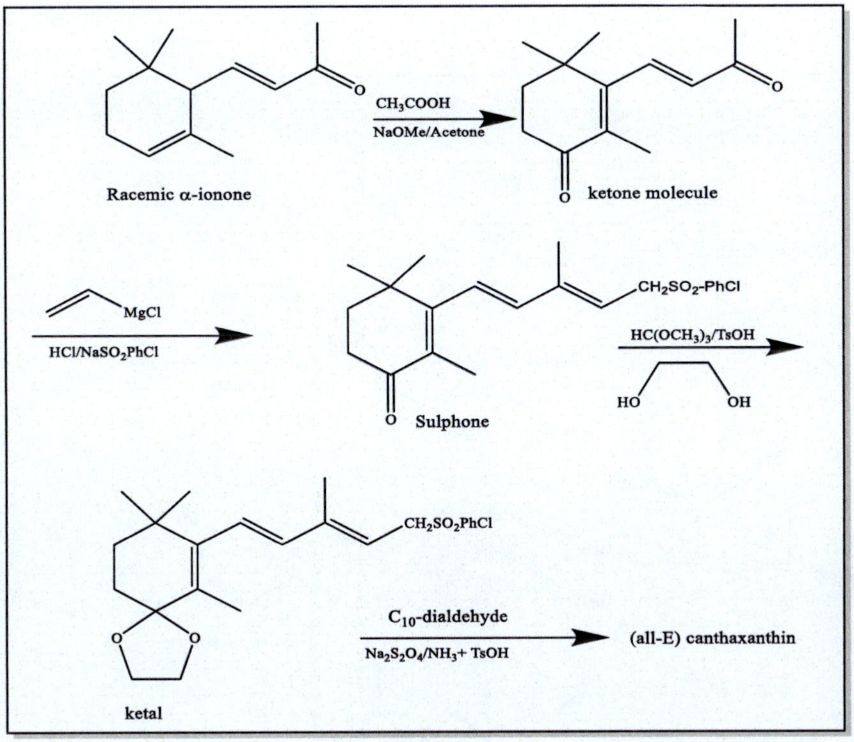

Fig. 3.26 Stepwise synthesis of (all-E) canthaxanthin

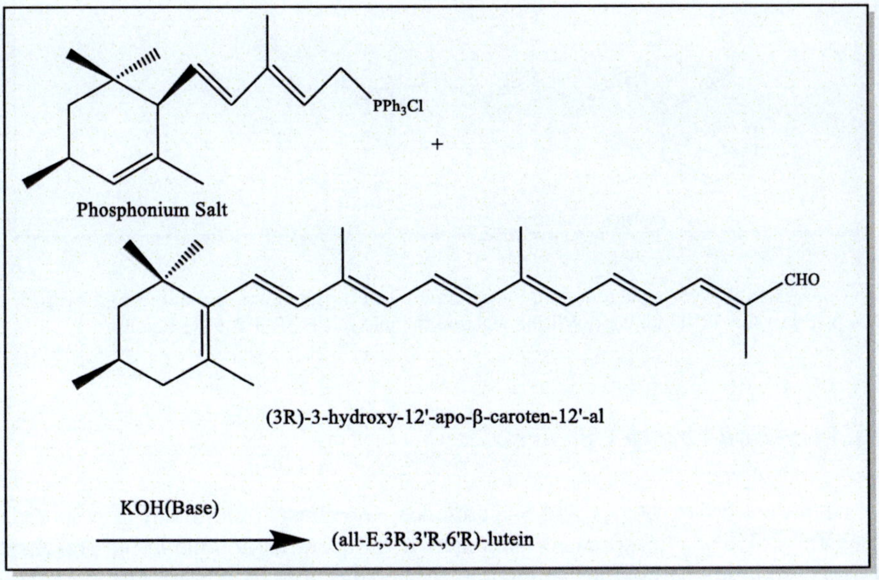

Fig. 3.27 Formation of (all-E,3R,3'R,6'R)-lutein via condensation mechanism

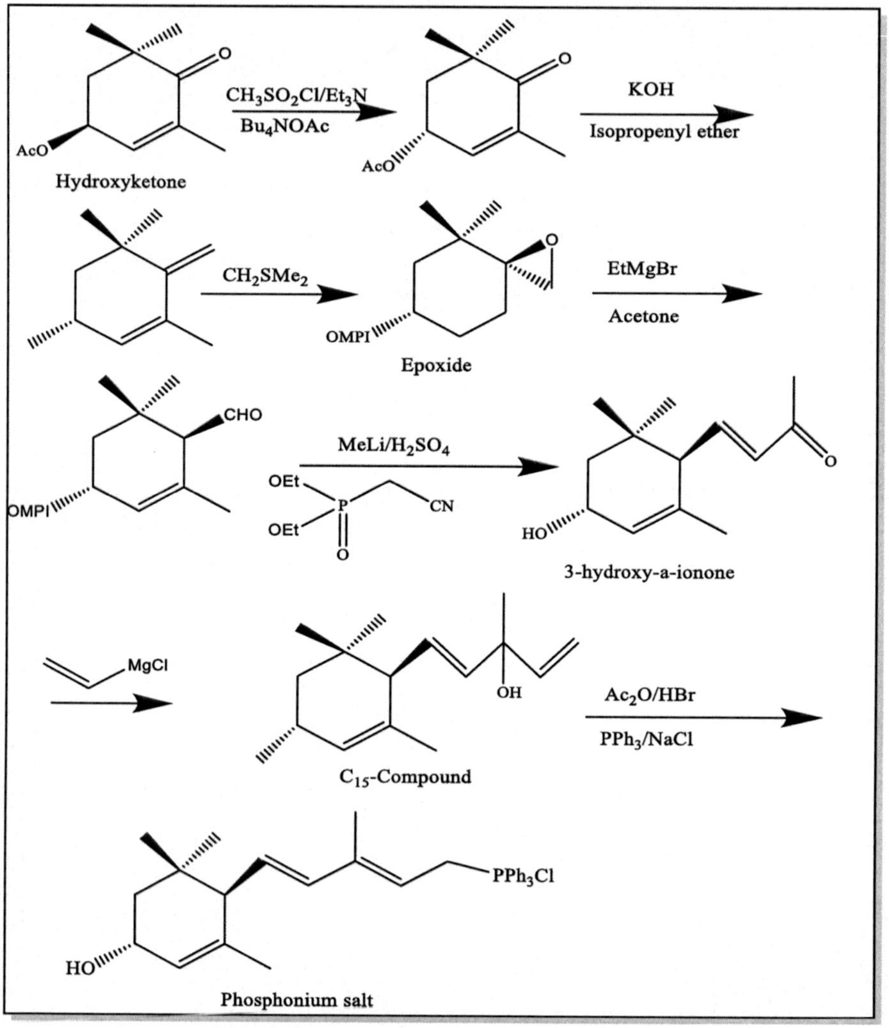

Fig: Conversion of hydroxyketone into phosphonium salt hosphonium salt and (3R)-3-hydroxy-12′-apo-β-caroten-12′-al undergo Witting reaction to yield (all-E,3R,3'R,6'R)-lutein

3.5.3 γ-End Group Carotenoids

To synthesize β,γ-carotenes and γ- carotenes, C_{15}-phosphonium salt is prepared, when ethyl diethyl phosphonoacetate reacts with racemic γ-ionone in the presence of base, sodium methoxide, following Homer-Emmons reaction. The obtained

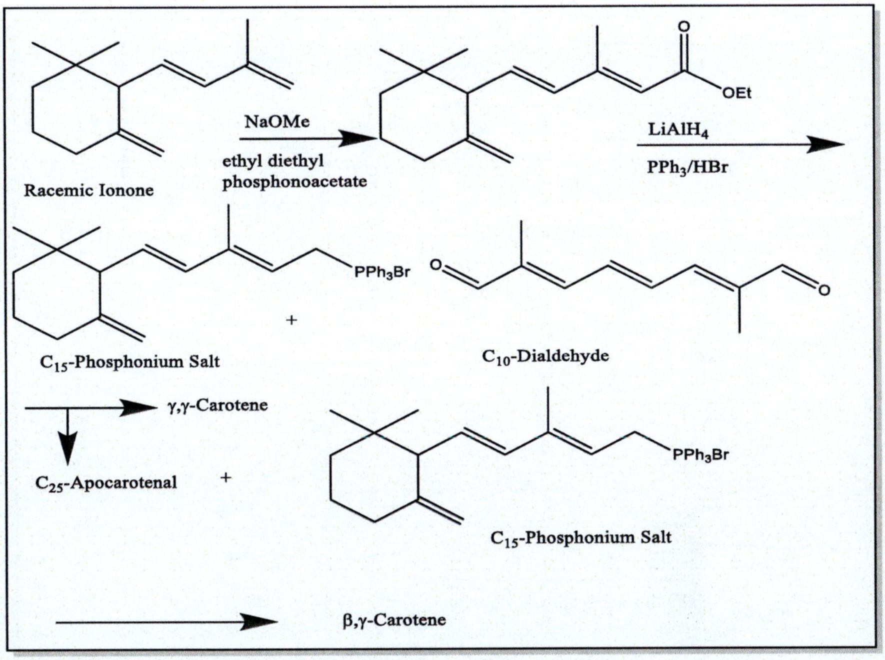

Fig. 3.28 Possible pathway for the synthesis of recemic mixture of β,γ-carotene

product is reduced with $LiAlH_4$ and triphenylphosphine hydrobromide to C_{15}-phosphonium salt (Fig. 3.28).

γ,γ-Carotene is synthesized through reaction C_{15}-phosphonium salt between C_{10}-dialdehyde, producing C_{25}-apocarotenal as by product, which with C_{25}-apocarotenal through Witting reaction produces β,γ-carotene in recemic form [25].

3.5.4 Carotenoids with κ-End Group

Capsorubin, capsanthin and cryptocapsin contain characteristic 3-hydroxy-κ end group. For their synthesis, trans-C_{10}-hydroxyketone is used as precursor. It is prepared by two ways: in former way, camphor undergoes oxidation, esterification withdimethyl sulphate, saponification and decarboxylation to form unsaturated ester. This ester is reduced, methylated and acetaliazed via(+)-diisopinocamphenylborane to form trans-C_{10}-hydroxyketone (Fig. 3.29).

In the other way, (3R)-3-hydroxy-p-cyclocitral (71) under multiple reactions and rearrangements to give trans-C_{10}-hydroxyketone (Fig. 3.30).

In the presence of base sodium hydride, trans-C_{10}-hydroxyketone is condensed with crocetinsialdehyde, following aldol condensation, to form capsorubin [26] (Fig. 3.31).

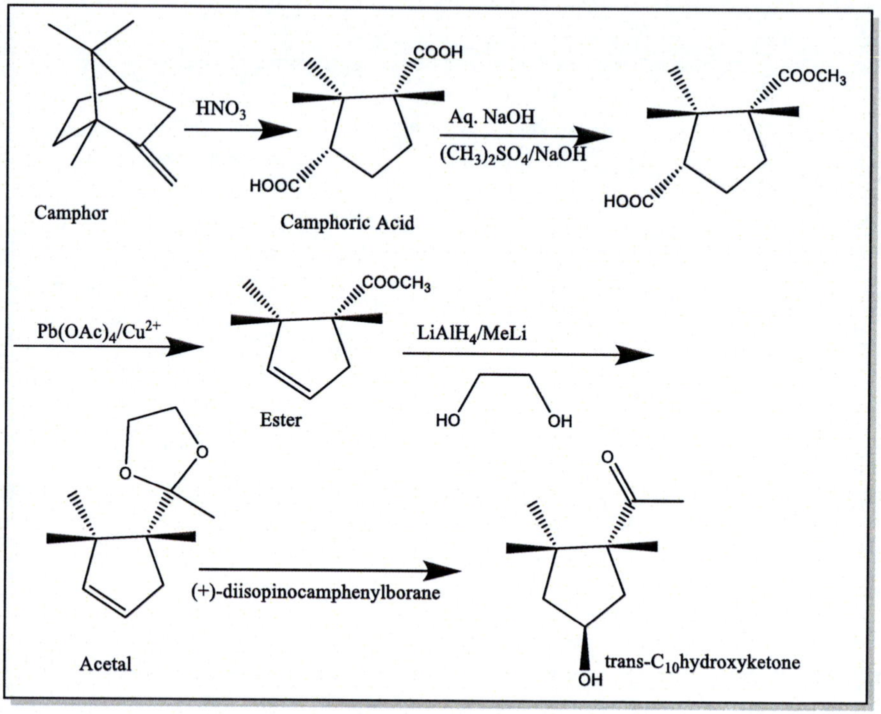

Fig. 3.29 Trans-C_{10}-hydroxyketone synthesis via oxidation, reduction and boration of camphor

3.5.5 *Carotenoids with Aromatic End Group [27]*

Symmetrical carotene isorenieratene and renieratene have aromatic end group. For their synthesis, there are two main methods:

1. 2,3,6-Trimethylbenzaldehyde, as a starting material, is condensed with acetone following aldol condensation to give C_{13}-compound. Its chain is elongated by propargyl bromide to C_{16}-alcohol. C_{16}-alcohol reacts with diketone in the presence of Grignard reagent and the obtained product undergoes a number of reactions: partially hydrogenation, dehydration and isomerization, as a result, isorenieratene is produced (Fig. 3.32).

2. 2,3,6-trimethylbenzaldehyde is also used as precursor; in this reaction, C_{16}-Alcohol is formed by reduction of aldehyde group 2,3,6-trimethylbenzaldehyde of in the presence of $LiAIH_4$ followed by formation of corresponding phosphonium salt which reacts with crocetindialdehyde under basic conditions of BuLi through Witting reaction (Fig. 3.33).

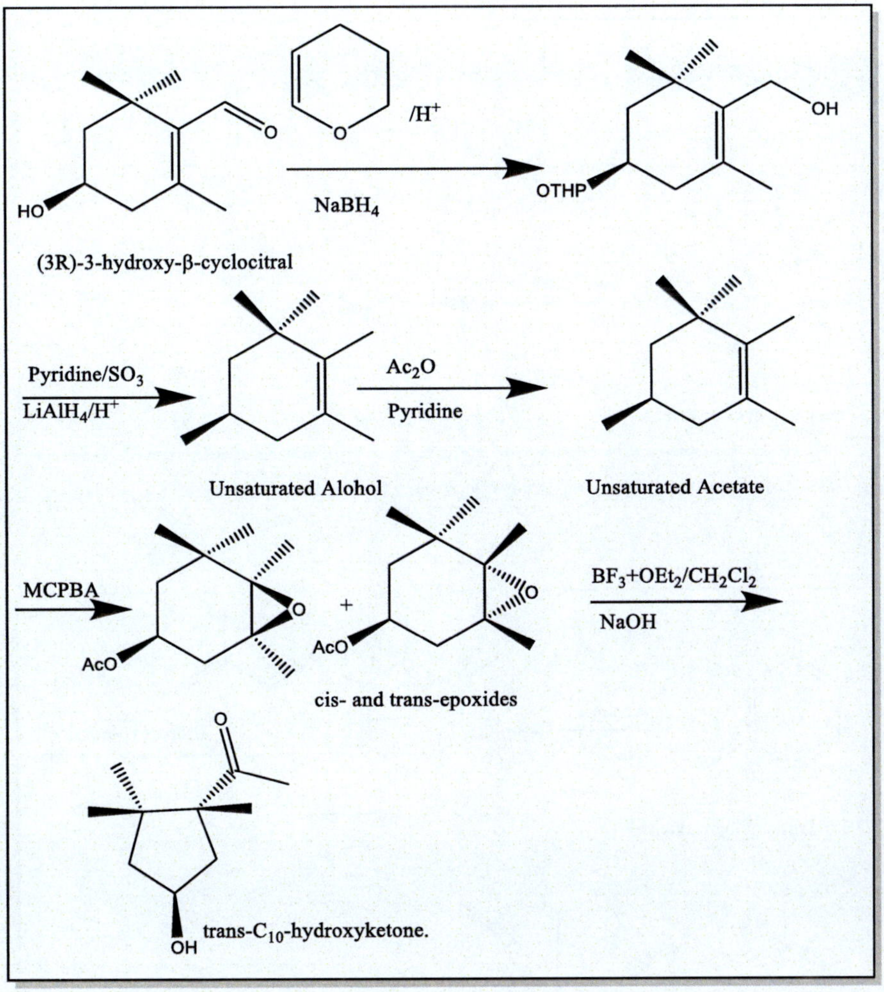

Fig. 3.30 Trans-C_{10}-hydroxyketone synthesis via multiple reaction mechanisms

3.6 Synthesis of Acetylenic Carotenoids

Acetylenic carotenoids have an acetylene group in the C(7)- position. Alloxanthin and xanthin isomers are examples of acetylenic carotenoids, which are synthesized by using an intermediate 3-methylpent-2en-4-yn-l-ol (It contains C_6-unit) (Fig. 3.34).

A diol specie is prepared through Grignard reaction between ketone and 3-methylpent-2en-4-yn-l-ol. This product is hydrolyzed, reduced in the presence of sodium borohydride, acetylate and dehydrated with $POCl_3$ into diacetate product.

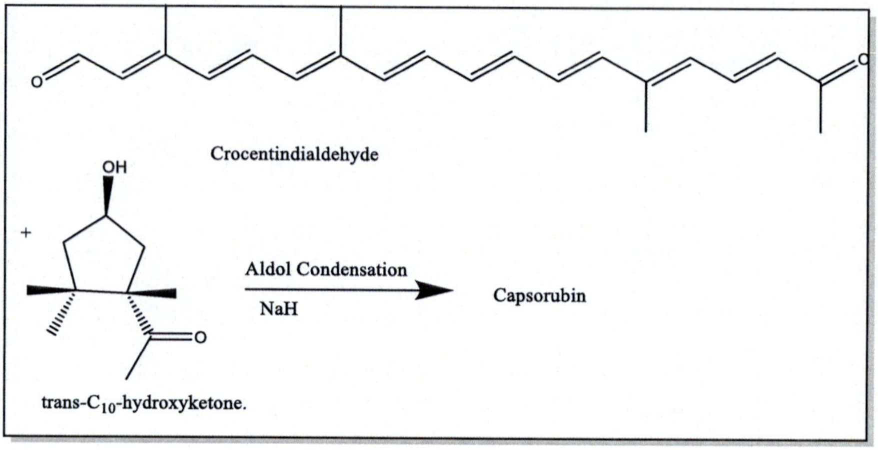

Fig. 3.31 Condensation of trans-C_{10}_hydroxyketone and crocentindialdehyde

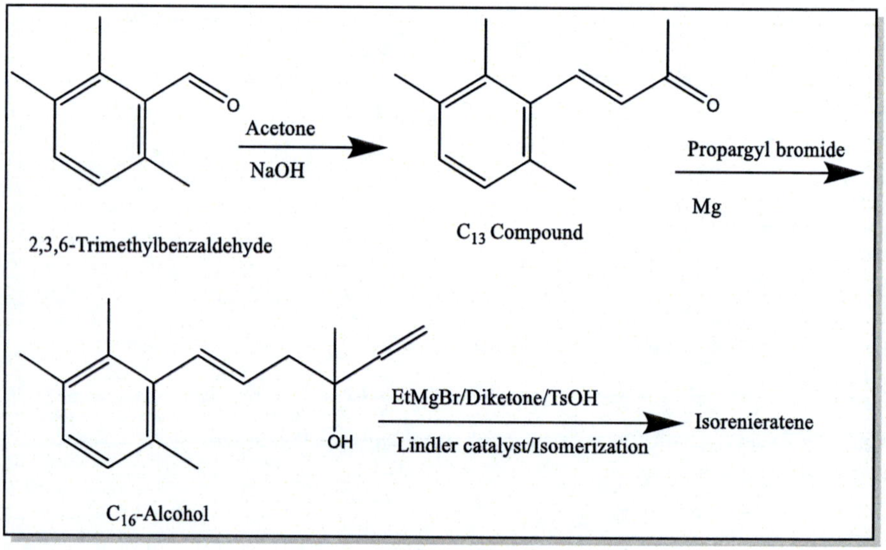

Fig. 3.32 Isorenieratene synthesis via condensation machanism

Phosphonium salt is obtained by hydrolysis and reaction of PPh_3 with diacetate product (Fig. 3.35).

Witting condensation reaction occurs between phosphonium salt and C_{10}-dialdehyde giving (all-E)acetylenic aldehyde. (all-E) acetylenic aldehyde further condenses slowly with phosphonium salt under basic conditions at 40 °C and forms

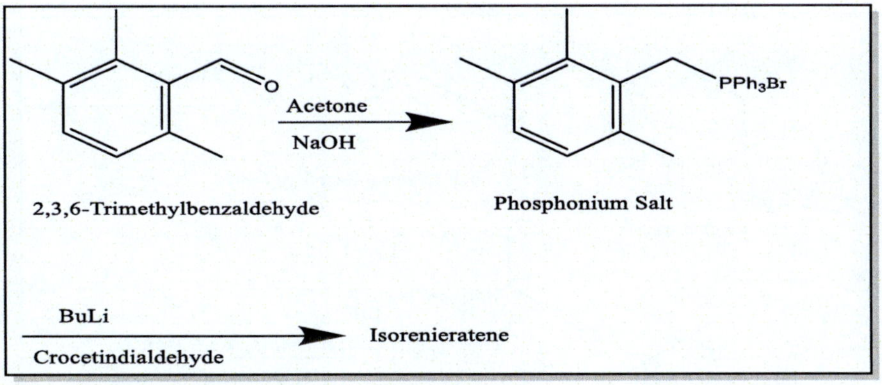

Fig. 3.33 Isorenieratene synthesis via reduction mechanism

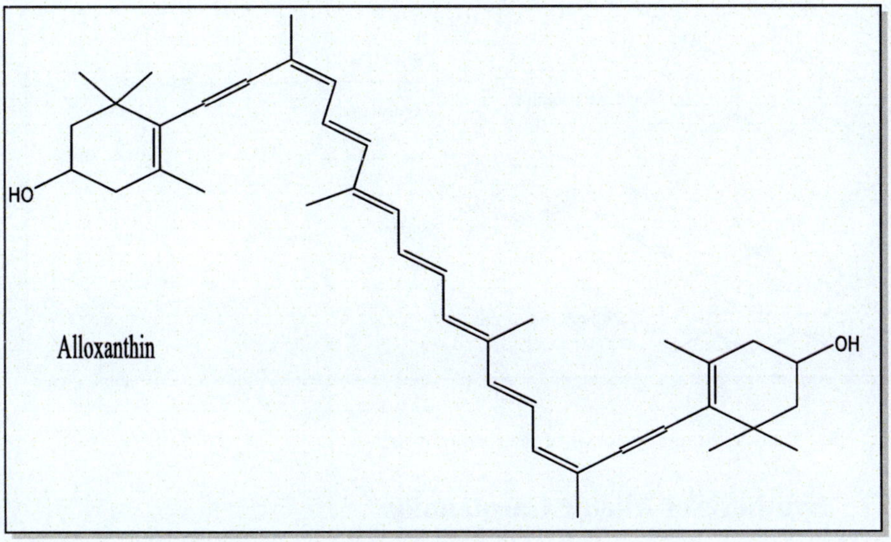

Fig. 3.34 Molecular structure of Alloxanthin

alloxanthin having (Z)-configuration. It is reflected that alloxanthin having (Z)-configuration are more stable than their (E)-forms (Fig. 3.36).

(all-E)-alloxanthin and its (9Z)-isomer can also be prepared by using acetoxy phosphonium salt withC_{10}-dialdehyde in the presence of KOH preferably, as NaOMe in propan-2-ol require low temperature of −30 °C [27] (Fig. 3.37).

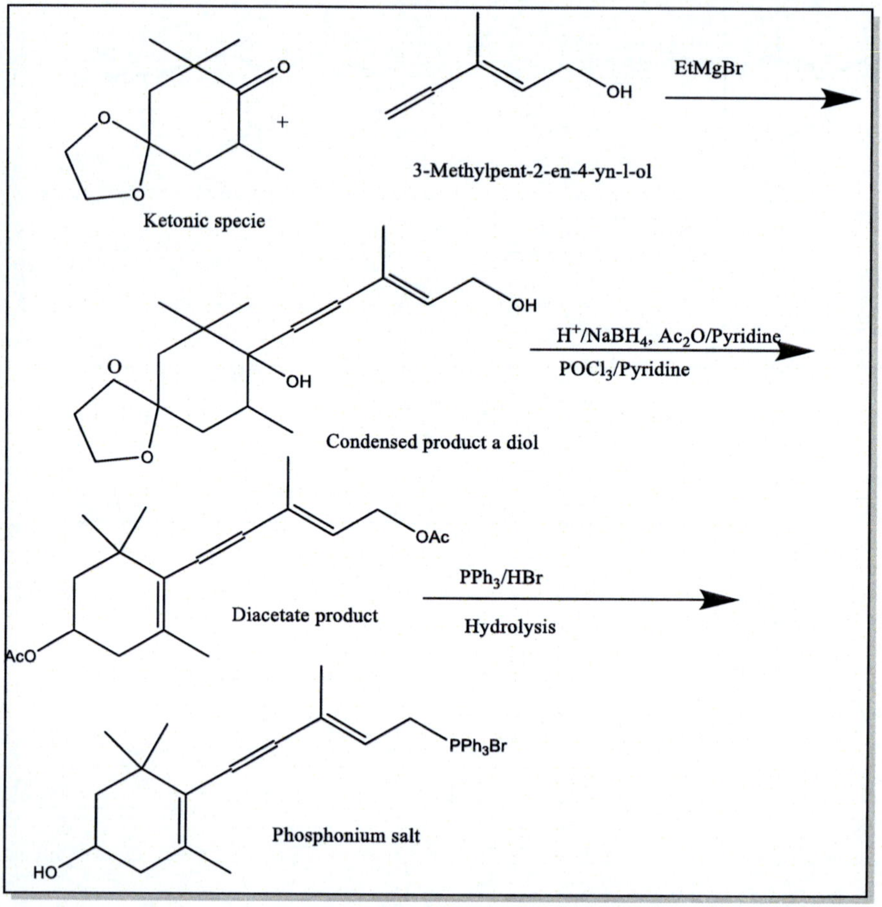

Fig. 3.35 Pathway for the synthesis of phosphonium salt

3.7 Synthesis of Allenic Carotenoids

Allenic carotenoids are synthesized in the following two ways: [28].

3.7.1 *Synthesis of C_{15}-Epoxyformyl Ester*

Palladium (II) catalyst enhances the process ofolefination of the vinyl triflate that is used for preparing C_{15} epoxyforrnyl ester. TBS ether reacts with phenyltrifluoromethanesulphonimide (Tf2NPh) in the presence of LDA to synthesize vinyl triflate. It forms diene ester by reaction withmethyl acrylate catalysed by palladium (II).

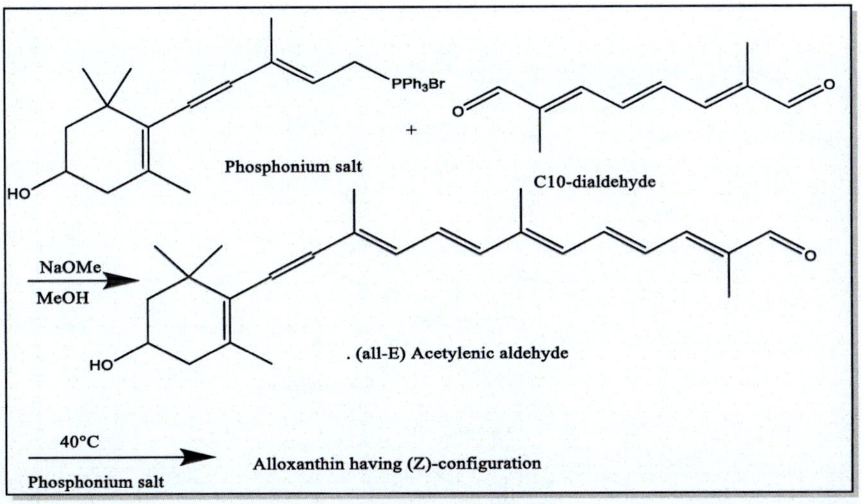

Fig. 3.36 Alloxanthin synthesis via condensation mechanism

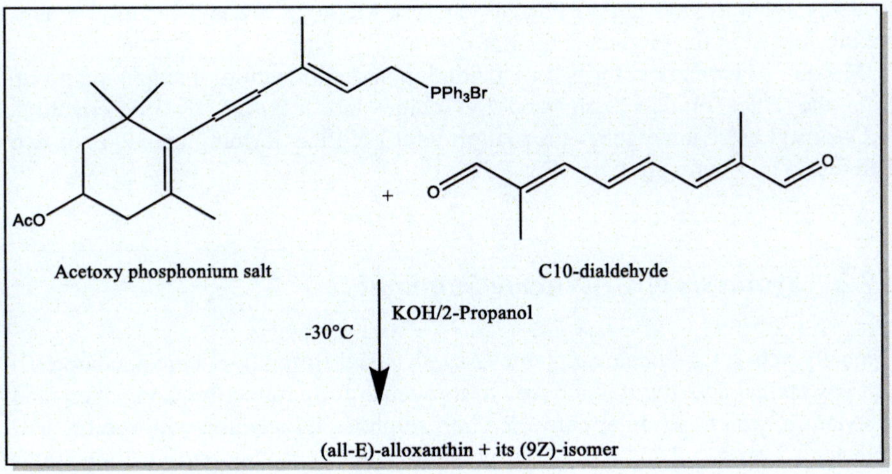

Fig. 3.37 Synthesis of (all-E)-alloxanthin and its isomers

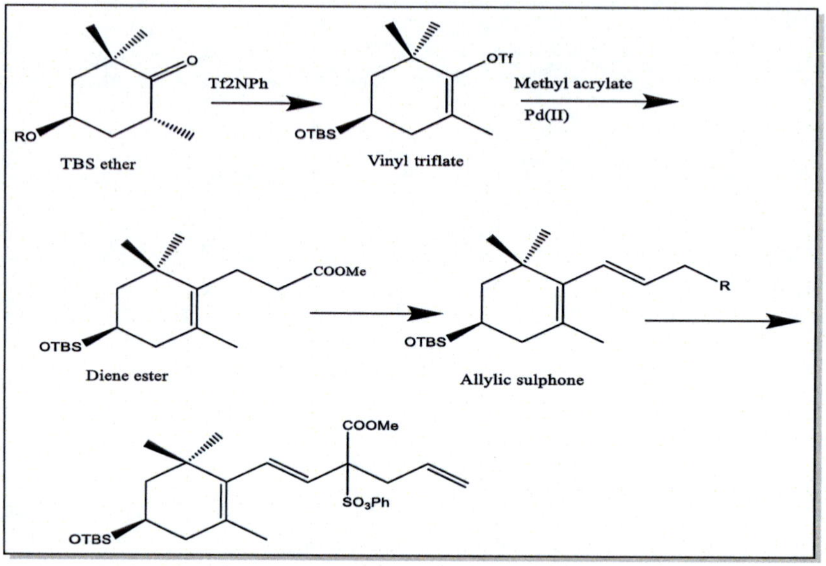

Fig. 3.38 Possible reaction for Synthesis of intermediate for C_{15}-epoxyformyl ester

Allylic sulphone is obtained by reaction in which sodium sulphinate is used as a reducing agent as well as $Pd(PPh_3)_4$ as catalyst. Methoxycarbonyl and allyl groups are introduced in the product (Fig. 3.38).

Mixture of isomers is formed by oxidation and elimination of sulphonic group from the latter product, catalyzed by iodine and followed by isomerization. (9Z)-formyl ester undergoes epoxidation with MCPBA forming epoxides of two different stereo-configurations (Fig. 3.39).

3.7.2 Synthesis of C_{22}-Allenic Sulphone

Optically active C_{22}-allenic sulphone is synthesized from C_{22}-allenic apocarotenal that has optical activity. It is reduced to its acetate form, under reducing agent and acetylation, which is further converted into sulphone by reaction with sodium sulphinate in 2-propanol and water. Phosphonium chloride having seven carbon atoms is condensed with C_{15} allenic aldehydethrough Witting condensation reaction forming C_{22}-allenic apocarotenal. C_{15} allenic aldehydeis prepared by TMS ether through different intermediates (Fig. 3.40).

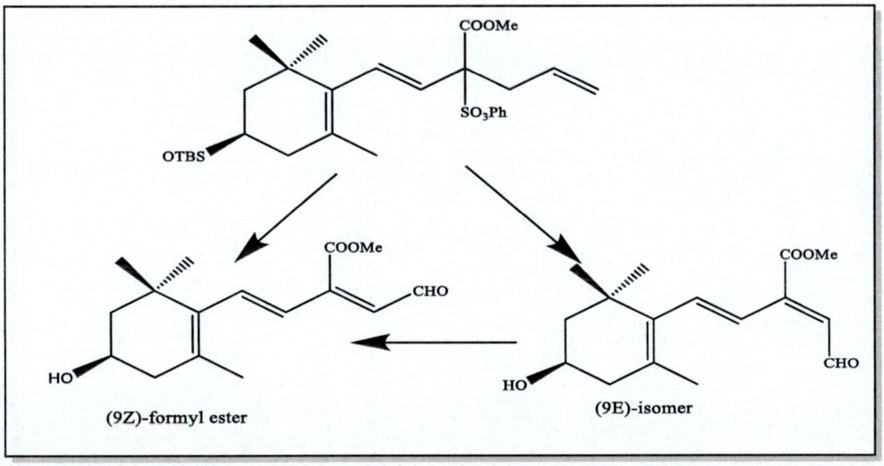

Fig. 3.39 Pathway for C_{15}-epoxyformyl ester intermediate for epoxidation

3.8 Synthesis of Butenolid Carotenoids

An important example of butenolide carotenoid is peridinin, a light-harvesting pigment that is used in photosynthesis by some species of algae. It has 4-alkylidenebutenolide ring structure that has an allene function in the main chain. It has conjugation at C(2)-position (Fig. 3.41).

The following method is used to prepare alkylidenebutenolide carotenoids, (±)-peridinin, (±) pyrrhoxanthin and optically active peridinin. In this method, mixture of (Z)- and (E)-isomers of conjugated alkylidenebutenolide is prepared by reaction of the conjugated formyl ester between allylic sulphones in the presence of LDA under extremely low temperature. Its isomers arise from the position of side chains about ylidene double bond. This is a nucleophilic addition reaction in which carbanion of allylic sulphone is added to the formyl ester followed by cyclization and elimination of the sulphone group (Fig. 3.42).

Isomers of optically active peridinin synthesized are separated through preparative HPLC in the dark [29].

3.9 Carotenoids and their Isomers

These are main factors that cause isomerization of the carotenoids: heat/drying, pH, light and structural rearrangements. Isomerization involves the process of oxidation as well as transformation [30–32]. Through oxidation, cis-trans isomerization occurs resulting in yielding epoxides, [33] as in case of **β-carotne**, oxidation takes place at terminal double bond of the conjugated double bond but in two step: at first,

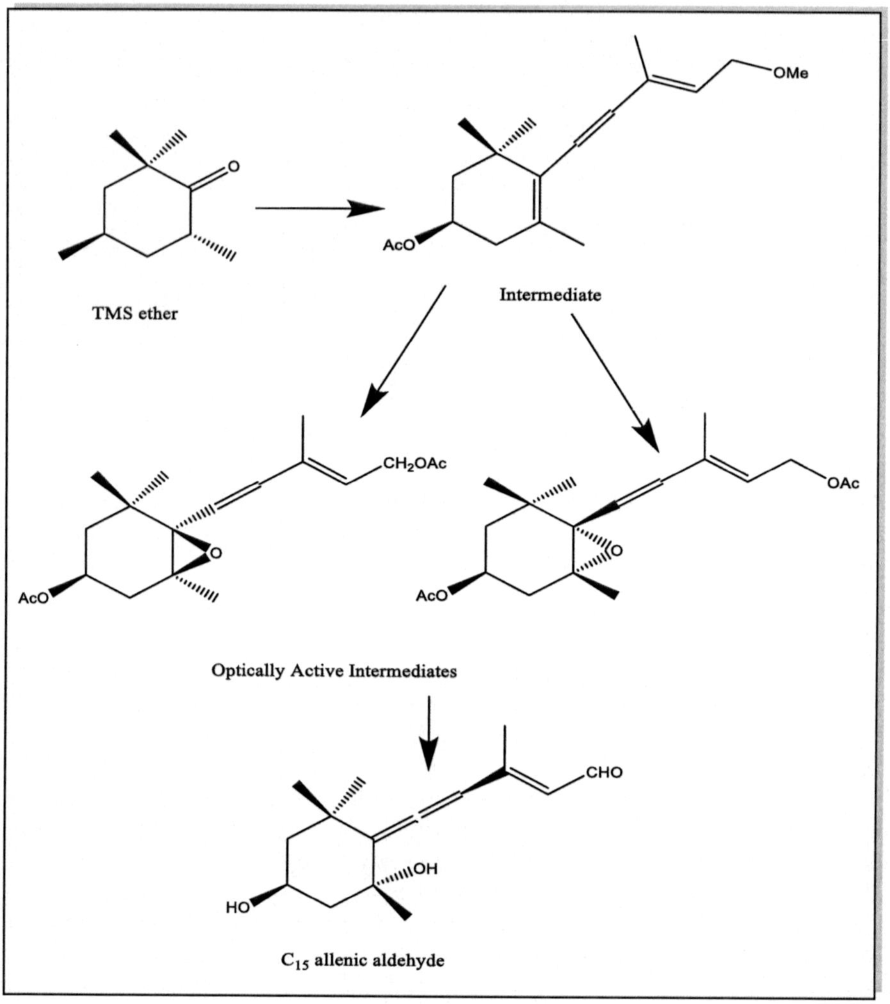

Fig. 3.40 C_{22}-Allenic Sulphone synthesis through reduction and acetylation

Fig. 3.41 Molecular structure of 4-alkylidenebutanolide

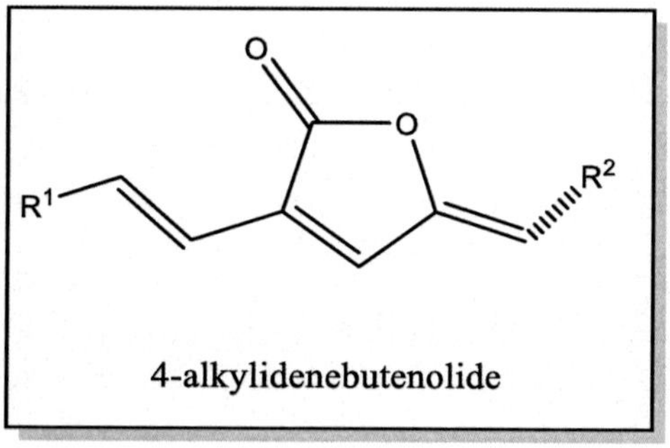

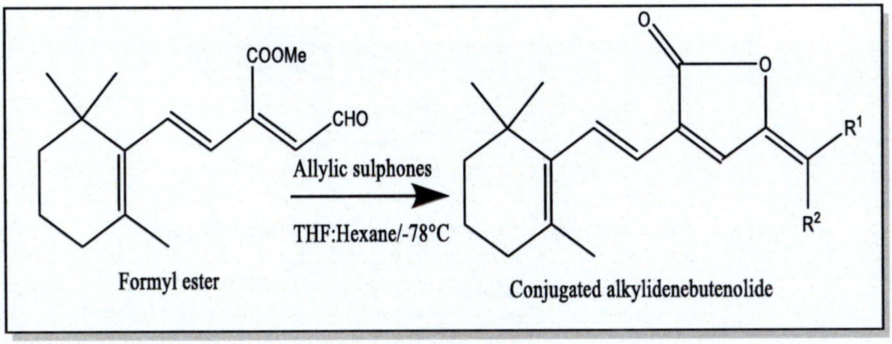

Fig. 3.42 Nucleophilic addition reaction of formyl ester

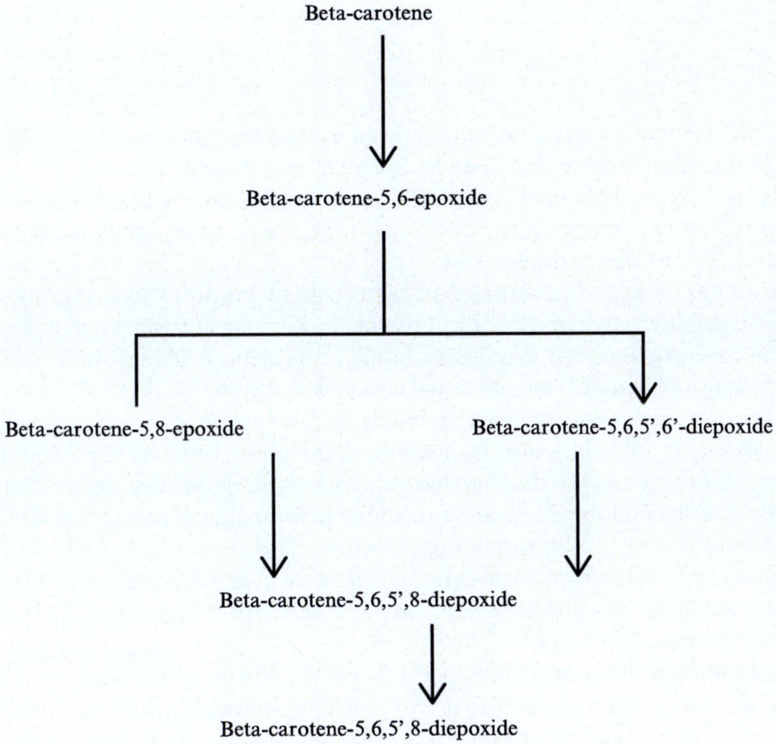

Fig. 3.43 Isomerization of β-carotenoid into epoxy carotenoids

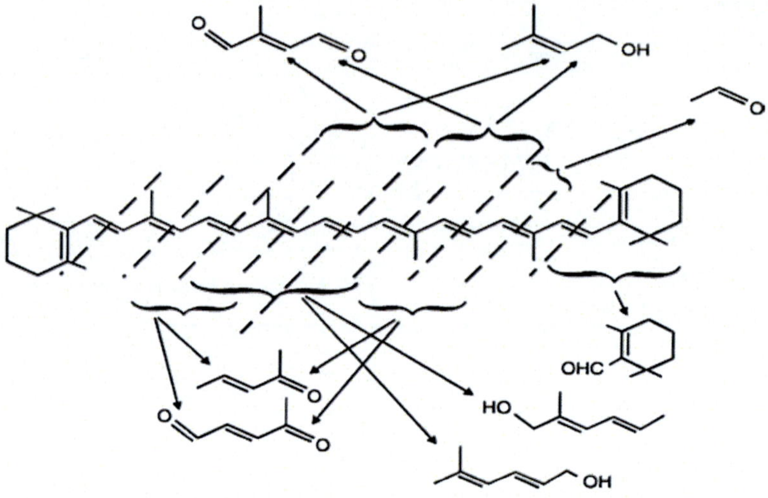

Fig. 3.44 Possible fragmentation products of β-carotene [35]

one side of the molecule is oxidized followed by oxidation at other side of the molecule [34], as described on Fig. 3.43 in case of Beta-carotene:

β-carotene is unstable in all trans form, but in least cis forms, that are isomerized into cis form due to rearrangement of bonds [35]. These are cleaved on addition of oxygen as well as shown in Fig. 3.44:

Lycopene consists of unsaturated straight chain hydrocarbon with11 conjugated and two unconjugated double bonds. It has no β-ionic ring at terminal, that is why it does not have provitamin A activity. Naturally, is present in all tans form. As it has seven conjugated double bond, it has red colour and absorbs UV-Vis light more than any other carotenoid [36–38]. Double bonds help it to isomerize into mono-cis- or poly-cis-isomers [39]. It is also isomerized in the presence of heat, pH and light. Heating effect may cause oxidation that takes place at terminal double bond as well as single double bond simultaneously, resulting in formation of complex products in large number [40, 41], as shown in Figs. 3.45 and 3.46:

Xanthophylls are formed when carotenes are oxidatively derivatized. Presence of hydroxyl groups makes it more polar than the other class of caroteniods [43]. It has chemical formula $C_{40}H_{56}O_2$ and yellow to red color [44, 45]. Lutein, zeaxanthin, and cryptoxanthin are the most common xanthophylls. Pigment that does not absorb energy, are converted into form that absorb energy through xanthophyll cycle in which violaxanthin, antheraxanthin and zeaxanthin take part [10]. Lutein is fat soluble and stable in the form of emulsion. Lutein and zeaxanthin are isomers [46, 47].

In plants, violaxanthin is present in high percentage and is transformed into its isomers on addition of epoxide [48] (Fig. 3.47).

For isomerization, iodine and diphenyl diselenide are also used in hexane, benzene or dichloromethane under inert conditions [49, 50].

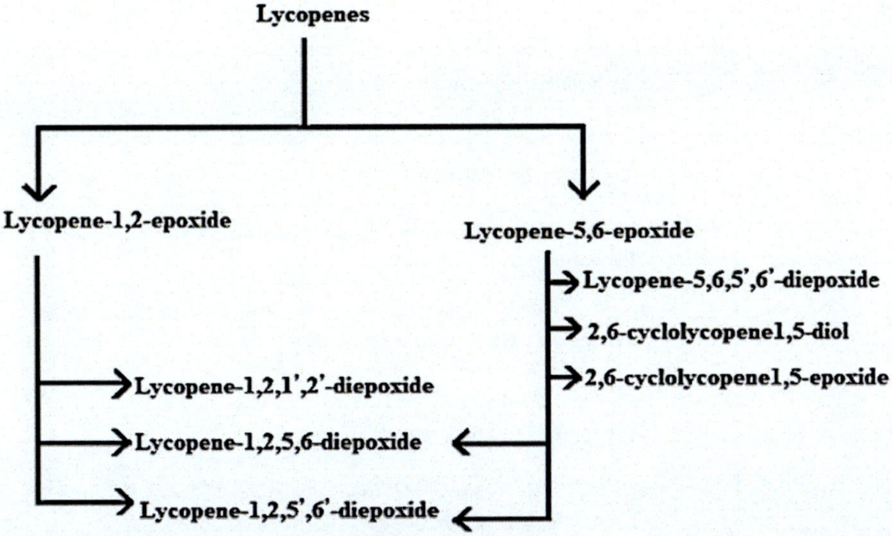

Fig. 3.45 Possible pathway for oxidation of lycopenes [42]

3.10 Synthesis of z-Isomers

As all carotenoids are synthesized in the form of stereoimers, so their synthesis has been discussed earlier in this chapter.

3.11 Effect of z-Isomerization on Physicochemical Properties of Carotenoids

As, almost all biological molecules including carotenoids exist in isomeric forms, so their bioavailability and functionality changes on conversion of one form to another. In case of carotenoids, Z-isomers are abundant in nature than E-isomers. Here, effect of Z-isomeriaztion on functions and their bioavailability is being discussed below:

Lycopene (molecular formula $C_{40}H_{56}$) has ability to reduce oxidative stress, scavenge radicals leading to lowering the probability of arteriosclerosis, atherogenesis, prostate and esophageal cancer. On Z-isomerization, it increases its positive effect on human health. Z-isomers of lycopenes are present in present abundantly in vegetables and fruits [5–7, 51, 52]. Test was conducted on the humans who consumed such food that contained high percentage of -isomers of lycopene, its increased amount was found in their bile acid micelles, human intestinal Caco-2 cells, lymph-cannulated ferret and blood than in the humans who did not consume [37, 53–58]. So, it should be taken into consideration that Z-isomers of lycopenes

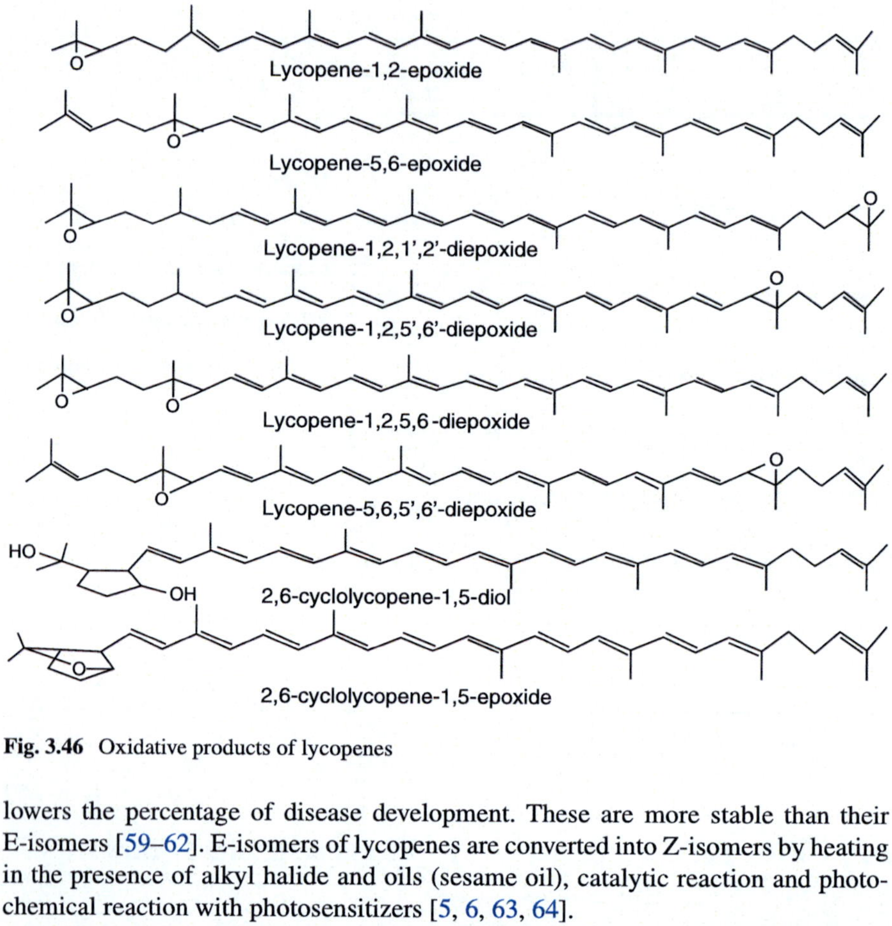

Fig. 3.46 Oxidative products of lycopenes

lowers the percentage of disease development. These are more stable than their E-isomers [59–62]. E-isomers of lycopenes are converted into Z-isomers by heating in the presence of alkyl halide and oils (sesame oil), catalytic reaction and photochemical reaction with photosensitizers [5, 6, 63, 64].

β-carotene has antioxidant activity and reduces risk of as cancer and atherogenesis. It can be converted into retinol rapidly. Z-isomerization lessens its bioavailability, while has both positive and negative effects on development of diseases [65, 66]. E-isomers of β-carotene are more available than Z-isomers, it was confirmed through in vivo tests Caco-2 cells, HSC-T6 cells, liver microsomes, ferrets and gerbils [67–69]. This study was negated, as high bioavaolability of Z-isomers of β-carotene in human intestinal Caco-2 cells and ferrets was found [70, 71]. β-carotene Z-isomers show increased antiatherogenic activity and antiatherosclerotic activity [72]. β-carotene in converted into molecules, (all-E)-retinoic acid and (9Z)-retinoic acid, involved in gene regulation. (all-E)-β-carotene acts as substrate for cleavage of vitamin A, this affects vision [73–78] (Fig. 3.48).

Astaxanthin, due to having strong antioxidant activity, lessens carcinogenic effects in human body, such as cancer, eye disease, and cardiovascular disease alongwith stress-induced suppression of tumor-fighting natural killer cells [79–82]. These help the micro-organisms to change color of their body [83]. Z-isomers of

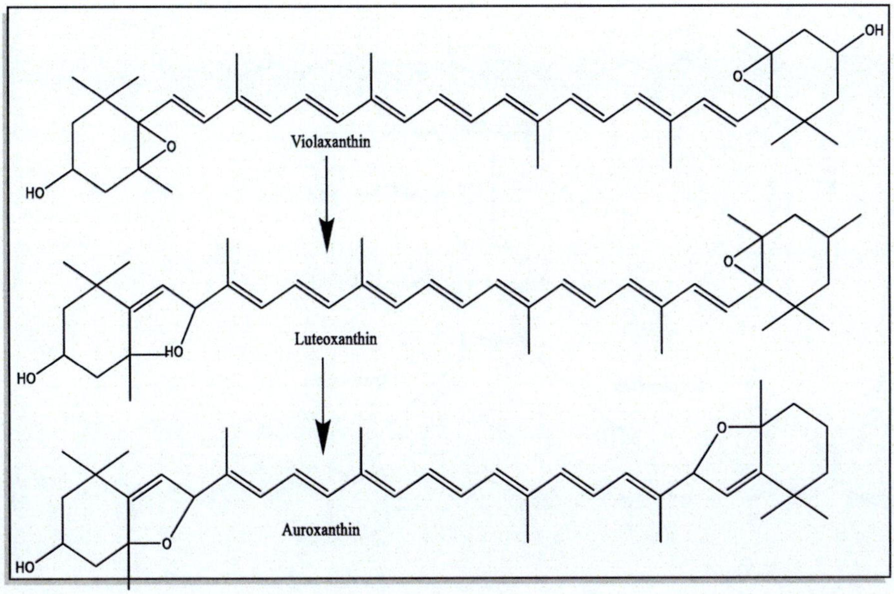

Fig. 3.47 Epoxidation of violaxanthin into its isomers

astaxanthin have more bioavailability and antioxidant activity than its E-isomers, it helps in transport of molecules to the specific position in body [84]. Its increased antioxidant ability was confirmed through enzyme activities, oxygen radical-absorption capacity (ORAC), cellular antioxidant activity (CAA) assay photochemiluminescence (PLC) and peroxidation [85–89]. So, all-E-isomer are changed into Z-isomers via heating as well as catalysis [90].

Canthaxanthin, a xanthophylls, lowers mammary tumor volumes, inflammation, reduces chances of development of neurodegenerative disorder and cancer by decreasing the level of IL-1, IL-6, and TNF-α alongwith enhancing the ability of GPX and catalase [91–94]. High antioxidant activity of Z-isomers than E-isomers has been measured via various assays, as DPPH radicalscavenging assay, superoxide radical-scavenging assay and fluorescence assay. Z-isomers also show enhanced activity of pro-apoptosis process [95–98]. Catalysts and heat are used to get Z-isomers from E-isomers [99].

Fucoxanthin, an allenic xanthophylls, has anticancer, antiangiogenic, antiobesity, antidiabetic and antioxidant effects [100–104]. It lessens number of Caco-2, HT-29, and DLD-1 cells in human colon. Order of antioxidant ability 13Z-isomer ≈ 13′Z-isomer > all-E-isomer >9′Z-isomer was measured by DPPH radical-scavenging and superoxide-detection assay [105, 106]. So, with the increase in percentage of Z-isomers, antioxidant effect of fucoxanthin decreases, but anticancer effect increases. All-E-isomer change into Z-isomers via light effect [107, 108].

Lutein, a xanthophylls, reduces the risk of macular degeneration, eye diseases and cardiovascular diseases. It has strong antioxidant effect that increases on

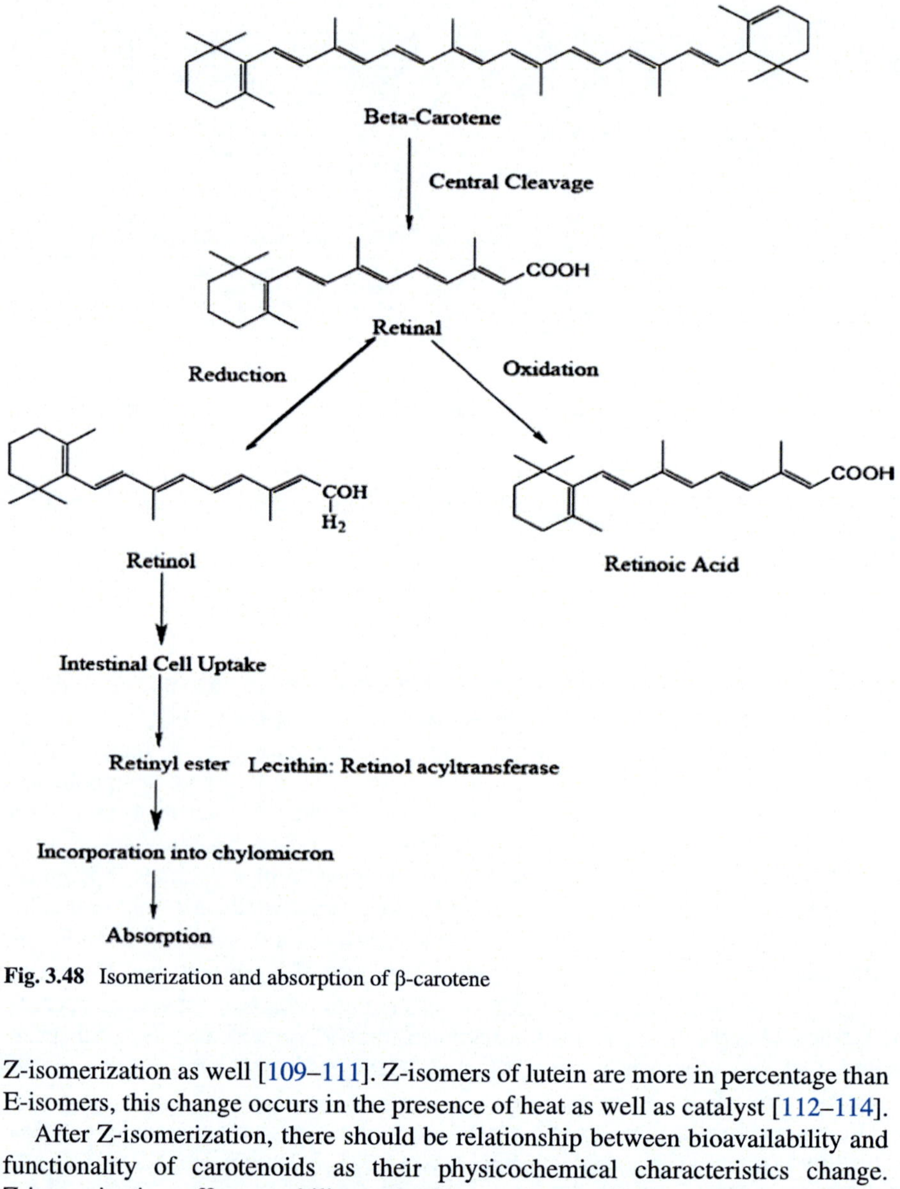

Fig. 3.48 Isomerization and absorption of β-carotene

Z-isomerization as well [109–111]. Z-isomers of lutein are more in percentage than E-isomers, this change occurs in the presence of heat as well as catalyst [112–114].

After Z-isomerization, there should be relationship between bioavailability and functionality of carotenoids as their physicochemical characteristics change. Z-isomerization affects stability, solubility, and crystallinity. Stability order of carotenoids is following: all-E-isomer ≈ 5Z-isomer >9Z-isomer >13Z-isomer >15Z-isomer >7Z-isomer ≈ 11Z-isomer for lycopene; all-E-isomer >9Z-isomer >13Z-isomer >15Z-isomer >7Z-isomer ≈ 11Z-isome for β-carotene. In terms of stability, Gibbs free energy play an important role, it can be measured by Gaussian program. Antioxidant effect changes remarkably with the change in Gibbs free energy. Same case has been observed in terms of solubility. As, solubility increase, the carotenoid uptake percentage increases with increasing the bioaccessibility of that molecule in the particular location. So, for the uptake purpose, carotenoid transport proteins also play an important role. Z-isomers mostly exist in amorphous form, while E-isomers in crystalline form. Crystallinity of carotenoids in confirmed by optical observations, differential scanning calorimetry, powder X-ray diffraction, and scanning electron microscopy analyses [115–119].

More than hundred types of carotenoids have been reported, but limited study has been done on the bioavailability and functionality of the carotenoids, that is summarized in Table 3.1 in the following:

Table 3.1 Summary of bioavailability and functions of some carotenoids

Carotenoids	Evaluation	Results	Effects
Lycopene [37, 53, 54, 56, 57, 120, 121]	Bioavailability/ bioaccessibility [diffusion model, bile acid micelles and lymph-cannulated ferrets, evaluated in Caco-2 cells and human oral-dosing tests]	Z-isomers > all-E-isomer	+ve
	Antioxidant activity [TEAC, FRAP, MbFeIII-LP assay]	Z-isomers > all-E-isomer	+ve/–ve rarely
β-Carotene [68, 69, 122–131]	Bioavailability/ bioaccessibility [Caco-2 cells, HSCT6 in cells, rat liver microsomes, small intestines of ferrets and in ferret, gerbil, human oral-dosing test]	All-E-isomer > Z isomers 9Z-isomer > all-E-isomers	+ve/–ve
	Antioxidant activity [LDL oxidation, TEAC and PSC assay]	9Z-isomer > all-E-isomer All-E-isomer >9Z-isomes	+ve/–ve
	Atherogenesis activity [in knockout mice]	9Z-isomer > all-E-isomers	+ve
	Atherosclerosis activity [in female LDLR and apoE-deficient mice]	9Z-isomer > all-E-isomer	+ve
Astaxanthin [83, 86, 87, 132–134]	Bioavailability/bioaccessibility [via digestion model, Caco-2 cells, in human oral-dosing test]	Z-isomers > all-E-isomers 13Z-isomer > all-E-isomer>9Z-isomers All-E-isomer > Z-isomers	+ve
	Antioxidant activity [in antioxidant enzyme-activity, DPPH, lipidperoxidation, ORAC, PLC and CAA assay]	Z-isomers > all-E-isomers 13Z-isomer > all-E-isomer >9Z-isomers	+ve
Canthaxanthin [95, 97]	Antioxidant activity [DPPH, superoxide radical-scavenging, and fluorescence assay]	9Z-isomer > all-E-isomers	+ve
	Pro-apoptotic activity [THP-1 macrophages]	9Z-isomer > all-E-isomers	+ve

(continued)

Table 3.1 (continued)

Carotenoids	Evaluation	Results	Effects
Fucoxanthin [108, 135]	Antioxidant activity [DPPH, superoxide-detection, ABTS, hydroxyl radical-scavenging, hydrogen peroxide-scavenging, superoxide anion, and reducing power assay]	13Z-isomer ≈ 13′Z-isomer > all-E-isomer >9′Z-isomer	+ve/–ve
		9′Z-isomer > all-E-isomer >13Z-isomer ≈ 13′Z-isomer	
	Anticancer activity [HL-60 cells and Caco-2 cells]	Z-isomers > all-E-isomer	+ve
Lutein [112, 136–138]	Bioavailability/ bioaccessibility [digestion model and Caco-2 cells]	Z-isomers > all-E-isomer	+ve/–ve
		All-E-isomer > Z-isomers	
		13Z-isomer > all-E-isomer	
	Antioxidant activity [FRAP, DPPH and ORAC assay]	Z-isomers > all-E-isomer	+ve
		13′Z-isomer > all-E-isomer ≈ 9Z-isomer	

3.12 Analysis of Isomers

These isomers are separated, identified and quantified by various analytical techniques, some of them are briefly discussed in the following:

3.12.1 *UV/Vis-Spectroscopy*

Cis-carotenoids absorb low wavelength with low absorption coefficients due to maximum hypso-chromical shift at absorption band, while *trans*-carotenoids have high absorption coefficients. In case of *cis*-carotenoids, larger peak is obtained that shows the position of double bond at the center of the molecule. Amount of *cis*-isomer is difficult to estimate, still now a few have been determined. The following data are reported for lycopene: (2% CH_2Cl_2 in hexane), as nm: (all-*E*) 470 (187); (5Z) 470 (184); (5Z,5'Z) 470 (182); (9Z,9'Z) 459 (168); (7Z) 469 (154); (7Z,7'Z) 466(128); (7Z,9Z) 444 (115); (7Z,9Z,7'Z,9'Z) 437 (105); (15Z) 468 (110). In the same way, for zeaxanthin, corresponding data are nm: (all-*E*) 452 (133) hexane; (7Z) 459 (115) chloroform; (9Z) 458 (95) chloroform; (13Z) 458 (86) chloroform; (15Z) 449 (77) hexane, showing a systematic reduction in absorption coefficient in the series (all-*E*) > (9Z) > (13Z) > (15Z) [139, 140].

3.12.2 *NMR Spectroscopy*

Structure of geometrical isomers is determined by ^{1}H NMR data through chemical shift differences of protons (in *cis*-isomers, protons on one side of the loop are-deshielded and those on the other side shielded, as compared to the *trans*isomer).

Structures of lycopene, fucoxanthin, peridinin, and neoxanthin have been determined through this technique. ^{13}H NMR spectra is also used for this purpose. Recently, LC-NMR is used for the direct determination of the stereoisomers [141].

3.12.3 IR and Resonance Raman Spectroscopy

It is employed for the confirmation of *cis/trans* double bonds as well as sterically hindered *cis* double bonds in the molecule [142].

3.12.4 Mass Spectrometry

It is used for the identification of the trans isomers while coupled with high performance liquid chromatography, position of double bind in cis-isomers is not confirmed through simple mass spectrometry technique [143].

3.13 Conclusion

Carotenoids are plant pigments present in fruits and vegetable. Humans, animals and micro-organisms get them through ingestion. These are present mostly in trans forms and can be classified depending upon end groups. Carotenoids can be synthesized in laboratory as well as on industrial scale through various pathways. *Trans* isomers are changed into cis isomers by heating, processing, light irradiation, oxidation, catalytic effect, degradation, isomerization and cleavage. These transformation processes are responsible to enhance the applications of carotenoids. Bioavailability, antioxidant effect, macular degeneration, eye diseases, cardiovascular diseases, tumor volume, inflammation, chances of development of neurodegenerative disorders decreases as percentage of *Z*-Isomers of carotenoids increases. Their effect may vary depending upon the type of carotenoid. These are analyzed by Infra-red and resonance Raman spectroscopy, UV-Vis spectroscopy, NMR spectroscopy, mass spectrometry and chromatography. Stability, solubility and crystallinity may vary with the change in the configuration of the isomer.

References

1. Maoka T (2011) Carotenoids in marine animals. Mar Drugs 9:278–293
2. Yabuzaki J (2017) Carotenoids database: structures, chemical fingerprints and distribution among organisms. Database 2017:1–11

3. Cooper DA (2004) Carotenoids in health and disease: recent scientific evaluations, research recommendations and the consumer. J Nutr 134:221–224
4. Young CYF, Yuan HQ, He ML, Zhang JY (2008) Carotenoids and prostate cancer risk. Med Chem 8:529–537
5. Krinsky NI, Johnson EJ (2005) Carotenoid actions and their relation to health and disease. Mol Asp Med 26:459–516
6. Xu XR, Zou ZY, Huang YM, Xiao X, Ma L, Lin XM (2012) Serum carotenoids in relation to risk factors for development of atherosclerosis. Clin Biochem 45:1357–1361
7. Ouchi A, Aizawa K, Iwasaki Y, Inakuma T, Terao J, Nagaoka S, Mukai K (2010) Kinetic study of the quenching reaction of singlet oxygen by carotenoids and food extracts in solution. Development of a singlet oxygen absorption capacity (SOAC) assay method. J Agric Food Chem 58:9967–9978
8. Dutta D, Chaudhuri UR, Chakraborty R (2005) Structure, health benefits, antioxidant property and processing and storage of carotenoids. Afr J Biotechnol 4:1510–1520
9. Takeungwongtrakul S, Benjakul S (2016) Astaxanthin degradation and lipid oxidation of Pacific white shrimp oil: kinetics study and stability as affected by storage conditions. Int Aquat Res 8(1):15–27
10. Shahidi F (2009) Nutraceuticals and functional foods.Whole versus processed foods. Trends Food Sci Technol 20:376–387
11. Krinsky NI, Landrum JT, Bone RA (2003) Biologic mechanisms of the protective role of lutein and zeaxanthin in the eye. Annu Rev Nutr 23:171–201
12. Holick CN, Michaud DS, Stolzenberg SR (2002) Dietary carotenoids, serum b-carotene, and retinol and risk of lung cancer in the a-tocopherol, b-carotene cohort study. Am J Epidemiol 156:536–547
13. Stahl W, Sies H (2007) Carotenoids and flavonoids contribute to nutritional protection against skin damage from sunlight. Mol Biotechnol 37:26–30
14. Bonnie TP, Choo YM (1999) Oxidation and thermal degradation of carotenoids. J Oil Palm Res 2(1):62–78
15. Clinton SK (1998) Lycopene chemistry, biology, and implications for human health and disease. Nutr Rev 56:35–51
16. Nguyen M, Francis D, Schwartz S (2001) Thermal isomerisation susceptibility of carotenoids in different tomato varieties. J Sci Food Agric 81(9):910–917
17. Sy C, Dangles O, Borel P, Caris-Veyrat C (2013) Iron-induced oxidation of (all-E)-β-carotene under model gastric conditions: kinetics, products, and mechanism. Free Radic Biol Med 63:195–206
18. Zepka LQ, Borsarelli CD, da Silva MAAP, Mercadante AZ (2009) Thermal degradation kinetics of carotenoids in a cashew apple juice model and its impact on the system color. J Agric Food Chem 57(17):7841–7845
19. Jin E (2003) Role of the reversible xanthophyll cycle in the photosystem II damage and repair cycle in Dunaliella salina. Plant Physiol 132(1):352–364
20. Hunter CN (1994) Introduction of new carotenoids into the bacterial photosynthetic apparatus by combining the carotenoid biosynthetic pahtways of Erwinia herbicola and Rhodobacter sphaeroides. J Bacteriol 176:3692–3697
21. Fukuoka S, Ajiki Y, Ohga T, Kawanami Y, Izumori K (2004) Production of dihydroxy C50-carotenoid by *Aureobacterium* sp. FERM P-18698. Biosci Biotech Bioch 68:2646–2648
22. Ito M, Yamano Y, Tode C, Wada A (2009) Carotenoid synthesis: retrospect and recent progress. Arch Biochem Biophys 15; 483(2):224–228
23. Bishop NI, Urbig T, Senger H (1995) Complete separation of the f3,e- and (3,(3-carotenoid biosynthetic pathways by a unique mutation of the lycopene cyclase in the green alga, Scenedesmus obliquus. FEBS Lett, 367:158–162
24. Pogson B, McDonald KA, Truong M, Britton G, DellaPenna D (1996) Arabidopsis carotenoid mutants demonstrate that lutein is not essential for photosynthesis in higher plants. Plant Cell 8(9):1627–1639

25. Taylor CB (1996) Control of cyclic carotenoid biosynthesis: no lutein, no problem! Plant Cell 8(9):1447–1451
26. Yamano Y, Ito M (2007) Total synthesis of capsanthin and capsorubin using Lewis acid-promoted regio- and stereoselective rearrangement of tetrasubsutituted epoxides. Org Biomol Chem 5(19):3207–3212
27. Yamano Y, Chary MV, Wada A (2012) Stereoselective total synthesis of the acetylenic carotenoids alloxanthin and triophaxanthin. Org Biomol Chem 10(20):4103–4107
28. Furuichi N, Hara H, Osaki T, Nakano M, Mori H, Katsumura S (2004) Stereocontrolled total synthesis of a polyfunctional carotenoid, peridinin. J Org Chem 69(23):7949–7959
29. Vaz B, Alvarez R, Brückner R, de Lera AR (2005) The stille reaction in the synthesis of carotenoid butenolides: synthesis of 6′-epi-peridinin. Org Lett 7(4):545–548
30. Chen BH, Tang YC (1998) Processing and stability of carotenoid powder from carrot pulp waste. J Agric Food Chem 46:2312–2318
31. Goula AM, Adamopoulos KG, Chatzitakis PC, Nikas VA (2006) Prediction of lycopene degradation during a drying process of tomato pulp. J Food Eng 74:37–46
32. Mordi RC, Walton JC, Burton GW, Hughes L, Keith IU, David LA, Douglas MJ (1993) Oxidative degradation of β-carotene and β-apo-8′-carotenal. Tetrahedron 49:911–928
33. Wachea Y, Bosser DAL, Lhuguenot JC, Belin JM (2003) Effect of cis/trans isomerism of β-carotene on the ratios of volatile compounds produced during oxidative degradation. J Agric Food Chem 51:1984–1987
34. Rodriguez E, Rodriguez AD (2007) Formation of apocarotenals and epoxycarotenoids from β-carotene by chemical reactions and by autoxidation in model systems and processed foods. Food Chem 101:563–572
35. Kuki M, Koyama Y, Nagae H (1991) Triplet-sensitized and thermal isomerization of all-trans, 7-cis, 9-cis, 13-cis and 15-cis isomers of β-carotene: configurational dependence of the quantum yield of isomerization via the T1 state. J Phys Chem 95:7171–7180
36. Rao AV, Rao LG (2007) Carotenoids and human health. Pharmacol Res 55:207–216
37. Boileau TWM, Boileau AC, Erdman JW Jr (2002) Bioavailability of all-trans and cis-isomers of lycopene. Exp Biol Med 227:914–919
38. Schulz H, Baranska M, Baranski R (2005) Potential of NIR-FT-Raman spectroscopy in natural carotenoid analysis. Biopolymers 77:212–221
39. Kong K-W, Khoo H-E, Prasad KN, Ismail A, Tan C-P, Rajab NF (2010) Revealing the power of the natural red pigment lycopene. Molecules 15:959–987
40. Khachik F, Pfander H, Traber B (1998) Proposed mechanisms for the formation of synthetic and naturally occurring metabolites of lycopene in tomato products and human serum. J Agric Food Chem 46:4885–4890
41. Khachik F, Steck A, Niggli UA, Pfander H (1998) Partial synthesis and structural elucidation of the oxidative metabolites of lycopene identified in tomato paste, tomato juice, and human serum. J Agric Food Chem 46:4874–4884
42. Rodriguez E, Rodriguez AD (2009) Lycopene epoxides and apo-lycopenals formed by chemical reactions and autoxidation in model systems and processed foods. J Food Sci 74:674–682
43. Matsuno T, Hirono T, Ikuno Y, Maoka T, Shimizu M, Komori T (1986) Isolation of three new carotenoids and proposed metabolic pathways of carotenoids in hen's egg yolk. Comp Biochem Physiol 84:477–481
44. Handelman GJ (2001) The evolving role of carotenoids in human biochemistry. Nutrition 17:818–822
45. Emenhiser C, Sande LC, Schwartza SJ (1995) Capability of a polymeric C30 stationary phase to resolve cis-trans carotenoid isomers in reversed-phase liquid chromatography. J Chromatogr 707:205–216
46. Holden JM, Eldridge AL, Beecher GR, Buzzard IM, Bhagwat AS, Davis CS, Douglass LW, Gebhardt ES, Haytowitz D, Schakel S (1999) Carotenoid content of U.S. foods. An update of the database. J Food Compos Anal 12:169–196

47. Tee E-S, Lim C-L (1991) Carotenoid composition and content of Malaysian vegetables and fruits by the AOAC and HPLC methods. Food Chem 41:309–339
48. Mercadante A, Rodriguez AD (1998) Influence of ripening, cultivar differences, and processing on the carotenoid composition of mango. J Agric Food Chem 46:128–130
49. Haugan JA, Englert G, Glinz E, Liaacn-Jensen S (1992) Algal carotenoids. 48. Structural assignments of geometrical. Acta Chem Scand 46:389–395
50. Glaser T, Lienau A, Zeeb D, Krucker M, Dachtler M, Albert K (2003) Qualitative and quantitative determination of carotenoid stereoisomers in a variety of spinach samples by use of MSPD before HPLC-UV, HPLC-APCI-MS, and HPLC-NMR on-line coupling. Chromatographia 57(S1):S19–S25
51. Amorim-Carrilho K, Cepeda A, Fente C, Regal P (2014) Review of methods for analysis of carotenoids. TrAC Trends Anal Chem 56:49–73
52. Honda M, Murakami K, Watanabe Y, Higashiura T, Fukaya T, Wahyudiono KH, Goto M (2017) The E/Z isomer ratio of lycopene in foods and effect of heating with edible oils and fats on isomerization of (all-E)-lycopene. Eur J Lipid Sci Technol 119, 1600389(1–9)
53. Boileau A, Merchen N, Wasson K, Atkinson C, Erdman JJ (1999) Cis-lycopene is more bioavailable than trans-lycopene in vitro and in vivo in lymph-cannulated ferrets. J Nutr 129:1176–1181
54. Failla M, Chitchumroonchokchai C, Ishida B (2008) In vitro micellarization and intestinal cell uptake of cis isomers of lycopene exceed those of all-trans lycopene. J Nutr 138:482–486
55. Stahl W, Sies H (1992) Uptake of lycopene and its geometrical isomers is greater from heat processed than from unprocessed tomato juice in humans. J Nutr 122:2161–2166
56. Unlu N, Bohn T, Francis D, Nagaraja H, Clinton S, Schwartz S (2007) Lycopene from heat-induced cis-isomer-rich tomato sauce is more bioavailable than from all-trans-rich tomato sauce in human subjects. Br J Nutr 98:140–146
57. Burri B, Burri B, Chapman M, Neidlinger T, Seo J, Ishida B (2009) Tangerine tomatoes increase total and tetra-cis-lycopene isomer concentrations more than red tomatoes in healthy adult humans. Int J Food Sci Nutr 60:116–122
58. Desmarchelier C, Borel P (2017) Overview of carotenoid bioavailability determinants: from dietary factors to host genetic variations. Trends Food Sci Technol 69:270–280
59. Muller L, Goupy P, Frohlich K, Dangles O, Caris-Veyrat C, Bohm V (2011) Comparative study on antioxidant activity of lycopene (Z)-isomers in different assay. J Agric Food Chem 59:4504–4511
60. Takehara M, Kuwa T, Inoue Y, Kitamura C, Honda M (2015) Isolation and characterization of (15Z)-lycopene thermally generated from a natural source. Biochem Biophys Res Commun 467:58–62
61. Honda M, Kudo T, Kuwa T, Higashiura T, Fukaya T, Inoue Y, Kitamura C, Takehara M (2017) Isolation and spectral characterization of thermally generated multi-Z-isomers of lycopene and the theoretically preferred pathway to di-Z-isomers. Biosci Biotechnol Biochem 81:365–371
62. Murakami K, Honda M, Takemura R, Fukaya T, Wahyudiono KH, Goto M (2018) Effect of thermal treatment and light irradiation on the stability of lycopene with high Z-isomers content. Food Chem 250:253–258
63. Honda M, Takahashi N, Kuwa T, Takehara M, Inoue Y, Kumagai T (2015) Spectral characterisation of Z-isomers of lycopene formed during heat treatment and solvent effects on the E/Z isomerisation process. Food Chem 171:323–329
64. Honda M, Horiuchi I, Hiramatsu H, Inoue Y, Kitamura C, Fukaya T, Takehara M (2016) Vegetable oil-mediated thermal isomerization of (all-E)-lycopene: facile and efficient production of Z-isomers. Eur J Lipid Sci Technol 118:1588–1592
65. Ouchi A, Aizawa K, Iwasaki Y, Inakuma T, Terao J (2010) Nagaoka S, Mukai K. Kinetic study of the quenching reaction of singlet oxygen by carotenoids and food extracts in solution. Development of a singlet oxygen absorption capacity (SOAC) assay method. J Agric Food Chem 58:9967–9978

66. Krinsky N, Russett M, Handelman G, Snodderly D (1990) Structural and geometrical isomers of carotenoids in human plasma. J Nutr 120:1654–1662
67. Hussain AM, Morel D, Harrison E (2002) Carotenoid uptake and secretion by CaCo-2 cells: β-carotene isomer selectivity and carotenoid interactions. J Lipid Res 43:1086–1095
68. Erdman JW, Thatcher AJ, Hofmann NE, Lederman JD, Block SS, Lee CM, Mokady S (1998) Alltrans β-carotene is absorbed preferentially to 9-cis β-carotene, but the latter accumulates in the tissues of domestic ferrets (Mustela putorius puro). J Nutr 128:2009–2013
69. Deming DM, Teixeira SR, Erdman JW (2002) All-trans β-carotene appears to be more bioavailable than 9-cis or 13-cis β-carotene in gerbils given single oral doses of each isomer. J Nutr 132:2700–2708
70. Ferruzzi M, Lumpkin J, Schwartz S, Failla M (2006) Digestive stability, micellarization, and uptake of β-carotene isomers by Caco-2 human intestinal cells. J Agric Food Chem 54:2780–2785
71. Hébuterne X, Wang X, Johnson E, Krinsky N, Russell R (1995) Intestinal absorption and metabolism of 9-cis-β-carotene in vivo: biosynthesis of 9-cis-retinoic acid. J Lipid Res 36:1264–1273
72. Harari A, Harats D, Marko D, Cohen H, Barshack I, Kamari Y, Gonen A, Gerber Y, BenAmotz A, Shaish A (2008) A 9-cis β-carotene–enriched diet inhibits atherogenesis and fatty liver formation in LDL receptor knockout mice. J Nutr 138:1923–1930
73. Nagao A, Olson J (1994) Enzymatic formation of 9-cis, 13-cis, and all-trans retinals from isomers of β-carotene. FASEB J 8:968–973
74. Wang X, Krinsky N, Benotti P, Rusell R (1994) Biosynthesis of 9-cis-retinoic acid from 9-cis-β-carotene in human intestinal mucosa in vitro. Arch Biochem Biophys 313:150–155
75. Heyman R, Mangelsdorf D, Dyck J, Stein R, Eichele G, Evans R, Thaller C (1992) 9-Cis retinoic acid is a high affinity ligand for the retinoid X receptor. Cell 68:397–406
76. Levin A, Sturzenbecker L, Kazmer S, Bosakowski T, Huselton C, Allenby G, Speck J, Kratzeisen C, Rosenberger M, Lovey E, Grippo J (1992) 9-Cis retinoic acid stereoisomer binds and activates the nuclear receptor RXRα. Nature 355:359–361
77. Minguez MM, Hornero MD, Perez GA (2002) Carotenoids and provitamin A in functional foods. In: Hurst WJ (ed) Methods of analysis for Func tional foods and nutraceuticals. Taylor & Francis, Oxford, pp 101–157
78. Schieber A, Carle R (2005) Occurrence of carotenoid cis-isomers in food: technological, analytical, and nutritional implications. Trends Food Sci Technol 16:416–422
79. Yuan J, Peng J, Yin K, Wang J (2011) Potential health-promoting effects of astaxanthin: a high-value carotenoid mostly from microalgae. Mol Nutr Food Res 55:150–165
80. Guerin M, Huntley M, Olaizola M (2003) Haematococcus astaxanthin: applications for human health and nutrition. Trends Biotechnol 21:210–216
81. Pashkow F, Watumull, D.,& Campbell C. (2008) Astaxanthin: a novel potential treatment for oxidative stress and inflammation in cardiovascular disease. Am J Cardiol 101:S58–S68
82. Ambati R, Phang S, Ravi S, Aswathanarayana R (2014) Astaxanthin: sources, extraction, stability, biological activities and its commercial applications—a review. Mar Drugs 12:128–152
83. Yang C, Zhang H, Liu R, Zhu H, Zhang L, Tsao R (2017) Bioaccessibility, cellular uptake, and transport of astaxanthin isomers and their antioxidative effects in human intestinal epithelial Caco-2 cells. J Agric Food Chem 65:10223–10232
84. Osterlie M, Bjerkeng B, Liaaen JS (2000) Plasma appearance and distribution of astaxanthin E/Z and R/S isomers in plasma lipoproteins of men after single dose administration of astaxanthin. J Nutr Biochem 11:482–490
85. Bjerkeng B, Folling M, Lagocki S, Storebakken T, Olli JJ, Alsted N (1997) Bioavailability of all-E-astaxanthin and Z-isomers of astaxanthin in rainbow trout (Oncorhynchus mykiss). Aquaculture 157:63–82
86. Østerlie M, Bjerkeng B, Liaaen JS (1999) Accumulation of astaxanthin all-E, 9Z and 13Z geometrical isomers and 3 and 3′ RS optical isomers in rainbow trout (Oncorhynchus mykiss) is selective. J Nutr 129:391–398

87. Liu X, Osawa T (2007) Cis astaxanthin and especially 9-cis astaxanthin exhibits a higher antioxidant activity in vitro compared to the all-trans isomer. Biochem Biophys Res Commun 357:187–193
88. Yang C, Zhang L, Zhang H, Sun Q, Liu R, Li J, Wu L, Tsao R (2017) Rapid and efficient conversion of all-E-astaxanthin to 9Z- and 13Z-isomers and assessment of their stability and antioxidant activities. J Agric Food Chem 65:818–826
89. Yuan J, Chen F (2011) Kinetics for the reversible isomerization reaction of trans-astaxanthin. Food Chem 73:131–137
90. Zhao L, Chen F, Zhao G, Wang Z, Liao X, Hu X (2005) Isomerization of trans-astaxanthin induced by copper (II) ion in ethanol. J Agric Food Chem 53:9620–9623
91. Cho J, Zhang Z, Kim I (2012) Effects of canthaxanthin on egg production, egg quality, and egg yolk color in laying hens. J Agric Sci 5:269–274
92. Malik K, Tokkas J, Goyal S (2012) Microbial pigments: a review. Int J Microb Resour Technol 1:361–365
93. Chew B, Park J, Wong M, Wong T (1999) A comparison of the anticancer activities of dietary β-carotene, canthaxanthin and astaxanthin in mice in vivo. Anticancer Res 19:1849–1853
94. Chan K, Mong M, Yin M (2009) Antioxidative and anti-inflammatory neuroprotective effects of astaxanthin and canthaxanthin in nerve growth factor differentiated PC12 cells. J Food Sci 74:H225–H231
95. Venugopalan V, Tripathi S, Nahar P, Saradhi P, Das R, Gautam H (2013) Characterization of canthaxanthin isomers isolated from a new soil Dietzia sp. and their antioxidant activities. J Microbiol Biotechnol 23:237–245
96. Li M, Rahman M, Wu B, Lin Y (2017) Effects of dietary canthaxanthin on growth and body colour of blood parrot cichlid Amphilophus citrinellus × Paraneetroplus synspilus. Aquac Int 25:705–713
97. Venugopalan V, Verma N, Gautam H, Saradhi P, Das R (2009) 9-cis-Canthaxanthin exhibits higher pro-apoptotic activity than all-trans-canthaxanthin isomer in THP-1 macrophage cells. Free Radic Res 43:100–105
98. Qiu D, Zhu W, Tang C, Shi L, Gao H (2014) Identification of the composition of isomeric canthaxanthin sample by NMR, HPLC, and mass spectrometry. Food Anal Methods 7:597–605
99. Sundquist A, Hanusch M, Stahl W, Sies H (1993) Cis/trans isomerization of carotenoids by the triplet carbonyl source 3-hydroxymethyl-3,4,4-trimethyl-1,2-dioxetane. Photochem Photobiol 57:785–791
100. Maeda H, Hosokawa M, Sashima T, Funayama K, Miyashita K (2005) Fucoxanthin from edible seaweed, Undaria pinnatifida, shows antiobesity effect through UCP1 expression in white adipose tissues. Biochem Biophys Res Commun 332:392–397
101. Sachindra N, Sato E, Maeda H, Hosokawa M, Niwano Y, Kohn M, Miyashita K (2007) Radical scavenging and singlet oxygen quenching activity of marine carotenoid fucoxanthin and its metabolites. J Agric Food Chem 55:8516–8522
102. Hosokawa M, Kudo M, Maeda H, Kohno H, Tanaka T, Miyashita K (2004) Fucoxanthin induces apoptosis and enhances the antiproliferative effect of the PPARγ ligand, troglitazone, on colon cancer cells. Biochimica et Biophysica Acta (BBA)—General Subject 1675:113–119
103. Sugawara T, Matsubara K, Akagi R, Mori M, Hirata T (2006) Antiangiogenic activity of brown algae fucoxanthin and its deacetylated product, fucoxanthinol. J Agric Food Chem 54:9805–9810
104. Maeda H (2015) Nutraceutical effects of fucoxanthin for obesity and diabetes therapy: a review. J Oleo Sci 64:125–132
105. Maeda H, Hosokawa M, Sashima T, Murakami F, K., & Miyashita, K. (2009) Anti-obesity and anti-diabetic effects of fucoxanthin on diet-induced obesity conditions in a murine model. Mol Med Rep 2:897–902
106. Maeda H, Kanno S, Kodate M, Hosokawa M, Miyashita K (2015) Fucoxanthinol metabolite of fucoxanthin, improves obesity-induced inflammation in adipocyte cells. Mar Drugs 13:4799–4813

107. Kawee-ai A, Kuntiya A, Kim S (2013) Anticholinesterase and antioxidant activities of fucoxanthin purified from the microalga Phaeodactylum tricornutum. Nat Prod Commun 8:1381–1386
108. Zhao D, Kim S, Pan C, Chung D (2014) Effects of heating, aerial exposure and illumination on stability of fucoxanthin in canola oil. Food Chem 145:505–513
109. Abdel AE, Akhtar H, Zaheer K, Ali R (2013) Dietary sources of lutein and zeaxanthin carotenoids and their role in eye health. Nutrients 5:1169–1185
110. Mares P, Millen J, Ficek AT, Hankinson S (2002) The body of evidence to support a protective role for lutein and zeaxanthin in delaying chronic disease. Overview. J Nutr 132:518S–524S
111. Mozaffarieh M, Sacu S, Wedrich A (2003) The role of the carotenoids, lutein and zeaxanthin, in protecting against age-related macular degeneration: a review based on controversial evidence. Nutr J 2:1–8
112. Rodrigues D, Mariutti, & L., Mercadante, A. (2016) An in vitro digestion method adapted for carotenoids and carotenoid esters: moving forward towards standardization. Food Funct 7:4992–5001
113. Subagio A, Morita N, Sawada S (1998) Thermal isomerization of all-trans-lutein in a benzene solution. Biosci Biotechnol Biochem 62:2453–2456
114. Updike A, Schwartz S (2003) Thermal processing of vegetables increases cis isomers of lutein and zeaxanthin. J Agric Food Chem 51:6184–6190
115. Guo W, Tu C, Hu C (2008) Cis-trans isomerizations of β-carotene and lycopene: a theoretical study. J Phys Chem 112:12158–12167
116. Gamlieli BI, Korin E, Cohen S (2002) Selective separation of cis-trans geometrical isomers of β-carotene via CO2 supercritical fluid extraction. Biotechnol Bioeng 80:169–174
117. Kaga K, Honda M, Adachi T, Honjo M, Wahyudiono KH, Goto M (2018) Nanoparticle formation of PVP/astaxanthin inclusion complex by solution-enhanced dispersion by supercritical fluids (SEDS). Effect of PVP and astaxanthin Z-isomer content. J Supercrit Fluids 36:44–51
118. Stahl W, Schwarz W, Sundquist A, Sies H (1992) cis-trans isomers of lycopene and β-carotene in human serum and tissues. Arch Biochem Biophys 294:173–177
119. Hempel J, Schadle C, Leptihn S, Carle R, Schweiggert R (2016) Structure related aggregation behavior of carotenoids and carotenoid esters. J Photochem Photobiol A Chem 317:161–174
120. Sun Q, Yang C, Li J, Raza H, Lycopene ZL (2016) Heterogeneous catalytic E/Z isomerization and in vitro bioaccessibility assessment using a diffusion model. J Food Sci 81:2381–2389
121. Stahl W, Sies H (1992) Uptake of lycopene and its geometrical isomers is greater from heat-processed than from unprocessed tomato juice in humans. J Nutr 122:2161–2166
122. During A, Hussain M, Morel D, Harrison E (2002) Carotenoid uptake and secretion by CaCo-2 cells: β-carotene isomer selectivity and carotenoid interactions. J Lipid Res 43:1086–1095
123. Lavy A, Amotz AB, Aviram M (1993) Preferential inhibition of LDL oxidation by the alltrans isomer of β-carotene in comparison with 9-cis β-carotene. Clin Chem Lab Med 31:83–90
124. Levin G, Mokady S (1994) Antioxidant activity of 9-cis compared to all-trans β-carotene in vitro. Free Radic Biol Med 17:77–82
125. Stahl W, Schwarz, W.,& Sies, H. (1993) Human serum concentrations of all-trans β-and α-carotene but not 9-cis β-carotene increase upon ingestion of a natural isomer mixture obtained from Dunaliella salina (Betatene). J Nutr 123:847–851
126. Stahl W, Schwarz W, Von LJ, Sies H (1995) All-trans β-carotene preferentially accumulates in human chylomicrons and very low density lipoproteins compared with the 9-cis geometrical isomer. J Nutr 125:2128–2133
127. Gaziano J, Johnson E, Russell R, Manson J, Stampfer M, Ridker P, Frei B, Hennekens, C.,& Krinsky, N. (1995) (1995). Discrimination in absorption or transport of β-carotene isomers after oral supplementation with either all-trans-or 9-cis-β-carotene. Am J Clin Nutr 61:1248–1252

128. Johnson E, Qin J, Krinsky N, Russell R (1997) β-Carotene isomers in human serum, breast milk and buccal mucosa cells after continuous oral doses of all-trans and 9-cis β-carotene. J Nutr 127:1993–1999
129. Johnson E, Krinsky, N.,& Russell, R. (1996) Serum response of all-trans and 9-cis isomers of β-carotene in humans. J Am Coll Nutr 15:620–624
130. Relevy N, Rühl R, Harari A, Grosskopf I, Barshack I, Ben AA, Nir U, Gottlieb H, Kamari Y, Harats D, Shaish A (2015) 9-cis β-carotene inhibits atherosclerosis development in female LDLR−/− mice. Funct Foods Health Dis 5:67–79
131. Harari A, Abecassis R, Relevi N, Levi Z, Ben AA, Kamari Y, Harats D, Shaish A (2013) Prevention of atherosclerosis progression by 9-cis-β-carotene rich alga Dunaliella in apoE-deficient mice. Biomed Res Int 169517:1–7
132. Coral HG, Ytrestøyl T, Ruyter B, Bjerkeng B (2014) Plasma appearance of unesterified astaxanthin geometrical E/Z and optical R/S isomers in men given single doses of a mixture of optical 3 and 3′R/S isomers of astaxanthin fatty acyl diesters. Comp Biochem Physiol Part C Toxicol Pharmacol 139:99–110
133. Østerlie M, Bjerkeng B, Liaaen JS (2000) Plasma appearance and distribution of astaxanthin E/Z and R/S isomers in plasma lipoproteins of men after single dose administration of astaxanthin. J Nutr Biochem 11:482–490
134. Bjerkeng B, Følling M, Lagocki S, Storebakken T, Olli J, Alsted N (1997) Bioavailability of all-E-astaxanthin and Z-isomers of astaxanthin in rainbow trout (Oncorhynchus mykiss). Aquaculture 157:63–82
135. Nakazawa Y, Sashima T, Hosokawa M, Miyashita K (2009) Comparative evaluation of growth inhibitory effect of stereoisomers of fucoxanthin in human cancer cell lines. J Funct Foods 1:88–97
136. Zhang Y, Fang H, Xie Q, Sun J, Liu R, Hong Z, Yi, R.,& Wu, H. (2014) Comparative evaluation of the radical-scavenging activities of fucoxanthin and its stereoisomers. Molecules 19:2100–2113
137. Yang C, Fischer M, Kirby C, Liu R, Zhu H, Zhang H, Chen Y, Sun Y, Zhang L, Tsao R (2018) Bioaccessibility, cellular uptake and transport of luteins and assessment of their antioxidant activities. Food Chem 249:66–76
138. Böhm V, Puspitasari NN, Ferruzzi M, Schwartz S (2002) Trolox equivalent antioxidant capacity of different geometrical isomers of α-carotene, β-carotene, lycopene, and zeaxanthin. J Agric Food Chem 50:221–226
139. Schweiggert U, Kammerer DR, Carle R, Schieber A (2005) Characterization of carotenoids and carotenoid esters in red pepper pods (Capsicum annuum L.) by high-performance liquid chromatography/atmospheric pressure chemical ionization mass spectrometry. Rapid Commun Mass Spectrom 19(18):2617–2628
140. Lichtenthaler HK, Buschmann C (2001) Chlorophylls and carotenoids: measurement and characterization by UV-VIS spectroscopy. Curr Protocol Food Anal Chem 1(1):F4.3.1–F4.3.8
141. Khoo H-E, Prasad KN, Kong K-W, Jiang Y, Ismail A (2011) Carotenoids and their isomers: color pigments in fruits and vegetables. Molecules 16(2):1710–1738
142. Berg CJ, LaFountain AM, Prum RO, Frank HA, Tauber MJ (2013) Vibrational and electronic spectroscopy of the retro-carotenoid rhodoxanthin in avian plumage, solid-state films, and solution. Arch Biochem Biophys 539(2):142–155
143. Prasain JK, Moore R, Hurst JS, Barnes S, van Kuijk FJGM (2005) Electrospray tandem mass spectrometric analysis of zeaxanthin and its oxidation products. J Mass Spectrom 40(7):916–923

Chapter 4
Apocarotenoids

Madiha Ilyas, Faraz Ali Rana, and Muhammad Riaz

4.1 Introduction

Apocarotenoids are organic compounds commonly found in living organisms. They are obtained from carotenoids with the help of bond breaking process and are catalyzed through enzyme named as carotenoid oxygenases. Due to the significant quantity of carotenoid precursors, deviations at the adjacent sequence in the oxidation site and functional changes, the wide variety of apocarotenoids are present in nature [1]. Sources are vitamin A and abscisic acid of the plant hormone. Retinol, retinal and retinoic acid, these all vitamin A are apocarotenoids. However, apocarotenoids are known for the pleasing odors and edible essences of many floras and fruits (for example rose, violet, tomato and raspberry) [2–4] and carotenoids also give rise to the optical beauty to blossoms and fruits. In the same way to the colored carotenoids these apocarotenoid fragrances attract pollinators and give rise to interactions between plants and insects [5].

Furthermore, particular apocarotenoids play their role like hormones. For instance in plant growth procedure, the plant growing hormone named as abscisic acid, also has several purposes including bud dormancy and reaction to environmental stress and plant pathogens [2]. Carotenoids and apocarotenoids have been broadly utilized in diet, feed, nutritious, therapeutic and particular care industries because of their pigment, fragrance, significant nutritional and health claims. There are rapidly increased consumer demands for carotenoids and apocarotenoids, as more and more clinical trials reveal numerous health and medicinal benefits [6].

M. Ilyas (✉)
Department of Nutrition, Government College Women University, Faisalabad, Pakistan

F. A. Rana
Department of Haematology, Armed Forces Institute of Pathology, Rawalpindi, Pakistan

M. Riaz
Department of Pharmacy, Shaheed Benazir Bhutto University, Sheringal Dir (U), Pakistan

© Springer Nature Switzerland AG 2021
M. Zia-Ul-Haq et al. (eds.), *Carotenoids: Structure and Function in the Human Body*, https://doi.org/10.1007/978-3-030-46459-2_4

By 2021, it is estimated that the worldwide carotenoid market will hit USD 1.53 billion. To overcome vitamin A deficiency, it is important to take diet with a high amount of carotenoids (for example beta-carotene) or retinoids on daily basis. Serious complications of Vitamin A deficiency are blindness, reduced immunity and in severe conditions it may lead to death. For the prevention of age-related macular deterioration, lutein and zeaxanthin both are good for eye health [7].

4.1.1 Classification of Apocarotenoids

Apocarotenoids are classified in to Abscisic acid, Apocarotenal, Bixin, Crocin, Crocetin, saffranal and Ionones. Strigolactones constitute additional essential sub class of apocarotenoids, which act as an inhibitor of shoot-branching; furthermore, it also stimulates the synthesis of symbiotic relationship between plants and fungi [8]. Astaxanthin another subclass, has few more advantages including effective antioxidant activities, stimulating immune reactions, decreasing eye weakness, improving functions of muscles and many others [9]. Alpha-ionone and beta-ionone are broadly utilized in beauty products such as scents as they have little extraordinary fragrance capability [10]. Another useful apocarotenoid which is known as crocin, and red color of saffron is due to crocin. Saffron is a highly valuable spice whose retail cost ranging between 2000 and 7000 Euros/kg.

To date, almost 1,117 natural occurring carotenoids and apocarotenoids have been identified, which comprise of C30, C40, C45 and C50 carotenoids [11]. Between them, the most abundant carotenoids are C40 carotenoids and their derivative apocarotenoids which have 1,093 separate structures. In this chapter, I will cover few of the apocarotenoids that will illustrate the antioxidant potential and health claims.

4.1.2 Apocarotenoid Functions

Apocarotenoids and their roles have become more and more evident in the last decade. These include the volatile C13 norisoprenoid; "β-ionone" and "β-damascenone" accountable for a typical raised fragrance [12], and the C13 or isoprenoids, which is responsible for the flowery and fruity features of grapevine of wine (*Vitisv inifera* L.) and berries [13]. Other examples showed characters for undesired, chemo appellants, progress simulants and blockers of apocarotenoids [14]. Many flora and vegetable tissues release changeable apocarotenoid to encourage interactions between insects and plants [15]. New and interesting roles have become evident in comparison to well-studied apocarotenoids linked to the taste, scent and fragrances. The hormones elaborated in herbal production and development like ABA and SLs are best recognized examples.

4.2 Biosynthesis of Apocarotenoids

Biosynthesis of apocarotenoids initiated from the precursors i.e. carotene, zeaxanthin and lycopene, lead up to the end products i.e. pigments (bixin), flavoring and fragrance compounds (crocin, picrocrocin), signal molecules (mycorradicin) and plant hormones (abscisic acid and strigolactones).

Apocarotenoids produced by the oxidative cleavage of carotenoids; a class of terpenoid compounds. The huge quantity of carotenoid precursors above than 750 to date have been recognized, distinctions in particular cleavage sites and adaptation after cleavage results in the abundant diversity of apocarotenoids [16]. The apocarotenoid absorbs detectable light depending on the mass of the chromophore and is thus effective in attracting pollinators as color stains and also act as seed dispersants. Throughout plants, the etioplast, leucoplast, chromoplast and other plant tissues such as flowers and roots are found to contain apocarotenoids. The carotenoids are converted into apocarotenoids using regionally defined carotenoid cleavage oxygenases (CCDs) that object various double-bonds in the polyene carotenoid chain. Lycopene, β-carotene and zeaxanthin are the key precursors of identified apocarotenoids. Because of their high commercial importance, two of the most significant apocarotenoids are bixin and crocin. Bixin is polyene produced as a result of the enzymatic breakdown of the 5–6 and 50–60 double bonds by the inner portion of C40 lycopene. Enzymes like bixin aldehyde dehydrogenase (BiADH) and Lycopene cleavage dioxygenase (LCD) cause this cleavage and activate norbixin, the C24 dicarboxylic acid.

Synthesis of bixin results from the catalyzed reaction of Norbixin methyl transferase (nBMT) amongst S-adenosyl L-methionine and norbixin, which is obtained from the Bixaorellana (annatto bush native to Central and South America) seeds. Crocin; the glycosylated Crocetin form (Figs. 4.1 and 4.2), accumulates in huge quantities (up to 8% by dry weight) in saffron stigmata. The red pigmentation of saffron stigmas is due to the both pigments, the sour taste of saffron is due to the presence of picrocrocin in saffron and the presence of saffranal (Fig. 4.3) produce fragrance in saffron [17].

The suggested biosynthetic process starts with a symmetric zeaxanthine cleavage at 7, 8/7′, 8′ positions by non-heme iron carotenoid cleavage dioxygenase (CCD2). The cleavage of two components, 3-OH—cycellocitral and crocetine dialdehyde (ALDH) are dehydrogenated by the glycosylated agent (UGT) and ALDH and transferred into the vacuoles [18] to produce picrocrocin and crocin respectively [19]. It is interesting to note that the CsCCD2 expression involves a regular

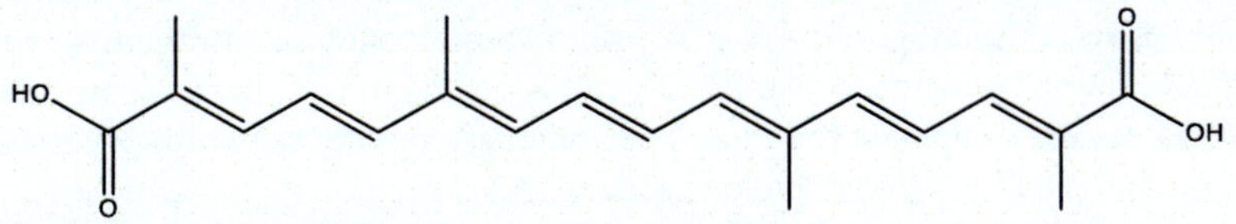

Fig. 4.1 Chemical structure of Crocin

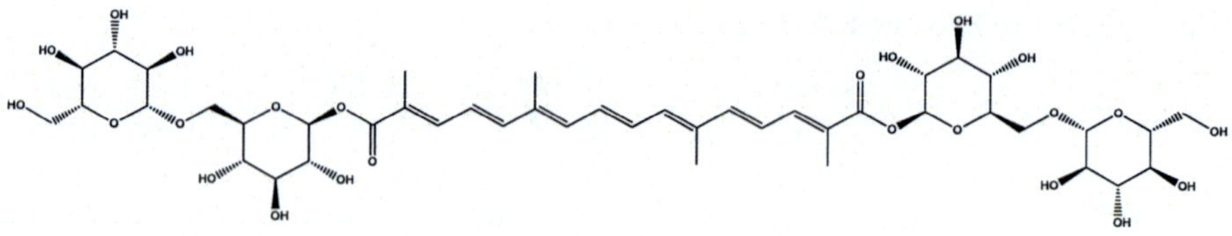

Fig. 4.2 Chemical structure of Crocetin

Fig. 4.3 Chemical structure of Safranal

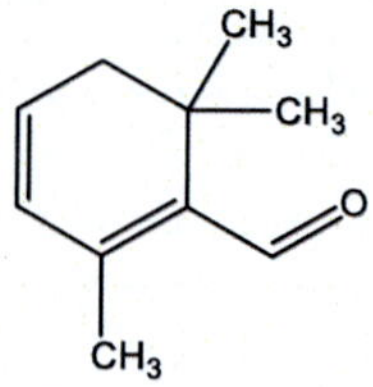

Fig. 4.4 Chemical structure of Abscisic acid

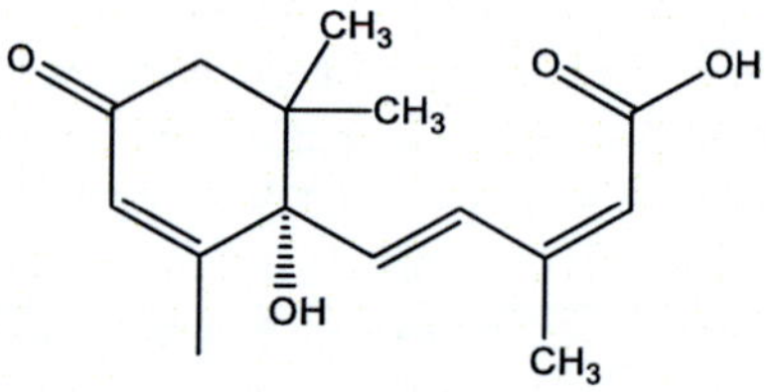

rhythmic modulation and "cyclocitral" induction, which indicates that the apocarotenoid act as a plastid to nucleus signaling molecule for the production of crocetin for stigmata of saffron [20].

From all of these, Abscisic acid (ABA) (Fig. 4.4) is an important factor in the control of the seed production, tolerance to drought, and sensing of sugar [21]. The initial stage which takes place in the plastid is particular to the abscisic acid biological production, the epoxidation of zeaxanthin and anther xanthin in to violet axanthin. An enzyme named zeaxanthin epoxidase (ZEP) which first time exposed its molecular identity in tobacco catalyzes that stage. Violaxanthin is transformed to 9-cys-epoxycarotenoid after a sequence of fundamental variations. The 9-cis-epoxycarotenoid dioxygenase (NCED) performed oxydative cleavage of the main epoxycarotenoid 9-cy-neoxanthin creates an intermediate C-15, called xanthoxin [22]. This phase is the fisrt stage performed in the abscisic acid biosynthesis process. The xanthoxin compound is then transported to the cytosol, where a two-step reaction converts the drug into ABA. A brief chain alcohol dehydrogensis / reductase (SDR) coded by the gene (AtABA2), catalyzes the initial phase of this reaction and produces ABA aldehyde [23–25]. The end phase on the route of biosynthesis is that, it is catalyzed by ABA aldehyde oxidase (AAO).

The plant Phytohormone Strigolactone (SL) explored in recent times has been found to engage in numerous physiological procedures i.e. shoot branching, plant production and plant growth synchronization and root morphology depending on supply of phosphate [26–28]. The root tissue of the synthesis of the strigolactone starts with a reversible "β-carotene" isomeration, catalytic of a D27 protein that yields 9-cis "β-carotene" (9-cis C40). The cleavage carotenoide dioxygenase CCD7 then splits the carotene into 9-cis "(carotene), forming the carotenes (9-cis C27), plus "β-ionone". After that the 9-cis C27 complex clustered with a carotenoid cleavage dioxygenase CCD to synthesize carlactone (CL), which is then transformed into strigolactone (SL) in the form of a tri-cyclic lactone structure (ABC ring) through the activity of MAX1 and probably other enzymes, which are still to be recognized. The symbiosis of arbuscular mycorrhizal (AM) is encouraged by the use of two other apocarotenoids i.e. C13′ ionol (previously identified as cyclohexenone) [29] and mycorradicine [30] (a C14 linear polyene dicarboxylic acid). These compounds are generated in roots when AM fungi are colonized specifically, but not in disease interfaces. The compounds C13 and C14 are both produced from simultaneous action of D27, CCD7 and CCD1 and are able together in 9 apocarotenoids at impartially higher levels. The compound C14 gives a macro-visible yellow coloration because of its yellow color [30], by following intense colonization with AM fungi to the roots of certain plants (for example maize). Both are glycosylated (C13) and undefined (C14) compounds [31].

4.2.1 *Production of Apocarotenoids*

CCDs or other oxygenases may further convert carotenoids into apocarotenoids. Retinol, or vitamin A, is the apocarotenoid having major importance to individual's health having good market quality. Retinol has an integral role in the development of vision, bones and act as antioxidants in stimulating skin health [32].The three more are aromatic compounds, alpha-ionone that is obviously present in raspberry, beta-ionone, exists in most of the fruits and blossoms like strawberries, osmanthus and violets and gamma-ionone is a part of Tamarindus indica (tamarind) and present in fruits (Fig. 4.5).The chemical production of the three molecules is not very complex and makes a significant difference to the existing market share (Fig. 4.6).

4.3 Apocarotenoids in Foods

Foods may also contain β-Apocarotenes which is produced through autoxidation, β-carotene thermal degradation, and throughout food processing. Natural autoxidation of toluene of ß-carotene solutions with 100% oxygen for 60 °C over a duration of 120 min culminated in the homologation of β-carotene carbonyl cleavage products such as retinaldehyde, β-apo-13-carotenone, ß-apo-14-carotenal, β-apo-12 carotenal and β-apo-10-carotenal. The inverse HPLC stage was used to detach these beta-carotenes, and with the help of an online diodes detector, specific peaks were

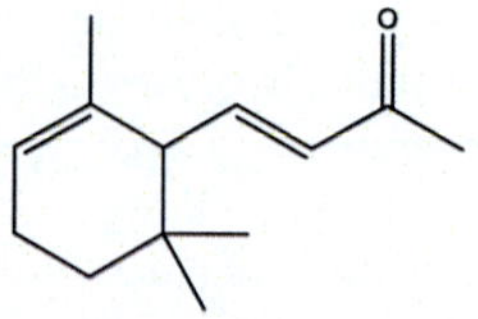

(3E)-4-(2,6,6-Trimethylcyclohex-2-en-1-yl)but-3-en-2-one

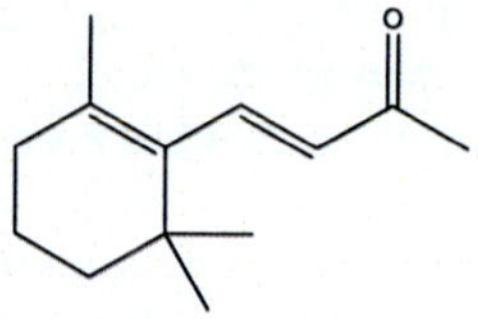

(3E)-4-(2,6,6-Trimethylcyclohex-1-en-1-yl)but-3-en-2-one

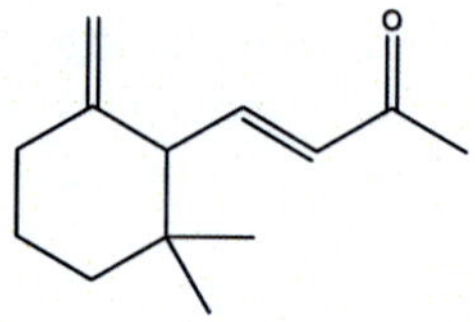

(3E)-4-(2,2-Dimethyl-6-methylenecyclohexyl)but-3-en-2-one

Fig. 4.5 chemical structures of Ionones (α+β+γ)

Fig. 4.6 Production of Apocarotenoids

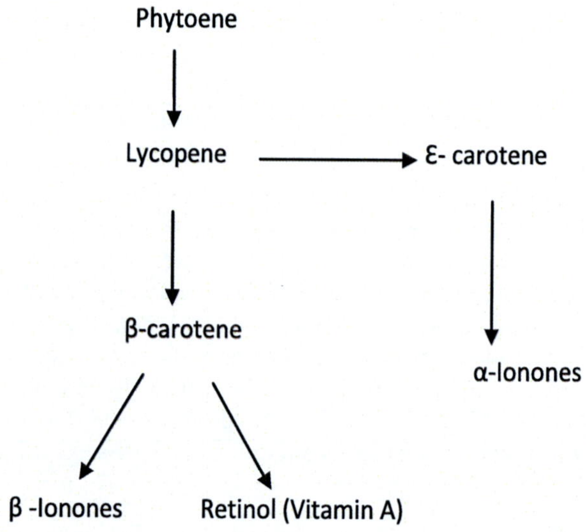

categorized [33]. In the simultaneous deodorization, β-apo-13-carotene, ß-apo-14-carotene and the retinaldehyde were separated in segments by utilizing a silicic acid column. Electromagnetic spectroscopy and mass spectrometer detections were recognized by the use of refined palm oil and β-carotene [34]. In a further study related to food, β-apo-8-carotenal, β-apo-10% carotenal, β-apo-12-carotenal, β-apo-14 carotenal and seoxidation products were originated to be 5% of the original β-carotene, synthesized by several heat treatments known as thermal degradation of beta carotene. The food also contains β-apocarotenosids [35].

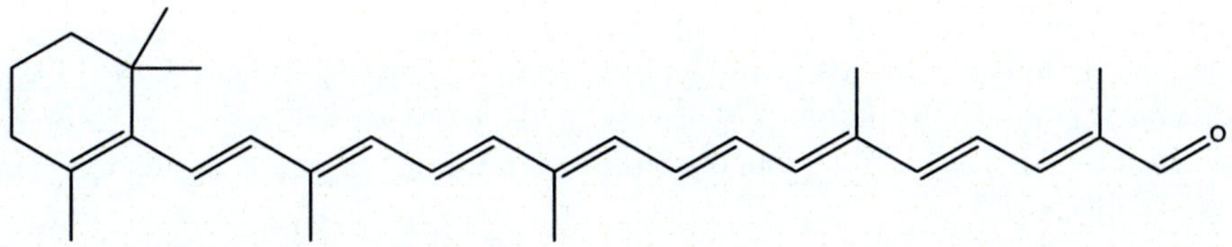

Fig. 4.7 Chemical structure of Apocarotenal

Thin layer chromatography, HPLC, visible absorption spectrums acquired from a photodiode array detector and mass spectra were studied to distilled mango juice, acerola juice, and dried apricots. The presence of β-apocarotenals (Fig. 4.7) including β-apo-8,-10-,-12-, and-14-β-carotenes has been tested in these juices and foods. Nothing was detected any presence of oxidation product in acerola or in the dried apricots, while mango juice containing only β-apo-12-carotenal were detected [36]. β-apocarotenoids are present in orange-fleshed melons which is an organic food. The ranks of β-apo-13-carotenone and β-apocarotenals comprising β-apo-8-, -10-,-12-, and -14-carotenal were detected in melons (cantaloupe and greenhouse-grown orange tube). The five β-apocarotenoids in both fruits were around 30 pmol/g wet in weight and in total around 1.5% of the β-carotene level in the melons [37].

Bixin and crocetin are dietary apocarotenoids present in foodstuffs at high levels. Bixin is a key part of annatto, a pigment extracted from the outer covering of the sensitive soft seed of Bixaorellana, a tiny tropical tree. Annatto has been utilized in foods (sausages, margarine, cheeses or vegetable oils) and in cosmetics for coloring for many hundreds of years [38–40]. Crocetin is the main Saffron carotenoid (*Crocus sativus* stigmas obtained by saffron spice). Since ancient times, Saffron has been utilized as a dye in food. Certain other apocarotenoids have also been shown to be available in food stuffs (including green leafy vegetables, green vegetables, cereal grains, berries and soft drinks) discovered by different studies [41, 42]. Though the effectiveness is slightly below than parent carotenoids from which they are extracted in concentrations e.g. in tomatoes, red grapefruits and watermelon products contain apo-6-, apo-8-, apo-10-, apo-12-, and apo-14-lycopenals have been detected in tomato, red grapefruit, and watermelon product [43–45].

4.4 Apocarotenoids in Mammals

Apocarotenoids are produced in the tissues of mammals. Retinoids are mainly formed in mammalian tissues through central β-carotene cleavage and regulated via cytosolic beta-carotene enzyme 15, 15-oxygenase (BCO1), resulting in two retinaldehyde molecules [46]. Vitamin A is essential for the maintenance of human health and mainly performed its functions through the transcriptional activity of retinoic acid, the ligand of the nuclear retinoic acid receptors (RARs)/retinoic X receptors (RXRs) which modulated the genes expression [47].

Retinaldehyde molecules can be then either oxidized to retinoic acid or diminished to retinol, which makes retinyl esters after esterification and in this form vitamin-A stores in the tissues [48]. Likewise, the retinaldehyde act as the chromophore of rhodopsin, form the foundation for the necessary function of vitamin A in eyes health [46].

4.5 Biological Functions of Apocarotenoids

In 1960s and beginning of the 1970s, numerous important biological roles of β-apocarotenoids has been illustrated [49, 50]. These studies revealed that β-carotene which includes β-apo-8-carotene; β-apo-10-carotene and β-apo-12-carotene obtained from apocarotenoids, can improve growth in vitamin A deficient quails, rats and chickens. Nevertheless, during these earlier studies, it was not obvious whether β-apocarotenoids could carry out this process as entirely retinoid precursors or as such. A number of influential findings on the enzyme of mammals and its cleavage products i.e. β-carotene, BCO2 has been carried out [51]. BCO2 cleaves β-carotene at double bond of C9 C10 and therefore only produces β-apo-10-carotene and β-ionone [51, 52]. It was therefore assumed that the lengths of β-apocarotene chain could be generated by a non enzymatic pathway [53].The researchers recommended that β-apocarotenoids produced via β-carotene might be responsible in biosyntheis of retinoic acid from β-carotene.

However, the mice without having sufficient amount of *Bco1* became vitamin A deficits in spite of BCO2; β-carotene was the sole supply of vitamin A in their food intakes, which indicated that the enzyme and its cleavage products had a probable differentiated physiological role [54]. Furthermore, *Bco2*$^{-/-}$ mice were explored, that were fed on a diet with xanthophylls as these carotenoids possibly will not be digested by BCO1, which was expressed in the *Bco2*$^{-/-}$ animals [55]. The researchers have described that Lutein and Zeaxanthin metabolites (i.e., 3-dehydro derivatives) are accumulated in the tissues of supplemented mice's. In addition, this kind of build-up causes mitochondrial improper function because it is evaluated through boosting in the dismutase of manganese superoxide and a lower Adenosine diphosphate dependent respiration. The researchers prove that the accumulation of carotenoids has directly impaired the electric transport chain regardless of the structural intact of mitochondria and they also reported the improved ROS synthesis and depolarization of mitochondrial membrane that was induced in HepG2 cells when treated with zeaxanthine, lutein, their derivatives and as well as β-carotene. This deficiency of mitochondrial function was attributed to cell signaling pathways induction and correlated with oxidative stress and disease. It is evident that β-carotene provoked the generation of ROS and reduced mitochondrial membrane depolarization in BCO2-dependent ways that contribute to apoptotic pathways activation [56].

After that, in vivo studies were carried out with single and double knock-out rats for *Bco1* and *Bco2*, to further develop BCO1 and BCO2 contributions to β-carotene metabolism. They showed that BCO1 in adult mouse tissue is the major β-carotene-metabolizing enzyme, and that BCO2 single does not make significant contributions to β-carotene homeostasis, β-apo-10-carotenoid (β-apo-10-carotenol) was produced by activity of β-carotene that is again confirmed in vivo. Yet β-carotene was appeared to be a weak BCO2 substrate in vivo [57].

The researchers demonstrated that, asymmetrical conversion of provitamin-A carotenoid, i.e. β-cryptoxanthin, in to retinods, could be important. It has been suggested by these and other facts that "the β-carotene enzymatic cleavage can take place in the test tube but is mostly lacking in live organisms." [58]. Moreover, the asymmetric BCO2 cleavage helps to protect mitochondrion from toxic effects of carotenoids (comprising β-carotene) which may cause oxidative stress. This concept was later on confirmed the relationship among carotenoids metabolism and disorders like various cancers, age-linked macular deterioration, and metabolic disorders [59–64].

It has been observed that, numerous chain lengths and shapes (aldehyde, acid and ketone) of beta-apocarotenoids provoked trans-retinoic acid mediated trans-activation of all three RAR isoforms at nano-molar concentrations, either by conflicting openly with retenoic acid or, for example in cases with β-apo-13-carotenone, for receptor binding by modifying the RXR oligomeric state [65, 66]. Likewise, β-apo-14-carotene efficiently blocked agonist induced RXRα, PPARα and PPARγ initiation, reducing adipogenesis. It is evident that β-carotene can produce RAR agonists (all-trans retinoic acids) by symmetric cleavage and powerful RAR antagonist (β-apocarotenoids), have been reported the adverse effects of β-carotene given for cancer prevention [66–68].

In another study, it has been found that adipocyte marker gene expression has been enhanced in 3T3-L1 cells with β-apo-13-carotenone or β-apo-10-caroteneic acid. On the other hand, β-apocarotenoid was failed to decrease the obstruction of 3T3 L1 cell differentiation for every-trans-retinoic therapy. That's why scientists proposed, however, that few β-apocarotenoids can also work by different pathways, e.g. by non-genomic activity as well as a transcriptional antagonist.

It has been validated in in-vivo study that, β-apocarotenoid's transcription regulatory activity by suggesting that a feed-forward controlling pathway is formed by ß-carotene metabolites to provide placental-fetal β-carotene transfers by regulation of placental lipoprotein biosynthesis. In particular, the researchers suggested that the placental transcription of Bco2 would get higher when β-carotene is present, hence, the production through asymmetrical cleavage of β-apo-10-carotenal from β-carotene also increased [69]. This metabolite enhances the transcription of Hnf4α that in result elevated the transcription and functions of microsomal triglyceride transfer protein and improving the transfer of intact β-canotene to the embryo by the regulator in interacellular movement and excretion of apo B-lipoproteins. It has not been observed that, nor Hnf4α or Mttp potentially are the primary targets of the action of β-apo-10-carotenoids [70].

4.5.1 Absorption of Apocarotenoids in Gastrointestinal Tract

There is partial evidence obtained on the intact absorption of apocarotenoids from diet in the intestine and this area needs further investigations. In one human clinical trial, the study participants were consumed rich amount of β-carotene (35 μg/day) tomato juice and lycopene tomato juice (120 μg/day) for about 1 month. Only β-apo-13-carotenone at a concentration of 0.5 nM was observed exceeding the limits (50 pM). It indicated that apocarotenoids are quickly metabolized or not absorbed by the intestine. The same finding is validated by initial in-vitro trial via human Caco-2 cells of intestine. The cells were subjected to 1–5 μM β-apo-8-carotenes, β-apo-10-carotene or β-apo-13 carotenone, revealed that, cells uptake these compounds rapidly and metabolized extensively [71].

4.6 Saffron- Apocarotenoid

Crocus, having species more than 85 and it belongs to the family iridaceae. The greatest significant species is *Crocus sativus*, the *C. sativus* is very costly spice that needs intensive work and time for its cultivation [72].

Saffron is a grown plant in different countries of the world, including Spain, Iran, India, Turkey and Greece. Saffron was cultivated from hundreds of years. The market for saffron has risen due to its high medicinal value. Its high demand in the markets is regarded as the key purpose for rising saffron growth in different areas of the world. Carotenoids, glycosides, mono-terpenes, anthocyanins, aldehydes, vitamins (specifically vitamin B1 & B2), flavonoids, amino acids, fats, starch, gums and minerals are the chief ingredients of saffron [73].

Moreover, the essential bioactive components are known to be found is apocarotenoids included crocetin, safranal (zeaxanthin bio-oxidative cleavage products) picrocrocin and crocin. Strong pigment is due to the presence of Crocin whereas aroma is due to safranal and picrocrocin causes sour taste. Saffron is regarded as an essential medicinal plant because of essential nutrients and bioactive compounds present in it. Crocin (digentiobiosyl 8, 8′-diapocarotene-8, 8μ-oat; C44H64O24) is regarded as one of the exceptional water soluble carotenoids present naturally, as a diester of gentiobiosis (disaccharide) and dicarboxylic acid crocetin [74]. Except crocin-1 all the other crocin derivatives are claimed to be present in cis-trans isomeric forms. The other bioactive element in saffron is Crocetin, which cannot dissolve in water and other organic solvents. It is a dicarboxylic acid, a carotenoid which present naturally. Crocetin glucosyl ester forms with 7 double bonds and having four methyl groups on the side chains. One, two or three glucose are esterified at the end [75]. Previous studies showed that, only 6 glycosides of crocetin are present in saffron [76]. Furthermore, crocetin esters and their cis/trans isomers have been separated with the help of the UV / spectrometry and UV/HPLC photodiode array [74]. Safranal (C10H14O7) accounts for 60–70% of the saffron varies portion and is the most liable factor for the fragrance of saffron [77]. Cyclic terpenic aldehyde, is an enzyme produced by picro-crocin and HTCC (4-hydroxy-2, 6, 6-trimethyl-1-cyclohexen-1-carboxaldehyde) and also act as thermal degradation component in the storage process [78].

4.6.1 *Saffron World Production*

Saffron production is estimated to about 200 tons per annum globally [79]. In Europe, especially, in Italy, Greece and Spain, saffron cultivated areas are reduced in the last century, due to rise in the prices of saffron. It dropped from 6000 ha in 1971 to 200 ha in Spain, 1600 in 1982 to 860 in Greece [73, 80] and it reduced from 300 ha in 1910 to 6 ha in central Italy (Abruzzo) few years back. Conversely, in Iran from last three decades eminent increase has been observed in the production. Currently, Greece, Iran and India are the main producer countries as shown in (Table 4.1). It has been reported that Iran has the highest cultivation of saffron i.e. 47,000 ha mostly grown in the province of Khorasan [81]. The major countries import the saffron is Switzerland, Germany, USA, Italy, United Kingdom and France [82]. Spain also imports huge quantities of saffron, especially from Greece, Iran, and Morocco for its internal market and re-export.

4.6.2 *Therapeutic Potential of Saffron*

In traditional drugs, saffron was used for the treatment and prevention of cardiovascular diseases, inflammatory disorders, respiratory problems, infectious fever, back pain, period pains and whooping cough [88]. During the last 20 years decaying carotenoid derivatives have acquired significance in contemporary medicinal trials

Table 4.1 Saffron world production [83]

Country	Cultivated area (ha)	Production (kg)	References
Iran	47,000	160,000	[81]
India	n.a.	8000–10,000	[73]
Greece	860	4000–6000	[73]
Morocco	500	1000	[84]
Spain	200	300–500	[73]
Italy	35	120	[79]
Turkey	n.a.	10	[85]
France	1	4	[86]
Switzerland	n.a.	0.4	[72]
Azerbaijan	675	n.a.	[87]

and showed many features, such as anti-inflammatory, antimicrobial, anxiolytic [89] and anti-neurodegenerative ailments [90].

4.6.3 *Antioxidant Activity of Saffron*

Production of ROS rates in a body contribute to ' oxidative stress,' which is thought to be a major reason of different ailments [91]. In cells, the formation by the ROS species is a common procedure. It is estimated that carotenoids as an antioxidant are highly effective for scavenging of free radicals to prevent oxidation [74–76].

4.6.4 *Bioactive Compounds of Saffron*

It is evident that phenolics and flavonoids rich spices act as antioxidants and often utilized as nutraceuticals [92, 93]. The role of *C. sativusas* an antioxidant can primarily be linked to its dynamic constituents, which include safranal, crocin, crocetin and carotene, with antioxidant characteristics [94]. Numerous in vivo and in vitro researches revealed that the synergistic anti-oxidation capacity of all the bioactive constituents of saffron is primarily responsible for the medicinal and biological properties of saffron particularly for its alcoholic condensed form [92].

4.7 Saffron Health Benefits

Saffron is beneficial for health and used for the management of various disorders (Fig. 4.8). Likewise, it is also used as coloring agent in foods with a particular societal consideration, by adding the brilliant yellow and orange colors in dishes.

4.7.1 *Bioactivities of Saffron Components*

Saffron has preventive and protective role on many organs, such as kidneys, lungs, pancreas, G I Tract, heart, nervous system etc as reported in Table 4.2.

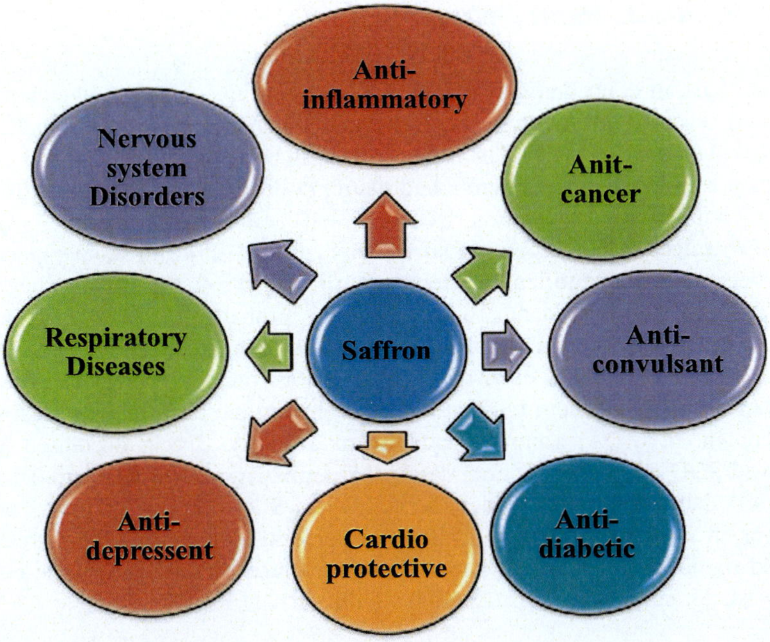

Fig. 4.8 Medicinal effects of saffron

Table 4.2 Biological functions of saffron [83]

Activity	Functions	References
Anti-inflammatory	In vivo antinociceptive effect	[95]
	Analgesic properties	[96]
	In vivo anti-inflammatory effect	[97]
Anticancer	Against lung cancer	[98]
	Skin carcinoma: a histopathological study	[99]
	in vivo antitumor effect	[100]
	Breast cancer	[101]
Antioxidant	Against lipoprotein oxidation	[102]
	Neuron oxidative stress	[103]
	Diabetic neuropathy	[104]
Central nervous system	Antidepressant	[105–109]
	Alzheimer's disease	[110–112]
	Epilepsy	[113]
	Memory enhancer	[114]
	Anticonvulsant	[115]
Cardio protective effect	Reducing the chances of heart attacks	[116, 117]
	Hypotensive effect	[118]
	Against atherosclerosis	[119]
	Calcium channel inhibitory	[120]

4.7.2 *Ischemic Heart Diseases*

Nervous System contributes to ischemia, stroke, Parkinson's & Alzheimer's disease, and several other neurodegenerative and psychiatric problems which occur due to oxidative stress [121]. The use of saffron in compounds has claimed to play important role in treating ischemia and neuronal disorders [122]. So, saffron has been recommended in cerebral ischemia and autoimmune encephalomyelitis (C57BL/6 mice) [123], hippocampal ischemia [124, 125] and renal reperfusion [126] because it is identified as effective curing agents for neuronal ailments as well. Saffron has antioxidant property which is probably the primary reason of decrease in cerebral ischemia which causes oxidative damage in rats [127]. Similarly, It has been observed that ethanolic extract of saffron increased the antioxidant potential in mice and there by increased the glutathione rates by suppressing the high content of MDA, aspartate and glutamate [128]. Crocin decreased MDA, improved SOD and GPx function, decreased peroxidation of lipids, and enhanced the antioxidant potential to reduce the activity of SOD, Na+K+-ATPase, catalase enzymes in ischemic stroke rats [129]. In cerebral ischaemia, crocin treatment reduced oxidizing reactions in micro vessels of mice and modulates the cortical microvascular endothelic cells (CMEC) significantly [130].

4.7.3 *Memory Loss*

Saffron consumption reduced the levels of GSH and MDA and also reduces the changes in monoamine oxidase (MAO) function induced by Al (aluminum). It also reduces the activity of AChE in cerebral tissues in Aluminum persuaded memory loss mice [131]. It was found that crocin is beneficial to fight against oxidative stress, in prevention of memory impairment and cognitive loss and restoration of memory in old age mice [132]. All of these progressive results of having saffron in diet can be linked to the antioxidants mechanism that induces the scavenging of free radicals which are involved in memory-impairment.

4.7.4 *Neurodegenerative Disorders*

Saffron is capable of inhibiting β-amyloid accumulation and prevents memory loss in brain, as well as being harmless and essential in minor to modest Alzheimer's disease (AD) depression [133]. Saffron has high antioxidant potential to inhibit the Aβ fibrillogenesis in a time dependent manner. It was observed that crocin amphiphilic properties make it most beneficial in the prevention of production of toxic amyloid structures [134]. Saffron plays an important role in the management of Alzheimer's disease by inhibiting the acetyl choline esterase (AChE), the enzyme which breakdown the acetylcholine. Saffron inhibits (30%) the activity of AChE, and considered beneficial for the treatment of AD [93].

In Parkinson's disease (PD), crocetin strengthen the antioxidant mechanism along with the reduction of thiobarbituric acid (TBARS), prevents the harmful effect of components such as the 6-hydroxy dopamine (involving in PD) and reduced the use of dopamine by tissue [135]. In hypertensive rats, which are at the risk of stroke, the antioxidant activity of crocetin on reactive species of oxygen inhibits the production of hydroxyl radical significantly [136].

4.7.5 *Anti-convulsant Role*

C. sativus have anti-convulsant activity in maximal electroshock seizure and in pentylene tetrazole (PTZ) [137]. The safranal also showed comparable results by minimizing generalized tonic colonic seizures and colonic seizures, decreasing the time and death rates of seizures and decreasing the occurrence of convulsions conversely to crocin which had no impacts [138]. The anticonvulsant and analgesic features of saffron and its impact on the morphine extraction are owing to the relation among GABA, saffron and opioid system [139].

4.8 Conclusions

Apocarotenoids are cleavage molecules of carotenoid, and generally obtained from β-carotene lycopene, and zeaxanthin. Mostly have important characteristics and play biological roles in plants, for example act as repellents, hormones, pigments and give aroma and color to the foods. From last decade several investigations showed the progress in the biosynthesis of apocarotenoid and its biological activities. Though, the studies in this direction are in its initial stages and having key issues that need to be explored further. In the last decade, apocarotenoids can be produced in the tissues of the mammals that have consumed or gained instantly from diet that are supplemented and are produced by the metabolic engineering of microbes; the biological availability of these molecules should be further understood.

Different clinical trials validated the claim that regular intake of carotenoids and apocarotenoids are useful for the prevention and treatment of various diseases. Functional foods and nutraceuticals are considered to be a sustainable solution for the management of such chronic diseases. Saffron is one such functional food which has gained popularity since many years.

Saffron-apocarotenoid practices now have a much extended list of positive health impacts and they are increasing day by day. These nutritional derivative components have been described to effect the growth of cancer, heart diseases, metabolic disorders and neurodegenerative diseases etc. Many chronic diseases such as cancers, inflammatory disorders, cardiovascular diseases, etc are associated with oxidative stress. Saffron and its components have good free radical scavenging activity and regulate the metabolic and cell signaling pathways. The information described in this chapter promoting the utilization of saffron for the management of disorders. Though, there is a dire need to explore the mechanism of saffron and its components, to estimate the toxic and upper intake levels. More clinical trials on human beings need to be carried out to validate the findings.

References

1. Alejandro FB, Pinoa MMHD, Arteagab JF, Delgadob JAG (2017) Occurrence and chemical synthesis of Apocarotenoids from mucorales: a review. Nat Prod Commun 7(5):733–741
2. Walter MH, Floss DS, Strack D (2010) Apocarotenoids: hormones, mycorrhizalmetabolite-sand aroma volatiles. Planta 232:1–17
3. Lalko J, Lapczynski A, McGinty D, Bhatia S, Letizia CS, Api AM (2007) Fragrance material review on beta-ionone. Food Chem Toxicol 45(Suppl 1):S241–S247. https://doi.org/10.1016/j.fct.2007.09.052
4. Larsen M, Poll L (1990) Odour thresholds of some important aroma compounds in raspberries. Z Lebensm Unters Forsch 191:129–131. https://doi.org/10.1007/bf01202638
5. McQuate GT, Peck SL (2001) Enhancement of attraction of alpha-ionol to male Bactrocera latifrons (Diptera: Tephritidae) by addition of a synergist, cade oil. J Econ Entomol 94:39–46
6. Bernstein PS, Li B, Vachali PP, Gorusupudi A, Shyam R, Henriksen BS, Nolan JM (2016) Lutein, zeaxanthin, and meso-zeaxanthin: the basic and clinical science underlying carotenoid based nutritional interventions against ocular disease. Prog Retin Eye Res 50:34–66
7. Akhtar S, Ahmed A, Randhawa MA, Atukorala S, Arlappa N, Ismail TT (2013) Prevalenceof vitamin A deficiency in South Asia: causes, outcomes, and possible remedies. J Health Popul Nutr 31:413–423
8. Gutjahr C, Parniske M (2013) Cell and developmental biology of arbuscular mycorrhiza symbiosis. Annu Rev Cell Dev Biol 29:593–617. https://doi.org/10.1146/annurev-cellbio-101512-122413
9. Kidd P (2011) Astaxanthin, cell membrane nutrient with diverse clinical benefits and antiaging potential. Altern Med Rev 16:355–364
10. Curtis T, Williams DG (1994) Introduction to perfumery. Ellis Horwood Limited, Hemel Hemstead
11. Yabuzaki J (2017) Carotenoids database: structures, chemical fingerprints and distribution among organisms. Database 2017:bax004. https://doi.org/10.1093/database/bax004
12. Huang FC, Molnar P, Schwab W (2009) Cloning and functional characterization of carotenoid cleavage dioxygenase 4 genes. J Exp Bot 60:3011–3022
13. Mendes-Pinto MM (2009) Carotenoid breakdown products the—norisoprenoids—in wine aroma. Arch Biochem Biophys 483:236–245
14. Bouvier F, Isner JC, Dogbo O, Camara B (2005) Oxidative tailoring of carotenoids: a prospect towards novel functions in plants. Trends Plant Sci 10:187–194
15. Azuma H, Toyota M, Asakawa Y, Takaso T, Tobe H (2002) Floral scent chemistry of mangroveplants. J Plant Res 115:0047–0053
16. Schwartz SH, Qin X, Zeevaart JA (2001) Characterization of a novel carotenoid cleavage dioxygenase from plants. J Biol Chem 276:25208–25211
17. Caballero OH, Miranda PR, Abdullaev FI (2005) HPLC quantification of major activecomponents from 11 different saffron (*Crocus sativus* L.) sources. Food Chem 100:1126–1131
18. Bouvier F, Dogbo O, Camara B (2003) Biosynthesis of the food and cosmetic plant pigment bixin (annatto). Science 300:2089–2091

19. Grilli-Caiola MG, Canini A (2004) Ultrastructure of chromoplasts and other plastids in Crocus sativus L. (Iridaceae). Plant Biosyst 138:9. https://doi.org/10.1080/11263500410001684116
20. Ahrazem O, Moraga RA, Picazo AJ, Castillo R, Gómez LG (2016) Intronretention and rhythmic diel pattern regulation of carotenoid cleavage dioxygenase 2 duringcrocetin biosynthesis in saffron. Plant Mol Biol 91:355–374
21. Schwartz SH, Tan BC, McCarty DR, Welch W, Zeevaart JA (2003) Substrate specificity and kinetics for VP14, a carotenoid cleavage dioxygenase in the ABA biosynthetic pathway. Biochim Biophys Acta 1619:9–14
22. Schwartz SH, Tan BC, Gage DA, Zeevaart JA, McCarty DR (1997) Specific oxidative cleavage of carotenoids by VP14 of maize. Science 276:1872–1874
23. Rook F, Corke F, Card R, Munz G, Smith C, Bevan MW (2001) Impaired sucrose-induction mutants reveal the modulation of sugar-induced starch biosynthetic gene expression by abscisic acid signalling. Plant J Cell Mol Biol 26:421–433
24. Cheng WH, Endo A, Zhou L, Penney J, Chen HC, Arroyo A, Leon P, Nambara E, Asami T, Seo M, Koshiba T, Sheen J (2002) A unique short-chain dehydrogenase/reductase in Arabidopsis glucose signaling and abscisic acid biosynthesis and functions. Plant Cell 14:2723–2743
25. Gonzalez-Guzman M, Apostolova N, Belles JM, Barrero JM, Piqueras P, Ponce MR, Micol JL, Serrano R, Rodriguez PL (2002) The short-chain alcohol dehydrogenase ABA2 catalyzes the conversion of xanthoxin to abscisic aldehyde. Plant Cell 14:1833–1846
26. Brewer PB, Koltai H, Beveridge CA (2013) Diverse roles of strigolactones in plant development. Mol Plant 6:18–28
27. Rasmussen A, Depuydt S, Goormachtig S, Geelen D (2013) Strigolactones fine-tune the root system. Planta 238:615–626
28. Ruyter-Spira C, Al-Babili S, vanderKrol S, Bouwmeester H (2013) The biology of strigolactones. Trends Plant Sci 18:72–83
29. Maier W, Peipp H, Schmidt J, Wray V, Strack D (1995) Levels of a terpenoid glycoside (blumenin) and cell wall-bound phenolics in some cereal mycorrhizas. Plant Physiol 109:465–470
30. Klingner A, Hundeshagen B, Kernebeck H, Bothe H (1995) Localization of the yellow pigment formed in roots of gramineous plants colonized by arbuscular fungi. Protoplasma 185:50–57
31. Fester T, Hause B, Schmidt D, Halfmann K, Schmidt J, Wray V, Hause G, Strack D (2002) Occurrence and localization of apocarotenoids in arbuscular mycorrhizal plantroots. Plant Cell Physiol 43:256–265
32. O'Byrne SM, Blaner WS (2013) Retinol and retinyl esters: biochemistry and physiology. J Lipid Res 54:1731–1743
33. Handelman GJ, van Kuijk FJ, Chatterjee A, Krinsky NI (1991) Characterization of products formed during the autoxidation of β-carotene. Free Radic Biol Med 10:427–437
34. Ouyang JM, Daun H, Chang SS, Ho C-T (1980) Formation of carbonyl compounds from P-carotene during palm oil deodorization. J Food Sci 45:1214–1217
35. Marty C, Berset C (1990) Factors affecting the thermal degradation of all-trans-β-carotene. J Agric Food Chem 38:1063–1067
36. Rodriguez EB, Rodriguez-Amaya DB (2007) Formation of apocarotenals and epoxycarotenoids from β-carotene by chemical reactions and by autoxidationin modelsystems and processedfoods. Food Chem 101:563–572
37. Fleshman MK, Lester GE, Riedl KM, Kopec RE, Narayanasamy S et al (2011) Carotene and novel apocarotenoid concentrations in orange-fleshed Cucumis melo melons: determinations of β-carotene bioaccessibility and bioavailability. J Agric Food Chem 59:4448–4454
38. Rios ADO, Borsarelli CD, Mercadante AZ (2005) J Agric Food Chem 53:2307
39. Rivera-Madrid R, Godoy-Hernandez G, Aguilar-Espinosa M, Cardenas-Conejo Y, Garza-Caligaris LE, Lugo-Cervantes EC (2017) Carotenoides En Agroalimentacion y Salud (Melendez Martınez AJ, ed). Editorial Terracota, Ciudad de Mexico, pp 260–279

40. Britton G, Liaaen-Jensen S, Fander HP (2017) Carotenoids: a colourful history, 1st edn. Carote Nature, Lupsingen
41. Kopec RE, Riedl KM, Harrison EH, Curley RW, Hruszkewycz DP, Clinton SK, Schwartz SJ (2010) J Agric Food Chem 58:3290
42. Schaub P, Wust F, Koschmieder J, Yu Q, Virk P, Tohme J, Beyer P (2017) J Agric Food Chem 65:6588
43. Alvarez R, Vaz B, Gronemeyer H, Lera ARA, Rosana A, Gronemeyer H, De Lera AR (2014) Chem Rev 114(1)
44. Sommer A (2008) Vitamin A deficiency and clinical disease: an historical overview. J Nutr 138:1835–1839
45. Britton G (2009) Carotenoids, vol 5: Nutrition. healing (Britton G, Liaaen-Jensen S, Pfander H, eds). Birkhauser, Basel, pp 173–190
46. von Lintig J (2012) Provitamin A metabolism and functions in mammalian biology. Am J Clin Nutr 96:1234S–1244S
47. Iskakova M, Karbyshev M, Piskunov A, Rochette-Egly C (2015) Nuclearandextranucleareffectsofvitamin A. Can J Physiol Pharmacol 93:1065–1075
48. Blaner WS, Li Y, Brun PJ, Yuen JJ, Lee SA, Clugston RD (2016) Vitamin A absorption, storage and mobilization. Subcell Biochem 81:95–125
49. Sharma RV, Mathur SN, Dmitrovskii AA, Das RC, Ganguly J (1976) Studies on the metabolism ofβ-carotene and apo-β-carotenoids in rats and chickens. Biochim Biophys Acta 486:183–194
50. Al-Hasani SM, Parrish DB (1972) Forms of vitamin A and of carotenoids in tissues, blood serum and yolkof eggs from *Coturnix coturnix japonica* fed β-apo-carotenals. J Nutr 102:1437–1440
51. Kiefer C, Sumser E, Wernet MF, von Linter J (2002) A class B scavenger receptor mediates the cellularuptake of carotenoids in *Drosophila.* PNAS 99:10581–10586
52. Dela SC, Sun J, Narayanasamy S, Riedl KM, Yuan Y et al (2016) Substrate specificity of purifiedrecombinant chicken β-carotene 9_,10_-oxygenase (BCO2). J Biol Chem 291:14609–14619
53. Wang XD, Russell RM, Liu C, Stickel F, Smith DE, Krinsky NI (1996) b-Oxidation in rabbit liver in vitro and in the perfused ferret liver contributes to retinoic acid biosynthesis from b-apocarotenoic acids. J Biol Chem 271:26490–26498
54. Hessel S, Eichinger A, Isken A, Amengual J, Hunzelmann S et al (2007) CMO1 deficiency abolishesvitamin A production from β-carotene and alters lipid metabolism in mice. J Biol Chem 282:33553–33561
55. Amengual J, Lobo GP, Golczak M, Li HN, Klimova T et al (2011) A mitochondrial enzyme degrades carotenoids and protects against oxidative stress. FASEB J 25:948–959
56. Lobo GP, Isken A, Hoff S, Babino D, von Linter J (2012) BCDO2 acts as a carotenoid scavenger andgatekeeper for the mitochondrial apoptotic pathway. Development 139:2966–2977
57. Amengual J, Widjaja-Adhi MA, Rodriguez-Santiago S, Hessel S, Golczak M et al (2013) Two carotenoidoxygenases contribute to mammalian provitamin A metabolism. *J Biol Chem* 288:34081–34096
58. Palczewski G, Amengual J, Hoppel CL, von Linter J (2014) Evidence for compartmentalization of mammaliancarotenoid metabolism. FASEB J 28:4457–4469
59. Babino D, Palczewski G, Widjaja-Adhi MA, Kiser PD, Golczak M, Von Linter J (2015) Characterization ofthe role of beta-carotene 9,10-dioxygenase in macular pigment metabolism. J Biol Chem 290:24844–24857
60. Guo X, Wu L, Lyu Y, Chowanadisai W, Clarke SL et al (2017) Ablation ofβ-carotene-9_,10_-oxygenase2 remodels the hypothalamic metabolome leading to metabolic disorders inmice. J Nutr Biochem 46:74–82
61. Ip BC, Liu C, Ausman LM, Von Linter J, Wang XD (2014) Lycopene attenuated hepatic tumorigenesis viadifferential mechanisms depending on carotenoid cleavage enzyme in mice. Cancer Prev Res 7:1219–1227

62. Li B, Vachali PP, Shen Z, Gorusupudi A, Nelson K et al (2017) Retinal accumulation of zeaxanthin,lutein, and beta-carotene in mice deficient in carotenoid cleavage enzymes. Exp Eye Res 159:123–131
63. Tan HL, Thomas-Ahner JM, Moran NE, Cooperstone JL, Erdman JWJ et al (2017) β-Carotene 9_,10_oxygenase modulates the anticancer activity of dietary tomato or lycopene on prostate carcinogenesis inthe TRAMP model. Cancer Prev Res 10:161–169
64. Wu L, Guo X, Hartson SD, Davis MA, He H et al (2017) Lack ofβ,β-carotene-9_,10_-oxygenase 2 leads tohepaticmitochondrial dysfunction and cellular oxidative stress inmice. Mol Nutr Food Res 61(5):1600576
65. Sun J, Narayanasamy S, Curley RWJ, Harrison EH (2014) β-Apo-13-carotenone regulates retinoid Xreceptor transcriptional activity through tetramerization of the receptor. J Biol Chem 289:33118–33124
66. Eroglu A, Hruszkewycz DP, Dela SC, Narayanasamy S, Riedl KM et al (2012) Naturally occurringeccentric cleavage products of provitamin A β-carotene function as antagonists of retinoic acid receptors. J Biol Chem 287:15886–15895
67. Russell RM (2004) The enigma of beta-carotene in carcinogenesis: what can be learned from animal studies. J Nutr 134:262S–268S
68. Wang XD, Russell RM (*1999*) Procarcinogenic and anticarcinogenic effects of beta-carotene. *Nutr Rev* 57:263–272
69. Costabile BK, Kim YK, Iqbal J, Zuccaro MV, Wassef L et al (2016) β-Apo-10_-carotenoids modulateplacental microsomal triglyceride transfer protein expression and function to optimize transport of intactβ-carotene to the embryo. J Biol Chem 291:18525–18535
70. Hussain MM (2014) Intestinal lipid absorption and lipoprotein formation. Curr Opin Lipidol 25:200–206
71. Cooperstone JL, Riedl K, Cichon MJ, Francis DM, Curley RWJ et al (2017) Carotenoids and apocarotenoidsin human plasma after continued consumption of high β-carotene or high lycopene tomatojuice. FASEB J 31:635.13
72. Negbi M (1999) Saffron cultivation: past, present and future prospects. In: Negbi M (ed) Saffron: medicinal and aromatic plants—industrial profiles, vol 8: Crocus sativus. Harwood Academic Publishers, Australia, pp 1–18
73. Fernández JA (2004) Biology, biotechnology and biomedicine of saffron. Recent Res Dev Plant Sci 2:127–159
74. Alonso GL, Salinas MR, Garijo J, Sanchez-Fernandez MA (2001) Composition ofcrocins and picrocrocin from Spanish saffron (Crocus sativus L.). J Food Qual 24:219–233. https://doi.org/10.1111/j.1745-4557.2001.tb00604.x
75. Tarantilis PA, Tsoupras G, Polissiou M (1995) Determination of saffron (Crocus sativusL.) components in crude plant extract using high-performance liquid chromatography-UV-visible photodiode-array detection-mass spectrometry. J Chromatogr A 699:107–118. https://doi.org/10.1016/0021-9673(95)00044-N
76. Pfander HP, Schurtenberger H (1982) Biosynthesis of C20-carotenoids in Crocus sativus. Phytochemistry 21:1039–1042. https://doi.org/10.1016/S0031-9422(00)82412-7
77. Alonso GL, Salinas MR, Esteban-Infantes FJ, Sánchez-Fernández MA (1996) Determination of safranal from saffron (Crocus sativus L.) by thermal desorption−gas chromatography. J Agric Food Chem 44:185–188. https://doi.org/10.1021/jf940665i
78. Rezaee R, Hosseinzadeh H (2013) Safranal: from an aroermatic natural product to a rewardingpharmacological agent, Iran. J Basic Med Sci 16:12–26
79. Gresta F, Lombardo GM, Siracusa L, Ruberto G (2008a) Saffron, an alternative crop for sustainable agricultural systems. A review. Agron Sustain Dev 28:95–112
80. Skrubis B (1990) The cultivation in Greece of Crocus sativus L. In: Tammaro F, Marra L (eds) Proceedings of the international conference on saffron (Crocus sativus L.). L'Aquila, pp 171–182

81. Ehsanzadeh P, Yadollahi AA, Maibodi AMM (2004) Productivity, growth and quality attributes of 10 Iranian saffron accession under climatic conditions of Chahar-Mahal Bakhtiari, central Iran. Acta Hortic 650:83–188
82. International Trade Centre (2006) UNCTAD/WTO world markets in the spice trade 2000–2004. ITC, Geneva, p 111
83. Siracusa L, Gresta F, Ruberto G (2011) Saffron (*Crocus sativus* L.) Apocarotenoids: a review of their biomolecular features and biological activity perspectives. In: Yamaguchi M (ed) Carotenoids: properties, effects and diseases. Nova Science Publishers, Inc. ISBN 978-1-61209-713-8
84. Ait-Oubahou A, El-Otmani M (1999) Saffron cultivation in Morocco. In: Negbi M (ed) Saffron: medicinal and aromatic plants—industrial profiles, vol. 8: Crocus sativus. Harwood Academic Publishers, Amsterdam, pp 87–94
85. Thiercelin J-M (2004) Room table: industrial perspectives for saffron. Acta Hortic 650:399–404
86. Girard N, Navarrete M (2005) Quelles synergies entre connaissances scientifiques et empiriques? L'exemple des cultures du safran et de la truffe. Natures Sciences Sociétés 13:33–44
87. Azizbekova NSH, Milyaeva EL (1999) Saffron in cultivation in Azerbaijan. In: Negbi M (ed) Saffron: medicinal and aromatic plants—industrial profiles, vol 8: Crocus sativus. Harwood Academic Publishers, Amsterdam, pp 63–71
88. Schmidt M, Betti G, Hensel A (2007) Saffron in phytotherapy: pharmacology andclinical uses. Wien Med Wochenschr 157:315–319. https://doi.org/10.1007/s10354-007-0428-4
89. Hosseinzadeh H, Noraei NB (2009) Anxiolytic and hypnotic effect of Crocus sativusaqueous extract and its constituents, crocin and safranal, in mice. Phyther Res 23:768–774. https://doi.org/10.1002/ptr.2597
90. Soeda S, Aritake K, Urade Y, Sato H, Shoyama Y (2016) Neuroprotective activities of saffron and crocin. Springer, Cham. https://doi.org/10.1007/978-3-319-28383-8_14
91. Cerutti PA (1994) Oxy-radicals and cancer. Lancet 344:862–863
92. Martínez-Tomé M, Jiménez AM, Ruggieri S, Frega N, Strabbioli R, Murcia MA (2001) Antioxidant properties of Mediterranean spices compared withcommon food additives. J Food Prot 64:1412–1419. https://doi.org/10.4315/0362-028X-64.9.1412
93. Ghadrdoost B, Vafaei AA, Rashidy-Pour A, Hajisoltani R, Bandegi AR, Motamedi F, Haghighi S, Sameni HR, Pahlvan S (2011) Protective effects of saffronextract and its active constituent crocin against oxidative stress and spatiallearning and memory deficits induced by chronic stress in rats. Eur J Pharmacol 667:222–229. https://doi.org/10.1016/j.ejphar.2011.05.012
94. Karimi E, Oskoueian E, Hendra R, Jaafar HZE (2010) Evaluation of Crocus sativus L.stigma phenolic and flavonoid compounds and its antioxidant activity. Molecules 15:6244–6256. https://doi.org/10.3390/molecules15096244
95. Tamaddonfard E, Hamzeh-Gooshchi N (2010) Effect of crocin on the morphine-induced antinociception in the formalin test in rats. Phytother Res 24:410–413
96. Zhu Y, Han T, Zhang QY, Qin LP (2008) The comparative studies on the analgesic and anti-inflammatory activities of different parts from Crocus sativus. J Pharm Pract 4:269–271
97. Hosseinzadeh H, Younesi HM (2002) Antinociceptive and anti-inflammatory effects of Crocus sativus L. stigma and petal extracts in mice. BMC Pharmacol 2:7. www.biomedcentral.com/1471-2210/2/7
98. Magesh V, Singh JPV, Selvendiran K, Ekambaram G, Sakthisekaran D (2006) Antitumour activity of crocetin in accordance to tumour incidence, antioxidant status, drug metabolizing enzymes and histopathological studies. Mol Cell Biochem 287:127–135
99. Das I, Das S, Saha T (2010) Saffron suppresses oxidative stress in DMBA-induced skin carcinoma: a histopathological study. Acta Histochem 112:317–327
100. Premkumar K, Thirunavuk karasu C, Abraham SK, Santhiya ST, Ramesh A (2006) Protective effect of saffron (Crocus sativus L.) aqueous extract against genetic damage induced by anti-tumor agents in mice. Hum Exp Toxicol 25:79–84

101. Chryssanthi DG, Lamari FN, Iatrou G, Pylara A, Karamanos NK, Cordopatis P (2007) Inhibition of breast cancer cell proliferation by style constituents of different Crocus species. Anticancer Res 27:357–362
102. Verma SK, Bordia A (1998) Antioxidant property of Saffron in man. Indian J Med Sci 52:205–207
103. Ochiai T, Ohno S, Soeda S, Tanaka H, Shoyama Y, Shimeno H (2004) Crocin prevents the death of rat pheochromyctoma (PC-12) cells by its antioxidant effects stronger than those of α-tocopherol. Neurosci Lett 362(1):61–64
104. Mousavi SH, Tayarami NZ, Parsaee H (2010) Protective effect of saffron extract and crocin on reactive oxygen species-mediated high glucose-induced toxicity in PC12 cells. Cell Mol Neurobiol 30:185–191
105. Akhondzadeh S, Fallah-Pour H, Afkham K, Jamshidi AH, Khalighi-Cigaroudi F, Miller LG (2005a) A comparative trial of Crocus sativus L. (saffron) and imipramine in mild to moderate depression. Focus Altern Complement Ther 10(1):22–23
106. Akhondzadeh S, Tahmacebi-Pour N, Noorbala AA, Amini H, Fallah-Pour H, Jamshidi AH, Khani M (2005b) Crocus sativus L. in the treatment of mild to moderate depression: a double-blind, randomized and placebo-controlled trial. Phytother Res 19:148–151
107. Karimi G, Hosseinzadeh H, Khaleghpanah P (2001) Study of antidepressant effect of aqueous and ethanolic extract of Crocus sativus in mice. Irn J Basic Med Sci 4:11–15
108. Hosseinzadeh H, Karimi G, Niapoor M (2004) Antidepressant effect of Crocus sativus L. stigma extracts and their constituents, crocin and safranal, in mice. Acta Hortic 650:435–445
109. Wang Y, Han T, Zhu Y, Zheng C-J, Ming Q-L, Rahman K, Qin L-P (2010) Antidepressant properties of bioactive fractions from the extract of Crocus sativus L. J Nat Med 64:24–30
110. Akhondzadeh S, Abbasi SH (2006) Herbal medicine in the treatment of Alzheimer's disease. Am J Alzheimers Dis Other Dement 21:113–118
111. Akhondzadeh S, Sabet MS, Harirchian MH, Togha M, Cheraghmakani H, Razeghi S, Hejazi SS, Yousefi MH, Alimardani R, Jamshidi AH, Zare F, Moradi A (2010) Saffron in the treatment of patients with mild to moderate Alzheimer's disease: a 16-week, randomized and placebo-controlled trial. J Clin Pharm Ther 35:581–588
112. Ebrahim-Habibi M-B, Amininasab M, Ebrahim-Habibi A, Sabbaghian M, Nemat-Gorgani S (2010) Fibrillation of α-lactalbumin: effect of crocin and safranal, two natural small molecules from Crocus sativus. Biopolymers 93(10):854–865
113. Pathan SA, Alam S, Jain GK, Zaidi SMA, Akhter S, Vohora D, Khar RK, Ahmad FJ (2010) Quantitative analysis of safranal in saffron extract and nanoparticle formulation by a validated high-performance thin layer chromatographic method. Phytochem Anal 21:219–223
114. Pitsikas N, Zisopoulou S, Tarantilis PA, Kankis CD, Polissiou MG, Sakellaridis N (2007) Effects of active constituents of Crocus sativus L., crocins on recognition and spatial rats' memory. Behav Brain Res 183:141–146
115. Hosseinzadeh H, Khosravan V (2002) Anticonvulsant effects of aqueous and ethanolic extracts of Crocus sativus L. stigmas in mice. Arch Intern Med 81:44–47
116. Giaccio M (2004) Crocetin from saffron: an active component of an ancient spice. Crit Rev Food Sci Nutr 44:155–172
117. Xu GL, Yu SQ, Gong ZN, Zhang SQ (2005) Study of the effect of crocin on rat experimental hyperlipemia and the underlying mechanisms. Zhongguo Zhong Yao Za Zhi 30:369–372
118. Imenshahidi M, Hosseinzadeh H, Javadpour Y (2010) Hypotensive effect of aqueous saffron extract (Crocus sativus L.) and its constituents, safranal and crocin, in normotensive and hypertensive rats. Phytother Res 24:990–994
119. He SY, Qian ZY, Tang FT, Wen N, Xu GL, Sheng L (2005) Effect of crocin on experimental atherosclerosis in quails and its mechanism. Life Sci 77:907–921
120. Boskabady MH, Shafei MN, Shakiba A, Sang SH (2008) Effect of aqueous-ethanol extract from Crocus sativus (Saffron) on Guinea-pig isolated heart. Phytother Res 22:330–334
121. Uttara B, Singh AV, Zamboni P, Mahajan RT (2009) Oxidative stress and neurodegenerative-diseases: a review of upstream and downstream antioxidant therapeutic options. Curr Neuro Pharmacol 7:65–74. https://doi.org/10.2174/157015909787602823

122. Saleem S, Ahmad M, Ahmad AS, Yousuf S, Ansari MA, Khan MB, Ishrat T, Islam F (2006) Effect of saffron (Crocus sativus) on neurobehavioral and neurochemical changes in cerebral ischemia in rats. J Med Food 9:246–253. https://doi.org/10.1089/jmf.2006.9.246
123. Vakili A, Einali MR, Bandegi AR (2014) Protective effect of crocin against cerebral ischemia in a dose-dependent manner in a rat model of ischemic stroke. J Stroke Cerebrovasc Dis 23:106–113. https://doi.org/10.1016/j.jstrokecerebrovasdis.2012
124. Lindahl T, Karran P, Wood RD (1997) DNA excision repair pathways. Curr Opin Genet Dev 7:158–169. https://doi.org/10.1016/S0959-437X(97)80124-4
125. Debatin K-M, Poncet D, Kroemer G (2002) Chemotherapy: targeting the mitochondrial cell death pathway. Oncogene 21:8786–8803. https://doi.org/10.1038/sj.onc.1206039
126. Hosseinzadeh H, Sadeghnia HR, Ziaee T, Danaee A (2005) Protective effect of aqueous saffron extract (Crocus sativus L.) and crocin, its active constituent, on renal ischemia-reperfusion-induced oxidative damage in rats. J Pharm Pharm Sci 8:387–393
127. Hosseinzadeh H, Sadeghnia HR (2005) Safranal, a constituent of Crocus sativus (saffron), attenuated cerebral ischemia induced oxidative damage in rat hippocampus. J Pharm Pharm Sci 8:394–399
128. Ghazavi A, Mosayebi G, Salehi H, Abtahi H (2009) Effect of ethanol extract of saffron(Crocus sativus L.) on the inhibition of experimental autoimmune encephalomyelitisin C57bl/6 mice. Pak J Biol Sci 12:690–695. http://www.ncbi.nlm.nih.gov/pubmed/19634472
129. Urrutia EC, Riverón-Negrete L, Abdullaev F, Del-Angel DS, Martínez NLH, Cruz MEG, Cruz VPD, Silva-Adaya D, González-Cortés C, Santamaría A (2007) Saffron extract ameliorates oxidative damage and mitochondrial dysfunction inthe rat brain. Acta Hortic:359–366
130. Zheng YQ, Liu JX, Wang JN, Xu L (2007) Effects of crocin on reperfusion-inducedoxidative/nitrative injury to cerebral microvessels after global cerebral ischemia. Brain Res 1138:86–94. https://doi.org/10.1016/j.brainres.2006.12.064
131. Linardaki ZI, Orkoula MG, Kokkosis AG, Lamari FN, Margarity M (2013) Investigation of the neuroprotective action of saffron (Crocus sativus L.) in aluminum-exposed adult mice through behavioral and neurobiochemical assessment. Food Chem Toxicol 52:163–170. https://doi.org/10.1016/j.fct.2012.11.016
132. Yang R, Vernon K, Thomas A, Morrison D, Qureshi N, Van Way CW 3rd. (2011) Crocetin reduces activation of hepatic apoptotic pathways and improves survivalin experimental hemorrhagic shock. JPEN J Parenter Enteral Nutr 35:107–113. https://doi.org/10.1177/0148607110374058
133. Ebrahim-Habibi MB, Amininasab M, Ebrahim-Habibi A, Sabbaghian M, Nemat-Gorgani N (2010) Fibrillation of α-lactalbumin: effect of crocin and safranal, twonatural small molecules from Crocus sativus. Biopolymers 93:854–865. https://doi.org/10.1002/bip.21477
134. Alavizadeh SH, Hosseinzadeh H (2014) Bioactivity assessment and toxicity of crocin: acomprehensive review. Food Chem Toxicol 64:65–80. https://doi.org/10.1016/j.fct.2013.11.016
135. Ahmad AS, Ansari MA, Ahmad M, Saleem S, Yousuf S, Hoda MN, Islam F (2005) Neuroprotection by crocetin in a hemi-parkinsonian rat model. Pharmacol Biochem Behav 81:805–813. https://doi.org/10.1016/j.pbb.2005.06.007
136. Yoshino F, Yoshida A, Umigai N, Kubo K, Lee MC (2011) Crocetin reduces the oxidativestress induced reactive oxygen species in the stroke-prone spontaneouslyhypertensive rats (SHRSPs) brain. J Clin Biochem Nutr 49:182–187. https://doi.org/10.3164/jcbn.11-01
137. Hosseinzadeh H, Khosravan V (2002) Anticonvulsant effects of aqueous and ethanolicextracts of Crocus sativus L. Stigmas in mice. Arch Iran Med 5:44–50
138. Hosseinzadeh H, Talebzadeh F (2005) Anticonvulsant evaluation of safranal and crocinfrom Crocus sativus in mice. Fitoterapia 76:722–724. https://doi.org/10.1016/j.fitote.2005.07.008
139. Khazdair MR, Boskabady MH, Hosseini M, Rezaee R, Tsatsakis AM (2015) The effectsof Crocus sativus (saffron) and its constituents on nervous system: a review. Avicenna J Phytomed 5:376–391

Chapter 5
Role of Carotenoids in Photosynthesis

Saima Zulfiqar, Shahzad Sharif, Maham Saeed, and Arifa Tahir

5.1 Introduction

Photosynthesis is the principal process that occurs naturally and plays an important role in the survival of life on the earth by converting solar light energy into chemical energy [1]. There are living organisms which undergo photosynthesis by absorbing sunlight through light harvesting complex, this excited energy is transferred to the reaction center where charge separation play an important role in the conversion of excited energy into electrochemical potential. Here, light-harvesting molecules chlorophylls and carotenoids act as accessory antenna molecules [2–8]. Carotenoids belong to the class of natural pigments, known as tetra-terpenoid, and are present in plants, algae, fungi, and bacteria. The color of fruits, flowers and leaves due to presence of carotenoids. Light in the blue-green region of visible light spectrum is captured by the carotenoids [9–11]. Besides functioning as light harvesting molecule, it absorbs reactive oxygen species, that have damaging effect towards photosynthetic machinery, that are produced when chlorophyll molecules are over-excited [12–15]. Phytohormones, abscisic acids (ABA) and strigolactones (SLs), are also synthesized via the help of precursors present in carotenoids molecules [16, 17].

These have been classified into two mainclasses: (1) carotenes and (2) xanthophylls. There is no functional group in the former class, while, oxygen in the form of functional group is present in the latter. There is special organ present in plants that contain carotenoids. These have complex structure in an organized form. In photosynthesis, carotenoids having more than seven conjugate double bonds, such as, carotenes and xanthophylls, play an important role by absorbing light energy.

S. Zulfiqar · S. Sharif (✉) · M. Saeed
Department of Chemistry, GC University Lahore, Lahore, Pakistan
e-mail: mssharif@gcu.edu.pk

A. Tahir
Environmental Science Department, Lahore College for Women University, Lahore, Pakistan

© Springer Nature Switzerland AG 2021

M. Zia-Ul-Haq et al. (eds.), *Carotenoids: Structure and Function in the Human Body*, https://doi.org/10.1007/978-3-030-46459-2_5

The electronic structure of the catorenoids helps to understand the formation of chemical energy from sunlight [18–22]. Is this chapter, we have explained, how carotenoids act as an accessory pigment in the process of photosynthesis, which mechanisms and factors plat an important role in this process.

5.2 Role of Carotenoids as Accessory Pigment

Their functional diversity makes them important part of photosynthetic mechanism of plants. Carotenoids play an important role as light harvesting agents and facilitate the electron transport chain in by transferring the energy required for excitation. That is the reason they are known as 'accessory Pigments'. Cartenoid molecules are located near chlorophyll molecules, having higher capacity to capture the light. Carotenoids also play a major role in the photo-protection, by dissipating the excessive energy that can cause damage to the tissues and reaction centers in leaves [23].

Carotenoids rely on their physical properties to perform functions like protection from oxidative damage and harvesting of sunlight. There are two important excited states in the carotenoids, one is high energy excited singlet state and other is low lying first excited triplet state. On the exposure of sunlight, free or bound carotenoid molecule excites from the ground state to the higher energy (S_2) singlet state, ($1B^*2u$). This excitation leads to a very quick, internal conversion process to a low lying, symmetry forbidden, (1A*g) S_1 singlet state. When it comes to harvesting of light, 1A*g acts as a donor state to transfer energy to the molecules of chlorophyll, without involving 1B*2u state. On the other hand, Low lying triplet state T*1 of the carotenoid molecule, is responsible to take energy from the excited triplet state of chlorophyll molecule. Recent studies show that, triplet state of zeaxanthin and xanthophyll interacts with the singlet state of chlorophyll, resulting in quenching of chlorophyll fluorescence [24].

5.2.1 Light Harvesting by Carotenoids

Carotenoids perform a great role in photosynthesis by harvesting the light in the region between 400 and 500 nm, which is unapproachable to chlorophyll molecules thus, enhancing the range to drive the photosynthesis. Carotenoid molecules then follow the singlet-singlet energy transfer mechanism for the efficient transfer of energy to chlorophyll molecules in the time scale of femto-seconds [25, 26].

They are composed of alternating C-C and C=C bonds in a linear chain. Difference in pi electron conjugation type, chain length, and functional group number make different groups of carotenoid molecules. There are carotenoid molecules exceeding 1000 but only 50 play their role in the light harvesting process of photosynthesis. Energy transfer between cholorophyll and carotenoid molecules is governed by three factors;

1. Coupling of electrons between carotenoids and chlorophyll molecules.
2. Overlapping between emission spectra of donor and absorption spectra of acceptor.
3. Lifetime of donor molecules, when no energy transfer takes place.

The energy transfer rate is determined by first two factors i.e. electron coupling and spectral overlap. The third factor determines the efficiency of energy transfer because there is competition between energy transfer and internal conversion process in donor carotenoid molecule. Light harvesting process of carotenoids is related to spectroscopic properties, which can be described through a "three states" model. This model consists of a ground state S_0, excited state S_1 and excited state S_2, as depicted in Fig. 5.1. S_0 and S_1 states are assigned A_g^- symmetry while S_2 has been given B_u^+ symmetry.

According to rule of quantum mechanics, transition of one photon between the electronic states of same symmetry is prohibited thus, transition from S_0 to S_1 is forbidden and S_0 to S_2 is the allowed transition with minimum energy [27]. Light is absorbed through S_0 to S_2 transition followed by relaxation of S_2 state to S_1 within few femtoseconds. The life-time of this transition varies from 1 to 200 picoseconds, depending on number of conjugated double bonds [28]. Due to short lifetime of states S_1 and S_2 as compared to other pigments like (bacterio)clorophylls, it is difficult to expect the light harvesting nature of carotenoids. For the energy transfer process, energy donor states S_1 and S_2 have to compete with very short relaxation time span. Moreover, S_1 state cannot be a part of Forster type dipole-dipole mediated energy transfer, due to forbidden transition between the ground and S_1 states which results in minute dipole moment. However, all photosynthetic organisms

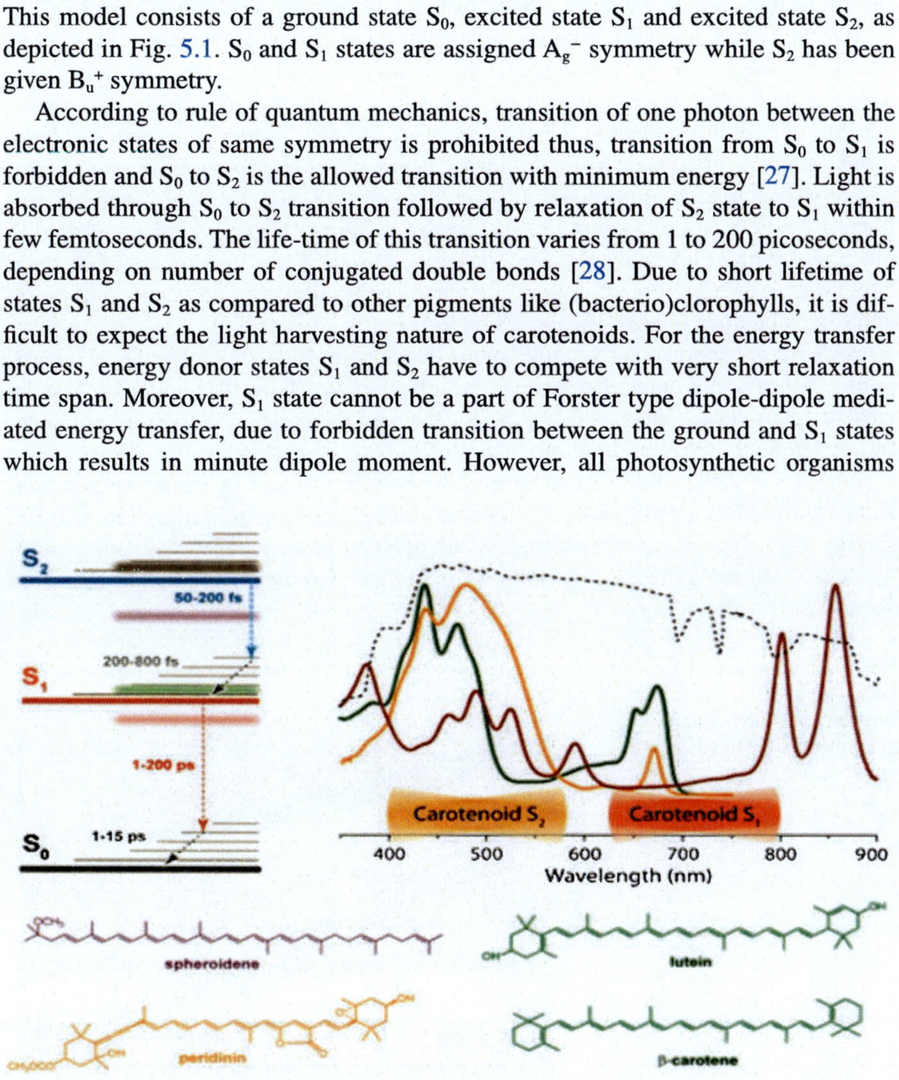

Fig. 5.1 Model and absorption spectra of carotenoids

depend on carotenoids for light harvesting and S_1 and S_2 states act as energy donors for chlorophylls [29].

Forster theory is based on the fact that excitation transfer is dependent on configuration, assuming the dipole-dipole moment can be related to interaction between the donor and acceptor. It is valid only when the distance between the interacting molecules is large as compared to their extension. In the energy transfer process between carotenoid and chlorophyll, distance between the centre of molecules is about 15 Å as small as the 20 Å extension of carotenoid molecules and larger as compared to distance between two closed atoms i.e. 3–6 Å. The distance between the atoms depends on the species involved and carotenoid-chlorophyll pair [30].

For the light harvesting process, efficiency of energy transfer process by carotenoid molecule depends on the optimized distance and geometry of donor and acceptor molecule. In most of the light harvesting complexes, this distance between acceptor and donor molecule is from 3 to 10 Å. When a proper geometry and large dipole moment of S_2 state are combined, energy transfer through S_2 state of carotenoid to (bacterio)chlorophyll takes place between 100 and 300 femtoseconds, following the Forster mechanism [31]. S_1 mediated energy transfer is also believed to proceed through Forster mechanism, but due to dipole forbidden nature of S_1 state, computation of 'Coulomb coupling' between energy donating and accepting molecules is mandatory to meet the experimental results.

Because the energies of S_1 and S_2 state depends on N of the molecule (N = no of double bonds in conjugated system) so it is the part which decides which will be the energy donor state. Optimized Forster mechanism depends on the maximum spectral overlapping between the donor state S_1/S_2 emission and acceptor absorption.

Figure 5.2 shows energy transfer process between S_2 and Q_x state of carotenoid and chlorophyll molecule is likely to occur in larger conjugation length molecules. Energy from S_2 state is transferred in all light harvesting carotenoid molecules, without considering the conjugation system in them. On the other hand, S_1 mediated

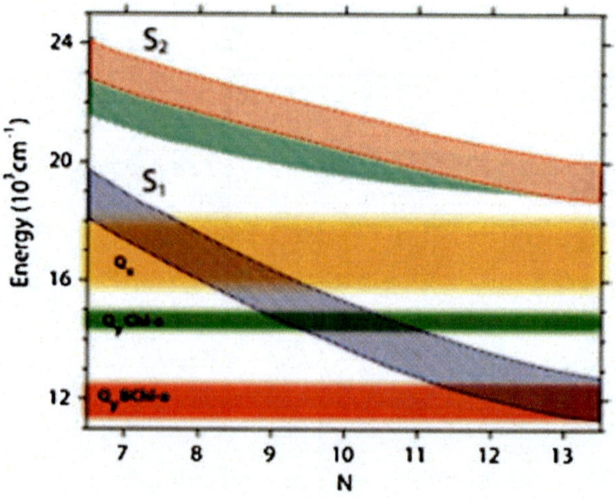

Fig. 5.2 Variation in the energy states of carotenoids on the number of conjugated double bonds

energy transfer takes place only when the acceptor molecule has comparatively lower energy e.g. (bacterio)Chl a in light-harvesting complexes. When acceptor molecule is Chl a, S_1 energy transfer will take place only when there is short pi-electron conjugation systems. Energy of S_2 state can be modified with the interacting proteins, while that of S_1 remain ineffective. Moreover, excited states energies and efficiency of conjugation system can be changed by the structural changes in the carotenoid molecules, induced by binding site of protein. Effective conjugation length occurs when the pi-electron in the conjugation are levelled with the linearity of chain or twisted for the complete separation of terminal pi bonds from the conjugation system. These all effects lead to diversity in the energy transfer process and efficient performance of light harvesting protein complexes [32].

In green plants there is complex network of light harvesting proteins complexes involving Chl a. While considering the energy diagram in above Fig. 5.2, it can be concluded that energy transfer will take place through S_2 state to the Q_x state of Chl a. S_1 state is the least involved in energy transfer process and S_2 state compensate this transfer in Chl a involving complexes. LHCII protein is widely found among these antenna complexes. In reaction centre of PS II, this protein complex is found on the outer periphery. LHCII complex is composed of Chl a, Chl b, a violaxanthin ($N = 9$), two luteins ($N = 10$), and a neoxanthin ($N = 9$) as shown in Fig. 5.3 below. Energy transfer is always preferred through S_2 state in this complex [33, 34].

Two lutein molecules are located at the centre of the complex with the purpose of stabilization. This role also makes carotenoids mandatory for the reformation of complex. There is Vander Waals interaction between lutein molecules and many chlorophylls, which is beneficial for light harvesting and photoprotective role of carotenoids [35].

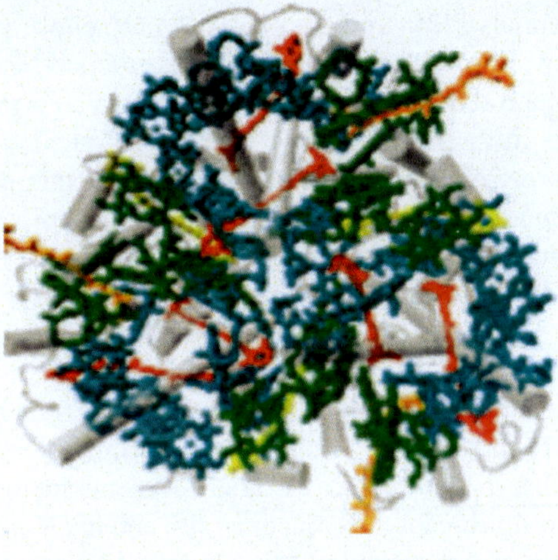

Fig. 5.3 LHII trimer utilizing carotenoids lutein (R), neoxanthin (Y), and violaxathin (O) in energy transfer to chlorophylls (a + b)

Here, reaction sequences are given for the energy transfer process from carotenoid to chlorophyll.

$$\text{Carotenoids} + h\nu \rightarrow \text{Carotenoids}^*$$

It is the absorption step to convert carotenoid into excited singlet state specie Car*.

$$\text{Chlorophyll} + \text{Carotenoids}^* \rightarrow \text{Carotenoids} + \text{Chlorophyll}^*$$

Singlet-singlet excitation occurs to transfer energy from carotenoid to ground state of chlorophyll.

$$\text{Chlorophyll}^* \rightarrow\rightarrow\rightarrow \text{Chlorophyll}_n^*$$

Singlet-singlet excitation energy transfer from one to other chlorophyll molecules.

$$\text{Chlorophyll}_n^* \rightarrow \text{Chlorophyll}_n^* + \text{Photochemistry}$$

Excitation energy is transferred to derive reactions like fixation of CO_2 and evolution of O_2 in the reaction centres but in the absence of reaction (4), chlorophyll molecule is de-excited with emission of fluorescence.

$$\text{Chlorophyll}_n^* \rightarrow \text{Chlorophyll}_n + \text{Fluorescence}$$

These all reactions show that analysis of light harvesting process efficiency can be done by comparison of different reaction rates involved in photosynthesis like fixation of CO_2 and evolution of O_2 and emission of energy.

At start, purple bacteria were used to study light harvesting function of carotenoids [13]. Spectra of donor carotenoid and acceptor chlorophyll showed minor overlap with the caotenoid absorbance between 400 and 550 nm and BChl soret band at 370 nm (Fig. 5.4a). Process of energy transfer from carotenoid to chlorophyll is different in different species of purple bacteria and thus different for different protein complexes in one species. For example, it was observed in *Rhodopseudomonas sphaeroides* that effeciency of light absorption by carotenoids is 90% and 100% for another protein complex B80 (Fig. 5.4b). While carotenoids are not efficient for light transfer activity in a specie *Rhodospirillum rubrum* [36, 37].

Red shifts in soret bands of Chl-a (435 nm), Chl-b (470 nm) and Chl-c (452 nm) result in spectral overlapping of carotenoid and chlorophyll molecules for higher plants and algae. Thus, presence of carotenoids is not clearly shown by absorbance spectra (Fig. 5.5a). In this situation, energy transfer efficiency is determined by absorbance spectra of other pigments involved. But it is also creates problems due to significant differences in situ and in an organic solvent measured absorption spectra of carotenoids [38]. Therefore, in case of intact tissues various studies have reported the approximated data to study the transfer efficiency For example,

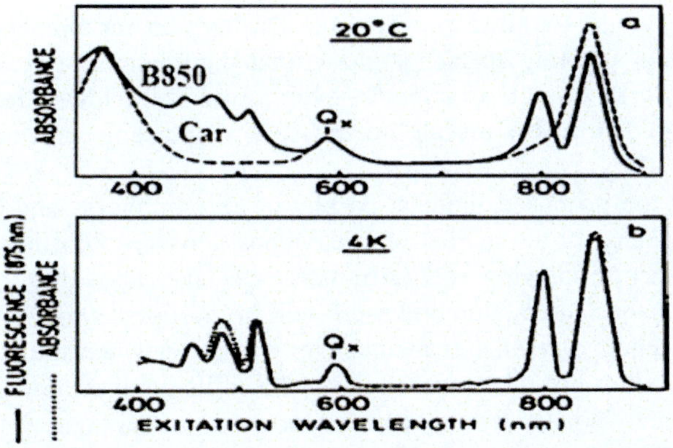

Fig. 5.4 Light harvesting function of carotenoids in the light harvesting pigment-protein complex of Rhodopseudomonas sphaeroides. (a) Comparison between absorption spectrum of B850 protein from carotenoids containing wild type and that from less mutant R26. (b) Comparison of excitation spectrum of chlorophyll b with absorption spectrum of B850 from carotenoid containing wild type

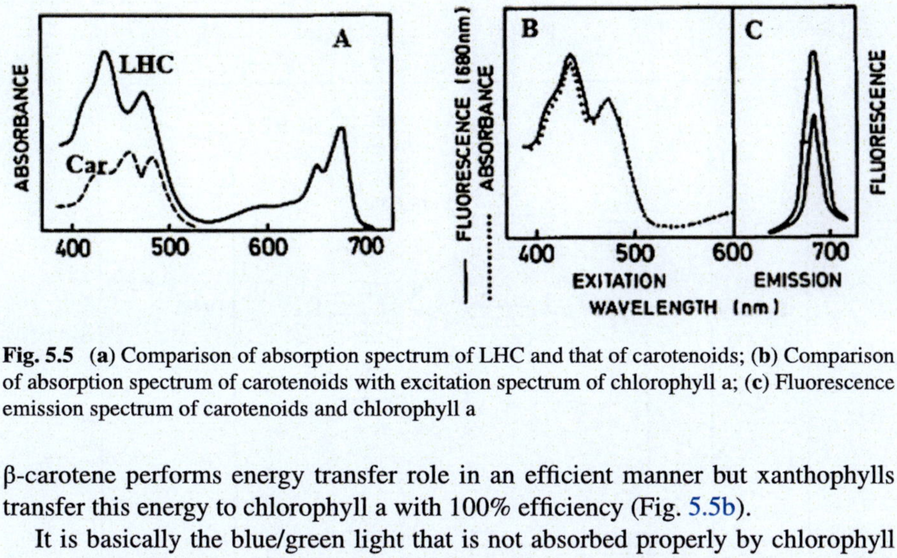

Fig. 5.5 (a) Comparison of absorption spectrum of LHC and that of carotenoids; (b) Comparison of absorption spectrum of carotenoids with excitation spectrum of chlorophyll a; (c) Fluorescence emission spectrum of carotenoids and chlorophyll a

β-carotene performs energy transfer role in an efficient manner but xanthophylls transfer this energy to chlorophyll a with 100% efficiency (Fig. 5.5b).

It is basically the blue/green light that is not absorbed properly by chlorophyll and carotenoids help in transfer of this light to them. In aquatic species carotenoids are certainly important as they may absorb blue/green light but not the red light. In terrestrial plants, low molecular weight carotenoids harvest blue light more efficiently than chlorophyll. Moreover, in Chl a/b lutein complex, carotenoids are 23% but assist in light absorption to 43% [39, 40].

Energy is transferred from carotenoids to chlorophyll molecules through singlet-singlet excitation mechanism. When there is no energy transfer, then excited carotenoid becomes de-excited within 1 picosecond, so transfer of energy to chlorophyll takes place during this limited time [41, 42]. During transfer of energy no diffusive displacements of reacting species suggests that pigments are orderly arranged in their complexes. Moreover, no effect in energy even at the lowest temperature like 4 Kelvin, (Fig. 5.5b) also reveals no diffusive displacement occurs during this mechanism.

In most of the proposed models, porphyrin chromophores and carotenoid are present at a distance <1 nm in their respective positions. On the other hand, (B)Chls cannot fluorescence showing energy transfer will take place through Dexter type exchange resonance interaction and there will be no Forster mechanism [43, 44]. Dexter type exchange resonance mechanism will occur when the reacting species are located at a short distance and it involves the collision of the complex with overlapped electronic clouds. It has not been properly understood that whether the excited state of carotenoid is energy donor state because it shows no similarity with dipole allowed $^1B_u^*$ excited state. Huge energy gap between excited state of (B)Chls and $^1B_u^*$ reveals that visible absorption spectra of various carotenoids is due to $^1B_u^*$ state. $^1Ag^*$ is the low lying state in which symmetry is forbidden and $^1B_u^*$ state is used to populate it. According to the Fig. 5.6, $^1Ag^*$ can act as a mediator when $^1B_u^*$ state transfers the energy to (B)Chls [45].

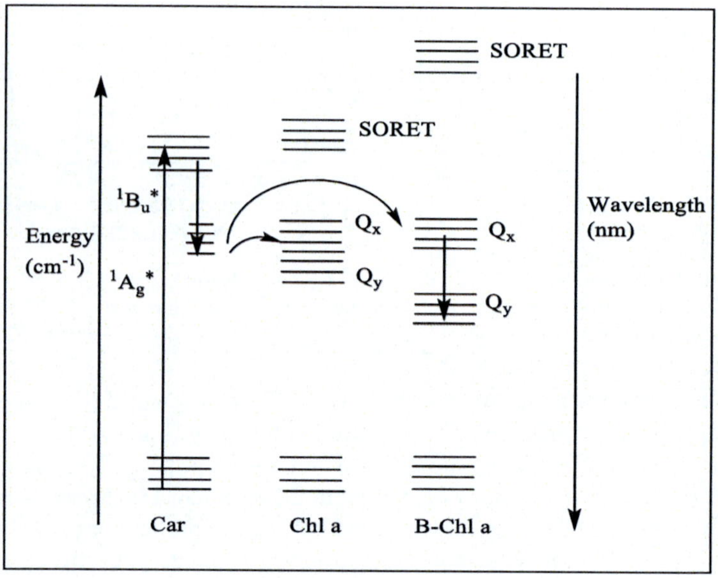

Fig. 5.6 Energy transfer from excited carotenoids to chlorophyll a and to B-chlorophyll a (Hypothetical scheme)

5.2.2 Photoprotection

Besides, light harvesting, carotenoids are also known for their photoprotective role. Their importance can be judged by illumination of organism by molecular oxygen, lacking carotenoids due to some mutations or inhibition in the synthesis. Organisms of wild-type also suffer from photodamage due to absence of carotenoids, which may lead to their death. This deterioration process is also known as "Photodynamic reaction" which has been seen in *Rhodobacter sphaeroides* specie of bacteria and plants like diflufenican and norflurazon.

It is believed that carotenoids protect against oxidative damage in the following ways;

1. They cause quenching of triplet state of Chl/ (B)Chl to prevent the singlet oxygen formation.
2. When a singlet oxygen is produced, it is scavenged by them [44].

Above Fig. 5.7 shows how carotenoids are involved in protection to oxidative damage. Actually, singlet oxygen is a strong oxidizing agent, which can destroy the components of cell, resulting in organism's death. For example in *Rhodobacter sphaeroides,* absence of carotenoid leads to illumination of B850 complexes and when carotenoid is unbound, bacteriochlorophyll a triplets form so rapidly but

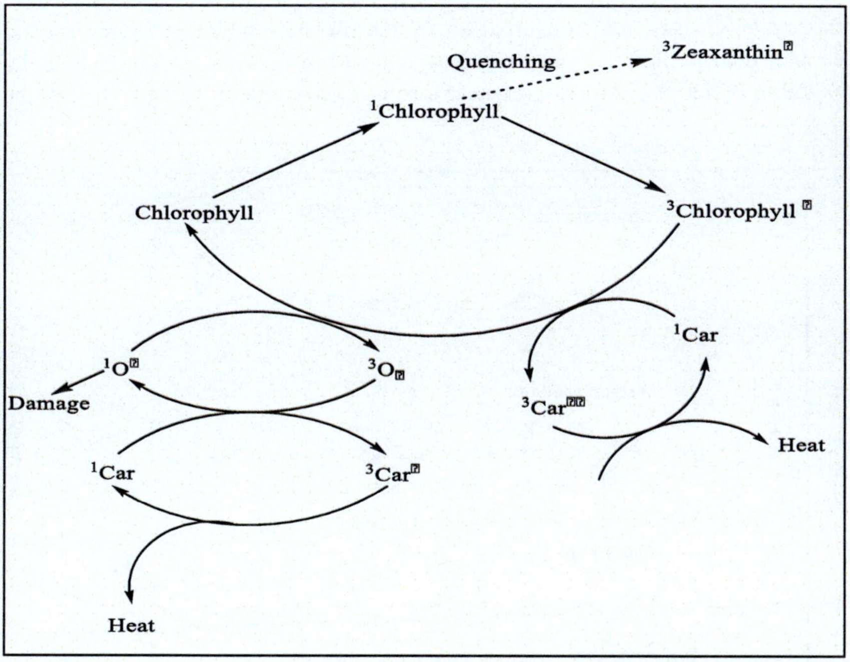

Fig. 5.7 Photo-sensitized production of singlet oxygen via triplet chlorophyll and involvement of carotenoids

decay between 10 and 100 ms. This triplet state on reacting with molecular oxygen, produces singlet oxygen. Thus, complexes destroy with no photo-protection. When carotenoids are present, life-time of triplet state becomes 1000 fold shorter and formation of singlet oxygen is prevented. Further studies have shown that carotenoids can also remove very reactive and toxic species of oxygen like OH radicals. However, it has also been revealed that membrane bound carotenoids are not good in scavenging these oxygen containing toxics as it depends how proximate a carotenoid and molecular oxygen are [46].

Sunlight is essential to derive the photosynthesis process, but at the same time, excessive radiation leads to overloading of electrons in electron transport chain which are to be transported to Photosystem I and Photosystem II in order to start the Calvin Benson cycle. This highly energised chain of electrons may give rise to charged and reactive molecules of oxygen which in turn react with protein molecules of the chloroplast, harming the tissue of leaf. Excited molecule of chlorophyll releases the excessive energy, which is accepted by carotenoid in order to dissipate it in form of heat. Dispersion of excess energy into heat is achieved through Xanthophyll cycle.

5.2.3 *Xanthophyll Cycle*

In this cycle, violaxanthin converts into an theraxanthin, which is an intermediate. Then this intermediate is converted into zeaxanthin as seen in (Fig. 5.8) below. Zeaxanthin helps in dissipation of heat through geometrical changes in light har-

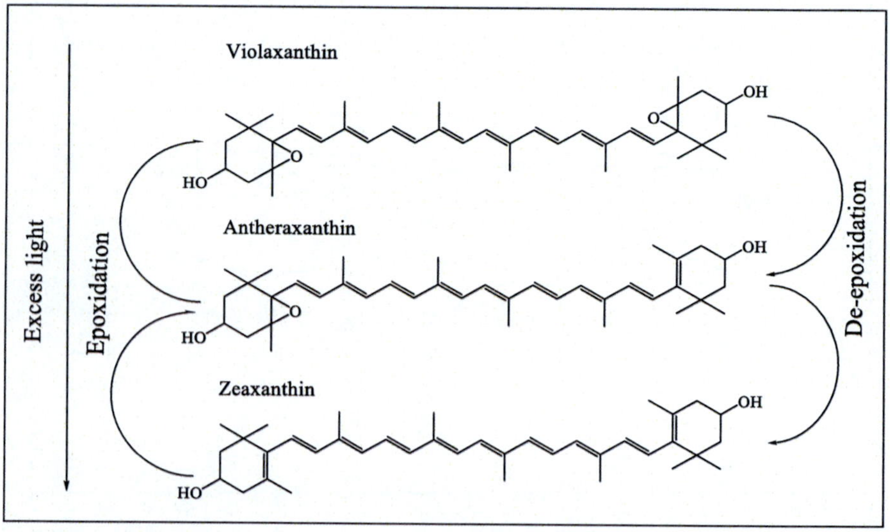

Fig. 5.8 Xanthophyll cycle

vesting complex. A process called de-epoxidation is responsible for such enzymatic conversions. It is activated by the decreased pH levels in thylakoid lumin. In this process, an oxygen molecule is released by six-sided two carbon rings. During low level of light, this process gets reversed. The time limit in which leaf stressing is continued, energy is deviated from chlorophyll through preservation of zeaxanthin in the chloroplast. When this stress is released by thermal dissipation, pH level becomes high, zeaxanthin is converted back into violaxanthin and whole process is called epoxidation. In this way, whole cycle gets reversed. Epoxidation and de-epoxidation both occurs at different locations of thylakoid. Xanthophyll cycle is responsible for optimized level of energy to be captured i.e. Releasing the energy in form of heat, when it exceeds the required amount but not when it is limited. Another hypothesis reveales that when antenna system of PS II captures the excessive light, acidification of lumen is triggered and de-epoxidation in xanthophylls cycle is activated. It may cause changes in geometry of antenna pigment-protein complexes so that they can interact with xanthophylls leading to non-phtochemical quenching NPQ. NPQ is a reversible process and is associated with formation of zeaxanthin in de-epoxidation. Carotenoids are present in plant leaves in all the seasons of growth and appear as yellow and orange pigments during autumn. When the seasonal growth is at the mid, chlorophyll molecules optically mask the presence of carotenoids. During senescence stage of leaves, incomplete breakdown of chlorophyll molecules occur before degradation of carotenoids. In this stage, carotenoids may also perform role in photoprotection and antioxidant activity in order to withdraw nutrients from leaves to the storage sections. They also control the level of species with very active oxygen, which keep accumulating in the period of senescence. Additionally, carotenoids are also believed to participate in formation of abscisic acid, which triggers the closing of stomata in leaves [47–49].

Excess energy is dissipated either by photochemical or non-photochemical ways. Photochemical processes include Cyt b_{559} induced electron transfer cycle or thermal dissipation in Photosystem II. Non-photochemical quenching (NPQ) occurs through three ways;

1. Protein phosphorylation controls the cross sectional area of photosystem II, in which light is aborbed.
2. When illumination occurs for a long time, non-radiative dissipation is related to zeaxanthin formation..
3. Non-radiative dissipation also depends on the pH gradient in the thylakoid [50].

Several studies have been done to find relationship between zeaxanthin levels in leaves and energy release. During dark situation, leaves only contain violaxanthin and there is no zeaxanthin but its formation can occur when in low light, leaves are exposed to 2% oxygen, 98% nitrogen and 0% carbodioxide. This illumination causes quenching of fluorescence which is related to the formation of zeaxanthin. Non-photochemical quenching of fluorescence occurs in leaves only with high content of zeaxanthin. Quenching of chlorophyll fluorescence during dark or low light level is called energy dependent quenching (qE) or high energy state. This photochemical quenching is directly related to trans-thylakoid pH gradient and non-radiative dissipation of excess energy [51].

To check the xanthophylls cycle role in chlorophyll fluorescence, in vitro studies have been performed in the isolated chloroplasts. With high zeaxanthin content, formation of qE takes place at low pH gradient in the illuminated leaves while in non-illuminated leaves, qE forms at high pH gradient. It was concluded from the experimental research that, in the presence of zeaxanthin, steady conditions and low pH gradient lead to high qE formation. On the other hand, in non-illuminated leaves or dithiothreitol illuminated leaves, large pH gradient is required to obtain same quenching level [52, 53].

Sulfhydral reagent dithiothreitol (DTT) inhibit the formation of zeaxanthin and thus de-epoxidation process. Influence of DTT has been seen on chlorophyll fluorescence by taking spinach leaves. Without zeaxanthin, reaction number in photosystem II is decreased. It reveals that zeaxanthin plays an important role in non-photochemical dissipation of heat [54].

5.3 Spectral Properties of Carotenoids

The characteristic of carotenoids to absorb between 300 and 550 nm of spectral region is due to long polyene chain with alternating C-C and C=C system [55]. Two double isoprene units are present at the centre of carotenoid molecule, which are combined from tail to tail. Open or ring structures are present at both sides of molecule. Basically, there are two main groups in carotenoids, one is carotene and the other is xanthophylls. Carotenes contain non-polar β- carotene while xanthophylls are polar with oxygen containing derivatives. These functional groups are present at the terminals of chain without changing double bonds in conjugated system [56].

Spectral properties of carotenoids can be investigated by different ways of transferring energy to excited states of carotenoids [57]. Absorption spectra of carotenoids are usually represented by three-peaks absorption or sometimes two-peaks as well and a high intensity band at the middle. Shape and position of absorption spectra depends on several factors;

1. No of conjugated double bonds
2. Length of chain
3. Chromophore group

When conjugated double bonds are large in a system, it leads to longer wavelength shifts in the spectrum. When hydroxyl groups are substituted, no spectral change is observed. The spectra of β-cryptoxanthin and zeaxanthin resembles to β-carotene, while zeinoxanthin and α-cryptoxanthin is similar to α-carotene. Actually, it's the type of solvent which affects the absorption spectra e.g. when more OH-groups are substituted in xanthophylls, their polarity is enhanced and solubility in organic solvents is reduced [58].

Carotenoids dissolve in organic solvents due to hydrophobic nature with little solubility in water. Specific coefficients of absorption and lambda max position are affected by solvent used and amount of water. Extraction of a specific carotenoid in a proper organic solvent and its quantitative analysis is difficult due to lacking knowledge. When dissolved in pure methanol, carotenoids show maximum absorption at a lower range of wavelength [59].

5.3.1 Absorption Spectra

Experiments were performed with healthy young and senescing leaves of maple. For the extraction of pigments, leaves were crushed in methanol and acetone and $CaCO_3$ was added to avoid pheophytininzation of chlorophyll. A homogenised mixture was obtained and was subjected to centrifugation for 3–4 min [6]. Extracts were taken and analysed by spectrophotometer. For further calculations, carotenoids weight was taken as 570 and absorbance was calculated by formula, log(100-R/T).

Moving from light green to dark green leaves, more than 90% of light is absorbed by leaves in the blue and red regions of spectra as shown in the Fig. 5.9. In spectral blue region appearance of peaks at 440 and 470 nm indicates absorbance of Chl a, Chl b with total carotenoids respectively [60, 61]. In the red region chlorophyll a shows absorption near 680 nm and a shoulder appears at 650 nm indicating absorption of chlorophyll b. In green region 90% of absorption can be seen. Yellow leaves show less absorption in green and red regions due to less amount of chlorophyll. On the other hand in blue region carotenoids show upto 90% of absorbance [62, 63].

No absorbance can be seen at 678 run for yellow leaves. It can be inferred that absorption of yellow leaves, that appeared in blue region is merely due to carotenoids. Due to similarity between absorption peaks of carotenoids and spectral characteristics at 420 nm, 460 nm and 475 nm, it can be concluded that carotenoids are the main absorbers of yellow leaves [60]. Rising chlorophyll content in leaves leads to strong and overlapped absorption bands and carotenoids band disappear. Hence, for analysis of carotenoids spectral characteristics, its necessary to minimize

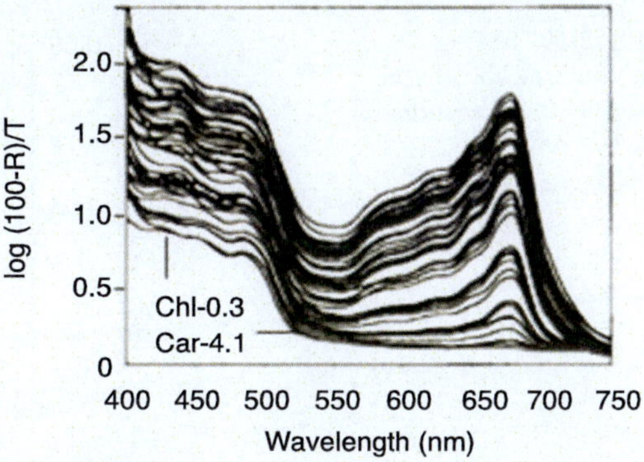

Fig. 5.9 Absorbance spectra of variety of pigment contents and composition of Maple leaves

the absorption effect due to chlorophyll [63]. It is done by normalization of absorption spectra at 678 nm in the red region. It is the main absorbance peak of Chl a and reduces the chlorophyll absorbance in the visible region [64].

Standard deviation of absorbance can be performed by normalization at 678 nm with three different classes of leaves. First class is of yellow to dark green leaves and chlorophyll level is 10 nmolcm^{-2}. Second class is of light green to dark green leaves and chlorophyll level is greater than 20. Third class is of green colored leaves to dark green leaves and chlorophyll level is greater than 30 nmolcm^{-2}. All STD spectra appeared similar. Normalization led to smaller STD at 440–460 nm (chlorophylls absorb strongly in this range) [65]. Weak absorbing regions of chlorophylls in spectra ranging from 550 to 620 nm and close to 700 nm showed nearly equal values of STD. A curve at 520 nm due to total chlorophylls can be seen in STD normalized spectra [66, 67].

Some troughs appeared at 440 nm and 460 nm (main absorbing region of chlorophylls) which is indication of decreased chlorophyll participation in total absorbance (Fig. 5.10). It also shows that area close to 520 nm is more sensitive to varying amount of total carotenoids.

Appearing of reflectance at 550 nm wavelength is relevant to the both chlorophyll present in green leaves. Measuring absorbance at the corresponding wavelength may help in measuring total chlorophylls also. Absorbance by carotenoids and chlorophyll affects the absorbance at 520 nm [65, 66]. Therefore, difference of absorbance between 520 and 550 nm would be highly influenced by absorbance of carotenoids relative to absorbance of cholorophyll.

Comparison of both "chlorophyll v/s carotenoids" and "A_{550} v/s A_{520}-A_{550}" lead to demonstration of contributed absorbance of total carotenoids and chlorophylls at 520 nm. Above figure shows the relation between chlorophylls and carotenoids. In the senescencing leaves, the relationship differs a lot especially at the ending stage when chlorophyll content is less than 10 nmolcm^{-2} prior to its level greater than 10 nmolcm^{-2} in green leaves. Similarity between A_{550} and A_{520}–A_{550} can be seen in the second figure. Similarity between both Fig. 5.11a and b indicate that important part is played by carotenoids to A_{520} [68].

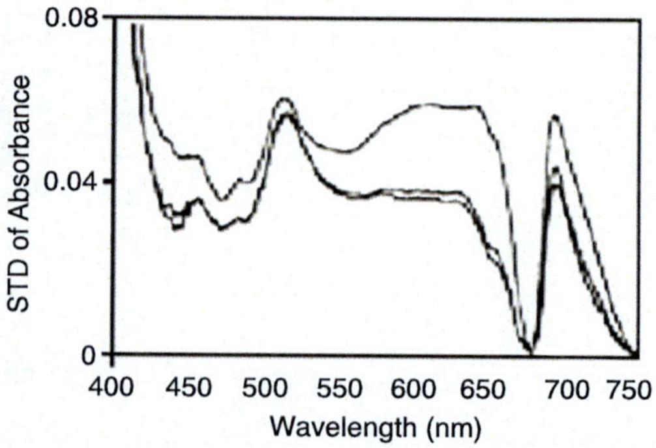

Fig. 5.10 STD of normalized absorbance spectra to the absorbance at 678 nm

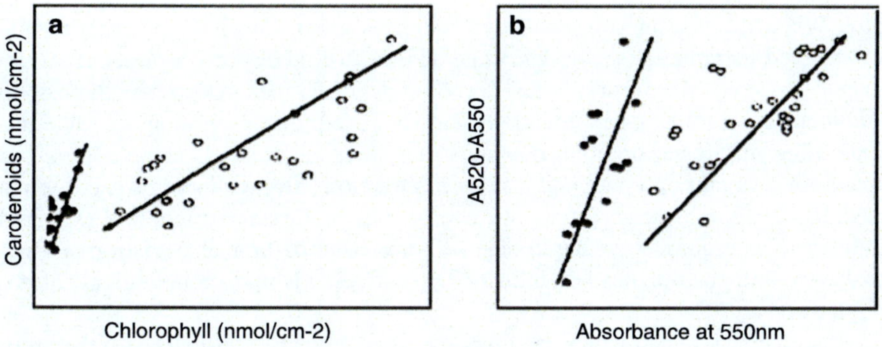

Fig. 5.11 (**a**) Carotenoids contents versus chlorophyll contents nmol/cm^{-2}; (**b**) Difference in Absorbance at 520 and 550 nm versus Absorbance at 550 nm

Fig. 5.12 Absorption spectra of β-carotene, lutein, vilaxanthin and leoxanthin in diethyl ether

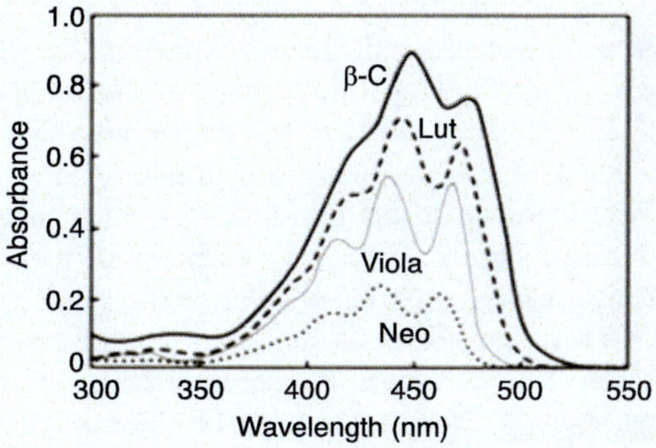

5.3.2 *Absorption Maxima*

Broad absorption of isolated carotenoids with three absorption maxima from 400 to 500 nm in the blue region can be seen in Fig. 5.12. Absorption maxima is dependent on two factors;

1. Slovent type to extract the pigments.
2. To some extent on the spectrophotometer.

Wavelength shifts in absorption maxima due to these factors are related to altering absorption coefficients which help in quantitative determination of carotenoids. Thus, it is necessary to measure the absorbance of extracted pigment at correct position of wavelength i.e. the absorption maxima of a carotenoid in a particular solvent [69]. For quantification of extracted pigments, solvent extinction coefficients must be considered when calculations are applied. Small difference in the λ_{max} can exist depending on the spectrophotometer used for measurement. Therefore, 1–1.5 nm deviation in position of wavelength can be seen.

Spectroscopic measurements of green leaves pigment extract in maximum absorption region can be performed by determination of maxima in the red region with spectrophotometer and then comparison of these measured values with literature. When the position of wavelength deviates greater than 1 nm, experimental determined maxima is used to calculate absorbance of respective pigment extract inspite of using values from literature. When the difference in wavelength position continues to 2 nm, same equations can be used for absorption calculations. Exceeding difference of 2 nm demands that wavelength needs to be adjusted in spectrophotometer or a pure and different extraction solvent should be used. When calculations have to be performed for carotenoids in the same solution of extracted pigments, position of wavelength may be adjusted to 470 nm as deviation of 1 nm in wavelength position has less influence on carotenoids than on individual chlorophyll a and b [70].

Absorbance range in which readings are taken, must be considered for the accurate spectroscopic measurements. It must be taken from 0.3 to 0.85. Less than 0.3 absorbance value in the red region cannot be used to calculate correct amount of pigment. Absorbance of 0.9 from leaves may be due to accuracy problems in detector as calculations of absorbance is done from the measured transmitted light by detector. Accuracy decreases exponentially when linear units of transmitted light are converted to logarithmic units of absorbance. When measured absorbance is less than 0.3, extracts need to be concentrated by evaporating solvent and adding more content of tissues. Many spectrophotometers can measure absorbance up-to 1.0. With them, an absorbance reading greater than 0.85 is not convenient and dilution must be performed in order to have fair measurements. Thus, for accurate absorbance readings volume of solution must be controlled carefully.

Therefore, for exact determination of carotenoids at 470 nm wavelength, it is essential to determine the chlorophyll b which also shows absorption at the same wavelength. Decreasing level of chlorophyll b enhances the level of total carotenoids. Redetermination of extinction coefficients allows the estimation of all carotenoids with chlorophyll a and chlorophyll b present in the same extract [71].

5.3.3 *Effect of Extraction Solvents on Spectral Characteristics of Carotenoids*

Absorbance is used for quantitative analysis of different carotenoids by using their stock solutions in different solvents. This analysis depends on molar extinction coefficient of each carotenoid in the respective solvent. Since solubility of carotenoids varies from polar to a-plor solvent, the characteristic spectra of carotenoids is changed in different solvents [72–74].

Usually, when polar solvents are used, three peak absorption spectra can be observed in the visible region. When a certain amount of water is added to the solutions made in ethanol, isopropanol, and acetone etc., characteristic changes in the absorption spectra are observed. Addition of water leads to appearance of a new peak in the UV-region, e.g. lutein show peak at 370 nm. Thus, 3 peak spectra starts disappearing as more amount of water is added. Several carotenoids like lutein, zeaxanthin, lycopene and violaxanthin etc. show this type of change in spectra. When excited thermally (45 °C) UV-peak diminishes and 3-peak spectra is re-observed. Water molecules may cause aggregation of carotenoid molecules which lead to respective changes. Polymer formation alters the light absorption range of carotenoids due to electronic distribution change in chromophore group.

Carotenoids extracted from spinach leaves give single spectral peak in visible region and strong absorbance in UV region. When these water soluble carotenoids are transferred to polar solvents, 3-peak characteristic spectra of carotenoids can be seen with. At the same time absorption maxima in UV region diminishes. It means membrane bound carotenoids give 4-peak spectra, which is same as of carotenoids in alcohol solutions with water described above [75]. Moreover, optical density of each carotenoid is different in each solvent thus, quantification is also changed if UV/Vis spectroscopy is used. Here, comparison of spectral characteristics of four important carotenoids lutein, zeaxanthin, β-carotene and lycopene in polar and non-polar solvents is described.

To make stock solutions, few crystals of carotenoids were taken and mixed with hexane. Their spectra were recorded and optical density from 1 to 2 was achieved by addition of hexane. Addition of butylated hydroxytoluene (BHT) in the stock solution made the final concentration till 10 mg/ml. Lutein, zeaxanthin, β-carotene and lycopene were added in the stock solutions and absorption spectra were determined from 350–500 nm by taking each 2 ml hexane solution. Samples were removed by using argon and 2 ml of methanol (MeOH) was added instantly. Spectra were again scanned from 350–550 nm after 30s vortexing. Due to formation of deposits, 1 ml of centrifuged lutein solution from the top was used for measurement. Re-extraction of methanol containing lycopene solution was done by 1 ml of phosphate buffered saline (7.4pH) and 1 ml hexane. 2 min of vortexing was done followed by centrifugation of sample for 3 min. Again top layer of hexane was taken to measure the absorption from 350 to 550 nm. Dilutions of each carotenoid (2–256 fold) were performed using hexane as a solvent and repetition of all steps was done as above.

UV/Vis spectra were recorded by Shimadzu UV-2101 PC double beam scanning spectrophotometer. Before measurement, each solvent base-line spectra were also recorded. Experiments were performed under low light conditions. Absorption maxima and UV/Vis spectra was not affected by the two solvents used, up-to 10 μM concentration for lutein. As seen in Fig. 5.13 below lutein spectra in hexane was similar to that of methanol except a minor difference in absorption maxima. Difference between heaxane and methanol spectra is related to concentration as decreasing concentration decreased this spectral difference. Inshort, with small molar extinction coefficient, lutein shows less absorption in hexane as compared to methanol [73, 74, 76].

Absorption spectra of zeaxanthin showed more dependency on the solvents used (Fig. 5.14a). Although, there was not a significant change in absorption maxima at increased concentration but when measured, absorption maxima in methanol was 2 folds greater than hexane. Absorption of zeaxanthin is directly propotional to the methanol concentration. On the other hand, in hexane absorption increased with

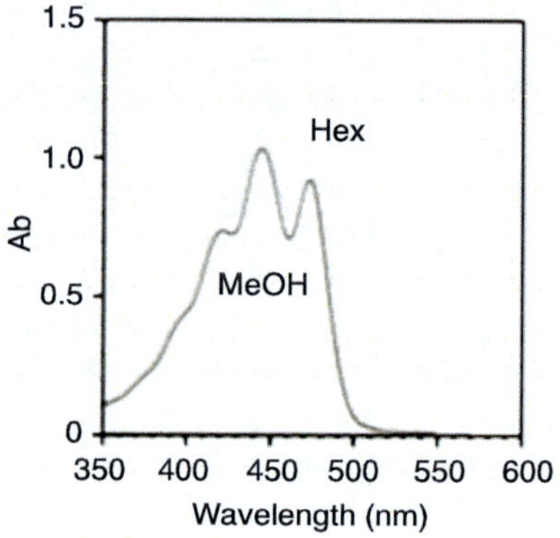

Fig. 5.13 UV-Vis spectra of lutein in hexane and methanol solvent.

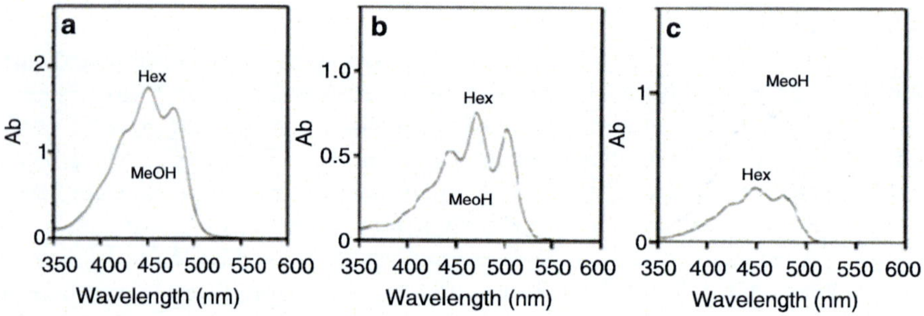

Fig. 5.14 Spectra of; (**a**) Zeasxanthin, (**b**) Lycopene, (**c**) β-carotene in hexane and methanol solvent in the presence of BHT

concentration range of upto 1.72 μM, after this it increased in a non-linear way. As the figure shows, the measured absorbances of zeaxanthin are 0.517 and 1.016 in methanol and 0.454, 0.463 in hexane at the concentrations of 3.5 μM and 7.0 μM respectively. This difference is not merely related to molar extinction coefficients, suggesting there might be formation of small polymers of zeaxanthin due to its poor solubility in hexane. To confirm this, both solutions were filtered through 0.2 μM pore sized membrane and absorbance was taken. In methanol it was same as before but for hexane, from 7 to 3 μM concentration it was decreased from 0.524 to 0.044 respectively. It indicated the formation of deposits which stayed in filter. Filter was rinsed with methanol to recover the left zeaxanthin. Formed crystal of zeaxanthin in hexane were made resoluble and rinsed with methanol. Solubility of lutein was found to be 20 mgL^{-1} in hexane and 200 mgL^{-1} in methanol while that of zeaxanthin was 2 mgL^{-1} in hexane. It was surprising with only an alteration in position of double bond, lutein and zeaxanthin showed much different solubilities in hexane. So, it was better to keep zeaxanthin in methanol solution.

Absorption maxima of lycopene in hexane is 4 times greater than in methanol, showing it is affected by solvent. Re-extraction of the sample in hexane gives recovering of sample absorbance in methanol. Figure 5.14b shows that absorbance is directly propotional to concentration, when measured in hexane but not in methanol, showing less solubility of lycopene in methanol. Solvent also affects the UV/Vis spectra of β-carotene that can be seen in Fig. 5.14c. Absorption maxima of hexane and methanol differs slightly and absorbance in methanol is smaller than hexane. Absorbance in hexane is directly propotional to the concentration. But in case of methanol, it varies linearly when concentration is less (upto 5 mM) and starts deviating at higher concentration. These changes in absorption due to concentration indicate the formation of sample crystals in methanol at 3 mgL^{-1} concentration, which is less as compared to reported data on solubility i.e. 10 mgL^{-1}.

Zeaxanthin is present as monomer in methanol but in form of micro-crystals in hexane. On the other hand, β-carotene and lycopene exist as monomers in hexane and formation of crystals can been seen at increased methanol concentrations. In-short, when stock solutions are to be made, that solvent should be chosen which completely dissolves the desired amount of carotenoid in it [76].

5.4 Excitation Energy Transfer: Sensitized Fluorescence and Photosynthesis

Carotenoids paly an important role in capturing and transferring of light energy to chlorophyll a. This is proved by following two methods:

- measurement of action spectrum of photosynthesis in the region carotenoids and Chls absorb and evaluation of the quantum efficiency of light absorbed by carotenoids in photosynthesis
- sensitized fluorescence method

In sensitized fluorescence method, action spectrum of Chlorophyll a is measured through fluorescence analysis followed by calculations of the quantum efficiency of excitation energy transfer from the carotenoids to Chlorophyll a. During fluorescence, fluorescent donor is excited followed by transfer of its fluorescence intensity to the receptor molecule; first discovered by Cario and Franck [77] in gases. Nowadays, this technique is used to measure the fluorescence of liquids, solids, proteins and photosynthetic systems [78, 79].

5.4.1 *Photosynthetic Yields in Different Wavelength Regions*

Engelmann measured the spectrum of green, red and brown algae and aerotactic motile bacteria through microscopic study, when there light of different wavelengths was used, to determine at which arte oxygen evolved [80]. It was concluded that accessory pigments absorb light and use it to drive the process of photosynthesis. In manometric method, to determine how oxygen is evolved from the green alga

Chlorella on exposure of light of different wavelengths, absolute quantum yield was measured. Carotenoids and Chlorophylls absorb blue wavelength light and show a suitable yield, but carotenoids due to absorbance of red light evolve oxygen slowly giving low absolute quantum yields [81]. It is followed by an experiment in which different photosynthetic pigments were used for absorption of light of different wavelengths and its was concluded that process of photosynthesis is driven by the light absorbed by a pigment fucoxanthin present in marine brown algae [82, 83]. The experiment, in which oxygen evolution by exposure of light on cyanobacterium Chroococcus and the green alga Chlorella was observed, is considered an accurate. This experiment was performed in Carnegie Institution of Washington through a monochromator that was fabricated by Emerson and Lewis themselves under 0.01 mm pressure. It was investigated that ten quantas of light were required for evolution of one molecule of oxygen [84, 85], as depicted in Fig. 5.15.

In an another experiment by Dutton and Manning, it was concluded that 905 oxygen was evolved from fucoxanthin present in diatom Nitzschia closterium; 40–50% oxygen evolved from carotenoids present in Chlorella; and 10% from Chroococcus [86]. Same results were reported by Wassink and Kersten [87]. A student of Emerson, Tanada, performed by using the same monochromator built up by Emerson and Lewis. In this experiment, it was concluded that fucoxanthin absorbed maximum light of 500 nm in blue-green region and presented high quantum yield of oxygen evolution and there were eight quantas of light required for the evolution of one molecule of oxygen [88], in Fig. 5.16.

In another study, it was observed that green alga Ulva had inactive carotenoids, so it had no photosynthesis [89]. The exact mechanism of transfer of energy required

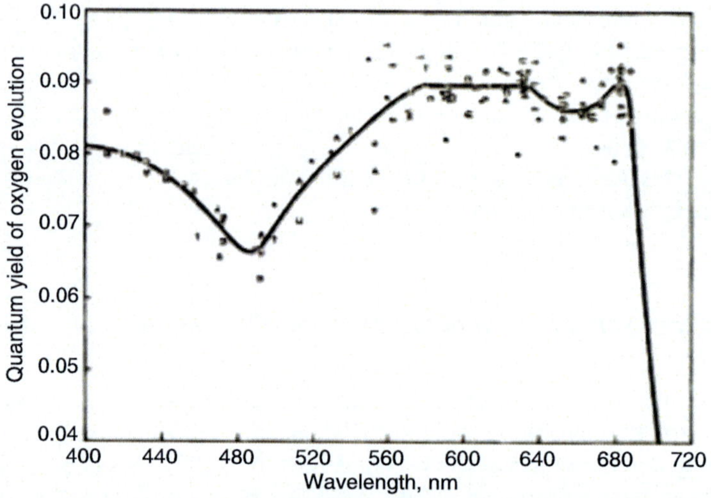

Fig. 5.15 Action spectra of oxygen evolution in cyanobacterium Chroococcus and the green alga Chlorella

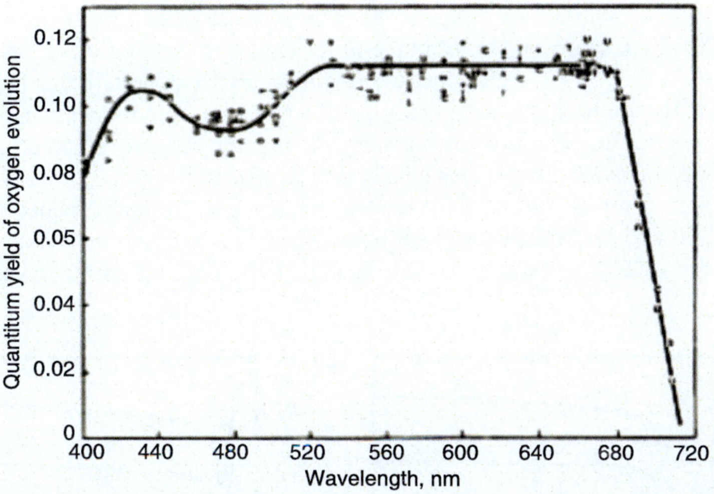

Fig. 5.16 Quantum yield action spectra representing oxygen evolution from fucoxanthin present in diatom Nitzschia closterium

to drive the process of photosynthesis, action spectra does not provide any sufficient data. This transfer can be studied through excitation energy transfer mechanism.

5.4.2 Sensitized Fluorescence: Excitation of Chl a Fluorescence by Different Wavelengths of Light Absorbed by Carotenoids

Transfer of energy from carotenoids to chlorophyll takes place in light harvesting complex 2. For this purpose, carotenoid components are changed by using reconstituted complexes in monomeric form. Light harvesting complex has three types of carotenoids; among them, two are tightly bound between two helices A and B, while third one has binding site adjacent to helix C. The former two carotenoids show high affinity towards lutin [90–92] Third one for neoxanthin. To understand energy transfer, two samples were used; one had mixture of lutein and neoxanthin, while the other lutein. The components of chlorophylls in both samples were the same. Difference resides in the molecular structures of lutein and neoxanthin, that cause the change with the spectra in only some chlorophylls. For example, when chlorophyll interacts with neoxanthin, there is difference in absorption at 652 nm, due to binding of Chlorophyll in LHCII [93]. For two intervals of excitation wavelengths (450–570 nm) and (610–730 nm), transition absorption kinetic was calculated in the absorption region of carotenoids. The former wavelength was showed in carotenoid S2 absorption and the latter one Qx/Qy absorption regions in the presence of very low absorption intensity. As a result, it was seen that excitation probability did not depend upon excitation wavelength. This was performed for LHCII-neolutein and the LHCII-lutein samples. Decay in the Soret region is more quick at 500 nm that 490 nm. In (Fig. 5.17a), to get spectrum of chlorophyll b in protein that has narrow band, LHCII samples containing chlorophyll a and chlorophyll b were used. Shape of this spectrum was the same as that of in the organic solvent. When acetone was used as solvent, same spectra was obtained as in protein [94].

The best soret region fit is obtained by using a complex containing two Chlorophyll a forms, three Clorophyll b forms and two lutein forms alongwith neoxanthin for LHCII-neolute, as shown in Fig. 5.18. Time of excitation at longer

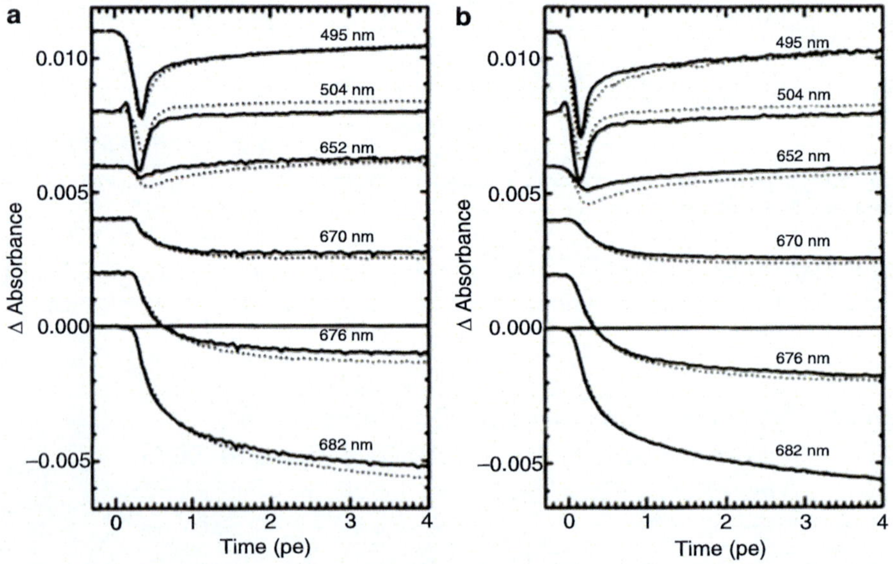

Fig. 5.17 Experimental transient absorption kinetics for the selected exciation/emission wavelength pairs for LHCII-Lutein (**a**) and LHCII-Neolutein (**b**)

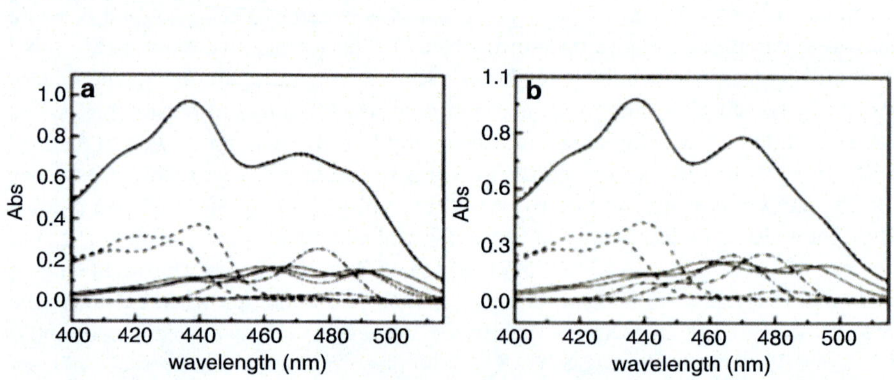

Fig. 5.18 SORET region of absorption spectra of LHCII-Neolutein (A) and LHCII-Lutein

wavelength of 490 and 500 nm was measured from width of the femto second excitation pulse plot.

500 nm wavelength was chosen for the red lutein. In case of combination with LHCII, at 490 nm, carotenoids decay takes place slow than at 500 nm, while there is quick bleaching effect in chlorophyll b. Both wavelengths showed a pronounced bleaching effect. Neoxanthin causes an increase in bleaching in case of LHCII-lutein sample. These transient spectra are obtained from lifetime density maps in the Chl absorption region for the two excitation wavelengths and both samples. Bleaching effect in LHCII-lutein sample is investigated in wavelength range of 640–660 nm within 60–80 fs. In this case, two maxima are observed at 650 and 660 nm. It takes place within 50 fs. Even though longer wavelength cause distinct increase in the excited state absorption, but with the same kinetic in both chlorophylls. Chlorophyll a and b in organic solvent separately do not show this behavior. Due to excitation of lutein into S_2 state, it can be said that these strong excited state absorptions are due to the presence of carotenoids.Transfer of energy from carotenoids to chlorophyll a takes place less time than 300 fs. Energy from chlorophyll b to chlorophyll a is transferred in different times such as, 170 fs, 350 fs, 1 ps, and10 ps. Both these lifetime ranges are different from each other visibly. In case of LHCII-neolutein sample, excited state absorption occurs at 650 nm [95, 96] The same result was obtained when sample containing LHCII neolutein was used. Time of 50–70 fs decays absorption based excited state while, in the same time, there is no transfer of energy from chlorophyll b to chlorophyll a. Formation of exciton coupling of chlorophyll a/b, chlorophyll a absorbs light energy and undergoes alteration in its working. On the other hand, in the presence of light of longer wavelength, spectrum of chlorophyll b shows flat small peak of absorption. It is the same effect when energy transfer from carotenoids to chlorophyll a takes place. This data does not explain how much energy has been transferred, so for this purpose there is a new strategy described in the following. As sample is excited, its excited state distribution is investigated to determine how much energy has been transferred from carotenoids to the chlorophyll a. For this purpose, individual spectrum of pigment is obtained from the absorption spectra of CP29 and LHCII [97], as described in Fig. 5.19:

Chlorophyll was dissolved in acetone to get absorption spectra, and carotenoids in 3-methylpentane [98] Their absorption spectra describing the stoichiometric ratio of the pigment alongwith extinction coefficients can be obtained via various spectroscopic techniques [99, 100] Fraction of excited pigments present in LHCII and CP29 has been listed in Table 5.1. For estimation of these fractions, carotenoids and

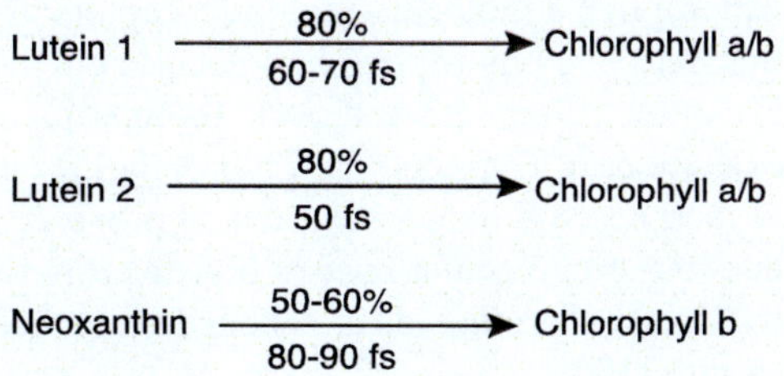

Fig. 5.19 Kinetic diagram of principal energy transfer pathway in LHCII

Table 5.1 Fractions of excited pigments in HCII and CP29

Saniple/Exc.	Lut	Neo	Vio	Chl-b	Chl-a
LHCII/506tun	0.46	0.04	0.15	0.25	0.10
LHCII/489 nm	0.28	0.13	0.04	0.50	0.05
CP29/506 nm	0.27	0.05	0.35	0.20	0.13
CP29/489 nm	0.22	0.16	0.17	0.40	0.05

chlorophylls are at lower limit and upper limit respectively at room temperature. The individual spectrum is sharp at lower temperature. Absorption of large amount of pump light in chlorophyll b is obvious. So, it is expected that CP29 is odd-excited and carotenoids even [101, 102].

5.4.2.1 Energy Transfer from S_2 State

In vitro studies investigated 200 fs lifetime of the carotenoids S_2 state, when there is no energy transfer takes place: this phenomenon is quite different when protein environment is used due to interaction of S_2 energy level to S_1 state. However, SE signals between 520 and 620 nm corresponds the decay of S_2 state [103]. It shows negative signal, which can be converted into positive signal through chlorophylls. This conversion was about 490–500 nm within 70–90 fs. When carotenoids in organic solvents undergo fluorescence analysis at room temperature, spectrum having separate excitation and emission peaks is obtained, on the other hand, when temperature is lowered to 77 K, this separation is more prominent. In this case, SE signal at lower wavelength of 500–510 nm is excluded. It can be assured that carotenoids to chlorophyll energy transfer takes place at 490 nm on bleaching of S_0.S_2 transition over IC to the S1 state. Area between 480 and 510 nm in the spectrum is used to measure the excited state density of carotenoids. This area shrinking within 70–90-fs to 35% indicates transfer of energy that is absorbed by the carotenoids bound to CP29 to chlorophylls through S_2 state. The amplitude as well as lifetime of the absorption decay does not depend upon the type of carotenoid but is determined by the chlorophyll environment, as explained in Table 5.1.

5.4.2.2 Xanthophylls in CP29 Transfer Excitations to Chlorophyll a

Chlorophyll a molecules are the receptors of energy in CP29 protein. Energy is not transferred to chlorophyll b molecules but in spectrum, it is present due to its direct excitation. (Figs. 5.3 and 5.4). To determine the amount of chlorophyll b in EET, excitations of Carotenoids, Chlorophyll b, and Chlorophyll a is correlated with that of their spectral decompositions. It proved that chlorophyll b molecule gets excited due to intramolecular energy transfer I no time (<<100 fs). Figure 5.3b It is noteworthy that carotenoids are excited more that the decomposition of their absorption spectra [104].

5.4.2.3 In LHCII Lut and Vio Transfer to Chl a and Neo Transfers to Chl b

LHCII molecules have different receptors which receive energy from S_2 state for different wavelengths. So, Lut and Neo present in both Chlorophyll a and Chlorophyll b absorb energy at 489 nm; Chl a 506 nm receive energy from Lut and Vio [105]. To understand this better, we extracted from the SADS determined by global analysis, the relative concentrations of excited pigments at each intermediate step is present. Chlorophyll b and Chlorophyll a are excited at 489 nm wavelength, in the meanwhile, caroteonids signal decrease because, they transfer their 65% energy to Chlorophyll a molecules containing Luts and 35% to Chlorophyll b molecules containing Neo. Neo is bound to Chlorophyll b adjacent to its helix C domain. The interaction of Chlorophyll b to Neo can be understood from the experiment in which LHCII excited at 650 nm at 77 K shows bleaching response at 485 nm instantaneously within 0.5–0.6 ps, depicting energy transfer from Chlorophyll b to Chlorophyll a [106, 107]. Many chlorophylls surround Luts in their middle section. Mostly Q_y transitions of receptors take place at 678 nm, but a few at 673 nm. Vio shows excitation at 506 nm representing its bond with blue chlorophyll a at different peripheral sites. In this case, there is specific loss when either monomeric LHCII consist of trimeric particles or detergent treatment of trimers under mild conditions [108].

5.4.2.4 Coupling Between Carotenoid S_2 State and Chlorophyll Excited States

Processes of transfer of energy from Carotenoid to Chlorophyll b EET is followed by Chlorophyll b to Chlorophyll a take place independent because, Carotenoid bleaching/SE takes much more time to decay than that of the Chlorophyll a bleaching/SE. Energy is transferred from the excited state of chlorophyll from the S_2 states of carotenoids in CP29 and LHCII. In this case, chlorophyll a occupies all four symmetrically arranged binding sites of chlorophyll, at the centre, which has 3D structure. Here, Coulombic interactions play an important role in energy transfer [109, 110]. So, energy is transferred from S_2 state to vibrant transitions of Chlorophyll a molecule. Average Coulombic coupling (200 cm^{-1}) between receptor molecules of chlorophyll to xanthophylls L1 and L2 causes excitation.

5.4.2.5 Energy Transfer from the S_1 State

Excitation based fluorescence was used to determine efficiency of transfer of energy from Carotenoid to Chlorophyll that was upto 90% for LHCII at room temperature and 80% for CP29 complex in the presence of temperature upto 77 K. As, 60–65% of the Carotenoid S_2 state excitation energy is shifted directly to Q_x and Q_y state, 15–20% energy is from optically forbidden S_1 excited state at 506 nm for both CP29 and LHCII; regardless of the fact that 489 nm is the wavelength at which maximum excited states are produced. However, decay of Carotenoids S_2 bleaching S_1-S_N spectrum takes place between 480–500 nm and 520–620 nm respectively within 0.2 ps. The in vitro S1 lifetimes are following: for Lut 14.6 ps, Neo 35 ps and Vio 23-25 ps [111, 112]. Although, S_1 lifetimes should have same values for in-vivo and in-vitro, but there are some interactions that may increase the excited state level, as at 506 nm, for LHCII, there is 30–35% and for CP29 10% increase in the excited state lifetime. Rapid transfer of the energy from S_1 state to the neighboring chlorophyll molecule is due to the fact that this phenomenon can be observed only in some chlorophylls; for this purpose, molecular properties and the binding site of the chlorophyll play an important role. In LHCII, after excitation of the sample, 1 ps time is required to transfer 50–60% of the S_1 excited state energy to the chlorophyll [113, 114]. So it is quite obvious that, in LHCII, for transfer of energy from S_1 excited state to chlorophyll a that has four receptors sites, Luts plays an important role, but CP29 has no binding site.

5.5 Light Absorption and Transfer of Resonance Energy to Reaction Center

Carotenoids absorb visible light and transfer it to the reaction center by following different transfer models as described below (Fig. 5.20):

5.5.1 Resonance Energy

Transfer of the resonance energy, a spectroscopic method, takes place through transition dipole-dipole interactions and radiation-less process between physically distinct atomic or molecular components. This process was discovered by Cario and Franck in gaseous phase, while in solid phase, by Gaviola and Pringsheim [115, 116].

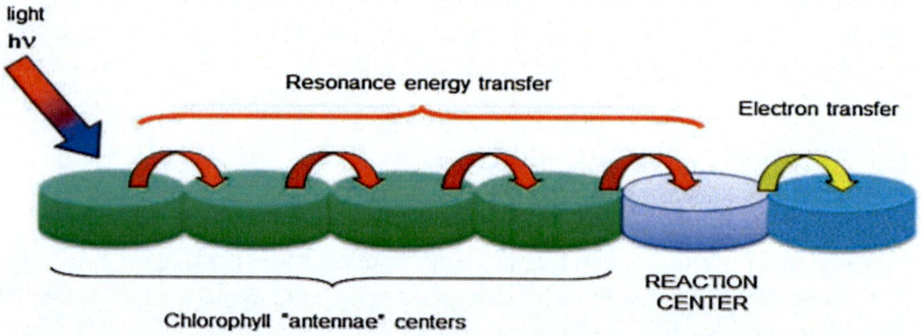

Fig. 5.20 Absorption and transfer of light to from carotenoids to reaction center

This process was explained theoretically by the Nobel laureate Perrin (1927); when an excited molecule and neighbouring molecule are in proximity, transfer of energy takes place via dipole interaction from excited to neighbouring molecule. He also investigated that radiation-less process took place within intermolecular distance of 1000 A°, this assumption was rejected later on [117].

Based upon Perrin's experiment, Forster investigated two main factors (spectrum overlap and intermolecular distance) that played an important role in the resonance energy transfer; he also reported that energy is transferred when there is proportionality between rate of energy transfer and dipole-dipole coupling and measured 10 and 100 A° distance between excited and acceptor molecule [118, 119].

In resonance energy transfer, a dielectric molecule on interaction with visible or UV light absorbs photon of energy resulting in the excitation of the molecule; that molecule is called chromophore. The absorption of energy is simultaneous with the loss of energy in the form of either fluorescence or dissipation of energy locally, from the lowest energy level of the excited state [120].

$$1D^{*} + 1A(RET) \rightarrow 1D + 1A^{*}$$

How fluorescence resonance energy is transferred from donor to acceptor [121] is shown in Fig. 5.21:

For the fluorescence resonance energy transfer, there should be fluorescent donor, that after excitation is converted into a dipole that interacts with that of acceptor through dipole-dipole interactions; the transfer takes place through vibrational energy states, the fluorescent state and resonance energy transfer [122], as explained in Fig. 5.22:

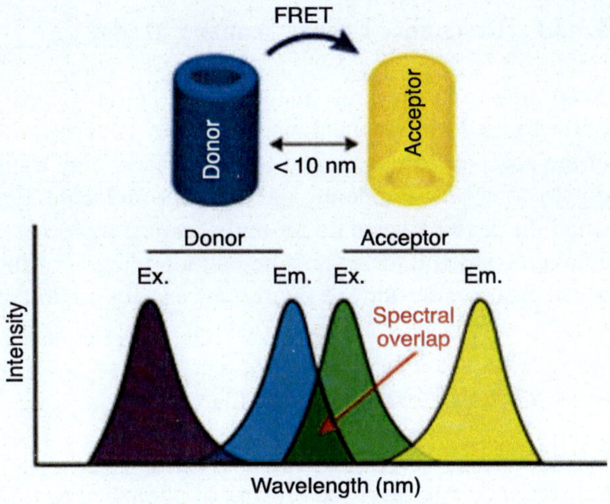

Fig. 5.21 Fluorescence resonance energy transfer

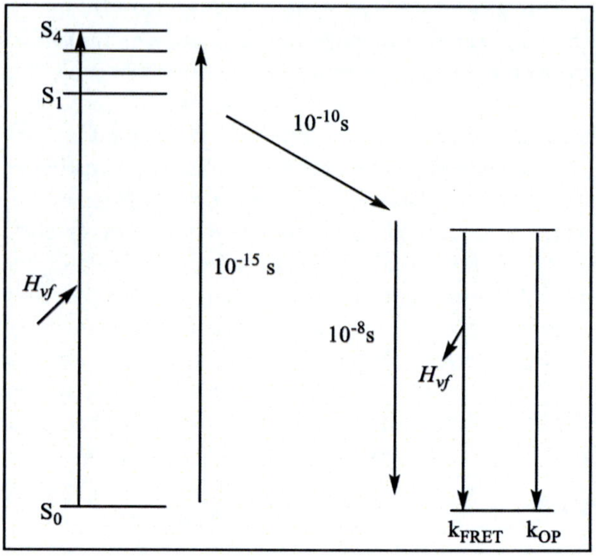

Fig. 5.22 Modified Jablonski diagram

5.5.2 *Models of Energy Transfer to Reaction Center*

As excitation energy is transferred from carotenoids (for example, fuxoxanthin, peridinin) to chlororphyll a; this mechanism was explained by William Arnold first time, where he used phycocyanin carotenoid. This mechanism was further extended by Oppenheimer. So, nowadays, there are two models as following:

5.5.2.1 Resonance Energy Transfer Model

According to Forster, resonance energy is transferred in two main steps; (1) decay of the excited state in the donor molecule, (2) upward transition of the ground state of the acceptor to the excited followed by energy transfer. It occurs due to dipole-dipole bond between donor and acceptor molecule. For this purpose, there must be small distance between donor and acceptor molecules, dipoles should have a suitable direction and there must be overlap integral of the absorption spectrum of the acceptor molecule and the fluorescence spectrum of the donor molecule [123–127].

Mechanism of Energy Transfers [128–130]

Spectroscopic analysis reveals that there are two excited states, S_1 and S_2, which transfer energy to drive the process of photosynthesis. There exists Ag symmetry of the ground state, S_0, and the first excited, S_1, of polyenes with C_{2h} point group, on the other hand, second excited state, S_2, has Bu symmetry. When transition S_0–S_1 takes place, there exists forbidden symmetry, while transition S_0–S_2 allows absorption as well as fluorescence in the visible region of t the spectrum. Carotenoids do not adhere strictly to C2h symmetry but have various spectral properties of the parent polyenes from which they were derived.

S_0–S_2 transition in carotenoids depends on polarizability of the solvent. It is represented through absorption spectrum which shows that carotenoids are tightly bound to the protein molecule of chlorophyll b due to a minor red shift in the emission spectrum. So, chlorophyll b is not well known for its light-harvesting properties.

When carotenoid molecules are excited by direct method, the excited singlet states are and changed into ground state via radiationless processes, at the same time, triplet states are empty leading to no intersystem crossing. So, it is required to excite the carotenoids into triplet excited states by transferring energy from the triplet states of chlorophylls through van der Waals interactions, and short proximal distance, between energy donor and acceptor molecules. In this way, there is very small lifetime of the triplet excited states; This can be achieved by using the carotenoid molecules which have high π-conjugation.

How energy from singlet excited state move back to the S_1 state, can be understood by the following equation.

$$[S_1]=[S_1]_0+k_{ic2}[S_2]_0\left(\text{epx}(-k_{k2t})-\exp(-k_{1t})/(k_1-k_2)\right)$$

where,

$$k_2 = k_{ET2} + k_{ic2} + k_{ic3}$$
$$k_1 = k_{ET1} + k_{ic1}$$

k_1 represents the single-exponential decay expression. k_{ET1} value can be obtained from kic_1 that can be derived from the values from the fit of the energy gap law to the dynamics for the decay of the S1 states of the carotenoid in solution. Lesser the value of k_{ET1}, more the extent of the π-electron conjugation of the carotenoids increases.

$$\epsilon = (k_{ET2}/k_{ET2} + k_{ic2}) + (k_{ic2}/k_{ET2} + k_{ic2})\,(k_{ET1}/k_{ET1} + k_{ic})$$

Equation 8 helps to calculate the values of k_{ET2}, representing how efficiently energy is transferred from the single excited state to chlorophylls. As number of double bonds in a molecule increase, S_2 excited state is preferred for the transfer of energy. It was observed that value of k_{ET1} decreases in much less time as compared to value of k_{ET2} with the increase in number of double bonds. This difference between two excited state, S_2 and S_1 explains well the mechanism of transfer of energy to the reaction center of the chlorophyll, in Fig. 5.23.

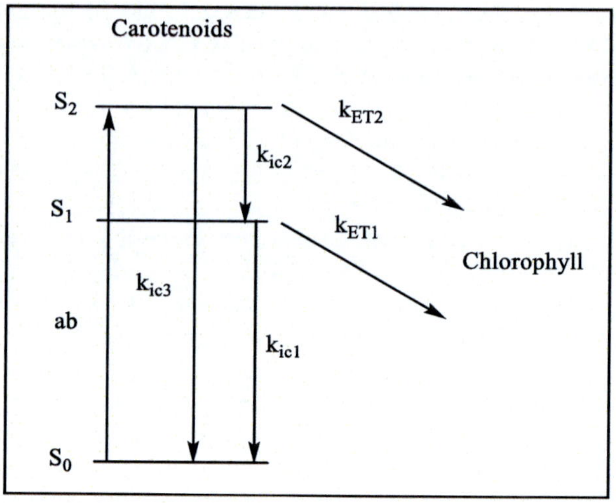

Fig. 5.23 kinetic diagram of transfer of carotenoids excitation energy to chlorophyll

5.5.2.2 Electron Exchange

Dexter explains that donor and acceptor molecules residing in the prixmity exchange electron simultaneously. According to Jablonski diagram of energy, second excited state S_2 of chlorophyll is at lower energy level than that of carotenoids. Due to short lifetime of second excited states (S_2), there is loss of heat as a result of de-excitation [131].

To understand the mechanism, consider a pair of donor (D) and acceptor (A) molecules having inter-distance R. Energy is transferred via dipole-dipole bond, the rate constant k_{ET} is given by the following equation:

$$k_{ET}\left(\text{dipole}\quad\text{dipole}\right) = Ak^2\Omega k^0{}_D\, R^{-6}$$

where, 'A' stands for acceptor molecule that has constant k. Value of k depends upon the orientation of both donor and acceptor molecules. The absorption spectra of acceptor and emission spectra of donor overlap due to:

$$\Omega = \int F_D(v)\epsilon A(v)\, v^{-4}\, dv$$

where F_D represent fluorescence spectrum of D on a wave number (v) scale, ϵ_A molar decadic extinction coefficient of A. Equation 1 elaborates the direct relationshil between, k_D and $k_{ET.}$

When $k \neq 0$ and k is substitute of Ak^2R^{-6}, the efficiency of the electron transfer can be calculated from:

$$\eta = k_{ET} / k_{ET} + k_D = \left(1 + 1/k\Omega \text{ł}_D\right)^{-1}$$

and the rate constant of electron-exchange-mediated energy transfer may be expressed as:

$$k_{ET}(\text{exchange}) = (2\pi / h)U^2 \square F_D(v)\beta_A(v)dv$$

where, β_A is the spectral distribution of the acceptor absorption on a wave number (v) scale.

In the thylakoid membrane, when chlorophyll and carotenoid molecule reside adjacent to each other, triplet-triplet energy transfer occurs:

$$\text{D}^\dagger(\text{Chlorophyll}) + \text{A}(\beta\ \text{Carotenoids}) \overset{\text{Electron exchange}}{\rightarrow} \text{D} + \text{A}^\dagger$$

There can take place the singlet-singlet electron transfer as well.

$$\text{D}^*(\beta\ \text{Carotenoids}) + \text{A}(\text{Chlorophyll}) \overset{\text{Electron exchange}}{\rightarrow} \text{D} + \text{A}^*$$

As, strong transfer access is forbidden excited state, there is strong electron coupling and exchange to the chlorophyll. Carotenoid symmetry plays an important role in this process. Naturally, electronic coupling is possible for the carotenoid molecules which have side group functional groups that play an important role in the breakage of the symmetry [132, 133].

Photosynthesis converts solar energy into chemical energy. It provides food and oxygen; and, in the future, it could directly provide bioenergy or renewable energy sources, such as bio-alcohol or hydrogen. To exploit such a highly efficient capture of energy requires an understanding of the fundamental physics. The process is initiated by photon absorption, followed by highly efficient and extremely rapid transfer and trapping of the excitation energy.

5.6 Dependencies of Carotenoids Assisted Photosynthesis in Plant Adaptive to Low Light Intensities

Naturally, UV light (280–350) and visible light (350–750 nm) encounter the plant [134]. UV-radiation, especially UV-B lowers te rate of photosynthesis by damaging the reaction centre of photosystem II (PSII), so there must be such mechanisms in the plant to reduce the damage of the reaction centre [135, 136]. When plant is exposed to light of high intensity, orientation and optical characteristics of leaves are changed. It affects the size of chloroplast, contents of chlorophylls and halts grana stacking present in the chloroplast as well. There is also lowering in concentration of light-harvesting complex II (LHCII) which plays an important role in PSII by getting bound to chlorophyll b present in in the thylakoid membrane. This leads to the less ratio of chlorophyll b and chlorophyll a [137–143]. On the other hand, visible light plays a significant role in the development and metabolism of the plant [144, 145], as well as for photosynthesis and photo-morphogenesis. There are so many factors, such as horticulture facilities, clouds, high intensity of visible light and low intensity of visible light, snow, etc., which cause disturbance in the light passage. Due to presence of low light, antioxidant properties, carbon and nitrogen fixation processes, photosynthetic processes along-with agronomic traits of plants are highly affected [146–149]. Plant also grows slowly, with thin leaves, less number of buds in the flower, low contents of sugar and starch [150–152], followed by change in color of plant and leaves and less time required to mature [153]. Chlorophyll takes up light, from the sun rays, which plays an important role during the process of photosynthesis [154]. If experiments are performed in the presence of low light, hybrid rice plant gets high contents of chlorophyll b [155]. It is due to low stomatal conductance as a result, rice leaves absorb high concentration of intercellular carbon dioxide [156, 157].

More components that play an important role in the process of photosynthesis are carotenoid and chlrorophylls [158]. Generally, it is thought that in the presence of low light intensity, content of chlorophyll lessens, but there are some contradictions in the following; According to Ma et al. [159], intensity of light does not affect the amount of chlorophyll a content, but there is an increase in contents of chlorophyll b; According to Lakshmi and Singh [160, 161], contents of both chlorophylls increase with the decrease in intensity of light, but Bell [162] proved effect of low light by performing an experiment on Agrostis stolonifera and concluded that when a plant is exposed to low intensity of light for a longer period of time, it gets dry, withered and etiolated. Bell s experiment has also been proved by Zhu by working on purple pak-choi. In this experiment, there was a significant decrease in contents of chlorophyll a when the plant was exposed to low light according to Table 5.2. On increasing the time of exposure, chlorophyll a was damaged, as a result there was damage of photosynthetic machinery, but contents of chlorophyll b decreased on 15th day depicting resistance to the low light stress [163].

On the other hand, light intensity also affects the contents of carotenoids present in the photosynthetic machinery [164], in one experiment it was showed that carotenoid contents decrease upto 0.048 mg g^{-1}, when light of 500 $\mu.mol.m^{-2}.g^{-1}$ was used. It also leads to change in color of the leaves [165] as depicted in Fig. 5.24,

So, it be said that in the presence of less intensity of light, plants are not able to grow and flourish because of damage of their photosynthetic machinery that is halted due to inactivity of the molecules which are responsible to capture sunlight and transmit it for driving the process.

Table 5.2 Treatment of low light on the plat seedlings

Category	Rate of light transmittance(%)	Treatment	Illuminescence intensity (μ mol m^{-2} s^{-1})
NL	100	Normal illumination	1000
TL-1	75	A layer of white guaze	750
TL-2	50	A layer of black shade	500
TL-3	25	A layer of black shade	250

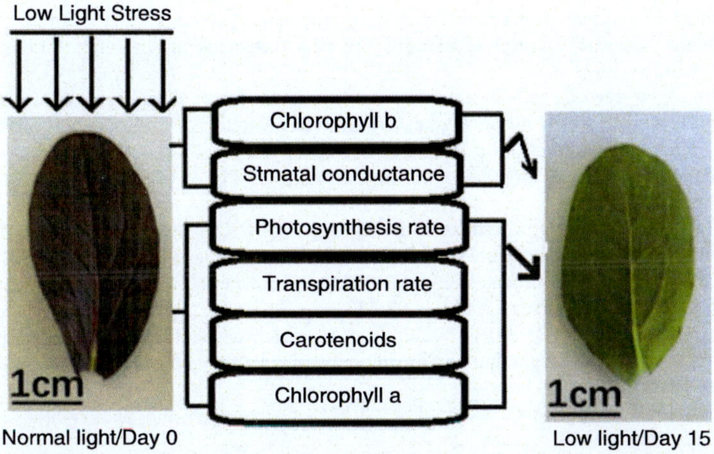

Fig. 5.24 Effect of low intensity light on the color of leaves

5.7 Conclusion

Photosynthesis takes place naturally in plants and micro-organisms, which have light harvesting molecules as chlorophyll and carotenoids which absorb sunlight and convert them it into chemical energy. Carotenoids absorb energy and get excited at different wavelengths followed by coupling with the chlorophyll molecules via different interactions and transfer energy to the reaction center of the chlorophyll. In this process, fluorescence resonance energy transfer is the most prominent way of energy transfer. There are also some factors which depend upon the rate of photosynthesis and health of the plant, such as low light intensity causes color change and rate change of process owing to damage of photosynthetic machinery. We conclude by examining the physical basis of some current models concerning the roles of coherent excitons and incoherent hopping in the exceptionally efficient transfer of energy into the reaction center.

References

1. Flores-Hidalgo M, Torres-Rivas F, Monzon-Bensojo J, Escobedo-Bretado M, Glossman Mitnik D, Barraza Jimenez D (2017) Electronic structure of carotenoids in natural and artificial photosynthesis. Carotenoids. https://doi.org/10.5772/67636
2. Blankenship RE (2014) Molecular mechanisms of photosynthesis. Wiley Blackwell, Oxford
3. Polıvka T, Frank HA (2010) Molecular factors controlling photosynthetic light harvesting by carotenoids. Acc Chem Res 43:1125–1134
4. Croce R, van Amerongen H (2014) Natural strategies for photosynthetic light harvesting. Nat Chem Biol 10(7):492–501

5. Fuciman M, Keşan G, LaFountain AM, Frank HA, Polívka T (2015) Tuning the spectroscopic properties of aryl carotenoids by slight changes in structure. J Phys Chem B 119(4):1457–1467
6. Berera R, Herrero C, van Stokkum IHM, Vengris M, Kodis G, Palacios RE, Kennis JTM (2006) A simple artificial light-harvesting dyad as a model for excess energy dissipation in oxygenic photosynthesis. Proc Natl Acad Sci 103(14):5343–5348
7. Cogdell RJ (1978) Carotenoids in photosynthesis. Philos Trans R Soc B Biol Sci 284(1002):569–579
8. Holt NE, Zigmantas D, Valkunas L, Li X-P, Niyogi KK, Fleming GR (2005) Carotenoid cation formation and the regulation of photosynthetic light harvesting. Science 307:433–436
9. Domonkos I, Kis M, Gombos Z, Ughy B (2013) Carotenoids, versatile components of oxygenic photosynthesis. Prog Lipid Res 52:539–561
10. Niyogi KK, Truong TB (2013) Evolution of flexible non-photochemical quenching mechanisms that regulate light harvesting in oxygenic photosynthesis. Curr Opin Plant Biol 16:307–314
11. Hashimoto H, Uragami C, Cogdell RJ (2016) Carotenoids and photosynthesis. Subcell Biochem 79:111–139
12. Krinsky NI (1989) Antioxidant function of carotenoids. Free Radic Biol Med 7:617–635
13. Havaux M (1998) Carotenoids as membrane stabilizers in chloroplasts. Trends Plant Sci 3:147–151
14. Joyard J, Ferro M, Masselon C, Seigneurin-Bernya D, Salvia D, Garin J et al (2009) Chloroplast proteomics and the compartmentation of plastidial isoprenoid biosynthetic pathways. Mol Plant 2:1154–1180
15. Murchie EH, Niyogi KK (2011) Manipulation of photoprotection to improve plant photosynthesis. Plant Physiol 155:86–92
16. Nambara E, Marion-Poll A (2005) Abscisic acid biosynthesis and catabolism. Annu Rev Plant Biol 56:165–185
17. Al-Babili S, Bouwmeester HJ (2015) Strigolactones, a novel carotenoid-derived plant hormone. Ann Rev Plant Biol 66:161–186
18. Namitha KK, Negi PS (2010) Chemistry and biotechnology of carotenoids. Crit Rev Food Sci Nutr 50(8):728–760
19. Lemmens L, Colle I, Van Buggenhout S, Palmero P, Van-Loey A, Hendrickx M (2014) Carotenoid bioaccessibility in fruit and vegetable based food products as affected by product (micro) structural characteristics and the presence of lipids: a review. Trends Food Sci Technol 38:125–135
20. Bartley GE, Scolnik PA (1995) Plant carotenoids: pigments for photoprotection, visual attraction, and human health. Plant Cell 7:1027–1038
21. Vishnevetsky M, Ovadis M, Vainstein A (1999) Carotenoid sequestration in plants: the role of carotenoid associated proteins. Trends Plant Sci 4:232–235
22. Haskell MJ (2013) Provitamin A carotenoids as a dietary source of vitamin A. In: Tanumihardjo SA (ed) Carotenoids and human health, 1st edn. Springer, New York, pp 249–260. https://doi.org/10.1007/978-1-62703203-2-15
23. Croft H, Chen JM (2018) Leaf pigment content. In: Comprehensive remote sensing. Elsevier, Oxford, pp 117–142
24. Schreiber U, Neubauer C (1990) O_2-dependent electron flow, membrane energization and the mechanism of non-photochemical quenching of chlorophyll fluorescence. Photosynth Res 25(3):279–293
25. Collini E (2019) Carotenoids in photosynthesis: the revenge of the "accessory" pigments. Chemistry 5(3):494–495
26. Frank H, Violette CA, Trautman JK, Shreve AP, Owens TG, Albrecht AC (1991) Photosynthetic carotenoids: structure and photochemistry. Pure Appl Chem 63:109–114
27. Hudson B, Kohler B (1974) Linear polyene electronic structure and spectroscopy. Annu Rev Phys Chem 25(1):437–460

28. Polívka T, Sundström V (2004) Ultrafast dynamics of carotenoid excited states– from solution to natural and artificial systems. Chem Rev 104(4):2021–2072
29. Krueger BP, Scholes GD, Jimenez R, Fleming GR (1998) Electronic excitation transfer from carotenoid to bacteriochlorophyll in the purple bacterium Rhodopseudomonas acidophila. J Phys Chem B 102(12):2284–2292
30. Ritz T, Damjanović A, Schulten K, Zhang JP, Koyama Y (2000) Efficient light harvesting through carotenoids. Photosynth Res 66(1–2):125–144
31. Scholes GD (2003) Long-range resonance energy transfer in molecular systems. Annu Rev Phys Chem 54(1):57–87
32. Frank HA, Brudvig GW (2004) Redox functions of carotenoids in photosynthesis†. Biochemistry 43(27):8607–8615
33. Liu Z, Yan H, Wang K, Kuang T, Zhang J, Gui L et al (2004) Crystal structure of spinach major light-harvesting complex at 2.72 Å resolution. Nature 428(6980):287–292
34. Peterman EJ, Dukker FM, van Grondelle R, van Amerongen H (1995) Chlorophyll a and carotenoid triplet states in light-harvesting complex II of higher plants. Biophys J 69(6):2670–2678
35. French CS (1937) The rate of CO2 assimilation by purple bacteria at various wave lengths of light. J Gen Physiol 21(1):71–87
36. Goedheer JC (1959) Energy transfer between carotenoids and bacteriochlorophyll in chromatophores of purple bacteria. Biochim Biophys Acta 35:1–8
37. Nuijs AM, van Grondelle R, Joppe HLP, van Bochove AC, Duysens LN (1985) Singlet and triplet excited carotenoid and antenna bacteriochlorophyll of the photosynthetic purple bacterium Rhodospirillum rubrum as studied by picosecond absorbance difference spectroscopy. Biochim Biophys Acta (BBA)-Bioenergetics 810(1):94–105
38. Haxo FT, Blinks LR (1950) Photosynthetic action spectra of marine algae. J Gen Physiol 33(4):389–422
39. Goedheer JC (1969) Energy transfer from carotenoids to chlorophyll in blue-green, red and green algae and greening bean leaves. Biochim Biophys Acta (BBA)-Bioenergetics 172(2):252–265
40. Siefermann-Harms D (1985) Carotenoids in photosynthesis. I. Location in photosynthetic membranes and light-harvesting function. Biochim Biophys Acta (BBA)- Bioenergetics 811(4):325–355
41. Dallinger RF, Woodruff WH, Rodgers MAJ (1981) The lifetime of the excited singlet state of β-carotene: consequences to photosynthetic light harvesting. Photochem Photobiol 33(2):275–277
42. Gonen O, Levanon H, Patterson LK (1981) Orientation and energy transfer in chlorophyll monolayers diluted with hexadecane. Fluorescence and sensitized fluorescence. Israel J Chem 21(4):271–276
43. Sineshchekov VA, Litvin FF, Das M (1972) Chlorophyll a and carotenoid aggregates and energy migration in monolayers and thin films. Photochem Photobiol 15(2):187–197
44. Moore AL, Dirks G, Gust D, Moore TA (1980) Energy transfer from carotenoid polyenes to porphyrins: a light-harvesting antenna. Photochem Photobiol 32(5):691–695
45. Siefermann-Harms D (1987) The light-harvesting and protective functions of carotenoids in photosynthetic membranes. Physiol Plant 69(3):561–568
46. Davidson E, Cogdell RJ (1981) Reconstitution of carotenoids into the light-harvesting pigment-protein complex from the carotenoidless mutant of Rhodopseudomonas sphaeroides R26. Biochim Biophys Acta (BBA)-Bioenergetics 635(2):295–303
47. Cogdell RJ, Frank HA (1987) How carotenoids function in photosynthetic bacteria. Biochim Biophys Acta (BBA)-Bioenergetics 895(2):63–79
48. Pizarro L, Stange C (2009) Light-dependent regulation of carotenoid biosynthesis in plants. Ciencia e Investigación Agraria 36(2):143–162
49. Bartley GE, Scolnik PA (1995) Plant carotenoids: pigments for photoprotection, visual attraction, and human health. Plant Cell 7(7):1027

50. Weis E, Berry JA (1987) Quantum efficiency of photosystem II in relation to 'energy'-dependent quenching of chlorophyll fluorescence. Biochim Biophys Acta (BBA)-Bioenergetics 894(2):198–208
51. Demmig-Adams B (1990) Carotenoids and photoprotection in plants: a role for the xanthophyll zeaxanthin. Biochim Biophys Acta (BBA)-Bioenergetics 1020(1):1–24
52. Rees D, Young A, Noctor G, Britton G, Horton P (1989) Enhancement of the ΔpH-dependent dissipation of excitation energy in spinach chloroplasts by light-activation: correlation with the synthesis of zeaxanthin. FEBS Lett 256(1–2):85–90
53. Noctor G, Rees D, Young A, Horton P (1991) The relationship between zeaxanthin, energy-dependent quenching of chlorophyll fluorescence, and trans-thylakoid pH gradient in isolated chloroplasts. Biochim Biophys Acta (BBA)-Bioenergetics 1057(3):320–330
54. Demmig-Adams B, Winter K, Krüger A, Czygan FC (1989) Light response of CO2 assimilation, dissipation of excess excitation energy, and zeaxanthin content of sun and shade leaves. Plant Physiol 90(3):881–886
55. Fraser PD, Bramley PM (2004) The biosynthesis and nutritional uses of carotenoids. Prog Lipid Res 43:228–265
56. Felt L, Pacakova V, Stulik K, Volka K (2005) Reliability of carotenoid analysis: a review. Curr Anal Chem 1:93–102
57. Polivka T, Sundstrom V (2004) Ultrafast dynamics of carotenoid excited states-from solution to natural and artificial systems. Chem Rev 104:2021–2072
58. Lichtenthaler HK (1987) Chlorophylls and carotenoids: pigments of photosynthetic biomembranes. Methods Enzymol 148:350–382
59. Andersson PO, Gillbro T, Ferguson L, Cogdell RJ (1991) Absorption spectral shifts of carotenopids related to medium polarizability. Photochem Photobiol 54(3):353–360
60. Lichtenthaler HK (1987) [34] chlorophylls and carotenoids: pigments of photosynthetic biomembranes. Plant Cell Membr 148:350–382
61. Merzlyak MN, Gitelson AA, Chivkunova OB, Rakitin VY (1999) Non destructive optical detection of pigment changes during leaf senescence and fruit ripening. Physiol Plant 106(1):135–141
62. Merzlyak MN, Gitelson A (1995) Why and what for the leaves are yellow in autumn? On the interpretation of optical spectra of senescing leaves (Acerplatanoides L.). J Plant Physiol 145(3):315–320
63. Datt B (1998) Remote sensing of chlorophyll a, chlorophyll b, chlorophyll a+ b, and total carotenoid content in eucalyptus leaves. Remote Sens Environ 66(2):111–121
64. Gitelson A, Merzlyak MN (1994) Quantitative estimation of chlorophyll-ausing reflectance spectra: experiments with autumn chestnut and maple leaves. J Photochem Photobiol B Biol 22(3):247–252
65. Gitelson AA, Kaufman YJ, Merzlyak MN (1996) Use of a green channel in remote sensing of global vegetation from EOS-MODIS. Remote Sens Environ 58(3):289–298
66. Gitelson AA, Merzlyak MN (1997) Remote estimation of chlorophyll content in higher plant leaves. Int J Remote Sens 18(12):2691–2697
67. Gitelson AA, Merzlyak MN (1996) Signature analysis of leaf reflectance spectra: algorithm development for remote sensing of chlorophyll. J Plant Physiol 148(3–4):494–500
68. Merzlyak MN, Gitelson AA, Chivkunova OB, Solovchenko AE, Pogosyan SI (2003) Russ J Plant Physiol 50(5):704–710
69. Arnon DI (1949) Copper enzymes in isolated chloroplasts. Polyphenoloxidase in Beta vulgaris. Plant Physiol 24(1):1
70. Lichtenthaler HK, Buschmann C (2001) Extraction of phtosynthetic tissues: chlorophylls and carotenoids. Curr Protocol Food Anal Chem 1(1):F4–F2
71. Lichtenthaler HK, Kuhn G, Prenzel U, Buschmann C, Meier D (1982) Adaptation of chloroplast-ultrastructure and of chlorophyll-protein levels to high-light and low-light growth conditions. Zeitschrift für Naturforschung C 37(5–6):464–475

72. Sommerburg O, Zang LY, Van Kuijk FJ (1997) Simultaneous detection of carotenoids and vitamin E in human plasma. J Chromatogr B Biomed Sci Appl 695(2):209–215
73. Handelman GJ, Snodderly DM, Krinsky NI, Russett MD, Adler AJ (1991) Biological control of primate macular pigment. Biochemical and densitometric studies. Invest Ophthalmol Vis Sci 32(2):257–267
74. Craft NE, Soares JH (1992) Relative solubility, stability, and absorptivity of lutein and. beta.-carotene in organic solvents. J Agric Food Chem 40(3):431–434
75. Hager A (1970) Ausbildung von Maxima im Absorptionsspektrum von Carotinoiden im Bereich um 370 nm; Folgen for die Interpretation bestimmter Wirkungsspektren. Planta 91(1):38–53
76. Buchwald M, Jencks WP (1968) Optical properties of astaxanthin solutions and aggregates. Biochemistry 7(2):834–843
77. Cario G, Franck J (1923) Uber sensibilisierte Fluoreszenz von Gasen. Zeitschrift Fur Physik 17(1):202–212
78. Frank HA, Cogdell RJ (1993) Photochemistry and functions of carotenoids. In: Young A, Britton G (eds) Carotenoids in photosynthesis. Springer, London, pp 252–326
79. Frank HA, Violette CA, Trautman JK, Shreve AP, Owens TG, Albrecht AC (1991) Carotenoids in photosynthesis: structure and photochemistry. Pure Appl Chem 63(1):109–114
80. Kamen MD (1986) On creativity of eye and ear: a commentary on the career of TW Engelmann. Proc Am Philos Soc 130:232–246
81. Warburg O, Negelein E (1923) Über den Einfluss der Wellenlänge auf den Energie Umsatz bei der Kohlensauereassimilation. Z Phys Chem 106:191–121
82. Montfort C (1936) Carotinoide, Photosynthese und Quantentheorie. *Jahrb Wiss Botan* 83:725–772
83. Montfort C (1941) Die Ausnutzung grünen Lichtes bei braunen Zellen im Hinblick auf den Energiegewinn durch den Fucoxanthin-Eiweisskomplex. Planta 32(1):118–120
84. Emerson R, Lewis CM (1942) The photosynthetic efficiency of phycocyanin in Chrococcus and the problem of carotenoid participation in photosynthesis. J Gen Physiol 25(4):579–595
85. Emerson R, Lewis CM (1943) The dependence of quantum yield of Chlorella photosynthesis on wavelength of light. Am J Bot 30:165–178
86. Dutton HJ, Manning (1941) Evidence for carotenoid-sensitized photosynthesis in the diatom Nitzschia closterium. *American Journal of Botany* 28:516–526
87. Wassink EC, Kersten JAH (1945) Photosynthesis and fluorescence of the chlorophylls of diatoms. Enzymologia 11:282–312
88. Tanada T (1951) The photosynthetic efficiency of carotenoid pigments in navicula minima. Am J Bot 38(4):276–283
89. Haxo F, Blinks LR (1950) Photosynthetic action spectra of marine algae. J Gen Physiol 33(4):389–422
90. Croce R, Remelli R, Varotto C, Breton J, Bassi R (1999) The neoxanthin binding site of the major light harvesting complex (LHCII) from higher plants. FEBS Lett 456:1–6
91. Croce R, Weiss S, Bassi R (1999) Carotenoid-binding sites of the major light-harvesting complex II of higher plants. J Biol Chem 274(42):29613–29623
92. Hobe S, Niemeier H, Bander A, Paulsen H (2000) Carotenoid binding sites in LHCIIb. Relative affinities towards major xanthophylls of higher plants. Eur J Biochem 267:616–624
93. Grabowski B, Cunningham FX, Gantt E (2001) Chlorophyll and carotenoid binding in a simple red algal light-harvesting complex crosses phylogenetic lines. Proc Natl Acad Sci 98(5):2911–2916
94. Croce R, Cinque G, Holzwarth AR, Bassi R (2000) The Soret absorption properties of carotenoids and chlorophylls in antenna complexes of higher plants. Photosynth Res 64(2–3):221–231
95. Valkunas L, Cervinskas V, Trinkunas G, Müller MG, Holzwarth AR (1999) Effects of excited state mixing on transient absorption spectra in dimers. Application to photosynthetic light-harvesting complex II. J Chem Phys 111:3121–3132

96. Valkunas L, Cervinskas V, Trinkunas G, Müller MG, Holzwarth AR (1998) Excited state mixing effects in transient absorption spectra of photosynthetic light-harvesting complex II. In: Garab G (ed) Photosynthesis: mechanism and effects/XI. International congress photosynthesis Budapest, Kluwer Academic Publishers, Dordrecht, The Netherlands, pp 281–284
97. Croce R, Muller MG, Bassi R, Holzwarth AR (2001) Carotenoid-to-chlorophyll energy transfer in recombinant major light-harvesting complex (LHCII) of higher plants. I. Femtosecond transient absorption measurements. Biophys J 80:901–915
98. Bondarev SL, Knyukshto VN (1994) Fluorescence from the S1 (2 1Ag) state of all-trans-β-carotene. Chem Phys Lett 225(4–6):346–350
99. Dexter DL (1953) A theory of sensitized luminescence in solids. J Chem Phys 21(5):836–850
100. Zurdo J, Fernandez-Cabrera C, Ramirez JM (1993) A structural role of the carotenoid in the light-harvesting II protein of Rhodobacter capsulatus. Biochem J 290(2):531–537
101. Gust D, Moore TA, Moore AL, Devadoss C, Liddell PA, Hermant R et al (1992) Triplet and singlet energy transfer in carotene-porphyrin dyads: role of the linkage bonds. J Am Chem Soc 114(10):3590–3603
102. Hartwich G, Scheer H, Aust V, Angerhofer A (1995) Absorption and ADMR studies on bacterial photosynthetic reaction centres with modified pigments. Biochim Biophys Acta (BBA)-Bioenergetics 1230(3):97–113
103. Christensen RL (1985) Electronic energy levels in long Polyenes: S2-So emission. J Am Chem Soc 107(14):4117–4122
104. Demmig-Adams B (1990) Carotenoids and photoprotection in plants: a role for the xanthophyll zeaxanthin. Biochim Biophys Acta (BBA)-Bioenergetics 1020(1):1–24
105. Grondelle R, Dekker JP, Gillbro T, Sundstrom V (1994) Energy transfer and trapping in photosynthesis. Biochim Biophys Acta (BBA)-Bioenergetics 1187(1):1–65
106. Frank HA, Cogdell RJ (1993) The photochemistry and function of carotenoids in photosynthesis. In: Carotenoids in photosynthesis. Springer, Dordrecht, pp 252–326
107. Fujiwara M, Hayashi H, Tasumi M, Kanaji M, Koyama Y, Satoh K (1987) Structural studies on a photosystem II reaction-center complex consisting of D-1 and D-2 polypeptides and cytochrome b-559 by resonance Raman spectroscopy and high-performance liquid chromatography. Chem Lett 16(10):2005–2008
108. Cogdell RJ, Hipkins MF, MacDonald W, Truscott TG (1981) Energy transfer between the carotenoid and the bacteriochlorophyll within the B-800–850 light-harvesting pigment-protein complex of Rhodopseudomonas sphaeroides. Biochim Biophys Acta (BBA)-Bioenergetics 634:191–202
109. Peterman EJ, Dukker FM, van Grondelle R, van Amerongen H (1995) Chlorophyll a and carotenoid triplet states in light-harvesting complex II of higher plants. Biophys J 69(6):2670–2678
110. Young AJ, Frank HA (1996) Energy transfer reactions involving carotenoids: quenching of chlorophyll fluorescence. J Photochem Photobiol B Biol 36(1):3–15
111. Bondarev SL, Knyukshto VN (1994) Fluorescence from the S1 (2 1Ag) state of all-trans-β-carotene. Chem Phys Lett 225(4–6):346–350
112. Frank HA, Farhoosh R, Aldema ML, DeCoster B, Christensen RL, Gebhard R, Lugtenburg J (1993) Carotenoid to bacteriochlorophyll singlet energy transfer in carotenoid incorporated b850 light harvesting complexes of Rhodobacter sphaeroides R–26.1. Photochem Photobiol 57(1):49–55
113. DeCoster B, Christensen RL, Gebhard R, Lugtenburg J, Farhoosh R, Frank HA (1992) Low-lying electronic states of carotenoids. Biochim Biophys Acta Protein Struct Mol Enzymol 1102(1):107–114
114. Harry AF, Farhoosh R, Gebhard R, Lugtenburg J, Gosztola D, Wasielewski MR (1993) The dynamics of the S1 excited states of carotenoids. Chem Phys Lett 207(1):88–92
115. Ronald KC, Craig SL, Julius CN, Rebek J (2000) Detection and mechanistic studies of multicomponent assembly by fluorescence resonance energy transfer. J Am Chem Soc 122(33):7876–7882

116. Cano-Raya C, Fernández-Ramos MD, Capitán-Vallvey LF (2006) Fluorescence resonance energy transfer disposable sensor for copper (II). Anal Chim Acta 555(2):299–307
117. Perrin J (1927) Fluorescence et induction moléculaire par résonance. C R Hebd Seances Acad Sci 184:1097–1100
118. Jang S (2007) Generalization of the Förster resonance energy transfer theory for quantum mechanical modulation of the donor-acceptor coupling. J Chem Phys 127(17):174710
119. Andrews DL, Rodríguez J (2007) Resonance energy transfer: spectral overlap, efficiency, and direction. J Chem Phys 127(8):084509
120. Latt SA, Cheung HT, Blout ER (1965) Energy transfer. A system with relatively fixed donor-acceptor separation. J Am Chem Soc 87(5):995–1003
121. Broussard JA, Green KJ (2017) Research techniques made simple: methodology and applications of Förster resonance energy transfer (FRET) microscopy. J Investig Dermatol 137(11):185–191
122. Dos Remedios CG, Miki M, Barden JA (1987) Fluorescence resonance energy transfer measurements of distances in actin and myosin. A critical evaluation. J Muscle Res Cell Motil 8(2):97–117
123. Förster T (1946) Energiewanderung und Fluoreszenz. Naturwissenschaften 33:166–175
124. Förster T (1948) Zwischenmolekulare Energiewanderung und Fluoreszenz. Ann Physik 6(2):55–75. (English translation 'Intermolecular energy migration and fluorescence' is available from Professor RS Knox, Department of Physics and Astronomy, University of Rochester, Rochester, New York)
125. Förster T (1965) Delocalized excitation and excitation transfer. In: Sinanoglu O (ed) Part II.B. 1 of modern quantum chemistry: Istanbul lectures. Part III. Action of light and organic crystals. Academic, New York, pp 93–137
126. Knox RS (1975) Excitation energy transfer and migration: theoretical considerations. In: Govindjee (ed) Bioenergetics of photosynthesis. Academic, New York, pp 183–221
127. Van-Grondelle R, Amesz J (1986) Excitation energy transfer in photosynthetic systems. In: Govindjee AJ, Fork DC (eds) Light emission by plants and Bacteria. Academic, Orlando, pp 191–223
128. Dale RE, Eisinger J, Blumberg WE (1979) The orientation freedom of molecular probes. The orientation factor in intramolecular energy transfer. Biophys J 26:161–194
129. Lewis FD, Zhang L, Zuo X (2005) Orientation control of fluorescence resonance energy transfer using DNA as a helical scaffold. J Am Chem Soc 127:10002–10003
130. Clegg RM (1992) Fluorescence resonance energy transfer and nucleic acids. Methods Enzymol 211:353–388
131. Jablonski AZ (1935) Über den Mechanismus des Photolumineszenz von Farbstoffphosphoren. Z Phys 94:38–46
132. Haas E, Katchalski-Katzir E, Steinberg IZ (1978) Effect of the orientation of donor and acceptor on the probability of energy transfer involving electronic transitions of mixed polarization. Biochemistry 17:5064–5070
133. Jimenez R, Dikshit SN, Bradforth SE, Fleming GR (1996) Electronic excitation transfer in the LH2 complex ofRhodobacter sphaeroides. J Phys Chem 100(16):6825–6834
134. Melis A (1999) Photosystem-II damage and repair cycle in chloroplasts: what modulates the rate of photodamage in vivo? Trends Plant Sci 4:130–135
135. Jordan BR (1996) The effects of ultraviolet-B radiation on plants: a molecular perspective. Adv Bot Res 22:97–162
136. Vass I (1997) Adverse effects of UV-B light on the structure and function of the photosynthetic apparatus. In: Pessarakli M (ed) Handbook of photosynthesis. Marcel Dekker, New York, pp 931–949
137. Björkman O, Demmig-Adams B (1994) Regulation of photosynthetic light energy capture, conversion, and dissipation in leaves of higher plants. In: Schulze E-D, Caldwell MM (eds) Ecophysiology of photosynthesis. Springer, Berlin, pp 17–47

138. Smith WK, Vogelmann TC, DeLucia EH, Bell DT, Shepherd KA (1997) Leaf form and photosynthesis. Do leaf structure and orientation interact to regulate internal light and carbon dioxide? Bioscience 47:785–793
139. Adams WW, Demmig-Adams B, Rosenstiel TN, Ebbert V (2001) Dependence of photosynthesis and energy dissipation activity upon growth form and light environment during the winter. Photosynth Res 67:51–62
140. Anderson JM, Chow WS, Park Y-I (1995) The grand design of photosynthesis: acclimation of the photosynthetic apparatus to environmental cues. Photosynth Res 46:129–139
141. Siefermann-Harms D (1985) Carotenoids in photosynthesis. I. Location in photosynthetic membranes and light-harvesting function. Biochim Biophys Acta 811:325–355
142. Melis A (1991) Dynamics of photosynthetic membrane composition and function. Biochim Biophys Acta 1058:87–106
143. Anderson JM, Chow WS, Park Y-I (1995) The grand design of photosynthesis: acclimation of the photosynthetic apparatus to environmental cues. Photosynth Res 46:129–139
144. Nagy F, Schafer E (2006) Photomorphogenesis in plants and bacteria. Springer, Dordrecht
145. Berenschot AS, Quecini V (2014) A reverse genetics approach identifies novel mutants in light responses and anthocyanin metabolism in petunia. Physiol Mol Biol Plants 20:1–13
146. Wang L, Deng F, Ren WJ, Yang WY (2012) Effects of shading on starch pasting characteristics of indica hybrid rice (*Oryza sativa* L.). PLoS One 8:e68220
147. Demao J, Xia L (2001) Cultivar differences in photosynthetic tolerance to Photooxidation and shading in Rice (*Oryza Sativa L.*). Photosynthetica 39:167–175
148. Janardhan KV, Murty KS (1980) Effect of low light during vegetative stage on photosynthesis and growth attributes in rice. Indian J Plant Physiol 23:156–162
149. Apel K, Hirt H (2004) Reactive oxygen species: metabolism, oxidative stress, and signal transduction. Annu Rev Plant Biol 55:728–749
150. Ishida K (1989) Influence of respiration rate and metabolic substances on nodal position of first flower bud of eggplant seedlings. Engei Gakkai Zasshi 58:657–664
151. Vol N (1996) Stress and stress recovering by grapevines. Botanica Helvatica 106:73–84
152. Callejonferre AJ, Manzanoagugliaro F, Dıazperez M, Carreñoortega A, Perezalonso J (2009) Effect of shading with aluminised screens on fruit production and quality in tomato (*Solanum lycopersicum* L.) under greenhouse conditions. Span J Agric Res 7:41–49
153. Proebsting P (1986) Effect of different artificial shading times and natural light intensities on the fruit quality of 'Bing' sweet cherry. J Am Soc Hortic Sci 111:360–363
154. Liu Q, Wu X, Chen B, Ma J, Gao J (2014) Effects of low light on agronomic and physiological characteristics of Rice including grain yield and quality. Rice Sci 21:243–251
155. Zhu P, Yang SM, Ma J, Li SX, Chen Y (2008) Effect of shading on the photosynthetic characteristics and yield at later growth stage of hybrid Rice combination. Acta Agron Sin 34:2003–2009
156. Meng L, Chen WF, Li LX, Xu ZJ, Liu LX, Sun JW (2002) Influence of low light on stomatal characters in Rice leaves. J Shenyang Agric Univer 33:87–89
157. Yang D, Duan LS, Xie HA, Li ZH, Huang TX (2011) Effect of pre-flowering light deficiency on biomass accumulation and physio-logical characteristics of rice. Chin J Eco-Agric 19:347–352
158. Yang FJ (2007) Effect of light stress on the content of chloroplast pigment and Taxol in the leaves of Taxus chinensis var.mairei. Bulletin of Botanical Research 27:556–558
159. Ma ZQ, Li SS, Zhang MJ, Jiang SH, Xiao YL (2010) Light intensity affects growth, photosynthetic capability, and total flavonoid accumulation of Anoectochilus plants. HortScience 45:863–867
160. Singh VP, Dey SK, Murty KS (1988) Effect of low light stress on growth and yield of rice. Indian J Plant Physiol 31:84–91
161. Liu Q, Wu X, Chen B, Ma J, Gao J (2014) Effects of low light on agronomic and physiological characteristics of Rice including grain yield and quality. Rice Sci 21(5):243–251

162. Li YC, Lin TC, Martin CE (2015) Leaf anthocyanin, photosynthetic light-use efficiency, and ecophysiology of the South African succulent Anacampseros rufescens (*Anacampserotaceae)*. S Afr J Bot 99:122–128
163. Sui X, Mao S, Wang L, Zhang B, Zhang Z (2012) Effect of low light on the characteristics of photosynthesis and chlorophyll a fluorescence during leaf development of sweet pepper. J Integr Agric 11(10):1633–1643
164. Bell GE, Danneberger TK (1999) Temporal shade on creeping Bentgrass Turf. Crop Sci 39(4):1142–1146
165. Zhu H, Li X, Zhai W, Liu Y, Gao Q, Liu J, Ren L, Chen H, Zhu Y (2017) Effects of low light on photosynthetic properties, antioxidant enzyme activity, and anthocyanin accumulation in purple pak-choi (*Brassica campestris ssp*. Chinensis Makino). PLoS One 12(6):1–17

Chapter 6
Carotenoids as Coloring Agents

Arnab Karmakar, Abhishek Kumar Das, Sumit Ghosh, and Parames C. Sil

6.1 Introduction

Pigments are the chemical compounds responsible for producing color on an organism and are present in photosynthetic plants and microbes including colored fruits, vegetables, leaves, flowers, bacterial colony, skin, eye and many more. Color is one of the essential features of foods, influencing its demand among consumers. The proper color increases its appeal, whereas inappropriate coloration leads to an impression that the food is ineligible of consumption [1]. Food colors can be of various categories: (i) *natural colors* are pigments raised by living organisms, (ii) industrially produced natural pigments are known as *nature identical colors*, (iii) *synthetic colors* are laboratory produced colors that are not found in nature, (iv) *inorganic colors* are obtained from inorganic salts and metallic compounds. The colorant is added to the food for various reasons like (i) enhancing or replacing the color lost during processing or storage, (ii) maintaining color uniformity due to seasonal (batch to batch) variation, (iii) coloring the uncolored food products, (iv) increasing the acceptability of food items to the consumers, etc. [2].

From ancient times, colorants are used for marketing purpose in better satisfactoriness of food products as well as cosmetics, textile and other kinds of stuff [3]. Although colored garments were found in the remnants of Mohenjodaro and Harappa civilization (3500 BC); the oldest written record of usage of natural dye was found in China dated 2600 BC. Subsequently uses of dyes were observed in other parts of the world including the Indian subcontinent, Egypt, Europe and Brazil [2]. The first synthetic organic dye was mauveine, discovered by Sir William Henry Perkin at 1856 [4]. Although the synthetic colorants gained huge popularity at that time, gradually the demand of natural colorants increased due to perspectives of toxicity, hygiene and environmental consciousness. Color additives used in food,

A. Karmakar · A. K. Das · S. Ghosh · P. C. Sil (✉)
Division of Molecular Medicine, Bose Institute, Kolkata, India
e-mail: parames@jcbose.ac.in

© Springer Nature Switzerland AG 2021

M. Zia-Ul-Haq et al. (eds.), *Carotenoids: Structure and Function in the Human Body*, https://doi.org/10.1007/978-3-030-46459-2_6

drug and cosmetics must be on a positive note as directed by different organizations like Food and Agricultural Organization (FAO), World Health Organization (WHO), Joint FAO/WHO Expert Committee on Food Additives (JECFA), etc. [5].

Based on the chemical structure, pigments are divided as tetraterpenoids (carotenoids), anthraquinones (carmine), flavonoids (anthocyanins) and tetrapyrroles (chlorophyll) [6]. Carotenoids are natural pigment synthesized by plants and microbes. These pigments may be red, yellow or orange depending on its necessity of fulfilling its physiological function [7]. Carotenoids are heavily used as a biocolorant. Large scale production of nature identical carotenoids has been flourished due to its demand.

6.2 Mechanism of Coloration by Carotenoids

Carotenoids are broadly characterized based on their molecular composition. Carotenoids made up of solely hydrogen and carbon is known as carotene (α-carotene, β-carotene, lycopene etc.) whereas molecules bearing oxygen atoms are known as xanthophylls (lutein, zeaxanthin, astaxanthin etc.) (Figs. 6.1 and 6.2) [7]. These molecules are tetraterpene derivatives, i.e., they are generated from 8 isoprene molecules and bear 40 carbon atoms. Usually, carotenoids absorb light in

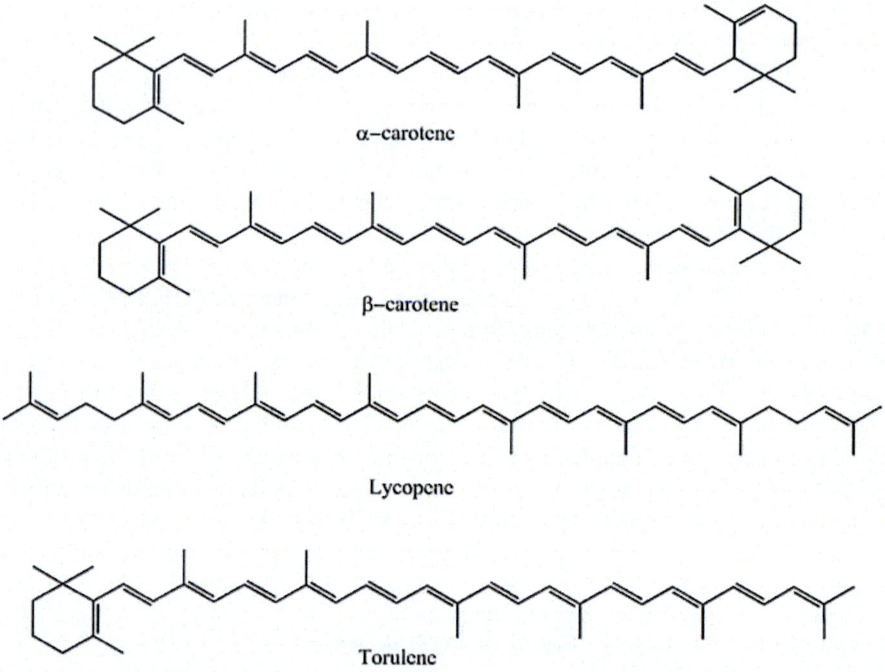

Fig. 6.1 Different types of carotenes

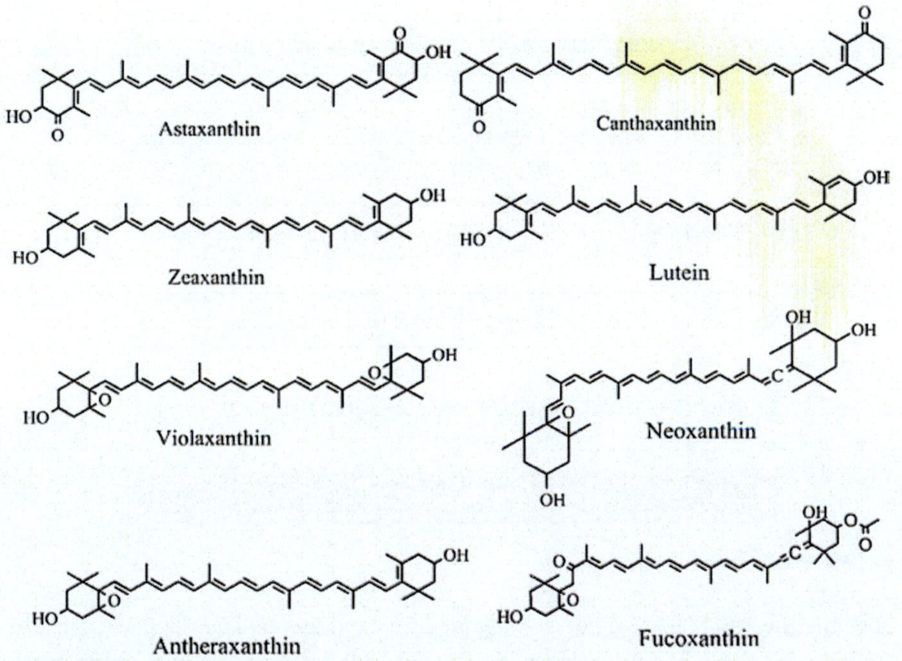

Fig. 6.2 Different types of xanthophylls

Fig. 6.3 Color of various carotenoids: Carotenoids can have different colors based on the number of double bonds present in its polyene tail

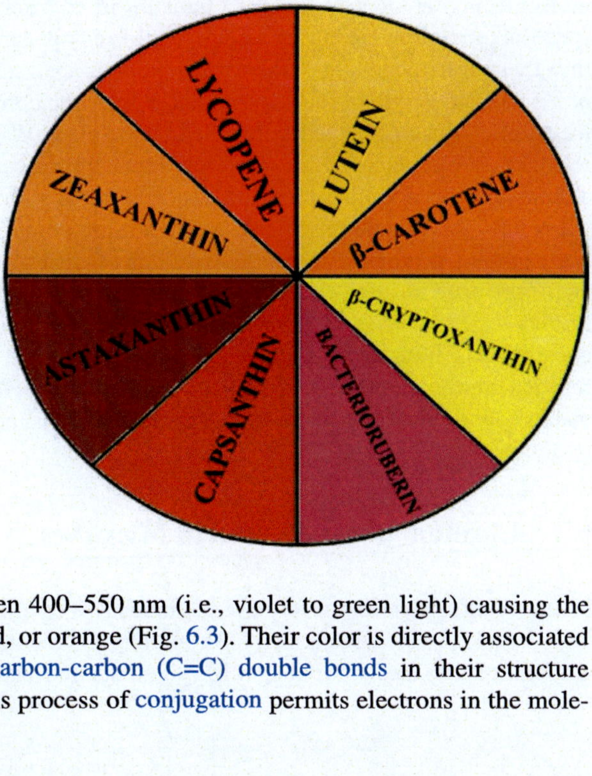

wavelengths ranging between 400–550 nm (i.e., violet to green light) causing the compounds to be yellow, red, or orange (Fig. 6.3). Their color is directly associated with their structure. The carbon-carbon (C=C) double bonds in their structure interact with each other. This process of conjugation permits electrons in the mole-

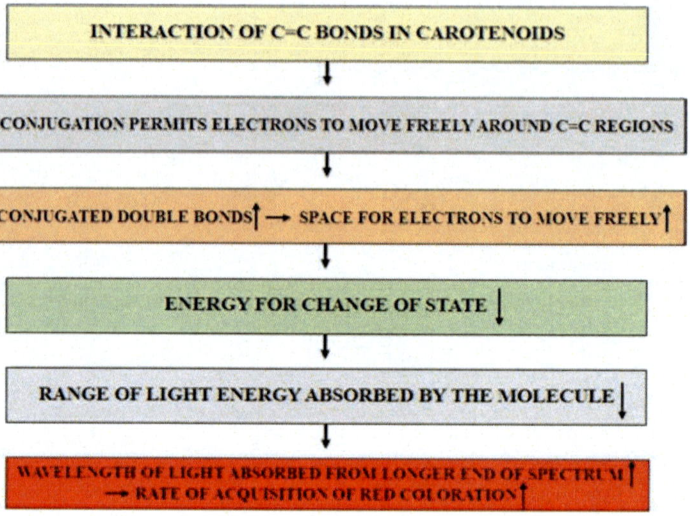

Fig. 6.4 Mechanism of color formation

cule to move freely around the C=C regions [8]. With the increase in the number of conjugated double bonds, electrons associated with the conjugation system experience more space to move freely and the energy requirement for the change of state decreases. With the resultant decrease in the range of light energy absorbed by the molecule, higher the wavelength of light absorbed from the longer end of the visible spectrum, higher is the rate of acquisition of red appearance by the compounds. On the other hand, the length of the carotenoids plays an essential role in the coloration of plants. The length of the polyene tail defines the chromophore, thus determining the wavelengths of light to be absorbed by the plant. Wavelengths, not absorbed, are reflected as the plant's color [9]. Therefore, different species of plants containing carotenoids of varying tail lengths allowing the absorption and reflection of different colors. Carotenoid based coloration is more complicated in animals (Fig. 6.4). For example, the bright feather colorations of birds can be due to pigments or by well-organized tissue. Reports reveal that yellow plumage color of American goldfinches is generated by both reflection of light from the white structural tissue and absorption of light by the carotenoids. Thus, structural components of feathers are linked with the carotenoid display of yellow color [10]. Several prominent carotenoids found on different classes of organisms are listed on Table 6.1.

6.3 Carotenoid Coloration in Microbes

Microorganisms are massively used in industries to produce various enzymes [11], organic acids [12], insecticides [13], antibiotics, recombinant proteins [14], bread and other beverages [15] along with pigments and natural colors. The large-scale

Table 6.1 **Major carotenoids** found in different classes of organism

Organism	Possible carotenoids found
Green algae	β-Carotene, lutein, astaxanthin, canthaxanthin, zeaxanthin
Micro algae	Fucoxanthin
Paracoccus	Astaxanthin, zeaxanthin
Fungi	β-Carotene, astaxanthin, neurosporaxanthin, etc.
Yeast	β-Carotene, torulene, torularhodin etc.
Plants	β-Carotene, xanthophyll, lycopene, capsanthin, lutein, violaxanthin, etc.
Sea sponge	Bastaxanthin
Corals	Diadinoxanthin, pyrrhoxanthin, peridinin, etc.
Sea anemones	Actinioerythrin, 2-nor astaxanthin
Arthropods	Canthaxanthin, astaxanthin, crustacyanin, 3-hydroxy echinenone, etc.
Molluscs	β-Carotene, astaxanthin, alloxanthin, fucoxanthin, pectenolone, 7,8-didehydro astaxanthin, etc.
Echinoderms	β-Carotene, 4-keto depoxy neoxanthin, etc.
Tunicates	Fucoxanthin, diatoxanthin
Fishes	Tunaxanthin, taraxanthin, eichinenone, β-carotene, lutein
Birds	β-Carotene, xanthophyll, etc.
Mammals	β-Carotene, lutein, astaxanthin, etc.

production of pigments from microorganisms is developing very fast due to easy fermentation, low cost production from agro waste products and relatively easy isolation process. Several critical factors are considered during pigment formation like optimum temperature, pH, carbon source, nitrogen source, fermentation time, etc [16].

6.3.1 Algae

Characteristically, algal carotenoid is present in a complex form inside chloroplasts. The accumulation of pigments is observed under different stress conditions like high light intensity, high temperature, nutrient starvation, etc. [17]. Among the chlorophyte algae (green algae), *Dunaliella bardawil* is the most notable green algae, acting as a source of a massive amount of β-carotene [18]. *Dunaliella salina* is rich in 9-cis-β-carotene and lutein [19]. Another green algae, *Haematococcus pluvialis,* is rich in astaxanthin [20], canthaxanthin and lutein [21]. *Scenedesmus* sp. contains lutein and β-carotene [22]. *Chlamydomonas reinhardti,* another type of green algae, produce zeaxanthin under stress conditions [23]. Other microalgae like *Isochrysis galbana (*haptophyte), *Mallomonas* sp., *Phaeodactylum tricornutum* (o*chrophyta) and Odontella aurita* (bacillariophyte) *contain*s fucoxanthin and diatoxanthin [24]. Rhodophytes or red algae contain α- and β-carotene derivatives [25].

6.3.2 Bacteria

Pigments in bacteria are mainly found as secondary metabolites [26]. Different species of *Brevundimonas* [27] and *Paracoccus* [28] are rich in astaxanthin. Zeaxanthin can be found on *Paracoccus zeaxanthinifaciens* [29], *Erwinia herbicola* [30] and *Synechocystis* sp. [31]. Another carotenoid Canthaxanthin is mostly found on *Bradyrhizobium* sp. which is used in salmon and poultry feed as a colorant [32]. Apart from the native carotenoid producing bacterial strains, several other E. coli strains are present with modified metabolic pathways to produce a more considerable amount of β-carotene and other carotenoids [33].

6.3.3 Fungus

The main three carotenoids, namely β-carotene, astaxanthin and neurosporaxanthin are abundant among fungal families. β-carotene can be found in numerous species of the order Mucorales like *Blakesleatri spora* [34] and *Phycomyces blakesleeanus* [35]. Different types of basidiomycetes like *Rhodosporidium* sp., *Sporidiobolus pararoseus, Sclerotinia sclerotiorum, Ustilagomaydis* sp. were found to exhibit β-carotene biosynthetic pathways [36]. Several ascomycetes like *Penicillium* sp., *Aschersonia aleyroides* and *Aspergillus giganteus* also contain β-carotene [36]. Astaxanthin can be found on *X. dendrorhous* which is used in the industrial purpose for large scale pigment production. Neurosporaxanthin was first discovered in *Neurospora crassa* but can also be found on *Fusarium fujikuroi* [36] and *Verticillium agaricinum* [37]. Different types of yeast species are of industrial importance for carotenoid production. The yeast *Phaffia rhodozyma* sp. is used as an industrial source of astaxanthin [38]. Carotenoid profiling of *Rhodotorula* sp. has shown the presence of β-carotene, torulene and torularhodin. The latter two carotenoids are also synthesized by *Sporidiobolus*, *Sporobolomyces* and *Rhodosporidium* sp. [39].

6.4 Carotenoid Coloration in Plants

In plants, the carotenoids are synthesized in both chromoplasts and chloroplasts. They impart color to photosynthetic tissues and also to fruits, flowers, storage organs, etc. [40]. Carotenoids found in mature chloroplasts mainly take part in photosynthesis [41, 42] whereas carotenoids found in chromoplasts primarily function as a coloring agent and lead to the attraction of some pollinators, some seed-distributing herbivores, etc. [43]. Chromoplast derived carotenoids are a rich source of antioxidants and pigments.

6.4.1 Leaves

Carotenoids are found in leaves and their occurrence varies according to the change of seasons. Tree leaves undergo senescence in autumn, which causes their color change from green to yellow.

During spring, some xanthophylls are found to be absent in tree leaves, whereas they come back during autumn. Earlier it was thought that some oxidation process in the plastid caused carotenoids formation, resulting in the phenomena. Later, detailed studies reported that the xanthophyll-carotene ratio increases during autumn, which causes a color change in leaves [44, 45].

6.4.2 Flowers

The flower coloration is solely dependent on the synthesis of carotenoids in chromoplast. Thus, carotenoids have a substantial impact on the flower industry mainly for aesthetic reasons as well as in the areas of traditional medicine because some carotenoids extracted from dried flowers are used as a food colorant and flavoring agents. The diversity and quantity of carotenoids differ widely in plants even within the same species. Flowers with white petals contain very few carotenoid molecules. In contrast, flowers with bright and dark colored petals consist up to 20-fold of the carotenoid content of leaves [46].

Marigold flowers are the major source of lutein, which is used commercially in flower and food industries. Lutein is responsible for the orange to yellow hues of the marigold petals [47]. Bright yellow colored flowers of Gentian *(Gentiana lutea)* are abundant in β-carotene and xanthophylls [48]. Development of such flowers shows upregulation of carotenoid synthesis as well as a shift in carotenoid profile from lutein to neoxanthin, zeaxanthin, and antheraxanthin [48, 49]. Another spectacular flower called Morning glory of genus *Ipomoea* is renowned for its diverse flower colors. Many species are rich in carotenoid due to their bright orange and yellow petals.

Interestingly, the Japanese morning glory (*Ipomoea nil*) lacks carotenoid accumulation since its petals are white. In a particular study, an attempt was made to determine the accumulation of carotenoids by comparing Japanese morning glory to two yellow-flowered species of Morning glory [50]. It was found that during early development of the flowers, all the species accumulated β-carotene, lutein and violaxanthin in the petals, the same carotenoids present in the leaves. During later development, however, the yellow flowers switched to chromoplast-derived carotenoid accumulation including zeaxanthin, β-cryptoxanthin and β-carotene [46, 51]. The white flowers did not accumulate any chromoplast derived carotenoid even in later developmental stages [50].

The Asiatic hybrid lily (*Lilium* sp.) is another commercially valuable ornamental plant which has flowers ranging from red, yellow to pink. Carotenoids impart the red, orange and yellow colorations whereas the pink color is due to anthocyanin. Most carotenoids found in yellow petals are violaxanthin, antheraxanthin, cis-lutein, etc. An unusual accumulation of capsanthin is also found in red petals [52, 53].

6.4.3 Fruits and Vegetables

There are mainly three types of carotenoids found in fruits and vegetables: β-carotene, xanthophyll and lycopene. They are responsible for the orange-yellow and red pigmentation in various vegetables and fruits [54]. Orange-yellow carotenoids include carotene and xanthophyll occurring in fruits and vegetables. β-carotene is orange in color while α-carotene is yellow [55], and the former being more abundant in fruits and vegetables than the latter. β-carotene is found less in yellow colored fruits and vegetables and more in quantity in orange colored ones. Xanthophylls like lutein, which appear yellow and zeaxanthin which appears orange-yellow, are mainly found in many fruits and vegetables with yellow-orange color [56].

Vegetables like carrot, tomato and sweet potato are good sources of β-carotene. Along with that, many green vegetables are also reported to have a decent amount of β-carotene [57]. Tomato *(Solanum lycopersicum)* is the widely researched plant model system for the study of carotenoid accumulation [58]. Green leafy vegetables from the genera of *Moringa, Brassica, Coriandrum, Solanum*, etc. are reported to contain lutein and zeaxanthin [59, 60].

Lycopene is abundant in fruits like pink grapefruit, watermelon, papaya, etc. [61]. The red color of ripe fruits shows the accumulation of lycopene. The red coloration is believed to be originated from lycopene, though contributions from anthocyanins [62, 63] and xanthophylls [64–66] are also reported. Similarly, orange colored fruits like mango, papaya, pink guava, watermelon, etc. are good sources of β-carotene. Commercially important kiwi fruit of genus *Actinidia* is an important model for studying carotenoid accumulation in fruits. This fruit comes in a range of colors including green, red, purple, orange and yellow. In ripe kiwi fruit, *A. deliciosa*, green flesh is a distinguishing feature and it is due to the retention of chlorophyll during ripening. The carotenoids occurring in this species are those related to the chlorophyll containing tissues [67] including 9′-cis-neoxanthin, lutein, violaxanthin and β-carotene. *A. chinensis* Cv. Hort16A is a recently commercialized selection ZESPRI Gold Kiwi fruit. It has bright yellow flesh and also the presence of green color with varying levels of brightness is seen. It has been found that its yellow color is due to the presence of lutein and violaxanthin, but there is an absence of chlorophyll-b which is present in *A. deliciosa*. Also, the bright orange color of *A. macrosperma* is due to the presence of high level of lutein and β-carotene. Concentrations of β-carotene are high at about ~20 mg/100 g of fresh fruit weight that provides a total carotenoid concentration higher than that of most other yellow colored fruits [68]. Another fruit of this genus, *A. polygama,* is light orange to yellow in color. Here, the concentration of xanthophylls was found to be higher than in other species. Moreover, it showed a high degree of β-carotene and zeaxanthin concentration [69].

6.5 Carotenoid Coloration in Animals

Carotenoids in animals are not synthesized *de novo*; instead, they are taken from food or modified metabolically [70, 71]. Various kinds of carotenoids with structural diversity are found in marine animals, birds, etc.

6.5.1 Poriferans

Phylum Porifera consists of a diverse group of sponges with brilliant colors because of the presence of carotenoids. Mostly aryl carotenoids are found in the sponges such as renierapurpurin, isorenieratene and renieratene [72]. Aryl carotenoids are also found in green sulfur bacteria other than sea sponges. So, it has been assumed that the aryl carotenoids in sponges originate from symbiotic bacteria. Bastaxanthins, a class of acetylenic carotenoids have been isolated from sea sponge *Ianthella basta* [73], which are thought to be metabolites of fucoxanthin, a carotenoid found in microalgae.

6.5.2 Coelenterates

In jellyfishes, the most dominant form of carotenoid found is astaxanthin, which comes from the zooplanktons that they feed on. Other forms like the diadinoxanthin, pyrrhoxanthin and peridinin in some corals [74] come from the symbiotic dinoflagellates. Sea anemones are another group of brightly colored coelenterates that possess carotenoids. For example, *Actinia equina* and *Tealia feline* [73] possess two unique carotenoids: 2-nor astaxanthin and actinioerythrin, respectively.

6.5.3 Arthropods

The crustaceans have carapace with bright color due to the presence of carotenoids, mostly astaxanthin. They feed on algae from where they acquire β-carotene and metabolize it to astaxanthin via other intermediates, namely echinenone, 3-hydroxyechinenone, canthaxanthin and adonirubin [70]. This astaxanthin exists in carapace in the form of carotenoproteins like crustacyanin and results in yellow, purple and blue color formation.

Accumulation of carotenoids by insects to acquire color in their body, eggs and even galls are seen widely. The green color in some insects and the purple-blue color in other arthropods are due to the presence of carotenoids. This alteration in color is due to the reaction with specific carotene protein or some of the chlorophyll degradation products like pterobilin [75]. Blue pigments like pterobilin combine with carotenoids to impart cryptic coloration in insects [76]. Studies have shown that carotenoids are responsible for the various intraspecific color morphs in aphids. Torulene imparts red coloration to the color morphs of aphid [77]. The bright red color of elytra in several ladybirds is also found to be derived from carotenoids [78]. Carotenoids are extracted from hair tuft and body of several lepidopterans [79] including the monarch butterflies where it confers the dramatic yellow stripes of caterpillar [80]. The green color of butterfly larva and stick insect (*Dixippus morosus*) is due to the combination of carotenes with other blue pigments [81]. Carotenoids, along with chlorophyll degraded products, impart green color to many lepidopteran larvae.

6.5.4 Molluscans

Chitons are molluscan herbivores, which mainly feed on algae. Carotenoids found in chitons are zeaxanthin, fucoxanthin, lutein and some of their metabolites [82]. Carotenoids found in sea snails like *Haliotis discus* and *Turbo cornutus* are zeaxanthin, fucoxanthin, lutein, α-carotene and β-carotene [74]. Apart from the herbivores, there are carnivores like the sea snails which feed on corals and zooplankton. The carotenoids found in them are mostly dependent on their diet as in the case of *Charoniasauliae,* which feeds on starfish, has been found to have 7,8,7′,8′-tetradehydroastaxanthin, 7,8-didehydroastaxanthin and astaxanthin which are some of the specific carotenoids of starfish. Diadinoxanthin and peridinin present in corals are also found in the sea snails (*Drupella fragum)* which prey upon these corals [74]. In spindle shells (*Fushinus perplexus)*, (3*S*)-adonirubin and (3*S*,3′*S*)-astaxanthin are the major carotenoids [83]. Sea hares and sea slugs are herbivores who feed on red and brown algae. Apocarotenoids are found in sea hares and sea slugs [75]. In *Aplysia kurodai*, apocarotenoid derived from zeaxanthin, lutein and β-carotene are found [84]. Bivalves modify the carotenoids they accumulate from dietary microalgae. Major carotenoids reported in bivalves are the metabolites of diadinoxanthin, diatoxanthin, alloxanthin and fucoxanthin [70]. Oxidative metabolites of alloxanthin and diatoxanthin like 4-hydroxyalloxanthin, 4-ketoalloxanthin, pectenolone and pectenol are widely distributed among ark shells and scallops [70, 71]. Bright red and orange colors are found to be present in some edible clams. Major carotenoids found in *Meretrix petechialis*, *Ruditapes philippinarum* and *Mactra chinensis* are mostly fucoxanthinol 3-ester and fucoxanthin [85]. In cuttlefish and octopus, major carotenoids found are astaxanthin and its esters.

6.5.5 Echinodermatans

Echinenone, an oxidative metabolite of β-carotene, is found in the gonads of sea urchin [70]. Starfishes mainly feeding on small crustaceans and bivalves have been reported to acquire astaxanthin, 7,8,7′,8′-didehydroastaxanthin and 7,8-didehydroactaxanthin. *Acanthaster planci* also called the crown-of-thorns starfish is a nocturnal sea star which preys on coral polyps. Four new carotenoids, namely 3′- epigobiusxanthin, 7,8-dihydrodiadinoxanthin, 4-keto-4′-hydroxydiatoxanthin and 4-ketodepoxyneoxanthin are found in *A. planci* as the minor components and 7,8-didehydroastaxanthin, peridininol and astaxanthin as significant components [86]. Ophioxanthin is reported in *Ophioderma longicaudum*, the brittle star. Astaxanthin and canthaxanthin are found in the gonads of sea cucumber as the major carotenoids [87]. Lutein, zeaxanthin and astaxanthin have been isolated from spiny sea star *Marthasterias glacialis* [88].

6.5.6 Protochordates (Tunicates)

Being filter feeders, phytoplankton and diatoms, are the main sources of food of tunicates. Carotenoids originating from diatoms, like the metabolites of fucoxanthin, alloxanthin and diatoxanthin are mostly found in the tunicates [70].

6.5.7 Pisces

Predominant carotenoids found in fishes are tunaxanthin (yellow), α-β-dordexanthin (yellow), β-carotene (orange), lutein (greenish yellow), canthaxanthin (orange-red), zeaxanthin (yellow-orange), taraxanthin (yellow), eichinenone (red) and astaxanthin (red). These carotenoids are the metabolites of other carotenoids which fishes accumulate in their body through diet. Astaxanthin is the primary carotenoid found in ornamental as well as exotic fishes which causes their red and pink coloration. Accumulation of carotenoids mainly occurs in the gonads and integuments of fishes. Interestingly, salmonids accumulate astaxanthin in muscles. Salmonidae and Perciformes fishes cannot synthesize astaxanthin. Hence, the astaxanthin present in their body comes solely from the dietary zooplanktons. Perciformes fishes are found to possess tunaxanthin which imparts the bright yellow color to the skin and fins of these marine fishes [70, 89]. Cyprinid fish oxidatively metabolize zeaxanthin and synthesize 3S,3’S-astaxanthin. Micropteroxanthins, some unique apocarotenoids, are found in the integuments of *Micropterus salmoides* [90].

6.5.8 Birds

Carotenoids impart the red, yellow and orange coloration to the plumage in the majority of birds [91, 92]. As birds cannot synthesize carotenoids *de novo*, they modify dietary carotenoids biochemically [92]. The main dietary carotenoids for most of the birds are β-carotene, zeaxanthin & β-cryptoxanthin and these are converted biochemically to red ketocarotenoids and yellow canary xanthophylls which get deposited in the feathers and some bare parts [92]. It has been found that wild type canaries deposit xanthophylls in the feathers that impart the yellow coloration unique to this species [93].

6.5.9 Mammals

Marine mammals like dolphins are reported to have β-carotene and lutein [94]. Whales accumulate astaxanthin as a result of their preferred feeding habit on krill, which are small crustaceans carrying this carotenoid type in their body.

6.6 Carotenoids as Food Colorants

The range of colors of carotenoids varies from yellow to dark red. β-carotene is the most widespread carotenoid which occurs naturally in many foods like egg yolk, fish, milk, butter, spinach, tomato, corn, oranges, pineapples, mangoes, etc. β-apo-8′ carotenal (apo-carotenal) is found in citrus fruits, spinach, marigold, grass, other green plants and animal tissues. Canthaxanthin (4,4′-diketo-β-carotene) is another carotenoid which is distributed in algae, crustaceans, sea trout, edible mushrooms, etc. Therefore, it is relevant that carotenoids are used as food coloring agents artificially as well [95, 96].

The β-carotene is used in several forms to color different food items. 30% β-carotene liquid suspension is used for coloring oil and fat products, cheese, margarine, frozen egg yolk, winter butter, etc. 24% β-carotene semisolid suspension is used in margarine and other fat based products. 22% β-carotene HSR liquid suspension is used for coloring cooking oil (heat stressed) and in fat based foods where the stability of carotene is required in a greater way. 3.6% β-carotene liquid emulsion is used for colouring fruit juice blends especially orange color drinks. Dry 10% β-carotene beadlets are used to color water-based foods, beverages and reconstituting dry products in warm water. Dry 2.4% β-carotene beadlets is used for coloration of beverages, reconstitution of water based liquid and dry foods in water [95, 96].

Similarly, apo-carotenoid is utilised in many forms for food coloration. Apo-carotenal is a good replacement for oleoresin of paprika in many dressing products. It not only improves the color stability but also offers a uniform composition as well as colour. French dressings are also coloured readily with apo-carotenal. 20% Apo-carotenal liquid suspension is used in fat and oil product coloration, cheese and salad dressing. 2% carotenal solution is also used for salad dressing and coloring some fat-based products. Moreover, it is also used for plating dry spices and other breading mixers [95, 96].

Canthaxanthin is also used as a food colorant. Roxanthin Red 10® dry canthaxanthin beadlets is used for the coloration of tomato products, meat products, dry or water based products which are reconstituted in warm water. 10% Canthaxanthin SD is also used for colouring water based products, tomato products and any other products for reconstitution in water [95, 96].

6.7 Carotenoids as Dyes

One of the most critical methods for the isolation of lipophilic pigments (e.g., lutein & zeaxanthin) from crude plant extracts involves high-speed counter-current chromatography (HSCCC). However, the pigments neoxanthin, violaxanthin, β-carotene, chlorophylls a and b can also be isolated by this process [97]. Another mode of extraction of carotenoids is by using magnetic stirring, which is an alternative to ultrasound extraction techniques [98].

Owing to their toxicity, nowadays, natural pigments are being explored as potential textile dyes. Fungi are significant sources of carotenoids. In a recent study, a fungal strain *Talaromyces verruculosus* from spoiled mango and capable of producing pigments suitable for textile dyeing, was isolated. The extracted pigment was applied to cotton fabric following a standard dyeing procedure and it exhibited adequate color yield. However, the exact nature and structure of the extracted pigment are yet to be investigated [99].

In another study, carotenoids extracted from orange peels were investigated of their potency as textile dyes. Effect of combination of different solvents like acetone, ethanol & hexane and their mixtures on the carotenoid yield from the orange peel wastes was studied. It was observed that acetone-hexane-ethanol (55–45-4%) mixture was able to extract more carotenoids than other mixtures. 3% alum of the net weight fraction of cotton and copper sulfate were used as mordants. Carotenoids dye isolated from orange peels were used as dyes for cotton fabrics with potent fastness properties [100].

6.8 Carotenoids in Industry

Many industrial sectors like food, textile and cosmetics are leaning towards biocolorants from synthetic artificial colors as they are more stable, less toxic and cheaper than the synthetic ones [101]. Pigments with microbial origin are gradually gaining special interest over synthetic dyes due to their higher yield, easy downstream processing and higher shelf life [102]. The current market share of carotenoids is not clearly known, but several market research reports have shown a clear increasing trend of carotenoid market growth. BBC research has stated that the global carotenoid market had reached \$1.5 billion in 2017 and it is expected to touch \$2.0 billion by 2022 (www.bccresearch.com). Mordor Intelligence (www.mordorintelligence.com) has predicted the total global feed carotenoid market to cross \$2100 million in 2020 in their report 'Feed Carotenoids Market - By Type, Animal Type and Geography-Trends and Forecasts (2017–2022). 60% of the total market share of carotenoids are made up of β-carotene, astaxanthin and lutein [103]. According to Grand View Research (www.grandviewresearch.com), the leading industrial source of β-carotene is from algae and fruit-vegetable. They have also presented a possible β-carotene market application showing dietary supplements and food & beverages as the prime consumption area.

6.9 Conclusion

Carotenoids act as important coloring agents. They are found to impart color in different types of microbes, plants and animals. The mechanism of color generation is dependent on the structure of different types of carotenoids. They impart colors in the range of yellow to red to orange, i.e., within a wavelength range of 440 to 500 nanometers. Due to growing health consequences, the use of nature identical colors is increasing day by day. The use of microbial pigments is one of the major developing industries in the food sector. The industrial production of natural pigments is relatively low to compensate for the vast number of synthetic dyes that are being used. So, it is necessary to discover novel and new natural pigments. For large scale productions optimising several factors like metabolic engineering, biotechnological manipulations and strain specific fermentation can make better result in quality, quantity and cheap production of the pigments. Nano-pigments could be a solution for carotenoids with low solubility or stability, increasing its shelf life.

References

1. Delgado-Vargas F, Jiménez AR, Paredes-López O (2000) Natural pigments: carotenoids, anthocyanins, and betalains—characteristics, biosynthesis, processing, and stability. Crit Rev Food Sci Nutr 40(3):173–289
2. Chattopadhyay P, Chatterjee S, Sen SK (2008) Biotechnological potential of natural food grade biocolorants. Afr J Biotechnol 7(17)
3. Rymbai H, Sharma RR, Srivastav M (2011) Bio-colorants and its implications in health and food industry–a review. Int J Pharm Tech Res 3(4):2228–2244
4. Burrows JD, A. (2009) Palette of our palates: a brief history of food coloring and its regulation. Compr Rev Food Sci Food Saf 8(4):394–408
5. Wrolstad RE, Culver CA (2012) Alternatives to those artificial FD&C food colorants. Annu Rev Food Sci Technol 3:59–77

6. Arimboor R, Natarajan RB, Menon KR, Chandrasekhar LP, Moorkoth V (2015) Red pepper (Capsicum annuum) carotenoids as a source of natural food colors: analysis and stability—a review. J Food Sci Technol 52(3):1258–1271
7. Alcaíno J, Baeza M, Cifuentes V (2016) Carotenoid distribution in nature. In: Carotenoids in nature. Springer, Cham, pp 3–33
8. Krinsky NI, Mayne ST, Sies H (2004) Carotenoids in health and disease. CRC Press
9. Vershinin A (1999) Biological functions of carotenoids-diversity and evolution. Biofactors 10(2–3):99–104
10. Shawkey MD, Hill GE (2005) Carotenoids need structural colours to shine. Biol Lett 1(2):121–124
11. Singh R, Kumar M, Mittal A, Mehta PK (2016) Microbial enzymes: industrial progress in 21st century. 3 Biotech *6*(2):174
12. Sauer M, Porro D, Mattanovich D, Branduardi P (2008) Microbial production of organic acids: expanding the markets. Trends Biotechnol 26(2):100–108
13. Weinzierl R, Henn T, Koehler PG, Tucker CL (1989) Microbial insecticides, Cooperative Extension Service, vol 1295. University of Illinois, Urbana-Champaign
14. Demain AL, Adrio JL (2008) Contributions of microorganisms to industrial biology. Mol Biotechnol *38*(1):41
15. Legras JL, Merdinoglu D, Cornuet JM, Karst F (2007) Bread, beer and wine: Saccharomyces cerevisiae diversity reflects human history. Mol Ecol 16(10):2091–2102
16. Tuli HS, Chaudhary P, Beniwal V, Sharma AK (2015) Microbial pigments as natural color sources: current trends and future perspectives. J Food Sci Technol 52(8):4669–4678
17. Minhas AK, Hodgson P, Barrow CJ, Adholeya A (2016) A review on the assessment of stress conditions for simultaneous production of microalgal lipids and carotenoids. Front Microbiol 7:546
18. Ben-Amotz A, Avron M (1983) On the factors which determine massive β-carotene accumulation in the halotolerant alga Dunaliellabardawil. Plant Physiol 72(3):593–597
19. García-González M, Moreno J, Manzano JC, Florencio FJ, Guerrero MG (2005) Production of Dunaliellasalina biomass rich in 9-cis-β-carotene and lutein in a closed tubular photobioreactor. J Biotechnol 115(1):81–90
20. Collins AM, Jones HD, Han D, Hu Q, Beechem TE, Timlin JA (2011) Carotenoid distribution in living cells of Haematococcuspluvialis (Chlorophyceae). PLoS One *6*(9):e24302
21. Choubert G, Heinrich O (1993) Carotenoid pigments of the green alga Haematococcuspluvialis: assay on rainbow trout, Oncorhynchusmykiss, pigmentation in comparison with synthetic astaxanthin and canthaxanthin. Aquaculture 112(2–3):217–226
22. Mehariya S, Iovine A, Di Sanzo G, Larocca V, Martino M, Leone GP, Casella P, Karatza D, Marino T, Musmarra D, Molino A (2019) Supercritical fluid extraction of lutein from Scenedesmusalmeriensis. Molecules *24*(7):1324
23. Baroli I, Do AD, Yamane T, Niyogi KK (2003) Zeaxanthin accumulation in the absence of a functional xanthophyll cycle protects Chlamydomonasreinhardtii from photooxidative stress. Plant Cell 15(4):992–1008
24. Lu X, Sun H, Zhao W, Cheng KW, Chen F, Liu B (2018) A hetero-photoautotrophic two-stage cultivation process for production of fucoxanthin by the marine diatom Nitzschialaevis. Marine Drugs *16*(7):219
25. Dufossé L, Galaup P, Yaron A, Arad SM, Blanc P, Murthy KNC, Ravishankar GA (2005) Microorganisms and microalgae as sources of pigments for food use: a scientific oddity or an industrial reality? Trends Food Sci Technol 16(9):389–406
26. Rao N, Prabhu M, Xiao M, Li WJ (2017) Fungal and bacterial pigments: secondary metabolites with wide applications. Front Microbiol 8:1113
27. Asker D, Awad TS, Beppu T, Ueda K (2012) Isolation, characterization, and diversity of novel radiotolerant carotenoid-producing bacteria. In: Microbial carotenoids from bacteria and microalgae. Humana Press, Totowa, pp 21–60

28. Osanjo GO, Muthike EW, Tsuma L, Okoth MW, Bulimo WD, Luuml H, Abraham WR, Dion M, Timmis KN, Golyshin PN, Mulaa FJ (2009) A salt lake extremophile, Paracoccusbogoriensis sp. nov., efficiently produces xanthophyll carotenoids. Afr J Microbiol Res 3(8):426–433
29. McDermott JC, Britton G, Goodwin TW (1973) Carotenoid biosynthesis in a Flavobacterium sp.: stereochemistry of hydrogen elimination in the desaturation of phytoene to lycopene, rubixanthin and zeaxanthin. Biochem J 134(4):1115–1117
30. Hundle BS, O'Brien DA, Beyer P, Kiemig H, Hearst JE (1993) In vitro expression and activity of lycopene cyclase and β-carotene hydroxylase from Erwiniaherbicola. FEBS Lett 315(3):329–334
31. Lagarde D, Beuf L, Vermaas W (2000) Increased production of zeaxanthin and other pigments by application of genetic engineering techniques to Synechocystis sp. strain PCC 6803. Appl Environ Microbiol *66*(1):64–72
32. Lorquin J, Molouba F, Dreyfus BL (1997) Identification of the carotenoid pigment canthaxanthin from photosynthetic bradyrhizobium strains. Appl Environ Microbiol 63(3):1151–1154
33. Albermann C, Beuttler H (2016) Synthesis of *β*-carotene and other important carotenoids with Bacteria. Ind Biotechnol Vitam Biopigments Antioxidants (eds. Vandamme EJ, Revuelta JL) 229–261 (John Wiley & Sons, 2016)
34. Avalos J, Cerdá-Olmedo E (2004) Fungal carotenoid production. In: Handbook of fungal biotechnology, vol 20. Marcel Dekker, New York, pp 367–378
35. Mehta BJ, Salgado LM, Bejarano ER, Cerdá-Olmedo E (1997) New mutants of Phycomycesblakesleeanus for (beta)-carotene production. Appl Environ Microbiol 63(9):3657–3661
36. Avalos J, Limón MC (2015) Biological roles of fungal carotenoids. Curr Genet 61(3):309–324
37. Valadon LRG, Mummery RS (1977) Natural beta apo 4′ carotenoic acid methyl ester in the fungus Verticilliumagaricinum. Phytochemistry
38. Calo P, Gonzalez T (1995) The yeast Phaffiarhodozyma as an industrial source of astaxanthin. Microbiologia (Madrid, Spain) 11(3):386–388
39. Buzzini P, Innocenti M, Turchetti B, Libkind D, van Broock M, Mulinacci N (2007) Carotenoid profiles of yeasts belonging to the genera Rhodotorula, Rhodosporidium, Sporobolomyces, and Sporidiobolus. Can J Microbiol 53(8):1024–1031
40. Britton G, Liaaen-Jensen S (2012) In: Pfander H (ed) Carotenoids: handbook. Birkhäuser, Basel
41. Frank HA, Cogdell RJ (1996) Carotenoids in photosynthesis. Photochem Photobiol 63(3):257–264
42. Demmig-Adams B, Adams WW (2002) Antioxidants in photosynthesis and human nutrition. Science 298(5601):2149–2153
43. Bartley GE, Scolnik PA (1995) Plant carotenoids: pigments for photoprotection, visual attraction, and human health. Plant Cell *7*(7):1027
44. Keresztes A, Faludi-Daniel A (1973) Ultrastructure, pigment content and photosynthetic activity of the normal and mutant chloroplasts in developing Tradescantia leaves. Actabiol Acad Sci Hung *24*(3):175
45. Sestak Z (1978) Photosynthetic characteristics during ontogenesis of leaves. 3. Carotenoids. Photosynthetica
46. Moehs CP, Tian L, Osteryoung KW, DellaPenna D (2001) Analysis of carotenoid biosynthetic gene expression during marigold petal development. Plant Mol Biol 45(3):281–293
47. Hina S, Mazhar S, Kalim I, Ijaz A, Zahra N, Masood S, Saeed MK, Asif M (2018) Stability of lutein content in color extracted from Marigold flower and its application in candies. Pak J Agric Res 31(1)
48. Zhu C, Yamamura S, Nishihara M, Koiwa H, Sandmann G (2003) cDNAs for the synthesis of cyclic carotenoids in petals of Gentianalutea and their regulation during flower development. Biochim Biophys Acta (BBA)-Gene Struct Expression 1625(3):305–308

49. Zhu C, Yamamura S, Koiwa H, Nishihara M, Sandmann G (2002) cDNA cloning and expression of carotenogenic genes during flower development in Gentianalutea. Plant Mol Biol 48(3):277–285
50. Yamamizo C, Kishimoto S, Ohmiya A (2009) Carotenoid composition and carotenogenic gene expression during Ipomoea petal development. J Exp Bot 61(3):709–719
51. Nakayama M, Tanikawa N, Morita Y, Ban Y (2012) Comprehensive analyses of anthocyanin and related compounds to understand flower color change in ion-beam mutants of cyclamen (Cyclamen spp.) and carnation (Dianthus caryophyllus). Plant Biotechnol:12–0102
52. Nakayama M, Tanikawa N, Morita Y, Ban Y (2012) Comprehensive analyses of anthocyanin and related compounds to understand flower color change in ion-beam mutants of cyclamen (Cyclamen spp.) and carnation (Dianthus caryophyllus). Plant Biotechnol:12–0102
53. Deli J, Molnár P, Matus Z, Tóth G, Steck A, Pfander H (1998) Isolation and characterization of 3, 5, 6-trihydroxy-carotenoids from petals ofLiliumtigrinum. Chromatographia 48(1–2):27–31
54. Khoo HE, Prasad KN, Kong KW, Jiang Y, Ismail A (2011) Carotenoids and their isomers: color pigments in fruits and vegetables. Molecules 16(2):1710–1738
55. Takyi EE (2001) Bioavailability of carotenoids from vegetables versus supplements. CRC Press LCC, Danvers, pp 19–31
56. Minguez-Mosquera MI, Gandul-Rojas B, Garrido-Fernandez J, Gallardo-Guerrero L (1990) Pigments present in virgin olive oil. J Am Oil Chem Soc 67(3):192–196
57. Rajyalakshmi P, Venkatalaxmi K, Venkatalakshmamma K, Jyothsna Y, Devi KB, Suneetha V (2001) Total carotenoid and beta-carotene contents of forest green leafy vegetables consumed by tribals of South India. Plant Foods Hum Nutr 56(3):225–238
58. Giuliano G, Bartley GE, Scolnik PA (1993) Regulation of carotenoid biosynthesis during tomato development. Plant Cell 5(4):379–387
59. Holden JM, Eldridge AL, Beecher GR, Buzzard IM, Bhagwat S, Davis CS, Douglass LW, Gebhardt S, Haytowitz D, Schakel S (1999) Carotenoid content of US foods: an update of the database. J Food Compos Anal 12(3):169–196
60. Bub A, Möseneder J, Wenzel G, Rechkemmer G, Briviba K (2008) Zeaxanthin is bioavailable from genetically modified zeaxanthin-rich potatoes. Eur J Nutr 47(2):99–103
61. Yano M, Kato M, Ikoma Y, Kawasaki A, Fukazawa Y, Sugiura M, Matsumoto H, Oohara Y, Nagao A, Ogawa K (2005) Quantitation of carotenoids in raw and processed fruits in Japan. Food Sci Technol Res 11(1):13–18
62. Tanaka Y (2006) Flower colour and cytochromes P450. Phytochem Rev 5(2–3):283–291
63. Bakker J, Timberlake CF (1997) Isolation, identification, and characterization of new color-stable anthocyanins occurring in some red wines. J Agric Food Chem 45(1):35–43
64. Camara B, Hugueney P, Bouvier F, Kuntz M, Monéger R, Kwang WJ, Jonathan J (1995) Biochemistry and molecular biology of chromoplast development. Int Rev Cytol 163:175–247
65. Bouvier F, d'Harlingue A, Hugueney P, Marin E, Marion-Poll A, Camara B (1996) Xanthophyll biosynthesis cloning, expression, functional reconstitution, and regulation of β-cyclohexenyl carotenoid epoxidase from pepper (Capsicum annuum). J Biol Chem 271(46):28861–28867
66. Prasad KN, Chew LY, Khoo HE, Yang B, Azlan A, Ismail A (2011) Carotenoids and antioxidant capacities from CanariumodontophyllumMiq. fruit. Food Chem 124(4):1549–1555
67. Khachik F, Beecher GR (1988) Separation of carotenol fatty acid esters by high-performance liquid chromatography. J Chromatogr A 449:119–133
68. Rodrigez-Amaya DB (2001) A guide to carotenoid analysis in food. ILSI Press, Washington, DC
69. McGhie TK, Ainge GD (2002) Color in fruit of the genus Actinidia: carotenoid and chlorophyll compositions. J Agric Food Chem 50(1):117–121
70. Liaaen-Jensen S (1991) Marine carotenoids: recent progress. Pure Appl Chem 63(1):1–12
71. Matsuno T (2001) Aquatic animal carotenoids. Fish Sci 67(5):771–783
72. Ackman RG (1989) Marine biogenic lipids, fats and oils, vol 2. CRC Press

73. Mercadante AZ, Egeland ES, Britton G, Liaaen-Jensen S, Pfander H (2004) In: Britton G, Liaaen-Jensen S, Pfander H (eds) Carotenoids handbook. Springer Basel, Basel
74. Maoka T, Akimoto N, Tsushima M, Komemushi S, Harada R, Sameshima N, Iwase F, Sakagami Y (2010) Carotenoids in corals and sea snails. Carotenoid Sci
75. Britton G (2008) Functions of intact carotenoids. In: Carotenoids. Birkhäuser, Basel, pp 189–212
76. Rothschild M, Mummery R (1985) Carotenoids and bile pigments in danaid and swallowtail butterflies. Biol J Linn Soc 24(1):1–14
77. Losey JE, Ives AR, Harmon J, Ballantyne F, Brown C (1997) Maintenance of an aphid color polymorphism through a balance of parasitism and predation. Nature (London) 388:269Ð272
78. Bezzerides AL, McGraw KJ, Parker RS, Husseini J (2007) Elytra color as a signal of chemical defense in the Asian ladybird beetle Harmoniaaxyridis. Behav Ecol Sociobiol 61(9):1401–1408
79. Sandre SL, Tammaru T, Esperk T, Julkunen-Tiitto R, Mappes J (2007) Carotenoid-based colourpolyphenism in a moth species: search for fitness correlates. Entomol Exp Appl 124(3):269–277
80. Rothschild M, Gardiner B, Mummery R (1978) The role of carotenoids in the "golden glance" of danaid pupae (Insecta: Lepidoptera). J Zool 186(3):351–358
81. Fox DL (1976) Animal biochromes and structural colours: physical, chemical, distributional & physiological features of coloured bodies in the animal world. Univ of California Press, Berkley
82. Tushima M, Maoka T, Matsuno T (1989) Comparative biochemical studies of carotenoids in marine invertebrates—the first positive identification of ϵ, ϵ-carotene derivatives and isolation of two new carotenoids from chitons. Comp Biochem Physiol, Part B: Comp Biochem 93(3):665–671
83. Tsushima M, Maoka T, Matsuno T (2001) Structures of carotenoids with 5, 6-dihydro-β-end groups from the spindle shell Fusinusperplexus. J Nat Prod 64(9):1139–1142
84. Yamashita E, Matsuno T (1990) A new apocarotenoid from the sea hare Aplysiakurodai. Comp Biochem Physiol, Part B: Comp Biochem 96(3):465–470
85. Maoka T, Akimoto N, Murakoshi M, Sugiyama K, Nishino H (2010) Carotenoids in clams, Ruditapesphilippinarum and Meretrixpetechialis. J Agric Food Chem 58(9):5784–5788
86. Maoka T, Akimoto N, Terada Y, Komemushi S, Harada R, Sameshima N, Sakagami Y (2010) Structure of minor carotenoids from crown-of-thorns starfish, Acanthaster planci. J Nat Prod 73:675–678
87. d'Auria MV, Riccio R, Minale L (1985) Ophioxanthin, a new marine carototenoid sulphate from the ophiuroid ophioderma longicaudum. Tetrahedron Lett 26(15):1871–1872
88. Ferreres F, Pereira DM, Gil-Izquierdo A, Valentão P, Botelho J, Mouga T, Andrade PB (2010) HPLC-PAD-atmospheric pressure chemical ionization-MS metabolite profiling of cytotoxic carotenoids from the echinoderm Marthasterias glacialis (spiny sea-star). J Sep Sci 33(15):2250–2257
89. Schiedt K (1998) Absorption and metabolism of carotenoids in birds, fish and crustaceans. Carotenoids Biosynthesis Metab 3:321–343
90. Yamashita E, Arai S, Matsuno T (1996) Metabolism of xanthophylls to vitamin A and new apocarotenoids in liver and skin of black bass, Micropterus salmoides. Comp Biochem Physiol B: Biochem Mol Biol 113(3):485–489
91. Goodwin TW (1984) The biochemistry of the carotenoids. II. Animals. Chapmann and Hall, London
92. Hill GE, Hill GE, McGraw KJ (eds) (2006) Bird coloration: mechanisms and measurements, vol 1. Harvard University Press
93. Saino N, Stradi R, Ninni P, Pini E, Møller AP (1999) Carotenoid plasma concentration, immune profile, and plumage ornamentation of male barn swallows (Hirundo rustica). Am Nat 154(4):441–448

94. Slifka KA, Bowen PE, Stacewicz-Sapuntzakis M, Crissey SD (1999) A survey of serum and dietary carotenoids in captive wild animals. J Nutr 129(2):380–390
95. Bauernfeind JC (1981) Carotenoids as colorants and vitamin A precursors; technological and nutritional applications. Elsevier Science, Saint Louis
96. Gordon HT, Bauernfeind JC, Furia TE (1983) Carotenoids as food colorants. CRC Crit Rev Food Sci Nutr 18(1):59–97
97. Aman R, Carle R, Conrad J, Beifuss U, Schieber A (2005) Isolation of carotenoids from plant materials and dietary supplements by high-speed counter-current chromatography. J Chromatogr A 1074(1–2):99–105
98. Almahy HA, Ali MA, Band A (2013) Extraction of carotenoids as natural dyes from the Daucus carota Linn (carrot) using ulturasound in Kingdoum of Saudi Arabia. Res J Chem Sci 3(1):63–66
99. Chadni Z, Rahaman MH, Jerin I, Hoque KMF, Reza MA (2017) Extraction and optimisation of red pigment production as secondary metabolites from Talaromyces verruculosus and its potential use in textile industries. Mycology 8(1):48–57
100. Baaka N, Mahfoudhi A, Mhenni MF (2017) Orange peels waste as a low cost cotton natural dye. Moroccan J Chem 5(2):5–2
101. Tuli HS, Chaudhary P, Beniwal V, Sharma AK (2015) Microbial pigments as natural color sources: current trends and future perspectives. J Food Sci Technol 52(8):4669–4678
102. Dufossé L, Galaup P, Yaron A, Arad SM, Blanc P, Murthy KNC, Ravishankar GA (2005) Microorganisms and microalgae as sources of pigments for food use: a scientific oddity or an industrial reality? Trends Food Sci Technol 16(9):389–406
103. Saini RK, Keum YS (2017) Progress in microbial carotenoids production. Indian J Microbiol 57(1):129–130

Chapter 7
Effect of Processing on Foods Containing Carotenoids

Matloob Ahmad, Sana Aslam, Madiha Rohi, and Saiqa Yaqoob

7.1 Introduction

Fruits and vegetables are basic dietary sources of carotenoids. They could be processed at peak harvest time to get the access of seasonal foods during other seasons. The processing enables the minimization of the post-harvest losses as well as it makes sure their distribution to far removed places from agricultural side of production. The proper understanding of nature and stability of carotenoids under specific processing conditions will be beneficial to get best and desirable results. Carotenoids undergo oxidation and isomerization during samples storage and/or during analysis and even during storage of foods. Moreover, they undergo structural rearrangements and degradation reactions as well. A good understanding about processing methodologies and their effects could lead to best desirable results. Carotenoids possess diverse applications in food and pharmaceutical industries leading the focus on processing of vegetables and fruits to get them. The natural profile of carotenoids is altered unavoidably during food processing due to their easily oxidizing nature, instability and less resistance to the processing stresses [1]. They are extensively processed in the form of fruit juices, jams, and jellies etc. The traditional way of food processing includes addition of sugar and/or chemicals, freezing, drying, juice

M. Ahmad
Department of Chemistry, Government College University, Faisalabad, Pakistan

S. Aslam (✉)
Department of Chemistry, Government College Women University Faisalabad, Faisalabad, Pakistan
e-mail: dr.sana@gcwuf.edu.pk

M. Rohi
Department of Food Science and Technology, Government College Women University Faisalabad, Faisalabad, Pakistan

S. Yaqoob
Department of Chemistry, University of Sargodha, Sargodha, Pakistan

© Springer Nature Switzerland AG 2021
M. Zia-Ul-Haq et al. (eds.), *Carotenoids: Structure and Function in the Human Body*, https://doi.org/10.1007/978-3-030-46459-2_7

extraction, pulping, and pasteurization etc. but the major drawbacks of these methodologies are significant loss and alteration of carotenoid contents.

7.2 Processing of Carotenoids

Food processing (thermal and mechanical) at both domestic as well as industrial level, affect the cellular matrix having phytochemicals. Food processing methodology, preparation and storage of variety food products rich in carotenoids depends upon various factors like variety, cultivar or stage of maturity, farming practices, geographic site or climate of production and the nature of plant part being utilized. Food processing and storage also have significant qualitative as well as quantitative effects on the contents, stability, bio-availability and bioaccessibility of carotenoids. Therefore to preserve the physicochemical stability and bioaccessibility of carotenoid contents, there is a need to determine the effective way for processing the food products. This work aims to update the current knowledge and provide useful insights on carotenoid's bioavailability and bioaccessibility in food matrices.

7.2.1 Home Processing of Carotenoid Based Foods

Home preparation of food products rich in carotenoids includes chopping, boiling, agitation and addition of salt, sugar and vegetable oil. But these operations lead to significant carotenoid losses. The quality of home products depends on many factors like temperature, cooking time, exposure to air and light, moisture etc. The stability of carotenoid contents is enhanced by dark and cold storage for minimal time duration, especially for green leafy vegetables. Research data indicates that the β-carotene contents increase with storage, due to continued enzymatic activity which stimulates the biosynthesis of β-carotene even in storage.

7.2.1.1 Boiling and Freezing

Boiling and freezing are the most commonly used methods for the preservation of food products rich in carotenoids. The processing of tomatoes has long been carried out by chopping, boiling, agitation and addition of vegetable oil and it is observed that the thermally induced isomerization of all-*E*-lutein and all-*E*-β-carotene into their respective *Z*-configurations occur. The physical ultrastructure changes provoked by heat treatment is confirmed by Electron microscopy technique [2]. Lycopene contents preserve up to some extent after baking and boiling of two tomato cultivars, while frying reduces the lycopene contents significantly [3].

Another study indicates significant losses of lycopene by an average of 48% were calculated by peeling of skin and seeds, assessed in three cultivars of tomato [4].

Boiling of food products adversely affect the carotenoid-compounds. Basically α- and β- carotene contents are predominantly present in its all-*trans* stereoisomeric form in raw foods. However, all-*trans* double bonds isomerizes partially or fully into its *cis* form as a result of processing like exposure to light, increased temperature (e.g. boiling, with blanching, etc.), and catalysts (e.g. acid) [5]. The *trans*- to *cis*- isomerization rate of α- and β-carotenes has been reported by boiling of both fresh and frozen carrots [6]. The β-carotenes are 1.9 times more susceptible to isomerization as compared to α-carotene during cooking and blanching processes [7]. Pressure also effects the carotenoid degradation. Moreover, α- and β-carotene contents are significantly lost as result of boiling the carrots under pressure for 50 s as compared to heating in saucepan without pressure for 19 min [8].

Boiling and grilling treatment of red jalapeño pepper slightly increase the all-*E*-capsanthin contents, while lutein contents slightly decrease in treated green jalapeño pepper [9]. There is a significant decrease in carotenoid compounds upto 31–53% in boiled non-pungent peppers as compared pungent peppers (i.e. 3–24%) [10].

The storage of orange fruits as a whole, in segments and peeled form at 4 °C for 12 days was studied and it was determined that the total carotenoid contents were increased significantly in case of a whole fruit storage than in segments or peeled fruits. It was proposed that the carotenogenic enzymes activity was conserved in whole fruit [11]. The individual carotenoids contents like α- and β-carotene, α- and β-cryptoxanthin, lutein and zeaxanthin enhance during storage at low temperature, showing more significant results in the whole fruit samples. Total carotenoids content remain unaffected when the fresh-cut fruits like watermelon cubes and kiwi slices are kept at 5 °C for 6 days, while the 10–15% total carotenoids content losses occur in cantaloupe cubes, strawberry slices and mango cubes. On the other hand, 25% loss in total carotenoids content occur in pineapple pieces [12]. The carotenoid level is reported to be increased in whole fruits i.e. pineapple and kiwi (up to 3 days), strawberry (during 9 days of storage), while total carotenoid content of mango as whole fruits, was first decreased (up to 3 days) but increased subsequently for 9 days of storage. The total β-carotene and lutein contents in the fresh-cut kiwi fruit after 9 days has been reported to be slightly higher than the whole and fresh-cut watermelon.

7.2.1.2 Chopping, Peeling and Juicing

Chopping, peeling and juicing of fruits and vegetables directly expose their inner tissues to the light and oxygen and easily prone to oxidation which is attributed to the significant loss of carotenoid concentration especially in leafy vegetables [13, 14].

7.2.1.3 Baking, Grilling and Frying

Processing of vegetables by baking, grilling and frying (including deep frying) may result into a significant loss of carotenoids contents. It was reported that the significant decrease of carotenoid concentration occurred in domestic deep fried broccoli, carrots and courgettes samples in peanut oil at 170 °C as compared to their respective steamed and boiled samples [15]. In a contrary study, it was investigated that the stability of carotenoids (i,e, 9′-*cis*-neoxanthin) contents were 7% more in boiled broccoli (5 min) as compared to stir-fried broccoli (4 min) [16]. In another study, total carotenoid contents were retained in both treated stir-fried (10 min) and boiled (5 min) green beans samples [17].

It was observed that all-*E*-β-carotene contents of tomato, sweet potato and sweet bell pepper were 2–27% decreased after stir-frying for 0.5–3 min. β-Carotene was found to be extracted into the frying oil, although in low amount after 3 min frying of sweet potato shreds [18].

Boiled leafy vegetables of Taiwan cultivar possess the same carotenoid concentration as in untreated raw leafy vegetables, while there was a significant decrease in fried and deep-fried samples, attributed due to their lipophilic nature which results into leaching of these hydrophobic compounds into oil. In addition, these (all-*E*)-isomers of carotenoids are unstable at the high temperatures (190 °C) [19]. Twenty different vegetables were processed by boiling, baking and grilling and their antioxidant activity were determined. It was evaluated that griddling, baking, and microwave cooking products showed good antioxidant activity as compared to boiled and pressure-cooked vegetables [20].

7.2.2 *Processing of Carotenoid Based Foods at Domestic and Industrial Level*

Fruits and vegetables are processed by different conventional processing methods like drying, freezing, pulping, dehydrated, pasteurization, juice extraction, addition of chemicals e.g. vinegar, salts and/or sugar and canned (as a whole or in pieces form). Industrial processing techniques are the main focus on commercial level in food industry.

7.2.2.1 Drying

Drying is considered as one of the oldest food preservation method [21]. Fruits and vegetables are processed by different dehydrating procedures like sun, hot air, freeze, spray, crossflow, drum, puff and microwave drying. Dehydrated carrot slices are used as an ingredient in different cooked foods like soups and healthy snack foods [22]. Drying of vegetables resulted in 10–20% carotenoids contents losses

[23], due to increased surface area of powdered or dried products which are more prone towards further environmental factors of degradation like autoxidation in light and air (Table 7.1).

Conventional sun drying is consider as a cheapest and accessible method for food preservation although causes significant carotenoid degradation [25]. Drying food under the shade also showed good preservation results. Another simple and inexpensive way of food drying is in a solar dryer with however a considerably decrease in carotenoid losses.

Savoy beet, fenugreek and amaranth leaves were dried under different dehydration conditions like shade, sun, solar drier, low temperature drier and cabinet drier. The best results for the stability of β-carotene contents were observed for processed product in low temperature drier [26]. In another research, the higher degradation rate of β-carotene contents of amaranth leaves and savoy beet was reported in case of solar drying than in cabinet drying [27]. It was determined that the all-*E*-β-carotene contents were partially degradated (i.e. 31% in Tommy Atkins mango and 7% in Kent mango),while xanthopylls contents were completely degradated in dried mango slices processed in laboratory overflow dryer at 75 °C for 3.5 h [28]. In another study two different varieties of Thai mangoes i.e. Kaew and Nam Dokmai were dried in sunlit tunnel dryer and solar dryer, led to significant carotenoid degradation upto 42% and 34% respectively. Eight Tanzanian green leafy vegetables i.e. sweet potato, cowpea, mgagani, amaranth, pumpkin, ngwiba nsonga and maimbe were dried by solar drying and open sun drying and quantity of all-*E*-β-carotene substantially reduced in open sun dried vegetables as compared to solar dried vegetables [29]. In another study different vegetables were dehydrated by solar drying and their β-carotene contents and dehydration rate were determined. It was reported that 2–20%, 4–6%, 49–67% β-carotene constituents were degradated in collard green, sweet potato and carrots respectively [30]. β-Carotene contents of different treated Cassava were determined i.e. 56%, 59%, and 72% preservative stability under boiling, shadow drying and oven drying treatments respectively [31]. Gari (i.e. processed Cassava, includes peeling, grating, pressing, toasting and packing in perforated sack), showed less preservation potential of β-carotene (34%). It was also reported that stability of dried cassava was more when stored as a chips as compared to flour.

Orange fleshed sweet potato slices were also processed by open air sun drying and solar drying, although resulted in the ≤10% decrease of moisture contents but also reported the all-*E*-β-carotene degradation upto 16% and 9% respectively [32].

Table 7.1 Percentage retention of all-*trans* β-carotene (%) dried by different drying methods [24]

Different forms of drying	Retention of all-*trans* β-carotene (%) in carrot dices as that in fresh carrot.
Blanched	95%
Freeze-dried	85%
Conventional air dried	80%
Explosive puff-dried	72%

All-*E*-β-carotene constituents of dried orange fleshed slices, processed in a forced air oven at 57°C for 10 hours, were reduced upto 12%. In another study all-*E*-β-carotene contents were determined in flour prepared from dried sweet potato pieces/chips by different drying treatments. Sun and solar drying process did not retained the β-carotene contents significantly. While the hot air cross flow drying showed good results as compared to sun drying. It was also observed that the crimped slices showed more preservation potential of β-carotene contents as compared to chips [33]. Raw shrimp was processed via cooking followed by direct sunlight drying and it was observed that astaxanthin constituents of the cooked shrimp were readily degraded upto 75% and 83% during sun drying and in storage respectively [34].

Conventionally dried food products usually decreases its quality as compared to the original foodstuff [35]. Continuous stream of hot air passed through the food produces dehydrated products with extended shelf life. The major drawback of this processing is a considerable shrinkage of food product due to cell collapse as a result of moisture loss, which leads to undesired changes in flavour, colour, texture and nutrients as well as poor rehydration of the dried product. In dehydrated products, water itself has a protective effect for carotenoid stability, determined as water activity (Aw). At water activity i.e. Aw = 0.43 for carotenoid, contents are most stable, so by adding different dehydrating agents like salt, sugar and sodium metabisulphite ($Na_2S_2O_5$) significantly reduces the rate of further carotenoid degradation [36]. Pre-treatment parameters and drying conditions determine the quality of final dehydrated product.

7.2.2.2 Freeze-Drying

Freeze-drying is also known as lyophilisation process, generally a drying process applied for preservation of heat sensitive foods like fruits and vegetables. Lyophilisation process starts with dehydration i.e. by freezing the water and/or the suspension medium into ice or respective crystallized state of medium and then removing the ice /crystals via sublimation directly into the vapours. It is reported that significant loss of carotenoids in processed summer fruits and vegetables takes place during lyophilisation process [37, 38] (Fig. 7.1).

Type of Foods Processed
Meat, fish vegetables, fruits, coffee, and juices are usually processed through this methodology.

Effect on Carotenoids
Carotenoid contents of fruits and vegetables decreases. Freezing at −20 °C seems to be more effective methodology for preservation as compared to lyophilisation. The stability of carotenoids at ambient temperature was limited in cassava biofortified with palm oil [40]. The fatty acids constituents of palm oil may protect the carotenoids skeleton, resultantly the shelf life of gari increased by the addition of red palm oil.

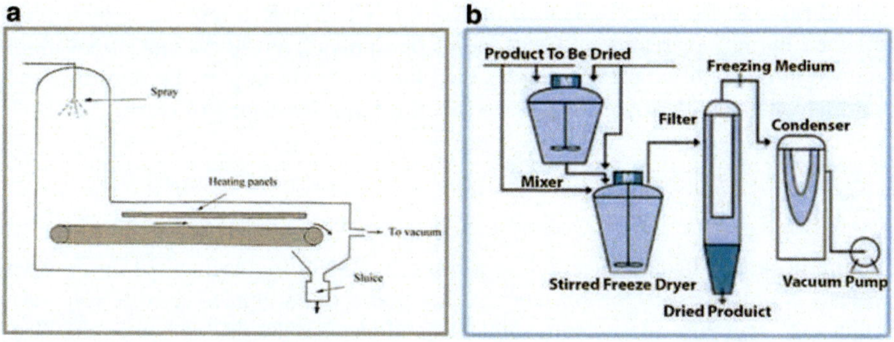

Fig. 7.1 Continuous freeze-drying [39]

Vacuum freeze-drying is a better choice for dehydration resulting in high quality products as compared to other drying methods [41, 42]. The process is based on sublimation of water from frozen product at low temperature. Freeze-drying retains the shape and primary structure of a product with minimum volume reduction. The major benefits achieved are high quality product with long shelf life and retard the microbiological reactions and deterioration of the nutrients. It was reported that the carotenoid stereoisomers were not formed during processing, although carotenoid contents were decreased in freeze-dried carrots [6].

7.2.2.3 Blanching

Principle of Operation
Blanching slows down the carotenoid degradation rate by inactivating the enzymes, like lipoxygenase and peroxidase. Blanching is classified into various types, depending upon its sources, including water, steam, hot-air, vacuum-steam, microwave and in-can. Water blanching takes place at 75–95 °C for 1–10 min either in the presence of sulphite (as antimicrobial agent as well as inhibitor of nitrosamine formation and enzymes).

Type of Foods Processed
Vegetables are usually processed through this methodology.

Effect on Carotenoids
Blanching is commonly employed due to its relatively low running cost. This technique decreases the degradation of vitamin C contents as well as carotenoids. The organoleptic properties (colour) of the vegetable products are not much affected. Blanching can cause 5–13% loss of carotenoids [43–45]. Carotenoids and polyacetylene (i.e. falcarindiol, falcarinol and falcarindiol-3-acetate) contents of freeze dried and hot air dried carrot discs were examined for the comparative study of blanching and ultrasound pretreatments. It was noted that stability of carotenoids and polyacetylene was significantly higher in case of pretreatment of dried carrot

with ultrasound followed by hot air drying (i.e. UPHD) as compared to blanching followed by hot air treatment. Carotenoids and polyacetylene contents were more retained in freeze dried carrot samples than hot air dried samples. Thus, ultrasound pretreatment is a potential substitute to the blanching for drying of carrots [46, 47].

7.2.2.4 Peeling

Peeling of raw fruits and vegetables, as a pre-treatment by mechanical or abrasive peelers, is processed in the presence of hot water or sodium hydroxide (NaOH). Fruits and vegetables like carrots, tomatoes and peaches are peeled before canning.

7.2.2.5 Coating

Another processing way is the coating of carrots, proceeded by steam blanching the carrot slices in 2.5% starch solution for 5 min. This processing method not only increases the shelf life but also prevents the carotenoid degradation of dehydrated carrots [48]. Coating with edible starch covers the surface of food products which protects their carotenoid contents against further autoxidation, decolouration, loss of activity, and development of off-flavours in spray-dried and air-dried foods.

7.2.3 Advanced Processing Techniques at Industrial Scale

7.2.3.1 High Pressure Processing (HPP)/ (Cold Pasteurization Technique)

Principle of Operation
HPP is a non-thermal, cold processing preservation technique carried out at high applied pressure i.e. 200–900 MPa. This preservation technology is based on the application of high hydrostatic pressure to the food in its final packaging stage, as a result inactivation of the food-borne microorganisms occurs which increases the shelf life of the product. This technology does not disturb the covalent linkages within the food product, as a result maintaining both the nutritional and sensory aspects of the food products [49].

Type of Foods Processed
Meat, sea food, dairy products, fruits, juices, dips and sauces.

Effect on Carotenoids
It is an emerging technique for food preservation. HPP preserves the nutritious value of the processed food with improved extractability of carotenoid under optimum conditions as compared to the thermally processed food. Its advantageous edge is increase in the overall carotenoid contents of vegetables. However, bioaccessibility of carotenoid contents decreases in fruit-milk beverages [50–52].

7.2.3.2 High Pressure Homogenization (HPH)

Principle of Operation

HPH is widely used food homogenizing technique to enhance the physical stability with improved textural properties of various food products. The samples are first forced through a special narrow gap of homogenizing valve where conditions of high shear and turbulence (pressure) combined with acceleration, compression and pressure drop are maintained. As a result, homogeneous product is formed via altering the sample structure by standardizing and reducing the particle size. Finally, uniform and homogeneous product with improved shelf life, taste and texture are obtained [53] (Fig. 7.2).

Type of Foods Processed

Liquid foods (e.g. fruit juices, milk, ice cream, beverages, baby foods and sauces)

Effect on Carotenoids

High pressure homogenization is attributed the improved bioaccessibility and microstructure of carotenoids by cell disruption. HPH processed carrot emulsions possess the increased bioaccessibility of carotenoids. However, bioaccessibility of tomato product (i.e. pulp) is disrupted by HPH. The effect of HPH on bioaccessibility of carotenoid depends on the nature of carotenoid as well as the fibrous network developed after processing [55–57]. Carotenoid profile of HHP treated purees of different commercial tomato varieties, like Black Prince Cerise, Maliniak, and Lima was done by photo-chemiluminescence technique (PCLACL), HPLC-DAD as well as cyclic voltammetry (CV). The ferric-reducing antioxidant power (i.e. FRAP) assay was performed to find out the antioxidant potential. It was reported that the HHP processing significantly enhanced the antioxidant potential of HHP-treated

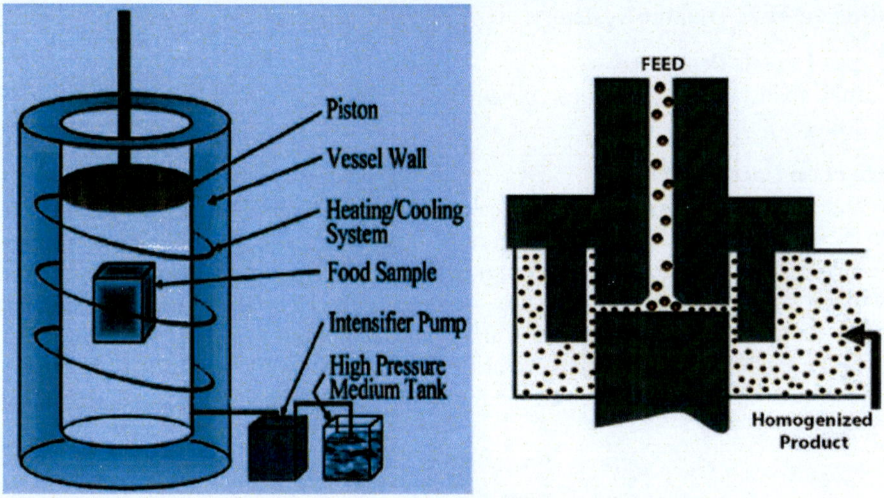

Fig. 7.2 Schematic setup for high pressure homogenization (HPH) [53, 54]

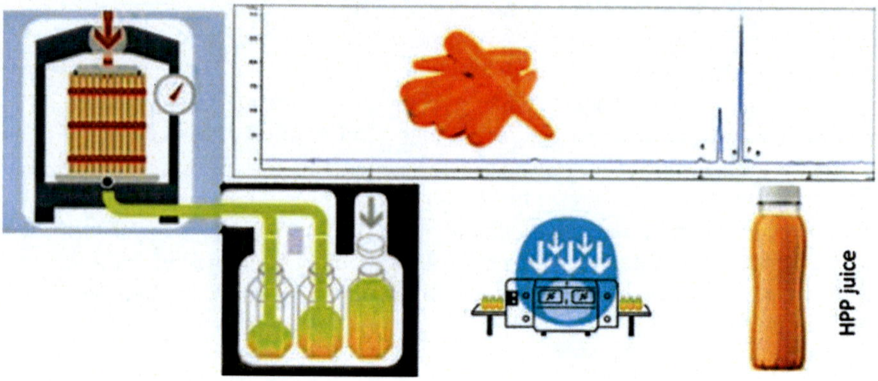

Fig. 7.3 Experimental Setup for high pressure homogenization (HPH) [59]

purees but their β-Carotene and lycopene contents were considerably reduced. It was also observed that lutein contents of the Black Prince and Cerise tomatoes puree extracts were slightly affected at 650 and 550 MPa/5 min respectively [58] (Fig. 7.3).

7.2.3.3 High Hydrostatic Pressure (HHP)

Principle of Operation
HHP processing is a 'mild technology' for food products. First of all the sample is closed in a Teflon tube, then the pressurization is carried out at different pressures (like 400, 500, 600 MPa) and at specific temperature for particular time duration by using the High Pressure System.

Type of Foods Processed
Liquid foods (i.e. fruit juices, puree, milk, ice cream, beverages, baby foods and sauces)

Effect on Carotenoids
Tomato puree was treated by HHP using 400 MPa/15 min hydrostatic pressure at 25 °C and an increase in carotenoid content upto 50% was observed as compared to untreated tomato puree [60]. An increase in total carotenoid content by more than twofold was noted when tomato puree was treated by HHP at 600 MPa/15 min and 20 °C temperature [61]. β-Carotene and lycopene contents of HHP treated tomato juice were unaffected on treatment at 600 MPa/5 min [62]. In another study, the puree was lyophilised and then stored for estimation of its reducing and antioxidant capacity (Fig. 7.4).

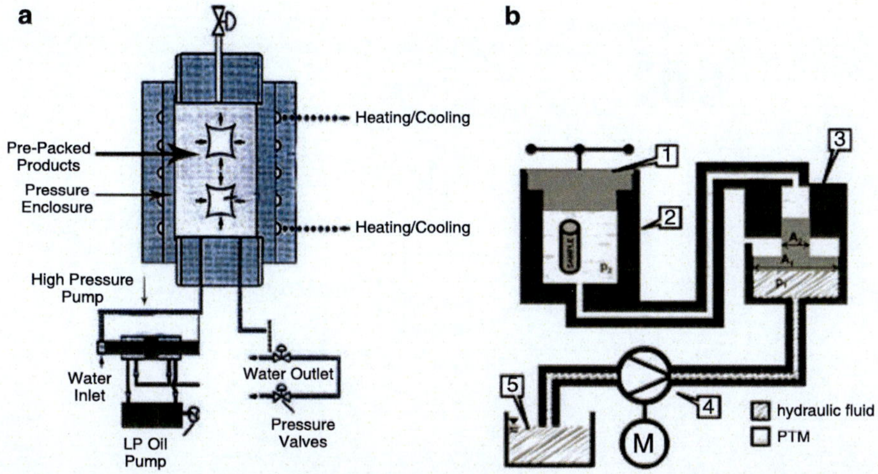

Fig. 7.4 Schematic Experimental System for High hydrostatic pressure (HHP) (screwable closure (1), reaction volume/pressure autoclave (2), tight pressure converter (3). controlled supercharger (4) [63, 64]

7.2.3.4 Pulsed Electric Fields (PEF)

Principle of Operation

PEF is a non-thermal technique in which food is first placed between two electrodes subjected to the high voltage (i.e. 20–80 kV/cm) at ambient temperature (25 °C) for less than 1s. This treatment method is considered superior as compared to the traditional heat treatment due to their effectiveness to inactivate the microbial growth without changing any unwanted structural profile of foods with no or minimum detrimental effect on its quality (Fig. 7.5).

Type of Foods Processed

Liquid and semi-liquid foods like milk, yoghurt, soups, juices and liquid eggs.

Effect on Carotenoids

Carotenoid contents decrease in PEF treated fruit-soymilk juice under processing conditions of 35 kV/cm pulsed electric fields with 4 μs bipolar pulses at 200 Hz for 800/1400 μs. The bioavailability of carotenoid contents of PEF treated tomato products increases. The effectiveness of PEF methodology on the carotenoid based bioactive compounds depends upon the intensity of pulsed electric fields as well as the nature of the food matrix. [65]. A comparative study of PEF treated orange juice–milk beverage at different conditions of PEF (i.e 15, 25, 35 and 40 kV/cm, for 40–700 μs) was studied. The concentration of extracted carotenoids slight increases at 15 kV/cm and decrease at 40 kV/cm [66].

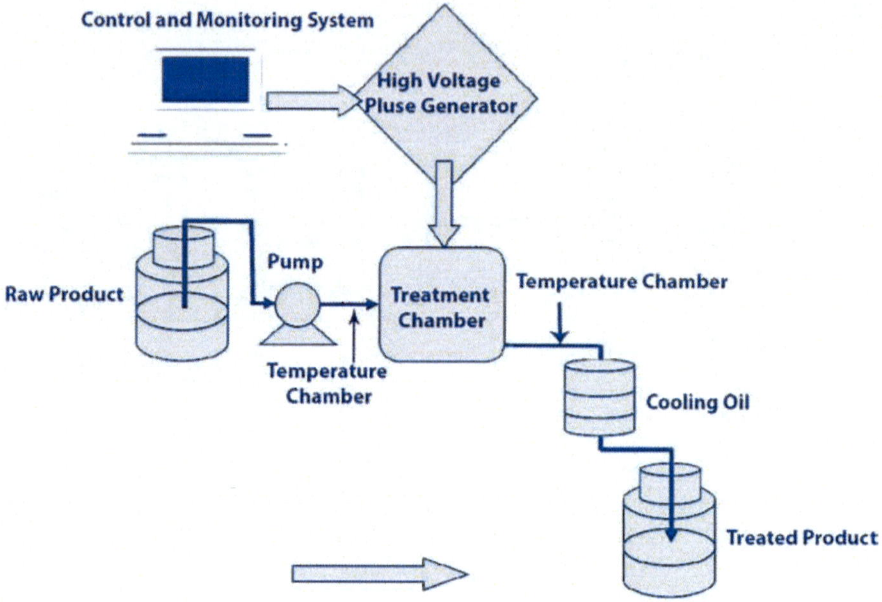

Fig. 7.5 Schematic experimental setup for pulsed electric fields (PEF) [65]

Moderate-Intensity Pulsed Electric Field (MIPEF)

The concentration of carotenoids contents i.e. α-carotene, β-carotene, lutein, 13-*Z*-lycopene 9-*Z*-lycopene and all-*E*-lycopene increases in tomato juice treated with MIPEF at 1 kV/cm with 16 monopolar pulses for 4 μs. This was attributed by the accumulation of secondary metabolites and enhanced metabolic activity induced by MIPEF [67]. The pretreated tomato juice samples with MIPEF, were further subjected to HIPEF (at 35 kV/cm for 1500 μs) as well as thermal (at 90 °C for 60 s) processing and then stored under refrigeration for 56 days. It was investigated that 10–20% of higher carotenoid levels was maintained in HIPEF-treated juice than in untreated and heat-treated juices. Thus, it was concluded that the combination treatment approach i.e. MIPEF and HIPEF has a great potential for processing of food products with high carotenoid content.

High-Intensity Pulsed Electric Field Processing (HIPEF)

A comparative study of HIPEF (i.e. at 40, 35, 30, and 25 kV/cm for 30–340 μs) and pasteurized (i.e. at 98 °C for 21 s) processed orange-carrot juice mixture was done and quantified 16 carotenoids in this juice mixture. The results showed best outcomes with significant increase in carotenoid contents of HIPEF treated juice mixture at 30 and 25 kV/cm as compared to pasteurized juice [68].

Orange juice was processed by HIPEF at 30 kV/cm for 100 μs and results showed that the stability of carotenoids contents improved significantly as compared to untreated orange juice [69]. Untreated orange juice and pasteurized processed orange juice at 90 °C for 20 s were refrigerated at 2 °C for 7 weeks and 10 °C for 6 weeks respectively and a significant reduction in total carotenoid contents were observed as compared to juice subjected to HIPEF. Further study showed that pasteurized orange juice at 90 °C for 20 s decreased total carotenoid content upto 13%, while only 8%, 6%, and 10% of total carotenoid contents were decreased at applied electric fields of 40, 30 and 15 kV/cm respectively during HIPEF processing [70].

Optimization conditions for HIPEF treatment of tomato juice were designed with reference to pulse polarity, pulse width and frequency. It was evaluated that maximum lycopene content was obtained at 250 Hz for 1 μs in bipolar mode. [71].

Watermelon juice was subjected to HIPEF treatment at 30–35 kV/cm (electric field strengths), 50–250 Hz (frequency), 1–7 μs (pulse widths), for 50–2050 μs and the best results for maximum lycopene retention was reported at 35 kV/cm and 200 Hz using 7 μs bipolar pulses for 50 μs [72]. It was also reported that the individual (i.e. β-carotene, lycopene, and phytofluene) and total carotenoids content and even reddish color of tomato juice was significantly enhanced under thermal pasteurization (at 90 °C for 30 and 60 s) and HIPEF (at 35 kV/cm and 100 Hz with 4 μs bipolar pulse for 1500 μs) treatments as compared to fresh juice. The results of HIPEF-processed juice predominates over pasteurized sample [73]. In a contrast study, it was reported that carotenoids contents (like β-cryptoxanthin, zeaxanthin and lutein) of different treated fruit juices (i.e. pineapple, kiwi and orange) and soymilk beverage by thermal (at 90 °C for 60 s) and HIPEF (at 35 kV/cm, and 200 Hz with 4 μs bipolar pulses for 800–1400 μs), were stored at 4 °C for 56 days. The decrease in total carotenoid content was less in the HIPEF-processed beverages than in heat-treated beverages [74] (Fig. 7.6).

7.2.3.5 Ultrasonication

Principle of Operation
Ultrasonication is a homogenization technique in which food product is subjected to ultrasonic waves of 20–40 kHz frequency, generated by either acoustic horns or sonotrodes using applied electrical energy (Fig. 7.7). Ultrasonic waves produce high mechanical effects and shear force. Cavitation technique, as an alternate of ultrasonication, is used for patterning of emulsions in food industry.

Type of foods Processed
Vegetables and fruits

Effect on Carotenoids
Ultrasonication is considered as a potential technique for food processing due to its various advantages like improved homogenization, better drying, enhanced heat

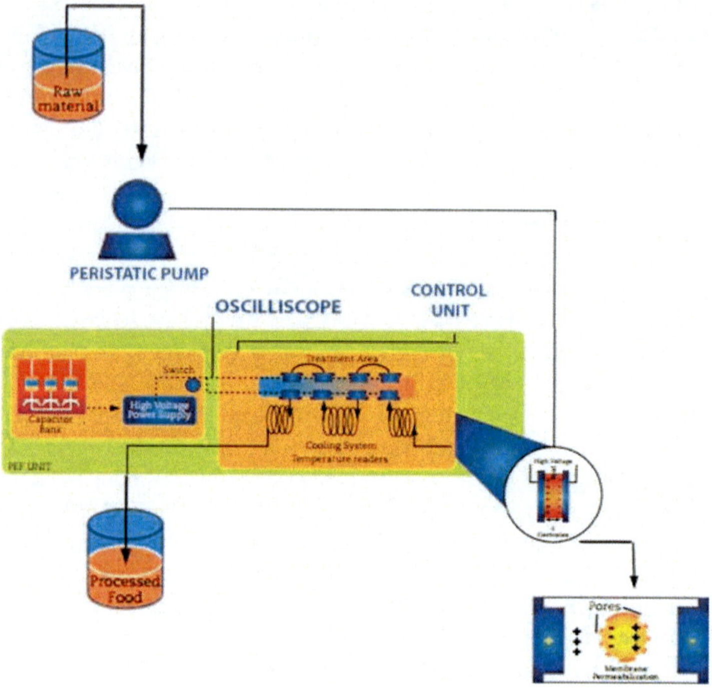

Fig. 7.6 Experimental setup for high-intensity pulsed electric field processing (HIPEF) [75]

Fig. 7.7 Schematic diagram for ultrasonication [53]

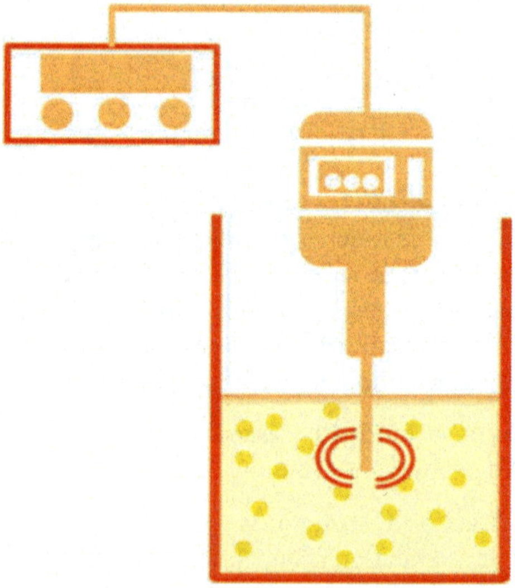

transfer in crystallization or freezing uniformly, better filtration speed, degassing of liquid foods and enhanced mass transfer during pickling or marinating.

Oil-in-water emulsions are also prepared by ultrasonication. Ultrasounds used in acoustic cavitation phenomenon, causes reduction of droplet size uniformly for emulsification.

The efficiency of ultrasound-assisted emulsion is effected by different parameters like biopolymer concentration, sonication power and sonication time [76]. The rate of droplet breakage increases in the dispersion medium at high intensity of ultrasonication and less emulsifier concentration.

Emulsifiers not only help in the formation of interfacial film which reduces the surface tension but also reduces the size of the droplets, assisted by ultrasound. Carotenoid fabricated on nanostructures via ultrasonication and the effects of this technique on its bioavailability in GI tract and functionality was reported [77]. Different vegetables and food products like carrots, cauliflower, brussels sprouts and mushrooms, have been processed by ultrasound assisted drying [78, 79]. In another study, blanching and ultrasound pretreatments were given to the freeze and hot air dried carrot discs and the effects of these pretreatments on carotenoid compounds and polyacetylene (i.e. falcarindiol-3-acetate, falcarindiol and falcarinol) was investigated. Ultrasound pretreatment showed higher stability of carotenoids and polyacetylenes contents of hot air dried carrot discs as compared to blanching pretreated hot air dried carrot samples [46].

7.2.3.6 Pasteurization

Pasteurization is a method in which food products are first heated below 100 °C and then rapidly cooled at 4 °C, as a results the microorganisms are killed. The heating can be generated by different means like dry heat, hot water, or electric current. On Industrial scale, usually two pasteurization approaches are employed:

1. At high temperature for a short processing time
2. At low temperature for a longer processing time.

Microwave pasteurization is another alternate for heating process of many food products, in which the temperature of food product is raised to 70 °C only just for 1.5–2 min and is found to be effective for inactivating the growth of microorganisms. Besides the conventional pasteurization techniques, many other approaches include pasteurization by blanching, via irradiation, microwave pasteurization, thermization, extrusion pasteurization and cold pasteurization etc (Fig. 7.8).

Type of Foods Processed

Pasteurization treatment of food products like milk, beer, and fruit juices are carried out at low temperatures i.e. ~ 62.8 °C for 30 min. These treatments are sufficient to kill the nonspore-forming pathogenic organisms e.g. Coxiella burnetti and Mycobacterium tuberculosis [81, 82].

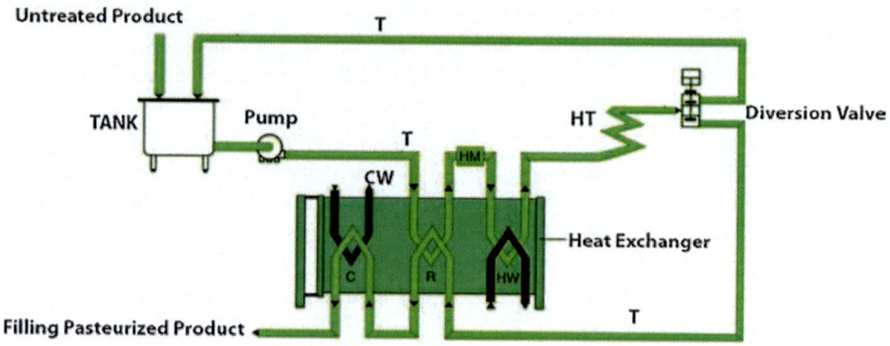

Fig. 7.8 Schematic experimental setup for continuous pasteurizer. C fields (PEF) cooler; *CW* cold water, *HM* homogenizer, *HT* holding tube, *HW* hot water, *R* regenerator [80].

Effect on Carotenoids

The pasteurization of orange juice processing was studied and it was observed that degradation rate of carotenoid and violaxanthin increases. Valencia orange juice undergo pasteurization process at 95–105 °C for 10 s and the percentage loss of constituents i.e. zeaxanthin (9%), β-carotene (11%), α-carotene (13%), ζ-carotene (14%), lutein (20%) and violaxanthin (38%) were reported [83]. In agreement with previous study, it was reported that during the concentration process of orange juice constituents losses were as follow; violaxanthin (31%), zeaxanthin (24%), β-carotene (5%), α-carotene (12%), ζ-carotene (29%), lutein (17%) and violaxanthin (38%) [84]. Orange juice and milk beverage were processed by pasteurization at 90 °C for 30 s and resulted in significant losses of zeaxanthin (22%), and lutein (23%) contents, while β-cryptoxanthin, α-carotene and β-carotene contents were retained potentially [66].

7.2.4 *Microencapsulation Method*

Microencapsulation is basically the encapsulation of tiny particles or droplets in a homogenous or heterogenous coating matrix [85]. Microencapsulation provides more stability to the bioactive compounds against oxidation in air, light and water, with it's controlled release [86]. Microencapsulation is considered as a suitable choice for processing, protecting and storage of sensitive food components e.g. vitamins, oleoresins and carotenoids. So different types of raw materials like isolated/ extracted natural or synthetic bioactive compounds and juices are encapsulated by using different coating materials. The carrier material attributed the physicochemical and thermodynamical barrier against processing conditions or external environmental like heat, light, oxygen, pH, water vapours or enzymes [87].

The most promising techniques include in microencapsulation are

- Spray-drying
- Spray Coating
- Centrifugal Extrusion
- Annular Jet
- Spinning Disk
- Prilling

7.2.4.1 Spray Drying

Spray drying consists of four steps; atomization, spray contact with drying medium, evaporation and finally product is separation of dried product by airflow (Fig. 7.9) [53].

Principle of Operation
Spray drying is performed by dispersing, dissolving, or emulsifying the bioactive ingredients in a solution of coating material prepared in aqueous medium followed by atomization under specific conditions of temperature and pressure. Finally, this mixture is sprayed into a drying chamber. The next step is solvent evaporation which results in the homogenized contacted form of droplets and drying medium (usually hot air). The final step is the separation of dried particles from the humid air in a cyclone and the final product is collected in powder form.

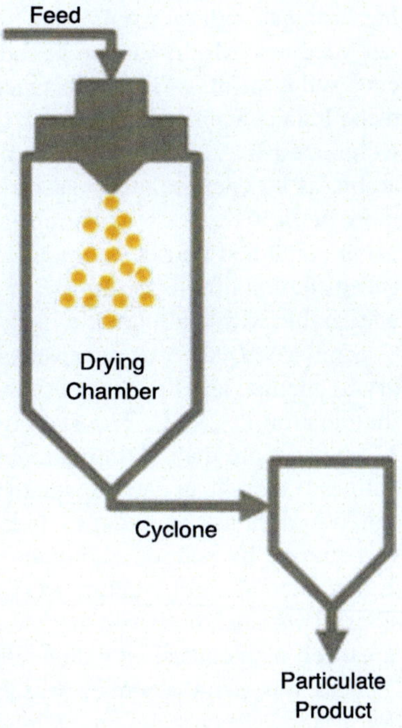

Fig. 7.9 Schematic experimental setup for spray drying [53]

A number of parameters control the efficiency of final encapsulated products like, feed flux, inlet and outlet air temperatures, speed of spray disc, type of atomizer, drying time and type of coated material.Primarily conditions like concentration of bioactive substances, nature of carrier material and drying conditions create limitation for this process. The final encapsulated products obtained with lot of dents and wrinkles on its surface as a result of shrinkage (i.e. loss of moisture) during spray drying process [88]. It was reported that the rate of water evaporation is slow at low inlet air temperature, as a result, particles got more time to shrink, however water evaporation increased at high inlet air temperature which also attribute to particle shrinkage [89].

Type of Foods Processed

Meat, fish vegetables, fruits, coffee, and juices are usually processed through this methodology.

Effect on Carotenoids

Spray drying is the most popular choice for processing the carotenoids rich food products, resulted in high yield with excellent encapsulation efficiency. Spray drying is considered a simple but versatile processing method due to miscellaneous choices for optimal changes in drying parameters and different carrier agents, moreover, carotenoids constituents itself possess uniform particle size, good thermal stability, less moisture content (5–8%) and low bulk density [90]. The most commonly used coated materials are modified starch, maltodextrin and gum arabic in the formation of spray-dried powders. The stable emulsions were prepared by gum arabic (using as wall materials), due to its high permeability and low viscosity in aqueous medium. Modified starches also possess low viscosity, which facilitates the encapsulation of various substances. Moreover, carotenoids losses decreases by using high concentration of starch coated material in the spray drying encapsulation technique [91, 92]. It was reported that the combination of maltodextrin and gum arabic as an encapsulated material, exhibited enhanced carotenoids retention and more stability against oxidation [93]. Astaxanthin oleoresin, onencapsulation by using combined carrier polymers i.e. gum arabic and whey protein, enhanced the encapsulation efficiency up to 70% as compared to the polymeric combination of gum arabic with maltodextrin [94].

Similarly, Bixin was encapsulated by maltodextrin or gum arabic using spray-drying method and their stability was determined in aqueous solution under dark or illumination at 21 °C. The stability of bixin-gum arabic microcapsules appeared 3–4 times more than bixin-maltodextrin combination. The stability of encapsulated bixin was more than unencapsulated bixin and also more stable when microcapsules stored in dark as compared to under illumination [95]. The encapsulation of Bixin was studied by sodium caseinate in hot aqueous ethanolic solution using spray-drying method and hydration of spray-dried powder resulted in transparent dispersions. It was also observed that the stability of encapsulated bixin was significantly enhanced with consistent yellow color [96].

Rosa mosqueta oleoresin was encapsulated by gelatin or starch under spray-drying processing technique. Oleoresin-gelatin microcapsules exhibited improved stability for carotenoids content especially all-*E*-β-carotene, with low degradation rate constants as well as longer half-lives. The degradation rate of all-*E*-β-carotene, all-*E*-lycopene, all-*E*-rubixanthin were same for oleoresin-starch microcapsules [97].

Paprika oleoresin was microencapsulated with soy protein isolate and gum arabic by spray-drying and HPH (high-pressure homogenization) processing methods. The maximum carotenoid content stability was observed for microencapsulated paprika oleoresin with A_w of 0.710 and 0.274 for soy protein and isolated gum arabic respectively [98].

The encapsulation of Gac oil by gum arabic and whey protein concentrate using spray-drying method and the spray-drying conditions resulted in enhancing the stability of encapsulated oil contents like β-carotene, lycopene and unsaturated fatty acids as well as appeared in attractive red-yellow colour. Encapsulated oil powder showed excellent stability against oxidation and light, so could be stored for longer time period. It was proposed that encapsulated oil powder could also be used as a natural food colorant and as nutrient supplements [99].

7.2.4.2 Spray Coating

Coating technologies are traditionally quick and economical way to prepare pharmaceutical capsules on macro scale. The performance efficiency and surface properties of engineering materials could be enhanced by thermal spray coatings. The common processes includes;

- Thermal Spray Coatings
- Air-Suspension Coating
- Wurster Coating or Fluidised Bed Coating

Principle of Operation

The process is started by spraying the fine semi-molten or molten particles of coating material onto the bioactive matrix or substrates to produce capsules. Air stream is permitted to rise up through a cylindrical drum having inserted perforated plate with holes of specific size and patterns, during the downward movement of active ingredients. As a result the coated particles fluidize on the surface. The air stream slows down at the top and diverges back downward to repeat the cycle. The particles pass many times through the inner cylinder in few minutes. Spraying of the coated material can be done by different ways depending upon the heating source of air stream [100].

- Plasma Spraying
- Flame Spraying
- High Velocity Oxy-Fuel Spraying
- Arc Spraying (Fig. 7.10)

The smooth coated microcapsules must have specific features like porosity, lamella structure and flexibility. The finishing procedures or post-treat of the coated

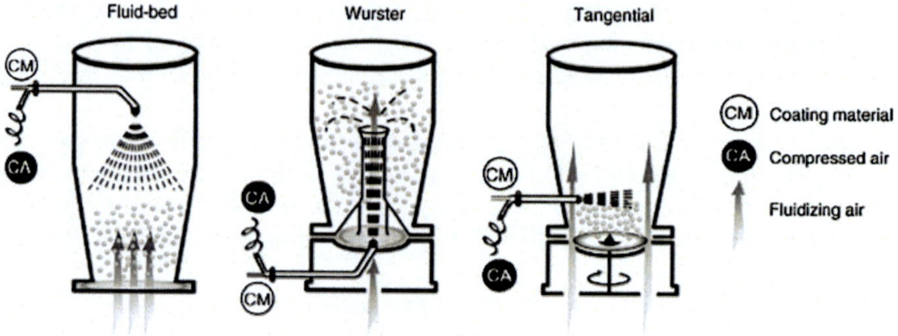

Fig. 7.10 Schematic diagram for spray coating [101]

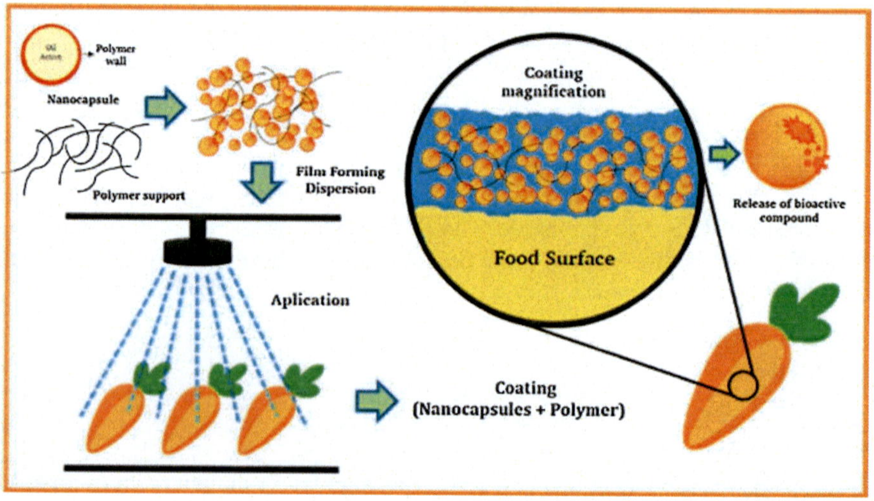

Fig. 7.11 Application of polymeric nanoparticles in edible coating [102]

microstructure is necessary to get the improved dimensional requirements with appropriate tolerances. The post-treat methods involve different mechanical, chemical, and thermal techniques. The key choice for procedure adopted depends upon many factors like colour match, coating thickness, ease of finishing, bond strength, surface profile after finishing and application cost (Fig. 7.11).

The Wurster's process has been used more commonly for encapsulation of minerals, vitamins, carotenoids and other functional food ingredients in the food industry. This process improves the shelf life, stability and mask, an undesirable smell or flavor. The suitable approach is to coat a hydrophilic or hydrophobic active ingredient with a hydrophobic or hydrophilic polymer as wall material respectively.

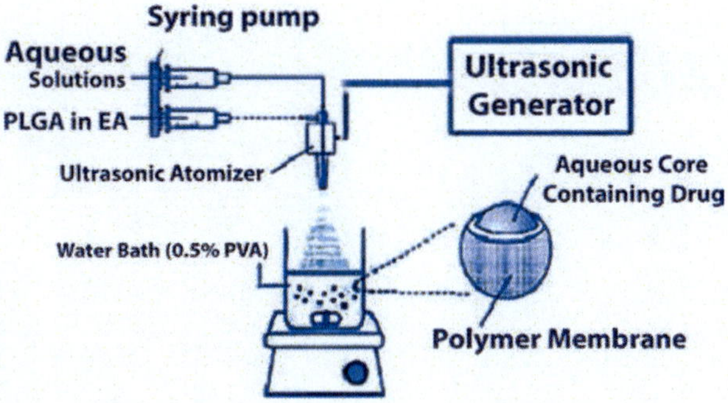

Fig. 7.12 Experimental setup for centrifugal extrusion [103]

7.2.4.3 Centrifugal Extrusion

Another potent encapsulation choice for protection of food ingredients is a liquid co-extrusion process. Various types of shell materials used for encapsulation are starches, gum arabic, fats, alginates, polyethylene glycol, gelatin, cellulose derivatives, waxes and carrageenan. Desirable characteristics could also be obtained by using blend of two or more polymers as a shell system. This approach results into the microencapsulated product having particle size in the range of 150–2000 μm with high loading capacity (20–80%). Flavor oils are efficiently encapsulated by this methodology. The centrifugal extrusion assembly consists of a rotating cylinder having nozzles with concentric orifices at its outer circumference. First the liquid core and wall materials are forced through the inner and outer orifices respectively. As a result, co-extruded rod of core substance coated with wall material is obtained. Then, the extruded rod finally breaks as the device rotates, into droplets to form microcapsules (Fig. 7.12).

7.2.4.4 Annular Jet

Annular Jet assembly consist of two concentric jets, the active ingredient and molten wall material are filled in the inner and outer jet respectively. The technique involves the atomization of both fluid streams into droplets and then, the solidification on exiting the jet forms the microcapsules (Fig. 7.13).

The uniform product and droplet size of microcapsule (sub-micron diameters sizes) could be maintained by controlling the vibrational rate of nozzle. The solidification of microcapsule droplets can be initiated by adding binder (e.g. in a slurry) or by gelation system (e.g. sol-gel processing, melt).

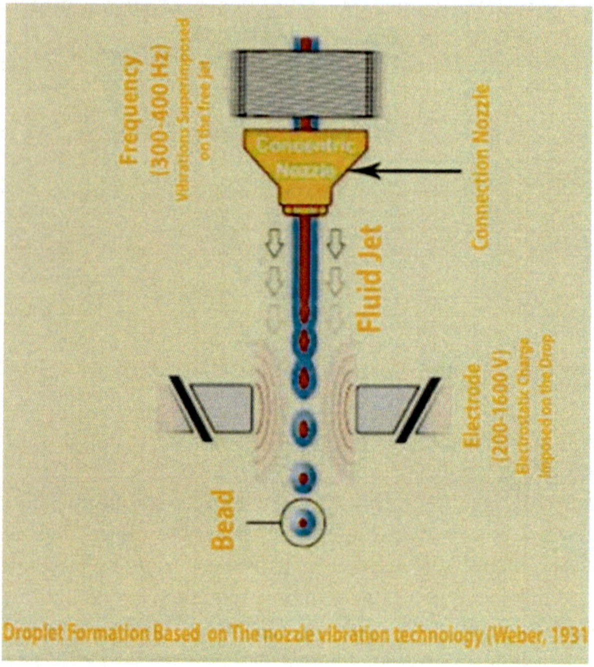

Fig. 7.13 Schematic experimental setup for annular jet [104, 105]

7.2.4.5 Spinning Disk

In this method, the rotational forces are generated using spinning disk system to form fine droplets. The basic operational principle is similarly to centrifugal extrusion.

The active substrate is first suspended in a coated material and then dropped onto the rotating disk. The droplets, then throws out by centrifugal rotational forces towards the circumference of disk. The coating material finally solidifies either by chilling or drying. The optimal narrow particles size (between i.e. 5 and 3000 microns) could be attained by controlling the spinning rate of disk (Fig. 7.14).

7.2.4.6 Prilling

Prilling technique is similar to spray drying, also known as spray cooling, spray chilling, or spray congealing. In this methodology, atomization of active ingredient in a liquid coating material is carried out and then, the microcapsules are formed by solidification with the help of chilled air (Fig. 7.15).

The optimal narrow particles size (between i.e. 10 and 400 microns) could be attained by this technique. A variety of shell materials used for microencapsulation e.g. wax, polymers and hydrogenated vegetable oils. This microencapsulation technique has wide range of applications especially in pharmaceutical, agrochemicals and food industry. Microcapsule technology has been used for preservation of fragrances and flavors.

Fig. 7.14 Spinning disk setup

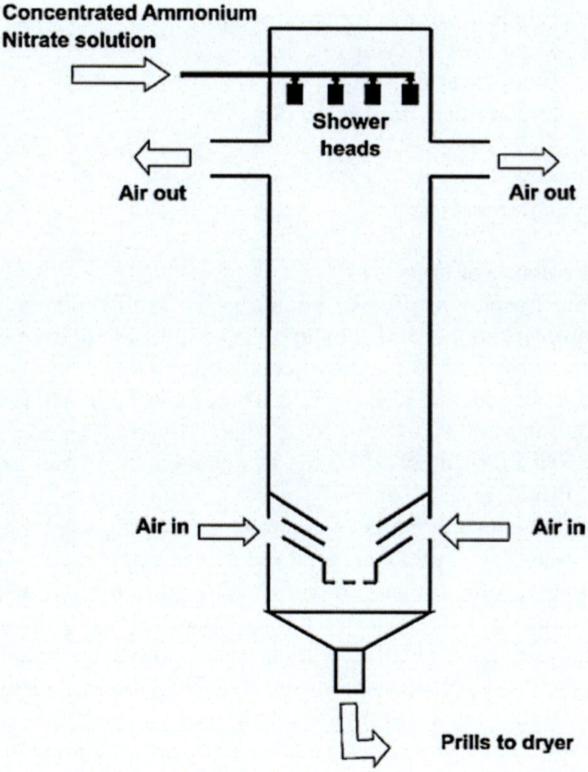

Fig. 7.15 Experimental setup for prilling tower and granulation [106]

7.2.5 *Nano-Encapsulation Techniques*

Nano-encapsulation techniques are generally more complex than microencapsulation techniques because it is difficult to produce capsules on nanoscale. Nano-encapsulation techniques are classified in to two types;

1. Top Down Nano-Encapsulation Technique
2. Bottom Up Nano-Encapsulation Technique

7.2.5.1 Top Down Nano-Encapsulation Technique

Top down method is more precise tool to get the desired structural shaping and size reduction of nanomaterial [107]. Top down approaches includes

- Electrospraying
- Extrusion Gelation Techniques
- Electrospinning
- Emulsification solvent evaporation
- Nano Spray Drying
- Emulsification
- Supercritical fluid extraction [108].

Electro-spraying

Principle of Operation

Electrospraying is also known as electrohydrodynamic spraying, generally fine droplets are generated by applying an electric field (electrostatically). This process starts by flowing the polymer solution through capillary nozzle which is subjected to high potential electric field. Then, the solution jet disrupt the flight of the droplets due to various sizes of these charged particles under the influence of coulombic explosions, followed by the evaporation of solvent and finally move towards the collector (Fig. 7.16).

Electrospraying is an emerging processing technology for carotenoids encapsulation. The particle diameter reduces by applying the electric field, which develops coulomb force that competes with cohesion force of particle. As Coulomb force dominates over Cohesive force, surface tension decreases and results in nanoparticle formation [110]. Electrospraying processing usually proceeded at room temperature. Lycopene was encapsulated by emulsion-electrospraying technique which enhances the encapsulation efficiency as compared to spray drying. The results showed that spray drying causes thermal degradation of lycopene [110]. β-Carotene encapsulation by whey protein concentrate and zein under electrospraying techniqueimproves stability against thermal degradation along with higher encapsulation efficiency of β-Carotene [111].

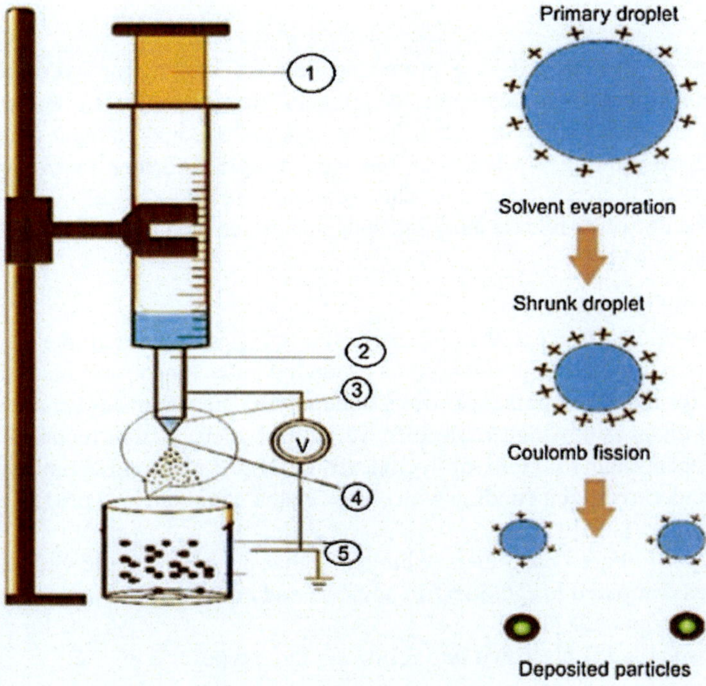

Fig. 7.16 Experimental setup for electrospraying method [109]

Extrusion Gelation Techniques

A biopolymer solution passes through a needle like nozzle (i.e. single or dual feed) into a gelling medium. Extrusion gelation is considered as a mild and suitable technique for encapsulation of both hydrophobic and hydrophilic bioactive compounds. Single or co-extrusion approach *i.e.* bioactive substance entrapped by a biopolymer (e.g. chitosan, pectins, alginates, or δ-carrageenan) exhibits the cross-linkages throughout the ionotropic gelation skeleton of fabricated microspheres.

Principle of Operation

In extrusion gelation, the biopolymer solution (coated material), first loaded into a syringe and passes through a needle into gelling condition to form gelation. The technique is also applicable on industrial scale and it provides large particle sizes [112]. The technical optimization of extrusion set-up devices mainly depends upon the atomization principle (i.e. stationary, vibrating or rotating) and the nozzle conformation [113].

Electrospinning

Nanofibers are considered as a promising delivery system for hydrophilic and hydrophobic bioactive compounds. Nanofibers are assembled by encapsulation technique i.e. electrospinning performed at high voltage and ambient temperature [114]. Electrospinning is a low cost, one step, simple, and time saving with high encapsulation efficiency technique. This technique attributed the electrospun nanofibers with controlled release and high specific surface area of encapsulated bioactive compounds [115].

Principle of Operation
In electro-spinning process, the polymer solution is passed through the electrically charged jet and then is deposited onto a grounded collector to form nano-fibers. In principle, for the development of strong bonding linkages or entanglement between fibers and chain needs high molecular weight polymers. It was reported that the highly concentrated solutions of cyclodextrin and phospholipids bonded through intermolecular hydrogen bonding with encapsulated compounds to produce electrospun nanofibers [116].

Cyclodextrin-based nanofibers exhibited higher porous structure with large surface area as compared to cyclodextrin powders and could be used for encapsulation of volatile compounds [117].

Electrospinning is classified into following three types;

1. Blend Electrospinning
2. Coaxial Electrospinning
3. Emulsion Electrospinning

Blend Electrospinning

The bioactive compounds are dissolved or dispersed in the solution. The homogeneous distribution of bioactive compounds in the electrospun nanofibers depends upon solution properties and their interaction with solvent molecules [118]. It was reported that the functional integrity of bioactive compounds e.g. peptides was lost due to interaction with solvent molecules. The bioactive compounds are bonded by charge repulsion on the surface of nanofibers which results with rapid and uncontrolled release [119].

Coaxial Electrospinning

In this method, nanofibers are produced with core-shell structures by using a concentric double needle [120]. Optimal parameters for coaxial electrospinning are the viscosity and interfacial tension of two polymers are required to get the best results [121]. Coaxial electrospinning processing provides more stability and protection to the bioactive compound by using separate injection (Fig. 7.17).

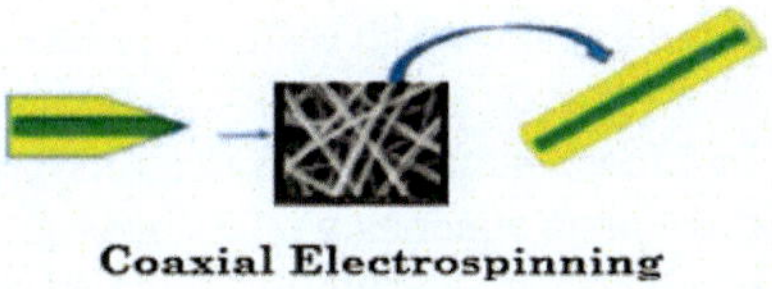

Fig. 7.17 Schematic layout for coaxial electrospinning [122]

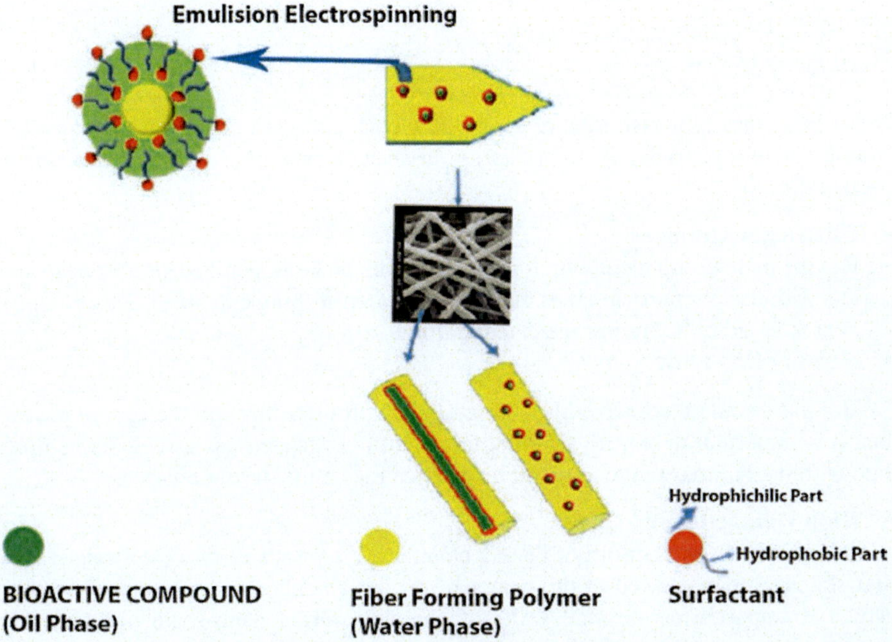

Fig. 7.18 Schematic layout for emulsion electrospinning

Emulsion Electrospinning

In this method, emulsification of bioactive compound is done in the presence of surfactant which is then subjected to elctrospinning to form nanofibers. Oil in water (O/W) or water in oil (W/O), both emulsions is formed by this method. Electrospun nanofibers have localized regions of dispersed phase or core-shell structure morphology [123]. Hydrophobic bioactive compounds were encapsulated more efficiently in core-shell structures by emulsion electrospinning as compared to the nanofibers formed by blend electrospinning. Another advantage of this processing is that the organic solvents are not used, resulting ineasily up-scalable and marketable in foods industry [120]. The major drawback of this technique is the lost of functionality of bioactive compounds between water and oil phases due to the interface tension [124] (Fig. 7.18).

Emulsification Solvent Evaporation

In this technique, emulsification of polymer solution having bioactive compound is done in an aqueous phase, subsequently solvent is vaporized to get the nanospheres of precipitated polymer. This technique has great potential to load lipophilic compounds more efficiently (Fig. 7.19).

Nano Spray Drying

Nano spray drying is a simple, relatively low cost and rapid process for encapsulation of nano-materials. A nano spray dryer consists of following functional parts: [126]

- **Ultrasonic Atomizer**
  It is the main component which is used to produce homogenized small size droplets with narrow particle-sizes that comes out from spray cap holes. This component is based on vibrating mesh technology.
- **Heating Gas Flow**

Hot gas circulate co-currently with atomized droplets through the drying chamber. It is important to control the inlet temperature of heated gas as well as its flow rate within a laminar range to avoid turbulence inside the drying chamber.

- **High Voltage Electric Field**
  It is present at the bottom of drying chamber. It consists of two electrodes i.e. a star shaped anode, placed at the center of drying chamber and a circular cathode which is imparted at the side walls of chamber. Dried nano-scale particles are charged by the central anode during their downward movement and are collected on the cathode surface. Finally, dried nanoparticles are gently scrapped from collector's surface and drying gas exits the drying chamber without carrying any particles [127]. This technology provides 99% separating and collecting efficiency of nanoparticles (Fig. 7.20).

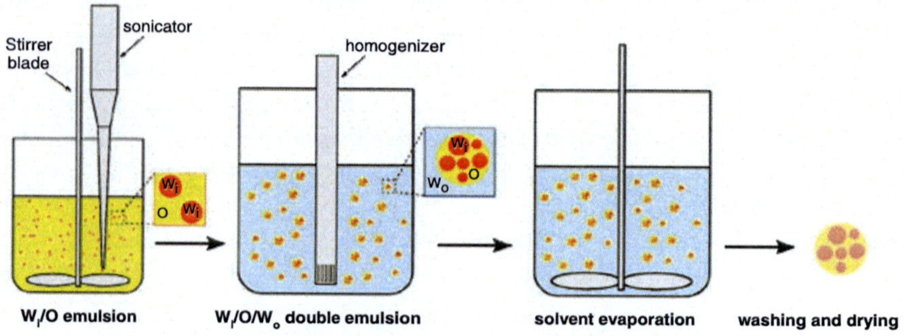

Fig. 7.19 Schematic diagram for emulsification solvent evaporation [125]

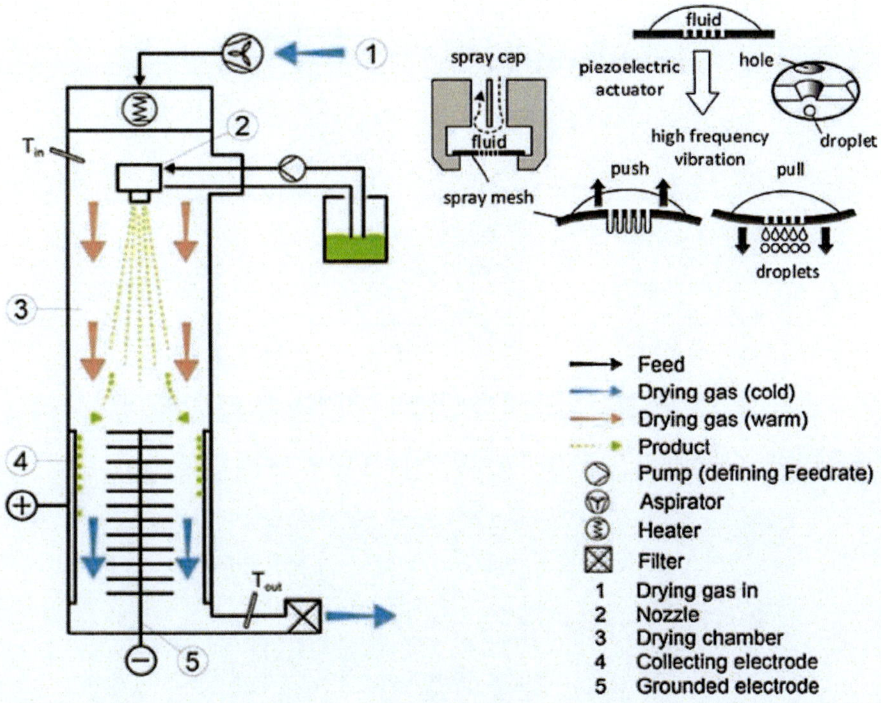

Fig. 7.20 Experimental setup for nano-spray dryer and its functional parts [128]

Emulsification

Emulsification is considered as thermodynamically stable isotropic system with high solubilization capacity. This approach is applicable for both hydrophobic and hydrophilic compounds.

A process in which homogenized single phase emulsion is prepared by mixing of two immiscible liquids in the presence of emulsifying agents (i.e., surfactant and co-surfactant). However, a large amount of surfactants is needed [129].

Two types of schemes are followed to form emulsions.

- Single emulsion (O/W and W/O)
- Double emulsion (W/O/W and O/W/O) (Figs. 7.21 and 7.22)

Supercritical Fluid Extraction

Supercritical fluid extraction is considered as green, nontoxic, scalable, and economical technology. It is a promising approach for food industry. SFE is a separation process, in which substances like phospholipid cholesterol, are dissolved in

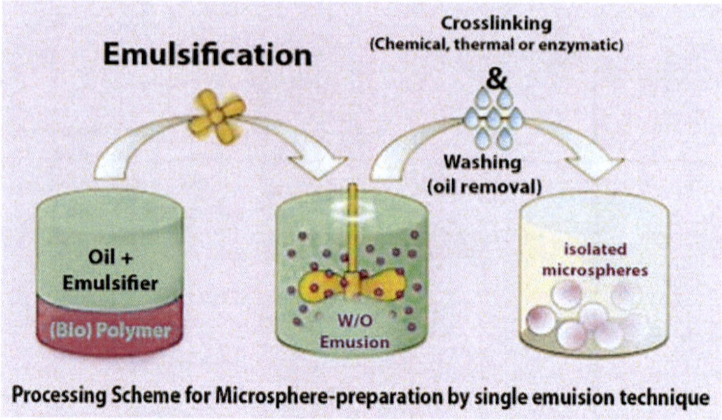

Fig. 7.21 Encapsulation by single emulsion (O/W and W/O) technique [130]

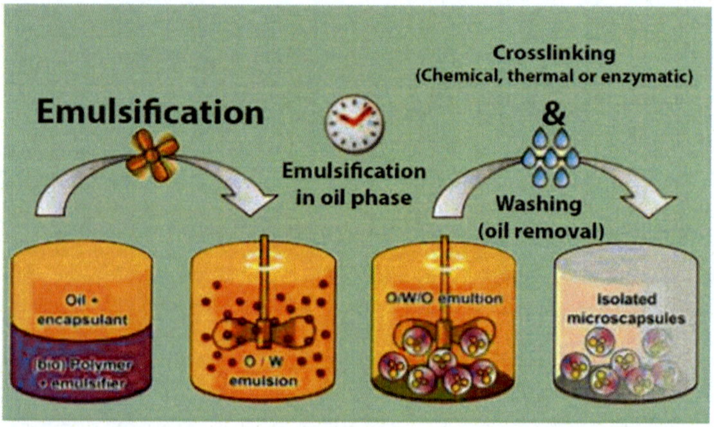

Fig. 7.22 Encapsulation by double emulsion (W/O/W and O/W/O) technique [131]

supercritical CO_2 followed by precipitation process to form ultrafine particles at above their critical pressure and temperature [132] (Fig. 7.23).

Extraction liquid or gases are used as solvents at above their critical pressure and temperature. CO_2 is considered as an ideal supercritical extraction fluid especially in the food industry, attributed to its distinctive properties, which are:

- Critical temperature = 31.06 °C
- Critical pressure = 73.83 bar
- Critical density = 0.46 g/cm^3

Lycopene extraction was proceeded by using carbon dioxide under supercritical fluid extraction and it was reported that lycopene degradation potential decreased considerably [134]. Kinetically SFE generates entropy-driven hydrophobic effect.

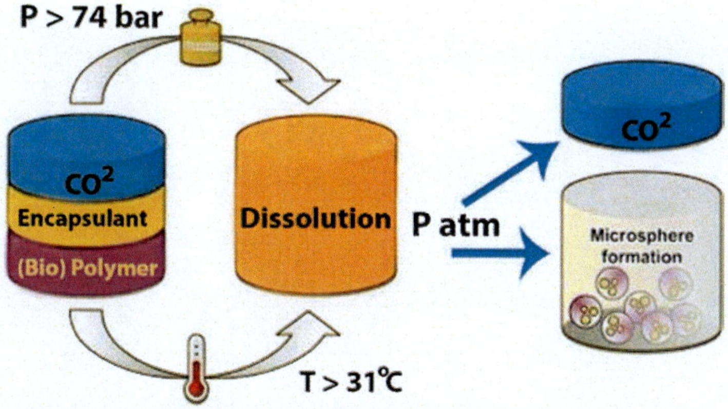

Fig. 7.23 Encapsulation by supercritical fluid extraction [133]

7.2.5.2 Bottom up Nano-Encapsulation Technique

Nano-encapsulated materials developed via self-organization and self-assembly of bioactive molecules [135], which depends upon many factors like pH, temperature, concentration, and ionic strength. Bottom up approaches include:

- Nanoprecipitation
- Layer by layer deposition
- Coacervation
- Inclusion Complexation

Nanoprecipitation

Nanoprecipitation is easily adaptable and relatively inexpensive with simple assembly methodology. The process is based on the spontaneous emulsification approach which consists of organic internal phase having bioactive compound and polymer in organic solvent with an aqueous external phase. Nanoprecipitation results multilayers to nanocapsules. It was reported that recovery of nanoparticles was difficult at low concentration of polymer [136–139] (Fig. 7.24).

Layer by Layer Deposition

It is easily adoptable method having low-cost and a simple assembly. It is extensively used technique for nanoencapsulation especially for lipophilic compounds. In this process, polyelectrolytes are deposited layer by layer around an oppositely charged template, as a result multilayer nanocarriers are developed. This technique has limited applications on industrial scale [140] (Fig. 7.25).

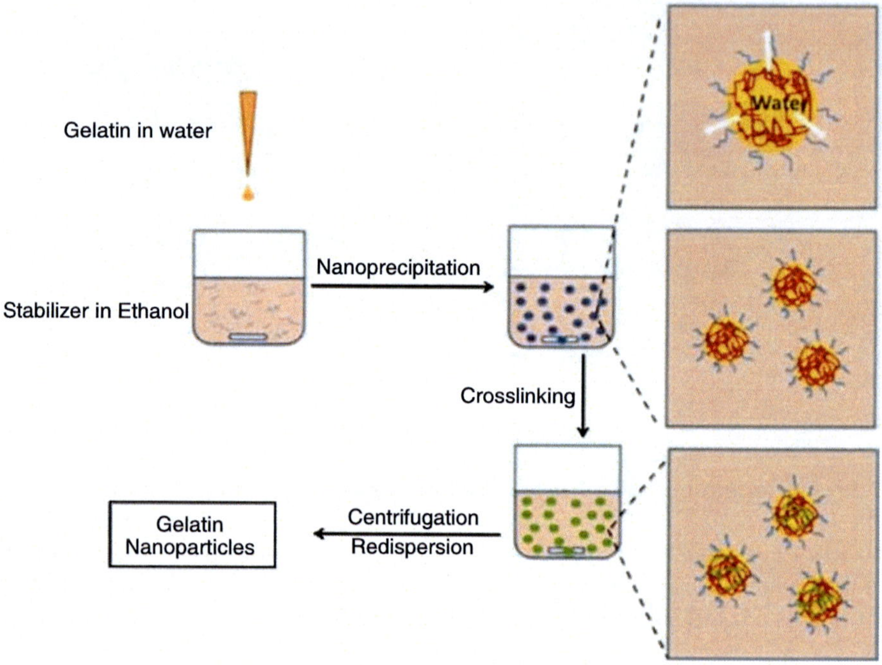

Fig. 7.24 Schematic layout for nanoprecipitation [139]

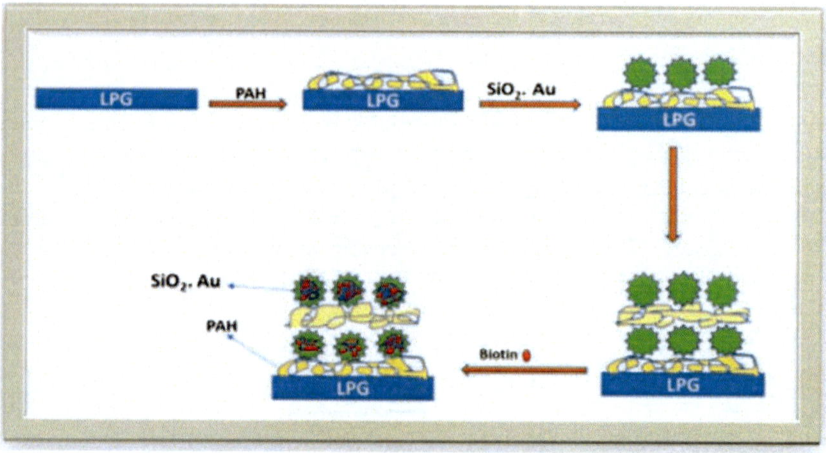

Fig. 7.25 Layer by layer deposition

Inclusion Complexation

It is a potential method for encapsulation of volatile organic compounds with high encapsulation efficiency. Generally, supra-molecules of encapsulation ingredient (a ligand) are entrapped into cavity-bearing substrate (i.e. shell material) via Van der Waals force or hydrogen bonding. Only a few specific bioactive compounds are successfully encapsulated by this method [108].

Coacervation

Coacervation is a chemical method with high encapsulation efficacy. This potent technique involves a separation of two liquid phases into single concentrated colloidal phase containing charged macro-ions for generating polymer droplets in suspension. Polymer (single or mixture of polyelectrolyte) separated from the supernatant solution and then the resulted coacervate, having the agglomerated colloidal polymer particles deposited around the immiscible active core to form the coacervate (simple or complex coacervate, respectively) [108, 141, 142] (Fig. 7.26).

Lycopene was encapsulated by gelatin and pectin using complex coacervation process and it was reported that pigment stability did not improve considerably due to disintegration during freezing as well as the initial loss of coacervate structure. However, the functionality of encapsulated carotenoids was effected significantly after post processing operations [145]. In another study, encapsulation by complex coacervation attributed to enhance ability to impart color to the food product, due to the structural compatibility and homogenous distribution of the pigment within the food matrix [146].

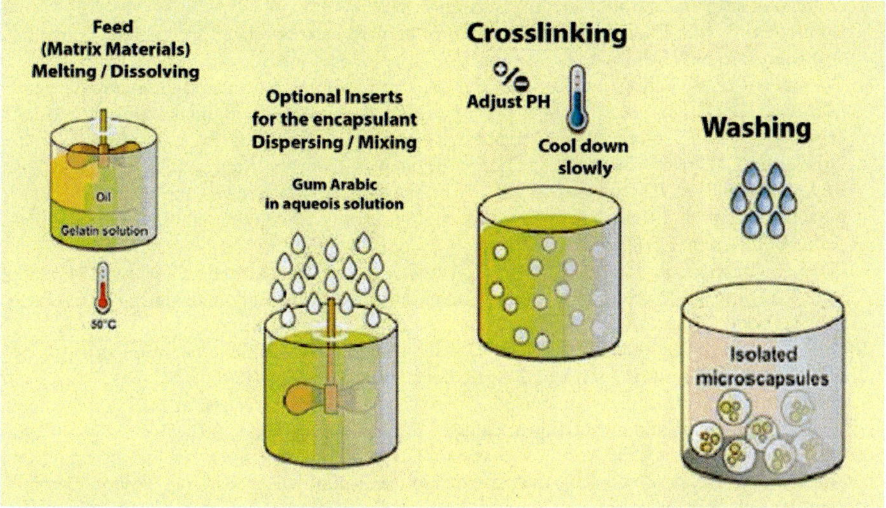

Fig. 7.26 Basic principle for coacervation [143, 144]

7.3 Conclusion

Food storage has been a topic of interest since the prehistoric times and thus, a number of home processing methods have been used, for example, boiling, baking and frying etc. A high contents of carotenoids is lost during these processes. Industrial processing techniques include drying, blanching, coating etc. Advanced industrial methods include high pressure processing, high pressure homogenization, high hydrostatic pressure, pulsed electric fields, ultrasonication, pasteurization, microencapsulation and nano-encapsulation etc. The advancements in the technology has resulted in maintaining a significant carotenoid contents and greater shelf life.

References

1. Chuyen HV, Roach PD, Golding JB, Parks SE, Nguyen MH (2019) Encapsulation of carotenoid -rich oil from Gac peel: optimisation of the encapsulating process using a spray drier and the storage stability of encapsulated powder. Powder Technol 344:373–379
2. Nguyen M, Francis D, Schwartz S (2001) Thermal isomerization susceptibility of carotenoids in different tomato varieties. J Sci Food Agric 81:910–917
3. Sahlin E, Savage GP, Lister CE (2004) Investigation of the antioxidant properties of tomatoes after processing. J Food Compos Anal 17:635–647
4. Toor RK, Savage GP (2005) Antioxidant activity in different fractions of tomatoes. Food Res Int J 38:487–494
5. Chandler LA, Schwartz SJ (1988) Isomerization and losses of trans beta-carotene in sweet potatoes as affected by processing treatment. J Agric Food Chem 36(1):129–133
6. Sweeney JP, Marsh AC (1971) Effect of processing on provitamin A in vegetables. J Am Diet Assoc 59(9):238–243
7. Baloch AK, Buckle KA, Edwards RA (1977) Stability of beta carotene in model systems containing sulphate. J Food Technol 12(3):309–316
8. Borchgrevink NC, Charley H (1966) Color of cooked carrots related to carotene content. J Am Diet Assoc 49(1):116–121
9. Cervantes Paz B, Yahia EM, Ornelas Paz JJ, Gardea-Béjar AA, Ibarra-Junquera V, Pérez Martínez JD (2012) Effect of heat processing on the profile of pigments and antioxidant capacity of green and red jalapeño peppers. J Agric Food Chem 60:10822–10833
10. Ornelas Paz JJ, Cira Chávez LA, Gardea Béjar AA, Guevara Arauza JC, Sepúlveda DR, Reyesb Hernández J, Ruiz Cruz S (2013) Effect of heat treatment on the content of some bioactive compounds and free radical scavenging activity in pungent and non-pungent peppers. Food Res Int J 50:519–525
11. Plaza L, Crespo I, de Pascual Teresa S, de Ancos B, Sánchez-Moreno C, Muñoz M, Cano MP (2011) Impact of minimal processing on orange bioactive compounds during refrigerated storage. Food Chem 124:646–651
12. Gil MI, Aguayo E, Kader AA (2006) Quality changes and nutrient retention in fresh-cut versus whole fruits during storage. J Agric Food Chem 54:4284–4296
13. de Azevedo CH, Rodriguez-Amaya DB (2005) Carotenoid composition of kale as influenced by maturity, season and minimal processing. J Sci Food Agric 85(4):591–597
14. de Azevedo-Meleiro CH, Rodriguez-Amaya DB (2005) Carotenoids of endive and New Zealand spinach as affected by maturity, season and minimal processing. J Food Compos Anal 18(8):845–855

15. Miglio C, Chiavaro E, Visconti A, Fogliano V, Pellegrini N (2008) Effects of different cooking methods on nutritional and physicochemical characteristics of selected vegetables. J Agric Food Chem 56:139–147
16. De Sa MC, Rodriguez-Amaya DB (2003) Carotenoid composition of cooked green vegetables from restaurants. Food Chem 83(4):595–600
17. De Sa MC, Rodriguez-Amaya DB (2004) Optimization of HPLC quantification of carotenoids in cooked green vegetables – comparison of analytical and calculated data. J Food Compos Anal 17(1):37–51
18. Kidmose U, Yang R-Y, Thilsted SH, Christensen LP, Brandt K (2006) Contents of carotenoids in commonly consumed Asian vegetables and stability and extractability during frying. J Food Compos Anal 19:562–571
19. Kao FJ, Chiu YS, Tsou MJ, Chiang WD (2012) Effects of Chinese domestic cooking methods on the carotenoid composition of vegetables in Taiwan. LWT Food Sci Technol 46(2):485–492
20. Jimenez-Monreal AM, Garcia-Diz L, Martinez-Tome M, Mariscal M, Murcia MA (2009) Influence of cooking methods on antioxidant activity of vegetables. J Food Sci 74(3):97–103
21. Pokorny J, Schmidt S (2001) Natural antioxidant functionality during food processing, antioxidants in food: practical applications. Woodhead Publishing, Cambridge, pp 331–354
22. Lin TM, Durance TD, Scaman CH (1998) Characterization of vacuum microwave, air and freeze dried carrot slices. Food Res Int J 31:111–117
23. Clydesdale FM, Ho C-T, Lee CY, Mondy NI, Shewfelt RL (1991) The effects of postharvest treatment and chemical interactions on the bioavailability of ascorbic acid, thiamin, vitamin A, carotenoids, and minerals. CRC Crit Rev Food Sci Nutr 30(6):599–638
24. Della-Monica ES, McDowell PE (1965) Comparison of β-carotene content of dried carrots prepared by three dehydration processes. Food Technol 19(10):141–143
25. Rodriguez Amaya DB (1997) Carotenoids and food preparation: the retention of provitamin A carotenoids in prepared, processed, and stored foods. Opportunities for Micronutrient Intervention, Arlington
26. Negi PS, Roy SK (2000) Effect of blanching and drying methods on β-carotene, ascorbic acid and chlorophyll retention of leafy vegetables. Lebens Wissen Technol 33:295–298
27. Negi PS, Roy SK (2001) Effect of drying conditions on quality of green leaves during long term storage. Food Res Int J 34:283–287
28. Pott I, Marx M, Neidhart S, Mühlbauer W, Carle R (2003) Quantitative determination of β-carotene stereoisomers in fresh, dried, and solar dried mangoes (Mangifera indica L.). J Agric Food Chem 51:4527–4531
29. Mulokozi G, Svanberg U (2003) Effect of traditional open sun drying and solar cabinet drying on carotene content and vitamin A activity of green leafy vegetables. Plant Food Hum Nutr 58:1–15
30. Mdziniso P, Hinds MJ, Bellmer DD, Brown B, Payton ME (2006) Physical quality and carotene content of solar dried green leafy and yellow succulent vegetables. Plant Food Human Nutr 61:13–21
31. Chávez AL, Sánchez T, Ceballos H, Rodriguez Amaya DB, Nestel P, Tohme J, Ishitani M (2007) Retention of carotenoids in cassava roots submitted to different processing methods. J Sci Food Agric 87:388–393
32. Bengtsson A, Namutebi A, Alminger ML, Svanberg U (2008) Effects of various traditional processing methods on the all-trans-β-carotene content of orange-fleshed sweet potato. J Food Compos Anal 21:134–143
33. Bechoff A, Dufour D, Dhuique Mayer C, Marouzé C, Reynes M, Westby A (2009) Effect of hot air, solar and sun drying treatments on provitamin A retention in orange fleshed sweet potato. J Food Eng 92:164–171
34. Hernández BJA, Ochoa FAA, Valerio AG, Soto RI, Rodríguez EMT, García HS (2014) Cholesterol oxidation and astaxanthin degradation in shrimp during sun drying and storage. Food Chem 145:832–839

35. Ratti C (2001) Hot air and freeze-drying of high-value foods: a review. J Food Eng 49:311–319
36. Mazza G (1983) Dehydration of carrots. Effect of pre-drying treatment on moisture transport and product quality. J Food Technol 18:113–123
37. Gross J (1991) Pigments in vegetables: chlorophylls and carotenoids. Van Nostrand Reinhold, New York
38. Leong SY, Oey I (2001) Effects of processing on anthocyanins, carotenoids and vitamin C in summer fruits and vegetables. Food Chem 133:1577–1587
39. Duroudier J-P (2016) Heat transfer in the chemical, food and pharmaceutical industries, pp 359–377
40. Aurelie B, Ugo C, Keith T, Pesila G, Paul I, Andrew W, Erick B (2015) Carotenoid stability during storage of yellow gari made from biofortified cassava or with palm oil. J Food Compos Anal 44:36–44
41. Genin N, René F (1995) Analyse du Rôle de la transition vitreuse dans les procédés de conservation agroalimentaires. J Food Eng 26:391–408
42. Irzyniec Z, Klimczak J, Michalowski S (1995) Freeze-drying of the black currant juice. Dry Technol 13:417–424
43. Bolin HR, Stafford AE (1974) Effect of processing and storage on provitamin A and vitamin C in apricots. J Food Sci 39(5):1034–1036
44. Baloch AK, Buckle KA, Edwards RA (1977) Effect of processing variables on quality of dehydrated carrot. II. Leaching losses and stability of carrot during dehydration and storage. J Food Technol 12(3):295–307
45. Guerrant NB, Vavich MG, Fardig OB, Ellenberger HA, Stern RM, Coonen NH (1947) Effect of duration and temperature of blanch on vitamin retention of certain vegetables. Ind Eng Chem 39:1000–1007
46. Rawson A, Tiwari BK, Tuohy MG, Donnell CPO, Brunton N (2011) Effect of ultrasound and blanching pretreatments on polyacetylene and carotenoid content of hot air and freeze dried carrot discs. Ultrason Sonochem 18(5):1172–1179
47. Ramesh MN, Wolf W, Tevini D, Bognar A (2002) Microwave blanching of vegetables. J Food Sci 67(1):390–398
48. Baloch AK, Buckle KA, Edwards RA (1986) Effect of coating with starch and nordihydroguaiaretic acid on the stability of carotenoid of dehydrated carrot. J Chem Soc Pak 8(1):59–62
49. Oey I, Lille M, Van Loey A, Hendrickx M (2008) Effect of high-pressure processing on colour, texture and flavour of fruit- and vegetable-based food products: a review. Trends Food Sci Technol 19(6):320–328
50. Mor-Mur M, Yuste J (2005) Microbiological aspects of high pressure processing. In: Sun DW (ed) Emerging technologies for food processing. Elsevier Academic Press, London, pp 47–66
51. Hogan E, Kelly A, Sun DW (2005) High pressure processing of foods: an overview. In: Sun DW (ed) Emerging technologies for food processing. Elsevier Academic Press, London, pp 3–32
52. Matser AM, Krebbers B, Berg RW, Bartels PV (2004) Advantages of high-pressure sterilisation on quality of food products. Trends Food Sci Technol 15(2):79–85
53. Nandita D, Shivendu R, Eric L. Environmental chemistry for a sustainable world, Environmental Nanotechnology, Volume 3, Springer, New York.
54. Naik L, Sharma R, Rajput YS, Manju G (2013) Application of high pressure processing technology for dairy food preservation – future perspective: a review. J Anim Products Adv 3(8):232–241
55. Svelander CA, Lopez-Sanchez P, Pudney PD, Schumm S, Alminger MA (2011) High pressure homogenization increases the in vitro bioaccessibility of α-and β-carotene in carrot emulsions but not of lycopene in tomato emulsions. J Food Sci 76(9):H215
56. Panozzo A, Lemmens L, Van Loey A, Manzocco L, Nicoli MC, Hendrickx M (2013) Microstructure and bioaccessibility of different carotenoid species as affected by high

pressure homogenisation: a case study on differently coloured tomatoes. Food Chem 141(4):4094–4100
57. Colle I, Van BS, Van LA, Hendrickx M (2010) High pressure homogenization followed by thermal processing of tomato pulp: influence on microstructure and lycopene in vitro bioaccessibility. Food Res Int 43(8):2193–2200
58. Maja J, Wioletta B, Danuta Z, Wiesław W, Ireneusz B (2020) Carotenoids and lipophilic antioxidant capacities of tomato purees as affected by high hydrostatic pressure processing. Int J Food Sci Technol 55:65–73
59. Carla MS, Justyna S, Krystian M, Carlos A, Pinto RS, Inácio PM-B, Francisco JB, Jose ML, Jorge AS, Antonio JM (2019) Effect of high-pressure processing on carotenoids profile, colour, microbial and enzymatic stability of cloudy carrot juice. Food Chem 299(30):112–125
60. Sánchez Moreno C, Plaza L, de Ancos B, Cano MP (2006) Impact of high pressure and traditional thermal processing of tomato purée on carotenoids, vitamin C and antioxidant activity. J Sci Food Agric 86:171–179
61. Patras A, Brunton NP, Da Pieve S, Butler F (2009) Impact of high pressure processing on total antioxidant activity, phenolic, ascorbic acid, anthocyanins content and colour of strawberry and blackberry purees. Innovative Food Sci Emerg Technol 10:308–313
62. Yan B, Martinez-Monteagudo SI, Cooperstone JL, Riedl KM, Schwartz SJ, Balasubramaniam VM (2017) Impact of thermal and pressure-based technologies on carotenoid retention and quality attributes in tomato juice. Food Bioprocess Technol 10:808–818
63. Benjamin F, Christina J, Nina E, Silke M, Eberhard S, Roland M-P, Rainer F, Martin H, Udo SG (2008) Cells under pressure – treatment of eukaryotic cells with high hydrostatic pressure, from physiologic aspects to pressure induced cell death. Curr Med Chem 15(23):2329–2336
64. Anon. High pressure processing of food. Avure.com/.../hpp-foodprocessing.asp. Accessed May 2008
65. Mohamed MEA, Amer Eissa AH (2012) Pulsed electric fields for food processing technology, Structure and Function of Food Engineering, Ayman Amer Eissa, IntechOpen, https://doi.org/: 10.5772/48678.
66. Zulueta A, Barba FJ, Esteve MJ, Frígola A (2010) Effects on the carotenoid pattern and vitamin A of a pulsed electric field-treated orange juice-milk beverage and behavior during storage. Eur Food Res Technol 231:525–534
67. Vallverdú Queralt A, Odriozola Serrano I, Oms Oliu G, Lamuela-Raventós RM, Elez Martínez P, Martín- Belloso O (2013) Impact of high- intensity pulsed electric fields on carotenoids profile of tomato juice made of moderate-intensity pulsed electric field treated tomatoes. Food Chem 141:3131–3138
68. Torregrosa F, Cortés C, Esteve MJ, Frígola A (2005) Effect of high-intensity pulsed electric fields processing and conventional heat treatment on orange carrot juice carotenoids. J Agric Food Chem 53:9519–9525
69. Cortés C, Torregrosa F, Esteve MJ, Frígola A (2006) Carotenoid profile modification during refrigerated storage in untreated and pasteurized orange juice and orange juice treated with high intensity pulsed electric fields. J Agric Food Chem 54:6247–6254
70. Cortés C, Esteve MJ, Rodrigo D, Torregrosa F, Frígola A (2006) Changes of colour and carotenoids contents during high intensity pulsed electric field treatment in orange juices. Food Chem Toxicol 44:1932–1939
71. Odriozola-Serrano I, Aguiló-Aguayo I, Soliva-Fortuny R, Gimeno-Añó V, Martín-Belloso O (2007) Lycopene, vitamin C, and antioxidant capacity of tomato juice as affected by high-intensity pulsed electric fields critical parameters. J Agric Food Chem 55:9036–9042
72. Oms Oliu G, Odriozola Serrano I, Soliva Fortuny R, Martín Belloso O (2009) Effects of high intensity pulsed electric field processing conditions on lycopene, vitamin C and antioxidant capacity of watermelon juice. Food Chem 115:1312–1319
73. Isabel O-S, Robert S-F, Teresa H-J, Olga M-B (2009) Carotenoid and phenolic profile of tomato juices processed by high intensity pulsed electric fields compared with conventional thermal treatments. Chemistry 112:258–266

74. Moraes M, Carvalho JMP, Silva CR, Cho S, Sola MR, Pinho SC (2013) Liposomes encapsulating beta-carotene produced by the proliposomes method: characterisation and shelf life of powders and phospholipid vesicles. Int J Food Sci Technol 48(2):274–282
75. Aguiló-Aguayo I (2013). Pulsed Electric Fields in Food Production, http://www.thebeefsite.com/articles/3710/pulsed-electric-fields-in-food-production/
76. Abbas S, Bashari M, Akhtar W, Li WW, Zhang X (2014) Process optimization of ultrasoundassisted curcumin nanoemulsions stabilized by OSA-modified starch. Ultrason Sonochem 21(4):1265–1274
77. Sari T, Mann B, Kumar R, Singh R, Sharma R, Bhardwaj M, Athira S (2013) Preparation and characterization of nanoemulsion encapsulating curcumin. Food Hydrocoll 43:540–546
78. Mason TJ, Riera E, Vercet A, Lopez-Bueza P (2005) Application of ultrasound. In: Sun DW (ed) Emerging technologies for food processing. Elsevier Ltd, Cambridge, MA, pp 323–351
79. Jambrak AR, Mason TJ, Paniwnyk L, Lelas V (2007) Accelerated drying of button mushrooms, brussels sprouts and cauliflower by applying power ultrasound and its rehydration properties. J Food Eng 81:88–97
80. Shahin R, Mohamed K. Sustainable Food Systems from Agriculture to Industry, 2018
81. Zeeshan AKM (2015) Protection of foods from microbes: a review on food preservation. Int J Pharm Biomed Res 2(1):13–18
82. Florin I, Irina G, Veronica L, Carmen C, Lia MD, Alexandru MG, Alina MH (2017) Nanostructurated materials for prolonged and safe food preservation 9:305–335
83. Gama JJT, de Sylos CM (2007) Effect of thermal pasteurization and concentration on carotenoid composition of Brazilian Valencia orange juice. Food Chem 100:1686–1690
84. Lee HS, Coates GA (2003) Effect of thermal pasteurization on Valencia orange juice color and pigments. Lebens-Wiss Technol 36:153–156
85. Gharsallaoui A, Roudaut G, Chambin O, Voilley A, Saurel R (2007) Applications of spray drying in microencapsulation of food ingredients: an overview. Food Res Int J 40:1107–1121
86. Shahidi F, Han XQ (1993) Encapsulation of food ingredients. Crit Rev Food Sci Nutr 33:501–547. Shi I, Le Maguer M (2000) Lycopene in tomato: chemical and physical properties affected by food processing. Crit Rev Biotechnol 42:293–334
87. Augustin MA, Sanguansri L (2008) Encapsulation of bioactives. In: Food materials science. Springer, New York, pp 577–601
88. Madhavi DL, Kagan DI (2008) Bioavailable carotenoid-cyclodextrin formulations for softgels and other encapsulation systems 7(446):101B1
89. Rajabi H, Ghorbani M, Jafari SM, Sadeghi Mahoonak A, Rajabzadeh G (2015) Retention of saffron bioactive components by spray drying encapsulation using maltodextrin, gum Arabic and gelatin as wall materials. Food Hydrocoll 51:327–337
90. Mahdavi SA, Jafari SM, Ghorbani M, Assadpoor E (2014) Spray-drying microencapsulation of anthocyanins by natural biopolymers: a review. Dry Technol 32(5):509–518
91. Nunes IL, Mercadante AZ (2007) Encapsulation of lycopene using spray-drying and molecular inclusion processes. Braz Arch Biol Technol 50:893–900
92. Özkan G, Bilek SE (2014) Microencapsulation of natural food colourants. Int J Nutr Food Sci 3(3):145–156
93. Kandansamy K, Somasundaram PD (2012) Microencapsulation of colors by spray drying-a review. Int J Food Eng 8(2):1–15
94. Bustos Garza C, Yáñez Fernández J, Barragán Huerta BE (2013) Thermal and pH stability of spray dried encapsulated astaxanthin oleoresin from Haematococcus pluvialis using several encapsulation wall materials. Food Res Int J 54:641–649
95. Barbosa MIMJ, Borsarelli CD, Mercadante AZ (2005) Light stability of spraydried bixin encapsulated with different edible polysaccharide preparations. Food Res Int J 38:989–994
96. Zhang Y, Zhong Q (2013) Encapsulation of bixin in sodium caseinate to deliver the colorant in transparent dispersions. Food Hydrocolloids J 33:1–9

97. Robert P, Carlsson RM, Romero N, Masson L (2003) Stability of spray dried encapsulated carotenoid pigments from rosa mosqueta (Rosa rubiginosa) oleoresin. J Am Oil Chem Soc 80:1115–1120
98. Rascón MP, Beristain CI, García HS, Salgado MA (2011) Carotenoid retention and storage stability of spray-dried encapsulated paprika oleoresin using gum Arabic and Soy protein isolate as wall materials. LWT Food Sci Technol 44:549–557
99. Kha TC, Nguyen MH, Roach PD, Stathopoulos CE (2014) Microencapsulation of Gac oil: optimization of spray drying conditions using response surface methodology. Powder Technol 264:298–309
100. Dorfman MR (2018) Handbook of environmental degradation of materials, 3rd edn. (2012) Handbook of environmental degradation of materials, 2nd edn
101. Claude PC, Patrick F (2007) Microencapsulation for the improved delivery of bioactive compounds into foods. Curr Opin Biotechnol 18:184–190
102. María L, Zambrano Z, Ricardo G-R, Néstor M-M, Verónica M-L, Tania F, Bernal C, Susana M-E, David Q-G (2018) Nanosystems in edible coatings: a novel strategy for food preservation. Int J Mol Sci 19:705
103. Aebischer P, Wahlberg L, Tresco PA, Winn SR (1991) Macroencapsulation of dopamine secreting cells by co-extrusion with an organic polymer solution. Biomaterials 12:50–56
104. Shimano K, Kondo O, Miwa A, Higashi Y, Koyama I, Yoshida T, Ito Y, Hirose J, Goto S (1995) Evaluation of uniform sized microcapsules using vibrational nozzle method. Drug Dev Ind Pharm 21(3):331–347
105. Zhang Y, Rochefort D (2010) Comparison of emulsion and vibration nozzle methods for microencapsulation of laccase and glucose oxidase by interfacial reticulation of poly(ethyleneimine). J Microencapsul 27(8):703–713
106. Saad N, Saleh SB (2016) Reduction of fine particle emission from a prilling tower using cfd simulation. Chem Eng Res Des 109:171–179
107. Suganya V, Anuradha V (2017) Microencapsulation and nanoencapsulation: a review. Int J Pharm Clin Res 9(3):233–239
108. Ezhilarasi PN, Karthik P, Chhanwal N, Anandharamakrishnan C (2013) Nanoencapsulation techniques for food bioactive components: a review. Food Bioprocess Technol 6:628–647
109. Afsaneh T, Seid MJ (2019) Nanostructures of gums for encapsulation of food ingredients, Jafari, S. M. Biopolymer nanostructures for food encapsulation purposes (First edition.). Academic Press, an imprint of Elsevier.
110. Pérez-Masiá R, López-Nicolás R, Periago MJ, Ros G, Lagaron JM, López-Rubio A (2015) Encapsulation of folic acid in food hydrocolloids through nanospray drying and electrospraying for nutraceutical applications. Food Chem 168:124–133
111. Gómez-Mascaraque LG, Perez-Masiá R, González-Barrio R, Periago MJ, López-Rubio A (2017) Potential of microencapsulation through emulsion-electrospraying to improve the bioaccessibility of β-carotene. Food Hydrocoll 73:1–12
112. Jia Z, Dumont MJ, Orsat V (2016) Encapsulation of phenolic compounds present in plants using protein matrices. Food Biosci 15:87–104
113. Zuidam NJ, Shimoni E (2010) Overview of microencapsulates for use in food products or processes and methods to make them. In: Encapsulation technologies for active food ingredients and food processing. Heidelberg, pp 3–29
114. Jafari SM (2017) Chapter 1 – an introduction to nanoencapsulation techniques for the food bioactive ingredients. In: Nanoencapsulation of food bioactive ingredients. Academic, pp 1–62
115. Rezaei A, Nasirpour A, Fathi M (2015) Application of cellulosic nanofibers in food science using electrospinning and its potential risk. Compr Rev Food Sci Food Saf 14:269–284
116. Celebioglu A, Uyar T (2011) Electrospinning of polymer-free nanofibers from cyclodextrin inclusion complexes. Langmuir 27:6218–6226
117. Celebioglu A, Sen HS, Durgun E, Uyar T (2016) Molecular entrapment of volatile organic compounds (Vocs) by electrospun cyclodextrin nanofibers. Chemosphere 144:736–744

118. Zamani M, Prabhakaran MP, Ramakrishna S (2013) Advances in drug delivery via electrospun and electrosprayed nanomaterials. Int J Nanomedicine 8:2997–3017
119. Szentivanyi A, Chakradeo T, Zernetsch H, Glasmacher B (2011) Electrospun cellular microenvironments: understanding controlled release and scaffold structure. Adv Drug Deliv Rev 63:209–220
120. Camerlo A, Vebert-Nardin C, Rossi RM, Popa AM (2013) Fragrance encapsulation in polymeric matrices by emulsion electrospinning. Eur Polym J 49(12):3806–3813
121. Yang Y, Li X, Qi M, Zhou S, Weng J (2008) Release pattern and structural integrity of lysozyme encapsulated in core-sheath structured poly(dl-lactide) ultrafine fibers prepared by emulsion electrospinning. Eur J Pharm Biopharm 69:106–116
122. Atefe R, Milad F, Seid MJ (2019) Nanoencapsulation of hydrophobic and low-soluble food bioactive compounds within different nanocarriers. Food Hydrocoll 88:146–162
123. Wang J, Windbergs M (2017) Functional electrospun fibers for the treatment of human skin wounds. Eur J Pharm Biopharm 119:283–299
124. Nikmaram N, Roohinejad S, Hashemi S, Koubaa M, Barba FJ, Abbaspourrad A (2017) Emulsion-based systems for fabrication of electrospun nanofibers: food, pharmaceutical and biomedical applications. RSC Adv 7:28951–28964
125. Mostafa Y, Ariene AL-K, Berdan A, Ana KB-R, Scott RW (2017) Encapsulation of grape seed extract in polylactide microcapsules for sustained bioactivity and time-dependent release in dental material applications. Dent Mater 33:630–636
126. Rehman A, Ahmad T, Aadil RM, Spotti MJ, Bakry AM, Khan IM, Tong Q (2019) Pectin polymers as wall materials for the nano-encapsulation of bioactive compounds. Trends Food Sci Technol 90:35–46
127. Assadpour E, Jafari SM (2019b) Advances in spray-drying encapsulation of food bioactive ingredients: from microcapsules to nanocapsules. Annu Rev Food Sci Technol 10:8–29
128. Arpagaus C, Collenberg A, Rütti D, Assadpour E, Jafari SM (2018) Nano spray drying for encapsulation of pharmaceuticals. Int J Pharm 546:194–214
129. Qian C, Decker EA, Xiao H, McClements DJ (2012a) Nanoemulsion delivery systems: influence of carrier oil on b-carotene bioaccessibility. Food Chem 135:1440–1447
130. Arshady R (1989) Microspheres and microcapsules: a survey of manufacturing techniques. Part I: suspension-crosslinking. Polym Eng Sci 29(24):1746–1758
131. Pavanetto F, Perugini P, Conti B, Modena T, Genta I (1996) Evaluation of process parameters involved in chitosan microsphere preparation by the o/w/o multiple emulsion method. J Microencapsul 130(6):679–688
132. Andrea C, Massimo EM, Massimo E, Maffei A, Occhipinti AO (2013) Supercritical fluid extraction of plant flavors and fragrances. Molecules 18(6):7194–7238
133. Krober H, Teipel U (2005) Microencapsulation of particles using supercritical CO_2. Chem Eng Process 44(2):215–219. Chiou AH, Cheng HC, Wang DP (2006) Micronization microencapsulation of felodipine by supercritical CO_2. J Microencapsul 23(3):265–276
134. Zuknik MH, Norulaini NAN, Omar AKM (2012) Supercritical carbon dioxide extraction of lycopene: a review. J Food Eng 112:253–262
135. Anandharamakrishnan C (2014) Techniques for nanoencapsulation of food ingredients. Springer, New York
136. Gou M, Men K, Shi H, Xiang M, Zhang J, Song J, Long J, Wan Y, Luo F, Zhao X, Qian Z (2011) Curcumin-loaded biodegradable polymeric micelles for colon cancer therapy in vitro and in vivo. Nanoscale 3:1558–1567
137. Suwannateep N, Banlunara W, Wanichwecharungruang SP, Chiablaem K, Lirdprapamongkol K, Svasti J (2011) Mucoadhesive curcumin nanospheres: biological activity, adhesion to stomach mucosa and release of curcumin into the circulation. J Control Release 151:176–182
138. Hari N, Francis S, Nair AGR, Nair AJ (2018) Synthesis, characterization and biological evaluation of chitosan film incorporated with β -Carotene loaded starch nanocrystals. Food Packag Shelf Life 16:69–76

139. Saeed AK, Marc S (2013) Improvement of nanoprecipitation technique for preparation of gelatin nanoparticles and potential macromolecular drug loading. Macromol Biosci 13(94):455–463
140. Pavlitschek T, Gretz M, Plank J (2013) Microcapsules prepared from a polycondensate based cement dispersant via layer by layer self assembly on melamine formaldehyde core templates. J Appl Polym Sci 127(5):3705–3711
141. Yakindra PT, Taiwo OA, Nauman K, Benu A, Barrow CJ (2019) Complex coacervation: principles, mechanisms and applications in microencapsulation. Int J Biol Macromol 121:1276–1286
142. Wieland -Berghausen S, Schote U, Frey M, Schmidt F (2002) Comparison of microencapsulation techniques for the water -soluble drugs nitenpyram and clomipramine HCl. J Control Release 85(1–3):35–43
143. Burgess DJ, Carless JF (1985) Manufacture of gelatin/gelatin coacervate microcapsules. Int J Pharm 27:61–70
144. Takenaka H, Kamashima Y, Lin SY (1980) Micromeritic properties of sulfamethoxazole microcapsules prepared by gelatin-acacia coacervation. J Pharm Sci 69(5):513–516
145. Silva DF, Favaro-Trindade CS, Rocha GA, Thomazini M (2012) Microencapsulation of lycopene by gelatin–pectin complex coacervation. J Food Process Preserv 36(2):185–190
146. Rocha-Selmi GA, Favaro-Trindade CS, Grosso CR (2013a) Morphology, stability, and application of lycopene microcapsules produced by complex coacervation. J Chem 2013:1–7

Chapter 8
Stability of Carotenoids

Sana Aslam, Matloob Ahmad, and Muhammad Riaz

8.1 Introduction

Carotenoids are natural lipophilic pigments synthesized by plants and some microorganisms, which play an important function in the photosynthetic apparatus [1, 2]. They are primarily categorized into two basic classes i.e. carotenes or hydrocarotenoids and xanthophylls or oxycarotenoids (oxygenated derivatives) [3]. Carotenoids are basically present in binding form with in the chloroplasts. The biosynthesis of carotenoids occur partially in the chloroplasts and further specific steps are completed within cytoplasm in the presence of phytoene synthase enzyme. The environmental stresses played a significant role in the expression of this enzyme [4]. These miscellaneous natural pigments impart an important role in the process of photosynthesis by protecting the chlorophyll against near-UV or visible light and also stabilizes the cell membrane by binding with free radicals [5, 6]. Carotenoids have been proven health-promoting effects like role as vitamin A precursor. Carotenoids showed potent therapeutic potential like anti-tumour [7, 8], antioxidant [9], anticardiac [10], anti-inflammatory [11] and anti-ageing [12] properties. Keeping in view its potential properties and diverse functions especially in food and pharmaceutical industries; many researchers focused on processing of carotenoids in vegetables and fruits. To understand the

S. Aslam
Department of Chemistry, Government College Women University, Faisalabad, Pakistan

M. Ahmad (✉)
Department of Chemistry, Government College University, Faisalabad, Pakistan
e-mail: Matloob.Ahmad@gcuf.edu.pk

M. Riaz
Department of Pharmacy, Shaheed Benazir Bhutto University, Sheringal, Pakistan

© Springer Nature Switzerland AG 2021

M. Zia-Ul-Haq et al. (eds.), *Carotenoids: Structure and Function in the Human Body*, https://doi.org/10.1007/978-3-030-46459-2_8

stability of carotenoids in food processing and storage methodologies one must have knowledge about their chemical structures and related isomerization, possible structural rearrangements and degradation mechanisms. The natural profile of carotenoids is altered unavoidably during food processing due to their easily oxidizing nature, chemical instabiliy and less resistance to the processing stresses. Additionally, various unfavourable conditions like side reactions induced by heat or light, acid preservatives, unfavourable pH, less water solubility, air oxidation as well as digestive environment during many processes are among the major reasons of their limited food processing [13, 14]. Therefore, in order to provide them as a functional food with good yield to humans and to increase their bioavailability as well as bioaccessibility and minimizing conversions or losses of carotenoids, there is a need of more realistic approach to meet the current challenges of food and pharmaceutical industries. They are extensively processed in the form of fruit juices, jams, and jellies etc. The traditional way of food processing includes addition of sugar and/or chemicals, freezing, drying, juice extraction, pulping, and pasteurization etc. but the major drawbacks of these methodologies are significant loss and alteration of carotenoids. In this regard, the role of nanocarriers in food technology and pharmaceutical industry is significantly enhanced in modern research era. Nanoencapsulation technique for preservation of carotenoids by using biopolymeric nanocarriers like polysaccharides and proteins as well as lipid supported nanocarriers are the most recent advances in the field of food technology.

8.2 Factors Effecting the Chemical Stability of Carotenoids in Foods

The physicochemical characteristics of carotenoids are directly related to their chemical backbone which mainly consist of conjugated polyene chain with extended delocalized π-electrons system. This chemical backbone is more susceptible to the degradation process especially when they are extracted from a biological source. This polyene chain is terminated with cyclic end-groups and may also be substituted with oxygen containing functional groups at the ring systems. The nature and position of these functional groups ultimately determines the polarity as well as biological potential of these natural tetraterpenes. Carotenoids are extremely hydrophobic in nature and usually present in the interior of membranes and other hydrophobic sites where they play a role as antioxidants and protect the plant tissues from damage induced by oxygen and light [15]. The understanding of oxidation mechanisms is mendatory to design the most suitable delivery system to protect these natural compounds in functional food products. The terminal double bonds are found to be more susceptible to the oxidation process. Carotenoids are easily degraded oxidation induced by light, heat, oxygen, acids, and transition metals etc. The processing techniques for carotenoids fortified the functional foods with best product quality.

The resulting products led the researchers to developed novel and improved technologies to optimize stability of carotenoid in functional foods.

8.2.1 *Oxidation of Carotenoids*

Oxidation process of carotenoids is relatively fast with atmospheric oxygen especially during processing of carotenoids in organic solvents. Once the oxidation reaction starts, many carotenoid skeleton based radical species generate which not only reacts with themselves but also with other chemical species present in the reacting medium. As a result, a numbers of by- products are formed which decrease the quality of fortified carotenoids food i.e. colour loss, rancidity etc. as well as loss of its bioactivity. General mechanism of carotenoids oxidation is summarized in Fig. 8.1.

8.2.2 *Autoxidation of β-Carotene*

Reaction of beta-carotene with atmospheric oxygen in dark at 30 °Cusing tetrachloromethane (CCl_4) or benzene as solvent medium, autoxidation starts within a period of one hour with productions of many active oxidative products and beta-carotene undergo oxidative damage completely within 30 hours. The free radical reaction mechanism was confirmed with the addition of free radical reaction initiator like AIBN (2,2-azo-bis-isobutyronitrile). As a result, the rate of reaction increases which was further verified by addition of antioxidant reagents like α-tocopherol and BHT (butylated hydroxytoluene) which reduces the oxidation rate. Spectroscopic (i.e., FT-IR, and GC-MS) and HPLC analysis identifies that more than 20 oxidative products are formed.

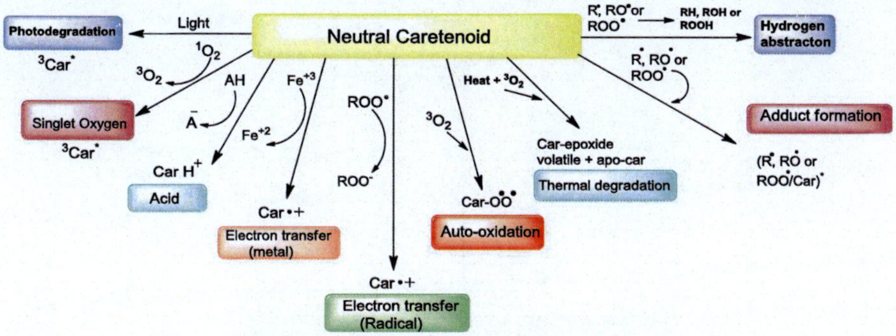

Fig. 8.1 Mechanism of carotenoid oxidation and their initial products

8.2.2.1 Mechanistic Pathways of Autoxidation of β-Carotene

Oxidative pathway starts with the formation of different structural isomers in which molecule twists with opposite electronic unpaired spin state followed by fast reaction with oxygen leading to the carbon-peroxyl triplet biradicals. The resultant biradical is either changed to endo-peroxides or it further reacts with another neutral molecule of beta-carotene to form corresponding epoxide and alkoxyl radical of carotene. These intermediates further undergo oxidation to forms different carbonyl derivatives, carboxylic acid and carbon dioxide [16] (Fig. 8.2).

Another research evidence of the oxidative degradation was observed by flowing oxygen-saturated water over the C_{18} solid phase supported beta-carotene. The degradation products include 9-*cis,* 13-*cis*, di-*cis* isomers, beta-carotene 5,8-endo-peroxide, beta-carotene 5,8-epoxide, beta-apo-13-carotenone and beta-apo-14-carotenal detected by electrospray LC-MS and UV-Vis spectroscopy [17] (Fig. 8.3).

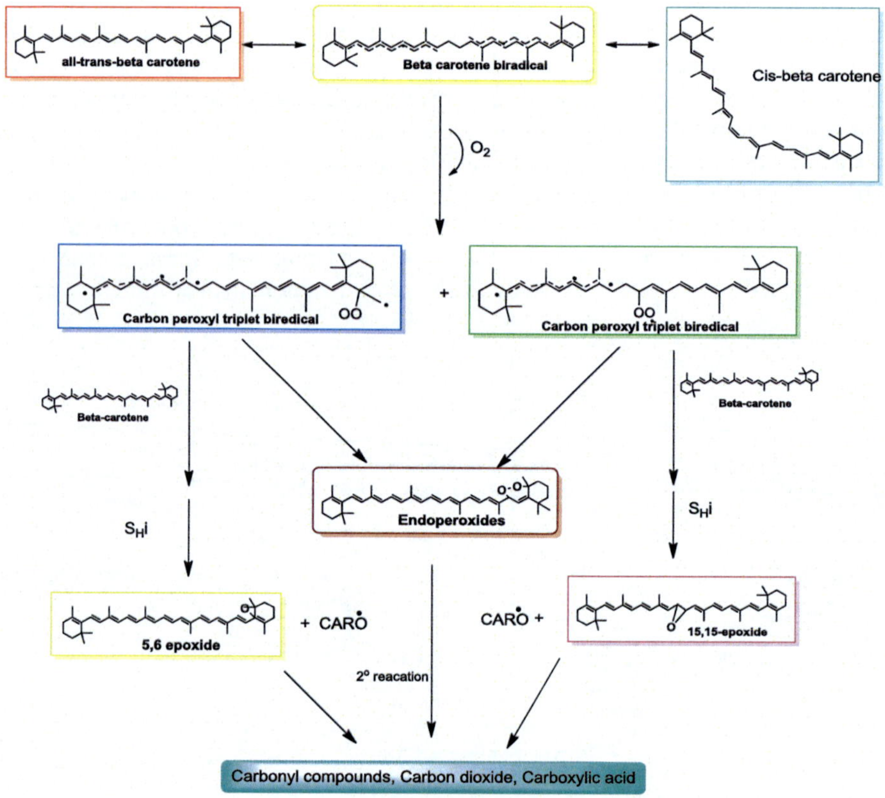

Fig. 8.2 Potential pathways of autoxidation of β-carotene [16]

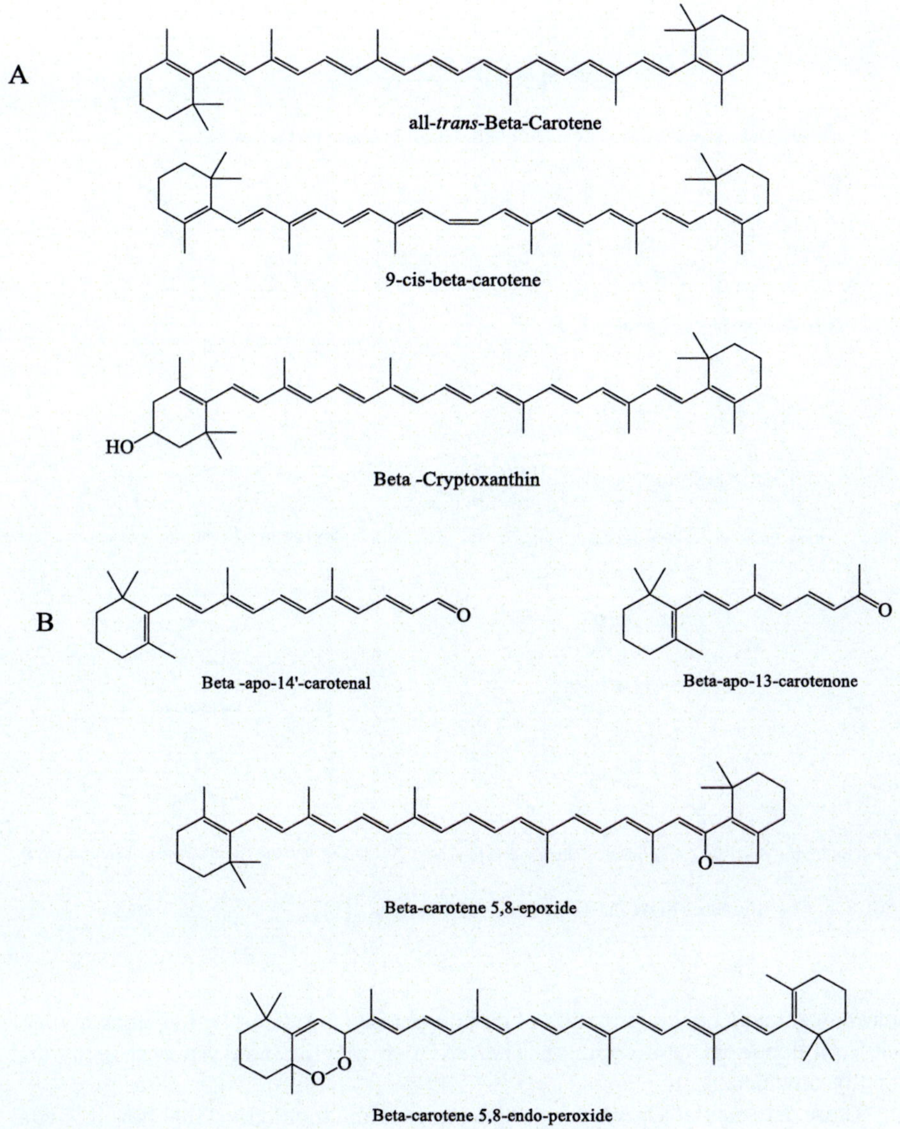

Fig. 8.3 Chemical structures of (**a**) carotenoids studied and (**b**) tentatively identified *β*-carotene degradation products [17]

Provitamin-A carotenoids derivative i.e. Retinoic acid, [18] directly reacts with oxygen in triplet state using 90% ethanol as reaction medium at 25 °C. The oxidative products result by direct addition of oxygen in the presence of BMP (2,6-di-tert-butyl-4-methylphenol) used as peroxyl radical scavenger. The structures of oxidative products are shown in Fig. 8.4 [19]. The proposed oxygenation reaction

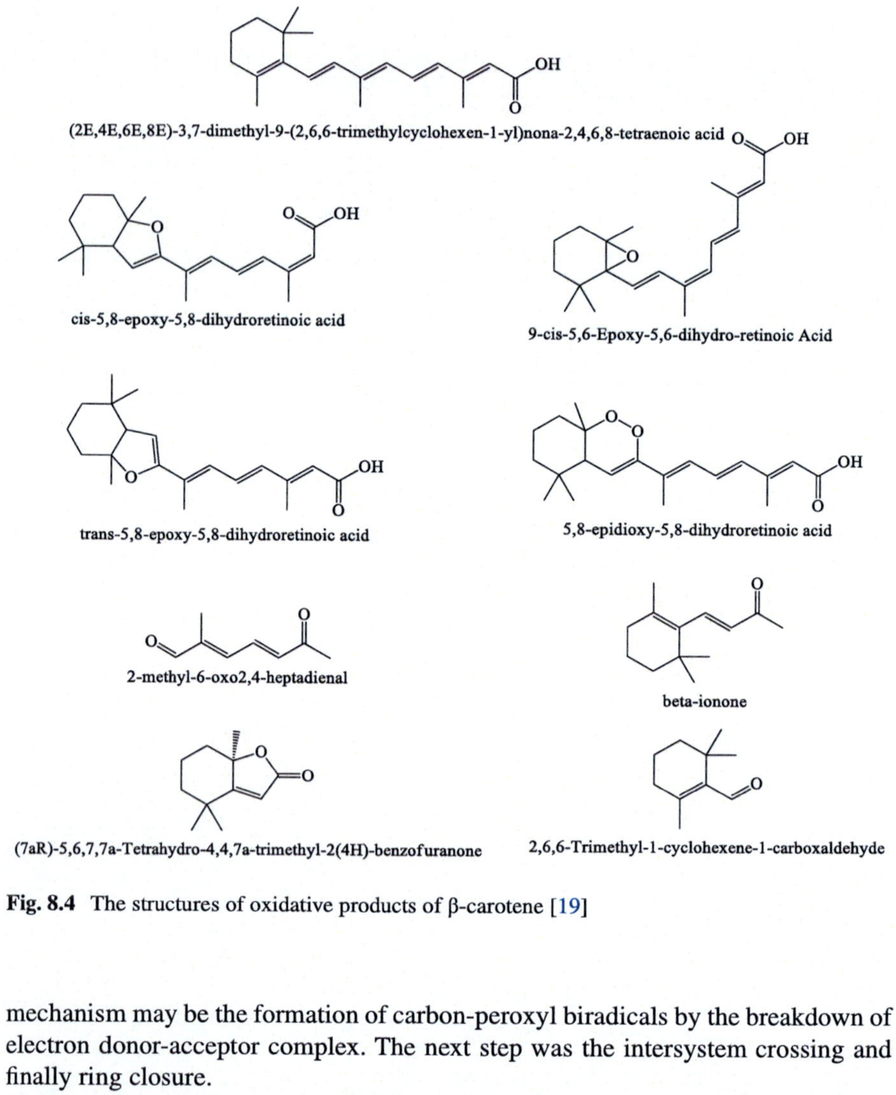

Fig. 8.4 The structures of oxidative products of β-carotene [19]

mechanism may be the formation of carbon-peroxyl biradicals by the breakdown of electron donor-acceptor complex. The next step was the intersystem crossing and finally ring closure.

These detrimental aspects of carotenoid oxidation highlighted that how it is vital to have a critical knowledge about the oxidation mechanisms so that we can be optimize the processing conditions with minimum degradation damage.

8.2.3 *Thermal Degradation of Carotenoids*

A large number of volatile and non-volatile compounds are formed during the heating of carotenoids in the presence of O_2 (air) [20]. Beta-carotene is thermally degradated at 97 °C in air, resulting in a variety of derivatives, as shown in the

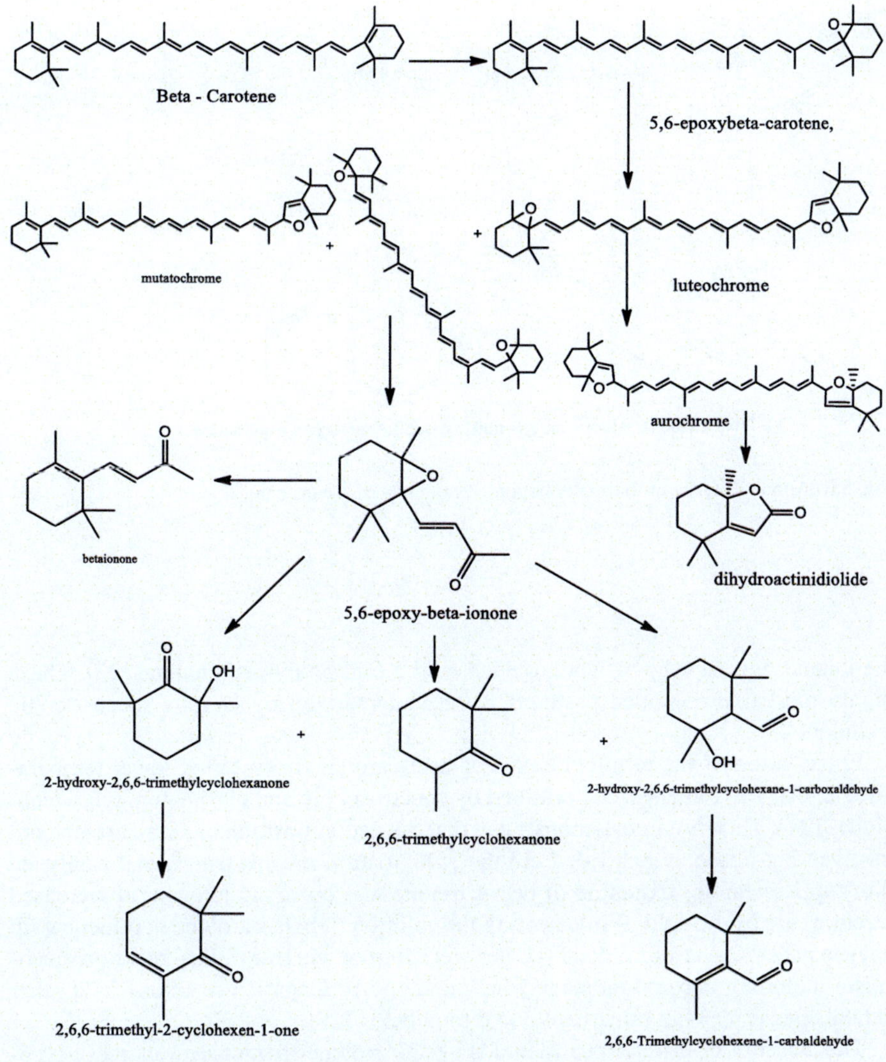

Fig. 8.5 Thermally degraded products of β-carotene [21]

Fig. 8.5 [21]. Heating the β-carotene at elevated temperature i.e. 180 °C for 2 hours, variety of *cis* isomers along with many oxidation products formed. The degradation mechanism of β-carotene molecule starts at its both terminal ends and moves towards centre [22].

The rate of oxidation increases as the amount of the air increased due to the direct interaction of beta-carotene molecules with oxygen. At higher temperature i.e. 240 °C crystalline beta-carotene under vacuum interacts with oxygen resulted in

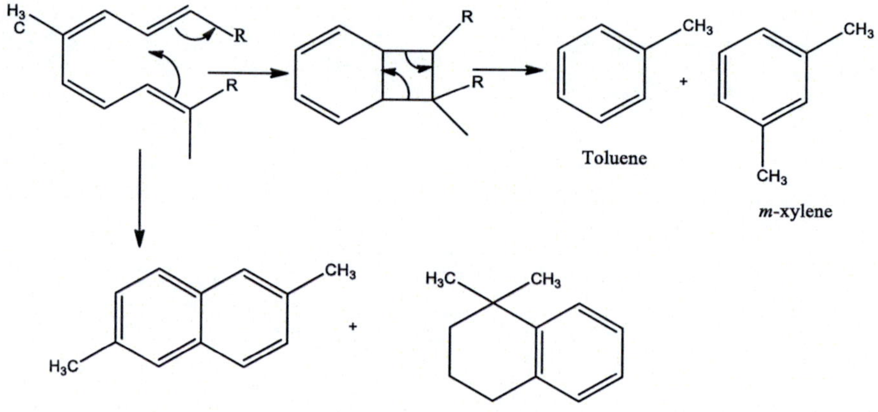

Fig. 8.6 Another layout of thermally degraded products of β-carotene [27]

the ionene, *m*-xylene, *p*-xylene, toluene, and 2,6-dimethylnaphthalene [23]. These highly oxidative degraded products were characterized by various spectroscopic techniques.

Epoxidation of the terminal bonds of beta-carotene occurred at lower temperature i.e. 60 °Cin toluene as determined by spectroscopic and chromatographic techniques [24]. This research demonstrated that the decomposition of beta-carotene not involved lag phase, which ruled out the autoxidation mechanism. Metal catalyzed (i.e. cupric stearate) oxidation of beta-carotene was observed to occur at increased reaction rate by 4.3 fold. While heating the solution of beta-carotene in toluene with oxygen at 60 °C, various radical species were formed via free radical reaction mechanism including peroxyl radicals, which undergo propagation reactions with other radical species or with other carotenoid molecules [25].

The stability of different carotenoids i.e. β-carotene (from carrot extract), lycopene (from tomato extract) and lutein (from Marigold extract) was studied by subjecting them to specific ranges of light and temperature. Prominent effect of light and temperature was observed on colour extracts. The marigold extract was the most stable under both light and temperature conditions followed by the tomato extract while carrot extract was the least stable [26]. Natural profile of carotenoid is inevitably destructed and altered at higher processing temperature with longer processing times (Figs. 8.6 and 8.7).

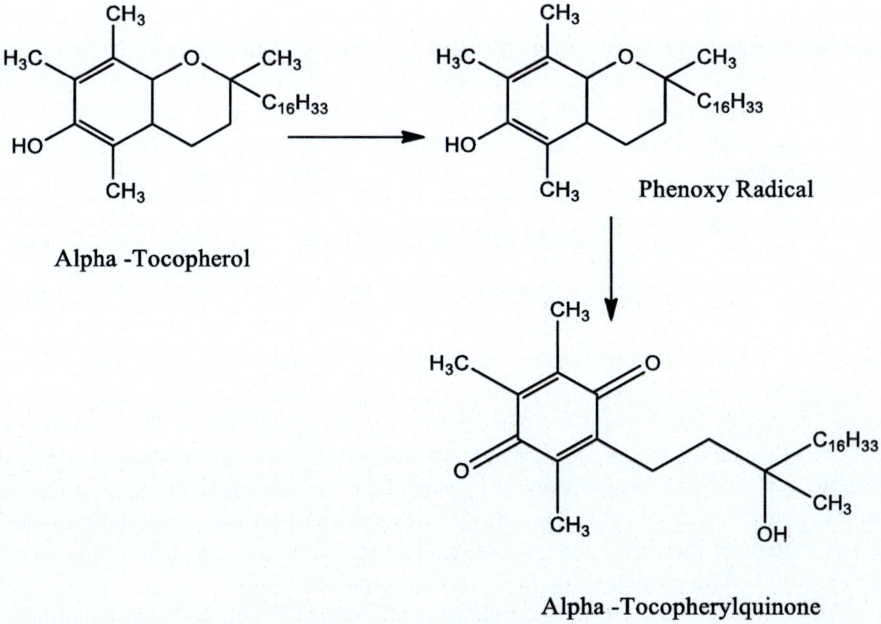

Fig. 8.7 Degradation mechanism of alpha-tocopherol [27]

8.2.4 *Photodegradation*

Carotenoids degrade easily by exposure to light. The photoactive carotenoids produces radical cations species [28]. Bleaching of beta-carotene in chloroform ($CHCl_3$) occurred by exciting the beta-carotene molecules, induced with Laser flash. Laser flash photolysis results when these excited molecules reacts directly with solvent molecule to produce either beta-carotene radical via intramolecular hydrogen abstraction or carotenoid-chloroform free radical adduct. This free radical adduct also reacts with the excited beta-carotene molecules on its return to ground state followed by the slow degradation process to give beta-carotene radical cationic species [29].

8.2.5 *Degradation by Singlet Oxygen and Free Radicals*

β-Carotene is a potent antioxidant and its mechanism of action is due to its potential to act either as a free-radicals scavenger or to quench singlet oxygen ($^1O_2$). β-Carotene quenches $^1O_2$ (singlet oxygen i.e. excited state of oxygen) mainly by following a physical type of mechanism [30], which starts with the excitation of β-carotene to triplet state by getting excitation energy of $^1O_2$ (i.e. within red frequency range of electromagnetic region) followed by its de-excitation into the ground state through dissipating this excess energy via rotational and vibrational interactions with the surrounding molecules of solvent [31–34]. The degree of quenching $^1O_2$ in benzene at 25 °C temperature was reported as 6.64–16, 12–12.6, 13 and 17 × 10^9 M^{-1} s^{-1} for lutein, zeaxanthin, β-carotene, and lycopene respectively [35, 36].

$$^1O_2 + CAR \rightarrow 3O_2 + 3CAR^* \tag{8.1}$$

$$3CAR* \rightarrow CAR + Heat \tag{8.2}$$

Moreover, a computational study by using DFT (density functional theory) has determined that the physical quenching route described in Eqs. (8.1 and 8.2), is the most favourable mechanism for carotenoids and singlet oxygen ($^1O_2$) interaction, On the other hand, the excited carotenoid molecules may also undergo chemical degradation pathway. This study suggested that the chemical reaction pathway might involved a direct attack of singlet oxygen ($^1O_2$) on the conjugated double bonds system of the carotenoid, which results in biradicals formation followed by the carbonyl chain cleavage that leads to final products [37].

Yamauchi et al. reported two products i.e. β-carotene 5,6-epoxide and β-carotene 5,8- endoperoxide as a result of chlorophyll sensitized photoxidation of β-carotene in methyl linoleate by $^1O_2$ [38] (Fig. 8.8).

β-Carotene 5,8-endo-peroxide has been proposed as the primary oxidative product. β-Carotene 5,6-epoxide is formed from oxygen-centered radical, generated during reaction with $^1O_2$, which then abstracts a hydrogen atom from the surrounding lipid medium followed by intramolecular hemolytic substitution reaction to give the final product. β-Carotene also scavenges free radicals by three basic mechanisms and produces different oxidized products (degradation) [39, 40].

In bioactive systems, like in food, lipid autoxidation may produce free radicals [34]. Carotenoids have a potential to generates both antioxidant as well as prooxidant effects in these systems comprising pre-formed radicals [41]. It was also reported that the mechanisms and rates of oxidation reactions were mainly dependent on the characteristics and stability of free radicals, oxygen concentration, nature and type of carotenoid used, as well as the polarity of reaction medium (solvent i.e. aqueous or lipid phase) [42, 43].

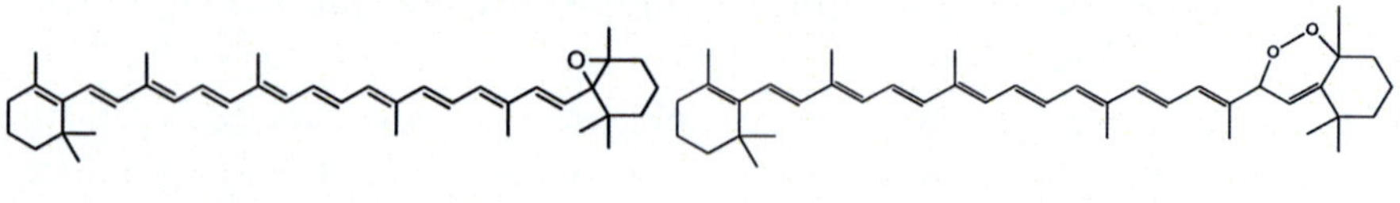

β-Carotene 5,6-epoxide β-Carotene 5,8- endo-peroxide

Fig. 8.8 Chlorophyll sensitized photoxidative products of β-carotene

In addition, β-carotene can also act as a prooxidant when the oxygen pressure (quantitatively) is sufficiently high [44, 45]. On the other hand, β-carotene is less likely to favour oxidation under biological conditions [46]. When the radicals are present in a system, carotenoids were interacted with these radicals by various pathways. These reactions include hydrogen abstraction, electron transfer as well as addition of radical species to form different carotenoid-radical adducts (Fig. 8.1) [46–48].

$$\mathrm{CAR} + \mathrm{ROO}* \rightarrow (\mathrm{ROO}-\mathrm{CAR})^{*}\,(\text{addition}) \tag{8.3}$$

$$\mathrm{CAR} + \mathrm{ROO}* \rightarrow \mathrm{CAR}*+ + \mathrm{ROO}^{-}\,(\text{electron transfer}) \tag{8.4}$$

$$\mathrm{CAR} + \mathrm{ROO}* \rightarrow \mathrm{CAR}* + \mathrm{ROOH}\,(\text{hydrogen abstraction}) \tag{8.5}$$

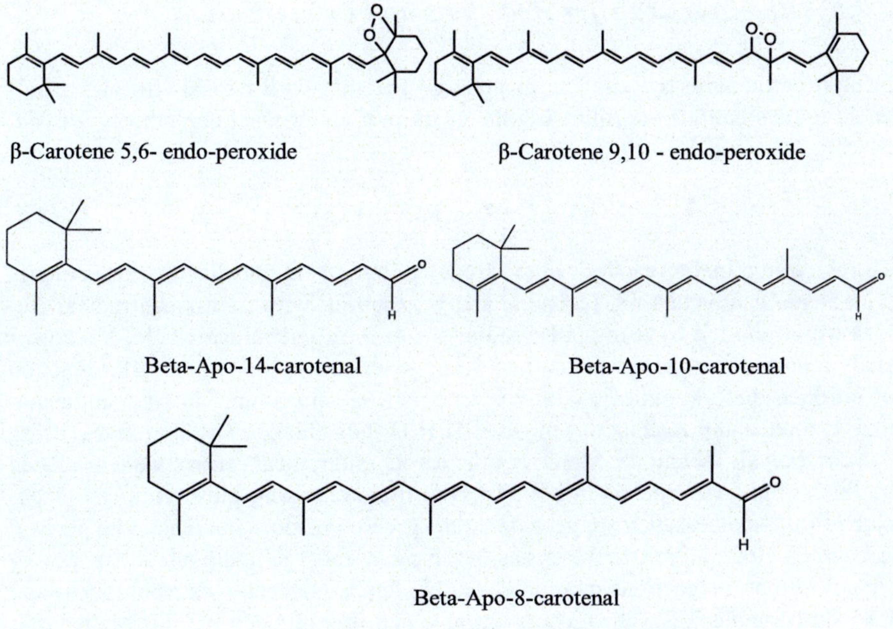

Fig. 8.9 Oxidative products formed by oxidation of β-carotene with $^{1}O_2$ [49]

The primary products of these reactions were further reacted with other reactive species by means of various reaction mechanisms e.g. with other radical interactions, to give variety of products.

Oxidation of β-carotene with $^1O_2$ was also reported to produce beta-apo-8-carotenal, beta-apo-10-carotenal, beta-apo-14-carotenal, beta-carotene 5,8-endoperoxide and beta-ionone, as identified by LCMS/MS, GC-MS, and HPLC techniques [49] (Fig. 8.9).

Chemical oxidative products including apo-6-lycopenal, and a combination of seven different short-chain oxygenated compounds like 2-methyl-2-hepten-6-one, were formed by irradiating the lycopene in the presence of methylene blue (sensitizer) and oxygen (air). The oxidative products were characterized by UV, IR, ^{1}H-NMR, ^{13}C-NMR and MS spectroscopic techniques. The mechanistic pathway led to the hypothesis that singlet oxygen was generated by methylene blue sensitizer which then reacted with lycopene to give cyclic peroxide at C5-C6. This cyclic peroxide is relatively unstable compound, could easily break cyclic structure to form the final products [50].

8.2.6 *Electron Transfer Reactions and Isomerization*

Neutral carotenoids are capable to undergo reaction with radicals to form carotenoid radical cations via following the electron transfer reaction pathway [46, 47, 51, 52].

$$ROO^{\bullet} + CAR \rightarrow ROO \;\; + CAR^{\bullet+} \tag{8.6}$$

The electron transfer reactions of different carotenoids in micelles like astaxanthin, β-carotene, canthaxanthin, lycopene and zeaxanthin, with tryptophan radical cations were reported by using pulse radiolysis to form radical cations [52]. Another study reported that the radical cations were generated by electron transfer reaction of nitrogen dioxide radicals with carotenoids (i.e. astaxanthin, lutein, canthaxanthin, lycopene and zeaxanthin) in *t*-BuOH/H_2O (tert-butanol/water) mixtures using pulse radiolysis technique. Moreover, the carotenoid radical cations were produced by thiyl-sulphonyl radicals in *t*-BuOH/H_2O mixtures using pulse radiolysis technique [48]. Another study suggested that the β-carotene does not react with peroxyl radical via electron transfer reactions, but instead undergo hydrogen abstraction or adduct formation reactions more likely, as evident from steady-state photolysis and laser flash studies [53]. The most favourable pathway of electron transfer reactions includes first the interaction of carotenoid with oxygen to give a carotenoid peroxyl radical cation (Eq. 8.7), followed by its reduction by reducing agents like ferrous iron or with another carotenoid molecule to form the respective carotenoid peroxide (Eqs. 8.8 and 8.9) [54].

$$CAR^{\bullet+} + O_2 \rightarrow CarO_2^{\bullet+} \tag{8.7}$$

$$CARO_2^{\bullet+} + Fe_2^{+} \rightarrow CARO_2 + Fe_3^{+} \tag{8.8}$$

$$CARO_2^{\bullet+} + CAR \rightarrow CARO_2 + CAR^{\bullet+} \tag{8.9}$$

The probability of electron transfer reactions may depend on the nature and type of radical and carotenoid used. The 5,8-endo-peroxides of both ethyl all-*trans*-8 -apo-β-caroten-8-oate and all-trans-β-carotenene were produce (~90% yield) by following these previously mentioned reaction pathways [54]. While, in the presence of iron, carotenoids are converted to the dimers.

$$CAR^{\bullet+} + Fe_3^{+} \rightarrow CAR_2^{+} + Fe_2^{+} \tag{8.10}$$

Carotenoid dimers of canthaxanthin was also produced by its bulk electrolysis with ferric chloride ($FeCl_3$) in dichloromethane at low temperature (i.e. –10 °C) followed by its irradiation with UV-Vis for about 1.5 min. The proposed mechanistic route for dimer formation first include the radical cation formation then its interaction with neutral carotenoids [51]. A study proposed the radical cations decay occurred during pulse radiolysis especially in a bimolecular process, while the nature of products formed were not indentified [48]. It was also reported that the carotenoid radical cations finally interact with reaction medium (solvent systems) via deprotonation reaction (Eq. 8.11). This deprotonated carotenoid radicals may either further react with oxygen (O_2) to form additional radicals (Eq. 8.12) [55] or with other deprotonated radicals to give didehydrodimers (Eq. 8.13) [54].

$$CAR^{\bullet+} \leftrightarrow *CAR^{\bullet} + H^{+} \tag{8.11}$$

$$*CAR^{\bullet} + O_2 \rightarrow *CAR - OO\bullet \tag{8.12}$$

$$*CAR^{\bullet} + *CAR^{\bullet} \rightarrow (*CAR)_2 \tag{8.13}$$

(* means Carotenoids with one less proton)

β-Carotene and canthaxanthin radical cations undergo deprotonation to give didehydrodimer in CH_2Cl_2, as determined by electrochemistry, MALDI-TOF, optical and EPR spectroscopic techniques [56]. In another study, it was confirmed by DFT that deprotonation of β-carotene radical cations occurred at 5 or 5 methyl group position on the cyclohexene ring to give more stable product [57]. β-Carotene radical cations also formed dications by undergoing electrochemical oxidation in THF (tetrahydrofuran), CH_2Cl_2 (dichloromethane) and $C_2H_4Cl_2$ (dichloroethane) solutions, as confirmed by EPR [58]. The structure and stability of the carotenoid determined. The dications were the predominant species in case of β-carotene and β-apo-8-carotenal, while radical cations produced predominantly in case of canthaxanthin, as confirmed by simultaneous electrochemical-EPR techniques [59]. It was further found that dications also undergo decay reactions either by reacting with neutral carotenoid to give two radical cations (Eq. 8.14) [60], or it may be further deprotonated with enhanced reaction rate in the water (Eq. 8.15) [51, 60–62].

$$CAR^{2+} + CAR \leftrightarrow 2CAR^{\bullet+} \tag{8.14}$$

$$CAR^{2+} \leftrightarrow *CAR^{+} + H^{+} \tag{8.15}$$

Isomerization of carotenoids thought to be mediated by carotenoid dications and radical cations, which were formed via electron transfer reactions of neutral carotenoids with either radicals or compounds or elements e.g. iron (Eqs. 8.16 and 8.17).

$$trans \quad CAR^{2+} \leftrightarrow cis \quad CAR^{2+} \tag{8.16}$$

$$trans \quad CAR^{\bullet+} \leftrightarrow cis \quad CAR^{\bullet+} \tag{8.17}$$

This is supported by the low energy barrier gaps for configural transformation of cationic species as compared to barrier gaps of neutral carotenoids [56, 63, 64]. The radical cations were first formed by oxidation of 8-apo-betacaroten-8-al and canthaxanthin followed by isomerization to form *cis*-isomers, as confirmed by optical spectroscopy and HPLC respectively [63]. Another study resulted that β-carotene and canthaxanthin gives radical cations by bulk electrolysis in CH_2Cl_2 as determined by absorption spectroscopy [64]. The further isomerization process of these *cis*- dications and radical cations occurred by reacting with neutral *trans*-carotenoids molecules to gives either *cis*-radical cations or neutral *cis*-carotenoid along with new *trans*-radical cations. These new *trans*-radical cations further isomerized to continue the reaction sequence (Eqs. 8.18 and 8.19) [63].

$$cis \quad CAR^{\bullet+} + trans \quad CAR \leftrightarrow cis \quad CAR + trans \quad CAR^{\bullet+} \tag{8.18}$$

$$cis \quad CAR^{2+} + trans \quad CAR \leftrightarrow cis \quad CAR^{\bullet+} + trans \quad CAR^{\bullet+} \tag{8.19}$$

β-Carotene and ethyl 8-apo-caroten-8-oate also isomerized to *cis*-dications and radical cations by oxidizing with ferric chloride ($FeCl_3$) which further reduced to the either *cis*-radical cation or neutral *cis*-carotenoid by ferrous iron (Fe^{2+}) and itself oxidized to ferric ion (Fe^{3+}) [54].

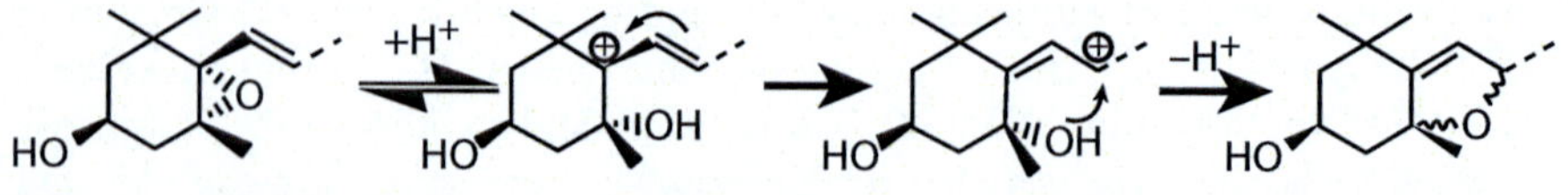

Fig. 8.10 Acid catalyzed rearrangement of a 5,6-epoxide derivative to its respective 5,8-furanoxide [67]

Isomerization of carotenoids also occur if exposed to acids. The *trans-cis* isomerization of carotenoid, radical cations formed during electron transfer reaction by acids occur via intermediate i.e. carotenoid carbocation ($CARH^+$). These carbocation intermediates were determined by calculating the rotation barriers through Austin Model 1 (AM1) and resulted that isomerization process was favourable due to lower energy barrier for rotation of radical cations than neutral carotenoids [65].

8.2.7 *Degradation of Carotenoid by Acid*

Degradation of carotenoid in the presence of acidic impurities produce ion-pairs followed by its dissociation to give respective carotenoid carbocation (Eq. 8.20) [65].

$$CAR + AH \leftrightarrow \left(CARH^{+} \ldots A-\right) \leftrightarrow CARH^{+} + A^{-} \tag{8.20}$$

β-Carotene, canthaxanthin and 8-apo-caroten-8-al produce respective carotenoid carbocations in the presences of CF_3COOH (in benzene), acetonitrile and CH_2Cl_2 solution, as confirmed by their optical spectra [65].

8-Apo-betacaroten-8-al and canthaxanthin entrapped into the sol-gels degraded by sulfuric acid (pH 3–3.5) [66]. 5,6-Epoxide derivative undergo acid catalyzed rearrangement into its respective 5,8-furanoxide Fig. 8.10 [67].

8.3 Stability of Carotenoids in Biofortified Foods

Nutritional density of different food products could be enhanced by modern biofortification process without sacrificing their properties and nutritional value, which includes conventional plant breeding, modern biotechnology and improved agronomic practices. Pro-vitamin A carotenoid based biofortification of different foods products like maize, cassava, rice, and sweet potato are underway of developmental stages [68, 69].

Biofortified maize was studied for the stability of carotenoids during storage and the results concluded that degradation rate of carotenoids was more in first week as compared to the later days [70]. The dramatic loss (80%) of carotenoids profile in biofortified orange-fleshed sweet potato (as dried chips) was observed within 4 months which is confirmed by loga-86 rithmic curve [71]. Carotenoid degradation rate is mainly effected by oxygen and temperature while the moisture had only a minor effect. A mathematical model is also developed to predict the degradation rate of *trans*-β-carotene under specific conditions of oxygen, temperature and humidity [72].

Gari is produced by grating followed by fermentation at acidic pH value and finally drying. The biofortified varieties of gari from White Cassava are very much similar in visible colour and rich in vitamin A, as prepared by addition of crude palm oil [73–75]. Abu et al. (2006) reported 57% loss of carotenoid within 4 months at ambient temperature [75]. Another research group also reported the significant loss of carotenoids i.e. about 25% and 50% after second and third week of storage at 28 °C respectively [75]. In contrast another study showed that the stability of carotenoids contents during processing of gari were retained [74]. These contradictory results demanded more detailed research to find the proper degradation and stability factors. However, the main disadvantage of adding palm oil in Cassava was the rancidity and it's not cost effective as well [76].

The comparison of the carotenoids stability at different temperature range in two varieties of cassava i.e. yellow cassava (BG) and the second one is white cassava biofortified with palm oil (RPG). They reported that 9-*cis*- and *trans*-β-carotene derivatives obey the first order kinetics (logarithmic) for degradation equation at the storage time of both varieties (i.e. BG and RPG). They studied that vitamin A activity and *trans*-β-carotene initial profile was almost double in BG as compared to RPG. The mathematical model i.e. Arrhenius and Eyring models successfully predicted the storage times between specific range of temperature. They finally concluded that the stability of carotenoids at ambient temperature was limited in cassava biofortified with palm oil [77]. The fatty acids constituents of palm oil may protect the carotenoids skeleton, So resultantly the shelf life of gari increased by the addition of red palm oil.

8.4 Encapsulation of Carotenoids

In the modern food technology, encapsulation of bioactive compounds using different types of micro/nano carrier is the most successful delivery approach for particular bioactive natural product [78]. Nano-encapsulation is considered as auspicious approach due to its significant benefits like improved bioavailability, enhanced stability, increased solubility, more loading capacity, hiding the unwanted flavours and potent efficiency of encapsulation [79]. A variety of nanocarriers like nanoliposomes, nanoemulsions, nano-hydrogels, solid lipid nanoparticles (SLNs) and nanostructured lipid carriers (NLCs) could be used for delivery and protection of carotenoids [80]. The dissolution rate is increased by nanocarriers (less than 100 nm), because a large surface area is provided to the H_2O molecules [81, 82]. The incorporation of carotenoids could be enhanced by specifying the improved parameters for the fabrication of nanocarriers. Moreover, the current state of art the encapsulation of various types of carotenoids by means of biopolymeric based nanocarriers have been tabulated and enclosed well. Nanoencapsulation has a wide range of diversity for the protection of carotenoids.

Table 8.1 Potential applications of biopolymeric nanocarriers for encapsulation of various carotenoids [88]

Nanocarriers	Encapsulating systems	Carotenoids	Size (nm)	Zeta potential (mV)	Encapsulation efficiency (%)	References
Whey protein	Nanoparticles	β-Carotene	96.2	Not reported	97	[89]
Zein-propylene glycol alginate	Composite nanoparticles	β-Carotene	~441–756	−19.63 to +34.67	37.10–69.37	[90]
Chitosan and alginate	Nanoparticles	Crocin	236	Not reported	38.16	[91]
Chitosan	Nano-structured lipid carriers	Fucoxanthin	250–400	+9.45 to −44.97	88.12–98.90	[92]
Chitosan	Nanoparticles	Astaxanthin	~92–237	~+29.3 to +32.2	Not reported	[93]
Chitosan and oligomerized (−) epigallocatechin-3-Ogallate	Nanoparticles	Lycopene	~152	~58.3	89	[94]
Carboxymethyl and chitosan	Nanoparticles	β-Carotene	~70.411552	−20.3 to −45.1	56.5–92.7	[95]
Sugar beet pectin, fish gelatin, and whey protein isolate	Nanoemulsions	Lutein	138–202	+26.5 to +50.2	Not reported	[96]
Sodium caseinate, whey protein concentrate, whey protein isolate, and whey protein hydrolysate	Nanodispersions	β-Carotene	13–1730		Not reported	[97]
OSA-modified starch	Nanosuspensions	β-Carotene	100–200	Not reported	90	[98]
Casein-graft dextran	Nanoparticles	β-Carotene	~70–200	0	4.7–54	[99]
Sodium caseinate	Nanoemulsions	Lutein	232	−37.75	93.7	[100]
Whey protein isolate	Nanoemulsions	β-Carotene and lutein	178–237	−4.08 to +13.43	Not reported	[101]
Whey protein concentrate	Nanocapsules	β-Carotene	~56.3–1764		~90	[102]
Sodium caseinate, whey protein isolate, and soybean protein isolate	Nanoparticles		78, 90, and 370	−36, −30, and − 38	99.1, 98.8, and 98.7	[103]

(continued)

Table 8.1 (continued)

Nanocarriers	Encapsulating systems	Carotenoids	Size (nm)	Zeta potential (mV)	Encapsulation efficiency (%)	References
Potato protein	Nanoparticles	Astaxanthin	<600	+33.65	85.34	[104]
Sodium alginate and calcium caseinate	Nanocapsules	β-Carotene	211	Not reported	~79.63	[105]
Whey protein isolate	Nanoemulsions	Lutein	68.8	−28.7	80.7	[106]
OSA-modified starch	Nanoemulsions	Lycopene	143–165	−19.73 to −20.77	Not reported	[107]
Whey protein isolate and quillaja saponins	Nanoemulsions	β-Carotene	140–160	~−50 to −67	Not reported	[108]
Whey protein isolate	Nanoparticles	Lycopene	100–350	−5.6	50–68.1	[109]
Sodium caseinate, whey protein isolate, and soy protein isolate	Solid lipid nanoparticles	β-Carotene	75–503	−30 to −38	98.7–99.1	[110]
Chitosan and dextran sulphate	Nanoparticles	Lutein	282–461	+31.06 to +54	60.71–77.17	[111]
Whey protein isolate	Solid lipid nanoparticles	β-Carotene	<220	−39 to +38	Not reported	[112]
Hydroxypropylmethyl cellulose phthalate	Nanocapsules	Lutein	163–219	Not reported	88.41	[113]
Octenyl succinic anhydride modified short glucan chains	Nanoparticles	Lutein	108–371	−17.42 to −38.14	~85	[114]
Chitosan	Nanoparticles	Lutein	200	Not reported	100	[115]

8.4.1 *Biopolymers Based Nanocarriers for Carotenoids Encapsulation*

Biopolymers are the best choice to be used as nanocarriers due to their significant technical and environment friendly advantages. Carbohydrates, proteins, lipids, and many other biodegradable polymers i.e. PLA, PGA etc. are commonly used natural food grade biopolymers. Their structural skeleton have close resemblance with basic human structural features and also have potent nutritional value as well, so efficiently used in manufacturing of food products [83]. They are mostly fabricated either as coating agent or wall material. The desired formulation for carotenoids encapsulation achieved by using different combination of polysaccharides with proteins via a range of nanoencapsulation strategies. The lipid supported nanocarriers are the most potent candidates for the encapsulation of carotenoids [84–87] (Table 8.1).

8.4.1.1 Polysaccharides as Nanocarriers

Polysaccharides are preferably used as a natural biomaterials due to their physiochemical profile, abundantly available in nature, cost effective, enhanced biocompatibility, high biodegradability, non-reactogeic, and non-toxic nature [116–118]. The most widely used polysaccharides based nano-materials include starches, chitosan, cellulose, alginate, pectin and gums etc. The successful application of these biopolymers for the carotenoids nanoencapsulation not only increases the stability with efficient loading or fabrication rate but also enhanced their delivery to the targeted site.

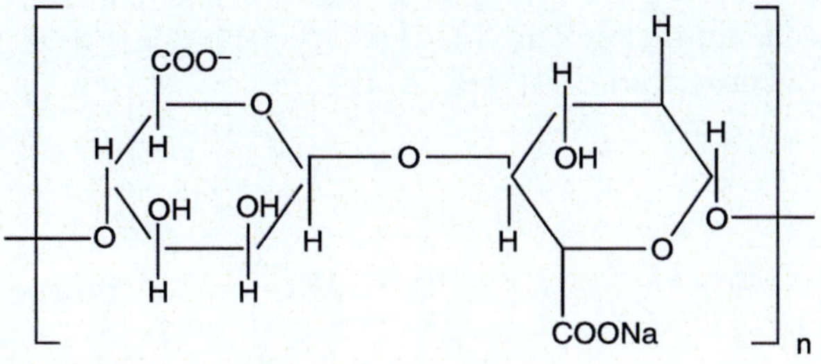

Fig. 8.11 General chemical structure of Alginate

Alginates

Alginate is a rectilinear anionic type of polysaccharide, having (1–4)-linked chemical structural backbone of α-L-guluronic (G) acid and β-D-mannuronate (M) in pyranosic conformation [119] (Fig. 8.11).

The alginates isomeric structures and their physiochemical properties depends upon the binding sequence of these polymer and β-D-mannuronate (M)/α-L-guluronic (G) acid ratio. The alginates isomeric structures are either flat ribbons shaped and buckled streamers in case of pure M and G polymer respectively, or in helix form for intermingled GM blocks. Alginate has extensive industrial applications due to its stabilising, viscosifying, gelling and water retention properties, resulting from the di-equatorial or di-axial linkages of M-blocks (flexible chain) with G-blocks (stiffest chain) which enhances the flexibility in three dimensional grid. The gelling properties could be enhanced by G chains elongation and by higher the molecular weights [120, 121].

Alginates derived from bacteria and brown seaweed i.e. Phaeophyceae but currently commercial alginates stems from algal sources e.g. marine brown algae. Alginates extracted in salt form e.g. sodium or calcium salt, using aqueous alkaline media [122]. These salt forms the non-toxic, stable, biocompatible and biodegradable gels in aqueous media which are highly compatible for encapsulation and improved delivery systems for bioactive compounds [123].

Crocin loaded nanoparticles were prepared by using chitosan and alginate. The results of this study pointed its enhanced bioavailability as well as excellent antioxidant properties [91]. β-Carotene bilayer emulsions were prepared by its fabrication on composites of chitosan and alginate. The results showed the significant improved stability against unfavourable conditions [95].

Cellulose

Cellulose is chemically repeated D-glucopyranose ring units, which are covalently bonded by β-1,4-glycosidic linkages i.e. inter-connected via C1 of one and C4 of the adjacent glucose ring. They are very stable due to their physiochemical and mechanical properties [124] (Fig. 8.12).

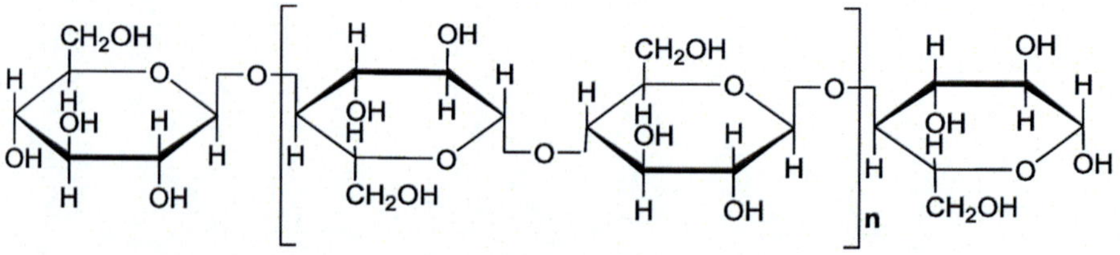

Fig. 8.12 General chemical structure of Cellulose

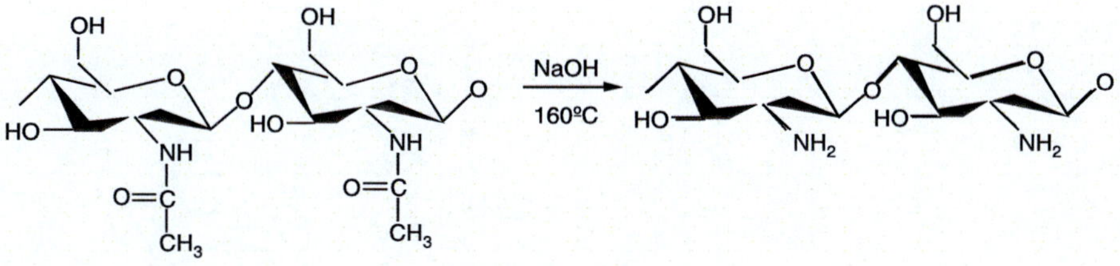

Fig. 8.13 General chemical synthesis of chitosan from chitin

Cellulose is cheap low toxic, biocompatible and biodegradable raw material for encapsulation and for the safe and suitable delivery of biomedical compounds, but the main hurdle to use as encapsulant is its less water solubility. So the big challenge is to modify it by different physiochemical and bio-chemical methods to enhance its utilization as a biopolymer encapsulant [125]. Cotton, flax, ramie, jute and wood are the most common sources of cellulose [126].

Carboxymethyl cellulose (CMC) is one of the water soluble, biocompatible and biodegradable modified cellulose that is available in market for nanoencapsulation. CMC is used for stabilizing, thickening, binding, encapsulation and making therapeutic tablets [127].

Hydroxyethyl cellulose-loaded-linoleic acid nanomicelles was used for encapsulation of β-carotene, which increases its encapsulation efficiency upto 84.67% (w/w) while its loading content upto 4.23%. *In-vitro* studies reported the continuous release of β-carotene from the cellulose-based nanomicelles in phosphate buffer for about 7 days. Therefore, nanomicelles were proved to be potential nanocarriers candidate for carotenoids [128].

Chitosan

Chitosan, a polycationic polymer, composed of D-glucosamine (deacetylated units) and *N*-acetyl-D-glucosamine (acetylated units) bonded by β (1,4) glycosidic linkage [129]. Chitosan derived from chitin via its alkaline deacetylation Fig. 8.13 [130].

The natural sources of chitin includes cuticle walls of insects and fungi and the exoskeleton of different marine arthropods. Chitosan has a wide applications not only in pharmaceuticals but also in food technology due to its potent therapeutic properties like anti-tumour [131], antioxidant [132], immune-enhancing [133], antifungal [134], anti-microbial [135], and wound healing properties [136]. Due to diversify properties, it is consider as an excellent nanocarrier for the delivery and encapsulation of different functional compounds like carotenoids. It is also considered as a potent biomedical compound due to unique features like low price, nontoxicity, biodegradability, biocompatibility and non-antigenic nature. Due to these marvellous characteristics, chitosan is an excellent option especially in food technology, drug delivery systems, protein binding and gene delivery [137, 138].

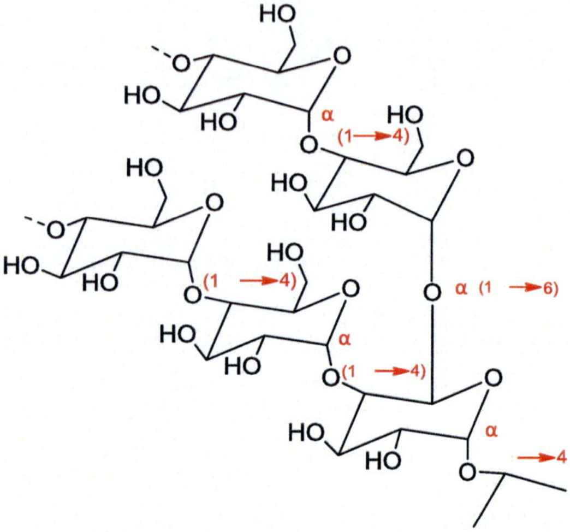

Fig. 8.14 General chemical structure of dextrins

Chitosan-coated liposomes were used for the encapsulation of four different types of carotenoids, which not only enhanced the stability and encapsulation efficiency of carotenoids but also inhibits the carotenoids leakage from this double layer [139]. Chitosan behave as a cationic polymer because of amine (NH_2) groups, thereby extensively used for the development of complex matrices with anionic polymers. These complex matrices not only enhanced the stability but also increases the efficiency of encapsulation system [93]. Marine nutraceutical carotenoid i.e. fucoxanthin with glycolipid nanogels were coated with chitosan and studied as a potent anticancer agent against human colon cancer cells. They induced apoptosis via ROS production which inhibited the BCl-2 protein linked with high level of Bax by caspase-3 cascade pathway and also inhibited the viability of Caco-2 cells [140]. The lutein bioavailability significantly enhanced by nanocapsulation with chitosan and recommended as a healthier dietary complex in food and pharmaceutical industry [141]. Another study reported that the solubility of nanoencapsulated lutein with poly-ϒ-glutamic acid and chitosan based nanoparticles significantly enhanced around about 12-times as compared to non encapsulated lutein [115].

The chitosan-ferulic acid conjugates was prepared by carbodiimide mediated coupling reaction. This conjugate have more water solubility with enhanced antioxidant activity. Then the β-carotene bilayer emulsions were designed by fabrication of CFC with propylene glycol alginate. Being oppositely charged, CFC and PGA shaped bilayer interface which covered the oil droplet of β-carotene emulsions via electrostatic force of attraction. β-Carotene exhibited enhanced stability and resistant against UV light or heat in bilayer emulsions [142].

Dextrins

Dextrins are modified starches, formed by hydrolysis. They are water soluble starches. However pyrodextrins are formed by dry heating and exists in brown, white and yellow colours [143] (Fig. 8.14).

They are used as excellent binders, coating agents and oral delivery carriers for biomedical compounds [144]. β-Carotene fabricated nanoemulsions were prepared by grafting on the whey protein isolate and dextrin conjugates. The dextran impart significant stability to β-carotene as compared to whey protein isolate and also avoids the flocculation and aggregate formation during storage at 25 and 50 °C for 30 days [145].

Octenyl Succinic Anhydride (OSA) Modified Starches

OSA treated native starch (hydrophilic) behave as a amphiphilic compound after esterfication with OSA, that offer excellent emulsifying, interfacial, gelling, encapsulating and filming characteristics [146]. Successful carotenoids loading was furnished on OSA modified starches as encapsulant [147, 148]. Recently, OSA modified starch (purity gum ultra) used to prepare the nanoemulsions of beta-carotene and apha-tocopherol, as a result, the stability profile of these biomedical compounds increase i.e. 42% for beta-carotene and 90% for apha-tocopherol as comparative to Tween 80 during 4 weeks of storage period [147]. Liang et al. 2013 reported the higher stabilization and retention of β-carotene in the form of nanoemulsions fabricated with OSA modified starch i.e. Hi-cap 100 in comparison with bulk oil [148]. While multilayer β-carotene-grafted nanoemulsions were prepared by its fabrication with Hi-cap 100 (OSA modified starch) and chitosan. This nanoemulsion showed a considerable stability and protection during 30 days of storage [149]. In another investigation, loading capacity and encapsulation efficiency of lycopene-loaded nanoemulsions were enhanced by fabrication of OSA modified starch [150].

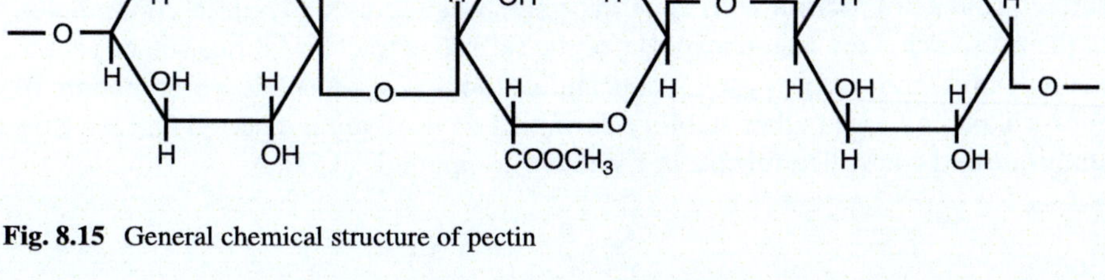

Fig. 8.15 General chemical structure of pectin

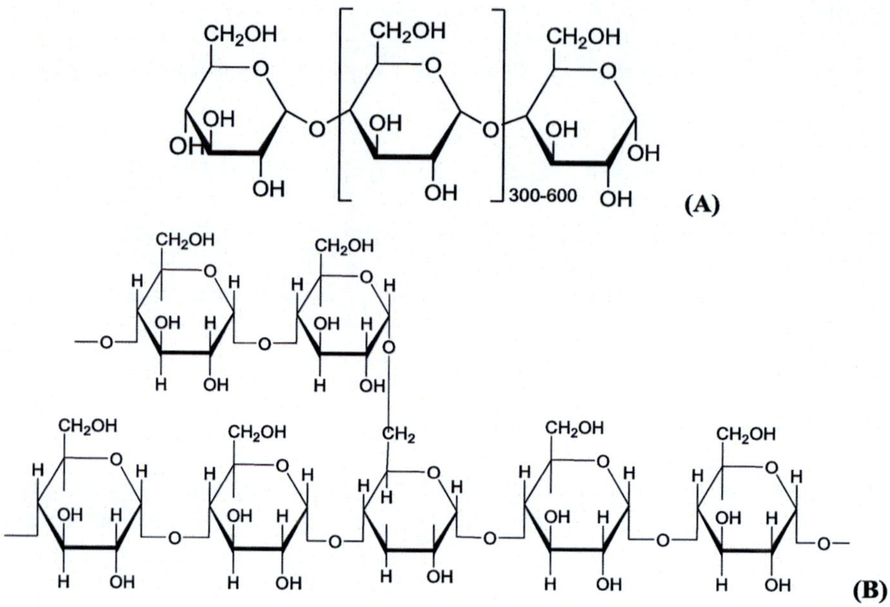

Fig. 8.16 General Chemical Structure of amylose (**a**) and amylopectin (**b**)

Pectins

Pectins are structurally methyl ester of poly-galacturonic acid, having at least 17 different monosaccharides [151]. Pectins are comprised of D-galacturonic acid and D-galactose as leading element which then tailed by L-arabinose [152, 153] (Fig. 8.15).

The main sources of pectin are apple pomace, citrus peel, sunflower and sugar beet [152]. They are classified as low and high methoxy pectin with degree of esterification (DE) i.e. DE $<$ 50% and DE $>$ 50% respectively [154]. Pectins are cheap, non-toxic, remarkably stability, due to significant functionalities pectins are consider as a potential candidates for encapsulation and for controlled release of bioactive compounds [82]. Moreover, whey protein isolate grafted on beet pectin to prepare stabilized emulsions, which showed controlled release means for β-carotene [155]. Double layered nanoemulsion was prepared by fabrication of whey protein concentrate and pectin on saffranal, crocin or picrocrocin. Owing to this nanoencapsulation system, efficient loading capacity (~97%), greater stability with minimum surface contact of saffron with apparent plane was reported [156]. In continuation of their research, they nanoencapsulated the saffron extract by fabrication of pectin, maltodextrin and whey protein concentrate combination. This combination of encapsulant provides higher stability during 22 days of storage period and *in-* vitro study showed controlled release of bioactive compounds [157].

Starches

Starches are the natural carbohydrates, having large number of glucose units bonded by glucosidic linkages [158]. They are classified into two main types, first one is the amylose and the second one is the amylopectin. More than 99% of amylose is naturally present as a linear polymer with α-1,4 -linkage and only 1% structurally branched having linked points are α-1,6-linkages. However 95% of amylopectin consist of branched structure with α-1,4 linked amorphous phase core and 5% comprised of α -1,6 linkages of 20–30 glucose units which make the exterior crystalline structural backbone of starch pellets [159]. The basic composition of starch is 75–80% of amylopectin and 20–25% of amylose [158] (Fig. 8.16).

Starch is also good and cheap choice for encapsulation of many bioactive compounds. But it's not a good emulsifier in its raw form so it first physicochemically modified by different methods like acetylation, hydroxypropylation, oxidation, partial hydrolysis, and via cross linkages [160–162]. Starches are also processed enzymatically or by biochemical methods [163], to used as a coating agent and emulsifier in food technology and pharmaceutical sciences. Starches modified by hydroxypropylation showed improved release of biomedical compounds as a result its resistant increase against the hydrolytic cleavage of pancreatin within GIT track [164]. Generally, oxidatively modified starches exhibited strong adhesivity, enhanced film forming capacity and excellent release functionality [165]. Cross-linked starches give better structural framework which not only enhanced its mechanical properties but also improved resistant against the environment stresses [166]. Enzymatic hydrolysis resistant, improved hydrophobicity and control delivery were reported in acetylated starches [167]. Partially hydrolysis of starches make their structural backbone more stable against oxidation and heat.

Carotenoid loaded microcapsules were successfully prepared by entrapping in modified starches and maltodextrins [168]. Carotenoid encapsulation efficiency were enhanced by optimizing the parameters like amylose quantity, ratio of 1-octenyl succinate as surface active functional group and considering thickening capacity of starch. β-Carotene was also encapsulated by acid modify native tapioca starch and it was reported that the chemical stability and retention timings of β-carotene were significantly enhanced by hydrogen bonded continuous network of amylose. This was also attributed the improved film forming capacity, as physical protection and stability against oxygen and heat as well as reduced deposit amount on its surface. In addition lycopene was encapsulated by Capsul® (i.e. waxy maize starch modified hydrophobically), which enhancing its stability during storage period at 10 and 20 °C [169].

8.4.1.2 Protein-Based Nanocarriers

Proteins are commonly obtained from animals, plants and microorganism [170]; animal protein sources include: gelatine, whey proteins, caseins, elastin, albumin, collagen, and silk whereas plant proteins are cereal proteins, soy proteins, gliadin, zein and pulse proteins [171]. Protein based nanocarriers are one of the best choice for the nanoencapsulation and for controlled and targeted delivery of biomedical compounds due to their nutritional, bio-based origin, abundant renewable sources, water binding capacity, emulsification, bio-degradable foaming, gelation, non-antigenic properties and easy to prepare. They provide enhanced bioaccessibility and protection to the nutrients from degradation [172, 173]. So successfully applied for the delivery of flavours, oils, fatty acids, fats and many other bioactive compounds [174]. Major type of proteins that have been used for carotenoid's encapsulation are discussed below.

Caseins

Caseins, are amphiphilic in nature. Caseins structurally comprised of four different types of proteins i.e. α-1, α-2, β-, and k-casein, having 38, 10, 34 and 15 percentage composition respectively [175]. It is also known as colloidal proteins due to its light scattering property via non-covalent attraction between the colloidal particles. These colloidal proteins form the casein micelles in the milk i.e. calcium and calcium phosphate insoluble protein, which give white colour to milk [176]. Casein micelles are consider as a suitable nanocarriers candidate because of their vast range of physiochemical properties like excellent solubilizing, water holding, emulsification, biocompatibility, acid and heat stability, gel forming, gelation, bio-degradable and binding properties with small ions and molecules [177–180].

Casein-graft-dextran copolymer were prepared which were then used for the nanoparticle encapsulation of β-carotene via hydrophobic interactions. Consequently β-carotene stored with enhanced stability against change in ionic strength and pH with improved release during hydrolysis of pepsin or trypsin [99]. Casein micelles were used as a natural nanocarriers for the delivery and protection of carotenoids [181].

A natural food colorant and apocarotenoid i.e. Bixin, was microcapsuled by casein micelle using spray dried technique (dissociation–reassembly mechanisms), which enhances its stability efficiently as well as improves dispersibility over a pH range 2–10 [182]. It was also reported that the soluble soy polysaccharides adsorbed on the casein under acidic conditions, which attributed the stability even at casein isoelectric point. The size of the bixin core-casein shell particles minimized (i.e.90 nm) upon rehydration, resulted in the carotenoid microcapsules degradation stability against light and heat.

Astaxanthin was microencapsulated with sodium caseinate and lactose using monodisperse droplet spray drying method, which attributed many technological features like narrow sized microparticles, flowability, oil surface coverage, residual moisture, agglomeration as well as controlled released in neutral pH buffer solution at 25 ° C temperature. It was also reported that the astaxanthin's radical scavenging activity was maintained despite of the high temperature (i.e. 105 °C) of drying chamber [183].

Casein micelles were also reported as protective agent for β-carotene against processing stressors i.e. high pressure and temperature and also provide heat-resistant especially in the bakery products [184]. These characteristics of casein micelles make them suitable candidates for the nano-delivery of bioactive food ingredients. β-carotene encapsulated within the casein via hydrophobic interactions that attributed the high storage stability and high encapsulation efficiency [185]. Casein-based emulsions were successfully recommended for lutein application in the beverage industry, which increases its chemical stability in photo-oxidation environment [186].

Cereal Proteins

Most abundant natural sources of cereal proteins are potato protein, wheat protein, amaranth, barley protein, and zein. They have been used as an encapsulant for many bioactive compounds because of owning many functional properties like elasticity, stability, water binding capacity, emulsifying capacity, foaming, and cohesiveness, [187]. Wheat proteins comprised on two basic proteins i.e. gluten and gliadin which have excellent film forming and gel forming functional properties [188], So could be used as encapsulant either alone or in combined form with other polymeric encapsulation systems [189]. Pickering emulsions having β-carotene ingredients were stabilized via complexes based on gluten nanoparticles xanthan gum and wheat gluten nanoparticles. This complex nanoparticles system not only provided chemical stability to β-carotene during one month of storage time at 25 and 37 °C but also enhanced the *in*-vitro bioaccessibility [190]. Barley comprised of 52% hordein, 23% glutelin and 20–25% globulins and albumins proteins [191]. β-Carotene was successfully nanoencapsulated by barley protein and reported as ideal choice for targeted delivery system development [192]. In another research, β-carotene was also fabricated via barley protein based nanoparticles which not only enhanced the stability, safety but also helped in the controlled release of lipophilic compounds [193].

Gelatin

Gelatin is an amphiphilic natural polymer with Arginine-Glycine-Aspartic structural sequences of amino acids. It is classified as gelatin A and B on the basis of isoelectric points i.e. acid and alkaline hydrolysis with isoelectric points 7–9 and 4.8–9 respectively [194]. It is mainly obtained from the bones, skin and connective tissues of different animals. Gelatin is suitable choice as a nanocarrier for functional compounds in food technology and for gene and drugs delivery in pharmaceuticals due to owing significant properties like low cost, readily available, non-toxic, low antigenicity, biocompatible and bio-degradable [195–198]. Gelatin based nanoparticles were prepared for the delivery and encapsulation of carotenoids. Nanoemulsions containing astaxanthin were stabilized with high emulsifying efficiency by fabrication with gelatin. The mixture of these surface active compounds also provide maximum zeta potential as well as minimum polydispersity index to astaxanthin nanodispersions [199].

Soy Proteins

Soy proteins consist of two globular proteins; the first one is β-conglycinin (7S) protein while the second one is glycinin (11S) protein [200]. Soy proteins showed potent nutritional, functional, and health beneficial properties like anti-carcinogenic and protective effects like for diabetes, obesity, heart and kidney diseases [201–203]. Additionally, they are cheap with remarkable properties e.g. antioxidant, water binding, emulsification, biocompatible, gel formation, fat absorption and biodegradable. So, they have a great potential as an ideal nanocarriers for carotenoids encapsulation [204, 205].

To protect the degradation of β-carotene along with its sustained release, soy glycinin was added as a food grade pickering stabilizer [206]. The stability, bioaccessibility and water dispersibility of β-carotene was enhanced remarkably via fabrication of nanocomplexes with soy protein isolates [207].

Whey Proteins

Whey proteins basically milk proteins which consist of two subunits of globular proteins i.e. β-lactoglobulin (50%) and α-lactalbumin (20%). β-Lactoglobulin a small globular protein (18.3 kDa), comprised of 162 amino acids having a free thiol (SH) group and two disulphide bridges on its structural backbone. It is present in stable dimeric form at slightly acidic or neutral pH at room temperature. While at acidic pH (i.e. pH = 2–3), it is dissociated into its subunits [208]. β-Lactoglobulin have marvellous properties like less molecular weight, low hydrophobicity, gel forming ability and high unfolding, which make it use as encapsulant for the efficient delivery of functional components in food and pharmaceutical industry [209].

Currently β-lactoglobulin is one of the best choice as an encapsulant for the transport of carotenoids. β-Carotene physical stability was significantly enhanced by fabricated with combination of whey protein isolates and pectin (as emulsifier), because of small particles size of nanoemulsion conjugates as compared to whey protein isolates alone [210]. In another study whey protein isolates and beet pectin conjugates were used to encapsulate the β-carotene, which enhanced its stability against degradation (25% loss) than whey protein isolates (60% loss) at pH 7 and 70 °C during 5 days of storage [211].

β-Carotene was also encapsulated by the combination tween 20 and whey protein concentrate. The results concluded the efficient retention of β-carotene as compared to mixture of tween 20 and bovine serum albumin [212]. A study reported that 90% efficiency of encapsulated β-carotene was enhanced by whey protein nanocarriers [102]. In another study, it was reported that β-carotene micellarization level (<3%) dramatically reduced by using lactoferrin stabilizer either emulsion having β-lactoglobulin or not. The results suggested that poor solubility of β-carotene was due to interactions between lactoferrin and β-carotene as well as with metabolic products of lactoferrin [213]. β-carotene loaded nano-emulsions were designed by using α-La-EGCG (i.e. combination of α-lactalbumin and (−)-epigallocatechin gallate) covalent complex at alkaline pH. The results reported the potent emulsifying as well as antioxidative properties than α-lactalbumin [214]. Moreover stability of β-carotene (6% loss) was significantly enhanced by α-La-EGCG system as compared to native α-lactalbumin (13% loss) after 20 days (25 °C) [215]. A comparative study on stability of β-carotene loaded nano-emulsions was made by forming conjugates of EGCG with various milk proteins (i.e., β-lactalbumin, α-lactoglobulin, sodium caseinate and lactoferrin). It was concluded that EGCG-protein complexes worked better as compared to individual proteins but the best stability of β-carotene results from the combination of EGCG with β-lactalbumin and sodium caseinate [216].

β-Carotene also fabricated by protein-polyphenol conjugates (i.e. chlorogenic acid or catechin), which showed improved stability [103, 215]. The ternary complex i.e. polyphenol-protein-polysaccharide have designed with enhanced stability of β-carotene (especially light stability) as compared to binary conjugates [217]. β-Lactoglobulin was fabricated to prepare the β-carotene nanoemulsions with enhanced stability and bio-accessibility. The results also showed the lipolysis inhibition as well as β-carotene leakage inhibition [145]. The physiochemical stability of Lutein-loaded nanoemulsions was enhanced via grafted with whey protein isolate at 4 °C. It was reported that only 4% lutein was degraded during storage time i.e. four weeks [218]. Whey protein isolate and lactoferrin were used as a encapsulant for lutein heteroaggregation and droplets of docosahexaenoic acid to prepare stabilize emulsions. The results indicated that the stability of nanoemulsions were enhanced via limited mobility of droplets and decreases the gap for Brownian motion [150].

8.4.1.3 Lipid-Based Nanocarriers

Lipids are abundantly present in nature with low toxicity and excellent emulsification functionalities [219]. Keeping these properties in view, both classes of lipids i.e. polar lipids e.g. monoglycerides, phospholipids etc. and non-polar lipids e.g. cholesterol, triacylglycerol etc. are consider as a potent choice as a nano-materials or an encapsulating agents [220]. Moreover, they are excellent carriers for bioactive compounds e.g., carotenoids within GIT due to their intermingled micelles formation ability which solubilize and targeted the passage of hydrophobic bioactive compounds [221]. Phospholipids (Polar lipids) are highly biocompatible because of their exceptional surface active properties along with enhanced functional proper-

Table 8.2 Applications of different carotenoids encapsulated by lipid-based nanocarriers [88]

Nanocarriers	Carotenoids	Surfactants	Size (nm)	Zeta Potential (Mv)	Encapsulation Efficiency (%)	Refrences
Nano-structured lipid carriers	Astaxanthin	Tween 80 and lecithin	85–138	−22 to −35	Not reported	[230]
	β-Carotene	Tween 20	8–309	Not reported	26.64–97.95	[231]
	Lycopene	Orange wax and Eumulgin sodium stearoyl glutamate	150–160	−73 to −75	100	[232]
	β-Carotene	Propylene glycol monostearate	~300	Not reported	Not reported	[233]
	Lutein	Span 60 and Tween 80	190 and 360	−28 to −34	68	[162]
	Lycopene	Orange wax and Eumulgin sodium stearoyl glutamate	150–160	−73 to −75	100	[232]
	Lycopene	Tween 80 and Poloxamer 188	~122	−29	84.50	[109]
	Lutein	Plantacare 810, cetyl palmitate, glyceryl tripalmitate, and carnauba wax	150–350	−40 to −63	Not reported	[234]
	Bixin	Soy and egg lecithin	135 to 353	−17.9 to −36.5	>99	[235]
	β-Carotene	Egg YolkEEeg	~160–2700	Not reported	40	[236]
	Lutein	Plantacare 810, cetyl palmitate, glyceryl tripalmitate, and carnauba wax	150–350	−40 to −63	Not reported	[234]
	Lutein	Carnauba wax and glycerol stearate	167–387	−34.2	88.5	[237]

Nanocarriers	Carotenoids	Surfactants	Size (nm)	Zeta Potential (Mv)	Encapsulation Efficiency (%)	Refrences
Solid lipid nanoparticles	Lycopene	Tween 80 or Poloxamer 407	~125	~ – 10.06	~98.4	[238]
	β-Carotene	Poloxamer 188 and Tween 20	~115	−35.9	>70	[239]
	β-Carotene	Tween 20	60–140	Not reported	Not reported	[240]
	Bixin	Poly-ε-caprolactone and sorbitan monostearate	168–222	±30	100	[241]
Nanoemulsions	β-Carotene	Tween 20	132–184	No reported	100	[242]
	Lutein	Phospholipon 85G	150	No reported	100	[243]
	Lycopene	OSA-modified starch	145–161	−19.73 to −20 to 77	Not reported	[107]
	Lycopene	Soy protein isolate	69	−28.7	80.7	[106]
	β-Carotene	Protein poly phenol conjugates	159 to 163	−57 to −61	No reported	[244]
	Lutein	Hyderogenated soya and phosphatidylcholine	200	Not reported	>90	[245]
	Lycopene	Tween 20	100–200	−33 to −42	51–65	[246]
	Lutein	Whey protein isolate and lactoferrin	317	22.6 to −37.4	Not reported	[107]
	Lycopene	Soya protein concentrate	346.90	−42.32	85.58	[247]
	β-Carotene	Soya lecithin and tween 80	119	No reported	85.63	[248]
	Lutein	Tween 80, whey protein isolate and cascinate	200	−55 to −62	Not reported	[249]
	Lutein	Bovine and caprine caseins	205–206	−35 to −37	Not reported	[186]

(continued)

Table 8.2 (continued)

Nanocarriers	Carotenoids	Surfactants	Size (nm)	Zeta Potential (Mv)	Encapsulation Efficiency (%)	Refrences
Nanoliposomes	Lycopene	Lecithin and cholesterol	~840	Not reported	71.65	[250]
	β-Carotene	Polyvinyl alcohol and polyethylene oxide	~250	Not reported	86.8	[251]
	β-Carotene	Xanthan and guar gums	~2000	−32 to −105	90	[252]
	Lutein	Cholesterol, lecithin and Tween 80	~264 to 367		92.93	[253]
	β-Carotene	Enzyme modified fluid lecithin, fatfree powder lecithin, standardized fluid lecithin	~115 to 460	−45 to −60	60	[254]
	Lycopene	Phospholipid	Not reported	Not reported	71.9	[255]
	Lutein	Soy lecithin	147–195	−54.5 to −61	56–97	[222]

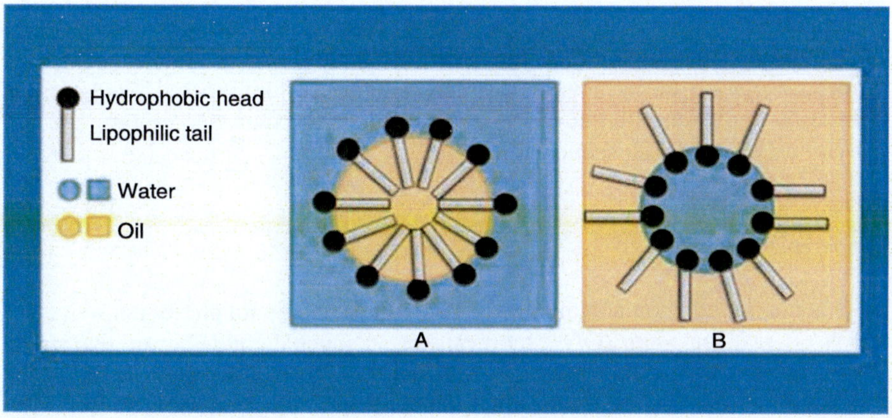

Fig. 8.17 Graphical representation of oil in water (O/W, A) and water in oil (W/O, B) emulsions

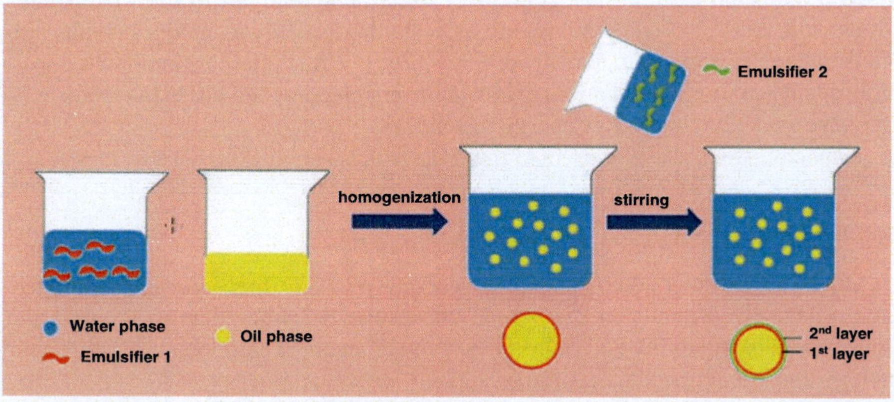

Fig. 8.18 Schematic diagram for the preparation of a double layers emulsion (i.e. multilayered emulsion) (emulsifier 1 and 2 are oppositely charged)

ties like, stability, protection, maintenance and specific delivery of bioactive compounds [220]. Phospholipids also have unique emulsifying capacity, self-assembly functionalities and wetting ability, So used as important nano-encapsulant for both hydrophilic and hydrophobic compounds [222–224]. Consequently, lipid-based nanocarriers are considered as the advance approach for carotenoids nanoencapsulation [225].

The bioaccessibility of β-carotene was significantly enhanced by adding phosphatidylcholine, (i.e. phospholipid in the membrane of human cells). The results concluded that phosphatidylcholine helps in mixed micelles formation along with high emulsifying capacity [226]. Lycopene was also incooperated with phosphatidylcholine and results found that the bioaccessibility was significantly enhanced [227].

The different approaches for lipid-based nanocarriers includes SLNs, NLCs nanoemulsions, nanoliposomes, niosomes and cubosomes [79, 228, 229]. The nanoencapsulation of carotenoids by these engineered lipid-based nano-carriers techniques, their formulation processes and potential controlled release are discussed briefly in the next sections (Table 8.2).

Nanoemulsions

Nanoemulsions are basically mixture of two immiscible liquid phases i.e. water -in-oil (W/O) or oil-in-water (O/W) by means of an suitable emulsifier to give single phase emulsion as shown in Fig. 8.17 [256]. They are also known as mini-emulsions and sub-micron emulsions due to their small mean radii (about 50–500 nm). The stability of nanoemulsions are directly related to the particles size i.e. emulsion are more stable as smaller the particles size [257]. Homogenization, ultrasonication and microfluidization are the most commonly used methodologies to form nanoemulsions [258, 259] First two methods are basically high energy techniques results into the smaller size and homogeneous droplets [260], While phase inversion composition, spontaneous emulsification, emulsion inversion point and phase inversion temperature are belonging to low energy techniques.

Nanoemulsions can be classified into two groups.

(a) Single emulsions (O/W or W/O)
(b) Double emulsions (W/O/W & O/W/O)

Double emulsions could be formed either by low or high energy techniques. These nano-sized emulsion used as an encapsulated nanocarriers to improve the solubility, bioavailability and stability of many biomedical compounds in food technology and in pharmaceuticals [261, 262].

Double emulsions heterogeneous system e.g. $W_1/O/W_2$, in which dispersed phase aW_1/O emulsion is continuously diffused in W_2 aqueous phase [156]. So W_1 in these emulsions, has ability to hold the hydrophilic bioactive compounds like aminoacids, polyphenols, vitamins, minerals and colours [260]. Moreover, both hydrophilic and hydrophobic bioactive compounds can be encapsulated with nanoemulsions [263, 264] (Fig. 8.18).

β-Carotene nanoemulsions were prepared by combination of either three glycerol monooleates (i.e. decaglycerol, hexaglycerol, tetraglycerol) or by three glycerol monolaurates. The results concluded that the β-carotene loaded emulsions enhanced stability of bioactives [265]. In addition, β-carotene was encapsulated with oil-in-water nanoemulsions via high pressure homogenization. The physical stability enhanced either by greater number of homogenization cycle (i.e. up to three cycles) and at high pressure (i.e. up to 100 MPa), while decreases at elevated temperature. β-carotene-loaded on nanoemulsion was degraded upto 25% at 4–25 °C during storage but more loss occurred at 25 °C [266]. In another study β-carotene-loaded nanoemulsion were prepared by using tween 20 (i.e. polyoxythylene sorbitan ester) and decaglycerol monolaurate (i.e. polyglycerol ester). The results concluded that higher concentration of β-carotene was retained in the nano-emulsion system of polyglycerol ester as compared to tween 20-stabilized nano-emulsion [267, 268].

β-Carotene was loaded on oil-in-water nanoemulsions which stabilized by β-lactoglobulin (globular protein). Degradation rate of β-carotene was considerably decreased in β-lactoglobulin stabilized nanoemulsions as compared to Tween 20-stabilized nanoemulsions. In addition colour fading of degraded β-carotene increases at high storage temperature i.e. 5–55 °C and acidic pH [205]. Another study reported that the β-carotene stability against degradation was significantly enhanced by using different water-soluble (i.e. EDTA and ascorbic acid) and oil-soluble (i.e. coenzyme Q10 and vitamin E acetate) antioxidants during storage time at high temperature (55 °C). It was also reported that EDTA showed good results than ascorbic acid while Coenzyme Q10 also showed better results as compared to vitamin E acetate. However, a mixture of vitamin E acetate and EDTA was not as much efficient as the individual compounds [269, 270].

β-Carotene light stability was significantly enhanced in oil-in-water nanoemulsion system with α-tocopherol as compared to ascorbyl palmitate and tertiary butyl hydroquinone [271, 272]. β-Carotene was also loaded on oil-in-water nanoemulsions using α-tocopherol or ascorbic acid as good antioxidant. The results concluded that β-carotene degradation was significantly decreased in case of α-tocopherol at storage temperature of 25 and 55 °C [106].

The comparative study of functional properties of different β-carotene-loaded nanodispersions with decaglycerol monolaurate (ML750), Tween 20, sodium caseinate (SC), and sucrose fatty acid ester (L1695), prepared via solvent displacement technique was reported. The most potent oxidative stability was exhibited by β-carotene-loaded SC-stabilized nanodispersion systems, owing due to their small definite particles surface area and the physical barrier of sodium caseinate (SC). The antioxidative activity was improved by combination of SC-L1695 and SC-ML750 nanodispersions. The results reported the enhanced stability of β-carotene due to the larger mean particle size as compared to L1695or ML750 nano-systems alone. In another study β-carotene loaded on different low (i.e. olive oil and corn) and high temperature melting lipids (i.e. hydrogenated coconut oil and cocoa butter), while Tween 80 was used as stabilizer. It was found that the stability of β-carotene enhanced along with *in*-vitro digestibility kinetics and bioaccessibility in both cocoa butter and corn oil LNPs (lipid nanoparticles) [273].

The physicochemical stability of SLN (solid lipid nanoparticle) encapsulated suspensions of β-carotene depends upon surfactant type. The synthetic layout of oil-in-water emulsions start by homogenizing the 90% w/w aqueous phase (i.e. surfactant and cosurfactant) and 10% w/w lipid phase (i.e. 1 mg/g of β-carotene in carrier lipid) at neutral pH and 75 °C followed by cooling to 20 °C. The aqueous phase comprise of 2.4% w/w low and high-melting (HM) lecithin each and 1.3% w/w Tween 80 or 1.3% w/w Tween 60. Moreover 0.6% taurodeoxycholate used as a cosurfactant. It was reported that solid particles were formed by using HM lipid e.g. tripalmitin while LM lipid e.g. medium chain triglycerides (MCT) exist in liquid droplets. In addition alpha-crystals were formed by using combination of Tween 60 and HM-lecithin in high fraction. The degradation rate of β-carotene was 91,100, 97, and 11% with HM lipid i.e. tripalmitin, while 90, 95, 21, and 16% with LM lipid i.e. MCT for Tween 60, Tween 80, LM-lecithin and HM-lecithin respectively during 21 days storage period. The enhanced chemical stability of β-Carotene was attributed due to alpha-crystals with marginal surfactant tails [274].

β-Carotene nanoemulsions were formulated via homogenizing the β-carotene (0.05% w/w) dispersed in HPKO (hydrogenated palm kernel oil, 10% w/w) with sucrose (30% w/w) as water phase and whey protein isolate (WPI 1%, 1.5%, 2%, and 3% w/w) at pH 7. The degradation rate of β-carotene decreases by increasing the amount of protein supported on solid lipid carrier and liquid dispersed phase [275]. β-Carotene nanoemulsions were prepared by homogenizing the lysolecithin and sucrose monoester as emulsifiers while combination of corn and lemon oil as lipid phase at high pressure. The functional properties like physical stability, bioaccessibility and microstructure of β-carotene nanoemulsions were influenced by the percentage composition of carrier oil. The β-carotene bioaccessibility and production of free fatty acid in small intestine increased via increasing the quantity of digestible oil in the droplets. So providing the effective means for β-carotene delivery systems [276]. Oil-in-water nanoemulsions of carotenoids were prepared by using nonionic surfactant i.e. Tween 20 as emulsifier while oil phase was long-chain triglyceride from corn oil. The digestion rate and bioaccessibility of β-carotene was investigated and it was suggested that the β-carotene bioaccessibility was much lower in high-fat group (i.e. 39%) as compared to low-fat group (i.e. 84%), due to incomplete hydrolyzation within GIT [277]. In another study stable nanoemulsions were prepared by using different carrier oils like coconut oil, corn oil fish oil, and palm oil. The nanoemulsion was comprised of aqueous solution (90% v/v), whey protein isolate (2% w/v) and dispersed oil (10% v/v). It was reported that the β-carotene bioaccessibility was significantly improved by corn oil = palm oil > fish oil > coconut oil in increasing order. The stability of β-carotene nanoemulsions were 49.58, 54.91, 63.81 and 69.36%, with fish oil, corn oil, coconut oil and palm oil respectively. The best combination reported with excellent stability and bioaccessibility was whey protein isolate-stabilized β-carotene nanoemulsion with carrier oil (palm oil) as a suitable delivery system [278].

β-Carotene (BC, 0.1 wt%) nanodispersions encapsulated in cocoa butter (15 wt%) were prepared as crystalline solid emulsions (i.e. SE and SE-BC) or undercooled liquid emulsions (i.e. LE and LE-BC) droplets at 25 °C. This work demonstrated that the bioaccessibility and *in*-vitro digestive lipolysis were depends upon the physical state of lipid crystallinity, like spherical morphology, beta-V polymorphism, thermal behaviour (peak melting = 30 °C), zeta potential (~ – 44 mV), particle size distributions ($D_{4,3}$ ~ 0.7 μm), and degradation of β-carotene in accelerated lighting conditions. The results showed that the β-carotene bioaccessibility and duodenal hydrolysis were lower in case of SE-BC [279].

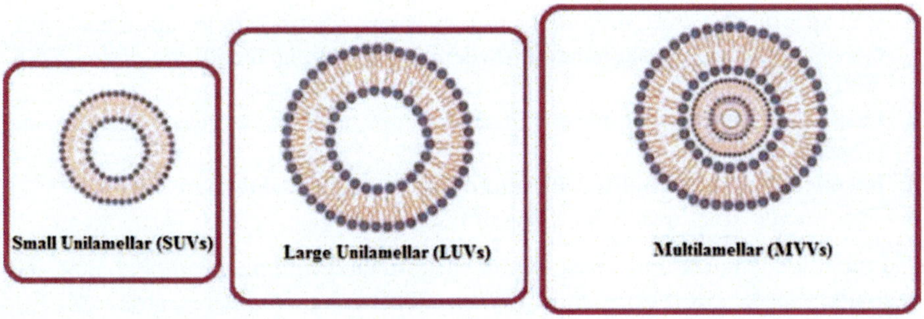

Fig. 8.19 Classification of Nanoliposomes

Nanoliposomes

Liposomes basically consist of one or two layers. Liposomes are spherical in shape with diameter less than 200 nm [280], comprised of internal core of hydrophilic molecules which are directed towards the aqueous phase and bilayer membrane constitute by the phospholipids lipophilic tails e.g. Soy lecithin [281, 282]. On the basis of lipid bilayer they are divided into two classes [259] (Fig. 8.19).

(a) Unilamellar (i.e. With a single lipid bilayer)
(b) Multilamellar (i.e. With two or more lipid bilayers)

Nanoliposomes preparation need high energy as compared to liposomes due to their larger surface area [279]. Various non-mechanical techniques for the fabrication of nanoliposomes includes reverse-phased evaporation, shrinking of mixed detergent-lipid micelles, freeze drying rehydration, injection method and freeze thawing. However mechanical approaches comprise of colloid mills, sonication, microfluidization, extrusion and high pressure homogenization. In addition more recent methodologies are cross-flow filtration technology, freeze-drying double method, dual asymmetric centrifugation, dense gas technique, supercritical fluid technology and membrane contractor technology [283].

The basic physicochemical properties of nanoliposomes includes biocompatibility, biodegradability and cell-specific targeting. So liposomes and nanoliposomes have potential applications in various fields like nanotherapy (including diagnosis, gene delivery, cancer therapy) [281], food technology, cosmetics and agriculture [284, 285]. Nanoliposomes have consider as a potent candidate for carrier vehicles of nutraceuticals, nutrients, food additives, enzymes, flavours and food antimicrobials in food technology. Consequently their working mechanism enhanced the solubility, stability, and bioavailability of bioactive agents. Both hydrophobic and hydrophilic bioactive compounds can encapsulated by nanoliposomes [77].

The functional characteristics includes

1. Encapsulation and emancipation of lipid soluble, water soluble, and amphiphilic bioactive compounds [286, 287].

2. Cell-specific targeting drug delivery by various biocompatible routes e.g. pinocytosis, without bio-degradation or demolition during blood circulation [285, 288].
3. Easily modifiable by simple chemical reactions to get the targeted functions and properties [289].
4. Excellent biocompatibility with improved metabolism and no toxicity [289].
5. The fasten carrier vesicle has form of fluidity [289].

Lutein-loaded nanoliposomes system encapsulated with phophatidylcholine and egg yolk, showed enhanced stability against different environmental stresses [290]. Electrospun nanofibers were incorporated with β-carotene-loaded nanoliposomes. The results reported the greater stability against UV-light as well as during electrospinning with enhanced delivery of β-carotene from nanofibers. In another study, lycopene encapsulated with nanoliposome i.e. phospholipids system exhibited improved therapeutic potential like antioxidant activity and efficient recovery against methotrexate-induced kidney injury [254].

Niosomes

Niosomes are microscopic nonionic surfactants vesicles. Niosomes are prepared by self-clustering of nonionic surfactants (i.e. polyoxyethylene alkyl esters or ethers and alkyl or dialkyl polyglycerol ether, which form the bilayers of niosomes) in aqueous media with or without cholesterol or other fats (Fig. 8.20).

Niosomes are classified into three groups on the basis of their particles size (Fig. 8.21):

1. Small unilamellar (particle diameters = 10–100 nm)
2. Large unilamellar (particles diameter = 100–3000 nm)
3. Multilamellar (comprised of more than one bilayer) [291].

In addition, different types of niosomes were also reported with similarities to liposomes like aspasome, discomes, elastic niosomes, polyhedral niosomes, surfactant ethosomes, and proniosome etc. [292]. The significant properties of niosomes include cost effective, non-toxic, biodegradable, highly stable, nonimmunogenic, better skin penetration, easy to handle, targeted delivery, longer storage time and with an osmotic activity [293–296]. Additive agents like cholesterol play an effective role for self-assembly of surfactants into niosomes, which gives essential vesicular properties like chemical stability, greater discharge rate, improved entrapment, and greater storage time [297]. Both lipophilic and hydrophilic biomedical compounds are encapsulated by means of niosomes, in which lipophilic compounds are bonded with lipophilic shell of vesicles while the hydrophilic components are in the aqueous core [298]. Because of their unique functional properties, they are potentially used in cosmeceuticals [292], pharmaceuticals [295], gene delivery [299], drug delivery [300] and foods industry [293].

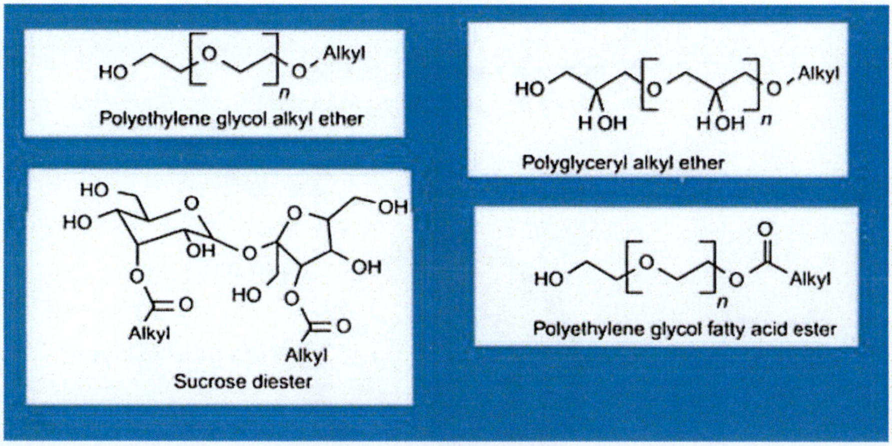

Fig. 8.20 Chemical Strucutres of different types of nonionic surfactants

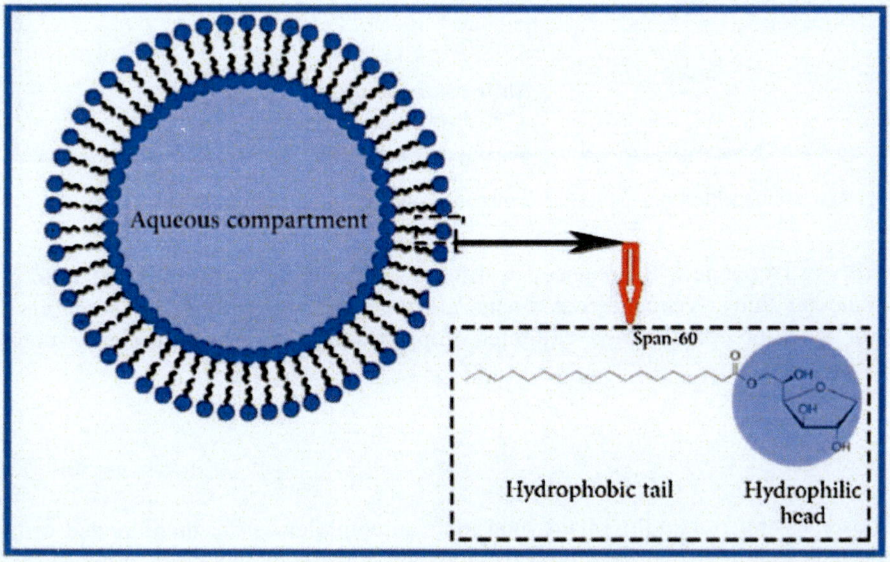

Fig. 8.21 General Strucutre of Niosomes prepared by Span-60(sorbitanmonostearate) (redrawn from Ref. [292])

β-Carotene-loaded niosomes were developed by means of different combination of surfactants i.e. Tween 20, 40, 60 and Spans 40, 60, 80. β-Carotene in nano-niosomes exhibited significant resistance against oxidative stresses caused by free radicals and also against high heat. They also showed enhanced stability in the culture medium upto 96 h [301]. Lycopene-loaded nano-niosomal vesicles were fabricated with Span 60 and cholesterol. The results showed the efficient bioavail-

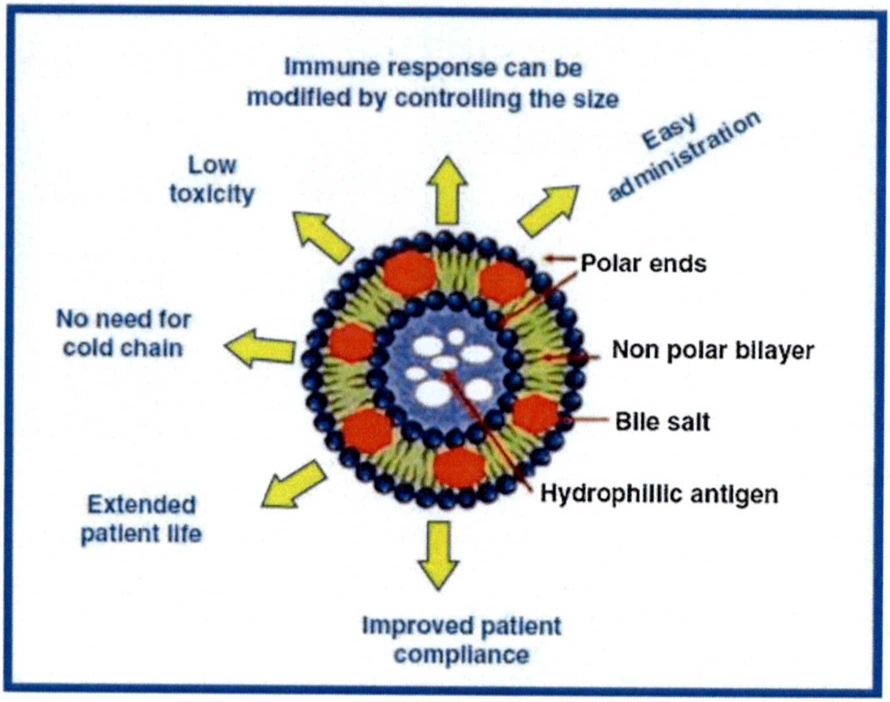

Fig. 8.22 Composition and main characteristics of bilosomes (redrawn from Ref. [309])

ability of lycopene with enhanced proliferative activity against cancer cells [302]. In another study, lycopene-coated nano-niosomes grafted on Span 60 and cholesterol were prepared which exhibited improved encapsulation efficiency upto 62.8% [303].

Bilosomes

Bilosomes are structurally bilayer non-ionic amphiphiles carrier incorporated with bile salts which enable them to protect the biomedical compounds from the acidic environment of the GIT [304]. Bilosomes are modified type of liposomes and niosomes [305]. Bilosomes are specific delivery carrier due to their biodegradable and biocompatible nature, thereby enabling them to used for oral delivery of biomedical compounds like vitamins, proteins, antibodies, DNA, and hormones [306]. Additionally, bilosomes also consist of deoxycholate or glycocholate which improved the bio-macromolecules oral delivery like cyclosporine A, calcitonin, insulin and salmon [307]. There are two main types for the preparation of bilosomes (Fig. 8.22)

1. Hot homogenization technique [308]

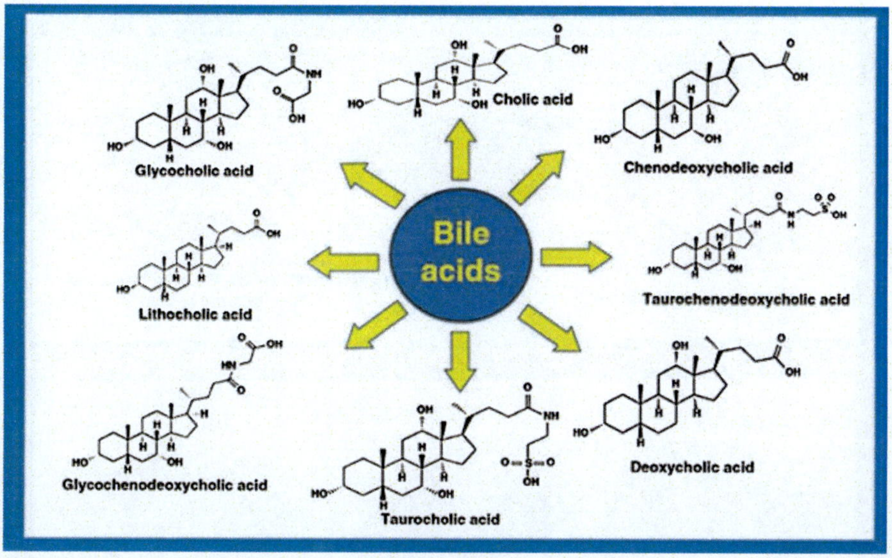

Fig. 8.23 Different bile acids used in oral drug delivery systems (redrawn from Ref. [309])

2. Thin film hydration.

The brief overview of functional characteristics of bilosomes are mentioned as follow (Fig. 8.23):

1. Continuous and stable release [310]
2. Enhanced mucosal penetration [311]
3. Bilosomes are more stable carriers for drug delivery with enhanced absorption and transmembrane delivery [312].
4. Better cellular uptake [313]
5. Enhanced bioavailability [314]
6. Tissue targeted delivery [315]
7. Highly stable in air from oxidation at high temperatures (~130 °C) [304]
8. Easy to prepare, low cost and passive to lyophilisation [316].

Bilosomes potently protect the vaccine antigen, So potentially used for the oral delivery of vaccines [317] with enhanced mucosal epithelial tissues penetration. [310]. They prevent the antigens from degradation [318]. In another study, the lipophilic components channels increases across biological membranes due to bile salts consequently oral bioavailability of bioactive compounds were significantly enhanced [319] (Fig. 8.24, Table 8.3).

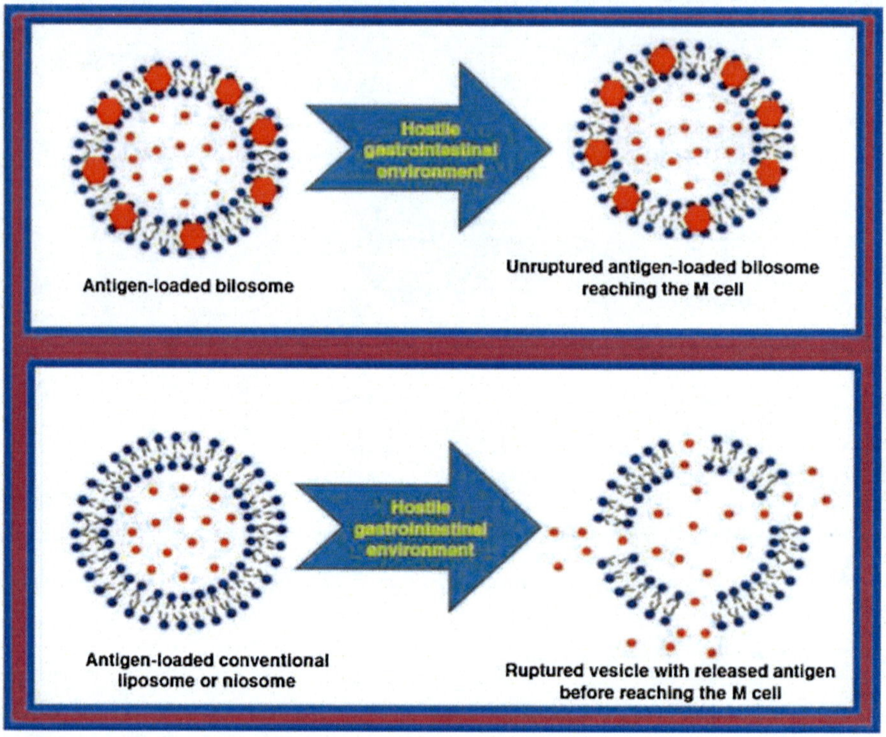

Fig. 8.24 Comparative stability of antigen loaded liposomes, bilosomes and niosomes in gut

Table 8.3 Comparative summary of liposomes, bilosomes and niosomes

Parameter	Liposomes	Bilosomes	Niosomes
Composition	Natural phospholipids, cholesterol	Non-ionic surfactant, bile salt and charge inducer	Non- ionic surfactant with cholesterol and charge inducer
GIT stability	Unstable	Stable	Unstable
Antigen dose	Comparatively high	Comparatively low	Comparatively high
Storage stability	Required liquid nitrogen for storage	Special conditions not required	Special conditions not required
Chemical stability	Phospholipids undergo the oxidative degradation	Stable, does not undergo oxidative degradation	Stable, does not undergo oxidative degradation
Storage and handling conditions	Required special conditions (e.g. liquid nitrogen storage)	Do not required special conditions	Do not required special conditions

Solid Lipid Nanoparticles (SLNs)

Solid lipid nanoparticles are considered as a new generation of lipid emulsions with submicron-sized, which comprised of solid lipid as an alternative of liquid lipid (oil), that are dispersed either in an aqueous surfactant or in water solution. Different types of solid lipids are used as nano-carriers e.g. triglycerides, steroids, partial glycerides, fatty acids and waxes etc. [320], while polysorbates [321], soybean and egg lecithin [322], and bile salts [323] etc. were used as surfactants for SLN's fabrication [324]. SLNs owing excellent properties like small size, easy large scale production, large surface area, bioavailability, non-biotoxicity, high drug loading, controlled release, better physical stability, easy biodegradability and interface interactions of phases [325, 326]. Keeping in view this diversity, SLNs are placed at forefront of advanced field of nanotechnology with many potential applications like in pharmaceuticals (especially drug delivery), clinical medicines, neutraceuticals and research in other varied sciences. SLNs as a colloidal carriers have ability to entrap both lipophilic as well as hydrophilic compounds.

β-Carotene was encapsulated by SLNs i.e. palmitic acid and corn oil while whey protein isolate was used as a stabilizer. The results reported that the physicochemical and oxidative stabilities of β-carotene were significantly enhanced [112]. A study reported the successful encapsulation of β-carotene by SLNs and the impact of lipid composition on their physicochemical properties. It was concluded that the stability for aggregation and degradation of β-carotene was improved for high concentration of saturated fatty acids [112]. Lycopene-loaded SLNs were prepared by fabrication of myristic acid, glyceryl behenate and glyseryl palmitostearate as lipid phases. The results showed enhanced loading capacity i.e. 98.4% as well as fine distribution of nano-particles [238]. Another research conducted on lycopene encapsulated by SLNs using glycerol distearate as a fabricated nanocarriers, which showed large smooth spherical surface and higher stability as compared to SLNs based on glycerol monostearate as solid lipid after 30 days of storage period at 6 and 25 °C [327].

Nano Structured Lipid Carriers (NLCs)

NLCs are formulated as a combination of both liquid and solid lipids as a core matrix along with emulsifiers in water which form the outer layer. NLCs are very similar to SLNs, the only difference is the substitution of solid phase (5–40%) to liquid phase in NLCs, which give them some unique encapsulation properties like enhanced loading capacity with improved and controlled release of therapeutics throughout the GIT [225, 328]. NLCs are formed via controlling the mixing rate of

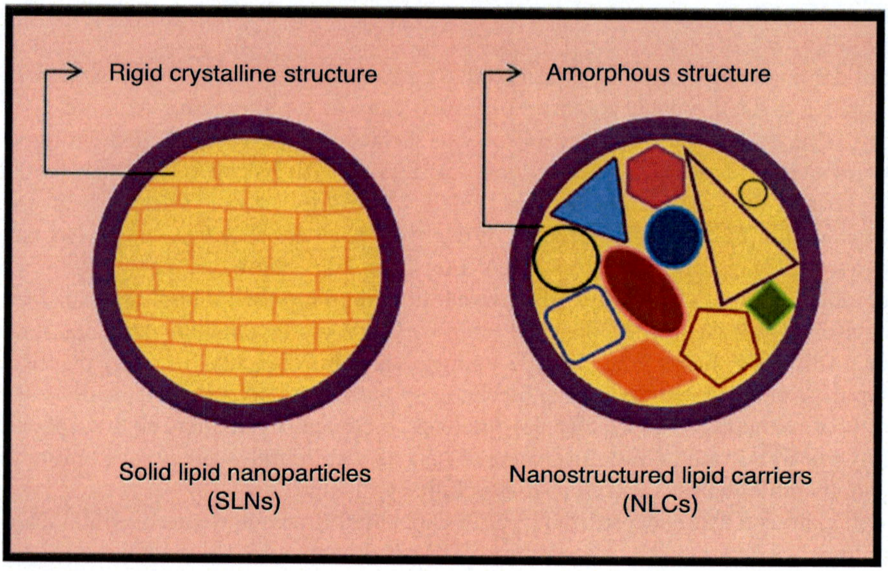

Fig. 8.25 General Structural Comparison of SLNs And NLCs

Table 8.4 Different types of important ingredients for NLCs Preparation and their sources

Essential components	Examples (Natural & Synthetic Sources)
Liquid lipids	Medium Chain Triglycerides, Oleic Acid, Vitamin E, 2-Octyl Dodecanol, Squalene, Paraffin Oil, Isopropyl Myristate, Transcutol® HP, Miglyol® 812, Labrafil Lipofile® WL 1349, Lauroglycol® FCC, Labrafac® PG, Capryol® 90
Solid lipids	Tristearin, cholesterol, cetyl palmitate, stearic acid, Precirol® ATO 5, Compritol® 888 ATO, Cutina® CP, Dynasan® 118, Dynasan® 116, Emulcire® 61, Gelot® 64, Geleol®, Imwitor® 900 P, Softisan® 154
Amphiphilic emulsifiers	Egg lecithin, phosphatidylethanolamines, phosphatidylcholines, soya lecithin, Gelucire® 50/13
Lipophilic emulsifiers	Span 60, Span 40, Span 20, Myverol® 18-04K,
Hydrophilic emulsifiers	Solutol® HS15, Pluronic® F68 (poloxamer 188), Tween 80, Tween 40, Tween 20, Pluronic® F127 (poloxamer 407), trehalose, sodium oleate, sodium glycocholate, sodium deoxycholate, polyglycerol methyl glucose distearate, polyvinyl alcohol

solid lipids within the liquid oil, resulting the formation of special network of nanostructures within the matrix (Fig. 8.25).

Different techniques were used to prepare NLCs like cold or hot homogenization and microemulsion, depending upon the hydrophobicity and heat stability of the bioactive compounds as well as the cloud point of emulsifier etc. This method has many advantages for the heat sensitive compounds but mostly carotenoids are

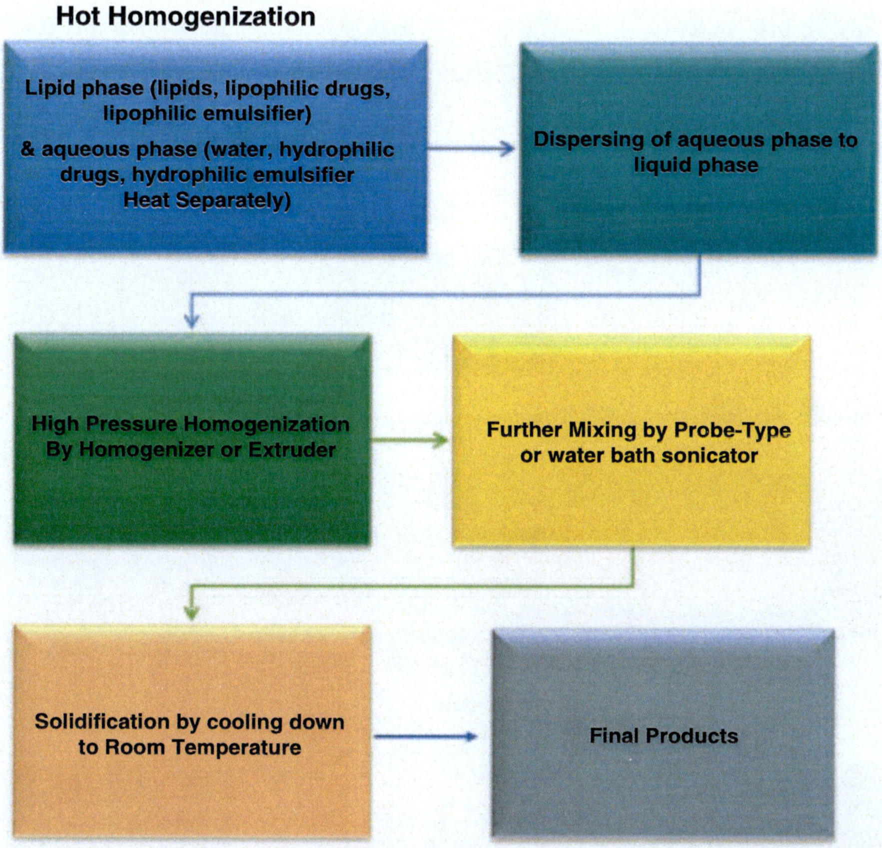

Fig. 8.26 Schematic Layout of Hot homogenization for NLCs Preparation

encapsulated by hot homogenization method because carotenoids are able to withstand heat [273]. Additionally, fat crystallization can be controlled through the selection of proper and suitable emulsification systems [329].

NLCs are classified on the basis of its composition i.e. Amorphous type, Multiple type [330] and Imperfect type [331] (Table 8.4).

Hot homogenization is carry out at temperatures above the lipids melting point. First lipid phase (i.e. liquid and solid lipids along with lipophilic emulsifiers) and aqueous phase (i.e. double distilled water as well as hydrophilic emulsifiers) are prepared separately. Then the mixture of two are homogenized or ultrasonicated at high temperature and pressure followed by cooling to get nanoparticle precipitation (Fig. 8.26).

While the cold homogenization starts with melting the solid lipid matrix followed by cooling in cold-mill using liquid nitrogen and finally ground to fine microparticles which are dispersed in a solution of cold emulsifier. This pre-suspension is then homogenized at low temperature ($\geq$25 °C) (Fig. 8.27).

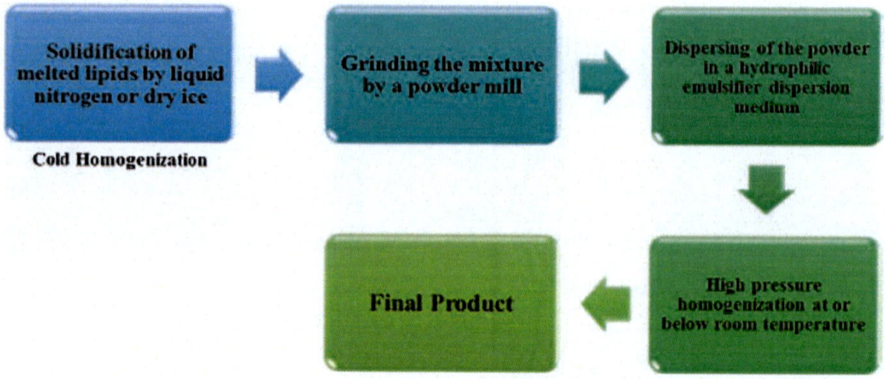

Fig. 8.27 Schematic layout of cold homogenization for NLCs preparation

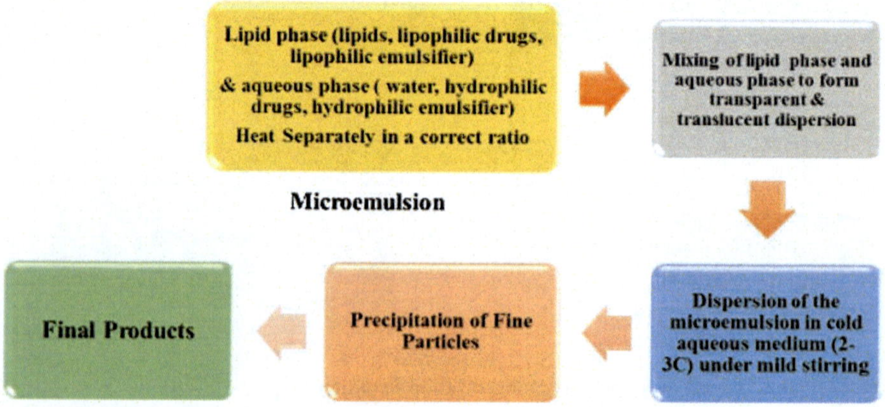

Fig. 8.28 Schematic Layout of microemulsion for NLCs Preparation

Third methodology i.e. microemulsion is discussed in flow sheet diagram (Fig. 8.28).

Keeping in view its unique properties NLCs were potently used in food technology like for entrapment of carotenoids. Hot homogenization method was mostly used for encapsulation of carotenoids because of their ability to withstand higher temperature [273].

β-Carotene and tocols (vitamin E) were loaded into NLCs with improved processing parameters i.e. lower-temperature condition (reduced from 85 °C to 60 °C). The results reported that the chemical stability was significantly enhanced of these heat-sensitive compounds (i.e. β-carotene and tocols) and no degradation was observed during 90 days of storage period as compared to conventional technique. While the physical stability was preserved throughout the study duration [332]. In another study β-carotene -loaded NLCs was prepared and reported that the NLC

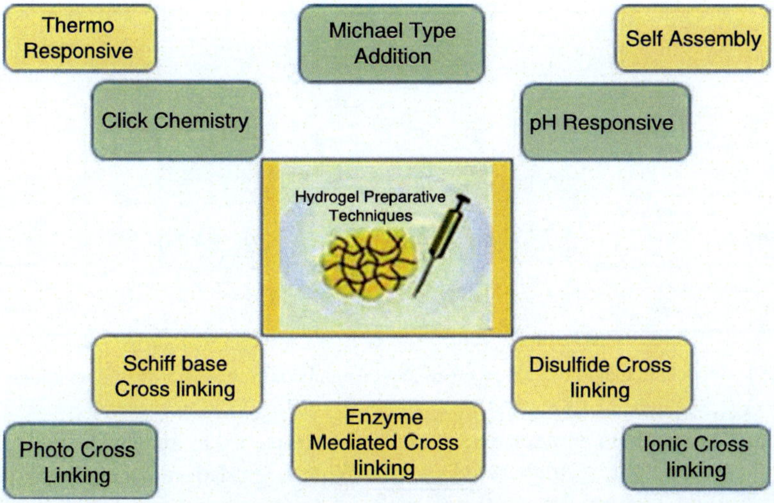

Fig. 8.29 Different Techniques for the preparation of Hydrogels

particles size were significantly affected by concentration ratio of emulsifier/lipid [246]. Lutein was nanoencapsulated with NLCs which leads to enhanced encapsulation efficiency and chemical stability. These nano-cargos also exhibited significant antioxidant activity against free radicals along with better controlled released as compared to lutein-loaded nanoemulsion [237]. β-Carotene was also entrapped within NLCs which were fabricated by Precirol ATO5 (solid lipid) and octyl octanate (liquid oil) while Poloxamer 407 used as surfactant. It was reported that encapsulation efficiency enhanced upto 97.7% and the β-carotene was protected from degradation during the storage period of 14 days at 25 °C. The results also concluded that smallest particles size were gained by 2% concentration of Poloxamer 407 as compared to other parameters [333].

8.4.2 Nano-Hydrogels

Hydrogels are basically three-dimensional network of polymeric molecules which have great capacity to hold definite amount of liquids within their three-dimensional system via osmosis, capillary forces, and polymer/liquid molecular interactions [334]. These factors played a important role in designing the structural as well as dimension model of the hydrogels [335].

Nanohydrogels have been prepared by shaping these hydrogels in to the self-assembly nano-structures [334], which developed by the electrostatic force of interactions between the unlike charged polymers. The synthetic mechanism of fluid phase gelling is controlled by crosslinking agents, temperature adjustment, multivalent ions addition or acidification [336] (Fig. 8.29).

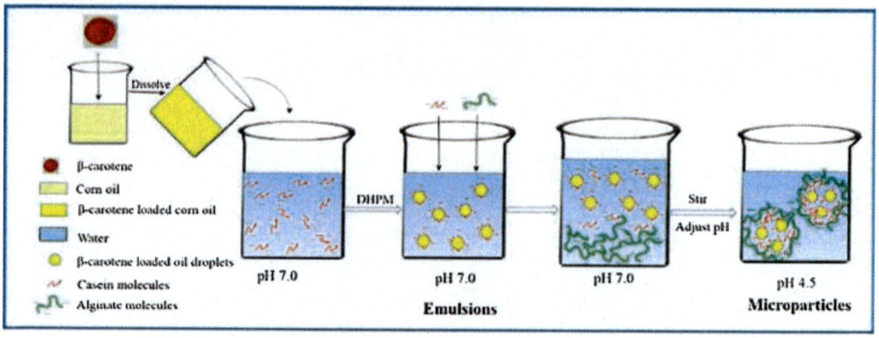

Fig. 8.30 Schematic diagram for β-carotene-loaded emulsions fabricated by microparticles [340]

Nano-hydrogels is another excellent choice to use as an encapsulant for many bioactive compounds, which could prevent from degradation either by providing lipophilic and hydrophilic shield as well as helps in effective targeted release [259, 337].

Crocin (i.e. saffron carotenoid) was encapsulated by means of chitosan and alginate based hydrogels using ionic gelation method. These crocin-loaded nanoparticles exhibited a potent anticancer and antioxidant properties with controlled release. Moreover they also showed enhanced chemical stability against harsh environmental conditions including pH, light and temperature. This study concluded that these nanohydrogels as a novel and favourable candidate for food industries [338].

β-carotene-loaded nanohydrogels have been prepared by alginate and whey protein isolate. These nanohydrogels exhibited a improved chemical stability at elevated temperatures during storage period and passing through the GIT as comparative to nanoemulsions. In another study, researcher found a lower bioaccessibility of these nanohydrogels formulation instead of nanoemulsions [339]. β-Carotene-loaded microparticles prepared by mixing the β-carotene-loaded emulsions with 2% sodium caseinate (biopolymer solution 1:1 ratio, at pH 7.0) then with 2% alginate solution (volume ratio of 1:2). Then the pH of resulting mixture was adjusted at 4.5 by adding 1 M citric acid. The biopolymer microparticles enhanced the chemical stability, water-solubility, and bioavailability β-carotene confirmed by *in* vitro digestion experiment. The results reported that 32% of carotenoids were retained during 42 days of storage period [340] (Fig. 8.30).

8.5 Conclusion

Carotenoids, natural products obtained from plant origin are well known in literature for their ability to act as anti-oxidants, anti-tumour and anti-ageing. They posses unique ability to absorb reactive oxygen species (ROS) and free radicals in the body. These characteristics have enabled them for their use in food industry to improve the quality of human life. This chapter deals with the stability of carotenoids under various conditions of food processing. Moreover, the literature is reviewed for the ability of the carotenoids to undergo oxidation reactions, thermal degradation and photo-degradation reactions. In-addition, literature dealing various encapsulation methods is also discussed in-detail.

References

1. Huang JJ, Lin S, Xu W, Cheung PCK (2017) Occurrence and biosynthesis of carotenoids in phytoplankton. Biotechnol Adv 35(5):597–618
2. Khalid M, Bilal M, Iqbal HM, Huang D (2018) Biosynthesis and biomedical perspectives of carotenoids with special reference to human health -related applications. Biocatal Agric Biotechnol 17:399–407
3. Bezerra PQM, de Matos MFR, Ramos IG, Magalhães Guedes KT, Druzian JI, Costa JAV, Nunes IL (2019) Innovative functional nanodispersion: combination of carotenoid from Spirulina and yellow passion fruit albedo. Food Chem 285:397–405
4. Gong M, Bassi A (2016) Carotenoids from microalgae: a review of recent developments. Biotechnol Adv 34(8):1396–1412
5. Solovchenko A, Merzlyak M (2008) Screening of visible and UV radiation as a photoprotective mechanism in plants. Russ J Plant Physiol 55(6):719
6. Ueki N, Ide T, Mochiji S, Kobayashi Y, Tokutsu R, Ohnishi N, Minagawa J (2016) Eyespotdependent determination of the phototactic sign in Chlamydomonas reinhardtii. Proc Natl Acad Sci 113(19):5299–5304
7. Rostamabadi H, Falsafi SR, Jafari SM (2019a) Nanoencapsulation of carotenoids within lipid -based nanocarriers. J Control Release 298:38–67
8. Saini RK, Moon SH, Gansukh E, Keum YS (2018) An efficient one -step scheme for the purification of major xanthophyll carotenoids from lettuce, and assessment of their comparative anticancer potential. Food Chem 266:56–65
9. Moreira MD, Melo MM, Coimbra JM, dos Reis KC, Schwan RF, Silva CF (2018) Solid coffee waste as alternative to produce carotenoids with antioxidant and antimicrobial activities. Waste Manag 82:93–99
10. Zeng J, Zhao J, Dong B, Cai X, Jiang J, Xue R, Liu C (2019) Lycopene protects against pressure overload -induced cardiac hypertrophy by attenuating oxidative stress. J Nutr Biochem 66:70–78
11. Kaulmann A, Bohn T (2014) Carotenoids, inflammation, and oxidative stress —implications of cellular signaling pathways and relation to chronic disease prevention. Nutr Res 34(11):907–929
12. Neville CE, Young IS, Gilchrist SE, McKinley MC, Gibson A, Edgar JD, Woodside JV (2013) Effect of increased fruit and vegetable consumption on physical function and muscle strength in older adults. Age 35(6):2409–2422
13. Chuyen HV, Roach PD, Golding JB, Parks SE, Nguyen MH (2019) Encapsulation of carotenoid -rich oil from Gac peel: optimisation of the encapsulating process using a spray drier and the storage stability of encapsulated powder. Powder Technol 344:373–379
14. Xianquan S, Shi J, Kakuda Y, Yueming J (2005) Stability of lycopene during food processing and storage. J Med Food 8(4):413–422
15. Britton G (1995) Structure and properties of carotenoids in relation to function. FASEB J 9:1551–1558
16. Mordi R, Walton JC, Burton GW, Hughes L, Ingold KU, Lindsay DA, Moffatt DJ (1993) Oxidative degradation of beta-carotene and beta-apo-8-carotenal. Tetrahedron 49(4):911–928

17. Henry LK, Puspitasari-Nienaber NL, Jaren-Galan M, van Breemen RB, Catignani GL, Schwartz SJ (2000) Effects of ozone and oxygen on the degradation of carotenoids in an aqueous model system. J Agric Food Chem 48:5008–5013
18. Byrne SM, Blaner WS (2005) Introduction to retinoids. In: Carotenoids and retinoids: molecular-aspects and health issue. Marcel Dekker, New York, pp 1–22
19. Clark KB, Howard JA, Oyler AR (1997) Retinoic acid oxidation at high oxygen pressures: evidence for spin-forbidden direct addition of triplet molecular oxygen. J Am Chem Soc 119:9560–9561
20. Bonnie TP, Choo YM (1999) Oxidation and thermal degradation of carotenoids. J Oil Palm Res 2(1):62–78
21. Kanasawud P, Crouzet JC (1990) Mechanism of formation of volatile compounds by thermal degradation of carotenoids in aqueous medium. 1. Beta-carotene degradation. J Agric Food Chem 38:237–243
22. Marty C, Berset C (1990) Factors affecting the thermal degradation of all-trans-beta-carotene. J Agric Food Chem 38:1063–1067
23. Mader I (1964) Beta-carotene: thermal degradation. Science 144:533–534
24. El-Tinay AH, Chichester CO (1970) Oxidation of beta-carotene. Site of initial attack. J Org Chem 35(7):2290–2293
25. Handelman G, van Kuijk F, Chatterjee A, Krinsky NI (1991) Characterization of products formed during the autoxidation of beta-carotene. Free Radic Biol Med 10:427–437
26. Kadian SS, Sharma A, Sood DR (2013) Effect of light and heat on stability of crude carotenoid extract from natural sources. Int J Pharm Sci Res 4(6):2415–2418
27. Hung LC, Basri M, Tejo BA, Ismail R, Nang HL, Abu Hassan H, May CY (2011) An improved method for the preparations of nanostructured lipid carriers containing heat-sensitive bioactives. Colloids Surf B Biointerfaces 87(1):180–186
28. Konovalova TA, Gao YL, Schad R, Kispert LD, Saylor CA, Brunei LC (2001) Photooxidation of carotenoids in mesoporous MCM-41, Ni-MCM-41 and Al-MCM-41 molecular sieves. J Phys Chem B 105(31):7459–7464
29. Mortensen A, Skibsted LH (1996) Kinetics of photobleaching of betacarotene in chloroform and formation of transient carotenoid species absorbing in the near infrared. Free Radic Res 25(4):355–368
30. Stahl W, Sies H (1992) Uptake of lycopene and its geometrical isomers is greater from heat-processed than from unprocessed tomato juice in humans. J Nutr 122:2161–2166
31. Stahl W, Sies H (1996) Lycopene: a biologically important carotenoid for humans. Arch Biochem Biophys 336(1):1–9
32. Jung MY, Min DB (1991) Effects of quenching mechanisms of carotenoids on the photosensitized oxidation of soybean oil. J Am Oil Chem Soc 68(9):653–658
33. Krinsky NI (1998) The antioxidant and biological properties of the carotenoids. Ann N Y Acad Sci 854:443–447
34. Choe E, Min DB (2006) Mechanisms and factors for edible oil oxidation. Compr Rev Food Sci Food Saf 5:169–186
35. Edge R, McGarvey DJ, Truscott TG (1997) The carotenoids as antioxidants: a review. J Photochem Photobiol B Biol 41:189–200
36. Conn PF, Schalch W, Truscott TG (1991) The singlet oxygen and carotenoid interaction. J Photochem Photobiol B Biol 11:41–47
37. Garavelli M, Bernardi F, Olivucci M, Robb MA (1998) DFT study of the reactions between singlet-oxygen and a carotenoid model. J Am Chem Soc 120:10210–10222
38. Yamauchi R, Miyake N, Inoue H, Kato K (1993) Products formed by peroxyl radical oxidation of beta-carotene. J Agric Food Chem 41:708–713
39. Edge R, McGarvey DJ, Truscott TG (1997) The carotenoids as antioxidants-a review. J Photochem Photobiol Biol 41:189–200
40. Rodriguez-Amaya DB (2015) Food carotenoids: chemistry, biology and technology. Wiley Blackwell, West Sussex

41. Haila KM, Nielsen BR, Heinonen MI, Skibsted LH (1997) Carotenoid reaction with free radicals in acetone and toluene at different oxygen partial pressures. Eur Food Res Technol 204(2):81–87
42. Everett SA, Dennis MF, Patel KB, Maddix S, Kundu SC, Willson RL (1996) Scavenging of nitrogen dioxide, thiyl, and sulfonyl free radicals by the nutritional antioxidant β-carotene. J Biol Chem 271:3988–3994
43. El-Agamey A, Lowe GM, McGarvey DJ, Mortensen A, Phillip DM, Truscott TG, Young AJ (2004) Carotenoid radical chemistry and antioxidant/pro-oxidant properties. Arch Biochem Biophys 430:37–48
44. Burton GW, Ingold KU (1984) Beta-carotene: an unusual type of lipid antioxidant. Science 224:569–573
45. Palozza P, Serini S, Nicuolo FD, Piccioni E, Calviello G (2003) Prooxidant effects of β-carotene in cultured cells. Mol Asp Med 24:353–362
46. Young AJ, Lowe GM (2001) Antioxidant and prooxidant properties of carotenoids. Arch Biochem Biophys 385(l):20–27
47. Britton G, Liaaen-Jensen S, Pfander H (1995) Carotenoids today and challenges for the future. In: Carotenoids. Isolation and analysis, vol 1A. Birkhäuser, Basel, pp 13–26
48. Mortensen A, Skibsted LH, Sampson J, Rice-Evans C, Everett SA (1997) Comparative mechanisms and rates of free radical scavenging by carotenoid antioxidants. FEBS Lett 41:91–97
49. Stratton SP, Schaefer WH, Liebler DC (1993) Isolation and identification of singlet oxygen oxidation products of beta-carotene. Chem Res Toxicol 6:542–547
50. Ukai N, Lu Y, Etoh H, Yagi A, Ina K, Oshima S, Ojima F, Sakamoto H, Ishiguro Y (1994) Photosensitized oxygenation of lycopene. Biosci, Biotechnol Biochem 58(9):1718–1719
51. Gao GQ, Deng Y, Kispert LD (1997) Photoactivated ferric chloride oxidation of carotenoids by near-UV to visible light. J Phys Chem Biol 101(39):7844–7849
52. Burke M, Edge R, Land EJ, McGarvey DJ, Truscott TG (2001) One-electron reduction potentials of dietary carotenoid radical cations in aqueous micellar environments. FEBS Lett 500(3):132–136
53. Mortensen A, Skibsted LH (1988) Reactivity of beta-carotene towards peroxyl radicals studied by laser flash and steady-state photolysis. FEBS Lett 426:392–396
54. Gao YL, Kispert LD (2003) Reaction of carotenoids and ferric chloride: equilibria, isomerization, and products. J Phys Chem B 107(22):5333–5338
55. Konovalova TA, Kispert LD, Polyakov NE, Leshina TV (2000) EPR spin trapping detection of carbon-centered carotenoid and beta-ionone radicals. Free Radic Biol Med 28(7):1030–1038
56. Gao YL, Webb S, Kispert LD (2003) Deprotonation of carotenoid radical cation and formation of a didehydrodimer. J Phys Chem B 107(47):13237–13240
57. Gao YL, Focsan AL, Kispert LD, Dixon DA (2006) Density functional theory study of the beta-carotene radical cation and deprotonated radicals. J Phys Chem B 110(48):24750–24756
58. Grant JL, Kramer VJ, Ding R, Kispert LD (1988) Carotenoid cation radicals: electrochemical, optical, and EPR study. J Am Chem Soc 110:2151–2157
59. Khaled M, Hadjipetrou A, Kispert LD (1991) Simultaneous electrochemical and electron paramagnetic resonance studies of carotenoid cation radicals and dications. J Phys Chem 95:2438–2442
60. Khaled M, Hadjipetrou A, Kispert L (1990) Electrochemical and electron paramagnetic resonance studies of carotenoid cation radicals and dications: effect of deuteration. J Phys Chem 94:5164–5169
61. Jeevarajan JA, Wei CC, Jeevarajan AS, Kispert LD (1996) Optical absorption spectra of dications of carotenoids. J Phys Chem 100:5637–5641
62. Deng Y, Gao GQ, He ZF, Kispert LD (2000) Effects of polyene chain length and acceptor substituents on the stability of carotenoid radical cations. J Phys Chem B 104(23):5651–5656
63. Wei C, Gao G, Kispert LD (1997) Selected cis/trans isomers of carotenoids formed by bulk electrolysis and iron(III) chloride oxidation. J Chem Soc Perkin Trans 2:783–786

64. Gao G, Wei CC, Jeevarajan AS, Kispert LD (1996) Geometrical isomerization of carotenoids mediated by cation radical/dication formation. J Phys Chem 100:5362–5366
65. Konovalov VV, Kispert LD (1999) AMI, INDO/S and optical studies of carbocations of carotenoid molecules. Acid induced isomerization. Journal of Chemical Society Perkin Transaction 2:901–909
66. He ZF, Kispert LD (2001) Carotenoids in sol-gels: incorporation, stability, and sensitivity to oxidant and acid. Chem Mater 13(1):227–231
67. Britton G, Liaaen-Jensen S, Pfander H (1994) Carotenoids, volume 1 a: isolation and analysis. Birkhäuser, Basel
68. Bouis HE, Hotz C, McClafferty B, Meenakshi JV, Pfeiffer WH (2011) Biofortification: a new tool to reduce micronutrient malnutrition. Food Nutr Bull 32(1 Suppl):S31–S40
69. Saltzman A, Birol B, Bouis HE, Boy E, De Moura FF, Islam Y, Pfeiffer WH (2013) Biofortification: progress toward a more nourishing future. Glob Food Secur 2:9–17
70. Mugode L, Ha B, Kaunda A, Sikombe T, Phiri S, Mutale R, De Moura FF (2014) Carotenoid retention of biofortified provitamin A maize (Zea mays L.) after Zambian traditional methods of milling cooking and storage. J Agric Food Chem 62:6317–6325
71. Bechoff A, Tomlins KI, Dhuique-Mayer C, Dove R, Westby A (2011) On-farm evaluation of the impact of drying and subsequent storage on the carotenoid content of orange-fleshed sweet potato. Int J Food Sci Technol 46:52–60
72. Bechoff A (2010) Investigating Carotenoid Losses After Drying and Storage of Orange-Fleshed Sweet Potato. (PhD thesis) University of Greenwich, UK
73. Abu JO, Badifu GIO, Akpapunam MA (2006) Effect of crude palm oil inclusion on some physic-chemical properties of gari – a fermented cassava food product. Nigererian Food J 24:73–79
74. Gouado I, Demasse Mawamba A, Meyimgo Ouambo RS, Some IT, Félicite TM (2008) Provitamin A carotenoid content of dried fermented cassava flour: the effect of palm oil addition during processing. Int J Food Eng 561(4):1556–1558
75. Uzomah A, Ubbaonu CN, Osuji CM (2006) Vitamin A retention in palm oil- gari during processing and storage. Nigerian Food J 23(1):69–73
76. Burri BJ (2012) Evaluating global barriers to the use of red palm oil as an intervention food to prevent vitamin A deficiency. Compr Rev Food Sci Food Saf 11(2):221–232
77. Bechoff A, Ugo Chijioke U, Tomlins K, Govinden P, Ilon P, Westby A, Boy E (2015) Carotenoid stability during storage of yellow gari made from biofortified cassava or with palm oil. J Food Compos Anal 44:36–44
78. Soukoulis C, Bohn T (2018) A comprehensive overview on the micro-and nano-technological encapsulation advances for enhancing the chemical stability and bioavailability of carotenoids. Crit Rev Food Sci Nutr 58(1):1–36
79. Assadpour E, Mahdi, Jafari S (2018) A systematic review on nanoencapsulation of food bioactive ingredients and nutraceuticals by various nanocarriers. Crit Rev Food Sci Nutr 59(1):1–47
80. Jafari S (2017) An introduction to nanoencapsulation techniques for the food bioactive ingredients. In: Nanoencapsulation of food bioactive ingredients. Taylor and Francis Group, Boca Ratón, pp 1–62
81. Medeiros AK, de Carvalho OC, Gomes C, de Araújo Amaral MLQ, de Medeiros LDG, Medeiros I, Porto DL, Passos TS (2019) Nanoencapsulation improved water solubility and color stability of carotenoids extracted from Cantaloupe melon (Cucumis melo L.). Food Chem 270:562–572
82. Rehman A, Ahmad T, Aadil RM, Spotti MJ, Bakry AM, Khan IM et al (2019) Pectin polymers as wall materials for the nano-encapsulation of bioactive compounds. Trends Food Sci Technol 90:35–46
83. McClements DJ, Gumus CE (2016) Natural emulsifiers biosurfactants, phospholipids, biopolymers, and colloidal particles: molecular and physicochemical basis of functional performance. Adv Colloid Interf Sci 234:3–26

84. Bhatt PC, Srivastava P, Pandey P, Khan W, Panda BP (2016) Nose to brain delivery of astaxanthin-loaded solid lipid nanoparticles: fabrication, radio labeling, optimization and biological studies. RSC Adv 6(12):10001–10010
85. Esposito E, Drechsler M, Mariani P, Panico AM, Cardile V, Crascì L, Puglia C (2017) Nanostructured lipid dispersions for topical administration of crocin, a potent antioxidant from saffron (Crocus sativus L.). Mater Sci Eng C 71:669–677
86. Xia S, Tan C, Zhang Y, Abbas S, Feng B, Zhang X, Qin F (2015) Modulating effect of lipid bilayer–carotenoid interactions on the property of liposome encapsulation. Colloids Surf B: Biointerfaces 128:172–180
87. Xia Z, McClements DJ, Xiao H (2015) Influence of physical state of β-carotene (crystallized versus solubilized) on bioaccessibility. J Agric Food Chem 63(3):990–997
88. Rehman A, Tong Q, Jafari SM, Assadpour E, Shehzad Q, Aadil RM, Iqbal MW (2020) Rashed, MA, Mushtaq BS. Ashraf W Carotenoid-loaded nanocarriers: A comprehensive review Advances in Colloid and Interface Science 275:102048
89. Salem A, Ramadan AR, Shoeib T (2018) Entrapment of β -carotene and zinc in whey protein nanoparticles using the pH cycle method: evidence of sustained release delivery in intestinal and gastric fluids. Food Biosci 26:161–168
90. Wei Y, Sun C, Dai L, Zhan X, Gao Y (2018) Structure, physicochemical stability and in vitro simulated gastrointestinal digestion properties of β-carotene loaded zein-propylene glycol alginate composite nanoparticles fabricated by emulsification-evaporation method. Food Hydrocoll 81:149–158
91. Rahaiee S, Hashemi M, Shojaosadati SA, Moini S, Razavi SH (2017) Nanoparticles based on crocin loaded chitosan-alginate biopolymers: antioxidant activities, bioavailability and anticancer properties. Int J Biol Macromol 99:401–408
92. Cordenonsi LM, Faccendini A, Catanzaro M, Bonferoni MC, Rossi S, Malavasi L et al (2019) The role of chitosan as coating for nanostructured lipid carrier for skin delivery of fucoxanthin. Int J Pharm 2019:118487
93. Wang W, Jung J, Zhao Y (2017) Chitosan-cellulose nanocrystal microencapsulation to improve encapsulation efficiency and stability of entrapped fruit anthocyanins. Carbohydr Polym 157:1246–1253
94. Li W, Yalcin M, Lin Q, Ardawi MSM, Mousa SA (2017) Self-assembly of green tea catechin derivatives in nanoparticles for oral lycopene delivery. J Control Release 248:117–124
95. Wang M, Fu Y, Chen G, Shi Y, Li X, Zhang H, Shen Y (2018) Fabrication and characterization of carboxymethyl chitosan and tea polyphenols coating on zein nanoparticles to encapsulate β-carotene by anti-solvent precipitation method. Food Hydrocoll 77:577–587
96. Beicht J, Zeeb B, Gibis M, Fischer L, Weiss J (2013) Influence of layer thickness and composition of cross -linked multilayered oil -in -water emulsions on the release behavior of lutein. Food Funct 4(10):1457–1467
97. Chu BS, Ichikawa S, Kanafusa S, Nakajima M (2007) Preparation and characterization of β –carotene nanodispersions prepared by solvent displacement technique. J Agric Food Chem 55(16):6754–6760
98. De Paz E, Martín Á, Estrella A, Rodríguez-Rojo S, Matias AA, Duarte CM, Cocero MJ (2012) Formulation of β-carotene by precipitation from pressurized ethyl acetate -on -water emulsions for application as natural colorant. Food Hydrocoll 26(1):17–27
99. Pan X, Yao P, Jiang M (2007) Simultaneous nanoparticle formation and encapsulation driven by hydrophobic interaction of casein-graft-dextran and β-carotene. J Colloid Interface Sci 315(2):456–463
100. Gumus CE, Davidov-Pardo G, McClements DJ (2016) Lutein -enriched emulsion -based delivery systems: impact of Maillard conjugation on physicochemical stability and gastrointestinal fate. Food Hydrocoll 60:38–49
101. Lim AS, Griffin C, Roos YH (2014) Stability and loss kinetics of lutein and β -carotene encapsulated in freeze -dried emulsions with layered interface and trehalose as glass former. Food Res Int 62:403–409

102. López-Rubio A, Lagaron JM (2012) Whey protein capsules obtained through electrospraying for the encapsulation of bioactives. Innovative Food Sci Emerg Technol 13:200–206
103. Yi J, Zhang Y, Liang R, Zhong F, Ma J (2015a) Beta-carotene chemical stability in nanoemulsions was improved by stabilized with beta-lactoglobulincatechin conjugates through free radical method. J Agric Food Chem 63:297–203
104. Edelman R, Engelberg S, Fahoum L, Meyron-Holtz EG, Livney YD (2019) Potato protein -based carriers for enhancing bioavailability of astaxanthin. Food Hydrocoll 96:72–80
105. Gupta SS, Ghosh M (2012) In vitro study of anti -oxidative effects of β -carotene and α -lipoic acid for nanocapsulated lipids. LWT-Food Sci Technol 49(1):131–138
106. Teo A, Lee SJ, Goh KK, Wolber FM (2017) Kinetic stability and cellular uptake of lutein in WPI -stabilised nanoemulsions and emulsions prepared by emulsification and solvent evaporation method. Food Chem 221:1269–1276
107. Li D, Li L, Xiao N, Li M, Xie X (2018) Physical properties of oil -in -water nanoemulsions stabilized by OSA -modified starch for the encapsulation of lycopene. Colloids Surf A Physicochem Eng Asp 552:59–66
108. Luo X, Zhou Y, Bai L, Liu F, Deng Y, McClements DJ (2017) Fabrication of β -carotene nanoemulsion -based delivery systems using dual -channel microfluidization: physical and chemical stability. J Colloid Interface Sci 490:328–335
109. Singh A, Neupane YR, Panda BP, Kohli K (2017) Lipid based nanoformulation of lycopene improves oral delivery: formulation optimization, ex vivo assessment and its efficacy against breast cancer. J Microencapsul 34(4):416–429
110. Yi J, Lam TI, Yokoyama W, Cheng LW, Zhong F (2014) Cellular uptake of β -carotene from protein stabilized solid lipid nanoparticles prepared by homogenization –evaporation method. J Agric Food Chem 62(5):1096–1104
111. Chaiyasan W, Srinivas SP, Tiyaboonchai W (2015) Crosslinked chitosan -dextran sulfate nanoparticle for improved topical ocular drug delivery. Mol Vis 21:1224
112. Mehrad B, Ravanfar R, Licker J, Regenstein JM, Abbaspourrad A (2018) Enhancing the physicochemical stability of β-carotene solid lipid nanoparticle (SLNP) using whey protein isolate. Food Res Int 105:962–969
113. Heyang J, Fei X, Jiang C, Yaping Z, Lin H (2009) Nanoencapsulation of lutein with hydroxypropylmethyl cellulose phthalate by supercritical antisolvent. Chin J Chem Eng 17(4):672–677
114. Chang R, Yang J, Ge S, Zhao M, Liang C, Xiong L, Sun Q (2017) Synthesis and self-assembly of octenyl succinic anhydride modified short glucan chains based amphiphilic biopolymer: micelles, ultrasmall micelles, vesicles, and lutein encapsulation/release. Food Hydrocoll 67:14–26
115. Lee JS, Lee HG (2016) Chitosan/poly – γ -glutamic acid nanoparticles improve the solubility of lutein. Int J Biol Macromol 85:9–15
116. Debele TA, Mekuria SL, Tsai HC (2016) Polysaccharide based nanogels in the drug delivery system: application as the carrier of pharmaceutical agents. Mater Sci Eng C 68:964–981
117. Kang B, Opatz T, Landfester K, Wurm FR (2015) Carbohydrate nanocarriers in biomedical applications: functionalization and construction. Chem Soc Rev 44(22):8301–8325
118. Yang X, Du H, Liu J, Zhai G (2015) Advanced nanocarriers based on heparin and its derivatives for cancer management. Biomacromolecules 16(2):423–436
119. rnivec IGO, Ulrih NP (2019) Nano-hydrogels of alginate for encapsulation of food ingredients. In: Biopolymer nanostructures for food encapsulation purposes. Academic Press, New York, pp 335–380
120. Lee KY, Mooney DJ (2012) Alginate: properties and biomedical applications. Prog Polym Sci 37(1):106–126
121. Peteiro C (2018) Alginate production from marine macroalgae, with emphasis on kelp farming. In: Alginates and their biomedical applications. Springer, Singapore, pp 27–66

122. Flórez-Fernández N, Domínguez H, Torres M (2019) A green approach for alginate extraction from Sargassum muticum brown seaweed using ultrasound-assisted technique. Int J Biol Macromol 124:451–459
123. Zhang JP, Wang Q, Xie XL, Li X, Wang AQ (2010) Preparation and swelling properties of pH sensitive sodium alginate/layered double hydroxides hybrid beads for controlled release of diclofenac sodium. J Biomed Mater Res Part B Appl Biomater Off J Soc Biomater Jpn Soc Biomater Austr Soc Biomater Kor Soc Biomater 92(1):205–214
124. Bajpai P (2019) Biobased polymers: properties and applications in packaging. Elsevier, Amsterdam
125. da Rosa ZE, Kringel DH, Dias ARG (2019) Nano -scale polysaccharide materials in food and agricultural applications. Food Appl Nanotechnol 88:85
126. Gopi S, Balakrishnan P, Chandradhara D, Poovathankandy D, Thomas S (2019) General scenarios of cellulose and its use in the biomedical field. Mater Today Chem 13:59–78
127. Rokhade AP, Agnihotri SA, Patil SA, Mallikarjuna NN, Kulkarni PV, Aminabhavi TM (2006) Semi -interpenetrating polymer network microspheres of gelatin and sodium carboxymethyl cellulose for controlled release of ketorolac tromethamine. Carbohydr Polym 65(3):243–252
128. Yang Y, Guo Y, Sun R, Wang X (2016) Self-assembly and β -carotene loading capacity of hydroxyethyl cellulose -graft -linoleic acid nanomicelles. Carbohydr Polym 145:56–63
129. Zimet P, Mombrú ÁW, Mombrú D, Castro A, Villanueva JP, Pardo H, Rufo C (2019) Physicochemical and antilisterial properties of nisin-incorporated chitosan/carboxymethyl chitosan films. Carbohydr Polym 219:334–343
130. Fan Z, Qin Y, Liu S, Xing R, Yu H, Chen X, Li P (2018) Synthesis, characterization, and antifungal evaluation of diethoxyphosphoryl polyaminoethyl chitosan derivatives. Carbohydr Polym 190:1–11
131. Zou P, Yuan S, Yang X, Guo Y, Li L, Xu C, Wang J (2019) Structural characterization and antitumor effects of chitosan oligosaccharides against orthotopic liver tumor via NF-κB signaling pathway. J Funct Foods 57:157–165
132. Rambabu K, Bharath G, Banat F, Show PL, Cocoletzi HH (2018) Mango leaf extract incorporated chitosan antioxidant film for active food packaging. Int J Biol Macromol 126:1234–1243
133. El-Naby FSA, Naiel MA, Al-Sagheer AA, Negm SS (2019) Dietary chitosan nanoparticles enhance the growth, production performance, and immunity in Oreochromis niloticus. Aquaculture 501:82–89
134. Kurniasih M, Cahyati T, Dewi RS (2018) Carboxymethyl chitosan as an antifungal agent on gauze. Int J Biol Macromol 119:166–171
135. Brink I, Šipailienė A, Leskauskaitė D (2019) Antimicrobial properties of chitosan and whey protein films applied on fresh cut turkey pieces. Int J Biol Macromol 130:810–817
136. Patrulea V, Laurent-Applegate LA, Ostafe V, Borchard G, Jordan O (2019) Polyelectrolyte nanocomplexes based on chitosan derivatives for wound healing application. Eur J Pharm Biopharm 140:100–108
137. Fazli Y, Shariatinia Z (2017) Controlled release of cefazolin sodium antibiotic drug from electrospun chitosan-polyethylene oxide nanofibrous mats. Mater Sci Eng C 71:641–652
138. Zhang L, Ma Y, Pan X, Chen S, Zhuang H, Wang S (2018) A composite hydrogel of chitosan/heparin/poly (γ-glutamic acid) loaded with superoxide dismutase for wound healing. Carbohydr Polym 180:168–174
139. Tan C, Feng B, Zhang X, Xia W, Xia S (2016) Biopolymer-coated liposomes by electrostatic adsorption of chitosan (chitosomes) as novel delivery systems for carotenoids. Food Hydrocoll 52:774–784
140. Ravi H, Kurrey N, Manabe Y, Sugawara T, Baskaran V (2018) Polymeric chitosan-glycolipid nanocarriers for an effective delivery of marine carotenoid fucoxanthin for induction of apoptosis in human colon cancer cells (Caco-2 cells). Mater Sci Eng C 91:785–795
141. Arunkumar R, Prashanth KVH, Baskaran V (2013) Promising interaction between nanoencapsulated lutein with low molecular weight chitosan: characterization and bioavailability of lutein in vitro and in vivo. Food Chem 141(1):327–337

142. Wang D, Mao L, Dai L, Yuan F, Gao Y (2018) Characterization of chitosan-ferulic acid conjugates and their application in the design of β-carotene bilayer emulsions with propylene glycol alginate. Food Hydrocoll 80:281–291
143. Bai Y, Shi YC (2016) Chemical structures in pyrodextrin determined by nuclear magnetic resonance spectroscopy. Carbohydr Polym 151:426–433
144. Alvani K, Qi X, Tester RF (2011) Use of carbohydrates, including dextrins, for oral delivery. Starch Stärke 63(7):424–431
145. Fan Y, Yi J, Zhang Y, Wen Z, Zhao L (2017) Physicochemical stability and in vitro bioaccessibility of β -carotene nanoemulsions stabilized with whey protein -dextran conjugates. Food Hydrocoll 63:256–264
146. Chang R, Xiong L, Li M, Chen H, Xiao J, Wang S, Sun Q (2019) Preparation of octenyl succinic anhydride -modified debranched starch vesicles for loading of hydrophilic functional ingredients. Food Hydrocoll 94:546–552
147. Sharif HR, Goff HD, Majeed H, Shamoon M, Liu F, Nsor-Atindana J, Zhong F (2017) Physicochemical properties of β -carotene and eugenol co-encapsulated flax seed oil powders using OSA starches as wall material. Food Hydrocoll 73:274–283
148. Liang R, Shoemaker CF, Yang X, Zhong F, Huang Q (2013) Stability and bioaccessibility of β – carotene in nanoemulsions stabilized by modified starches. J Agric Food Chem 61(6):1249–1257
149. Fang S, Zhao X, Liu Y, Liang X, Yang Y (2019) Fabricating multilayer emulsions by using OSA starch and chitosan suitable for spray drying: application in the encapsulation of β -carotene. Food Hydrocoll 93:102–110
150. Li X, Wang X, Xu D, Cao Y, Wang S, Wang B, Gao Y (2018) Enhancing physicochemical properties of emulsions by heteroaggregation of oppositely charged lactoferrin coated lutein droplets and whey protein isolate coated DHA droplets. Food Chem 239:75–85
151. Yabe T (2018) New understanding of pectin as a bioactive dietary fiber. J Food Bioactives 3:95–100
152. Roy MC, Alam M, Saeid A, Das BC, Mia MB, Rahman MA, Ahmed M (2018) Extraction and characterization of pectin from pomelo peel and its impact on nutritional properties of carrot jam during storage. J Food Process Preserv 42(1):13411
153. Yapo BM (2011) Pectic substances: from simple pectic polysaccharides to complex pectins. A new hypothetical model. Carbohydr Polym 86(2):373–385
154. Marić M, Grassino AN, Zhu Z, Barba FJ, Brnčić M, Brnčić SR (2018) An overview of the traditional and innovative approaches for pectin extraction from plant food wastes and by-products: ultrasound-, microwaves-, and enzyme-assisted extraction. Trends Food Sci Technol 76:28–37
155. Xu D, Yuan F, Gao Y, Panya A, McClements DJ, Decker EA (2014) Influence of whey protein–beet pectin conjugate on the properties and digestibility of β-carotene emulsion during in vitro digestion. Food Chem 156:374–379
156. Esfanjani AF, Jafari SM, Assadpoor E, Mohammadi A (2015) Nano-encapsulation of saffron extract through double-layered multiple emulsions of pectin and whey protein concentrate. J Food Eng 165:149–155
157. Esfanjani AF, Jafari SM, Assadpour E (2017) Preparation of a multiple emulsion based on pectin when protein complex for encapsulation of saffron extract nanodroplets. Food Chem 221:1962–1969
158. Li H, Qi Y, Zhao Y, Chi J, Cheng S (2019) Starch and its derivatives for paper coatings: a review. Prog Org Coat 135:213–227
159. Xie F, Pollet E, Halley PJ, Averous L (2013) Starch -based nano -biocomposites. Prog Polym Sci 38(10–11):1590–1628
160. Bravo-Núñez Á, Pando V, Gómez M (2019) Physically and chemically modified starches as texturisers of low -fat milk gels. Int Dairy J 92:21–27
161. Cai C, Wei B, Tian Y, Ma R, Chen L, Qiu L, Jin Z (2019) Structural changes of chemically modified rice starch by one-step reactive extrusion. Food Chem 288:354–360

162. Liu J, Wang B, Lin L, Zhang J, Liu W, Xie J, Ding Y (2014) Functional, physicochemical properties and structure of cross -linked oxidized maize starch. Food Hydrocoll 36:45–52
163. Leyva-López R, Palma-Rodríguez HM, López-Torres A, Capataz-Tafur J, Bello-Pérez LA, Vargas Torres A (2019) Use of enzymatically modified starch in the microencapsulation of ascorbic acid: microcapsule characterization, release behavior and in vitro digestion. Food Hydrocoll 96:259–266
164. Onofre F, Wang YJ (2010) Hydroxypropylated starches of varying amylose contents as sustained release matrices in tablets. Int J Pharm 385(1–2):104–112
165. Wu CN, Lai HM (2019) Novel pH -responsive granules with tunable volumes from oxidized corn starches. Carbohydr Polym 208:201–212
166. Balakrishnan P, Sreekala M, Geethamma V, Kalarikkal N, Kokol V, Volova T, Thomas S (2019) Physicochemical, mechanical, barrier and antibacterial properties of starch nanocomposites crosslinked with pre -oxidised sucrose. Ind Crop Prod 130:398–408
167. Xiao H, Yang T, Lin Q, Liu GQ, Zhang L, Yu F, Chen Y (2016) Acetylated starch nanocrystals: preparation and antitumor drug delivery study. Int J Biol Macromol 89:456–464
168. Loksuwan J (2007) Characteristics of microencapsulated b-carotene formed by spray drying with modified tapioca starch, native tapioca starch and maltodextrin. Food Hydrocoll 21:928–935
169. Rocha GA, Fívaro-Trindade CSL, Grosso CRF (2012) Microencapsulation of lycopene by spray drying: characterization, stability and application of microcapsules. Food Bioprod Process 90:37–42
170. Nehete JY, Bhambar RS, Narkhede MR, Gawali SR (2013) Natural proteins: sources, isolation, characterization and applications. Pharmacogn Rev 7(14):107
171. Tarhini M, Greige-Gerges H, Elaissari A (2017) Protein-based nanoparticles: from preparation to encapsulation of active molecules. Int J Pharm 522(1–2):172–197
172. Elzoghby AO, Samy WM, Elgindy NA (2012) Albumin-based nanoparticles as potential controlled release drug delivery systems. J Control Release 157(2):168–182
173. Madalena DA, Pereira RN, Pinheiro AC, Teixeira JA, Vicente AA, Ramos ÓL (2017) Design of β-lactoglobulin nanostructures for encapsulation and controlled release of riboflavin in the gastrointestinal tract
174. Pathakoti K, Manubolu M, Hwang HM (2017) Nanostructures: current uses and future applications in food science. J Food Drug Anal 25(2):245–225
175. Ranadheera C, Liyanaarachchi W, Chandrapala J, Dissanayake M, Vasiljevic T (2016) Utilizing unique properties of caseins and the casein micelle for delivery of sensitive food ingredients and bioactives. Trends Food Sci Technol 57:178–187
176. Rehan F, Ahemad N, Gupta M (2019) Casein Nanomicelle as an emerging biomaterial–a comprehensive review. Colloids Surf B: Biointerfaces. https://doi.org/10.1016/j.colsurfb.2019.03.051
177. Chen M, Sala G, Meinders MB, van Valenberg HJ, van der Linden E, Sagis LM (2017) Interfacial properties, thin film stability and foam stability of casein micelle dispersions. Colloids Surf B: Biointerfaces 149:56–63
178. Nogueira MH, Tavares GM, Silva NFN, Casanova F, Stringheta PC, Gaucheron F, Carvalho AF (2019) Physico-chemical stability of casein micelles cross-linked by transglutaminase as a function of acidic pH. Food Struct 19:100–103
179. Gandhi S, Roy I (2019) Doxorubicin-loaded casein nanoparticles for drug delivery: preparation, characterization and in vitro evaluation. Int J Biol Macromol 121:6–12
180. Ghayour N, Hosseini SMH, Eskandari MH, Esteghlal S, Nekoei AR, Gahruie HH, Naghibalhossaini F (2019) Nanoencapsulation of quercetin and curcumin in casein-based delivery systems. Food Hydrocoll 87:394–403
181. Cornacchia L, Roos YH (2011) Stability of β-carotene in protein-stabilized oil-in-water delivery systems. J Agric Food Chem 59(13):7013–7020
182. Zhang Y, Zhong Q (2013) Encapsulation of bixin in sodium caseinate to deliver the colorant in transparent dispersions. Food Hydrocoll J 33:1–9

183. Dai Q, You X, Che L, Yu F, Selomulya C, Chen XD (2013) An investigation in microencapsulating astaxanthin using a monodisperse droplet spray dryer. Dry Technol 31:1562–1569
184. Sáiz-Abajo MJ, González-Ferrero C, Moreno-Ruiz A, Romo-Hualde A, González-Navarro CJ (2013) Thermal protection of β-carotene in re-assembled casein micelles during different processing technologies applied in food industry. Food Chem 138(2–3):1581–1587
185. Jarunglumlert T, Nakagawa K, Adachi S (2015) Influence of aggregate structure of casein on the encapsulation efficiency of β-carotene entrapped via hydrophobic interaction. Food Struct 5:42–50
186. Mora-Gutierrez A, Attaie R, de González MN, Jung Y, Woldesenbet S, Marquez S (2018) Complexes of lutein with bovine and caprine caseins and their impact on lutein chemical stability in emulsion systems: effect of arabinogalactan. J Dairy Sci 101(1):18–27
187. Wang C, Tian Z, Chen L, Temelli F, Liu H, Wang Y (2010) Functionality of barley proteins extracted and fractionated by alkaline and alcohol methods. Cereal Chem 87(6):597–606
188. Arte E, Huang X, Nordlund E, Katina K (2019) Biochemical characterization and techno-functional properties of bioprocessed wheat bran protein isolates. Food Chem 289:103–111
189. Liao L, Luo Y, Zhao M, Wang Q (2012) Preparation and characterization of succinic acid deamidated wheat gluten microspheres for encapsulation of fish oil. Colloids Surf B: Biointerfaces 92:305–314
190. Fu D, Deng S, McClements DJ, Zhou L, Zou L, Yi J, Liu W (2019) Encapsulation of β-carotene in wheat gluten nanoparticle-xanthan gum-stabilized Pickering emulsions: enhancement of carotenoid stability and bioaccessibility. Food Hydrocoll 89:80–89
191. Sullivan P, Arendt E, Gallagher E (2013) The increasing use of barley and barley by-products in the production of healthier baked goods. Trends Food Sci Technol 29(2):124–134
192. Wang R, Tian Z, Chen L (2011) Nano-encapsulations liberated from barley protein microparticles for oral delivery of bioactive compounds. Int J Pharm 406(1–2):153–162
193. Yang J, Zhou Y, Chen L (2014) Elaboration and characterization of barley protein nanoparticles as an oral delivery system for lipophilic bioactive compounds. Food Funct 5(1):92–101
194. Elzoghby AO (2013) Gelatin-based nanoparticles as drug and gene delivery systems: reviewing three decades of research. J Control Release 172(3):1075–1091
195. Karthikeyan S, Prasad NR, Ganamani A, Balamurugan E (2013) Anticancer activity of resveratrolloaded gelatin nanoparticles on NCI-H460 non-small cell lung cancer cells. Biomed Prev Nutr 3(1):64–73
196. Pal A, Bajpai J, Bajpai A (2018) Poly (acrylic acid) grafted gelatin nanocarriers as swelling controlled drug delivery system for optimized release of paclitaxel from modified gelatin. J Drug Deliv Sci Technol 45:323–333
197. Vafania B, Fathi M, Soleimanian-Zad S (2019) Nanoencapsulation of thyme essential oil in chitosangelatin nanofibers by nozzle-less electrospinning and their application to reduce nitrite in sausages. Food Bioprod Process 116(28):240–248
198. Zillies JC, Zwiorek K, Hoffmann F, Vollmar A, Anchordoquy TJ, Winter G, Coester C (2008) Formulation development of freeze-dried oligonucleotide-loaded gelatin nanoparticles. Eur J Pharm Biopharm 70(2):514–521
199. Anarjan N, Nehdi I, Sbihi H, Al-Resayes S, Malmiri H, Tan C (2014) Preparation of astaxanthin nanodispersions using gelatin-based stabilizer systems. Molecules 19(9):14257–14265
200. Miraghajani M, Zaghian N, Mirlohi M, Feizi A, Ghiasvand R (2017) The impact of probiotic soy milk consumption on oxidative stress among type 2 diabetic kidney disease patients: a randomized controlled clinical trial. J Ren Nutr 27(5):317–324
201. Chen J, Cheng YQ, Yamaki K, Li LT (2007) Anti-α-glucosidase activity of Chinese traditionally fermented soybean (douchi). Food Chem 103(4):1091–1096
202. Roberts CK, Hevener AL, Barnard RJ (2013) Metabolic syndrome and insulin resistance: underlying causes and modification by exercise training. Compr Physiol 3(1):1–58
203. Zhang X, Shu XO, Gao YT, Yang G, Li Q, Li H, Zheng W (2003) Soy food consumption is associated with lower risk of coronary heart disease in Chinese women. J Nutr 133(9):2874–2878

204. Prusty K, Biswal A, Biswal SB, Swain SK (2019) Synthesis of soy protein/polyacrylamide nanocomposite hydrogels for delivery of ciprofloxacin drug. Mater Chem Phys 234:378–389
205. Yin B, Deng W, Xu K, Huang L, Yao P (2012) Stable nano-sized emulsions produced from soy protein and soy polysaccharide complexes. J Colloid Interface Sci 380(1):51–59
206. Liu F, Tang CH (2016) Soy glycinin as food-grade Pickering stabilizers: part. III. Fabrication of gel like emulsions and their potential as sustained-release delivery systems for β-carotene. Food Hydrocoll 56:434–444
207. Deng XX, Zhang N, Tang CH (2017) Soy protein isolate as a nanocarrier for enhanced water dispersibility, stability and bioaccessibility of β- carotene. J Sci Food Agric 97(7):2230–2237
208. Elzoghby AO, Elgohary MM, Kamel NM (2015) Implications of protein-and Peptide-based nanoparticles as potential vehicles for anticancer drugs. Adv Protein Chem Struct Biol 98:169–221
209. Lohcharoenkal W, Wang L, Chen YC, Rojanasakul Y (2014) Protein nanoparticles as drug delivery carriers for cancer therapy. Biomed Res Int 2014:180549
210. Xu D, Yuan F, Wang X, Li X, Hou Z, Gao Y (2011) The effect of whey protein isolate-dextran conjugates on the free-thaw stability of oil-in-water emulsion. J Dispers Sci Technol 32:77–83
211. Xu D, Wang X, Jiang J, Yuan F, Gao Y (2012) Impact of whey protein-beet pectin conjugation on the physicochemical stability of β-carotene emulsions. Food Hydrocoll 28:258–266
212. Trentin A, De Lamo S, Güell C, López F, Ferrando M (2011) Proteinstabilized emulsions containing beta-carotene produced by premix membrane emulsification. J Food Eng 106:267–274
213. Tokle T, Mao Y, McClement DJ (2013) Potential biological fate of emulsion-based delivery systems: lipid particles nanolaminated with lactoferrin and βlactoglobulin coatings. Pharm Res 30:3200–3213
214. Wang X, Zhang J, Lei F, Liang C, Yuan F, Gao Y (2014b) Covalent complexation and functional evaluation of (−)-epigallocatechin gallate and α-lactalbumin. Food Chem 150:341–347
215. Wang X, Liu F, Liu L, Wei Z, Yuan F, Gao Y (2015) Physicochemical characterisation of β-carotene emulsion stabilised by covalent complexes of α-lactalbumin with (−)-epigallocatechin gallate or chlorogenic acid. Food Chem 173:564–568
216. Wei Z, Yang W, Fan R, Yuan F, Gao Y (2015) Evaluation of structural and functional properties of protein–EGCG complexes and their ability of stabilizing a model β-carotene emulsion. Food Hydrocoll 45:337–350
217. Liu F, Ma C, McClements DJ, Gao Y (2016c) Development of polyphenolprotein-polysaccharide ternary complexes as emulsifiers for nutraceutical emulsions: impact on formation, stability, and bioaccessibility of β-carotene emulsions. Food Hydrocoll 61:578–588
218. Zhao C, Shen X, Guo M (2018) Stability of lutein encapsulated whey protein nano-emulsion during storage. PLoS One 513(2):e0192511
219. Fathi M, Mozafari MR, Mohebbi M (2012) Nanoencapsulation of food ingredients using lipid based delivery systems. Trends Food Sci Technol 23(1):13–27
220. Đorđević V, Paraskevopoulou A, Mantzouridou F, Lalou S, Pantić M, Bugarski B, Nedović V (2016) Encapsulation technologies for food industry. In: Emerging and traditional technologies for safe, healthy and quality food, pp 329–382. https://doi.org/10.1007/978-3-319-24040-4
221. Fathi M, Varshosaz J, Mohebbi M, Shahidi F (2013) Hesperetin-loaded solid lipid nanoparticles and nanostructure lipid carriers for food fortification: preparation, characterization, and modeling. Food Bioprocess Technol 6(6):1464–1475
222. Zhao L, Temelli F, Chen L (2017) Encapsulation of anthocyanin in liposomes using supercritical carbon dioxide: effects of anthocyanin and sterol concentrations. J Funct Foods 34:159–167
223. Lyu S, Chen X, Shah S, Wu X (2019) Experimental study of influence of natural surfactant soybean phospholipid on wettability of high-rank coal. Fuel 239:1–12
224. Ismail M, Ling L, Du Y, Yao C, Li X (2018) Liposomes of dimeric artesunate phospholipid: a combination of dimerization and self-assembly to combat malaria. Biomaterials 163:76–87

225. Katouzian I, Esfanjani AF, Jafari SM, Akhavan S (2017) Formulation and application of a new generation of lipid nano-carriers for the food bioactive ingredients. Trends Food Sci Technol 68:14–25
226. Verrijssen TAJ, Smeets KHG, Christiaens S, Palmers S, van Loey AM, Hendrickx ME (2015) Relation between in vitro lipid digestion and β-carotene bioaccessibility in β-carotene-enriched emulsions with different concentrations of L-αphosphatidylcholine. Food Res Int 67:60–66
227. Nishimukai M, Hara H (2004) Enteral administration of soybean phosphatidylcholine enhances the lymphatic absorption of lycopene, but reduces that of alpha-tocopherol in rats. J Nutr 134:1862–1866
228. Jafari SM (2017b) An overview of nanoencapsulation techniques and their classification Nanoencapsulation technologies for the food and nutraceutical industries. Elsevier, London, pp 1–34
229. Rafiee Z, Jafari SM (2019) Application of lipid nanocarriers for the food industry. In: Bioactive molecules in food. Springer, Cham, pp 623–665
230. Tamjidi F, Shahedi M, Varshosaz J, Nasirpour A (2014) Design and characterization of astaxanthin – loaded nanostructured lipid carriers. Innovative Food Sci Emerg Technol 26:366–374
231. Hejri A, Khosravi A, Gharanjig K, Hejazi M (2013) Optimisation of the formulation of β -carotene loaded nanostructured lipid carriers prepared by solvent diffusion method. Food Chem 141(1):117–123
232. Okonogi S, Riangjanapatee P (2015) Physicochemical characterization of lycopene -loaded nanostructured lipid carrier formulations for topical administration. Int J Pharm 478(2):726–735
233. Hentschel A, Gramdorf S, Müller R, Kurz T (2008) β-Carotene loaded nanostructured lipid carriers. J Food Sci 73(2):N1–N6
234. Mitri K, Shegokar R, Gohla S, Anselmi C, Müller RH (2011) Lipid nanocarriers for dermal delivery of lutein: preparation, characterization, stability and performance. Int J Pharm 414(1–2):267–275
235. Rao MP, Manjunath K, Bhagawati ST, Thippeswamy BS (2014) Bixin loaded solid lipid nanoparticles for enhanced hepatoprotection –preparation, characterisation and in vivo evaluation. Int J Pharm 473(1–2):485–492
236. Triplett MD, Rathman JF (2009) Optimization of β -carotene loaded solid lipid nanoparticles preparation using a high shear homogenization technique. J Nanopart Res 11(3):601–614
237. Lacatusu I, Mitrea E, Badea N, Stan R, Oprea O, Meghea A (2013) Lipid nanoparticles based on omega −3 fatty acids as effective carriers for lutein delivery. Preparation and in vitro characterization studies. J Funct Foods 5(3):1260–1269
238. Nazemiyeh E, Eskandani M, Sheikhloie H, Nazemiyeh H (2016) Formulation and physicochemical characterization of lycopene-loaded solid lipid nanoparticles. Adv Pharm Bull 6(2):235
239. Nik AM, Langmaid S, Wright AJ (2012) Digestibility and β -carotene release from lipid nanodispersions depend on dispersed phase crystallinity and interfacial properties. Food Funct 3(3):234–245
240. Tan CP, Nakajima M (2005) β -carotene nanodispersions: preparation, characterization and stability evaluation. Food Chem 92(4):661–671
241. de Sousa LKB, Paese K, Forgearini JC, Guterres SS, Jablonski A, de Oliveira RA (2013) Characterisation and stability evaluation of bixin nanocapsules. Food Chem 141(4):3906–3912
242. Yuan Y, Gao Y, Mao L, Zhao J (2008) Optimisation of conditions for the preparation of β -carotene nanoemulsions using response surface methodology. Food Chem 107(3):1300–1306
243. Vishwanathan R, Wilson TA, Nicolosi RJ (2009) Bioavailability of a nanoemulsion of lutein is greater than a lutein supplement. Nano Biomed Eng 1(1):38–49
244. Yi J, Fan Y, Zhang Y, Zhao L (2016) Characterization of catechin-α-lactalbumin conjugates and the improvement in β-carotene retention in an oil-in-water nanoemulsion. Food Chem 205:73–80

245. Xia F, Hu D, Jin H, Zhao Y, Liang J (2012) Preparation of lutein proliposomes by supercritical anti -solvent technique. Food Hydrocoll 26(2):456–463
246. Ha TVA, Kim S, Choi Y, Kwak HS, Lee SJ, Wen J, Ko S (2015) Antioxidant activity and bioaccessibility of size -different nanoemulsions for lycopene -enriched tomato extract. Food Chem 178:115–121
247. Dhakane JP, Kar A, Patel A, Khan I (2017) Effect of soy proteins and emulsification -evaporation process on physical stability of lycopene emulsions. Int J Chem Stud 5(5):1354–1358
248. Mehmood T, Ahmed A, Ahmad A, Ahmad MS, Sandhu MA (2018) Optimization of mixed surfactants -based β -carotene nanoemulsions using response surface methodology: an ultrasonic homogenization approach. Food Chem 253:179–184
249. Weigel F, Weiss J, Decker EA, McClements DJ (2018) Lutein -enriched emulsion -based delivery systems: influence of emulsifiers and antioxidants on physical and chemical stability. Food Chem 242:395–403
250. Ou CF, Ye SY, Zhang J, Han X (2012) Preparation of lycopene liposome using membrane – homogeneous method. J South Agric 2012:4
251. de Freitas ZRH, López-Rubio A, de Pinho SC, Lagaron JM, de la Torre LG (2015) Hybrid encapsulation structures based on β -carotene -loaded nanoliposomes within electrospun fibers. Colloids Surf B: Biointerfaces 134:475–482
252. Toniazzo T, Berbel IF, Cho S, Fávaro-Trindade CS, Moraes IC, Pinho SC (2014) β -carotene – loaded liposome dispersions stabilized with xanthan and guar gums: Physico -chemical stability and feasibility of application in yogurt. LWT-Food Sci Technol 59(2):1265–1273
253. Jiao Y, Li D, Liu C, Chang Y, Song J, Xiao Y (2018) Polypeptide –decorated nanoliposomes as novel delivery systems for lutein. RSC Adv 8(55):31372–31381
254. Michelon M, Mantovani RA, Sinigaglia-Coimbra R, de la Torre LG, Cunha RL (2016) Structural characterization of β -carotene -incorporated nanovesicles produced with non -purified phospholipids. Food Res Int 79:95–105
255. Stojiljkovic N, Ilic S, Jakovljevic V, Stojanovic N, Stojnev S, Kocic H, Kocic G (2018) The encapsulation of lycopene in nanoliposomes enhances its protective potential in methotrexate induced kidney injury model. Oxidative Med Cell Longev 2018:1–11
256. Jafari SM (2017a) An overview of nanoencapsulation techniques and their classification. In: Nanoencapsulation technologies for the food and nutraceutical industries, pp 1–34. https://doi.org/10.1016/C2015-0-04253-8
257. Purkayastha MD, Manhar AK (2016) Nanotechnological applications in food packaging, sensors and bioactive delivery systems. Nanoscience in Food and Agriculture 2:59–128
258. Gharehbeglou P, Jafari SM, Hamishekar H, Homayouni A, Mirzaei H (2019) Pectin-whey protein complexes vs. small molecule surfactants for stabilization of double nano-emulsions as novel bioactive delivery systems. J Food Eng 245:139–148
259. Shishir MRI, Xie L, Sun C, Zheng X, Chen W (2018) Advances in micro and nano-encapsulation of bioactive compounds using biopolymer and lipid-based transporters. Trends Food Sci Technol 78:34–60
260. Gharehbeglou P, Jafari SM, Homayouni A, Hamishekar H, Mirzaei H (2019) Fabrication of double W1/O/W2 nano-emulsions loaded with oleuropein in the internal phase (W1) and evaluation of their release rate. Food Hydrocoll 89:44–55
261. Donsì F, Sessa M, Mediouni H, Mgaidi A, Ferrari G (2011) Encapsulation of bioactive compounds in nanoemulsion-based delivery systems. Proc Food Sci 1:1666–1671
262. Ezhilarasi P, Karthik P, Chhanwal N, Anandharamakrishnan C (2013) Nanoencapsulation techniques for food bioactive components: a review. Food Bioprocess Technol 6(3):628–647
263. Gumus CE, Decker EA, McClements DJ (2017) Gastrointestinal fate of emulsion -based ω −3 oil delivery systems stabilized by plant proteins: lentil, pea, and faba bean proteins. J Food Eng 207:90–98
264. Jia Z, Dumont MJ, Orsat V (2016) Encapsulation of phenolic compounds present in plants using protein matrices. Food Biosci 15:87–104

265. Tan CP, Nakajima M (2005b) Effect of polyglycerol esters of fatty acids on physicochemical properties and stability of β-carotene nanodispersions prepared by emulsification/evaporation method. J Sci Food Agr 85:121–126
266. Yuan Y, Yanxiang G, Jian X, Like M (2008) Characterization and stability evaluation of β-carotene nanoemulsions prepared by high pressure homogenization under various emulsifying conditions. Food Res Int 41(1):61–68
267. Mao L, Xu D, Yang J, Yuan F, Gao Y, Zhao J (2009) Effects of small and large molecule emulsifiers on the characteristics of β-carotene nanoemulsions prepared by high pressure homogenization. Food Technol Biotechnol 47:336–342
268. Yin LJ, Chu BS, Kobayashi I, Nakajima M (2009) Performance of selected emulsifiers and their combinations in the preparation of β-carotene nanodispersions. Food Hydrocoll 23:1617–1622
269. Qian C, Decker EA, Xiao H, McClements DJ (2012a) Inhibition of β-carotene degradation in oil-in-water nanoemulsions: influence of oil-soluble and water-soluble antioxidants. Food Chem 135:1036–1043
270. Qian C, Decker EA, Xiao H, McClements DJ (2012) Physical and chemical stability of beta-carotene-enriched nanoemulsions: influence of pH, ionic strength, temperature, and emulsifier type. Food Chem 132(3:1221–1229
271. Liu Y, Hou Z, Yang J, Gao Y (2015) Effects of antioxidants on the stability of β-carotene in O/W emulsions stabilized by gum Arabic. J Food Sci Technol 52:3300–3311
272. Liu Y, Hou Z, Lei F, Chang Y, Gao Y (2012) Investigation into the bioaccessibility and microstructure changes of β-carotene emulsions during *in vitro* digestion. Innovative Food Sci Emerg Technol 15:86–95
273. Salvia-Trujillo L, Verkempinck S, Rijal SK, Van Loey A, Grauwet T, Hendrickx M (2019) Lipid nanoparticles with fats or oils containing β-carotene: storage stability and in vitro digestibility kinetics. Food Chem 25(278):396–405
274. Helgason T, Awad TS, Kristbergsson K, Decker EA, McClements DJ, Weiss J (2009) Impact of surfactant properties on oxidative stability of beta-carotene encapsulated within solid lipid nanoparticles. J Agric Food Chem 57(17):8033–8040
275. Cornacchia L, Roos YH (2011) State of dispersed lipid carrier and interface composition as determinants of beta-carotene stability in oil-in-water emulsions. J Food Sci 76(8):C1211–C1218
276. Rao J, Decker EA, Xiao H, McClements DJ (2013) Nutraceutical nanoemulsions: influence of carrier oil composition (digestible versus indigestible oil) on β-carotene bioavailability. J Sci Food Agric 93(13):3175–3183
277. Xia Z, McClements DJ, Xiao H (2017) Influence of lipid content in a corn oil preparation on the bioaccessibility of β-carotene: a comparison of low-fat and high-fat samples. J Food Sci 82(2):373–379
278. Zhou X, Wang H, Wang C, Zhao C, Peng Q, Zhang T, Zhao C (2018) Stability and in vitro digestibility of beta-carotene in nanoemulsions fabricated with different carrier oils. Food Sci Nutr 6(8):2537–2544
279. Hart SM, Lin XL, Thilakarathna SH, Wright AJ (2018) Emulsion droplet crystallinity attenuates early in vitro digestive lipolysis and beta-carotene bioaccessibility. Food Chem 15(260):145–151
280. Jafari S, McClements D (2017) Nanotechnology approaches for increasing nutrient bioavailability. Adv Food Nutr Res 81:1–30
281. Niaz T, Shabbir S, Noor T, Rahman A, Bokhari H, Imran M (2018) Potential of polymer stabilized nano-liposomes to enhance antimicrobial activity of nisin Z against foodborne pathogens. LWT 96:98–110
282. Ruiz CJ, Velderrain RGR, Domínguez AJA, Preciado SAM, Alvarez PE, Villegas OMA, González AGA (2018) Encapsulation to protect different bioactives to be used as nutraceuticals and food ingredients. In: Bioactive molecules in food. Springer, Cham, pp 1–20

283. Rafiee Z, Nejatian M, Daeihamed M, Jafari SM (2018) Application of different nanocarriers for encapsulation of curcumin. Crit Rev Food Sci Nutr 59:1–30
284. Ghorbanzade T, Jafari SM, Akhavan S, Hadavi R (2017) Nano -encapsulation of fish oil in nano – liposomes and its application in fortification of yogurt. Food Chem 216:146–152
285. Khorasani S, Danaei M, Mozafari M (2018) Nanoliposome technology for the food and nutraceutical industries. Trends Food Sci Technol 79:106–115
286. Cui H, Li W, Li C, Vittayapadung S, Lin L (2016) Liposome containing cinnamon oil with antibacterial activity against methicillin -resistant Staphylococcus aureus biofilm. Biofouling 32(2):215–225
287. Lin L, Cui H, Zhou H, Zhang X, Bortolini C, Chen M, Dong M (2015) Nanoliposomes containing Eucalyptus citriodora as antibiotic with specific antimicrobial activity. Chem Commun 51(13):2653–2655
288. Mozafari MR (2010) Nanoliposomes: preparation and analysis. Methods Mol Biol 605:29–50
289. Li H, Li X, Shi X, Li Z, Sun Y (2019) Effects of magnetic dihydroartemisinin nano-liposome in inhibiting the proliferation of head and neck squamous cell carcinomas. Phytomedicine 6:215–228
290. Salvia-Trujillo L, Qian C, Martín-Belloso O, McClements D (2013) Influence of particle size on lipid digestion and β-carotene bioaccessibility in emulsions and nanoemulsions. Food Chem 141(2):1472–1480
291. Kaur IP, Garg A, Singla AK, Aggarwal D (2004) Vesicular systems in ocular drug delivery: an overview. Int J Pharm 269(1):1–14
292. Moghassemi S, Hadjizadeh A (2014) Nano-niosomes as nanoscale drug delivery systems: an illustrated review. J Control Release 185:22–36
293. Basiri L, Rajabzadeh G, Bostan A (2017) α-Tocopherol-loaded niosome prepared by heating method and its release behavior. Food Chem 221:620–628
294. Ghadi ZS, Dinarvand R, Asemi N, Amiri FT, Ebrahimnejad P (2019) Preparation, characterization and in vivo evaluation of novel hyaluronan containing niosomes tailored by Box-Behnken design to co-encapsulate curcumin and quercetin. Eur J Pharm Sci 130:234–246
295. Gugleva V, Titeva S, Rangelov S, Momekova D (2019) Design and in vitro evaluation of doxycycline hyclate niosomes as a potential ocular delivery system. Int J Pharm 567:118431
296. Fidan-Yardimci M, Akay S, Sharifi F, Sevimli-Gur C, Ongen G, Yesil-Celiktas O (2019) A novel niosome formulation for encapsulation of anthocyanins and modelling intestinal transport. Food Chem 293:57–65
297. Ravaghi M, Razavi SH, Mousavi SM, Sinico C, Fadda AM (2016) Stabilization of natural canthaxanthin produced by Dietzia natronolimnaea HS-1 by encapsulation in niosomes. LWT 73:498–504
298. Attia N, Mashal M, Grijalvo S, Eritja R, Zárate J, Puras G, Pedraz JL (2018) Stem cell-based gene delivery mediated by cationic niosomes for bone regeneration. Nanomed: Nanotechnol, Biol Med 14(2):521–531
299. Barani M, Nematollahi MH, Zaboli M, Mirzaei M, Torkzadeh-Mahani M, Pardakhty A, Karam GA (2019) In silico and in vitro study of magnetic niosomes for gene delivery: the effect of ergosterol and cholesterol. Mater Sci Eng C 94:234–246
300. Takzare A, Ghafoor DD, Siddiqi AF, Ravali S, Shalbaf M, Bakhtiar M (2019) Trachyspermum copticum essential oil incorporated niosome for cancer treatment. J Drug Deliv Sci Technol 52:818–824
301. Palozza P, Muzzalupo R, Trombino S, Valdannini A, Picci N (2006) Solubilization and stabilization of β-carotene in niosomes: delivery to cultured cells. Chem Phys Lipids 139(1):32–42
302. Sharma P, Saxena P, Jaswanth A, Chalamaiah M, Tekade K, Balasubramaniam A (2016) Novel encapsulation of lycopene in niosomes and assessment of its anticancer activity. J Bioequival Bioavail 8(5):224–232
303. Sharma P, Saxena P, Jaswanth A, Chalamaiah M, Balasubramaniam A (2017) Anti-diabetic activity of lycopene niosomes: experimental observation. J Pharm Drug Dev 4(1):103

304. Shukla A, Katare O, Singh B, Vyas SP (2010) M-cell targeted delivery of recombinant hepatitis B surface antigen using cholera toxin B subunit conjugated bilosomes. Int J Pharm 385(1–2):47–52
305. D'Elia RV, Woods S, Butcher W, McGahon J, Khadke S, Perrie Y, Roberts CW (2019) Exploitation of the bilosome platform technology to formulate antibiotics and enhance efficacy of melioidosis treatments. J Control Release 298:202–212
306. Aburahma MH (2016) Bile salts-containing vesicles: promising pharmaceutical carriers for oral delivery of poorly water-soluble drugs and peptide/protein-based therapeutics or vaccines. Drug Deliv 23(6):1847–1867
307. Guan P, Lu Y, Qi J, Wu W (2016) Readily restoring freeze-dried probilosomes as potential nanocarriers for enhancing oral delivery of cyclosporine a. Colloids Surf B: Biointerfaces 144:143–151
308. Wilkhu JS, McNeil SE, Anderson DE, Perrie Y (2013) Characterization and optimization of bilosomes for oral vaccine delivery. J Drug Target 21(3):291–299
309. Shukla A, Mishra V, Kesharwani P (2016) Bilosomes in the context of oral immunization: development, challenges and opportunities. Drug Discov Today 21(6):888–899
310. Kaur IP, Aggarwal D, Singh H, Kakkar S (2010) Improved ocular absorption kinetics of timolol maleate loaded into a bioadhesive niosomal delivery system. Graefes Arch Clin Exp Ophthalmol 248(10):1467–1472
311. Shukla A, Singh B, Katare O (2011) Significant systemic and mucosal immune response induced on oral delivery of diphtheria toxoid using nano bilosomes. Br J Pharmacol 164(2b):820–827
312. Guan P, Lu Y, Qi J, Niu M, Lian R, Hu F, Wu W (2011) Enhanced oral bioavailability of cyclosporine A by liposomes containing a bile salt. Int J Nanomedicine 6:965
313. Al-mahallawi AM, Abdelbary AA, Aburahma MH (2015) Investigating the potential of employing bilosomes as a novel vesicular carrier for transdermal delivery of tenoxicam. Int J Pharm 485(1–2):329–340
314. Ahad A, Raish M, Ahmad A, Al-Jenoobi FI, Al-Mohizea AM (2018) Eprosartan mesylate loaded bilosomes as potential nano-carriers against diabetic nephropathy in streptozotocin-induced diabetic rats. Eur J Pharm Sci 111:409–417
315. Elnaggar YS, Omran S, Hazzah HA, Abdallah OY (2019) Anionic versus cationic bilosomes as oral nanocarriers for enhanced delivery of the hydrophilic drug risedronate. Int J Pharm 564:410–425
316. Abdelkader H, Alani AW, Alany RG (2014) Recent advances in non-ionic surfactant vesicles (niosomes): self-assembly, fabrication, characterization, drug delivery applications and limitations. Drug Deliv 21(2):87–100
317. Csaba N, Garcia-Fuentes M, Alonso MJ (2006) The performance of nanocarriers for transmucosal drug delivery. Expert Opin Drug Deliv 3(4):463–478
318. Mann JF, Ferro VA, Mullen AB, Tetley L, Mullen M (2004) Optimisation of a lipid based oral delivery system containing A/Panama influenza haemagglutinin. Vaccine 22:2425–2429
319. Holm R, Müllertz A, Mu H (2013) Bile salts and their importance for drug absorption. Int J Pharm 453(1):44–55
320. Rizvi SZH, Shah FA, Khan N, Muhammad I, Ali KH, Ansari MM, Choe Y-H (2019) Simvastatin-loaded solid lipid nanoparticles for enhanced anti-hyperlipidemic activity in hyperlipidemia animal model. Int J Pharm 560:136–143
321. Mohtar N, Khan NA, Darwis Y (2015) Solid lipid nanoparticles of atovaquone based on 24 fullfactorial design. Iran J Pharm Res: IJPR 14(4):989
322. Shi F, Zhao J-H, Liu Y, Wang Z, Zhang Y-T, Feng N-P (2012) Preparation and characterization of solid lipid nanoparticles loaded with frankincense and myrrh oil. Int J Nanomed 7:2033
323. Cavalli R, Bocca C, Miglietta A, Caputo OMR, Gasco MR (1999) Albumin adsorption on stealth and non-stealth solid lipid nanoparticles. STP Pharma Sci 1999:183

324. Jain S, Patel N, Shah MK, Khatri P, Vora N (2017) Recent advances in lipid-based vesicles and particulate carriers for topical and transdermal application. J Pharm Sci 106(2):423–445
325. Gupta B, Yong CS, Kim JO (2017) Solid matrix-based lipid nanoplatforms as carriers for combinational therapeutics in cancer. J Pharm Investig 47(6):461–473
326. Trucillo P, Campardelli R (2019) Production of solid lipid nanoparticles with a supercritical fluid assisted process. J Supercrit Fluids 143:16–23
327. Zardini AA, Mohebbi M, Farhoosh R, Bolurian S (2018) Production and characterization of nanostructured lipid carriers and solid lipid nanoparticles containing lycopene for food fortification. J Food Sci Technol 55(1):287–298
328. Wang T, Xue J, Hu Q, Zhou M, Luo Y (2017) Preparation of lipid nanoparticles with high loading capacity and exceptional gastrointestinal stability for potential oral delivery applications. J Colloid Interface Sci 507:119–130
329. Gutiérrez FJ, Albillos SM, Casas-Sanz E, Cruz Z, García-Estrada C, García-Guerra A, Gonzalez-Ferrero C (2013) Methods for the nanoencapsulation of β-carotene in the food sector. Trends Food Sci Technol 32(2):73–83
330. Tamjidi F, Shahedi M, Varshosaz J, Nasirpour A (2013) Nanostructured lipid carriers (NLC): a potential delivery system for bioactive food molecules. Innovative Food Sci Emerg Technol 19:29–43
331. Yang Y, Corona A III, Schubert B, Reeder R, Henson MA (2014) The effect of oil type on the aggregation stability of nanostructured lipid carriers. J Colloid Interface Sci 1418:261–272
332. Hung LC, Basri M, Tejo BA, Ismail R, Nang HLL, Hassan HA, May CY (2011) An improved method for the preparations of nanostructured lipid carriers containing heat-sensitive bioactives. Colloid Surf B 87:180–186
333. Pezeshki A, Hamishehkar H, Ghanbarzadeh B, Fathollahy I, Nahr FK, Heshmati MK, Mohammadi M (2019) Nanostructured lipid carriers as a favorable delivery system for β-carotene. Food Biosci 27:11–17
334. Abaee A, Mohammadian M, Jafari SM (2017) Whey and soy protein-based hydrogels and nanohydrogels as bioactive delivery systems. Trends Food Sci Technol 70:69–81
335. Slavutsky AM, Bertuzzi MA (2019) Formulation and characterization of hydrogel based on pectin and brea gum. Int J Biol Macromol 123:784–791
336. Ahmed EM (2015) Hydrogel: preparation, characterization, and applications: a review. J Adv Res 6(2):105–121
337. Zhang Z, Zhang R, Chen L, Tong Q, McClements DJ (2015) Designing hydrogel particles for controlled or targeted release of lipophilic bioactive agents in the gastrointestinal tract. Eur Polym J 72:698–716
338. Rahaiee S, Shojaosadati SA, Hashemi M, Moini S, Razavi SH (2015) Improvement of crocin stability by biodegradeble nanoparticles of chitosan-alginate. Int J Biol Macromol 79:423–432
339. Zhang Z, Zhang R, McClements DJ (2016) Encapsulation of β-carotene in alginate-based hydrogel beads: impact on physicochemical stability and bioaccessibility. Food Hydrocoll 61:1–10
340. Liu W, Wanga J, McClements DJ, Zoua L (2018) Encapsulation of β-carotene-loaded oil droplets in caseinate/alginate microparticles: enhancement of carotenoid stability and bioaccessibility. J Funct Foods 40:527–535

Chapter 9
Analysis of Carotenoids

Saqib Mahmood, Muhammad Azeem, Sadia Javed, Beenish Afzal, and Muhammad Zia-Ul-Haq

9.1 Introduction

Eye catching visibility of carotenoids had attracted researchers since long starting from 1826 (Wachenroder in 1826 as cited by Palmer [1]. Carotenoids have always been investigated and utilized to good biomarker for fruit and vegetable intake, leading towards a pool of knowledge about their isolation to characterization. Now it is evident that they are diverse with more than 700 types [2, 3], that vary in quality and quantity with specie and even from variety to variety. Commercial production of carotenoids has been well established now a days still their extraction from natural sources has eminent place [4].

Their analysis and preservation became complicated task due to their interconversion from cis to trans isomers and oxidation potential. Collection of reliable quantifiable information about food carotenoids is beneficial in different fields of life. In agriculture this data helps in selection of varieties and to optimize yield management after and before harvesting; In food sciences assisting in selection of raw constituents, monitoring and control of degradability of processed and stored items; In nutrition and public health it is helping to asses adequacy of in taken diet along with guidance about diet formulation; in the field of clinical medicine helps to establish link between diet components and disease resistance. .

Detailed documentation about carotenoid related issues including their composition in raw food source, food processing/storage related problems and about their

S. Mahmood (✉) · M. Azeem · B. Afzal
Department of Botany, Government College University, Faisalabad, Pakistan

S. Javed (✉)
Department of Biochemistry, Government College University, Faisalabad, Pakistan

M. Zia-Ul-Haq
Office of Research, Innovation and Commercialization, Lahore College for Women University, Lahore, Pakistan

© Springer Nature Switzerland AG 2021
M. Zia-Ul-Haq et al. (eds.), *Carotenoids: Structure and Function in the Human Body*, https://doi.org/10.1007/978-3-030-46459-2_9

pharmaceutical/pharmacological aspects is available. Where, a move from classics to modernization has an associated shift towards advanced methodologies. Still to plan out some research related to the analysis of carotenoids is confusing. In this context it is aimed to review relevant literature arranging them with respect to the technical availabilities and the metabolite of interest to make it simple, easier and least confusing for new researchers.

Chemical reactions responsible for carotenoid identification and isolation depend upon functional group involved in reaction. Before discussing analytical approaches used by biological researchers (coping with their structural variability, lower concentration and instability related issues), we will discuss prerequisites for the successful assay of carotenoids.

9.2 Prerequisites for Carotenoid Analysis

For assured, authentic outcomes careful selection of analytical tools, sampling, validation of analytical methods, training of researcher, supporting staff and appropriate facilitation is compulsory. There are some key prerequisites that must be fulfilled for precise and reliable results here we have grouped major reactions with their functional groups and expected products.

- Understanding physical and chemical properties of carotenoids (Fig. 9.1)
- Acquisition of method validity and quality assurance
- Basic understanding of technique
- Carotenoids with variable quality/quantity, matrix type, could be converted into their other possible isomeric forms while being assayed/stored exists in nature [5]. That inherits difficulty during analysis with a chain of errors. Therefore, precision of carotenoid analysis will be lost if following sources of errors will be ignored. Hence proper understanding of possible sources of errors has been summarized in Table 9.1.
- Chances of error can be minimized if carefully cautionary instruction are adopted. Here is the list of the things needed to be done and to be avoided (Table 9.2)

9.3 Appropriate Selection of Analytical Technique

Major objective of this chapter is to provide comprehensive review of previous carotenoid studies. Additionally it is aimed to simplify the selection of appropriate technique. It would facilitate researcher to decide about their experimentation keeping in view their own laboratory facilities and technical expertise. There is simplified grouping of analytical tools available in literature on the basis of the interest of researcher (Fig. 9.2).

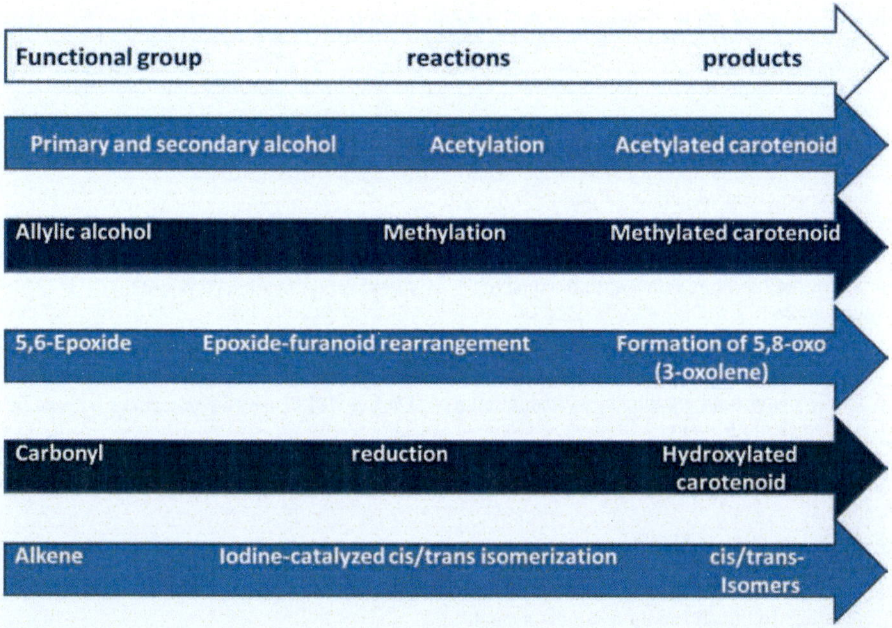

Fig. 9.1 Chemical reactions used for carotenoid identification

Table 9.1 General sources of errors experienced in carotenoid analysis

Sources of errors
1. Inadequate sampling, so far samples would not represent whole food lot
2. If volume reduced in homogenizing activity,sample could not represent analyte
3. Enzymatic oxidation/ release of acids of carotenoids during homogenization possibly promote geometric isomerization
4. Possibility of incomplete extraction
5. It the transfer of solvent would not be complete/there would be emulsion formation
6. Degradation or artifactformation during saponification
7. Possible damage of carotenoids
8. Possible degradation and/or adherence to the wall of container particularly when carotenoid are being dried
9. Coelution, overlapping peaks, low recovery from column
10. Erroneous identification, unaccounted impurity and degradation of standards, calibration errors
11. Calculation errors/ misinterpretation

9.4 Quantitative Analysis

An extra detail of quantitative analysis is beyond the scope of this book but brief summary of quantification of carotenoid will be discussed here. For carotenoid assays quantitative analysis are of two types

- Total carotenoids
- Individual carotenoid components

Table 9.2 List of possible cautionary measures for carotenoid analysis

Do's	Don'ts
1. Fulfill prerequisites mentioned above	1. Storage in dichloromethane/ethyl ether, or acetone degrade carotenoids. Use petroleum ether/ hexane or dry under vacuum or nitrogen
2. Go through some training under supervision of some experienced person (if possible)	2. Avoid prolonged storage
3. Clean the glassware properly to get rid of contamination	3. Don't use plastic blenders to avoid corrosion by acetone and adherence of carotenoids
4. Survey of literature relevant to the type of your sample	4. Avoid saponification (if possible)
5. Work in separate lab or part of lab with low light	5. During partition; avoid shaking separatory funnel to stop emulsification.
6. Cover vessels and equipment with aluminum foil or black cloth	6. During HPLC; overlapping peaks will not be quantified
7. Protect yourself from solvent fumes by work in proper ventilated lab preferably to protect yourself 8. Made complete extraction 9. During partitioning be careful to avoid loss of carotenoids 10. During drying add sodium sulfate in small quantity till few crystals appear loose 11. During concentration (rotary evaporator) set temperature not more than 40 °C avoid complete drying of carotenoids (if necessary then evaporate/ use nitrogen) 12. For OCC use small quantity of sample to minimize band broadening 13. For spectrophotometry, calibrate first with potassium dichromate with concentrations ranging absorbance 0.2–1.0. 14. For HPLC; use compatible injection solvent to evade distorted/ splitted peak. 15. Follow manufacturer's instructions about column. 16. Use guard columns with metal-free frits to avoid carotenoid –metal reactions 17. Check the purity of carotenoid standards on arrival 18. Divide pure standards into several portions. Avoid prolonged storage (sealed under nitrogen or argon at −20 °C) 19. For quantitative HPLC; prepare standard curve with 3–5 concentrations of standard 20. Use minimal volume of chemicals to make it economic and lower ecological problems of wastes	

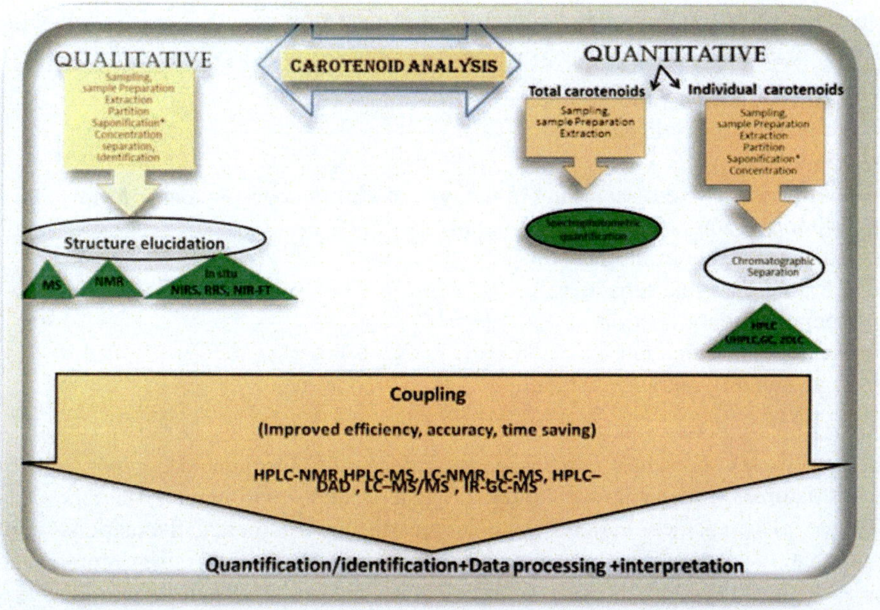

Fig. 9.2 Summary of analytical techniques used for carotenoid of biological samples based upon research objectives

9.5 Total Carotenoid Content (UV and Visible Spectroscopy)

Simplest total carotenoids can be analyzed by using visible absorption at the λmax. It has limitations but still is in use due to its simplicity [6]. A spectrophotometer is employed to measure the amount of light that a sample absorbs. For its operation beam of light is passed from sample and intensity of light that reach detector is measured. In this technique absorbance is directly proportional to the concentration therefore they can be quantified by using formula drawn by standard graph trend. Spectrophotometric determination of individual carotenoids is possible only if it is employed after chromatographic separation of carotenoids.

9.6 Quantification of Individual Carotenoids

To assess the nutritive and medicinal benefits of some samples information about its individual components along with total composition is needed. Each individual carotenoid varies for its polarity, stability, bioavailability and antioxidant activity. Therefore, they have differential health related activities in routine diet and medicinal plants. Additionally, variability of sample matrices made it impossible to adopt single analytical method for each plant sample and each fraction [5–7] For quantitative estimation of individual component from sample matrices following steps are involved

- Careful sampling and its storage
- To prepare analytical sample,
- Extraction,
- Partition of solvent according to the type of chromatography being followed
- Saponification and washing (optional)
- To concentrate/ evaporate solvent,
- Chromatographic separation,
- Technical identification,
- Precise quantification
- Data handing out and elucidation

Sampling
Sample must represent complete ration under study. Carotenoid compositions of plant samples depend upon number of genetic and environmental factors. That include plant species, variety, stage of growth, environmental fluctuations, geographical factors and plant organ under study. Similarly, in food during storage and processing it vary conditionally. Samples of animal (plasma/ serum) also show differential pattern of accumulation and structural pattern. Therefore, sample number is recommended to be large, randomly collected and homogenized. Finally mean values will be used in data (accompanied with standard deviation).

Storage of Samples
For ideal results period of storage must be short to avoid changes in the carotenoid composition due to gain/loss of moisture content. Therefore, lyophilization (freeze-drying) is considered best to preserve samples till analysis.

Preparation of the Analytical Sample
Objective of sampling is to make sure that small part of sample is being homogenized, resulting into such a small volume that is representative of whole sample. It is further based on the purpose of the assessment, type of experimental population under study, type of material used to be measured, variability of composition faced by unstable constituents in the course preparation.

Extraction
Extraction involves disturbing tissue and removal of unwanted components. There is no generalized/ standard method to extract carotenoid in labs. Though there is one thing common that is the release of desired components from matrices (Fig. 9.3).

Till date most frequently practiced extraction methods are as following (Table 9.3)

- Agitation, Homogenization and Shaking
- Soxhlet extraction
- ultrasonic extraction
- Electro-plasmolysis (EP)
- water-bath
- protein precipitation (PPT)

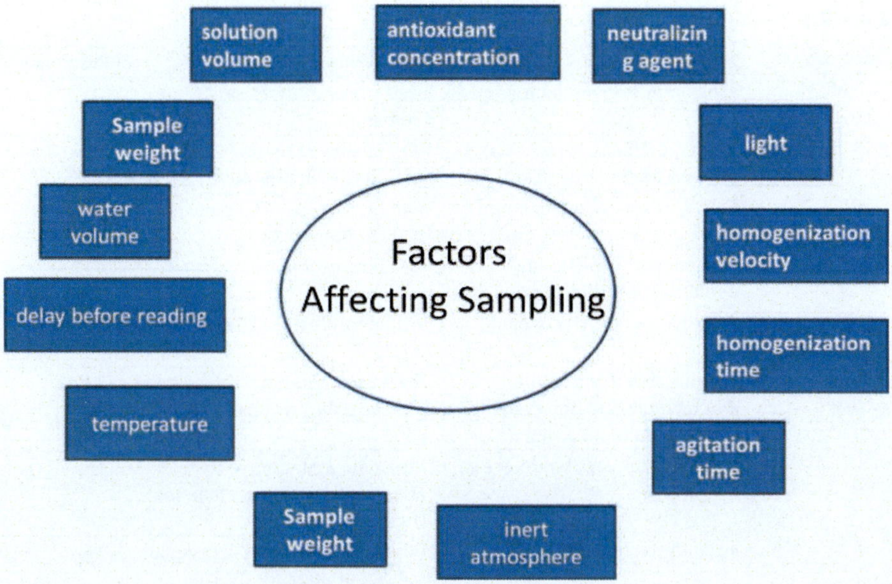

Fig. 9.3 Factors that affect accuracy of extraction procedure

- microwave-assisted extraction (MAE)
- supercritical fluid extraction (SFE)
- pressurized-liquid extraction (PLE)
- matrix solid phase dispersion (MSPD)
- Accelerated solvent extraction (ASE)
- Pressurized hot-water extraction (PHWF)
- liquid–liquid extraction (LLE)
- solid-phase extraction (SPE)
- Supercritical Fluid extraction
- Enzymatic pretreatment

9.7 Agitation, Homogenization and Shaking

At low pressure conditions carotenoids can be extracted by mechanical options. Solvent and matrix material is left for some time with each other. Agitation, homogenization and shaking may be used to facilitate separation carotenoids from samples. Later on it needs proper isolation of analyte extract from other material of sample. Heat can facilities this type of extraction but it is usually avoided due to β-carotene like carotenoids.

Main advantage of this method is its simplicity. Disadvantages of this method include time consumption, solvents left at the end, its need of concentration step

Table 9.3 Most frequently practiced extraction methods

Method	Sample	Extraction conditions	Reference
Soxhlet extraction	Marjoram (Origanum marojana L	n-hexane and EtOH, 50 1C and 450 bar	[8]
	Palm oil (Elaeis guineensis) pressed fiber (carotenes)	With hexane at76–80 °C for 8 h, humidity fresh (35–40%), dried	[9]
	Paprika (cap- sicum annuum)	With hexane, 0.4–0.6 mm and 10–11% humidity	[10]
	Red pepper (cap- sicum annuum L.), pelletized	With hexane 4%	[11]
	Tomato pomace (lycopene)	With chloroform, 125 (v/w) for 7 h	[12]
Ultrasound assisted extraction	Red paprika (Capsanthin, Zeaxanthin, Cucurbitaxanthin, Capsorubin & β-carotene	Ultrasonic agitation with acetone, methanol, diethyl ether	[14]
	Beans (lutein &&β-carotene)	Ultrasonic agitation, centrifugation with acetone for 30 s	[15]
	Tomato (lycopene)	Ultrasonic with ethyl acetate, at 86.4 °C, 40 kHz for 29.1 min, Ultrasound/microwave assisted extraction (UMAE); eith ethyl acetate, 10.6:1 (v/w) for 367 s at 98 W	[16]
	Spinach and tomatoes (β-carotene, lycopene)	Electrical-field treatment (at 80 V/cm for 4 s)	[17]
	Food complements (free lutein lutein esters lycopene b-carotene)	Extraction with ethyl acetate +0.1% BHT in ultrasonic bath for 30 min centrifuged, supernatant filtere	[18]
	Elaeis guineensis Jacq (b-carotene, lutein, and zeaxanthin)	L-ascorbic acid (in ethanol), for b-carotene (30.14 °C, 37.11 min, and 23.18 mL/g), lutein (30.00 °C, 39.09 min, and 19.24 mL/g), and zeaxanthin (30.09 °C, 36.76 min, and 22.38 mL/g)	[19]
	Tomato juice	Olive oil	[20]
	Rapeseed	Petroleum ether/acetone (v/v = 1/1), temperature 49.6C, liquid to material ratio 41.4 mL/g, duration 48.5 min, ultrasonic power 252.9 W	[21]
	Lemon balmand peppermint leaves	35 kHz/140 W, 15 min at 46.0 °C, 20 min at 54.2 °C and 25 min at 54.7 °C	[22]

(continued)

Table 9.3 (continued)

Method	Sample	Extraction conditions	Reference
Electro-plasmolysis	Tomato peels (lycopene, phytoene, phytofluene, β-carotene, *cis*-lycopene and lutein)		[23]
	Pomegranate peels, vegetable oils as solvents, sunflower oil	T = 30 min, 51.5 °C, peels/solvent ratio, 0.10; amplitude level, 58.8%	[24]
Water-bath extraction	Spinach and tomatoes		[17]
Enzyme assisted extraction	Spinach and tomatoes (b-carotene,lycopene)	At 80 °C	[25]
MAE	Tomato paste	Solvent ethyl acetate; solvent to feed ratio = 10.6:1 (v/w), 367 s, 98 W	[16]
Matrix solid phase dispersion	Medical food (β-carotene)	Sorbent C18, elution agent isopropanol; methylene chloride/ethyl acetate/hexane (3:3:4, v/v/v) with 0.5% (v/v) isopropanol, stationary phase Si 60, mobile phase hexane or isopropanol	[44]
Centrifugal	β-Carotene (48.7 wb), (cis + trans)-β-carotene (130 wb) cis-trans-carotene (8.9 wb)	Solvent tetrahydrofuran: Methanol (1:1 v/v) at 0 °C	[45]
	Orange (Citrus sinesis) juice (cis-violaxanthin, β-carotene lutein)	Solvent zinc sulphate + monohydrate potassium ferrocyanide; at 25 °C	[46]

(continued)

Table 9.3 (continued)

Method	Sample	Extraction conditions	Reference
Liquid-liquid extraction	Vegetables (α- and β-carotene)	Solvent pentane	[47]
	Five carotenoids in human serum	n-hexane	[42]
	Five carotenoids in human serum	Ethanol (etoh)	[48]
	Corn (xanthophyll)		[49]
	Human serum	Ethyl acetate (Barua	[50]
	Grape berries	Diethylether/hexane	[51]
	Tomato	Tetrahydrofuran (THF)	[52]
	Seaweed (cooked)	Methanol, hexane and dichloromethane (50:25:25, v/v/v) and a second extraction with acetone.	[53]
	Human serum	Ethanol +1 mL ethyl acetate and 3 mL hexane	[54]
	Tomato	Ethyl acetate for b-carotene dichloromethane for others	[55]
	Maize	Methanol with other less polar solvents	[56]
	Alga	Acetone + methanol/ ethanol (7:3)	[57]
	Tomato processing waste	Hexane, acetone and ethanol (50:25:25, v/v/v)	[17]
	Tomato	N-hexane	[58]
	Microalgae	Ethanol and 2-methyltetrahydrofuran (MTHF)	[59]
Enzymatic extraction	Tomato(total carotenoid and lycopene)	Pectinase and cellulase followed by high-pressure extraction	[60]
	Momordica cochinchinensis Spreng, and (β-carotene) for pretreatment before air-drying	Pectinase at 0.1% (w/w)	[61]
	Alga	31 °C, modifier; 15% MeOH/ EtOH	[64]

(continued)

Table 9.3 (continued)

Method	Sample	Extraction conditions	Reference
	Tomato processing byproducts (lycopene and β-carotene)	Solvent CO_2 with ethanol (100:0–8:15, w/w) @ 150 (w/w) at 35–65 °C and 20–30 MPa pressure with 6% humuduty for 1–4 hours	[65]
	Origanum majorana L	Solvent CO_2,50 °C and 450 bar	[8]
	Tomato skin and seed (lycopene and β-carotene)	Solvent CO2 at 32–86 °C and 13.78–48.26 MPa pressure for 20 min with 48.4% humidity	[66]
	Spirulina Pacifica algae (Zeaxanthin, β - cryptoxanthin, β -carotene)	CO2 with 15% EtOH modifier at 80 °C (for zeaxanthin), 76 °C (for β - cryptoxanthin), 60 °C (for β -carotene), at 350 bar ressure with flow-rate of 2 mL min21 for 60 min	[67]
	Tomato skin and pulp	At (8.7 MPa, 9.3 MPa and 28.1 MPa ppressure 30 min each at 40 °C	[68]
	Carrot (α-and β-carotene)	Canola oil, moisture (0.8–84.6%), particle size (0.25–2 mm), temperature (40–70, C), pressure (276–551 Bar), concentration of canola oil (0–5%), and CO2 flow rate (0.5–2 L/min).	[69]
	Rose fruit (R. canina) Totalcarotenoidcontent, lycopene, 췌-carotene, and lutein	CO_2, 15–45 MPa, 40–80 ∘C	[70]
	Pitanga fruit (E. uniflora L.), Totalcarotenoidcontent, lycopene, rubixanthin and 췌-cryptoxanthin	CO_2,10–40 MPa, 40 and 60 C at pressure (10, 15, 20, 25, 30, 35, and 40 MPa)	[71]
	Haematococcus pluvialis	Solvent CO_2, ethanol, soybean oil, or olive oil as cosolvents, 70 °C and 40 MPa	[72]
Pressurized liquid extraction	Common sample preparation methods	High pressure	[73]

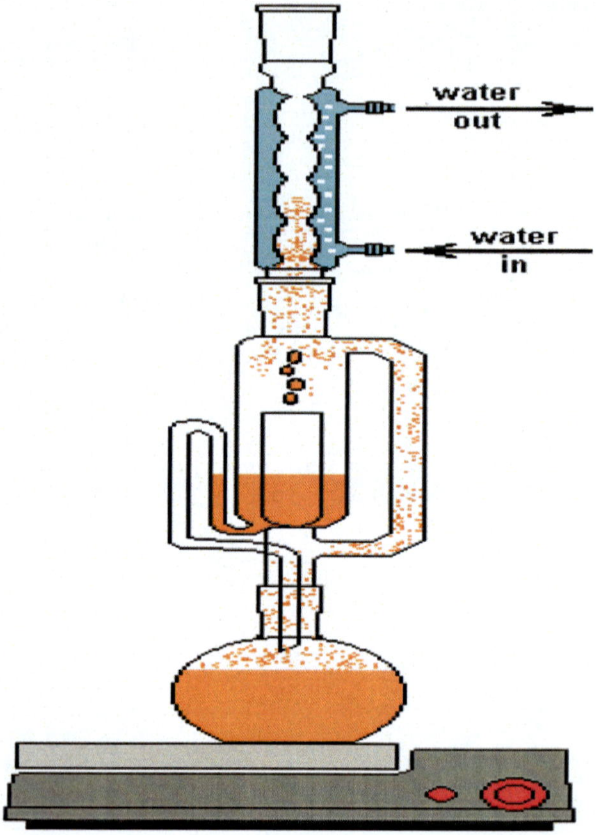

Fig. 9.4 Schematic diagram of apparatus used for soxhlet extraction

that leads toward some cleanup issues, reduction of product quality and expected degradation of compound of interest like vitamins (Fig. 9.4).

9.8 Soxhlet Extraction

During this type of extraction first task is to convert solvent into vapors then moved towards a chamber via distillation arm. Afterwards with the help of condenser it is allowed to cool down and droped in the chamber. Some of required compound become part of warm solvent. After complete filling of soxhlet chamber it is used to empty. Whole procedure is repeated in cycles followed by concentration of desired material in. It is plus point of this method that it makes possible recycling of single batch of solvent.

Major advantage of this extraction is that it needs no filtration step. Disadvantage include that it is time consuming and need greater volume of solvent and samples.

Moreover, it may face thermal decomposition of the compound of interest. It is being adopted by many researchers for carotenoid analysis [8–12].

9.9 Ultrasound Assisted Extraction

For some samples simple solvent extraction failed to extract carotenoids then UAE is considered helpful for enhancement of cell disruption and agitation. It ends with improved efficiency, uniformity and minimum chemical-degradation (expected due to alkalinity during saponification (Fig. 9.5).

It differs from conventional methods and involves variation of cell structure with the help of sound waves (higher than 20 kHz) [13]. Improved extraction with ultrasound is because of acoustic cavitation. High energy of sound waves produce bubbles in solvent, when these bubble could not absorb more energy they violently burst. It results in disruption of cell wall responsible for enhancement of solvent infiltration in cell and release of matrix components. The cavities with bubbles act as hydrophobic surfaces resulting into a rise in total hydrophobic properties of medium. That facilitates the extraction of polar analytes with otherwise hydrophilic

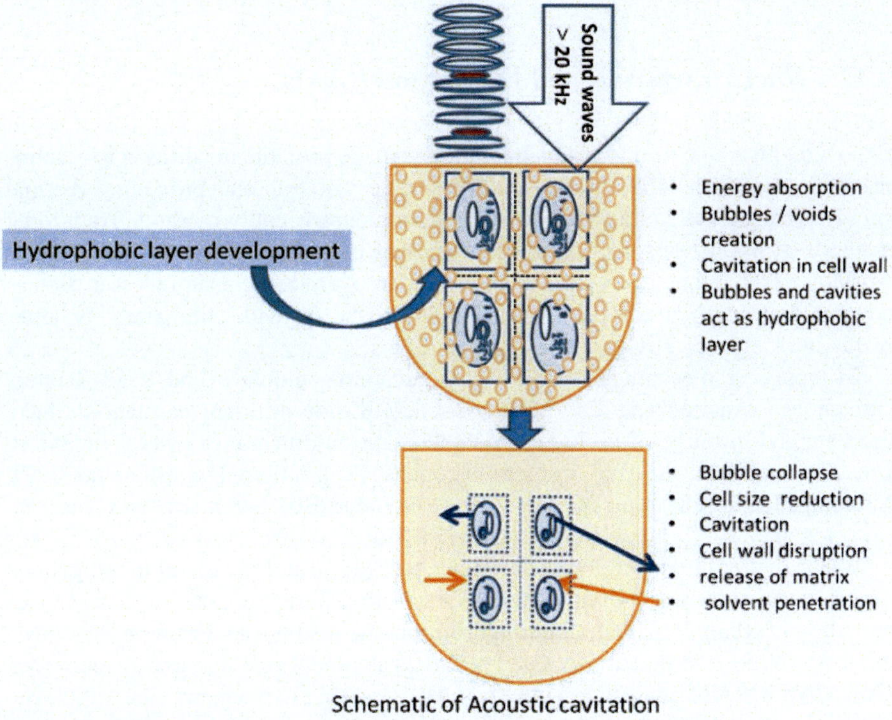

Fig. 9.5 Schematic diagram of apparatus used for ultrasound assisted extraction

medium. That is why it needs no other hydrophobic/ strong polar medium to improve efficiency. UAE frequency, pressure, temperature and sonication time can be monitored as per interest [14–22].

9.10 Electro-Plasmolysis (EP)

As plasmolysis means cell wall loosening resulted in the presence of comparatively hypertonic solution. Number of researchers have adopted this phenomenon in assistance with electricity [23, 24]. Aktas and Yildiz extracted carotenoids from spinach and tomato and for lycopene estimation from tomato they found EP better than water bath extraction [17].

9.11 Water-Bath Extraction

For spinach as compared to EP better extraction yield was observed by water-bath process at 80 °C [17].

9.12 Microwave-Assisted Extraction (MAE)

Exposure of sample to microwave radiation is made possible in addition to electromagnetic spectrum. That induce polar molecules rotation and producing thermal energy. That makes sample hot. It is an environment friendly method. Traditional methods of extractions need great quantities of organic solvents. In MAE major advantage is that they don't have such limit moreover, they consume lesser time as compared to lengthy traditional methods. Further they provide comparatively better extraction yield, selectivity and reproducibility.

By applying mechanical forces like sonication, autoclave and bead- beaters carotenoid extraction can also be improved. Till date different carotenoids have been studied in range of biological samples with modification in MAE including *Capsicum annuum* L. [26], tomato lycopene [16], microalgal pigments [27], Microalga lutein [28], food samples [29], b- carotene [30], b-carotene and lycopene [31], vegetables [32] lycopene in tomato [33].

Where, better lutein extraction was made by bead-beater treatment in indigenous microalga *Scenedesmus obliquus* [28]. Water with different organic modifiers (acetone/dioxin/ ethanol/ methanol/tetrahydrofuran) are also in use. From paprika powder it was optimized that at 120 s and 50 W carotenoids were maximally recovered [26]. MAE was also coupled to UAE (UMAE), resulting in lycopene yield of 97.4%, higher than the yield achieved by UAE only (89.4% of total lycopene (Fig. 9.6).

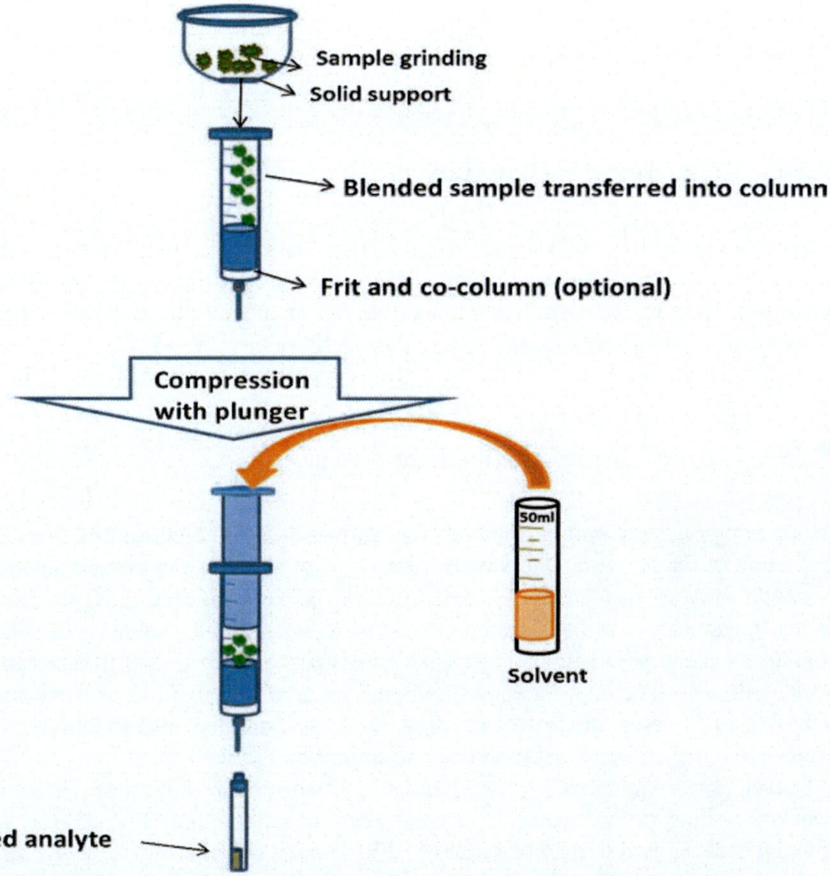

Fig. 9.6 Schematic diagram of apparatus used for matrix solid phase dispersion extraction

9.13 Matrix Solid Phase Dispersion (MSPD)

It is highly efficient technique developed in 1989 and patented in 1993 [34] in which samples are placed in mortar having in it some suitable solid supporting material like octadecylsilyl -derivatized silica (C18). Typical derivatives solid support material C18; supply alkyl chains that protect carotenoids from environment with its hydrophobic nature. Silica facilitates breaking of plant vesicular structure. Resultantly, negligible isomerization/degradation of carotenoids makes it one of the successful extraction techniques. After manual blending (in presence of internal standard) sample is allowed to be compressed in a column followed by addition of appropriate solvent. Where compressing creates compact column bed and with the help of polar solvents polar impurities are removed. Optionally co-columns with some other solid phase may also be added for assistance. Finally analyte is eluted out from column and used for analysis. This extraction method has been used by different biological researchers for carotenoid studies [35–44].

9.14 Centrifugal Extraction

It employs separation with centrifugation. Raw material is placed along with solvent in a vessel subjected to centrifugation followed by filtration (to get rid of raw material). Its major drawback is the requirement of special filtration operation unit. Therefore rarely adopted and have rare data in literature [32, 45, 46].

9.15 Liquid-Liquid Extraction

Most adopted preparation method for carotenoids is liquid-liquid extraction. Selection of liquid as solvent is based upon type of sample, analyte and techniques. Susceptibility of carotenoids to oxidation will be kept in mind during extraction. When present in matrices carotenoids are somewhat stable whereas, in solutions manifest sensitivity to light, heat, acid or oxygen exposure. Moist samples with water miscible (acetone, methanol, ethanol, or mixtures) organic solvent and dry samples with water-immiscible solvents are best extracted. But if they rehydrate water-miscible solvents extraction can be improved (Table 9.2).

Extraction solvents are preferred on basis of carotenoid of interest solvents with too low boiling points are best like petroleum ether fraction (B.P. 35–60 °C). In general acetone and tetrahydrofuran (THF) are most popular. For aqueous sample; water-miscible organic solvent (like acetone, methanol, ethanol, or their mixtures) and for dry samples; water-immiscible solvents will be used.

Tzouganaki et al. [40] compared liquid–liquid extraction (using n-hexane) with solid-phase extraction (using an ISOLUTE C18 cartridge). For isolating lycopene liquid–liquid extraction was found better with 33% greater recovery. Depending upon nature of carotenoids solvent is selected. For non-polar carotenoids (carotenes) some of non-polar solvent would be used lie hexane. Contractedly, for polar ones like xanthophylls ethanol, acetone and other polar solvent are used. Along with different organic solvents vegetable oils are also used for carotenoid extractions where vegetable oils got edge as they provide protection to carotenoids against possible oxidation. Further it can be processed without its removal. Because oil/extract with carotenoids have not any threat of thermal degradation (as it is in case of extraction with organic solvent at high temperature) [4]. A brief review of extraction choices opted for carotenoid analysis from human and plant samples has been shown in Table 9.3 [48–58].

Note Plant samples are directly extracted whereas, serum and plasma samples need de-proteinization prior to extraction [59–60].

9.16 Enzymatic Pretreatment

Some of researchers are pretreating samples with enzymes and got successful results [60, 61].

9.17 Supercritical Fluid (SFE)

Carotenoids are also being extracted with the help of supercritical technology, using some supercritical compound (a compound that is at its pressure and temperature above critical level). Most preferably used supercritical compound is carbon dioxide (Fig. 9.7).

SFE has proved its superiority than conventional extraction methods due to following characteristics of supercritical solvents,

- free of toxic waste
- free of post-processing removal of solvent
- there is not any risk of thermal degradation
- prevents oxidation (as oxygen is absent in column)
- It shows flexibility for adjustment during entire process for solvent solvation. Solubility of analyte can be enhanced monitoring pressure at particular temperature varying solvent density though at high pressure there is some fall of selectivity.
- Less viscous than liquid

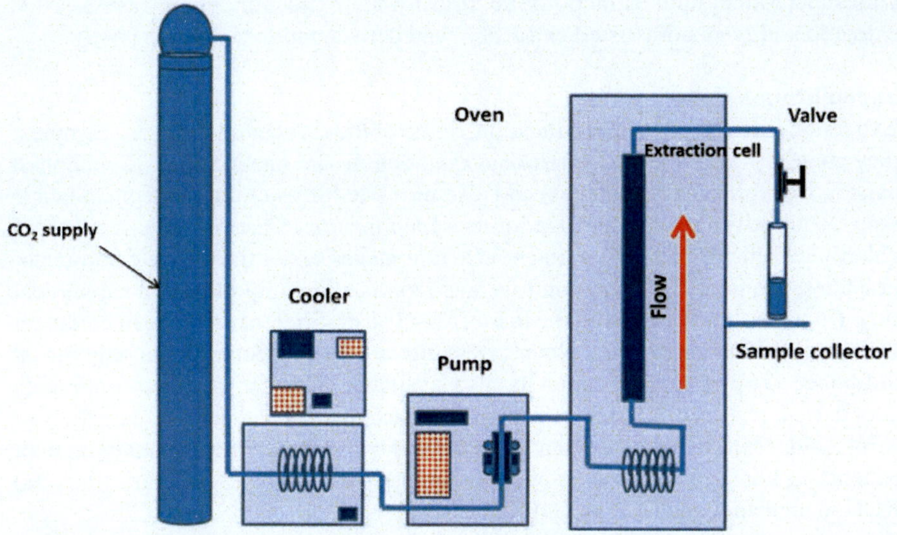

Fig. 9.7 Schematic diagram of SFE providing oxygen free environment to samples

- Easy and greater diffusion through solids
- Better transport potential as compared to liquids results in time saving extraction. Above all they can modify fluid density varying pressure and/or temperature.
- Their use is considered safe
- Owes better efficiency in terms of yields
- and lower extraction times, and the possibility of direct coupling with analytical chromato- graphic techniques as supercritical fluid chromatography (SFC)
- Co-solvents /modifiers can be used along with supercritical used (most preferred CO_2) improving efficiency
- Above all it is Speedy & economic

In a number of biological sample analysis this method is in use [62, 63] Relevant studies have been reviewed in Table 9.3 [64–72].

9.18 Pressurized Liquid Extraction

.Jaime et al. [73] opted Pressurized liquids as an alternative process to antioxidant carotenoids extraction from Haematococcus pluvialis microalgae with successful outcomes.

On-line Coupling of Extraction to Liquid Chromatography Coupling of extraction method with advanced technique ensure automation of the analytical procedure, lowering time span, possibility of contamination, and loss of analyte resulting into improved, precise and accurate methodology. These include coupling of LC with supercritical fluid extraction, subcritical water extraction, microwave-assisted extraction, ultrasound-assisted extraction, and derivatization technique [74].

Saponification

Extraction is followed by saponification. Some of the carotenoids are very complex, they possess great structural variations, xanthophylls are one of those. Xanthophyll exist in free form (like carotenes) and also in stable forms of their esters. To tackle their complexity they are analyzed after saponification. That involves removal of chlorophylls and lipids from carotenoids. That makes easier the ultimate chromatographic separation. There are controversial reports about importance of saponification. Divya et al. [75] observed the loss of 20–50% of different carotene in coriander during saponification process. Nonetheless, literature about better or non-significant difference [76] of saponification is also available. That is why it is optionally adopted by researchers. Saponification is considered beneficial for detection of carotenoids from food and human samples, and consequently for the study on their benefits to human health. Most commonly used saponifying agents are (0.02 M) KOH in methanol and (2.5%) KOH [77].

Homogenization Tool
One of the requirements for accurate findings for analytical techniques is the maintenance of homogenate in the extracts. Typically used homogenizing tools are as follows

- Mechanical blender (seed like samples)
- Mortar and pestle (leaves and roots)
- Polytron homogenizer
- Vortexing (for finely ground sample)

Some additive materials are often used to facilitate extraction like Polytron homogenizer, Celite/ Hyflosupercel, $MgCO_3$ (to neutralize and prevent isomerization and degradation). Nitrogen/dry ice (to prevent oxidation) and speedy work will also help.

Filtration Extraction (with or without saponification) is followed by filtration for that purpose sintered glass funnel (with pore size 20–30 μm) or buchner funnel are used. Repeated filtration and extraction is preferred till colorlessness.

9.19 Identification and Quantification

Till date various methods have been used to identify and quantify carotenoids that are enlisted below

- Colorimetry
- Spectrophotometry
- Fluorometry
- Paper, open-column and thin-layer chromatography, high performance liquid chromatography (HPLC)
- Capillary electrophoresis
- Nuclear magnetic resonance spectroscopy (NMR)
- Mass spectrometry
- Infra red spectroscopy
- Raman spectroscopy
- Combination of all above like HPLC-NMR, LC-NMR, GC-MS, LC-MS and positive mode atmospheric pressure chemical ionization (APCI+ mode) are in use. Most preferred chromatographic techniques have been reviewed in this chapter (Table 9.4).

Chromatographic Separation
A plant scientist Mikhail Tswett [78] explored that plant pigments can be parted into variable color zones on different material columns. Later, the term "chromatography" (writing with color) was coined for this technique. With the course of time chromatography travelled from simpler to advanced ones including paper,

open-column, thin-layer chromatography, high performance liquid chromatography (HPLC) and UHPLC.

Generally carotenoids exhibit conjugated double bonds (ranging from three to eleven) additional to their non-conjugated double bonds. That results into their cis and trans isomerization. Lycopene a famous carotenoid may produce more than thousand geometric isomers [79]. C18 stationary phases was employed in earlier work. They have greater selectivity for isomers monomeric columns. In 1980–90, a C30 stationary phase was introduced and even was used to say carotenoid column. It can efficiently part various carotenoid along with their isomers. Now it is well established that for same carotenoid C18 and C30 differ for its retention time resulting into different potential of separation [80].

Chromatography made possible the study amongst and within classes of carotenoids. Literature about successful outcomes of carotenoid analysis has been tabulated in Table 9.4 for animals and plant samples [42, 43, 57, 77, 81–113] where options of stationary and mobile phase vary sample to sample.

Here is brief background description of the chromatographic terminologies used the chapter (Fig. 9.8).

9.20 Planar (Paper/Thin Layer) Chromatography

It uses differential potential of migration rates (on paper) possessed by different components present in a solution. Sample is spotted on sheet of filter paper along with appropriate solvent to act as stationary liquid phase). One of the paper edge is placed in another solvent (capable of differently dissolving components at varying degrees) that penetrates in paper and passes over the sample spot, carrying along components of the sample. Difference of velocities and solubility make separation of the components.

Classically, TLC is considered as speedy, simple and low- cost useful technique. Planar chromatography (paper/ thin layer) experienced limitation for carotenoid analysis because they are thermally instable and have limitation of volatility. TLC showed limited application due to lower resolution and greater sample concentration. As TLC operates in an open system it leads towards oxidation/degradation of carotenoids. That was later improved as HPTLC (with smaller sized particles) and 2 dimensional TLC to improve resolution potential [18].

HPTLC was performed with high performance technique using silica gel plates. It is a direct method used to quantify biochemical effectively. Densitometer (370–700 nm) is used to record the spectra (in situ) of every band .that make possible the recording with maximum absorption wavelength. Quantification is further made with the help of standards curve developed with standard.

Table 9.4 Chromatographic techniques review

Source/analyte	Pre-injection steps	Chromatographic specification	Reference
Serum(9carotenoids)	200 µL of serum +400 µL of THF. Vortex mixing for 1 min. Settled for 10 min. Centrifuged for 6 min at 85 g. supernatant transfer to auto sampler vial. Solid phase extraction (SPE).	LC-PDA; C18 (100 mm × 3.9 mm, 4 lm) column. Mobile phase; (A) ACN-water (90:10) and (B) THF gradient elution	[42]
Tomato (lutein Lycopene b-carotene lycopene b-carotene isomers)	0.1 g (spiked with 0.01 to 1 lL/mL for b-carotene 0.01 to 4 µL/ mL for lutein) + 2 mL of ethanol +0.1 g For ascorbic acid. Vortexed for 1 min, water bath for 5 min + 100 µL (KOH 80% in water), vortexed for 20s, in water bath for 15 min and mixed every 5 min. Supernatant removed, residue re-extracted With 2 mL of ethanol by sonication for 1 min, Centrifuged. Extraction repeated one more time With 1 mL ethanol. Combined extract applied to SPE Cartridge (OASIS HBL) conditioned with 3 mL MeOH and 3 mL of water. Extracts 8 + 2 mL water +2 mL DCM. Evaporation under N2 at 30 °C reconstituted with 200 µL of methanolic solution of a-tocopherol acetate (IS, 50 µm/mL)	LC-PDA C18 (250 mm _ 4.6 mm, 5 lm) column Mobile phase: MeOH/ACN/isopropanol Gradient elution	[43]

(continued)

Table 9.4 (continued)

Source/analyte	Pre-injection steps	Chromatographic specification	Reference
Serum (30 carotenoids)	1 mL of blood into a 10 mL brown vial +1 mL of 0.01% Vit. C in ethanol +1 mL ethyl acetate and 3 mL hexane. Vortex (30 s), shaking (10 min at 200 rpm), centrifuging (20 min at 3000 rpm and 4 °C). To the residue, 3 mL of hexane was added and the procedure repeated three times	LC-PDA-APCI; (+)/MS C30 column. Mobile phase; (A) MeOH/ACN/water (84:14:4, v/v/v) and (B) DCM gradient elution LC-PDA	[57]
25 vegetables lutein Zeaxanthin, b-carotene trans/cis (E/Z)-b-carotene	10 g of raw or cooked +20 mL cold acetone 10 mL of internal standard b-apo-80carotenal, 6 Mg/200 mL hexane; homogenized for 60s; centrifuged at 1500 _ g for 5 min. Precipitate reconstituted in 20 mL acetone and homogenized 60s; centrifuged. Acetone extraction repeated until colorless. Supernatants combined and dried under vacuum at a temperature below 30_C. residue dissolved in acetone by ultrasonic agitation to a final volume of 20 mL. (Ambersep 900 OH, 0.1 g) + acetone extract (20 mL)	LC-PDA C30 (4.6 mm _ 250 mm, 5 lm) carotenoid column Mobile phase: (A) MeOH/MTBE (85:15, v/v), (B) MeOH/MTBE (6:94, v/v)	[77]
Serum (lutein Zeaxanthin lycopene b-carotene)	500 µL serum +500 µL internal standard (8-apo-b- carotenal) + 500 µL ethanol pippeted into 5 mL disposable tubes. Vortex mixed for 1 min. Mixture +700 µL hexane (three times). Upper layer collected and evaporated under N2 at 4 °C. residue dissolved in solvent A (MeOH: ACN, 40:60, v/v	LC-UV-vis; C30 (4.6 mm? 25 mm, 5 lm) column, guard C30 column. Mobile phase; (A) MeOH: ACN (40:60, v/v) (B) MeOH: ACN (25:75, v/v) and (C) MTBE gradient elution	[81]

Women with type 1 diabetes and preeclampsia (serum carotenoids)	200 μL + 200 μL ethanol-butylated hydroxytoluene containing 50 μL internal standard cocktail. Vortexed (60 s), extracted (1000 μL n-hexane), dried (N2, 10 min). Reconstituted in 200 μL ethanol-butylated hydroxytoluene solution	LC-PDA; C18 ODS 4.6 mm column Mobile phase: MeOH 60%; ACN, 20%; and DCM, 20%, isocratic elution	[82]
Plasma (lutein Zeaxanthin b-carotene lycopene)	0.5 mL + 0.5 mL ethanol (0.1% BHT) + 2,5 mL of HEAT (hexane/ethanol/acetone/toluene, 10:6:7:7, v/v/v/v)	LC-PDA; C30 (150 mm × 4.96 mm, 5μlm) column Mobile phase; (A) MeOH/MTBE/ H2O/2% aqueous ammonium acetate solution, 88:5:5:2, v/v/v/v)	[83]
Plasma (lutein/Zeaxanthin lycopene a-carotene b-carotene b-cryptoxanthin)	200 μL plasma extracted twice with 1 mL of hexane containing 0.01% BHT 8 0.167 μL/mL of echinone. Dried under N2; reconstituted in 100 μL of mixture CH3Cl:MeOH:CH3CN (30:35:35)	LC-PDA; C18 column. Mobile phase: (A) 0.0125% ammonium acetate in MeOH, (B) 100% CHCl3 and (C) CH3CN with 0.1% triethylamine gradient elution	[84]
Plasma Analyte lutein Zeaxanthin b-criptoxanthin a-carotene b-carotene lycopene Phytoene Phytofluene 36 different carotenoid isomers	4 mL of centrifuged blood +5 ml of saline and Ethanol (1:2 ratios). Extraction twice with 5 mL of Hexane and ether (1:1 ratio). Vortexing for 1 min. Centrifuging at 2000 g for 10 min at 4 degrees. Collecting upper layer. Pooled extractions Concentrated to dryness under nitrogen, Reconstituted in 100 μL of ethyl acetate	LC-PDA C30 column (3 lm, 4,6, 150 mm) Mobile phase: MeOH (containing 0.1% Ammonium acetate) and MTBE Gradient elution	[85]
Plasma(Z-퀘-carotene isomers)	800 μL of plasma +800μL of ethanol +2 mL of hexane/BHT (100 mg/L). Then addition of 300 μL of stock solution followed by a vortex-mixing (1 minute) & centrifuged (4∘C).	*HPLC-DAD-RP* *HPLC; C30 (5μm,250 mm × 4.6 mm),mobile phase of methanol + methyl tert-butyl ether + water* DAD; detector adjusted at 450 nm	[86]

(continued)

Table 9.4 (continued)

Source/analyte	Pre-injection steps	Chromatographic specification	Reference
Serum (lycopene a-carotene b-carotene)	200 μL + 5 mL of hexane +1 mL of ethanol. Hexane layer evaporated under N2. Residue dissolved in 200 μL of the mobile phase (ACN-MeOH-chloroform 47:47, v/v/v)	LC-PAD C18 ODS-2 (100? 4.6 mm, 3 lm) column. Mobile phase: ACN/MeOH/chloroform 47:47:6, v/ v/v Isocratic elution	[87]
Serum (lutein Zeaxanthin lycopene a-carotene b-carotene b-cryptoxanthin)	H_2O + ethanol containing b-apo- 80carotenal and extracted into hexane. Organic layer was removed, evaporated till drying & resolvation in chloroform/ethanol (1:19)	LC-PAD RP C18 Mobile phase: MeOH/THF/H2O(94:5:1)	[88]
Human plasma (b-carotene)	Oral dose of all-trans [10,10′,11,11′-(14) C]-beta-carotene (1.01 nmol; 100 nCi) for healthy man. Fate measurement in serial plasma, feces and. Selected plasma samples were spiked with reference standards of retinol, beta-apo-12′-carotenal, beta-apo-8′-carotenal, 13-cis-retinoic acid, all-trans-retinoic acid, beta-carotene-5,6-epoxide, all-trans-beta-carotene, and retinyl palmitate.	RP- HPLC At 14 °C with the use of accelerator mass spectrometry	[98]
Plasma (lutein and Zeaxanthin	100 μL plasma +1100 μL 20% mixture of n-hexane and chloroform +100 μL water +200 μL ethanol. Centrifugation. 800 μL supernatant dried under N2 and reconstituted in 200 μL of n-hexane and acetone; 19% by vol)	LC-PDA Si60 (250 mm × 4 mm, 5 μm) column Mobile phase: n-hexane and acetone (19% by vol)	[90]

Vegetable leaves (lutein, α-carotene, Zeaxanthin,β-criptoxanthin, All-trans-carotene 13- cis-carotene, 9-cis-β-carotene)	Ethanol+ BHT, saponification (80% KOH), mixing (85 °C/10 m. icing +3 mL deionized water. b-Apo-80 carotenyl decanoate added postsaponification extraction (hexane), drying under argon, reconstitution (50:50) MeOH: DCM	LC-PDA C30 (4.6, 250 mm, 3 lm) column Mobile phase: (A) MeOH:Water (92:8, v/v) +10 mmol/L ammonium acetate and (B) MTBE	[91]
Glabrous canary seeds (lutein, Zeaxanthin, b-caroten)	Acetone+ethanol+BHT. Centrifugation (at 1500 for 15 min). Volume upto 100 mL	LC-PDA YMC carotenoid column Mobile phase: (A) MTBE (B) 1% water in MeOH	[92]
Serum (lutein Zeaxanthin)	400 μL + 300 μL ethanol containing 0.25 g/L BHT 200 μL a-tocopherol acetate (0.25 g/L) + 500 μL heptane Vortexed for 2 min; centrifuged at 2000 rpm for 5 min second heptane extraction; combined heptane layers evaporated under N_2 and reconstituted in 200 μL MeOH containing 0.25 g/L BHT	LC-PDA RP (250 mm × 4.6, 5 μm) column connected with an in-line guard-column Mobile phase: 97% MeOH, 3% THF isocratic elution	[93]
Plasma (lutein Zeaxanthin lycopene b-cryptoxanthin b-carotene a-carotene,)	100 μL plasma diluted with 100 μLofH2O + 200 μL of ethanol. Vortex mixed for 30 s, extracted twice with 1 ml n-hexane stabilized with 0.05% BHT; vortex mixed for 10 min and centrifuged for 10 min at 1500 g Supernatants pooled, evaporated under N2 and reconstituted	HPLC-RP C30 (250 × 4.6 mm, 5 μm) column in line with a C18 pre-column Mobile phase: (A) MeOH/water (90:10, v/v, with 0.4 g/l ammonium acetate in H2O) (B) MeOH/ MTBE/water (8:90:2 v/v/v, with 0.1 g/l ammonium acetate in H2O).	[94]

(continued)

Table 9.4 (continued)

Source/analyte	Pre-injection steps	Chromatographic specification	Reference
Commercially available plant extracts	Saponified with an equal volume of 10% potassium hydroxide in methanol (under nitrogen in the dark with stirring) for 1 h at room temperature. Extracted from the KOH/ methanolic phase by shaking BHT + 20 ml 10% sodium chloride. The residue was redissolved by ultrasonic agitation in 10 ml of the mobile phase	HPLC-RP DAD column ((250 × 4.6 mm i.d., 5 mm particle size), Mobile phase; 0.1% BHT + 0.05% TEA. Detection at: 450 nm for caroteneS, 472 nm. Gradient elution of (I) 75:25 (v/v) acetone-water, (II) 75:25 (v/v) acetone-methanol. Gradient elution: 0% (I) to 65% (II) in 10 min. To 80% (II) B in 30 min to 20% (II) B in 60 min flow rate: 1.5 ml/min. Detection at 450 nm.	[95]
Dietary carotenoid Isomers	0.5 mL standard +0.5 mL ethanol/10 mg/L BHA. Mixing +1.5 mL of hexane, mixing and centrifugation (4000 rpm/10 min), re-extraction, evaporation under N_2. Re-dissolved in 0.25 mL of mobile phase.	HPLC-ECD; Column: C30, 5 μm, 4.6 × 250 mm. Mobile Phase:Methanol – Methyl-Ter t-butyl Ether (MTBE) – 1.0 M Ammonium Acetate, pH 4.4 (63:35:2) (v/v/v). Flow Rate; 1.0 mL/min, Temperature: 28 °C, Injection. Volume; 10 μL. Electrochemical Detector: Model 5600A, CoulArray,Applied Potentials:100, 160, 220, 280, 340, 400, 460 and 520 mV vs. Pd. Detector Wavelength;450 nm	[96]
Melon (Lutein,a-carotene b-carotene,Phytoene Phytofluene)	10 g sample + 125 mL THF, 5 g anhydrous sodium sulfate, and 1 g $MgCO_3$, blending, vacuum- filter, re-extraction, 500-mL volume with THF, evaporation/40 °C. re-dissolved in 10 mL THF with sonication.	LC-PDA C18 (200 × 4.6 mm, 5 μm) + guard column Mobile phase: (A) ACN/dicloromethane/ethyl acetate 40:25:35, v/v/v (B) ACN Gradient elution	[97]

Cereal grains (Lutein,Zeaxanthin,b-cryptoxanthin,13-cis lutein,9-cis lutein,90-cis lutein)	Water saturated butanol, vortex mixing, centrifugation, homogenizing, filteration and storage at 20 °C	LC-PDA S-3 (4.6 × 100 mm, 3 μm) carotenoid column Mobile phase: (A) MeOH/MTBE/Milli Q water (81:15:4,v/v/v), (B) MTBE/MeOH (90:19, v/v), gradient elution	[98]
Berries (lutein Zeaxanthin, lycopene b-carotene, b-cryptoxanthin a-carotene)	Extraction (magnesium carbonate +MeOH/ tetrahydrofuran) with BHT, reduction under N2 and re-dissolved in ACN/MeOH/ethyl acetate	LC-PDA,RP C18 (150 × 4.6 mm, 5 μm) with guard column,Mobile phase, with BHT(0.1%) + TEA(0.05%),(A) ACN/MeOH (B) ACN/MeOH/ethyl acetate (60: 20:20)	[99]
Orange juice (lutein Zeaxanthin, b-cryptoxanthin a-cryptoxanthin bcarotene, a-carotene)	THF + BHT till colorless + APO, addition of diethyl ether + salt water. Organic layer dried with anhydrous sodium sulfate reduced to 30 mL + KOH/ 16 h + NaCl soln.Diethyl ether, washing and evaporation +2 mL DCM.	RP -LC-PDA; C18 (250× 4.6 mm, 5 μm) column, Mobile phase: (A) MeOH/water (75:25), (B)ACN/DCM/MeOH (70:5:25)	[100]
Carrot b-carotene Carotenes, Xantophylls	Extraction (n-hexane/ethanol 96% (1:1) till colorless and preserved at 20 °C.	RP -LC-PDA; C18, 250 mm column,Mobile phase: (A): 1% water in MeOH (B) MeOH,(C) 10% n-hexane with CAN	[101]
Carrot(b-carotene, carotenes Xantophylls	10 g extracted with n-hexane/ethanol 96% (1:1, v/v) Until colorless; kept at? 20?C and analyzed within 72 h	RP-LC-PDA; C18, 250 mm, Mobile phase: (A): 1% water in MeOH (B) MeOH and(C) 10% n-hexane with ACN	[102]
Citrus (Zeaxanthin lutein b-cryptoxanthin, b-citraurin, b-carotene, a-carotene, Violaxanthin, (9z)-violaxanthin, Mutatoxanthin)	Extraction (MeOH, diethyl ether), →evaporation and saponification (ether with 30% KOH/ MeOH)	LC-PDA C18 (250 _ 4.6 mm, 5 lm) column Mobile phase: (A) 12% (v/v) H2O in MeOH, (B) MeOH, (C) 30% (v/v) DCM in MeOH	[103]

(continued)

Table 9.4 (continued)

Source/analyte	Pre-injection steps	Chromatographic specification	Reference
Watermelon (Lycopene Lutein, Violaxanthin Neurosporene Zeacarotene, Phytofluene Phytoene)	Acetone extracts + hexane (50 mL) + 300 mL water. N_2 gass passing,reconstituted with 0.5 mL acetone	LC-PDA OSD2 (250 mm _ 4.6 mm, 5 lm) column with a guard column, Mobile phase: (A) ethylacetate (B) ACN/water (9:1, v/v) both containing 0.1% triethylamine	[104]
Soybean (Lutein)	EtOH, 0.1% TBHQ, vortexing (85 °C) + KOH, vortexing +3 NaCl +3 mL hexane, vortexing; centrifugation (1000xg /5 min at 4 °C). Hexane extraction, washing with Na_2CO_3; centrifuged at 1000 g, x5 (4°C).Washed supernatants with 5 mL ultrapure water;evaporated under N_2 and dissolved in 250 mL isopropyl alcohol.	LC-PDA Phenyl (150 mm _ 3.9 mm, 5 lm) column Connected with a guard column,Ternary isocratic solvent system: ACN, MeOH and water (48:22.5:29.5, v/v/v). Isocratic conditions for 40 min, followed by a 4 min linear gradient to 100% MeOH for 7 min, subsequent 4 min linear gradient to initial conditions	[105]
chicen	10 mL of ethanol with 0.1% BHT was added per g. homogenised for 120 s in total (6 × 20 s pulses). Transferred to 250 mL polypropylene centrifuge bottles, vortexed for 10 s, sonicated at 24 oC for 2 min and vortexed again for 10 s. centrifuged at 4700 rpm at 25 oC	HPLC-DAD; column 100 × 10 mm (L, D) S-5 µm. Mobile phase; hexane and isopropyl alcohol (90:10, v/v). The flow rate was set at 2 mL min − 1, at 35 oC. Isocratic elution with hexane and isopropanol (90:10, v/v) and a flow rate of 0.5 mL min − 1. at 25 oC	[106]

Dried peppers (Violaxanthin b-cryptoxanthin,b-carotene)	1 g of ground pepper +20 mL cold acetone (5 °C); vacuum filtered. Filtrate transferred to a separation funnel together with petroleum ether and water. The upper ether layer washed several times with water; added anhydrous sodium sulfate. Ether phase adjusted to 10 mL with petroleum ether; concentrated with N_2 and re-suspended in 1 mL acetone)	LC-PDA ODS (25 cm _ 4.6 mm, 5 lm) column Isocratic, mobile phase: ACN/MeOH/ tetrahydrofuran (THF) (58:35:7)	[107]
Sweet potato (b-carotene)	2 g + 5 mL acetone + acetone/petroleum ether (20:80); filtration. Procedure repeated until colorless. Filtrate + anhydrous sodium sulfate. Concentrated in a rotary vacuum evaporator max. 35 °C until 2 mL remained, dried under N_2 and re-suspended in 5 mL petroleum ether	RP -LC-UV-Vis C18–18 (4.6 mm _ 200 mm, 5 lm) column Mobile phase: ACN/MeOH/ethylacetate/ triethylamine (79:10:20:0.05)	[108]
Tomato (Lutein,Lycopene,b-Carotene,Lycopene,b-Carotene isomers)	100 mg + 3 mL ethyl acetate containing BHT (100 mg/L); vortex mixing for 1 min; centrifuged for 5 min Upper hexane removed. Extraction with 2 mL of ethyl acetate (3 or 4_) until colorless. Ethyl acetate fractions combined, evaporated in a vacuum evaporator. Residue re-suspended in 25 µL of diethyl ether and 75 µL of MeOH/ACN/THF (50:45:5, v/v/v) Total individual carotenoid	LC-PDA C18 (3.9 × 150 mm, 5 µm) column Mobile phase: MeOH/ACN and THF (50:45:5, v/v/v),Lycopene and b-Carotene isomers: LC-PDA,C18 column,Mobile phase: MeOH/ACN/2-propanol (54:44:2, v/ v/v)	[109]

(continued)

Table 9.4 (continued)

Source/analyte	Pre-injection steps	Chromatographic specification	Reference
Sweet potato (b-Carotene and Isomers)	5 g extracted thrice with 50 mL of cold acetone. Extracts pooled and transferred to a separation funnel and 40 mL PE added. Washed thrice with 200 mL of ultra-pure water. Upper PE-phase collected, dried with anhydrous sodium sulfate. Sample filtered and concentrated to dryness in vacuum (30 °C) and under N2. Re-suspended in 2 mL of acetone	LC-PDA C30 (250 _ 4.6 mm, 5 lm) carotenoid column Isocratic elution with a mobile phase of MeOH/ MTBE (80:20)	[110]
Einkorn, emmer and spring wheat (Lutein,Zeaxanthin, b-Carotene)	3 g + 15 mL of ethanol/acetone/hexane (1:1:2). Overnight extraction Ultrasonic bath for 10 min. Samples filtered. Filter cake washed until colorless with 3 mL of extraction mixture. Filtrates transferred into 25 mL flasks and made up with extraction mixture.Aliquots (20 mL) evaporated on a rotary evaporator and the residue dissolved in 2–3 mL ethanol/acetone (6:4) with 0.2% BHT	LC-PDA C30 (3.0 _ 150 mm, 5 lm) Carotenoid column Mobile phase: MeOH/water/TBME	[111]
Water spinach (b-carotene Lutein, Violaxanthin)	Extraction with a mixture of MeOH, acetone and petroleum ether (1:1:1, v/v/v) until colorless. Saponified by 30% (w/v) KOH in MeOH for 16 h. Extract washed until all KOH removed Concentrated in a rotary evaporator (30 °C). Resuspended in acetone	TLC and HPLC-PDA Purified by TLC silica gel (activated for 2 hours at 110C) with n-hexane: hylacetate:acetone: MeOH (27:4:2:2, v/v/v/v),YMC C30 column (150 mm _ 4.6 mm, 3 lm), Mobile phase: (A) MeOH/ water (9:1,v/v), (B) MTBE	[112]

spinach and sweet corn (xanthophylls, lutein, zeaxanthin) spinach (violaxanthin, neoxanthin, β-carotene	(hexane/acetone (1:1) + BHT (100 mg/l) + BHA (100 mg/l) vigorous shaking of sample with the two phase system. hexane/ethanol/water, 6:5:1.3, v/v) Shaking (2 min), separation and crude extraction	HPLC –RP C30 column (250 mm, 10 mm i.d.). particle size 5 lm, system controller LC-CaDI 22–14, two solvent delivery modules, compact pump 2250, UV–VIS detector, monitored at 445 nm. 350 lL xanthophyll isomers eluted with acetone/ water (89:11, v/v) at ambient temperature for 25 min. The flow rate was set at 4 mL/ min. Eluent A; Xanthophylls elution with acetone/ water (82/18, v/v), 20 °C, 45 min. Flow rate = 1 mL/min. Eluent B; metha-nol/tert-butyl methyl ether (MTBE)/water (92:4:4, v/v, 20 °C,flow rate 1 mL/ min	[113]

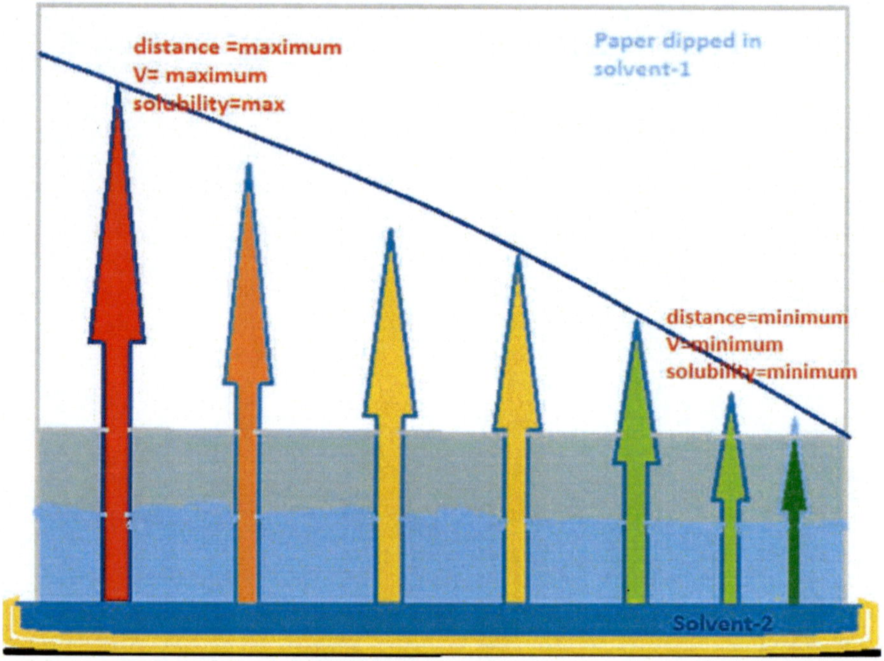

Fig. 9.8 Schematic diagram of paper chromatography

9.21 Open-Column Chromatography

For separation and isolation of biochemical components open column chromatography (on MgO: Hyflosupercel) can be employed with better results even as compared to commercial standards [7, 114–116].

9.22 Gas Chromatography

It is well established fact that customary gas chromatography could not prove itself appropriate for carotenoid assay because they are instable with lower volatility. Therefore, it can be employed for carotenoid analysis if coupled with other techniques like GC-MS. Schmidt, et al. [117] studied stilbene from Arabidopsis thaliana using this coupling. Recently, in complementary analyses of esters of carotenoids with fatty acids this coupling has been used for fatty acid determination followed by coupling of HPLC–DAD along with LC–MS/MS spectra for identification of the carotenoids found in the form of esters [118]. Gas chromatographic coupling with mass spectroscopy (GC-MS) has been adopted to cope with problems raised by isotope ratios. Successful combination of GC-MS for ^{13}C-carotenoids and $^{13}CO_2$

(isotope ratio-gas chromatography-mass spectrometry, IR-GC-MS) for isotope ratios in tracer experiments has been noted [119].

9.23 Liquid Chromatography

Liquid chromatography separates individual components of sample based upon differential pattern of interaction between mobile and stationary phases. Most adopted LC technique is HPLC (Fig. 9.9).

9.24 High-Performance Liquid Chromatography

HPLC facilitates extra discrete parting of carotenoids [120]. It uses a pump that passes pressurized liquid solvent having sample mixed with it, from a column already filled with a material that has potential to adsorb. Individual components of sample differ for their adsorbent potential. That results into differential flow rate of each component in adsorbent solid stationary phase. Ultimately it results into variable time of elution. This difference of the time of retention in column and the elution time help to separate each component from each other. Carotenoids with greater polarity (like E-lutein, E-zeaxanthin, E-canthaxanthin elute first as compared to the

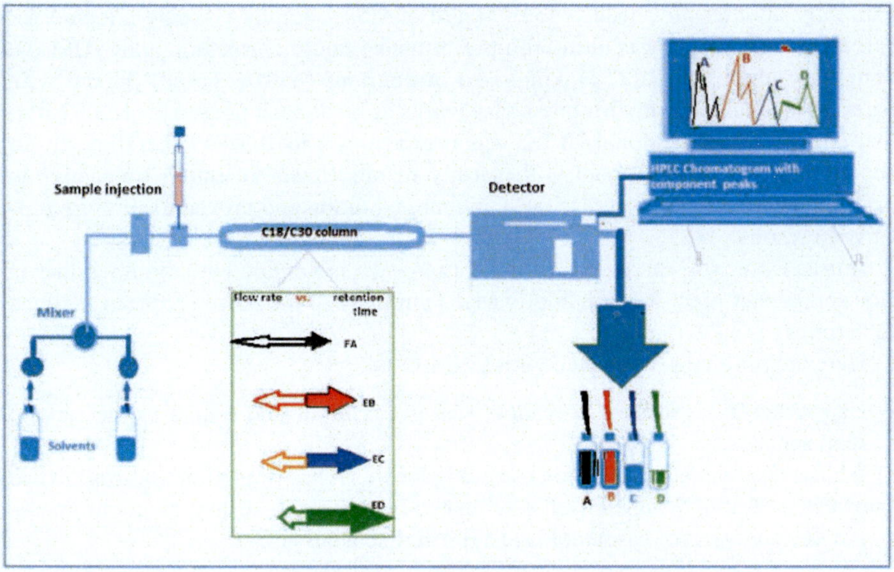

Fig. 9.9 Schematic diagram of high performance liquid chromatography

carotenoids exhibiting lesser polarity (like E- b-carotene) followed by nonpolar carotenoids (like E-lycopene).

In general, carotenoids are with 3–11 conjugated double bonds along with their non-conjugated double bonds. That facilitates the possibility of cis and trans isomerization. As Lycopene can exist in more than thousand geometric isomeric forms [79].

Previously C18 stationary phases was employed whereas now C30 carbon bound stationary phase (C30 column) is preferred as discussed above.

9.25 2D LC

As progression of HPLC, two-dimensional liquid chromatography (LC × LC) in different modes is under use for carotenoid assays. It may be in normal phase (NP) or it may be performed with reverse phase (RP) [121–123]. Where primary column is used along with secondary one for combined separation orthogonal to other one. Effluent of column one is used to inject in second column facilitated by some special switchable valve [124].

9.26 Ultra High-Performance Liquid Chromatography (UHPLC)

Currently another advanced type of liquid chromatography is being adopted for carotenoid analysis that is ultra-high-performance liquid chromatography (UPLC). Stationary phase of UPLC is comprised of columns (narrow bored) packed with particles size under 2 μm. Mobile phase works at great back pressures (103.5 MPa) that in case of conventional HPLC was in range of 35e40 MPa. Therefore, it got superiority over conventional with speedy results, lower retention time, narrow peaks and better signal-to-noise ratio, higher resolution and proved itself more sensitive in results.

It uses lesser solvent and greener technology. It is flexible with minimal dispersion and greater precision that improves resolution (2–4 times) and reduce run times (2–8 times).

Here are problems associated with UHPLC

- Experiencing a problem in readout just get printout and e-mail to the service representative.
- It has a disadvantage that is not easy to ignore, as it has limitation towards the use of C30 column. Whereas well established benefits of C30 regarding chromatographic analysis for carotenoids [125] had discussed above.

With minor modifications a number of researchers had used this tool for carotenoid analysis. Here is brief survey of literature available for recent utilization of UHPLC in Table 9.5.

9.27 Countercurrent Chromatography (CCC)

It is a category of LC with biphasic liquid to liquid separating analysis. It employs two immiscible liquid phases in the absence of any solid support. Liquid stationary phase is used for separation, identification, and quantification placed with the help of centrifugal force. One of liquids will be pumped via column/chambers having both phases. Separation of components will be manifested according to their solubility in both liquids.

Upon the basis of force that help to retain one of the liquids as stationary phase without any solid support CCC analysis are divided into two types. One type relies upon gravity while the other one upon centrifugal force in the column followed by mobile phase entry facilitated by pumps. CCC can be further sub grouped as following.

Gravity/Droplet CCC (DCCC)

Gravity method is also known as droplet counter current chromatography (DCCC). Gravity helps to transfer the mobile phase into the stationary phase, that has been detained in an elongated vertically arranged tubes serially linked with each other. There are further two categories under these analysis (Fig. 9.10).

- Descending mode; where denser mobile phase along with sample poured through the columns having comparatively lighter stationary phase, whole facilitation is by the gravity.
- Ascending mode; mobile phase is lesser dense there force is used to rise up in stationary phase
- Its efficiency vary with number of columns. DCCC provided good results with natural product partitioning. But mostly replaced with high-speed countercurrent chromatography. As it faces major limitation of lower flow rates and particularly for binary solvent sets it does not show good mixing

Centrifugal Method

This assay uses centrifugal force. Overall, two different ways are opted to retain stationary phase via centrifugal force (Fig. 9.11).

Table 9.5 A review of generally used Ultra High-Performance Liquid Chromatographic approaches for carotenoid analysis

Sample/analyte	Technical specifications	Refrences
Lutein, lycopene and -carotene	Solvent A: Millipore water containing 0.1% TFA and solvent B: methanol/acetonitrile/isopropyl alcohol (54:44:2, v/v/v). Gradient elution: 85% B to 95% B in 1 min, 95% B to 99% B in 1 min, 99% B to 99% B in 3 min, and 99% B to 95% B in 1 min BEH C18 (2.1 mm × 100 mm); 1.7 μm; NRb BEH Flow rate 0.6 mL/min Run time; 6 min	[126]
Antheraxanthin, violaxanthin, neoxanthin, astaxanthin, adonixanthin, zeaxanthin, lutein, −apo-8?-carotenal, 3-hydroxyechinenone, and-cryptoxanthin, echinenone, lycopene, carotene, phytofluene and phytoene	Solvent A: acetonitrile:methanol (70:30, v/v) and solvent B: 100% water. Gradient elution: 85% A in 2 min, 85% A to 100% A in 1 min, 100% A in 8.6 min, 100% A to 85% A in 1 min and 85% A in 2.4 min BEH C18 (2.1 mm × 100 mm); 1.7 μm; 32oC Flow rate 0.4 and 0.6 mL/min Run time; 15 min	[127]
Neoxanthin, violaxanthin, lutein 5,6 epoxide, antheraxanthin, lutein, zeaxanthin, cryptoxanthin, echinenone, all-trans-carotene, 9-cis-carotene, 13-cis-carotene	Solvent A: acetonitrile: dichloromethane:methanol (75:10:15, v/v/v) and solvent B: water containing 0.05 M acetate ammonium. Gradient elution: 75% A in 20 min, 75% A to 100% A in 1 min, 100% A to 98% A in 9 min, 98% A to end HSS T3 (2.1 mm × 150 mm); 1.8 μm; 35oC Flow rate 0.4 mL/min Run time; 46 min	[128]
Cryptoxanthin, lycopene, and carotene Lutein,	Solvent A: acetonitrile containing 0.1% formic acid and solvent B: TBME. Gradient elution: 2.5% B in 0.1 min, 2.5% B to 7.5% B in 0.3 min, 7.5% B to 10% B in 0.1 min, 10% B to 12.5% in 1 min and 2.5% B in 1.5 min BEH C18 (2.1 mm × 50 mm); 1.7 μm; 32oC Flow rate 0.45 mL/min Run time; 3 min	[129]
Lutein, zeaxanthin, ˇ-cryptoxanthin, lycopene, and carotene	Solvent A: acetonitrile: methanol (85:15, v/v) and solvent B: 2-propanol Gradient elution: 95% A in 0.8 min, 95% A to 50% A in 2 min. At min 3.5 the system is returned to initial conditions and maintained for 1 min HSS T3 (2.1 mm × 100 mm); 1.8 μm; 35oC Flow rate 0.5 mL/min Run time; 4.5 min	[130]

(continued)

Table 9.5 (continued)

Sample/analyte	Technical specifications	Refrences
//20 tomato cultivars (geometric isomers)	acetone/ethanol, Shaking (150 rpm for 1 h), centrifugation, re-extraction, volume upto 10 mL UPLC-PDA C18 (100 mm × 2.1 mm, 1.7 μm) + guard column Mobile phase: (A) MeOH/MTBE/water (90:5:5, v/v/ v) and (B) MeOH/MTBE/water (90:5:5, v/v/v)	[131]
Capsicum annuum L. (free carotenoids and carotenoid esters)	NP-LC _ RP-UHPLC with PDA and IT-TOF (APCI ionization) MS MeOH/ethyl acetate/ petroleum ether (1:1:1, v/v/v), C18 column (30 mm _ 4.6 mm, 2.7 lm), 2D Mobile phase: 1D (A) n-hexane (B) n-hexane/butylacetate/ acetone (80:15:5, v/v/v); 2D (A) water/CAN (10:90, v/v) (B) isopropanol.	[121]
Capsanthin, capsorubin, antheraxanthin, zeaxanthin, ˇ-cryptoxanthin and ˇ-carotene	Solvent A: 10% isopropanol and solvent B: 100% acetonitrile. Gradient elution: 75% B in 3 min, 75% B to 95% B in 0.2 min and 95% B to 100% B in 7.8 min HSS C18 (2.1 mm × 100 mm); 1.8 μm; NRb Flow rate 0.75 mL/min and Run time; 11 min	[132]

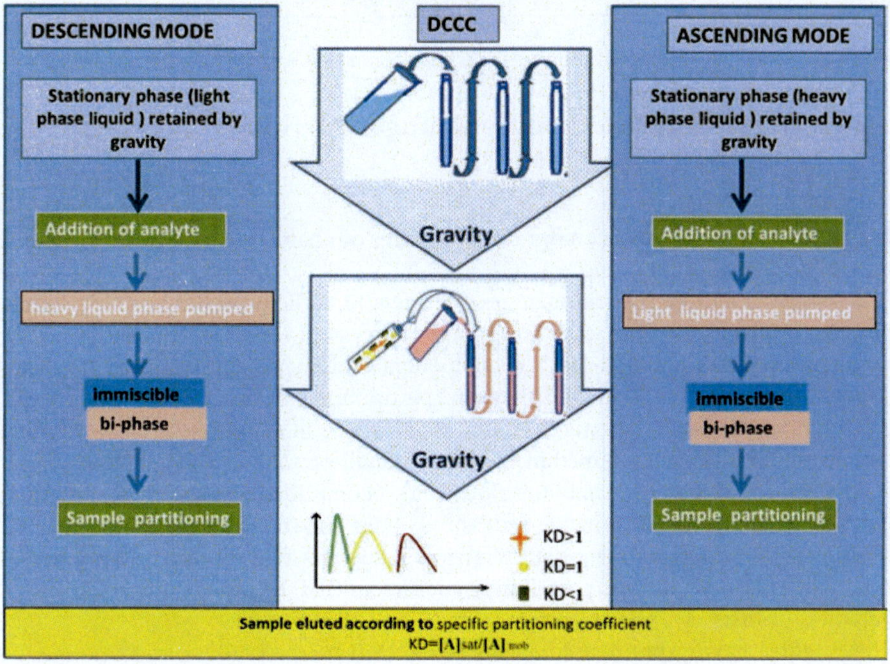

Fig. 9.10 Schematic diagram of droplet gravity counter current chromatography

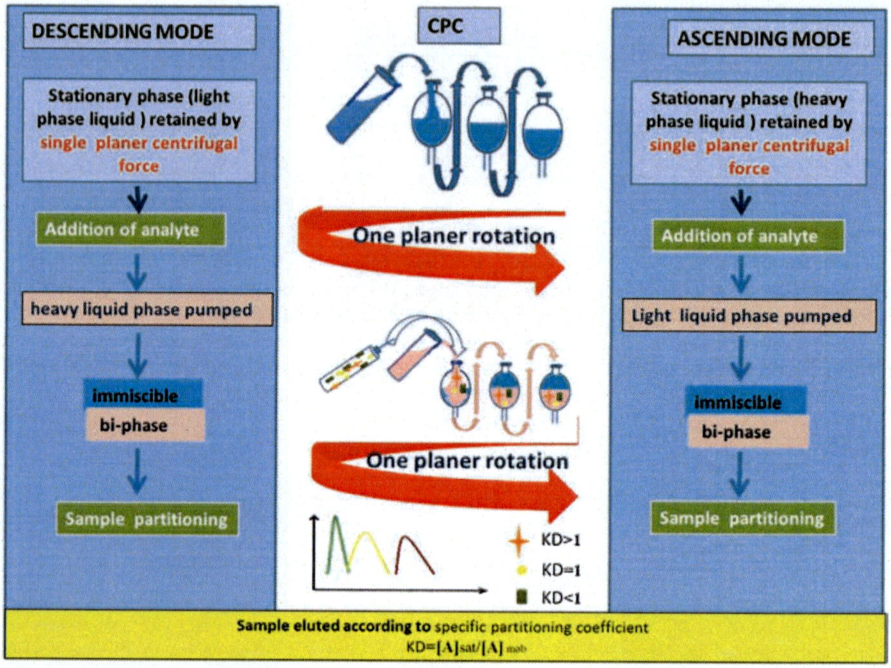

Fig. 9.11 Schematic diagram of centrifugal partition chromatography

9.28 Hydrostatic Method/Centrifugal Partition Chromatography (CPC)

In this procedure column is comprised of a series of chambers linked with channels that rotate around a central axis [133].

It is fast liquid-liquid chromatographic method. Where two immiscible liquid phases are made into contact forming biphasic system together. It is followed by multiple separations afterwards. The component analyte present in sample are separated on the basis of individual differential partitioning coefficients in both phases.

One liquid phases (stationary) is used to introduce in column (the rotor rotating around single axis) and retained in stationary condition in the rotor with the help of centrifugal force. Later on analyte solutes and second liquid (in spinning condition induced by moderate rotational speed) is fed under pressure with the help of pump. Molecular exchange amongst both phases is made possible by their mixing. Each solute differs for its specific partitioning coefficient (Kd) that made differential separation of solutes eluting from each outlet (Fig. 9.12).

All elutes having purified solutes are collected for particular period (minutes-hours). Centrifugal Partition Chromatography is beneficial in a number of ways

- It has not requirement of costly packed solid phase
- Solvent being consumed is ten to twenty five percent only

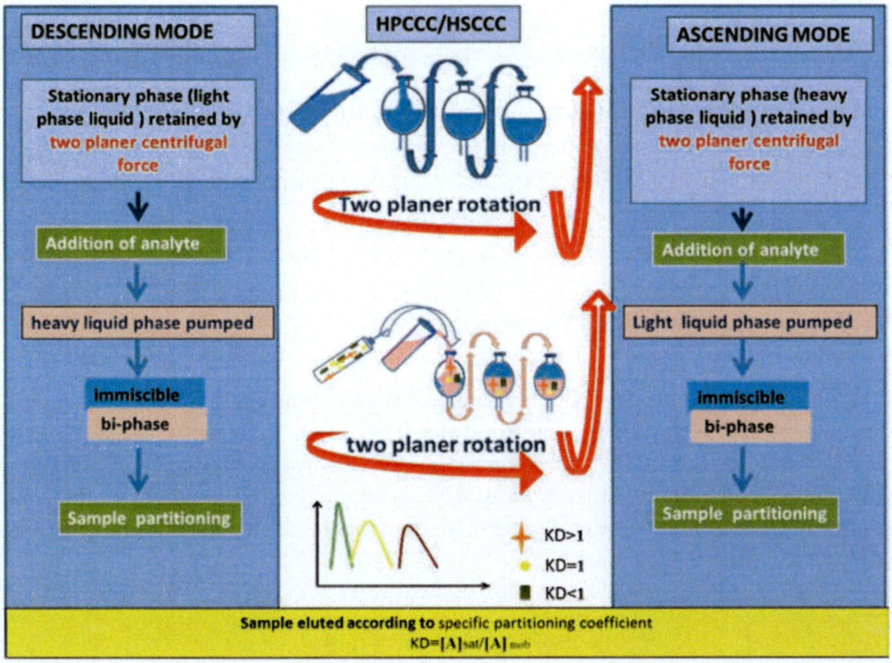

Fig. 9.12 Schematic diagram of Hydrodynamic/high-speed CCC (HSCCC and HPCCC)

- Single rotor for various partitions
- It is soft due to good sample recovery
- No possibility of interaction between silica and molecules denaturing them
- Good efficiency with almost 99.9% purity
- It is speedy (3–5 times) as compared to other options
- Good flexibility for adopting great variety of naturally and synthetically occurring biomolecules and can adapt wider range of polarities
- Easy scale up study as sample ranges from mili grams to many kilo grams

9.29 Hydrodynamic (HSCCC and HPCCC)/ High-Speed CCC

It is also termed as high-speed /high-performance countercurrent chromatography. There is Archimedes' screw force of helical coil that made possible the maintenance of liquid as stationary phase in the column. In this case stationary phase in column is retained by centrifugal force with planetary rotation movement around two rotating axes.

HSCCC is extensively used for preparation of partitioned samples from natural sources. Along with its potential to separate better than HPLC, HSCCC has the edge due to its capability of direct application using crude extracts. Moreover it shows good recovery of the analytes. In general there are limited lipophilic approaches for the separation of carotenoids.

Dual-mode countercurrent chromatography (CCC) has been successfully applied for different carotenoids (β-carotene, α-carotene and lutein) assay from carrot Englert et al. (2015) [134]. Other studies regarding utilization of CCC for carotenoid analysis also available in literature [135, 136].

9.30 Electrochemical Detection

For the samples equipped with minute quantity of carotenoids electrochemical detection (ECD) is preferably employed [96, 137–140].

Basic principal of ECD is based on separation of carotenoids on specific analytical column that generate current positively correlated (proportional) with the concentration of carotenoids generally there are two kind of ECD i.e. amperometric and coulometric (Fig. 9.13).

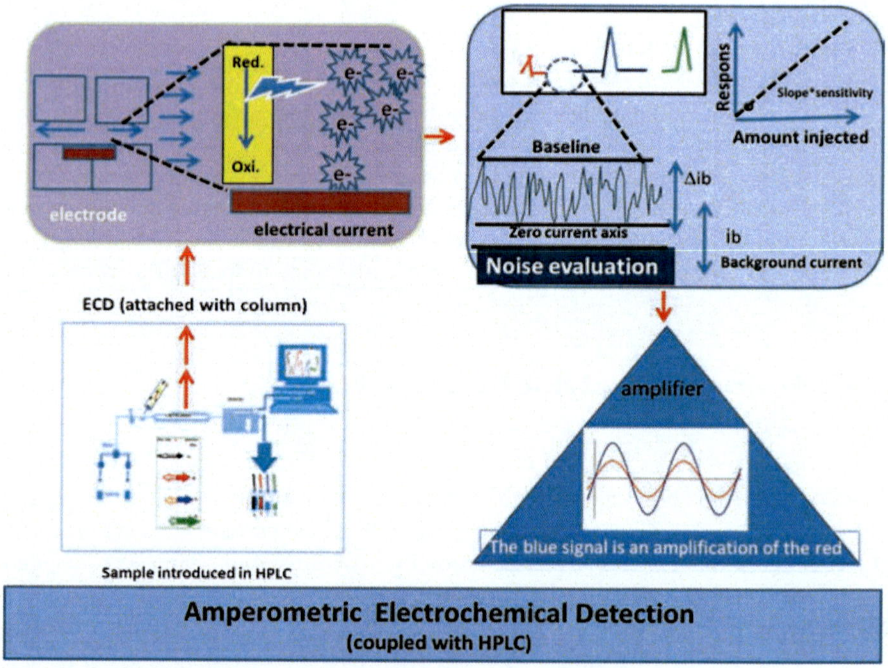

Fig. 9.13 Schematic diagram of an amperometric electrochemical detection system

9.31 Amperometric-ECD

The amperometric electrochemical detection measure the electricity in the form of current. That is produced from oxidation/reduction reactions. Here first of all sample is injected in HPLC for separation via chromatography of column. Next came an ECD cell and an electrochemical sensor followed by electrode. After elution from column the electrochemically active substances experience electrochemical reaction, due to which electrons formulate an electrical current. As the electrodes have been further connected with an electric source having powerful (low noise) amplifier capable of converting a pico- or nanoampere current in a signal in the range of ±1 Volt. That is usually required for data acquisition (Fig. 9.14).

9.32 Coulometric-ECD

In this type of ECD it is electrolyzed (100%) with the help of porous electrode structure and greater surface area after application of a sufficient voltage. Whereas in amperometry there was smooth electrode electrolyzing limited components.

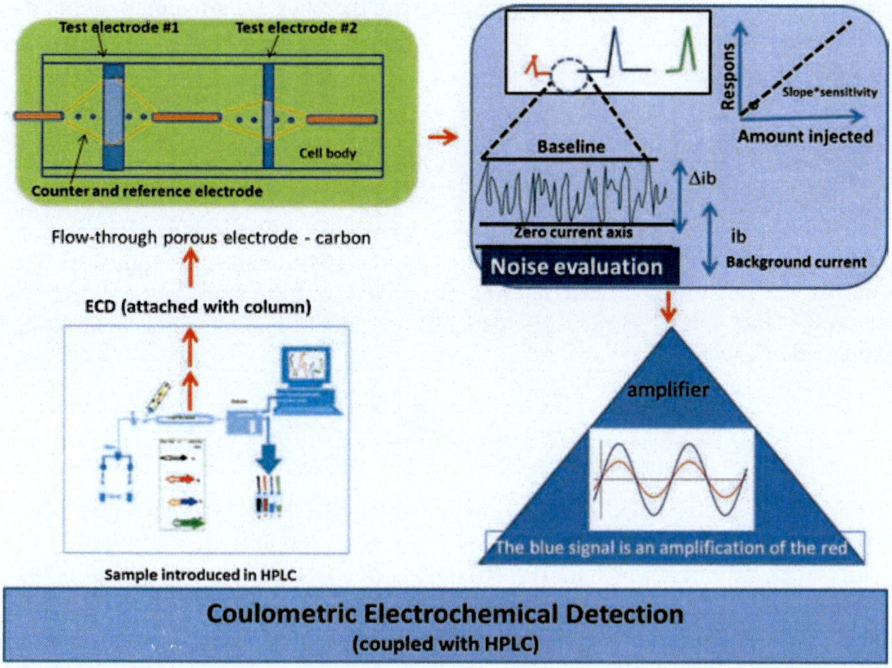

Fig. 9.14 Schematic diagram of Coulometric-ECD

Gamache et al. [96] successfully assayed carotenoids using HPLC-ECD. Still limited data is available regarding carotenoid analysis using ECD. Currently, Nagy et al. had used ECD approach for carotenoid glycoside isolation and identification from *cyanobacterium Cylindrospermopsis raciborskii* [141].

9.33 Carotenoid Analysis by Metabolomics Tools

Metabolomics is considered as the peak of analysis for metabolites. It provides complete set of information about metabolites present in analyte at particular stage. Amongst metabolomics tools MS- and NMR are most frequently opted.

9.34 NMR

Nuclear magnetic resonance (NMR) is considered as a authoritative approach for the structural elucidation of carotenoids. It can help to explore about stereo-isomerization. One- and two-dimensional NMR positively estimated the carotenoids from different biological samples. In this case comparatively larger amounts of sample are needed than mass spectroscopy. For successful out comes of NMR as an analytical tool, it is compulsory to understand the physical principles behind the technique (Fig. 9.15).

9.34.1 Background

Each nucleus of isotopes has its own specific spin (+/−). As a result of their spinning movement an internal magnetic field is generated (with magnetic moment = **μ** ∝ spin). In the same time an external field is applied to these nucleus the difference between + and - state energy (ΔE) depends on the external magnetic field too as expressed below.

$$\Delta E = \frac{\mu Bx}{1}$$

For NMR spectroscopy this ΔE is expressed generally in MHz (10^6 Hz) (20 to 900 Mz). Most frequently used isotopes in literature include ^{1}H, ^{13}C, ^{19}F and ^{31}P (with I = ½). Amongst which most frequently used isotope is of hydrogen atom (the proton).

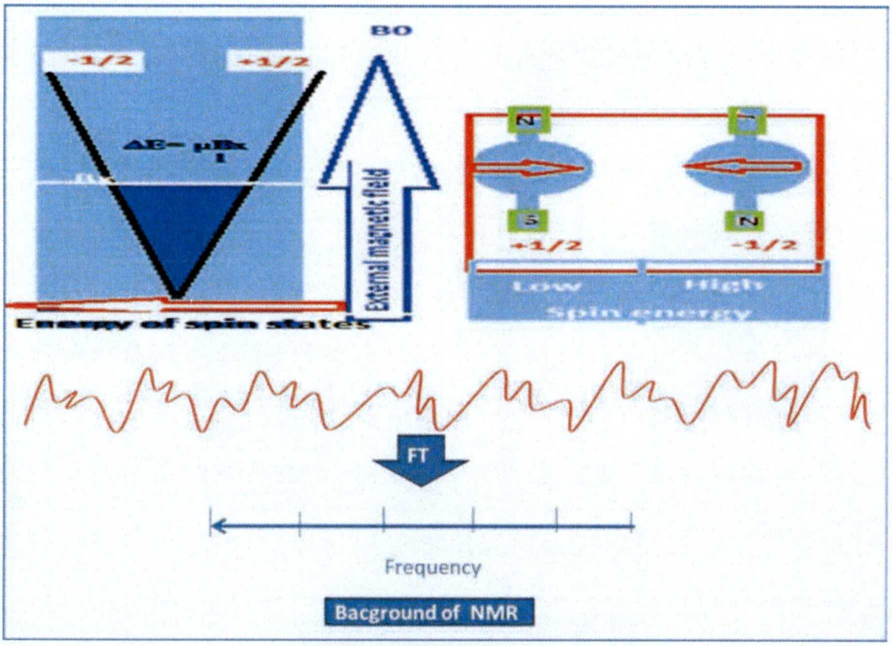

Fig. 9.15 Schematic diagram of NMR background

9.34.2 Proton NMR Spectroscopy

A classical CW (**continuous wave)** spectrometer is shown above (Fig. 9.16). Sample solution is positioned in a glass tube (5 mm) between magnetic poles and allowed to spin. To make it more simple and understandable red colored arrows have been used to mention step wise travel of energy. First of all radio frequency radiation is generated and passed towards the antenna coil (red arrow #1). Secondly from these antenna coils it is transmitted into the sample (red arrow # 2). Thirdly from sample it is received by receiver coil surrounding the sample tube (red arrow #3) that finally transmitted towards RF receiver. Here come the task of appropriate monitoring by electronic devices and a computer.. Next step is to collect overlapped resonance signals by a computer and used for Fourier transform (FT) mathematical assay. As shown in (Figs. 9.15 and 9.16) FT converts emitted signals into the frequency. Proton chemical shifts vary with functional group, intensity of signals molar concentration of analyte and number of hydrogen.

For identification of specified carotenoid structures of geometric isomers from peak shown in chromatogram nuclear magnetic resonance (NMR) is employed with promising efficacy.

NMR is non-destructive, and give pretty information about samples even smaller than mili grams without destructing samples.

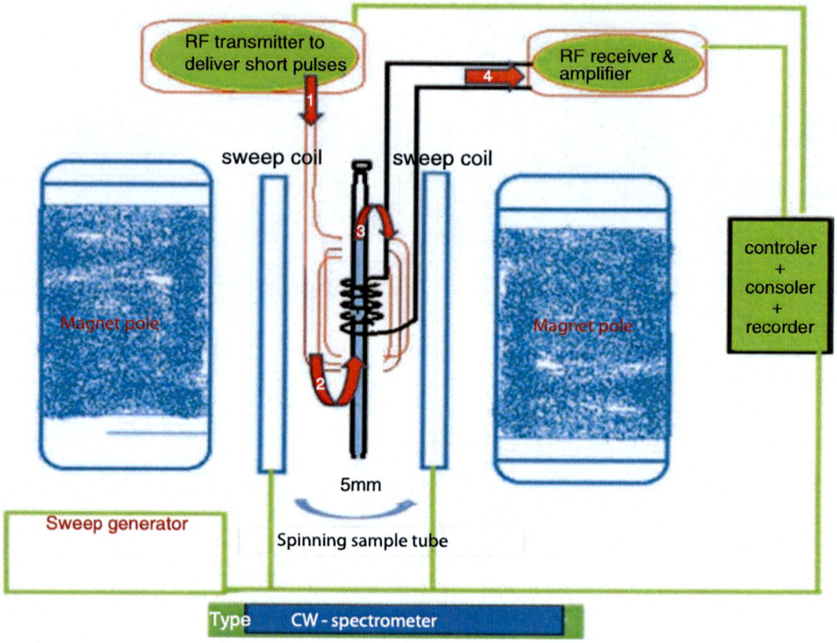

Fig. 9.16 Schematic diagram of NMR spectrometer

Though literature shows a number of successful studies of carotenoids using NMR (Table 9.6) [38, 141–144] still data regarding carotenoid is not very extensive with independent NMR technique (Fig. 9.17).

9.35 NMR Coupling with Chromatography

This is merging of NMR technique chromatographic approach to magnify its potential of quantification and qualification. The chromatographic run provides peak maxima of carotenoids that is shifted and detected by NMR probe to compare NMR spectra of each partitioned component.

Carotenoids are bioactive and naturally exist in isomeric forms too. Some carotenoids naturally occur in the form of optical (R/S) isomers. Such carotenoids like Lutein and zeaxanthin need more advanced technique for separation of their products of optical isomerization including HPLC-NMR and HPLC-MS enhancing carotenoid profiling. These techniques along with sensitivity and accuracy of results need extra care while preparing sample, technical expertise while interpreting data otherwise it would lead towards extreme errors HPLC-NMR On-line capillary HPLC–NMR. By this coupling all cis and all-trans isomer of carotenoids have been elucidated with successful isomer profiling [145] with high sensitivity (nano-liter).

Table 9.6 A review of generally used NMR specifications for carotenoid analysis

Sample/analyte	NMR specification	References
polar carotenoids (all-*E*) lutein, (all-*E*) zeaxanthin, and (all-*E*) canthaxanthin (all-*E*) b-carotene and nonpolar (all-*E*) lycopene,	1D 1H NMR, Capillary pump (with 50 mm capillaries), HPLC pump, the UV detector, and the NMR flow probe. C30 capillary Column, solvent gradient of acetone: water = 80:20 (v/v) to 99:1 (v/v) and a flow rate of 5 mL min − 1	[38]
cyanobacterium Cylindrospermopsis raciborskii	13C NMR; CDC13 with spectrometer (500.12/125.4 MHz for 1H/13C, respectively) at 25 °C. The ^{13}C and ^{1}H NMR on the basis of 1D (1H, 13C APT) and 2D (COSY, HSQC, HMBC) experiments. Standard Bruker software. The 1D and 2D data were processed using the programs MestReC 4.9.9.6. and ACD/NMR Processor 12.01	[141]
fruit of red mamey (Pouteria sapota) (neoxanthin and (9'Z)-neoxanthin)	1H- and 13C NMR; 600 MHz Varian DDR 5 mm IDPFG, Standard pulse sequences VnmrJ 3.2C/ Chempack 5.1,direct 1H–13C, long-range 1H–13C, and scalar spin–spin connectivities using 1D 1H, 13C, 1H–1H gCOSY, 1H–1H TOCSY, 1H–1H NOESY, 1H–1H ROESY, 1H–13C gHSQCAD (J = 140 Hz), 1H–13C gHMBCAD (J = 8 Hz) HSQC and HMBC spectra aided, selective 1D TOCSY, at 298 K and standard 5 mm NMR tubes. The 1H and 13C chemical shifts, TMS (0.00 ppm). Chloroform through column (10 cm), Al2O3 prior to the dissolution. CDC13 (99.8 atom% D)	[142]
Vegetable (beta-carotene)	1H HR-MAS NMR. (solid aqueous samples)	[143]
crystalline -carotene complexes (with cyclodextrin and micronizate)	13C CP-MAS NMR (solid state)	[144]

For structural elucidation on-line HPLC-NMR has been successfully implemented by Dachtler, [36, 37]. In studies of carotenoid isomer that exist in very low (ng/mol) quantity like lutein and zeaxanthin simple NMR was with limited precision. Thereafter to solve this issue of HPLC-NMR after wards, modification in magnets and the probes has made possible detection of carotenoids in nanogram range.

LC-NMR deals in one dimensional and two dimensional manner for NMR spectra of partitioned components (HPLC based). Higher magnetic fields made it sensitive and most adopted. Were NMR helps to explore conformational geometry for structural elucidation. From food and biological samples like other metabolites carotenoids also been characterized by this technique [146].

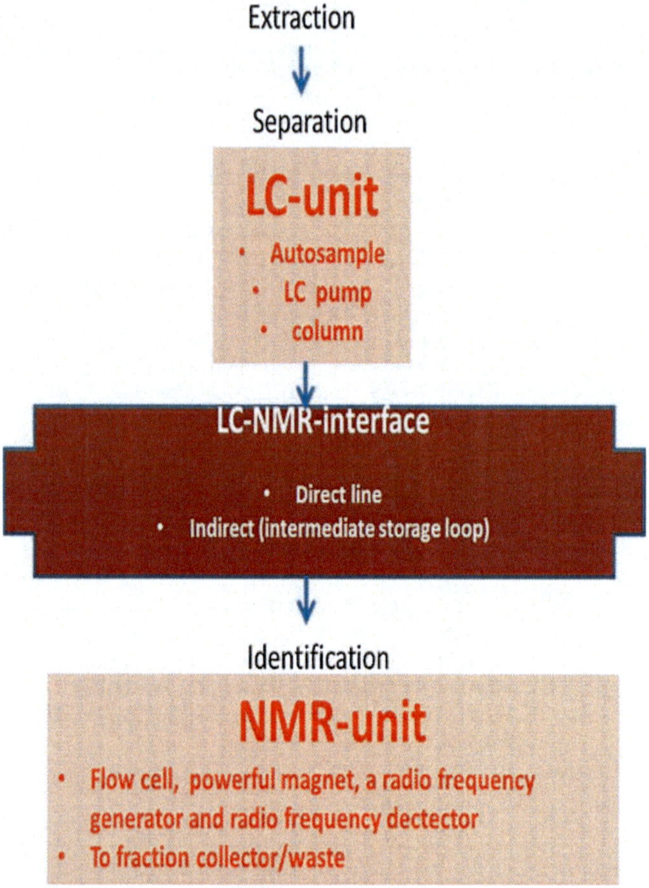

Fig. 9.17 NMR analysis coupled with LC/HPLC

9.36 Mass Spectrometry

Liquid chromatography (PDA/ UV-Vis detector) provides similar qualitative and quantitative spectra for many carotenoids that pointed out towards the need of some others detectors. Moreover it cannot explore molecular structure of unknown sample having complex matrix. Finally MS instrumentation was explored that helps to reduce spectral interferences resulting with better sensitivity during complex mixture analysis and data collection for structural elucidation based on molecular mass and fragmentation pattern (Fig. 9.18).

For the sake of the measurement of features of individual molecules, this instrument converts molecules into their ionic forms. That facilitate their movement in response to the external electric and magnetic fields. Basic steps of its functioning are being summarized below

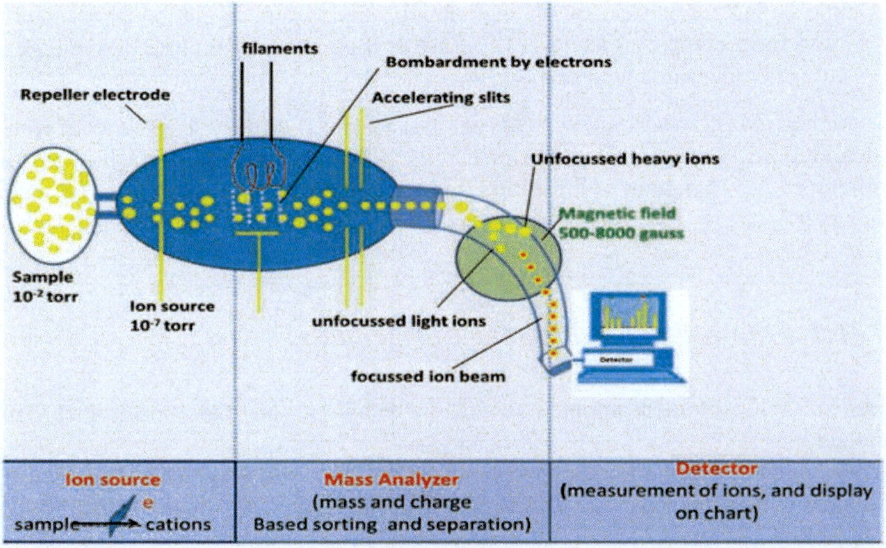

Fig. 9.18 Schematic diagram of a mass spectrometer

- Sample are converted into ions in ion source being reactive these ions are formed and manipulated in vacuum (at 10^{-5} to 10^{-8} torr pressure). As electrons have negligible molecular mass its removal during ionization do not impose detectable change in mass of analyte
- Stable isotopes (usually labeled with ^{13}C or ^{2}H) are used as internal standards
- Electron bombardment from heated filament facilitate ionization
- Slits converts them into a beam
- Separation of ions (mass and charge based) is facilitated by external magnetic field
- Detector detects and display chart on attached screen
- A mass spectrum is generally expressed as bars (varying with mass, charge and abundance)
- MS fragment produced by carotenoid will be compared with previously identified carotenoids. If it differs from literature it would be identified by number of steps
  (a) if in sufficient quantity it would be confirmed by NMR
  (b) if found in smaller fractions it would be followed by synthesis of putative "standards" from intermediates or by oxidation of parent carotenoids. Then they are classified using other tools like UV/Vis, IR, NMR, HPLC, GC, and MS. Number of carotenoids have been successfully identified in this way (Table 9.7)
- In general this technique is considered advantageous because there is no limitation of any pre-infusion chromatographic separation of carotenoids they can be directly infused in system. It promises accuracy, reproducibility, sensitivity and

time saving. Drawbacks of MS include ion suppression, co-elution and possibility to form complex spectra [147]. Further the decisive structure of some ions still is not identified with cert.

Modern Mass spectrometric analysis of carotenoids deals with an array of technical options with variable operational principles and background science. Major difference lies in pattern of ionization, mass analyzing, power resolution and mass accuracy.

9.36.1 Ionization Methods for Carotenoids

Here is a list of most adopted ionization techniques in mass spectrometry of carotenoids.

Presently atmospheric pressure photoionization (APPI) and APCI is preferably adopted. Other methods include EI, ESI, FAB, IMS, SIMS, APCI, APPI, ASAPP, and MALDI (Table 9.7).

9.37 Chemical Facilitation to Improve Ionization

To enhance ionization efficiency different chemicals are also being used by researchers Ammonium acetate, acetic acid, halogen-containing eluents were used by Rivera, et al. [59] to improve efficiency of ESI technique. However Mertz et al. [170] used 20 n*M* ammonium acetate, methanol/20 nM ammonium acetate and MTBE. Similar work was done for APPI by Rivera et al. [125] using dopants (acetone, toluene, anisole and chlorobenzene) and by Crupi, et al. [171] with methanol and tert-butyl methyl ether.

MS Couplings

Tandem mass spectrometry (MS-MS)

For the sake of identification of structures of carotenoid constituents Tandem mass spectrometry (MS-MS) is being adopted by a number of researchers with successful outcomes. It provides information about substitutions (e.g. fatty acid esters of xanthophylls). . Due to its triple quadrat pole instrumentation it helps to quantify components via selected reaction monitoring (SRM). As lycopene and g-carotene were parted and identified from a-carotene and b-carotene (at 467 developed by collision- induced dissociation). Successful results of this technique have been published by Fang et al. [172] and Kopec et al. [41].

LC–MS/MS This coupling provides additional specificity to the system as compared to LC–MS. It is comparatively selective for detection due to its reduced interference of impurities that leads towards chain of advantages including better sample

Table 9.7 List of ionization methods adopted in literature for carotenoids analysis by MS

Ionization methods	Schematic presentation	References
EI Sample is volatilized directly (source placed vacuum system linked with analyzer). Gaseous molecules bombarded by an electrons beam developed via heating filament comparatively –ve voltage (−70 volts than sample). radical ion is produced	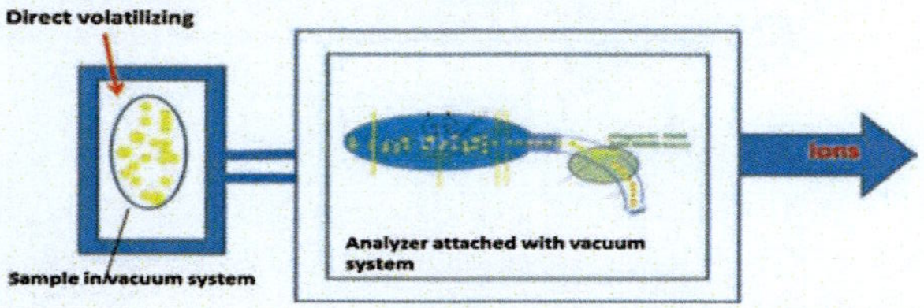	[145]
ESI Electrospray created (putting a voltage on the spray) and directed to an opening in the vacuum system, where the droplets are de-solvated into gas. Finally the ions are emitted from the droplets and enter in analyzer by voltages. Its derivatives includes nanospray, static nanospray or picospray	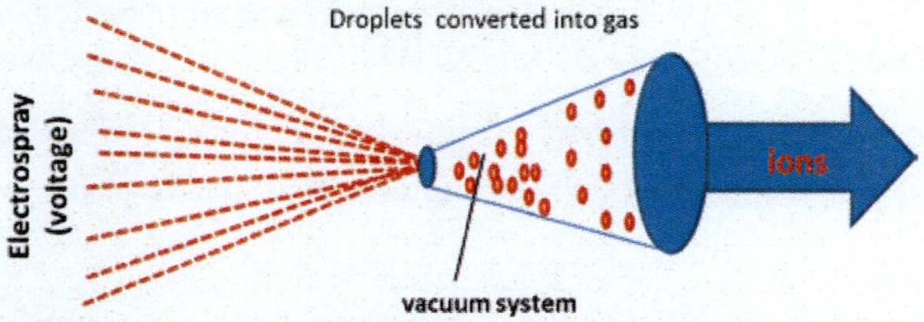	56, 148, 149
FAB Ionization by bombarding a sample (mixed in a matrix) in a vacuum with a beam of atoms, typically Ar or Xe (Kilovolt) The sample was typically mixed in a matrix. The ions formed by FAB were adducts to the molecule, where the adducts could be protons, sodium ions, potassium ions or ammonium ions.	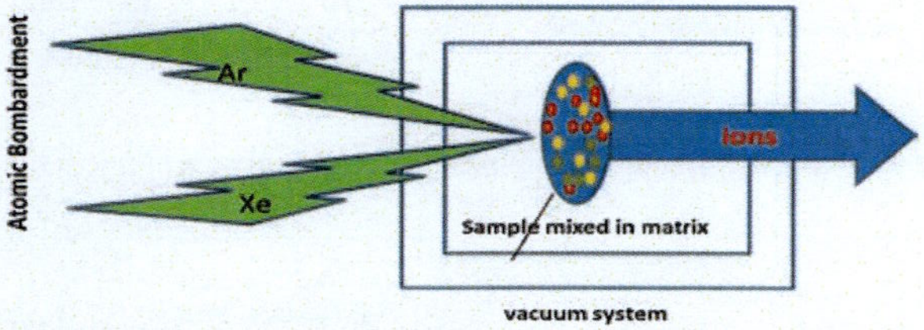	[149]
IMS Ion mobility mass spectrometry (IMS) is the solvent-free gas-phase separation of ions in an electric field on the basis of size and shape. Ions are separated based on size and shape using an electric field. Separates cis/trans isomers. (some cis isomers cannot be resolved by using IMS)		[157]
SIMS Modification of FAB replacing atom beam with ions (cesium) which was called secondary ion mass spectrometry (SIMS).		[150]

(continued)

Table 9.7 (continued)

Ionization methods	Schematic presentation	References
APPI variation of ESI with photo ionization, solvent is vaporized in a heated nebulizer and the gaseous analytes are then ionized with photons from a lamp	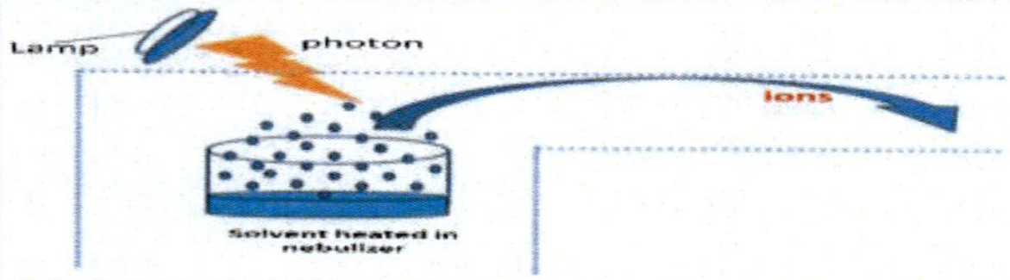	[56]
ASAPP Modification of APCI, samples not subjected to flow of liquid but placed on a probe placed directly into the corona discharged.		[56]
MALDI analyte co-crystallized with matrix, bombarded with a laser (pulsed), matrix absorbs the laser radiation and transfer a proton to the sample. The laser is always pulsed, and typically in a vacuum	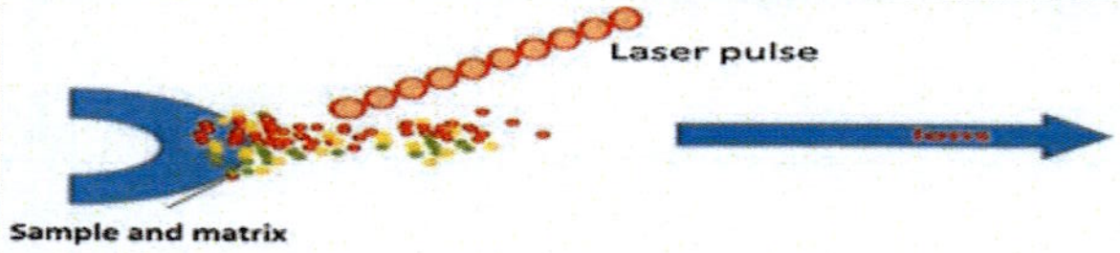	[56, 151]
LAESI It is one of derivation of ESI method. Usually opted for carotenoids study. It involves excitation of OH vibrations in H_2O molecules with mid-IR laser beam		[152]
TOF Time-of-Flight In addition, a TOF chamber (interfaced with MALDI or an ESI ion source) is employed to target metabolomics analysis of precise mass of a mixture of biochemical. It is capable of differentiating carotenoids from other ions		[41, 153, 154]
MALDI- TOF Coupling of Matrix-assisted laser desorption ionization with time-of-flight (MALDI/TOF). That is highly accurate and (84–26 ppm). MALDI-TOF-MS helps in direct analysis without requirement of chromatographic separation. MALDI needs HPLC as prerequisite.		[151, 154–156]
APCI Source like ESI but voltage is placed on a needle creating corona discharged that creates ions. The sample is injected into the discharge by a spray (by flow of liquid + hot gas) that volatilizes sample (by proton transfered from the H3O+/ water clusters). Ions + electrospray extracted into same opening vacuum.	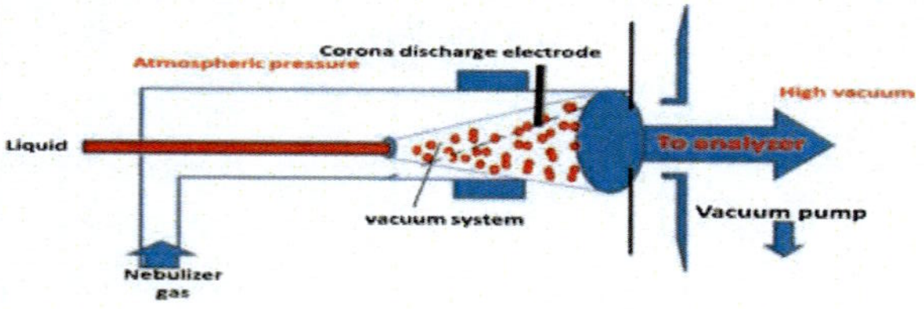	41, 77, 158–169

output (less clean-up), distinction of carotenoids from co-elute, info of structural isomers and time saving.

Fang et al. [172] distinguished between lycopene and its structural isomers with LC–MS/MS.

2D LC-MS This has been successfully opted for carotenoid analysis, particularly in case of complete elaboration in biological samples. Cacciola et al. [121] studied intact carotenoid in red chili peppers.

HPLC-MS Carotenoids have been detected and quantified with hundred times better sensitivity as compared to PDA [127] with this coupling.

GC-MS

Scarce data is available about its use for carotenoids analysis. As it could not distinguish amongst stereoisomers which yield similar mass spectra and fragmentation pattern. Though PDA detection can help to elucidate stereoisomer but it could not deliver sufficient information to determine entire structure.

Accelerator Mass Spectrometry (AMS) Is one of advanced MS where ions are accelerated to their extreme high kinetic energies before any mass analysis. Using this technology various characteristics of carotenoid (absorption, distribution, and metabolism) have been successfully studied and reviewed by Dueker et al. [173] Where radio-labeled carotenoid (normally with 14C) were subjected, processed, collected and analyzed (directly /by HPLC).All fractions that have supposed isotopes labeled like parent carotenoid were collected. Afterwards that collection was oxidized producing carbon dioxide that was later on changed into graphite [174]. The sample was made ionized that after acceleration got high kinetic energy following routine MS. With AMS pharmacokinetics of β-carotene [175], absorption of b-carotene, vitamin A metabolism [176], kinetics of lutein metabolism [177], lycopene bioavailability and metabolic activities [178] have been explored in biological samples.

OrbitrapMS It is comparatively new high-resolution mass spectrometric method providing greater resolution for carotenoids analysis. Here detectors (Orbitrap) are used that screen molecule and generate mass spectrum with mass accuracy between 2 ppm. Thus facilitating recognition of unidentified fragments to elucidate proper fragmentation pattern Bijttebier et al. [125] first time used orbitrap technology to elucidate fragmentation pattern of carotenoids. Currently orbitrap-MS has been used to study carotenoids in tomato [58].

Note During mass spectrometric studies of carotenoids, background matrix pose challenging conditions particularly in samples having sufficient lipid. Competition induced by these lipids for ionization enhance possible background noise reducing precision and sensitivity of results. That problem is resolved by removal of lipophilic background / HPLC-based separation/addition of internal standards [166].

Literature about carotenoid analysis using different MS techniques has been reviewed here (Table 9.8).

Table 9.8 Review of carotenoid analysis using different MS techniques

Sample	Analyte	Extractions	Specification	References
Plant samples				
Papaya	b-carotene b-cryptoxanthin Lycopene	Extraction (hexane:DCM (1:1), centrifugation (9000 g /10 min at 5 °C) + methanolic KOH 40% (1:1)/1 h at 50 °C at 100 rpm, saponification, 10 mL of 10% sodium sulfate darkness/1 hr. and evaporation at 30 °C + 2 mL acetone	LC-PDA-APCI(+)/ MS RP C30 (150 mm × 4.6 mm, 3 μm) column Mobile phase: (A) MeOH (B) MTBE	[159]
Saffron	(Trans and cis-isomers,of crocins)	Extraction (→MeOH) sonication/15 min storage (60 °C /15 min)	LC-PDA-ESI (+)/ MS SB-C18 (4.6 × 150 mm, 3.5 μm) + guard column Mobile phase: (A) 0.15% formic acid in water (B) 0.15% formic acid in MeOH	[148]
Chinese herb	Taraxacum formosanum 25 carotenoids	Extracted with 30 mL hexane/ ethanol/acetone/ toluene (10:6:7:7, v/v/v/v). Shaken for one hour. 2 mL of 40% methanolic potassium hydroxide for saponification for 16 hours under N2.15mL hexane-shaken for 10 min. 15 mL 10% sodium sulfate solution-shaken for 1 min. Supernatant +15 mL of hexane shaken for 10 min (repeated 4 times until colorless). Supernatants evaporated to dryness, dissolved in 5 mL of MeOH:methylene chloride (1:1, v/v)	LC-PDA-APCI(+)-MS C30 (250 mm? 4.6 mm, 5 lm) column Mobile phase: (A) MeOH/ ACN/water (79:14:7, v/v/ v) and (B) methylene chloride	[77]
Enriched fruits juice	Retinol Retinyl acetate Retinyl palmitate b-carotene Tamarillo	DLLMEC (dispersive liquid-liquid micro extraction) 0.1–2 mL + 10 mL water +2 mL MeOH containing 150 lL of carbon tetrachloride. Manually shaken for several seconds. Centrifugation for 2 min at 3000 rpm. Sediment phase evaporated under N2 and reconstituted with 50 lL of MeOH	LC-UV-APCI (+)/ MS RP C8 (15 cm? 46 mm, 5 lm) column Mobile phase: mixture of MeOH/water Gradient elution	[160]

(continued)

Table 9.8 (continued)

Sample	Analyte	Extractions	Specification	References
Arazá fruits	15 carotenoid	0.25 g lyophilized +5 mL MeOH, homogenized at 5000 rpm for 2 min, centrifuged at 300 g for 10 min. MeOH extract separated. 5 mL hexane/ acetone (1:1) + pellet, homogenized, centrifuged; hexane/ acetone extract + MeOH extract. Pulp-hexane/ acetone twice; for peel, once more. 1 mL sodium chloride +5 mL water + pooled hexane/ acetone/ MeOH extract. Upper hexane removed, dried over sodium sulfate. A fraction was dried under N2, reconstituted in MeOH/MTBE (MTBE) (1:1, v/v)	LC-PDA-APCI (−)/ MS C30 S-3 column (2 mm? 150 mm, 3 lm) Mobile phase: (A) MeOH/water/2% aqueous ammonium acetate, 80:18:2 (B) MeOH/ MTBE/2% aqueous ammonium, 20:78:2 Gradient elution	[161]
Amazonian fruits Extruded	60 carotenoids	Pulps extracted with acetone, transferred to petroleum ether (30-70 °C)/diethyl ether and saponified overnight at room temperature with 10% Methanolic KOH. Prior to ether transference, the carotenoid extracts were kept in the freezer (−18 °C) for 2 h, followed by filtration and washing with cold acetone	LC-PDA-APCI (+)/ MS C30 (250? 4.6 mm, 3 lm) column Mobile phase: MeOH with 0.1% Triethylamine/ MTBE Gradient elution	[162]
Mango	All-trans-violaxanthin 9-cis-violaxanthin All-trans-b-Carotene	6 g + 0.2 g calcium carbonate +15 mL MeOH. Homogenates filtered by adding MeOH until colorless. Methanolic extract +50 mL of hexane- acetone (1:1, v/v) containing 0.1% BHT. Vigorous stirring +40 mL of 10% sodium sulfate. Upper layer separated, washed with water and evaporated in a rotary evaporator at 35?C. Residue dissolved in diethyl ether (30 mL) + 0.2 mL of 40% KOH in MeOH (16 h in the dark). Extract washed with water, evaporated and dissolved in 2-propanol (2 mL) 2	HPLC-PDA-APCI(+)/MS C30 RP (4.6 mm? 150 mm, 3 lm) column Mobile phase: (A) water (B) MeOH (C) MTBE	[163]

(continued)

Table 9.8 (continued)

Sample	Analyte	Extractions	Specification	References
Crocus sativus	Lutein diesters	2 g + 30 mL cold acetone, stirred for 3 h. Extract filtered and residue re-extracted with 30 mL cold n- hexane for one hour. Filtrates centrifuged at 2200 g for 10 min at 4?C. Supernatants concentrated in a rotary evaporated. Residue dissolved in 20 mL acetone. 1 mL acetone extract evaporated under argon, dissolved in 50μLofCH2Cl2 and completed to 500 μL with CH3OH Saponification: 2 mL acetone extract evaporated under argon, dissolved in 2 mL n-hexane with 2 mL of CH3OH/ KOH (20% v/w), stirred, centrifuged at 2200 × g for 5 min at 4 °C. Upper hexane layer washed with CH3OH and water. Lower methanolic phase washed with diethyl ether, same volume water and washed again with water Extracts evaporated and dissolved in 50 μLof CH_2Cl_2 and completed with CH3OH	UPLC PDA-APCI(+)/MS UPLC C18 (150 mm × 2.1 mm, 1.8 lm) column Mobile phase: (A) CH3OH/H2O (80:20,v/v) (B) ethyl acetate	[164]
Capsicum	52 carotenoids	10 g + 1 g sodium bicarbonate. Extract with acetone until colorless. Combined acetone extracts evaporated under vacuum at 35?C until 50 mL. Concentrated into a separator funnel with 100 mL of diethyl ether, shacken and left to settle. NaCl solution (10%) + several times with Na2SO4 solution (2%). Ether phase evaporated to dryness at 30 °C Dry residue dissolved in MeOH/MTBE (1:1, v/v)	HPLC-PDA-APCI(+/−)/MS C30 (250 × 4.6 mm, 5 lm) column Mobile phase: (A) MeOH/ MTBE/water (82:16:2, v/ v/v) (B) MeOH/MTBE/water (10:88:2, v/v/v)	[165]

(continued)

Table 9.8 (continued)

Sample	Analyte	Extractions	Specification	References
Spinach and collard greens Lutein Mandarin	Lutein	Spinach and collard greens: 500 mg of purée vegetables +10 mL MeOH for 1 h in a shaking incubator at 120 rpm. Homogenized for 30 s in an ice bath and washed with MeOH. Centrifuged at 3000 rpm for 5 min. MeOH layer collected. Extraction repeated four times with THF; vortexed, centrifuged. THF layers combined with MeOH layers. 1 mL dried under N2 and re-suspended in 1 mL of ethanol	LC-APCI(+)/MS RP C30 (150 mm × 4.6 mm, 3 lm) Mobile phase: (A) MeOH/TBME/water (83:15:2, v/v/v) and (B) MeOH/TBME/water (8:90:2,v/v/v)	[166]
Cabbage	Lutein b-carotene Phytoene Prolycopene	5 g + 100 mL DCM:MeOH solution (2:1, v/v). Centrifuged at 17000 g. Extraction repeated twice with 100 mL of DCM. Filtrated. Added 100 mL of hexane and 100 mL of 5% NaCl. Supernatant evaporated to dryness. Precipitated dissolved in 50 mL of diethyl ether +5 mL of 60% (w/v) KOH. Equal volume (50 mL) of hexane: diethyl ether solution (1:1, v/v) and 5% NaCl added to saponified solution Organic solvent collected and treated with 5% NaCl, evaporated to dryness and dissolved in diethyl ether	LC-PDA-APCI (+)/MS 5C18 ODS (4.6 × 250 mm) Mobile phase: ACN/ethanol (8:2,v/v) LC-PDA	[167]
Red chili peppers (Capsicum annuum	Free carotenoids and carotenoid esters (>30)	200 g of homogenate extracted three times with 300 mL of MeOH/ethyl acetate/petroleum ether (1:1:1, v/v/v) Combined extracts dissolved in 4 mL of MeOH/tert- butyl methyl ether (1:1, v/v) + filtration 0.45 lm acrodisc nylon membrane	NP-LC × RP-UHPLC with PDA and IT-TOF (APCI ionization) MS Micro-bore cyano column (250 mm? 1.0 mm, 5 lm) for 1D C18 column (30 mm? 4.6 mm, 2.7 lm) packed with fused-core particles for 2D Mobile phase: 1D (A) n-hexane (B) n-hexane/butyl-acetate/acetone (80:15:5, v/v/v); 2D (A) water/ACN (10:90, v/v) (B) isopropanol	[168]

(continued)

Table 9.8 (continued)

Sample	Analyte	Extractions	Specification	References
Animal samples				
Serum	(30 carotenoids)	1 mL of blood into a 10 mL brown vial +1 mL of 0.01% Vit. C in ethanol +1 mL ethyl acetate and 3 mL hexane. Vortex (30 s), shaking (10 min at 200 rpm), centrifuging (20 min at 3000 rpm and 4 °C). To the residue, 3 mL of hexane was added and the procedure repeated three times	LC-PDA-APCI (+)/ MS C30 column Mobile phase: (A) MeOH/ACN/water (84:14:4, v/ v/v) and (B) DCM Gradient elution	[54]
Red blood cells	Lutein Zeaxanthin α-carotene β-carotene Lycopene β-cryptoxanthin	2,5 mL + 2,5 water +5 ml pyrogallol (0–400 mmol/L in ethanol) + 1 mL aqueous potassium hydroxide (0–7 mmol/L) and 40 lL of echinone (1 lmol/L in ethanol); sonicated for 5 min; vortexed for 2 min, incubated at various temp. (20–70) for different time periods (0–24 h). Mixed; extracted with 1.25 mL of 0.1 mol/L sodium dodecyl sulfate aqueous solution and 15 ml of hexane/dichloromethane (5:1, v/v) containing 1.2 mmol/L BHT; sonicated, vortexed, centrifuged at 1000 g for 10 min. Supernatants collected	LC-PDA-APCI(+) / MS C30 (250 mm × 4,6 mm I.D, 5 lm) column Mobile phase: (A) MeOH/MTBE/water (83:15:2, v/ v/v) containing 3,9 mmol/L ammonium acetate (B) MeOH/MTBE/ water (8:90:2, v/v/v) containing 2,6 mmol/L ammonium acetate	[169]
plasma	Apo-lycopenals	2 ml + HEAT (hexane/ethanol/ acetone/toluene, 10:6:7:7, v/v/v/v) + equal volume of ethanol containing 0.1% butylated hydroxytoluene +0.5 ml of a saturated NaCl solution +10 ml of HEAT. Vortex, centrifuged for 5 min at 300xg. Upper non-polar layer removed. Remaining aqueous plasma extracted two more times. Non-polar layer was combined and dried under argon.	UPLC-PDA-APCI(−)/MS C30 (150 mm × 4.96 mm, 5 lm) column Mobile phase: (A) MeOH/0.1% aqueous formic acid solution (80:20) (B) MTBE /MeOH/0.1% aq. formic acid solution (78:20:2)	[41]

(continued)

Table 9.8 (continued)

Sample	Analyte	Extractions	Specification	References
Serum	Lutein	Serum: 500 ml spiked with 375 μL of echinone (1.25 lM) + 4 ml chloroform/MeOH (2:1, v/ v) + 0.5 ml saline solution (0.85%). Vortexed, centrifuged for 10 min at 4?C and 3000 rpm. Chloroform layer removed. 5 ml of hexane added. Vortexed, centrifuged. Extract combined and dried in a water bath at 40 °C and dissolved in 300 μL	LC-APCI (+)/MS RP C30 (150 mm x4.6 mm, 3 lm) Mobile phase: (A) MeOH/tert-butyl methyl ether/ water (83:15:2, v/v/v) and (B) MeOH/MTBE/ water (8:90:2, v/v/v) Gradient	[166]
plasma	Astaxanthin	0.5 ml + 0.5 ml of water +1 ml of ethanol +3 ml hcxane with 4 lmol/L of an echinone ethanol solution (50 μL). Vortexed for 30 s, centrifuged at 1000 g for 10 min at 4 °C. Upper hexane collected. Extraction repeated. Combined hexane layers dried under N2 and re-dissolved in 100 lL of MeOH/MTBE (2:3, v/v)	HPLC-APCI(+)/MS C30 (4.6 × 250 mm, 5 lm) carotenoid column Mobile phase: (A) MeOH/ MTBE/water (8:90:2) containing 2.6 mmol/L of ammonium acetate	[158]

Non-Destructive Carotenoid Study

In general to get exact information, sampling is recommended to follow extraction as discussed above. In certain condition it is not possible for biological samples to destruct by extraction. Geared by the motivation of time based importance of carotenoid studies for different health issues of human patients. Advancement of analytical approaches is going on for better qualification and quantification. For example in case of the study of human there is limitation of above mentioned extractions because of the compartmentation of carotenoids in different living tissues. Further in the case study of intact plants majority of techniques almost failed. So there is need of such techniques that could facilitate carotenoid assays without destruction and the task was achieved by followings

- Infrared spectroscopy (NIRS/MIRS)
- Raman Spectroscopy

9.38 Infrared Spectroscopy (Near and Mid)

Infrared Spectroscopy is now well known for its speed, sensitivity and reliability for biochemical estimation without sample preparations. Use of this technology (for on–/inline measurements) is getting more and more importance and preference.

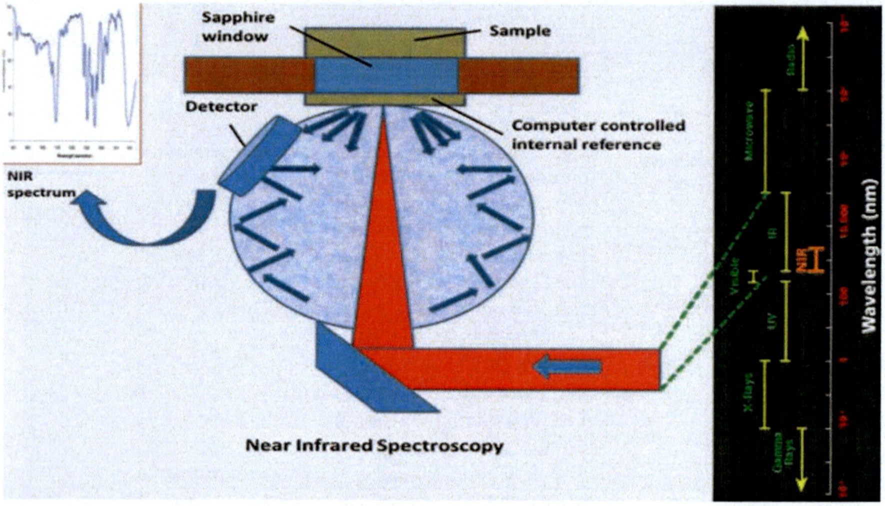

Fig. 9.19 Schematic diagram for Near infrared (NIR) spectroscopy

Chemo metrics is needed for data analysis and interpretation, both quantitative and qualitative analysis are possible with this tool [179] (Fig. 9.19).

Near infrared (NIR) and mid infrared (MIR) spectroscopy were used to study β-carotene with similar results for apricot [180] and maize [181]. Unfortunately the researchers concluded that both of techniques were unsuitable for carotenoids analysis. Probable justification was their low concentrations and lower intensity of absorption bands.

But afterwards it has been declared as a rapid simple, safe, economic and non-destructive technique. It proved itself rapid and successful for carotenoids quantification particularly for lycopene [182–184]. Rubio-Diaz et al. [185] successfully studied carotenoid without destruction of tomato using infrared spectroscopy. Table 9.9 enlisted some studies of carotenoid sample using IR spectroscopy.

9.39 Raman Spectroscopy

Generalized components of Raman spectroscopy instrument are these

- Laser source to provide directional monochromatic lights
- Sample chamber with lens, filters and sample holders)
- Determination system
- Data analysis system

Strong bands are formed vibrating nonpolar components groups and determine the single (C-C) and double (C=C) bonds of nonpolar components. Then bonds are observed possible in range of 1150 cm^{-1} to 1170 cm^{-1} minima and 1500 cm^{-1} to

Table 9.9 Summery of infrared spectroscopy applications for carotenoid analysis from biological samples

Sample	Carotenoiud	Measurement range	Measurement mode	Wavelength range	Data pretreatment	Regression method	Regression performance	References
Banana and planta pulp	α-carotene	0–110 nmol g^{-1} DW	reflectance	367–2388	1st deriv	PLS	RMSECV = 18.76 nmol g^{-1} DW RMSEP = 26.00 nmol g^{-1} DW r2 = 0.82 RPD = 1.68	[182]
	β- carotene	00–144.8 nmol g^{-1} DW	Reflectance	//	1st derive	PLS	RMSECV = 16.16 nmol g^{-1} DW RMSEP = 17.07 nmol g^{-1} DW r2 = 0.92 RPD = 2.74	
	13 cis β carotene	0–14 nmol g^{-1} DW	Reflectance	//	1st derive	PLS	RMSECV = 2.82 nmol g^{-1} DW RMSEP = 2.25 nmol g^{-1} DW r2 = 0.80 RPD = 1.96	
	Leutene	//	Reflectance	//	1st deriv	PLS	RMSECV = 3.10 nmol g^{-1} DW RMSEP = 4.75 nmol g^{-1} DW r2 = 0.56 RPD = 1.16	
Carrot root	carotene	2.08–1929 mg/100 ml	Reflectance	1200–4000 cm^{-1}	NA	PLS	RMSECV = 1.54 mg 100 ml^{-1} r2 = 0.8	[186]
Cucurbita pepo fruits	total carotenoid,	67.1–451.2 (flesh) mg kg^{-1} DW 85.0–1822 (skin) mg kg^{-1} DW	reflectance	1st step (400 to 2500 nm) 1100–2378 nm	2nd deriv	mPLS	RPD =4.31 (flesh) 4.55 (skin)	[187]

(continued)

Table 9.9 (continued)

Sample	Carotenoiud	Measurement range	Measurement mode	Wavelength range	Data pretreatment	Regression method	Regression performance	References
Fresh cheese	β-Carotene	0.00–2.55 mg kg – 1	Reflectance	400–2498 nm	2nd deriv	mPLS	SECV = 0.20 mg kg^{-1} SEP = 0.21 mg kg^{-1} r2 = 0.92 RPD = 3.43	[188]
	Xanthophylls	10.0–160.0 μg kg^{-1}	Reflectance	400–2498 nm	SNV-DT and ist derive.	mPLS	SECV = 27.0 μg kg^{-1} SEP = 24.0 μg kg^{-1} r2 = 0.60 RPD = 1.58	
Freeze-dried cheese	β-Carotene	0.00–3.78 mg kg^{-1}	Reflectance	400–2498 nm	1st deriv	mPLS	SECV = 0.21 mg kg^{-1} SEP = 0.46 mg kg^{-1} r2 = 0.88 RPD = 2.3	
	Xanthophylls	20.0–270.0 μg kg^{-1} Xanthophylls	Reflectance	//	SNV-DT and first deriv	mPLS	SECV = 50.2 μg kg^{-1} SEP = 47.0 μg kg^{-1} r2 = 0.32 RPD = 1.19	
	lutein	50.3–434.3 (flesh) mg kg^{-1} DW, 78.4–1529 (skin) mg kg^{-1} DW	Reflectance	950 nm to 1650 nm			RPD =4.81 2.32 (flesh) 4.79 4.89 (skin)	
	β-carotene	0–24(flesh) mg kg^{-1} DW 0–194 (skin) mg kg^{-1} DW	Reflectance				RPD = R0.76 4.79 (flesh) 4.89 (skin)	
Triticum spp.			reflectance	1088–1697, 1972–2355	2nd derive	PL:S	RMSECV = 1.13 R^2 = 87.75RM SEE = 1.05	[189]
Tomato	lycopene				MSC and second derive.	PLS-1	RMSEP =21.5779, *r*values 0.9996	[190]
	β-carotene				MSC and second derive.	PLS-1	RMSEP = 0.7296, *r* values = 0.9981	[190]

Sample	Carotenoiud	Measurement range	Measurement mode	Wavelength range	Data pretreatment	Regression method	Regression performance	References
Maize	violaxanthin	2.38 ± 3.88 mg kg^{-1} (Cal: mean ± SD) 21.60	reflectance	400–2500 nm	2nd derive and SNV-DT	mPLS	SECV = 0.13 mg g^{-1} SEP = 0.14 mg g^{-1} r2 = 0.82 RPD = 2.3	[191]
	Zea xanthn	21.60 ± 11.79 mg kg^{-1} (Cal: mean ± SD) Constituent	reflectance	400–2500 nm	2nd derive. and SNV-DT	mPLS	SECV = 0.01 mg g^{-1} SEP = 0.01 mg g^{-1} r2 = .74 RPD = 1.7	[191]
	Total carotenoid	47.49 ± 35.09 mg kg^{-1} (Cal: mean ± SD	reflectance	400–2500 nm	2nd derive. and SNV-DT	mPLS	SECV = 0.14 mg g^{-1} SEP = 0.17 mg g^{-1} r2 = 0.81 RPD = 2.0	[191]
passion fruit	β-carotene	0–0.287 mg g − 1DW	reflectance	MIR 4000–1000 cm^{-1} NIR 2000–800 nm	1st derive. and smoothing	PLSR	RMSECV = 0.043 for MIR RMSECV = 0.045 for NIR	[192]
four tomato hybrids	All-trans lycopene	31.5–103.9	reflectance	950–1650	1st derive. + MSC	PLS (2)	RMSECV = 6.88 μg g^{-1}	[193]
	Cis-lycopene diepoxide	0.38–3.28	reflectance	950–1650 nm	1st derive. + MSC	PLS (8)	RMSECV = 0.269 μg g–[1]	
	Lycoxanthin	0.54–2.32	reflectance	950–1650 nm	1st derive. + MSC	PLS (7)	RMSECV = 0.281 μg g^{-1}	
	Zeaxanthin	0.21–0.94	reflectance	950–1650 nm	1st derive. + MSC	PLS(7)	RMSECV = 0.099 μg g^{-1}	
	β-carotene	0.49–1.47	reflectance	950–1300 nm	2nd derive.	PLS (4)	RMSECV =0.174 μg g^{-1}	

1550 cm^{-1} maxima. Polyene chain having methyl groups can generate Raman shift in the range of 1000 cm^{-1} to 1020 cm^{-1}.

Deficiencies of in vivo conventional Raman analysis

- Strong fluorescence
- Could not distinguish some carotenoids
- Could not observe the structure at such resolution level
- fail to detect internal Raman spectrum of solids

following modified advanced Raman spectroscopic determination.

are now in practice for carotenoid noninvasive analysis.

1. Fourier transform Raman spectroscopy (FT-Raman)
2. Resonance Raman spectroscopy (RRS)
3. Raman microspectroscopy
4. Spatially offset Raman spectroscopy (SORS)
5. Surface-enhanced Raman spectroscopy (SERS)

Amongst these only four mostly used types of Raman Spectroscopy for carotenoids (noninvasive) has been compared in Table 9.10.

9.40 Overview of Carotenoid Studies in 2019

Literature survey for the year 2019 showed that most preferred techniques during this year were as following. Visible and near-infrared Vis/NIR spectroscopy along with HPLC [209], visible to near-infrared spectroscopy and chemometrics [210], reflectance spectrometer using remote sensing portable spectroradiometer RS-3500 [211], visible and near-infrared spectroscopy coupled with chemometric method of principal component analysis (PCA) [212]. Raman spectroscopy was employed for direct assay of paprika powder with laser excited at 785 nm for the first time [213]. Microscopic confocal Raman (MCR) [214], resonant Raman spectroscopy, and surface-enhanced Raman scattering employing laser excitation at 532 nm [215] were also used for carotenoid determinations from living samples.

Currently for the study of metabolites in depth of several millimeters to turbid medium Spatially offset Raman spectroscopy (SORS) has been adopted too. Yu et al. [216] prepared modular inverse spatially offset Raman spectroscopy (Inverse SORS) system, to get metabolic data from biological tissues present at depth. Proving that the Inverse SORS helped for data collection on surface and in depth helped in detection of Raman signal from paracetamol powders for 8 mm deep tissue. Coupling of Raman spectroscopy and chemometrics tool was successfully opted for olive oil analysis [217] in same year.

Table 9.10 Comparison of four mostly used types of Raman Spectroscopy for carotenoids (noninvasive)

Raman	Schematic diagram of Raman spectroscopy experimental setup	Analysis	Reference
FT- R S	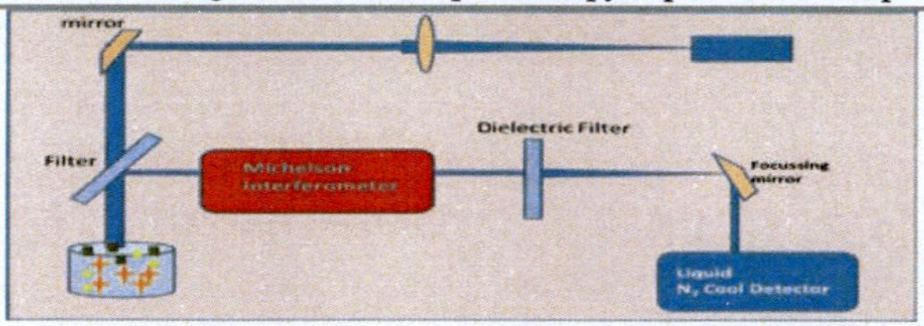 Monochromatic light (1064 nm) emits by the laser, collimated by lens, reflected by mirror and irradiated on the sample. Raman signal collected by Michelson interfer- ometer (filtered by dichroic filter). Improved signal-to-noise ratio. Scattered light pass from filter then focused to liquid N2 cooled detector.	conjugated carotenoids	[194]
		bands of C-C and C=C stretching and C-CH3 bending	[195]
		lycopene (R2 = 0.91 and SECV = 74.34) and beta-carotene (R2 = 0.89 and SECV = 0.34)	[196]
		spatial distribution of carotenoids in intact carrot roots of	[197]
		tissue specific accumulation of different carotenoids of carrot root	[198]
RRS	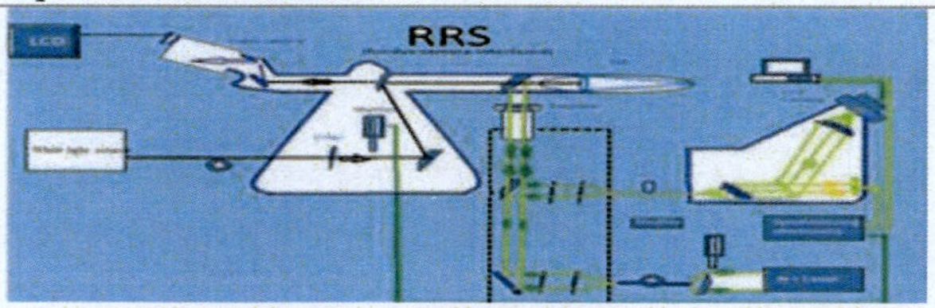 Exciting beam filtered by a BPF (band pass filter) that focuson sample, with dichroic filter and a LPF (low pass filter) to eliminating the laser. The band of BPF and LPF are according to the sample under test.	lycopene owns special peak position at 514 nm	[199]
		diadinoxanthin and diatoxanthin	[200]
		RRS - tristimulus colorimetry; Mango beta-carotene at 1008, 1158 and 1523 cm^{-1}. R2 = 0.962	[201]
		lycopene and beta-carotene, second harmonic of stretching vibration of C=C	[202]
		9-cis-beta-carotene existance in cytochrome b6f complex of spinach	[203]

(continued)

Table 9.10 (continued)

Raman	Schematic diagram of Raman spectroscopy experimental setup	Analysis	Reference
Raman microspectroscopy	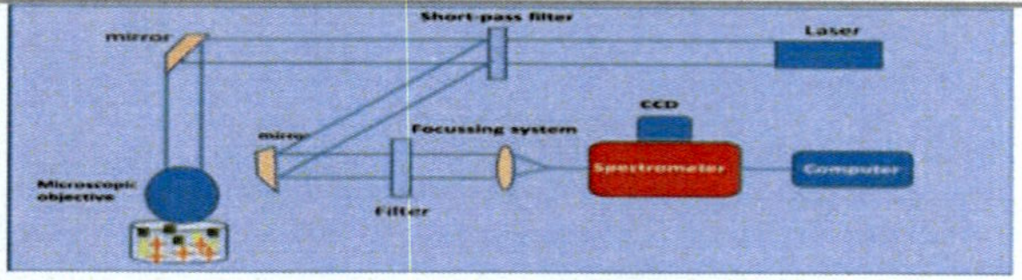 Measurement at micro-scale with an optical microscope, excitation laser, a sensitive detector and some other optical devices.	Lycopene in ripened and immature tomato	[204]
		microstructure of carotenes in tomato and carrot emulsions under varying degrees of high pressure homogenization	[205]
		Effect of thermal treatment on carrots carotenoids	[206]
SORS	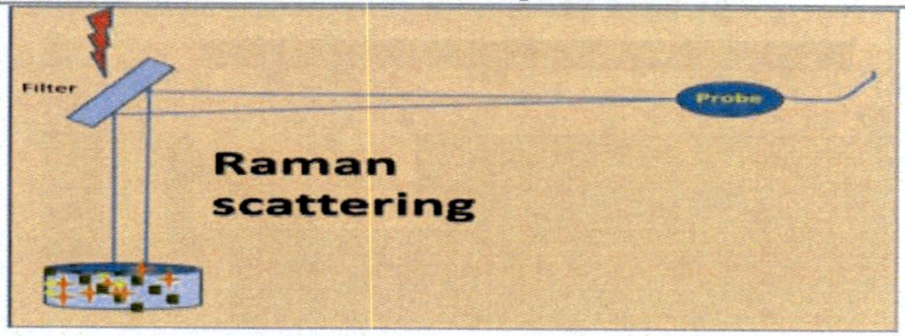 Raman spectra collected from the spatial regions away from the place of illumination, direction of laser irradiation is contrasting to the probe	lycopene during postharvest ripening of tomatoes	[207]
		carotenoids of salmon through the skin	[208]

9.41 Future Prospects

The journey of carotenoid analysis towards better precision has not been stopped. With the course of time even improvement of existing techniques or the invent of new ones is expected.

References

1. Palmer LS (1971) In: Isler O (ed) Carotenoids and related pigments: the chromolipoids 1922. American Chemical Society, New York
2. Isler O (1971) Introduction. In: Isler O (ed) Carotenoids. Birkhfiuser, Basel, pp 11–27
3. Amorim-Carrilho KT, Cepeda A, Fente C, Regal P (2014) Review of methods for analysis of carotenoids. TrAC Trends Anal Chem 56:49–73
4. Mezzomo N, Ferreira SRS (2016) Carotenoids functionality, sources , and processing by supercritical technology: a review. https://doi.org/10.1155/2016/3164312
5. Rodriguez-Amaya, D., 1999. Changes in carotenoids during processing and storage of foods. Arch Latinoam Nutr n 49, 38–47
6. Mieko Kimura CNKBR-A (2007) Screening and HPLC methods for carotenoids in sweetpotato, cassava and maize for plant breeding trials. Food Chem 100(4):1734–1746
7. Rodriguez-Amaya DB, Kimura M (2004) HarvestPlus handbook for carotenoid analysis, vol 2. International Food Policy Research Institute (IFPRI), Washington, DC
8. Vagi E, Simándi B, Daood HG, Deák A, Sawinsky J (2002) Recovery of pigments from Origanummajorana L. by extraction with supercritical carbon dioxide. J Agric Food Chem 50:2297–2301
9. Chuah HLNL, Yuen MCNMH (2006) Quality of residual oil from palm-pressed mesocarp fiber (Elaeis guineensis) using supercritical CO2 with and without ethanol. J Am Oil Chem 83(10):893–898
10. Daood HG, Illés V, Gnayfeed MH, Mészáros B, Horváth G, Biacs PA (2002) Extraction of pungent spice paprika by supercritical carbon dioxide and subcritical propane. J Supercrit Fluids 23(2):143–152
11. Uquiche E, del Valle JM, Ortiz J (2004) Supercritical carbon dioxide extraction of red pepper (Capsicum annuum L.) oleoresin. J Food Eng 65(1):55–66
12. Huang W, Li Z, Niu H, Li D, Zhang J (2008) Optimization of operating parameters for supercritical carbon dioxide extraction of lycopene by response surface methodology. J Food Eng 89(3):298–302
13. Takeuchi TM, Pereira CG, Braga MEM, Maróstica MR, Leal PF, Meireles MAA (2009) Low-pressure solvent extraction (solid-liquid extraction, microwave assisted, and ultrasound assisted) from condimentary plants. In: Extracting bioactive compounds for food products. CRC Press, Nueva York, pp 137–218
14. Deli J, Molnar P, Matus Z, Tóth G (2001) Carotenoid composition in the fruits of red paprika (Capsicum annuum var. lycopersiciforme rubrum) during ripening; biosynthesis of carotenoids in red paprika. J Agric Food Chem 49(3):1517–1523
15. Edelenbos M, Christensen LP, Grevsen K (2001) HPLC determination of chlorophyll and carotenoid pigments in processed green pea cultivars (Pisum sativum L.). J Agric Food Chem 49(10):4768–4774
16. Lianfu Z, Zelong L (2008) Optimization and comparison of ultrasound/microwave assisted extraction (UMAE) and ultrasonic assisted extraction (UAE) of lycopene from tomatoes. Ultrason Sonochem 15(5):731–737

17. Aktas ET, Yildiz H (2011) Effects of electroplasmolysis treatment on chlorophyll and carotenoid extraction yield from spinach and tomato. J Food Eng 106(4):339–346
18. Rodić Z, Simonovska B, Albreht A, Vovk I (2012) Determination of lutein by high-performance thin-layer chromatography using densitometry and screening of major dietary carotenoids in food supplements. J Chromatogr A 1231:59–65
19. Ofori-Boateng C, Lee KT (2013) Response surface optimization of ultrasonic-assisted extraction of carotenoids from oil palm (Elaeis guineensis Jacq.) fronds.f. Food Sci Nutr 1(3):209–221
20. Arranz S, Martínez-Huélamo M, Vallverdu-Queralt A, Valderas-Martinez P, Illán M, Sacanella E et al (2015) Influence of olive oil on carotenoid absorption from tomato juice and effects on postprandial lipemia. Food Chem 168:203–210
21. Yan F, Fan K, He J, Gao M (2015) Ultrasonic-assisted solvent extraction of carotenoids from rapeseed meal: optimization using response surface methodology. J Food Qual 38(6):377–386
22. Zlabur JS, Voca S, Dobricevic N, Pliestic S, Galic A, Boricevic A et al (2016) Ultrasound-assisted extraction of bioactive compounds from lemon balm and peppermint leaves. Int Agrophys 30(1):95
23. Suslick KS, Didenko Y, Fang MM, Hyeon T, Kolbeck KJ, McNamara WB et al (1999) Acoustic cavitation and its chemical consequences. Philos Trans R Soc Lond A 357(1751):335–353
24. Heffernan N, Smyth TJ, FitzGerald RJ, Vila-Soler A, Mendiola J, Ibáñez E et al (2016) Comparison of extraction methods for selected carotenoids from macroalgae and the assessment of their seasonal/spatial variation. Innovative Food Sci Emerg Technol 37:221–228
25. Ranveer RC, Patil SN, Sahoo AK (2013) Effect of different parameters on enzyme-assisted extraction of lycopene from tomato processing waste. Food Bioprod Process 91(4):370–375
26. Kiss GAC, Forgacs E, Cserhati T, Mota T, Morais H, Ramos A (2000) Optimisation of the microwave-assisted extraction of pigments from paprika (Capsicum annuum L.) powders. J Chromatogr A 889(1):41–49
27. Pasquet V, Chérouvrier JR, Farhat F, Thiéry V, Piot JM, Bérard JB et al (2011) Study on the microalgal pigments extraction process: Performance of microwave assisted extraction. Process Biochem 46(1):59–67
28. Chan MC, Ho SH, Lee DJ, Chen CY, Huang CC, Chang JS (2013) Characterization, extraction and purification of lutein produced by an indigenous microalga Scenedesmus obliquus CNW-N. Biochem Eng J 78:24–31
29. Xiao X, Song W, Wang J, Li G (2012) Microwave-assisted extraction performed in low temperature and in vacuo for the extraction of labile compounds in food samples. Anal Chim Acta 712:85–93
30. Hiranvarachat B, Devahastin S, Chiewchan N, Raghavan GV (2013) Structural modification by different pretreatment methods to enhance microwave-assisted extraction of β-carotene from carrots. J Food Eng 115(2):190–197
31. Tuyen CK, Nguyen MH, Roach PD, Stathopoulos CE (2013) Effects of Gac aril microwave processing conditions on oil extraction efficiency, and β-carotene and lycopene contents. J Food Eng 117(4):486–491
32. Prado MJ, Veggi C, P., & Meireles, A. A. (2014) Extraction methods for obtaining carotenoids from vegetables-review. Curr Anal Chem 10(1):29–66
33. Ho KKHY, Ferruzzi MG, Liceaga AM, San Martín-González MF (2015) Microwave-assisted extraction of lycopene in tomato peels: Effect of extraction conditions on all-trans and cis-isomer yields. LWT-Food Sci Technol 62(1):160–168
34. Barker SA (2007) Matrix solid phase dispersion (MSPD). J Biochem Biophys Methods 70(2):151–162
35. Fernandez-Alba AR, Martinez ML, Contreras M (2001) Analysis-evaluation of the relationship between carotenoids and colour measures by chromatographic and colorimetric based procedures: application to fruits and vegetables. Spec Publ Royal Soc Chem 269:280–284

36. Dachtler M, Glaser T, Kohler K, Albert K (2001) Combined HPLC– MS and HPLC– NMR on-line coupling for the separation and determination of lutein and zeaxanthin stereoisomers in spinach and in retina. Anal Chem 73(3):667–674
37. Glaser T, Lienau A, Zeeb D, Krucker M, Dachtler M, Albert K (2003) Qualitative and quantitative determination of carotenoid stereoisomers in a variety of spinach samples by use of MSPD before HPLC-UV, HPLC-APCI-MS, and HPLC-NMR on-line coupling. Chromatographia 57:S19–S25
38. Putzbach K, Krucker M, Grynbaum MD, Hentschel P, Webb AG, Albert K (2005) Hyphenation of capillary high-performance liquid chromatography to microcoil magnetic resonance spectroscopy—determination of various carotenoids in a small-sized spinach sample. J Pharm Biomed Anal 38(5):910–917
39. Mateos R, García-Mesa JA (2006) Rapid and quantitative extraction method for the determination of chlorophylls and carotenoids in olive oil by high-performance liquid chromatography. Anal Bioanal Chem 385(7):1247–1254
40. Tzouganaki ZD, Atta-Politou J, Koupparis MA (2002) Development and validation of liquid chromatographic method for the determination of lycopene in plasma. Anal Chim Acta 467(1):115–123
41. Kopec RE, Riedl KM, Harrison EH, Curley RW Jr, Hruszkewycz DP, Clinton SK et al (2010) Identification and quantification of apo-lycopenals in fruits, vegetables, and human plasma. J Agric Food Chem 58(6):3290–3296
42. Ferreiro-Vera C, Mata-Granados JM, Gómez JQ, de Castro ML (2011) On-line coupling of automatic solid-phase extraction and HPLC for determination of carotenoids in serum. Talanta 85(4):1842–1847
43. Irakli MN, Samanidou VF, Papadoyannis IN (2011) Development and validation of an HPLC method for the simultaneous determination of tocopherols, tocotrienols and carotenoids in cereals after solid phase extraction. J Sep Sci 34(12):1375–1382
44. Chase GW, Eitenmiller RR, Long AR (1999) Liquid chromatographic analysis of vitamin K~ 1 in milk-based infant formula with matrix sollid-phase dispersion. J-AOAC Int 82:1140–1145
45. Konings EJ, Roomans HH (1997) Evaluation and validation of an LC method for the analysis of carotenoids in vegetables and fruit. Food Chem 59(4):599–603
46. Mouly PP, Gaydou EM, Lapierre L, Corsetti J (1999) Differentiation of several geographical origins in single-strength Valencia orange juices using quantitative comparison of carotenoid profiles. J Agric Food Chem 47(10):4038–4045
47. Marsili R, Callahan D (1993) Comparison of a liquid solvent extraction technique and supercritical fluid extraction for the determination of α-and β-carotene in vegetables. J Chromatogr Sci 31(10):422–428
48. Gueguen S, Herbeth B, Siest G, Leroy P (2002) An isocratic liquid chromatographic method with diode-array detection for the simultaneous determination of μ-tocopherol, retinol, and five carotenoids in human serum. J Chromatogr Sci 40(2):69–76
49. Moros EE, Darnoko D, Cheryan M, Perkins EG, Jerrell J (2002) Analysis of xanthophylls in corn by HPLC. J Agric Food Chem 50(21):5787–5790
50. Thibeault D, Su H, MacNamara E, Schipper HM (2009) Isocratic rapid liquid chromatographic method for simultaneous determination of carotenoids, retinol, and tocopherols in human serum. J Chromatogr B 877(11):1077–1083
51. Kamffer Z, Bindon KA, Oberholster A (2010) Optimization of a method for the extraction and quantification of carotenoids and chlorophylls during ripening in grape berries (Vitis vinifera cv. Merlot). J Agric Food Chem 58(11):6578–6586
52. Zanfini A, Corbini G, La Rosa C, Dreassi E (2010) Antioxidant activity of tomato lipophilic extracts and interactions between carotenoids and α-tocopherol in synthetic mixtures. LWT-Food Sci Technol 43(1):67–72

53. Amorim K, Lage-Yusty MA, López-Hernández J (2012) Changes in bioactive compounds content and antioxidant activity of seaweed after cooking processing. CyTA-J Food 10(4):321–324
54. Hsu BY, Pu YS, Inbaraj BS, Chen BH (2012) An improved high performance liquid chromatography–diode array detection–mass spectrometry method for determination of carotenoids and their precursors phytoene and phytofluene in human serum. J Chromatogr B 899:36–45
55. Lucini L, Pellizzoni M, Baffi C, Molinari GP (2012) Rapid determination of lycopene and β-carotene in tomato by liquid chromatography/electrospray tandem mass spectrometry. J Sci Food Agric 92(6):1297–1303
56. Rivera S, Canela R (2012) Influence of sample processing on the analysis of carotenoids in maize. Molecules 17(9):11255–11268
57. Sarkar CR, Das L, Bhagawati B, Goswami BC (2012) A comparative study of carotenoid extraction from algae in different solvent systems. Asian J Plant Sci Res 2:546–549
58. Van Meulebroek L, Vanden Bussche J, Steppe K, Vanhaecke L (2014) Unravelling the phytohormonal status and carotenoid profile of tomato plants by using metabolomics, including generic extraction and UHPLC-Orbitrap-MS. In: 13th International symposium on Hyphenated Techniques in Chromatography and separation technology (HTC-13); 3rd International symposium on Hyphenated Techniques for Sample Preparation (HTSP-3)
59. Refardt D (2017) Extraction of carotenoids from Chlorella vulgaris using green solvents and syngas production from residual biomass. Algal Res 25:488–495
60. Strati IF, Gogou E, Oreopoulou V (2015) Enzyme and high pressure assisted extraction of carotenoids from tomato waste. Food Bioprod Process 94:668–674
61. Kha TC, Phan H, Nguyen MH (2014) Effects of pre- treatments on the yield and carotenoid content of Gac oil using supercritical carbon dioxide extraction. J Food Eng 120(1):44–49
62. Díaz-Reinoso B, Moure A, Domínguez H, Parajó JC (2006) Supercritical CO2 extraction and purification of compounds with antioxidant activity. J Agric Food Chem 54(7):2441–2469
63. Reverchon E, De Marco I (2006) Supercritical fluid extraction and fractionation of natural matter. J Supercrit Fluids 38(2):146–166
64. Herrero M, Cifuentes A, Ibañez E (2006) Sub-and supercritical fluid extraction of functional ingredients from different natural sources: Plants, food-by-products, algae and microalgae: A review. Food Chem 98(1):136–148
65. Baysal T, Ersus S, Starmans DAJ (2000) Supercritical CO2 extraction of β-carotene and lycopene from tomato paste waste. J Agric Food Chem 48(11):5507–5511
66. Rozzi NL, Singh RK, Vierling RA, Watkins BA (2002) Supercritical fluid extraction of lycopene from tomato processing byproducts. J Agric Food Chem 50(9):2638–2643
67. Careri M, Furlattini L, Mangia A, Musci M, Anklam E, Theobald A et al (2001) Supercritical fluid extraction for liquid chromatographic determination of carotenoids in Spirulina Pacifica algae: a chemometric approach. J Chromatogr A 912(1):61–71
68. Gómez-Prieto MS, Caja MM, Herraiz M, Santa-María G (2003) Supercritical fluid extraction of all-trans-lycopene from tomato. J Agric Food Chem 51(1):3–7
69. Sun M, Temelli F (2006) Supercritical carbon dioxide extraction of carotenoids from carrot using canola oil as a continuous co-solvent. J Supercrit Fluids 37:397–408
70. Machmudah S, Kawahito Y, Sasaki M, Goyo M (2008) Pro- cess optimization and extraction rate analysis of carotenoids extraction from rosehip fruit using SC-CO2. J Supercrit Fluids 44(3):308–314
71. Genival Filho L, De Rosso VV, Meireles MAA, Rosa PT, Oliveira AL et al (2008) Supercritical CO2 extraction of carotenoids from pitanga fruits (Eugenia uniflora L.). J Supercrit Fluids 46(1):33–39
72. Krichnavaruk S, Shotipruk A, Goto M, Pavasant P (2008) Supercritical carbon dioxide extraction of astaxanthin from Haematococcus pluvialis with vegetable oils as co-solvent. Bioresour Technol 99(13):5556–5560

73. Jaime L, Rodríguez-Meizoso I, Cifuentes A, Santoyo S, Suarez S, Ibáñez E, Señorans FJ (2010) Pressurized liquids as an alternative process to antioxidant carotenoids' extraction from Haematococcus pluvialis microalgae. LWT-Food Sci Technol 43(1):105–112
74. Chen L, Wang H, Zeng Q, Xu Y, Sun L, Xu H et al (2009) On-line coupling of solid-phase extraction to liquid chromatography—a review. J Chromatogr Sci 47(8):614–623
75. Divya P, Puthusseri B, Neelwarne B (2012) Carotenoid content, its stability during drying and the antioxidant activity of commercial coriander (Coriandrum sativum L.) varieties. Food Res Int 45(1):342–350
76. Watanabe M, Musumi K, Ayugase J (2011) Carotenoid pigment composition, polyphenol content, and antioxidant activities of extracts from orange-colored Chinese cabbage. LWT-Food Sci Technol 44(9):1971–1975
77. Kao FJ, Chiu YS, Tsou MJ, Chiang WD (2012) Effects of Chinese domestic cooking methods on the carotenoid composition of vegetables in Taiwan. LWT-Food Sci Technol 46(2):485–492
78. Tswett M (1906) Adsorptionsanalyse und Chromatogra-phische Methode. Anwendung auf die Chemie des Chlorophylls. Ber Deut Bot Ges 24:384–393
79. Damodaran S, Parkin KL (2017) Fennema's food chemistry. CRC Press, Boca Raton
80. Sander LC, Sharpless KE, Pursch M (2000) C 30 stationary phases for the analysis of food by liquid chromatography. J Chromatogr A 880(1):189–202
81. Xu XR, Zou ZY, Huang YM, Xiao X, Ma L, Lin XM (2012) Serum carotenoids in relation to risk factors for development of atherosclerosis. Clin Biochem 45(16):1357–1361
82. Azar M, Basu A, Jenkins AJ, Nankervis AJ, Hanssen KF, Scholz H et al (2011) Serum carotenoids and fat-soluble vitamins in women with type 1 diabetes and preeclampsia. Diabetes Care 34(6):1258–1264
83. Barona J, Jones JJ, Kopec RE, Comperatore M, Andersen C, Schwartz SJ et al (2012) A Mediterranean-style low-glycemic-load diet increases plasma carotenoids and decreases LDL oxidation in women with metabolic syndrome. J Nutr Biochem 23(6):609–615
84. Hodge A, Cunningham J, Maple-Brown L, Dunbar T, O'Dea K (2011) Plasma carotenoids are associated with socioeconomic status in an urban Indigenous population: an observational study. BMC Public Health 11(1):76
85. Melendez-Martinez AJ, Stinco CM, Liu C, Wang XD (2013) A simple HPLC method for the comprehensive analysis of cis/trans (Z/E) geometrical isomers of carotenoids for nutritional studies. Food Chem 138(2):1341–1350
86. Colmán-Martínez M, Martínez-Huélamo M, Miralles E, Estruch R, Lamuela-Raventós RM (2016) A new method to simultaneously quantify the antioxidants: Carotenes, xanthophylls, and vitamin A in human plasma. Oxidative Med Cell Longev 2016
87. Karppi J, Kurl S, Ronkainen K, Kauhanen J, Laukkanen JA (2013) Serum carotenoids reduce progression of early atherosclerosis in the carotid artery wall among Eastern Finnish men. PLoS One 8(5):e64107
88. Sugiura M, Nakamura M, Ogawa K, Ikoma Y, Yano M (2012) High serum carotenoids associated with lower risk for bone loss and osteoporosis in post-menopausal Japanese female subjects: prospective cohort study. PLoS One 7(12):e52643
89. Ho CC, de Moura FF, Kim SH, Clifford AJ (2007) Excentral cleavage of β-carotene in vivo in a healthy man. Am J Clin Nutr 85(3):770–777
90. Renzi LM, Hammond BR, Dengler M, Roberts R (2012) The relation between serum lipids and lutein and zeaxanthin in the serum and retina: results from cross-sectional, case-control and case study designs. Lipids Health Dis 11(1):33
91. Djuikwo VN, Ejoh RA, Gouado I, Mbofung CM, Tanumihardjo SA (2011) Determination of major carotenoids in processed tropical leafy vegetables indigenous to Africa. Food Nutr Sci 2(8):793
92. Li W, Beta T (2012) An evaluation of carotenoid levels and composition of glabrous canaryseed. Food Chem 133(3):782–786

93. Connolly EE, Beatty S, Thurnham DI, Loughman J, Howard AN, Stack J et al (2010) Augmentation of macular pigment following supplementation with all three macular carotenoids: an exploratory study. Curr Eye Res 35(4):335–351
94. Meinke MC, Darvin ME, Vollert H, Lademann J (2010) Bioavailability of natural carotenoids in human skin compared to blood. Eur J Pharm Biopharm 76(2):269–274
95. Varzakas T, Kiokias S (2016) HPLC analysis and determination of carotenoid pigments in commercially available plant extracts. Curr ResNut Food Sci J 4:01–14
96. Gamache P, Acworth I Measurement of dietary carotenoid isomers using HPLC-ECD. Thermo Fisher Scientific, Chelmsford
97. Condurso C, Verzera A, Dima G, Tripodi G, Crinò P, Paratore A et al (2012) Effects of different rootstocks on aroma volatile compounds and carotenoid content of melon fruits. Sci Hortic 148:9–16
98. Ndolo VU, Beta T (2013) Distribution of carotenoids in endosperm, germ, and aleurone fractions of cereal grain kernels. Food Chem 139(1):663–671
99. Marinova D, Ribarova F (2007) HPLC determination of carotenoids in Bulgarian berries. J Food Compos Anal 20(5):370–374
100. Plaza L, Sánchez-Moreno C, De Ancos B, Elez-Martínez P, Martín-Belloso O, Cano MP (2011) Carotenoid and flavanone content during refrigerated storage of orange juice processed by high-pressure, pulsed electric fields and low pasteurization. LWT-Food Sci Technol *44*(4):834–839
101. Mech-Nowak A, Swiderski A, Kruczek M, Luczak I, Kostecka-Gugała A (2012) Content of carotenoids in roots of seventeen cultivars of Daucus carota L. Acta Biochim Pol 59(1)
102. Andersson SC, Rumpunen K, Johansson E, Olsson ME (2011) Carotenoid content and composition in rose hips (Rosa spp.) during ripening, determination of suitable maturity marker and implications for health promoting food products. Food Chem 128(3):689–696
103. Agócs A, Nagy V, Szabó Z, Márk L, Ohmacht R, Deli J (2007) Comparative study on the carotenoid composition of the peel and the pulp of different citrus species. Innovative Food Sci Emerg Technol 8(3):390–394
104. Yoo KS, Bang H, Lee EJ, Crosby K, Patil BS (2012) Variation of carotenoid, sugar, and ascorbic acid concentrations in watermelon genotypes and genetic analysis. Hortic Environ Biotechnol 53(6):552–560
105. Slavin M, Yu LL (2012) A single extraction and HPLC procedure for simultaneous analysis of phytosterols, tocopherols and lutein in soybeans. Food Chem 135(4):2789–2795
106. Phelan D, Prado-Cabrero A, Nolan JM (2018) Analysis of lutein, zeaxanthin, and meso-zeaxanthin in the organs of carotenoid-supplemented chickens. Foods 7(2):20
107. Hernández-Ortega M, Ortiz-Moreno A, Hernández-Navarro MD, Chamorro-Cevallos G, Dorantes-Alvarez L, Necoechea-Mondragón H (2012) Antioxidant, antinociceptive, and anti-inflammatory effects of carotenoids extracted from dried pepper (Capsicum annuum L.). Biomed Res Int 2012:524019
108. Wu X, Sun C, Yang L, Zeng G, Liu Z, Li Y (2008) β-carotene content in sweet potato varieties from China and the effect of preparation on β-carotene retention in the Yanshu No. 5. Innovative Food Sci Emerg Technol 9(4):581–586
109. Hwang ES, Stacewicz-Sapuntzakis M, Bowen PE (2012) Effects of heat treatment on the carotenoid and tocopherol composition of tomato. J Food Sci 77(10)
110. Kidmose U, Christensen LP, Agili SM, Thilsted SH (2007) Effect of home preparation practices on the content of provitamin A carotenoids in coloured sweet potato varieties (Ipomoea batatas Lam.) from Kenya. Innovative Food Sci Emerg Technol 8(3):399–406
111. Lachman J, Hejtmánková K, Kotíková Z (2013) Tocols and carotenoids of einkorn, emmer and spring wheat varieties: Selection for breeding and production. J Cereal Sci 57(2):207–214
112. Fu H, Xie B, Ma S, Zhu X, Fan G, Pan S (2011) Evaluation of antioxidant activities of principal carotenoids available in water spinach (Ipomoea aquatica). J Food Compos Anal 24(2):288–297

113. Aman R, Carle R, Conrad J, Beifuss U, Schieber A (2005) Isolation of carotenoids from plant materials and dietary supplements by high-speed counter-current chromatography. J Chromatogr A 1074(1):99–105
114. Kimura M, Rodriguez-Amaya DB (2002) A scheme for obtaining standards and HPLC quantification of leafy vegetable carotenoids. Food Chem 78(3):389–398
115. Gama JJT, de Sylos CM (2007) Effect of thermal pasteurization and concentration on carotenoid composition of Brazilian Valencia orange juice. Food Chem 100(4):1686–1690
116. Griffiths DW, Dale MFB, Morris WL, Ramsay G (2007) Effects of season and postharvest storage on the carotenoid content of Solanum phureja potato tubers. J Agric Food Chem 55(2):379–385
117. Schmidt H, Kurtzer R, Eisenreich W, Schwab W (2006) The carotenase AtCCD1 from Arabidopsis thaliana is a dioxygenase. J Biol Chem 281(15):9845–9851
118. García-de Blas E, Mateo R, Viñuela J, Alonso-Álvarez C (2011) Identification of carotenoid pigments and their fatty acid esters in an avian integument combining HPLC–DAD and LC–MS analyses. J Chromatogr B 879(5):341–348
119. Parker RS, Swanson JE, Marmor B, Goodman KJ, Spielman AB, Brenna JT et al (1993) Study of β-carotene metabolism in humans using 13Cβ carotene and high precision isotope ratio mass spectrometry. Ann N Y Acad Sci 691(1):86–95
120. Hart DJ, Scott KJ (1995) Development and evaluation of an HPLC method for the analysis of carotenoids in foods, and the measurement of the carotenoid content of vegetables and fruits commonly consumed in the UK. Food Chem 54(1):101–111
121. Cacciola F, Donato P, Giuffrida D, Torre G, Dugo P, Mondello L (2012) Ultra high pressure in the second dimension of a comprehensive two-dimensional liquid chromatographic system for carotenoid separation in red chili peppers. J Chromatogr A 1255:244–251
122. Dugo P, Škeříková V, Kumm T, Trozzi A, Jandera P, Mondello L (2006) Elucidation of carotenoid patterns in citrus products by means of comprehensive normal-phase× reversed-phase liquid chromatography. Anal Chem 78(22):7743–7750
123. Dugo P, Herrero M, Kumm T, Giuffrida D, Dugo G, Mondello L (2008) Comprehensive normal-phase× reversed-phase liquid chromatography coupled to photodiode array and mass spectrometry detection for the analysis of free carotenoids and carotenoid esters from mandarin. J Chromatogr A 1189(1):196–206
124. Saini RK, Nile SH, Park SW (2015) Carotenoids from fruits and vegetables: chemistry, analysis, occurrence, bioavailability and biological activities. Food Res Int 76:735–750
125. Bijttebier S, D'Hondt E, Noten B, Hermans N, Apers S, Voorspoels S (2014) Ultra high performance liquid chromatography versus high performance liquid chromatography: stationary phase selectivity for generic carotenoid screening. J Chromatogr A 1332:46–56
126. Van Hung P, Hatcher DW (2011) Ultra-performance liquid chromatography (UPLC) quantification of carotenoids in durum wheat: Influence of genotype and environment in relation to the colour of yellow alkaline noodles (YAN). Food Chem 125(4):1510–1516
127. Rivera S, Vilaró F, Canela R (2011) Determination of carotenoids by liquid chromatography/mass spectrometry: effect of several dopants. Anal Bioanal Chem 400(5):1339–1346
128. Chauveau-Duriot B, Doreau M, Noziere P, Graulet B (2010) Simultaneous quantification of carotenoids, retinol, and tocopherols in forages, bovine plasma, and milk: validation of a novel UPLC method. Anal Bioanal Chem 397(2):777–790
129. Epriliati I, Kerven G, D'Arcy B, Gidley MJ (2010) Chromatographic analysis of diverse fruit components using HPLC and UPLC. Anal Methods 2(10):1606–1613
130. Granado-Lorencio F, Herrero-Barbudo C, Blanco-Navarro I, Pérez-Sacristán B (2010) Suitability of ultra-high performance liquid chromatography for the determination of fat-soluble nutritional status (vitamins A, E, D, and individual carotenoids). Anal Bioanal Chem 397(3):1389–1393
131. Li H, Deng Z, Liu R, Loewen S, Tsao R (2012) Ultra-performance liquid chromatographic separation of geometric isomers of carotenoids and antioxidant activities of 20 tomato cultivars and breeding lines. Food Chem 132(1):508–517

132. Foucault AP, Chevolot L (1998) Counter-current chromatography: instrumentation, solvent selection and some recent applications to natural product purification. J Chromatogr A 808(1–2):3–22
133. Foucault AP (1994) Centrifugal partition chromatography, Chromatographic Science Series, vol 68. CRC Press. ISBN 978-0-8247-9257-2
134. Englert M, Hammann S, Vetter W (2015) Isolation of β-carotene, α-carotene and lutein from carrots by countercurrent chromatography with the solvent system modifier benzotrifluoride. J Chromatogr A 1388:119–125
135. Li HB, Chen F, Zhang TY, Yang FQ, Xu GQ (2001) Preparative isolation and purification of lutein from the microalga Chlorella vulgaris by high-speed counter-current chromatography. J Chromatogr A 905(1–2):151–155
136. Xiao X, Si X, Yuan Z, Xu X, Li G (2012) Isolation of fucoxanthin from edible brown algae by microwave-assisted extraction coupled with high-speed countercurrent chromatography. J Sep Sci 35(17):2313–2317
137. Ferruzzi MG, Sander LC, Rock CL, Schwartz SJ (1998) Carotenoid determination in biological microsamples using liquid chromatography with a coulometric electrochemical array detector. Anal Biochem 256(1):74–81
138. Ferruzzi MG, Nguyen ML, Sander LC, Rock CL, Schwartz SJ (2001) Analysis of lycopene geometrical isomers in biological microsamples by liquid chromatography with coulometric array detection. J Chromatogr B Biomed Sci Appl 760(2):289–299
139. Allen CM, Schwartz SJ, Craft NE, Giovannucci EL, De Groff VL, Clinton SK (2003) Changes in plasma and oral mucosal lycopene isomer concentrations in healthy adults consuming standard servings of processed tomato products. Nutr Cancer 47(1):48–56
140. Unlu NZ, Bohn T, Francis D, Clinton SK, Schwartz SJ (2007) Carotenoid absorption in humans consuming tomato sauces obtained from tangerine or high-β-carotene varieties of tomatoes. J Agric Food Chem 55(4):1597–1603
141. Nagy V, Agócs A, Deli J, Gulyás-Fekete G, Illyés TZ, Kurtán T et al (2018) Carotenoid glycoside isolated and identified from cyanobacterium Cylindrospermopsis raciborskii. J Food Compos Anal 65:6–10
142. Agócs A, Murillo E, Turcsi E, Béni S, Darcsi A, Szappanos Á et al (2018) Isolation of allene carotenoids from mamey. J Food Compos Anal 65:1–5
143. Miglietta ML, Lamanna R (2006) 1H HR-MAS NMR of carotenoids in aqueous samples and raw vegetables. Magn Reson Chem 44(7):675–685
144. Kolodziejski W, Kasprzycka-Gutman T (1998) 13C CP/MAS NMR in trans-β-carotene. Solid State Nucl Magn Reson 11(3–4):177–180
145. Rehbein J, Dietrich B, Grynbaum MD, Hentschel P, Holtin K, Kuehnle M, Albert K (2007) Characterization of bixin by LC-MS and LC-NMR. J Sep Sci 30(15):2382–2390
146. Tode C, Sugiura M (2012) LC-NMR for the analysis of caroten- oids in foods. In: Preedy VR (ed) Vitamin A and carotenoids chemistry, analysis, function and effects. RSC, Cambridge, pp 269–281. 282–302
147. Draper J, Lloyd AJ, Goodacre R, Beckmann M (2013) Flow infusion electrospray ionisation mass spectrometry for high throughput, non-targeted metabolite fingerprinting: a review. Metabolomics 9(1):4–29
148. Lech K, Witowska-Jarosz J, Jarosz M (2009) Saffron yellow: characterization of carotenoids by high performance liquid chromatography with electrospray mass spectrometric detection. J Mass Spectrom 44(12):1661–1667
149. Van Breemen RB, Dong L, Pajkovic ND (2012) Atmospheric pressure chemical ionization tandem mass spectrometry of carotenoids. Int J Mass Spectrom 312:163–172
150. Felicissimo MP, Bittencourt C, Houssiau L, Pireaux JJ (2004) Time-of-flight secondary ion mass spectrometry and X-ray photoelectron spectroscopy analyses of Bixa orellana seeds. J Agric Food Chem 52(7):1810–1814
151. Manikandan M, Hasan N, Wu HF (2013) Rapid detection of haloarchaeal carotenoids via liquid–liquid microextraction enabled direct TLC MALDI-MS. Talanta 107:167–175

152. Takats Z, Wiseman JM, Gologan B, Cooks RG (2004) Mass spectrometry sampling under ambient conditions with desorption electrospray ionization. Science 306(5695):471–473
153. Lakshminarayana R, Aruna G, Sangeetha RK, Bhaskar N, Divakar S, Baskaran V (2008) Possible degradation/biotransformation of lutein in vitro and in vivo: isolation and structural elucidation of lutein metabolites by HPLC and LC-MS (atmospheric pressure chemical ionization). Free Radic Biol Med 45(7):982–993
154. Fraser PD, Enfissi E, Goodfellow M, Eguchi T, Bramley PM (2007) Metabolite profiling of plant carotenoids using the matrix-assisted laser desorption ionization time-of-flight mass spectrometry. Plant J 49(3):552–564
155. Sommerburg O, Siems W, Kraemer K (eds) (2013) Carotenoids and vitamin A in translational medicine. CRC Press
156. Yoshida K, Ueda S, Maeda I (2009) Carotenoid production in Bacillus subtilis achieved by metabolic engineering. Biotechnol Lett 31(11):1789
157. Dong L, Shion H, Davis RG, Terry-Penak B, Castro-Perez J, Van Breemen RB (2010) Collision cross-section determination and tandem mass spectrometric analysis of isomeric carotenoids using electrospray ion mobility time-of-flight mass spectrometry. Anal Chem 82(21):9014–9021
158. Miyazawa T, Nakagawa K, Kimura F, Satoh A, Miyazawa T (2011) Plasma carotenoid concentrations before and after supplementation with astaxanthin in middle-aged and senior subjects. Biosci Biotechnol Biochem 75(9):1856–1858
159. Sancho LEGG, Yahia EM, González-Aguilar GA (2011) Identification and quantification of phenols, carotenoids, and vitamin C from papaya (Carica papaya L., cv. Maradol) fruit determined by HPLC-DAD-MS/MS-ESI. Food Res Int 44(5):1284–1291
160. Viñas P, Bravo-Bravo M, López-García I, Hernández-Córdoba M (2013) Quantification of β-carotene, retinol, retinyl acetate and retinyl palmitate in enriched fruit juices using dispersive liquid–liquid microextraction coupled to liquid chromatography with fluorescence detection and atmospheric pressure chemical ionization. J Chromatogr A 1275:1–8
161. Garzón GA, Narváez-Cuenca CE, Kopec RE, Barry AM (2012) Mass spectrometry. J Chromatogr A 1275:1–8; Riedl KM, Schwartz SJ, Determination of carotenoids, total phenolic content, and antioxidant activity of Arazá (Eugenia stipitata McVaugh), an Amazonian fruit. J Agric Food Chem 60:4709–4717
162. de Rosso VV, Mercadante AZ (2007) HPLC–PDA–MS/MS of anthocyanins and carotenoids from dovyalis and tamarillo fruits. J Agric Food Chem 55(22):9135–9141
163. de Jesús Ornelas-Paz J, Yahia EM, Gardea AA (2008) Changes in external and internal color during postharvest ripening of 'Manila'and 'Ataulfo'mango fruit and relationship with carotenoid content determined by liquid chromatography–APcI+–time-of-flight mass spectrometry. Postharvest Biol Technol 50(2–3):145–152
164. Goupy P, Vian MA, Chemat F, Caris-Veyrat C (2013) Identification and quantification of flavonols, anthocyanins and lutein diesters in tepals of Crocus sativus by ultra performance liquid chromatography coupled to diode array and ion trap mass spectrometry detections. Ind Crop Prod 44:496–510
165. Giuffrida D, Dugo P, Torre G, Bignardi C, Cavazza A, Corradini C (2013) G Dugo. Food Chem 140(4):794–802
166. Lienau A, Glaser T, Tang G, Dolnikowski GG, Grusak MA, Albert K (2003) Bioavailability of lutein in humans from intrinsically labeled vegetables determined by LC-APCI-MS1. J Nutr Biochem 14(11):663–670
167. Watanabe M, Musumi K, Ayugase J (2011) Carotenoid pigment composition, polyphenol content, and antioxidant activities of extracts from orange-colored Chinese cabbage. LWT-Food Sci Technol 44(9):1971–1975
168. Kirkland JJ, Schuster SA, Johnson WL, Boyes BE (2013) Fused-core particle technology in high-performance liquid chromatography: an overview. J Pharm Anal 3(5):303–312

169. Nakagawa K, Kiko T, Hatade K, Asai A, Kimura F, Sookwong P et al (2008) Development of a high-performance liquid chromatography-based assay for carotenoids in human red blood cells: application to clinical studies. Anal Biochem 381(1):129–134
170. Mertz C, Brat P, Caris-Veyrat C, Gunata Z (2010) Characterization and thermal lability of carotenoids and vitamin C of tamarillo fruit (Solanum betaceum Cav.). Food Chem 119(2):653–659
171. Crupi P, Milella RA, Antonacci D (2010) Simultaneous HPLC-DAD-MS (ESI+) determination of structural and geometrical isomers of carotenoids in mature grapes. J Mass Spectrom 45(9):971–980
172. Fang L, Pajkovic N, Wang Y, Gu C, Van Breemen RB (2003) Quantitative analysis of lycopene isomers in human plasma using high-performance liquid chromatography– tandem mass spectrometry. Anal Chem 75(4):812–817
173. Dueker SR, Lin Y, Buchholz BA, Schneider PD, Lamé MW, Segall HJ, Clifford AJ (2000) Long-term kinetic study of β-carotene, using accelerator mass spectrometry in an adult volunteer. J Lipid Res 41(11):1790–1800
174. Getachew G, Kim SH, Burri BJ, Kelly PB, Haack KW, Ognibene TJ, Clifford AJ (2006) How to convert biological carbon into graphite for AMS. Radiocarbon 48(3):325–336
175. Van Lieshout M, West CE, Van Breemen RB (2003) Isotopic tracer techniques for studying the bioavailability and bioefficacy of dietary carotenoids, particularly β-carotene, in humans: a review. Am J Clin Nutr 77(1):12–28
176. Dueker SR, Lin Y, Buchholz BA, Schneider PD, Lamé MW, Segall HJ et al (2000) Long-term kinetic study of β-carotene, using accelerator mass spectrometry in an adult volunteer. J Lipid Res 41(11):1790–1800
177. Moura FF, Ho CC, Getachew G, Hickenbottom S, Clifford AJ (2005) Kinetics of 14C distribution after tracer dose of 14C-lutein in an adult woman. Lipids 40(10):1069–1073
178. Ross AB, Vuong LT, Ruckle J, Synal HA, Schulze-König T, Wertz K, Bourgeois A (2011) Lycopene bioavailability and metabolism in humans: an accelerator mass spectrometry study. Am J Clin Nutr 93(6):1263–1273
179. Bittner K, L., A Schonbichler, S., K Bonn, G., & W Huck, C. (2013) Near infrared spectroscopy (NIRS) as a tool to analyze phenolic compounds in plants. Curr Anal Chem 9(3):417–423
180. Ruiz D, Reich M, Bureau S, Renard CM, Audergon JM (2008) Application of reflectance colorimeter measurements and infrared spectroscopy methods to rapid and nondestructive evaluation of carotenoids content in apricot (Prunus armeniaca L.). J Agric Food Chem 56(13):4916–4922
181. Brenna OV, Berardo N (2004) Application of near-infrared reflectance spectroscopy (NIRS) to the evaluation of carotenoids content in maize. J Agric Food Chem 52(18):5577–5582
182. Davey MW, Saeys W, Hof E, Ramon H, Swennen RL, Keulemans J (2009) Application of visible and near-infrared reflectance spectroscopy (Vis/NIRS) to determine carotenoid contents in banana (Musa spp.) fruit pulp. J Agric Food Chem 57(5):1742–1751
183. de Oliveira GA, Bureau S, Renard CMGC, Pereira-Netto AB, de Castilhos F (2014) Comparison of NIRS approach for prediction of internal quality traits in three fruit species. Food Chem 143:223–230
184. Saad AG, Jaiswal P, Jha SN (2014) Non-destructive quality evaluation of intact tomato using VIS-NIR spectroscopy. Int J 2(12):632–639
185. Rubio-Diaz DE, De Nardo T, Santos A, de Jesus S, Francis D, Rodriguez-Saona LE (2010) Profiling of nutritionally important carotenoids from genetically-diverse tomatoes by infrared spectroscopy. Food Chem 120(1):282–289
186. Baranska M, Baranski R, Schulz H, Nothnagel T (2006) Tissue-specific accumulation of carotenoids in carrot roots. Planta 224(5):1028–1037
187. Azevedo-Meleiro CH, Rodriguez-Amaya DB (2007) Qualitative and quantitative differences in carotenoid composition among Cucurbita moschata, Cucurbita maxima, and Cucurbita pepo. J Agric Food Chem 55(10):4027–4403

188. Lucas A, Andueza D, Rock E, Martin B (2008) Prediction of dry matter, fat, pH, vitamins, minerals, carotenoids, total antioxidant capacity, and color in fresh and freeze-dried cheeses by visible-near-infrared reflectance spectroscopy. J Agric Food Chem 56(16):6801–6808
189. Geleta N, Grausgruber H, Vollmann J (2014) Near-Infrared Spectroscopy of Yellow Endosperm Pigments in Wheat. Int J Agric For 4(6):446–450
190. OUYANG SL, ZHOU M, CAO B (2009) Lycopene and β-Carotene content in tomato analyzed by the second harmonic. Spectrosc Spectr Anal 29(12):3362–3364
191. Brenna OV, Berardo N (2004) Application of near-infrared reflectance spectroscopy (NIRS) to the evaluation of carotenoids content in maize. J Agric Food Chem 52(18):5577–5582
192. de Oliveira GA, de Castilhos F, Renard CMGC, Bureau S (2014) Comparison of NIR and MIR spectroscopic methods for determination of individual sugars, organic acids and carotenoids in passion fruit. Food Res Int 60:154–162
193. Deák K, Szigedi T, Pék Z, Baranowski P, Helyes L (2015) Carotenoid determination in tomato juice using near infrared spectroscopy. Int Agrophys 29(3):275–282
194. Schulz H, Baranska M, Baranski R (2005) Potential of NIR-FT-Raman spectroscopy in natural carotenoid analysis. Biopolymers 77(4):212–221
195. de Oliveira VE, Castro HV, Edwards HG, de Oliveira LFC (2010) Carotenes and carotenoids in natural biological samples: a Raman spectroscopic analysis. J Raman Spectrosc 41(6):642–650
196. Baranska M, Schütze W, Schulz H (2006) Determination of lycopene and β-carotene content in tomato fruits and related products: comparison of FT-Raman, ATR-IR, and NIR spectroscopy. Anal Chem 78(24):8456–8461
197. Baranski R, Baranska M, Schulz H (2005) Changes in carotenoid content and distribution in living plant tissue can be observed and mapped in situ using NIR-FT-Raman spectroscopy. Planta 222(3):448–457
198. Baranska M, Baranski R, Schulz H, Nothnagel T (2006) Tissue-specific accumulation of carotenoids in carrot roots. Planta 224(5):1028–1037
199. Darvin ME, Gersonde I, Meinke M, Sterry W, Lademann J (2005) Non-invasive in vivo determination of the carotenoids beta-carotene and lycopene concentrations in the human skin using the Raman spectroscopic method. J Phys D Appl Phys 38(15):2696
200. Alexandre MT, Gundermann K, Pascal AA, van Grondelle R, Büchel C, Robert B (2014) Probing the carotenoid content of intact Cyclotella cells by resonance Raman spectroscopy. Photosynth Res 119(3):273–281
201. Bicanic D, Dimitrovski D, Luterotti S, Twisk C, Buijnsters JG, Dóka O (2010) Estimating rapidly and precisely the concentration of beta carotene in mango homogenates by measuring the amplitude of optothermal signals, chromaticity indices and the intensities of Raman peaks. Food Chem 121(3):832–838
202. OUYANG SL, ZHOU M, CAO B (2009) Lycopene and β-Carotene content in tomato analyzed by the second harmonic. Spectrosc Spectr Anal 29(12):3362–3364
203. Yan J, Liu Y, Mao D, Li L, Kuang T (2001) The presence of 9-cis-β-carotene in cytochrome b6f complex from spinach. Biochim Biophys Acta (BBA)-Bioenerg 1506(3):182–188
204. Trebolazabala J, Maguregui M, Morillas H, de Diego A, Madariaga JM (2013) Use of portable devices and confocal Raman spectrometers at different wavelength to obtain the spectral information of the main organic components in tomato (Solanum lycopersicum) fruits. Spectrochim Acta A Mol Biomol Spectrosc 105:391–399
205. Svelander CA, Lopez-Sanchez P, Pudney PD, Schumm S, Alminger MA (2011) High pressure homogenization increases the in vitro bioaccessibility of α-and β-carotene in carrot emulsions but not of lycopene in tomato emulsions. J Food Sci 76(9)
206. Camorani P, Chiavaro E, Cristofolini L, Paciulli M, Zaupa M, Visconti A et al (2015) Raman spectroscopy application in frozen carrot cooked in different ways and the relationship with carotenoids. J Sci Food Agric 95(11):2185–2191
207. Qin J, Chao K, Kim MS (2011) Investigation of Raman chemical imaging for detection of lycopene changes in tomatoes during postharvest ripening. J Food Eng 107(3–4):277–288

208. Afseth NK, Bloomfield M, Wold JP, Matousek P (2014) A novel approach for subsurface through-skin analysis of salmon using spatially offset Raman spectroscopy (SORS). Appl Spectrosc 68(2):255–262
209. Saad AG, Pék Z, Szuvandzsiev P, Gehad DH, Helyes L (2019) Determination of carotenoids in tomato products using Vis/NIR spectroscopy. J Microbiol Biotechnol Food Sci 2019:27–31
210. Olarewaju OO, Magwaza LS, Nieuwoudt H, Poblete-Echeverria C, Fawole OA, Tesfay SZ, Opara UL (2019) Model development for non-destructive determination of rind biochemical properties of 'Marsh' grapefruit using visible to near-infrared spectroscopy and chemometrics. Spectrochim Acta A Mol Biomol Spectrosc 209:62–69
211. Dubova L, Erdberga, I., Duma, M., Rakutko, S., Avotiņš, A., &Alsina, I. (2019). Comparison of lycopene and β-carotene content in tomatoes determined with chemical and non-destructive methods Agron Res, v17 n2: 343-348
212. Cheng W, Sørensen KM, Mongi RJ, Ndabikunze BK, Chove BE, Sun DW, Engelsen SB (2019) A comparative study of mango solar drying methods by visible and near-infrared spectroscopy coupled with ANOVA-simultaneous component analysis (ASCA). LWT 112:108214
213. Monago-Maraña O, Eskildsen CE, Afseth NK, Galeano-Díaz T, de la Peña AM, Wold JP (2019) Non-destructive Raman spectroscopy as a tool for measuring ASTA color values and Sudan I content in paprika powder. Food Chem 274:187–193
214. Wang K, Liao Y, Meng Y, Jiao X, Huang W, Liu TCY (2019) The early, rapid, and non-destructive detection of citrus Huanglongbing (HLB) based on microscopic confocal Raman. Food Anal Methods 12(11):2500–2508
215. Bocsa CD, Moisoiu V, Stefancu A, Leopold LF, Leopold N, Fodor D (2019) Knee osteoarthritis grading by resonant Raman and surface-enhanced Raman scattering (SERS) analysis of synovial fluid. Nanomedicine 20:102012
216. Yu F, Li H, Song D, Li J, He Q, Wang S (2019) Inverse spatially offest Raman spectroscopy and its applications sub-surface material detections. In: Eleventh international Conference on Information Optics and Photonics (CIOP 2019), vol 11209. International Society for Optics and Photonics, p 1120928
217. Portarena S, Anselmi C, Zadra C, Farinelli D, Famiani F, Baldacchini C, Brugnoli E (2019) Cultivar discrimination, fatty acid profile and carotenoid characterization of monovarietal olive oils by Raman spectroscopy at a single glance. Food Control 96:137–145

Chapter 10
Carotenoids Fortification

Muhammad Riaz, Muhammad Zia-Ul-Haq, and Najm-ur-Rahman

10.1 Introduction

According to World Health Organization and The Food and Agriculture Organization of the United States "the practice of deliberately increasing the content of an essential micronutrient, i.e. vitamins and minerals (including trace elements) in food, so as to improve the nutritional quality of the food supply and to provide a public health benefit with minimal risk to health", whereas **enrichment** is defined as "synonymous with fortification and refers to the addition of micronutrients to food which is lost during processing." Fortification is the process of increasing the content of the given natural source more than the usual occurrence. Fortified foods include energy drinks or sports drinks or bars that hold positions between traditional food and dietary supplement. The number of active ingredients or targeted ingredients in fortified food is lesser than the corresponding dietary supplement, while more than the corresponding traditional food [1].

Fortification of foods is one of the World Health Organization strategies to eradicate malnutrition at a global level. The common foods that were targeted to be fortified food are cereals and its products, milk and its products, other necessary food items, tea and beverages, and baby milk formulas.

The importance of carotenoids in daily life cannot be overlooked; however, the yield of carotenoids containing the natural sources is of attention to researchers to maintain a healthier life through fortified carotenoid food. The fortification of carotenoids foods needs to pass legislative requirements in some countries; however, in other countries, carotenoids are permitted to be used as food additive colorants only.

M. Riaz (✉) · Najm-ur-Rahman
Department of Pharmacy, Shaheed Benazir Bhutto University, Sheringal, Pakistan

M. Zia-Ul-Haq
Office of the Research Innovation and Commercialization, Lahore College for Women University, Lahore, Pakistan

© Springer Nature Switzerland AG 2021

M. Zia-Ul-Haq et al. (eds.), *Carotenoids: Structure and Function in the Human Body*, https://doi.org/10.1007/978-3-030-46459-2_10

Fortification is an important strategy to avoid frequent and bulk intake of the non-fortified food; however, there may be a risk of an overdose of the targeted ingredient, and toxicity may occur. The cost of the fortified food is also higher. With fortification, the colour of carotene may not be acceptable to consumers [1]. Children and women of childbearing age in the world, particularly Africa, North America, and Asia face a deficit in micronutrients including pro-vitamin A carotenoids [2]. The concentration of micronutrients and carotenoids is low in their sources, so biofortification is the strategy to overcome this deficiency. Vitamin A fortified margarine was feed to pre-school children as 27 g per day for 6 months to manage the low serum retinol level [3] Similarly, Vitamin A fortified wheat flour bun were feed to school children (6–13 years) in Philippine to treat low serum retinol level [4].

The source of carotenoids are plants, algae, bacteria, animals. However, most of the animals and humans depend on their diet source for carotenoids. There are 40–50 dietary carotenoids that have been identified in human blood and tissues. Carrot is one of the foods, which is used as a source of pro-vitamin A (β-carotene). Various food preparations like juices have been fortified with β-carotene. The most common in the dietary carotenoids are α-carotene, β-carotene, β-cryptoxanthin, lycopene, lutein, and zeaxanthin. The first, three of the mentioned carotenoids are termed as provitamin A carotenoids as they are converted by the body to retinol. The rest of the three carotenoids are playing a beneficial role in human health like eyesight, skin, and reduction of inflammation. The carotenoids are also in use as approved food colorant e.g., β-Carotene, lutein, lycopene in Europe listed as E 160a–f, E 161b, and E 161 g. Due to the beneficial effects on human health, the interest in fortified carotenoids food or derived products has been increasing. The fortified food or supplements can help in reducing the frequency of intake comparative to non-fortified.

Various methods are used to fortify the carotenoid content; one is convention plant breeding techniques improvement, the second is genetic modification, and the third one is the addition of micronutrients to other food.

In the first method, various parameters in agriculture techniques are needed to be practised targeting the enrichment of carotenoids. While biofortification or genetic engineering involves changes in caroto-gene makeup to fortify the content but this needs skill and technology, which is unfortunately difficult to exercise in developing countries. The third method is already in use, as various fortified foods exist in markets that are fortified by the addition of micronutrients (Fig. 10.1).

10.2 Agriculture Techniques

10.2.1 Breeding Methods

Selection, consumption, need, the current status of the carotenoids in the targeted food or plants, and interest/capacity or collaboration of the region for the fortification of the nutrients is very much crucial. Currently, Pakistan, India, and Bangladesh

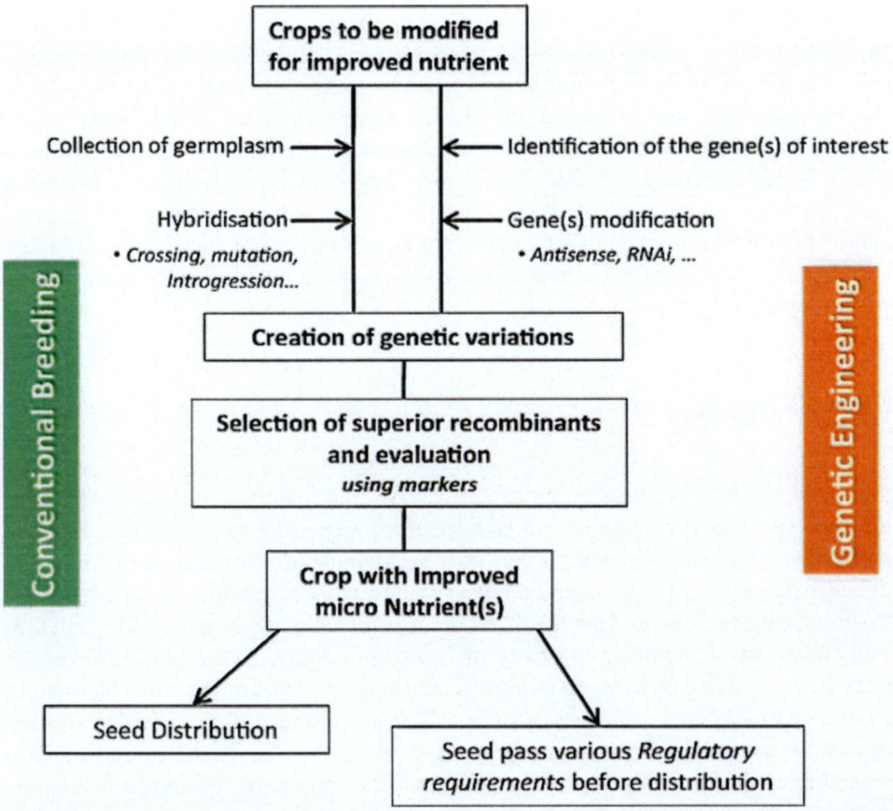

Fig. 10.1 The interconnectivity of conventional breeding and genetic engineering in the development of crops with enhanced micronutrients [5]

are the leading consumer of wheat, which has a trace amount of pro-vitamin A, that need to be fortified [6]. The organic farming has shown a very positive correlation with the total content of carotenoid fortification in Sweet Peppers, cv. Almuden and mandarin orange juice, cv. Hernandina [7, 8].

10.2.2 Ontogenic Factors/Ontogeny

Ontogeny is the complete sequence of events involved in the development of an organism, i.e., from seed to senescence. Various stages are involved during this entire life cycle; the quantity of carotenoids vary with the stage. The growing stage effects the carotenoid content by selecting the final stage of maturity will help to choose the best carotenoid source with enhanced content. In Tommy Atkins mangoes β-carotene, violaxanthin and *cis*-violaxanthin content increase with the stage of maturity (mature green, partially ripe and ripe). α-Carotene and β- carotene increased dramatically during maturation in *Cucurbita moschata* cultivar Menina Verde [9].

Similarly, the β-carotene content of mature leaves of lettuce was about three times those of the young leaves [10]. Carotenogenesis in the plant is augmented in most cases by a rise in temperature and prolonged exposure to sunlight. Papayas of the cultivar Formosa produced in the hot environment of Bahia gave higher β-carotene, β-cryptoxanthin, and lycopene contents with those produced in the temperate Southeastern state of São Paulo. Keitt mangoes and acerola, produced in a hot environment, gave higher β-carotene [9]. It was found that even in the same stage of maturity, samples of kale that were grown in organic fields showed higher carotenoid content. Some herbicides have been reported to inhibit the biosynthesis of carotenoids [11].

10.2.3 Grafting

Agriculture practices like grafting can be used to fortify the carotenoid content. The effect on carotenoid content in grafted vegetable crops is dependent on the rootstock and implant, e.g., the lycopene content of watermelon and total carotenoid, increased by 20% via grafting [12]. Other studies in watermelon by grafting of different rootstock improved carotenoid profile (α-carotene, β-carotene, lycopene, and lutein [13, 14]. Helyes et al. reported lowering of lycopene content in tomato; however, an increase in the yield of the crop minimizes this impact [15]. Apart from watermelon, other fruits like peach [16], bell pepper [17], and oriental melon [18] also showed enhanced carotenoid content using grafting procedures. The grafting procedure not only improves the bioactive composition but also produces resistance to various environmental factors like salinity, infection, and drought, etc.

10.2.4 Environmental Factors

The environmental factors may be biotic or abiotic that affect the carotenoid content of the source species. Plants of the same species but grown in a different environment will produce different carotenoid content. Considering these factors will undoubtedly help to collect the best source of carotenoids with higher content. Biotic stress is the stress induced by other living organisms that may include plants and micro-organisms etc. while abiotic stress is influenced by non-living factors or things like light, water, chemicals, and fertilizers, etc.

10.2.5 Biotic Factors

Various living factors might change the content of carotenoid, a study carried out across multiple carrot genotypes, only Bolero showed higher content of β-carotene, α-carotene and lutein stress induced by *Alternaria dauci* [19]. *Alternaria dauci* infection and water stresses combined caused the lowering of carotenoid content both in roots and leaves of carrot [20].

10.2.6 Abiotic Factors

These are nonliving parts of the environment that affect the ecosystem that includes rain, temperature, humidity, salinity, acidity, and soil composition, etc.

10.2.7 Cultivar and Climatic or Geographic Effects

10.2.7.1 Seasonal Variations

The geography and cultivar of a specie effect the carotenoids content, so choosing the best geographical conditions and selection of cultivar that has a maximum yield of carotenoids, will help in fortification via agricultural way. *Carica papaya* of Tailandia, produced in the Brazilian State of Bahia, has maximum lycopene and β-cryptoxanthin than other cultivars [9].

There are varieties of carrot, white, yellow, orange, pink, and purple. The quantity and types of carotenoids vary with the kind of carrot. Orange color carrot has been reported for the highest total carotenoid content compared to other types, followed by a pink color carrot. However, the quantity of individual carotenoid has great variation among the types of carrot, e.g., lycopene has been reported only in pink carrot. Lutein, α and β -carotene have been reported in all models except white carrot [21]. The carrot was reported for yearly variation of 24% for β-carotene content [22]. β-carotene content in melons remained constant throughout the growing season. However, the content varied significantly with sowing time and cultivar [23].

But not all species are capable of being chosen as favorable cultivar for future breeding. So, one has to go for genetic modification of the species. The petals of *Adonis aestivalis* produce astaxanthin; the content of astaxanthin was fortified by developing varieties [24]. Various strains of bacteria and types of algae have been discovered and developed into an enriched source of carotenoids at commercial levels [25].

Yellow fleshed potato grown at Colorado, and white-fleshed developed at Texas showed a difference in their total carotenoids level and individual carotenoid profile like neoxanthin, lutein, zeaxanthin, and violaxanthin. It was higher in variety grown at Texas [26]. Similar carotenoids difference in genotypes grown in New Zealand and the Netherlands was reported [27].

10.2.7.2 Cultivation System

Different cultivation system was used to study the effect of plants secondary metabolites. In a study, it was found that organically farmed plants have shown a greater concentration of secondary metabolites compared to conventionally grown plants. However, the carotene content was not affected by the cropping system in a carrot or seemed to decrease [22, 28].

There are various abiotic factors, e.g., light, temperature, fertilizers, salts, and drought, that regulate the biosynthesis of carotenoids.

10.2.7.3 Light

The exposure of carrot to more light or for a longer period causes a significant increase in carotene content [23, 29]. Stage et al. determined in *Daucus carota* that light affects differentially several carotenoid genes and, thus, the accumulation of carotenoids. The intensity of light alters the content of carotenoids in the plant, for example, *Raphanus sativus* grown under a shade has a higher content of carotenoids usually β-carotene, lycopene, lutein, and antheraxanthin [30, 31]. Direct sunlight exposure four-fold increases zeaxanthin content in spinach and rocket [32]. Similarly, the lycopene content can also be increased by direct sunlight exposure in tomato [33]. Photo-stress in leaves of oranges can enhance the content of carotenoids in nearby fruits [34].

The α-carotene content in carrot was increased with a decrease in plant density [22]. This may show a relation to light exposure that has been reported to increase the carotenoid content.

10.2.7.4 Temperature

The temperature has been found to affect the carotenoid content; in tomato, a higher temperature is correlated to the decrease in lycopene content; the optimum temperature is 20–40 °C [35, 36]. Carrot has been reported for lower β-carotene content when grown at a lower temperature [22] (Fig. 10.2).

10.2.7.5 Chemical Stress

Minerals/Fertilizers

Fertilizers based on their composition are also linked to the content of carotenoids. N fertilizers have a negative effect on carotenoids, e.g., lycopene in tomatoes [37] and apples [38]. In contrast, P and K fertilizers have a positive impact on their content [35, 39, 40]; however, the level of nitrogen does not affect the carotene content in carrot [29]. A study reported that reduction in nitrogen fertilization from 240 to

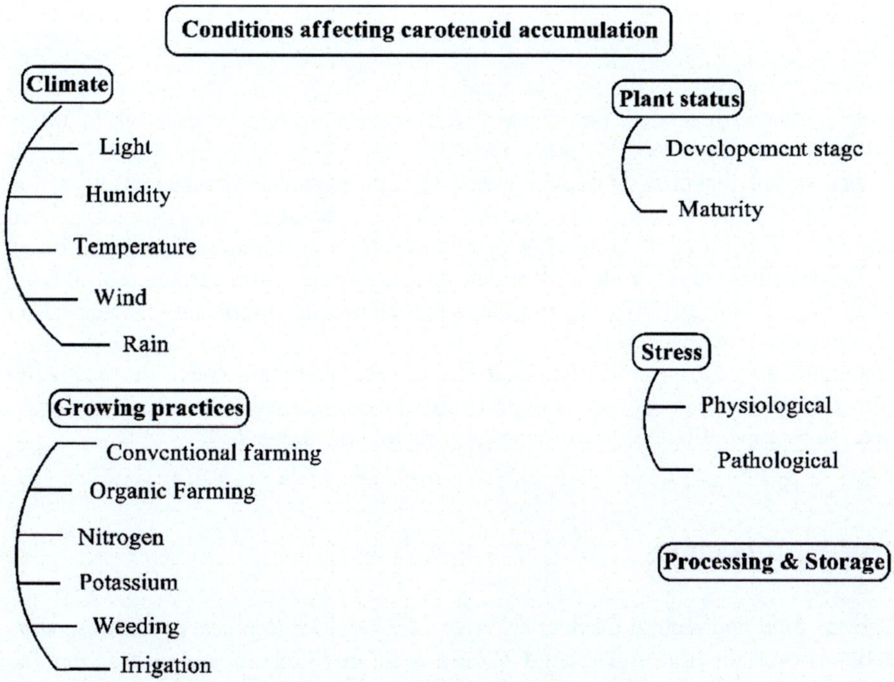

Fig. 10.2 Environmental conditions affecting carotenoid accumulation

60 kg N ha^{-1} resulted in a 12% decrease in β-carotene content [41]; a similar effect in carrot has been reported for reduction of molybdenum, selenium, and zinc in fertilization [42]. Another study published in kale that N-type fertilizer caused an increase in lutein, β-carotene, and chlorophyll content. The effect was further enhanced with NO^{3-} application compared to NH^{4+} [43]. The same enhanced carotenoid effect was also observed in lettuces [44]. Spent coffee grounds caused the enrichment of carotenoids up to 90% in *Lactuca sativa* [45]. Organic and mineral fertilizers have been reported for a significant increase in carotenoid content of sweet potato variety (TIB-440060) [46]. Karin, Agria and Marabel varieties of potato were reported for higher carotenoids in response to mineral fertilizers and polypropylene fiber exposure and other climatic conditions [47] When the culture of *Haematococcus pluvialis* is subjected to stress i.e., lack of nutrients cause the fortification of astaxzanthin [25]. Potassium does not affect the content of β-carotene [48]. The potassium containing fertilizers when applied weekly to melons in the fruiting stage, the increases β-carotene in fruit at harvest [49]. No change in carotenoids content was observed for this NPK nutrition in potato [50].

Salt and Droughts

Various genes responsible for the regulation of carotenoids, the up-regulation of genes is factor dependent one factor will up-regulate a type of gene while not by others. In rice third, the PSY gene has been reported to be up-regulated by stress induce via salt and drought treatment, but the light has no effect on its up-regulation [51] similar stress-induced ABA accumulation in roots of Zea mays have also been reported Li et al. [52]. Increasing sodium chloride concentration during growth causes fortification of carotenoids in buckwheat sprouts; however, the morphology of plants was lower [53]. The carotenoids content was significantly increased by drought stress in cherry tomatoes [54], while no such effect was reported in more abundant tomato fruits [55]. It was found in orange-colored carrot that irrigation showed a negative impact on carotenoid content comparative to rainfed crop; however, the composition of the carotenoids remained the same [56].

10.2.8 Radiations

Various food and natural product are irradiated in order to preserve it during storage; however, the time and strength should be of specified unit and dose so that the irradiated food or herb remain safe and healthy to the consumer. Exposure to radiation the carotenoids content shows variations, or sometimes no significant effect in carotenoid content has been reported (Table 10.1).

The seedling of lettuce was exposed to gamma radiations (2–30 grey). Increase carotenoid content in their leaves; the content showed a decrease when exposed to 70 Gray [24]. This experiment is a sort of activating the carotenoid biosynthetic cascade in seed to produce the effect in leaves there should be an optimum range of radiations and time of exposure to fortify the carotenoid content to its maximum level.

10.2.9 Post-harvest Handling/Processing and Storage

Various articles have been written about the loss or retention of carotenoids during processing or storage. However, Rodriguez-Amaya et al. have reported some conclusive comments about the post-harvest handling of carotenoid sources. The tropical conditions favor the enhanced content of carotenoids in plants, but the same situation may lead to loss or destruction of carotenoids during storage. The biosynthesis of carotenoids might work during storage if the conditions favor the existence of enzymes. Cutting or milling the plant parts during processing usually results in degradation of carotenoids because of its exposure to oxidations. Sun-drying or high-temperature drying of maize grain will increase carotenoid degradation, as will fine-milling and storing flour in translucent packaging [6]. The carotenoid con-

Table 10.1 Irradiation effect carotenoid content

Carotenoid source	Carotenoids	Radiation type	Effect on Carotenoids	References
Peppers (*Capsicum annum* L. cv. Zafiro)	Carotenoids	7 kJm^{-2} UV-C	Decrease in phenolic content and no significant effect on the carotenoids content	[57]
Artichoke (*Cynara scolymus L.*)	β-carotene	Gamma (0, 10, 20, 30 kGy)	slight decrease in β-carotene content	[58]
Fresh-cut mangoes	β-carotene	UV-C, for 0, 10, 20, and 30 min	decrease in β-carotene content	[59]
Mango (*Mangifera indica L.*)	Carotenoids	Electron beam (1-3.1 kGy)	No significant changes in carotenoid profile	[60]
Sweet basil (Ocimum basilicum L.)	β-carotene	Gamma, 30 kGy (10 intervals),	somewhat reduced the β-carotene content of the extracts	[58]
cinnamon, oregano, parsley, rosemary, bird pepper, and sage	Carotenoids	γ-radiation (10 kGy),	decrease in Carotenoids	[61]
Spinach cultivar Lazzio	Neoxanthin, violaxanthin, lutein/zeaxanthin and β-carotene	irradiation dose (1.5 KGy)	Increase in carotenoids	[62]
Spinach cultivar Samish	Neoxanthin, violaxanthin, lutein/zeaxanthin and β-carotene	irradiation dose (1, 1.5, 2 KGy)	gradually decrease in the carotenoid content	[62]
Tobacco leaves	lutein, β-carotene,	UV-B exposure (+20.76 mW/cm^2)	rapid increase in carotenoid	[63]

tent is negatively affected by the low storage temperature (−1 °C) and positively by higher (25 °C) storage temperature [64].

The stability of carotenoids is source dependent even at the same conditions of processing and storage. The stability may vary with change in processing conditions. In a home, the processing of carotenoids is directly linked to energy and processing time. The order of carotenoids degradations are microwaving < steaming < boiling < sautéing. It is estimated 37% of β-carotene is lost during maize porridge preparation following traditional Ghanaian cooking methods, more than 50% of β-carotene is lost during preparation of bread from fortified maize flour [2]. Most of the pro-vitamin A in Zambian bio-fortified maize hybrid occurs during storage compared to milling or cooking [65]. The low temperature (4 °C) during storage increases β-carotene up to 23% content in carrot comparative to the harvesting time [66]. However, the increase in storage temperature caused the lowering of β-carotene content 46% (7.5–8.5 °C), 51% (17–21 °C), and 70% (22–37.5 ° C) [67]. Dehydrated and unpacked carrots showed a decrease in β-carotene content [67, 68]. The β-carotene showed a 900% variation among the cultivars in carrot and 40% variation due to climate [22].

The carotenoid content of most of the plants at the mature fruit stage is at maximum, e.g., β- and δ-carotene in tomatoes and apricots [69, 70], lycopene-β-cyclase play its role to increase the content of lycopene in tomato [71]. However, if the plants are harvested before maturity, then the content will vary during the storage with temperature [72]. The storage time affects carotenoids content; invariably, the storage time has a direct relation with the carotenoid content in spinach leaves; however, there is the indirect relationship of the carotenoid content in potatoes [50, 73]. The storage time and temperature play a key role in affecting the carotenoid content. Three days or four weeks storage of carrot at 4 °C and 20 °C for 14 days caused an increase in β-carotene content [73, 74]; however, a decrease in β-carotene content at 7.5 °C in store for 8 days have been reported [67]. The loss of carotenoids during storage decreased with packing as reported packed broccoli florets in polypropylene micro-perforated bags, and packed carrots in polyethylene bags lost significantly less β-carotene [22, 75]. The leafy vegetable loses its carotenoids content during cold storage, as reported by Bunea et al. [76] for cutting spinach and shredded kale [77]. β-carotene content in peas and carrots remained the same at freezing for up to 6 months; however, a decline in content was observed after 12 months of storage freezing in processed carrots [78]. Decreasing the temperature to −25 °C for 8 weeks reduced the β-carotene in carrots [22].

10.3 Bio-fortification

When Genetic modification and plant breeding techniques are used to fortify the nutrient quality or fortify the carotenoid content, the process is termed as bio-fortification. An example of the method is Golden rice that produces carotene [1]. Bio-fortification or genetic engineering required sound knowledge of carotenoid pathways in the targeted plant need to be fortified; the second most crucial step is to know about the source or availability of cloned genes or enzyme encoded genes in pathways, the third important is the availability of suitable promoter to express the transgene, and the fourth one is the knowledge of tissues or compartments where the metabolites pathways are regulated.

Carotenoids are biosynthesized in plants via condensation reaction of the isomeric precursor's isopentenyl diphosphate (IPP) and dimethylallyl diphosphate (DMAPP) de novo within plastids; the pathway is given in Fig. 10.3. The carotenoids content can be fortified by modulation in these biosynthetic pathways in individual plants via providing an elevated levels of precursors. This strategy works well in tomato carotenoid fortification and chlorophyll, tocopherols fortification in Arabidopsis [79].

Another approach is to introduce new functionality in plants that has no capability to produce carotenoids via bioengineering as phytoene was produced in rice endosperm [80].

Modifying the storage capacity in plants provide another approach to fortify carotenoids content. Carotenoids accumulate in specialized structures made of sequestering lipoproteins in chromoplasts [81], leucoplasts, and amyloplasts [82, 83]. Mutation in the cauliflower Orange (Or) gene caused an increase in the accretion of carotenoids in chromoplasts [84, 85].

10.3.1 Phytoene Synthase

Phytoene synthase (PSY, as shown in Fig. 10.3) atalyzes the Geranylgeranyl diphosphate to form the first member of the biosynthetic chain carotenoids called phytoene. Multiple gene families are responsible for most of the plants except for Arabidopsis for the programming of the PSY. These genes expressed their character differentially in tissues or organs and affected by environmental stimuli [51, 52, 86, 87] (Table 10.2).

As shown in Fig. 10.3, the enzymes involved in various steps play an important role in the production of carotenoids. The genes responsible for the regulation of

Table 10.2 Phytoene synthase affects carotenoid content

Stress/or genetic engineering	Gene	Source	References
Salt and drought	PSY3	Increase carotenoid content in maize	[52]
Light	PSY1 and PSY2	In rice	[51]
ABA	PSY3	In rice	[51]
Seed-specific overexpression	PSY	50-fold increase *Brassica napus*	[88]
Overexpression	Daffodil PSY gene	Rice endosperm	[80]
Overexpression	Daffodil PSY gene	Higher carotenoid in rice endosperm japonica rice model variety Taipei 309	[89]
Genetic engineered plant, overexpression	Bacterial phytoene synthase	Two- to five-fold increase in carotenoids carrot roots	[90]
Overexpression	PSY, CaMV 35S promoter	Increased the carotenoid levels in tomato	[91]
Overexpression	Lycopene *β*-cyclase genes from the daffodil	Increase in β carotene in the fruits and xanthophyll in the leaves tomato	[92]
Downregulation using RNAi	Lycopene ε-cyclase	In *Brassica napus* increased carotenoids	[93]

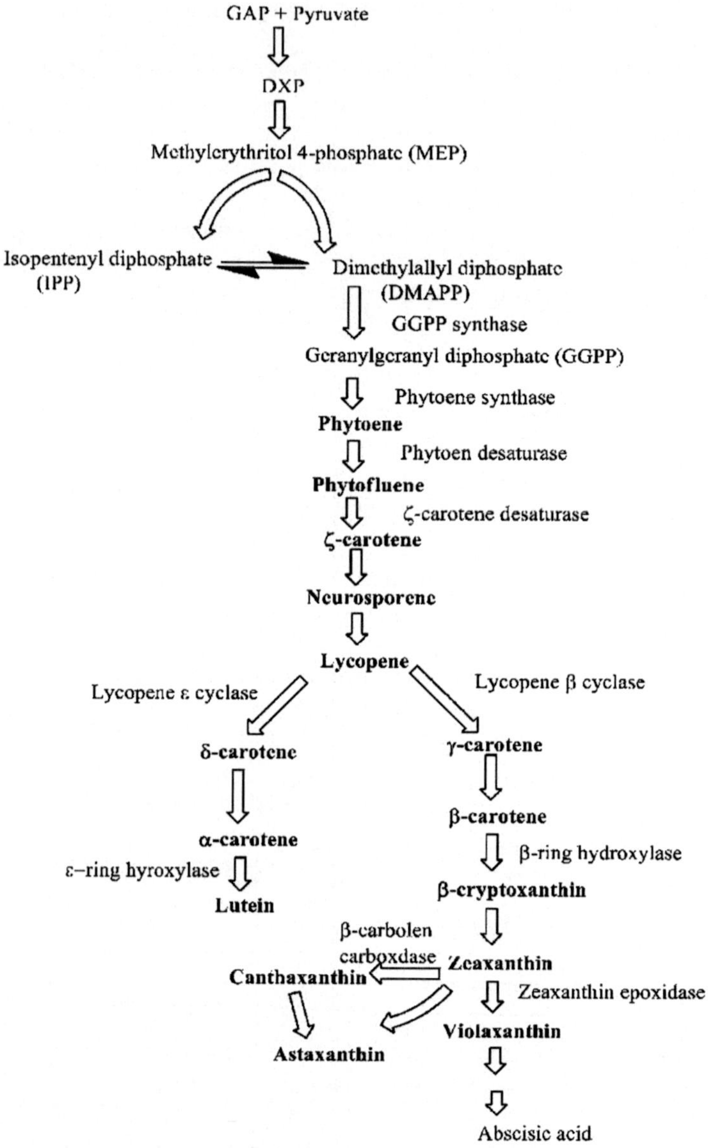

Fig. 10.3 Biosynthetic pathway of carotenoids; the transcriptional activation and suppression are a major regulatory mechanism for carotenoid biosynthesis, Key genes have been discovered in this pathway that if appropriately overexpressed will lead to enhanced carotenoid content

enzymes like Isomerases provide another regulatory step in carotenoid biosynthesis. Thus, selecting different lycopene ε-cyclase maize varieties may help in different isomers content [94].

10.3.2 The Xanthophyll Cycle (Fig. 10.4)

10.3.3 Genetics Changes in Crops Effect Level of Carotenoids

Genetic engineering has brought various changes in plant morphology and chemistry. The following crops were used mostly for genetic experiments to know the effect of the carotenoids profile.

10.3.3.1 Rice

Rice is an important essential food that is consumed in more than half of the population of the world. Being a vital food crop attracted researchers to fortify this crop by conventional agricultural techniques or genetically. The content of provitamins is significantly low in rice to be used as food for vitamin A deficient population that results in millions of cases of avertible impaired vision every year in developing countries. Thus, 'Golden Rice' the first of the genetic engineering mission to address

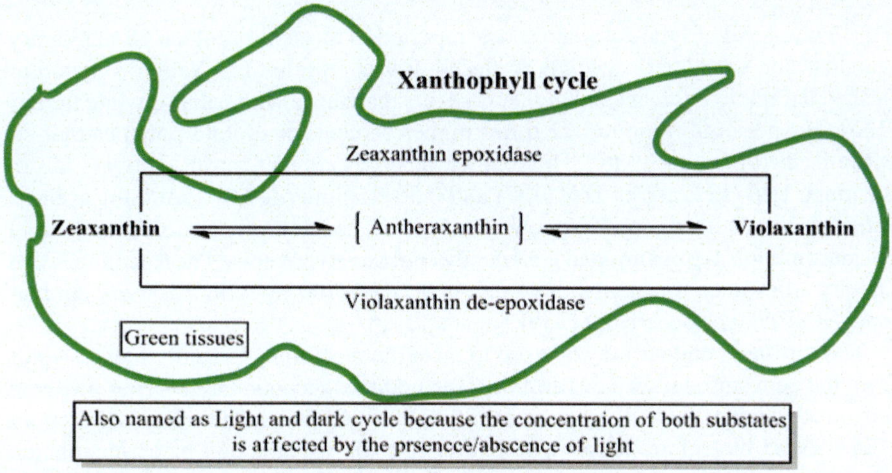

Fig. 10.4 Xanthophyll cycle; controlled by light and dark reactions [95]

vitamin A deficiency [96]. The rest of the experiments that resulted in the elevated carotenoid level in rice is given in Table 10.3.

10.3.3.2 Tomato

Tomato, when get matured, contains lycopene, the chief carotenoids, and lower level of β-carotene. Several publications have been published related to the fortification of lycopene, lutein, and β-carotene in tomato as tomato is among the essential economic crops. Significant variation in carotenoid profiles has been reported; the enrichment of carotenoids varies with the type of promoter and type of transgenic genes, for example, lutein 50-fold increase, 8-fold for β-carotene for more details can be read in Table 10.3. The adding of carotenogenic genes e.g., crtE, crtB, crtI, and crtY enables the de novo synthesis of lycopene, β-carotene and zeaxanthin [131, 132] while the addition of crtZ and crtW enables the synthesis of astaxanthin [133]. The up-regulation of PSY gene and downregulation of LCYB and LCYE at the onset of maturity is a key regulatory factor for carotenoid enrichment [105].

10.3.3.3 Carrots

One of the vegetables that consumed as raw and cooked. Several techniques may be used to fortify carotenoids content in carrot root. The elevation of Phytoene Synthase Protein Levels caused the increased in carotenoid levels in carrot roots to 858 g/g dry weight [134]

10.3.3.4 Potatoes

The higher level of total carotenoid was reported in the skin of tubers 28 mg/kg dry weight while 9 mg/kg dry weight in fleshy. It was reported that Yellow-skinned or fleshed tetraploid cultivars had higher contents than those with paler or white tissues [135]. Haynes et al. reported 22 times higher content for diploid potatoes than in white-fleshed potatoes [136]. The total carotenoids content for all cultivars fall in the range 1.10–12.2 mg/kg DW [137] and 0.50–15.5 mg/kg DW [138], the average value of about 4.35 mg/kg DW; however, 50% of the RDA of vitamin A can be met by consuming 250 g of enriched genetically engineered potatoes. The Recommended Dietary Allowance for men and women is 900 and 700 μg retinol activity equivalents (RAE)/day, respectively [139].

The fortified potatoes may be used to meet the recommended daily intake. Apart from the conventional way of fortifying the potatoes, genetic engineering involved the introduction of Or allele mutant cauliflower in potatoes with fortified carotenoid [124]. These biotechnological approaches, like the introduction of carotenogenic genes or suppression of endogenous enzymes, are used to improve carotenoid levels in potato tubers. Table 10.3 may be consulted for related studies.

Table 10.3 Genetically altered crops and carotenoids

Inserted gene/cDNA	Promoter	Variety	Carotenoid phenotype	References
Rice				
Daffodil Psy cDNA	CaMV 35S	Jt 309	Phytoene in endosperm (0.3 μg/g FW)	[80]
Daffodil Psy cDNA	Glutelin	Jt 309	Phytoene in endosperm (0.6 μg/g FW)	[80]
E. uredovora crtI + Psy + Lcy-b (daffodil)	CaMV 35S Glutelin	Jt 309	β-Carotene in endosperm (1.6 μg/g FW)	[97]
crtI + Psy + Lcy-b CaMV 35S	Glutelin	Indica varieties	β-Carotene in endosperm (1.6 μg/g FW)	[98]
crtI + Psy	Glutelin	Indica	β-Carotene (6.8 μg/g FW)	[99]
PSY (*Zea mays*) and crtI (*Pantoea ananatis*)	Glutelin Glut01	Asanohikari	Increase in carotenoids 23-times in seeds compared to Golden Rice 1	[89]
PSY (*Z. mays*)	Glutenin	White corn M37W	Zeaxanthin	[100]
CrtI (*P. ananatis*)	Barley hordein	White corn M37W	Slight increase in total carotenoids	[100]
PSY (*Zea mays*) crtI (*P. ananatis*)	Rice prolamin	White corn M37W	Rise in lycopene	[100]
PSY (*Z. mays*), crtI (*P. ananatis*), Lycb (*Gentiana lutea*),	Rice glutelin-1	White corn M37W	Rise in β-carotene	[100]
PSY (Z. mays), crtI (*Pantoea ananatis*), CBHX and crtW (*Paracoccus* spp), LYCB (*Gentiana lutea*)	Maize γ-zein	White corn M37W	Rise in carotenoids and de novo synthesis of ketocarotenoids	[100]
crtB and crtI (*Pantoea ananatis)*	Super gamma-zein	Hi-II corn	Increase in carotenoids 34 times	[101]
Tomato				
Tomato Psy-1 antisense	CaMV 35S	AC	Decrease in carotenoids up to 100 times	[102]

(continued)

Table 10.3 (continued)

Inserted gene/ cDNA	Promoter	Variety	Carotenoid phenotype	References
E. uredovora, crtI	CaMV 35S	AC	Increase in β-carotene 2–4 times, reduction in lycopene	[103]
E. uredovora, crtB	CaMV 35S	AC	Increases in phytoene, lycopene and β-carotene 2–3 times	[104]
Tomato Psy-1 sense	PG	AC	Early lycopene buildup	[105]
Yeast ySAMdc	E8	VF 36	β-carotene increase proportion wise, lycopene increases 3 times	[106]
Tomato Lcy-b sense and antisense	Pds	Moneymaker	β-carotene increases up to 7 times	[107]
Tomato Lcy-b + pepper CrtR-b	Pds	Moneymaker	Lycopene increases up to 30%; β-cryptoxanthin: 5 μg/g FW; zeaxanthin 13 μg/gFW	[108]
Paracoccus crtW + crtZ	CaMV 35S	AC	Ketocarotenoids in leaf	[109]
Tomato *Cry-2*	CaMV 35S	Moneymaker	Hp phenotype	[110]
E. coli Dxps	CaMV 35S and fibrillin	AC	Carotenoids increase	[111]
Tomato *Det-1*	RNAi + fruit specific promoters	Moneymaker	Increase in carotenoids, β-carotene 8-fold to 130 g/g dry weight	[112]
Tomato lcy-b cDNA	CaMV 35S	Red setter	Converting all the lycopene into β-carotene, increasing the total carotenoid content	[113]
Canola				
E. uredovora, crtB	Napin	Cv 212/86 Quantum	Carotenoids increase 50 times	[88, 114]
Carrot				

(continued)

Table 10.3 (continued)

Inserted gene/ cDNA	Promoter	Variety	Carotenoid phenotype	References
E. herbicola, *crt* genes	CaMV 35S	–	Carotenoids increase 2–5 times	[90]
alga Haematococcus pluvialis gene	Double CaMV 35S, *Arabidopsis-ubiquitin*, and RolD from *Agrobacterium rhizogenes*	–	Total carotenoids up to 70% were converted to novel ketocarotenoids	[115]
Potato				
Tobacco *Zep*, antisense and sense	GBSS	Freya and Baltica	Zeaxanthin increase 130 times; total carotenoid 5–7 times	[116]
E. uredovora, *crtB*	Patatin	Desiree	Increase in carotenoids 7 times	[117]
		Mayan gold	Increase in carotenoids 5 times	
E. coli Dxps	Patatin	Desiree	Increase in carotenoids 2 times	[118]
Potato *e-Lcy*, antisense	Patatin	Desiree	Increase in β-carotene 14 times; increase in total carotenoids 2.5 times	[119]
Algal *Bkt-1*	Patatin	*S. tuberosum*, *S. phureja*	Accumulation of ketocarotenoids	[120]
Erwinia crtB, *crtI* and *crtY*	CaMV 35S or Patatin	Desiree	Increase in β-carotene 20 times	[121]
Synechocystis crtO + antisense *crtZ*	CaMV 35S	Desiree	Rise in astaxanthin	[122]
Antisense *Chy-1* and *Chy-2*	Patatin	Desiree	Increase in β-carotene 38 times	[123]
Cauliflower or gene	GBSS	Desiree	Resulted in orange tuber flesh, tenfold the normal level of β-carotene	[124]
RNA interference to silence bch gene	GBSS CaMV 35S	Yema de Huevo, 91E22, and 'Desiree	Increase in Beta carotene 3.31 g/g dry weight and lutein	[125]

(continued)

Table 10.3 (continued)

Inserted gene/ cDNA	Promoter	Variety	Carotenoid phenotype	References
Wheat				
PSY (*Z. mays*) and crtI (*P. ananatis*)	CaMV 35S	Chinese elite wheat	Increase in carotenoid 10.8 times	[126]
crtB (*P. ananatis*)	CaMV35S,Glutenin,TP, Ubiquitin-1	Bobwhite	Slightly increase carotenoid content	[127]
crtI (*P. ananatis*)	CaMV35S, Ubiquitin-1, glutenin,TP	Bobwhite	Slightly increase carotenoid content	[127]
crtB and crtI (*P. ananatis*)	CaMV35S, TP, CaMV35S, glutenin, Ubiquitin-1	Bobwhite	Increase in provitamin A content 65 times	[127]
Sorghum				
PSY (*Z. mays*) and crtI (*P. ananatis*)	α-kafirin promoter β-kafirin promoter	TX430	A three to fourfold increase in carotenoid content	[128]
Soybean				
PSY (Capsicum) and crtI (*P. ananatis*)	β-conglycinin (β) or CaMV-35S (35S)	Kwangan	Increase 62 times in β-carotene	[129])
crtB (*P. ananatis*)	Lectin	Jack	Increase 1500 times in β-carotene	[130]

Abbreviations: *AC* Ailsa Craig, *bch* b-carotene hydroxylase gene, *Bkt-1* β-carotene 4-oxygenase, *CaMV 35S* cauliflower mosaic virus promoter, *Chy 1 and 2* β-carotene hydroxylases, *crtB* phytoene synthase, *crtI* phytoene desaturase, *crtO* carotenoid 4-oxygenase, *crtW* β-carotene 4-oxygenase, *crtY* lycopene β-cyclase, *crtZ*, β-carotene hydroxylase, *Det-1* de-etiolated 1, *Dxps* 1-deoxy-D-xylulose 5-phosphate synthase, *FW* fresh weight, *GBSS* granule bound starch synthase, *Jt* Japonica Taipei, *Pds* phytoene desaturase, *PG* polygalacturonase, *Psy* phytoene synthase, *TP* transit peptide from pea Rubisco small subunit, *ySAMdc* yeast S-adenosyl methionine decarboxylase, *Zep* zeaxanthin epoxidase, *β-Lcy* lycopene β-cyclase, *ε-Lcy* lycopene ε-cyclase

10.3.3.5 Maize

Maize is an outstanding staple food consumed mostly by the indirect way in corn starch and syrup. It contains vitamin A, β-carotene, lutein, and zeaxanthin. However, the amount of these carotenoids is quite less than the daily recommended allowance. It is a valued model for carotenoid investigation due to diverse gene pool, its pliability, and display of clear phenotypes. The level of carotenoids in corn kernels varies, that may provide conventional breeding may be used for fortification purpose [140].

10.3.3.6 Canola

Brassica napus L., commonly called canola, is an oil crop that contains carotenoids (18–26 g/g dry weight). Consuming the oil with sufficient carotenoids may fulfill the deficient dietary carotenoid. The plant may be a valued target to be used for carotenoid fortification research.

10.4 Fortification of Carotenoids and Health

Carotenoids play their health-protective role due to antioxidant and reactive oxygen scavenging potentials. α, β-carotene act as provitamin A, Lutein and zeaxanthin play an important role in eye health [21], while lycopene in prostate health. The consumption of lycopene enriched foods is associated with reduced risk of prostate cancer [141]. The Lutein and zeaxanthin prevent age-related macular degeneration, which was confirmed in a trial conducted in the US with the name of Age-Related Eye Diseases Study II [142]. β-carotene and apo-carotenal both showed excellent stability in fortified drinks, but the latter was less stable on exposure to sunlight. Two aspects were discussed in the chapter on the fortification in a source, and the second one is to add the micronutrients to the food product. The fortification in the source was discussed in detail; however, the addition technique of fortification required to be elaborated. The dietary carotenoids are added in required quantities to various food products for example β- carotene is used to fortify food that may be used as a provitamin A source. 2 μg of β-carotene supplement, 12 μg of dietary β-carotene, 24 μg of dietary α- carotene and 24 μg of dietary β- cryptoxanthin is sufficient to be converted by the body to 1 μg of retinol [143]. Apart from vitamin A source, it is also added as coloring agent (E 160a) to various foods. The usual products are butter, cheese, margarine, oils, milk-derived products, ice creams, soups, sauces, drinks (powder), bakery products, chewing gums, juices, jams, and egg-derived products. Lycopene is also added to various food products both for fortification or for coloring purposes (E 160d). As the beneficial health effects are well known so fortified food products, weather drinks/beverages or juices or oils/fats or confectionary, are consumed for the said purpose. Lutein is also added to various food products to enhance the eye-protective effects and other health benefits due to oxidative stress prevention.

10.5 Conclusion

Fortification is used to overcome micronutrient deficiency in various parts of the world; however, in relation to carotenoids, only pro-vitamin A carotenoids have been extensively considered as it is part of the micronutrients consumed daily. The sources of carotenoids have been studied in biofortification using various biotechnological techniques; however, the use of fortified carotenoids is limited to few like zeaxanthin, lutein, β-carotene, α-carotene, and lycopene even the number of food carotenoids is more than 50. This area needs to be investigated as the role of carotenoids in health cannot be overlooked.

References

1. Britton G, Liaaen-Jensen S, Pfander H (eds) (2009) Carotenoids. Nutrition and health. Birkhäuser Verlag, Basel/Boston/Berlin
2. Bouis HE, Welch RM (2010) Biofortification—a sustainable agricultural strategy for reducing micronutrient malnutrition in the global south. Crop Sci 50(Supplement_1):S-20–S-32
3. Dary O, Hurrell R (2006) Guidelines on food fortification with micronutrients. Switzerland World Health Organization, Food and Agricultural Organization of the United Nations, Geneva
4. Solon FS, Klemm RD, Sanchez L, Darnton-Hill I, Craft NE, Christian P et al (2000) Efficacy of a vitamin A–fortified wheat-flour bun on the vitamin A status of Filipino schoolchildren. Am J Clin Nutr 72(3):738–744
5. Badejo AA (2018) Elevated carotenoids in staple crops: the biosynthesis, challenges and measures for target delivery. J Genet Eng Biotechnol 16(2):553–562
6. Ortiz-Monasterio JI, Palacios-Rojas N, Meng E, Pixley K, Trethowan R, Pena R (2007) Enhancing the mineral and vitamin content of wheat and maize through plant breeding. J Cereal Sci 46(3):293–307
7. Pérez-López AJ, López-Nicolas JM, Núñez-Delicado E, FMD A, Carbonell-Barrachina ÁA (2007) Effects of agricultural practices on color, carotenoids composition, and minerals contents of sweet peppers, cv. Almuden. J Agric Food Chem 55(20):8158–8164
8. Beltrán-González F, Pérez-López AJ, López-Nicolás JM, Carbonell-Barrachina ÁA (2008) Effects of agricultural practices on instrumental colour, mineral content, carotenoid composition, and sensory quality of mandarin orange juice, cv. Hernandina. J Sci Food Agric 88(10):1731–1738
9. Rodriguez-Amaya DB (ed) (2003) Enhancing the carotenoid levels of foods through agriculture and food technology. FoodAfrica, Internet Forum; 2003: Citeseer
10. Ramos D, Rodriguezamaya D (1987) Determination of the vitamin-a value of common Brazilian leafy vegetables. J Micronutr Anal 3:147
11. Mercadante AZ, Rodríguez-Amaya DB (1991) Carotenoid composition of a leafy vegetable in relation to some agricultural variables. J Agric Food Chem 39(6):1094–1097
12. Davis AR, Perkins-Veazie P (2005) Rootstock effects on plant vigor and watermelon fruit quality. Rpt-Cucurbit Genet Coop 28:39
13. Condurso C, Verzera A, Dima G, Tripodi G, Crinò P, Paratore A et al (2012) Effects of different rootstocks on aroma volatile compounds and carotenoid content of melon fruits. Sci Hortic 148:9–16
14. Candir E, Yetisir H, Karaca F, Üstun D (2013) Phytochemical characteristics of grafted watermelon on different bottle gourds (Lagenaria siceraria) collected from Mediterranean region of Turkey. Turk J Agric For 37(4):443–456
15. Helyes L, Lugasi A, Pogonyi A, Pék Z (2008) Effect of variety and grafting on lycopene content of tomato (Lycopersicon lycopersicum L. Karsten) fruit. Acta Aliment 38(1):27–34
16. Remorini D, Tavarini S, Degl'Innocenti E, Loreti F, Massai R, Guidi L (2008) Effect of rootstocks and harvesting time on the nutritional quality of peel and flesh of peach fruits. Food Chem 110(2):361–367
17. Chávez-Mendoza C, Sanchez E, Muñoz-Marquez E, Sida-Arreola JP, Flores-Cordova MA (2015) Bioactive compounds and antioxidant activity in different grafted varieties of bell pepper. Antioxidants 4(2):427–446

18. Zhou X, Wu Y, Chen S, Chen Y, Zhang W, Sun X et al (2014) Using Cucurbita rootstocks to reduce fusarium wilt incidence and increase fruit yield and carotenoid content in oriental melons. HortScience 49(11):1365–1369
19. Emmanuel G, Florent P, Cécile D-L, Sébastien H, Anita S, Valérie L C, Mathilde B, Didier P (2017) Effects of biotic and abiotic stress on carotenoid content in carrot. 38th International Carrot Conference Bakersfield, 20–22 March 2017; IRHS, Agrocampus Ouest, INRA, Université Angers, Angers
20. Perrin F, Dubois-Laurent C, Gibon Y, Citerne S, Huet S, Suel A et al (2017) Combined Alternaria dauci infection and water stresses impact carotenoid content of carrot leaves and roots. Environ Exp Bot 143:125–134
21. Geoffriau E, Charpentier T, Huet S, Hägnefelt A, Lopes V, Nothnagel T et al (2018) CarrotDiverse: understanding variation in a wild relative of carrot. II International Symposium on Carrot and Other Apiaceae 1264
22. Seljåsen R, Kristensen HL, Lauridsen C, Wyss GS, Kretzschmar U, Birlouez-Aragone I et al (2013) Quality of carrots as affected by pre-and postharvest factors and processing. J Sci Food Agric 93(11):2611–2626
23. Wolbang CM, Singh DP, Sykes SR, McInerney JK, Bird AR, Treeby MT (2010) Influence of pre-and postharvest factors on β-carotene content, its in vitro bioaccessibility, and antioxidant capacity in melons. J Agric Food Chem 58(3):1732–1740
24. Mawson R (1995) Astaxanthin from flowers of the genus Adonis. Google Patents
25. Ausich RL (1997) Commercial opportunities for carotenoid production by biotechnology. Pure Appl Chem 69(10):2169–2174
26. Al-Saikhan MS (2000) Antioxidant, proteins and carotenoids in potato (Solanum tubureosum L.). Texas A & M University, College Station
27. Othman R (2009) Biochemistry and genetics of carotenoid composition in potato tubers. Lincoln University, Christchurch
28. Rashed A (2009) Variation of bioactive constituents, sensory qualities and yield in carrots grown with different types of crop management. Newcastle University
29. Northolt M, van der Burgt G-J, Buisman T, Vanden Bogaerde A (2004) Parameters for carrot quality: and the development of the inner quality concept
30. Stange C, Fuentes P, Handford M, Pizarro L (2008) Daucus carota as a novel model to evaluate the effect of light on carotenogenic gene expression. Biol Res 41(3):289–301
31. Grumbach KH, Lichtenthaler HK (1982) Chloroplast pigments and their biosynthesis in relation to light intensity. Photochem Photobiol 35(2):209–212
32. Esteban R, Fleta-Soriano E, Buezo J, Míguez F, Becerril JM, García-Plazaola JI (2014) Enhancement of zeaxanthin in two-steps by environmental stress induction in rocket and spinach. Food Res Int 65:207–214
33. Helyes L, Lugasi A, Pék Z (2007) Effect of natural light on surface temperature and lycopene content of vine ripened tomato fruit. Can J Plant Sci 87(4):927–929
34. Poiroux-Gonord F, Santini J, Fanciullino AL, Lopez-Lauri F, Giannettini J, Sallanon H et al (2013) Metabolism in orange fruits is driven by photooxidative stress in the leaves. Physiol Plant 149(2):175–187
35. Dumas Y, Dadomo M, Di Lucca G, Grolier P (2003) Effects of environmental factors and agricultural techniques on antioxidantcontent of tomatoes. J Sci Food Agric 83(5):369–382
36. Krumbein A, Schwarz D, Kläring H-P (2012) Effects of environmental factors on carotenoid content in tomato (Lycopersicon esculentum (L.) Mill.) grown in a greenhouse. J Appl Bot Food Qual 80(2):160–164
37. Dorais M (ed) (2005) Effect of cultural management on tomato fruit health qualities. I International symposium on Human Health Effects of Fruits and Vegetables 744
38. Reay PF, Fletcher RH, Thomas V (1998) Chlorophylls, carotenoids and anthocyanin concentrations in the skin of 'Gala'apples during maturation and the influence of foliar applications of nitrogen and magnesium. J Sci Food Agric 76(1):63–71

39. Dorais M, Ehret DL, Papadopoulos AP (2008) Tomato (Solanum lycopersicum) health components: from the seed to the consumer. Phytochem Rev 7(2):231
40. Kopsell DE, Kopsell DA, Randle WM, Coolong TW, Sams CE, Curran-Celentano J (2003) Kale carotenoids remain stable while flavor compounds respond to changes in sulfur fertility. J Agric Food Chem 51(18):5319–5325
41. Sorensen J (1998) Nitrogen effects on vegetable crop production and chemical composition. International workshop on Ecological Aspects of Vegetable Fertilization in Integrated Crop Production 506
42. Biacs PA, Daood HG, Kadar I (1995) Effect of Mo, Se, Zn, and Cr treatments on the yield, element concentration, and carotenoid content of carrot. J Agric Food Chem 43(3):589–591
43. Kopsell DA, Kopsell DE, Curran-Celentano J (2007) Carotenoid pigments in kale are influenced by nitrogen concentration and form. J Sci Food Agric 87(5):900–907
44. Mozafar A (1994) Plant vitamins: agronomic, physiological and nutritional aspects. CRC Press, Boca Raton
45. Cruz R, Baptista P, Cunha S, Pereira JA, Casal S (2012) Carotenoids of lettuce (Lactuca sativa L.) grown on soil enriched with spent coffee grounds. Molecules 17(2):1535–1547
46. Koala M, Hema A, Somé K, Palé E, Sérémé A, Belem J et al (2013) Effects of organic and mineral fertilizers on total antioxidant, polyphenolic and carotenoid contents of Orange fleshed sweetpotato tubers. J Nat Sci Res 3(6):23–30
47. Kotikova Z, Hejtmankova A, Lachman J, Hamouz K, Trnkova E, Dvorak P (2007) Effect of selected factors on total carotenoid content in potato tubers (Solanum tuberosum L.). Plant Soil Environ 53(8):355
48. Hochmuth GJ, Brecht JK, Bassett MJ (2006) Fresh-market carrot yield and quality did not respond to potassium fertilization on a sandy soil validated by Mehlich-1 soil test. HortTechnology 16(2):270–276
49. Lester GE, Jifon JL, Makus DJ (2010) Impact of potassium nutrition on postharvest fruit quality: Melon (Cucumis melo L) case study. Plant Soil 335(1–2):117–131
50. Ezekiel R, Singh N, Sharma S, Kaur A (2013) Beneficial phytochemicals in potato—a review. Food Res Int 50(2):487–496
51. Welsch R, Wüst F, Bär C, Al-Babili S, Beyer P (2008) A third phytoene synthase is devoted to abiotic stress-induced abscisic acid formation in rice and defines functional diversification of phytoene synthase genes. Plant Physiol 147(1):367–380
52. Li F, Vallabhaneni R, Yu J, Rocheford T, Wurtzel ET (2008) The maize phytoene synthase gene family: overlapping roles for carotenogenesis in endosperm, photomorphogenesis, and thermal stress tolerance. Plant Physiol 147(3):1334–1346
53. Lim J-H, Park K-J, Kim B-K, Jeong J-W, Kim H-J (2012) Effect of salinity stress on phenolic compounds and carotenoids in buckwheat (Fagopyrum esculentum M.) sprout. Food Chem 135(3):1065–1070
54. Matsuzoe N, Zushi K, Johjima T (1998) Effect of soil water deficit on coloring and carotene formation in fruits of red, pink, and yellow type cherry tomatoes. J Jpn Soc Hortic Sci 67(4):600–606
55. Zushi K, Matsuzoe N (1998) Effect of soil water deficit on vitamin C, sugar, organic acid, amino acid and carotene contents of large-fruited tomatoes. J Jpn Soc Hortic Sci 67(6):927–933
56. Ombódi A, Daood HG, Helyes L (2014) Carotenoid and tocopherol composition of an orange-colored carrot as affected by water supply. HortScience 49(6):729–733
57. Vicente AR, Pineda C, Lemoine L, Civello PM, Martinez GA, Chaves AR (2005) UV-C treatments reduce decay, retain quality and alleviate chilling injury in pepper. Postharvest Biol Technol 35(1):69–78
58. Koseki PM, Villavicencio ALC, Brito MS, Nahme LC, Sebastião KI, Rela PR et al (2002) Effects of irradiation in medicinal and eatable herbs. Radiat Phys Chem 63(3–6):681–684

59. González-Aguilar GA, Villegas-Ochoa MA, Martínez-Téllez M, Gardea A, Ayala-Zavala JF (2007) Improving antioxidant capacity of fresh-cut mangoes treated with UV-C. J Food Sci 72(3):S197–S202
60. Reyes LF, Cisneros-Zevallos L (2007) Electron-beam ionizing radiation stress effects on mango fruit (Mangifera indica L.) antioxidant constituents before and during postharvest storage. J Agric Food Chem 55(15):6132–6139
61. Calucci L, Pinzino C, Zandomeneghi M, Capocchi A, Ghiringhelli S, Saviozzi F et al (2003) Effects of γ-irradiation on the free radical and antioxidant contents in nine aromatic herbs and spices. J Agric Food Chem 51(4):927–934
62. Lester GE, Hallman GJ, Pérez JA (2010) γ-Irradiation dose: effects on baby-leaf spinach ascorbic acid, carotenoids, folate, α-tocopherol, and phylloquinone concentrations. J Agric Food Chem 58(8):4901–4906
63. Shen J, Jiang C, Yan Y, Liu B, Zu C (2017) Effect of increased UV-B radiation on carotenoid accumulation and total antioxidant capacity in tobacco (Nicotiana tabacum L.) leaves. Genet Mol Res 16(1)
64. Rivera-Pastrana DM, Yahia EM, González-Aguilar GA (2010) Phenolic and carotenoid profiles of papaya fruit (Carica papaya L.) and their contents under low temperature storage. J Sci Food Agric 90(14):2358–2365
65. Mugode L, Ha B, Kaunda A, Sikombe T, Phiri S, Mutale R et al (2014) Carotenoid retention of biofortified provitamin A maize (Zea mays L.) after Zambian traditional methods of milling, cooking and storage. J Agric Food Chem 62(27):6317–6325
66. Berger M, Küchler T, Maaßen A, Busch-Stockfisch M, Steinhart H (2008) Correlations of carotene with sensory attributes in carrots under different storage conditions. Food Chem 106(1):235–240
67. Negi PS, Roy SK (2000) Effect of low-cost storage and packaging on quality and nutritive value of fresh and dehydrated carrots. J Sci Food Agric 80(15):2169–2175
68. Soria AC, Olano A, Frías J, Peñas E, Villamiel M (2009) 2-Furoylmethyl amino acids, hydroxymethylfurfural, carbohydrates and β-carotene as quality markers of dehydrated carrots. J Sci Food Agric 89(2):267–273
69. Carrillo-Lopez A, Yahia E (eds) (2009) Qualitative and quantitative changes in carotenoids and phenolic compounds in tomato fruit during ripening. VI International Postharvest Symposium 877
70. Campbell OE, Merwin IA, Padilla-Zakour OI (2013) Characterization and the effect of maturity at harvest on the phenolic and carotenoid content of Northeast USA apricot (Prunus armeniaca) varieties. J Agric Food Chem 61(51):12700–12710
71. Bravo S, García-Alonso J, Martín-Pozuelo G, Gómez V, García-Valverde V, Navarro-González I et al (2013) Effects of postharvest UV-C treatment on carotenoids and phenolic compounds of vine-ripe tomatoes. Int J Food Sci Technol 48(8):1744–1749
72. Kalt W (2005) Effects of production and processing factors on major fruit and vegetable antioxidants. J Food Sci 70(1):R11–RR9
73. Bergquist SÅ, Gertsson UE, Olsson ME (2006) Influence of growth stage and postharvest storage on ascorbic acid and carotenoid content and visual quality of baby spinach (Spinacia oleracea L.). J Sci Food Agric 86(3):346–355
74. Imsic M, Winkler S, Tomkins B, Jones R (2010) Effect of storage and cooking on β-carotene isomers in carrots (Daucus carota L. cv.'Stefano'). J Agric Food Chem 58(8):5109–5113
75. Nath A, Bagchi B, Misra L, Deka BC (2011) Changes in post-harvest phytochemical qualities of broccoli florets during ambient and refrigerated storage. Food Chem 127(4):1510–1514
76. Bunea A, Andjelkovic M, Socaciu C, Bobis O, Neacsu M, Verhé R et al (2008) Total and individual carotenoids and phenolic acids content in fresh, refrigerated and processed spinach (Spinacia oleracea L.). Food Chem 108(2):649–656
77. de Azevedo CH, Rodriguez-Amaya DB (2005) Carotenoid composition of kale as influenced by maturity, season and minimal processing. J Sci Food Agric 85(4):591–597

78. Puupponen-Pimiä R, Häkkinen ST, Aarni M, Suortti T, Lampi AM, Eurola M et al (2003) Blanching and long-term freezing affect various bioactive compounds of vegetables in different ways. J Sci Food Agric 83(14):1389–1402
79. Farré G, Sanahuja G, Naqvi S, Bai C, Capell T, Zhu C et al (2010) Travel advice on the road to carotenoids in plants. Plant Sci 179(1–2):28–48
80. Burkhardt PK, Beyer P, Wünn J, Klöti A, Armstrong GA, Schledz M et al (1997) Transgenic rice (Oryza sativa) endosperm expressing daffodil (Narcissus pseudonarcissus) phytoene synthase accumulates phytoene, a key intermediate of provitamin A biosynthesis. Plant J 11(5):1071–1078
81. Goodwin TW (1988) Distribution and analysis of carotenoids. In: Plant pigments. Academic, San Diego, pp 61–132
82. Marano MR, Serra EC, Orellano EG, Carrillo N (1993) The path of chromoplast development in fruits and flowers. J Plant Sci 94(1–2):1–17
83. Bartley GE, Scolnik PA (1995) Plant carotenoids: pigments for photoprotection, visual attraction, and human health. Plant Cell 7(7):1027
84. Li L, Paolillo DJ, Parthasarathy MV, DiMuzio EM, Garvin DF (2001) A novel gene mutation that confers abnormal patterns of β-carotene accumulation in cauliflower (Brassica oleracea var. botrytis). Plant J 26(1):59–67
85. Lu S, Van Eck J, Zhou X, Lopez AB, O'Halloran DM, Cosman KM et al (2006) The cauliflower or gene encodes a DnaJ cysteine-rich domain-containing protein that mediates high levels of β-carotene accumulation. Plant Cell 18(12):3594–3605
86. Li F, Vallabhaneni R, Wurtzel ET (2008) PSY3, a new member of the phytoene synthase gene family conserved in the Poaceae and regulator of abiotic stress-induced root carotenogenesis. Plant Physiol 146(3):1333–1345
87. Arango J, Wüst F, Beyer P, Welsch R (2010) Characterization of phytoene synthases from cassava and their involvement in abiotic stress-mediated responses. Planta 232(5):1251–1262
88. Shewmaker CK, Sheehy JA, Daley M, Colburn S, Ke DY (1999) Seed-specific overexpression of phytoene synthase: increase in carotenoids and other metabolic effects. Plant J 20(4):401–412
89. Paine JA, Shipton CA, Chaggar S, Howells RM, Kennedy MJ, Vernon G et al (2005) Improving the nutritional value of Golden Rice through increased pro-vitamin A content. Nat Biotechnol 23(4):482
90. Hauptmann R, Eschenfeldt WH, English J, Brinkhaus FL (1997) Enhanced carotenoid accumulation in storage organs of genetically engineered plants. Google Patents
91. Fray RG, Wallace A, Fraser PD, Valero D, Hedden P, Bramley PM et al (1995) Constitutive expression of a fruit phytoene synthase gene in transgenic tomatoes causes dwarfism by redirecting metabolites from the gibberellin pathway. Plant J 8(5):693–701
92. Apel W, Bock R (2009) Enhancement of carotenoid biosynthesis in transplastomic tomatoes by induced lycopene-to-provitamin A conversion. Plant Physiol 151(1):59–66
93. Yu B, Lydiate DJ, Young LW, Schäfer UA, Hannoufa A (2008) Enhancing the carotenoid content of Brassica napus seeds by downregulating lycopene epsilon cyclase. Transgenic Res 17(4):573–585
94. Harjes CE, Rocheford TR, Bai L, Brutnell TP, Kandianis CB, Sowinski SG et al (2008) Natural genetic variation in lycopene epsilon cyclase tapped for maize biofortification. Science 319(5861):330–333
95. Demmig-Adams B, Adams WW (2002) Antioxidants in photosynthesis and human nutrition. Science 298(5601):2149–2153
96. Dubock A (2017) An overview of agriculture, nutrition and fortification, supplementation and biofortification: Golden Rice as an example for enhancing micronutrient intake. Agriculture & Food Security 6(1):59. https://doi.org/:10.1186/s40066-017-0135-3.
97. Ye X, Al-Babili S, Klöti A, Zhang J, Lucca P, Beyer P et al (2000) Engineering the provitamin A (β-carotene) biosynthetic pathway into (carotenoid-free) rice endosperm. Science 287(5451):303–305

98. Datta K, Baisakh N, Oliva N, Torrizo L, Abrigo E, Tan J et al (2003) Bioengineered 'golden'indica rice cultivars with β-carotene metabolism in the endosperm with hygromycin and mannose selection systems. Plant Biotechnol J 1(2):81–90
99. Al-Babili S, Hoa TTC, Schaub P (2006) Exploring the potential of the bacterial carotene desaturase CrtI to increase the β-carotene content in Golden Rice. J Exp Bot 57(4):1007–1014
100. Zhu C, Naqvi S, Breitenbach J, Sandmann G, Christou P, Capell T (2008) Combinatorial genetic transformation generates a library of metabolic phenotypes for the carotenoid pathway in maize. Proceedings of the National Academy of Sciences. 2008:pnas. 0809737105
101. Aluru M, Xu Y, Guo R, Wang Z, Li S, White W et al (2008) Generation of transgenic maize with enhanced provitamin A content. J Exp Bot 59(13):3551–3562
102. Bird CR, Ray JA, Fletcher JD, Boniwell JM, Bird AS, Teulieres C et al (1991) Using antisense RNA to study gene function: inhibition of carotenoid biosynthesis in transgenic tomatoes. Nat Biotechnol 9(7):635
103. Römer S, Fraser PD, Kiano JW, Shipton CA, Misawa N, Schuch W et al (2000) Elevation of the provitamin A content of transgenic tomato plants. Nat Biotechnol 18(6):666
104. Fraser PD, Romer S, Shipton CA, Mills PB, Kiano JW, Misawa N et al (2002) Evaluation of transgenic tomato plants expressing an additional phytoene synthase in a fruit-specific manner. Proc Natl Acad Sci 99(2):1092–1097
105. Fraser PD, Bramley PM (2009) Genetic manipulation of carotenoid content and composition in crop plants. In: Carotenoids. Springer, Birkhäuser Verlag Basel, pp 99–114
106. Mehta RA, Cassol T, Li N, Ali N, Handa AK, Mattoo AK (2002) Engineered polyamine accumulation in tomato enhances phytonutrient content, juice quality, and vine life. Nat Biotechnol 20(6):613
107. Rosati C, Aquilani R, Dharmapuri S, Pallara P, Marusic C, Tavazza R et al (2000) Metabolic engineering of beta-carotene and lycopene content in tomato fruit. Plant J 24(3):413–420
108. Dharmapuri S, Rosati C, Pallara P, Aquilani R, Bouvier F, Camara B et al (2002) Metabolic engineering of xanthophyll content in tomato fruits. FEBS Lett 519(1–3):30–34
109. Ralley L, Enfissi EM, Misawa N, Schuch W, Bramley PM, Fraser PD (2004) Metabolic engineering of ketocarotenoid formation in higher plants. Plant J 39(4):477–486
110. Giliberto L, Perrotta G, Pallara P, Weller JL, Fraser PD, Bramley PM et al (2005) Manipulation of the blue light photoreceptor cryptochrome 2 in tomato affects vegetative development, flowering time, and fruit antioxidant content. Plant Physiol 137(1):199–208
111. Enfissi EM, Fraser PD, Lois LM, Boronat A, Schuch W, Bramley PM (2005) Metabolic engineering of the mevalonate and non-mevalonate isopentenyl diphosphate-forming pathways for the production of health-promoting isoprenoids in tomato. Plant Biotechnol J 3(1):17–27
112. Davuluri GR, Van Tuinen A, Fraser PD, Manfredonia A, Newman R, Burgess D et al (2005) Fruit-specific RNAi-mediated suppression of DET1 enhances carotenoid and flavonoid content in tomatoes. Nat Biotechnol 23(7):890
113. D'Ambrosio C, Giorio G, Marino I, Merendino A, Petrozza A, Salfi L et al (2004) Virtually complete conversion of lycopene into β-carotene in fruits of tomato plants transformed with the tomato lycopene β-cyclase (tlcy-b) cDNA. Plant Sci 166(1):207–214
114. Ravanello MP, Ke D, Alvarez J, Huang B, Shewmaker CK (2003) Coordinate expression of multiple bacterial carotenoid genes in canola leading to altered carotenoid production. Metab Eng 5(4):255–263
115. Jayaraj J, Devlin R, Punja Z (2008) Metabolic engineering of novel ketocarotenoid production in carrot plants. Transgenic Res 17(4):489–501
116. Römer S, Lübeck J, Kauder F, Steiger S, Adomat C, Sandmann G (2002) Genetic engineering of a zeaxanthin-rich potato by antisense inactivation and co-suppression of carotenoid epoxidation. Metab Eng 4(4):263–272
117. Ducreux LJ, Morris WL, Hedley PE, Shepherd T, Davies HV, Millam S et al (2004) Metabolic engineering of high carotenoid potato tubers containing enhanced levels of β-carotene and lutein. J Exp Bot 56(409):81–89

118. Morris WL, Ducreux LJ, Hedden P, Millam S, Taylor MA (2006) Overexpression of a bacterial 1-deoxy-D-xylulose 5-phosphate synthase gene in potato tubers perturbs the isoprenoid metabolic network: implications for the control of the tuber life cycle. J Exp Bot 57(12):3007–3018
119. Diretto G, Tavazza R, Welsch R, Pizzichini D, Mourgues F, Papacchioli V et al (2006) Metabolic engineering of potato tuber carotenoids through tuber-specific silencing of lycopene epsilon cyclase. BMC Plant Biol 6(1):13
120. Morris WL, Ducreux LJ, Fraser PD, Millam S, Taylor MA (2006) Engineering ketocarotenoid biosynthesis in potato tubers. Metab Eng 8(3):253–263
121. Diretto G, Al-Babili S, Tavazza R, Papacchioli V, Beyer P, Giuliano G (2007) Metabolic engineering of potato carotenoid content through tuber-specific overexpression of a bacterial mini-pathway. PLoS One 2(4):e350
122. Gerjets T, Sandmann G (2006) Ketocarotenoid formation in transgenic potato. J Exp Bot 57(14):3639–3645
123. Diretto G, Welsch R, Tavazza R, Mourgues F, Pizzichini D, Beyer P et al (2007) Silencing of beta-carotene hydroxylase increases total carotenoid and beta-carotene levels in potato tubers. BMC Plant Biol 7(1):11
124. Lopez AB, Van Eck J, Conlin BJ, Paolillo DJ, O'neill J, Li L (2008) Effect of the cauliflower or transgene on carotenoid accumulation and chromoplast formation in transgenic potato tubers. J Exp Bot 59(2):213–223
125. Van Eck J, Conlin B, Garvin D, Mason H, Navarre D, Brown C (2007) Enhancing beta-carotene content in potato by RNAi-mediated silencing of the beta-carotene hydroxylase gene. Am J Potato Res 84(4):331
126. Cong L, Wang C, Chen L, Liu H, Yang G, He G (2009) Expression of phytoene synthase1 and carotene desaturase crtI genes result in an increase in the total carotenoids content in transgenic elite wheat (Triticum aestivum L.). J Agric Food Chem 57(18):8652–8660
127. Wang C, Zeng J, Li Y, Hu W, Chen L, Miao Y et al (2014) Enrichment of provitamin A content in wheat (Triticum aestivum L.) by introduction of the bacterial carotenoid biosynthetic genes CrtB and CrtI. J Exp Bot 65(9):2545–2556
128. Lipkie TE, De Moura FF, Zhao Z-Y, Albertsen MC, Che P, Glassman K et al (2013) Bioaccessibility of carotenoids from transgenic provitamin A biofortified sorghum. J Agric Food Chem 61(24):5764–5771
129. Kim M-J, Kim JK, Kim HJ, Pak JH, Lee J-H, Kim D-H et al (2012) Genetic modification of the soybean to enhance the β-carotene content through seed-specific expression. PLoS One 7(10):e48287
130. Schmidt MA, Parrott WA, Hildebrand DF, Berg RH, Cooksey A, Pendarvis K et al (2015) Transgenic soya bean seeds accumulating β-carotene exhibit the collateral enhancements of oleate and protein content traits. Plant Biotechnol J 13(4):590–600
131. Misawa N, Nakagawa M, Kobayashi K, Yamano S, Izawa Y, Nakamura K et al (1990) Elucidation of the Erwinia uredovora carotenoid biosynthetic pathway by functional analysis of gene products expressed in Escherichia coli. J Bacteriol 172(12):6704–6712
132. Hundle B, Alberti M, Nievelstein V, Beyer P, Kleinig H, Armstrong G et al (1994) Functional assignment of Erwinia herbicola Eho10 carotenoid genes expressed in Escherichia coli. Mol Gen Genet MGG 245(4):406–416
133. Misawa N, Satomi Y, Kondo K, Yokoyama A, Kajiwara S, Saito T et al (1995) Structure and functional analysis of a marine bacterial carotenoid biosynthesis gene cluster and astaxanthin biosynthetic pathway proposed at the gene level. J Bacteriol 177(22):6575–6584
134. Maass D, Arango J, Wüst F, Beyer P, Welsch R (2009) Carotenoid crystal formation in Arabidopsis and carrot roots caused by increased phytoene synthase protein levels. PLoS One 4(7):e6373
135. Valcarcel J, Reilly K, Gaffney M, O'Brien N (2015) Total carotenoids and l-ascorbic acid content in 60 varieties of potato (Solanum tuberosum L.) grown in Ireland. Potato Res 58(1):29–41

136. Haynes KG, Clevidence BA, Rao D, Vinyard BT (2011) Inheritance of carotenoid content in tetraploid× diploid potato crosses. J Am Soc Hortic Sci 136(4):265–272
137. Hamouz K, Pazderů K, Lachman J, Čepl J, Kotíková Z (2016) Effect of cultivar, flesh colour, locality and year on carotenoid content in potato tubers. Plant Soil Environ 62(2):86–91
138. Fernandez-Orozco R, Gallardo-Guerrero L, Hornero-Méndez D (2013) Carotenoid profiling in tubers of different potato (Solanum sp) cultivars: accumulation of carotenoids mediated by xanthophyll esterification. Food Chem 141(3):2864–2872
139. Russell R, Beard JL, Cousins RJ, Dunn JT, Ferland G, Hambidge K et al (2001) Dietary reference intakes for vitamin A, vitamin K, arsenic, boron, chromium, copper, iodine, iron, manganese, molybdenum, nickel, silicon, vanadium, and zinc. A Report of the Panel on Micronutrients, Subcommittees on Upper Reference Levels of Nutrients and of Interpretation and Uses of Dietary Reference Intakes, and the Standing Committee on the Scientific Evaluation of Dietary Reference Intakes Food and Nutrition Board Institute of Medicine
140. Kurilich AC, Juvik JA (1999) Quantification of carotenoid and tocopherol antioxidants in Zea m ays. J Agric Food Chem 47(5):1948–1955
141. Siler U, Herzog A, Spitzer V, Seifert N, Denelavas A, Buchwald H et al (2005) Promises and perils of lycopene/tomato supplementation and cancer prevention. J Nutr 135:2050–2052
142. Group A-REDSR (1999) The age-related eye disease study (AREDS): design implications AREDS report no. 1. Control Clin Trials 20(6):573
143. Linus Pauling Institute's Micronutrient Information Center (2020) Carotenoids. Oregon State University, 307 Linus Pauling Science Center Corvallis, Oregon. https://lpi.oregonstate.edu/mic/dietary-factors/phytochemicals/carotenoids. Accessed 17 Jan 2020

Chapter 11
Metabolism of Carotenoids

Huma Umbreen, Madiha Javid, Muhammad Riaz, and Mehar-un-Nisa

11.1 Introduction

Carotenoids had been a part of human diet throughout the evolution and were familiar to the body biochemically. For the first time carotenoids were isolated from paprika and crocin in 1817 and 1818 respectively. With evolution in technology the structure of carotenoids was elucidated between 1930 and 1950. This was the era when starting substance and intermediates for synthesis of carotenoids were recognized. In the years1954–1968 commercial manufacture of five different carotenoids as coloring substances was introduced [1]. Most of the knowledge that is available today about carotenoids especially β-carotene was reported in 1960's, where small dose of radioactive substance was used. The study demonstrated that retinyl esters are major product of carotene metabolism. Further studies in 1960's also showed the biosynthetic pathway of carotenoids but it was unsuccessful to purify and characterize the enzymes related to carotenoids. However, carotenogenesis was understood in 1990, when the genes of carotene metabolism were cloned, thus explaining carotenogenesis at molecular level [2].Color is of great importance among the sensory attributes as it is determinant of quality and appearance of the food. As described in earlier chapters in fruits and vegetable, color is responsible for

H. Umbreen (✉)
Institute of Home and Food Sciences, Government College University Faisalabad, Faisalabad, Pakistan

M. Javid
Department of Home Economics, Government College Women University Faisalabad, Faisalabad, Pakistan

M. Riaz
Department of Pharmacy, Shaheed Benazir Bhutto University Sheringal, Sheringal, Pakistan

Mehar-un-Nisa
Institute of Home and Food Sciences, Govt. College University, Faisalabad, Faisalabad, Pakistan

© Springer Nature Switzerland AG 2021

M. Zia-Ul-Haq et al. (eds.), *Carotenoids: Structure and Function in the Human Body*, https://doi.org/10.1007/978-3-030-46459-2_11

acceptability but also these are major bioactive compounds that improve health and provide ability to fight against diseases [3]. Among various coloring compounds, carotenoids are lipophilic in nature consisting of isoprenoid units. These are synthesized in plant and certain micro-organisms and there are almost 500–600 different carotenoids which have been studied. In many diets, carotenoids are precursors of vitamin A and may protect from development of degenerative diseases such as cancer, muscular degeneration and heart diseases [4]. However, human body cannot synthesize carotenoids, therefore dietary provision of these compounds are mandatory for metabolic processes. Only a few carotenoids have been detected in human plasma including α-carotene, β-carotene, β-cryptoxanthin, zeaxanthin, lutein and lycopene, their cis-isomers and some degradation products being the most abundant [5].

Due to lipophilic nature, carotenoids depend upon lipid metabolic pathways for their digestion, absorption and metabolism. Their absorption from the intestine is facilitated by proteins and transport through the circulation is carried by lipoproteins [6]. Furthermore, carotenoids metabolism in the body is mediated by oxygenases (enzymes) to be converted into apocarotenal and retinoids [7]. Carotenoids have unsaturated double bonds in their structure which are responsible for isomerization, siglet oxygen scavenging and enzymatic cleavage for formation of retinoids. All such factors from food bioavailability to metabolism of the carotenoids will be discussed later in this chapter.

11.2 Food Sources and Bioavailability of Carotenoid

Carotenoids are highly unsaturated organic pigments which are widespread in nature. The synthesis of carotenoids takes place mainly in plants and in some micro-organisms such as bacteria, algae and fungi where they are stored in photosynthetic parts. In animals, carotenoids are accumulated in the tissues which they take in through their diet. The photosynthetic material can produce carotenoids using fats and other organic metabolic building blocks. Around 600 carotenoids have been identified and isolated from natural sources; 40 among them are commonly found in fruits and vegetables. Tomato juice has the richest concentration of the carotenoids. Other sources are kale, collard greens, spinach, sweet potato, card, water melon, carrots and pumpkin. Many carotenoids, such as carotenes, are used in the formation of vitamin A in the body, whereas others, including lycopene and xanthophyll, show no vitamin A activity [8]. Carotenoids chemically comprise eight isoprene units with a symmetrical skeleton of 40 carbon atoms forming a long chain of conjugated carbon double bonds that is the reason for the characteristic yellow to red color of the carotenoids and their antioxidant activities [9].

11.3 Bioaccessibility and Intestinal Absorption

The amount of a consumed nutrient or dietary constituent available for utilization in metabolism, normal physiological functioning and storage is referred to as its bioaccessibility or bioavailability. The bioavailability of carotenoids is the proportion of an ingested carotenoid that is absorbed by the intestinal enterocytes and transported in the bloodstream [10].

Physical and mechanical actions like heating, processing and chewing, and chemical actions of gastric acid and digestive enzymes help carotenoids to release from the food matrix which are then solubilized into micelles; absorbed, packaged into chylomicrons and finally transported and stored in various tissues. The uptake of ingested carotenoids by the intestinal mucosal cells is termed as 'entry' whereas the term 'absorption' refers to the movement of carotenoids from the mucosal cells into the lymphatic blood system. The absorption of carotenoids takes place in the similar manner as that of dietary lipids and the cholesterol transport inhibitors have been found to be the carotenoid transport inhibitor as well [8].

Solubilized in the lipid globules, carotenoids become accessible to the enterocytes where they are taken up via passive diffusion after the lipids have been digested in the small intestine. From the enterocytes, carotenoids are carried in the chylomicrons to the lymphatic system either to be released into the circulation or stored in the liver [11].

11.3.1 Release of Carotenoid from Food Matrix

The first step in the bioavailability of carotenoids is their release from the food matrix which can happen on the mechanical destruction of food matrix. Heating and processing of food also enhance the release of carotenoids. Carotenoids particularly in vegetables require more effort to release because of rigid cell walls of the vegetables. For instance, a study showing 2.5-fold higher concentration of lycopene in human plasma after ingestion of tomato paste than that of fresh tomato [12] provides evidence in this context (Fig. 11.1).

11.3.2 Distribution in the Digestive Tract

When released from the food matrices, carotenoids in the stomach interact with the gastric juices, where mixing movements of the stomach cause a lipid emulsion to form [13]. The transfer of carotenoids from the food to the lipid globules is initiated by the stomach activity. Different carotenoids may differ in their location within the lipid globule [14], e.g. xanthophylls is located near the surface and hence can be easily incorporated into the lipid droplets while the carotenes have to penetrate into

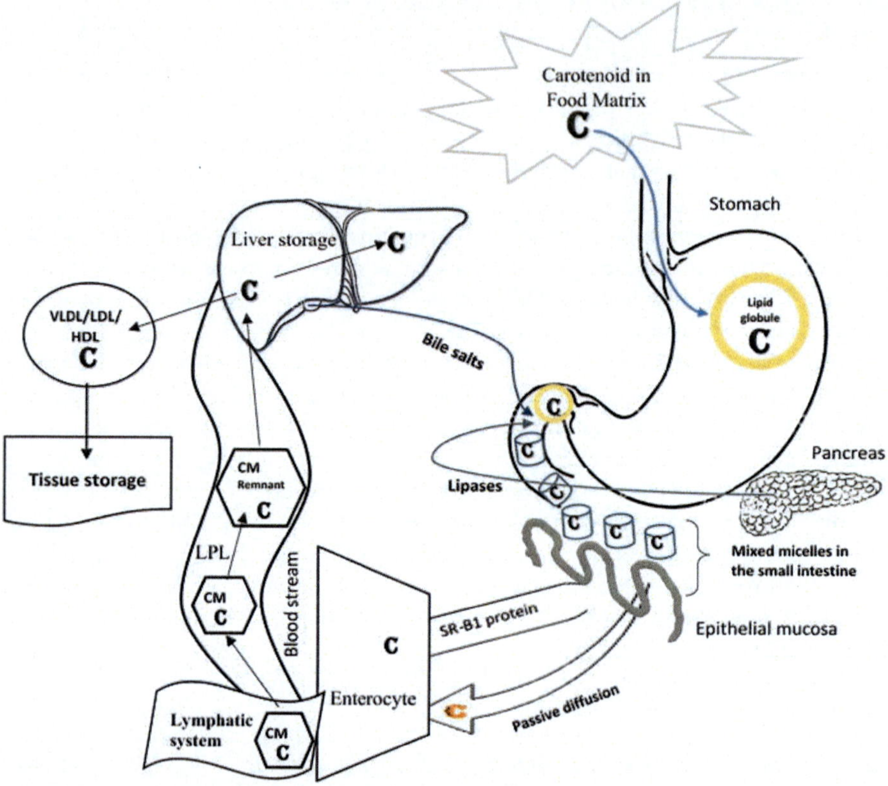

Fig. 11.1 Diagram illustrating the pathway of carotenoid molecules (C) from the food matrix to the bloodstream and liver and tissue storage

the lipid core, have less chances of incorporation [15]. This localization influences the transfer of carotenoids into the mixed micelles [16], ultimately affecting the carotenoid bioavailability. The entrance of lipid-carotenoid emulsion in the duodenum stimulates the secretion of bile acids (from the gall bladder) and lipases (from the pancreas). Due to the lipophilic nature of carotenoids, their emulsification and subsequent dispersion co-occur with the dietary lipid emulsification and dispersion under the action of phospholipids and bile acids present in the bile [11].

11.3.3 Solublization in the Mixed Micelles

The next step after the hydrolysis of biliary and dietary lipids, is the solubilization of carotenoids in the mixed micelles. The mixed micelles are the disc-like particles, 40–200 nm in diameter, consisting of cholesterol, monoacylglycerol, phospholipids, fatty acids and bile acids. Only the carotenoids which are solubilized in the micelles are considered to be available for absorption through the epithelial cells of the small intestine. Solubilized in the mixed micelles, carotenoids are able to get an access to the epithelial cells of the jejunum, where they are taken up by the epithelial cells through passive diffusion or via scavenger receptor class B type I protein (SR-BI) to be incorporated in the chylomicrons and then secreted to lymph [11, 17]. Hence, the ratio of the solubilized carotenoids in the mixed micelles to the amount present in food ingested represents the bioavailability [11]. Thus, the bioavailability of carotenoids is affected largely by their solubilization in the gastrointestinal tract and absorption into the small intestine which in turn is influenced by the dietary fats and oils [17].

The polarity of the carotenoid and the fatty acid composition along with the chain length and level of saturation of the fat may greatly affect the extent of carotenoids solubilization into the micelles. Pancreatic enzymes as well as bile salts are essential for carotenoids to be efficiently incorporated into micelles which provide a vehicle for the lipid-soluble carotenoids to diffuse across the water layer. The effects of lipids on carotenoid bioavailability are described in more detail later in the chapter.

11.4 Dietary Factors Affecting Bioavailability

Carotenoids have been known to be associated with a number of biological functions, including modulation of detoxifying enzymes, upregulating cell-to-cell communication, antioxidant activity, enhancing the immune system, regulating gene expression and upregulating cell-to-cell communication [18]. The bioavailability of carotenoid may be defined as a continuum ranging from the lowest bioavailability in raw whole-food sources, to slightly better in mildly processed foods, and the highest bioavailability in the purified food sources available in water-dispersible or oily preparations [19, 20]. The bioavailability of carotenoids is influenced by a number of dietary factors; some of these factors are described in detail as under:

11.4.1 Food Matrix

The food matrix is often the major factor that determines bioavailability. The release of carotenoids from this matrix is considered to be the first step needed to facilitate absorption. The more the food matrix is disrupted the greater the possibility of carotenoid absorption [21]. Many studies have shown that the bioavailability of carotenoids from commercial supplements is greater than that from food sources [11]. The bioavailability of β–carotene from vegetables in particular has been shown to be low (14% from mixed vegetables) compared with that of purified β-carotene added to a simple matrix (e.g. salad dressing), whereas for lutein, the difference is much smaller (relative bioavailability of 67% from mixed vegetables) [22].

11.4.1.1 Carotenoids in Fruits and Vegetables

The type of food matrix in which carotenoids are stored is a major bioavailability determining factor [22]. The amount of carotenoids present in raw vegetables and fruits depend on a number of variables including location and growing season, outdoor versus indoor growth, maturity at harvest, harvest/storage conditions and variety or cultivar [23]. The absolute amount of bioavailable carotenoids from fruits is considerably greater than that from vegetables. Conclusion have been drawn from the results of many well-designed studies that orange fruits contain higher amounts of bioavailable carotenoids than the green leafy vegetables. An equal amount of β-carotene from orange fruits such as mango, papaya and squash pumpkin can provide double vitamin A activity than that of green-leafy vegetables [24]. Moreover, mean four times greater serum response for β-carotene was found for subjects eating fruits than vegetables [25].

11.4.1.2 Location of Carotenoids

The bioavailability of carotenoids can be influenced by their location in the plant tissues and their physical state in which they are stored in the tissues [26]. The reason why fruit sources are better sources of absorbable carotenoid than vegetables may be the differences between the cellular structures in fruits and vegetables that are responsible for sequestering carotenoids [25]. Lutein, β-carotene and zeaxanthin are located in the chloroplasts, together with chlorophyll, where they form pigment-protein complexes in green leafy vegetables [27] and in crystalline form in dark green vegetables [28, 29]. In carrot, the carotenes are found in crystalline form and enclosed in membranous sheets of large proteins [26]. Carotenoids are found dissolved in oil droplets within the chromoplast structure in orange and yellow fruits as well as sweet potatoes which have more chances of solubilization in the digestive environment. However, carotenoids contained within chloroplasts in crystalline sheets or those in the form of protein-complexes may not be as accessible to the digestive processes to make these compounds bioavailable [28–30].

11.4.2 Food Cooking and Processing

Different observations are available about the effect of cooking on availability of carotenoid. Some scientists claim that heat involved with the process of cooking may deteriorate the carotenoids in vegetables, while other describe that moderate level of cooking make them more available due to release from the cell matrix [18]. Cooking processes such as microwave, pan cooking, boiling, steaming and scrambling affect differently (both negatively and positively) to the various carotenoids in foods of animal and plant origin (Table 11.1).

Table 11.1 Factors affecting the bioaccessibility of carotenoids

Factor	Effect	Type of carotenoid studied	References
1. Food matrix			
a. Chromoplast	During ripening carotenoids move into the chromplast and bioavailability is dependent upon the shape, size and physicochemical properties of the chromoplast.	Carotene, leutin, lycopene, β-cryptoxanthin	[21, 109]
b. Cell wall	Cell wall is a natural barrier that prevents the release of carotenoids. Processing methods that disrupt the cell wall and chromoplast membrane help to release carotenoids	β-Carotene, lycopene	[110–112]
c. Pectin	Pectin type, content in cell wall and degree of methyl esterification affect the bioaccessibility of carotenoids.	β-Carotene, β-cryptoxanthin	[113, 114]
2. Cooking style			
a. Boiling	Boiling has both positive (carrots and eggs) and negative (yellow peppers) effects on bioaccessibility of carotenoids due to disruption of cell wall and penetration of oil droplets into cell.	β-Carotene, zeaxanthin	[115, 116]
b. Scrambling	Higher content of polyunsaturated fatty acids in scramling decreases the bioavailability of carotenoids	Zeaxanthin, lutein	[117]
c. Baking, pan frying, steaming	Adversely affect the bioaccessibility of the carotenoids due to trapping in denatured muscle proteins.	Astaxanthin, canthaxanthin	[48, 118]
3. Processing			
a. Drying	Results in increased bioavailability of the carotenoid with higher lipophilicity	Lutein, capsanthin, β-carotene, β-cryptoxanthin, zeaxanthin	[119]
b. Heating	Heating through different method has different effects on carotenoids content. Thermal treatment decreases carotenoid availability in tomatoes, while microwave heating makes carotenoids more available in presence of lipids	Lycopene, β-carotene	[32]
c. Pasteurization	Pasteurization of juices increases bioavailability of carotenoids due to lower content of pectin.	β-Cryptoxanthin, lutein, α and β-carotene	[120]
d. Ultrasound	Due to increased networking of fiber, ultrasound decrease the bioaccessibility of carotenoids	Lycopene	[33]

(continued)

Table 11.1 (continued)

Factor	Effect	Type of carotenoid studied	References
4. Dietary fat			
a. Presence of fats	Consumption of fats and oils promote the absorption of carotenoids due to secretion of lipases,bile salts, phospholipids and delayed gastric emptying rate.	Xanthophyll, β-carotene, lycopene	[121, 122]
b. Saturated fats	Uptake via brush border is reduced in presence of saturated fats as butter	Xanthophyl, β-carotene, lutein	[123]
c. Unsaturated fats	Increased secretion of carotenoids with micelle rich in unsaturated fatty acids with larger number of chylomicrons. Carotenoid absorption is increased in case with monounsaturated fatty acids compared to polyunsaturated fatty acids	Lutein, zeaxanthin and Xanthophyl	[124, 125]
5. Carotenoids and fat soluble vitamins			
a. Different carotenoids	Competition is there between different carotenoids, xanthophyll are polar and easily up taken by enterocytes	Xanthophyl, leutin, zeaxanthin	[126]
b. Hydrolysis of carotenoids	Hydrolysis of carotenoids increases the bioavailability of carotenoids via greater uptake by epithelial cells.	Lutein, capsanthin, zeaxanthin, and β-cryptoxanthin	[48]
c. Fat soluble vitamins	Fat soluble vitamins compete for absorption sites in enterocytes, vitamin E is found to decrease the absorption of carotenoids, while effect of vitamin K and D has not yet been established.	Xanthophyll canthaxanthin	[127]

Similarly, many research studies are accomplished to know about the relation present between the type and degree of food processing (heat treatment and mechanical homogenization) and its effect on availability of carotenoids [31]. It is argued that trivial processing may disrupt the cell walls and thus helps to set free carotenoids from intracellular organelles. Furthermore, the processing may also break the carotenoid-protein complexes, may cause the reduction in particle size resulting in higher absorptions of carotenoids in the body. It is also proposed that processing may even deactivate oxidizing enzymes involved in degradation of carotenoids (Table 11.1). However, excessive heat processing may form isomers of carotenoids and makes it unavailable as studied with lycopene from tomatoes [32]. Similarly, Anese et al. [33] described that ultrasound processing of lycopene may reduce its absorption due to increased networking of fiber content. Contrarily, van Het Hof et al. [22] projected that homogenization or heating may enhance carotenoid bioavailability as much as six times.

11.4.3 Dietary Fat

As discussed earlier that uptake of carotenoids from the intestine is dependent upon the same pathway as of dietary fat, therefore its importance is quite obvious and intake of higher doses of carotenoids without dietary fats may result in lower plasma concentration [18]. Similarly, the other fat soluble compounds may compete for the absorption with carotenoids and thus may lower its bioavailability [22]. Moreover, fats are needed to stimulate gall bladder for bile production and release into the small intestine. Carotenoids solubilize with bile salt micelle, so it is proposed that fats should be present in the same meal containing carotenoids or some fat from previous meal may also facilitate the process [34]. Carotenoid may absorb properly even if only 3–5 g of fat in a meal is available [35]. Similarly, Brown et al. [36] described that full fat salad dressing showed better absorption of carotenoids from carotenoid containing foods as compared to low fat dressing. Moreover, a study showed that astaxanthin (carotenoid of seafood) showed higher absorption rate when was provided with lipids formulation [37]. Similarly, another research work by Deming et al. [38] proved that the Gerbils which consumed higher fat content (30%) as source of energy showed more conversion of β-carotene into vitamin A in post-absorptive state compared to those which consumed lower fat content (10%).

Not only the amount of fat but also the type of fat may alter the rate of bioavailability of carotenoids and also influence its efficiency but still more research work is needed to clear the effect of type of the fat [39]. While describing the role of monounsaturated fatty acids, Unlu et al. [40] used avocado oil as dressing to salsa and salads and observed increased absorption of carotenoids compared to fat free dressing. Similarly, mono-unsaturated fats from oleic acid also increased the rate of β-carotene absorption in a study conducted by Hollander and Ruble [41], however, intake of linoleic and linolenic acid (containing PUFA) resulted in decreased uptake of β-carotene. Contrarily, Fuller et al. [42] opposed the importance of fat in carotenoid bioavailability in human subjects and observed that β-carotene supplements provided with oil base showed less absorption compared to β-carotene beadlets mixed with water.

11.4.4 Dietary Fiber

Dietary fiber especially soluble dietary fiber (pectin, gaur and alginate) may alter the absorption and availability of carotenoids from the lumen [39]. Rock and Swendseid [43] observed the effect of pectin rich diet (12 g citrus pectin) consumption on supplementation with β-carotene (25 mg) instantly after meal in human subjects. They observed the effect after 30 h of treatment and found an increased level of plasma β-carotene in both groups but the concentration in low fiber diet was higher (141%) and in pectin rich diet was lower (60%) compared to the baseline values. In opposite to results of the most of the other studies, Unlu et al. [40], described no significant inhibitory effect of dietary fiber from avocado fruit on absorption of carotenoids and this study was unable to be compared with others. Whereas, Castenmiller et al. [25] also demonstrated no inhibitory effect of dietary fiber added to liquefied spinach. It was attributed to disrupted cell wall which has little effect on carotenoid metabolism.

11.4.5 Alcohol Consumption

Consumption of alcohol may interfere with formation of vitamin A through β-carotene. Studies have shown that heavy alcohol consumption not only inhibits the formation of vitamin A (in spite of presence of β-carotene in liver and plasma), but also is associated with its depletion. This may be associated with the liver injury caused by alcohol consumption [44]. Further with decreased level of conversion to vitamin A, carotene content in plasma is increased and may also show clearance issues (with lower tissue uptake). Studies about the impact of low to moderate level of alcohol consumption on plasma carotenoid concentration are scarce. But Rock et al. [45] proposed that metabolism and plasma clearance of carotenoids is also affected by even low to moderate level of alcohol consumption and that it is not limited to heavy consumers. Furthermore, Forman et al. [46] studied women consuming two drinks of alcohol every day for 3 months and observed an increased level of α and β-carotene (19 and 13% respectively), however, lutein and zeaxanthin ratio was found to reduce by 17%.

11.4.6 Interactions Among Carotenoids

In general, the higher intake of one of the carotenoid may alter the availability and absorption of other carotenoids. This effect may be of various type as has been shown by different studies as under;

- ↑ amount of one carotenoid may ↑ absorption of the other
- ↑ amount of one carotenoid may ↓ absorption of another due to competition for absorption site
- ↑ amount of one carotenoid may spare another carotenoid
- ↑ amount of one carotenoid may change the metabolic rate of other carotenoids [18]

Hypothetically, higher doses of various carotenoids administered together may show hindrance in absorption, transport and metabolism of each other in accordance with the existing serum content of that particular carotenoid. Though, in case the carotenoid source is a food it may show different results depending upon biological importance of the required carotenoid. However, the fight of carotenoids for receptors and carriers, required in uptake by enterocytes and transport by chylomicron is a debatable topic about carotenoid bioavailability [47]. According to serum demand for type of carotenoid following reaction may be observed;

- Carotenoids in intestinal mucosa may obstruct or enhance the activity of the carotenoid cleavage enzymes
- Plasma lipoprotein may also exchange the already circulating carotenoids
- One carotenoid can ↑or ↓ the uptake and accessibility of another [10, 22].

11.4.7 Bioavailability of Carotenoids from Whole Foods and Supplements

Whole foods, as ample quantity of carotenoid-rich vegetables and fruits may add to carotenoid content of blood in healthy subjects. As described earlier in this chapter the availability and quantity of carotenoids also varies with type of plant matrix so, different plants vary in their content and bioavailability of carotenoids. Furthermore, there may be some inhibitory factors in plants which can hinder carotenoid absorption [18]. β-carotene may be actively available from supplements due to some added factors however, harsh processing of foods may result in loss of carotenoids due to exposure to high temperature, oxygen and long storage time, so maintenance of biological activity of carotenoids is a major issue of food industry (Table 11.1). Micro and nano encapsulation of these carotenoids with amorphous sugars, polyols, polysaccharides and protein may enhance the bioavailability [31].

11.4.8 Carotenoid Structure and Isomers

Structure of carotenoids is an important factor to change its bioavailability. The studies using HPLC investigation of ferret lymph showed that absorption of zeaxanthin and lutein is higher as compared to β-carotene and lycopene [20]. The reason behind the difference may be the more polar nature of xanthophylls that helps it to absorb in micelle, so are efficiently entered into the enterocytes [17]. Similarly, carotenoids and their acyl esters may also show differences and conflicting results in bioavailability. The esters of hydroxycarotenoids including zeaxanthin, lutein, β-cryptoxanthin and caspanthin are hydrolyzed by carboxyl ester lipase present in exocrine pancreatic discharge. This hydrolysis can be helpful in uptake of carotenoid in enterocytes as depicted in case of zeaxanthin esters, which are hydrolyzed by carboxyl ester lipase resulting in increased bioavailability of free zeaxanthin in micelles and higher uptake by epithelial cell [48]. However, some other researches have depicted no significant relation between the bioavailability of lutein esters and free lutein [49].

11.4.9 Host Related Factors

Many studies have demonstrated the effect of host on bioavailability of carotenoids such as gender is important as females have higher blood concentration of carotenoids that might be due to difference in food consumption, blood volume and metabolism [50]. Similarly, the absorption of carotenoids may vary with age due to deterioration of gastro intestinal tract. However, Cardinault et al. [51] did not find any difference in absorption of lutein, α and β-carotene and lycopene in young and older adults. Any disease which is related to the malfunction of intestine may alter the absorption of carotenoids as blood of patients with cystic fibrosis show lowered leutin and zeaxanthin concentration compared to normal subjects [52]. Another study showed 37% deceased level of macular carotenoids in patient suffering from celiac disease and crohn's disease [53]. Furthermore, parasites and dysbiosis may also result in alteration in pattern of nutrient absorption, as was observed in case of Indonesian children suffering from helimenths [35].

11.5 Post-Absorption Metabolism of Carotenoid

11.5.1 In Circulation

Carotenoid are transported via chylomicrons same as fats and are secreted into the lymphatic vessel of the mesentery from where these are directed to venous blood through thoracic duct [54]. The lipids from chylomicrons first pass through hepatic metabolism and then carotenoids and its products in form of retinyl esters can be transported to extrahepatic tissues for further use [47]. In the circulation chylomicron pass through the process of hydrolysis and are metabolized to chylomicron remnants (CRs), with loss of triacyl glycerols (TAG) [55]. The hydrolysis process is catalyzed by lipoprotein lipase following the acquisition of apo E (apolipiprotein E) that also points the hepatic clearance. At the same time retinyl esters from chylomicrons are also hydrolyzed to retinol and are taken up by peripheral tissues with the help of lipoprotein lipase (Fig. 11.2) [54].

In the circulation, with the loss of triglyceride content, chylomicrons convert into chylomicron remnants (CRs). This process is accomplished through hydrolysis and is catalyzed by lipoprotein lipase (LPL). This results in hepatic clearance through the acquisition of apo E [56]. Along with this process retinyl esters also undergo hydrolysis with LPL and causes the breakdown of triglycerides so the resulting retinol can be taken up by peripheral tissues. This postprandial hydrolysis of retinyl esters are of much significance in building extrahepatic stores of retinol and most of these ester (25%) contribute to concentration of peripheral tissues [57].

Before the clearance of CRs from the blood, they are metabolized further to smaller size through triglyceride hydrolysis. CRs have ApoE on the surface which binds to LDL receptors on hepatocytes, playing a part in entry of these remnants into the liver [58]. From the liver carotenoids can be transported into the blood

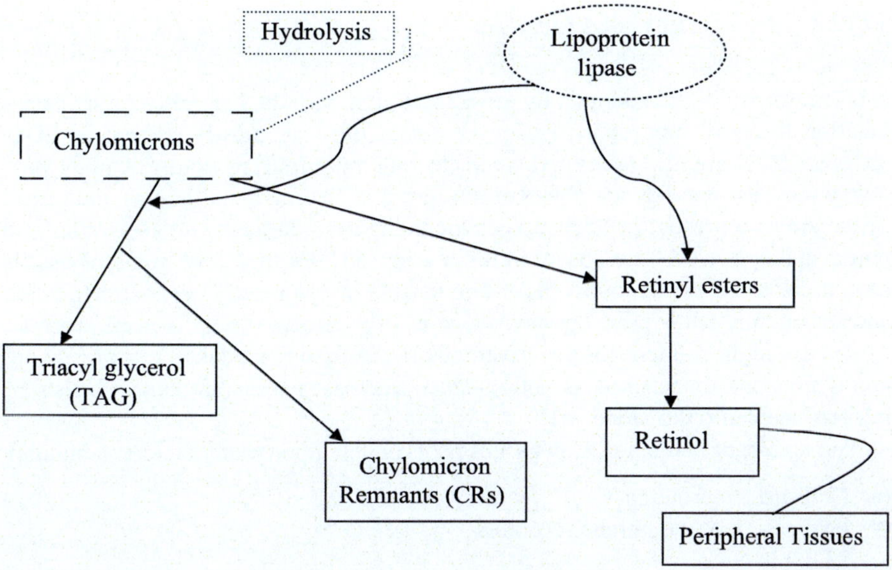

Fig. 11.2 Transport of retinyl esters along with triacyglycerols

Fig. 11.3 Amount of different carotenoids carried by lipoproteins

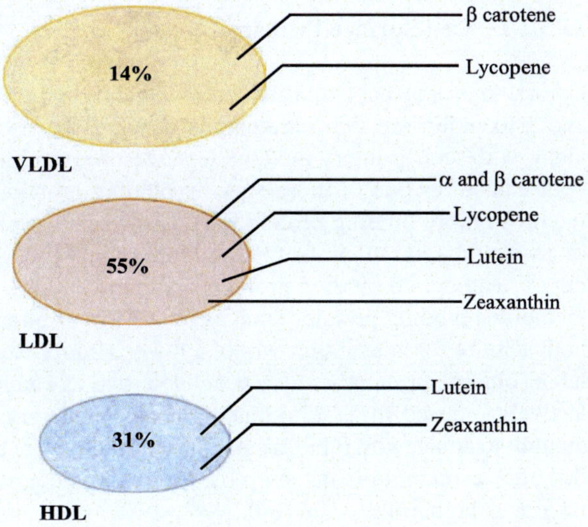

through various lipoprotein classes (LDL, VLDL and HDL). The various studies have demonstrated that in fasting state the lipoprotein of a normal subject may carry carotenoids (Fig. 11.3) as 14% in VLDL consisting mainly of β-carotene and lyopene, 55% in LDL mainly containing α and β carotene, lycopene, zeaxanthin and lutein whereas HDL has 31% of carotenoids consisting mainly of lutein and zeaxanthin [47, 59, 60].

11.5.2 In Tissues and Organs

Blood carotenoids are taken up by different tissues with the help of enzymatic degradation through lipoprotein lipase. As carotenoids are mostly present in LDL, therefore these are also taken up by cells having more LDL receptors. Among various tissues, carotenoids are abundantly present in liver, adrenal glands and testes. These are also present in other tissues including eyes, kidneys, lungs, ovaries and pineal gland, however are absent in brain stem [54]. Furthermore, some of tissues have preferential carotenoids uptake like macula of eye mostly concentrates lutein and zeaxanthin, while pineal gland has stores of p-carotene [60], likewise prostrate tissues are high in lutein [54]. Carotenoids in the tissues are taken up for retinoid production or can be used as antioxidants whereas; excess amount can also be released back into the blood [61].

The uptake of carotenoids by tissues is dependent upon many factors including:

(a) Cell surface proteins
(b) Expression of membrane receptors
(c) LDL receptors
(d) Skin Cells

11.5.2.1 Cell Surface Proteins

Caroteinoids may act as antioxidants when bright light is incident on macula. Lutein and zeaxanthin are two carotenoids that are specifically accumulated in macula along with their primary metabolites (Mesozeaxanthin) [62]. This highly selective uptake requires one or more specific binding proteins as retinoid uptake in retina requires retinol binding protein [63]. Likewise, it is also proposed that for transfer of zeaxanthin and lutein from blood into retinal tissues, specific pathways consisting of xanthophyll binding proteins are involved [64]. In a study, it was shown that membrane binding proteins such as pi isoform of glutathione S-transferase showed high affinity for zeaxanthin while, a lower affinity was observed for lutein [65]. A lutein-binding protein has also been identified as a member of a protein family collectively referred to as teroidogenic acute regulatory protein. Moreover, lutein is thought to attach with tubulin's paclitaxel binding site but it has been further studied that after carotenoid is taken up by specific binding proteins these are deposited at tubulin as high affinity site [66].

11.5.2.2 Expression of Membrane Receptors

Recent researchers have evidenced that in transportation of carotenoids and retinoids through different membranes and into the cells various facilitators are involved including

- ATP-binding cassette transporter (ABCA4)
- Scavenger receptor class B type 1
- Channel-like membrane protein (encoded by STRA6) [67].

Carotene, lycopene, and lutein are mainly present in adipose tissue, and the proposed pathway to adipocytes through CD36 by a facilitated-uptake process [61]. The concentration of abdominal adipose carotenoids is related to their serum level and dietary intake and it is thought to be consistent marker of carotenoid status in body [49].

The liver also has a large capacity for carotenoid accumulation and carotenoids reach the liver through lipoprotein remnant particles, which are taken up via endocytosis into hepatocytes for accumulation. Carotenoinds dissociate from the lipoprotein remnants and may be transferred to hepatic stellate cells (HSCs). Various studies show that HSCs can also accumulate β-carotene although at lower concentrations as compared to hepatocytes [68]. Furthermore, it has also been proposed that HSCs accumulate all-*trans*-lycopene in the lipid droplets and convert these into isomeric form (cis conformation) [69, 70]. Along with the transfer through cell surface protein as described in previous section SR-BI also appear to be involved in preferential uptake of zeaxanthin by macula of retina [8, 71]. Conversely, genetic changes in CD36are also linked with change in plasma and retina concentrations of lutein [72]. Furthermore, another study has proposed that CD36 is also linked with the uptake of lycopene and lutein by adipose tissues [61].

11.5.2.3 LDL Receptors

A number of studies conducted on humans and animal models have proposed that β-carotene and other carotenoids are associated with all types of lipoproteins, but as described earlier in this chapter, low-density lipoprotein (LDL) is major transporter of carotenoids [73]. LDL and VLDL (ApoB and apoE-containing lipoproteins) are required by the cell and enter into the cell by the process of endocytosis using LDL receptor (LDLr) [74]. In fact, in adult cells, cholesterol concentration is negatively affected if LDLr is not present and it is also confirmed by animal models that absence of this receptor results in unusually high concentrations of serum LDL-cholesterol and less clearance of lipoproteins from the blood [75]. LDLr are dominantly present in the liver as well as in the placenta and the embryo and molecular systems of carotenoid uptake by the liver or developing tissues is still unclear. Moreover, a study conducted by Shete et al. [75] investigated the role played by LDLr in β-carotene uptake by maternal liver, placenta and embryo and elucidated that LDLr significantly contributes to β-carotene acquisition in the adult mouse liver. Contrarily, LDLr did not show any role to facilitate the uptake by the placental–fetal unit.

11.5.2.4 Skin

Human skin has higher concentration of carotenoids to protect it from pro-oxidative damage and cytotoxic effects caused by UV light, air pollutants, cigarette smoke and microbes [76]. Raman spectroscopy has indicated that epidermis has uneven distribution of carotenoids across, with evidently higher content at the skin surface [77]. The mechanism behind it is that just as topical use of carotenoids causes the penetration into the epidermis layer of the skin, the serum concentration also has direct connection with retention in the skin, so serum content is involved to protect from photo-damages [78]. Furthermore, it has been elucidated that the level of carotenoids in the epidermis is more in parts of the body where larger number of sweat glands are present, including the forehead, the soles and the palms, being most abundant in hands [77]. Moreover, carotenoid supplementation also increases the carotenoid content in the skin however its content is affected by drop in serum concentration. Among the carotenoids metabolites xanthoplyll esters with fatty acids are abundantly present in the skin, which use the same transporters (SR-B1 and CD 36) for absorption as has been described previously [79].

11.5.3 At Cellular Level

Carotenoids have unique structure due to which it has the ability to turn into isoforms, to quench singlet oxygen species and to transfer electron during oxidation and reduction processes [59]. Carotenoid molecule has different point where different oxidative and non-oxidative enzymes can act to cleave such as

- Cyclohexene rings
- Double bonds
- Methyl groups [60].

Due to highly reactive nature of conjugated double bonds, carotenoids can form various metabolic products through enzymatic and non-enzymatic chemical transformations [80]. The reactive oxygen species produced as a result of different metabolic and physiological processes of the body, react with various site of carotenoid to form various metabolic products and are eliminated from the body after modification. Extensive studies on *in vitro* metabolic products of carotenoids have been performed to show their effect as antioxidants during oxidative stresses. The true picture about the definite processes involved in enzymatic reactions of carotenoids is yet not clear due to involvement of various reaction sites and still needs lot of research to be done in this regard [81]. To explain in further detail, the metabolism of carotenoids has been discussed under two major areas according to the function of carotenoids i.e.

11.5.3.1. Pro-vitamin A carotenoids
11.5.3.2. Non pro-vitamin A carotenoids

11.5.3.1 Pro-Vitamin A Carotenoids

Carotenoids can be converted to apocarotenoids such as retinoids, which include all natural and synthetic derivatives of vitamin A (all-*trans*-retinol) [82]. The studies have shown that among cellular metabolism of provitamin A carotenoids (α-carotene, β-carotene and β-cryptoxanthin) major focus has been on β-carotene. Knowing the metabolic importance of retinoids, β-carotene was first to be studied for conversion into retinoid using animal tissues [83]. For the first time John Glover in 1960 suggested two possible alternatives for formation of retinal from β-carotene that were named as central fission and asymmetric cleavage (eccentric cleavage). Additionally, it was revealed by two independent research groups [84, 85] that homogenates from small intestine of rat can cleave β-carotene enzymatically at the central 15–15′-carbon double bond. This process yields two aldehyde molecules of vitamin A i.e. retinal and retinaldehyde. It was also suggested that provitamin A carotenoids may also be enzymatically oxidized at the central double bond to form two molecules of retinal. This pathway of oxidation process was observed in intestinal epithelium and is catalyzed by 15,15-dioxygenase [86]. Moreover, retinal is reduced to form retinol that then converts to retinyl esters that are transported to the liver via chylomicron (Described later in this section).

The enzyme required for the purpose (earlier called as dioxygenase, then renamed as monooxygenase) is present in several tissues except in intestine and is involved in production of vitamin A from carotenoids stores present in the peripheral tissues. Besides the central cleavage pathway described earlier, eccentric cleavage yields p-apocarotenoids that has also been shown in human intestinal tissue [9]. The discussion about the central and eccentric cleavage of β-carotene continued till the cloning and characterization of enzymes required for the purpose [87]. These enzymes are termed as:

- β-carotene-15,15′-monooxygenase, (BCMO1 or BCO1) responsible for central cleavage
- β-carotene-9′10′-monooxygenase, (BCMO2 or BCO2), responsible for eccentric cleavage

β-Carotene are mostly present as all-trans β-carotene in nature which upon central cleavage with β-carotene-15,15′-monooxygenase, (BCMO1) produces 2 molecules of retinaldehyde [88]. This retinaldehyde can be oxidized to all trans-retinoic acid (active form of vitamin A) with the help of enzyme retinaldehyde dehydrogenase [89]. Retinoic acid is involved with transcription regulation (as retinoic acid receptor RAR) and has its predominant role for several hundred target genes. If the production of retinoic acid outstrips a specific level, it may go through oxidative degeneration to 4-hydroxy retinoic acid (relatively less active form) with the help of enzyme belonging to family of cytochrome P450 [90]. On the other hand, retinaldehyde with the help of retinoldehydrogenase enzyme may be converted into retinol (alcohol form of vitamin A) [91]. Furthermore, β-carotene can also produce 1 molecule of retinaldehyde through eccentric cleavage into apocarotenals with the help

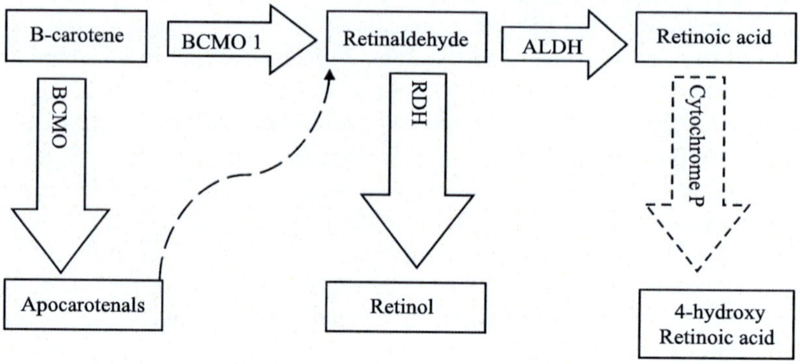

Fig. 11.4 Conversion of β-carotene through BCMO 1 and BCMO 2

of enzyme β-carotene-9′10′-monooxygenase (BCMO2) [92] as has been described through Fig. 11.4.

11.5.3.2 Non-Provitamin A Carotenoids

As discussed previously, BCO1 is efficiently involved with the cleavage of pro-vitamin A carotenoids as compared to non–provitamin A carotenoids, the enzymatic activity of BCO2 is more shown toward non–provitamin A carotenoidsas substrate, including lycopene, lutein, and zeaxanthin [93]. As evidence, it is observed that mutation in BCO2 gene causes the accumulation of xanthoplyll as lutein and zeaxanthin in adipose tissues and skin [94]. It is also proposed through many animal studies that BCO2 is an enzyme to cleave xanthophyll carotenoid that can split carotenoids eccentrically at the 9–10 or 9′-10′carbon–carbon double bonds into C13- and C27-apocarotenoids and further into C14-dialdehyde [95]. BCO2 is present in mitochondria and is highly expressed in liver and testis whereas it shows a lower activity in brain, kidneys, heart, prostrate, spleen, lungs, stomach, intestine and colon [96]. The studies demonstrate that the recombinant ferret BCO2 is more involved with the eccentric cleavage of cis-lycopene isomers than all-trans lycopene, at the 9 and10 double bond. The exact reason behind this preference is unknown however, it may be postulated that the chemical structure of cis-lycopene could mimic the ring structure of the β-carotene molecule and thus may form the substrate-enzyme complex [97].

Further studies on mice have confirmed that fucoxanthin and lutein are transformed to keto-carotenoids through oxidation with resulting products as fucoxanthinol and amarouciaxanthin A in mammals [98]. In intestinal tract fucoxanthin is hydrolyzed to fucoxanthinol in which is further converted oxidatively to amarouciaxanthin A. Such trans formation of fucoxanthinol into amarouciaxanthin A has also been observed in human hepatoma HepG2 cells [99]. Additionally, the metabolic oxidative conversion of carotenoids is mediated in liver of mammals and

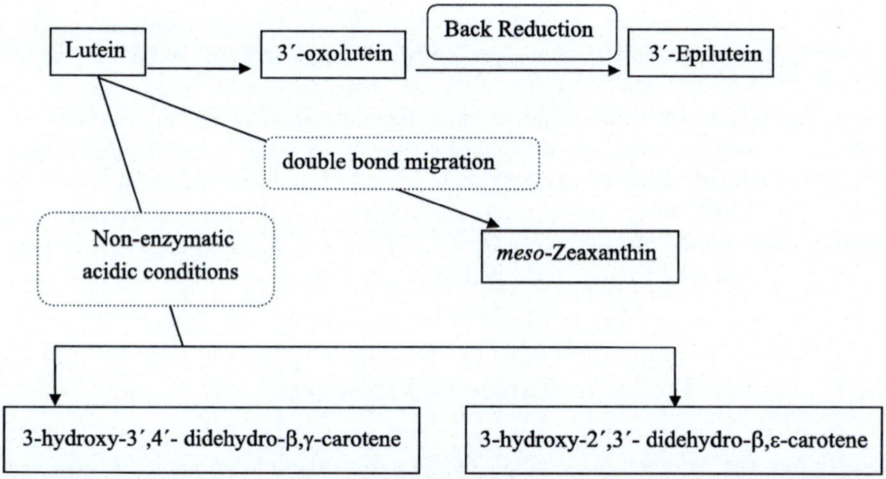

Fig. 11.5 Metabolites of lutein identified in human subjects. (Derived from [98, 101–103])

requires NAD^+ as a cofactor (most probably through NAD-dependent dehydrogenases). It also demonstrates the metabolic conversion xanthophyll (at 3-hydroxyl end group) at enzyme level reaction in various animals [100].

In accordance with the animal studies about lutein (in mice), other xanthophylls also showed metabolic oxidation even in human tissues where various metabolites of lutein were identified. Some of the examples of these metabolites are given in Fig. 11.5 along with their precursor.

11.6 Regulation of Carotenoid Metabolism

The regulation of carotenoids bioavailability and metabolism is directly related with regulation and physiological functions of the carotenoid oxygenases. Recent research has demonstrated that a negative feedback mechanism is present in intestine to control carotenoids absorption and β-carotene conversion. Takitani et al. [105] described that inclusion of retinyl acetate, 9-*cis*-retinoic acid and all-*trans*-retinoic acid in diet resulted in reduced activity of BCMO1 enzyme in intestinal tissues (down regulation). Contrarily, diet deficient in retinoids resulted in greater expression of BCMO1 gene (Up-regulation). This may be caused by regulation of intestine specific transcription factor i.e. intestine specific homebox (ISX) as it has been proved that BCMO1 and SR-B1 expression in intestine is enhanced by ISX deficient mice [106]. Furthermore, Zaripheh et al. [107] described that BCMO1 gene expression can also be regulated by other carotenoids, for instance dietary intake of lycopene can down-regulate expression of BCMO1 in kidneys and adrenal glands of rats.

Deficiency of the genes related to enzyme BCMO2 (Bcmo2) results in alteration in carotenoids metabolism and specificity of substrate for BCMO2 enzyme (Carotene and xanthophyll). While at tissue level, lack of BCMO2 activity in a tissue-specific manner caused no adverse developmental or health consequences and did not affect serum and liver retinoid levels. Bcmo2 deficient mice display altered carotenoid homeostasis and carotenoids accumulate excessively in several tissues, including blood, heart, liver, and adipose [107]. Little is known about Bcmo2 gene regulation. Luvizotto et al. [108] reported that contrary to Bcmo1, Bcmo2 expression is not much affected by retinoid status.

11.7 Future Trends in Carotenoid Research

Insight into the carotenoids metabolism has opened new horizons to be explored. There is need for deep and more elaborative studies into the genes regulation of the enzymes related to carotenoids metabolism. More thoughtful and thorough research is required to explore the role of receptors or involvement of hormones in regulation of carotenoids uptake, physiological functions and metabolism. Furthermore, the uptake, accumulation and metabolism at cellular level need to be addressed at molecular point.

11.8 Conclusion

On the basis of data reviewed from different research papers about carotenoids metabolism, it can be concluded that carotenoids bioaccessibility is dependent upon type of food, its processing, intake, digestion, absorptions and other host related factors. Furthermore, different enzymes and receptors regulate the metabolism of carotenoids. Enough information is available about mechanism of action of carotenoids as vitamin A and antioxidants but needs further exploration at molecular and genetic level.

References

1. Isler O (1979) History and industrial applications of carotenoids and vitamin A. Pure Appl Chem 51:447–461
2. Rodroguez-Amaya DB (2016) Food carotenoids. IFT Press, Wiley Blackwell, Chichester
3. During A, Harrison EH (2004) Intestinal absorption and metabolism of carotenoids: insights from cell culture. Arch Biochem Biophys 430(1):77–88
4. Arscott SA (2013) Food sources of carotenoids. In: Tanumihardjo SA (ed) Carotenoids and human health. Humana Press/Springer, London/Totowa

5. Faulks RM, Southon S (2005) Challenges to understanding and measuring carotenoid bioavailability. Biochim Biophys Acta 1740(2):95–100
6. van Bennekum AM, Fisher EA, Blaner WS, Harrison EH (2000) Hydrolysis of retinyl esters by pancreatic triglyceride lipase. Biochemistry 39(16):4900–4906
7. During A, Doraiswamy S, Harrison EH (2008) Xanthophylls are preferentially taken up compared with beta-carotene by retinal cells via a SRBI-dependent mechanism. J Lipid Res 49(8):1715–1724
8. Nagao A (2009) Absorption and function of dietary carotenoids. In: Yoshikawa T (ed) Food factors for health promotion, Forum nutrition. S. Karger, Basel, pp 55–63
9. Ferruzzi MG, Lumpkin JL, Schwartz SJ, Failla M (2006) Digestive stability, micellarization, and uptake of β-carotene isomers by Caco-2 human intestinal cells. J Agric Food Chem 54:2780–2785
10. Canene-Adams K, Erdman JW (2009) Absorption, transport, distribution in tissues and bioavailability. In: Britton G, Pfander H, Liaaen-Jensen S (eds) Carotenoids, vol 5. Birkhäuser, Basel
11. Gartner C, Stahl W, Sies H (1997) Lycopene is more bioavailable from tomato paste than from fresh tomatoes. Am J Clin Nutr 66:116
12. Yonekura L, Nagao A (2007) Intestinal absorption of dietary carotenoids. Mol Nutr Food Res 51(1):107–115
13. Karger B, von Lintig J (2010) Colors with functions: elucidating the biochemical and molecular basis of carotenoid metabolism. Annu Rev Nutr 30:35–56
14. Borel P, Grolier P, Armand M, Partier A, Lafont H, Lairon D et al (1996) Carotenoids in biological emulsions: solubility, surface-to-core distribution, and release from lipid droplets. J Lipid Res 37:250–261
15. Scita G (1992) Stability of β-carotene under different laboratory conditions. In: Packer L (ed) Methods in enzymology, carotenoids part A, chemistry, separation, quantitaion, and antioxidation. Academic, San Diego
16. Borel P (2012) Genetic variations involved in inter individual variability in carotenoid status. Mol Nutr Food Res 56:175–185
17. Deming DM, Erdman JW (1999) Mammalian carotenoid absorption and metabolism. Pure Appl Chem 71(12):2213–2223
18. Clevidence B, Paetau I, Smith JC (2000) Bioavailability of carotenoids from vegetables. HortScience 35(4):585–588
19. Yeum KJ, Russell RM (2002) Carotenoid bioavailability and bioconversion. Annu Rev Nutr 22:483–504
20. Boileau TWM, Moore AC, Erdman JW Jr (1999) Antioxidant status. In: Pappas AM (ed) Diet, nutrition, and health. CRC Press, New York, pp 133–158
21. Schweiggert RM, Carle R (2017) Carotenoid deposition in plant and animal foods and its impact on bioavailability. Crit Rev Food Sci Nutr 57(9):1807–1830
22. van het Hof KH, West CE, Weststrate JA, Hautvast J, J GA (2000) Dietary factors that affect the bioavailability of carotenoids. J Nutr 130(3):503–506
23. Nisar N, Li L, Shan L, ChiKhin N, Pogson BJ (2015) Carotenoid metabolism in plants. Mol Plant 8(1):68–82
24. de Pee S, West CE, Permaesih D, Martuti S, Muhilal, Hautvast JGAJ (1998a) Orange fruit is more effective than are dark-green, leafy vegetables in increasing serum concentrations of retinol and betacarotene in schoolchildren in Indonesia. Am J Clin Nutr 68:1058–1067
25. Castenmiller JJM, West CE, Linssen JPH, van het Hof KH, Voragen AGJ (1999) The food matrix of spinach is a limiting factor in determining the bioavailability of b-carotene and to a lesser extent of lutein in humans. J Nutr 129(2):349–355
26. Frey-Wyssling A, Schwegler F (1965) Ultrastructure of the chromoplasts in the carrot root. J Ultrastruct Res 13(5–6):543–559
27. Tracewell CA, Vrettos JS, Bautista JA, Frank HA, Brudvig GW (2001) Carotenoid photooxidation in photosystem II. Arch Biochem Biophys 385(1):61–69

28. Zhou JR, Gugger ET, Erdman JW Jr (1994) Isolation and partial characterization of an 18 kDa carotenoid-protein complex from carrot roots. J Agric Food Chem 42(11):2386–2390
29. Ben-Shaul Y, Treffry T, Klein S (1968) Fine structure studies of carotene body development. J Microsc (Paris) 7:265–274
30. Desobry SA, Netto FM, Labuza TP (1998) Preservation of β-carotene from carrots. Crit Rev Food Sci Nutr 38:381–396
31. Soukoulis C, Bohn T (2018) A comprehensive overview on micro and nano technological advances for enhancing the chemical stability and bioavailability of carotenoids. Crit Rev Food Sci Nutr 58(1):1–36
32. Palmero P, Lemmens L, Hendrickx M, Van Loey A (2014) Role of carotenoid type on the effect of thermal processing on bioaccessibility. Food Chem 157:275–282
33. Anese M, Mirolo G, Beraldo P, Lippe G (2013) Effect of ultrasound treatments of tomato pulp on microstructure and lycopene in vitro bioaccessibility. Food Chem 136(2):458–463
34. Ribaya-Mercado JD (2002) Influence of dietary fat on beta-carotene absorption and bioconversion into vitamin A. Nutr Rev 60(4):104–110
35. Jalal F, Nesheim MC, Agus Z, Sanjur D, Habicht JP (1998) Serum retinol concentrations in children are affected by food sources of beta-carotene, fat intake, and anthelmintic drug treatment. Am J Clin Nutr 68(3):623–629
36. Brown MJ, Ferruzzi MG, Nguyen ML, Coope DAR, Eldridge AL, Schwartz SJ et al (2004) Carotenoid bioavailability is higher from salads ingested with full-fat than with fat-reduced salad dressings as measured with electrochemical detection. Am J Clin Nutr 80:396–403
37. Odeberg MJ, Lignell A, Pettersson A, Höglund P (2003) Oral bioavailability of the antioxidant astaxanthin in humans is enhanced by incorporation of lipid based formulations. Eur J Pharm Sci 19(4):299–304
38. Deming DM, Baker DH, Erdman JW Jr (2002) The relative vitamin A value of 9-*cis*-β-carotene is less and that of 13-*cis*-β-carotene may be greater than the accepted 50% that of all-*trans*-β-carotene in gerbils. J Nutr 132:2709–2712
39. Priyadarshani MB (2017) A review on factors influencing bioaccessibility and bioefficacy of carotenoids. Crit Rev Food Sci Nutr 57(8):1710–1717
40. Unlu NZ, Boh T, Clinton SK, Schwartz SJ (2005) Carotenoid absorption from salad and salsa by humans is enhanced by the addition of avocado or avocado oil. J Nutr 135(3):431–436
41. Hollander D, Ruble PEJ (1978) Beta-carotene intestinal absorption: bile, fatty acid, pH, and flow rate effects on transport. Am J Phys 235(6):E686–E691
42. Fuller RC (2001) Patterns in male breeding behaviors in the blue fin killifish, *Lucania goodei*: a field study (cyprinodontiformes: Fundulidae). Copeia 3:823–828
43. Rock CL, Swendseid ME (1992) Plasma β-carotene response in humans after meals supplemented with dietary pectin. Am J Clin Nutr 55:96–99
44. Stice CP, Wang X (2013) Carotenoids and alcoholic liver disease. Hepatobiliary Surg Nutr 2(5):244–247
45. Rock CL (1997) Carotenoids: biology and treatment. Pharmacol Ther 75(3):185–197
46. Forman MR, Beecher GR, Lanza E, Reichman ME, Graubard BI, Campbell WS et al (1995) Effect of alcohol consumption on plasma carotenoid concentrations in premenopausal women: a controlled dietary study. Am J Clin Nutr 62(1):131–135
47. Tyssandier V, Choubert G, Grolier P, Borel P (2002) Carotenoids, mostly the xanthophylls, exchange between plasma lipoproteins. Int J Vitam Nutr Res 72(5):300–308
48. Chitchumroonchokchai C, Failla ML (2017) Bioaccessibility and intestinal cell uptake of astaxanthin from salmon and commercial supplements. Food Res Int 99(2):936–943
49. Chung HY, Ferreira AL, Epstein S, Paiva SA, Castaneda-Sceppa C, Johnson EJ (2009) Site-specific concentrations of carotenoids in adipose tissue: relations with dietary and serum carotenoid concentrations in healthy adults. Am J Clin Nutr 90(3):533–539
50. Desmarchelier C, Borel P (2017) Overview of carotenoid bioavailability determinants: from dietary factors to host genetic variations. Trends Food Sci Technol 69:270–280

51. Cardinault N, Tyssandier V, Grolier P, Winklhofer-Roob BM, Ribalta J, Bouteloup-Demange C et al (2003) Comparison of the post prandial chylomicron carotenoid responses in young and older subjects. Eur J Nutr 42:315–323
52. Schupp C, Olano-Martin E, Gerth C, Morrissey BM, Cross CE, Werner JS (2004) Lutein, zeaxanthin, macular pigment, and visual function in adult cystic fibrosis patients. Am J Clin Nutr 79:1049–1052
53. Ward MS, Zhao DY, Bernstein PS (2008) Macular and serum carotenoid concentrations in patients with malabsorption syndromes. J Ocul Biol Dis Inform 1:12–18
54. Smidt CR, Bruke DS (2004) Nutritional significance and measurement of carotenoids. Curr Top Nutraceutical Res 2(2):79–91
55. Choi MY, Romer AI, Hu M, Lepourcelet M, Mechoor A, Yasilaltay A et al (2006) A dynamic expression survey identifies transcription factors relevant in mouse digestive tract development. Development 133(20):4119–4129
56. van Bennekum AM, Kako Y, Weinstock PH, Harrison EH, Deckelbaum RJ, Goldberg IJ et al (1999) Lipoprotein lipase expression level influences tissue clearance of chylomicron retinyl ester. J Lipid Res 40(3):565–574
57. Blaner WS, Obunike JC, Kurlandsky SB, al-Haideri M, Piantedosi R, Deckelbaum RJ et al (1994) Lipoprotein lipase hydrolysis of retinyl ester. Possible implications for retinoid uptake by cells. J Biol Chem 269(24):16559–16565
58. Cooper AD (1997) Hepatic uptake of chylomicron remnants. J Lipid Res 38(11):2173–2192
59. Krinsky NI, Johnson EJ (2005) Carotenoid actions and their relation to health and disease. Mol Asp Med 26(6):459–516
60. Olson JA (1994) Absorption, transport and metabolism of carotenoids in human. Pure Appl Chem 66(5):1011–1016
61. Moussa M, Gouranton E, Gleize B, Yazidi CE, Niot I, Besnard P et al (2011) CD36 is involved in lycopene and lutein uptake by adipocytes and adipose tissue cultures. Mol Nutr Food Res 55(4):578–584
62. Landrum JT, Bone RA (2001) Lutein, zeaxanthin, and the macular pigment. Arch Biochem Biophys 385(1):28–40
63. Krinsky NI, Landrum JT, Bone RA (2003) Biologic mechanisms of the protective role of lutein and zeaxanthin in the eye. Annu Rev Nutr 23:171–201
64. Li B, Vachali P, Bernstein PS (2010) Human ocular carotenoid-binding proteins. Photochem Photobiol Sci 9(11):1418–1425
65. Bhosale P, Larson AJ, Frederick JM, Southwick K, Thulin CD, Bernstein PS (2004) Identification and characterization of a Pi isoform of glutathione S-transferase (GSTP1) as a zeaxanthin-binding protein in the macula of the human eye. J Biol Chem 279(47):49447–49454
66. Li B, Vachali P, Frederick JM, Bernstein PS (2011) Identification of StARD3 as a lutein-binding protein in the macula of the primate retina. Biochemistry 50(13):2541–2549
67. Kelly M, von Lintig J (2015) STRA6: role in cellular retinol uptake and efflux. Hepatobiliary Surg Nutr 4(4):229–242
68. Shmarakov I, Fleshman MK, D'Ambrosio DN, Piantedosi R, Riedl KM, Schwartz SJ et al (2010) Hepatic stellate cells are an important cellular site for beta carotene conversion to retinoid. Arch Biochem Biophys 504(1):3–10
69. Teodoro AJ, Perrone D, Martucci RB, Borojevic R (2009) Lycopene isomerisation and storage in an *in vitro* model ofmurine hepatic stellate cells. Eur J Nutr 48(5):261–268
70. Blaner WS, O'Byrne SM, Wongsiriroj N, Kluwe J, D'Ambrosio D, Jiang D et al (2009) Hepatic stellate cell lipid droplets: a specialized lipid droplet for retinoid storage. Biochem Biophys Acta 1791(6):467–473
71. Wang Y, Illingworth DR, Connor SL, Barton Duell P, Connor WE (2010) Competitive inhibition of carotenoid transport and tissue concentrations by high dose supplements of lutein, zeaxanthin and beta-carotene. Eur J Nutr 49(6):327–336
72. Borel P, de Edelenyi FS, Vincent-Baudry S, Malezet-Desmoulin C, Margotat A, Lyan B et al (2011) Genetic variants in BCMO1 and CD36 are associated with plasma lutein concentrations and macular pigment optical density in humans. Ann Med 43(1):47–59

73. Wassef L, Shete V, Costabile B, Rodas R, Quadro L (2015) High preformed vitamin A intake during pregnancy prevents embryonic accumulation of intact β-carotene from the maternal circulation in mice. J Nutr 145(7):1408–1414
74. Ramasamy I (2014) Recent advances in physiological lipoprotein metabolism. Clin Chem Lab Med 52(12):1695–1727
75. Shete V, Costabile BK, Kim Y, Quadro L (2016) Low-density lipoprotein receptor contributes to β-carotene uptake in the maternal liver. Nutrients 8(12):765
76. Antille C, Tran C, Sorg O, Saurat JH (2004) Topical beta-carotene is converted to retinyl esters in human skin ex vivo and mouse skin in vivo. Exp Dermatol 13(9):558–561
77. Lademann J, Meinke MC, Sterry W, Darvin ME (2011) Carotenoids in human skin. Exp Dermatol 20(5):377–382
78. Terao J, Minami Y, Bando N (2011) Singlet molecular oxygen-quenching activity of carotenoids: relevance to protection of the skin from photo aging. J Clin Biochem Nutr 48(1):57–62
79. Richelle M, Sabatier M, Steiling H, Williamson G (2006) Skin bioavailability of dietary vitamin E, carotenoids, polyphenols, vitamin C, zinc and selenium. Br J Nutr 96(2):227–238
80. Nagao A (2004) Oxidative conversion of carotenoids to retinoids and other products. J Nutr 134(1):237S–240S
81. Siems W, Salerno C, Crifo C, Sommerburg O, Wiswedel I (2009) Beta-carotene degradation products—formation, toxicity and prevention of toxicity. Forum Nutr 61:75–86
82. von Lintig J (2010) Colors with function: elucidating the biochemical and molecular basis of carotenoid metabolism. Annu Rev Nutr 30:35–56
83. Grune T, Lietz G, Palou A, Ross AC, Stahl W, Tang G et al (2010) Beta-carotene is an important vitamin A source for humans. J Nutr 140(12):2268S–2285S
84. Olson JA, Hayaishi O (1965) The enzymatic cleavage of beta-carotene into vitamin A by soluble enzymes of rat liver and intestine. Proc Natl Acad Sci U S A 54(5):1364–1370
85. Goodman DS, Huang HS (1965) Biosynthesis of vitamin A with rat intestinal enzymes. Science 149(3686):879–880
86. Nagao A, During A, Hoshino C, Terao J, Olson JA (1996) Stoichiometric conversion of all trans-beta-carotene to retinal by pig intestinal extract. Arch Biochem Biophys 328(1):57–63
87. Wyss A (2004) Carotene oxygenases: a new family of double bond cleavage enzymes. J Nutr 134(1):246S–250S
88. Shete V, Quadro L (2013) Mammalian metabolism of β-carotene: gaps in knowledge. Nutrients 5(12):4849–4868
89. Al Tanoury Z, Piskunov A, Rochette-Egly C (2013) Vitamin A and retinoid signaling: genomic and nongenomic effects. J Lipid Res 54(7):1761–1775
90. Abu-Abed S, Dolle P, Metzger D, Beckett B, Chambon P, Petkovich M (2001) The retinoic acid-metabolizing enzyme, CYP26A1, is essential for normal hindbrain patterning, vertebral identity, and development of posterior structures. Genes Dev 15(2):226–240
91. D'Ambrosio DN, Clugston RD, Blaner WS (2011) Vitamin A metabolism: an update. Nutrients 3(1):63–103
92. Pares X, Farres J, Kedishvili N, Duester G (2008) Medium- and short-chain dehydrogenase/reductase gene and protein families: medium-chain and short-chain dehydrogenases/reductases in retinoid metabolism. Cell Mol Life Sci 65(24):3936–3949
93. Mein JR, Dolnikowski GG, Ernst H, Russell RM, Wang XD (2011) Enzymatic formation of apo-carotenoids from the xanthophyll carotenoids lutein, zeaxanthin and beta-cryptoxanthin by ferret carotene-9′, 10′-monooxygenase. Arch Biochem Biophys 506(1):109–121
94. Vage DI, Boman IA (2010) A nonsense mutation in the beta-carotene oxygenase 2 (BCO2) gene is tightly associated with accumulation of carotenoids in adipose tissue in sheep (*Ovis aries*). BMC Genet 11:10
95. Li B, Vachalia PP, Gorusupudia A, Shena Z, Sharifzadeha H, Bescha BM et al (2014) Inactivity of human β, β-carotene-9′, 10′-dioxygenase (BCO2) underlies retinal accumulation of the human macular carotenoid pigment. PNAS 111(28):10173–10178

96. Hu KQ, Liu C, Ernst H, Krinsky NI, Russell RM, Wang XD (2006) The biochemical characterization of ferret carotene-9′, 10′-monooxygenase catalyzing cleavage of carotenoids in vitro and in vivo. J Biol Chem 281(28):19327–19338
97. Liu C, Lian F, Smith DE, Russell RM, Wang XD (2003) Lycopene supplementation inhibits lung squamous metaplasia and induces apoptosis via up-regulating insulin-like growth factor-binding protein 3 in cigarette smoke-exposed ferrets. Cancer Res 63(12):3138–3144
98. Kotake-Nara E, Nagao A (2011) Absorption and metabolism of xanthophylls. Mar Drugs 9(6):1024–1037
99. Asai A, Sugawara T, Ono H, Nagao A (2004) Biotransformation of fucoxanthinol into amarouciaxanthin A in mice and HepG2 cells. Drug Metab Dispos 32(2):205–211
100. Yonekura L, Kobayashi M, Terasaki M, Nagao A (2010) Keto-carotenoids are the major metabolites of dietary lutein and fucoxanthin in mouse tissues. J Nutr 140(10):1824–1831
101. Khachik F, Spangler CJ, Smith JC Jr, Canfield LM, Steck A, Pfander H (1997) Identification, quantification, and relative concentrations of carotenoids and their metabolites in human milk and serum. Ann Chem 69(10):1873–1881
102. Khachik F, de Moura FF, Zhao DY, Aebischer CP, Bernstein PS (2002) Transformations of selected carotenoids in plasma, liver, and ocular tissues of humans and in nonprimate animal models. Invest Ophthalmol Vis Sci 43(11):3383–3392
103. Khachik F (2003) An efficient conversion of (3*R*,3′*R*,6′*R*)-lutein to (3*R*,3′*S*,6′*R*)-lutein (3′-epilutein) and (3*R*,3′*R*)-zeaxanthin. J Nat Prod 66(1):67–72
104. Eroglu A, Harrison EH (2013) Carotenoid metabolism in mammals, including man: formation, occurrence, and function of apocarotenoids. J Lipid Res 54(7):1719–1730
105. Takitani K, Zhu CL, Inoue A, Tamai H (2006) Molecular cloning of the rat beta-carotene 15,15 ¢ -monooxygenase gene and its regulation by retinoic acid. Eur J Nutr 45(6):320–326
106. Seino Y, Miki T, Kiyonari H, Abe T, Fujimoto W, Kimura K et al (2008) Isx participates in the maintenance of vitamin A metabolism by regulation of beta-carotene 15,15′-monooxygenase (Bcmo1) expression. J Biol Chem 283(8):4905–4911
107. Zaripheh S, Nara TY, Nakamura MT, Erdman JW Jr (2006) Dietary lycopene downregulates carotenoid 15,15 ¢ -monooxygenase and PPAR-gamma in selected rat tissues. J Nutr 136(4):932–938
108. Luvizotto RA, Nascimento AF, Veeramachaneni S, Liu C, Wang XD (2010) Chronic alcohol intake upregulates hepatic expression of carotenoid cleavage enzymes and PPAR in rats. J Nutr 140(10):1808–1814
109. Schweiggert RM, Kopec RE, Villalobos-Gutierrez MG, Högel J, Quesada S, Esquivel P et al (2014) Carotenoids are more bioavailable from papaya than from tomato and carrot in humans: a randomised cross-over study. Br J Nutr 111:490–498
110. Salvia-Trujillo L, Qian C, Martín-Belloso O, McClements DJ (2013) Influence of particle size on lipid digestion and β-carotene bioaccessibility in emulsions and nanoemulsions. Food Chem 141:1472–1480
111. Lemmens L, Colle I, Van Buggenhout S, Palmero P, Van Loey A, Hendrickx M (2014) Carotenoid bioaccessibility in fruit- and vegetable-based food products as affected by product (micro) structural characteristics and the presence of lipids: a review. Trends Food Sci Technol 38:125–135
112. Palmero P, Colle I, Lemmens L, Panozzo A, Nguyen TTM, Hendrickx M et al (2016) Enzymatic cell wall degradation of high-pressure-homogenized tomato puree and its effect on lycopene bioaccessibility. J Sci Food Agric 96(1):254–261
113. Sriwichai W, Berger J, Picq C, Avallone S (2016) Determining factors of lipophilic micronutrient bioaccessibility in several leafy vegetables. J Agric Food Chem 64:1695–1701
114. Verrijssen TAJ, Christiaens S, Verkempinck SHE, Boeve J, Grauwet T, Van Loey AM et al (2016) In vitro β-carotene bioaccessibility and lipid digestion in emulsions: influence of pectin type and degree of methyl-esterification. J Food Sci 81:C2327–C2336

115. Zhang R, Zhang Z, Zou L, Xiao H, Zhang G, Decker EA et al (2015) Enhancing nutraceutical bioavailability from raw and cooked vegetables using excipient emulsions: influence of lipid type on carotenoid bioaccessibility from carrots. J Agric Food Chem 63:10508–10517
116. Liu X, Bi J, Xiao H, McClements DJ (2015) Increasing carotenoid bioaccessibility from yellow peppers using excipient emulsions: impact of lipid type and thermal processing. J Agric Food Chem 63:8534–8543
117. Nimalaratne C, Savard P, Gauthier SF, Schieber A, Wu J (2015) Bioaccessibility and digestive stability of carotenoids in cooked eggs studied using a dynamic in vitro gastrointestinal model. J Agric Food Chem 63:2956–2962
118. Chacón-Ordóñez T, Esquivel P, Jiménez VM, Carle R, Schweiggert RM (2016) Deposition form and bioaccessibility of keto-carotenoids from mamey sapote (*Pouteria sapota*), red bell pepper (*Capsicum annuum*), and sockeye salmon (*Oncorhynchus nerka*) filet. J Agric Food Chem 64:1989–1998
119. Pugliese A, O'Callaghan Y, Tundis R, Galvin K, Menichini F, O'Brien N et al (2014) In vitro assessment of the bioaccessibility of carotenoids from sun-driedchilli peppers. Plant Food Hum Nutr 69(1):8–17
120. Aschoff JK, Kaufmann S, Kalkan O, Neidhart S, Carle R, Schweiggert RM (2015) In vitro bioaccessibility of carotenoids, flavonoids, and vitamin C from differently processed oranges and orange juices *Citrus sinensis* (L.) Osbeck. J Agric Food Chem 63:578–587
121. Xavier AAO, Mercadante AZ, Garrido-Fernández J, Pérez-Gálvez A (2014) Fat content affects bioaccessibility and efficiency of enzymatic hydrolysis of lutein esters added to milk and yogurt. Food Res Int 65:171–176
122. McClements DJ (2015) Nanoscale nutrient delivery systems for food applications: improving bioactive dispersibility, stability, and bioavailability. J Food Sci 80(7):N1602–N1611
123. Failla ML, Chitchumronchokchai C, Ferruzzi MG, Goltz SR, Campbell WW (2014) Unsaturated fatty acids promote bioaccessibility and basolateral secretion of carotenoids and α-tocopherol by Caco-2 cells. Food Funct 5:1101–1112
124. Nidhi B, Mamatha BS, Baskaran V (2014) Olive oil improves the intestinal absorption and bioavailability of lutein in lutein-deficient mice. Eur J Nutr 53(1):117–126
125. Goltz SR, Campbell WW, Chitchumroonchokchai C, Failla ML, Ferruzzi MG (2012) Meal triacylglycerol profile modulates postprandial absorption of carotenoids in humans. Mol Nutr Food Res 56(6):866–877
126. Reboul E, Thap S, Tourniaire F, André M, Juhel C, Morange S et al (2007) Differential effect of dietary antioxidant classes (carotenoids, polyphenols, vitamins C and E) on lutein absorption. Br J Nutr 97(3):440–446
127. Goncalves A, Roi S, Nowicki M, Dhaussy A, Huertas A, Amiot MJ et al (2015) Fat-soluble vitamin intestinal absorption: absorption sites in the intestine and interactions for absorption. Food Chem 172:155–160

Chapter 12
Carotenoids as Antioxidants

Saikat Dewanjee, Niloy Bhattacharjee, Pratik Chakraborty, and Simanta Bhattacharjee

12.1 Introduction

Carotenoids are in-charge of red, orange and yellow shading of plant leaves, fruits, and flowers. These are also responsible for the shades of different animal species, such as birds, insects, fish, and crustaceans. Carotenoids are principally produced by plants, animals, and microbes for fulfilling their vital physiological activities [1]. In photosynthetic living beings, carotenoids are fundamental for photosynthesis and photo-protection; while in non-photosynthetic life forms, these take an interest in lightening photo-oxidative harm. Considering the properties, carotenoids find different mechanical applications as colours, and due to their different beneficial impacts for wellbeing, these have subsequently been utilised by the pharmaceutical and nutraceutical companies. In 2004, a report revealed that the pharmaceutical business gained a yearly production of more than 100 million tons of carotenoids [2].

More than 700 carotenoids have been reported [3], around 40 carotenoids are ingested by human through diet. However, fewer than 20 including both polar xanthophylls and nonpolar carotenes with both cyclic and acyclic structures have been found in plasma and tissues [4]. Carotenoids in the leaves are uncovered before the leaves pass on and tumble to the ground. Plants seem to create carotenoids to shield their stems and leaves from the vitality of the sun. Bright-UV wavelengths can create free radicals by the expulsion of an electron from an atom, which have the potential to harm living cells. As antioxidants, carotenoids restrain free radical harm by giving electrons to quench, or kill, the oxidant radicals.

Various free radicals are generating within the body during metabolism of food items, xenobiotics whether in normal conditions or in disease state or exposure to radiations. These free radicals and reactive oxygen species are neutralized by

S. Dewanjee (✉) · N. Bhattacharjee · P. Chakraborty · S. Bhattacharjee
Advanced Pharmacognosy Research Laboratory, Department of Pharmaceutical Technology, Jadavpur University, Kolkata, West Bengal, India
e-mail: saikat.dewanjee@jadavpuruniversity.in

© Springer Nature Switzerland AG 2021
M. Zia-Ul-Haq et al. (eds.), *Carotenoids: Structure and Function in the Human Body*, https://doi.org/10.1007/978-3-030-46459-2_12

carotenoids when consumed. These free radicals are produced tens of thousands in 1 s. The free radical generation is a chain process within the system that is mediated by the transfer of electrons. These free radicals are quenched by the body's own defence system; if not properly quenched the DNA or other body proteins will be damaged. The body antioxidant system consists of several enzymes like superoxide dismutase (SOD), catalase (CAT), glutathione peroxidase (GPx) etc. that diminish the free radicals. Apart from these enzyme generations various dietary supplements play its role in free radical nullification. Activities of this antioxidant enzymes and carotenoids are described in Fig. 12.1.

Chemically, carotenoids consist of 8 isoprene units or 40 carbon atoms. They may be cyclic or acylic but majority of the carotenoids are acyclic. The general molecular formula is $C_{40}H_{56}$. The central conjugated polyene system is responsible for light absorption from the visible range of the electromagnetic spectrum (400–500 nm). There are variations in the central structure of carotenoids, which lead to changes in their physical and chemical properties. These variations include unsaturation, cyclization at one or both ends, and the addition of oxygen-containing functional groups, such as hydroxy, epoxy, oxo groups etc. [5]. carotenoid system containing oxygen functional groups are termed as xanthophylls [6]. These oxygens containing functional groups affects their polarity and solubility, therefore, different types of separation techniques are applied for their separation.

Carotenoids play a major role in plant, algae and some microorganism by harvesting light energy during the photosynthesis. Usually, xanthophylls are more important in light harvesting and are essential for the organisms living in the environments with less availability of light. Xanthophylls are located in specific pigment-protein complexes in the cell membrane [7]. Carotenoinds also prevent the

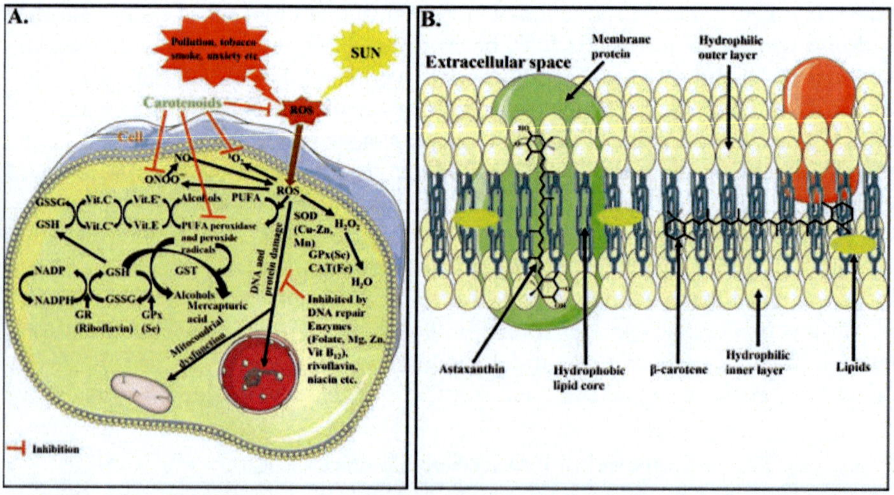

Fig. 12.1 Schamatic diagram A indicates normal antioxidant defence mechanism of human body and involvement of carotenoids on it and diagram B indicates the arrangement of two types of carotenoids (xanthophylls like astaxanthin and carotene like β-carotene) in cell membrane

formation of reactive free radical production due to high light energy and prevent plant cells from oxidative damage [8]. Vibrant colour of carotenoids also attracts insects and helps in the pollination of the plant species [9]. Carotenoids are present in the animals as carotenoid-protein complexes termed as carotenoproteins [10]. Denaturation of these complexes release free carotenoids [10]. Carotenoids are essential for the good health but unfortunately most of the animals cannot synthesize these of their own and essentially depend upon exogenous carotenoids [11].

In human being, carotenoids are essential for good health and fitness and absence of these may lead to chronic health problems. Until now only 20 of carotenoids have been found within the human body and 3 of these are well abundant in functional foods [12]. It has been observed that carotenes and xanthophylls possess several beneficial functions in human body. α, β-carotene and β-cryptoxanthin are pro-vitamin A carotenoids that act as precursor of vitamin A, which is useful for good ocular health and its deficiency leads to night blindness. Not all carotenoids meet the structural requirements to produce vitamin A. Carotenoids which contain same type ring free from oxygen and has a polyene chain with at least 11 carbon atoms have been observed to be potential precursors of vitamin A. Considering this structural requirement, it has been observed that only 10% of the existing carotenoids exhibit the activity of pro-vitamin A [13]. Zeaxanthin and lutein were also observed as two major components of the yellow spot (fovea centralis) in retina, which help in vision and their deficiency may cause complete blindness [14]. Furthermore, all carotenoids have powerful antioxidant property and also help in disease prevention like cancer, diabetes, immune-deficiency, cardiovascular diseases, and cataract formation. It has been observed that astaxanthin is stronger antioxidant than any of the known antioxidants [15].

It has been already revealed that carotenoids are the good quenchers of singlet oxygen produced during photosynthesis in plant species [16]. Carotenoids have been revealed as most powerful singlet oxygen quenchers and some of these were found to be more potent than that of α-tocopherol and other biological antioxidants [17]. One molecule of β-carotene has the capability to quench up to 1000 singlet oxygen via irreversible reaction resulting the end of the oxidation [18]. Young and co-worker observed that carotenoids could inhibit the lipid peroxidation under some circumstances but in humans, the action of carotenoids is more complex [19].

The major focus of this chapter is to describe the importance of carotenoids as antioxidants and its action in removal of oxidative stress.

12.2 Importance of Carotenoids as Antioxidants

Although the antioxidant action of carotenoids provide a mechanism and firm base for other biological actions but still it has not been able to attract noteworthy intrigue and discussion than other biological activities. We will briefly describe the significant features of the antioxidant activity and capacity, which are different in concept but they are generally utilized without distinction [20]. The antioxidant activity is related to the in vitro oxidative procedures, while antioxidant capacity is related to the in vivo oxidative procedures and truly includes a biological activity.

12.2.1 Antioxidant Activity and Its Mechanism

Antioxidant activity is process of chemical reaction between antioxidant and free radicals. In the human body, a range of reactive oxygen species (ROS) are produced, like $_1O^2$ (singlet oxygen), $OH^•$ (hydroxyl radical), $O_2^{•-}$(superoxide radicals), and H_2O_2 (hydrogen peroxide) etc. [16]. The mechanisms and the rate of scavenging of free radicals by carotenoids in solution strongly depend upon the nature of the ROS [21]. Carotenoids, such as β-carotene are very reactive to peroxyl radicals but less reactive to $OH^•$. Burton and Ingold first observed the reaction between β-carotene and peroxyl radicals, where β-carotene undergoes an addition reaction and provided evidence that β-carotene can function as an effective chain-breaking antioxidant at low partial pressures of oxygen [22]. The amount of ROS within the cells is very critical. Carotenoids are very effective in absorbing energy from oxygen free radicals and make other free radicals non- functional [23].

Carotenoids, in their structures, possess highly reactive, electron enriched polyene chains. Conjugative double bonds present in the polyene chains are susceptible to electrophilic attack to form stabilized radicals (R·) [24]. Oxidation mediated heterolytic cleavage of carotenoid pigments leads to generation of carbonyl fragments and dialkyl peroxides resulting in consumption of two peroxyl radicals producing an antioxidant effect [25]. Carotenoids' lowest excited triplet state ($^3Car^*$) can convert the excess energy within oxygen free radical into heat, thus leading to deactivation of $^1O_2$. However, the possibility of damages from excited carotenoids can be ignored owing to their low energy and short lifespan [26]. Carotenoids are also able to quench $^1O_2$ chemically by oxidation or oxygenation [27]. Antioxidants react with radical species mainly via three different mechanisms, such as radical addition, radical cations production (electron transfer), and hydrogen atom abstraction [28–30] (Fig. 12.2). These mechanisms may occur simultaneously on the basis of the structural attributes of carotenoids, such as cyclic or acyclic terminals, presence of polar or nonpolar groups etc., the factors related to both the reaction environment, such as aqueous or lipidic, and the natures of radical species.

Carotenoids such as lycopene and β-carotene are lipophilic and embedded deep within the cell membrane [5]. In this case, the nonpolar environment would not support charge separation, thus electron transfer would not be the preferred mechanism. So, quite clearly electron transfer is possible only for hydrophillic carotenoids e.g. zeaxanthin, astaxanthin etc.

Carotenoid radical cations can absorb light at the near-infrared region (820–950 nm) [31]. It has also been observed that the radical adducts absorb energy at visible region (400–500 nm) and often absorb in parent ground-state absorption [32]. Radical species with high redox potential tends to produce carotenoid radical cation through electron transfer, whereas reduction of the carotenoid molecule takes

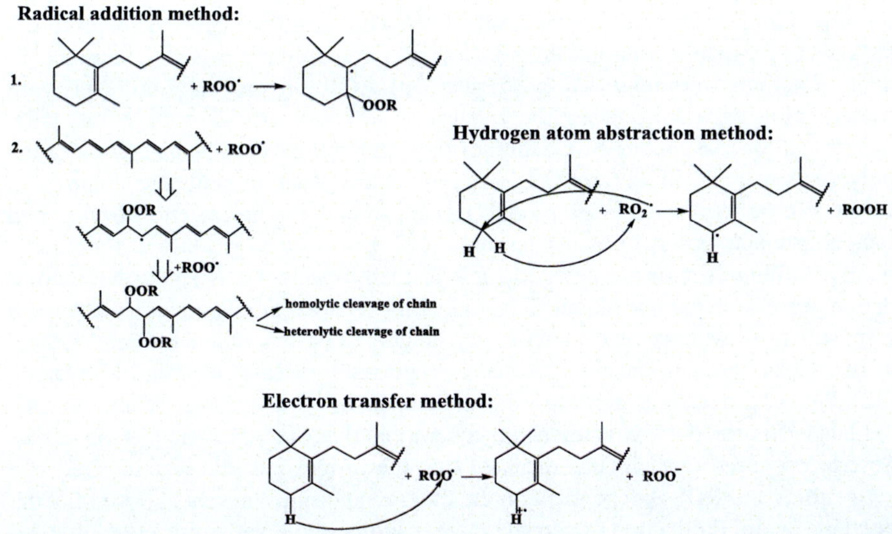

Fig. 12.2 Different reaction mechanisms between free radicals and carotenoids

place; as a result corresponding anions are formed [33]. Only the allylic hydrogens can act via hydrogen abstraction mechanism by reacting with peroxy radicals to produce a resonance-stabilized radical. This, in turn may continue the radical propagation chain.

Carotenoids are lipophilic antioxidants and show activity against free radical species. Different types of free radicals are produced like superoxide anions, hydroxyl, per-hydroxyl radical and nitrogen derived radicals, which are probably the target of carotenoids. Although, interactions between carotenoids and reactive nitrogen species (RNS) or nitrogen derived radicals are not completely understood, the capacity of carotenoids to scavenge per-oxynitrite in vitro has been studied systematically [34, 35]. β-carotene scavenge the per-oxynitrite radicals more efficiently than lutein and zeaxanthin in both in vivo and in vitro systems. However, the activity of the lutein and zeaxanthin was comparable to that of glutathione.

Figure 12.1 indicates that the carotenoids, which are hydrocarbon in nature, such as lycopene and β-carotene are very lipophilic and embedded deep within the cell membrane [5] so more efficient charge separation will not occur due to lipophilic environment however, xanthophyllic carotenoids prepare hydrophilic environment that ease the electron transfer more efficiently. Carotenoid radical cations absorb light in the near-infrared region and give peaks in a range of 820–950 nm (7,70-dihydro β-carotene to lycopene in methanol) of infrared but it varies with environment [31]. It has also been observed that the radical adducts absorb light energy at the visible region (400–500 nm) and often may absorb in parent ground-state absorption [31, 32]. It has been investigated that carotenoid redicals and oxygen molecules are not reactive [36]. Species with higher redox potentials produced carotenoid radical cations [33], while carotenoids undergo reduction to produce corresponding carotenoid radical anions [37]. These radicals cause damage to tissues or cells due to oxidation [38, 39]. Neutral carotenoid radicals and their mechanism of actions were reviewed by [5] and [30]. Allylic hydrogen atom reacts with peroxyl radicals and produce radical that may continue radical propagation chain. Additional process includes oxidation of the carotenoids via radical addition to the polyenic chain or in the ring. If reaction takes place in polyenic chain then carotenoid-peroxyl adducts are produced that can undergo heterolytic cleavage and radical propagation chain is discontinued [18]. But, if it takes place at the double bond position of the ring, a new radical is formed, which undergoes further autoxidation process. In the line of study, the fate of the neutral carotenoid radicals derived respectively from other two mechanisms, such as H-atom abstraction and radical addition has not been clearly revealed as reported by El Agamey and McGarvey [32]. Fig. 12.2 describes the reaction mechanisms discussed above. When the antioxidants capture the free radical at generation then autooxidation is observed. Several proposed mechanisms revealed that the progress of the reaction between carotenoids and ROS is dependent on the functional groups present at the end of the polyenic chain. Production of peroxyl radical and its stabilization via electron delocalization depend on these functional groups. It has been observed that these functional groups can prevent further oxidation of carotenoids through one of the aforementioned reaction mechanisms and reduce the chances of autoxidation of the carotenoid [40].

During past three decades, researchers are involved in determining the antioxidant potentials of different natural and synthetic substances including plant extracts, food ingredients, seeds, and even complete food systems [41]. But there are lack of information regarding the chemistry behind these methods [42, 43]. These methods include oxygen radical absorbance capacity, total radical trapping antioxidant parameter, trolox equivalent antioxidant capacity, total oxyradical scavenging capacity and peroxyl radical scavenging capacity etc. The process to determine the antioxidant activity of carotenoids still follow the methods described by [44] and the TRAP method by Bartosz and co-workers (1998) [45], which are based on the reactions in the lipid autoxidation process.

12.2.2 *Antioxidant Capacity*

Antioxidant capacity may be described as the 'capacity of a compound to protect biological systems, such as tissues, biological samples such as plasma and cell cultures from the potentially harmful effects of the processes or reactions involving ROS and RNS'. TEAC method is generally employed to determine the antioxidant capacity of carotenoids [54–59]. Additionally, DPPH scavenging method [60–62] and ex vivo oxidation of LDLs assay [63] are also used to determine the antioxidant capacity. However, these methods can only predict the antioxidant capacity of carotenoids. Lipids are the major constituents of cell membrane and most of the cellular structure of a biological system. Therefore, oxidation of lipids has a negative impact on the functioning of the membrane and its integrity. Carotenoids can accommodate in the lipid portion of these structures, and able to delay radical propagation chain during lipid peroxidation. Therefore, carotenoids together with tocopherols are known as membrane antioxidants [46]. Thus, in vivo antioxidant effect of carotenoids observed in LDLs and cellular membranes follows the reaction mechanisms described earlier in this chapter. Despite several reports revealed the beneficial roles of dietary carotenoids in protecting cellular structures from oxidative stress. However, at the same time harmful effects of the carotenoid autoxidation products cannot be ignored. Most studies revealed that the carotenoids, such as β-carotene, zeaxanthin, lutein and lycopene are effective antioxidants [47]. Contrarily, some reports depicted the ineffectiveness of carotenoids in protecting LDLs oxidation [48].

It has been observed that few factors can modulate antioxidant and pro-oxidant properties of carotenoids in biological systems. Factors, such as chemical properties (size, shape, nature, position and number of substituent groups, unsaturation etc.), physical forms (aggregated, monomeric, isomer, configuration etc.), location/ site of action within the cells, potential for interactions with other antioxidants, concentration, and partial pressure of oxygen can influence the antioxidant and pro-oxidant properties of carotenoids [19]. Solubility of the carotenoids is generally very poor in aqueous environment except xanthophylls, which contain polar groups. Therefore, distribution of carotenoids in biological membrane varies depending on their types, such as carotenes or xanthophylls [49]. The cellular locations of carotenoids also influences their capacity of scavenging free radicals. Reactions of carotenoids with different ROS depend on the reaction environment, such as hydrophilic or lipophilic, which causes difference in the mode of action in different regions in the cell [50]. This controversy claims to an acute screening of the factors that might affect the affinity of LDLs to oxidation. Factors, such as study population, design of the intervention trial, pharmacokinetic properties and bioavailability of carotenoids, efficiency of assimilation of carotenoids, possible biotransformation of carotenoids into less active compounds, and generation of new free radicals are important aspects in this regard [51].

12.3 Antioxidant Protection of Carotenoids

On the basis of several researches and practices we can say that the carotenoids enriched diet can protect from numerous health problems. Carotenoids can inhibit the onset of various diseases like cancer, diabetes, and blindness etc. due to their antioxidant activity and it is known that carotenoids exert their anti-oxidative role by scavenging radicals and singlet oxygen. Several carotenoids with promising antioxidant potential have been summarized in this chapter. In order to deactivate ROS and stop oxidative reactions, carotenoids usually withdraw or donate electrons or protons depending upon the radical species and the structural aspects of carotenoids

[52]. Some of these may act inside cell, while other exhibit their antioxidant property through chain breaking mechanism based on their structural diversity [53].

On further transformation of the carotenoid radical products, a diverse group of secondary carotenoid derivatives can be formed with different reactivity. In many instances, these secondary species may act as potential pro-oxidant agents rather than acting as anti-oxidants. Furthermore, the free radical scavenging ability of carotenoids depends largely on their redox potentials [54]. So, the relative antioxidant ability can be modified by the presence or absence of salts. Astaxanthin loses its antioxidant ability in presence of salts due to decrease in oxidation potential and reduced stability of astaxanthin radical cations [55].

Carotenoids with oxygenic substituents (xanthophylls) can easily donate protons to the reactive species and exhibit better antioxidant activity than non-oxygenated carotenoids (carotenes) [56]. In this contrast, fucoxanthin, an oxygenated carotenoid, has been reported as a potent ROS scavenger, which significantly reduces oxidative stress-mediated cellular damages [57]. Due to high antioxidant activity, fucoxanthin can regulate certain genes involved in the cellular metabolism [58]. Fucoxanthin from marine algae has been revealed to regulate lipid markers in the blood and liver in various animal models [59]. In addition, it could reduce white adipose tissue count via modulating lipid oxidation [59]. It was also observed that fucoxanthin can scavenge HOCl [60]. Astaxanthin possesses the highest antioxidant activity among the carotenoids, namely β-carotene, zeaxanthin and canthaxanthin and its ${}_1O^2$ quenching activity was found to be higher than that of vitamin E [61]. In the line of study, Kurashige and co-workers [62] observed that astaxanthin can prevent lipid peroxidation better than α-tocopherol. Astaxanthin can scavenge few other free radicals, such as peroxynitrite radical. Astaxanthin in low concentration can protect mitochondria of dopaminergic neurons by inhibiting lipid peroxidation and thereby, prevents the progression of Parkinson's disease [63]. In biological systems, exposure of light can lead to tissue damage by generating ROS [64]. Junghans and co-workers suggested that carotenoids, specially lutein and zeaxanthin have the ability to filter blue light effectively from liposomal formulations [64]. β-carotene and α-tocopherol have also been shown to prevent and/or protect from photo-oxidative damages [64].

Reactive oxygen species react nitric oxide and form per-oxynitrite, which can react with carotenoids to form nitro-carotenoids [65], thus carotenoids stop the reactive process. These Per-oxynitrite reacts with amino acids lipids and some fragments of DNA [66]. For example, Fucoxanthin stop this nitration of tyrosine by forming Z-nitro-fucoxanthin. This new product also showed anti-tumor and anti-proliferative properties. Fucoxanthin has less anti-carcinogenic potentials than nitro-fucoxanthin [65] similarly, reactive nitrogen species are scavenged by other carotenoids like astaxanthin and lutein to corresponding 15 Z-nitro-carotenoids [67]. However, astaxanthin possesses higher antioxidant activity as a ${}_1O^2$ quencher than nitro-astaxanthin and β-carotene. Lutein and nitro-lutein possess similar ${}_1O^2$ quenching activities and the quenching potentials were found to be higher than that of β-carotene [67]. Astaxanthin (<16.75 μM) did not show any sign of cytotoxicity in normal or tumor cells [68]. In addition, astaxanthin has not been mentioned to exhibit any considerable side effect [69]. The antioxidant properties of carotenoids are given in Table 12.1.

12.4 Importance of Carotenoids as Antioxidants in Disease Treatment

12.4.1 Cancer

According to Word Health Organization the death rate due to tumour is increasing yearly and estimated to be raised over 11 million in the year 2030. Free radicals and reactive oxygen species are one the causes that results in tumour formation. Therefore, certain steps are required to minimize the exposure to reactive oxygens species or free radicals or these species need to be scavenged or quenched before causing damage [105]. Different searches have shown that consumption of carotenoid enriched fruits and vegetables can decrease the risk and development of several types of cancer [106]. It was already reported that β-carotene supplementation in non-smoking adults can prevent the chances of lung cancer [107]. In past few years, some statements were published regarding the risk of lung cancer and intakes of food sources enriched with carotenoids, such as α-carotene, β-carotene, lutein, lycopene, and β-cryptoxanthin [108]. We know that DNA oxidation can lead to the mutagenesis or carcinogenesis. Several reports suggested that the carotenoids may impede oxidative DNA damage. The potential role of the carotenoids in preventing DNA oxidation has been reviewed by Collins [109]. At the same time, some evidences showed that carotenoid molecules may increase DNA damage at high concentrations via pro-oxidant mechanism. An enhancement of H_2O_2-induced oxidative DNA damage by β-carotene was observed in HepG2 cells [70]. Some evidences showed that β-carotene and other carotenoids, such as lycopene failed to protect HT29 cells against free-radical-induced DNA damage [110, 111]. It has been observed that oxidised β-carotene can highly increase DNA oxidation in human Hs68 fibroblasts. Although, a large number of epidemiological studies generally support the idea that several carotenoids, as well as carotenoid enriched foods, could be involved in the reduction of the risk of prostate cancer [112]. It has been observed that 8-hydroxy-2′-deoxyguanosine (8-OHdG) is a good biomarker for oxidative DNA damage caused by smoke. In this aspect, some researches revealed that TAR obtained from smoke caused increased levels of 8-OHdG even in presence of β-carotene [113]. Sometimes, pro-oxidant activity of carotenoids is helpful in the treatment of certain types of cancer. β-carotene can activate caspase-3 in some cancer cells, which is a key component in programmed cell death. Activation of caspase-3 leads to the apoptotic death of cancer cells. β-carotene can also inhibit Bcl-2 protein expression in cancer cells. Bcl-2 prevents programmed cell death by increasing the formation of ROS and lipid peroxidation products in cancer cells (Fig. 12.3). So, activation of caspase-3 and decrease in Bcl-2 expression may lead to the death

Table 12.1 Some important carotenoids and their antioxidant properties

Name	Related literature	Antioxidant activity	References
β-Carotene (**1**)	It serves as an antioxidant inhibiting free radical production. It also serves as pro-oxidant propagating free radical-induced reactions depending on its concentration, intrinsic properties, and on the redox potential of the biological environment in which it acts. β-carotene is able to modify ROS production in human colon adenocarcinoma cells and human leukemia cells. β-Carotene can act as an endogenous generator of ROS through its ability to induce various P450 isoforms. β-Carotene (10–20 μM) decreases intracellular concentration of GSH and a concomitant increase in the levels of oxidized glutathione. It can induce apoptosis and arrest cell cycle progression. β-carotene can induce oxidative DNA damage.	$_1O^2$ quencher; radical scavenger; inhibits Na^+K^+-ATPase pump; stimulates CAT and GST.	[35, 70–75]
α-Carotene (2)	α-Carotene is a better antioxidant than β-carotene tested in phosphatidyl choline vesicles. Therefore, it is useful in prevention of free radical mediated peroxidative damage against membrane phospholipids in vivo.	$_1O^2$ quencher; radical scavenger; inhibits lipid peroxidation.	[76, 77]
Lycopene (3) Phytoene (4) Phytofluene (5) ζ-Carotene (6)	All these compounds showed significant antioxidant potential against peroxyl radical. Lycopene inhibits lipid peroxidation tested in egg yolk liposome. The antioxidant capacity of lycopene and other carotenoids offers protection against gamma-radiation induced damage to cells. Pre-treatment with lycopene resulted in a significant reduction of DNA damage and lipid peroxidation from the radicals produced from gamma irradiation. Watermelon lycopene has significantly higher antioxidant activities than those of tomato lycopene and has comparable superoxide anion scavenging activity to that of Trolox.	Scavenge trichloromethyl peroxyl radical (CCl_3O_2.) and other radicals; peroxyl radical scavenger; $_1O^2$ quencher; scavenge $O_2^{\bullet-}$	[78–82]

(continued)

Table 12.1 (continued)

Name	Related literature	Antioxidant activity	References
Astaxanthin (7)	Astaxanthin quenches singlet oxygen in a membrane model and also protects membrane lipids from free radical-mediated damage. It can inhibit RNS-induced inflammation. Astaxanthin on reaction with peroxynitrite produces nitroastaxanthins which is more powerful antioxidant than astaxanthin. Astaxanthin shows the highest antioxidant activity toward peroxyl radicals. It can also inhibit ROS mediated IκB kinase-dependent NF-κB activation.	RNS quencher; chain-breaking AO; $_1O^2$ quencher; lipid peroxidation inhibitor; peroxyl radical scavenger.	[67, 76, 83–85]
Lutein (8)	Lutein can scavenge peroxynitrite radical in vitro and reaction with peroxynitrite leads to the formation of nitrolutein which is also an antioxidant. Lutein decreases the malondialdehyde content and increases the oxygen radical absorbance capacity level, reduced glutathione concentration, vitamin C contents and total SOD and GPx activities. Lutein also increases expressions of copper-zinc SOD, manganese SOD and GPx mRNA and can promote antioxidant activity in *BALB/c* mice. Lutein protects unsaturated lipids more effectively in membranes made of raft-forming mixtures than in homogeneous membranes. It can prevent macula from photo-oxidative damage.	RNS scavenger; chain-breaking antioxidant; peroxyl radical scavenger; stimulates antioxidant enzymes; lipid peroxidation inhibitor.	[6, 67, 86–88]
Canthaxanthin (9)	It increases resistance to lipid peroxidation primarily by enhancing membrane α-tocopherol level and secondarily by providing weak direct antioxidant activity. Canthaxanthin inhibits radical-initiated lipid peroxidation in rat liver microsomes.	ROS and RNS scavenger; chain-breaking antioxidant.	[84, 89, 90]

(continued)

Table 12.1 (continued)

Name	Related literature	Antioxidant activity	References
Fucoxanthin (10)	Fucoxanthin enhances binding activities of nuclear Nrf2 with ARE and increased mRNA and protein expressions of HO-1 and NQO1 and exert its antioxidant activity. Fucoxanthin can reduce lipid hydroperoxide levels in liver and abdominal white adipose tissue of obese/diabetes KK-A^y mice. It can also scavenge peroxynitrite radical and reduced its harmful effects.	$_1O^2$ quencher; inhibits Na^+K^+-ATPase; RNS quencher; stimulates catalase and glutathione transferase; radical scavenger; scavenge $O_2^{\bullet-}$, $OH^{\bullet}$.	[91–93]
Zeaxanthin (11)	Zeaxanthin protects thylakoid membrane lipids from photo-oxidation and the effect is noticeably higher than that of all other xanthophylls of *Arabidopsis* leaves. It protects the retinal pigment epithelium against a photo-oxidative damage initiated in part by light absorption. Zeaxanthin shows synergistic antioxidant role with GSTP1 in egg yolk phosphatidylcholine (EYPC) liposomes and prevents lipid peroxidation.	ROS scavenger; scavenge $O_2^{\bullet-}$, $OH^{\bullet}$. Lipid peroxidation inhibitor.	[6, 86, 94, 95]
Violaxanthin (12)	It produces antioxidant activity. Violaxanthin is useful in light harvesting and photo-protection.	ROS scavenger; inhibits photo-oxidation.	[96]
Echinenone (13)	It provides protection from ROS related disorders.	ROS scavenger.	[97–99]
Halocyanthiaxanthin (14)	Halocyanthiaxanthin is the best suppressor of $O_2^{\bullet-}$ among about 20 carotenoids present in human body.	$O_2^{\bullet-}$ scavenger; RNS scavenger.	[100]
Bixin (15)	Bixin can increase GR, thioredoxin reductase, SOD, and serum NOx levels representing potent antioxidant and protective effects. As well it can inhibit lipid peroxidation.	Stimulates antioxidant enzymes; ROS scavenger	[101, 102]
Crocin (16)	Crocin can reduce lipid peroxidation and xanthine oxidase activity and can increase GSH in the kidney of the diabetic rats. It protects oxidative stress-induced neuropathy.	Stimulates antioxidant enzymes; ROS scavenger.	[103, 104]

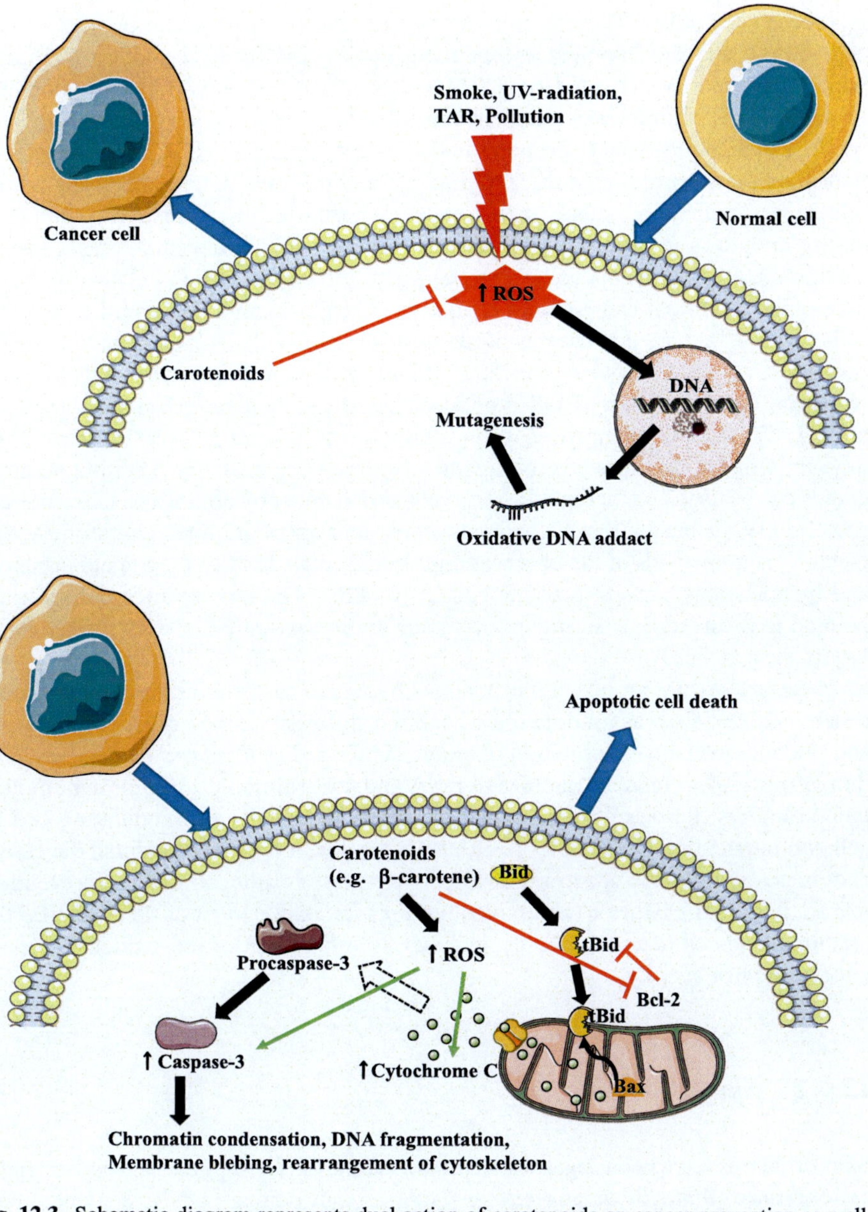

Fig. 12.3 Schematic diagram represents dual action of carotenoids on cancer prevention as well as treatment. Carotenoids prevents oxidative DNA damage and mutagenesis of normal cells and on the other hand it produces apoptotic response in cancer cells via its prooxidant function

of cancer cells [114]. Therefore, it can be said that β-carotene can prevent cancer cell growth and development by inducing apoptosis.NF-κB is another mediator, which is primarily activated by ROS and on activation it carries pro-apoptotic response [115]. Carotenoid treatment can enhance the production of NF-κB in cancer cells, which may lead to programmed cell death of cancer cells [116]. Treatment with β-carotene can alter mitochondrial membrane potential, which leads to the release of another -apoptotic mediator, cytochrome C. β-carotene was found to induce apoptosis in different tumour cells, such as human leukemia, colon adenocarcinoma, and melanoma cells [71]. Schwartz and co-workers [117] observed that β-carotene treatment can enhance oxidative stress in human epidermal carcinoma cells, which in turn activates heat shock proteins, such as hsp70 and/or hsp90 involved in apoptosis. Normal cells are highly resistant than cancer cells towards induction of apoptosis by ROS produced due to pro-oxidant effect of β-carotene [117, 118]. β-carotene improves its antioxidant efficiency at low partial pressure of oxygen, while it becomes a pro-oxidant at high oxygen pressure. Therefore, carotenoids can be used as a therapeutic intervention in cancer by both ways. Giovannucci (2002) [119] discussed that lycopene is a potential agent to impede tumor development. A meta-analysis of the observational studies revealed that the tomato products and lycopene can prevent prostate cancer [6, 120]. The human intervention trials pointed to β-carotene as an important factor in the prevention of oral, pharynx and larynx cancers [121].

There are several proposed mechanisms by which carotenoids can produce anticancer activity, such as modulation of cellular differentiation and proliferation via antioxidant mechanism, induction of apoptosis via pro-oxidant mechanism, prevention of free radical induced damage to DNA and other biomolecules, immunomodulation, and restriction of clonal expansion of initiated cells via modulating cell to cell communication [122]. After an extensive review, it would be said that the carotenoids possess diverse spectrum of effects on cancer pathogenesis and prevention due to their dual character (antioxidant and pro-oxidant). Most of the scientific literature evidenced that carotenoids, whether as antioxidant or pro-oxidant can produce antitumor activity.

12.4.2 Skin Problems

Skin problems have been regarded as one of the serious complications by free radical exposure. Sunlight or any strong radiation can damage DNA, enzymes, and collagen in the skin, which lead to the appearance of deep grooves and bumps on/within the skin [123]. Prolong exposure to sunlight and radiation may produce various skin problems, such as sunburn, pigmentation, hyperplasia, immunosuppression, DNA damage, and even can cause skin cancer [124]. Several studies claimed that β-carotene can protect against photo-oxidative damage and burns of skin. It was observed that β-carotene can reduce lipid peroxidation and can quench singlet oxygen produced due to photochemical reactions in rodent's skin. In cultured skin cells, it decreases photo-inactivation of catalase (CAT), superoxide dismutase (SOD) and protein cross linking. It was also revealed that, β-carotene can reciprocate UV-A-induced CAT deactivation and lipid peroxidation in the kidney fibroblasts of rats [125]. Bando et al. observed that the dietary accumulation of β-carotene in mouse skin provide protection against UV-A induced oxidative damage. An LC-MS analysis of mouse skin lipids supported the aforementioned claims [126]. The formation of β-carotene 5,8-endoperoxide observed via LC-MS analysis is a clear evidence for the singlet oxygen chemical reaction. These observations suggest that β-carotene is a potential singlet oxygen quencher in the skin [126]. Considering the effects of increased UV fraction of solar radiation on earth, it has become necessary to evaluate more possibilities with carotenoids as photo-protectors. Some carotenoids, such as β-carotene and lycopene have already been justified and proven for their good photoprotective activity [125].

12.4.3 Eye Problems

Carotenoids are also helpful in eye related disorders due to their ROS scavenging property. Several researches indicate that lutein and zeaxanthin are accumulated in the macula lutea, the central part of the eye retina, at extremely high concentrations [127]. Several reports revealed the retino-protective mechanisms of xanthophylls. Removal of oxidative stress, interaction with signalling molecules, regulation in the expressions of inflammatory genes, such as connexin or inflammation-related genes, and interactions with membrane lipids and proteins are thought to be the major protective mechanisms of xanthophylls [128].

Many research articles stated that xanthophylls can protect light induced damage by absorbing high energy photons before the formation of ROS and quenching of generated ROS [129]. In addition, xanthophylls can efficiently quench excited triplet states of photosensitizers. Lutein, zeaxanthin and meso-zeaxanthin, an isomer of zeaxanthin and a metabolite of lutein, were reported to be accumulated in the Henle's fiber layer of photoreceptor axons and the membranes of photoreceptor outer segments. However, the position of xanthophylls in the lipid part of the membranes is not clear yet. Some reports suggested that lutein and zeaxanthin are directly incorporated in the lipid part of retinal cell membranes [130], while, other stated that macular xanthophylls are attached through the membrane associated xanthophyll-binding proteins [131].

Although, the results of various studies on model membranes indicated that the location of macular xanthophylls within the lipid part of the membrane, it cannot be excluded that protein-carotenoids interactions are comparably significant in the selective accumulation of lutein and zeaxanthin in the protein part of retinal cells. Bhosale et al. have been involved in searching of specific xanthophyll binding proteins, which could be involved in the uptake, stabilization, and also antioxidant functions of lutein and zeaxanthin [132]. First, Pi isoform of glutathione S-transferase

(GSTP1) has been identified as a zeaxanthin-binding protein in human macula exhibiting high affinity toward both dietary zeaxanthin and non-dietary meso-zeaxanthin [132]. Additionally, it has been shown that GSTP1 can act synergistically with zeaxanthin in protecting the model membranes against oxidative damage [86]. Few years later, the same group identified one of the steroidogenic acute regulatory domain proteins (StARD3) as a lutein-binding protein [133]. Interactions of macular xanthophylls and their binding proteins have been investigated by surface plasmon resonance (SPR) and the pigment binding constants have been determined [133]. Based on high affinity of GSTP1 for zeaxanthin and StARD3 for lutein, it has been hypothesized that xanthophyll binding proteins can be responsible for capturing these xanthophylls by the retina and their selective accumulation within the tissue [130, 133].

12.4.4 Diabetes

Carotenoids are also useful in the treatment of diabetes and its complications. Recent advances in carotenoid research predicted that carotenoids can treat or ameliorate diabetes and its complications. Several studies proposed that carotenoids can reduce the risk of type 2 diabetes in human beings [134]. It was observed that, carotenoids can reduce fasting plasma glucose concentrations, insulin resistance, and glycated haemoglobin level [134]. Different reports revealed that carotenoids can produce its antidiabetic activity through its antioxidant property or via directly modulating oxidative stress-provoked signalling pathways [134]. The following reports supported that the carotenoids are efficient in neutralizing diabetes related metabolic and molecular abnormalities, partly via their antioxidant mechanism. It has already been mentioned that the antioxidant activity of astaxanthin is higher than that of β-carotene or α-tocopherol [76]. Uchiyama and colleagues observed that astaxanthin can decrease glucose tolerance, enhance serum insulin levels and reduce blood glucose levels in db/db mice by protecting pancreatic β cells from oxidative damage [135]. In another study, astaxanthin has been found to ameliorate diabetic nephropathy due to its antioxidant properties [136]. They reported that high glucose can trigger the production of ROS in mitochondria, which disturb mitochondrial redox status and activate stress related signalling molecules, such as MCP-1, TGF-β, COX-2, NF-κB and AP-1 [136]. Astaxanthin was found to neutralize the mitochondrial oxidative stress and pathological signal transduction [136]. Astaxanthin can suppress the generation of superoxide, nitric oxide, and peroxynitrite in high glucose induced proximal tubular epithelial cells. Astaxanthin was found to reduce ROS production and endoplasmic reticulum stress in the liver of high fructose and high fat diet-fed mice [137]. Neutrophils are another source of ROS triggered by hyperglycaemia. Astaxanthin can partially prevent the increase in ROS and RNS accumulation and restored phagocytic capacity of the neutrophils under diabetic condition [138]. Lycopene is known for its antioxidant properties, which has been found to trigger CAT, SOD, and GPx in diabetic rats in both mRNA and protein levels [139]. Thereby, lycopene can prevent the extent of hydrogen peroxide accumulation, impede lipid peroxidation [140], and reduce NO level in plasma [141]. Bixin, another carotenoid, has been reported to possess antidiabetic property [142]. In addition, bixin was found to increase GR, thioredoxin reductase, SOD, and serum NOX levels and thereby, can reciprocate oxidative stress in diabetic rats [101]. Crocin was found to exhibit prophylactic effect to the organs of diabetic rats via antihyperglycemic and antioxidant mechanisms [143]. These findings revealed that the oxidative stress increases in hyperglycemic condition. Oxidative stress can further impair insulin secretion and increases the resistance to insulin. Carotenoids such as bixin, astaxanthin, crocin, and lycopene can prevent oxidative stress-induced complications in diabetes.

12.4.5 Coronary Artery Disease

The effects of carotenoids on coronary artery disease have been a subject of disagreement. In some case-control studies, a clear relationship between serum carotenoids content and risk of myocardial infarction was not found. Moreover in some cases, an inverse association between serum β-carotene level and the risk of cardiovascular diseases has been noticed [144]. Dietary lycopene has not been found to be strongly associated with the risk of cardiovascular diseases in women; however, a probable inverse association between the higher levels of tomato-based products intake and cardiovascular disease has been reported. In another report, a higher plasma concentration of lycopene was found to reduce the risk of cardiovascular diseases in women [89]. Plasma levels of lutein, lycopene, α-carotene, and β-carotene were found to be significantly lower in the patients with acute ischemic stroke and it may be correlated to that of increase in the oxidative stress [89].

12.5 Conclusion

It is quite clear that the carotenoids are the natural antioxidants, which can inhibit the oxidation of vital biological macromolecules, such as lipids, proteins, and nucleic acids. Due to the fact that oxidative damage is involved in the pathogenesis of majority of the human ailments, carotenoids seem to impart beneficial effects to the human health. Carotenoids attribute antioxidant effect principally through scavenging of ROS, inhibiting ROS generation, triggering endogenous redox defence mechanism, modifications of membrane properties etc. So far, a vast number of data were obtained from the epidemiological, interventional, and clinical studies with different carotenoids (Table 12.1 and Fig. 12.4), which generally support the beneficial roles of the carotenoids against the diseases predominantly through redox defence and antioxidant mechanisms. It is the fact that some of the results remain inconsistent; however, the antioxidant properties of the dietary carotenoids cannot

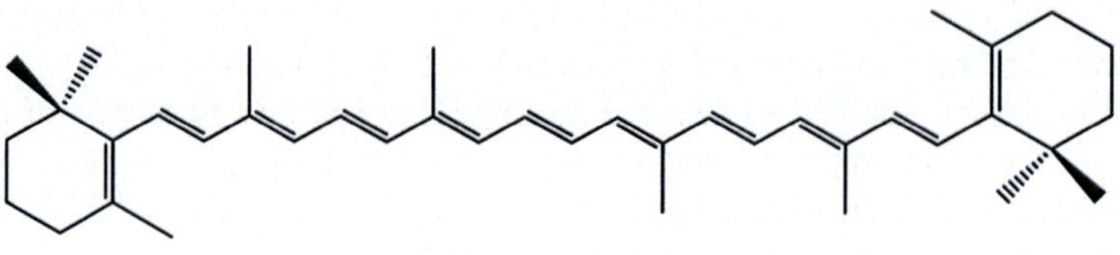

(1)

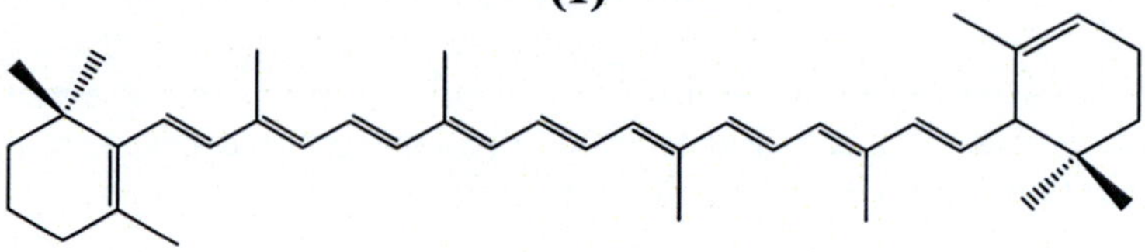

(2)

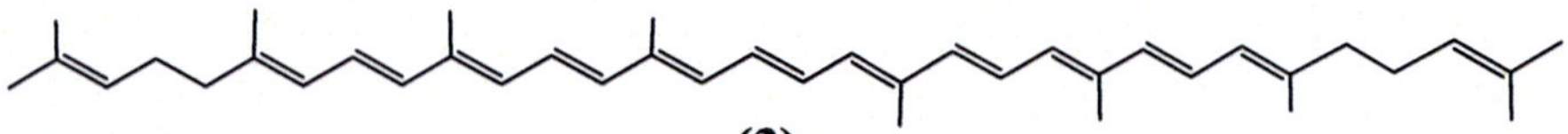

(3)

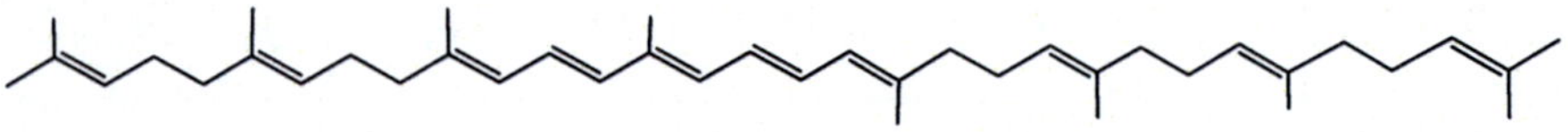

(4)

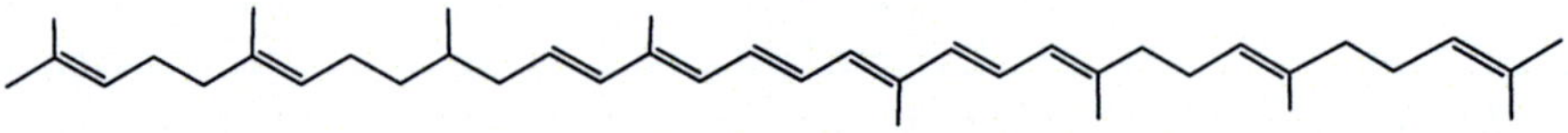

(5)

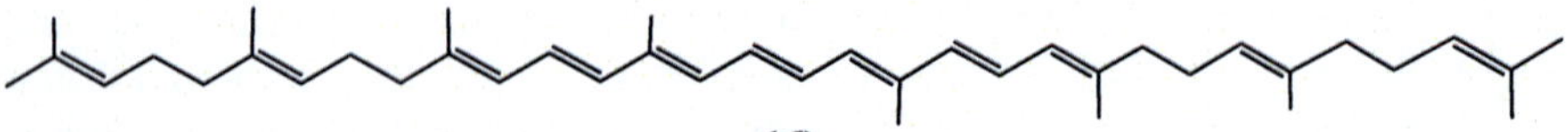

(6)

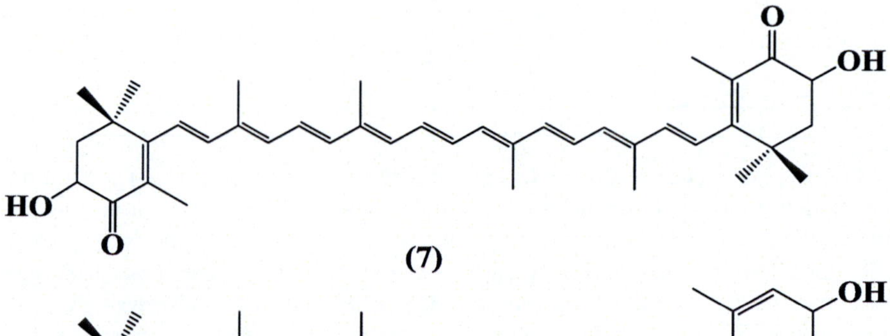

(7)

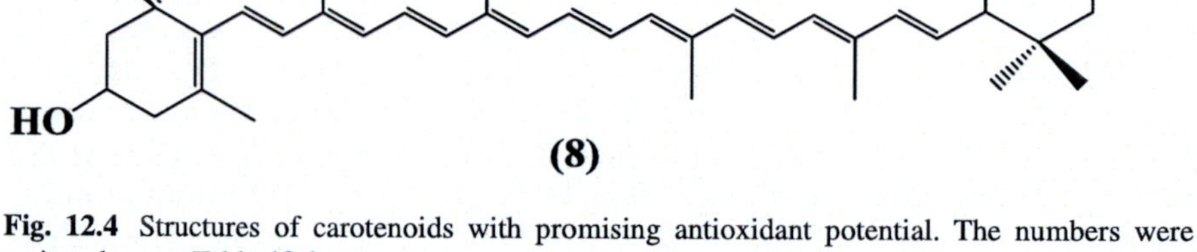

(8)

Fig. 12.4 Structures of carotenoids with promising antioxidant potential. The numbers were assigned as per Table 12.1

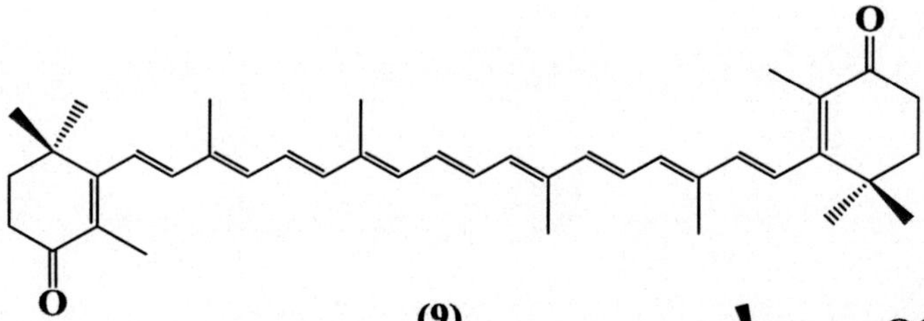

(9)

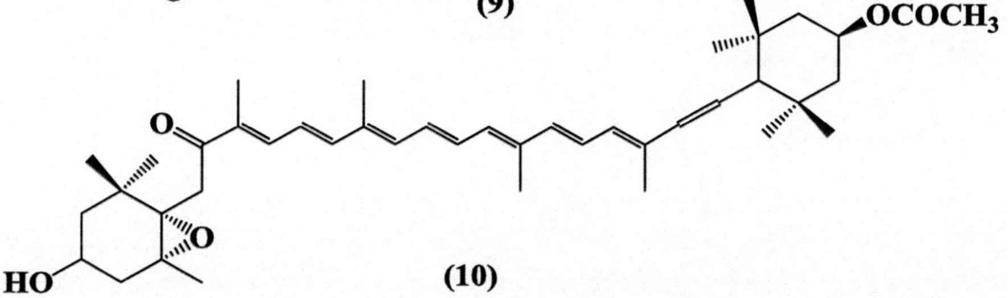

(10)

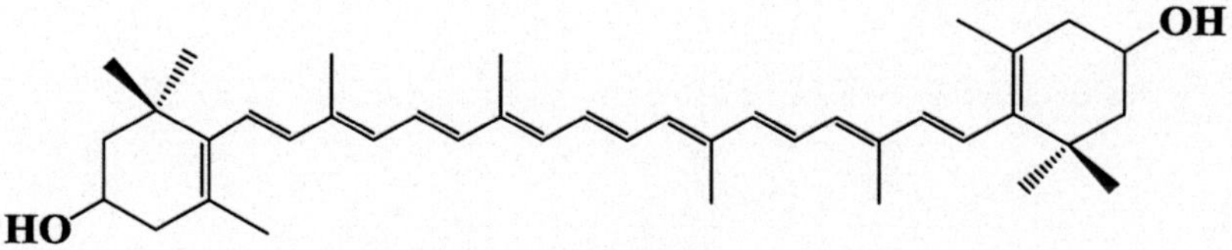

(11)

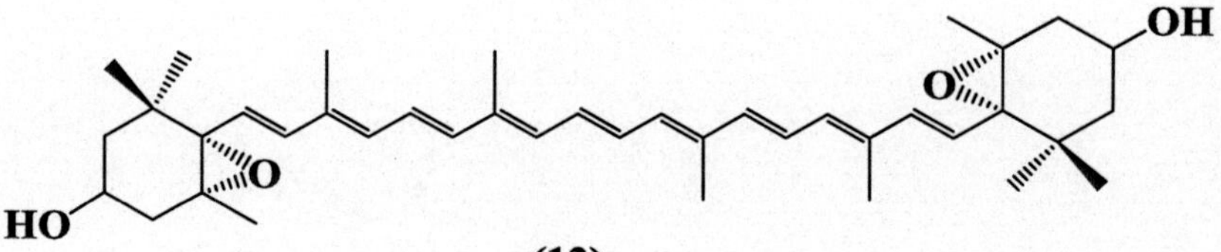

(12)

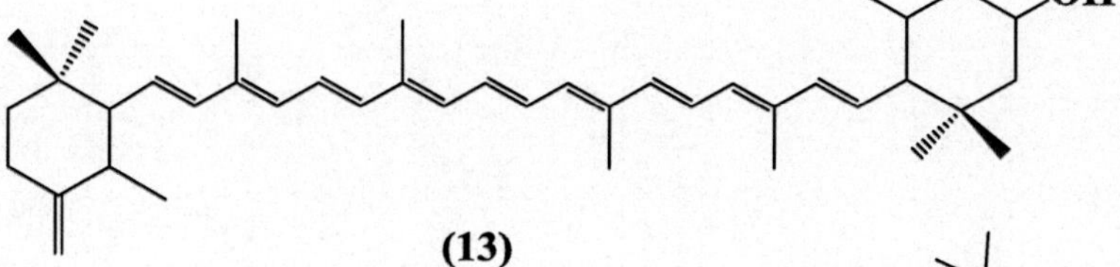

(13)

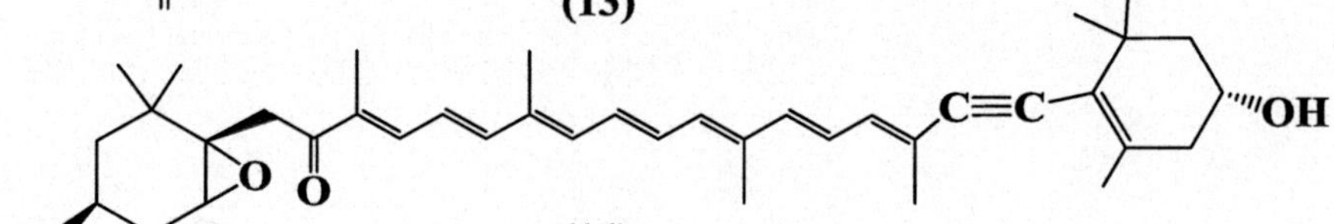

(14)

Fig. 12.4 (continued)

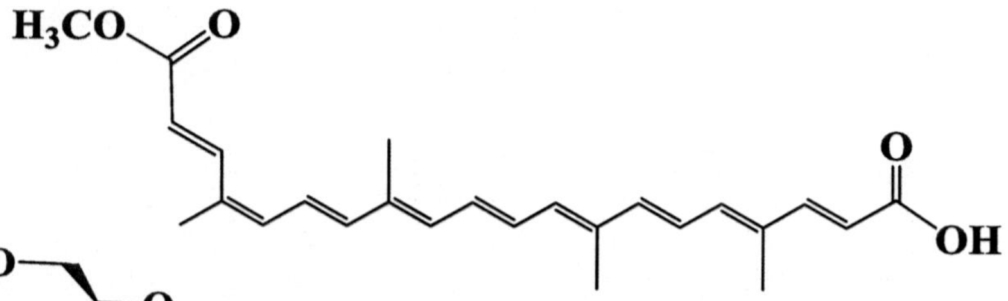

(15)

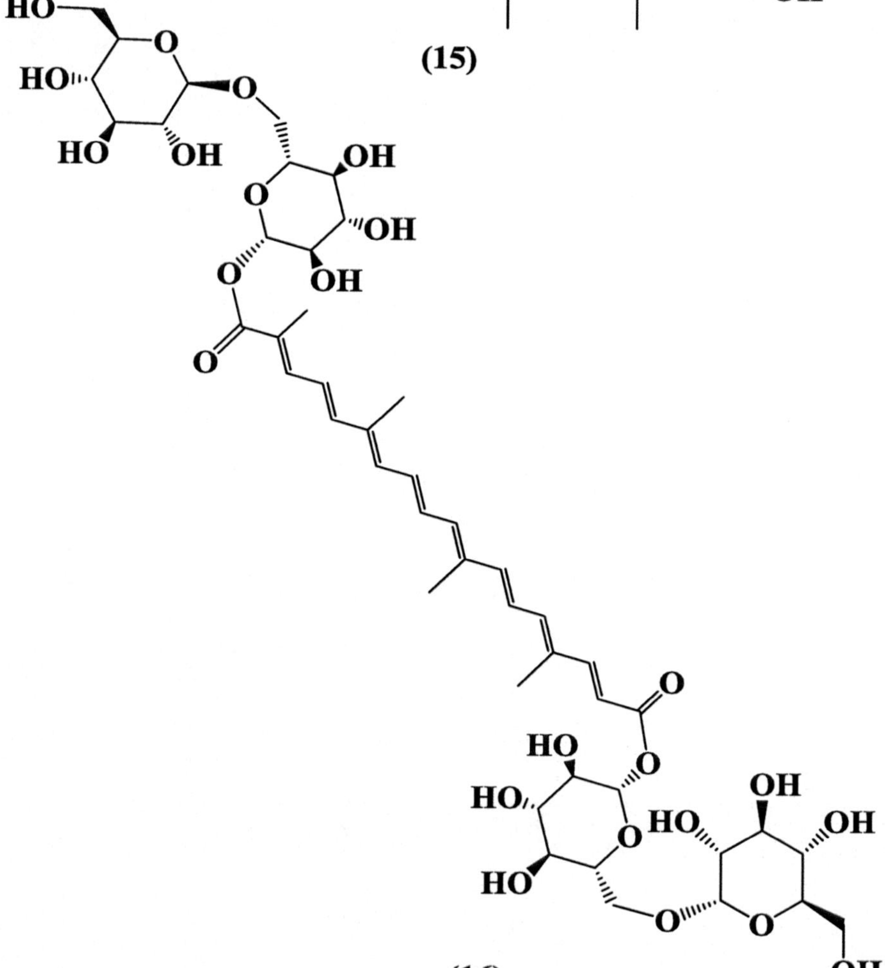

(16)

Fig. 12.4 (continued)

be ignored. Therefore, more studies are needed for the complete understanding of the exact antioxidative mechanism of carotenoids and their disease prevention capabilities.

References

1. Stahl W, Sies H (2005) Bioactivity and protective effects of natural carotenoids. Biochim Biophys Acta Mol basis Dis 1740(2):101–107
2. Fraser PD, Bramley PM (2004) The biosynthesis and nutritional uses of carotenoids. Prog Lipid Res 43(3):228–265
3. Takaichi S (2011) Carotenoids in algae: distributions, biosyntheses and functions. Mar Drugs 9(6):1101–1118
4. Khachik F, Spangler CJ, Smith JC, Canfield LM, Steck A, Pfander H (1997) Identification, quantification, and relative concentrations of carotenoids and their metabolites in human milk and serum. Anal Chem 69(10):1873–1881
5. Britton G (1995) Structure and properties of carotenoids in relation to function. FASEB J 9(15):1551–1558
6. Bhosale P, Bernstein PS (2005) Microbial xanthophylls. Appl Microbiol Biotechnol 68(4):445–455
7. Semba RD, Lauretani F, Ferrucci L (2007) Carotenoids as protection against sarcopenia in older adults. Arch Biochem Biophys 458(2):141–145
8. Britton G (2008) Functions of intact carotenoids. In: Carotenoids. Springer, Dordrecht, pp 189–212
9. Lu S, Li L (2008) Carotenoid metabolism: biosynthesis, regulation, and beyond. J Integr Plant Biol 50(7):778–785
10. Schiedt K (1998) Absorption and metabolism of carotenoids in birds, fish and crustaceans. Carotenoids Biosynth Metab 3:321–343
11. Goodwin T (1984) Carotenoid—protein complexes. In: The biochemistry of the carotenoids. Springer, Dordrecht, pp 1–21
12. Roberts RL, Green J, Lewis B (2009) Lutein and zeaxanthin in eye and skin health. Clin Dermatol 27(2):195–201
13. Minguez-Mosquera MI, Hornero-Mendez D (1997) Changes in provitamin A during paprika processing. J Food Prot 60(7):853–857
14. Abdel-Aal E-S, Akhtar H, Zaheer K, Ali R (2013) Dietary sources of lutein and zeaxanthin carotenoids and their role in eye health. Nutrients 5(4):1169–1185
15. Miki W (1991) Biological functions and activities of animal carotenoids. Pure Appl Chem 63(1):141–146
16. Halliwell B, Gutteridge JM (2015) Free radicals in biology and medicine. Oxford University Press, Oxford
17. Foote CS, Chang YC, Denny RW (1970) Chemistry of singlet oxygen. X. Carotenoid quenching parallels biological protection. J Am Chem Soc 92(17):5216–5218
18. Liebler DC (1993) Antioxidant reactions of carotenoids. Ann N Y Acad Sci 691(1):20–31
19. Young AJ, Lowe GM (2001) Antioxidant and prooxidant properties of carotenoids. Arch Biochem Biophys 385(1):20–27
20. Chantrell SJ, McAuliffe CA, Munn RW, Pratt AC, Land EJ (1977) Excited states of protoporphyrin IX dimethyl ester: reaction on the triplet with carotenoids. J Chem Soc Faraday Trans 1 73:858–865
21. Mortensen A, Skibsted LH, Sampson J, Rice-Evans C, Everett SA (1997) Comparative mechanisms and rates of free radical scavenging by carotenoid antioxidants. FEBS Lett 418(1–2):91–97

22. Burton GW, Ingold K (1984) Beta-carotene: an unusual type of lipid antioxidant. Science 224(4649):569–573
23. Kiokias S, Gordon MH (2004) Antioxidant properties of carotenoids in vitro and in vivo. Food Rev Int 20(2):99–121
24. Mordi RC, Walton JC, Burton GW, Hughes L, Keith IU, David LA et al (1993) Oxidative degradation of β-carotene and β-apo-8′-carotenal. Tetrahedron 49(4):911–928
25. Liebler DC, McClure TD (1996) Antioxidant reactions of β-carotene: identification of carotenoid− radical adducts. Chem Res Toxicol 9(1):8–11
26. Fiedor J, Fiedor L, Haeßner R, Scheer H (2005) Cyclic endoperoxides of β-carotene, potential pro-oxidants, as products of chemical quenching of singlet oxygen. Biochim Biophys Acta Bioenerg 1709(1):1–4
27. Stratton SP, Schaefer WH, Liebler DC (1993) Isolation and identification of singlet oxygen oxidation products of. beta.-carotene. Chem Res Toxicol 6(4):542–547
28. Polyakov NE, Kruppa AI, Leshina TV, Konovalova TA, Kispert LD (2001) Carotenoids as antioxidants: spin trapping EPR andoptical study. Free Radic Biol Med 31(1):43–52
29. Edge R, Land EJ, McGarvey D, Mulroy L, Truscott TG (1998) Relative one-electron reduction potentials of carotenoid radical cations and the interactions of carotenoids with the vitamin E radical cation. J Am Chem Soc 120(17):4087–4090
30. Edge R, Truscott TG (1999) Carotenoid radicals and the interaction of carotenoids with active oxygen species. In: The photochemistry of carotenoids. Springer, Dordrecht, pp 223–234
31. Mortensen A (2001) Scavenging of acetylperoxyl radicals and quenching of triplet diacetyl by β-carotene: mechanisms and kinetics. J Photochem Photobiol B Biol 61(1–2):62–67
32. El-Agamey A, McGarvey DJ (2003) Evidence for a lack of reactivity of carotenoid addition radicals towards oxygen: a laser flash photolysis study of the reactions of carotenoids with acylperoxyl radicals in polar and non-polar solvents. J Am Chem Soc 125(11):3330–3340
33. Jomová K, Kysel O, Madden J, Morris H, Enoch S, Budzak S et al (2009) Electron transfer from all-trans β-carotene to the t-butyl peroxyl radical at low oxygen pressure (an EPR spectroscopy and computational study). Chem Phys Lett 478(4–6):266–270
34. Panasenko OM, Sharov VS, Briviba K, Sies H (2000) Interaction of peroxynitrite with carotenoids in human low density lipoproteins. Arch Biochem Biophys 373(1):302–305
35. Rodrigues E, Mariutti LR, Mercadante AZ (2012) Scavenging capacity of marine carotenoids against reactive oxygen and nitrogen species in a membrane-mimicking system. Mar Drugs 10(8):1784–1798
36. Simic MG (1992) [42] carotenoid free radicals. In: Methods in enzymology. Elsevier, Amsterdam, pp 444–453
37. Bobrowski K, Das P (1985) Transient phenomena in the pulse radiolysis of retinyl polyenes. 3. Radical cations. J Phys Chem 89(23):5079–5085
38. Everett SA, Dennis MF, Patel KB, Maddix S, Kundu SC, Willson RL (1996) Scavenging of nitrogen dioxide, thiyl, and sulfonyl free radicals by the nutritional antioxidant-carotene. J Biol Chem 271(8):3988–3994
39. Miller NJ, Sampson J, Candeias LP, Bramley PM, Rice-Evans CA (1996) Antioxidant activities of carotenes and xanthophylls. FEBS Lett 384(3):240–242
40. Pérez-Gálvez A, Mínguez-Mosquera MI (2002) Degradation of non-esterified and esterified xanthophylls by free radicals. Biochim Biophys Acta Gen Subj 1569(1–3):31–34
41. Cao G, Verdon CP, Wu A, Wang H, Prior RL (1995) Automated assay of oxygen radical absorbance capacity with the COBAS FARA II. Clin Chem 41(12):1738–1744
42. Huang D, Ou B, Prior RL (2005) The chemistry behind antioxidant capacity assays. J Agric Food Chem 53(6):1841–1856
43. Prior RL, Wu X, Schaich K (2005) Standardized methods for the determination of antioxidant capacity and phenolics in foods and dietary supplements. J Agric Food Chem 53(10):4290–4302
44. Terao J (1989) Antioxidant activity of β-carotene-related carotenoids in solution. Lipids 24(7):659–661

45. Bartosz G, Janaszewska A, Ertel D, Bartosz M (1998) Simple determination of peroxyl radical-trapping capacity. IUBMB Life 46(3):519–528
46. Ringer TV, DeLoof M, Winterrowd G, Francom S, Gaylor S, Ryan J et al (1991) Beta-carotene's effects on serum lipoproteins and immunologic indices in humans. Am J Clin Nutr 53(3):688–694
47. Dugas TR, Morel DW, Harrison EH (1998) Impact of LDL carotenoid and α-tocopherol content on LDL oxidation by endothelial cells in culture. J Lipid Res 39(5):999–1007
48. Carroll Y, Corridan B, Morrissey P (2000) Lipoprotein carotenoid profiles and the susceptibility of low density lipoprotein to oxidative modification in healthy elderly volunteers. Eur J Clin Nutr 54(6):500
49. Šesták Z, Frank HA, Young AJ, Britton G, Cogdell RJ (eds) (2003) The photochemistry of carotenoids. Photosynthetica 41(2):190
50. Mortensen A (2000) Mechanism and kinetics of scavenging of the phenylthiyl radical by carotenoids: a laser flash photolysis study. Asian Chem Lett 4(1–2):135–143
51. Wright AJ, Southon S, Chopra M, Meyer-Wenger A, Moser U, Granado F et al (2002) Comparison of LDL fatty acid and carotenoid concentrations and oxidative resistance of LDL in volunteers from countries with different rates of cardiovascular disease. Br J Nutr 87(1):21–29
52. Nomura T, Kikuchi M, Kubodera A, Kawakami Y (1997) Proton-donative antioxidant activity of fucoxanthin with 1, 1-diphenyl-2-picrylhydrazyl (DPPH). IUBMB Life 42(2):361–370
53. Paravicini TM, Touyz RM (2008) NADPH oxidases, reactive oxygen species, and hypertension: clinical implications and therapeutic possibilities. Diabetes Care 31(Supplement 2):S170–SS180
54. Burke M, Edge R, Land EJ, McGarvey DJ, Truscott TG (2001) One-electron reduction potentials of dietary carotenoid radical cations in aqueous micellar environments. FEBS Lett 500(3):132–136
55. Polyakov NE, Focsan AL, Bowman MK, Kispert LD (2010) Free radical formation in novel carotenoid metal ion complexes of astaxanthin. J Phys Chem B 114(50):16968–16977
56. Shimidzu N, Goto M, Miki W (1996) Carotenoids as singlet oxygen quenchers in marine organisms. Fish Sci 62(1):134–137
57. Heo S-J, Ko S-C, Kang S-M, Kang H-S, Kim J-P, Kim S-H et al (2008) Cytoprotective effect of fucoxanthin isolated from brown algae Sargassum siliquastrum against H 2 O 2-induced cell damage. Eur Food Res Technol 228(1):145–151
58. Miyashita K (2009) Function of marine carotenoids. In: Food factors for health promotion. Karger Publishers, New York, pp 136–146
59. Hosokawa M, Miyashita T, Nishikawa S, Emi S, Tsukui T, Beppu F et al (2010) Fucoxanthin regulates adipocytokine mRNA expression in white adipose tissue of diabetic/obese KK-Ay mice. Arch Biochem Biophys 504(1):17–25
60. Kalam S, Gul MZ, Singh R, Ankati S (2015) Free radicals: implications in etiology of chronic diseases and their amelioration through nutraceuticals. Pharmacologia 6(1):11–20
61. Pashkow FJ, Watumull DG, Campbell CL (2008) Astaxanthin: a novel potential treatment for oxidative stress and inflammation in cardiovascular disease. Am J Cardiol 101(10):S58–S68
62. Kurashige M, Okimasu E, Inoue M, Utsumi K (1990) Inhibition of oxidative injury of biological membranes by astaxanthin. Physiol Chem Phys Med NMR 22(1):27–38
63. Liu X, Shibata T, Hisaka S, Osawa T (2009) Astaxanthin inhibits reactive oxygen species-mediated cellular toxicity in dopaminergic SH-SY5Y cells via mitochondria-targeted protective mechanism. Brain Res 1254:18–27
64. Junghans A, Sies H, Stahl W (2001) Macular pigments lutein and zeaxanthin as blue light filters studied in liposomes. Arch Biochem Biophys 391(2):160–164
65. Tsuboi M, Etoh H, Kato K, Nakatugawa H, Kato H, Maejima Y et al (2011) Nitrocapsanthin and nitrofucoxanthin, respective products of capsanthin and fucoxanthin reaction with peroxynitrite. J Agric Food Chem 59(19):10572–10578

66. Scheidegger R, Pande A, Bounds PL, Koppenol W (1998) The reaction of peroxynitrite with zeaxanthin. Nitric Oxide 2(1):8–16
67. Maoka T, Tokuda H, Suzuki N, Kato H, Etoh H (2012) Anti-oxidative, anti-tumor-promoting, and anti-carcinogensis activities of nitroastaxanthin and nitrolutein, the reaction products of astaxanthin and lutein with peroxynitrite. Mar Drugs 10(6):1391–1399
68. Régnier P, Bastias J, Rodriguez-Ruiz V, Caballero-Casero N, Caballo C, Sicilia D et al (2015) Astaxanthin from Haematococcus pluvialis prevents oxidative stress on human endothelial cells without toxicity. Mar Drugs 13(5):2857–2874
69. Fassett RG, Coombes JS (2011) Astaxanthin: a potential therapeutic agent in cardiovascular disease. Mar Drugs 9(3):447–465
70. Woods JA, Bilton RF, Young AJ (1999) β-Carotene enhances hydrogen peroxide-induced DNA damage in human hepatocellular HepG2 cells. FEBS Lett 449(2–3):255–258
71. Palozza P, Serini S, Torsello A, Di Nicuolo F, Piccioni E, Ubaldi V et al (2003) β-Carotene regulates NF-κB DNA-binding activity by a redox mechanism in human leukemia and colon adenocarcinoma cells. J Nutr 133(2):381–388
72. Palozza P, Serini S, Di Nicuolo F, Boninsegna A, Torsello A, Maggiano N et al (2004) β-Carotene exacerbates DNA oxidative damage and modifies p53-related pathways of cell proliferation and apoptosis in cultured cells exposed to tobacco smoke condensate. Carcinogenesis 25(8):1315–1325
73. Sachindra NM, Sato E, Maeda H, Hosokawa M, Niwano Y, Kohno M et al (2007) Radical scavenging and singlet oxygen quenching activity of marine carotenoid fucoxanthin and its metabolites. J Agric Food Chem 55(21):8516–8522
74. Perocco P, Paolini M, Mazzullo M, Biagi G, Cantelli-Forti G (1999) β-Carotene as enhancer of cell transforming activity of powerful carcinogens and cigarette-smoke condensate on BALB/c 3T3 cells in vitro. Mutat Res Genet Toxicol Environ Mutagen 440(1):83–90
75. Kikugawa K, Hiramoto K, Tomiyama S, Asano Y (1997) β-Carotene effectively scavenges toxic nitrogen oxides: nitrogen dioxide and peroxynitrous acid. FEBS Lett 404(2–3):175–178
76. Naguib YM (2000) Antioxidant activities of astaxanthin and related carotenoids. J Agric Food Chem 48(4):1150–1154
77. Farombi EO, Britton G (1999) Antioxidant activity of palm oil carotenes in peroxyl radical-mediatedperoxidation of phosphatidyl choline liposomes. Redox Rep 4(1–2):61–68
78. Yaping Z, Suping Q, Wenli Y, Zheng X, Hong S, Side Y et al (2002) Antioxidant activity of lycopene extracted from tomato paste towards trichloromethyl peroxyl radical CCl3O2·. Food Chem 77(2):209–212
79. Klebanov G, Kapitanov A, Teselkin Y, Babenkova I, Zhambalova B, Lyubitsky O et al (1998) The antioxidant properties of lycopene. Membr Cell Biol 12(2):287–300
80. Srinivasan M, Sudheer AR, Pillai KR, Kumar PR, Sudhakaran P, Menon V (2007) Lycopene as a natural protector against γ-radiation induced DNA damage, lipid peroxidation and antioxidant status in primary culture of isolated rat hepatocytes in vitro. Biochim Biophys Acta Gen Subj 1770(4):659–665
81. Takeoka GR, Dao L, Flessa S, Gillespie DM, Jewell WT, Huebner B et al (2001) Processing effects on lycopene content and antioxidant activity of tomatoes. J Agric Food Chem 49(8):3713–3717
82. Kim CH, Park MK, Kim SK, Cho YH (2014) Antioxidant capacity and anti-inflammatory activity of lycopene in watermelon. Int J Food Sci Technol 49(9):2083–2091
83. Seon-Jin L (2003) Astaxanthin inhibits nitric oxide production and inflammatory gene expression by suppressing IkB kinase-dependent NF-kB activation. Mol Cell 16:97–105
84. Palozza P, Krinsky NI (1992) Astaxanthin and canthaxanthin are potent antioxidants in a membrane model. Arch Biochem Biophys 297(2):291–295
85. Khan SK, Malinski T, Mason RP, Kubant R, Jacob RF, Fujioka K et al (2010) Novel astaxanthin prodrug (CDX-085) attenuates thrombosis in a mouse model. Thromb Res 126(4):299–305

86. Bhosale P, Bernstein PS (2005) Synergistic effects of zeaxanthin and its binding protein in the prevention of lipid membrane oxidation. Biochim Biophys Acta Mol basis Dis 1740(2):116–121
87. Tsuboi M, Etoh H, Yomoda Y, Kato K, Kato H, Kulkarni A et al (2010) Nitration reaction of lutein with peroxynitrite. Tetrahedron Lett 51(4):676–678
88. Wisniewska-Becker A, Nawrocki G, Duda M, Subczynski WK (2012) Structural aspects of the antioxidant activity of lutein in a model of photoreceptor membranes. Acta Biochim Pol 59(1):119–124
89. Krinsky NI, Johnson EJ (2005) Carotenoid actions and their relation to health and disease. Mol Asp Med 26(6):459–516
90. Mayne ST, Parker RS (1989) Antioxidant activity of dietary canthaxanthin. Nutr Cancer 12(3):225–236
91. Zaragozá M, López D, Sáiz MP, Poquet M, Pérez J, Puig-Parellada P et al (2008) Toxicity and antioxidant activity in vitro and in vivo of two Fucus vesiculosus extracts. J Agric Food Chem 56(17):7773–7780
92. Liu C-L, Chiu Y-T, Hu M-L (2011) Fucoxanthin enhances HO-1 and NQO1 expression in murine hepatic BNL CL. 2 cells through activation of the Nrf2/ARE system partially by its pro-oxidant activity. J Agric Food Chem 59(20):11344–11351
93. Iwasaki S, Widjaja-Adhi MAK, Koide A, Kaga T, Nakano S, Beppu F et al (2012) In vivo antioxidant activity of fucoxanthin on obese/diabetes KK-A y mice. Food and Nutr Sci 3(11):1491–1499
94. Havaux M, Dall'Osto L, Bassi R (2007) Zeaxanthin has enhanced antioxidant capacity with respect to all other xanthophylls in Arabidopsis leaves and functions independent of binding to PSII antennae. Plant Physiol 145(4):1506–1520
95. Broekmans WM, Berendschot TT, Klöpping-Ketelaars IA, de Vries AJ, Goldbohm RA, Tijburg LB et al (2002) Macular pigment density in relation to serum and adipose tissue concentrations of lutein and serum concentrations of zeaxanthin. Am J Clin Nutr 76(3):595–603
96. Cha KH, Koo SY, Lee D-U (2008) Antiproliferative effects of carotenoids extracted from Chlorella ellipsoidea and Chlorella vulgaris on human colon cancer cells. J Agric Food Chem 56(22):10521–10526
97. Kuwahara R, Hatate H, Yuki T, Murata H, Tanaka R, Hama Y (2009) Antioxidant property of polyhydroxylated naphthoquinone pigments from shells of purple sea urchin Anthocidaris crassispina. LWT-Food Sci Technol 42(7):1296–1300
98. Haug T, Kjuul AK, Styrvold OB, Sandsdalen E, Olsen ØM, Stensvåg K (2002) Antibacterial activity in Strongylocentrotus droebachiensis (Echinoidea), Cucumaria frondosa (Holothuroidea), and Asterias rubens (Asteroidea). J Invertebr Pathol 81(2):94–102
99. Lennikov A, Kitaichi N, Noda K, Mizuuchi K, Ando R, Dong Z et al (2014) Amelioration of endotoxin-induced uveitis treated with the sea urchin pigment echinochrome in rats. Mol Vis 20:171–177
100. Murakami A, Nakashima M, Koshiba T, Maoka T, Nishino H, Yano M et al (2000) Modifying effects of carotenoids on superoxide and nitric oxide generation from stimulated leukocytes. Cancer Lett 149(1–2):115–123
101. Roehrs M, Figueiredo CG, Zanchi MM, Bochi GV, Moresco RN, Quatrin A et al (2014) Bixin and norbixin have opposite effects on glycemia, lipidemia, and oxidative stress in streptozotocin-induced diabetic rats. Int J Endocrinol 2014:839095 https://doi.org/10.1155/2014/839095
102. Furukawa S, Fujita T, Shimabukuro M, Iwaki M, Yamada Y, Nakajima Y et al (2017) Increased oxidative stress in obesity and its impact on metabolic syndrome. J Clin Invest 114(12):1752–1761
103. Altinoz E, Oner Z, Elbe H, Cigremis Y, Turkoz Y (2015) Protective effects of saffron (its active constituent, crocin) on nephropathy in streptozotocin-induced diabetic rats. Hum Exp Toxicol 34(2):127–134

104. Tamaddonfard E, Farshid AA, Asri-Rezaee S, Javadi S, Khosravi V, Rahman B et al (2013) Crocin improved learning and memory impairments in streptozotocin-induced diabetic rats. Iran J Basic Med Sci 16(1):91–100
105. Organization WH. Cancer key statistics. Avenue Appia 20, 1211 Geneva. 2020. https://www.who.int/cancer/resources/keyfacts/en/
106. Voorrips LE, Goldbohm RA, Brants HA, van Poppel GA, Sturmans F, Hermus RJ et al (2000) A prospective cohort study on antioxidant and folate intake and male lung cancer risk. Cancer Epidemiol Prev Biomark 9(4):357–365
107. Mayne ST, Janerich DT, Greenwald P, Chorost S, Tucci C, Zaman MB et al (1994) Dietary beta carotene and lung cancer risk in US nonsmokers. JNCI: J Natl Cancer Inst 86(1):33–38
108. Michaud DS, Feskanich D, Rimm EB, Colditz GA, Speizer FE, Willett WC et al (2000) Intake of specific carotenoids and risk of lung cancer in 2 prospective US cohorts. Am J Clin Nutr 72(4):990–997
109. Collins AR (2001) Carotenoids and genomic stability. Mutat Res Fundam Mol Mech Mutagen 475(1–2):21–28
110. Lowe GM, Booth LA, Young AJ, Bilton RF (1999) Lycopene and β-carotene protect against oxidative damage in HT29 cells at low concentrations but rapidly lose this capacity at higher doses. Free Radic Res 30(2):141–151
111. Yeh S-L, Hu M-L (2001) Induction of oxidative DNA damage in human foreskin fibroblast Hs68 cells by oxidized β-carotene and lycopene. Free Radic Res 35(2):203–213
112. Chan JM, Gann PH, Giovannucci EL (2005) Role of diet in prostate cancer development and progression. J Clin Oncol 23(32):8152–8160
113. Kim KB, Lee BM (1997) Oxidative stress to DNA, protein, and antioxidant enzymes (superoxide dismutase and catalase) in rats treated with benzo (a) pyrene. Cancer Lett 113(1–2):205–212
114. Chlichlia K, Peter ME, Rocha M, Scaffidi C, Bucur M, Krammer PH et al (1998) Caspase activation is required for nitric oxide–mediated, CD95 (APO-1/Fas)–dependent and independent apoptosis in human neoplastic lymphoid cells. Blood 91(11):4311–4320
115. Bowie A, O'Neill LA (2000) Oxidative stress and nuclear factor-κB activation: a reassessment of the evidence in the light of recent discoveries. Biochem Pharmacol 59(1):13–23
116. La Rosa F, Pierce JW, Sonenshein GE (1994) Differential regulation of the c-myc oncogene promoter by the NF-kappa B rel family of transcription factors. Mol Cell Biol 14(2):1039–1044
117. Schwartz JL, Singh RP, Teicher B, Wright JE, Trites DH, Shklar G (1990) Induction of a 70KD protein associated with the selective cytotoxicity of beta-carotene in human epidermal carcinoma. Biochem Biophys Res Commun 169(3):941–946
118. Ortmann EK, Mayerhofer T, Getoff N, Kodym R (2004) Effect of antioxidant vitamins on radiation-induced apoptosis in cells of a human lymphoblastic cell line. Radiat Res 161(1):48–55
119. Giovannucci E, Rimm EB, Liu Y, Stampfer MJ, Willett WC (2002) A prospective study of tomato products, lycopene, and prostate cancer risk. J Natl Cancer Inst 94(5):391–398
120. Etminan M, Takkouche B, Caamaño-Isorna F (2004) The role of tomato products and lycopene in the prevention of prostate cancer: a meta-analysis of observational studies. Cancer Epidemiol Prev Biomark 13(3):340–345
121. Mayne S, Goodwin W Jr (1993) Chemoprevention of head and neck cancer. Curr Opin Otolaryngol Head Neck Surg 1:126–132
122. Serpeloni JM, Barcelos GRM, Angeli JPF, Mercadante AZ, Bianchi MLP, Antunes LMG (2012) Dietary carotenoid lutein protects against DNA damage and alterations of the redox status induced by cisplatin in human derived HepG2 cells. Toxicol In Vitro 26(2):288–294
123. Wlaschek M, Tantcheva-Poór I, Naderi L, Ma W, Schneider LA, Razi-Wolf Z et al (2001) Solar UV irradiation and dermal photoaging. J Photochem Photobiol B Biol 63(1–3):41–51
124. Ouhtit A, Ananthaswamy HN (2001) A model for UV-induction of skin cancer. Biomed Res Int 1(1):5–6

125. Biesalski HK, Obermueller-Jevic UC (2001) UV light, beta-carotene and human skin—beneficial and potentially harmful effects. Arch Biochem Biophys 389(1):1–6
126. Bando N, Hayashi H, Wakamatsu S, Inakuma T, Miyoshi M, Nagao A et al (2004) Participation of singlet oxygen in ultraviolet-a-induced lipid peroxidation in mouse skin and its inhibition by dietary β-carotene: an ex vivo study. Free Radic Biol Med 37(11):1854–1863
127. Bone R, Landrum J, Fernandez L, Tarsis S (1988) Analysis of the macular pigment by HPLC: retinal distribution and age study. Invest Ophthalmol Vis Sci 29(6):843–849
128. SanGiovanni JP, Neuringer M (2012) The putative role of lutein and zeaxanthin as protective agents against age-related macular degeneration: promise of molecular genetics for guiding mechanistic and translational research in the field. Am J Clin Nutr 96(5):1223S–1233S
129. Wooten BR, Hammond BR (2002) Macular pigment: influences on visual acuity and visibility. Prog Retin Eye Res 21(2):225–240
130. Loane E, Kelliher C, Beatty S, Nolan JM (2008) The rationale and evidence base for a protective role of macular pigment in age-related maculopathy. Br J Ophthalmol 92(9):1163–1168
131. Li B, Vachali P, Frederick JM, Bernstein PS (2011) Identification of StARD3 as a lutein-binding protein in the macula of the primate retina. Biochemistry 50(13):2541–2549
132. Bhosale P, Larson AJ, Frederick JM, Southwick K, Thulin CD, Bernstein PS (2004) Identification and characterization of a Pi isoform of glutathione S-transferase (GSTP1) as a zeaxanthin-binding protein in the macula of the human eye. J Biol Chem 279(47):49447–49454
133. Vachali P, Bhosale P, Bernstein PS (2012) Microbial carotenoids. In: Microbial carotenoids from fungi. Springer, New York, pp 41–59
134. Suzuki K, Ito Y, Nakamura S, Ochiai J, Aoki K (2002) Relationship between serum carotenoids and hyperglycemia: a population-based cross-sectional study. J Epidemiol 12(5):357–366
135. Uchiyama K, Naito Y, Hasegawa G, Nakamura N, Takahashi J, Yoshikawa T (2002) Astaxanthin protects β-cells against glucose toxicity in diabetic db/db mice. Redox Rep 7(5):290–293
136. Manabe E, Handa O, Naito Y, Mizushima K, Akagiri S, Adachi S et al (2008) Astaxanthin protects mesangial cells from hyperglycemia-induced oxidative signaling. J Cell Biochem 103(6):1925–1937
137. Bhuvaneswari S, Yogalakshmi B, Sreeja S, Anuradha CV (2014) Astaxanthin reduces hepatic endoplasmic reticulum stress and nuclear factor-κB-mediated inflammation in high fructose and high fat diet-fed mice. Cell Stress Chaperones 19(2):183–191
138. Guerra BA, Otton R (2011) Impact of the carotenoid astaxanthin on phagocytic capacity and ROS/RNS production of human neutrophils treated with free fatty acids and high glucose. Int Immunopharmacol 11(12):2220–2226
139. Aydin M, Celik S (2012) Effects of lycopene on plasma glucose, insulin levels, oxidative stress, and body weights of streptozotocin-induced diabetic rats. Turk J Med Sci 42(Sup. 2):1406–1413
140. Ali MM, Agha FG (2009) Amelioration of streptozotocin-induced diabetes mellitus, oxidative stress and dyslipidemia in rats by tomato extract lycopene. Scand J Clin Lab Invest 69(3):371–379
141. Duzguner V, Kucukgul A, Erdogan S, Celik S, Sahin K (2008) Effect of lycopene administration on plasma glucose, oxidative stress and body weight in streptozotocin diabetic rats. J Appl Anim Res 33(1):17–20
142. Kianbakht S, Hajiaghaee R (2011) Anti-hyperglycemic effects of saffron and its active constituents, crocin and safranal, in alloxan-induced diabetic rats. J Med Plants 3(39):82–89
143. Rajaei Z, Hadjzadeh M-A-R, Nemati H, Hosseini M, Ahmadi M, Shafiee S (2013) Antihyperglycemic and antioxidant activity of crocin in streptozotocin-induced diabetic rats. J Med Food 16(3):206–210
144. Van de Vijver L, Kardinaal A, Grobbee D, Princen H, Van Poppel G (1997) Lipoprotein oxidation, antioxidants and cardiovascular risk: epidemiologic evidence. Prostaglandins Leukot Essent Fat Acids 57(4–5):479–487

Chapter 13
Carotenoids as Anticancer Agents

Saikat Dewanjee, Sonjit Das, Swarnalata Joardar, Simanta Bhattacharjee, and Pratik Chakraborty

13.1 Introduction

Cancer is a lethal non-communicable disease. Incidence of this disease is increasing alarmingly both in the developed as well as in the developing countries. It can ruin the quality of life physically, psychologically, and financially. Compared to other diseases, cancer attributes much higher rate of death. Breast, colon, lung, and skin cancers have been categorized as the top four common forms cancer accounting 358,967, 342,137, 309,589, and 82,075, respectively in the diagnosed cases in the European Union in 2012 [1]. D'Souza et al. reported that 0.44 million cancer patients expired in 2011, while 1.11 million were at the borderline and are expected to die between 2016 and 2021 [2]. It has been projected that the mortality of cancer patients may reach up to 0.70 million by the year 2026 [2, 3]. At present, chemotherapy is the principal therapeutic intervention in cancer treatment, mainly targets multi-steps in carcinogenesis. Pharmacological modulation of various regulatory pathways that are involved in the carcinogenesis can be the best therapeutic approach. A number of epidemiologic studies revealed that the consumption of carotenoid enriched vegetables and fruits can reduce the risk of cancer [4, 5]. Several reports revealed the cancer inhibitory effects of carotenoids; while, β-carotene and fucoxanthinol have been found to possess both inhibitory and promoting effects in cancer [6–11]. Carotenoids have been found to exhibit anticancer effect via regulating various molecular processes, such as free radical production, Wnt signaling, Insulin-like growth factor 1 (IGF-1) signaling, autophagy, nuclear factor-κB (NF-κB) signaling, phosphatidylinositol-3 kinase (PI3K)/protein kinase B (AKT) signaling, cyclin-dependent kinases (CDK), peroxisome proliferator-activated receptors (PPARs)-mediated signal transduction, etc. This chapter deals with the role of

S. Dewanjee (✉) · S. Das · S. Joardar · S. Bhattacharjee · P. Chakraborty
Advanced Pharmacognosy Research Laboratory, Department of Pharmaceutical Technology, Jadavpur University, Kolkata, India
e-mail: saikat.dewanjee@jadavpuruniversity.in

© Springer Nature Switzerland AG 2021

M. Zia-Ul-Haq et al. (eds.), *Carotenoids: Structure and Function in the Human Body*, https://doi.org/10.1007/978-3-030-46459-2_13

different biochemical and molecular factors in cancer development and the pharmacotherapeutic target/s of carotenoids in cancer prevention.

13.2 Free Radicals in Cancer Pathogenesis and Chemo-preventive Role of Carotenoids

13.2.1 Free Radicals and Cancer Pathogenesis

Oxidative free radicals can oxidize other molecules through electron transferring [12]. Generally, cancerous cells exhibit higher level of oxidative radicals than that of normal cells caused by the abnormal oxygen metabolism in mitochondria [13]. These free radicals facilitate promotion and progression of different cancerous cells, such as melanoma, hepatoma, breast, pancreas, bladder, lung, colon, and prostate cancers [14]. The level of endogenous antioxidants reduces with increase in oxidative free radicals [15]. These reactive species can attack cellular macromolecules, such as nucleic acids, proteins, and lipids depending upon their concentration, cellular environment, and time of exposure. Free radicals provoke DNA damage that either up regulate or down-regulate the oncogenes [16–20]. The level of 8-hydroxydeoxy guanosine (8-OHdG) increases during oxidative DNA damage [16–20]. Increased level of H_2O_2 was reported to endorse various molecular events, which are necessary for the cell invasion and metastasis [21]. Most noticeably, oxidative free radicals are known to trigger cell proliferation via regulating several signal transduction pathways like IGF-1, IRS-1, PI3K, AKT, CDK, PPAR, etc., which are some of signal proteins affecting cell proliferation and are regulated by the free radicals [21–23]. Oxidative radicals can also activate c-Jun and c-Fos subunits of activator protein-1 (AP-1) and the active nuclear transcription factor, thereby activate genes involved in the cellular proliferation [24]. In addition, these free radicals have been proposed to promote cancer via modulation of cytosolic sulfhydryl bearing proteins, such as thioredoxin reductase (TRXR), thioredoxin (TRX), and peroxiredoxins [13]. The carcinogenic role of oxidative free radicals has been depicted in Fig. 13.1.

13.2.2 Chemo-preventive Role of Carotenoids by Targeting Free Radicals

Carotenoids are most common natural pigments and have achieved significant interest due to their antioxidant role. Over 600 different carotenoids have been extracted from plants, animals, and microbes [25]. Carotenoids exhibited antioxidant property due their capacity to delocalize unpaired electrons and efficiently deactivate sensitizer molecules [26–31]. Thus, β-carotene can quench singlet oxygen without

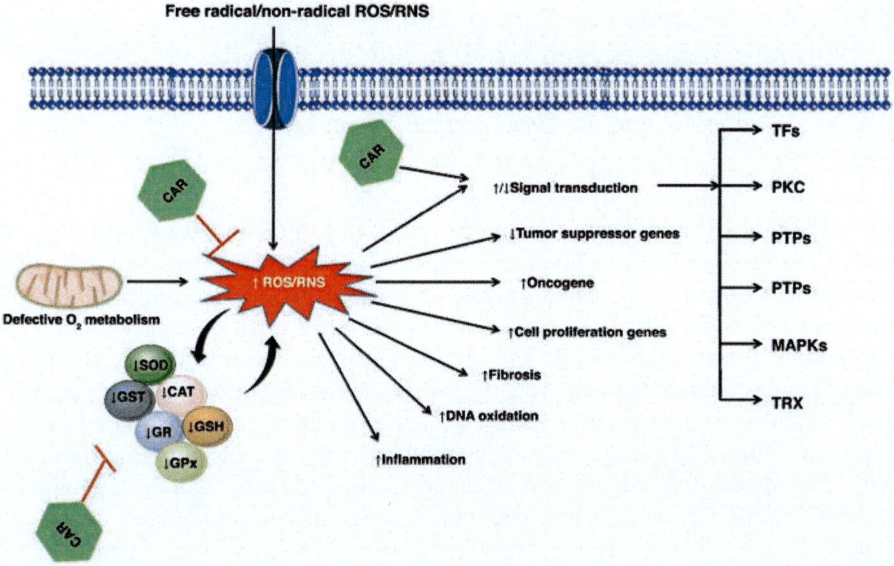

Fig. 13.1 Different roles of ROS/RNS in cancer and pharmacotherapeutic targets of carotenoids for ROS/RNS. *CAR* carotenoids, *MAPKs* Mitogen activated protein kinases (JNKs, Erks), *PKCs* Protein kinase C isoforms, *PTKs* non-receptor protein tyrosine kinases (Src, Abi, JAK), *PTPs* Protein phosphatases (PTEN), *TFs* Transcriptional factors (NF-κB, HIF-1α, AP-1, Nrf-2). Black arrows (→) indicate down-stream events. Red lines (T) show the restriction/inhibition

degradation [26, 27]. Lycopene has been reported to neutralize H_2O_2 and $ONO^{•-}$ [28–31]. Crocin exhibited chemo-preventive effect via down-regulating reactive oxygen species (ROS) production and oxidative DNA fragmentation in the human myeloma cells [32]. β-carotene can reduce ROS production at lower concentration [33]. However, at higher concentration, it has been reported to increase ROS production and simultaneously increased DNA fragmentation, lactic acid dehydrogenase (LDH) release in the undifferentiated and differentiated human leukemia cells [33, 34]. In addition, it can increase the transcriptions of c-Jun and c-Fos genes in ferrets [7]. Fucoxanthin has been confirmed to initiate DNA fragmentation in HL-60 cell [35]. Astaxanthin was found to inhibit lipid peroxidation and DNA cleavage in the preclinical assays [36, 37]. Furthermore, astaxanthin has been found to reciprocate the levels of SOD and GSH and inhibit DNA damage in the human skin fibroblasts, human intestinal CaCo-2, and human melanocytes cells [38]. Lutein has shown down-regulation in ROS production in human liver cells [39]. The carotenoids, such as neoxanthin, fucoxanthinol, and halocynthiaxanthin have been reported to induce DNA fragmentations in human leukemia, breast cancer, colon, and prostate cancer cells [40, 41]. Bixin was claimed to induce ROS production and to reduce TRX and thioredoxin reductase (TRXR1) activities in the myelomas [42]. Pharmaco-therapeutic targets of carotenoids in free radical-provoked carcinogenicity have been shown in Fig. 13.1.

13.3 Wnt Signaling in Cancer Pathogenesis and Chemo-preventive Role of Carotenoids

13.3.1 Wnt Pathway in Cancer Pathogenesis

The Wnt is a family of protein ligands, which effects several cellular processes [43]. The Wnt signaling pathway includes two intracellular pathways [44]. One of which executes β-catenin, while other is β-catenin-independent pathway [44, 45]. The upregulation of Wnt/β-catenin signaling in cancer cells have been confirmed in various studies [46–48]. However, its activation relies upon the concentration of cytoplasmic β-catenin. Normally, this protein is kept at a low level in the cytoplasm due to its ubiquitin-dependent proteolytic degradation, for details of Wnt pathway read ref. [44, 45]. Data from prostate cancer patients confirmed mutation in β-catenin [46, 47]. Up-regulation of β-catenin at mRNA and protein levels have been demonstrated in human cervical cancer CaSki cells [48]. The overall impression of Wnt/β-catenin signaling pathway in carcinogenicity has been shown in Fig. 13.2.

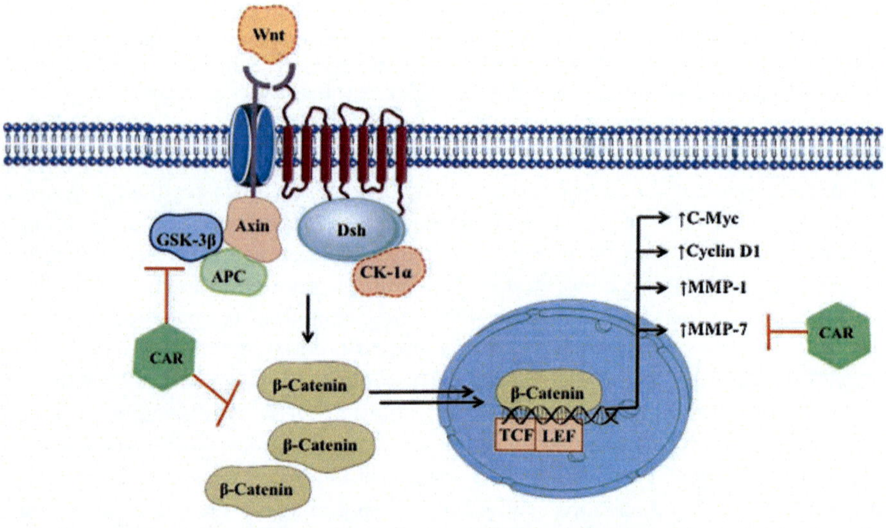

Fig. 13.2 Roles of Wnt/β-catenin signalling in cancer and pharmacotherapeutic targets of carotenoids in Wnt/β-catenin signalling. *APC* Adenomatous polyposis coli, *CAR* carotenoids, *CK-1α* Casein kinase 1α, *GSK-3β* Glycogen synthase kinase 3β, *LER* Lymphocyte enhancer factor, *MMP* Matrix metalloproteinase, *TCF* T-cell factor. Black arrows (→) indicate down-stream events. Red lines (T) show the restriction/inhibition

13.3.2 Chemo-preventive Role of Carotenoids by Targeting Wnt Signaling

Astaxanthin suppress Wnt/β-catenin signaling in human hepatocellular carcinoma cells and thus proliferation is inhibited and apoptosis is induced [49]. It was found to down-regulate the expressions GSK-3β and β-catenin, GSK-3β and β-catenin phosphorylation, and nuclear translocation of β-catenin in human hepatoma cell lines, such as LM3 and SMMC-7721 [49]. Kavitha et al. reported that astaxanthin can inhibit of Wnt/β-catenin signaling through down-regulation of GSK-3β in hamster oral cancer [50]. Lycopene was found to inhibit the expression of β-catenin protein in human colon cancer cells [51]. Furthermore, lycopene has been reported to inhibit leptin-mediated cell invasion, MMP-7 expression, AKT phosphorylation, GSK-3β expression, and extracellular signal-regulated kinases 1/2 (ERK 1/2) expression in human colon cancer cells [52]. Pharmaco-therapeutic targets of carotenoids in Wnt/β-catenin signal transduction have been shown in Fig. 13.2.

13.4 IGF-1R/IGR-1R Signaling in Cancer Pathogenesis and Chemo-preventive Role of Carotenoids

13.4.1 IGF-1R/IGR-1R Pathway in Cancer Pathogenesis

Under normal physiological state, liver cells discharge IGF-1 protein into blood, which is the essential source of circulating IGF-1. Generally, IGF- 1 level within our system is regulated by different insulin-like growth factor-binding proteins (IGFBPs). IGF-1 protein is involved in the embryonic development of pancreas and regulates the endocrine pancreas in postnatal life. It maintains the physiological and energy homeostasis of numerous tissues. Epidemiological investigations established the correlation between the risk of cancers and levels of circulating IGF-1 and IGFBP-3 [53]. IGF-1 stimulate cell proliferation and protect against initiation of apoptosis in primary mesenchymal and epithelial cells, and various human carcinoma cells [54]. It has been reported that autocrine activation of the IGF-1R system regulates the expression of vascular endothelial growth factor (VEGF) and angiogenesis in human pancreatic cancer [55]. Up-regulated IGF-1 ligand level and overexpression of IGF-1R have been observed in pancreatic cancer cells. Kopantzev et al. reported increased secretion of IGF-1 in cancer-associated fibroblasts and M2 macrophages [56]. A schematic overview of IGF-1R/IGR-1R signaling in cancer pathogenesis has been shown in Fig. 13.3.

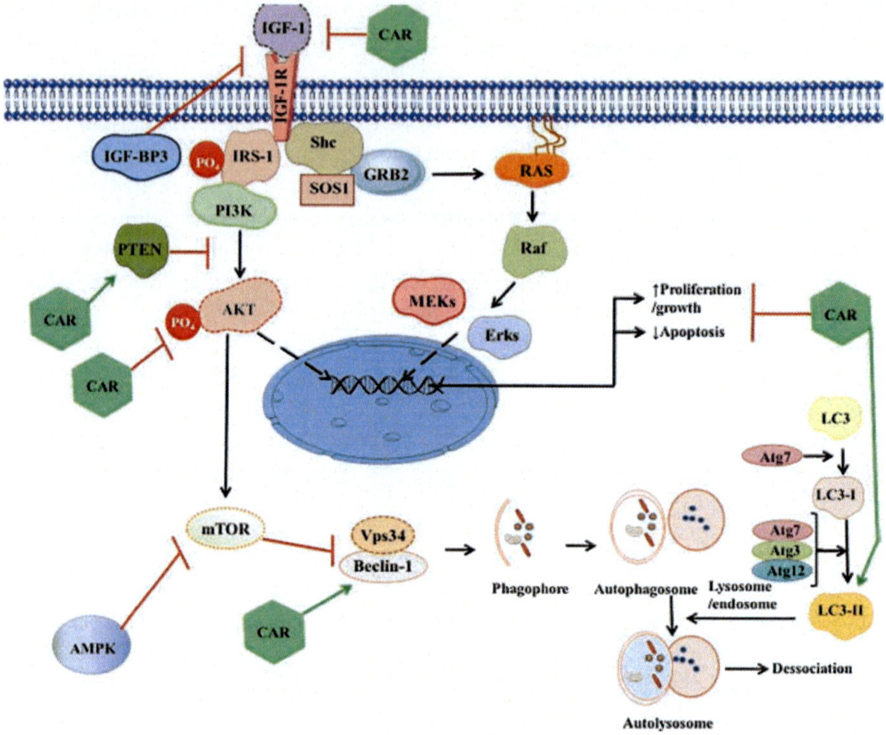

Fig. 13.3 Roles of IGF/IRS-1/PI3K/AKT/AMPK/mTOR signalling in cancer and pharmacotherapeutic targets of carotenoids. *AKT* Serine/threonine-specific protein kinase or protein kinase B, *AMPK* 5′ Adenosine monophosphate-activated protein kinase, *Atg* Autophagy related, *CAR* Carotenoids, *ERK* Extracellular signal-regulated kinases, *GRB2* Growth factor receptor-bound protein 2, *IGF-1* Insulin-like growth factor 1, *IRS-1* Insulin receptor substrate 1, *LC3* Microtubule-associated protein 1 A/1B-light chain 3, *LC3-I* Cytosolic form of LC3, *LC3-II* Microtubule associated protein 1 A/1B-light chain 3-phosphatidylethanolamine conjugate, *MEK* Mitogen-activated protein kinase kinase, *mTOR* Mechanistic target of rapamycin, *PI3K* Phosphatidylinositol-3 kinase, *Rafl* Rapidly accelerating fibrosarcoma kinase protein 1, *She* Src-homology and collagen homology, *SOS1* Son of seven less homolog 1, and *Vsp34* Vacuolar protein sorting 34. Black arrows (→) indicate down-stream events. Red lines (T) show the restriction/inhibition. Green arrows (→) indicate activation

13.4.2 Pharmacotherapeutic Targets of Carotenoid in the IGF-1R/IGR-1R Signaling

It has been revealed that the activation of IGF-1/IGF-1R can trigger the growth of pancreatic tumors [53, 56–58]. It could also regulate the sensitivity of the cancer cells toward chemotherapy agents. Therefore, IGF-1/ IGF-1R would be a potential target in the chemotherapy of pancreatic or other cancers. Lycopene has been reported to impair IGF-1R activation via suppressing IGF-1 activation and triggering the expression and secretion of IGFBP-3 in prostate cancer cells [57]. Through

IGF-1/IGF-1R inhibition, lycopene subsequently down-regulates downstream AKT kinase and survivin activities resulting induction of apoptosis [57]. Lycopene supplementation has been found to potentiate the chemotherapeutic effect of docetaxel in castration-resistant prostate cancer patients [57]. In addition, lycopene was found to arrest the proliferation in human endometrial, mammary, and lung cancer cells by reducing IGF-1R expression [58]. α- and β-carotenes can also exhibit similar growth suppressing effects like lycopene to human endometrial cancer cells, however, much higher concentrations of α- and β-carotenes are required [57]. Pharmacotherapeutic targets of carotenoids in IGF-1R/IGR-1R signal transduction have been shown in Fig. 13.3.

13.5 Autophagy in Cancer Pathogenesis and Chemo-preventive Role of Carotenoids

13.5.1 Autophagy in Cancer Pathogenesis

Autophagy has been proposed to constrain tumor initiation, promotion and maintenance [59]. Under normal physiological state, autophagy exerts cytoprotective role. In addition, it could arrest genomic damage which can induce abnormal mutation resulting in carcinogenesis. Therefore, it has been regarded as a tumor suppressive mechanism [11, 60]. Significant reduction in the beclin-1, autophagy-related gene, expression revealed defective autophagy in several cancer cells [61–63]. The monoallelic deletion of the beclin-1 can spontaneously develop tumors [64]. Allelic loss of beclin-1 was found to endorse the development of breast, ovarian, and prostate cancers [64]. Deletion of Atg5 and Atg7 has been reported to develop benign liver adenomas in mice by inducing mitochondrial damage and oxidative stress [65]. In contrast, suppression of beclin-1 gene expression has been reported in stage IIIB colon cancer and non-Hodgkin lymphomas, which could support the dual roles of autophagy in carcinogenesis [64, 66]. Another study demonstrated a reduction in the growth of mammary tumor after deletion of autophagic gene, FIP200, in mice [67]. A schematic impression of autophagy in cancer pathogenesis has been shown in Fig. 13.3.

13.5.2 Pharmacotherapeutic Targets of Carotenoid in the Autophagy

Considering the dual roles of autophagy as promoter or suppressor in tumor growth, it would be a delicate target in the chemoprevention of cancers. Preclinical data revealed that stress-induced autophagy in tumor cells is principally cytoprotective, and thus, inhibition of autophagy pathway can augment cell death by different anticancer agents [68]. In contrast, it has also been reported that several anticancer agents exhibit chemotherapeutic effect via inducing autophagic cell death [69]. Fucoxanthin has been found to induce autophagy-provoked cytotoxic effect in human epithelial cervical cancer cells via increasing LC3 II and beclin-1 expressions through inhibition of AKT/mTOR signaling [69]. Crocin has been reported to induce inefficient autophagy in p53-null colorectal cancer cells [70]. In addition, crocin can arrest the proliferation of wild and p53-null colorectal cancer cells via arresting cell cycle at G1 and G2, respectively [70]. Pharmaco-therapeutic targets of carotenoids in autophagy have been depicted in Fig. 13.3.

13.6 PI3K/AKT Signaling in Cancer Pathogenesis and Chemo-preventive Role of Carotenoids

13.6.1 PI3K/AKT Signaling and Cancer

PI3K is a major signal protein that catalyzes the production of phosphatidylinositol-3,4,5-triphosphate (PIP3), a lipid second messenger, at the cell membrane. PIP3 subsequently recruits and activates different downstream targets, such as AKT [71]. Activation of PI3K/AKT implicated to the survival, proliferation, and growth of cancer cells. Tumors have abnormal activation of PI3K/AKT signaling [72, 73]. PI3K/AKT signaling is antagonized by the tumor suppressor phosphatase and tensin homolog (PTEN) and thereby, PTEN serves as tumor suppressor factor [73]. The rational confirmation of tumor suppressive role of PTEN has been revealed to be mediated through PI3K/AKT inhibition [71]. Germline mutation of PTEN was found to induce autosomal dominant cancer syndromes [71]. In addition, the tissue specific deletion of PTEN has been proposed to promote tumorogenesis, proliferation, and cell survival by hyperactivating PI3K/AKT signaling [71]. A report implicated over-expressions of PI3K, AKT1 and AKT2 proteins in human ovarian cancer cells [74]. Earlier observations revealed that amplification of AKT2 oncogene can trigger tumorogenicity in the breast, ovarian, and pancreatic cells [75, 76]. An activation of the PI3K-AKT signaling was found to endorse cell invasiveness and trigger the progression of prostate cancer [77]. The overall role of PI3K-AKT signaling in carcinogenesis has been depicted in Fig. 13.3.

13.6.2 Pharmacotherapeutic Targets of Carotenoid in the PI3K/AKT Pathway

Components in the PI3K/AKT pathway have been regarded to be the efficient targets in cancer chemotherapy. Lycopene has been proposed to inhibit the proliferation of human colon, prostate, prostatic, and colorectal cancer cells through suppressing AKT signaling pathway [51, 57]. Fucoxanthin was found to increase the expression of PTEN resulting the decrease in AKT phosphorylation and suppression of its downstream p53/p70S6K/mTOR signaling in human epithelial cervical cancer cells [69]. Astaxanthin was claimed to inhibit PI3K/AKT signal transduction in the human hepatoma cells [49].

13.7 PPARs in Cancer Pathogenesis and Chemo-preventive Role of Carotenoids

13.7.1 PPARs in Cancer Pathogenesis

PPARs regulate gene transcription [78]. Three isoforms of PPAR, such as PPARα, PPARβ/δ, and PPARγ have been identified, which share high degree of homology but vary in their tissue distribution and ligand specificity [78]. In nucleus, PPARs bind with retinoid X receptor (RXR) to form a heterodimer and interact with cofactors to trigger the transcription of target genes [78]. Fatty acids and eicosanoids were recognized as the natural ligands for the PPARs [78]. After binding with ligands, PPARs endure conformational changes and interact with co-repressors and co-activators to regulate their functions [78]. PPARs are claimed to exhibit dual roles in carcinogenesis. Some reports revealed protective role of PPARs in cancer; in contrast, others implicated for the role of PPARs in cancer promotion and development [78]. PPARγ ligands have been claimed to impart anti-proliferative effects through its capacity to help differentiation in human tumors [78, 79]. It has also been claimed that PPARγ ligands can offer chemoprevention via inducing apoptosis and regulating cell cycle [78, 79]. In contrast, PPARγ activation has been claimed to develop the polyps in the colon and increase the chance of developing colorectal cancer in mice [78]. Similar observations have been documented for PPARα and PPARβ/δ in case of liver cancer [78]. Initially, PPARγ has been regarded as a key regulator in adipocyte differentiation, and the metabolism of carbohydrates and fats. However, several reports revealed the chemotherapeutic role of PPARγ agaist different types of cancers [78]. Significant down-regulation of PPARγ expression was seen in the colorectal cancer cells [80]. Other studies also reported significant reduction of PPARγ expression in esophageal and cervical cancer cells as compared to normal cells [81, 82]. In contrast, Zaytseva and co-workers claimed that down-regulation of PPARγ1 expression can induce apoptosis and impair the growth and proliferation of human breast cancer cells, MCF-7 [83]. The role of PPARs signaling in carcinogenesis has been represented in Fig. 13.4.

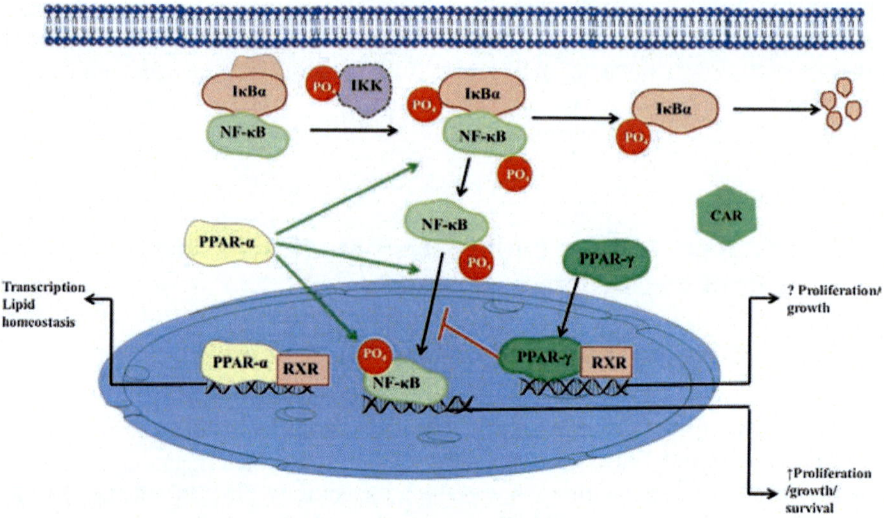

Fig. 13.4 Roles of NF-κB and PPARs in cancer pathogenesis and pharmacotherapeutic targets of carotenoids in IGF pathway. *CAR* Carotenoids, *IκBα* Nuclear factor of kappa light polypeptide gene enhancer in B-cells inhibitor, alpha, NF-κB Nuclear factor kappa-light-chain-enhancer of activated B cells, PPAR Peroxisome proliferator-activated receptor, RXR Retinoid X receptor. Black arrows (→) indicate down-stream events. Red lines (T) show the restriction/inhibition

13.7.2 Pharmacotherapeutic Targets of Carotenoids by Regulating PPARs

Studies have reported the chemo-preventive role of PPARγ, so it has been considered as a potential target in cancer chemotherapy [84]. β-carotene, astaxanthin, capsanthin, and bixin have been reported to up-regulate expression of PPARγ protein in the leukemia cells [84, 85]. Fucoxanthinol has been reported to activate PPARγ via integrin-provoked suppression of focal adhesion kinase in the colorectal cancer cells, however, the activity has been found to be recovered at 24 h [11].

13.8 NF-κB Signaling in Cancer Pathogenesis and Chemo-preventive Role of Carotenoids

13.8.1 NF-κB in Cancer Pathogenesis

NF-κB is a family of transcription factors consisting five subfamilies, such as RelA/p65, c-Rel, RelB, NF-κB (p50/p105), and NF-κB2 (p52/p100) [86]. NF-κB signaling is mainly regulated by nuclear factor of κ light polypeptide gene enhancer in B-cells inhibitor (IκB) proteins. Some of the human IκB proteins are IκBα, IκBβ,

IκBε, and IκBζ. NF-κB is normally present in cytoplasm of the un-stimulated cells in association with IκBs. IκBs contain two conserved serines, which are phosphorylated by the IkB kinases (IKKs). Cellular stress and activation of other signalling pathways can stimulate NF-κB [87]. NF-κB signalling has been found to be regulated by other transcription factors, such as Ras, MAPK, PI3K, AKT, Notch-1, STAT3, β-catenin, p53 etc. [87]. NF-κB has been shown to regulate cancer progression via delicate regulation of anti- and pro-apoptotic balance [86, 87]. NF-κB can also regulate expressions of gene products connected to the proliferation of tumors [86, 87]. Furthermore, NF-κB can control the expressions of MMPs, adhesion molecules, VEGF/TWIST, C-X-C chemokine receptor type 4, which are associated with invasion, angiogenesis, and metastasis in cancer [88]. A report revealed that NF-κB activation in prostate cancer cells can contribute to bone metastatic formation via promoting osteoclastogenesis [89]. It has been claimed to be the invariably main reason of developing osseous metastasis in the patients with advanced prostate cancer [89]. It has been reported that the level of NF-κB (p50) is higher in cancer cells [90, 91]. Over-expression of endogenous NF-κB (p65) was found to down-regulate PTEN expression in human lung, and thyroid cancer cells, which is a tumor suppressor gene [92]. However, NF-κB (p50) did not show any PTEN inhibitory effect [92]. The cross-talk between NF-κB and other transcription proteins have been implicated to be the critical factor of carcinogenicity as observed in different malignant cells [93]. The role of NF-κB signaling in carcinogenesis has been represented in Fig. 13.4.

13.8.2 Pharmacotherapeutic Targets of Carotenoid in NF-κB Pathway

Considering the role of NF-κB signaling in cancer pathogenesis, it has been regarded as a potential therapeutic strategy in tumor chemotherapy. Dietary astaxanthin has been reported to inhibit proliferation and to induce apoptosis in colitis-associated colon cancer in mice via down-regulating the NF-κB-triggered activation of pro-inflammatory cytokines, such as TNF-α and IL-1β and COX-2 [94]. Astaxanthin was also found to induce apoptosis in hepatocellular carcinoma cells via impeding NF-κB expression [49]. Kavitha and co-workers reported that astaxanthin can produce chemo-preventive effect in hamster's oral cancer by down-regulating NF-κB mediated through the inhibition of its up-stream MAPK/PI3K/AKT signaling and its key regulatory enzyme, IKKβ [50]. Peridinin has been found to impair proliferation and survival of human T-cell leukemia virus type 1 infected T cells via suppressing nuclear translocation of NF-κB through inhibition of phosphorylation IκBα, RelA, AKT and p70 S6 kinase [95]. It also imparted chemo-preventive effect via arresting cell cycle, suppressing anti-apoptotic proteins, and activating cleavage of caspases through NF-κB/AKT-dependent pathway [95].

13.9 CDKs in Cancer Pathogenesis and Chemo-preventive Role of Carotenoids

13.9.1 CDKs in Cancer Pathogenesis

The cell cycle is a series of cellular events during cell division, which leads to division and duplication of DNA to produce two identical cells [96]. Mammalian cell cycle consists of four phases. S phase involves DNA replication, G2 phase involves preparing for mitosis, M phase involves division of DNA and cellular components to produce 2 daughter cells, and G1 phase involves committing and preparing for next cycle [97]. Cell cycle is functionally regulated by the proteins, cyclin (Cyc)s, and their associated cyclin-dependent kinases (CDKs) [98]. CDK families includes 20 mammalian proteins of which, 5 members, such as CDK1, CDK2, CDK3, CDK4, and CDK6 have the partners designated as cyclins. Association between CDKs and Cycs integrally regulate mammalian cell cycle [98]. There are different members in Cyc family, which play the key roles in different phases of cell cycle [98]. Generally, CycD is activated from mid to late G1 phase to inactivates their regulatory functions of pRb and cell cycle inhibitor proteins via phosphorylation and allow the cell to enter into S phase [98]. At the boundary of G1/S, CycA is present, which activates CDK2 and CDK1, promoting progression through the G2 interval [98]. CycB1 and CDK1 association triggers mitosis [98]. Hyperphosphorylation of the RNA polymerase II at carboxyl-terminal domain (CTD) and phosphorylation of mitosis-specific transcription activators, such as Sp1, Myc, Myb can communicate transcriptional inhibition of mitosis [98]. In addition, general transcription factors, such as TFIID and TFIIH are involved in promoter clearance and regulation of transcription during mitosis [98]. Several CTD kinases, such as CDK7, CDK8, and CDK 9 are found to be the members of the CDK superfamily and associated with transcription initiation complexes [96]. A new kind of cell cycle regulator has been identified as CDK inhibitor. The major role of these proteins is to bind and inhibit the action of cyclin/CDK complexes. There are two families of CDK inhibitors: for details read Ref. [98]. The role of these CDKs in up-regulation and overexpression has been observed in in various tumors [99]. Cyclin A up-regulation has been found to be associated with carcinogenesis and metastasis in colorectal cancers [100]. Blocking of CDK4/6 activity has been claimed to reciprocate the resistance of a proto-oncogene, B-Raf, in melanoma [99]. Activation of the expressions of CDK7 and its cofactors, such as CycH and a mating-type protein (MAT1) has been reported over 900 breast cancer samples [99]. CDK9 has been proposed to regulate the expressions of the genes which are associated with drug resistance [99]. The role of CDKs in cell cycle has been represented in Fig. 13.5.

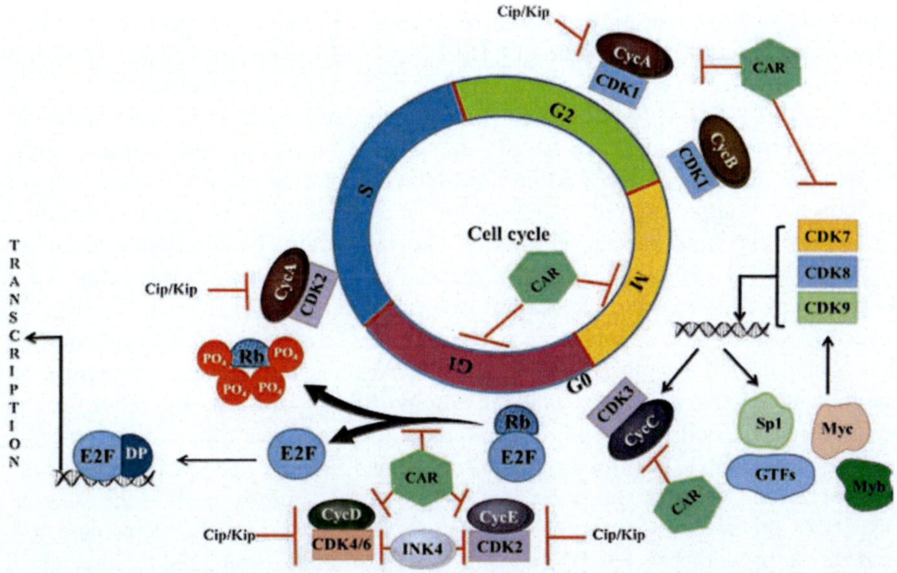

Fig. 13.5 Roles of CDK signalling in cancer and pharmacotherapeutic targets of carotenoids in CDK pathway. *CAR* Carotenoids,*CDK* Cyclin-dependent kinases, *Cip* CDK interacting protein, *Cyc* Cyclin, *DP* Dimerization partner, *E2F* E2F family of DNA-binding transcription factors, *GTFs* General transcription factors, *INK4* Family of cyclin-dependent kinase inhibitors, *Kip* Kinase inhibitory protein, *Myb* MYB Proto-Oncogene, *Myc* MYC Proto-Oncogene, *Rb* Tumor suppressor protein, *Spl* Specificity protein 1. Black arrows (→) indicate down-stream events. Red lines (T) show the rest restriction/inhibition

13.9.2 Pharmacotherapeutic Targets of Carotenoids by Regulating CDKs

Considering the role of CDKs in cell proliferation, these may be a potential target in cancer chemotherapy (Fig. 13.5). Lycopene has been reported to suppress the growth and expansion of prostate cancer cells via down-regulating the CDK-7 and CDK-9 expressions [101]. It was found to inhibit cell proliferation in human colon cancer cells through impairing the promoter activity and protein expression of CycD1 via activation of nuclear CDK inhibitor, p27Kip [51]. Lycopene was also found to induce the nuclear localization of p27Kip and to indirectly block pRb phosphorylation in human colon cancer cells [51]. Lycopene has been claimed to decrease the risk of gastric cancer in ferret via down-regulating the protein expressions of total p53, phosphorylated p53, CycD1, proliferating cellular nuclear antigen [102]. It was also found to up-regulate the expression of p21waf1/cip1, a CDK inhibitor, in the gastric mucosa of ferret during the process of carcinogenesis in stomach [102]. Lycopene was found to inhibit cell cycle progression at the G1 phase and thereby prevent the growth of human breast and endometrial cancer cells [103]. Lycopene treatment was able to decrease serum-induced phosphorylation of the pRb and related pocket proteins, such as p130 and p107, which has been found to be associated with the inhibition of CDK2 and CDK4 genes (not protein) [103]. It was also reported to suppress CycD1 and CycD3 expressions and to retain p27 in CycE-CDK2 [103]. It has been reported to induce cell cycle arrest through down-regulation of CycD1 and increase p21, p27 and p53 levels in prostate cancer cells [104]. Crocin has been reported to block G0/G1 phase in cell cycle and to trigger apoptosis via suppression of anti-apoptotic, and activation of pro-apoptotic factors in prostate and lung cancer cells [105]. Crocin has also been reported to down-regulate CycD1 expression in the carcinoma cells in mouse bladder [106]. Hire et al. reported that crocin can inhibit cell proliferation in breast cancer cells by affecting microtubule depolarizing potency [107]. Treatment with β-cryptoxanthin has shown to reduce pre-neoplastic and neoplastic lesions in urinary bladder in carcinogen-provoked urinary bladder carcinogenesis of mice by reducing the CycD1-positive cell ratio [108]. β-carotene at low dose was found to increase the expression of AP-1, proliferating cellular nuclear antigen, and cycD genes in the lung of ferrets during the process of smoke-provoked carcinogenesis in the lung tissue [108]. β-carotene was found to decrease CycA expression and to increase p21 and p27 expressions in HL-60 leukemia cells [33]. The carotenoids, astaxanthin, capsanthin, and bixin were also reported to down-regulate CycD1 and to up-regulate p21 expressions in K562 cancer cells [109]. Peridineine has been claimed to induce cell cycle arrest at G1 phase by suppressing G1 cell cycle regulators, such as CycD1, CycD2, CDK4, and CDK6 in human T-cell leukemia virus type 1-infected T cells [95].

13.10 Other Molecules/Receptors/Signaling Pathways in Cancer Pathogenesis and Chemo-preventive Role of Carotenoids

13.10.1 Other Molecules/Receptors/Signaling Pathways in Cancer Pathogenesis

Apart from aforementioned factors, several other molecules and/or signaling pathways have been mentioned to be involved in the process of carcinogenesis. Human multidrug resistance-associated protein (MRP) family, members of ATP-binding cassette (ABC) transporter family, consists of seven members and their presence has been confirmed in different types of cancer cells [110]. Ca^{+2}/calmodulin-dependent protein kinase IV (CaMKIV) has been reported to regulate various cellular processes affecting cell proliferation. The activation of CaMKIV has been revealed in small cell lung and hepatocellular carcinomas [111]. Prostate-specific antigen (PSA), is an important marker to diagnose prostate cancer and its magnitude/stage

[112]. PSA concentration remains low in serum under normal condition; however, elevated PSA level in serum accurately predict the existence of prostate cancer [112]. Focal adhesion kinase (FAK) in micro-vascular endothelial cells has been regarded to endorse angiogenesis. FAK has been found to be over-expressed in advanced cancers, which has been correlated to the cancer progression and metastatic transformation [113]. Caspases play a crucial role in the process of apoptosis. Lack of caspase-3 expression has been observed in breast cancer cells due to frame shift mutation within the exon 3 of the caspase-3 gene [114]. Caspase-8 suppression has been seen in a number of different types of tumor cells [115]. Mutation/alteration in the expression of pro-apoptotic and/or anti-apoptotic proteins under Bcl-2 family has been observed in the cancer cells [116, 117]. Up-regulation of Bcl-2 protein, an anti-apoptotic factor, has been observed in myeloid leukemia cells [116]. Anti-apoptotic to pro-apoptotic factor expression ratio has been observed high in children with acute lymphoblastic leukemia cells as compared to normal cells, which endorses survival of cancer cells via resistance towards apoptosis [117].

13.10.2 Pharmacotherapeutic Targets of Carotenoids

Crocin has been reported to suppress drug resistance by down-regulating MRP1 and MRP2 gene expression in cisplatin-resistant human ovarian cancer cells, A2780/RCIS [110]. β-carotene has been found to exhibit chemotherapeutic role in cancers associated with the abnormal CAMKIV expression [118]. Naz and co-workers claimed that β-carotene can inhibit cell proliferation, arrest the cell cycle, and induce apoptosis via inhibition of CaMKIV in HeLa, HuH7and MCF-7 cells [118]. Lycopene has been reported to reduce PSA and VEGF levels in the sera and to increase connexin 43 gene and Bax protein expressions in the prostate cancer patients [8, 119–125]. Fucoxanthinol has been demonstrated to down-regulate FAK expression in human colorectal cancer cells [11]. Carotenoids were found to induce apoptosis in different types of cancer cells. Lycopene has been claimed to decrease Bcl-2 and Bcl-2 L1 expressions, and to increase Bax expression in the human prostate cancer and benign prostate hyperplasia cells [101, 126]. Lycopene was found to induce apoptosis by inhibiting phosphorylation of Bad, and increasing Bax-1, and cleaved caspase 3 protein levels in the lung of ferrets during the process of smoke-provoked carcinogenesis in the lung tissue [102, 127]. It has also been reported to increase p53 level and Bax to Bcl-2 ratio in the prostate cancer cells [104]. Treatment with crocin was claimed to trigger apoptosis via down-regulation of Bcl-2, and up-regulation of p53 and Bax expressions in prostate cancer, gastric carcinoma, and lung adenocarcinoma cells [106, 128]. Crocin was also found to increase caspase activation in the cancer cells [128, 129]. β-carotene has been reported to endorse apoptosis by suppressing the expression of Bcl-2, and activating the expressions of Bax and caspase-3 human esophageal squamous cell carcinoma and HL-60 leukemia cells [33, 130]. Fucoxanthin has been reported to up-regulate caspases-3 and caspases-7 genes, and down-regulate poly-ADP-ribose polymerase (PARP) and Bcl-xL levels in human bladder cancer and leukemia cells [131, 132]. Neoxanthin and fucoxanthin have been demonstrated to induce apoptosis by reducing the expressions of Bax and Bcl-2 and increasing the cleavages of caspase-3 and PARP in the human prostate cancer cells [133]. Astaxanthin has been reported to down-regulate Bcl-2, phosphorylated Bad, survivin expressions and to up-regulate Bax and Bad expressions in oral cancer cells of hamsters [50]. Bi et al. reported that zeaxanthin can induce apoptosis via mitochondria dependent pathway through down-regulation of anti-apoptotic factors, up-regulation of pro-apoptotic proteins, cytochrome C release to the cytosol, and cleavage of caspases in human uveal melanoma cells [134]. Fucoxanthinol and halocynthiaxanthin was reported to decrease the expression of anti-apoptotic-Bcl-2 protein in human leukemia, breast, and colon cancer cells [41]. Deinoxanthin exhibited a decrease in Bcl-2 expression, and increased the expression of Bax protein in human hepatoma, prostate, and colon cancer cells [135]. Siphonaxanthin has been revealed to decrease the expression of Bcl-2, and to increase the activation of caspase-3 in the human leukemia cells [136] (Table 13.1).

13.11 Conclusion

Cancer is a life-threatening disease. Multiple signaling pathways are involved in the pathogenesis of cancer; however exact molecular mechanism is yet to be clearly understood. Chemotherapy is an important therapeutic strategy in cancer and plant-derived chemotherapeutic agents have contributed immensely in the development of useful chemotherapeutic agents. Dietary consumption of carotenoid-enriched fruits and vegetables have been proposed to reduce the risk of cancer. Carotenoids were found to impart their anticancer activity through the regulation of various signaling pathways. Carotenoids are reported to down-regulate the (NF)-κB, mTOR, Wnt/β--catenin, autophagy, and PI3K/AKT pathway in different types of cancer cells. Carotenoids have also been proposed to induce apotosis in cancer cells. Some of the carotenoids were found to arrest cell cycle via modulation of Cycs and CDKs activities. Therefore, it may be possible to develop a potential chemotherapeutic agent from the carotenoid family. The roles of different signaling events in cancer pathogenesis and the chemo-preventive roles of different carotenoids have been detailed in this chapter. An outline (Table 13.1) has been incorporated to show the chemotherapeutic roles of carotenoids along with therapeutic doses, preclinical/clinical pharmacological outcomes. However, more research is needed to develop novel anti-cancer carotenoid to be useful against different types of cancer.

Table 13.1 Anticancer activity of carotenoids

Compounds	Therapeutic mechanisms	Treatment	Observations	References
1. Lycopene	PSA inhibitor, antioxidant, proliferation inhibitor, apoptosis inducer, CDK inhibitor, growth inhibition, MMP-7 inhibitor.	Prostatic cancer patients (2 mg twice daily, 2 years, p. o.)	Serum PSA l ↓, primary tumor ↓, secondary tumors ↓.	[8, 51, 52, 57, 58, 101–103, 119–127]
		Patients with high prostate cancer risk (30 mg/day, 4 months, p. o.)	Serum PSA level ↓.	
		Recurring prostate cancer patients (25 mg/day in diet, 4 weeks)	Serum lycopene ↑, serum PSA ↓, serum vascular endothelial growth factor ↓.	
		Prostate cancer patients (15 mg twice daily, 3 weeks, p. o.)	Serum PSA ↓, connexin 43 gene ↑, Bax ↑.	
		High-grade prostatic intra-epithelial neoplasia patients (4 mg twice a day, 1 year, p.o)	Serum PSA ↓, HGPIN ↓.	
		Benign prostate hyperplasia patients (15 mg/day, 6 months, p. o.)	Serum PSA ↓, plasma lycopene ↑, prostate enlargement ↓.	
		Prostate cancer patients (15 mg twice a day, 6 months, p.o.)	Serum PSA ↓.	
		Prostate cancer or benign prostate hyperplasia patient (30 mg/day, 21 days)	Plasma lycopene ↑, prostate tissue lycopene ↑.	
		IC50 = 1–2 μM agaisnt endometrial (Ishikawa), mammary (MCF-7), and lung (NCI-H226) cancer cells.	Proliferation ↓, IGF-1 ↓.	

(continued)

Table 13.1 (continued)

Compounds	Therapeutic mechanisms	Treatment	Observations	References
		Hormone-refractory human prostate cancer (PC-3) cells at 50–100 μM.	Proliferation ↓.	
		Lycopene (25 μM)	Cell cycle ↓, TGF-β2 ↓, IGF-1R ↓, EGFR ↓, CDK-7 ↓, CDK-9 ↓, BRCA1 ↓, Bcl-2 L1 ↓, Bcl-2 ↓.	
		Lycopene (25 μM) + PPARγ agonists	Proliferation ↓, apoptosis ↑.	
		Lycopene (25 μM) + chemotherapeutic agents	Apoptosis ↑.	
		Human prostate cancer (PCa) and benign prostate higher plasia cells (BPH)	Cell viability (PCa) ↓.	
		2.5 μM	G0/G1 (PCa) ↑, G2/M (PCa) ↓, G0/G1 (BPH) ↓, G2/M (BPH) ↑, PPARγ (PCa, BPH) ↑, RXR (PCa, BPH) ↑, Tp53 (PCa, BPH) ↑, Bax (PCa, BPH) ↑, Bcl-2 (PCa) ↓, Bcl-2 (BPH) ↑.	
		5–10 μM		
		10 μM	Apoptosis (PCa) ↑.	
		Human prostate cancer cells (PCa)	Cell viability ↓ (IC50 = 5.1, 15, 36 16, and 50 μM, repectively), IGF-1 ↓, IGF-1R ↓, IGF-BP3 ↑, AKT kinase activity ↓, survivin ↓, apoptosis ↑.	
		Prostate tumor (DU145)-bearing mice (15 mg/kg per day, 3 weeks, p. o.)	Tumor growth ↓, survivin ↓, apoptosis ↑.	
		Human colon cancer cells (HT-29); IC50 = 10 μM	Cell proliferation ↓, phospho-AKT ↓, β-catenin ↓, phospho-β-catenin ↑, CycD1 ↓, nuclear p27kip ↑, Phospho-Rb ↓.	

		Leptin-stimulated human colon cancer cells (HT-29); 1–2 μM	Cell invasion ↓, MMP-7 ↓, phospho-AKT ↓, GSK-3β ↓, Erk 1/2 ↓.	
		Smoke-initiated lung cancer bearing ferrets (1.1 mg and 4.3 mg/kg/day, 9 weeks, p.o.)	Plasma lycopene ↑, lung lycopene ↑, squamous metaplasia ↓, PCNA ↓, phospho-bad ↓, IGF-BP3 ↑, apoptosis ↑.	
		Human mammary cancer cells (MCF-7, T-47D) and human endometrial cancer cells (ECC-1); 2–3 μM	G1 ↓, Phospho-Rb ↓, P107 ↓, P130 ↓, CDK2 gene ↓, CDK4 gene ↓, CycD1 ↓, CycD3 ↓, p27 ↑.	
		Smoke-initiated gastric cancer bearing ferrets (1.1 mg and 4.3 mg/kg/day in diet, 9 weeks)	Gastric total p53 ↓, phospho-p53 ↓, CycD1 ↓, PCNA ↓, CycD1 ↓, p21 (waf1/cip1) ↑, Bax-1 ↑, cleaved caspase 3 ↑, apoptosis ↑.	
		Human prostate prostatic carcinoma cells (LNCaP) (2.5–10 mM)	Total cholesterol ↓, Ras ↓, HMG-CoA reductase ↓, NF-κB ↓, CycD1 ↓, phospho-AKT ↓, p21 ↑, p27 ↑, p53 ↑, Bax ↑, Bcl-2 ↓, apoptosis ↑.	
2. Crocin	Proliferation inhibitor, apoptosis inducer, MRP inhibitor, growth inhibitor.	Colorectal cancer cells (HCT-116, SW-480, and HT-2); 1.0 mM	Cells proliferation ↓.	[9, 32, 70, 104, 106, 107, 128, 129]
		p53-null colorectal cancer cells; 10 mM	Cell proliferation ↓, G1 ↓, G2 ↓, LC3-II ↑, beclin-1 ↓, Atg7 ↓.	
		Prostate Cancer cell lines (LAPC-4, CWR22, LnCaP, 22rv1, C4-2B, DU145 and PC3) IC50 = 0.26 and 0.95 mM/ml	Cell proliferation ↓, G0/G1 ↓, Bcl-2 ↓, Bax ↑, caspase-9 ↑, apoptosis ↑.	
		Human gastric carcinoma cells (AGS); 2.5–3 mg/ml; IC50 = 3, 2.7, and 2.5 mg/ml at 24, 48, and 72 h, respectively	Cell growth ↓, sub-G1 population ↑, apoptosis ↑, Bax ↑, Bcl-2 ↓, caspase-9 ↑.	

(continued)

Table 13.1 (continued)

Compounds	Therapeutic mechanisms	Treatment	Observations	References
		Cisplatin-resistant human ovarian cancer cells (A2780/RCI); 80–100 μM	Cell proliferation ↓, cell viability ↓, multidrug-resistance ↓, MRP1 ↓, MRP2 ↓.	
		Human myeloma cells (U266); 250 and 500 μM	Cell proliferation ↓, cell viability ↓, ROS production ↓, DNA fragmentation ↑, apoptosis ↑,	
		Human bladder cancer cells (T24) and BALB/c nude mice bearing T24 cells (100 mM)	Cell growth ↓, cells in G0/G1 phase ↑, apoptosis ↑, Bcl-2 ↓, CycD1 ↓, survivin ↓, Bax ↑.	
		Breast cancer cell line (MCF-7); 2.5–4.5 mg/ml; IC50 = 3.5 mg/ml	Cell viability ↓, cell survival ↓, apoptosis ↑, caspase-7 ↑, caspase-9 ↑, p53 ↑, PARP-1 ↑.	
		Cancer cells (HCC70, HCC1806, HeLa and CCD1059sk); IC50 = 275 ± 8, 380 ± 7 and 660 ± 23 nM, respectively	Cell proliferation ↓, depolarization of microtubule ↑, multipolar spindles ↑, mitosis ↓.	
3. α-Carotene	Carcinogenesis inhibitor.	C3H/He mice having a high incidence of spontaneous liver tumor development (0.005% or 0.05% in drinking water, 40 weeks)	Number of hepatomas ↓, spontaneous liver carcinogenesis ↓, tumor formation ↓.	[6]
		4NQO-initiated lung cancer bearing mice (0.05% in drinking water, 30 weeks)	Number of lung tumors ↓, lung carcinogenesis ↓.	
		DMBA+TPA-initiated skin cancer bearing mice (200 and 400 mmol, topically, 20 weeks)	Number of skin tumor (papilloma) ↓, skin carcinogenesis ↓, tumor progression ↓, tumor initiation ↓.	
4. β-Cryptoxanthin	Carcinogenesis inhibitor.	OH-BBN-induced urinary bladder cancer bearing mice (1, 5 and 25 ppm in diet, 24 weeks)	Preneoplastic bladder lesions ↓, neoplastic bladder lesions ↓, bladder carcinoma ↓, CycD1-positive cell ratio ↓.	[108]

5. β-carotene	Growth inhibitor, proliferation inhibitor, CAMKIV inhibitor, apoptosis inducer, pro-oxidant, PPARγ inhibitor.	WAZ-2 T tumor cell-initiated mammary tumor bearing BALB/c mice (0.1%, 45 days, p.o)	Tumor growth ↓, tumor volume ↓.	[10, 33, 34, 37, 109, 118, 137]
		Human leukemia cells (U937, and HL-60); 20 μM	Cell number ↓, cell viability ↓, apoptosis ↑, DNA fragmentation ↑, LDH release ↑, G1 shift ↑.	
		Human cervical cancer (HeLa), breast cancer (MCF-7), hepatoma (HuH7) cells. 25–200 μM	Cell proliferation ↓, cell cycle arrest ↑, apoptosis ↑, CAMKIV ↓.	
		Smoke-initiated lung cancer bearing ferret (0.43 mg/kg/day, 6 months, p.o.)	RAR-β ↓, macrophage proliferation ↓, CycD1 ↓, keratinized squamous metaplasia ↓.	
		H2O2 or B[a]P-provoked human epithelial lung carcinoma (A549, BEAS-2B); 5 μM; and B[a]P-initiated lung cancer bearing ferrets (0.8 and 3.2 mg/kg/day, 6 months)	Hydroxyl free radical formation ↑, carbon centered radical formation ↓, 8-oxo-dG↑, M1dG ↓.	
		Undifferentiated human leukemia cells (HL-60)	Apoptosis ↑, ROS ↑, GSSG ↑, cell growth ↓, CycA ↓, Bcl-2 ↓, p21 ↑, p27 ↑.	
		Human leukemia cells (K562); 10–50 μM	Cell proliferation ↓, cell viability ↓, apoptosis ↑, PPARγ ↑, p21 ↑, CycD1 ↓, Nrf2 ↑.	
6. Fucoxanthin	Growth inhibitor, proliferation inhibitor, apoptosis inducer, AKT/mTOR inhibitor, autophagy promoter.	Human prostate cancer cells (PC-3, DU 145, and LNCaP); 20 μM	Cell viability ↓, apoptosis ↑, DNA fragmentation ↑.	[35, 40, 41, 69, 131–133]

(continued)

Table 13.1 (continued)

Compounds	Therapeutic mechanisms	Treatment	Observations	References
		Human prostate cancer cells (PC-3); 20 μM	Cell viability ↓, apoptosis ↑, cleaved caspase-3 ↑, cleaved PARP ↑, Bcl-2 ↓, ROS, DNA fragmentation ↑.	
		Human leukemia cells (HL-60); 11.3, 22.6 and 45.2 μM	Cell proliferation ↓, cell viability ↓, apoptosis ↑, DNA fragmentation ↑.	
		Human leukemia cells (HL-60); 15 μM	Cell proliferation ↓, ROS production ↑, cleaved caspase-3 ↑, cleaved caspase-7 ↑, cleaved PARP ↑, Bcl-xL ↓, DNA fragmentation ↑.	
		Human bladder cancer cells (EJ-1); 20 μM	Cell viability ↓, apoptosis ↑, hypodiploid cells ↑, caspase-3 ↑, DNA fragmentation ↑.	
		Human epithelial cervical cancer cell line (HeLa); 10, 20, 40 μM; IC50: 55.1 μM at 48 h	Cytotoxicity ↓, G0/G1 ↓, LC3 II ↑, Beclin 1 ↑, phospho-AKT ↓, p53 ↓, p70S6K ↓, mTOR ↓, PTEN ↑.	
		Human leukemia cells (HL-60); 6.25 and 12.5 μM	Cell viability ↓, cell growth ↓, induced DNA fragmentation ↑.	
7. Astaxanthin	Growth inhibitor, NF-κB inhibitor, Wnt/β-catenin inhibitor, carcinogenesis inhibitor, PPARγ inhibitor.	WAZ-2 T tumor cell-initiated mammary tumor bearing BALB/c mice (0.1–0.4%, 45 days, p.o)	Tumor growth ↓, tumor volume ↓, lipid peroxidation ↓.	[36, 37, 49, 50, 94]
		Colitis-associated azoxymethane-initiated colon cancer bearing male ICR mice (50, 100, and 200 ppm in diet, 17 weeks)	Colonic mucosal ulcers ↓, dysplastic crypts ↓, colonic adenocarcinoma ↓, NF-κB ↓, TNF-α ↓, IL-1β ↓, IL-6 ↓, COX-2 ↓, cell proliferation ↓, apoptosis ↑.	

		Human hepatoma cells (LM3 and SMMC-7721); 100, 200, 300 μM	Apoptosis ↑, phospho-NF-κB (p65) ↓, IκBα ↑, phospho-IKK-α/β ↓, phospho-GSK-3β ↓, β-catenin ↓, phospho-Erk ↓, PI3K ↓, phospho-AKT ↓, PCNA ↓.	
		DMBA-initiated oral cancer bearing hamster (15 mg/kg in diet, 14 weeks)	Nuclear NF-κB (p65) ↓, cytosolic NF-κB (p65) ↑, phospho-NF-κB ↓, IκBα ↑, phospho-IKK-β ↓, IKK-β ↓, GSK-3β ↓, phospho- GSK-3β ↓, Erk ↓, phospho-Erk ↓, AKT ↓, phospho-AKT ↓, PI3K ↓, PDK-1 ↓, Bax ↑, Bcl-2 ↓, phospho-Bad ↓, nuclear survivin ↑, cytosolic survivin ↓, cytosolic cytochrome C ↑, mitochondrial cytochrome C ↓, cleaved caspase-3 ↑, cleaved caspase-9 ↑, cleaved PARP ↑.	
		Aflatoxin B1-initiated liver carcinogenesis bearing rat (300 mg/kg in diet)	Hepatic preneoplastic foci ↓, DNA single-strand breaks ↓, aflatoxin B1 binding to liver DNA ↓, plasma albumin ↓, aflatoxin B1 metabolism ↑.	
		Human leukemia cells (K562); 10–50 μM	Cell proliferation ↓, cell viability ↓, apoptosis ↑, PPARγ ↑, p21 ↑, CycD1 ↓, Nrf2 ↑.	

(continued)

Table 13.1 (continued)

Compounds	Therapeutic mechanisms	Treatment	Observations	References
8. Lutein	Growth inhibitor, proliferation inhibitor, apoptosis inducer, cell cycle regulator, antioxidant, pro-oxidant, PI3K/AKT inhibitor.	Human breast cancer cells (MCF-7, MDA-MB-468); 0.5–2.0 μM	Cell growth ↓, cell viability ↓, ROS production ↑,anaphase promoting complex subunit 2 ↑, aurora kinase B ↑, CycF ↑, cell division cycle 25 homolog A ↑, cyclin- CDK6 ↑, p21 ↑, Cip1 ↑, p27 ↑, Kip1 ↑, E2F1 ↑, E2F4 ↑, CycD1 ↓, CycD3 ↓, CDK2 CycD1 ↓, CDK4, CycD1 ↓, Bax ↑, Bcl-2 ↓.	[138–141]
		WAZ-2 T cell-initiated mammary tumor bearing female BALB/c mice (0.002% in diet)	Tumor development ↓, tumor growth ↓, tumor volume ↓, lipid peroxidation ↓.	
		Rat prostate carcinoma cells (AT3), 2.0 μM	Cell growth ↓, cell viability ↓.	
		Human non-small-cell lung cancer cells (A549)	Cell viability ↓, cell invasion ↓, cell migration ↓, apoptosis ↑, PI3K ↓, AKT ↓, mTOR ↓.	
9. Zeaxanthin	Growth inhibitor, proliferation inhibitor, apoptosis inducer,	Human uveal melanoma cells (SP6.5 and C918); 10–100 μM; IC50 = 40.8 and 28.7, respectively	Cell viability ↓, apoptosis ↑, Bcl-2 ↓, Bcl-xL ↓, Bak ↑, Bax ↑, bad ↑, cytosolic cytochrome C ↑, caspase-3 ↑, caspase-9 ↑.	[134, 142]
		C918-inititated ocular tumor bearing mice (57, 114 μM, intravitreous injection, 3 weeks)	Tumor volume ↓, tumor mass ↓, tumor invasion, melanoma cells ↓.	
10. Neoxanthin	Apoptosis inducer,	Human prostate cancer cells (PC-3, DU 145 and LNCaP); 20 μM	Cell viability ↓, apoptosis ↑, DNA fragmentation ↑.	[40, 133]
		Human prostate cancer cells (PC-3); 20 μM	Cell viability ↓, apoptosis ↑, cleaved caspase-3 ↑, cleaved PARP ↑, Bcl-2 ↓, ROS, DNA fragmentation ↑.	

11. Fucoxanthinol	Anoikis inducer, proliferation inhibitor, apoptosis inducer.	Human colorectal cancer cells (HT-29, HCT116 and DLD-1); 10.0 mM	Cell proliferation ↓, apoptosis ↑, inhibit FAK activation ↓, integrin β1 ↑, PPARγ ↓, anoikis ↑, phospho-AKT ↓.	[11, 41]
		Human leukemia (HL-60), breast (MCF-7), and colon cancer cells (Caco-2); 6.25 and 12.5 μM	Cell viability ↓, cell growth ↓, apoptosis ↑, induced DNA fragmentation ↑, Bcl-2 ↓.	
12. Halocynthiaxanthin	Apoptosis inducer, proliferation inhibitor, N-myc inhibitor.	Human leukemia (HL-60), breast (MCF-7), and colon cancer cells (Caco-2); 6.25 and 12.5 μM	Cell viability ↓, cell growth ↓, apoptosis ↑, induced DNA fragmentation ↑, Bcl-2 ↓.	[41, 143, 144]
		Human neuroblastoma cells (GOTO); 1, 2 and 5 μg/ml	Cell proliferation ↓, cell growth ↓, tumor initiation ↓, n-Myc gene ↓, membrane-bound Ca2+ release ↓.	
		Human colon carcinoma cells (DLD-1), 40 μM	Cell proliferation ↓, apoptosis ↑, death receptor 5 ↑.	
13. Peridinin	Proliferation inhibitor, apoptosis inducer, PI3K/AKT inhibitor, NF-κB inhibitor.	Human T-cell leukemia virus type 1 infected T cells (MT-2); 10 μM	Cell proliferation ↓, cell viability ↓, G1 phase ↓, apoptosis ↑, cyclin D1 ↓, cyclin D2 ↓, CDK4 ↓, CDK6 ↓, c-Myc ↓, survivin, XIAP ↓, Bcl-2 ↓, cleaved caspase-3 ↑, cleaved caspase-8 ↑, cleaved caspase-9 ↑, cleaved PARP ↑, NF-κB nuclear translocation ↓, phospho-IκBα ↓, phospho-RelA ↓, phospho-AKT ↓, phospho-p70 S6 kinase ↓, PDK1 ↓.	[95, 144, 145]
		HUT-102 ATL xenograft mice tumor (8.5 mg/kg, 15 doses, 3 weeks, i. p.)	Tumor genesis ↓, tumor growth ↓, serum sIL-2R ↓, serum sCD30 ↓.	
		Human colon carcinoma cells (DLD-1), 40 μM	Cell proliferation ↓, apoptosis ↑, death receptor 5 ↑.	
		Human neuroblastoma cells (GOTO)	Cell proliferation ↓.	

(continued)

Table 13.1 (continued)

Compounds	Therapeutic mechanisms	Treatment	Observations	References
14. Capsanthin	PPARγ inhibiting agent, MDR-reversing agent, apoptosis inducer.	Human leukemia cells (K562); 10–50 μM	Cell proliferation ↓, cell viability ↓, apoptosis ↑, PPARγ ↑, p21 ↑, CycD1 ↓, Nrf2 ↑.	[109, 146]
		MDR-1 gene-transfected mouse lymphoma cells (L1210); 20 μM	MDR-reversal ↑, apoptosis ↑.	
15. Capsorubin	Cytostatic cell inhibitor, proliferation inhibitor, cell cycle arrest.	Human lung cancer cells (A549); 30 μM	Cell growth ↓, p27 ↑, cyclin A ↓, CDK2 ↓, CDC2 ↓.	[147, 148]
		Human lung cancer cells (A549); 10, 20, and 35 μM	Cell growth ↓, CMV-IE tumor antigen ↓, capsorubin-nuclear accumulation ↑.	
16. Bixin	Pro-oxidant, PPARγ inhibitor, MDR-reversing agent, apoptosis inducer, cell cycle arrest.	Human osteosarcoma (U2OS), prostate (PC3), colon (HCT-116), breast (MCF7), anaplastic thyroid (DRO), papillary thyroid (BHP5-16), and lung cancer (A549) cells; IC50 = 10–50 μM	Cell viability ↓, apoptosis.	[42, 109, 147, 149]
		Patients' bone marrow-derived myeloma leukocytes (CD138+); 50–400 μM	Cell viability ↓, apoptosis ↑, cleaved PARP ↑, ROS ↑, oxidative stress ↑, thioredoxin reductase ↓. Thioredoxin ↓.	
		Doxorubicin sensitive or resistant myeloma cells (RPMI-8226 or RPMI-8226R); 50–400 μM and dexamethasone sensitive or resistant myeloma cells (MM1RL or MM1); 100 μM.	Cell viability ↓, apoptosis ↑, MDR-reversal ↑.	
		Human leukemia cells (K562); 10–50 μM	Cell proliferation ↓, cell viability ↓, apoptosis ↑, PPARγ ↑, p21 ↑, CycD1 ↓, Nrf2 ↑.	

		Human melanoma cells (A2058); 50 μM	Cell viability (IC50 = 40.5 μM) ↓, cell migration ↓, apoptosis ↑, ROS ↑, oxidative stress ↑, G0/G1 ↓, G2/M ↑	
		Human hepatocellular carcinoma cells (Hep3B); 23 μg/ml (IC50 at 24 h)	Cell proliferation (10–50 μg/ml at 24 h; 5–50 μg/ml at 48 h) ↓, ROS production ↑, DNA damage ↑, apoptosis ↑, Bax ↑, cleaved caspase-9 ↑, cleaved caspase-8 ↑, cleaved caspase-3 ↑, FASL ↑, G2/M cell population ↑, G1/S cell population ↓.	
17. Deinoxanthin	Apoptosis inducer	Human hepatoma (HepG2), prostatecancer (PC-3), and colon cancer (HT-29) cells; 60, 80, and 60 μM respectively, 24 h	Cell viability (20–100 μM; IC50 = 59, 77, 61 μM, respectively) ↓, ROS production ↑, apoptosis ↑, cleaved caspase-3 ↑, Bcl-2 ↓, Bax ↑.	[135]
18. Siphonaxanthin	Proliferation inhibitor	Human leukemia cells (HD-60); 20 μM for 6 h.	Cell viability ↓, apoptosis ↑, chromatin condensation ↑, Bcl-2 ↓, cleaved caspase-3 ↑, GADD45α ↑, death receptor 5 ↑.	[136]
19. Capsanthin esters (capsanthin 3′-ester and capsanthin 3,3′-diester)	Proliferation inhibitor.	Burkitt's lymphoma-derived Epstein-Barr virus genome-carrying Raji cells; 500 and 1000 mol ratio/TPA	Cell viability ↓.	[150]
		DMBA (initiator) + TPA (promotor)-initiated skin papilloma bearing mouse (85 nmol, 2 doses/week, 20 weeks, topically)	Papilloma formation ↓, papilloma number ↓.	
20. Canthaxanthin	Proliferation inhibitor, apoptosis inducer.	Human colon adenocarcinoma (WiDr) and melanoma (SK-MEL-2) cells; 1 and 10 μM	Cell viability ↓, apoptosis ↑, nuclear chromatin condensation ↑, nuclear chromatin fragmentation ↑.	[37, 151–156]

(continued)

Table 13.1 (continued)

Compounds	Therapeutic mechanisms	Treatment	Observations	References
		WAZ-2 T tumor cell-initiated mammary tumor bearing BALB/c mice (0.1%, 45 days, p.o)	Tumor growth ↓, tumor volume ↓.	
		Rat ascites hepatoma cells (AH109A); 5–20 μM	Cell invasion ↓.	
		OH-BBN-initiated urinary bladder cancer bearing male mice (50 ppm in drinking water, 20 weeks)	Cell proliferation ↓, pre-neoplastic lesion ↓, neoplastic lesion ↓	
		4-NQO-initiated oral cancer bearing rats (0.01% in diet, 22 weeks)	Cell proliferation ↓, pre-neoplastic lesion ↓, tongue neoplasm ↓, tongue lesion ↓, polyamine level ↓.	
		Azoxymethane-initiated colon cancer bearing rats (500 ppm in diet, 34 weeks)	Cell proliferation ↓, intestinal neoplasm ↓, blood polyamine level ↓.	
		Skin papilloma bearing mice (200 mg/kg/day, 14 days, p.o)	Papilloma ↓, papilloma growth ↓, c-Myc ↓.	
21. Heteroxanthin	Proliferation inhibitor.	Burkitt's lymphoma-derived Epstein-Barr virus genome-carrying Raji cells; 500 and 1000 mol ratio/TPA	Cell viability ↓.	[157]
22. Violaxanthin	Proliferation inhibitor, apoptosis inducer, MDR-reversing agent,	Human breast cancer cells (MCF-7); 8 and 20 μg/ml	Cell viability ↓, apoptosis ↑.	[40, 158–160]
		Human breast cancer cells (MCF-7); 4 and 40 μg/ml	MDR ↓.	
		Human prostate cancer cells (PC-3, DU 145 and LNCaP); 5, 10, and 20 μmol/l	Cell viability ↓.	
		PPARγ2 CALUX reporter cells; 1, 10, and 100 μM	PPARγ ↑.	

23. Diatoxanthin	Proliferation inhibitor	Human malignant melanoma cells (A2058); 100 μg/ml for 72 h	Cell growth ↓, cell proliferation ↓.	[161]
24. Lactucaxanthin	Proliferation inhibitor.	Burkitt's lymphoma-derived Epstein-Barr virus genome-carrying Raji cells; 500 and 1000 mol ratio/TPA	Cell viability ↓.	[157]
25. Dinochrome A	Proliferation inhibitor.	Neuroblastoma (GOTO), osteosarcoma (OST), and cervical cancer (HeLa) cells; 5 μg/ml for 3 days	Cell proliferation ↓.	[162]
26. Dinoxanthin	Proliferation inhibitor.	Human malignant melanoma cells (A2058); 100 μg/ml for 72 h	Cell growth ↓, cell proliferation ↓.	[161]
27. 19-hexanoyloxymytiloxanthin	Cytotoxicity inducer.	Mouse leukemia cells (P388)	Cytotoxicity ↑.	[163]
28. 19-butanoyloxymytiloxanthin	Cytotoxicity inducer.	Mouse leukemia cells (P388)	Cytotoxicity ↑.	[163]

4-NQO 4-Nitroquinoline 1-oxide, *8-oxo-dG* 8-Oxo-2′-deoxyguanosine, *AKT* Protein kinase B, *Atg* Autophagy related, *B[a]P* Benzo[a]pyrene, *Bad* Bcl-2-associated death promoter, *Bax* BCL2 Associated X, *Bcl-2* B-cell lymphoma 2, *Bcl-xL* B-cell lymphoma-extra large, *BRCA1* Breast cancer 1, *CAMKIV* Calcium/calmodulin-dependent protein kinase type IV, *CDK* Cyclin dependent kinase, *CMV-IE* Cytomegalovirus immediate-early, *COX-2* Cyclooxygenase-2, *Cyc* Cyclin, *DMBA* 7,12-Dimethylbenz[a]anthracene, *DNA* Deoxyribonucleic acid, *EGFR* Epidermal growth factor receptor, *ERK* Extracellular regulated kinases, *GADD45α* Growth arrest and DNA damage-45 α, *GSK-3β* Glycogen synthase kinase-3β, *GSSG* Glutathione disulfide, *HGPIN* High-grade prostatic intraepithelial neoplasia, *HMG-CoA* β-Hydroxy β-methylglutaryl-CoA, *IGF* Insulin-like growth factor, *IGF-1R* Insulin-like growth factor 1 receptor, *IGF-BP3* Insulin-like growth factor binding protein 3, *IκBα* Inhibitor of κB, *IKK* IκB kinase, *IL* Interleukin, *i. p.* Intraperitoneal, *LDH* Lactate dehydrogenase, *MDR* Multi-drug resistance, *MMP* Matrix metalloproteinase, *MRP* Multidrug resistance-associated protein, *mTOR* Mammalian target of rapamycin, *NF-κB* Nuclear factor kappa-light-chain-enhancer of activated B cells, *n-Myc* N-myc proto-oncogene, *Nrf2* Nuclear factor erythroid 2-related factor 2, *OH-BBN* N-butyl-N-(4-hydroxybutyl)nitrosamine, *p. o.* Orally, *PARP* Poly (ADP-ribose) polymerase, *PCNA* Proliferating cell nuclear antigen, *PDK* Pyruvate dehydrogenase kinase, *PI3K* Phosphoinositide 3-kinase, *PPARγ* Peroxisome proliferator-activated receptor γ, *PSA* Prostate-specific antigen, *PTEN* Phosphatase and tensin homolog, *RAR-β* Retinoic acid receptor-β, *Rb* Retinoblastoma protein, *ROS* Reactive oxygen species, *sIL-2R* Soluble IL-2 receptor, *TGF-β* Transforming growth factor-β, *TNF-α* Tumor necrosis factor-α, *TPA* 12-O-Tetradecanoylphorbol-13-acetate

References

1. Martínez Andrade KA, Lauritano C, Romano G, Ianora A (2018) Marine microalgae with anti-cancer properties. Mar Drugs 16(5):165
2. D'Souza N, Murthy N, Aras R (2013) Projection of burden of cancer mortality for India, 2011–2026. Asian Pac J Cancer Prev 14(7):4387–4392
3. Bhandari PR (2015) Crocus sativus L.(saffron) for cancer chemoprevention: a mini review. J Tradit Complement Med 5(2):81–87
4. Tanaka T, Shnimizu M, Moriwaki H (2012) Cancer chemoprevention by carotenoids. Molecules 17(3):3202–3242
5. Ziegler RG (1989) A review of epidemiologic evidence that carotenoids reduce the risk of cancer. J Nutr 119(1):116–122
6. Murakoshi M, Nishino H, Satomi Y, Takayasu J, Hasegawa T, Tokuda H et al (1992) Potent preventive action of α-carotene against carcinogenesis: spontaneous liver carcinogenesis and promoting stage of lung and skin carcinogenesis in mice are suppressed more effectively by α-carotene than by β-carotene. Cancer Res 52(23):6583–6587
7. Wang X-D, Liu C, Bronson RT, Smith DE, Krinsky NI, Russell RM (1999) Retinoid signaling and activator protein-1 expression in ferrets given β-carotene supplements and exposed to tobacco smoke. J Natl Cancer Inst 91(1):60–66
8. Ansari M, Gupta N (2003) A comparison of lycopene and orchidectomy vs orchidectomy alone in the management of advanced prostate cancer. BJU Int 92(4):375–378
9. Aung H, Wang C, Ni M, Fishbein A, Mehendale S, Xie J et al (2007) Crocin from Crocus sativus possesses significant anti-proliferation effects on human colorectal cancer cells. Exp Oncol 29(3):175
10. Van Helden YG, Keijer J, Heil SG, Picó C, Palou A, Oliver P et al (2009) Beta-carotene affects oxidative stress-related DNA damage in lung epithelial cells and in ferret lung. Carcinogenesis 30(12):2070–2076
11. Terasaki M, Maeda H, Miyashita K, Mutoh M (2017) Induction of anoikis in human colorectal cancer cells by fucoxanthinol. Nutr Cancer 69(7):1043–1052
12. Liou G-Y, Storz P (2010) Reactive oxygen species in cancer. Free Radic Res 44(5):479–496
13. Tafani M, Sansone L, Limana F, Arcangeli T, De Santis E, Polese M et al (2016) The interplay of reactive oxygen species, hypoxia, inflammation, and sirtuins in cancer initiation and progression. Oxidative Med Cell Longev 2016:3907147. https://doi.org/10.1155/2016/3907147.
14. Prasad S, Gupta SC, Tyagi AK (2017) Reactive oxygen species (ROS) and cancer: role of antioxidative nutraceuticals. Cancer Lett 387:95–105
15. Cook JA, Gius D, Wink DA, Krishna MC, Russo A, Mitchell JB (2004) Oxidative stress, redox, and the tumor microenvironment. Semin Radiat Oncol 14:259–266.
16. Gottschling BC, Maronpot RR, Hailey JR, Peddada S, Moomaw CR, Klaunig JE et al (2001) The role of oxidative stress in indium phosphide-induced lung carcinogenesis in rats. Toxicol Sci 64(1):28–40
17. Miyake H, Hara I, Kamidono S, Eto H (2004) Oxidative DNA damage in patients with prostate cancer and its response to treatment. J Urol 171(4):1533–1536
18. Weiss J, Goode EL, Ladiges W, Ulrich CM (2005) Polymorphic variation in hOGG1 and risk of cancer: a review of the functional and epidemiologic literature. Mol Carcinogen 42(3):127–141. Published in cooperation with the University of Texas MD Anderson Cancer Center
19. Diakowska D, Lewandowski A, Kopeć W, Diakowski W, Chrzanowska T (2007) Oxidative DNA damage and total antioxidant status in serum of patients with esophageal squamous cell carcinoma. Hepato-Gastroenterology 54(78):1701–1704
20. Muguruma M, Unami A, Kanki M, Kuroiwa Y, Nishimura J, Dewa Y et al (2007) Possible involvement of oxidative stress in piperonyl butoxide induced hepatocarcinogenesis in rats. Toxicology 236(1–2):61–75

21. Galadari S, Rahman A, Pallichankandy S, Thayyullathil F (2017) Reactive oxygen species and cancer paradox: to promote or to suppress? Free Radic Biol Med 104:144–164
22. Vara JÁF, Casado E, de Castro J, Cejas P, Belda-Iniesta C, González-Barón M (2004) PI3K/Akt signalling pathway and cancer. Cancer Treat Rev 30(2):193–204
23. Klaunig JE, Kamendulis LM, Hocevar BA (2010) Oxidative stress and oxidative damage in carcinogenesis. Toxicol Pathol 38(1):96–109
24. Waris G, Ahsan H (2006) Reactive oxygen species: role in the development of cancer and various chronic conditions. J Carcinogen 5:14
25. Liu RH (2004) Potential synergy of phytochemicals in cancer prevention: mechanism of action. J Nutr 134(12):3479S–3485S
26. Young AJ, Lowe GM (2001) Antioxidant and prooxidant properties of carotenoids. Arch Biochem Biophys 385(1):20–27
27. Valko M, Izakovic M, Mazur M, Rhodes CJ, Telser J (2004) Role of oxygen radicals in DNA damage and cancer incidence. Mol Cell Biochem 266(1–2):37–56
28. Tinkler JH, Böhm F, Schalch W, Truscott TG (1994) Dietary carotenoids protect human cells from damage. J Photochem Photobiol B Biol 26(3):283–285
29. Böhm F, Edge R, Burke M, Truscott T (2001) Dietary uptake of lycopene protects human cells from singlet oxygen and nitrogen dioxide–ROS components from cigarette smoke. J Photochem Photobiol B Biol 64(2–3):176–178
30. Handelman GJ, Packer L, Cross CE (1996) Destruction of tocopherols, carotenoids, and retinol in human plasma by cigarette smoke. Am J Clin Nutr 63(4):559–565
31. Mortensen A, Skibsted LH, Sampson J, Rice-Evans C, Everett SA (1997) Comparative mechanisms and rates of free radical scavenging by carotenoid antioxidants. FEBS Lett 418(1–2):91–97
32. Rezaee R, Jamialahmadi K, Zanjani BR, Mahmoudi M, Abnous K, Rabe SZT et al (2014) Crocin effects on human myeloma cells regarding intracellular redox state, DNA fragmentation, and apoptosis or necrosis profile. Jundishapur J Nat Pharm Prod 9(4):e20131
33. Palozza P, Serini S, Torsello A, Boninsegna A, Covacci V, Maggiano N et al (2002) Regulation of cell cycle progression and apoptosis by β-carotene in undifferentiated and differentiated HL-60 leukemia cells: possible involvement of a redox mechanism. Int J Cancer 97(5):593–600
34. Upadhyaya K, Radha K, Madhyastha H (2007) Cell cycle regulation and induction of apoptosis by β-carotene in U937 and HL-60 leukemia cells. BMB Rep 40(6):1009–1015
35. Hosokawa M, Wanezaki S, Miyauchi K, Kurihara H, Kohno H, Kawabata J et al (1999) Apoptosis-inducing effect of fucoxanthin on human leukemia cell line HL-60. Food Sci Technol Res 5(3):243–246
36. Gradelet S, Le Bon A, Berges R, Suschetet M, Astorg P (1998) Dietary carotenoids inhibit aflatoxin B1-induced liver preneoplastic foci and DNA damage in the rat: role of the modulation of aflatoxin B1 metabolism. Carcinogenesis 19(3):403–411
37. Chew B, Park J, Wong M, Wong T (1999) A comparison of the anticancer activities of dietary beta-carotene, canthaxanthin and astaxanthin in mice in vivo. Anticancer Res 19(3A):1849–1853
38. Lyons NM, O'Brien NM (2002) Modulatory effects of an algal extract containing astaxanthin on UVA-irradiated cells in culture. J Dermatol Sci 30(1):73–84
39. Serpeloni JM, Barcelos GRM, Angeli JPF, Mercadante AZ, Bianchi MLP, Antunes LMG (2012) Dietary carotenoid lutein protects against DNA damage and alterations of the redox status induced by cisplatin in human derived HepG2 cells. Toxicol In Vitro 26(2):288–294
40. Kotake-Nara E, Kushiro M, Zhang H, Sugawara T, Miyashita K, Nagao A (2001) Carotenoids affect proliferation of human prostate cancer cells. J Nutr 131(12):3303–3306
41. Konishi I, Hosokawa M, Sashima T, Kobayashi H, Miyashita K (2006) Halocynthiaxanthin and fucoxanthinol isolated from Halocynthia roretzi induce apoptosis in human leukemia, breast and colon cancer cells. Comp Biochem Physiol C Toxicol Pharmacol 142(1–2):53–59

42. Tibodeau JD, Isham CR, Bible KC (2010) Annatto constituent cis-bixin has selective antimyeloma effects mediated by oxidative stress and associated with inhibition of thioredoxin and thioredoxin reductase. Antioxid Redox Signal 13(7):987–997
43. Dewanjee S, Das S, Das AK, Bhattacharjee N, Dihingia A, Dua TK et al (2018) Molecular mechanism of diabetic neuropathy and its pharmacotherapeutic targets. Eur J Pharmacol 833:472–523
44. Hu X-Y, Hou P-F, Li T-T, Quan H-Y, Li M-L, Lin T et al (2018) The roles of Wnt/β-catenin signaling pathway related lncRNAs in cancer. Int J Biol Sci 14(14):2003
45. Ochoa-Hernández AB, Juárez-Vázquez CI, Rosales-Reynoso MA, Barros-Núñez P (2012) Wnt-β-catenin signaling pathway and its relationship with cancer. Cir Cir 80:362–370
46. Voeller HJ, Truica CI, Gelmann EP (1998) β-Catenin mutations in human prostate cancer. Cancer Res 58(12):2520–2523
47. Chesire DR, Ewing CM, Sauvageot J, Bova GS, Isaacs WB (2000) Detection and analysis of β-catenin mutations in prostate cancer. Prostate 45(4):323–334
48. Sun X, Liu Y (2017) Activation of the Wnt/β-catenin signaling pathway may contribute to cervical cancer pathogenesis via upregulation of Twist. Oncol Lett 14(4):4841–4844
49. Li J, Dai W, Xia Y, Chen K, Li S, Liu T et al (2015) Astaxanthin inhibits proliferation and induces apoptosis of human hepatocellular carcinoma cells via inhibition of NF-κB P65 and Wnt/β-catenin in vitro. Mar Drugs 13(10):6064–6081
50. Kavitha K, Kowshik J, Kishore TKK, Baba AB, Nagini S (2013) Astaxanthin inhibits NF-κB and Wnt/β-catenin signaling pathways via inactivation of Erk/MAPK and PI3K/Akt to induce intrinsic apoptosis in a hamster model of oral cancer. Biochim Biophys Acta 1830(10):4433–4444
51. Tang FY, Shih CJ, Cheng LH, Ho HJ, Chen HJ (2008) Lycopene inhibits growth of human colon cancer cells via suppression of the Akt signaling pathway. Mol Nutr Food Res 52(6):646–654
52. Lin M-C, Wang F-Y, Kuo Y-H, Tang F-Y (2011) Cancer chemopreventive effects of lycopene: suppression of MMP-7 expression and cell invasion in human colon cancer cells. J Agric Food Chem 59(20):11304–11318
53. Renehan AG, Zwahlen M, Minder C, O'Dwyer ST, Shalet SM, Egger M (2004) Insulin-like growth factor (IGF)-I, IGF binding protein-3, and cancer risk: systematic review and meta-regression analysis. Lancet 363(9418):1346–1353
54. Fürstenberger G, Senn H-J (2002) Insulin-like growth factors and cancer. Lancet Oncol 3(5):298–302
55. Stoeltzing O, Liu W, Reinmuth N, Fan F, Parikh AA, Bucana CD et al (2003) Regulation of hypoxia-inducible factor-1α, vascular endothelial growth factor, and angiogenesis by an insulin-like growth factor-I receptor autocrine loop in human pancreatic cancer. Am J Pathol 163(3):1001–1011
56. Kopantzev E, Grankina E, Kopantseva M, Sverdlov E (2017) The IGF-I/IGF-IR signaling system and pancreatic Cancer. Mol Genet Microbiol Virol 32(3):131–136
57. Tang Y, Parmakhtiar B, Simoneau AR, Xie J, Fruehauf J, Lilly M et al (2011) Lycopene enhances docetaxel's effect in castration-resistant prostate cancer associated with insulin-like growth factor I receptor levels. Neoplasia (New York, NY) 13(2):108
58. Levy J, Bosin E, Feldman B, Giat Y, Miinster A, Danilenko M et al (1995) Lycopene is a more potent inhibitor of human cancer cell proliferation than either α-carotene or β-carotene. Nutr Cancer 24(3):257–266
59. Santana-Codina N, Mancias JD, Kimmelman AC (2017) The role of autophagy in cancer. Annu Rev Cancer Biol 1:19–39
60. Kung C-P, Budina A, Balaburski G, Bergenstock MK, Murphy M (2011) Autophagy in tumor suppression and cancer therapy. Crit Rev Eukaryot Gene Expr 21(1):145
61. Zhang Z, Shao Z, Xiong L, Che B, Deng C, Xu W (2009) Expression of Beclin1 in osteosarcoma and the effects of down-regulation of autophagy on the chemotherapeutic sensitivity. J Huazhong Univ Sci Technol [Med Sci] 29(6):737

62. Cai M, Hu Z, Liu J, Gao J, Liu C, Liu D et al (2014) Beclin 1 expression in ovarian tissues and its effects on ovarian cancer prognosis. Int J Mol Sci 15(4):5292–5303
63. Qiu D-M, Wang G-L, Chen L, Xu Y-Y, He S, Cao X-L et al (2014) The expression of beclin-1, an autophagic gene, in hepatocellular carcinoma associated with clinical pathological and prognostic significance. BMC Cancer 14(1):327
64. Singh SS, Vats S, Chia AY-Q, Tan TZ, Deng S, Ong MS et al (2018) Dual role of autophagy in hallmarks of cancer. Oncogene 37(9):1142–1158
65. Takamura A, Komatsu M, Hara T, Sakamoto A, Kishi C, Waguri S et al (2011) Autophagy-deficient mice develop multiple liver tumors. Genes Dev 25(8):795–800
66. Nicotra G, Mercalli F, Peracchio C, Castino R, Follo C, Valente G et al (2010) Autophagy-active beclin-1 correlates with favourable clinical outcome in non-Hodgkin lymphomas. Mod Pathol 23(7):937–950
67. Wei H, Wei S, Gan B, Peng X, Zou W, Guan J-L (2011) Suppression of autophagy by FIP200 deletion inhibits mammary tumorigenesis. Genes Dev 25(14):1510–1527
68. Chen N, Karantza-Wadsworth V (2009) Role and regulation of autophagy in cancer. Biochim Biophys Acta Mol Cell Res 1793(9):1516–1523
69. Hou L-L, Gao C, Chen L, Hu G-Q, Xie S-Q (2013) Essential role of autophagy in fucoxanthin-induced cytotoxicity to human epithelial cervical cancer HeLa cells. Acta Pharmacol Sin 34(11):1403–1410
70. Amin A, Bajbouj K, Koch A, Gandesiri M, Schneider-Stock R (2015) Defective autophagosome formation in p53-null colorectal cancer reinforces crocin-induced apoptosis. Int J Mol Sci 16(1):1544–1561
71. Luo J, Manning BD, Cantley LC (2003) Targeting the PI3K-Akt pathway in human cancer: rationale and promise. Cancer Cell 4(4):257–262
72. Vivanco I, Sawyers CL (2002) The phosphatidylinositol 3-kinase–AKT pathway in human cancer. Nat Rev Cancer 2(7):489–501
73. Martini M, De Santis MC, Braccini L, Gulluni F, Hirsch E (2014) PI3K/AKT signaling pathway and cancer: an updated review. Ann Med 46(6):372–383
74. Shayesteh L, Lu Y, Kuo W-L, Baldocchi R, Godfrey T, Collins C et al (1999) PIK3CA is implicated as an oncogene in ovarian cancer. Nat Genet 21(1):99–102
75. Bellacosa A, De Feo D, Godwin AK, Bell DW, Cheng JQ, Altomare DA et al (1995) Molecular alterations of the AKT2 oncogene in ovarian and breast carcinomas. Int J Cancer 64(4):280–285
76. Cheng JQ, Ruggeri B, Klein WM, Sonoda G, Altomare DA, Watson DK et al (1996) Amplification of AKT2 in human pancreatic cells and inhibition of AKT2 expression and tumorigenicity by antisense RNA. Proc Natl Acad Sci 93(8):3636–3641
77. Shukla S, MacLennan GT, Hartman DJ, Fu P, Resnick MI, Gupta S (2007) Activation of PI3K-Akt signaling pathway promotes prostate cancer cell invasion. Int J Cancer 121(7):1424–1432
78. Youssef J, Badr M (2011) Peroxisome proliferator-activated receptors and cancer: challenges and opportunities. Br J Pharmacol 164(1):68–82
79. Reddy RC, Srirangam A, Reddy K, Chen J, Gangireddy S, Kalemkerian GP et al (2008) Chemotherapeutic drugs induce PPAR-γ expression and show sequence-specific synergy with PPAR-γ ligands in inhibition of non-small cell lung cancer. Neoplasia (New York, NY) 10(6):597
80. Motawi T, Shaker O, Ismail M, Sayed N (2017) Peroxisome proliferator-activated receptor gamma in obesity and colorectal cancer: the role of epigenetics. Sci Rep 7(1):1–8
81. Terashita Y, Sasaki H, Haruki N, Nishiwaki T, Ishiguro H, Shibata Y et al (2002) Decreased peroxisome proliferator-activated receptor gamma gene expression is correlated with poor prognosis in patients with esophageal cancer. Jpn J Clin Oncol 32(7):238–243
82. Jung T-I, Baek W-K, Suh S-I, Jang B-C, Song D-K, Bae J-H et al (2005) Down-regulation of peroxisome proliferator-activated receptor gamma in human cervical carcinoma. Gynecol Oncol 97(2):365–373

83. Zaytseva YY, Wang X, Southard RC, Wallis NK, Kilgore MW (2008) Down-regulation of PPARgamma1 suppresses cell growth and induces apoptosis in MCF-7 breast cancer cells. Mol Cancer 7(1):90
84. Zhao W-E, Shi G, Gu H, Nguyen BN (2016) Role of PPARγ in the nutritional and pharmacological actions of carotenoids. Res Rep Biochem 6:13
85. Semple RK, Chatterjee VKK, O'Rahilly S (2006) PPARγ and human metabolic disease. J Clin Invest 116(3):581–589
86. Xia L, Tan S, Zhou Y, Lin J, Wang H, Oyang L et al (2018) Role of the NFκB-signaling pathway in cancer. Onco Targets Ther 11:2063–2073
87. Dolcet X, Llobet D, Pallares J, Matias-Guiu X (2005) NF-kB in development and progression of human cancer. Virchows Arch 446(5):475–482
88. Aggarwal BB, Gehlot P (2009) Inflammation and cancer: how friendly is the relationship for cancer patients? Curr Opin Pharmacol 9(4):351–369
89. Jin R, Sterling JA, Edwards JR, DeGraff DJ, Lee C, Park SI et al (2013) Activation of NF-kappa B signaling promotes growth of prostate cancer cells in bone. PloS one 8(4):e60983
90. Mukhopadhyay T, Roth JA, Maxwell SA (1995) Altered expression of the p50 subunit of the NF-kappa B transcription factor complex in non-small cell lung carcinoma. Oncogene 11(5):999–1003
91. Sovak MA, Bellas RE, Kim DW, Zanieski GJ, Rogers AE, Traish AM et al (1997) Aberrant nuclear factor-kappaB/Rel expression and the pathogenesis of breast cancer. J Clin Invest 100(12):2952–2960
92. Vasudevan KM, Gurumurthy S, Rangnekar VM (2004) Suppression of PTEN expression by NF-κB prevents apoptosis. Mol Cell Biol 24(3):1007–1021
93. Chaturvedi M, Sung B, Yadav V, Kannappan R, Aggarwal BB (2011) NF-κB addiction and its role in cancer:'one size does not fit all'. Oncogene 30(14):1615–1630
94. Yasui Y, Hosokawa M, Mikami N, Miyashita K, Tanaka T (2011) Dietary astaxanthin inhibits colitis and colitis-associated colon carcinogenesis in mice via modulation of the inflammatory cytokines. Chem Biol Interact 193(1):79–87
95. Ishikawa C, Jomori T, Tanaka J, Senba M, Mori N (2016) Peridinin, a carotenoid, inhibits proliferation and survival of HTLV-1-infected T-cell lines. Int J Oncol 49(4):1713–1721
96. Canavese M, Santo L, Raje N (2012) Cyclin dependent kinases in cancer: potential for therapeutic intervention. Cancer Biol Ther 13(7):451–457
97. Geleta B, Makonnen E, Abay SM (2016) Cyclic dependent kinase (CDK): role in cancer pathogenesis and as drug target in cancer therapeutics. J Cancer Sci Ther 8(6):160–167
98. Malumbres M, Barbacid M (2005) Mammalian cyclin-dependent kinases. Trends Biochem Sci 30(11):630–641
99. Biggs J, Kraft A (1995) Inhibitors of cyclin-dependent kinase and cancer. J Mol Med 73(10):509–514
100. Peyressatre M, Prével C, Pellerano M, Morris MC (2015) Targeting cyclin-dependent kinases in human cancers: from small molecules to peptide inhibitors. Cancers 7(1):179–237
101. Rafi MM, Kanakasabai S, Reyes MD, Bright JJ (2013) Lycopene modulates growth and survival associated genes in prostate cancer. J Nutr Biochem 24(10):1724–1734
102. Liu C, Russell RM, Wang X-D (2006) Lycopene supplementation prevents smoke-induced changes in p53, p53 phosphorylation, cell proliferation, and apoptosis in the gastric mucosa of ferrets. J Nutr 136(1):106–111
103. Nahum A, Hirsch K, Danilenko M, Watts CK, Prall OW, Levy J et al (2001) Lycopene inhibition of cell cycle progression in breast and endometrial cancer cells is associated with reduction in cyclin D levels and retention of p27 Kip1 in the cyclin E–cdk2 complexes. Oncogene 20(26):3428–3436
104. Palozza P, Colangelo M, Simone R, Catalano A, Boninsegna A, Lanza P et al (2010) Lycopene induces cell growth inhibition by altering mevalonate pathway and Ras signaling in cancer cell lines. Carcinogenesis 31(10):1813–1821

105. D'Alessandro AM, Mancini A, Lizzi AR, De Simone A, Marroccella CE, Gravina GL et al (2013) Crocus sativus stigma extract and its major constituent crocin possess significant anti-proliferative properties against human prostate cancer. Nutr Cancer 65(6):930–942
106. Zhao P, Luo C, Wu X, Hu H, Lv C, Ji H (2008) Proliferation apoptotic influence of crocin on human bladder cancer T24 cell line. Zhongguo Zhong yao za zhi= Zhongguo zhongyao zazhi=. China J Chinese Materia Medica 33(15):1869–1873
107. Hire RR, Srivastava S, Davis MB, Konreddy AK, Panda D (2017) Antiproliferative activity of crocin involves targeting of microtubules in breast cancer cells. Sci Rep 7:44984
108. Miyazawa K, Miyamoto S, Suzuki R, Yasui Y, Ikeda R, Kohno H et al (2007) Dietary β-cryptoxanthin inhibits N-butyl-N-(4-hydroxybutyl) nitrosamine-induced urinary bladder carcinogenesis in male ICR mice. Oncol Rep 17(2):297–304
109. Zhang X, Zhao W-E, Hu L, Zhao L, Huang J (2011) Carotenoids inhibit proliferation and regulate expression of peroxisome proliferators-activated receptor gamma (PPARγ) in K562 cancer cells. Arch Biochem Biophys 512(1):96–106
110. Mahdizadeh S, Karimi G, Behravan J, Arabzadeh S, Lage H, Kalalinia F (2016) Crocin suppresses multidrug resistance in MRP overexpressing ovarian cancer cell line. DARU J Pharm Sci 24(1):17
111. Williams CL, Phelps SH, Porter RA (1996) Expression of Ca2+/calmodulin-dependent protein kinase types II and IV, and reduced DNA synthesis due to the Ca2+/calmodulin-dependent protein kinase inhibitor KN-62 (1-[N, O-Bis (5-isoquinolinesulfonyl)-N-methyl-L-tyrosyl]-4-phenylpiperazine) in small cell lung carcinoma. Biochem Pharmacol 51(5):707–715
112. Gretzer MB, Partin AW (2002) PSA levels and the probability of prostate cancer on biopsy. Eur Urol Suppl 1(6):21–27
113. Yoon H, Dehart JP, Murphy JM, Lim S-TS (2015) Understanding the roles of FAK in cancer: inhibitors, genetic models, and new insights. J Histochem Cytochem 63(2):114–128
114. Jänicke RU, Sprengart ML, Wati MR, Porter AG (1998) Caspase-3 is required for DNA fragmentation and morphological changes associated with apoptosis. J Biol Chem 273(16):9357–9360
115. Debatin K-M (2004) Apoptosis pathways in cancer and cancer therapy. Cancer Immunol Immunother 53(3):153–159
116. Campos L, Rouault J-P, Sabido O, Oriol P, Roubi N, Vasselon C et al (1993) High expression of bcl-2 protein in acute myeloid leukemia cells is associated with poor response to chemotherapy. Blood 81(11):3091–3096
117. Bargou RC, Daniel PT, Mapara MY, Bommert K, Wagener C, Kallinich B et al (1995) Expression of the bcl-2 gene family in normal and malignant breast tissue: low bax-α expression in tumor cells correlates with resistance towards apoptosis. Int J Cancer 60(6):854–859
118. Naz H, Khan P, Tarique M, Rahman S, Meena A, Ahamad S et al (2017) Binding studies and biological evaluation of β-carotene as a potential inhibitor of human calcium/calmodulin-dependent protein kinase IV. Int J Biol Macromol 96:161–170
119. Bunker CH, McDonald AC, Evans RW, De La Rosa N, Boumosleh JM, Patrick AL (2007) A randomized trial of lycopene supplementation in Tobago men with high prostate cancer risk. Nutr Cancer 57(2):130–137
120. Grainger EM, Schwartz SJ, Wang S, Unlu NZ, Boileau TW-M, Ferketich AK et al (2008) A combination of tomato and soy products for men with recurring prostate cancer and rising prostate specific antigen. Nutr Cancer 60(2):145–154
121. Kucuk O, Sarkar FH, Sakr W, Djuric Z, Pollak MN, Khachik F et al (2001) Phase II randomized clinical trial of lycopene supplementation before radical prostatectomy. Cancer Epidemiol Prev Biomarkers 10(8):861–868
122. Mohanty NK, Saxena S, Singh UP, Goyal NK, Arora RP (2005) Lycopene as a chemopreventive agent in the treatment of high-grade prostate intraepithelial neoplasia. Urol Oncol 23:383–385.
123. Vaishampayan U, Hussain M, Banerjee M, Seren S, Sarkar FH, Fontana J et al (2007) Lycopene and soy isoflavones in the treatment of prostate cancer. Nutr Cancer 59(1):1–7

124. Schwarz S, Obermüller-Jevic UC, Hellmis E, Koch W, Jacobi G, Biesalski H-K (2008) Lycopene inhibits disease progression in patients with benign prostate hyperplasia. J Nutr 138(1):49–53
125. van Breemen RB, Sharifi R, Viana M, Pajkovic N, Zhu D, Yuan L et al (2011) Antioxidant effects of lycopene in African American men with prostate cancer or benign prostate hyperplasia: a randomized, controlled trial. Cancer Prev Res 4(5):711–718
126. Soares NCP, Teodoro AJ, Oliveira FL, Takiya CM, Junior AP, Nasciutti LE et al (2014) Lycopene induce apoptosis in human prostate cells and alters the expression of Bax and Bcl-2 genes. LWT-Food Sci Technol 59(2):1290–1297
127. Liu C, Lian F, Smith DE, Russell RM, Wang X-D (2003) Lycopene supplementation inhibits its lung squamous metaplasia and induces apoptosis via up-regulating insulin-like growth factor-binding protein 3 in cigarette smoke-exposed ferrets. Cancer Res 63(12):3138–3144
128. Hoshyar R, Bathaie SZ, Sadeghizadeh M (2013) Crocin triggers the apoptosis through increasing the Bax/Bcl-2 ratio and caspase activation in human gastric adenocarcinoma, AGS, cells. DNA Cell Biol 32(2):50–57
129. Vali F, Changizi V, Safa M (2015) Synergistic apoptotic effect of crocin and paclitaxel or crocin and radiation on MCF-7 cells, a type of breast cancer cell line. Int J Breast Cancer 2015
130. Zhang Y, Zhu X, Huang T, Chen L, Liu Y, Li Q et al (2016) β-Carotene synergistically enhances the anti-tumor effect of 5-fluorouracil on esophageal squamous cell carcinoma in vivo and in vitro. Toxicol Lett 261:49–58
131. Kim K-N, Heo S-J, Kang S-M, Ahn G, Jeon Y-J (2010) Fucoxanthin induces apoptosis in human leukemia HL-60 cells through a ROS-mediated Bcl-xL pathway. Toxicol In Vitro 24(6):1648–1654
132. Zhang Z, Zhang P, Hamada M, Takahashi S, Xing G, Liu J et al (2008) Potential chemoprevention effect of dietary fucoxanthin on urinary bladder cancer EJ-1 cell line. Oncol Rep 20(5):1099–1103
133. Kotake-Nara E, Asai A, Nagao A (2005) Neoxanthin and fucoxanthin induce apoptosis in PC-3 human prostate cancer cells. Cancer Lett 220(1):75–84
134. Bi M-C, Rosen R, Zha R-Y, McCormick SA, Song E, Hu D-N (2013) Zeaxanthin induces apoptosis in human uveal melanoma cells through Bcl-2 family proteins and intrinsic apoptosis pathway. Evid Based Complement Alternat Med 2013:205082. https://doi.org/10.1155/2013/205082.
135. Choi Y-J, Hur J-M, Lim S, Jo M, Kim DH, Choi J-I (2014) Induction of apoptosis by deinoxanthin in human cancer cells. Anticancer Res 34(4):1829–1835
136. Ganesan P, Noda K, Manabe Y, Ohkubo T, Tanaka Y, Maoka T et al (2011) Siphonaxanthin, a marine carotenoid from green algae, effectively induces apoptosis in human leukemia (HL-60) cells. Biochim Biophys Acta General Subjects 1810(5):497–503
137. Liu C, Wang X-D, Bronson RT, Smith DE, Krinsky NI, Russell RM (2000) Effects of physiological versus pharmacological β-carotene supplementation on cell proliferation and histopathological changes in the lungs of cigarette smoke-exposed ferrets. Carcinogenesis 21(12):2245–2253
138. Pan L, Chai H-B, Kinghorn AD (2012) Discovery of new anticancer agents from higher plants. Front Biosci (Scholar edition) 4:142
139. Gong X, Smith JR, Swanson HM, Rubin LP (2018) Carotenoid lutein selectively inhibits breast cancer cell growth and potentiates the effect of chemotherapeutic agents through ROS-mediated mechanisms. Molecules 23(4):905
140. Park JS, Chew BP, Wong TS (1998) Dietary lutein from marigold extract inhibits mammary tumor development in BALB/c mice. J Nutr 128(10):1650–1656
141. Gunasekera RS, Sewgobind K, Desai S, Dunn L, Black HS, McKeehan WL et al (2007) Lycopene and lutein inhibit proliferation in rat prostate carcinoma cells. HNUC 58(2):171–177
142. Zhang W-L, Zhao Y-N, Shi Z-Z, Cong D, Bai Y-S (2018) Lutein inhibits cell growth and activates apoptosis via the PI3K/AKT/mTOR signaling pathway in A549 human non-small-cell lung cancer cells. J Environ Pathol Toxicol Oncol 37(4):341

143. Xu XL, Hu D-N, Iacob C, Jordan A, Gandhi S, Gierhart DL et al (2015) Effects of zeaxanthin on growth and invasion of human uveal melanoma in nude mouse model. J Ophthalmol 57:4104. https://doi.org/10.1155/2015/392305.
144. Nishino H, Tsushima M, Matsuno T, Tanaka Y, Okuzumi J, Murakoshi M et al (1992) Antineoplastic effect of halocynthiaxanthin, a metabolite of fucoxanthin. Anti-Cancer Drugs 3(5):493–497
145. Yoshida T, Maoka T, Das SK, Kanazawa K, Horinaka M, Wakada M et al (2007) Halocynthiaxanthin and peridinin sensitize colon cancer cell lines to tumor necrosis factor–related apoptosis-inducing ligand. Mol Cancer Res 5(6):615–625
146. Nishino H (1991) Anti-tumor activity of peridinin and its structurelly related butenolide compounds. J Kyoto Pref Univ Med 100:831–835
147. Molnar J, Serly J, Pusztai R, Vincze I, Molnar P, Horváth G et al (2012) Putative supramolecular complexes formed by carotenoids and xanthophylls with ascorbic acid to reverse multidrug resistance in cancer cells. Anticancer Res 32(2):507–517
148. Tokuda H, Enjo F, Maoka T, Mochida K, Kuchide M, Ogata M et al (2004) Regulation of the cell cycle progression of A549 human lung cancer cell treatment by major paprica carotenoids elements capsanthin, capsorubin and beta-carotene. Cancer Res 64:1374.
149. de Oliveira Júnior RG, Bonnet A, Braconnier E, Groult H, Prunier G, Beaugeard L et al (2019) Bixin, an apocarotenoid isolated from Bixa orellana L., sensitizes human melanoma cells to dacarbazine-induced apoptosis through ROS-mediated cytotoxicity. Food Chem Toxicol 125:549–561
150. Kumar Y, Phaniendra A, Periyasamy L (2018) Bixin triggers apoptosis of human Hep3B hepatocellular carcinoma cells: an insight to molecular and in silico approach. Nutr Cancer 70(6):971–983
151. Maoka T, Mochida K, Kozuka M, Ito Y, Fujiwara Y, Hashimoto K et al (2001) Cancer chemopreventive activity of carotenoids in the fruits of red paprika Capsicum annuum L. Cancer Lett 172(2):103–109
152. Palozza P, Maggiano N, Calviello G, Lanza P, Piccioni E, Ranelletti FO et al (1998) Canthaxanthin induces apoptosis in human cancer cell lines. Carcinogenesis 19(2):373–376
153. Kozuki Y, Miura Y, Yagasaki K (2000) Inhibitory effects of carotenoids on the invasion of rat ascites hepatoma cells in culture. Cancer Lett 151(1):111–115
154. Tanaka T, Morishita Y, Suzui M, Kojima T, Okumura A, Mori H (1994) Chemoprevention of mouse urinary bladder carcinogenesis by the naturally occurring carotenoid astaxanthin. Carcinogenesis 15(1):15–19
155. Tanaka T, Makita H, Ohnishi M, Mori H, Satoh K, Hara A (1995) Chemoprevention of rat oral carcinogenesis by naturally occurring xanthophylls, astaxanthin and canthaxanthin. Cancer Res 55(18):4059–4064
156. Tanaka T, Kawamori T, Ohnishi M, Makita H, Mori H, Satoh K et al (1995) Suppression of azoxymethane-induced rat colon carcinogenesis by dietary administration of naturally occurring xanthophylls astaxanthin and canthaxanthin during the postinitiation phase. Carcinogenesis 16(12):2957–2963
157. Katsumura N, Okuno M, Onogi N, Moriwaki H, Muto Y, Kojima S (1996) Suppression of mouse skin papilloma by canthaxanthin and β-carotene in vivo: possibility of the regression of tumorigenesis by carotenoids without conversion to retinoic acid. Nutr Cancer 26(2):203–208
158. Tsushima M, Maoka T, Katsuyama M, Kozuka M, Matsuno T, Tokuda H et al (1995) Inhibitory effect of natural carotenoids on Epstein-Barr virus activation activity of a tumor promoter in Raji cells. A screening study for anti-tumor promoters. Biol Pharm Bull 18(2):227–233
159. Pasquet V, Morisset P, Ihammouine S, Chepied A, Aumailley L, Berard J-B et al (2011) Antiproliferative activity of violaxanthin isolated from bioguided fractionation of Dunaliella tertiolecta extracts. Mar Drugs 9(5):819–831

160. Gyémánt N, Tanaka M, Molnár P, Deli J, Mándoky L, Molnár J (2006) Reversal of multidrug resistance of cancer cells in vitro: modification of drug resistance by selected carotenoids. Anticancer Res 26(1A):367–374
161. Gijsbers L, van Eekelen HD, de Haan LH, Swier JM, Heijink NL, Kloet SK et al (2013) Induction of peroxisome proliferator-activated receptor γ (PPARγ)-mediated gene expression by tomato (Solanum lycopersicum L.) extracts. J Agric Food Chem 61(14):3419–3427
162. Haguet Q, Bonnet A, Bérard J-B, Goldberg J, Joguet N, Fleury A et al (2017) Antimelanoma activity of Heterocapsa triquetra pigments. Algal Res 25:207–215
163. Maoka T, Tsushima M, Nishino H (2002) Isolation and characterization of dinochrome A and B, anti-carcinogenic active carotenoids from the fresh water red tide Peridinium bipes. Chem Pharm Bull 50(12):1630–1633

Chapter 14
Carotenoids as Antidiabetic Agents

Ranabir Sahu and Saikat Dewanjee

14.1 Introduction

Diabetes mellitus (DM), a life-threatening chronic syndrome, is characterized by hyperglycemia brought about by the deficit in insulin production and/or setting up of a deficiency in insulin sensitivity. There are two types of DM, namely type 1 DM or "insulin-dependent diabetes mellitus" (IDDM) or "juvenile diabetes", where adequate insulin production is hampered due to the failure of pancreas and type 2 or "noninsulin-dependent diabetes mellitus" (NIDDM) or "adult-onset diabetes", which initiates with the insulin resistance i.e. cells refuse to react to insulin appropriately [1]. There is another DM called "gestational diabetes" referred as a condition in which a pregnant woman without diabetes suddenly develops high blood sugar levels. Insulin management has been revealed to be main therapeutic option in type 1 DM, while type 2 DM can be treated with medication and/or insulin. As stated by the World Health Organization (WHO), total 415 million people are experiencing DM where about 90% were diagnosed as type 2 DM patients. WHO reported that around 1.5 million deaths come about due to DM worldwide and it has been alarmingly increasing [2]. Obesity due to fat-enriched foods and poor manner of living has been regarded to expand the danger of type 2 DM (Fig. 14.1). Hyperglycemia and insulin resistance were revealed to develop oxidative stress via generating excess of oxidative free radicals, which participate a dominant role in the pathogenesis of type 2 DM (Fig. 14.2).

Antioxidants can normalize or deactivate free radicals by various mechanisms before imparting toxic manifestations to the cells [3]. The mechanisms include the

R. Sahu
Department of Pharmaceutical Technology, University of North Bengal, Darjeeling, India

S. Dewanjee (✉)
Advanced Pharmacognosy Research Laboratory, Department of Pharmaceutical Technology, Jadavpur University, Kolkata, India
e-mail: saikat.dewanjee@jadavpuruniversity.in

© Springer Nature Switzerland AG 2021

M. Zia-Ul-Haq et al. (eds.), *Carotenoids: Structure and Function in the Human Body*, https://doi.org/10.1007/978-3-030-46459-2_14

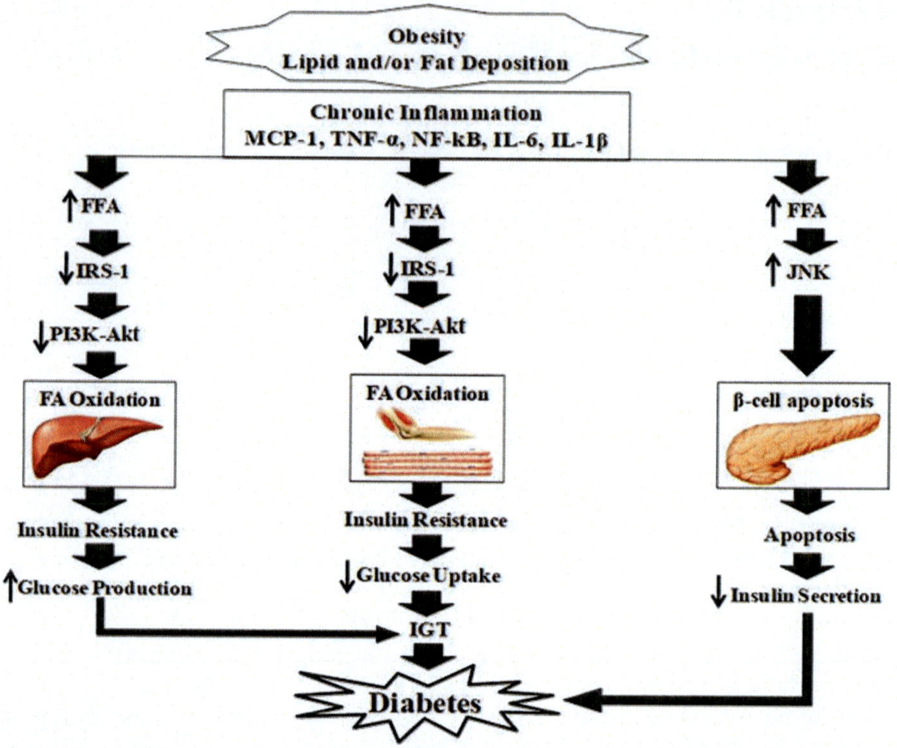

Fig. 14.1 The obesity can increase the risk of diabetes via involving several cellular and molecular events in skeletal muscle, liver, and pancreas in the development of diabetes. *Akt* serine/threonine kinase, *FA* Fatty acid, *FFA* Free fatty acid, *IGT* Impaired glucose tolerance, *IL* Interleukin, *IRS1* Insulin receptor substrate 1, *JNK* c-Jun N-terminal kinase, MCP-1 Monocyte-chemo-attractant protein-1, *NF-κB* Nuclear factor kappa-light-chain-enhancer of activated B cells, *PI3K* Phosphatidylinositol 3-kinase, *TNF-α* Tumor necrosis factor α. (↓) decrease and (↑) increase

direct scavenging of free radicals, boosting the production of cellular redox defense molecules, and/or regulating redox sensitive signaling events. Both endogenous and exogenous antioxidants can work synergistically [4]. Though endogenous antioxidant is important for maintaining cellular balance, it is not sufficient under the oxidative stress. Exogenous antioxidants can be supplied through diet or dietary supplements [5–7].

Fig. 14.2 (continued) growth factor, *GSH* reduced glutathione, *iNOS* inducible nitric oxide synthase, *IRS* insulin receptor substrate, *IRS ser/thr* Serine/ threonine phosphorylation of IRS-1, *MAPK* Mitogen-activated protein kinase, *NADPHox* Nicotinamide adenine dinucleotide phosphate oxidase, *NADPH* Nicotinamide adenine dinucleotide phosphate, *NF-κB* Nuclear factor kappa-light-chain-enhancer of activated B cells, *PKC* Protein kinase C, *RNS* Reactive nitrogen species, *ROS* Reactive oxygen species, *SOD* Superoxide dismutase, *TGF-β* Transforming growth factor beta, *VEGF* Vascular endothelial growth factor. (↓) decrease and (↑) increase

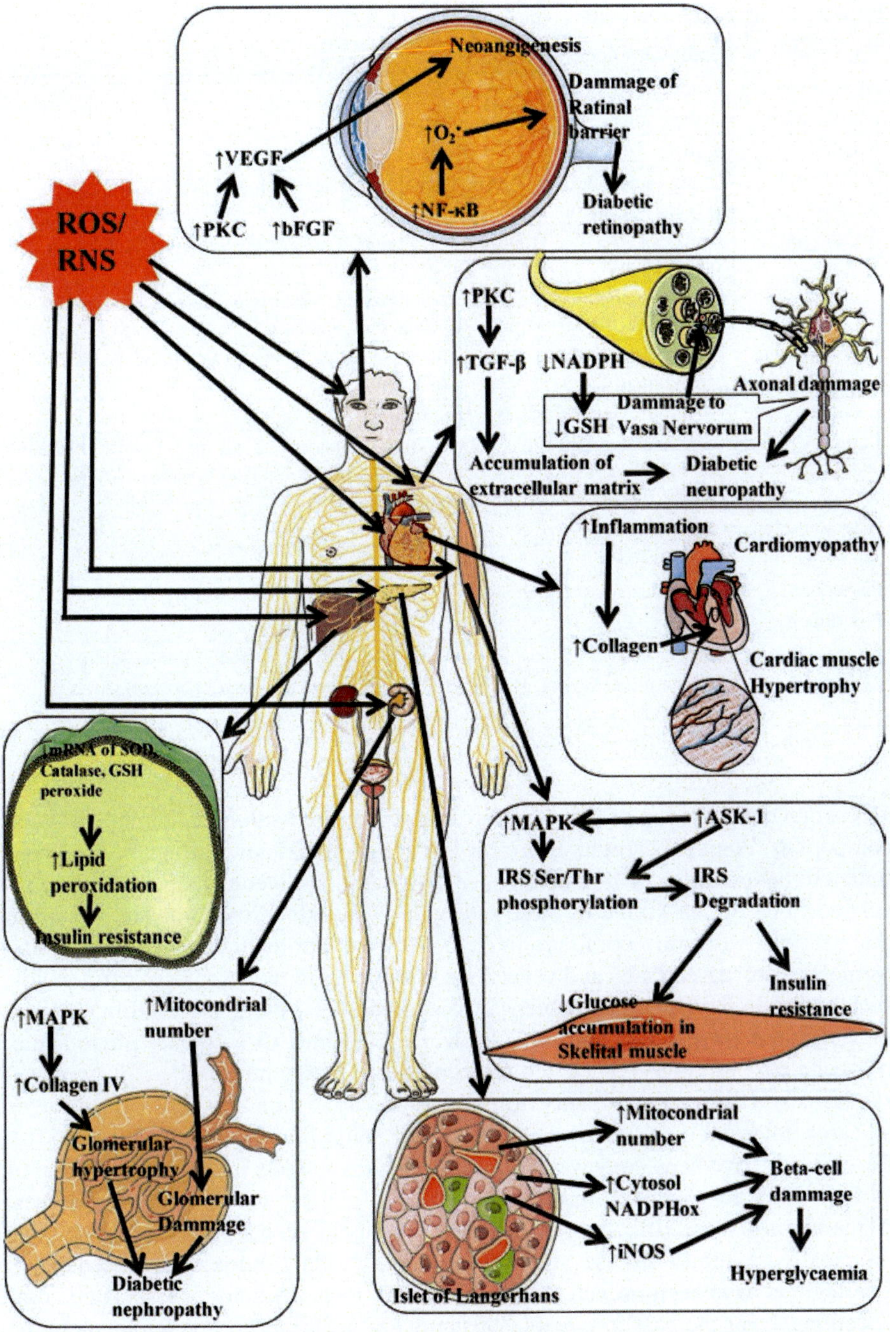

Fig. 14.2 Role of oxidative stress in insulin resistance and diabetic complications like retinopathy, nephropathy and neuropathy. *ASK1* Apoptosis signal-regulating kinase 1, *bFGF* basic fibroblast

Table 14.1 List of some common sources of carotenoids [9]

Carotenoids	Common sources
Antheraxanthin	Orange juice, Virgin olive oil, Capsicum
Auroxanthin	Orange juice
Crocins	Saffron
α-Carotene	Orange juice, Melon, Carrot, Spinach, Berries, Fish and Sea Food, *Citrus* Species, Capsicum
β-Carotene	Tomatoes, Berries, Orange Juice, Melon, Tea, Carrot, Spinach, Grapes, Berries, Table Olives
	Papaya, Fish and Sea Food, *Citrus* Species, *Opuntia ficus-indica,* Sweet Potato Flours, Dried Peppers, Mango, Capsicum, cabbage
Cryptoxanthin	Orange juice, Wheat, Barley, Berries, Papaya, Fish and Sea Food, *Opuntia ficus-indica*, Dried Peppers, Capsicum
Lycopene	Tomatoes, Berries, Papaya, Fish and Sea Food
Lutein	Tomatoes, Orange Juice, Wheat, Barley, Melon, Tea, Grapes, Berries, Fish and Sea Food, *Citrus* Species, *Opuntia ficus-indica,* Virgin Olive Oil, Soybean, *Crocus sativus,* Capsicum, Cabbage
Luteoxanthin	Orange juice
Mutatoxanthin	*Citrus* species
Phytoene	Melon, Capsicum, Cabbage
Phytofluene	Melon
Violaxanthin	Carrot, Spinach, *Citrus* Species, Virgin Olive Oil, Dried Peppers, Mango
Zeaxanthin	Orange juice, Wheat, Barley, Berries, Fish and Sea Food, *Citrus* Species, Capsicum
Zeinoxanthin	Orange juice

Carotenoids, a group of natural antioxidants, are lipid-soluble natural pigmented compounds. Plentiful carotenoids occur in the nature and new carotenoids are continued to be identified. Carotenoids have a 40-carbon polyene chain, which is considered to be the backbone of these molecules. Though carotenoids are present in the nature as all-*trans* configuration, 10–30% of carotenoids are found as *cis*-isomers in processed fruits and vegetables [8]. There are two groups of carotenoids, such as xanthophylls and carotenes [9]. Xanthophylls (lutein, astaxanthin, zeaxanthin etc.) are oxygenated carotenoids; while carotenes (β-carotene, phytofluene, lycopene etc.) are non-oxygenated molecules, [9]. Carotenoids take part in photosynthesis and they protect plants from photo-oxidative damage. The major sources of carotenoids include dietary vegetables and fruits (Tables 14.1 and 14.2) [10]. Carotenoids have been conveyed to possess valuable effects in the counteraction of various ailments, such as metabolic disorders, cancer, cardiovascular diseases, Alzheimer's disease, osteoporosis, etc. However, it has been argued that carotenoids principally exhibit disease prevention ability via antioxidant mechanism. Carotenoids have been proven to be beneficial to treat DM and its associated complications. Over the past few years, substantial effort has been given to explore the antidiabetic mechanisms of dietary carotenoids and carotenoids proved their antidiabetic effect, which is more than an antioxidant. This chapter focused on the present understanding of the potential efficacy of carotenoids on DM.

Table 14.2 Some recent works on carotenoids on diabetes

Type of Diabetes	Subject	Carotenoids	References
Type 2	Human	α/β-carotene, lycopene, α-tocopherol and, γ-tocopherol	[11]
Type 2	Human	β-carotene	[12]
Type 2	Human	α/β-carotene, β-cryptoxanthin, lycopene, and lutein.	[13]
Type 2	Human	α/β-carotene, β-cryptoxanthin, and zeaxanthin	[14]
Type 2	Human	α/β-carotene and lutein	[15]
Type 2	Human	α/β-carotene, lycopene, β-cryptoxanthin, lutein, and zeaxanthin	[16]
Type 2	Human	β-carotene	[17]
Type 2	Human	β-carotene	[18]
Type 1	STZ-Wistar rats	Astaxanthin and lutein	[19]
Type 1	STZ-Wistar rats	Bixin	[20]
Type 1	Alloxan-Wistar rats	Crocin and crocetin	[21]
Type 1	STZ-Wistar rats	Crocin	[22]
Type 1	Human	β-carotene	[23]
Type 2	Human	β-carotene	[24]
Type 2	Human	α/β-carotene,lutein, zeaxanthin, β-cryptoxanthin, and lycopene	[25]
Type 1	STZ-Wistar rats	Lutein	[26]
Type 2	STZ-Wistar rats	Crocin	[27]
Type 1	STZ-Wistar rats	Astaxanthin and crocin	[28]
Type 1	STZ-Wistar rats	Lycopene	[29]
Type 1	Alloxan-Wistar rats	Astaxanthin	[30]
Type 2	STZ-Wistar rats	Crocin	[31]

14.2 Carotenoids as Antidiabetic Agents

14.2.1 Astaxanthin

Astaxanthin is a carotenoid, which exhibits several pharmacological actions including antidiabetic activity.

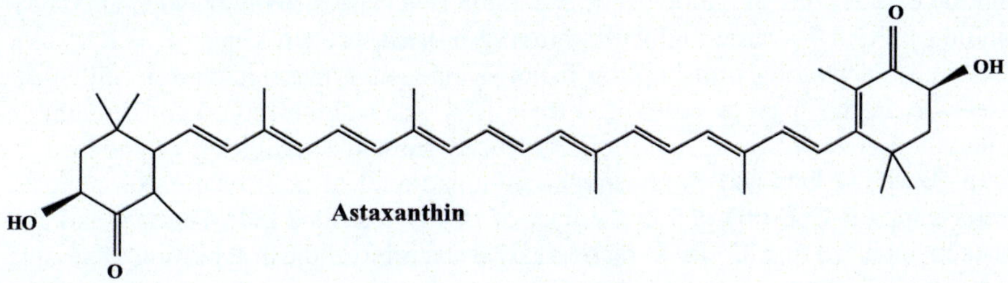

Astaxanthin

In addition, it simultaneously possesses antioxidative, immunomodulatory, anticancer, and anti-inflammatory activities [32]. Various studies revealed that astaxanthin is a potent antioxidant [33]. The antioxidant property of astaxanthin is quite greater than other naturally occurring antioxidants, such as Trolox, α-tocopherol, zeaxanthin, lycopene, α/β-carotene, lutein etc. [19, 34]. The roles of astaxanthin in diabetes prevention have been mentioned in many reports. Uchiyama and co-workers evaluated the therapeutic role of astaxanthin in C57BL/KsJ-db/db diabetic mice. The treatment (1.0 mg/mouse/day, orally) was started at 6 weeks of age and continued up to 18 weeks [35]. The levels of blood glucose in fasting condition were assessed at the specific intervals [35]. Astaxanthin demonstrated significant anti-hyperglycemic effect in type 2 diabetic mice [35]. Intraperitoneal glucose tolerance has been evaluated at 18 weeks of age [35]. Astaxanthin has been accounted for improving serum insulin levels and decreased glucose resilience in type 2 db/db mice [35]. In search of mechanistic insight, astaxanthin was found to protect pancreatic β-cells from oxidative injury [35]. Generally, diabetic patients suffer from hyperglycemia-provoked oxidative tissue injury as a secondary toxic manifestation, which could be prevented by astaxanthin. Astaxanthin (2 mg/kg/day for 4 weeks) treatment has been stated to display significant anti-hyperglycemic effect in high fructose + fat diet (HFFD)-fed mice [36]. Astaxanthin was observed to suppress fasting blood glucose, body weight, and homeostatic model assessment (HOMA) index and simultaneously increased serum astaxanthin and quantitative insulin sensitivity check index (QUICKI) [36]. In addition, astaxanthin treatment significantly increased hepatic glucose utilizing enzymes, such as hexokinase, pyruvate kinase, and hepatic glycogen content in diabetic mice [36]; while, it can inhibit the levels of hepatic gluconeogenic enzymes, for example, glucose-6-phosphatase, glycogen phosphorylase, and fructose-1,6-bisphosphatase [36]. In search of mechanistic insight, astaxanthin was found to activate insulin receptor downstream signaling, which subsequently increased the insulin sensitivity. Astaxantin can trigger insulin-stimulated tyrosine phosphorylation insulin receptor (IR)-β and insulin receptor substrate (IRS)-1/2 activation in the HFFD murine liver [36]. However, astaxanthin can decrease the phosphorylation of IRS-1/2 in serine residues. Consequently, astaxanthin endorses phosphatidylinositol 3-kinase (PI3K) triggering and protein kinase B (Akt) phosphorylation [36]. In addition, it could significantly suppress the expression of c-jun N-terminal kinase (JNK)-1 and extracellular signal-regulated kinases (ERK)-1 in the liver of diabetic mice [36]. In another study by the same group, astaxanthin (2 mg/kg/day from day 16 onwards for the next 45 days, orally) was found to impede ROS production, lipid accumulation, caspase 12 activation, and the expressions of endoplasmic reticulum (ER) stress markers, such as X-box binding protein 1, immunoglobulin-binding protein, protein kinase R (PKR)-like ER kinase, activating transcription factor 6, phosphorylated eukaryotic initiation factor 2α in the liver of HFFD-fed mice [37]. Astaxanthin could further inhibit inflammation in diabetic condition evidenced from the inhibition of inhibitory-kappa B kinase beta (IκBβ) phosphorylation and nuclear translocation of necrosis factor-kappa B (NF-κB) p65 in the liver of HFFD-fed mice [37]. Hussein and co-workers reported that 22-week treatment of astaxanthin at the oral portion of 50 mg/

kg/day can altogether lessen fasting blood glucose, serum triglycerides, serum non-esterified fatty acids, homeostasis catalog of insulin resistance and can improve insulin sensitivity in the transgenic human apolipoprotein B bearing SHR/NDmcr-cp rats [38]. Astaxanthin treatment can also reduce adiponectin and high-density lipoprotein (HDL) cholesterol level in experimental rats [38]. In addition, it can significantly reduce arterial blood pressure [38].

Peroxisome proliferator-activated receptor gamma (PPAR-γ) has a dominant role in the metabolism of carbohydrates, which has been revealed to be a principal target for astaxanthin [39]. Inoue and co-workers mentioned that astaxanthin can bind with PPAR-γ and trigger its interactions with transcriptional intermediary factor 2 (TIF2) and steroid receptor coactivator-1 (SRC-1) [39]. Thereby, it could regulate the transcriptions of PPAR-γ target genes. However, astaxanthin alone cannot stimulate adipocyte differentiation [39]. Eighteen day treatment of astaxanthin (50 mg/kg/day) was revealed to suppress the elevated glucose (liver and sera) concentration in streptozotocin (STZ)-evoked type 1 diabetic rats [28]. In addition, it could attenuate hyperglycemia-provoked hepatic inflammation via reciprocating the expressions of inflammation-related proteins, such as inducible nitric oxide synthase (iNOS), cyclooxygenase-2 (COX-2), NF-κB, monocyte chemoattractant protein-1 (MCP-1), and intercellular adhesion molecule 1 (ICAM-1) in the liver of type 1 diabetic rats (Fig. 14.3) [28]. In the same study, astaxanthin was found to attenuate the levels of advanced glycation end products (AGE), reactive oxygen species (ROS), and hepatic lipid peroxidation; however, no change was observed either in the levels of cellular antioxidants or in the expressions of ROS-related proteins [28].

Earlier evidences revealed that astaxanthin would exhibit protective effect in diabetic complications. Astaxanthin was found to cut the generation of absolute reactive species, nitric oxide, superoxide, and peroxynitrite in high glucose-evoked proximal tubular epithelial cells [40]. In addition, it can suppress high-glucose-induced lipid peroxidation, apoptosis, and expression of inflammatory factors, such as iNOS, COX-2, and NF-κB in the renal cells in vitro [40]. Manabe and colleagues claimed that astaxanthin can inhibit high glucose-elicited activation of mitochondrial ROS production in the normal human mesangial cells [41]. It has been found to reciprocate the expressions of inflammatory and pro-fibrotic signal proteins, such as MCP-1, COX-2, NF-κB, transforming growth factor β (TGF-β), and activator protein-1 (AP-1) in the high glucose-exposed normal human mesangial cells [41]. Oral treatment of astaxanthin (20 mg/kg/day) for 30 days was found to increase glutathione peroxidase (GPx) activity without offering any significant change in superoxide dismutase (SOD) and glutathione reductase (GR) levels in the pulp tissue of the teeth of alloxan-induced type 1 diabetic rats [30]. The impacts of astaxanthin (20 mg/kg/day) for 30 days in the redox status in the lymphocytes of alloxan-incited diabetic rodents have been assessed. In addition to hypoglycemic and hypolipidemic effects, astaxanthin can reciprocate the redox imbalance in the diabetic lymphocytes via regulating ROS/RNS metabolism [42]. Astaxanthin can also promote the activities of endogenous antioxidant enzymes, such as catalase (CAT), SOD, GPx, and GR in the lymphocytes of type 1 diabetic rats [42]. Naito and co-researchers investigated the effect of astaxanthin (0.02% in diet) on

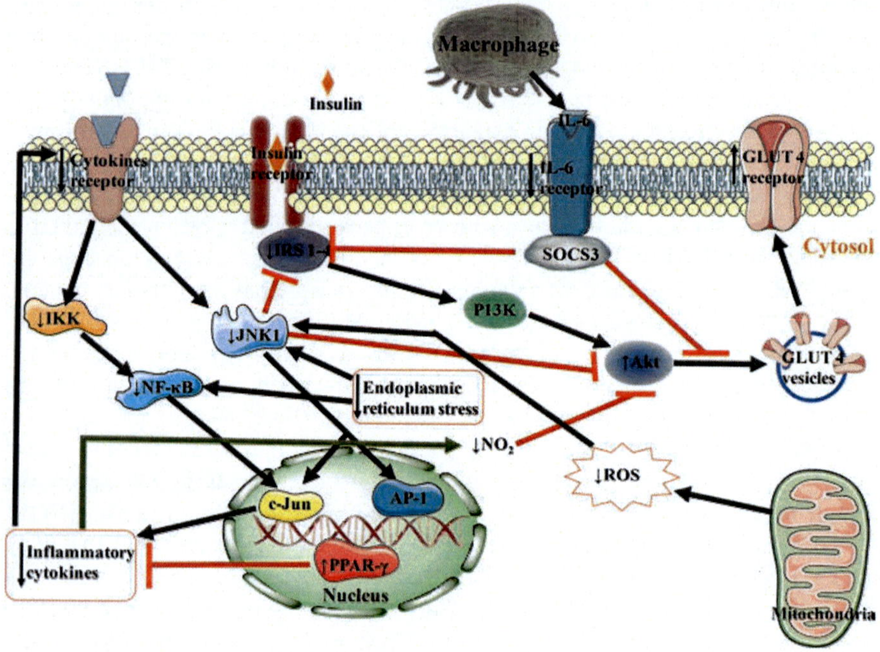

Fig. 14.3 The antidiabetic mechanisms of carotenoids. *Akt* Serine/threonine kinase, *AP-1* Activator protein 1, *GLUT4* Glucose transporter type 4, *IKK* Inhibitor of nuclear factor kappa-light-chain-enhancer of activated B cells kinase, *IL* Interleukin, *IRS* Insulin receptor substrate, *JNK1* c-Jun N-terminal kinase 1, *NF-κB* Nuclear factor kappa-light-chain-enhancer of activated B cells, *PI3K* Phosphatidylinositol 3-kinase, *PPAR-γ* Peroxisome proliferator-activated receptor gamma, *SOCS3* Suppressor of cytokine signaling 3. (↓) activation and (↑) suppression

diabetic-provoked renal complications in female db/db mice [32]. In their study, astaxanthin was found to attenuate the urinary albumin and 8-hydroxydeoxyguanosine levels within 12 weeks in type 2 diabetic mice, which suggested the reno-protective effect of astaxanthin [32, 43]. It has been claimed that astaxanthin imparted reno-protective effect via suppressing oxidative stress and preventing cell damage in mouse kidney [32]. Combination of α-tocopherol (100 mg/kg) and astaxanthin (100 mg/kg) for 20 weeks has been proven to be beneficial against hyperglycemia-induced nephrotoxicity in STZ-induced diabetic Osteogenic Disorder Shionogi rats [44]. Hyperglycemia-provoked inflammation and apoptosis in the central nervous system can develop diabetic neuropathy [45]. Astaxanthin (10, 20, and 40 mg/kg) for 5 days was found to hinder the hyperglycemia-provoked cognitive deficiencies and neurotoxicity in STZ-diabetic rodents [46]. Astaxanthin also prevented diabetes-provoked redox stress, apoptosis, and inflammatory disturbances in rat brain [46]. The increases in the expressions of NOS, NF-κB (p65), tumor necrosis factor (TNF-α), pro-inflammatory interleukin (IL)s, and intrinsic apoptotic caspases have been perceived in the cerebral cortex and hippocampus of the type 1 diabetic rats [46]. As opposed to, astaxanthin can significantly reciprocate the expression of aforementioned pro-inflammatory and pro-apoptotic signal proteins [46], which unveils the prophylactic mechanisms of astaxanthin in diabetic neuropathy (Fig. 14.3). Depression-like behavior has also been noticed in diabetic animals [47]. Astaxanthin (25 mg/kg/day) for 10 weeks has been found to exhibit the anti-depressant effect via reducing the counts of glial fibrillary acidic protein (GFAP)-responsive cells, suppressing caspase 3 activation, and inhibiting the levels of pro-inflammatory mediators, such as IL-6, IL-1β, and COX-2 in the hippocampus and hypothalamus of STZ-diabetic animals [48]. The aforementioned statements revealed that astaxanthin can be a impending therapeutic agent in DM and DM-associated complication [45].

14.2.2 Zeaxanthin

Zeaxanthin, a natural pigment, possessing potent antioxidant function [49–51]. Zeaxanthin exists in the retina and lens.

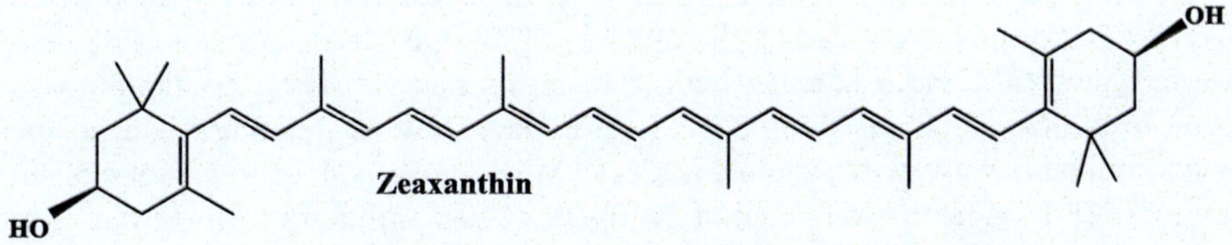

It protects photoreceptor in lens form toxic blue light. Scientists have unveiled the protective role of zeaxanthin in diabetic retinopathy [52, 53]. Persistent hyperglycemia was regarded to accelerate the age-linked degeneration in macula, which is considered to be a major etiology in diabetic retinopathy. Low level of zeaxanthin was found in the patients suffering from diabetic retinopathy [54]. Zeaxanthin (0.5 mg/day) for 3 months was found to improve macular edema, visual perception, and contrast sensitivity in the patients (n = 33) with non-propagative diabetic retinopathy [54]. Hyperglycemia-provoked oxidative stress to the retina due to high concentration of oxidative radicals has been regarded to be a key contributor in diabetic retinopathy [55]. Inflammation mediated through the activation of NF-κB signaling can also contribute in the progression of diabetic retinopathy. Vascular endothelial growth factor is a perilous angiogenic factor, which endorses vascular permeability and angiogenesis via up-regulating ICAM-1 expression in the eye [55]. Zeaxanthin (0.02% or 0.1% in diet) has been claimed to reinstate oxidative stress, angiogenesis, vascular permeability, and inflammation via VEGF and ICAM-1 inhibition in the retina of STZ-diabetic rats [55, 56]. Zeaxanthin can also reverse DM-induced augmentation of lipid peroxidation, DNA oxidation, iNOS, nitrotyrosine, and mitochondrial SOD without offering glycemic control [55, 56]. Zeaxanthin has also been reported to reduce the hazard of cataract advancement in DM [57, 58].

It has been reported that the cognitive impairment and memory function can be improved by zeaxanthin treatment. Zeaxanthin (50 mg/kg) for 12 weeks has been found to improve cognitive deficit via glycemic control, protection of redox defense, protection of neural cells, PI3K/Akt activation, inhibition of caspase-3 cleavage, and impediment of nuclear relocation of NF-κB in hippocampus of high-sugar+ high fat diet and single low-dose STZ-induced type 2 diabetic rats (Fig. 14.3) [59]. Four-week treatment of zeaxanthin (200 and 400 mg/kg) has been reported to normalized the body mass and hypergycemia in high-sugar + high-fat + single low-dose STZ-produced diabetic rats [60]. It also exhibited significant anti-hyperlipidemic activities via modulating of serum levels of triglycerides, low-density lipoprotein (LDL) cholesterol, HDL cholesterol, and whole cholesterol in type 2 diabetic rats [60]. In addition, zeaxanthin exhibited prophylactic role in diabetic nephropathy by restoring the kidney physiology, urine levels of n-acetyl-β-d-glucosaminidase and albumin, and serum levels of blood urea nitrogen to near normal status [60]. Zeaxanthin can significantly attenuate hyperglycemia-induced inflammation via reducing the level of inflammatory factors, such as IL-2, IL-6, TNF-α, and NF-κB [60]. Zeaxanthin also ensured significant antioxidant effect in type 2 diabetic rats by augmenting the level of endogenous antioxidant molecules in the sera, such as SOD, CAT, GPx and methane dicarboxylic aldehyde [60]. Obesity is a major complication in type 2 DM and it can also reduce insulin sensitivity [61]. Zeaxanthin was found to inhibit adipogenesis in 3 T3-L1 adipocytes by modulating 5′ adenosine monophosphate activated protein kinase (AMPK)-governed energy metabolism in vitro [62]. Intragastric treatment of 20 mg/kg of zeaxanthin for 4 weeks has been reported to possess anti-obesity activity via triggering AMPK activation, decreasing adipocytes size, reducing adipose weight, decreasing intracellular lipid contents, and inhibiting lipogenesis in high-fat-diet-provoked overweight mice [62]. It can also suppress sterol regulatory element-binding protein (SREBP)-1, fatty acid synthase, and PPAR-γ in adipocytes [62].

14.2.3 Bixin

Bixin is also known for its potential antidiabetic effect. It can normalize hyperglycemia in type 1 diabetic rats [20].

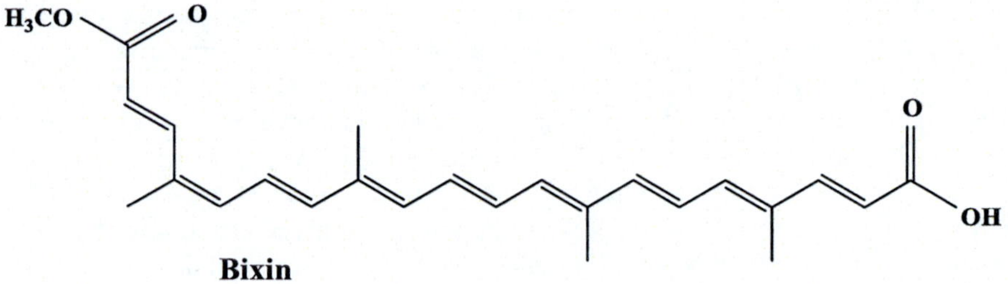

Bixin

Bixin has been reported to be a potent PPAR-γ gene regulator, which has a foremost role in carbohydrate metabolism [20, 63]. In addition, bixin can increase insulin sensibility via regulating PPAR-γ expression (Fig. 14.3) [63].

Annatto extract, a pigment from *Bixa orellana* L., contains 80% bixin, which has been informed to lessen the blood glucose level in normal and STZ-induced diabetic dogs [64]. *Bixa* extract was accounted to increase then insulin-to-glucose proportion in normoglycemic dogs after glucose loading [64]. In addition, it can prevent the postdiluvial upswing in blood glucose level following glucose feeding in normal dogs [64]. The *Bixa* extract has been claimed to improve peripheral glucose utilization to exhibit anti-hyperglycemic effect [64]. Assis and co-workers claimed that oral administration of bixin (5.5 mg/kg bixin in yoghurt) for 50 days can correct glycemic and lipid status in STZ-diabetic rats [65]. In addition, it can reduce lipid peroxidation and trigger endogenous redox defense molecules [65]. However, the effect is more pronounced when bixin is administered in combination with curcumin (90 mg/kg) [65]. In another study, 30 day oral treatment of bixin (10 and 100 mg/kg) has been reported to lower blood glucose, fructosamine, LDL cholesterol, and triglyceride levels in STZ-induced diabetic rats [20]. It can simultaneously improve HDL cholesterol level in type 1 diabetic rats [20]. Additionally, bixin treatment can prevent advanced protein oxidation and NO production and can restore SOD level in STZ-diabetic rats [20]. Bixin at the higher dose (100 mg/kg) was also found to increase serum levels of GR and thioredoxin reductase in STZ-diabetic rats [20].

14.2.4 β-Carotene

β-Carotene is a red-orange color pigment, which is especially abundant in orange-coloured fruits and vegetables, such as cantaloupe, mangoes, pumpkin, papayas, carrots, sweet potatoes etc. [66]. It is a potent antioxidant agent [66].

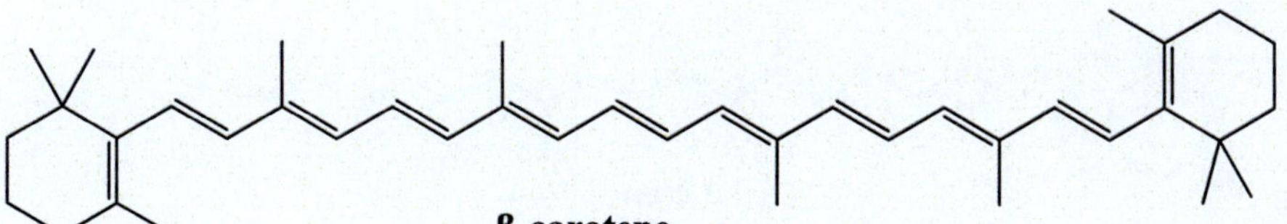

β-carotene

β-carotene has been found to lessen the threat of diabetes development and progression in both men and women [11]. There is inverse connection between the plasma β-carotene and plasma glucose during fasting conditions and the insulin resistance [11]. Another study revealed that a considerable inverse correlation between the serum hemoglobin A1c (HbA1c) and serum β-carotene concentrations in type 1 diabetic subject [67]. In the same experiment, Hozumi et al. also reported that there is an inverse association between serum fructosamine and serum β-carotene in type 1 diabetic children [67]. Therefore, low β-carotene concentration in blood would be demonstrated as impaired insulin sensitivity and vice versa [24]. Several studies supported that the inverse association between β-carotene intake and the risk of type 2 DM [25, 68]. In a STZ model of type 1 diabetes, β-carotene (10 mg/kg) treatment for about 2 weeks was found to attenuate hyperglycemia and augmented oxidative stress in the major organs, such as liver, kidney and heart of diabetic rats [69]. It has been found to exhibit an organ-specific of antioxidant responses in type 1 diabetic rats [69]. Diabetic retinopathy is a major microvascular complication in DM, which lead to visual impairment. Retinopathy is mostly induced by persistent hyperglycemia. The β-carotene has both important roles in the visual percept and restoring the macular pigments [70, 71]. β-carotene acts as provitamin A and is present in ocular tissue [70, 71]. Diabetic patients have the tendency to develop cataract, which causes decrease in the vision and lead to complete blindness [72]. β-carotene was found to inhibit development and progression of cataract via antioxidant mechanism [73]. In a population-based study, Dherani and co-workers revealed a significant reverse association between blood β-carotene concentration and the cataract formation in the people (n = 1112, ≥50 years) of 11 villages in North India [74]. However, β-carotene exhibited an inverse association ($P = 0.01$) with body mass index [74]. In addition, β-Carotene can offer an additional advantage due to its ability to produce vitamin A which is a major functional component in retina [71].

14.2.5 Lutein

Lutein is mainly present in the green and leafy vegetables. Lutein was reported to attenuate type 2 DM. Ylönen and co-worker claimed that the diabetic subject exhibited reduced level of serum lutein [11].

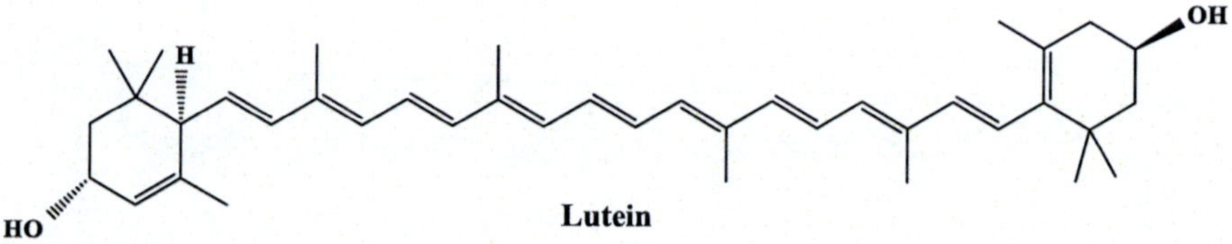

Lutein

In contrast, Muriach and co-workers demonstrated that lutein (0.5 mg/kg) treatment for 30 day could not suppress the elevated blood glucose and HbA1c levels in STZ-induced diabetic rats [75]. However, lutein can constrict high glucose-incited oxidative stress and inflammation both in vitro and in vivo in immune system cells [75]. In addition, lutein can prevent high glucose-induced loss in the viability of immune system cells [75]. It has been found to reciprocate hyperglycemia-provoked enhancement in lipid peroxidation, reduction of cellular level of GSH, and activation of NF-κB signaling in high glucose-induced U937 cells and rat lymphocytes

[75]. Thus, lutein can treat oxidative stress- and inflammation-provoked immune system impairment in diabetic condition without lowering hyperglycemia [75]. Redox insult has been implicated in the emergence and progression of diabetic retinopathy [76]. Lutein treatment was found to inhibit the extent of lipid peroxidation significantly in blood and can significantly enhance the level of GPx in the retinal tissue of alloxan-induced type 1 diabetic mice [76]. In addition, lutein was found to restore the amplitude in serial electroretinogram to near normal status [76]. However, it could not put any impact in the glycemic status of type 1 diabetic mice [76]. In addition to NF-κB inhibition, lutein can also down-regulate the inflammatory molecules, such as ICAM-1, MCP-1, and FKN [19]. Hu and co-workers revealed an inverse association between lutein burden and the development of diabetic retinopathy [54]. Lutein (6 mg/day) for 3 months was found to improve visual acuity, contrast sensitivity, and macular edema in the patients (n = 33) with non-proliferative diabetic retinopathy [54]. Lutein intake can also reduce the risk of cataract in type 2 diabetic patients [57, 58]. Arnal and co-workers showed that lutein alone or in combination with insulin can decrease lipid peroxidation and increase GSH levels in the lens of type 1 diabetic rats [72]. Therefore, lutein can be effective against diabetic retinopathy by ensuring redox defense and inhibiting retinal inflammation by inhibiting MCP-1, FKN, ICAM-1, and NF-κB [19]. Muriach and co-workers reported that lutein at very low dose can suppress lipid peroxidation and enhanced GSH, GPx, and B-wave amplitude of the electroretinogram in alloxan-induced diabetic mice without affecting glycemic status [77]. In addition, it could prevent NF-κB activation in the retina of type 1 diabetic mice [77]. Thus, lutein can serve as an adjuvant therapy in diabetic retinopathy via ensuring redox defense and inhibiting inflammation in retina [77]. Suzuki et al. reported that lutein can protect diabetic kidney via oxidative defense mechanism [68]. Lutein has been uncovered to assume a basic job in synaptic associations by improving the availability of cells in the nervous system, which led to their direct or indirect effects on binding protein gene expression. It has been accounted for that oxidative pressure and inflammation indispensably contribute in the pathological changes in brain during the DM [77]. Lutein (0.2 mg/day) treatment for 10 days has been found to suppress lipid peroxidation and NF-κB signaling in alloxan-induced diabetic mice without affecting glycemic status, GPx, and GSH levels in the hippocampus of type 1 diabetic mice [77]. 12-week treatment of lutein at the dose of 0.5 mg/day can significantly arrest lipid peroxidation, and enhanced the level of GSH and GPx in the cerebral cortex of STZ-diabetic rats without affecting glycemic status and HbA1c [78]. However, the statistical analyses do not seem convincing in the report [78]. 4-HNE immunofluorescence analysis in the cerebral cortex from type 1 diabetic rats supported the protective role of lutein against diabetic encephalopathy [78]. Combining all, it would be said that lutein can serve as an adjuvant therapy in diabetic microvascular complications.

14.2.6 Lycopene

A lipophilic carotenoid with linear structure and a major component of tomato or its products.

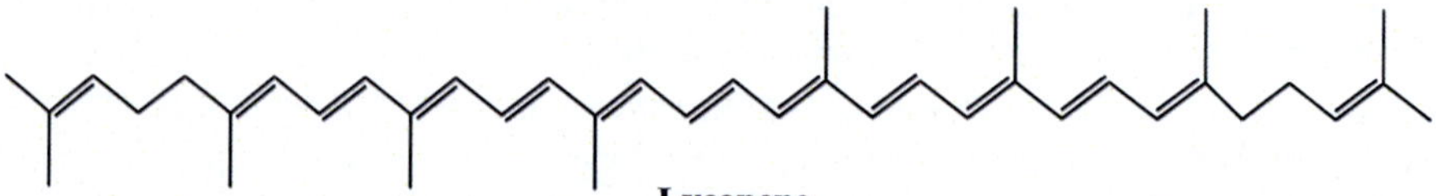

Lycopene

Lycopene exhibit prophylactic role in type 2 DM in an indirect mechanism [79]. However, an association between serum concentration of lycopene and DM has been seen. Ylonen et al. disclosed that serum concentration of lycopene remained significantly lower in case of hyperglycemia [11]. Ford and co-workers claimed that lycopene may improve the insulin sensitivity [80]. Oral administration of lycopene (30, 60 or 90 mg/kg/day) for 30 days can mitigate hyperglycemia and improved the levels of circulatory and pancreatic insulin in STZ-diabetic rats [29]. In addition, it can counteract with hyperglycemia-provoked oxidative stress [29]. Lycopene treatment was found to reduce hydrogen peroxide, lipid peroxidation marker, and nitric oxide level in the sera; while, it improved total antioxidants (serum), SOD (erythrocytes), CAT (erythrocytes), and GPx (erythrocytes) levels in type 1 diabetic rats [29]. Lycopene treatment can also mitigate dyslipidemia via restoring the serum concentrations of total lipids, triglycerides, cholesterols to close normal levels in the STZ-diabetic rats [29]. The highest effect has been observed at the dose of 90 mg/kg of lycopene [29]. In another preclinical assay, lycopene (4 mg/kg/day) for 8 weeks has been reported to reduce fasting blood glucose levels and improved the levels of circulatory insulin in STZ-diabetic rats [81]. Lycopene treatment significantly reduced extent of lipid peroxidation and nitric oxide level in the sera and improved serum levels of GSH in type 1 diabetic rats [81]. It simultaneously improved GSH levels and the degree of lipid peroxidation in the brain of STZ-diabetic rats [81]. In addition, lycopene was found to trigger both the transcriptions and the translations of redox stabilizing enzymes in the brain of type 1 diabetic rats [81]. Lycopene (4 mg/kg/day) treatment for 3 weeks has been found to correct glycemic status and disallowed the loss of body weight in STZ-diabetic rats [82]. Lycopene was found to suppress plasma nitric oxide level and the extent of lipid peroxidation in type 1 diabetic rats [82]. Despite some studies revealed potential anti-inflammatory effect of lycopene mediated through inhibition of mitogen-activated protein kinase (MAPK) activation, NF-κB signaling, and nitric oxide formation, but it failed to inhibit the activities of pro-inflammatory mediators in diabetic animals [83, 84].

There are very few studies related to lycopene and diabetic nephropathy. Diabetic rats receiving lycopene (20 mg/kg/day, orally) were shown to decrease in the levels of glucose, blood urea nitrogen, serum creatinine, and urea protein in the plasma [85]. Lycopene treatment can also mitigate dyslipidemia via reducing the serum concentration of total lipids, cholesterol, triglycerides, and LDL cholesterol and increasing serum HDL cholesterol in the STZ-diabetic rats [85]. Lycopene treatment also reduced redox stress via suppressing lipid peroxidation and increasing SOD level in the renal tissue of type 1 diabetic rats [85]. In search of mechanistic insight, lycopene has been found to prevent apoptosis and fibrosis in the diabetic kidney via triggering the phosphorylation of Akt and decreasing connective tissue growth factor (CTGF) signaling in the type 1 diabetic rats [85].

According to Reske-Nielsen and Lundbaek, patients of type 1 and 2 DM with diabetic encephalopathy and diabetic neuropathy have shown lowered cognitive and visual perception performance [86–88]. Oxidative stress has been regarded as the principal mechanism in DM-provoked nerve damage [45]. Lycopene (1, 2, and 4 mg/kg/day) treatment for 5 weeks was found to prevent neuropathic pain in STZ-diabetic rats via reducing serum levels of TNF-α and NO [89]. In this study, Kuhad and co-workers claimed that lycopene reversed thermal hyperalgesia, which is analogous to that of stage 1 diabetic neuropathy in human [89]. In another study, lycopene (4 mg/kg/day) treatment for 10 weeks corrected hyperglycemia, serum insulin level, and homeostasis catalog of insulin resistance in fructose-fed hyperglycemic rats [90]. In addition, it can correct cognitive impairments in diabetes milieu of via triggering antioxidant, anti-inflammatory, and cholinergic effects [90]. Lycopene treatment can fundamentally weaken the levels of ROS, lipid peroxides, and carbonylated proteins in the hippocampus and cerebral cortex of hyperglycemic rats; while, it can simultaneously endorse SOD, CAT and GPx activities in the brain tissues of fructose-fed rats [90]. In addition, lycopene treatment attenuated neuoinflammation by means of stifling pro-inflammatory factors and endorsed cholinergic activity via inhibiting cholinesterase activity in the hippocampus and cerebral cortex of fructose-fed insulin resistant rats [90]. In search of the role of lycopene on signaling events in diabetic brain, it has been found to trigger IR/IRS-1R/PI3K/Akt and PPAR-γ signaling events in the neocortex and hippocampus of hyperglycemic rats [90]. The increases in acetylcholinesterase activity and the oxidative stress in brain were proposed to be the major pathways of diabetes-provoked cognitive deficit [91]. However, lycopene was found to attenuate diabetes-provoked cognitive deficit through decreasing acetylcholinesterase activity in the murine brain [91].

14.3 Conclusion

Carotenoids are a class of coloured compounds possessing beneficial effects against DM and its associated microvascular complications via multiple mechanisms. From the earlier research, the mechanistic insights of individual carotenoid in DM have been explored; however, antioxidant is the principal disease preventing mechanism of carotenoids. Considering the contribution of hyperglycemia-triggered oxidative insult in the diabetes pathogenesis, it could be claimed that carotenoids would be effective against microvascular complications in diabetes. Among the carotenoids, lutein, zeaxanthin, β-carotene, and lycopene possess prophylactic role in diabetic retinopathy. These can also reduce the risk of cataract. Lycopene and astaxanthin induce prophylactic role in diabetic nephropathy by decreasing oxidative damages and inflammatory insult in the diabetic kidney. Astaxanthin and lycopene have also affected diabetic neuropathy by improving neurobehavioral abnormalities, preventing the defect in sensory awareness, and reinstating the learning and memory activities. Insulin imbalance has been said as secondary effect of oxidative stress. Therefore, carotenoid would reverse insulin imbalance and retard the risk of developing DM. The role of carotenoids against DM has been detailed in this chapter. We discussed the protective mechanisms of different carotenoids in DM and its associated microvascular complications. More research has been continuing to discover novel carotenoids with better therapeutic effects to combat against DM and its associated pathogenesis.

References

1. Pitocco D, Tesauro M, Alessandro R, Ghirlanda G, Cardillo C (2013) Oxidative stress in diabetes: implications for vascular and other complications. Int J Mol Sci 14(11):21525–21550
2. Patil PD, Mahajan UB, Patil KR, Chaudhari S, Patil CR, Agrawal YO et al (2017) Past and current perspective on new therapeutic targets for type-II diabetes. Drug Des Devel Ther 11:1567
3. Bajaj S, Khan A (2012) Antioxidants and diabetes. Indian J Endocrinol Metab 16(8):267
4. De Beer D, Joubert E, Gelderblom W, Manley M (2017) Phenolic compounds: a review of their possible role as in vivo antioxidants of wine. S Afr J Enol Vitic 23(2):48–61
5. Ames BN, Shigenaga MK, Hagen TM (1993) Oxidants, antioxidants, and the degenerative diseases of aging. Proc Natl Acad Sci 90(17):7915–7922
6. Halliwell B (1994) Free radicals, antioxidants, and human disease: curiosity, cause, or consequence? Lancet 344(8924):721–724
7. Sies H (1997) Oxidative stress: oxidants and antioxidants. Exp Physiol 82(2):291–295
8. Britton G, Khachik F (2009) Carotenoids in food. In: Carotenoids. vol 5, Springer, pp 45–66
9. Amorim-Carrilho K, Cepeda A, Fente C, Regal P (2014) Review of methods for analysis of carotenoids. TrAC Trends Anal Chem 56:49–73
10. Rodriguez-Amaya DB (2001) A guide to carotenoid analysis in foods. ILSI press, Washington, DC
11. Ylönen K, Alfthan G, Groop L, Saloranta C, Aro A, Virtanen SM et al (2003) Dietary intakes and plasma concentrations of carotenoids and tocopherols in relation to glucose metabolism in subjects at high risk of type 2 diabetes: the Botnia Dietary Study. Am J Clin Nutr 77(6):1434–1441
12. Sluijs I, Cadier E, Beulens J, Spijkerman A, van der Schouw Y (2015) Dietary intake of carotenoids and risk of type 2 diabetes. Nutr Metab Cardiovasc Dis 25(4):376–381
13. Brazionis L, Rowley K, Itsiopoulos C, O'Dea K (2008) Plasma carotenoids and diabetic retinopathy. Br J Nutr 101(2):270–277
14. Sugiura M, Nakamura M, Ogawa K, Ikoma Y, Yano M (2015) High-serum carotenoids associated with lower risk for developing type 2 diabetes among Japanese subjects: Mikkabi cohort study. BMJ Open Diabetes Res Care 3(1):e000147
15. Cooper AJ, Sharp SJ, Luben RN, Khaw K-T, Wareham NJ, Forouhi NG (2015) The association between a biomarker score for fruit and vegetable intake and incident type 2 diabetes: the EPIC-Norfolk study. Eur J Clin Nutr 69(4):449–454
16. Wang L, Liu S, Pradhan AD, Manson JE, Buring JE, Gaziano JM et al (2006) Plasma lycopene, other carotenoids, and the risk of type 2 diabetes in women. Am J Epidemiol 164(6):576–585

17. Liu S, Ajani U, Chae C, Hennekens C, Buring JE, Manson JE (1999) Long-term β-carotene supplementation and risk of type 2 diabetes mellitus: a randomized controlled trial. JAMA 282(11):1073–1075
18. Kataja-Tuomola M, Sundell J, Männistö S, Virtanen M, Kontto J, Albanes D et al (2008) Effect of α-tocopherol and β-carotene supplementation on the incidence of type 2 diabetes. Diabetologia 51(1):47–53
19. Yeh P-T, Huang H-W, Yang C-M, Yang W-S, Yang C-H (2016) Astaxanthin inhibits expression of retinal oxidative stress and inflammatory mediators in streptozotocin-induced diabetic rats. PLoS One 11(1):e0146438
20. Roehrs M, Figueiredo CG, Zanchi MM, Bochi GV, Moresco RN, Quatrin A et al (2014) Bixin and norbixin have opposite effects on glycemia, lipidemia, and oxidative stress in streptozotocin-induced diabetic rats. Int J Endocrinol 2014:1–10
21. Kianbakht S, Hajiaghaee R (2011) Anti-hyperglycemic effects of saffron and its active constituents, crocin and safranal, in alloxan-induced diabetic rats. J Med Plants 3(39):82–89
22. Rajaei Z, Hadjzadeh M-A-R, Nemati H, Hosseini M, Ahmadi M, Shafiee S (2013) Antihyperglycemic and antioxidant activity of crocin in streptozotocin-induced diabetic rats. J Med Food 16(3):206–210
23. Hozumi M, Murata T, Morinobu T, Manago M, Kuno T, Tokuda M et al (1998) Plasma β-carotene, retinol, and α-tocopherol levels in relation to glycemic control of children with insulin-dependent diabetes mellitus. J Nutr Sci Vitaminol 44(1):1–9
24. Ärnlöv J, Zethelius B, Risérus U, Basu S, Berne C, Vessby B et al (2009) Serum and dietary β-carotene and α-tocopherol and incidence of type 2 diabetes mellitus in a community-based study of Swedish men: report from the Uppsala Longitudinal Study of Adult Men (ULSAM) study. Diabetologia 52(1):97–105
25. Coyne T, Ibiebele TI, Baade PD, Dobson A, McClintock C, Dunn S et al (2005) Diabetes mellitus and serum carotenoids: findings of a population-based study in Queensland, Australia. Am J Clin Nutr 82(3):685–693
26. Katyal T, Singh G, Budhiraja R, Sachdeva M (2013) Effect of lutein in development of experimental diabetic nephropathy in rats. Afr J Pharm Pharmacol 7(47):3004–3010
27. Shirali S, Zahra Bathaie S, Nakhjavani M (2013) Effect of crocin on the insulin resistance and lipid profile of streptozotocin-induced diabetic rats. Phytother Res 27(7):1042–1047
28. Park CH, Xu FH, Roh S-S, Song YO, Uebaba K, Noh JS et al (2015) Astaxanthin and Corni Fructus protect against diabetes-induced oxidative stress, inflammation, and advanced glycation end product in livers of streptozotocin-induced diabetic rats. J Med Food 18(3):337–344
29. Ali MM, Agha FG (2009) Amelioration of streptozotocin-induced diabetes mellitus, oxidative stress and dyslipidemia in rats by tomato extract lycopene. Scand J Clin Lab Invest 69(3):371–379
30. Leite M, De Lima A, Massuyama M, Otton R (2010) In vivo astaxanthin treatment partially prevents antioxidant alterations in dental pulp from alloxan-induced diabetic rats. Int Endod J 43(11):959–967
31. Altinoz E, Oner Z, Elbe H, Cigremis Y, Turkoz Y (2015) Protective effects of saffron (its active constituent, crocin) on nephropathy in streptozotocin-induced diabetic rats. Hum Exp Toxicol 34(2):127–134
32. Naito Y, Uchiyama K, Aoi W, Hasegawa G, Nakamura N, Yoshida N et al (2004) Prevention of diabetic nephropathy by treatment with astaxanthin in diabetic db/db mice. Biofactors 20(1):49–59
33. Ambati RR, Phang S-M, Ravi S, Aswathanarayana RG (2014) Astaxanthin: sources, extraction, stability, biological activities and its commercial applications—a review. Mar Drugs 12(1):128–152
34. Naguib YM (2000) Antioxidant activities of astaxanthin and related carotenoids. J Agric Food Chem 48(4):1150–1154
35. Uchiyama K, Naito Y, Hasegawa G, Nakamura N, Takahashi J, Yoshikawa T (2002) Astaxanthin protects β-cells against glucose toxicity in diabetic db/db mice. Redox Rep 7(5):290–293

36. Bhuvaneswari S, Anuradha CV (2012) Astaxanthin prevents loss of insulin signalling and improves glucose metabolism in liver of insulin resistant mice. Can J Physiol Pharmacol 90(11):1544–1552
37. Bhuvaneswari S, Yogalakshmi B, Sreeja S, Anuradha CV (2014) Astaxanthin reduces hepatic endoplasmic reticulum stress and nuclear factor-κB-mediated inflammation in high fructose and high fat diet-fed mice. Cell Stress Chaperones 19(2):183–191
38. Hussein G, Nakagawa T, Goto H, Shimada Y, Matsumoto K, Sankawa U et al (2007) Astaxanthin ameliorates features of metabolic syndrome in SHR/NDmcr-cp. Life Sci 80(6):522–529
39. Inoue M, Tanabe H, Matsumoto A, Takagi M, Umegaki K, Amagaya S et al (2012) Astaxanthin functions differently as a selective peroxisome proliferator-activated receptor γ modulator in adipocytes and macrophages. Biochem Pharmacol 84(5):692–700
40. Kim YJ, Kim YA, Yokozawa T (2009) Protection against oxidative stress, inflammation, and apoptosis of high-glucose-exposed proximal tubular epithelial cells by astaxanthin. J Agric Food Chem 57(19):8793–8797
41. Manabe E, Handa O, Naito Y, Mizushima K, Akagiri S, Adachi S et al (2008) Astaxanthin protects mesangial cells from hyperglycemia-induced oxidative signalling. J Cell Biochem 103(6):1925–1937
42. Otton R, Marin DP, Bolin AP (2010) dos Santos RdCM, Polotow TG, Sampaio SC et al. Astaxanthin ameliorates the redox imbalance in lymphocytes of experimental diabetic rats. Chem Biol Interact 186(3):306–315
43. Wu LL, Chiou C-C, Chang P-Y, Wu JT (2004) Urinary 8-OHdG: a marker of oxidative stress to DNA and a risk factor for cancer, atherosclerosis and diabetics. Clin Chim Acta 339(1):1–9
44. Nakano M, Onodera A, Saito E, Tanabe M, Yajima K, Takahashi J et al (2008) Effect of astaxanthin in combination with α-tocopherol or ascorbic acid against oxidative damage in diabetic ODS rats. J Nutr Sci Vitaminol 54(4):329–334
45. Dewanjee S, Das S, Das AK, Bhattacharjee N, Dihingia A, Dua TK et al (2018) Molecular mechanism of diabetic neuropathy and its pharmacotherapeutic targets. Eur J Pharmacol 833:472–523
46. Xu L, Zhu J, Yin W, Ding X (2015) Astaxanthin improves cognitive deficits from oxidative stress, nitric oxide synthase and inflammation through upregulation of PI3K/Akt in diabetes rat. Int J Clin Exp Pathol 8(6):6083
47. Wang S-h, Sun Z-l, Guo Y-j, Yuan Y, Yang B-q (2009) Diabetes impairs hippocampal function via advanced glycation end product mediated new neuron generation in animals with diabetes-related depression. Toxicol Sci 111(1):72–79
48. Zhou X-y, Zhang F, Hu X-t, Chen J, Tang R-x, Zheng K-y et al (2017) Depression can be prevented by astaxanthin through inhibition of hippocampal inflammation in diabetic mice. Brain Res 1657:262–268
49. Johnson EJ (2014) Role of lutein and zeaxanthin in visual and cognitive function throughout the lifespan. Nutr Rev 72(9):605–612
50. Xiao J, Wang J, Xing F, Han T, Jiao R, Liong EC et al (2014) Zeaxanthin dipalmitate therapeutically improves hepatic functions in an alcoholic fatty liver disease model through modulating MAPK pathway. PLoS One 9(4):e95214
51. Sen A, Marsche G, Freudenberger P, Schallert M, Toeglhofer AM, Nagl C et al (2014) Association between higher plasma lutein, zeaxanthin, and vitamin C concentrations and longer telomere length: results of the Austrian Stroke Prevention Study. J Am Geriatr Soc 62(2):222–229
52. Wu J, Cho E, Willett WC, Sastry SM, Schaumberg DA (2015) Intakes of lutein, zeaxanthin, and other carotenoids and age-related macular degeneration during 2 decades of prospective follow-up. JAMA Ophthalmol 133(12):1415–1424
53. Moshetova L, Vorob'eva I, Alekseev I, Mikhaleva L (2014) Results of the use of antioxidant and angioprotective agents in type 2 diabetes patients with diabetic retinopathy and age-related macular degeneration. Vestn oftalmol 131(3):34–40, 2–4
54. Hu B-J, Hu Y-N, Lin S, Ma W-J, Li X-R (2011) Application of lutein and zeaxanthin in nonproliferative diabetic retinopathy. Int J Ophthalmol 4(3):303

55. Aiello LP, Wong J-S (2000) Role of vascular endothelial growth factor in diabetic vascular complications. Kidney Int 58:S113–S1S9
56. Kowluru RA, Menon B, Gierhart DL (2008) Beneficial effect of zeaxanthin on retinal metabolic abnormalities in diabetic rats. Invest Ophthalmol Vis Sci 49(4):1645–1651
57. Chasan-Taber L, Willett WC, Seddon JM, Stampfer MJ, Rosner B, Colditz GA et al (1999) A prospective study of carotenoid and vitamin A intakes and risk of cataract extraction in US women. Am J Clin Nutr 70(4):509–516
58. Christen WG, Liu S, Glynn RJ, Gaziano JM, Buring JE (2008) Dietary carotenoids, vitamins C and E, and risk of cataract in women: a prospective study. Arch Ophthalmol 126(1):102–109
59. Zhou X, Wang S, Ding X, Qin L, Mao Y, Chen L et al (2017) Zeaxanthin improves diabetes-induced cognitive deficit in rats through activiting PI3K/AKT signalling pathway. Brain Res Bull 132:190–198
60. Kou L, Du M, Zhang C, Dai Z, Li X, Zhang B (2017) The hypoglycemic, hypolipidemic, and anti-diabetic nephritic activities of zeaxanthin in diet-streptozotocin-induced diabetic Sprague dawley rats. Appl Biochem Biotechnol 182(3):1–12
61. Sharma D, Bhattacharya P, Kalia K, Tiwari V (2017) Diabetic nephropathy: new insights into established therapeutic paradigms and novel molecular targets. Diabetes Res Clin Pract 128:91–108
62. Liu M, Liu H, Xie J, Xu Q, Pan C, Wang J et al (2017) Anti-obesity effects of zeaxanthin on 3T3-L1 preadipocyte and high fat induced obese mice. Food Funct 8(9):3327–3338
63. Takahashi N, Goto T, Taimatsu A, Egawa K, Katoh S, Kusudo T et al (2009) Bixin regulates mRNA expression involved in adipogenesis and enhances insulin sensitivity in 3T3-L1 adipocytes through PPAR-γ activation. Biochem Biophys Res Commun 390(4):1372–1376
64. Russell K, Omoruyi F, Pascoe K, Morrison ESA (2008) Hypoglycaemic activity of Bixa orellana extract in the dog. Methods Find Exp Clin Pharmacol 30(4):301
65. Assis RP, Arcaro CA, Gutierres VO, Oliveira JO, Costa PI, Baviera AM et al (2017) Combined effects of curcumin and lycopene or bixin in yoghurt on inhibition of LDL oxidation and increases in HDL and Paraoxonase levels in streptozotocin-diabetic rats. Int J Mol Sci 18(4):332
66. Bonrath W, Bruins M, Mair P, Medlock J, Netscher T, Schütz J et al (2000) Vitamin A. In: Kirk-Othmer encyclopedia of chemical technology. Wiley, New York
67. Hozumi M, Murata T, Morinobu T, MANAGO M, TOKUDA M, KONISHI K et al (1998) Plasma β-carotene, retinol, and α-tocopherol levels in relation to glycemic control of children with insulin-dependent diabetes mellitus. J Nutr Sci Vitaminol 44(1):1–9
68. Suzuki K, Ito Y, Inoue T, Hamajima N (2011) Inverse association of serum carotenoids with prevalence of metabolic syndrome among Japanese. Clin Nutr 30(3):369–375
69. Maritim A, Dene BA, Sanders RA, Watkins JB (2002) Effects of β-carotene on oxidative stress in normal and diabetic rats. J Biochem Mol Toxicol 16(4):203–208
70. Chistiakov DA (2011) Diabetic retinopathy: pathogenic mechanisms and current treatments. Diabetes Metab Syndr Clin Res Rev 5(3):165–172
71. Krinsky NI, Johnson EJ (2005) Carotenoid actions and their relation to health and disease. Mol Asp Med 26(6):459–516
72. Arnal E, Miranda M, Almansa I, Muriach M, Barcia JM, Romero FJ et al (2009) Lutein prevents cataract development and progression in diabetic rats. Graefes Arch Clin Exp Ophthalmol 247(1):115–120
73. Thiagarajan R, Manikandan R (2013) Antioxidants and cataract. Free Radic Res 47(5):337–345
74. Dherani M, Murthy GV, Gupta SK, Young IS, Maraini G, Camparini M et al (2008) Blood levels of vitamin C, carotenoids and retinol are inversely associated with cataract in a North Indian population. Invest Ophthalmol Vis Sci 49(8):3328–3335
75. Muriach M, Bosch-Morell F, Arnal E, Alexander G, Blomhoff R, Romero F (2008) Lutein prevents the effect of high glucose levels on immune system cells in vivo and in vitro. J Physiol Biochem 64(2):149–157

76. Miranda M, Muriach M, Johnsen S, Bosch-Morell F, Araiz J, Roma J et al (2004) Oxidative stress in a model for experimental diabetic retinopathy: treatment with antioxidants. Arch Soc Esp Oftalmol 79(6):289–294
77. Muriach M, Bosch-Morell F, Alexander G, Blomhoff R, Barcia J, Arnal E et al (2006) Lutein effect on retina and hippocampus of diabetic mice. Free Radic Biol Med 41(6):979–984
78. Arnal E, Miranda M, Barcia J, Bosch-Morell F, Romero F (2010) Lutein and docosahexaenoic acid prevent cortex lipid peroxidation in streptozotocin-induced diabetic rat cerebral cortex. Neuroscience 166(1):271–278
79. Higuera-Ciapara I, Felix-Valenzuela L, Goycoolea F (2006) Astaxanthin: a review of its chemistry and applications. Crit Rev Food Sci Nutr 46(2):185–196
80. Ford ES, Will JC, Bowman BA, Narayan KV (1999) Diabetes mellitus and serum carotenoids: findings from the Third National Health and Nutrition Examination Survey. Am J Epidemiol 149(2):168–176
81. Aydin M, Celik S (2012) Effects of lycopene on plasma glucose, insulin levels, oxidative stress, and body weights of streptozotocin-induced diabetic rats. Turk J Med Sci 42(Sup. 2):1406–1413
82. Duzguner V, Kucukgul A, Erdogan S, Celik S, Sahin K (2008) Effect of lycopene administration on plasma glucose, oxidative stress and body weight in streptozotocin diabetic rats. J Appl Anim Res 33(1):17–20
83. Kim GY, Kim JH, Ahn SC, Lee HJ, Moon DO, Lee CM et al (2004) Lycopene suppresses the lipopolysaccharide-induced phenotypic and functional maturation of murine dendritic cells through inhibition of mitogen-activated protein kinases and nuclear factor-κB. Immunology 113(2):203–211
84. Yegin SÇ, Yur F, Ceylan E (2013) Effect of lycopene application in rats with experimental diabetes using lipoprotein, paraoxonase and cytokines. J Membr Biol 246(8):621–626
85. Li W, Wang G, Lu X, Jiang Y, Xu L, Zhao X (2014) Lycopene ameliorates renal function in rats with streptozotocin-induced diabetes. Int J Clin Exp Pathol 7(8):5008
86. Reske-Nielsen E, Lundbaek K (1963) Diabetic encephalopathy. Diffuse and focal lesions of the brain in long-term diabetes. Acta Neurol Scand Suppl 39(SUPPL4):273
87. Brands AM, Biessels GJ, De Haan EH, Kappelle LJ, Kessels RP (2005) The effects of type 1 diabetes on cognitive performance. Diabetes Care 28(3):726–735
88. Abbott CA, Malik RA, van Ross ER, Kulkarni J, Boulton AJ (2011) Prevalence and characteristics of painful diabetic neuropathy in a large community-based diabetic population in the UK. Diabetes Care 34(10):2220–2224
89. Kuhad A, Sharma S, Chopra K (2008) Lycopene attenuates thermal hyperalgesia in a diabetic mouse model of neuropathic pain. Eur J Pain 12(5):624–632
90. Yin Q, Ma Y, Hong Y, Hou X, Chen J, Shen C et al (2014) Lycopene attenuates insulin signalling deficits, oxidative stress, neuroinflammation, and cognitive impairment in fructose-drinking insulin resistant rats. Neuropharmacology 86:389–396
91. Kuhad A, Sethi R, Chopra K (2008) Lycopene attenuates diabetes-associated cognitive decline in rats. Life Sci 83(3):128–134

Chapter 15
Carotenoids as Antiparkinson Agents

Saikat Dewanjee, Muhammad Zia-Ul-Haq, Muhammad Riaz, Shounak Sarkhel, Pratik Chakraborty, and Sagheer Ahmed

15.1 Introduction

Parkinson's disease (PD) leads to movement disorders. Recently its incidences has amplified exponentially [1]. A recent survey reported that the number of Parkinson's cases in 1990 was ~2.5 (95% uncertainty interval 2.0–3.0) million, which has been climbed to ~6.1 (95% uncertainty interval 5.0–7.3) million by 2016. This number has been projected to reach to Σ12 million by 2040 [1, 2]. Over the past few decades, several therapeutic targets were identified, and different therapeutic agents have been developed, such as monoamine oxidase B inhibitors, catechol-O-methyltransferase inhibitors and dopamine agonists [3]. In addition, several cell-based approaches, including stem cell-strategies, are underway with varying grades of success; however, these cell-transplantation grafts convey a high possibility of tumorigenesis to the patients [3]. To date, levodopa remains the drug of choice for the symptomatic relief in PD [4, 5]. Hence, there is an dire need to discover potential therapeutic agents, which can ensure both slow progression and neuroprotection in PD without producing considerable adverse effects [6].

S. Dewanjee (✉) · S. Sarkhel · P. Chakraborty
Advanced Pharmacognosy Research Laboratory, Department of Pharmaceutical Technology, Jadavpur University, Kolkata, India
e-mail: saikat.dewanjee@jadavpuruniversity.in

M. Zia-Ul-Haq
Office of Research, Innovation and Commercialization, Lahore College for Women University, Lahore, Pakistan

M. Riaz
Department of Pharmacy, Shaheed Benazir Bhutto University, Sheringal, Pakistan

S. Ahmed
Shifa College of Pharmaceutical Sciences, Shifa Tameer-e-Millat University, Islamabad, Pakistan

© Springer Nature Switzerland AG 2021

M. Zia-Ul-Haq et al. (eds.), *Carotenoids: Structure and Function in the Human Body*, https://doi.org/10.1007/978-3-030-46459-2_15

Emerging evidence revealed that nutrients might play important roles in regulating the pathological events in PD [7–14]. They were found to impart neuroprotective effect via suppressing oxidative stress, reducing inflammation, restoring mitochondrial homeostasis, preventing apoptosis, and modulating signal transduction [9, 13, 15].

Carotenoids are a class of naturally occurring colouring agents having significant nutrition and disease-preventing capacity [16–20]. Dietary carotenoids are bio-accumulated to various organs and tissues; however, the tissue burden has been found to be positively linked with the number of low-density lipoprotein receptors [20]. Bio-accumulated carotenoids perform a vital functional role in restoring the normal physiological status of the tissues or organs [16, 21, 22]. Carotenoids suppress and prevent several pathological conditions, like cardiovascular complications, cancers, neurological disorders, aging, and photosensitive or eye-related disorders [11, 17, 23–25]. Emerging evidence revealed that carotenoids can decrease the possibility of neurodegenerative ailments, like PD [11]. Carotenoids prevent neurodegeneration and massive neuronal loss via scavenging free radicals, endorsing redox defense, suppressing inflammation, inhibiting amyloidosis, and resulting several signaling events. Epidemiological studies revealed that carotenoid-enriched food can decline the possibility of PD [26–31]. In contrast, few studies found the positive association between carotenoids ingestion and the risk of PD [31–34]. This chapter emphasized an overview of the therapeutic role of carotenoids in PD. Special attention has been given to clinical evidence.

15.2 Overview of PD

PD is characterised by a large number of the motor and non-motor symptoms [35]. The common Parkinsonian motor symptoms include movement disorders, such as slowness in movement, hypokinesia (decrease in bodily movement), and stiffness on passive movement [35–37]. The non-motor symptoms include psychiatric disorders, sleep disorders, hyposmia (poor sense of smell), gastrointestinal dysfunction, and cognitive impairment [37]. These non-motor symptoms always appear before Parkinsonian motor symptoms and can often be difficult to treat (Fig. 15.1) [37, 38]. The loss of dopaminergic neurons endorses motor dysfunction, while the accumulation of insoluble aggregates of α-synuclein has been regarded to lead non-motor symptoms in PD [38, 39].

The cell-specific factors include neurotransmitter, hyperpolarized membrane, Ca binding protein, L-type Ca channels, axonal branching etc. [40]. The non-specific cellular factors include genes associated with mitochondrial functions, environmental toxins, inflammation, and viral or prion-like infection [40]. Mutations of the genes, such as Parkinson disease protein 7 (PARK7), Parkin and phosphatase and tensin homolog-induced putative kinase 1 (PINK1), and Parkin can cause mitochondrial dysfunctions [40, 41]. These factors work together to trigger mitochondrial dysfunction resulting in proteostatic decline and cell death.

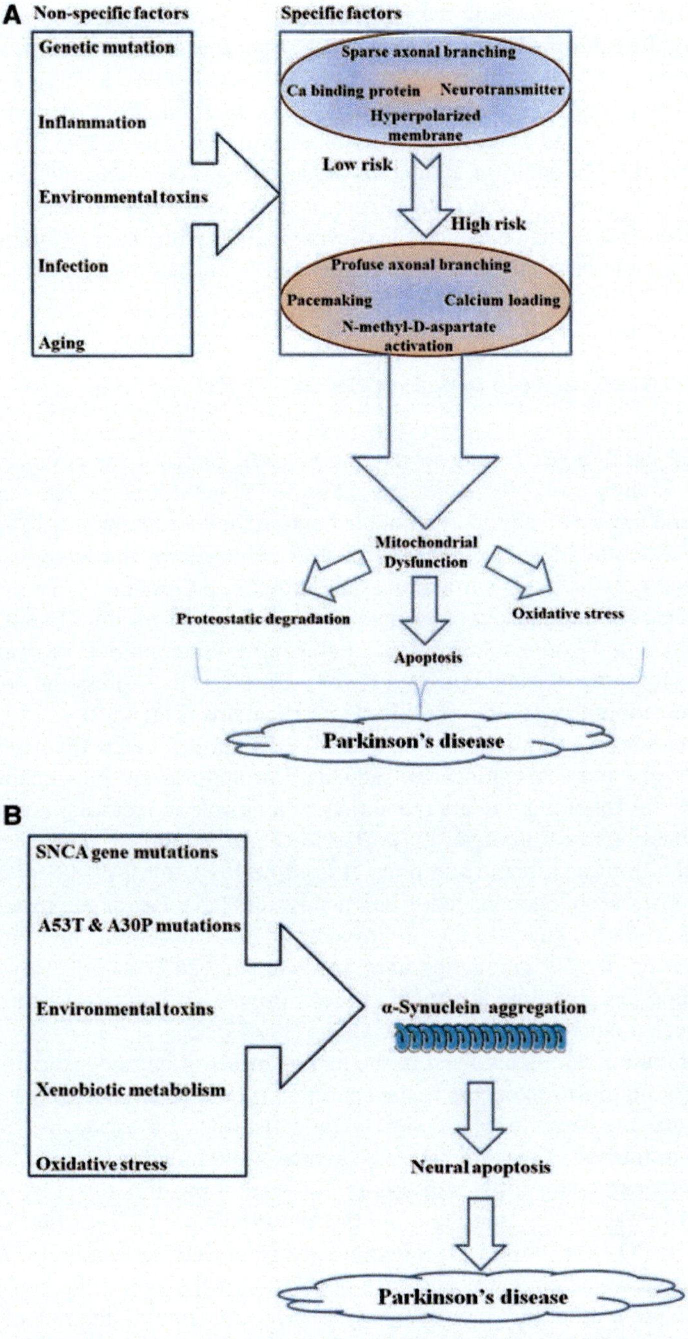

Fig. 15.1 Proposed pathogenesis in the PD. (**a**). Damage of dopaminergic neurons due to interplay between specific and non-specific cellular factors and (**b**). Neuronal loss attributes in PD due to α-synuclein aggregation

α-Synuclein, a presynaptic neuronal protein, is pathologically connected with PD. Accumulation of α-synuclein in β-sheet structures is implicated in the pathological lesions [38, 42]. Various alterations in the α-synuclein encoding SNCA gene on chromosome 4q21–23 are involved in the early onset familial Parkinsonism [42]. In addition, A53T and A30P mutations in α-synuclein can facilitate α-synuclein aggregation [43]. In addition, oxidative stress, aging, xenobiotic metabolism, and environmental factors play a critical role in α-synuclein accumulation [42]. The aforementioned two major pathological events along with other pathological proceedings, complement each other and implicated in the pathogenesis of PD.

15.3 Protective Role of Carotenoids in PD

Chemically carotenoids belong to terpenoid (C40) family comprising 8 isoprene units [11]. In most cases, the isoprenes are linked to yield a linear and symmetrical structure and having an alternating double bond in their structure [11, 18]. Based on polarity, carotenoids have been categorized as carotenes or non-polar carotenoids and xanthophylls or polar carotenoids. α-Carotene, β-carotene, γ-carotene, lycopene etc. belong to the group of carotenes; while β-cryptoxanthin, lutein, astaxanthin, violaxanthin, neoxanthin, flavoxanthin, β-cryptoxanthin, zeaxanthin etc. belong to the group of xanthophylls [11, 44]. Carotenoids supress the commencement and development of PD via multiple mechanisms (Fig. 15.2).

The characteristic structural features of carotenoids make them capable of absorbing light and scavenging free radicals. Carotenoids are bio-accumulated in various organs, including the brain and play crucial role in restoring normal physiological functions of this organ [11]. Due to the lipophilic structure, carotenoids reside in the lipid membrane and prevent lipid peroxidation [45–48]. Carotenoids can also boost the endogenous redox defence system [11]. Bearing in mind the critical role of oxidative stress in the progression of PD, carotenoids may impart prophylactic role in PD via antioxidant mechanism [26, 49–54]. However, the concentration of carotenoid is critical as it can impart oxidative stress through pro-oxidant mechanisms at higher concentrations [11, 23, 55].

Neuroinflammation is involved in the pathogenesis of neurodegeneration in PD [56]. Neuroinflammation occurs in the central nervous system due to the release of inflammatory mediators [56]. A variety of cells, such as astrocytes, microglia, peripheral immune cells etc. are involved in neuroinflammation [11, 57]. Activation of these cells can endorse the transcription of several pro-inflammatory genes and different factors which contribute in the amplification of neuro-inflammatory responses in PD [57]. Prolonged inflammatory responses, activation of glial cells, and accumulation of immune cells have been recognized as the characteristic features of PD brain after post-mortem analysis [57]. Considering the role of neuroinflammation in PD pathogenesis, it would not be unrealistic to claim that the suppression of neuro-inflammation can serve as an efficient therapeutic option in

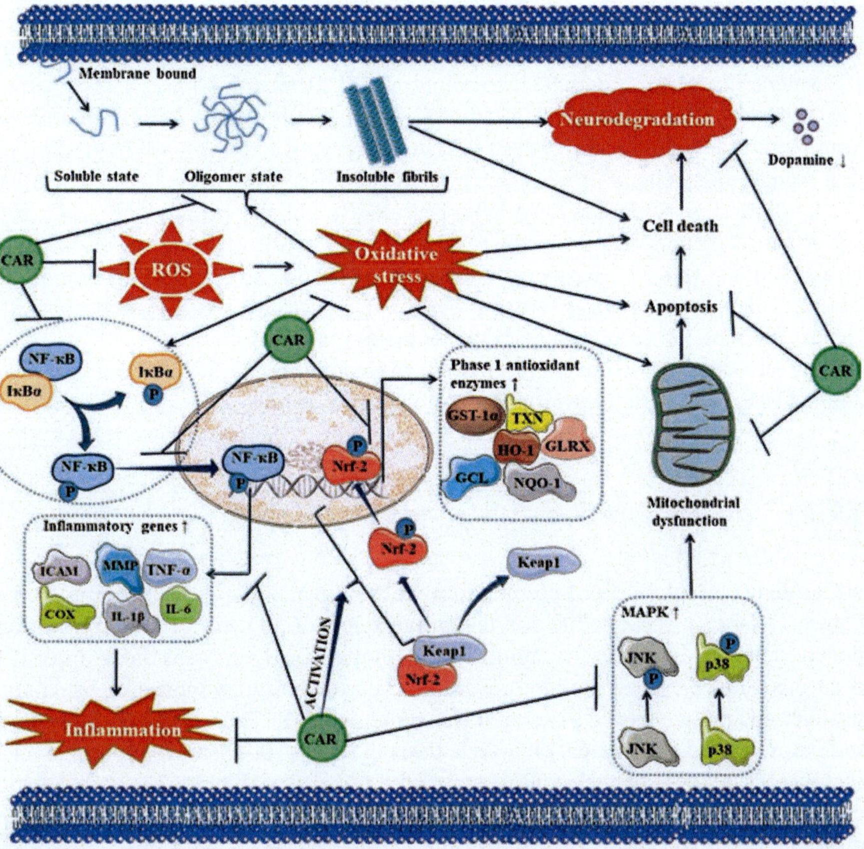

Fig. 15.2 The cellular targets of carotenoid in the management of PD. Blue arrows indicate downstream cellular events. Blue lines (T) indicate inhibition. *CAR* carotenoids, *COX* cyclooxygenase, *GCL* glutamate-cysteine ligase, *GLRX* glutaredoxin-1, *GST-1α* glutathione S-tranferase 1α, *HO-1* heme oxygenase-1, *ICAM* intercellular adhesion molecule, *IL* interleukin, *JNK* c-Jun N-terminal kinase, *Keap1* kelch-like ECH-associated protein 1, *MMP* matrix metalloproteinases, *NF-κB* nuclear factor kappa-light-chain-enhancer of activated B cells, *NQO-1* NAD(P)H dehydrogenase (quinone) 1, *Nrf-2* nuclear factor erythroid 2 related factor 2, *TNF-α* tumour necrosis factor-α, *TXN* thioredoxin

PD. Numerous carotenoids attenuate neuroinflammation [11]. Fucoxanthin [58, 59], astaxanthin [60–62], lycopene [63], lutein [64], crocin [65–67], crocetin [65, 68, 69] etc. were revealed to attenuate neuroinflammation mainly via suppressing nuclear factor kappa-light-chain-enhancer of activated B cells-provoked transcription of inflammatory genes and endorsing nuclear factor erythroid 2 related factor 2 signal transduction.

Amyloidosis featuring aggregation and accumulation of amyloid fibrils of α-synuclein is involved in the progression of PD [70]. Amyloid fibrils of α-synuclein with a cross-β structure in the brain endorse the propagation of PD in experimental models [70]. Carotenoids were found to possess an anti-amyloidogenic and amyloidolytic effect [11, 54, 71]. Lycopene [72], lutein [71], β-cryptoxanthin [71], β-carotene [71], α-carotene [71], zeaxanthin [71], astaxanthin [73, 74], fucoxanthin [75], crocin [76] etc. can prevent amyloid fibril accumulation in the brain by preventing their formation and/or destabilizing the preformed fibrils via regulating several signalling events.

In addition, carotenoids have been revealed to possess prophylactic role in PD via suppressing mitochondrial dysfunction [25, 77–79], inhibiting acetylcholinesterase activity [80–83], regulating autophagy [11, 84–86], inhibiting monoamine oxidases [87–89], restoring tyrosine hydroxylase activity [90–94], and alleviating the toxic effects of xenobiotics or environmental factors [95–99].

The subsequent section highlights the protective roles of individual carotenoid against PD along with their proposed mechanistic insights.

15.3.1 α-Carotene

α-Carotene can inhibit lipid peroxidation in the substantia nigra [29]. Lipid peroxidation is largely responsible for the progression of PD via promoting neuronal degeneration in the specific region (pars compacta) of the substantia nigra [29]. α-carotene has been revealed to prevent lipid peroxidation by inhibiting free radical production and quenching generated free radicals [100]. Thus, it can prevent peroxy radical-mediated free radical chain reactions in brain [100]. Earlier reports revealed that there is a link between α-carotene and the severity of PD [26, 30, 101]. Kim and co-workers included patients (n = 104 idiopathic Parkinson cases and 52 control) to establish the relationship between serum carotenoid levels with possibility of PD [33]. The patients with advanced phase of PD exhibit much lower level of serum α-carotene level relative to the patients with earlier phases of the disease and established inverse correlation with Hoehn and Yahr stage and unified PD rating scale motor score [33]. In contrast, in a population-based case-control study in Japan, Miyake and co-workers could not establish any association between dietary intake of α-carotene and the risk and progression of PD. In another cohort analysis, Hughes et al. [53] could not observe any link between risk and progression of PD and ingestion of α-carotene. Moreover, in a case-control study (n = 57 male Parkinson cases and 50 control receiving long-term treatment with dietary antioxidants), Scheider and co-workers reported that higher intakes of α-carotene can enhance the risks of PD [30].

15.3.2 *β-Carotene*

Epidemiological studies revealed that dietary consumption of β-carotene can mitigate the possibility and development of PD [8, 102]. The preclinical study revealed that β-carotene could partially alleviate xenobiotic-induced dopaminergic neuronal loss in the substantia nigra of mice, possibly via antioxidant mechanism [103]; however, the same was not observed in marmosets [104]. Accumulation of α-synuclein fibrils in brain are involved in the pathological events of PD [105]. β-carotene has been proposed to attribute a neuroprotective effect via inhibiting lipid peroxidation [8]. In addition, β-carotene was proved helpful in preventing PD via inhibiting α-synuclein formation and destabilizing the preformed of α-synuclein fibril [105]. Due to hydrophobic and antioxidant properties, β-carotene can bind to α-synuclein and/or β-amyloid fibrils, inhibits the filamentous accumulation of α-synuclein fibrils during PD [105]. β-carotene can influence the biological processes independently, along with acting as vitamin A precursor [106]. Kim et al. [33] revealed an opposite association between serum β-carotene levels and the risks and the development of PD. In a multicentre hospital-based study (n = 249 Parkinson cases and 368 control) in Japan, Miyake et al. [107] revealed a contrary relationship between β-carotene ingestion and the possibility of PD. β-carotene (~ 4 mg/day) exhibited the multivariate odd's ratio between extreme quartiles of 0.56 (95% confidence interval: 0.33–0.97, P for trend = 0.03) [107]. In a cohort study, dietary β-carotene reciprocated the risks of PD; however, the level of significance has been quite low [101]. In a German case-control study, β-carotene ingestion was contrariwise linked with the possibility of PD (p_{trend} = 0.06) [108]. In contrast, in a hospital-based case-control study (n = 61 Parkinson cases), Jiménez-Jiménez and co-researchers [29] failed to establish any correlation between serum β-carotene concentration and the possibility of having PD. A similar observation has been mentioned by Foy et al. [109] who did not find any depletion in serum β-carotene level in the patients (n = 18) with PD and dementia. Furthermore, higher intake of β-carotene was reported to enhance the risks of PD [30].

15.3.3 *Retinol*

Retinol, a component of vitamin A (vitamin A_1), has been regarded to be an essential micronutrient for developing and restoring the activities of the brain. However, it may impart neurotoxic effect on the individuals without vitamin A deficiency and/or with hereditary neurodegenerative diseases [110]. Epidemiological evidence indicates that some carotenoids may prevent PD through their participation in retinol synthesis and metabolism [111]. During embryonic development, retinol plays a central part in neuronal differentiation and regeneration [112]. In the adult, it not only regulates synaptic plasticity but also controls hippocampal functions [113, 114]. Oral supplementation of retinol could not prevent xenobiotic-induced dopaminergic denervation in the substantia nigra in rats; however, it has been found to stimulate astrocyte reactivity without affecting systemic parameters [114]. Retinoic acid, a metabolite of retinol, is involved in the development and maintenance of meso-diencephalic dopaminergic neurons, thus can inversely impact the possibility of PD [115]. Several studies mentioned an inverse correlation between ingestion of retinol and risk of PD [30, 116]. While others could not associate any connection between consumption of retinol and possibility of PD in the adult human population [26, 33, 117, 118]. In contrast, Anderson and co-workers [119] reported that vitamin A supplement can increase the risk and development of PD. Published results from multiple cohort studies contradict each-other to draw firm conclusions in this regard [28].

15.3.4 Lycopene

Lycopene has been revealed to possess powerful neuroprotective effects mediated through the reticence of oxidative stress, neuro-inflammation, neuronal apoptosis, and mitochondrial dysfunction [120, 121]. In addition, it can impart neuroprotective effect via endorsing nuclear factor erythroid 2-related factor, brain-derived neurotrophic factor, and intracellular Ca^{2+} homeostasis [121]. Lycopene attributes neuroprotective effect in inhibiting and postponing PD [11, 122]. Lycopene was found to attenuate rotenone-induced PD in mice via suppressing α-synuclein, microtubule-associated protein 3 light chain positive neurons, neurobehavioral abnormalities, and oxidative stress in the substantia nigra with subsequent growth in the activity of tyrosine hydroxylase [91]. In another study, it has been found to alleviate oxidative stress, apoptosis, neurochemical deficits, and physiological abnormalities in 1-methyl-4-phenyl-1,2,3,6-tetrahydropyridine-induced PD in mice [120]. In addition, it can control PD via improving the populations of dopaminergic and γ-aminobutyric acid neurons in hippocampus and substantia nigra in rats [123]. In a rodent model, lycopene-enriched tomato powder has succeeded in preventing the loss of dopaminergic neurons, which is an inherent reason for PD [124]. In addition, lycopene can protect from DNA damage, induce autophagy, and inflammation-provoked cognitive impairment [11]. Several reports are compiled to find the possible role of lycopene in PD. However, the reports contradict each-other to obtain a specific conclusion. In a case-control study (n = 57 male Parkinson cases and 50 control), Scheider and co-workers [30] concluded an inverse correlation between lycopene and possibility of PD. In a high-population case-control study (n = 104 idiopathic Parkinson patients and 52 control), it has been found that serum lycopene levels remained significantly low in the advanced stage PD patients as compared to early PD patients [33]. Thus, it could be said that the lycopene concentration is contrariwise linked with the progression of PD [33]. In another case-control study (n = 18 patients with PD and dementia and 50 control), plasma lycopene level remained considerably small in the patients with PD and dementia compared to control; however, no significant change in lycopene level are found in only PD patients (n = 41) [109]. Thus, it may be concluded that lycopene play critical role in reversing cognitive impairment in PD [109]. In a hospital-based case-control study (n = 61 Parkinson cases), Jiménez-Jiménez et al. [29] did not discover any connection between lycopene consumption and possibility of PD in adult human population. Zhang et al. [101] could not establish any association between lycopene ingestion and possibility of PD. In a good-quality case-control study, Powers and co-workers [32] observed a significantly ($p = 0.046$) positive link between lycopene and PD.

15.3.5 Fucoxanthin

Fucoxanthin is reported to exhibit a protecting influence against PD via monoamine oxidase (MAO) inhibition in vitro [89]. It can also possess neuroprotective effect by preventing β-amyloid assembly and β-amyloid fibrils-mediated cognitive dysfunctions [74, 75]. Fucoxanthin has been proposed to attenuate β-amyloid oligomer-provoked neurotoxicity by endorsing phosphoinositide 3-kinase/protein kinase B signaling and suppressing extracellular signal-regulated kinases activation [125]. Fucoxanthin ensures neuroprotection via antioxidant, anti-excitatory, anti-apoptotic, and anti-inflammatory mechanisms [126]. Fucoxanthin was found to play a vital part as an agonist for dopamine D3 and D4 receptors [126]. In an experimental model of PD in mice, fucoxanthin reciprocated 1-methyl-4-phenyl-1,2,3,6-tetrahydropyridine-induced decline in dopaminergic neurons and tyrosine hydroxylase activity with simultaneous suppression of α-synuclein accumulation, redox stress, and gliosis [94]. Thus, fucoxanthin can serve as a potential drug in PD [89, 126].

15.3.6 Zeaxanthin

Zeaxanthin is known to endorse glutathione (GSH) production via activating phase II antioxidant enzyme, γ-glutamyl-cysteine ligase, and can mitigate neurodegenerative diseases via executing GSH [127]. In addition, zeaxanthin can increase the systemic levels of brain-derived neural growth factors compared to untreated subjects [128]. Population-based case studies revealed a positive association between blood zeaxanthin concentration and cognitive functions in older population (n = 4076, $\geq$ 50 years) [34]. A cohort investigation indicated an opposite (non-significant) link between zeaxanthin ingestion and the possibility of PD as compared with 5th vs. the 1st quintile of distribution of the nutrient (hazard ratio 0.78, 95% confidence interval: 0.56–1.08 [101]. However, Kim et al. [33] could not establish any association between serum zeaxanthin levels and the risk of PD, when they compared serum zeaxanthin levels of 104 patients' idiopathic Parkinson cases with respect to 52 healthy controls.

15.3.7 Astaxanthin

Astaxanthin has been proposed to exhibit anti-neurotoxic effect via enhancing mitochondrial function, triggering redox defence, suppressing free radical production, and inhibiting apoptosis [61, 62, 92, 129]. It has been reported that the astaxanthin bio-accumulation in the cerebral cortex is critical in maintaining and improving cognitive performances [130]. It has been regarded to endorse nerve cell regeneration and the activation of the glial fibrillary acidic protein, microtubule-associated protein 2, brain-derived neurotrophic factor, and growth-associated protein 43 [97]. It has been proposed to suppress free radical- and 1-methyl-4-phenylpyridinium ion-provoked toxicity in dopaminergic nerve cells via inhibiting intrinsic apoptotic signalling, decreasing α-synuclein activation, and endorsing superoxide dismutase, catalase, and tyrosine hydroxylase [90, 131, 132]. It can further inhibit 6-hydroxydopamine-induced cellular apoptosis by interfering with the MAPK pathway [133]. Astaxanthin can alleviate xenobiotic-provoked brain aging in rats by reducing oxidative stress, inflammation, and apoptosis, and enhancing the expression of brain-derived neurotrophic factor [97]. In an experimental model, astaxanthin can prevent 1-methyl-4-phenyl-1,2,3,6-tetrahydropyridine-triggered apoptosis to neurons in the substantia nigra in mice [90]. It can reciprocate 1-methyl-4-phenyl-1,2,3,6-tetrahydropyridine-induced decrease in the population of tyrosine hydroxylase-positive neurons and an increase in the argyrophilic neuron counts in mice [90]. It can suppress neurotoxicity in the CNS in the experimental model of PD in young mice [92]. However, it has been found to be less effective in aged mice [92, 129]. In aged animal astaxanthin can only preserve neurons in the substantia nigra, but cannot restore tyrosine hydroxylase expression [92, 129]. Astaxanthin exhibited neuroprotective effect in rat primary hippocampal neurons through inhibition of homocysteine-induced neural apoptosis, possibly via repairing mitochondrial dysfunction and normalizing mitogen-activated protein kinase and phosphoinositide 3-kinase/protein kinase B signalling [134]. Astaxanthin-enriched *Haematococcus pluvialis* extract has been found to be safe and can improve cognitive functions in aged individuals ($n = 96$) in a randomised double-blind placebo-controlled study [135]. Astaxanthin-enriched *Paracoccus carotinifaciens* extract improved cognitive functions in the middle-aged (< 55 years) and older (≥ 55 years) individuals ($n = 28$) in a placebo-controlled study [136]. Combining all, it is concluded that astaxanthin can act as a potential drug against PD.

15.3.8 Canthaxanthin

Canthaxanthin, a keto-carotenoid, inhibits free radical accumulation maintains redox homeostasis [99, 137, 138]. It can suppress hydrogen peroxide- and 1-methyl-4-phenylpyridinium ion-provoked toxicity in nerve growth factor-differentiated dopaminergic nerve (PC12) cells via antioxidant and anti-inflammatory mechanisms [138]. It can also restore normal mitochondrial functions and Na^+-K^+-ATPase activity in nerve cells [138]. Canthaxanthin has been reported to offer partial neuro-protection against β-amyloid peptide (25–35)-induced injury in undifferentiated PC12 cells via restoring cell viability, neutralizing free radicals, suppressing free radical production, and reducing Ca^{2+} influx [139]. Other reports also supported that canthaxanthin exhibits neuroprotective effect and can serve as a potential therapeutic negotiator against neurodegenerative disorder, such as PD [140, 141].

15.3.9 β-Cryptoxanthin

β-Cryptoxanthin, is found in the human brain, and it is regarded to shield the brain from oxidative stress-provoked complications, such as neurodegenerative diseases [11]. β-Cryptoxanthin has been reported to prevent amyloid β protein accumulation via inhibiting the β-amyloid formation and destabilizing preformed β-amyloid fibrils [71]. It can also prevent age-related cognitive impairment, which has been proposed to be connected with increased accumulation of β-cryptoxanthin in the mouse brain [142]. β-Cryptoxanthin accumulation in the brain was found to prevent DNA oxidation in the cerebral cortex, suppress oxidative stress in the brain, and improve learning memory [142]. In addition, β-cryptoxanthin has been revealed to reduce neuro-inflammation significantly, as observed in a population-based study comprising 10 men and 28 women [143]. 1% increase in plasma β-cryptoxanthin has been found to cause 0.33% suppression in the level of IL-6 in cerebrospinal fluid [143]. The dopaminergic system has been recognized as one of the main targets of pro-vitamin A, including β-cryptoxanthin, thus increased intake of β-cryptoxanthin has been postulated to decrease the possibility of PD [111]. However, the clinical data could not find any connection between dietary β-cryptoxanthin and risk of PD. Kim et al. [33] reported an insignificant inverse relationship between serum levels of β-cryptoxanthin and advanced-stage PD patients (n = 57 patients and 52 control). Two large prospective cohort investigations suggested that β-cryptoxanthin ingesting does not significantly affect the risk of PD [53]. In another case-control study (n = 57 male Parkinson cases and 50 control receiving dietary antioxidants for long-term), Scheider et al. [30] could not establish any link between dietary consumption of β-cryptoxanthin and the risk of PD. In another large population-based, case-control study, Zhang and co-workers [101] did not find any association between β-cryptoxanthin and PD. In a multicentre hospital-based case-control study (n = 249 Parkinson cases and 368 control), Miyake and co-workers [107] could not find any link between β-cryptoxanthin intake and PD.

15.3.10 Lutein

Lutein, a carotenoid of xanthophyll family, is a potential neuroprotective agent. It is one of the dominant carotenoids in the human brain [33]. Epidemiological evidence revealed a positive correlation between brain lutein level and cognitive functions in young people [144]. In addition, it has been found to suppress neuronal damage in mice via reducing oxidative stress, apoptosis, and inflammation via regulating various signalling events [145]. Lutein has been reported to restore synaptic functionality in mitochondrial complex I deficiency by regulating the transcriptions of several genes. Thus, it can be effective in treating neurodegenerative diseases [146]. Lutein exhibits the potential inhibitory effect on β-amyloid fibril formation and fibril-destabilizing effects [71]. In an experimental murine model of PD, lutein protected dopaminergic neurons against 1-methyl-4-phenyl-1,2,3,6-tetrahydropyridine-provoked oxidative stress, apoptosis, and mitochondrial dysfunction [147]. In the same study, it can also prevent the exhaustion of striatal dopamine and ameliorate behavioural damages [147]. In another study, lutein was found to prevent dopaminergic neurodegeneration in an experimental model of PD in rats [148]. In this study, lutein significantly reciprocated rotenone-triggered behavioural abnormalities and the reduction in the quantity of dopamine, norepinephrine, and serotonin in brain [148]. In addition, lutein supplement has been found to potentiate the effect with levodopa [148]. The preclinical data conclude the possible protective effect of lutein against PD. However, clinical evidence did not fetch any conclusive evidence regarding the anti-Parkinson effect of lutein. In a long-term longitudinal cohort study ($n = 706$), a higher intake of lutein is found to retard the development of PD in adults [149]. In an occupational cohort study, Zhang and co-workers [101] found a non-significant negative link between serum lutein level and the risk of PD. Other studies reported either significant [30, 118] or borderline [32] positive associations between serum lutein level and PD possibility. Kim et al. [33] reported non-significant but positive connection between serum lutein level and risk of PD in the patients with both early ($n = 47$) and advanced ($n = 57$) PD compared to control ($n = 52$).

15.3.11 Crocin

Crocin, has been found to possess neuroprotective effects through antioxidant and anti-inflammatory mechanisms [150]. It has also been revealed to prevent memory deficit and cognitive impairment [150]. Crocin was reported to shield dopaminergic PC12 cells against 1-methyl-4-phenylpyridinium ion-provoked neurotoxicity through suppression of mitochondrial dysfunction and endoplasmic reticulum stress [151]. Crocin can interact with the α-synuclein in the fibrillation pathway and can block its conversion into fibrils, which is a key process in developing PD [76, 152]. It can simultaneously promote the dissociation of pre-formed fibrils [152]. Crocin can also inhibit the formation of apo-α-lactalbumin fibrillar assemblies under amyloidogenic conditions via formatting soluble oligomers [153]. Crocin has been proposed to up-regulate brain-derived neurotrophic factor expression, thus ensures the survival and protection of dopaminergic neurons in the brain [154]. In a Drosophila model of PD, crocin attenuated rotenone-induced neurodegenerative conditions via triggering redox defence, inhibiting mitochondrial dysfunction, suppressing acetylcholinesterase activity, enhancing dopamine level, and delaying locomotor deficit [82]. Rajaei and co-workers [155] reported that crocin can prevent hippocampal oxidative stress and memory shortfalls in 6-hydroxydopamine-induced PD in rats. Crocin was found alleviate rotenone-induced neurodegeneration in rats, which, which has been witnessed through increase in tyrosine hydroxylase and dopamine levels with concomitant decrease in α-synuclein level [156]. It has been revealed that crocin imparted neuroprotective effect via endorsing phosphoinositide 3-kinase/protein kinase B/mammalian target of rapamycin signalling, suppressing glycogen synthase kinase-3β/forkhead box transcription factor of the O class 3a/caspase-9 signalling, and activating microRNA-7 and microRNA-221 [156]. In another study, crocin prevented 1-methyl-4-phenyl-1,2,3,6-tetrahydropyridine-provoked dopaminergic neuron damage and PD via improving the number of tyrosine-hydroxylase-expressing neurons and preventing cell death in the substantia nigra in mice [93]. Abd Al-Fattah and Abo Zeid [157] reported that crocin could prevent Parkinson-mimic behaviour via reducing lipid peroxidation, DNA oxidation, inflammation in the brain, and improving glutathione level and dopamine level in the brain of rotenone-treated rats. Epidemiological studies revealed that pesticide and herbicide exposure could develop PD [158, 159]. Crocin can prevent malathion (an organophosphate pesticide)-triggered neurobehavioral impairments via suppressing oxidative stress, inflammation, and acetylcholinesterase activity [159]. In another report, Shahidani and co-workers [160] reported that crocin pre-treatment, along with exercise, could be protective in restoring normal motor and memory functions in hemiparkinsonian rats. Crocin pre-treatment, along with exercise, was found to ameliorate 6-hydroxydopamine-induced motor and memory deficits via attenuating striatal inflammation, suppressing hippocampal lipid peroxidation, and activating total thiol quantity in the hippocampus of experimental rats [160]. Combining all, it may be said that crocin has an opposite association with the risk of PD.

15.3.12 Crocetin

Crocetin, has been regarded as the potential neuroprotective agent in neurodegenerative diseases [161]. Crocetin has been found to rescue H_2O_2-induced neurotoxicity via repressing free radical production and decreasing apoptosis [162]. Neural oxidative stress is regarded to play a critical part in inducing neurotoxicity in PD. Crocetin has been reported to prevent β-amyloid$_{1-42}$-induced neurotoxicity in mouse hippocampal neuronal cells (HT-11) [163]. In an experimental model of rat Parkinsonism, crocetin pre-treatment attenuates 6-hydroxydopamine-induced neurotoxicity [164]. Crocetin pre-treatment protected dopaminergic neurones, reduced oxidative stress, and reciprocated motor and cognitive deficits in hemi-parkinsonian rats [164]. Abnormal aggregation of β-amyloid in nerve tissue has been regarded as a critical pathological event in PD [152, 165]. Crocetin can prevent β-amyloid aggregation by inhibiting α-synuclein aggregation and promoting pre-formed fibril dissociation in vitro. Thus, it can be useful to treat PD [152]. Crocetin nano-formulation was found to be efficient in averting 6-hydroxydopamine-induced memory dysfunction and depressive behaviour via triggering redox defence and improving dopamine and its metabolite level in the striatum of mice [166]. Thus, crocetin may be a potential drug in attenuating neurodegenerative diseases, like PD.

15.4 Conclusion

Oxidative stress, inflammation, and amyloidosis have been regarded as significant pathological events for the nigral loss in both familial and sporadic forms of PD [167–169]. Carotenoids are natural antioxidants, which can simultaneously exhibit anti-inflammatory, and anti-amyloidogenic effects via multiple mechanisms [11, 23, 49, 54, 58–62, 71, 152]. Thus, carotenoids may be a potential therapeutic option against Parkinsonism. However, only a few dietary carotenoids were assayed to evaluate their beneficial role against PD. Thus, there is an option to develop novel anti-Parkison's agents from the rest of the carotenoid members. In this regard, special attention may be given to the animal and marine carotenoids. [170]. Thus, the combinatorial effects of carotenoids, along with existing therapeutic agents/options, can also open a new prospect in the management of PD. However, substantial effort is required to develop carotenoid-based potential therapeutic options in PD.

This chapter includes up-to-date information about the protective/therapeutic effects of carotenoids against PD. Emerging evidence revealed that carotenoids could protect neurons in the context of PD. The preclinical assays exhibited the most promising results and explained the possible mechanistic insights. Carotenoids were found to decrease the risk of PD via scavenging free radicals, triggering redox defense, suppressing neuroinflammation, repairing mitochondrial dysfunctions, restoring cell viability, and regulating intercellular signaling events. Among different carotenoids, α-carotene, β-carotene, retinol, lycopene, fucoxanthin, lutein, zeaxanthin, astaxanthin, canthaxanthin, crocin, β-cryptoxanthin, and crocetin exhibit significant anti-Parkinson's effect. However, population-based case studies reported heterogeneity in observations to draw any specific conclusion. Some literature suggested that there is an opposite relationship between carotenoid intake/plasma carotenoids and the danger of PD in humans. However, significant literature could not substantiate any definite conclusion or even reported a positive link between carotenoid intake/plasma carotenoids and the risk of PD. The lack of reporting of measures, heterogeneity in measures, publication-biased interpretation, and insufficient data may be the reasons behind this contradiction. Thus, more clinical studies are required, considering the factors above to establish conclusive evidence regarding the relationship between carotenoids and the danger of PD.

References

1. Dorsey E, Sherer T, Okun MS, Bloem BR (2018) The emerging evidence of the Parkinson pandemic. J Parkinsons Dis 8(s1):S3–S8
2. Feigin VL, Nichols E, Alam T, Bannick MS, Beghi E, Blake N et al (2019) Global, regional, and national burden of neurological disorders, 1990–2016: a systematic analysis for the Global Burden of Disease Study 2016. Lancet Neurol 18(5):459–480
3. Stoker TB, Torsney KM, Barker RA (2018) Emerging treatment approaches for PD. Front Neurosci 12:693
4. Jankovic J, Aguilar LG (2008) Current approaches to the treatment of PD. Neuropsychiatr Dis Treat 4(4):743
5. Rizek P, Kumar N, Jog MS (2016) An update on the diagnosis and treatment of Parkinson disease. CMAJ 188(16):1157–1165
6. DeMaagd G, Philip A (2015) PD and its management: part 1: disease entity, risk factors, pathophysiology, clinical presentation, and diagnosis. Pharm Ther 40(8):504
7. Chao J, Leung Y, Wang M, RC-C C (2012) Nutraceuticals and their preventive or potential therapeutic value in PD. Nutr Rev 70(7):373–386
8. Seidl SE, Santiago JA, Bilyk H, Potashkin JA (2014) The emerging role of nutrition in PD. Front Aging Neurosci 6:36
9. Hang L, Basil AH, Lim K-L (2016) Nutraceuticals in PD. Neuromol Med 18(3):306–321
10. Maiti P, Manna J, Dunbar GL (2017) Current understanding of the molecular mechanisms in PD: targets for potential treatments. Transl Neurodegener 6(1):28
11. Cho KS, Shin M, Kim S, Lee SB (2018) Recent advances in studies on the therapeutic potential of dietary carotenoids in neurodegenerative diseases. Oxidative Med Cell Longev 2018:4120458
12. Zhao X, Zhang M, Li C, Jiang X, Su Y, Zhang Y (2019) Benefits of vitamins in the treatment of PD. Oxidative Med Cell Longev 2019
13. Ciulla M, Marinelli L, Cacciatore I, Stefano AD (2019) Role of dietary supplements in the management of PD. Biomol Ther 9(7):271
14. Muralidhara YS, Yenisetti S (2017) Current understanding on the beneficial role of nutrition in PD –an overview. J Aging Sci 5:1):1–1)10
15. Tan BL, Norhaizan ME, Liew W-P-P, Sulaiman Rahman H (2018) Antioxidant and oxidative stress: a mutual interplay in age-related diseases. Front Pharmacol 9:1162
16. Johnson EJ (2002) The role of carotenoids in human health. Nutr Clin Care 5(2):56–65
17. Tapiero H, Townsend D, Tew K (2004) The role of carotenoids in the prevention of human pathologies. Biomed Pharmacother 58(2):100–110
18. Mezzomo N, Ferreira SR (2016) Carotenoids functionality, sources, and processing by supercritical technology: a review. J Chem 2016:3164312. https://doi.org/10.1155/2016/3164312
19. Elvira-Torales LI, García-Alonso J, Periago-Castón MJ (2019) Nutritional importance of carotenoids and their effect on liver health: a review. Antioxidants 8(7):229
20. Maria AG, Graziano R, Nicolantonio DO (2015) Carotenoids: potential allies of cardiovascular health? Food Nutr Res 59(1):26762
21. Perera CO, Yen GM (2007) Functional properties of carotenoids in human health. Int J Food Prop 10(2):201–230
22. Galasso C, Corinaldesi C, Sansone C (2017) Carotenoids from marine organisms: biological functions and industrial applications. Antioxidants 6(4):96
23. Fiedor J, Burda K (2014) Potential role of carotenoids as antioxidants in human health and disease. Nutrients 6(2):466–488
24. Milani A, Basirnejad M, Shahbazi S, Bolhassani A (2017) Carotenoids: biochemistry, pharmacology and treatment. Br J Pharmacol 174(11):1290–1324
25. Tan BL, Norhaizan ME (2019) Carotenoids: how effective are they to prevent age-related diseases? Molecules 24(9):1801

26. Takeda A, Nyssen OP, Syed A, Jansen E, Bueno-de-Mesquita B, Gallo V (2014) Vitamin A and carotenoids and the risk of PD: a systematic review and meta-analysis. Neuroepidemiology 42(1):25–38
27. Agim ZS, Cannon JR (2015) Dietary factors in the etiology of PD. Biomed Res Int 2015:672838
28. Lange KW, Nakamura Y, Chen N, Guo J, Kanaya S, Lange KM et al (2019) Diet and medical foods in PD. Food Sci Human Wellness 8:83–95
29. Jiménez-Jiménez FJ, Molina J, Fernández-Calle P, Vázquez A, Cabrera-Valdivia F, Catalán MJ et al (1993) Serum levels of β-carotene and other carotenoids in PD. Neurosci Lett 157(1):103–106
30. Scheider W, Hershey L, Vena J, Holmlund T, Marshall J, Freudenheim J (1997) Dietary antioxidants and other dietary factors in the etiology of PD. Mov Disord 12(2):190–196
31. Sherzai AZ, Tagliati M, Park K, Pezeshkian S, Sherzai D (2016) Micronutrients and risk of PD: a systematic review. Gerontol Geriatr Med 2:2333721416644286
32. Powers K, Smith-Weller T, Franklin G, Longstreth W, Swanson P, Checkoway H (2003) PD risks associated with dietary iron, manganese, and other nutrient intakes. Neurology 60(11):1761–1766
33. Kim JH, Hwang J, Shim E, Chung E-J, Jang SH, Koh S-B (2017) Association of serum carotenoid, retinol, and tocopherol concentrations with the progression of PD. Nutr Res Pract 11(2):114–120
34. Feeney J, O'Leary N, Moran R, O'Halloran AM, Nolan JM, Beatty S et al (2017) Plasma lutein and zeaxanthin are associated with better cognitive function across multiple domains in a large population-based sample of older adults: findings from the Irish longitudinal study on aging. J Gerontol Ser A Biomed Sci Med Sci 72(10):1431–1436
35. Radhakrishnan DM, Goyal VPD (2018) Parkinson's Disease: a review. Neurol India 66(7):26–35
36. Clarke C (2007) Clinical review-PD. BMJ 335(7617):441–444
37. Chen H, Burton EA, Ross GW, Huang X, Savica R, Abbott RD et al (2013) Research on the premotor symptoms of PD: clinical and etiological implications. Environ Health Perspect 121(11–12):1245–1252
38. Stefanis L (2012) α-Synuclein in PD. Cold Spring Harb Perspect Med:2, a009399.2012
39. Xu L, Pu J (2016) Alpha-synuclein in PD: from pathogenetic dysfunction to potential clinical application. J Parkinsons Dis 2016:1720621. https://doi.org/10.1155/2016/1720621
40. Surmeier DJ, Guzmán JN, Sánchez-Padilla J, Goldberg JA (2010) What causes the death of dopaminergic neurons in PD? Progress in brain research. Elsevier, Netherlands, Vol 183, pp 59–77
41. Cookson MR (2012) Parkinsonism due to mutations in PINK1, parkin, and DJ-1 and oxidative stress and mitochondrial pathways. Cold Spring Harb Perspect Med 2(9):a009415
42. Fink AL (2006) Factors affecting the fibrillation of α-synuclein, a natively unfolded protein. In: Misbehaving proteins. Springer, New York, pp 265–285
43. Narhi L, Wood SJ, Steavenson S, Jiang Y, Wu GM, Anafi D et al (1999) Both familial PD mutations accelerate α-synuclein aggregation. J Biol Chem 274(14):9843–9846
44. Maoka T (2019) Carotenoids as natural functional pigments. J Nat Med 74:1–16
45. Augustynska D, Jemioła-Rzemińska M, Burda K, Strzałka K (2015) Influence of polar and nonpolar carotenoids on structural and adhesive properties of model membranes. Chem Biol Interact 239:19–25
46. Grudzinski W, Nierzwicki L, Welc R, Reszczynska E, Luchowski R, Czub J et al (2017) Localization and orientation of xanthophylls in a lipid bilayer. Sci Rep 7(1):1–10
47. Widomska J, Welc R, Gruszecki WI (2019) The effect of carotenoids on the concentration of singlet oxygen in lipid membranes. Biochimica et Biophysica Acta (BBA)-Biomembranes 1861(4):845–851
48. Kopsell DA, Kopsell DE (2006) Accumulation and bioavailability of dietary carotenoids in vegetable crops. Trends Plant Sci 11(10):499–507

49. Ben-Dor A, Steiner M, Gheber L, Danilenko M, Dubi N, Linnewiel K et al (2005) Carotenoids activate the antioxidant response element transcription system. Mol Cancer Ther 4(1):177–186
50. Etminan M, Gill SS, Samii A (2005) Intake of vitamin E, vitamin C, and carotenoids and the risk of PD: a meta-analysis. Lancet Neurol 4(6):362–365
51. Dias V, Junn E, Mouradian MM (2013) The role of oxidative stress in PD. J Parkinsons Dis 3(4):461–491
52. Manoharan S, Guillemin GJ, Abiramasundari RS, Essa MM, Akbar M, Akbar MD (2016) The role of reactive oxygen species in the pathogenesis of Alzheimer's disease, PD, and Huntington's disease: a mini review. Oxidative Med Cell Longev 2016:8590578
53. Hughes KC, Gao X, Kim IY, Rimm EB, Wang M, Weisskopf MG et al (2016) Intake of antioxidant vitamins and risk of PD. Mov Disord 31(12):1909–1914
54. Lakey-Beitia J, Kumar D, Hegde ML, Rao K (2019) Carotenoids as novel therapeutic molecules against neurodegenerative disorders: chemistry and molecular docking analysis. Int J Mol Sci 20(22):5553
55. Lowe GM, Booth LA, Young AJ, Bilton RF (1999) Lycopene and β-carotene protect against oxidative damage in HT29 cells at low concentrations but rapidly lose this capacity at higher doses. Free Radic Res 30(2):141–151
56. Troncoso-Escudero P, Parra A, Nassif M, Vidal RL (2018) Outside in: unraveling the role of neuroinflammation in the progression of PD. Front Neurol 9:860
57. Wang Q, Liu Y, Zhou J (2015) Neuroinflammation in PD and its potential as therapeutic target. Transl Neurodegener 4(1):19
58. Pangestuti R, Vo T-S, Ngo D-H, Kim S-K (2013) Fucoxanthin ameliorates inflammation and oxidative reponses in microglia. J Agric Food Chem 61(16):3876–3883
59. Zhao D, Kwon S-H, Chun YS, Gu M-Y, Yang HO (2017) Anti-neuroinflammatory effects of fucoxanthin via inhibition of Akt/NF-κB and MAPKs/AP-1 pathways and activation of PKA/CREB pathway in lipopolysaccharide-activated BV-2 microglial cells. Neurochem Res 42(2):667–677
60. Zhang X-S, Zhang X, Wu Q, Li W, Wang C-X, Xie G-B et al (2014) Astaxanthin offers neuroprotection and reduces neuroinflammation in experimental subarachnoid hemorrhage. J Surg Res 192(1):206–213
61. Galasso C, Orefice I, Pellone P, Cirino P, Miele R, Ianora A et al (2018) On the neuroprotective role of astaxanthin: new perspectives? Mar Drugs 16(8):247
62. Fakhri S, Yosifova Aneva I, Farzaei MH, Sobarzo-Sánchez E (2019) The Neuroprotective effects of Astaxanthin: therapeutic targets and clinical perspective. Molecules 24(14):2640
63. Zhao B, Ren B, Guo R, Zhang W, Ma S, Yao Y et al (2017) Supplementation of lycopene attenuates oxidative stress induced neuroinflammation and cognitive impairment via Nrf2/NF-κB transcriptional pathway. Food Chem Toxicol 109:505–516
64. Wu W, Li Y, Wu Y, Zhang Y, Wang Z, Liu X (2015) Lutein suppresses inflammatory responses through Nrf2 activation and NF-κB inactivation in lipopolysaccharide-stimulated BV-2 microglia. Mol Nutr Food Res 59(9):1663–1673
65. Nam KN, Park Y-M, Jung H-J, Lee JY, Min BD, Park S-U et al (2010) Anti-inflammatory effects of crocin and crocetin in rat brain microglial cells. Eur J Pharmacol 648(1–3):110–116
66. Deslauriers AM, Afkhami-Goli A, Paul AM, Bhat RK, Acharjee S, Ellestad KK et al (2011) Neuroinflammation and endoplasmic reticulum stress are coregulated by crocin to prevent demyelination and neurodegeneration. J Immunol 187(9):4788–4799
67. Xiao Q, Xiong Z, Yu C, Zhou J, Shen Q, Wang L et al (2019) Antidepressant activity of crocin-I is associated with amelioration of neuroinflammation and attenuates oxidative damage induced by corticosterone in mice. Physiol Behav 212:112699
68. Wang X, Zhang G, Qiao Y, Feng C, Zhao X (2017) Crocetin attenuates spared nerve injury-induced neuropathic pain in mice. J Pharmacol Sci 135(4):141–147
69. Zhang J, Wang Y, Dong X, Liu J (2018) Crocetin attenuates inflammation and amyloid-β accumulation in APPsw transgenic mice. Immun Ageing 15(1):24

70. Araki K, Yagi N, Aoyama K, Choong C-J, Hayakawa H, Fujimura H et al (2019) PD is a type of amyloidosis featuring accumulation of amyloid fibrils of α-synuclein. Proc Natl Acad Sci 116(36):17963–17969
71. Katayama S, Ogawa H, Nakamura S (2011) Apricot carotenoids possess potent anti-amyloidogenic activity in vitro. J Agric Food Chem 59(23):12691–12696
72. Sachdeva AK, Chopra K (2015) Lycopene abrogates Aβ (1–42)-mediated neuroinflammatory cascade in an experimental model of Alzheimer's disease. J Nutr Biochem 26(7):736–744
73. Lobos P, Bruna B, Cordova A, Barattini P, Galáz JL, Adasme T et al (2016) Astaxanthin protects primary hippocampal neurons against noxious effects of Aβ-oligomers. Neural Plast 2016:3456783
74. Alghazwi M, Smid S, Musgrave I, Zhang W (2019) In vitro studies of the neuroprotective activities of astaxanthin and fucoxanthin against amyloid beta (Aβ1-42) toxicity and aggregation. Neurochem Int 124:215–224
75. Xiang S, Liu F, Lin J, Chen H, Huang C, Chen L et al (2017) Fucoxanthin inhibits β-amyloid assembly and attenuates β-amyloid oligomer-induced cognitive impairments. J Agric Food Chem 65(20):4092–4102
76. Ghahghaei A, Bathaie SZ, Kheirkhah H, Bahraminejad E (2013) The protective effect of crocin on the amyloid fibril formation of Aβ42 peptide in vitro. Cell Mol Biol Lett 18(3):328–339
77. Jin H, Kanthasamy A, Ghosh A, Anantharam V, Kalyanaraman B, Kanthasamy AG (2014) Mitochondria-targeted antioxidants for treatment of PD: preclinical and clinical outcomes. Biochimica et Biophysica Acta (BBA)-Molecular Basis of Disease 1842(8):1282–1294
78. Sheriff SA, Shaik Ibrahim S, Devaki T, Chakraborty S, Agarwal S, Pérez-Sánchez H (2017) Lycopene prevents mitochondrial dysfunction during d-galactosamine/lipopolysaccharide-induced fulminant hepatic failure in albino rats. J Proteome Res 16(9):3190–3199
79. Sztretye M, Dienes B, Gönczi M, Czirják T, Csernoch L, Dux L et al (2019) Astaxanthin: a potential mitochondrial-targeted antioxidant treatment in diseases and with aging. Oxidative Med Cell Longev 2019:3849692
80. Geromichalos GD, Lamari FN, Papandreou MA, Trafalis DT, Margarity M, Papageorgiou A et al (2012) Saffron as a source of novel acetylcholinesterase inhibitors: molecular docking and in vitro enzymatic studies. J Agric Food Chem 60(24):6131–6138
81. Ghanam K, Deshpande J, Juturu V (2015) Lutein and zeaxanthin isomers inhibit cholinesterase and modulate the expression of inflammation-related genes: in vitro models. Int J Ophthalmol Clin Res 2:4
82. Rao SV, Yenisetti SC, Rajini PS (2016) Evidence of neuroprotective effects of saffron and crocin in a Drosophila model of parkinsonism. Neurotoxicology 52:230–242
83. Hira S, Saleem U, Anwar F, Sohail MF, Raza Z, Ahmad B (2019) β-Carotene: a natural compound improves cognitive impairment and oxidative stress in a mouse model of Streptozotocin-induced Alzheimer's disease. Biomol Ther 9(9):441
84. Zeng C, Li H, Fan Z, Zhong L, Guo Z, Guo Y et al (2016) Crocin-elicited autophagy rescues myocardial ischemia/reperfusion injury via paradoxical mechanisms. Am J Chin Med 44(03):515–530
85. Chang C-J, Lin J-F, Hsiao C-Y, Chang H-H, Li H-J, Chang H-H et al (2017) Lutein induces autophagy via beclin-1 upregulation in IEC-6 rat intestinal epithelial cells. Am J Chin Med 45(06):1273–1291
86. Rzajew J, Radzik T, Rebas E (2020) Calcium-involved action of phytochemicals: carotenoids and monoterpenes in the brain. Int J Mol Sci 21(4):1428
87. Seif-El-Nasr M, Atia AS, Abdelsalam RM (2008) Effect of MAO-B inhibition against ischemia-induced oxidative stress in the rat brain. Arzneimittelforschung 58(04):160–167
88. Jiang X, Zhu K, Xu Q, Wang G, Zhang J, Cao R et al (2017) The antidepressant-like effect of trans-astaxanthin involves the serotonergic system. Oncotarget 8(15):25552
89. Jung HA, Roy A, Choi JS (2017) In vitro monoamine oxidase A and B inhibitory activity and molecular docking simulations of fucoxanthin. Fish Sci 83(1):123–132

90. Lee D-H, Kim C-S, Lee YJ (2011) Astaxanthin protects against MPTP/MPP+-induced mitochondrial dysfunction and ROS production in vivo and in vitro. Food Chem Toxicol 49(1):271–280
91. Liu C, Wang R, Pan H, Ding Q, Lu F (2013) Effect of lycopene on oxidative stress and behavioral deficits in rotenone induced model of PD. Zhongguo ying yong sheng li xue za zhi= Zhongguo yingyong shenglixue zazhi= Chin J Appl Physiol 29(4):380–384
92. Grimmig B, Daly L, Hudson C, Nash K, Bickford P (2017) Astaxanthin attenuates neurotoxicity in a mouse model of PD. Funct Foods Health Dis 7(8):562–576
93. Haeri P, Mohammadipour A, Heidari Z, Ebrahimzadeh-bideskan A (2019) Neuroprotective effect of crocin on substantia nigra in MPTP-induced PD model of mice. Anat Sci Int 94(1):119–127
94. Sun G, Xin T, Zhang R, Liu C, Pang Q (2020) Fucoxanthin attenuates behavior deficits and neuroinflammatory response in 1-methyl-4-phenyl-1, 2, 3, 6-tetrahydropyridine-induced PD in mice. Pharmacogn Mag 16(67):51
95. El-Missiry M, Shalaby F (2000) Role of β-carotene in ameliorating the cadmium-induced oxidative stress in rat brain and testis. J Biochem Mol Toxicol 14(5):238–243
96. Qu M, Nan X, Gao Z, Guo B, Liu B, Chen Z (2013) Protective effects of lycopene against methylmercury-induced neurotoxicity in cultured rat cerebellar granule neurons. Brain Res 1540:92–102
97. Wu Q, Zhang X-S, Wang H-D, Zhang X, Yu Q, Li W et al (2014) Astaxanthin activates nuclear factor erythroid-related factor 2 and the antioxidant responsive element (Nrf2-ARE) pathway in the brain after subarachnoid hemorrhage in rats and attenuates early brain injury. Mar Drugs 12(12):6125–6141
98. Zhang F, Xing S, Li Z (2017) Antagonistic effects of lycopene on cadmium-induced hippocampal dysfunctions in autophagy, calcium homeostatis and redox. Oncotarget 8(27):44720
99. Sandmann G (2019) Antioxidant protection from UV-and light-stress related to carotenoid structures. Antioxidants 8(7):219
100. Farombi EO, Britton G (1999) Antioxidant activity of palm oil carotenes in peroxyl radical-mediatedperoxidation of phosphatidyl choline liposomes. Redox Rep 4(1–2):61–68
101. Zhang S, Hernan M, Chen H, Spiegelman D, Willett W, Ascherio A (2002) Intakes of vitamins E and C, carotenoids, vitamin supplements, and PD risk. Neurology 59(8):1161–1169
102. Perlmutter JS (1988) New insights into the pathophysiology of PD: the challenge of positron emission tomography. Trends Neurosci 11(5):203–208
103. Perry TL, Yong VW, Clavier RM, Jones K, Wright JM, Foulks JG et al (1985) Partial protection from the dopaminergic neurotoxin N-methyl-4-phenyl-1, 2, 3, 6-tetrahydropyridine by four different antioxidants in the mouse. Neurosci Lett 60(2):109–114
104. Perry TL, Yong VW, Hansen S, Jones K, Bergeron C, Foulks JG et al (1987) α-Tocopherol and β-carotene do not protect marmosets against the dopaminergic neurotoxicity of N-methyl-4-phenyl-1, 2, 3, 6-tetrahydropyridine. J Neurol Sci 81(2–3):321–331
105. Ono K, Yamada M (2007) Vitamin A potently destabilizes preformed α-synuclein fibrils in vitro: implications for Lewy body diseases. Neurobiol Dis 25(2):446–454
106. Sies H, Stahl W, Sundquist AR (1992) Antioxidant functions of vitamins: vitamins E and C, beta-carotene, and other carotenoids A. Ann N Y Acad Sci 669(1):7–20
107. Miyake Y, Fukushima W, Tanaka K, Sasaki S, Kiyohara C, Tsuboi Y et al (2011) Dietary intake of antioxidant vitamins and risk of PD: a case–control study in Japan. Eur J Neurol 18(1):106–113
108. Hellenbrand W, Boeing H, Robra B-P, Seidler A, Vieregge P, Nischan P et al (1996) Diet and PD II: a possible role for the past intake of specific nutrients: results from a self-administered food-frequency questionnaire in a case-control study. Neurology 47(3):644–650
109. Foy C, Passmore A, Vahidassr M, Young I, Lawson J (1999) Plasma chain-breaking antioxidants in Alzheimer's disease, vascular dementia and PD. QJM 92(1):39–45
110. Oliveira MRD (2015) The neurotoxic effects of vitamin a and retinoids. An Acad Bras Cienc 87(2):1361–1373

111. Guest J, Grant R (2016) Carotenoids and neurobiological health. In: The benefits of natural products for neurodegenerative diseases. Springer, Cham, pp 199–228
112. Maden M (2007) Retinoic acid in the development, regeneration and maintenance of the nervous system. Nat Rev Neurosci 8(10):755–765
113. McCaffery P, Zhang J, Crandall JE (2006) Retinoic acid signaling and function in the adult hippocampus. J Neurobiol 66(7):780–791
114. Kunzler A, Ribeiro CT, Gasparotto J, Petiz LL, da Rosa Silva HT, da Silva JD Jr et al (2019) The effects of retinol oral supplementation in 6-hydroxydopamine dopaminergic denervation model in Wistar rats. Neurochem Int 125:25–34
115. Jacobs FM, Smits SM, Noorlander CW, von Oerthel L, van der Linden AJ, Burbach JPH et al (2007) Retinoic acid counteracts developmental defects in the substantia nigra caused by Pitx3 deficiency. Development 134(14):2673–2684
116. Cerhan J, Wallace R, Folsom A (eds) (1994) Antioxidant intake and risk of parkinsons-disease (PD) In: Older women. Am J Epidemiol; Johns Hopkins University School Hygiene Public Health 111 Market Place, STE 840
117. Jiménez-Jiménez FJ, Molina J, Fernández-Calle P, Vázquez A, Pondal M, del Ser T et al (1992) Serum levels of vitamin a in PD. J Neurol Sci 111(1):73–76
118. Johnson C, Gorell J, Rybicki B, Sanders K, Peterson E (1999) Adult nutrient intake as a risk factor for PD. Int J Epidemiol 28(6):1102–1109
119. Anderson C, Checkoway H, Franklin GM, Beresford S, Smith-Weller T, Swanson PD (1999) Dietary factors in PD: the role of food groups and specific foods. Mov Disord 14(1):21–27
120. Prema A, Janakiraman U, Manivasagam T, Thenmozhi AJ (2015) Neuroprotective effect of lycopene against MPTP induced experimental PD in mice. Neurosci Lett 599:12–19
121. Chen D, Huang C, Chen Z (2019) A review for the pharmacological effect of lycopene in central nervous system disorders. Biomed Pharmacother 111:791–801
122. Przybylska S (2020) Lycopene–a bioactive carotenoid offering multiple health benefits: a review. Int J Food Sci Technol 55(1):11–32
123. Zabihollah K, Mohammad F (2015) The effect of lycopene on dopaminergic receptors and GABA neurons in Hippocampus and Substantia Nigra areas in Rat with PD. J Entomol Zool Stud 3(4):239–244
124. di Matteo V, Pierucci M, Di Giovanni G, Dragani LK, Murzilli S, Poggi A et al (2009) Intake of tomato-enriched diet protects from 6-hydroxydopamine-induced degeneration of rat nigral dopaminergic neurons. In: Birth, life and death of dopaminergic neurons in the Substantia Nigra. Springer, Wein, pp 333–341
125. Lin J, Yu J, Zhao J, Zhang K, Zheng J, Wang J et al (2017) Fucoxanthin, a marine carotenoid, attenuates β-amyloid oligomer-induced neurotoxicity possibly via regulating the PI3K/Akt and the ERK pathways in SH-SY5Y cells. Oxidative Med Cell Longev 2017:1–10
126. Paudel P, Seong SH, Jung HA, Choi JS (2019) Characterizing fucoxanthin as a selective dopamine D3/D4 receptor agonist: relevance to PD. Chem Biol Interact 310:108757
127. Zou X, Gao J, Zheng Y, Wang X, Chen C, Cao K et al (2014) Zeaxanthin induces Nrf2-mediated phase II enzymes in protection of cell death. Cell Death & Disease 5(5):e1218
128. Hammond BR Jr, Miller LS, Bello MO, Lindbergh CA, Mewborn C, Renzi-Hammond LM (2017) Effects of lutein/zeaxanthin supplementation on the cognitive function of community dwelling older adults: a randomized, double-masked, placebo-controlled trial. Front Aging Neurosci 9:254
129. Grimmig B, Daly L, Subbarayan M, Hudson C, Williamson R, Nash K et al (2018) Astaxanthin is neuroprotective in an aged mouse model of PD. Oncotarget 9(12):10388
130. Manabe Y, Komatsu T, Seki S, Sugawara T (2018) Dietary astaxanthin can accumulate in the brain of rats. Biosci Biotechnol Biochem 82(8):1433–1436
131. Liu X, Shibata T, Hisaka S, Osawa T (2009) Astaxanthin inhibits reactive oxygen species-mediated cellular toxicity in dopaminergic SH-SY5Y cells via mitochondria-targeted protective mechanism. Brain Res 1254:18–27
132. Ye Q, Huang B, Zhang X, Zhu Y, Chen X (2012) Astaxanthin protects against MPP+-induced oxidative stress in PC12 cells via the HO-1/NOX2 axis. BMC Neurosci 13(1):156

133. Ikeda R, Iwashita K-i, Sumizawa T, Beppu S-i, Tabata S, Tajitsu Y et al (2008) Hyperosmotic stress up-regulates the expression of major vault protein in SW620 human colon cancer cells. Exp Cell Res 314(16):3017–3026
134. X-j W, Chen W, X-t F, Ma J-k, Wang M-h, Y-j H et al (2018) Reversal of homocysteine-induced neurotoxicity in rat hippocampal neurons by astaxanthin: evidences for mitochondrial dysfunction and signaling crosstalk. Cell Death Dis 4:1):1–1)11
135. Katagiri M, Satoh A, Tsuji S, Shirasawa T (2011) Effects of astaxanthin-rich Haematococcus pluvialis extract on cognitive function: a randomised, double-blind, placebo-controlled study. J Clin Biochem Nutr 51:1203270150
136. Hayashi M, Ishibashi T, Maoka T (2018) Effect of astaxanthin-rich extract derived from Paracoccus carotinifaciens on cognitive function in middle-aged and older individuals. J Clin Biochem Nutr 62(2):195–205
137. Lim BP, Nagao A, Terao J, Tanaka K, Suzuki T, Takama K (1992) Antioxidant activity of xanthophylls on peroxyl radical-mediated phospholipid peroxidation. Biochimica et Biophysica Acta Lipids Lipid Metab 1126(2):178–184
138. Kc C, Mc M, Mc Y (2009) Antioxidative and anti-inflammatory neuroprotective effects of astaxanthin and canthaxanthin in nerve growth factor differentiated PC12 cells. J Food Sci 74(7):H225–HH31
139. Chang C-S, Chang C-L, Lai G-H (2013) Reactive oxygen species scavenging activities in a chemiluminescence model and neuroprotection in rat pheochromocytoma cells by astaxanthin, beta-carotene, and canthaxanthin. Kaohsiung J Med Sci 29(8):412–421
140. Guest JA, Grant RS (2012) Effects of dietary derived antioxidants on the central nervous system. Int J Nutr Pharmacol Neurol Dis 2(3):185
141. Zare Dorahi Z, Ilani M, Jamhiri M, Daneshparvar A, Zare S, Zare M et al (2019) The investigation of the induction of Diketocarotenoids senescence in SHSY-5Y cells. Rep Health Care 5(1):37–44
142. Unno K, Sugiura M, Ogawa K, Takabayashi F, Toda M, Sakuma M et al (2011) Beta-cryptoxanthin, plentiful in Japanese mandarin orange, prevents age-related cognitive dysfunction and oxidative damage in senescence-accelerated mouse brain. Biol Pharm Bull 34(3):311–317
143. Guest J, Grant R, Garg M, Mori TA, Croft KD, Bilgin A (2014) Cerebrospinal fluid levels of inflammation, oxidative stress and NAD+ are linked to differences in plasma carotenoid concentrations. J Neuroinflammation 11(1):117
144. Johnson EJ (2014) Role of lutein and zeaxanthin in visual and cognitive function throughout the lifespan. Nutr Rev 72(9):605–612
145. Li S-Y, Yang D, Fu ZJ, Woo T, Wong D, Lo ACY (2012) Lutein enhances survival and reduces neuronal damage in a mouse model of ischemic stroke. Neurobiol Dis 45(1):624–632
146. Maglioni S, Schiavi A, Melcher M, Brinkmann V, Luo Z, Raimundo N et al (2020) Lutein restores synaptic functionality in a C. elegans model for mitochondrial complex I deficiency. bioRxiv
147. Nataraj J, Manivasagam T, Thenmozhi AJ, Essa MM (2016) Lutein protects dopaminergic neurons against MPTP-induced apoptotic death and motor dysfunction by ameliorating mitochondrial disruption and oxidative stress. Nutr Neurosci 19(6):237–246
148. Makhija D, Jagtap A, Deshpande J, Shankaranarayanan J, Kulkarni S (2014) Nutritional supplement of lutein attenuates dopaminergic neurodegeneration induced by rotenone in rats. PharmaNutrition 2(3):95–96
149. Agarwal P, Wang Y, Buchman A, Bennett D, Morris MC (2018) Association of Carotenoids, vitamin E and vitamin C intake with parkinsonism and progression of Parkinsonian signs in older adults. Neurology 90:S4.006
150. Farkhondeh T, Samarghandian S, Yazdi HS, Samini F (2018) The protective effects of crocin in the management of neurodegenerative diseases: a review. Am J Neurodegener Dis 7:1):1
151. Zhang G-F, Zhang Y, Zhao G (2015) Crocin protects PC12 cells against MPP+-induced injury through inhibition of mitochondrial dysfunction and ER stress. Neurochem Int 89:101–110

152. Inoue E, Shimizu Y, Masui R, Hayakawa T, Tsubonoya T, Hori S et al (2018) Effects of saffron and its constituents, crocin-1, crocin-2, and crocetin on α-synuclein fibrils. J Nat Med 72(1):274–279
153. Ebrahim-Habibi MB, Amininasab M, Ebrahim-Habibi A, Sabbaghian M, Nemat-Gorgani M (2010) Fibrillation of α-lactalbumin: effect of crocin and safranal, two natural small molecules from Crocus sativus. Biopolymers 93(10):854–865
154. Rezai M, Mahmoodi M, Kaeidi A, Karimabad MN, Khoshdel A, Hajizadeh MR (2018) Effect of crocin carotenoid on BDNF and CREB gene expression in brain ventral tegmental area of morphine treated rats. Asian Pac J Trop Biomed 8(8):387–393
155. Rajaei Z, Hosseini M, Alaei H (2016) Effects of crocin on brain oxidative damage and aversive memory in a 6-OHDA model of PD. Arq Neuropsiquiatr 74(9):723–729
156. Salama RM, Abdel-Latif GA, Abbas SS, Hekmat M, Schaalan MF (2020) Neuroprotective effect of crocin against rotenone-induced PD in rats: Interplay between PI3K/Akt/mTOR signaling pathway and enhanced expression of miRNA-7 and miRNA-221. Neuropharmacology 164:107900
157. Abdalfattah AA, Abo Zeid A (2019) Possible neuroprotective effects of crocin against motor and neurochemical changes in rotenone induced animal model of PD. Bull Egypt Soc Physiol Sci 39(2):216–230
158. Tanner CM, Kamel F, Ross GW, Hoppin JA, Goldman SM, Korell M et al (2011) Rotenone, paraquat, and PD. Environ Health Perspect 119(6):866–872
159. Mohammadzadeh L, Hosseinzadeh H, Abnous K, Razavi BM (2018) Neuroprotective potential of crocin against malathion-induced motor deficit and neurochemical alterations in rats. Environ Sci Pollut Res 25(5):4904–4914
160. Shahidani S, Rajaei Z, Alaei H (2019) Pretreatment with crocin along with treadmill exercise ameliorates motor and memory deficits in hemiparkinsonian rats by anti-inflammatory and antioxidant mechanisms. Metab Brain Dis 34(2):459–468
161. Gupta S (2019) A review on therapeutic potentials of Crocetin-A carotenoid derived from saffron. Curr Res Pharm Sci 9:54–62
162. Papandreou MA, Tsachaki M, Efthimiopoulos S, Cordopatis P, Lamari FN, Margarity M (2011) Memory enhancing effects of saffron in aged mice are correlated with antioxidant protection. Behav Brain Res 219(2):197–204
163. Kong Y, Kong L-P, Luo T, Li G-W, Jiang W, Li S et al (2014) The protective effects of crocetin on aβ1-42-induced toxicity in Ht22 cells. CNS Neurol Disord Drug Targets 13(9):1627–1632
164. Ahmad AS, Ansari MA, Ahmad M, Saleem S, Yousuf S, Hoda MN et al (2005) Neuroprotection by crocetin in a hemi-parkinsonian rat model. Pharmacol Biochem Behav 81(4):805–813
165. Roberts HL, Schneider BL, Brown DR (2017) α-Synuclein increases β-amyloid secretion by promoting β-/γ-secretase processing of APP. PLoS One 12(2):e0171925
166. Sahai V, Kumar V (2018) Development of nano-formulation containing rutin for the protective and beneficial effect against 6-hydroxydopamine induced PD model via altered the genetic backgrounds. Parkinsonism Relat Disord 46:e43
167. Sutachan JJ, Casas Z, Albarracin SL, Stab BR, Samudio I, Gonzalez J et al (2012) Cellular and molecular mechanisms of antioxidants in PD. Nutr Neurosci 15(3):120–126
168. Filograna R, Beltramini M, Bubacco L, Bisaglia M (2016) Anti-oxidants in PD therapy: a critical point of view. Curr Neuropharmacol 14(3):260–271
169. Yang F, Wolk A, Håkansson N, Pedersen NL, Wirdefeldt K (2017) Dietary antioxidants and risk of PD in two population-based cohorts. Mov Disord 32(11):1631–1636
170. Fahn S (1992) A pilot trial of high-dose alpha-tocopherol and ascorbate in early PD. Ann Neurol 32(S1):S128–SS32

Chapter 16
Role of Carotenoids in Neurological Diseases

Sagheer Ahmed, Sidrah Tariq Khan, Aiman Aziz, Saima Gul, Lavinia Buvnariu, and Muhammad Zia-Ul-Haq

16.1 Introduction

Carotenoids are a large group of pigments that are naturally found in plants. Amongst this large group of pigments, β-carotene, β-cryptoxanthin, α-carotene, zeaxanthin, lycopene and lutein are found to be the most prominent dietary pigments. These pigments form an integral part of the plant tissue and due to their potent anti-inflammatory and anti-oxidant effects, their consumption has been associated with numerous health benefits, including the effects on the nervous system that lead to neuronal differentiation, neuronal patterning and motor axon growth [1]. In this chapter, we will describe and discuss the neuroprotective roles played by carotenoids in neurological diseases.

S. Ahmed (✉) · S. T. Khan · A. Aziz
Shifa College of Pharmaceutical Sciences, Shifa Tameer-e-Millat University, Islamabad, Pakistan
e-mail: sagheer.scps@stmu.edu.pk

S. Gul
Department of Physical Therapy, Shifa Tameer-e-Millat University, Islamabad, Pakistan

L. Buvnariu
Faculty of Medicine, Transilvania University of Brasov, Brasov, Romania

M. Zia-Ul-Haq
Office of Research, Innovation and Commercialization, Lahore College for Women University, Lahore, Pakistan

© Springer Nature Switzerland AG 2021

M. Zia-Ul-Haq et al. (eds.), *Carotenoids: Structure and Function in the Human Body*, https://doi.org/10.1007/978-3-030-46459-2_16

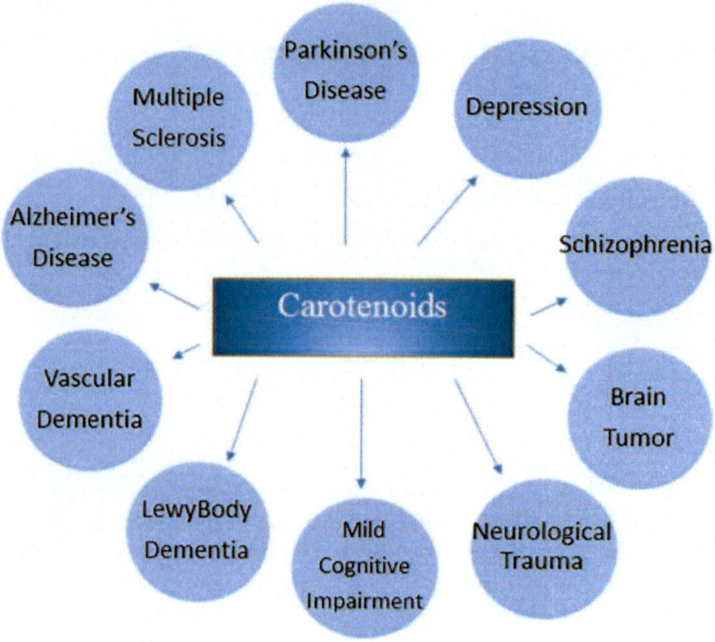

16.2 Alzheimer's Disease

Alzheimer's disease (AD) is one of the infamous neurodegenerative diseases, it is characterized by the formation of neurofibrillary tangles and β-amyloid plaques in certain parts of the brain, the most important consequence of which is the decline in cognitive function of the human mind [2]. Several theories about the pathogenesis of Alzheimer's have been put forth by scientists, one of the mechanisms through which these plaques are produced has been attributed to the production of reactive oxygen species (ROS) in the brain. It has been suggested through various studies that there is a fall in the serum levels of vitamin A (retinoids) and β-carotene in patients suffering with the disease [3]. Carotenoids have been found to possess certain neuroprotective properties such as inhibiting the synthesis and accumulation of β-amyloid plaques amongst other anti-neurotransmission, and anti-inflammatory properties [4, 5]. Experimental results have also proposed that there is a strong association between carotenoids and decreased rate of mortality associated with AD [6]. The antioxidant effect of carotenoids by scavenging toxic oxygen radicals has led to the targeting of these specific nutrients in order to prevent cognitive decline [7]. The application of carotenoids may in fact lead to lowering β-amyloid levels, thereby reducing the risk of AD [8].

Numerous studies are being carried out to evaluate the anti-inflammatory and anti-oxidant roles of carotenoids in the prevention and inhibition of neurodegenerative factors involved in AD [9]. Experiments revealed that carotenoids not only reduced β-amyloid secretion in SH-SY5Y cells but also protected the cell from copper and H202 induced oxidative stress, imparting nueroprotective effects to these cells [10]. This effect was also observed in cognitively impaired male Wister rats who upon administration of carotenoids show a reduction in β-amyloid induced mitochondrial damage [11]. A number of studies have emphasized on the use of dietary nutrient supplements containing carotenoids for the management of the disease. However, these findings also warrant further studies on animal models to unravel their mechanism of neuroprotection [4].

Accumulation of β-amyloid also leads to the down regulation of certain carotenoid signaling pathways such as those involving the retinoid X and the retinoic acid receptors [12]. These signaling pathways are involved in the inhibition of AD mediators, thereby decreasing the progression of AD and maintaining memory function in adult brains as the levels of β-amyloids in the RBCs increase with age [13, 14]. The activation of these signaling pathways also reduce β-amyloid accumulation and tau hyperphosphorylation [5]. Another mechanism by which these pathways increase clearance of β-amyloid plaques is through enhancing the activity of neprilysin and insulin degrading enzymes in the neurons and also in the microglial cells of the brain [15]. Recently, studies have suggested that targeting the oligomerization of β-amyloid (40 & 42) using carotenoids could serve as a preventive measure or even a therapeutic tool in AD by improving spatial learning and memory [15]. Recent studies have also implied that the antioxidant defenses of the body are compromised in AD, these findings underline the need for improved understanding of balancing the upregulation of exogenous antioxidants in response to dietary intake in further studies based on nutrition [16].

Experimental results further show that oxidized states of RBC, which are a characteristic of AD, contain higher levels of phospholipids hydroperoxides (PLOOH) and upon raising the levels of carotenoids there is a marked reduction in PLOOH accumulation [14], which reveals that there must be an inverse relation between PLOOH accumulation and carotenoids. Furthermore, the activity of phospholipase A2 in the neurons and glial cells, stimulated by retinoic acid plays a significant role in arachidonic acid redistribution, which through its downstream signaling is proposed to be involved in schizophrenia and AD [17]. This suggests that the development of new retinoic acid analogs could be proven to be significant in establishing new pharmacological interventions for the management of these neurodegenerative diseases. A recent study also provided evidence that cryptocapsin, cryptocapsin-5,6-epoxide, and zeaxanthin possess some potential to decrease the β amyloid plaques and may prove to be useful in AD treatment and prophylaxis [18]. Collectively, these findings reveal the burgeoning role of carotenoids in AD pathology and treatment.

16.3 Vascular Dementia

Vascular dementia (VaD) is another type of dementia that is characterized by the reduced blood supply to the brain. VaD progresses in a step-like manner unlike AD that progresses in a continuous manner, but both are the two extremes of one disease i.e. dementia [19]. Low antioxidant levels, particularly carotenoids in conjunction with high lipid peroxidation and protein oxidation levels may be a reason behind the accumulation of free radicals, cognitive impairment and memory loss [20]. Studies revealed that plasma levels of carotenoids are in fact low in patients with VaD, whereas free radicals are significantly increased [21]. Dementia patients have shown increased levels of DNA damage repair products like 8-oxoguanine in their urine and also CSF. Conversely, the levels of antioxidants (ascorbic acid, carotenoids) in these people are significantly reduced [22]. Spatial learning and memory tests done on rat models revealed that treatment with corcetin (a carotenoid) could protect the neurons of the hippocampus and the cerebro-cortex from ischemia bringing about improvement memory and spatial learning following cerebral hypoperfusion in rats [23].

Population based studies executed to explore the relation between dementia and plasma levels of carotenoids, have shown that maintaining higher plasma concentrations of carotenoids caused a moderate reduction in VaD risk [24]. Raising the levels of antioxidants via supplementation may prove pharmacologically beneficial in the treatment of VaD. Carotenoids, particularly astaxanthins have shown significant nueroprotective effects in ischemic mice and pretreating these mice with astaxanthin has shown to improve memory function in VaD, owing to their antioxidant properties [25]. However, several other studies have also revealed that the use of dietary carotenoids is not associated with reducing the risk of dementia or even improved cognitive performance [26]. Moreover, midlife use of dietary carotenoids/antioxidants doesn't result in reduced risk of late-life dementia or any of its other forms [26]. Therefore, the role of carotenoids in VaD is not conclusive at the moment.

16.4 Lewy Body Dementia

Lewy body dementia is the second most prevalent dementia but commonly misdiagnosed because of its close resemblance to Parkinson's disease. Clumps of α-synuclein fibrils are present and a hallmark of the disease. These clumps are formed due to misfolding of the α-synuclein proteins leading to its accumulation. Experiments revealed that antioxidants like carotenoids are useful in inhibiting the formation of α-synuclein fibrils, and therefore, could be helpful in the prevention and treatment of Lewy body dementia [27].

16.5 Mild Cognitive Impairment

Dementia is preceded by mild cognitive impairment and is sometimes considered as an early stage of dementia. It could be amnestic causing a reduction in memory or non-amnestic causing a reduction in cognitive function [28]. Despite of its type, cognitive impairment can occur due to a variety of reason like oxidative stress, which is now a well-established cause behind cognitive decline along with neuroinflammation [29]. In some cases, MCI indicates an early stage of AD [28].

Dose-dependent incorporation of carotenoids can lead to improvement in cognitive performance, due to their antioxidant properties [30]. Cognitive function is closely associated with the presence of lutein and zeaxanthin found in the macular pigment of the eye. In people with mild cognitive impairment, carotenoid supplementation may be beneficial in enhancing learning and memory as these patients have relatively low levels of carotenoids [31]. Better cognitive performance, especially in the elderly has been known to be linked to high plasma carotenoid levels [32]. In a recent study aimed to find out relationship between concentration of carotenoids in the plasma and cognitive performance in the elderly, it has been observed that low plasma carotenoid levels were associated with lower cognitive functioning [33]. Furthermore, low levels of the macular pigments have also been linked with a suppression of prospective memory [34]. Infant brain has significant quantities of carotenoids, most abundant of which is lutein, accounting for 59% of the total carotenoid levels and is mostly concentrated in regions involved in memory and overall cognitive functioning, which reveals that it may also be involved in early neuronal development and in verbal recognition memory [35]. Studies also suggest that these pigments are associated with cognitive enhancement in people in whom the cognitive function is already deteriorating [36, 37].

Use of carotenoids for longer duration may help in cognitive improvement. Gender-based randomized clinical trials on two groups of men, one receiving treatment with 50 mg β-carotene on alternate days, for 18 years and the other receiving treatment for only 1 year, showed that the group that received treatment for a longer duration showed higher global cognitive memory scores [38]. This study affirms that in order to improve cognitive memory, longer duration of treatments with carotenoids may be required. Similarly, studies carried out to elucidate the effects of dietary intake of docosahexaenoic acid (DHA) and carotenoids on cognitive performance, in elderly women revealed that following supplementation of both DHA and carotenoids, general cognitive performance improved significantly [39]. This suggests that supplementation with DHAs and carotenoids can in fact prove beneficial in certain aspects of cognitive decline. The results of a case control study by Rinaldi and his colleagues reveal that increasing the intake of carotenoids as well as other antioxidants could be beneficial in lowering the risk of cognitive impairment and other neurodegenerative diseases [40]. Although carotenoids can delay or even prevent cognitive decline under certain circumstances, other investigations suggest that they are not as such beneficial in treating cognitive decline associated with cardiovascular diseases [41].

Verbal fluency has been shown to be associated with serum lutein and zeaxanthin concentrations [42]. Administration of both lutein and zeaxanthin combined is found to improve not only learning efficiency and learning rate but enhances memory scores as well [31]. Retinal health is strongly influenced by lutein and zeaxanthin concentrations in the eye and this effect was found only in humans and some other higher primates [43]. Both lutein and zeaxanthin have anti-inflammatory and anti-oxidant properties which are most likely responsible for the observed beneficial effects of these carotenoids. In addition, increased gap junction communications and enhanced structural stability of the membranes by lutein and zeaxanthin which they do by being integrated into cell membranes, also play pivotal roles [44].

The prevalent view is that carotenoids improve cognition and other related neuroelectric indicators and hence protect brain in early or middle childhood [45]. Thus, it is thought that carotenoids afford brain protection only in early age decades before the aging process starts affecting brain. However, Hammaond and colleagues showed that lutein and zeaxanthin supplementation is also beneficial later in life as evident by their investigation which shows carotenoid consumption later in life is associated with enhanced cognition and cerebral perfusion [46]. In a randomized, double-blind, placebo-controlled study, cognitive function in community-dwelling older men and women was enhanced by lutein and zeaxanthin consumption [46]. Other interventional studies in adults indicate that omega-3 fatty acid and docosahexaenoic acid may also potentiate the positive effects of lutein on cognition [47].

16.6 Nuerological Trauma

Any injury to the brain, spinal cord or nerves can result in nuerological trauma that may lead to several complications like loss of motor control or even cognitive functioning. These injuries can result from motor vehicle accidents, falls, sports activities and a variety of other traumas. Carotenoids perform pivotal roles in the physiological neuronal development. Carotenoids and its derivatives protect neuronal tissue and help its regeneration by mediating through various signaling pathways [48]. Post-trauma therapy with carotenoids leads to the inhibition of neuronal apoptosis and increased levels of certain growth factors such as serum response factor and vascular endothelial growth factor receptor 2, as evident from experiments carried out on rat models [49]. The mechanism by which carotenoids inhibit apoptosis remains undiscovered. Certain hypotheses have been put forth regarding their anti-apoptotic activity. One such hypothesis based on experiments done on rat models with sub-arachanoid hemorrhage, suggests that carotenoids prevent apoptosis by modulating the PI3-kinase-Akt pathway. Enhancing the activity of this signaling pathway leads to decreased apoptosis. Moreover, carotenoids also decrease caspase-3 activity in the cerebral cortex [50]. It has been reported that functional recovery of the spinal cord injury is accelerated by astaxanthin by reducing pathological damage and inhibiting neuronal apoptosis [51].

Hypoxic conditions in the brain may also lead to cerebral ischemia, which could become a cause of neurological damage. Carotenoids play a major role in attenuating this damage by increasing the levels of neurogranin, which is a kinase C substrate found in the brain. This improves cerebral functioning and decreasing infarct volume [52]. In the brain, blood brain barrier is disrupted by damage to spinal cord or brain which further aggravates the neuronal damage by exacerbating inflammatory response. Therapy with carotenoids has shown to improve functional recovery in rat models by preserving the integrity of these barriers and diminishing the loss of adhesion molecules like P120, β-catenin, occludin and claudin5 of the tight junctions. Carotenoids have also shown to preserve the microvasculature of the brain [53]. Therapy with the carotenoid crocin, after a spinal cord injury has shown to improve mechanical recovery and also decrease chronic pain [54]. Another carotenoid, astaxanthin has shown to have some neuroprotective role in brain injuries owing to its anti-oxidant and anti-inflammatory properties [55]. These reports indicate that carotenoid therapy may be important in reducing neurological damage after trauma [56].

16.7 Brain Tumor

The ineffectiveness of present brain tumor therapies leaves the need for new and effective therapeutic strategies to be discovered. Many studies report that carotenoids have been useful in the treatment of brain tumor. It has been suggested that combining the chemo and immuno therapeutic effects of retinoic acid and gamma interferons, can prove beneficial in inducing apoptosis, cellular differentiation and immunity in different glioblastoma cell lines [57] and thus may be helpful in treating brain tumors.

Many studies have revealed that the levels of antioxidants such as carotenoids were significantly decreased in patients with brain tumor. The more malignant the tumor was, the lower were the levels of carotenoids [58]. This suggests that the levels of carotenoids may be beneficial in evaluating the severity of malignancy in brain tumors. In vitro studies in a rat models show that administration of carotenoids reduced the incidence of gliomas and also decreased its aggressiveness. The levels of tumor biomarkers like growth factors such as insulin growth factor binding protein and antioxidant enzymes such as superoxide dismutase, were greatly reduced after therapy with antioxidants including carotenoids [59]. These findings advocate the role of carotenoids in preventing tumor progression and also reducing its aggressiveness.

The homeobox transcription factor OTX2 is usually silenced after child birth but in certain cases due to genetic mutations, it can be overexpressed as evident in medulloblastomas. Treatment with carotenoids (esp. retinoic acid) has shown to suppress this transcription factor thereby inhibiting tumor growth [60]. This effect of retinoic acid is sometimes resisted by certain mediators like fibroblast growth factors resulting in decreased effectiveness [60]. Therefore, therapy with retinoic acid combined with targeted inhibition of FGF may prove to be pharmacologically beneficial in inhibiting tumor growth.

Another mechanism by which retinoic acid reduces medulloblastomas is by inhibiting the CyclinD1 and C-myc (involved in regulating cell cycle transitions) resulting in cell growth arrest and ultimately decreasing tumor progression [61]. Other clinical studies have proposed that retinoic acid promotes dopamine receptor subtype 2 expression in human pituitary adenomas, thereby decreasing hormone secretion. Retinoic acid was found to elevate pro-apoptotic mediators in growth hormone producing GH3 cell lines [62]. Retinoic acid also increases the expression of bone morphogenetic protein 4 (BMP 4), leading to decreased hormone secretion and cellular proliferation in pituitary adenomas [63]. These activities of retinoic acid can serve to be potential therapeutic targets in suppressing pituitary tumor growth.

16.8 Schizophrenia

The etiology of schizophrenia has been defined as the enlargement of cerebral ventricular resulting in abnormalities in the limbic structures of the brain. The disease results in cognitive decline, hallucinations, delusional thinking and motor deficits in certain cases, mostly occurring between the ages 16 and 30. Genetico-organic or environmental factors have shown some relationship with the onset of this disease [64].

Retinoic acid receptors are scattered throughout the brain and its related signaling pathways are crucial for normal neurological development. Any alterations in these pathways can result in serious neurodevelopmental diseases like schizophrenia, as evident from certain case control studies [65]. The neurotransmitter dopamine plays a pivotal role in controlling certain important neurophysiological functions [66]. Studies have proposed that transcriptional activation of the dopamine D2 receptor is down regulated in schizophrenia. Normally, the activation of this receptor is brought about by retinoic acid and alterations in the levels of D2R receptors can be reversed by the use of retinoids [65].

16.9 Depression

Depression is a chronic psychiatric disorder affecting how a person thinks or feels. The exact causes of depression are not known but stress along with low levels of serotonin neurotransmitter in the brain are thought to be implicated. Case controls studies revealed that these patients have significantly low levels of antioxidants including carotenoids and dietary supplementation of these antioxidants shows a marked reduction in depression [67]. Furthermore, depressive patients have very low levels of serotonin and brain derived neurotrophic factor in their brains [67]. In a recent study, the administration of β-carotene significantly raised the levels of serotonin and BDNF in patients suffering from depression [68]. These findings propose that increasing carotenoid levels by supplementation can lead to a decrease in depression symptoms.

16.10 Parkinson's Disease

Parkinson Disease is progressive neurodegenerative disorder of the motor neurons. The disease is caused by the destruction of the dopaminergic neurons in substania nigra reducing the action of dopamine in corpus striatum, involved in motor activity. PD is characterized by the tremors, muscular rigidity, bradykinesia, poor balance and coordination. Therapies being used today to treat PD provide only temporary relief and do not execute the complete reversal of the disease. Oxidative stress has shown to play some role in neuronal death and loss of dopaminergic neurons in PD. Serum levels of PD patients show a significant deficiency of antioxidants like carotenoids, resulting in increased iron levels and hence increased oxidative stress. Elevating levels of these antioxidants can prove to be beneficial in reducing this stress by scavenging reactive oxygen and nitrogen species as well as inhibiting synphilin-1 and alpha-synuclein aggregation [69]. Other case control studies have also showed low antioxidant and consequently high oxidative stress levels in patients suffering from PD [21]. Experiments carried out on rat models to elucidate the effects of retinoic acid on dopaminergic neurons revealed that retinoic acid shows neuroprotective effects on dopaminergic neurons by increasing dopamine secretion in the striatum [70]. However, certain studies also show that serum levels of carotenoids are unrelated to Parkinson's disease [3].

16.11 Multiple Sclerosis

Multiple sclerosis (MS) is an autoimmune, neurodegenerative disease mediated by T-cells (CD-4) and characterized by neuroinflammation and decreased insulation of the neurons due to disruption of the protective myelin sheath surrounding the axon [71]. The levels of Th17 helper cells are increased in MS whereas that of regulatory T cells are greatly reduced [72]. Recent studies suggest that enhancing retinoic acid levels can be helpful in balancing the levels of these proinflammatory and immunoprotective cells resulting in increased tolerance to autoimmunity and regeneration in neurons [73].

Case control studies reveal that astrocytes increase the synthesis of retinoid acid by modulating an enzyme involved in its synthesis-retinaldehyde dehydrogenase 2. This increase in retinoic acid synthesis by astrocytes protects blood brain barrier and is a representation of an endogenous response to neuroinflammation [74].

Retinoic acid has also shown to downregulate the expression of pro-inflammatory mediators like IL-17 and RORγt gene [2]. Moreover, certain studies have revealed that carotenoids can prove to be potential targets for treating MS [74].

16.12 Conclusions

Carotenoids are known to be very efficient quenchers of oxidants. Being effective as antioxidants, they are effective in the prevention of diseases with oxidative stress as their underlying mechanism. In neurological diseases there is a debate as whether oxidative stress is causative or consequence of various neurological diseases. Regardless, a large body of evidence supports the neuroprotective role of carotenoids in various neurological diseases. The beneficial effects of carotenoids in neurological diseases have been provided in a number of studies. Nevertheless, some studies suggest inconclusive data and therefore further studies are required to confirm their role. Furthermore, there is a scarcity of good animal models to study certain neurological diseases. Therefore, there is a significant need to make new and effective animal models to study the effects of carotenoids in neurological diseases such as in AD. This will help understand the molecular mechanisms of carotenoid functions *in vivo* in these diseases.

References

1. Maden M (2007) Retinoic acid in the development, regeneration and maintenance of the nervous system. Nat Rev Neurosci 8(10):755–765
2. Honarvar NM, Saedisomeolia A, Abdolahi M, Shayeganrad A, Sangsari GT, Rad BH et al (2017) Molecular anti-inflammatory mechanisms of retinoids and carotenoids in Alzheimer's disease: a review of current evidence. J Mol Neurosci 61(3):289–304
3. Jiménez-Jiménez FJ, Molina J, Fernández-Calle P, Vázquez A, Cabrera-Valdivia F, Catalán MJ et al (1993) Serum levels of β-carotene and other carotenoids in Parkinson's disease. Neurosci Lett 157(1):103–106
4. Obulesu M, Dowlathabad MR, Bramhachari P (2011) Carotenoids and Alzheimer's disease: an insight into therapeutic role of retinoids in animal models. Neurochem Int 59(5):535–541
5. Sodhi RK, Singh N (2014) Retinoids as potential targets for Alzheimer's disease. Pharmacol Biochem Behav 120:117–123
6. Min J-y, Min K-b (2014) Serum lycopene, lutein and zeaxanthin, and the risk of Alzheimer's disease mortality in older adults. Dement Geriatr Cogn Disord 37(3–4):246–256
7. Wang W, Shinto L, Connor WE, Quinn JF (2008) Nutritional biomarkers in Alzheimer's disease: the association between carotenoids, n-3 fatty acids, and dementia severity. J Alzheimers Dis 13(1):31–38
8. Kiko T, Nakagawa K, Tsuduki T, Suzuki T, Arai H, Miyazawa T (2012) Significance of lutein in red blood cells of Alzheimer's disease patients. J Alzheimers Dis 28(3):593–600
9. Carratu MR, Marasco C, Signorile A, Scuderi C, Steardo L (2012) Are retinoids a promise for Alzheimer's disease management? Curr Med Chem 19(36):6119–6125

10. Chen W, Mao L, Xing H, Xu L, Fu X, Huang L et al (2015) Lycopene attenuates Aβ1–42 secretion and its toxicity in human cell and Caenorhabditis elegans models of Alzheimer disease. Neurosci Lett 608:28–33
11. Prakash A, Kumar A (2014) Implicating the role of lycopene in restoration of mitochondrial enzymes and BDNF levels in β-amyloid induced Alzheimer's disease. Eur J Pharmacol 741:104–111
12. Goncalves MB, Clarke E, Hobbs C, Malmqvist T, Deacon R, Jack J et al (2013) Amyloid β inhibits retinoic acid synthesis exacerbating A lzheimer disease pathology which can be attenuated by an retinoic acid receptor α agonist. Eur J Neurosci 37(7):1182–1192
13. Chakrabarti M, McDonald AJ, Will Reed J, Moss MA, Das BC, Ray SK (2016) Molecular signaling mechanisms of natural and synthetic retinoids for inhibition of pathogenesis in Alzheimer's disease. J Alzheimers Dis 50(2):335–352
14. Kiko T, Nakagawa K, Satoh A, Tsuduki T, Furukawa K, Arai H et al (2012) Amyloid β levels in human red blood cells. PloS One 7(11):e49620
15. Ono K, Yamada M (2012) Vitamin A and Alzheimer's disease. Geriatr Gerontol Int 12(2):180–188
16. Mullan K, Williams MA, Cardwell CR, McGuinness B, Passmore P, Silvestri G et al (2017) Serum concentrations of vitamin E and carotenoids are altered in Alzheimer's disease: a case-control study. Alzheimers Dement Transl Res Clin Interv 3(3):432–439
17. Farooqui AA, Antony P, Ong W-Y, Horrocks LA, Freysz L (2004) Retinoic acid-mediated phospholipase A2 signaling in the nucleus. Brain Res Rev 45(3):179–195
18. Lakey-Beitia J, Doens D, Kumar DJ, Murillo E, Fernandez PL, Rao K et al (2017) Anti-amyloid aggregation activity of novel carotenoids: implications for Alzheimer's drug discovery. Clin Interv Aging 12:815
19. Bennett S, Grant MM, Aldred S (2009) Oxidative stress in vascular dementia and Alzheimer's disease: a common pathology. J Alzheimers Dis 17(2):245–257
20. Polidori MC, Mattioli P, Aldred S, Cecchetti R, Stahl W, Griffiths H et al (2004) Plasma antioxidant status, immunoglobulin g oxidation and lipid peroxidation in demented patients: relevance to Alzheimer disease and vascular dementia. Dement Geriatr Cogn Disord 18(3–4):265–270
21. Foy C, Passmore A, Vahidassr M, Young I, Lawson J (1999) Plasma chain-breaking antioxidants in Alzheimer's disease, vascular dementia and Parkinson's disease. QJM 92(1):39–45
22. Gackowski D, Rozalski R, Siomek A, Dziaman T, Nicpon K, Klimarczyk M et al (2008) Oxidative stress and oxidative DNA damage is characteristic for mixed Alzheimer disease/vascular dementia. J Neurol Sci 266(1–2):57–62
23. Tashakori-Sabzevar F, Hosseinzadeh H, Sadat Motamedshariaty V, Reza Movassaghi A, Ahmad Mohajeri S (2013) Crocetin attenuates spatial learning dysfunction and hippocampal injury in a model of vascular dementia. Curr Neurovasc Res 10(4):325–334
24. Feart C, Letenneur L, Helmer C, Samieri C, Schalch W, Etheve S et al (2016) Plasma carotenoids are inversely associated with dementia risk in an elderly French cohort. J Gerontol A Biol Sci Med Sci 71(5):683–688
25. Hussein G, Nakamura M, Zhao Q, Iguchi T, Goto H, Sankawa U et al (2005) Antihypertensive and neuroprotective effects of astaxanthin in experimental animals. Biol Pharm Bull 28(1):47–52
26. Laurin D, Masaki KH, Foley DJ, White LR, Launer LJ (2004) Midlife dietary intake of antioxidants and risk of late-life incident dementia: the Honolulu-Asia Aging Study. Am J Epidemiol 159(10):959–967
27. Ono K, Yamada M (2007) Vitamin A potently destabilizes preformed α-synuclein fibrils in vitro: implications for Lewy body diseases. Neurobiol Dis 25(2):446–454
28. Mecocci P (2007) The multifaceted aspects of Alzheimer's disease: from social to molecular problems. J Alzheimers Dis 12(1):1
29. Wyss-Coray T (2006) Inflammation in Alzheimer disease: driving force, bystander or beneficial response? Nat Med 12(9):1005–1015

30. Unno K, Sugiura M, Ogawa K, Takabayashi F, Toda M, Sakuma M et al (2011) Beta-cryptoxanthin, plentiful in Japanese mandarin orange, prevents age-related cognitive dysfunction and oxidative damage in senescence-accelerated mouse brain. Biol Pharm Bull 34(3):311–317
31. Johnson EJ (2012) A possible role for lutein and zeaxanthin in cognitive function in the elderly. Am J Clin Nutr 96(5):1161S–1165S
32. Perrig WJ, Perrig P, Stähelin HB (1997) The relation between antioxidants and memory performance in the old and very old. J Am Geriatr Soc 45(6):718–724
33. Akbaraly NT, Faure H, Gourlet V, Favier A, Berr C (2007) Plasma carotenoid levels and cognitive performance in an elderly population: results of the EVA Study. J Gerontol Ser A Biol Med Sci 62(3):308–316
34. Feeney J, Finucane C, Savva GM, Cronin H, Beatty S, Nolan JM et al (2013) Low macular pigment optical density is associated with lower cognitive performance in a large, population-based sample of older adults. Neurobiol Aging 34(11):2449–2456
35. Vishwanathan R, Kuchan MJ, Sen S, Johnson EJ (2014) Lutein and preterm infants with decreased concentrations of brain carotenoids. J Pediatr Gastroenterol Nutr 59(5):659–665
36. Renzi LM, Dengler MJ, Puente A, Miller LS, Hammond BR Jr (2014) Relationships between macular pigment optical density and cognitive function in unimpaired and mildly cognitively impaired older adults. Neurobiol Aging 35(7):1695–1699
37. Kelly D, Coen RF, Akuffo KO, Beatty S, Dennison J, Moran R et al (2015) Cognitive function and its relationship with macular pigment optical density and serum concentrations of its constituent carotenoids. J Alzheimers Dis 48(1):261–277
38. Grodstein F, Kang JH, Glynn RJ, Cook NR, Gaziano JM (2007) A randomized trial of beta carotene supplementation and cognitive function in men: the Physicians' Health Study II. Arch Intern Med 167(20):2184–2190
39. Johnson EJ, McDonald K, Caldarella SM, Chung H-y, Troen AM, Snodderly DM (2008) Cognitive findings of an exploratory trial of docosahexaenoic acid and lutein supplementation in older women. Nutr Neurosci 11(2):75–83
40. Rinaldi P, Polidori MC, Metastasio A, Mariani E, Mattioli P, Cherubini A et al (2003) Plasma antioxidants are similarly depleted in mild cognitive impairment and in Alzheimer's disease. Neurobiol Aging 24(7):915–919
41. Kang JHCN, Manson JE, Buring JE, Albert CM, Grodstein F (2009) Vitamin E, vitamin C, beta carotene, and cognitive function among women with or at risk of cardiovascular disease: the women's antioxidant and cardiovascular study. Circulation 119(21):2772–2780. https://doi.org/10.1161/CIRCULATIONAHA.108.816900
42. Vishwanathan R, Iannaccone A, Scott TM, Kritchevsky SB, Jennings BJ, Carboni G et al (2014) Macular pigment optical density is related to cognitive function in older people. Age Ageing 43(2):271–275
43. Erdman JW, Smith JW, Kuchan MJ, Mohn ES, Johnson EJ, Rubakhin SS et al (2015) Lutein and brain function. Foods 4(4):547–564
44. Maci S, Fonseca B, Zhu Y (2016) The role of lutein in brain health and function. Forum Nutr 15:179–188
45. Walk AM, Edwards CG, Baumgartner NW, Chojnacki MR, Covello AR, Reeser GE et al (2017) The role of retinal carotenoids and age on neuroelectric indices of attentional control among early to middle-aged adults. Front Aging Neurosci 9:183
46. Hammond BR Jr, Miller LS, Bello MO, Lindbergh CA, Mewborn C, Renzi-Hammond LM (2017) Effects of lutein/zeaxanthin supplementation on the cognitive function of community dwelling older adults: a randomized, double-masked, placebo-controlled trial. Front Aging Neurosci 9:254
47. Mohn ES, Johnson EJ (2017) Lutein and cognition across the lifespan. Nutr Today 52(4):183–189
48. Malaspina A, Michael-Titus AT (2008) Is the modulation of retinoid and retinoid-associated signaling a future therapeutic strategy in neurological trauma and neurodegeneration? J Neurochem 104(3):584–595

49. Bie X, Chen Y, Zheng X, Dai H (2011) The role of crocetin in protection following cerebral contusion and in the enhancement of angiogenesis in rats. Fitoterapia 82(7):997–1002
50. Zhang X-S, Zhang X, Wu Q, Li W, Zhang Q-R, Wang C-X et al (2014) Astaxanthin alleviates early brain injury following subarachnoid hemorrhage in rats: possible involvement of Akt/bad signaling. Mar Drugs 12(8):4291–4310
51. Masoudi A, Dargahi L, Abbaszadeh F, Pourgholami MH, Asgari A, Manoochehri M et al (2017) Neuroprotective effects of astaxanthin in a rat model of spinal cord injury. Behav Brain Res 329:104–110
52. Li L, Li Y, Ji X, Zhang B, Wei H, Luo Y (2008) The effects of retinoic acid on the expression of neurogranin after experimental cerebral ischemia. Brain Res 1226:234–240
53. Zhou Y, Zheng B, Ye L, Zhang H, Zhu S, Zheng X et al (2016) Retinoic acid prevents disruption of blood-spinal cord barrier by inducing autophagic flux after spinal cord injury. Neurochem Res 41(4):813–825
54. Karami M, Bathaie SZ, Tiraihi T, Habibi-Rezaei M, Arabkheradmand J, Faghihzadeh S (2013) Crocin improved locomotor function and mechanical behavior in the rat model of contused spinal cord injury through decreasing calcitonin gene related peptide (CGRP). Phytomedicine 21(1):62–67
55. Wu H, Niu H, Shao A, Wu C, Dixon BJ, Zhang J et al (2015) Astaxanthin as a potential neuroprotective agent for neurological diseases. Mar Drugs 13(9):5750–5766
56. Orellano EA, Rivera OJ, Chevres M, Chorna NE, González FA (2010) Inhibition of neuronal cell death after retinoic acid-induced down-regulation of P2X 7 nucleotide receptor expression. Mol Cell Biochem 337(1–2):83–99
57. Haque A, Banik NL, Ray SK (2007) Emerging role of combination of all-trans retinoic acid and interferon-gamma as chemoimmunotherapy in the management of human glioblastoma. Neurochem Res 32(12):2203–2209
58. Aggarwal S, Subberwal M, Sharma M (2006) Brain tumor and role of β-carotene, a-tocopherol, superoxide dismutase and glutathione peroxidase. J Cancer Res Ther 2(1):24
59. Hervouet E, Staehlin O, Pouliquen D, Debien E, Cartron P, Menanteau J et al (2013) Antioxidants delay clinical signs and systemic effects of ENU induced brain tumors in rats. Nutr Cancer 65(5):686–694
60. Bai R, Siu I-M, Tyler BM, Staedtke V, Gallia GL, Riggins GJ (2010) Evaluation of retinoic acid therapy for OTX2-positive medulloblastomas. Neuro-Oncology 12(7):655–663
61. Chang Q, Chen Z, You J, McNutt MA, Zhang T, Han Z et al (2007) All-trans-retinoic acid induces cell growth arrest in a human medulloblastoma cell line. J Neuro-Oncol 84(3):263–267
62. Grossrubatscher E, Veronese S, Dalino Ciaramella P, Pugliese R, Boniardi M, De Carlis L et al (2008) High expression of dopamine receptor subtype 2 in a large series of neuroendocrine tumors. Cancer Biol Ther 7(12):1970–1978
63. Giacomini D, Paez-Pereda M, Theodoropoulou M, Labeur M, Refojo D, Gerez J et al (2006) Bone morphogenetic protein-4 inhibits corticotroph tumor cells: involvement in the retinoic acid inhibitory action. Endocrinology 147(1):247–256
64. Ferreira AJ (1961) The etiology of schizophrenia—a review. Calif Med 94(6):369
65. Goodman AB (1998) Three independent lines of evidence suggest retinoids as causal to schizophrenia. Proc Natl Acad Sci 95(13):7240–7244
66. Samad TA, Krezel W, Chambon P, Borrelli E (1997) Regulation of dopaminergic pathways by retinoids: activation of the D2 receptor promoter by members of the retinoic acid receptor–retinoid X receptor family. Proc Natl Acad Sci 94(26):14349–14354
67. Gautam M, Agrawal M, Gautam M, Sharma P, Gautam AS, Gautam S (2012) Role of antioxidants in generalised anxiety disorder and depression. Indian J Psychiatry 54(3):244
68. Kim N-R, Kim H-Y, Kim M-H, Kim H-M, Jeong H-J (2016) Improvement of depressive behavior by Sweetme Sweet Pumpkin™ and its active compound, β-carotene. Life Sci 147:39–45
69. Sutachan JJ, Casas Z, Albarracin SL, Stab BR, Samudio I, Gonzalez J et al (2012) Cellular and molecular mechanisms of antioxidants in Parkinson's disease. Nutr Neurosci 15(3):120–126

70. Yin L-H, Shen H, Diaz-Ruiz O, Bäckman CM, Bae E, Yu S-J et al (2012) Early post-treatment with 9-cis retinoic acid reduces neurodegeneration of dopaminergic neurons in a rat model of Parkinson's disease. BMC Neurosci 13(1):120
71. Peterson LK, Fujinami RS (2007) Inflammation, demyelination, neurodegeneration and neuroprotection in the pathogenesis of multiple sclerosis. J Neuroimmunol 184(1–2):37–44
72. Abdolahi M, Yavari P, Honarvar NM, Bitarafan S, Mahmoudi M, Saboor-Yaraghi AA (2015) Molecular mechanisms of the action of vitamin A in Th17/Treg axis in multiple sclerosis. J Mol Neurosci 57(4):605–613
73. Dorosty-Motlagh AR, Honarvar NM, Sedighiyan M, Abdolahi M (2016) The molecular mechanisms of vitamin A deficiency in multiple sclerosis. J Mol Neurosci 60(1):82–90
74. Mirshafiey A, Mohsenzadegan M (2009) Antioxidant therapy in multiple sclerosis. Immunopharmacol Immunotoxicol 31(1):13–29

Chapter 17
Carotenoids as Antiobesity Agents

Muhammad Riaz, Rizwan Ahmad, and Muhammad Zia-Ul-Haq

17.1 The Role of Carotenoids in Obesity

Obesity and overweight are the abnormal or unnecessary fat buildup in body adipocytes that may impair human health [1]. The energy in the form of triglycerides gets stored in adipose tissues made of adipocytes (white cells) which are distributedthroughout the body, in the subcutaneous layer of skin, around vital organs, muscles and bone marrow. These tissues are highly vascularized in nature hence nutrients, hormones and enzymes can easily be delivered [2]. The global prevalence forobesity witnessed an annual mortality of 2.8 M [1]. Several strategies are used to control obesity, among which dietary intervention palys a promising role. Fruits and vegetables are rich sources of micronutrients including carotenoids which can inhibit and treat obesity [3]. Animal model studies involving carotenoids indicate prevention and cure of body-weight gain [4–8]. Several studies have established a linkage between carotenoids supplementation and improved health status [9].

Obesity is mediated by adiponectin, cytokines and leptin that have been associated to increase oxidative stress and inflammation which may pose health risk at world level due to cancer, hyperglycemia, diabetes, and other vital organs diseases. More specifically obesity is the leading metabolic disorders that has been treated conventionally using anti-obesity drugs and surgery that produces a continuing unwanted effects Therefore, functional food or nutraceuticals of natural origin may play a vital role to combat obesity and its comorbidities [10].

M. Riaz (✉)
Department of Pharmacy, Shaheed Benazir Bhutto University, Sheringal, Pakistan

R. Ahmad
Natural Products & Alternative Medicines, College of Clinical Pharmacy, Imam Abdulrahman Bin Faisal University, Dammam, Eastern Province, Kingdom of Saudi Arabia

M. Zia-Ul-Haq
Office of Research, Innovation, and Commercialization (ORIC), Lahore College for Women University, Lahore, Pakistan

© Springer Nature Switzerland AG 2021

M. Zia-Ul-Haq et al. (eds.), *Carotenoids: Structure and Function in the Human Body*, https://doi.org/10.1007/978-3-030-46459-2_17

17.2 Anti-obesity Effect of Pro-vitamins A Carotenoids and Retinoic Acid

There are several carotenoids that exert action when metabolized to vitamin [11]. It is significant to discern the carotenoids metabolism in animal models. Animals with variation in metabolizing carotenoids can produce confusing results. Looking to this variation some scientists has used BCO1 or BCO2-deficient mice to learn about carotenoids accumulation. Physiological amounts of dietary β-carotene reduced obesity in wild-type mice via downregulation of adipogenic genes [12], however, *Bco1*⁻/⁻ deficient mice did not indicate any important variations despite storing big quantities of β-carotene in adipose tissue. It was confirmed that β-carotene decreases obesity only when it is transformed to vitamin A [13].

Lazar et al. have found that retinoic acid prevents adipocyte differentiation in cell cultures via suppressing the key adipokines [14], however, if retinoic acid was added to fully differentiated adipocytes then lipolysis is favored via initiation of UCP1 and brown/beige adipocyte markers [15, 16]. β-carotene, exercises analogous properties to retinoic acid in mature brown adipocytes [17] and other cell types able to store lipids e.g. macrophages [18]. β-carotene is converted to retinoic acid by the action of β-carotene oxygenase 1(BCO1). β-carotene and vitamin A get stored in adipocytes to serve as a physiological source for use during adipocyte differentiation where BCO1 is observed to be upregulated that simply mean more retinoic acid (vitamin A) from β-carotene and intracellular lipid storing capacity is also increased. It was also noted that adipocytes exposed to β-carotene showed reduction in lipolysis and downregulation of peroxisome proliferator-activated receptors gamma (PPARγ) [12]. β-carotene concentration were found lesser in obese adipocytes comparative to lean individuals [19] (Fig. 17.1).

In most of the studies conducted in mice confirmed that that adipogenesis, oxidation of fatty acid and browning of AT are affected by retinoic acid [16, 20–22]. These effects of retinoic acid have been observed in liver, muscle cell culture studies [23, 24] and animal models [25, 26]. Studies in rodents demonstrated that Vitamin A dosage during adipogenesis advances adipocyte expansion in infant creatures, yet whenever enhanced with retinyl esters and oppressed high-fat eating regimen subsequent to weaning time for about 4 months offered an expanded heftiness list and raised leptin in WAT. A similar multiplication influence was accounted for by Lazar et al. in refined preadipocytes. Here, the job of presentation to retinoic corrosive is significant. On the off chance that an excess of nutrient An is regulated during the expansion phase of fat tissue, this could prompt heftiness sometime down the road, while if retinoic corrosive is enhanced to develop adipocytes, it will prompt weight reduction [14, 27, 28] (Tables 17.1 and 17.2; Fig. 17.2).

It is affirmed that all-trans-retinoic corrosive has portion subordinate impact on adipogenesis, the supraphysoiological focus (usually used in experiments 1 to 10 μM) smother the adipogenesis while at physiological fixation (1 pM to 10 nM go) it acts like intense adipogenic hormone [44]. Besides, foundational microorganism duty into adipocyte genealogy needs a period characterized treatment with

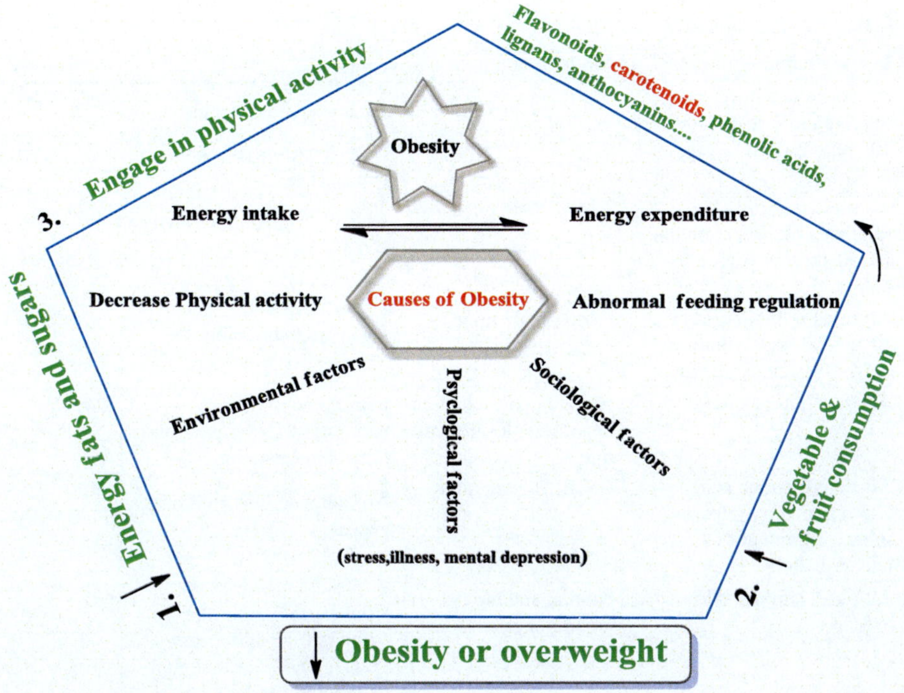

Fig. 17.1 Causes and reducing obesity

atRA [56, 57], and endogenous RA creation seems, to be needed for proficient adipogenesis of 3 T3-L1 fibroblast cell lines [39].

The link of insulin resistance and inflammation has been established with the raised adipokines profile such as resistin, leptin and RBP4 [58–63]. All-trans-retinoic acid suppresses adipocyte production of leptin [64], resistin [30] and retinol binding protein 4 (RBP4) [65].

β-carotene (10 mM) through antioxidant mechanism at cellular level suppress the ROS production in 3 T3-L1 adipocytes [35, 66].

BCO1 enzyme coverts β-carotene to retenoids which otherwise accumulated in BCO1 deficient mice WATs thus fat storage is altered [13]. Studies in rodents confirmed the anti-obesity potential of all trans retinoic acids [15, 67] (Fig. 17.3).

17.3 Cryptoxanthin

Oral intake of β-cryptoxanthin (0.8 mg/kg bw/day, 2 months) reduced body weight, visceral adipose tissue mass, adipocyte hypertrophy and serum lipid contents in mouse model [4]. Anti-obesity properties of mango (*Mangifera indica* L.) pulp, a rich source of β-cryptoxanthin are reported [68, 69]. The anti-adiposity results of

Table 17.1 Pro-vitamin A and Vitamin A effects on obesity

Pro or vitamin A	Model of study	Effect	References
Subcutaneous injection of all trans-retinoic acid (10–100 μg/g bw/day 4 days)	NMRI male mice	↑ adiposity and leptin mRNA expression, ↓ UCP1 expression	[29]
All trans-retinoic acid subcutaneous injection daily 10, 50 and 100 mg/kg bw for 4 days	NMRI male mice	↓ UCP1, UCP2 and PPARγ	[15]
All-trans-retinoic acid (10, 50, or 100 mg/kg body wt. for 4 days sub cutaneous)	NMRI male mice	↑ resistin expression	[30]
All-trans-retinoic acid (10 μM)	HepG2 cells, a human hepatoma-derived cell line	initiation of carnitine palmitoyl transferase 1 and fatty acid oxidation	[24]
All-trans-retinoic acid s (7.5 mg, 60-day release subacute implant of slow release pellet	C57BL/6Ntac mice	↓ of adipose PPARb/d target genes	[20]
All-trans-retinoic acid s (up to 1μM)	Human and mouse preadipocytes	↓ the expression of the adipogenesis inhibitors Pref-1, Sox9, and Kruppel-like factor 2 (KLF2)	[20]
All-trans-retinoic acid (1 μM, 10 μM)	Human adipocyte cell lines and primary human white adipocytes	Lack of UCP1 induction	[31, 32]
All-trans-retinoic acid (1 μM, 10 μM)	Mouse embryo fibroblasts and mature C3H10T1/2 adipocytes	↓ of UCP1	[31, 32]
All-trans-retinoic acid (2 μM)	Mature 3 T3-L1 adipocytes	Initiates oxidative phosphorylation and mitochondria biogenesis in adipocytes	[33]
All-trans-retinoic acid (1 μM, 10 μM)	Mature 3 T3-L1 adipocytes	↓ acquisition of BAT features in WAT, reduced intracellular lipid content	[21]
β-carotene (35 mg/kg bw/day, 14 weeks)	Wild-type mice	Suppression of PPARg in WAT	[13, 34]
β-carotene (2 μM)	Mature 3 T3-L1 adipocytes	Decreased lipid content and the expression of PPARg	[12]
β-carotene (20 μM)	3 T3-L1 adipocytes	Induced adiponectin expression	[35]

(continued)

Table 17.1 (continued)

Pro or vitamin A	Model of study	Effect	References
β-carotene (3.2 mg/kg bw/day for 6 months)	Ferret	↓ in adiposity and cell lipid accumulation ↑ UCP1 content	[36, 37]
β-Carotene (50 μ-c	Murine 3 T3-L1 preadipocytes	Inhibited the adipose conversion	[38]
β-Carotene metabolites (5 mM–50 mM	3 T3-L1 cells	↑ adipogenesis	[38]
β-carotene, a-carotene and lutein (up to 10 μM)	Mice brown adipocytes	↓ UCP1 expression	[17]
Ratinaldehyde (10–100 nM)	Murine 3 T3-L1 preadipocytes, *Aldh1a1*-deficient adipocytes	↑ *PPARγ* controlled by zinc figer protein 423 (ZEP423) dependent mechanism,	[39]
Ratinaldehyde (10–100 nM)	*Aldh1a1*$^{-/-}$ mice and C57/BL6	70% reduction in expression of ZEP423, *PPARγ* and Fabp4	[39]
Retinoic acid (>1 μM)	3 T3-F442A cells	Inhibited the adipose conversion	[40]
Retinoic acid (0.16 mg/day for 3 weeks)	Mice (C57BL/6Ntac mice)	↓ of PPARβ/δ and RAR target genes	[41]
Retinoic acid (0·25 and 25 mg/kg per d)	Ferret	Minor reduction in adiposity, reduced cell lipid ↓ growth and UCP1 quantity	[37]
Retinoic acid (1 μM)	3 T3-L2 fibroblasts	Inhibited the adipose conversion	[42]
Retinoic acid (10 μM)	3 T3-L1 preadipocytes	Inhibited the adipose conversion	[43]
Retinoic acid (1pM-10 nM)	Rat preadipocytes	Inhibited adipose differentiation	[44]
Retinaldehyde (all-trans-Rald, vitamin A or all-trans-retinoic acid, all 500 nM)	C57/BL6 mice, ob/ob mice	Repress obesity	[45]
Retinaldehyde (10 μM)	Murine MEFs-derived adipocytes	↓ I of UCP1	[32]
Retinaldehyde (1μM)	3 T3-L1 cells	Suppresses adipogenesis	[45]
Retinaldehyde (1μM)	White adipocytes	↓ of UCP1 via binding and activation of RAR	[46]
Retinoic acid (100 mg/kg per day, subcutaneously)	Rodent (diet based obesity model)	↑ PPARγ2 expression in WAT	[47, 48]

(continued)

Table 17.1 (continued)

Pro or vitamin A	Model of study	Effect	References
Vitamin A adjusted according to rat milk	Wistar rats	Reduction of expression of PPARγ and lipoprotein lipase	[27]
Vitamin A enriched diet (8 mg and 320 mg retinyl palmitate/kg diet, respectively) for 4.2 months	NMRI male mice	↑ resisting expression	[30]
Vitamin A for 8 weeks (27.3 IU/g)	Male Wistar rats	↓ of PPARγ and RXRα expressions	[49]
Vitamin A supplemented diet (2200 IU/Kg of DM)	Angus-cross steers	↓ intramuscular fat	[50]
Vitamin A-rich food (129 mg/kg diet)	WNIN/Ob lean and obese rats	↑ 11β-HSD1 activity, contribute to visceral fat loss	[51]
Vitamin A supplemented diet (129 mg retinol/kg diet) for 2 months	Rats	↓ UCP1 gene expression and reducing serum leptin profile	[52]

Abbreviations: (increases, upregulate, promote), (Downregulate, decrease, suppress) *UCP* uncoupling protein, *WAT*white adipose tissue

Table 17.2 Mechanism of antiobeisty effect (carotenoids)

Carotenoids	Mechanism	References
All trans retinoic acid	Upstream C/EBPβ, induction of specific proteins e.g. Pref-1, Sox9 and KLF2	[20]
All trans retinoic acid	Retinoylation of CRM1, which revokes the transfer of MEK1	[53]
Retinaldehyde and b-apo-14-carotenal	Suppress PPARg- and RXR-mediated responses by RAR- independent tools and following their direct binding to PPARg and RXR	[45, 54]
All-tans-retinoic acid	Repression of PPARγ, interference of liganded RAR via early transcription factor C/EBPβ	[14, 55]

β-cryptoxanthin in rodents are in line with studies in differentiating 3 T3-L1 adipocytes displaying decline of lipid growth following contact with b-cryptoxanthin (1–10 μM) [70, 71]; although there are inconsistent results [68]. Cryptoxanthin influence on adipogenesis involve RAR stimulation followed by PPARγ down-regulation [70]. Cryptoxanthin can effectively bind RARs (but not PPARγ), indicating that it acts *per se* as a RAR agonist to down-regulate PPARγ in adipocytes [70].

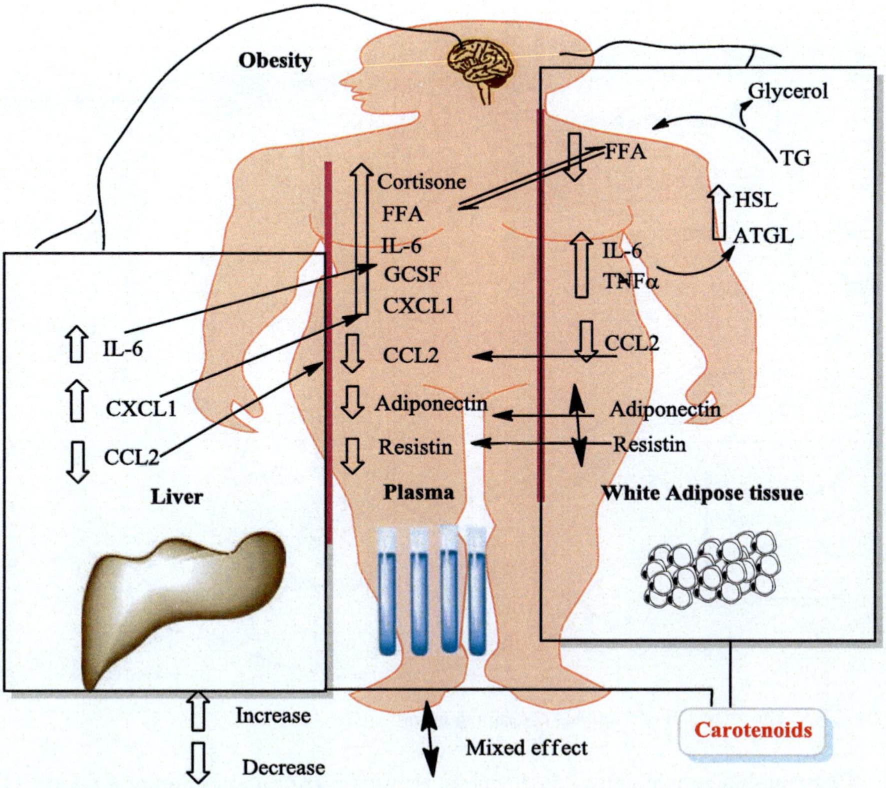

Fig. 17.2 Carotenoids, obesity and mediaters

17.4 Astaxanthin

Astaxanthin (6–30 mg/kg bw/day) prohibited visceral fat addition, and insulin confrontation prompted by fat rich food in mice and rats [6, 69], and decresed oxidative stress markers in adipose tissue and skeletal muscle of the fat rich food-fed mice [69]. The anti-adiposity effect was due to increased systemic fatty acid consumption, as shown by decreased respiratory quotient [6]. Astaxanthin prevents rosiglitazone (a PPAR" ligand)-induced adipogenesis of 3 T3-L1 cells by antagonizing PPAR" transcriptional activity [72] read Table 17.3 for studies.

17.5 Anti-obesity Effect of Fucoxanthin

Many studies indicate the anti-obesity effect of carotenoids [3, 94] especially fucoxanthin in numerous animal model studies, read Table 17.3 for details [95–101].

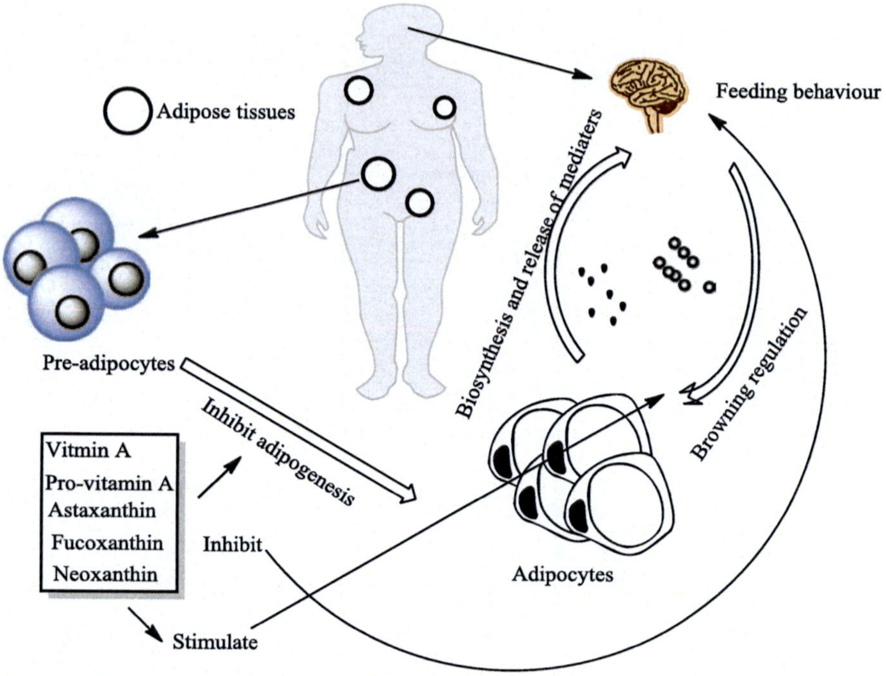

Fig. 17.3 The effects of carotenoids on adipogenesis

Fucoxanthin supplements cause weight reduction and suppression of fat deposition which is significant for the safe use of fucoxanthin therapy in humans.

17.6 Conclusion

Obesity is a global problem now. All antioxidants bioactive molecules are being explored as natural remdey due to underlying oxidative stress induced mechanism. Carotenoids modulate positively different enzymes, transcription factors and cytokines affiliated with obesity validating their antiobesity effects. Although metabolic fate of carotenoids and its metabolites vary in animal and human studies, it will be too eraly to suggest them as antiobesity remedy. Fucoxanthin and astxanthin show promising biological effects on obesity and can be explored as potential candidates for further research in this domain. The studies especially performed in last 2 decades indicate direct relationship between carotenoids consumption and decreased obesity and related disorders like insulin resistance, metabolic inflamamtin and liver steatosis. Although recent investigation tools havedeciphered molecular mechanism in a more better way, still many underlying mechanism are unclear. To complement

Table 17.3 The antiobesity effects of fucoxanthin or its preparations

Source	Study model	Action/Mechanism	References
Undaria pinnatifida (10% f FC)	Rats and mice	↓ UCP-1 in WAT	[73]
Medium-chain triacylglycerols (MCT) and FC 0.1%	KK-Ay obese mouse	↓ UCP-1 in WAT, MCT ↓ in absorption rate of FC	[74]
FC and fish oil	Diabetic/obese KK-A^y mice	↑ Leptin and tumor necrosis factor (TNFα) mRNA expression in WAT	[75]
FC	1. KK-A^y mice 2. Lean C57BL/6 J mice	Regulates mRNA expression of inflammatory adipocytokines in 1 but no effect on 2	[76]
0.2% FC	1. Hyperleptinemic KK-A^y mice 2. Leptin-deficient *ob/ob* mice	↑ SCD1 expression via leptin signaling in 1 but not in 2	[77]
FC and fucoxanthinol	3 T3-L1 cells	Prevent the adipocyte maturation of 3 T3-L1 cells via ↑ PPARAγ	[78]
Neoxanthin	3 T3-L1 cells	Suppression of adipocyte maturation via CCAAT/C/EBPffer PPART/C/EB	[68]
FC enriched seaweed	Diet induced obese murine model	↑ mRNA expression of Adrb3 in WAT and GLUT4 mRNA in skeletal muscle	[79]
Amarouciaxanthin A	3 T3-L1 cells	↑ adipocyte differentiation via decrease expression level of CCAAT/C/EBPe ex PPART/C/EB	[80]
0.083 and 0.167 mg/kg/bw FC intake in HFD	Male Sprague Dawley Rat	Liver & plasma triglyceride contents were ↑	[81]
0.2% FC diet	Obese KK-ay mice	↓ HDL and non-HDL cholesterol contents & ↑ liver cholesterol uptake	[67]
0.1 or 0.2% FC supplemented diet	KK-Ay mice and B6. V-Lepob/J (ob/ob) mice	Inhibited body weight, visceral WAT mass, and dropped serum leptin quantity	[82]
Petalonia binghamiae extract (FC enriched) 150 mg/kg/day	Diet induced obesity mice model	↓ β-oxidation and reducing lipogenesis	[83]
0.2% FC	C57BL/6 J and KK-Ay mice	↑ body weight increase and WAT weight	[76]

(continued)

Table 17.3 (continued)

Source	Study model	Action/Mechanism	References
Undaria pinnatifida (2.4 mg/kg/day)	Obese female volunteers	↓ energy expenditure in the body leading to significant weight decrease after 4 months	[84]
2% seaweed lipids (FC 16–21 mg/g)	Female KK-Ay mice	Substantial reduction in liver lipid hydroperoxide levels and abdominal WAT weight	[80]
0.05% and 0.2% FC in diet, w/w	C57BL/6 N mice	Recovers plasma and hepatic lipid metabolism and blood glucose level	[85]
0.05 or 0.2% FC supplemented diet	C57BL/6 J mice	Controlled plasma and hepatic lipid metabolism by increasing fecal lipid secretion	[86]
P. binghamiae extract (oral 150 mg/kg/day for 10 weeks)	30 male 4-week-old C57BL/6 mice	↑ body weight gain, adipose tissue weight, adipose cell size, serum triglyceride levels	[83]
P. binghamiae 10 mM treatment	3 T3-L1 adipocyte	↓ the phosphorylation of LKB1, AMPK, and ACC; and prevented the expression of PPAR, C/EBP_, and SREBP1c	[83]
0.2% FC powder in feed	Male Sprague-Dawley rats	↑ hepatic lipids cholesterols and triglycerides level	[87]
5% seaweeds powder supplemented in HFD	Sprague-Dawley rats	↑ body weight gain and plasma lipid peroxidation	[88]
0.1 or 0.2% FC rich food	KK-Ay mice and B6 V-Lepob/J (ob/ob) mice	Repressed body weight, visceral WAT mass, and dropped serum leptin quantity	[82]
0.05 or 0.2% FC rich food	C57BL/6 N mice	Controlled plasma and hepatic lipid metabolism; improved fecal lipid elimination	[86]
Haematococcus pluvialis 6, 12, and 18 mg/day of ASX	Human	↓ serum lipid profile and HDL-cholesterol	[89]
HF diet supplemented 0.003, 0.01 and 0.03% of ASX	Male C57BL/6 J mice	↑ triacylglycerol concentrations	[90]
0.03% ASX	Male apoE knockout (apoE)−/− mice	Improves cholesterol and lipid metabolism	[91]
6 mg/kg/day in olive oil for 2 months	Male Swiss albino mice	↑ lipid storage and oxidative stress and adipose tissue weight	[69]
5 mg and 20 mg soft ASX capsule	Human; overweight and obese young adults	Improved lipid metabolism	[92]

(continued)

Table 17.3 (continued)

Source	Study model	Action/Mechanism	References
6 mg/kg/day in olive oil for 2 months	Male mus musculus albino mice	Regulates weight gain, ↓ insulin sensitivity and reinstated liver lipid levels	[93]

Abbreviations: ↓ (increases, upregulate, promote), ↑ (Downregulate, decrease, suppress) *UCP* uncoupling protein, *WAT* white adipose tissue, *DCF1* Stearoyl-coenzyme A desaturase-1, *PPARAγ* peroxisome proliferator-activated receptor γ, *C/EBP* enhancer binding protein ct Adrb3, β3-adrenergic receptor, *GLUT4* glucose transporter 4, *Fc* Fucoxanthin, *ASX* Astaxanthin

observational and preclinical studies, pure carotenoids should be used in randomized clinical trials.

References

1. Organization WH (2017) 10 facts on obesity. WHO. https://www.who.int/features/factfiles/obesity/en/. Accessed 25 Mar 2020
2. Spalding KL, Arner E, Westermark PO, Bernard S, Buchholz BA, Bergmann O et al (2008) Dynamics of fat cell turnover in humans. Nature 453(7196):783–787. https://doi.org/10.1038/nature06902
3. Bonet ML, Canas JA, Ribot J, Palou A (2016) Carotenoids in adipose tissue biology and obesity. In: Carotenoids in nature. Springer, Cham, pp 377–414
4. Takayanagi K, Morimoto S-i, Shirakura Y, Mukai K, Sugiyama T, Tokuji Y et al (2011) Mechanism of visceral fat reduction in Tsumura Suzuki obese, diabetes (TSOD) mice orally administered β-cryptoxanthin from Satsuma mandarin oranges (Citrus unshiu Marc). J Agric Food Chem 59(23):12342–12351
5. Fenni S, Hammou H, Astier J, Bonnet L, Karkeni E, Couturier C et al (2017) Lycopene and tomato powder supplementation similarly inhibit high-fat diet induced obesity, inflammatory response, and associated metabolic disorders. Mol Nutr Food Res 61(9):1601083
6. Ikeuchi M, Koyama T, Takahashi J, Yazawa K (2007) Effects of astaxanthin in obese mice fed a high-fat diet. Biosci Biotechnol Biochem 71(4):893–899
7. Liu M, Liu H, Xie J, Xu Q, Pan C, Wang J et al (2017) Anti-obesity effects of zeaxanthin on 3T3-L1 preadipocyte and high fat induced obese mice. Food Funct 8(9):3327–3338
8. Seca AM, Pinto DC (2018) Overview on the antihypertensive and anti-obesity effects of secondary metabolites from seaweeds. Mar Drugs 16(7):237
9. Asemi Z, Alizadeh S-A, Ahmad K, Goli M, Esmaillzadeh A (2016) Effects of beta-carotene fortified synbiotic food on metabolic control of patients with type 2 diabetes mellitus: a double-blind randomized cross-over controlled clinical trial. Clin Nutr 35(4):819–825
10. Ojulari OV, Lee SG, Nam J-O (2020) Therapeutic effect of seaweed derived Xanthophyl carotenoid on obesity management; overview of the last decade. Int J Mol Sci 21(7):2502
11. Alvarez R, de Andrés J, Yubero P, Viñas O, Mampel T, Iglesias R et al (1995) A novel regulatory pathway of brown fat thermogenesis retinoic acid is a transcriptional activator of the mitochondrial uncoupling protein gene. J Biol Chem 270(10):5666–5673
12. Lobo GP, Amengual J, Li HNM, Golczak M, Bonet ML, Palczewski K et al (2010) β, β-carotene decreases peroxisome proliferator receptor γ activity and reduces lipid storage capacity of adipocytes in a β, β-carotene oxygenase 1-dependent manner. J Biol Chem 285(36):27891–27899

13. Amengual J, Gouranton E, van Helden YG, Hessel S, Ribot J, Kramer E et al (2011) Beta-carotene reduces body adiposity of mice via BCMO1. PloS One 6(6):e20644
14. Schwarz EJ, Reginato MJ, Shao D, Krakow SL, Lazar MA (1997) Retinoic acid blocks adipogenesis by inhibiting C/EBPbeta-mediated transcription. Mol Cell Biol 17(3):1552–1561
15. Mercader J, Ribot J, Murano I, Felipe F, Cinti S, Bonet ML et al (2006) Remodeling of white adipose tissue after retinoic acid administration in mice. Endocrinology 147(11):5325–5332
16. Wang B, Fu X, Liang X, Deavila JM, Wang Z, Zhao L et al (2017) Retinoic acid induces white adipose tissue browning by increasing adipose vascularity and inducing beige adipogenesis of PDGFRα+ adipose progenitors. Cell Discov 3(1):1–14
17. Serra F, Bonet M, Puigserver P, Oliver J, Palou A (1999) Stimulation of uncoupling protein 1 expression in brown adipocytes by naturally occurring carotenoids. Int J Obes 23(6):650–655
18. Relevy NZ, Bechor S, Harari A, Ben-Amotz A, Kamari Y, Harats D et al (2015) The inhibition of macrophage foam cell formation by 9-cis β-carotene is driven by BCMO1 activity. PLoS One 10(1):e0115272
19. Östh M, Öst A, Kjolhede P, Strålfors P (2014) The concentration of β-carotene in human adipocytes, but not the whole-body adipocyte stores, is reduced in obesity. PLoS One 9(1):e85610
20. Berry DC, DeSantis D, Soltanian H, Croniger CM, Noy N (2012) Retinoic acid upregulates preadipocyte genes to block adipogenesis and suppress diet-induced obesity. Diabetes 61(5):1112–1121
21. Mercader J, Madsen L, Felipe F, Palou A, Kristiansen K, Bonet L (2007) All-trans retinoic acid increases oxidative metabolism in mature adipocytes. Cell Physiol Biochem 20(6):1061–1072
22. Bonet M, Ribot J, Felipe F, Palou A (2003) Vitamin A and the regulation of fat reserves. Cell Mol Life Sci 60(7):1311–1321
23. Amengual J, García-Carrizo FJ, Arreguín A, Mušinović H, Granados N, Palou A et al (2018) Retinoic acid increases fatty acid oxidation and irisin expression in skeletal muscle cells and impacts irisin in vivo. Cell Physiol Biochem 46(1):187–202
24. Amengual J, Petrov P, Bonet ML, Ribot J, Palou A (2012) Induction of carnitine palmitoyl transferase 1 and fatty acid oxidation by retinoic acid in HepG2 cells. Int J Biochem Cell Biol 44(11):2019–2027
25. Amengual J, Ribot J, Bonet ML, Palou A (2010) Retinoic acid treatment enhances lipid oxidation and inhibits lipid biosynthesis capacities in the liver of mice. Cell Physiol Biochem 25(6):657–666
26. Amengual J, Ribot J, Bonet ML, Palou A (2008) Retinoic acid treatment increases lipid oxidation capacity in skeletal muscle of mice. Obesity 16(3):585–591
27. Granados N, Amengual J, Ribot J, Musinovic H, Ceresi E, Von Lintig J et al (2013) Vitamin A supplementation in early life affects later response to an obesogenic diet in rats. Int J Obes 37(9):1169–1176
28. Bonet ML, Ribot J, Palou A (2012) Lipid metabolism in mammalian tissues and its control by retinoic acid. Biochim Biophys Acta Mol Cell Biol Lipids 1821(1):177–189
29. Bonet M, Oliver J, Pico C, Felipe F, Ribot J, Cinti S et al (2000) Opposite effects of feeding a vitamin A-deficient diet and retinoic acid treatment on brown adipose tissue uncoupling protein 1 (UCP1), UCP2 and leptin expression. J Endocrinol 166(3):511–517
30. Felipe F, Bonet ML, Ribot J, Palou A (2004) Modulation of resistin expression by retinoic acid and vitamin A status. Diabetes 53(4):882–889
31. Murholm M, Isidor MS, Basse AL, Winther S, Sørensen C, Skovgaard-Petersen J et al (2013) Retinoic acid has different effects on UCP1 expression in mouse and human adipocytes. BMC Cell Biol 14(1):41
32. Mercader J, Palou A, Bonet ML (2010) Induction of uncoupling protein-1 in mouse embryonic fibroblast-derived adipocytes by retinoic acid. Obesity 18(4):655–662

33. Tourniaire F, Musinovic H, Gouranton E, Astier J, Marcotorchino J, Arreguin A et al (2015) All-trans retinoic acid induces oxidative phosphorylation and mitochondria biogenesis in adipocytes. J Lipid Res 56(6):1100–1109
34. van Helden YG, Godschalk RW, von Lintig J, Lietz G, Landrier JF, Luisa Bonet M et al (2011) Gene expression response of mouse lung, liver and white adipose tissue to β-carotene supplementation, knockout of Bcmo1 and sex. Mol Nutr Food Res 55(10):1466–1474
35. Kameji H, Mochizuki K, Miyoshi N, Goda T (2010) β-Carotene accumulation in 3T3-L1 adipocytes inhibits the elevation of reactive oxygen species and the suppression of genes related to insulin sensitivity induced by tumor necrosis factor-α. Nutrition 26(11–12):1151–1156
36. Murano I, Morroni M, Zingaretti MC, Oliver P, Sánchez J, Fuster A et al (2005) Morphology of ferret subcutaneous adipose tissue after 6-month daily supplementation with oral beta-carotene. Biochim Biophys Acta Mol Cell Biol Lipids 1740(2):305–312
37. Sánchez J, Fuster A, Oliver P, Palou A, Picó C (2009) Effects of β-carotene supplementation on adipose tissue thermogenic capacity in ferrets (Mustela putorius furo). Br J Nutr 102(11):1686–1694
38. Kawada T, Kamei Y, Fujita A, Hida Y, Takahashi N, Sugimoto E et al (2000) Carotenoids and retinoids as suppressors on adipocyte differentiation via nuclear receptors. Biofactors 13(1–4):103–109
39. Reichert B, Yasmeen R, Jeyakumar SM, Yang F, Thomou T, Alder H et al (2011) Concerted action of aldehyde dehydrogenases influences depot-specific fat formation. Mol Endocrinol 25(5):799–809
40. Kuri-Harcuch W (1982) Differentiation of 3T3-F442A cells into adipocytes is inhibited by retinoic acid. Differentiation 23(1–3):164–169
41. Berry DC, Noy N (2009) All-trans-retinoic acid represses obesity and insulin resistance by activating both peroxisome proliferation-activated receptor β/δ and retinoic acid receptor. Mol Cell Biol 29(12):3286–3296
42. Murray T, Russell TR (1980) Inhibition of adipose conversion in 3T3-L2 cells by retinoic acid. J Supramol Struct 14(2):255–266
43. Xue J-C, Schwarz EJ, Chawla A, Lazar MA (1996) Distinct stages in adipogenesis revealed by retinoid inhibition of differentiation after induction of PPARgamma. Mol Cell Biol 16(4):1567–1575
44. Safonova I, Darimont C, Amri E-Z, Grimaldi P, Ailhaud G, Reichert U et al (1994) Retinoids are positive effectors of adipose cell differentiation. Mol Cell Endocrinol 104(2):201–211
45. Ziouzenkova O, Orasanu G, Sharlach M, Akiyama TE, Berger JP, Viereck J et al (2007) Retinaldehyde represses adipogenesis and diet-induced obesity. Nat Med 13(6):695–702
46. Kiefer FW, Vernochet C, O'brien P, Spoerl S, Brown JD, Nallamshetty S et al (2012) Retinaldehyde dehydrogenase 1 regulates a thermogenic program in white adipose tissue. Nat Med 18(6):918
47. Ribot J, Felipe F, Bonet ML, Palou A (2001) Changes of adiposity in response to vitamin A status correlate with changes of PPARγ2 expression. Obes Res 9(8):500–509
48. Esteban-Pretel G, Marín MP, Cabezuelo F, Moreno V, Renau-Piqueras J, Timoneda J et al (2010) Vitamin A deficiency increases protein catabolism and induces urea cycle enzymes in rats. J Nutr 140(4):792–798
49. Bairras C, Menard L, Redonnet A, Ferrand C, Delage B, Noel-Suberville C et al (2005) Effect of vitamin A content in cafeteria diet on the expression of nuclear receptors in rat subcutaneous adipose tissue. J Physiol Biochem 61(2):353
50. Gorocica-Buenfil M, Fluharty F, Bohn T, Schwartz S, Loerch SC (2007) Effect of low vitamin A diets with high-moisture or dry corn on marbling and adipose tissue fatty acid composition of beef steers. J Anim Sci 85(12):3355–3366
51. Sakamuri VPS, Ananthathmakula P, Veettil GN, Ayyalasomayajula V (2011) Vitamin A decreases pre-receptor amplification of glucocorticoids in obesity: study on the effect of vitamin A on 11beta-hydroxysteroid dehydrogenase type 1 activity in liver and visceral fat of WNIN/Ob obese rats. Nutr J 10(1):70

52. Kumar MV, Sunvold GD, Scarpace PJ (1999) Dietary vitamin A supplementation in rats: suppression of leptin and induction of UCP1 mRNA. J Lipid Res 40(5):824–829
53. Dave S, Nanduri R, Dkhar HK, Bhagyaraj E, Rao A, Gupta P (2014) Nuclear MEK1 sequesters PPARγ and bisects MEK1/ERK signaling: a non-canonical pathway of retinoic acid inhibition of adipocyte differentiation. PloS One 9(6):e100862
54. Ziouzenkova O, Orasanu G, Sukhova G, Lau E, Berger JP, Tang G et al (2007) Asymmetric cleavage of β-carotene yields a transcriptional repressor of retinoid X receptor and peroxisome proliferator-activated receptor responses. Mol Endocrinol 21(1):77–88
55. Marchildon F, St-Louis C, Akter R, Roodman V, Wiper-Bergeron NL (2010) Transcription factor Smad3 is required for the inhibition of adipogenesis by retinoic acid. J Biol Chem 285(17):13274–13284
56. Dani C, Smith A, Dessolin S, Leroy P, Staccini L, Villageois P et al (1997) Differentiation of embryonic stem cells into adipocytes in vitro. J Cell Sci 110(11):1279–1285
57. Bost F, Caron L, Marchetti I, Dani C, Marchand-Brustel YL, Binétruy B (2002) Retinoic acid activation of the ERK pathway is required for embryonic stem cell commitment into the adipocyte lineage. Biochem J 361(3):621–627
58. Steppan CM, Bailey ST, Bhat S, Brown EJ, Banerjee RR, Wright CM et al (2001) The hormone resistin links obesity to diabetes. Nature 409(6818):307–312
59. Stofkova A (2010) Resistin and visfatin: regulators of insulin sensitivity, inflammation and immunity. Endocr Regul 44(1):25–36
60. Stofkova A (2009) Leptin and adiponectin: from energy and metabolic dysbalance to inflammation and autoimmunity. Endocr Regul 43(4):157–168
61. Huang K, Lin R, Kormas N, Lee L, Chen C, Gill T et al (2004) Plasma leptin is associated with insulin resistance independent of age, body mass index, fat mass, lipids, and pubertal development in nondiabetic adolescents. Int J Obes 28(4):470–475
62. Yang Q, Graham TE, Mody N, Preitner F, Peroni OD, Zabolotny JM et al (2005) Serum retinol binding protein 4 contributes to insulin resistance in obesity and type 2 diabetes. Nature 436(7049):356–362
63. Esteve E, Ricart W, Fernández-Real JM (2009) Adipocytokines and insulin resistance: the possible role of lipocalin-2, retinol binding protein-4, and adiponectin. Diabetes Care 32(suppl 2):S362–S3S7
64. Hollung K, Rise CP, Drevon CA, Reseland JE (2004) Tissue-specific regulation of leptin expression and secretion by all-trans retinoic acid. J Cell Biochem 92(2):307–315
65. Mercader J, Granados N, Bonet L, Palou A (2008) All-trans retinoic acid decreases murine adipose retinol binding protein 4 production. Cell Physiol Biochem 22(1–4):363–372
66. Sliwa A, Góralska J, Czech U, Gruca A, Polus A, Zapała B et al (2012) Modulation of the human preadipocyte mitochondrial activity by beta-carotene. Acta Biochim Pol 59(1):39–41
67. Beppu F, Hosokawa M, Niwano Y, Miyashita K (2012) Effects of dietary fucoxanthin on cholesterol metabolism in diabetic/obese KK-A y mice. Lipids Health Dis 11(1):112
68. Okada T, Nakai M, Maeda H, Hosokawa M, Sashima T, Miyashita K (2008) Suppressive effect of neoxanthin on the differentiation of 3T3-L1 adipose cells. J Oleo Sci 57(6):345–351
69. Arunkumar E, Bhuvaneswari S, Anuradha CV (2012) An intervention study in obese mice with astaxanthin, a marine carotenoid–effects on insulin signaling and pro-inflammatory cytokines. Food Funct 3(2):120–126
70. Shirakura Y, Takayanagi K, Mukai K, Tanabe H, Inoue M (2011) β-Cryptoxanthin suppresses the adipogenesis of 3T3-L1 cells via RAR activation. J Nutr Sci Vitaminol 57(6):426–431
71. Goto T, Kim YI, Takahashi N, Kawada T (2013) Natural compounds regulate energy metabolism by the modulating the activity of lipid-sensing nuclear receptors. Mol Nutr Food Res 57(1):20–33
72. Inoue M, Tanabe H, Matsumoto A, Takagi M, Umegaki K, Amagaya S et al (2012) Astaxanthin functions differently as a selective peroxisome proliferator-activated receptor γ modulator in adipocytes and macrophages. Biochem Pharmacol 84(5):692–700

73. Maeda H, Hosokawa M, Sashima T, Funayama K, Miyashita K (2005) Fucoxanthin from edible seaweed, Undaria pinnatifida, shows antiobesity effect through UCP1 expression in white adipose tissues. Biochem Biophys Res Commun 332(2):392–397
74. Maeda H, Hosokawa M, Sashima T, Funayama K, Miyashita K (2007) Effect of medium-chain triacylglycerols on anti-obesity effect of fucoxanthin. J Oleo Sci 56(12):615–621
75. Maeda H, Hosokawa M, Sashima T, Miyashita K (2007) Dietary combination of fucoxanthin and fish oil attenuates the weight gain of white adipose tissue and decreases blood glucose in obese/diabetic KK-Ay mice. J Agric Food Chem 55(19):7701–7706
76. Hosokawa M, Miyashita T, Nishikawa S, Emi S, Tsukui T, Beppu F et al (2010) Fucoxanthin regulates adipocytokine mRNA expression in white adipose tissue of diabetic/obese KK-Ay mice. Arch Biochem Biophys 504(1):17–25
77. Beppu F, Hosokawa M, Yim M-J, Shinoda T, Miyashita K (2013) Down-regulation of hepatic stearoyl-CoA desaturase-1 expression by fucoxanthin via leptin signaling in diabetic/obese KK-Ay mice. Lipids 48(5):449–455. https://doi.org/10.1007/s11745-013-3784-4
78. Maeda H, Hosokawa M, Sashima T, Takahashi N, Kawada T, Miyashita K (2006) Fucoxanthin and its metabolite, fucoxanthinol, suppress adipocyte differentiation in 3T3-L1 cells. Int J Mol Med 18(1):147–152
79. Maeda H, Hosokawa M, Sashima T, Murakami-Funayama K, Miyashita K (2009) Anti-obesity and anti-diabetic effects of fucoxanthin on diet-induced obesity conditions in a murine model. Mol Med Rep 2(6):897–902
80. Airanthi MW-A, Sasaki N, Iwasaki S, Baba N, Abe M, Hosokawa M et al (2011) Effect of brown seaweed lipids on fatty acid composition and lipid hydroperoxide levels of mouse liver. J Agric Food Chem 59(8):4156–4163
81. Hu X, Li Y, Li C, Fu Y, Cai F, Chen Q et al (2012) Combination of fucoxanthin and conjugated linoleic acid attenuates body weight gain and improves lipid metabolism in high-fat diet-induced obese rats. Arch Biochem Biophys 519(1):59–65
82. Beppu F, Hosokawa M, Yim M-J, Shinoda T, Miyashita K (2013) Down-regulation of hepatic stearoyl-CoA desaturase-1 expression by fucoxanthin via leptin signaling in diabetic/obese KK-A y mice. Lipids 48(5):449–455
83. Kang S-I, Shin H-S, Kim H-M, Yoon S-A, Kang S-W, Kim J-H et al (2012) Petalonia binghamiae extract and its constituent fucoxanthin ameliorate high-fat diet-induced obesity by activating AMP-activated protein kinase. J Agric Food Chem 60(13):3389–3395
84. Abidov M, Ramazanov Z, Seifulla R, Grachev S (2010) The effects of Xanthigen™ in the weight management of obese premenopausal women with non-alcoholic fatty liver disease and normal liver fat. Diabetes Obes Metab 12(1):72–81
85. Woo M-N, Jeon S-M, Kim H-J, Lee M-K, Shin S-K, Shin YC et al (2010) Fucoxanthin supplementation improves plasma and hepatic lipid metabolism and blood glucose concentration in high-fat fed C57BL/6N mice. Chem Biol Interact 186(3):316–322
86. Jeon SM, Kim HJ, Woo MN, Lee MK, Shin YC, Park YB et al (2010) Fucoxanthin-rich seaweed extract suppresses body weight gain and improves lipid metabolism in high-fat-fed C57BL/6J mice. Biotechnol J 5(9):961–969
87. Ha AW, Kim WK (2013) The effect of fucoxanthin rich power on the lipid metabolism in rats with a high fat diet. Nutr Res Pract 7(4):287–293
88. Matanjun P, Mohamed S, Muhammad K, Mustapha NM (2010) Comparison of cardiovascular protective effects of tropical seaweeds, Kappaphycus alvarezii, Caulerpa lentillifera, and Sargassum polycystum, on high-cholesterol/high-fat diet in rats. J Med Food 13(4):792–800
89. Yoshida H, Yanai H, Ito K, Tomono Y, Koikeda T, Tsukahara H et al (2010) Administration of natural astaxanthin increases serum HDL-cholesterol and adiponectin in subjects with mild hyperlipidemia. Atherosclerosis 209(2):520–523
90. Yang Y, Pham TX, Wegner CJ, Kim B, Ku CS, Park Y-K et al (2014) Astaxanthin lowers plasma TAG concentrations and increases hepatic antioxidant gene expression in diet-induced obesity mice. Br J Nutr 112(11):1797–1804

91. Yang Y, Seo JM, Nguyen A, Pham TX, Park HJ, Park Y et al (2011) Astaxanthin-rich extract from the green alga Haematococcus pluvialis lowers plasma lipid concentrations and enhances antioxidant defense in apolipoprotein E knockout mice. J Nutr 141(9):1611–1617
92. Choi HD, Kim JH, Chang MJ, Kyu-Youn Y, Shin WG (2011) Effects of astaxanthin on oxidative stress in overweight and obese adults. Phytother Res 25(12):1813–1818
93. Bhuvaneswari S, Arunkumar E, Viswanathan P, Anuradha CV (2010) Astaxanthin restricts weight gain, promotes insulin sensitivity and curtails fatty liver disease in mice fed a obesity-promoting diet. Process Biochem 45(8):1406–1414
94. Chuyen HV, Eun J-B (2017) Marine carotenoids: bioactivities and potential benefits to human health. Crit Rev Food Sci Nutr 57(12):2600–2610
95. Miyashita K, Hosokawa M (2017) Fucoxanthin in the management of obesity and its related disorders. J Funct Foods 36:195–202
96. Peng J, Yuan J-P, Wu C-F, Wang J-H (2011) Fucoxanthin, a marine carotenoid present in brown seaweeds and diatoms: metabolism and bioactivities relevant to human health. Mar Drugs 9(10):1806–1828
97. Riccioni G, D'Orazio N, Franceschelli S, Speranza L (2011) Marine carotenoids and cardiovascular risk markers. Mar Drugs 9(7):1166–1175
98. Gammone MA, D'Orazio N (2015) Anti-obesity activity of the marine carotenoid fucoxanthin. Mar Drugs 13(4):2196–2214
99. Maeda H (2015) Nutraceutical effects of fucoxanthin for obesity and diabetes therapy: a review. J Oleo Sci 64(2):ess14226
100. Muradian K, Vaiserman A, Min K-J, Fraifeld V (2015) Fucoxanthin and lipid metabolism: a minireview. Nutr Metab Cardiovasc Dis 25(10):891–897
101. Hu X, Tao N, Wang X, Xiao J, Wang M (2016) Marine-derived bioactive compounds with anti-obesity effect: a review. J Funct Foods 21:372–387

Chapter 18
Carotenoids in Liver and Lung Diseases

Naheed Bano and Imran Imran

18.1 Introduction

There are different species of plants and microscopic organisms (alga, fungi and bacteria) which can synthesize carotenoids. Nearly seven hundred verities of carotenoids have been reported but only fifty are used by humans in diet. There are two groups of carotenoids: one is pro-vitamin A and other is non-provitamin A. Both these types of carotenes work against oxidative damaging effects. Researchers and health authorities have interest in role of pro-vitamin A against chronic liver diseases [1, 2]. Carotenoids are edible and have important role in vision, morphogenesis, against cancer, aging and atherosclerosis by suppression of oxidative stress. Oxidative stress and reactive oxygen species by damaging biomolecules cause chorionic liver diseases (CLDs) [3]. Carotenoids act as dietary prophylactic antioxidant molecules by inactivating reactive oxygen species [2, 4]. As fruits and vegetables have phytochemicals and carotenoids, a regular intake of fruits and vegetables reduces the chances of CLDs. There is increase in human death rate due to CLDs, including alcoholic liver disease, hepatitis B, hepatitis C, cirrhosis, fibrosis, non-alcoholic fatty liver disease and hepatocellular carcinoma [5].

Hepatocellular carcinoma is famous, as it is third most mortality causing cancer. Nearly 700,000 people globally died each year due to liver cancer [6]. Major reason for liver cirrhosis and fibrosis is oxidative stress. In liver cell, the major source of reactive oxygen species is mitochondria [7]. As hepatocytes have several mitochondria, reactive oxygen species are also produced in large number [8]. This imbalance leads to pathophysiological modifications in liver cells e.g. hepatic stellate cells

N. Bano (✉)
Faculty of Veterinary and Animal Sciences (FVAS), MNS University of Agriculture, Multan, Pakistan

I. Imran
Department of Pharmacology, Faculty of Pharmacy, Bahauddin Zakariya University, Multan, Pakistan

© Springer Nature Switzerland AG 2021

M. Zia-Ul-Haq et al. (eds.), *Carotenoids: Structure and Function in the Human Body*, https://doi.org/10.1007/978-3-030-46459-2_18

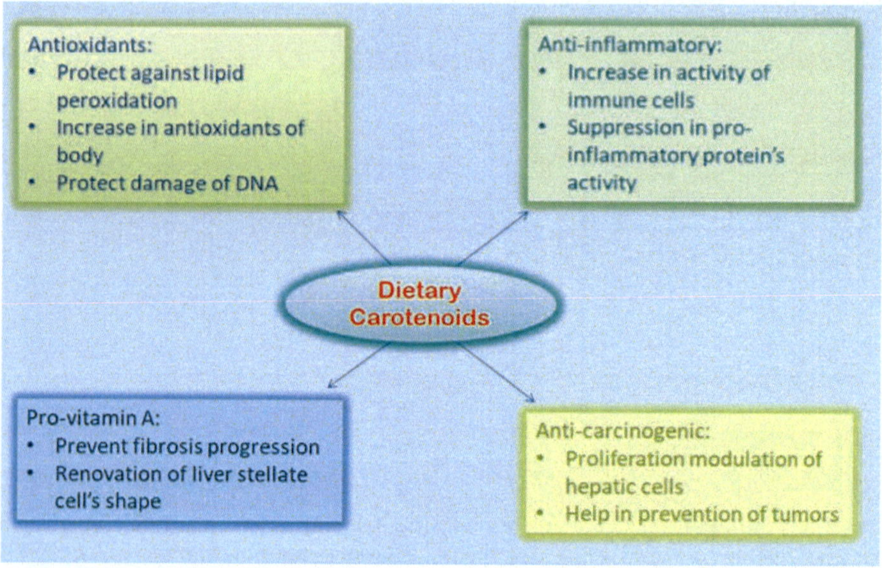

Fig. 18.1 Role of dietary carotenoids against chronic liver diseases

activation, initiation of collagen formation and proliferative progressions [6, 9]. In a healthy human, hepatic somatic cells stores up to 90% of vitamin A in the form of esters. Any chronic injuries to liver leads in activation of these cells and there is loss of storage capacity of vitamin A, proliferation and pro inflammation, controlling fibro genesis [10]. Fibrosis must be controlled as if remain untreated [11, 12]; its end result will be the full damage to liver leading to liver cancer (Fig. 18.1).

Alteration in lipid, carbohydrate metabolism, alcohol, viruses and xenobiotic are responsible for the induction of reactive oxygen species which leads to hepatic damage [13]. In this prospective, carotenoids are antecedent of vitamin A and essential part of antioxidant resistance system against CLDs [11, 14]. Currently, substitute medicines and therapies having least cost and side effects are getting attention globally. This chapter is focused on the carotenoids against liver diseases.

18.2 Carotenoids Against Liver Disease

18.2.1 Pathogenesis of NAFLD

Non-alcoholic fatty liver disease (NAFLD) consists of non-alcoholic steato-hepatitis (NASH), fatty liver, cirrhosis and fibrosis (Fig. 18.2). NAFLD may lead to hepatic cancer, second largest death causing cancer in the world [15, 16]. There is two-hit hypothesis for the pathogenesis of NAFLD. According to this hypothesis, firstly fat is accumulated leading to second hit (rout is oxidative stress to Inflammation to

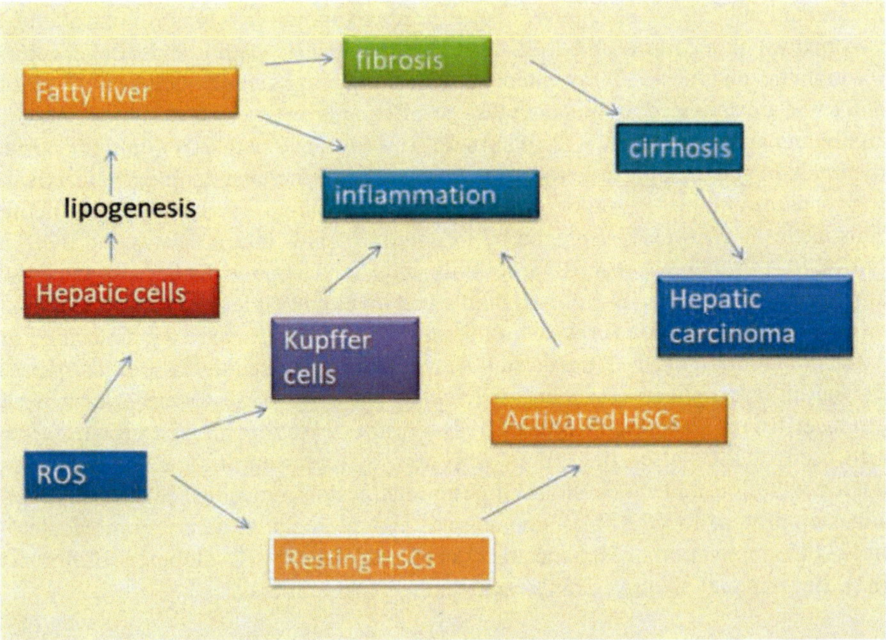

Fig. 18.2 pathogenesis of NAFLD

lipotoxicity to apoptosis and activation of hepatic injury). The progress of disease may be prevented by using inhibitory factors like xanthophyll and carotene. As carotene have antioxidant chattels and anti-inflammatory properties. For that reason dietary consumption of carotenoids may inhibit the growth of NAFLD [17, 18].

Hepatic toxicity can be induced by the accumulation of lipids and fatty acid. Lipotoxicity initiates the production of heat-shock proteins, trigger macrophages and uric acid production leading to NAFLD [19]. Toxic lipid intermediates like lysophosphatidyl choline, ceramides, diacyl glycerol and acylcarnitines can initiate insulin resistance, oxidative stress and endoplasmic reticulum stress [20]. These stresses can lead to apoptosis and liver cell damage and dysfunction. Any imbalance between elimination and production of reactive oxygen species (ROS) lead to hepatic oxidative stress [19, 21]. Mitochondria, endoplasmic reticulum and peroxisomes are primary sites of ROS production. Antioxidants like catalase, superoxide dismutase, carotene, ascorbate and vitamin E regulate ROS level inside cell [22]. Any damage in the electron transport chain affects the production of ROS e.g. antimycin can inhibit complex III which lead to hydrogen peroxide production in mitochondria [23].

NAFLD increases oxidative stress, lipid peroxidation products and 8-isoprotane and decreases antioxidants [24]. Lipid peroxides affects the cell membrane activity and damage proteins leading to apoptosis and nucleic acids. Reactive oxygen species stimulate the production of chemokines and cytokines which lead to hepatic inflammation [25]. NAFLD advancement leads to apoptosis which is controlled by two pathways: extrinsic and intrinsic [26]. In extrinsic pathway death receptors (proteolytic enzymes, TNF-related apoptosis-inducing ligand receptors, TNFR1 and FAS) are activated. In this path way Kupffer cells are important as they produce Fas ligand and TNFR1 [27]. In other path way (intrinsic path way), mitochondrial dysfunction and endoplasmic reticulum stress initiate the liver apoptosis by activating caspases and pro-apoptotic protein (B-cell lymphoma-2-associated X protein). This intrinsic pathway is triggered by FFAs, toxic lipids intermediates and ROS in NASH [28]. In hepatocytes, ROS not only activate pro apoptosis but also decrease anti-apoptotic proteins (B-cell lymphoma and myeloid cell leukemia sequence 1).

Advancement in NASH results in liver fibrosis, which is due to the deposition of extracellular matrix [29]. Hepatic stellate cells (HSCs) in healthy liver remain inactive but in NASH liver, HSCs become active. These active HSCs are primary scar producers in the liver and proliferative as a result of platelet-derived growth factor [30]. Active HSCs also affect immune system by recruiting immune cells by producing adhesion molecules. Some important adhesion molecules are vascular cell adhesion protein 1 (VCAM-1), chemokines and intercellular adhesion molecule 1. In NAFLD, activation of Hepatic stellate cells escalates by development of fibrosis [31]. But removal of active HSCs restricts the hepatic fibrosis.

18.2.2 Important Carotenoids

There are many important carotenoids which humans get from different dietary source like mangoes, sweet potato, papaya, carrot, pumpkin, spinach and cantaloupe. Some important carotenoids are β-Carotene, lutein, lycopene, β-cryptoxanthin, zeaxanthin, carotene and fucoxanthin [3, 4, 32].

18.2.2.1 β-Carotene

One of the important carotenoids essential precursor of vitamin A. it is essentially present in human diet is β-Carotene. It acts as antioxidants and decreases the level of reactive oxygen species by quenching per-oxy radicals [33, 34]. β-Carotene have protective liver potential, as it increases vitamin C and glutathione which acts as scavenger of free radicals and also acts against toxicity of aflatoxin as antioxidants. In hepatocytes, oxidative stress and inflammation is augmented by HBV and HCV [22]. Moreover, hepatocellular carcinoma HCC and cirrhosis are result of hepatic fibrosis. Normal resting hepatic stellate cells are store house of vitamin A, but in chronic liver disease condition, they become active [35, 36]. Furthermore, there is loss of retinol and production of extracellular matrix which leads to fibrosis [22, 37]. The activity of β-Carotene as provitamin A and inhibition of reactive oxygen species, depreciate the advancement and growth of HBV, HCV and HCC [38, 39].

In patients, having CHC showed high oxidative stress and decreased level of carotene, lycopene, retinol, α-tocopherol and β-cryptoxanthin. Beside this, β-Carotene prohibited liver damage by ethanol through delaying liver apoptosis by controlling Bcl-xL, caspase-3 and caspase-9 expressions and preventing lipid peroxidation and secretions of TNF-α [40, 41].

18.2.2.2 Lycopene

Lycopene is considered as antioxidant, de-toxicant and anti-cardiovascular diseases capacity [24, 42]. Carcinogenic activity of Aflatoxin-B1 (AFB1) can be decreased by lycopene as it ceases AFB1-N7-guanine formation and DNA destruction. Moreover it sacks liver injuries by blocking AFB enzymes like 3A4, 2A6, and 1A2 and improves antioxidant and de-toxicities by activating Nrf2 [20, 43]. It was found to be effective for alcoholic liver damage, cirrhosis, liver cancer, non- alcoholic steto-hepatitis and hepatitis [15, 37, 44]. Due to regulating properties of lipid metabolism, lycopene is considered useful NAFLD treating compound.

Naturally occurring compound such as lycopene, possess anti-carcinogenic and other beneficial properties, are referred to as chemo-preventers, high-fat diet-induced HCC, suppressors of oncogenic signs like B-cantoning protein and methionine mRNA. So lycopene not only prevents liver cancer but also decreases risk of cancer in NAFLD patients [45, 46].

18.2.2.3 Lutein

Lutein is famous for its antiviral activity and commercially it is prepared from *Tagetes erecta* L. which contains nearly 1–2% lutein. It blocks the transcription of virus (Hepatitis-B) and act as strong protective agent against ethanol-induced hepatic damage [47]. This also affects the immune system by decreasing hydroxyl proline and increasing the antioxidant enzymes i.e., catalase, peroxidase, glutathione, superoxide dismutase and glutathione peroxidase. As well as, it regulates the inflammation and oxidative stress by regulating antioxidant enzymes, cytokines and inflammatory proteins [48, 49]. Sometimes high fat diets can induce insulin resistance and lipid accumulation in liver, in that condition also, lutein supplementation can control this situation by activating peroxisome proliferator activated receptor (PPAR) through expressing sirtuin-1.

18.2.3 β-Cryptoxanthin

It can repress the gene expression interconnected with inflammation, controls non-alcoholic steto-hepatitis by inhibiting the expression of TNF and LPS genes. It increases the macrophage and immune cells as well as decreases the reactive oxygen species, which results into control of fibrosis and inflammation [31, 50]. This carotenoid prevents and converses not only the insulin resistance in NASH but also steato-hepatitus by stimulating the kupffer cells or macrophages. Carotenoid level in plasma is very important as the level of different carotenoids like β-cryptoxanthin, lutein, β-carotene and lycopene decreases in NASH cases [10, 51]. For such patients of NASH/NAFLD, antioxidant supplementation with β-cryptoxanthin hinders its advancement [52–54].

18.2.3.1 Other Carotenoids

Some other carotenoids like α-carotene, zeaxanthin and fucoxanthin are also important for the treatment of chronic liver diseases [1, 55, 56]. These can significantly decrease hepatomas, inhibit liver cancer and obesity related deformities like fatty liver, weight gain and liver functions [57, 58]. These carotenoids are also important in cases of nonalcoholic steatohepatitis, alcoholic liver disease and NASH as are reported to decrease oxidative stresses and fibrosis [53]. The way of action for carotenoids is lowering the expression of cytochrome (CYP2E1) and reduction in kappa (NF-κB) [59].

18.2.4 Liver Therapy by Carotenoids

There are emerging trends in using the carotenoids for the treatment of liver diseases. Liver therapy by carotenoids is found to reduce the oxidative stress, lipid peroxidation, inflammation and LPS levels [60, 61]. But carotenoids had not extensively used as treatment for NASH. The mechanisms of action are still unclear in NASH/NAFLD treatment but it is suggested that they may work through macrophage polarity regulation [62, 63]. Mechanism for prevention of chronic liver diseases by carotenoids can be understood easily (Fig. 18.3). Carotenoids have inverse effect on lipid peroxidation and oxidative DNA damage [54]. Recommended daily dose for different carotenoids is different (Table 18.1).

Serum studies showed that carotenoids especially cryptoxanthin have inverse relation with insulin resistance risk and increased levels of glutamyltransferase in non-diabetic as well as alcoholic persons [69, 70]. These also regulate the immunity by inducing changes in macrophage functioning, activating stellate cells and cellular redox status. Genetic expression study shows that carotenes can block the expression of genes responsible to convoy NASH. Among carotenes, cryptoxanthin may regulate the inflammation, infiltration, leukocytes, macrophage activation, cell death and free radicals by regulating the gene expressions [71, 72]. Moreover, lipid growth and peroxidation is inhibited by its robust anti-oxidant properties. It without affecting regulates the macrophages and T cells in liver.

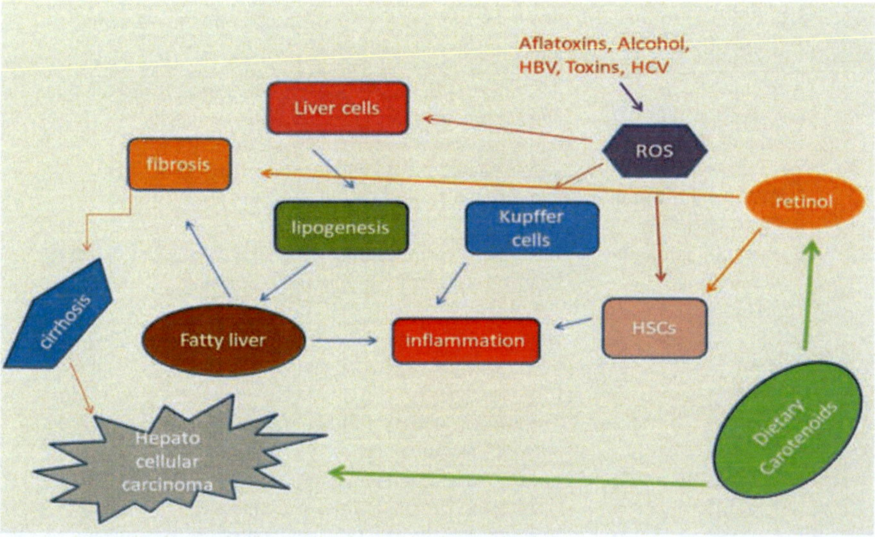

Fig. 18.3 Mechanism for prevention of chronic liver diseases by carotenoids

Table 18.1 Daily dose of different carotenoids

Sr.no	Recommended dose	Carotenoids	References
1	35 mg	Lycopene	[64]
2	7 mg	β-carotene	[65]
3	3 mg	β-cryptoxanthin	[66]
4	1–4 mg	Lutein	[67]
5	10 mg	Lutein	[68]

NAFLD is considered as one of the chronic liver diseases in the world. It is associated with some other diseases like metabolic syndrome, obesity, insulin resistance, type 2 diabetes mellitus and advancement to cirrhosis [73–75]. Pathogenesis of this chronic liver disease is very complex as many mechanisms are involved. For cure, therapeutic and preventive strategies, a detail understanding of mechanism is required [42]. And treatment involves reduction in oxidative stress, cardiovascular and dyslipidemia. Carotenoid therapy is proved as important treatment for NAFLD prevention and progression [12, 76]. Although the complete way of action of these carotenoids is still not completely known but it is suggested to act in three directions: the first suggested act is improvement in the liver antioxidants and secondly help in the production of vitamin A which will ultimately improve the retinoid signaling and the third action is production of apocarotenoids which may act as regulators in signaling pathway [34, 76]. Some of carotenoids used in therapy are listed in Table 18.2.

Table 18.2 Effect of carotenoids therapy on chronic liver diseases in cell line and human

S. No	Carotenoids	Effects	References
1	α- tocopherol, α-carotene Retinol, lycopene, lutein, γ-tocopherol, β-cryptoxanthin, and β- carotene	The effect of these was observed in human liver and serum and there was increase in oxidative stress in patients of chronic hepatitis C. very severe depletion of antioxidants in liver tissue and serum	[64, 66, 77–80]
2	Zeaxanthin, lutein, α- carotene and β-Carotene	In humans, serum carotenoid level was found to be inversely accompanying with the risk of NAFLD	[67, 81, 82]
3	Dietary carotenes and vitamin A	Dietary consumption of retinol, carotenes, and total vitamin A decreases the risk of primary liver cancer risk	[83, 84]
4	Lutein	The activity of HBV full-length promoter was inhibited in Human hepatoblastoma cells	[85, 86]
5	β-Cryptoxanthin	In NAFLD affected human liver, antioxidant and anti-inflammatory activities were induced. Macrophages directly targeted, contribute to liver homeostasis by regulating the polarization of M1/M2 macrophages/KCs	[66, 87]
6	Fucoxanthin	There was improvement in liver function with reduction in fat contents of obsessed body and liver	[65, 67, 79]
7	Astaxanthin	Prevent lipid per oxidation, contribute to liver homeostasis by regulating the polarization of M1/M2 macrophages/KCs. It act as hepato-protective and act against inflammation, ulcers, neurodegeneration, diabetes, and cardiovascular disease. It also improves insulin resistance by protecting cells from oxidative	[6, 88, 89]

18.3 Lung Diseases and Carotenoids

18.3.1 Factors Associated with Health of Lungs

There are many factors associated with health of lungs. The development of lungs and immune system is fast in childhood and after the age of 70 years, pulmonary function declines [90, 91]. With aging other factors associated with the health of lungs are inheritance, occupation, environment, exposure to chemicals and diet.

18.3.1.1 Dietary Factors

Unbalance diet and diet poor in antioxidant are biological factors that can enhance the oxidative stress, aging of lungs, inflammation and pathophysiology [90, 92]. The self-reported data on fruit and vegetable intake shows their association with chronic disruptive pulmonary infection. Many reports from japan and US presented reduction in risk of chronic disruptive pulmonary infection by the increase in fruit and vegetable intake. Other dietary factors associated with lung health are use of alcohol, tea, tobacco, calorie sources like protein, fat and cholesterol in daily life. Among fatty acids, marine source of omega-3 and omega-6 are important [93, 94]. The balanced diet for lung health and function may include vitamin A, C, E, α-carotein, lutein, β-carotein, β-cryptoxanthin, zeaxanthin and lycopene. Use of carotenoids in diet may protect against asthma and chronic bronchitis. Although carotenoids are directly associated with antioxidant defense in lung, variations in carotenoid content especially lycopene [64, 95]. The bioavailability of carotenoids contents are affected by different factors like quality of food, cooking techniques and time.

Environmental Factors

Environment is an important factor among other factors including inheritance, occupation, dietary habits, affecting lung health [89, 96]. All factors are influenced by the exposure to carcinogens. The source of carcinogens may be tobacco/ cigarette smoking which can lead to lung cancer. Pulmonary diseases can be observed in adults and especially after the age of 65 as there is slow decline in pulmonary function in adults and rate of decline increases after the age of 70. Slow decline in pulmonary function is assessed by forced expired volume per second and is associated with chronic pulmonary diseases like fibrotic lung disease and asthma [90]. As during childhood, there is rapid growth of immune system and lungs, so with the developmental factors, genetic tendency and environmental factors are very important in this age especially in the case of pathogenesis of infancy asthma [93, 97]. Lungs can be damaged by the induction of oxidative damage by smoke. This oxidative stress can lead to develop in active TB and source of smoke may be cigarette. This oxidative damage can be observed in active smokers as well as passive smokers. Air pollution is associated with oxidative stress and particles in polluted air include carbon nanoparticles and metals. Some persons work with metals like welders are directly exposed to metals present in welding fumes like iron and copper [98]. Carotenoids and antioxidants vitamin can neutralize the effects and are defensive agents against damage produced by free radicals in oxidative stress.

18.3.2 Different Carotenoids and Risk of Lung Diseases

18.3.2.1 Asthma

The asthma problem is reported to be increasing worldwide and the prevalence is more in children under the age of 18 years. Different studies reveal correlation between diet and lung health with providing the list of nutrients which can prevent respiratory diseases like bronchial hyper activity, asthmatic inflammation and asthma [93, 97]. Results from human are not easy to interpret because of alterations in background and demography of population. Carotenoids may be useful but further studies on relation between carotenoids and asthma needed. Some of studies are based on the quantity of carotenoids present in diet rather than monitoring nevertheless other studies are associated with the antioxidant activity of carotenoids in children [98, 99]. Beside these, instead of nutritional assessment, spectroscopy can be used for detection of skin carotenoids [100]. Raman spectroscopy is highly sensitive, precise and specific for measuring carotenoids level. Carotenoids studied and found important in asthma therapy include α-carotene, β-carotene, β-cryptoxanthin, zeaxanthin lycopene and lutein [94, 101].

There is increase in inflammation of lung cells in smokers. This increase can be observed in cells responsible for oxidants production like mast cells and eosinophils in both atopic and non-atopic asthma condition [99, 102]. But some cells like neutrophils can increase in number only in non-atopic asthma condition. This increase in number of neutrophil and eosinophil is harmful for host tissues and can cause inflammatory injury [103, 104]. The number of neutrophils also Increase in lung tissues in sjogren syndrome.

18.3.2.2 Tuberculosis

The world health organization described tuberculosis as one of top death causing infectious disease with ten million incidence cases and nearly two million death cases in 2015. The cases are mostly reported in low income countries whereas decrease in incidence and mortality observed in Europe and North America [97, 105]. This decrease in tuberculosis cases in developed countries is either due to effective chemotherapy, improvement in economic status and availability of vaccine (BCG). These differences of incidences between under developed and developed countries are associated with socioeconomic status which is intervened by nutrition [106]. Different studies on human support the ideology that tuberculosis and undernutrition are interlinked. As well as in vitro studies also deliver understanding in micronutrients mediated anti-bacterial activity [107]. From all these studies we can conclude that nutritional status and micronutrient deficiency studies are helpful in determining the progress of tuberculosis infection. Tuberculosis progression is linked with carotenoids and vitamin levels in diet [108, 109]. Carotenoids have been reported as in reverse connected with the progress of tuberculosis. An interesting study reported a decrease risk of tuberculosis among smokers utilizing large amount of fruits and vegetables in diet [96, 110].

18.3.2.3 Lung Cancer

There were reported only harmful effects or no effects of dietary carotenoids on lung cancer before the availability of database on food composition especially about a specific carotenoid [111, 112]. But now data base are available for specific carotenoid and we can relate different levels of carotenoids with the risk of lung cancer [112]. If we look at the leading factor causing lung cancer, the cigarette smoke is number one and other factors like inheritance, occupation, diet and environment can influence the effect of carcinogen released from tobacco [97, 113]. Nearly 15% of smokers ultimately are diagnosed with lung cancer. Different fat soluble pigments, carotenoids with and without vitamin A can help in control of lung cancer. Carotenoids which show activity with vitamin A in human's plasma are β-cryptoxanthin, α- carotene, β-carotene, and some carotenoids which show activity without vitamin A are lutein, lycopene and zeaxanthin [114, 115].

Some of the trials show that the pharmacological dose of carotene unassumingly improved the risk of lung cancer in asbestos workers and heavy smokers. Different experiments may not prove an association between lung cancer and carotenoids but indicator showing these carotenoids as bioactive pigments [91, 94]. Although some studies suggest carotenoids (α- carotene, β-carotene) are not linked with risk of lung cancer but β-cryptoxanthin intake was proved to have negative effect on lung cancer. The problem associated with these types of studies is the diagnosis of lung cancer at late stage so there may be a different dietary behavior in undiagnosed cancer [96, 116]. Lycopene, lutein, cryptoxanthin, α- and β carotene are proved to be significant in lowering the lung cancer risk. Some researchers also worked by adjusting smoking effect and after which among carotenoids only cryptoxanthin remain connected with lung cancer. Humans working in mines were selected for serum cryptoxanthin studies and results showed positive effect on risk of lung cancer. In most of studies, primary data was analysed instead of published literature review [95, 115]. Consequently, same carotenoid groups were created for the removal of heterogeneity. One of problem associated with smokers and non-smokers lung cancer is because of differences in feeding habits. Removal of confounding effect by smoking in analysis is also difficult during lung cancer studies.

There are additional factors with active smoking which may affect the risk of lung cancer are inhalation patterns, passive smoking, intensity of previous smoking or time since quitted smoking, type of tobacco, type of cigarette and pipe [97]. Effect of high and low dose of carotenoids was also checked by some scientists to compare the protective effect of different carotenoids. Although it was proved that carotenoids are biological antioxidants that defend body against infectious diseases like cancer but also the most active attainment against this disease is prevention of smoking [117, 118]. Therapeutic use of supplemental retinol, lutein, carotene and lycopene during the age of 51–77 years can significantly affect the lung cancer. This study also recommends short term use of carotenoids particularly for the prevention of lung cancer among smokers as increased mortality is reported by other studies [115, 119]. These findings are also very important as these revealed oxidative stress associated with smoking. We can conclude that carotenoids might be harmful for the organs and tissues experiencing excessive oxidative stress [120]. Moreover, these harmful effects of carotenoids may be experienced by the oxidative products of carotenoids. Hence it is also recommended taking into consideration the adverse effects of cancer chemoprevention [121]. Similarly one of carotenoid known as beta carotene previously recommended for cure of cardiac disease and lung cancer is no more recommended for therapy. Oxidative stress caused by smoking and asbestos is not only linked with lung cancer but carbon nanotubes, metals, fullerenes, ultrafine particulate pollutant and transition metals like copper or iron present in welding fumes are all environmental factors which are associated with lung health [122].

18.3.2.4 Oxidative DNA Damage in Lung Epithelial Cells

Oxidative stress may damage DNA and epithelial cells in lungs. This damage may occur due to reactive oxygen species which devastate the antioxidants [123]. The first defense line against oxidative DNA damage is the fluid present between epithelial cells lining and external environment. This fluid of respiratory tract contains antioxidants and metal binding protein which make it first defense line [124]. These enzymes and antioxidants include superoxide dismutase, catalase, glutathione peroxidase, urate, glutathione, ascorbate and proteins are transferrin, lactoferrin and ceruloplasmin [125]. Carotenoids are among the antioxidants of low molecular weight which are important in respiratory tract functioning as antioxidants. Among different carotenoids β-carotene is the one reported to protect lung against oxidative DNA damage [126]. Cryptoxanthin has been proved as an important antioxidant in preventing oxidative DNA damage. Moreover, this carotenoid has been proved for the reduction of DNA oxidation damage marker (8-OHDG cell number) with reduction in oxidized base group (8-oxo-7,8-dihydroguanine) derived from radiations and half-life of DNA breaks induced by hydrogen per oxide in HeLa cells [127].

18.4 Risk Factors of Liver Diseases

18.4.1 Insulin Resistance

Insulin resistance is fully understood as a risk factor for liver diseases. The insulin resistance mechanism of action for fatty liver disease is simply linked by oxidative stress and lipid metabolism [33]. Insulin resistance will lead into lipid overload which will lead to fatty liver disease. That will lead to lipogenesis in liver which is associated with metabolic syndrome. So insulin resistance is associated with metabolism syndrome, and will stimulate the accumulation of free fatty acid which will also start the anabolism [53]. These free fatty acids will be stored as droplets in liver cells, resulting into steatosis which will affect the immune system by inducing innate immune response. The nutritional status of carotenoids and vitamin is directly linked with liver health and obesity [71]. In humans it was observed that insulin resistance and carotene level are associated with each other and there is an inverse relation between carotenoids and class three obesity [69, 87]. Lutein is one of important carotenoid found to be effective against insulin resistance.

Insulin resistance is induced by high fat content in diet which also inhibits the expression of peroxisome proliferator activated receptors [85]. Lutein was proved to be effective also for the expression of peroxisome proliferator activated receptor. Another study proved cryptoxanthin an effective carotene that can reverse inflammation, progress in fibrosis and steatosis [79]. The adipose tissues in case of insulin sensitivity, increases uptake of fatty acids by activating lipase lipoprotein and fatty acid transporter protein induction whereas it repress lipolysis by preventing the lipase activity [93]. However, in insulin resistance, lipolysis becomes deregulated which releases free fatty acids from adipocytes. If insulin resistance exists, the adipose tissues can derive free fatty acids which may lead to non-alcoholic fatty liver disease. These free fatty acids produced by insulin resistance may trigger the formation of toxic intermediates like acyl carnitines, diacylglycerol, lysophosphatidyl choline and ceramides which in turn may produce stresses like oxidative and endoplasmic reticulum stress [101]. So in short we can conclude that insulin resistance can result in apoptosis and dysfunction of liver cells. And in humans suffering from obesity have low serum carotene levels. Moreover, there is significant relation between insulin resistance and carotenoids (especially B-carotene), which give us confidence in saying that b-carotene help in improving liver health.

18.4.2 Increases Oxidative Stress

Oxidative stress and inflammation are very important and observed in many studies related to liver diseases. Nonalcoholic fatty acid liver disease is always followed by inflammation and oxidative stress. The treatment of liver disease starts with the change in diet with the addition of carotenoids in diet [115]. As the carotenoids have the ability to reduce the oxidative stress by regulating lipid metabolism and modifying some genes. Most of people suffering from oxidative stress are smokers which can be treated by dietary carotenoids and the level in skin may also be increased by this treatment. Similarly, the increased levels of these antioxidants have a positive effect on liver metabolism [119]. To overcome the problem of oxidative stress and liver diseases, daily dose of 7 mg carotenoids (e.g. lycopene) is recommended for human and it is notable that cooking methods adopted in home cannot cause destruction of these pigments. Fibrogenosis is caused by the combined action of oxidative stress and inflammation [52]. Oxidative stress may also induce the production of pro inflammatory cytokines in liver cells (Kupffer cells), these cytokines are also known as tumor necrosis a factor (TNF-α) all this process is mediated by inflammation which may results in death or damage of cell [63]. However, carotenoids are well known antioxidants which work as defender against oxidative stress by delaying the process of oxidation, in this process these neutralize the free radicals and inhibit the process of steatohepatitis [42]. Effects of different carotenoids have been checked in the rat liver and significant results were observed for carotene, lutein, cryptoxanthin and lycopene.

Carotenoids as precursor of vit-A shows direct effect on reactive oxygen species and by fighting against them protect the body from oxidative stress [76]. Protective and preventive effects of carotenoids have been observed by recent researchers against apoptosis, fibrosis, hepatic steatosis, inflammation and oxidative stress. This also acts as hormone as through metabolism it is transformed into retinoic acid which works as a ligand and regulate the gene expression during metabolic progression.

Ozturk et al. [50] concluded that carotenoids may reduce the hepatic steatosis and effects of carbon tetrachloride. As carbon tetra chloride can alter the oxidative stress markers like glutathione, malondialdehyde, superoxide dismutase and catalase. But steatosis and liver damage were found to be improved by carotenoid uptake. Similar results were observed in cell culture system by Liu et al. [20] as they observed a decrease in hepatosteatosis by inhibition of RNA replication in hepatitis C. moreover carotenoids inhibit the development of cancer in hepatic cells by inhibition of reactive oxygen species. Among other carotenoids, carotene has been proved to be useful in cases of oxidative stress, inflammation, fibrosis and improve the expression of the gene fork head box 01 [72]. All these improvements may be due to accent of transcription factor. Like carotene, lycopene reduce oxidative stress and so hepatocarcinogenesis. In addition, it also regulates miRNA-21 by inhibiting fatty acid binding protein and reduces oxidation stress by decreasing isoprostane in urine. Moreover lycopene may also reverse the oxidative stress related hyoerhomo cysteinemia. Through the work of Xu et al. [90] it was proved that carotenoids improve hepatic oxidation by inproving liver injuries produced by aflatoxin B1. This property is based on the usage of free radicals and expression of peroxisome receptor. Oxidative stress, insulin resistance and endoplasmic stress are induced by lipid and fatty acid accumulation by induction of lipid intermediates like ceramides, diacylglycerols and acylcarnitines.

As oxidative stress is the imbalance between reactive oxygen species production and removal. The large removal and production of reactive oxygen species may lead to trigger peroxidation of lipid by free radical chain reaction. This lipid peroxide may damage protein, DNA and interrupt cell membrane which leads to necrosis and apoptosis [88]. Astaxanthin is a type of carotenoid which acts as antioxidant and protects cell membrane, protein, DNA and lipid from oxidative damage. It activates antioxidant enzymes expression which reduces oxidative stress of biomarkers like nitic oxide, products of protein oxidation and malondialdehyde [115].

In oxidative stress biological molecules are damaged. Oxidative stress is universally considered as responsible for all foremost diseases including liver and heart diseases.

Oxidative stress is associated with occupation, environment, smoking and alcohol consumption. But a study has proved that alcoholic oxidation stress cannot be improved by antioxidants (carotenoids). However, Paolini et al. [128] found 15 times increase in cytochrome p450 and oxygen radicals in liver. Oxidative stress in liver causes production of apo-8-carotenal which influences cytochrome P450 expression and radicals of oxygen. In addition to antioxidants (carotenoids), physical exercises regularly are recommended.

18.5 Case Studies (Table 18.3)

Table 18.3 Various studies on effect of carotenoids

various studies on effect of carotenoids						
Sample specification	Carotene	Duration	Assessment	Aim	Results	Reference
Mouse	Lactoferrin				Prevent liver steatosis	[129]
Mice	Astaxanthin (6 mg/kg)	60 days			Reduce hepatic steatosis, TGF1 protein and obesity	[130]
Mice	Astaxanthin				Reduce lipid formation, fibrosis and infiltration	[131]
Brain of mouse	Astaxanthin				Increased activities of antioxidant. Reduced biomarkers of oxidative stress,	[132]
Liver of mice	Astaxanthin				Reduce the pro-inflammatory M1-type macrophages	[133]
Skeletal muscle of mice	Astaxanthin				Stimulation of CPT1 and acyl-coenzyme A oxidase 1	[134]
Human cell line	Astaxanthin				Prevent the expression of profibrogenic genes persuaded by TGF1	[135]
Mice	Lycopene				Decrease in TG and NEFA	[136]
Rat liver	Lycopene				Gathering of pro-inflammatory cytokines (TNF & IL-1)	[137, 138]
Human cell line	Lycopene				Inhibit B kinase,	[139]
	B- Carotene				*prevent* oxidative damage in HepG2 cells	[68]
Mice	B- Carotene				Reduced body and serum TG levels	[140]
	B- cryptoxanthin (37.5 μg/kg/d)				Reduce DNA damage	[141]
HeLa cells	β-Cryptoxanthin (1-4 μM)				Reduced DNA damage	[142]

(continued)

Table 18.3 (continued)

various studies on effect of carotenoids						
Sample specification	Carotene	Duration	Assessment	Aim	Results	Reference
Liver cells	β-Cryptoxanthin				Reduce oxidative damage	[143]
Human	β-cryptoxanthin				Increase SOD level	[65]
Mice	β-cryptoxanthin				Inhibit liver lipid accumulation	[144]
	β-cryptoxanthin				Expression of TNFα	[144]
Human female	Fucoxanthin (2.4 mg)	16 week			Reduce liver fat	[79]
Mice	Fucoxanthin				Prevent insulin resistance	[145]
	2% fucoxanthin				Decrease white adipose tissues	[146]
Rat	Crocetin				Prevent DNA oxidative damage	[147]
Mice	Crocetin				Improves antioxidant enzymes	[148]
Rat	Crocetin				Increase in glucose tolerance	[149]
Mouse	β-cryptoxanthin				Prevent fibrosis and trigger stellate cells	[144]
Human liver	Retinol				Lower risk of cancer	[105]
Human	Retinol (1000 μg)				Suppress cancer	[103]
Human	Lutein				Inhibit HBV activity	[150]
Rat	Lutein				Restoration of peroxidase, catalase and glutathione	[151]
Rat liver	Lutein				Decrease cholesterol and increase liver insulin signaling	[152]
Rat	Lutein				Increase detoxification and prevent cytochrome p450	[151]
Human liver	Fucoxanthin				Reduce fat and improve liver functions	[79]

18.6 Conclusion

Carotenoids have protective actions which can be used in liver therapy and against lung diseases like tuberculosis, asthma and lung cancer. Some famous carotenoids useful in blocking the development of lung cancer and NAFLD are crocetin, lycopene, fucoxanthin, carotene, astaxanthin, lutein and cryptoxanthin. They act as antioxidants, anti-inflammatory, anti-fibrosis, lipid lowering and insulin-sensitizing agent. As there is lack of approved exercises, drug and food administration for the treatment of NAFLD, dietary carotenoids may be recommended because of their caring paraphernalia on development of NAFLD.

References

1. Bohn T (2008) Bioavailability of non-provitamin A carotenoids. Curr Nutr Food Sci 4:240–258
2. Ruhl R (2013) Non-pro-vitamin A and pro-vitamin A carotenoids in atopy development. Int Arch Allergy Immunol 161:99–115
3. Britton G (1995) Structure and properties of carotenoids in relation to function. FASEB J 9:1551–1558
4. Woodside JV, McGrath AJ, McKinley MC (2015) Carotenoids and health in older people. Maturitas 80:63–68
5. Gonzalez-Barrio R, Periago MJ, Luna-Recio C, Garcia-Alonso FJ, Navarro-Gonzalez I (2018) Chemical composition of the edible flowers, pansy (*Viola wittrockiana*) and snapdragon (*Antirrhinum majus*) as new sources of bioactive compounds. Food Chem 252:373–380
6. Maiani G, Periago Caston MJ, Catasta G, Toti E, Cambrodon IG, Bysted A, Bohm V (2009) Carotenoids: actual knowledge on food sources, intakes, stability and bioavailability and their protective role in humans. Mol Nutr Food Res 53:S194–S218
7. Sugiura M (2013) Carotenoids: liver diseases prevention. In: Watson RR, Preedy VR (eds) Bioactive foods as dietary interventions for liver and gastrointestinal disease, 1st edn. Academic, San Diego, pp 421–436
8. Coyne T, Ibiebele TI, Baade PD, McClintock CS, Shaw JE (2009) Metabolic syndrome and serum carotenoids: findings of a cross-sectional study in Queensland, Australia. Br J Nutr 102:1668–1677
9. Sharoni Y, Linnewiel-Hermoni K, Khanin M, Salman H, Veprik A, Danilenko M, Levy J (2012) Carotenoids and apocarotenoids in cellular signaling related to cancer: a review. Mol Nutr Food Res 56:259–269
10. Mikolasevic I, Milic S, Turk Wensveen T, Grgic I, Jakopcic I, Stimac D, Orlic L (2016) Nonalcoholic fatty liver disease-a multisystem disease. World J Gastroenterol 22:9488–9505
11. Azzam H, Malnick S (2015) Non-alcoholic fatty liver disease-the heart of the matter. World J Hepatol 7:1369–1376
12. Elvira-torales LI, Garcia-Alonso J, Periago-Caston MJ (2019) Nutritional importance of carotenoids and their effect on liver health: a review. Antioxidants 8:229
13. Krizhanovsky V, Yon M, Dickins RA, Hearn S, Simon J, Miething C et al (2008) Senescence of activated stellate cells limits liver fibrosis. Cell 134:657–667
14. Fabbrini E, Sullivin S, Klein S (2010) Obesity and nonalcoholic fatty liver disease: biochemical, metabolic and clinical implications. Hepatology 51:679–689
15. Singhal S, Baker SS, Baker RD, Zhu L (2018) Role of paraoxonase 1 as an antioxidant in nonalcoholic steatohepatitis. In: Patel VB, Rajendram R, Preedy VR (eds) The liver, 1st edn. Academic, London, pp 15–20

16. Younossi Z, Anstee QM, Marietti M, Hardy T, Henry L, Eslam M, Bugianesi E (2018) Global burden of NAFLD and NASH: trends, predictions, risk factors and prevention. Nat Rev Gastroenterol Hepatol 15:11–20
17. Calzadilla Bertot L, Adams LA (2016) The natural course of non-alcoholic fatty liver disease. Int J Mol Sci 17:774
18. Mencin AA, Lavine JE (2016) Nonalcoholic fatty liver disease in children. Curr Opin Clin Nutr Metab Care 14:151–157
19. Sayiner M, Koenig A, Henry L, Younossi ZM (2016) Epidemiology of nonalcoholic fatty liver disease and nonalcoholic steatohepatitis in the United States and the rest of the world. Clin Liver Dis 20:205–214
20. Liu W, Baker SS, Baker RD, Zhu L (2015) Antioxidant mechanisms in nonalcoholic fatty liver disease. Curr Drug Targets 16:1301–1314
21. Rolo AP, Teodoro JS, Palmeira CM (2012) Role of oxidative stress in the pathogenesis of nonalcoholic steatohepatitis. Free Radic Boil Med 52:59–69
22. Grune T, Lietz G, Palou A, Ross AC, Stahl W, Tang G, Biesalski HK (2010) Beta-carotene is an important vitamin A source for humans. Nutrition 140:2268S–2285S
23. Sanchez-moreno C, Cano MP, De Ancos B, Plaza L, Olmedilla B, Granado F, Martin A (2006) Mediterranean vegetable soup consumption increases plasma vitamin C and decreases F2-isoprostanes, prostaglandin E2 and monocyte chemotactic protein-1 in healthy humans. J Nutr Biochem 17:183–189
24. Heath E, Seren S, Sahin K, Kucuk O (2007) The role of tomato lycopene in the treatment of prostate cancer. In: Rao AV (ed) Tomatoes, lycopene and human health: preventing chronic diseases, 1st edn. Caledonian Science Press, Scotland, pp 127–140. ISBN:9780955356506
25. Huang YM, Dou HL, Huang FF, Xu XR, Zou ZY, Lin XM (2015) Effect of supplemental lutein and zeaxanthin on serum, macular pigmentation, and visual performance in patients with early age-related macular degeneration. Biomed Res Int 2015:564738
26. Ferramosca A, Di Giacomo M, Zara V (2017) Antioxidant dietary approach in treatment of fatty liver: new insights and updates. World J Gastroenterol 23:4146–4157
27. Murillo AG, diMarco DM, Fernandez ML (2016) The potential of non provitamin A carotenoids for the prevention and treatment of nonalcoholic fatty liver disease. Biology 5:42
28. Yilmaz B, Sahin K, Bilen H, Bahcecioglu IH, Bilir B, Ashraf S, Kucuk O (2015) Carotenoids and non-alcoholic fatty liver disease. Hepato Surg Nutr 4:161–171
29. Christensen K, Lawler T, Mares J (2019) Dietary carotenoids and non-alcoholic fatty liver disease among US adults, NHANES 2003–2014. Nutrients 11:1101
30. Jiang W, Guo MH, Hai X (2016) Hepatoprotective and antioxidant effect of lycopene on non-alcoholic fatty liver disease in rat. World J Gastroenterol 22:10180–10188
31. Matsuura B, Miyake T, Yamamoto S, Furukawa S, Hiasa Y (2016) Usefulness of beta-cryptoxanthin for nonalcoholic fatty liver diseases. J Food Nutr Disord 5:3
32. Elvira-Torales LI, Martin-Pozuelo G, Gonzalez-Barrio R, Navarro-Gonzalez I, Pallares FJ, Santaella M, Periago-Caston MJ (2019) Ameliorative effect of spinach on non-alcoholic fatty liver disease induced in rats by a high-fat diet. Int J Mole Sci 20:1662
33. Kitade H, Chen G, Ni Y, Ota T (2017) Nonalcoholic fatty liver disease and insulin resistance: new insights and potential new treatments. Nutrients 9:387
34. Clugston RD (2020) Carotenoids and fatty liver disease: current knowledge and research gaps. BBA-Mol Cell Biol Lipids. https://doi.org/10.1016/j.bbalip.2019.158597
35. Milani A, Basirnejad M, Shahbazi S, Bolhassani A (2017) Carotenoids: biochemistry, pharmacology and treatment. Br J Pharm 174:1290–1324
36. Johnson EJ (2002) The role of carotenoids in human health. Nutr Clin Care 5:56–65
37. Shete V, Quadro L (2013) Mammalian metabolism of α-carotene: gaps in knowledge. Nutrients 5:4849–4868
38. Rao AV, Rao LG (2007) Carotenoids and human health. Pharm Res 55:207–216
39. Stahl W, Sies H (2003) Antioxidant activity of carotenoids. Mol Asp Med 24:345–351
40. Stahl W, Sies H (2005) Bioactivity and protective effects of natural carotenoids. Biochem Biophys Acta 1740:101–107

41. Siani RK, Nile SH, Park SW (2015) Carotenoids from fruits and vegitables: chemistry, analysis, occurrence, bioavailability and biological activities. Food Res Int 76:735–750
42. Chen H, Tappel A (1996) A protection by multiple antioxidants against lipid peroxidation in rat liver homogenate. Lipids 31:47–50
43. Bohn T (2008) Bioavailability of non-provitamin A carotenoids. Curr Nutr Food Sci 4:240–258
44. Eggersdorfer M, Wyss A (2018) Carotenoids in human nutrition and health. Arch Biochem Biophys 652:18–26
45. Rodriguez-Amaya DB, Kimura M, Godoy HT, Amaya-Farfan J (2008) Updated Brazilian database on food carotenoids: factors affecting carotenoids composition. J Food Compos Anal 21:445–463
46. Latif U, Ahmad R (2018) Role of dietary carotenoids in different etiologies of chronic liver diseases. In: Valero Diaz A, Garcia-Gimeno RM (eds) Descriptive food science, 1st edn. Intech open, London, pp 93–112
47. Patel V, Sail S (2006) A-carotene protects the physiological antioxidants against aflatoxin-B1 induced carcinogenesis in albino rats. Pak J Biol Sci 9:1104–1111
48. Kheir Eldin AA, Motawi TM, Sadik NA (2008) Effect of some natural antioxidants on aflatoxin B1-induced hepatic toxicity. EXCLI J 7:119–131
49. Harari A, Harats D, Marko D, Cohen H, Barshack I, Kamari Y, Shaish A (2008) A9-cis beta-carotene-enriched diet inhibits atherogenesis and fatty liver formation in LDL receptor knockout mice. J Nutr 138:1923–1930
50. Ozturk F, Gul M, Ates B, Ozturk IC, Cetin A, Vardi N, Yilmaz I (2009) Protective effect of apricot (*Prunus armeniaca* L.) on hepatic steatosis and damage induced by carbon tetrachloride in Wister rats. Br J Nutr 102:1767–1775
51. Liu Q, Bengmark S, Qu S (2010) Nutrigenomics therapy of hepatitis C virus induced virus induced-hepatosteatosis. BMC Gastroenterol 10:49
52. Yadav D, Hertan HI, Schweitzer P, Norkus EP, Pitchumoni CS (2002) Serum and liver micronutrient antioxidants and serum oxidative stress in patients with chronic hepatitis C. Am J Gastroenterol 97:2634–2639
53. Younossi Z, Anstee QM, Marietti M, Hardy T, Henry L, Eslam M, Bugianesi E (2018) Global burden of NAFLD and NASH: trends, predictions, risk factors and prevention. Nat Rev Gastroenterol Hepatol 15:11–20
54. Shiota G, Tsuchiya H (2006) Pathophysiology of NASH: insulin resistance, free fatty acids and oxidative stress. J Clin Biochem Nutr 38:127–132
55. Gyamfi D, Patel V (2017) Liver metabolism: biochemical and molecular regulations. In: Preedy VR, Lakshman R, Srirajaskanthan R, Watson RR (eds) Nutrition, diet therapy, and the liver, 1st edn. CRC Press, Boca Raton, pp 3–15
56. Cabre Galada E, Pena Quintana L, Virgili Casas N (2017) Nutricion en las enfermedades hepatobiliares. In: Gil A (ed) Tratado de Nutricion: Nutricion y Enfermendad, 3rd edn. Editorial Medica Panamericana, Madrid, pp 865–906
57. Fabbrini E, Sullivan S, Klein S (2010) Obesity and nonalcoholic fatty liver disease: biochemical, metabolic, and clinical implications. Hepatology 51:679–689
58. Singhal S, Baker SS, Baker RD, Zhu L (2018) Role of paraoxonase 1 as an antioxidant in nonalcoholic steatohepatitis. In: Patel VB, Rajendram R, Preedy VR (eds) The liver, 1st edn. Academic, London, pp 15–20
59. Calzadilla Bertot L, Adams LA (2016) The natural course of non-alcoholic fatty liver disease. Int J Mol Sci 17:774
60. Mencin AA, Lavine JE (2011) Nonalcoholic fatty liver disease in children. Curr Opin Clin Nutr Metab Care 14:151–157
61. Sayiner m, Koenig A, Henry L, Younossi ZM (2016) Epidemiology of nonalcoholic fatty liver disease and nonalcoholic steatohepatitis in the United States and the rest of the world. Clin Liver Dis 20:205–214
62. Liu W, Baker SS, Baker RD, Zhu L (2015) Antioxidant mechanisms in nonalcoholic fatty liver disease. Curr Drug Targets 16:1301–1314

63. Rolo AP, Teodoro JS, Palmerira CM (2012) Role of oxidative stress in the pathogenesis of nonalcoholic steatohepatitis. Free Radic Biol Med 52:59–69
64. Rao AV, Agarwal S (2000) Role of antioxidant lycopene in cancer and heart disease. J Am Coll Nutr 19:563–569
65. Matsuura B, Miyake T, Yamamoto S, Furukawa S, Hiasa Y (2016) Usefulness of beta-cryptoxanthin for nonalcoholic fatty liver diseases. J Food Nutr Disord 5:3
66. Grune T, Lietz G, Palou A, Ross AC, Stahl W, Tang G et al (2010) B-carotene is an important vitamin A source for humans-3. J Nutr 140:2268S–2285S
67. Mares-Perlman JA, Millen AE, Ficek TL, Hankinson SE (2002) The body of evidence to support a protective role for lutein and zeaxanthin in delaying chronic disease. Overview. J Nutr 132:518S–524S
68. Martin kR, Failla ML, Smith JC Jr (1996) B-carotene and lutein protect hepg2 human liver cells against oxidant-induced damage. J Nutr 126:2098–2106
69. Ota T, Takamura T, Kurita S, Matsuzawa N, Kita Y, Uno M, Nakanuma Y (2007) Insulin resistance accelerates a dietary rat model of nonalcoholic steatohepatitis. Gastroenterology 132:282–293
70. Tilg H, Moschen AR (2010) Evolution of inflammation in nonalcoholic fatty liver disease: the multiple parallel hits hypothesis. Hepatology 52:1836–1846
71. Cusi K (2012) Role of obesity and lipotoxicity in the development of nonalcoholic steatohepatitis: pathophysiology and clinical implications. Gastroenterology 142:711–725
72. Karadeniz G, Acikgoz S, Tekin IQ, Tascylar O, Gun BD, Comert M (2008) Oxidized low-density-lipoprotein accumulation is associated with liver fibrosis in experimental cholestasis. Clinics 63:531–540
73. Fon Tacer K, Rozman D (2011) Nonalcoholic fatty liver disease: focus on lipoprotein and lipid deregulation. J Lipids 2011:783976
74. Werman MJ, Ben-Amotz A, Mokady S (1999) Availability and antiperoxidative effects of beta-carotene from Dunaliella bardawil in alcohol-drinking rats. J Nutr Biochem 10:449–454
75. Lee Y, Hu S, Park Y, Lee JI (2019) Health benefits of carotenoids: a role of carotenoids in the prevention of non-alcoholic fatty liver disease. Prev Nutr Food Sci 24:103–113
76. Whittaker P, Wamer WG, Chanderbhan RF, Dunkel VC (1996) Effects of alpha-tocopherol and beta-carotene on hepatic lipid peroxidation and blood lipids in rats with dietary iron overload. Nutr Cancer 25:119–128
77. Senoo H, Yoshikawa K, Morii m, Miura M, Imai K, Mezaki Y (2010) Hepatic stellate cell (vitamin A-storing cell) and its relative-past, present and future. Cell Biol Int 34:1247–1272
78. Xiao ML, Chen GD, Zeng FF, Qiu R, Shi WQ, Lin JS, Chen YM (2019) Higher serum carotenoids associated with improvement of non-alcoholic fatty liver disease in adults: a prospective study. Eur J Nutr 58:721–730
79. Abidov M, Ramazanov Z, Seifulla R, Grachev S (2010) The effects of Xanthigen ™ in the weight management of obese promenopausal women with non-alcoholic fatty liver disease and normal liver fat. Diabetes Obes Metab 12:72–81
80. Wang L, Ding C, Zeng F, Zhu H (2019) How levels of serum α-carotene and α-carotene/retinol ratio are associated with histological severity in nonalcoholic fatty liver disease patients. Ann Nutr Metab 74:156–164
81. Burri BJ (2002) Lycopene and human health. In: Meskin MS, Bidlack WR, Davies AJ, Omaye ST (eds) Phytochemicals in nutrition and health, 1st edn. CRC Press, Boca Raton, pp 157–172
82. Sarni RO, Suano de Souza FI, Ramalho RA, Schoeps Dde O, Kochi C, Catherino P, Colugnati FB (2005) Serum retinol and total carotene concentrations in obese pro-school children. Med Sci Monit 11:CR510–CR514
83. Seif El-Din SH, El-Lakkany NM, El-Naggar AA, Hammam OA (2015) Effects of rosuvastatin and/or α-carotene on non-alcoholic fatty liver in rats. Res Pharm Sci 10:275–287
84. Baybutt RC, Molteni A (1999) Dietary beta-carotene protects lung and liver parenchyma of rats treated with monocrotaline. Toxicology 137:69–80

85. Tainaka T, Shimada Y, Kuroyanagi J, Zang L, Oka T, Nishmura Y, Tanaka T (2011) Transcriptome analysis of anti-fatty liver action by Campari tomato using a zebrafish diet-induced obesity model. Nutr Metab 8:88
86. Xiao J, Liong EC, Ching YP, Chang RCC, Fung ML, Xu AM, Tipoe GL (2013) Lyceum barbarum polysaccharides protect rat liver from non-alcoholic steatohepatitis-induced injury. Nutr Diabetes 3:ne81
87. Villaca Chaves G, Pereira SE, Saboya CJ, Ramalho A (2008) A non-alcoholic fatty liver disease and its relationship with the nutritional status of vitamin A in individuals with class III obesity. Obes Surg 18:378–385
88. Gradelet S, Le Bon AM, Berges R, Suschetet M, Astorg P (1998) Dietary carotenoids inhibit aflatoxin B1-induced liver preneoplastic foci and DNA damage in rat: role of the modulation of aflatoxin B1 metabolism. Carcinogenesis 19:403–411
89. Yilmaz S, Kaya E, Kisacam MA (2017) The effect on oxidative stress of Aflatoxin and protective effect of lycopene on aflatoxin damage. In: Aflatoxin-control, analysis, detection and health risks. Croatia InTech, Rijeka, pp 67–90
90. Xu F, Yu K, Yu H, Wang P, Song M, Xiu C et al (2017) Lycopene relieves AFB 1-induced liver injury through enhancing hepatic antioxidation and detoxification potential with Nrf2 activation. J Funct Foods 39:215–224
91. Grievink L, Smit HA, Ocke MC, Vant Veer P, Kromhout D (1998) Dietary intake of antioxidant (pro)-vitamins, respiratory systems and pulmonary function: the MORGEN study. Thorax 53:166–171
92. Guenegou A, Leynaert B, Pin I, Le Moel G, Zureik M, Neukirch F (2006) Serum carotenoids, vitamin A and E, and 8 year lung function decline in a general population. Thorax 61:320–326
93. Tainaka T, Shimada Y, Kuroyangi J, Zang L, Oka T, Nishimura Y et al (2011) Transcriptome analysis of anti-fatty liver action by Campari tomato using a zebrafish diet-induced obesity model. Nutr Metab 8:88
94. Grievink L, Smit HA, Veer P, Brunekreef B, Kromhout D (1999) Plasma concentrations of the antioxidants beta-carotene and alpha-tocopherol in relation to lung function. Eur J Clin Nutr 53:813–817
95. Sharma g, Goodwin J (2006) Effect of aging on respiratory system physiology and immunology. Clin Interv Aging 1:253–260
96. Simpson CF, Punjabi NM, Wolfenden L et al (2005) Relationship between lung function and physical performance in disabled older women. J Gerontol A Biol Sci Med Sci 60:350–354
97. Sin DD, Wu LL, Man SFP (2005) The relationship between reduced lung function and cardiovascular mortality: a population-based study and a systematic review of the literature. Chest 127:1952–1959
98. Cao Y, Wang C, Liu J, Liu ZM, Ling WH, Chen YM (2015) Greater serum carotenoid levels associated with lower prevalence of nonalcoholic fatty liver disease in chinese adults. Sci Rep 5:12951
99. Crowle AJ, Ross EJ (1989) Inhibition by retinoic acid of multiplication of virulent tubercle bacilli in cultured human macrophages. Infect Immun 57:840–844
100. Wheelwright M, Kim EW, Inkeles MS et al (2014) All-trans retinoic acid-triggered antimicrobial activity against mycobacterium tuberculosis is dependent on NPC2. J Immunol 192:2280–2290
101. Chaves GV, Pereira SE, Saboya CJ, Ramalho A (2008) Non-alcoholic fatty liver disease and its relationship with the nutritional status of vitamin A in individuals with class III obesity. Obes Surg 18:378–385
102. Chattopadhyay MB, Kanna PS, Ray RS, Roy S, Chatterjee M (2004) Combined supplementation of vanadium and beta-carotene suppresses placental glutathione S-transferase-positive foci and enhances antioxidant functions during the inhibition of diethlnitrosamine induced rat liver carcinogenesis. J Gastroenterol Hepatol 19:683–693
103. Matsushima-Nishiwaki R, Okuno M, Takano Y, Kojima S, Friedman SL, Moriwaki H (2003) Molecular mechanism for growth suppression of human hepatocellular carcinoma cells by acyclic retinod. Carcinogenesis 24:1353–1359

104. Cegielski JP, McMurray DN (2004) The relationship between malnutrition and tuberculosis: evidence from studies in humans and experimented animals. Int J Tuberc Lung Dis 8:378–298
105. Lan QY, Zhang YJ, Liao GC, Zhou RF, Zhou ZG, Chen YM (2016) The association between dietary vitamin A and carotenes and the risk of primary liver cancer: a case-control study. Nutrients 8:624
106. Gerster H (1997) The potential role of lycopene for human health. J Am Coll Nutr 16:109–126
107. Kim CH (2011) Retinalic acid, immunity and inflammation. Vitam Horm 86:83–101
108. Kavanaugh CJ, Trumbo PR, Ellwood KC (2007) The US food and drug administration's evidence-based review for qualified health claims: tomatoes, lycopene, and cancer. J Natl Cancer Inst 99:1074–1085
109. Rahman I, Biswas SK, Kode A (2006) Oxidant and antioxidant balance in the airways and airway diseases. Eur J Pharm 533:222–239
110. Islamian JP, Mehrali H (2015) Lycopene as a carotenoid provides radioprotectant and antioxidant effects by quenching radiation-induced free radical singlet oxygen: an overview. Cell J (Yakhteh) 16:386–391
111. Muller L, Caris-Veyrat C, Lowe G, Bohm V (2016) Lycopene and its antioxidant role in the prevention of cardiovascular diseases-acritical review. Crit Rev Food Sci Nutr 56:1868–1879
112. Bates MN, Khalakdina A, Pai M et al (2007) Risk of tuberculosis from exposure to tobacco smoke: a systematic review and meta-analysis. Arch Intern Med 167:335–342
113. Villaca Chaves G, Pereira SE, Saboya CJ, Ramalho A (2008) Nonalcoholic fatty liver disease and its relationship with the nutritional status of vitamin A in individuals with class III obesity. Obes Surg 18:378–385
114. Pina-Zentella RM, Rosado JL, Gallegos-Corona MA, Madrigal-Perez LA, Garcia OP, Ramos-Gomez M (2016) Lycopene improves diet-mediated recuperation in rat model of nonalcoholic fatty liver disease. J Med Food 19:607–614
115. Van Helden YGJ, Keijer J, Heil SG, Pico C, Palou A, Oliver P, Munnia A, Briede JJ, Peluso M, Franssen-Van Hall NL, Van Schooten FJ, Godschalk RW (2009) Beta-carotene affects oxidative stress related DNA damage in lung epithelial cells and in ferret lung. Carcinogenesis 30:2070–2076
116. Brown MJ, Ferruzi MG, Nguyen ML, Cooper DA, Eldridge AL, Schwartz SJ, White WS (2004) Carotenoid bioavailability is higher from salads ingested with full-fat than with fat-reduced salad dressings as measured with electrochemical detection. Am J Clin Nutr 80:396–403
117. Tabak C, Smit HA, Heederik D, Ocke MC, Kromhout D (2001) Diet and chronic obstructive pulmonary disease: independent beneficial effects of fruits, whole grains and alcohol (the MORGEN study). Clin Exp Allergy 31:747–755
118. Ford ES, Mannino DM, Redd SC (2004) Serum antioxidant concentrations among U.S. adults with self-reported asthma. J Asthma 41:179–187
119. Wood LG, Gibson PG, Garg ML (2003) Biomarkers of lipid peroxidation, airway inflammation and asthma. Eur Resp J 21:177–186
120. Varreso R, Fung TT, Barr RG, Hu FB, Willett W, Camargo CA Jr (2007) Prospective study of dietary patterns and chronic obstructive pulmonary disease among US women. Am J Clin Nutr 86:488–495
121. Soh AZ, Chee CB, Wang YT et al (2016) Dietary cholesterol increases the risk whereas PUFAs reduce the risk of active tuberculosis in Singapore Chinese. J Nutr 146:1093–1100
122. Chandra RK, Kumari S (1994) Nutrition and immunity: an overview. J Nutr 124:1433s–1435s
123. Katona P, Katona-Apte J (2008) The interaction between nutrition and infection. Clin Infect Dis 46:1582–1588
124. Cegielski JP, McMurray DN (2004) The relationship between malnutrition and tuberculosis: evidence from studies in human and experimental animals. Int J Tuberc Lung Dis 8:286–298
125. Cegielski JP, McMurray DN (2004) The relationship between malnutrition and tuberculosis: evidence from studies in humans and experimental animals. Int J Tuberc Lung Dis 8:286–298
126. Lonnroth K, Williams BG, Cegieiski P et al (2010) A consistent log-linear relationship between tuberculosis incidence and body mass index. Int J Epidemiol 39:149–155

127. Robin K, Butler LM, Wang R et al (2013) Genetic and environmental predictors of serum 25-hydroxyvitamin D concentrations among middle-aged and elderly Chinese in Singapore. Br J Nutr 109:493–502
128. Paolini M, Antelli A, Pozzetti L, Spetlova D, Perocco P, Valgimigli L, Pedulli GF, Cantelli GF (2001) Induction of cytochrome P450 enzymes and over-generation of oxygen radicals in beta-carotene supplemented rats. Carcinogenesis 22:1483–1495
129. Lee S, Son B, Jeon J, Park G, Kim H et al (2018) Decreased hepatic lactotransferrin induces hepatic steatosis in chronic non-alcoholic fatty liver disease model. Cell Physiol Biochem 47:2233–2249
130. Bhuvaneswari S, Arunkumar E, Viswanathan P, Anuradha CV (2010) Astaxanthin restricts weight gain, promotes insulin sensitivity and curtails fatty liver disease in mice fed a obesity-promoting diet. Process Biochem 45:1406–1414
131. Kim B, Farruggia C, Ku CS, Pham TX, Yang Y, Bae M et al (2017) Astaxanthin inhibits inflammation and fibrosis in the liver and adipose tissue of mouse models of diet-induced obesity and nonalcoholic steatohepatitis. J Nutr Biochem 43:27–35
132. Al-Amin MM, Akhter S, Hasan AT, Alam T, Nageeb HSM, Saifullah ARM et al (2015) The antioxidant effect of astaxanthin is higher in young mice than aged: a region specific study on brain. Metab Brain Dis 30:1237–1246
133. Ni Y, Nagashimada M, Zhuge F, Zhan L, Nagata N, Tsutsui A et al (2015) Astaxanthin prevents and reverses diet-induced insulin resistance and steatohepatitis in mice: a comparison with vitamin E. Sci Rep 5:17192
134. Kim B, Farruggia C, Ku CS, Pham TX, Yang Y, Bae M et al (2017) Astaxanthin inhibits inflammation and fibrosis in the liver and adipose tissue of mouse models of diet-induced obesity and nonalcoholic steatohepatitis. J Nutr Biochem 43:27–35
135. Yang Y, Kim B, Park YK, Koo SI, Lee JY (2015) Astaxanthin prevents TGFβ1-induced pro-fibrogenic gene expression by inhibiting Smad3 activation in hepatic stellate cells. Biochim Biophys Acta 1850:178–185
136. Fenni S, Hammou H, Astier J, Bonnet L, Karkeni E, Counturier C et al (2017) Lycopene and tomato powder supplementation similarly inhibit high fat diet induced obesity, inflammatory response, and associated metabolic disorders. Mol Nutr Food Res 61:160–183
137. Jiang W, Guo MH, Hai X (2016) Hepatoprotective and antioxidant effects of lycopene on non-alcoholic fatty liver disease in rat. World J Gastroenterol 22:10180–10188
138. Yucel Y, Oguz E, Kocarslan S, Tatli F, Gozeneli O, Seker A et al (2017) The effects of lycopene on methotrexate-induced liver injury in rats. Britisl Lek Listy 118:212–216
139. Assar EA, Chopra M, Hafizi S (2015) Lycopene suppresses the NFκB signaling pathway through inhibition of IκB kinase in human prostate cancer cells. In: Proceedings of the 106th annual meeting of the American Association for cancer research. Apr 18–22. Philadelphia, PA, USA. Abstract LB-101
140. Mercader J, Ribot J, Murano I, Felip F, Cinti S, Bonet ML et al (2006) Remodeling of white adipose tissue after retinoic acid administration in mice. Endocrinology 147:5325–5332
141. Liu C, Bronson RT, Russell RM, Wang XD (2011) β-cryptoxanthin supplementation prevents cigarette smoke-induced lung inflammation, oxidative damage, and squamous metaplasia in ferrets. Cancer Prev Res 4:1255–1266
142. Lorenzo Y, Azqueta A, Luna L, Bonilla F, Dominguez G, Collins AR (2009) The carotenoid β-cryptoxanthin stimulates the repair of DNA oxidation damage in addition to acting as an antioxidant in human cells. Carcinogenesis 30:308–314
143. Katsuura S, Imamura T, Bando N, Yamanishi R (2009) β-carotene and β-cryptoxanthin but not lutein evoke redox and immune changes in RAW264 murine macrophages. Mol Nutr Food Res 53:1396–1405
144. Kobori M, Ni Y, Takahashi Y, Watanabe N, Sugiura M, Ogawa K et al (2014) β-cryptoxanthin alleviates diet-induced nonalcoholic steatohepatitis by suppressing inflammatory gene expression in mice. PLoS One 9:e98294
145. Park HJ, Lee MK, Park YB, Shin YC, Choi MS (2011) Beneficial effects of *Undaria pinnatifida* ethanol extract on diet-induced-insulin resistance in C57BL/6J mice. Food Chem Toxicol 49:727–733

146. Maeda H, Hosokawa M, Sashima T, Funayama K, Miyashita K (2005) Fucoxanthin from edible seaweed, *Undaria pinnatifida*, shows antiobesity effect through UCPI expression in white adipose tissues. Biochem Biophys Res Commun 332:392–397
147. Tseng TH, Chu CY, Huang JM, Shiow SJ, Wang CJ (1995) Crocetin protects against oxidative damage in rat primary hepatocytes. Cancer Lett 97:61–67
148. Chen P, Chen y, Wang Y, Cai S, Deng L, Liu J et al (2016) Comparative evaluation of hepatoprotective activities of geniposide, crocins and crocetin by CCl_4-induced liver injury in mice. Biomol Ther 24:156–162
149. Xi L, Qian Z, Shen X, Wen N, Zhang Y (2005) Crocetin prevents dexamethasone induced insulin resistance in rats. Planta Med 71:917–922
150. Pang R, Tao JY, Zhang SL, Zhao L, Yue X, Wang YG (2010) In vitro antiviral activity of lutein against hepatitis B virus. Phytother Res 24:1627–1630
151. Sindhu ER, Firdous AP, Preethi KC, Kuttan R (2013) Carotenoid lutein protects rats from paracetamol, carbon tetrachloride and ethanol induced hepatic damage. J Pharm Pharmacol 62:1054–1060
152. Qiu X, Gao DH, Xiang X, Xiong YF, Zhu TS, Liu LG, Hao LP (2015) Ameliorative effects of lutein on non-alcoholic fatty liver disease in rats. World J Gastroenterol 21:8061–8072

Chapter 19
Eye Sight and Carotenoids

Shagufta Kamal, Muhammad Junaid, Arslan Ejaz, Ismat Bibi, and Nicu Bigiu

19.1 Introduction

Approximately 600 carotenoids have been identified to date yet only 50 to 60 carotenoids have been found to have a significant role in human diet, 15 in human serum while only two *meso*-zeaxanthin and zeaxanthin are exclusively found in eye macula [1–3]. The yellow coloration of retina is sign of carotenoids presence therefore also known as "macular pigments" (MPs) [4, 5]. It has been established from the research of decades that MPs are potent formidable anti-oxidants due to their ability to filter high energy blue light [6–9]. Scientists suggested may be zeaxanthin and lutein can curtail the risk of numerous eye diseases, especially late silhouette of AMD and also oxidative damage [10, 11]. It is well-known now that consumption of carotenoids rich diet is related with lower occurrence of cataract formation and age related macular degeneration [12] and a glitch of carotenoids shows clinical signs of corneal aberrations including, xerophthalmia, night blindness, corneal ulceration, scarring, keratomalacia, resultant conjunctiva and irreversible blindness [13] Increased mortality due to adaptive immunity and weakened innate immunity is due to carotenoids deficiency (pro-vitamin A) [14]. Table 19.1 indicates the chronological sequence of importance of carotenoids in eye related disorders.

Little intake of carotenoids in developing countries is a matter of concern where deficiency leads to high eye disease and mortality rates. This chapter may

S. Kamal (✉) · M. Junaid · A. Ejaz
Department of Biochemistry, Govt. College University, Faisalabad, Pakistan
e-mail: shaguftakamal@gcuf.edu.pk

I. Bibi
Department of Chemistry, The Islamia University of Bahawalpur, Bahawalpur, Punjab, Pakistan

N. Bigiu
Faculty of Medicine, Transilvania University of Brasov, Brașov, Romania

© Springer Nature Switzerland AG 2021
M. Zia-Ul-Haq et al. (eds.), *Carotenoids: Structure and Function in the Human Body*, https://doi.org/10.1007/978-3-030-46459-2_19

Table 19.1 Historical review of carotenoids in different eye diseases

Sr. no.	Year	Successive development	References
1	2019	Combination of zinc with β-carotene reduces 25% reduction in eye dis-orders	[13]
2	2018	Generation of new therapies to develop neural connection between visual system and eyes	[14]
3	2016	It was studied that lutein possesses preventive and/or protective role against cataracts	[15]
4	2015	It was found that higher consumption of bioavailable lutein/zeaxanthin is related with a long-term decreased threat of advanced AMD	[16]
5	2014	The risk of cataract can be decreased by diets that are optimized for vitamin C, lutein/zeaxanthin, vitamin B complex, multivitamins, omega-3 fatty acids, and carbohydrates: Prescribed levels of micronutrients are salutary	[17]
6	2007	Carotenoids and other antioxidants intake retard the development of envelopment of wet neo-vascular AMD and early stages of age-related macular degeneration (AMD)	[18]
7	2004	A study was done in which AMD patient were supplemented with numerous antioxidants including 10 mg lutein that resulted in positive effects in visual function, as well as improved contrast sensitivity, glare recovery and Snellen acuity	[19]
8	2003	Recently it suggests that lutein supplementation can improve visual function in AMD patients	[20]
9	2003	Relationship between increasing age and hypertension with AMD was studied. Scientists have found a direct association	[21]
10	2002	Major study was done by addition with β- carotene, vitamin C, E and α-tocopherol for 3 years in controlled patients of cataract. It produced a small decrease in development of age-related cataract	[22]
11	2001	It was found through studies that high doses of antioxidants including β- carotene, zinc, vitamin A and vitamin E had significantly reduced the risk of AMD.	[23]
12	1999	Studies confirmed that higher intakes of lutein and zeaxanthin lowers the possibility of cataract extraction in men	[24]
13	1998	It was found that green or yellow color foods like egg yolk and corn, along with green vegetables contain a higher lutein.	[25]
14	1997	It was found that concentration of zeaxanthin decreases rapidly while concentration of lutein decreases slowly toward the periphery of the retina	[26]
15	1997	Human studies were done which were suggested that accumulation of carotenoids could be due to dietary intake and consequently provide protection against retinal diseases. Therefore, it may protect against retinal degeneration	[27]
16	1997	Khachik found that carotenoids act as antioxidants to minimize the oxidative stress of the tissue that is caused by light and metabolism	[27]
17	1995	It was proved experimentally that Serum Lycopene protects from AMD	[24]

(continued)

Table 19.1 (continued)

Sr. no.	Year	Successive development	References
18	1995	Association between vitamin A malnutrition and severe retinal diseases were studied	[28]
19	1995	It was discovered that zeaxanthin and lutein, and its metabolites, these are only carotenoids present in the lens.	[29]
20	1995	It was found that lutein and zeaxanthin are optimally found to minimize risk from blue light	[30]
21	1994	Inverse relationship was determined between green vegetables intake and AMD risk	[31]
22	1994	The relationship b/w intake of carotenoids and vitamins E, C, A, and and the risk of AMD was evaluated.	[32]
23	1993	Evidences were obtained to suggest that protection from AMD can be attained from lutein and zeaxanthin.	[33]
24	1993	It was found that lutein & zeaxanthin and their metabolites are the solitary carotenoids found in the retina	[34]
25	1991	Cataract risk in persons with low plasma carotenoid levels was observed first time	[35]
26	1986	First large-scale randomized field experiments of the effect of vitamin A supplementation was completed to study its association with retinal diseases and mortality rate	[36]
27	1984	Macular pigment hypothesis was proposed. According to this hypothesis, macular pigments are responsible for the filtration of blue light which is very dangerous for the retinal epithelium and photoreceptors.	[37]
28	1983	Treatment of cataract was studied with respect to effect of β-carotene, α-tocopherol for the first time.	[38]
29	1982	Night blindness and small foamy white spots on the outer features of the conjunctiva association was studied first time	[39]
30	1960	Hubbenet utilize the term "Bitot's spots"for classical lesions in retinal disorders for the first time	[40]
31	1890–1950	No prominent research was found in the field of retinal diseases and carotenoids	[41]
32	1896	First time officially reporting of xerophthalmia cases	[42]

serve as gestalt of food sources of carotenoids and role of carotenoids in vision & abnormalities of carotenoids.

19.2 Macular Carotenoids

Highest concentration of zeaxanthin (Z), meso-zeaxanthin (MZ) and lutein (L) is found in human fovea with the ratio of 1:1:1 [44]. Approximately 100 folds decline in the concentration of macular carotenoids is observed a few mm away from the middle of fovea.

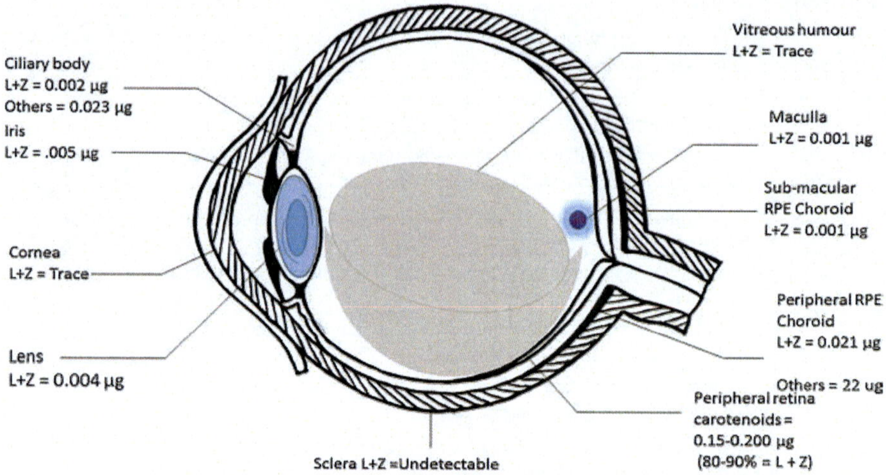

Fig. 19.1 Location of different carotenoids in different sections of internal macula

Cross-section of retinal macula (Fig. 19.1) revealed that these carotenoids are deposited in the internal plexiform layers of parafovea and in henle fiber layer of fovea [45–50]. Lutein or its isomers ensue vertical as well as horizontal orientations to the plane of the membrane while Zeaxanthin appears to be perpendicular [51, 52].

19.2.1 Advantage of Polyene Chain in Macula

Terminal hydroxyl groups and conjugated polyene chain of macular pigments upsurges its suitability as an antioxidant in polyunsaturated phospholipid rich membranes [53, 54]. These terminal OH^- groups allow Z, L and their isomeric forms to cross blood ocular and blood brain barriers [55]. β-carotene and lycopene due to deficiency of hydroxyl groups cannot cross ocular barriers and blood brain barriers [56–60]. Macular carotenoids are capable of satisfying singlet O_2 and trio state photosensitizers, scavenging ROS, plummeting lipofuscin development, impeding membrane phospholipids peroxidation and very prominent antioxidants [85]. These indispensable functions of macular pigment decrease oxidative stress in retina and improve vision in both normal retinas and abnormal retinas [61–65].

19.2.2 Chemical Nature of Macular Pigments

Hydroxyl groups of L, Z and MZ (Fig. 19.2) are attached with terminal ionone rings at the 3 and 3′ positions [83]. The position of double bond in L is at 4′, 5′ while in Z and MZ at 5′, 6′ position. The translocation of double bond position in Z & MZ and

Fig. 19.2 Position of OH in zeaxanthin and lutein

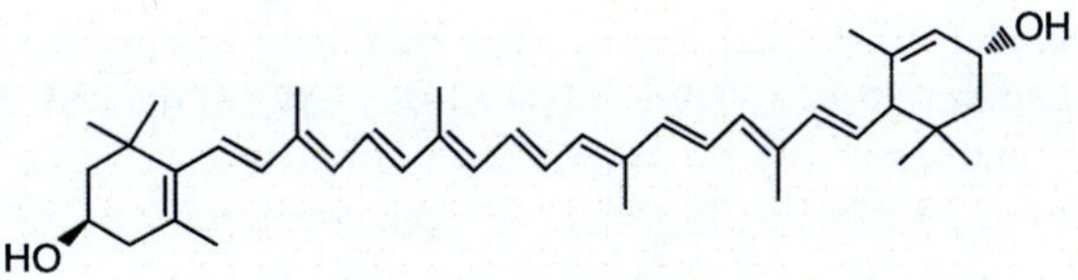

all trans-Lutein

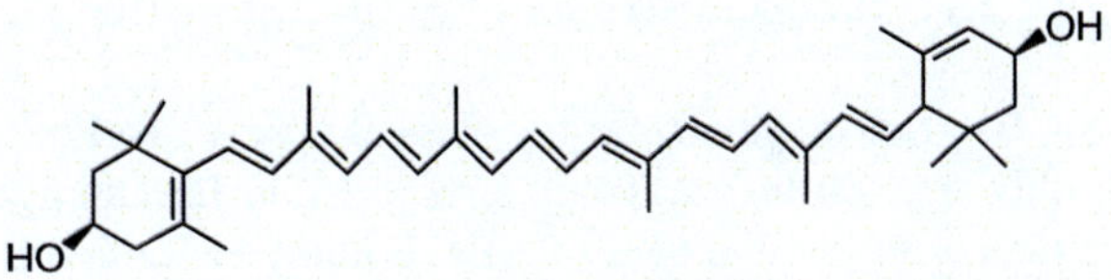

all trans-Zeaanthin

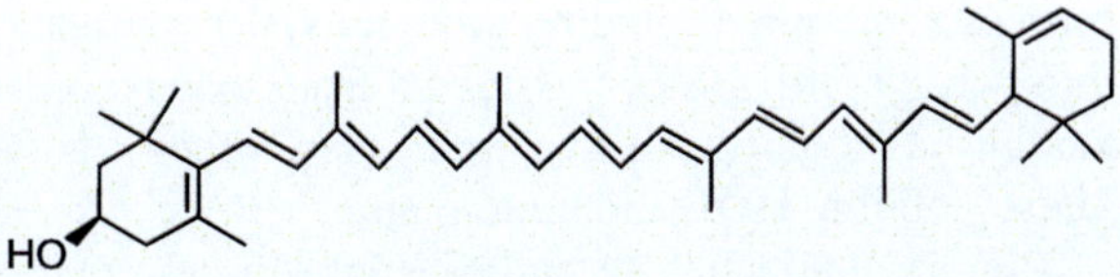

all trans- Beta- Cryptoxanthin

extra conjugated double bonds make them more stable and excellent antioxidant as compared to L [45, 58]. Z and L being chief carotenoids in macula and lens play idiosyncratic part against light-initiated oxidative damage [56, 66–70]. Significant level of L and Z are also observed in the outer membrane of rod and cortex of the lens where they shield these tissues from oxidative stress. Z and L exert beneficial effects by lowering lipofuscin, A2E formation and lipofuscin induced decrease in photo oxidative damage of RPE cells [71, 72]. Xanthophyll carotenoids also efficiently decrease the oxygen diffusion-concentration products, help to preserve membrane structure by preventing fatty acid peroxidation and control the rate of oxygen involved chemical reactions. Macular carotenoids due to their orientation in the membrane are also referred as "molecular rivets" and this property makes them more potent anti-oxidant to protect eye macula from oxidative damage than non-polar carotenoids such as lycopene or β-carotene [73, 74]. Macular carotenoids with protrude OH groups also increase the rigidity of membrane lipid bilayer [75].

Presence of non-dietary oxidative metabolites of carotenoids in the retina confirmed the anti-oxidant role of L and Z in the eye [76]. L and Z due to absorption of blue-light and destruction of reactive oxygen species (ROS) make them most potent anti-oxidant having superior filtering efficacy in model membrane system than β-carotene and lycopene [77]. While ordered orientation of high concentration of xanthophylls make them ideal optical filters on the other hand orthogonal orientations of two lutein molecule help to absorb light from all directions [78–80]. In accordance the diet, L dominates over Z in adult serum whilst this ratio is slightly variable and depends on distinct features such as life style, genetics etc. Hammond et al. [30] studied the supplementation of lutein or zeaxanthin with MP in primates and vertebrates. They reported that MP influenced the stability of Z or L as they remained stable for significant time periods (may be from many months to years) even after discontinuation of supplementation [81–85].

19.2.2.1 Biochemistry and Pharmacokinetics.

SR-B1 (Scavenger Receptor class B type-1) and NPC1L1 (cholesterol membrane influx transporter) are partly responsible for the uptake of lutein from intestinal lumen with involvement of ABC binding cassette transporters [86–88]. Member of steroidogenic acute regulatory domain "$StARD_3$" in human retina is determined as lutein binding protein. Glutathione S-transferase in p-isoform (GSTP1) "a special form of xanthophyll binding protein (XBP)" initiated the retinal capture of zeaxanthin [81, 89, 90]. GSTP1 extracted from macula presented highest affinity for zeaxanthin as compared to lutein than other xanthophyll carrier proteins like blactoglobulin, HDL and tubulin etc.

Dietary lutein before entering into the circulation via posterior vena cava is attached to chylomicrons consequently after being absorbed from enterocytes to serosal surface (Fig. 19.4). Lutein enters into circulatory system after absorbing into HDL from hepatocyte intake [91]. Prior to uptake esters of dietary lutein and zeaxanthin are hydrolyzed by esterases or lipases in the enterocytes or in the gut, yet esterification doesn't limit their bioavailability [92]. 17-day clinical studies of gelatin coated free and esterified lutein formulations reported that instead of esterification, bioavailability of lutein depends on solubilization and distribution in micelles [93]. Generally, bioavailability of carotenoids depends on dietary fiber contents, processing methods such as bio-encapsulation, heating etc., characteristics of food matrix, structural barriers [94]. Recent studies reported that dietary fiber decreases the bioavailability of lutein and zeaxanthin, whereas pasteurization increases the bioavailability of lutein without influencing the zeaxanthin [95]. Non-dietary factors which affect the availability and absorption of lutein and zeaxanthin consist of age, gender, body composition, alcohol intake, smoking, malabsorption of fats, and kidney or liver problem. Absorption of lutein is almost half from spinach or other vegetable sources than that from a purified crystalline lutein supplement [96–100].

Lutein and zeaxanthin are different from each other only in a double bond location in any of one hydroxyl group (Fig. 19.3). This hydroxyl group is the basis for the biochemical functions of these xanthophylls [101]. Dietary Lutein is sometimes converted into a non-dietary form called as meso-zeaxanthin. Due to less effective lutein conversion in infants there is less formation of meso-zeaxanthin. Anti-oxidant role of meso-zeoxanthin is more effective in central macula than lutein. Zeaxanthin appear to be taken up by cones of macula while lutein goes to rods and peripheral retina [102].

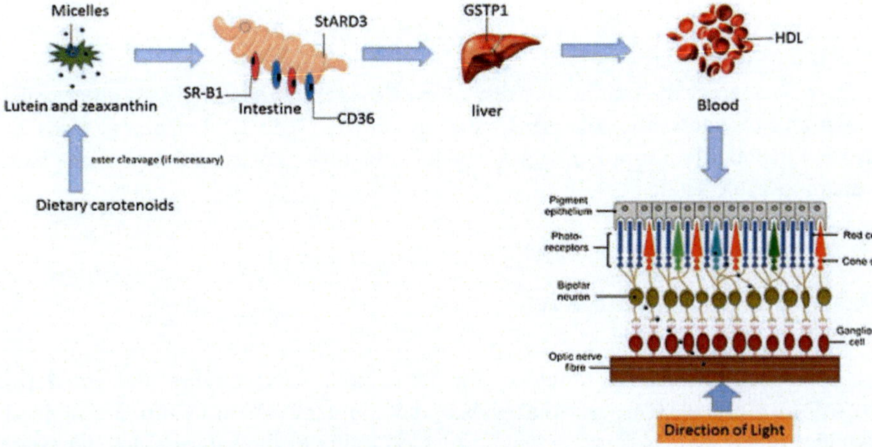

Fig. 19.3 Transport, uptake and absorption of carotenoids in eye (retina)

Interaction with Other Nutrients or Drugs

Red pigment in tomatoes is associated with decreased incidences of atherosclerosis, multiple sclerosis, age-related macular degeneration (ARMD), cancer and may other diseases [103]. Protection of lycopene against ARMD is surprising as lycopene is not present in significant concentrations; zeaxanthin, lutein and xanthophyll are prominent β-carotenoids [104]. Hydroxy-carotenoids and lycopene shared three type of reaction (Scheme 19.1) depending on nature such as zeaxanthin having polar chain, *type-(i)* reaction are feasible.

The strong ability of lycopene quenches radicals of astaxanthin indicated the link of lycopene with ARMD although not present in eyes (Eq. iv).

$$XAN^{\circ+} + LYC \rightarrow XAN + LYC^{\circ+} \qquad \text{(iv)}$$

Zeaxanthin and lutein interlink protect protein from the formation of protein cross-links and amino acid dimers (Eq. v).

$$CAR + ROO^{\circ} \rightarrow CAR^{+\circ} + ROO^{-} \ldots\ldots\ldots\ldots\ldots\ldots (i)$$

$$CAR + ROO^{\circ} \rightarrow CAR^{\circ} + ROOH \ldots\ldots\ldots\ldots\ldots\ldots (ii)$$

$$CAR + ROO^{\circ} \rightarrow ROO - CAR^{\circ} \ldots\ldots\ldots\ldots\ldots\ldots (iii)$$

Scheme 19.1 Three types of carotenoids reaction (i) carotenoid (CAR) radical cation formation (ii) neutral carotenoid formation (iii) addition of neutral carotenoid

$$CAR^{\circ+} + CySH \rightarrow CAR + CyS^{\circ} + H^{+} \quad \text{(v)}$$

They counteract the oxidative stress of eyes by communicating with other well-known anti-oxidants like vitamin C and vitamin E [105, 106]. Clinical trials of Zeaxanthin and lutein supplementation in different phases of eye disorders are given in detail in Table 19.2.

19.3 Anatomy of the Eye

Eye is "a complex sense organ" made from both mesenchyme and ectoderm (Fig. 19.4). Corneal, lens, lacrimal glands and conjunctival epithelium are molded from the surface of ectoderm on head side. Cell lining of the anterior chamber, part of vitreous body, iris, choroid, sclera and cornea are formed from mesenchyme. Protective structures of eye is composed of sclera, lids and the orbit, also christened sockets, are located in the front of skull. Orbital cavity protects eye ball and gives tough bony structure for ocular movement to six extrinsic muscles [107]. Cornea is a unique, transparent, flattened dome shaped structure located opposite to the pupil and iris has only nerve fibers without blood vessels [108]. Obviously, cornea serves as a filter having approximately 11 mm (0.43 inch) in diameter, 500 μm thickness in the center, and 700 μm in the periphery [109]. Edema results in increased light sprinkling and cloudy cornea. Complete opacity of the cornea will be produced in extreme conditions of swelled stroma [85]. An opaque sclera, white in adults, bluish tinge in children and yellowish in neonates forms posterior of eye ball [110]. Iris gives color to eye and blocks the entry of excess light in the eye. A magnifying glass "the lens" which focuses light is situated behind iris [111, 112]. Most important part of vision "the retina" with very sensitive photoreceptors having rod and cone cells, nerves and supporting cell is the brain sight center whereas macula is functional central of retina and also serves for color vision [113, 114]. Photoreceptors are chiefly involved with formation of electrical signals from light photon [115].

Cone pigments absorb light from yellow, blue, and green parts of spectrum and adopt with bright lights to resolve fine color vision while rod pigments absorb light from blue-green part of the spectrum [115, 116]. The density of rods and cones varies in different parts of the retina as fovea is deficient of rods but with high concentration of cones however huge numbers of rod cells are present in the periphery with very limited cone cells. Shapes of both rods and cones are long slender [117, 118]. Retinal pigment epithelium (RPE) composed of single layer cubical cells is the outer most layer of the retina in between choroid cramped with light absorbing capacity [119, 120]. Melanin, a black pigment in RPE is responsible for spattering and reflection of absorbed light. RPE is also involved with processing and elimination of photoreceptor waste products. The ability of RPE to cope with waste diminishes with the passage of time resulting in macular degeneration [121].

Table 19.2 Commercially available synthetic carotenoids

Sr.	Chemical nature	Trade name	Formulation	Concentration	References
1	Qoxy, health supplements	Lycopene, Methylcobalamine, carotenoids with multivitamin capsules	Capsules	Lycopene 6% (6000 mcg)	[128–134]
2	β- CAROTENE	Provitamins A	Capsules	15 mg	[135]
3	Dry vitamin A	Vitamin A and vegetable magnesium stearate	Tablets	5000 IU	[136–139]
4	Vitamin A	Vitamin A (as palmitate)	Capsules	25,000 IU	[140]
5	Vitamin A	Vitamin A	Soft gel	10,000 IU	[141]
6	Beta carotene	Provitamin A, carotenoids,	Soft gel	25,000 IU	[142]
7	Boots dry eyes vitamin A+ E eye drops	Vitamin A D + E	Drops	15 ml/1000 ml	[143, 144]
8	Vision essential gold	Lutein, zeaxanthin with Vit A	Capsules	40 mg lutein, 5000 IU Vit A	[145]
9	EyesightR$_x$	Carotenoids & Vit C	Tablets	50 mg	[146, 147]
10	Lutein	Lutein & Zeaxanthin	Capsules	6 mg	[148]
11	Vision health with lutein	Vit A+ lutein+ bilberry extract	Tablets	700 IU+ 10 mg + 100 mg	[152, 153]
12	Ultra-zeaxanthin	Lutein zeaxanthin & Vit C	Capsules	6 mg	[154]
13	Optein	Lutein, Vit C & E	Capsules	6 mg + 60 mg + 10 mg	[155]
14	Lutein with flora glow	Lutein & zeaxanthin	Soft gels	20 mg	[156–161]
15	Lutein with zeaxanthin	Lutein & zeaxanthin	Soft gels	10 mg	[162–166]
16	Ocuvite	Lutein & zeaxanthin	Capsules	6 mg	[167, 168]
17	Lutein	Lutein & zeaxanthin	Softgel capsules	40 mg	[169–172]
18	Advanced vision support	Lutein & Vit A	Capsules	25,000 IU	[86]
19	Blackcurrant + lutein	Lutein, blackcurrant	Capsules	200 mg + 10 mg	[173–176]
20	Bilberry Ginkgo eyebright complex	Vitamin A, C aND bilberry	Capsules	300 μg,300 mg, 20 mg	[177, 178]
21	Perfect vision	Beta Carotene & lutein	Capsules	12,500 IU, 10 mg,	[176, 179]
22	Lutein 20 mg	Lutein & zeaxanthin	Soft gel	20 mg	[180–182]

(continued)

Table 19.2 (continued)

Sr.	Chemical nature	Trade name	Formulation	Concentration	References
23	Eye pro MD	Lutein + xanthine+ zeaxanthin	Capsules	10 mg + 2 mg + 1 mg	[183]
24	Lycopene Lutein multivitamins multimineral capsule	Lycopene +lutein	Capsules	23 mg	[176]
25	Bio –Caroteen	β carotene + Vit E	Capsules	6 mg + 10 mg	[82]

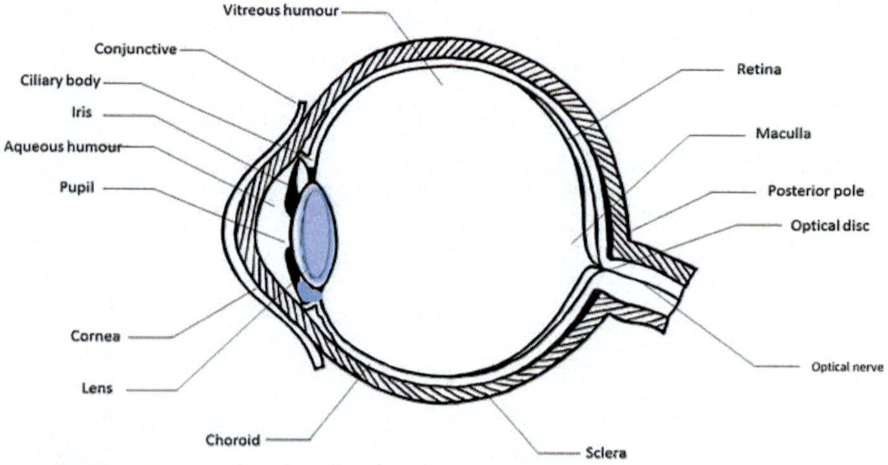

Fig. 19.4 Anatomy of eye

19.4 Commercially Available Synthetic Carotenoids for Eye Protection

Non-essential nutrient supplements are extensively accessible. Dietary supplements usage has become so popular that they can be easily fetched from supermarkets. Carotenoids are pervasive in fruits and vegetables rich diets [122]. The purpose of carotenoid supplements is either to enhance the intake of carotenoids in the individuals with previously well spent carotenoids from diet, or to supply carotenoids to those whose diet contains very limited amounts. Lutein, β-carotene, zeaxanthin and lycopene are four most important commercially available dietary supplements (Table 19.3). These carotenoids are also abundantly present in human serum [123]. Dietary supplements in different formulations e.g., capsule, sachets, liquids and tablets are available in market [124]. Oil suspension or oleoresin is the basic formulation. To increase the stability of the micronized synthetic carotenoid as compared with the pure crystalline form, it's usually dissolved in vegetable oils. Supplements are usually viscous yet pourable liquid due to presence of 20–30% carotenoids [124]. A two-piece carotenoid hard gel capsule, still flexible gelatin are oleoresins and oil suspension are traditionally used as powders. Starch or cellulose based

Table 19.3 Epidemiological studies examining lutein & zeaxanthin and age-related macular degeneration

Study year	Follow-up	Study population	Supplement dose	Sample size	Results	References
Huang et al., 2008	180 days	Both males & females, US	10 mg lutein & 2 mg zeaxanthin	40	↑ serum levels of lutein, zeaxanthin, MPOD	[228]
Landrum et al., 1997	140 days	Males, US	30 mg lutein	2	↑ serum lutein level and MPOD	[229]
Parisi et al., 2008	365 days	Both males and females, Italy	10 mg lutein, 1 mg Zeaxanthin plus antioxidants; placebo	27	↑ multifocal electro-retinogram N1–P1 response mplitude densities of R1	[230]
LUNA, Trieschmann et al., 2007	182 days	Both males and females, Germany	12 mg lutein and 1 mg zeaxanthin	136	↑ serum level lutein and MPOD	[231]
TOZAL, Cangemi et al., 2007	182 days	Both males and females, US	0.4 mg zeaxanthin, 8 mg lutein, + antioxidants	37	↑ visual acuity	[82]
Rosenthal et al., 2006	182 days	Both males and females, US	17.5 mg lutein	45	↑ serum lutein level without any effect on visual activity	[232]
LAST, Richer et al., 2004	365 days	Both males and females, US	20 mg lutein + antioxidants; placebo	90	↑ MPOD, visual acuity, contrast sensitivity, and glare recovery	[233]
Richer et al. (2004)	365 days	Men and women, US	Randomized, controlled 10 mg/d *all-trans* lutein	59	Patients experienced signiWcant 40% ↑ in MPOD; No signiWcant diVerence in adverse eVects vs. Placebo	[24]

(continued)

Table 19.3 (continued)

Study year	Follow-up	Study population	Supplement dose	Sample size	Results	References
Aleman et al. (2001)	182 days	Men and women, US	20 mg/d *all-trans* lutein; randomized, controlled	29	↑ macular pigment optical, density (MPOD); without any adverse eVects reported	[234]

carotenoids capsule for vegetarians are also available [124, 125]. Small spherical particles comprising of the carotenoid encapsulated in a gelatin-sucrose matrix are preferred for carotenoid tablets [126]. Mostly, carotenoid supplements (Table 19.3) contain multi-vitamin with multi-minerals as active ingredients, yet single dietary supplements e.g., lycopene or lutein are also available in the market [122].

Generally, these supplements contain few mg to 20 mg per dose carotenoid contents. A supplement ranges from a few mg up to around 20 mg per dose is the typical carotenoid content of dietary [127]. Doses >10,000 IU/d of vitamin A (both retinyl esters and retinol) are considered as safe and beyond these values malformations is observed in an epidemiologic study [127].

19.4.1 Complete Vision Cycle

ROL (Retinol) should be converted in chromophore to maintain and sustain vision after absorption by vertebrate eyes. Rhodopsin is an important protein present in the disc membranes of ROS (rod outer segments), and the chromophore and is covalently linked through Schiff base linkage. Isomerization of protein-bound chromophore from cis-to-trans is initiated by light that cause photo transduction [185]. RAL (Retinal) photoproduct is liberated by water entering from cytoplasmic side and causing hydrolysis of the Schiff base linkages. ABCA4 (ATP-binding cassette transporter) transfer the part of RAL to cytosol from disc lumen [186]. Retinol dehydrogenase (RDHs) catalyzes the reduction of RAL to ROL during first step of visual cycle. There are two short-chain dehydrogenase/reductase enzymes, one is RDH8 in photoreceptors outer segments and other one is RDH12 in photoreceptors inner segments which provide NADPH as a cofactor, are responsible for catalyzing this reaction in photoreceptors of mouse [187]. Under bright light conditions, large enzymatic capacity causes the redundancy that convert the reactive aldehyde group of photoproducts to alcohol (Fig. 19.5). Photoproduct can be present in millimolar concentration with in the cells after the bleaching of rhodopsin by bright light. Adducts can form by aldehyde group of photoproducts with primary amino group

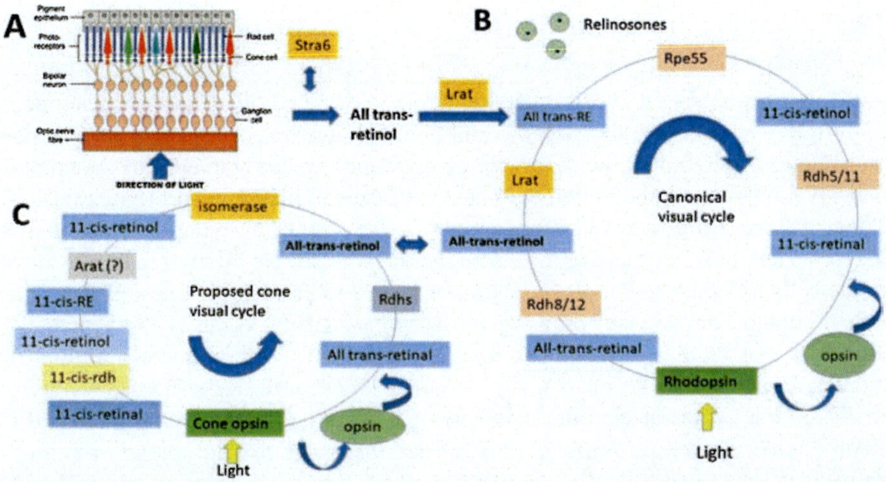

Fig. 19.5 Complete vision cycle

including lipids, protein and ribonucleotides present in many cellular molecules. Two molecules of RAL condense with the lipid phosphatidyl ethanolamine to form aberrant reaction side that is documented by the presence of retinoid A2E (bisretinoid N-retinylidene-N-retinylethanolamine). Redox reaction cause by the Ocular accumulation of A2E has been caught up in the pathology of eye diseases such as age-related macular degeneration [188]. ROL formed in ROS is facilitated to the RPE, where it undergoes esterification mediated by two retinoid-binding proteins, and CRBP1 (cellular retinol-binding protein-1), located within RPE cell. LRAT is the major ester synthase in RPE. In ocular retinoid metabolism LRAT play an important role to clarify the ROL from ROS and for the uptake of ROL from the blood. All trans-REs store vitamin A within internal membrane and retinosomes oil droplet-like structure due to their high hydrophobicity. However trans-RE also acts as substrate for RPE65 which catalyze the endothermic transformation of all trans-retinoid to its 11-cis conformation. 11-cis-retinol is the product of this isomerization which is then oxidized to 11-cis-retinal by the dehydrogenase/reductase activities of enzymes such as RDH5, RDH10 and RDH11 in the final catalytic step of visual cycle [189]. However, 11-cis-RDHs may also participate with RPE. Binding of cellular retinaldehyde-binding protein (CRALBP) to newly synthesize 11-cis-retinal leads to its protection, then initiate the transport, back to photoreceptors ROS, opsin couples with chromophore by completing the visual cycle. Any disturbance in enzymatic reaction during regeneration of chromophore in the RPE, especially in LRAT and RPE65 may affect retinal health. This deficiency of chromophore causes the slow death of rods cell and will contribute to activation visual phototransduction by un-liganded opsin, also causes the degeneration of cone [190].

19.5 Vision Abnormalities Related to Carotenoids Imbalance

In the past few decades, erudition about carotenoids metabolism has been exponentially improved. Regulatory aspects and aberrant side reactions of these are also unraveled by metabolic flow of targeted compounds via body in eyes of model animals. Animals have excogitated delineate pathways to maintain and sustain vision. These pathways are involved absorption of these dietary chromophore precursors like pro-vitamin A, all trans-retinol from the intestine, transportation in blood, entry in the cells and their metabolism. Requirement of vitamin A for tissue differentiation, immune competence, organogenesis and visual cycle clearly indicated the importance of vitamin A as an essential nutrient [191]. Approximately million or more cases of superfluous death and blindness every year throughout the developing world are the causes of vitamin A deficiency. β-Carotene an important, yet insufficient, source of vitamin A in developing nations, leads to widespread vitamin A deficiency. In many developing countries of Asia and Africa, approximately 20 million pregnant or lactating females and about 190 million pre-school children have severe vitamin A deficiency [192]. Mortality rate of VAD is higher in Pakistan than its counter parts [191]. According to Pakistan Institute of Community Ophthalmology [192] reports, extensively huge mass of Pakistani children below the age 15 years are seriously facing the consequences of xerophthalmia while 61% population (36% of females and 64% of males) have sever keratomalacia, corneal ulcer, blinding xerophthalmia, and corneal xerosis.

The prevalence of VAD is gradually increasing as in 2001 victims of VAD were only 5% females but recent data indicated that 42% females were facing the ramifications of vitamin A deficiency [94]. Therefore, dire need of time is to combat the hidden problem of hunger in Pakistan for the improvement of health status. Multiple interventions e.g., diversifying the dietary foods, supplementation through vaccines or capsules, bio fortification of food crops and industrial fortification of dietary food items to overcome the prevailing hazards of vitamin A deficiency have been declared by WHO [192].

Low innate retinal levels of zeaxanthin and lutein have adverse relation with risk of cataract, 1S2 glaucoma, age-related macular disease (AMD), various retinal diseases like pigmentosa (RP), diabetic retinopathy (DR) and blindness around the globe. Pathophysiology or etiology of all these diseases share some conventional mechanistic pathways and interestingly associated with aging. These mechanistic pathways comprising of apoptotic factors, inflammation and oxidative stress gives acumen for potentially convenient areas. In fact, many cases of ocular cell death due to oxidative stress of reactive oxygen or nitrogen species, lipid peroxidation in different eye disease have been reported [193]. Oxidative stress produces reactive oxygen species, which after interaction with the mitochondria, activate JNK pathway causing apoptosis. Surprisingly all these pathways are interconnected with the mechanism of action of several botanic compounds like zeaxanthin and lutein as glaucoma, DR, RP and AMD have a significant impact on worldwide populations [194].

19.5.1 *Xerophthalmia*

Xerophthalmia is a serious medical condition in which eyes are unable to produce tears. One of its major causes is the deficiency of vitamin A [195–197]. Conjunctiva in this situation develops wrinkled, dry and thick. Corneal ulceration which ultimately leads to blindness as a consequence of corneal damage if remain untreated. Systematic changes e.g., night blindness with conjunctival xerosis leads to corneal xerosis may occur in the eye, where Xerophthalmia earmarked as syndrome, with gradual worsening of vitamin A status. This reflects that retinol plays an essential role in the construction of rhodopsin [176, 179, 180]. According to WHO, signs of different eye diseases due to deficiency of vitamin A in children are.,

I. Nyctalopia (XN)
II. Conjunctival xerosis (X1A)
III. Bitot's spots (X1B)
IV. keratomalacia (X3B) also known as corneal ulcer covering 1/3 of the cornea
V. Corneal scarring (XS)

19.5.1.1 Nyctalopia

Children, pregnant and lactating women are more prone to nyctalopia. The prevalence of VAD is presented by local names e.g. night blindness. It is very difficult to find out night blindness, as it may be present at the time of birth or may be due to malnutrition. Persons infected with night blindness have not only poor vision in night, but also, they require spare time for their eyes to amend from luminously ignited zones to dim ones. Loss of peripheral vision due to presence of more rod cells results in night blindness. The child of 4 to 6 age groups will become less active and may be fearful of moving around [198]. Hemeralopia is another very rare and opposite problem of night blindness in which persons are unable to see in bright.

The characteristics of vitamin-A deficiency are usually restricted to the bulbar conjunctiva, but occasionally in long-standing cases, the conjunctiva of the lower lid and adjacent lower fornix may be rough and wrinkled [199].

19.5.2 Conjunctival Xerosis (Plate IA, D and Plate IIA, B, C)

This may be widespread throughout the visible part of bulbar conjunctiva or may be localized to a small part [200]. The literal meaning of "xerosis" is dryness which could be judged by lack of the normal luster or brilliance of the bulbar conjunctiva. The appearance has been likened to that of wax or dry paint. This occurs irrespective of the presence or absence of tears. Squares of xerosis emerge from their surroundings like "sandbars at retreating wave" when the child stops crying. This probably results from the disturbance of the steadiness of the pre-conjunctival film by the xerotic process in the epithelium [201]. Ability of the conjunctiva to transmit light is impaired, leading to reduced visibility of the conjunctival vessels [201]. On inspection with the slit-lamp, the translucent conjunctiva, which normally looks transparent (like an aquarium) and crossed by blood vessels, appears to be milky owed to fine precipitations. Soon the vascular pattern, apart from the large arterioles, becomes obscured. There is a tendency to generalized thickening and stiffness of the conjunctiva [202]. There are small, more or less vertical folds in the conjunctiva best demonstrated by rucking up the loose temporal conjunctiva against the outer canthus on maximal lateral movement of the eyeball [203].

Dark-skinned people have fine, diffuse smoky pigmentation, identical to the coarser and patchy pigmentation that is frequently seen in healthy objects of these races [204]. In sustained xerosis, the lower fornix first becomes yellowish, then light grey and finally dark brown owing to the presence of chromatophores in the basal cell layer of the epithelium. This characteristic "gutter" pigmentation responds slowly, over a period of weeks or months, to treatment [201].

19.5.3 Bitot's Spot (Plate IB, Plate IIA, B, C, and Plate IV)

The usual form taken by a Bitot's spot is a small plaque of a silvery-grey hue with a foamy surface. It is quite superficial and is raised above the general level of the conjunctiva; it is more or less readily removed by manipulation of the lids or direct wiping, revealing a xerotic conjunctival bed with a rough surface. Bitot's spot is invariably situated on the bulbar conjunctiva, frequently bilateral and temporal and less commonly nasal, and is usually confined to the interpalpebral fissure close to the limbus. This typical location of the spot would seem to be explained by the protection of the material here from the wiping movements of the lids, close to the protruding limbus. The shape varies considerably, being often irregularly circular or oval with the long axis horizontal. The classical triangular form with the base to the limbus is less common. Exceptionally, the following variations may be found. Bitot's spot material may be scattered widely over the conjunctiva, sometimes having a vertically corrugated arrangement [205].

Not all spots are foamy; some have a cheese-like or grease-like surface. Some accumulations are quite exuberant and not flat like a plaque. The significance of these differences in appearance is not known. The spots may be black in children whose eyelids are smeared with mascara (e.g., Kajal in India, a mixture of carbon and grease). If an unusual part of the conjunctiva is permanently exposed, as in strabismus, coloboma of the eyelid, or ectropion, a Bitot's spot may develop in relation to such an area, illustrating the etiological importance of exposure. Bitot's spot may or may not be associated with generalized conjunctival xerosis [206]. When so associated, the subjects are usually young children and may also have night blindness. These spots, together with the accompanying generalized xerosis, usually respond to vitamin-A therapy. Bitot's spots are also encountered in some parts of the world without generalized xerosis or evidence of retinal dysfunction, usually in older children and adults [207].

These are often minimal lesions; evidence of vitamin-A deficiency may be lacking (Plate IID), and there may be no response to therapy (Darby et al., 1960; Paton & McLaren, 1960). Accumulation of debris (Plate IC and Plate IIIA) in some cases of advanced xerosis, debris accumulates on the surface of the bulbar conjunctiva and may spread on to the adjacent part of the cornea. This material is creamy white, glistening, non-foamy and easily becomes detached to lie in the canthi, the lower fornix, or on the eyelid borders [208].

This is a quite unusual appearance, Bitot's spot being much more common. The material of the latter adheres more tenaciously to the eye. Any one, or even several, of these appearances may not be taken as diagnostic of conjunctival xerosis due to vitamin-A deficiency. The presence of most or all of them is highly suggestive of such a diagnosis, which will be confirmed by a return to the normal appearance under adequate therapy. Bitot's spot is very convenient gauge of vitamin-A deficiency, especially in young children, but it is non-pathognomonic. Corneal changes active stage these lead on from the appearances of conjunctival xerosis, which are evident with effect of cornea. Both corneas usually show changes but to widely varying degrees. Photophobia and inflammatory changes are not regarded as essential features. In the uncomplicated case, the mildness of congestion in the eye is truly remarkable [209].

19.5.4 Corneal Xerosis

Reversible changes (Plate ID) are categorized as general corneal xerosis, dryness, "un-wettability" and transparency loss, which lead to an early haziness of the cornea. These appearances may be demonstrated by holding the lids apart for 15 seconds. Slit-lamp examination at this early stage may reveal an increase of fine pigment in the para-limbal portions of the cornea, although it must be remembered that pigment in this area is common in healthy members of darkly pigmented races [210]. There may also be a loss of continuity of the surface epithelium and diminished tactile sensitivity. Later, cellular infiltration of the corneal stroma contributes to the intensity of the haziness of the cornea, which frequently has a cerulean, creamy appearance, mostly marked in the lower central part. In some cases, there is a cellular exudate in the lower part of the anterior chamber. Irreversible changes are characterized by the following signs [211]

19.5.5 Ulceration (Plate IE)

Ulceration involves a loss of substance of a part or the whole of the corneal thickness. This phenomenon is designated "ulceration" due to lack of more precise term, but it is characterized by mild signs of reaction or inflammation. Advanced degrees of stromal loss result in descemetocele and complete perforation with iris prolapse. These lesions are more common in the lower cornea [212].

19.5.6 Keratomalacia (Plate IF and Plate IIIB)

It is presented with a characteristic softening (colliquative necrosis) of the entire thick or more often the whole cornea, consistently leads to the destruction or deformation of the eyeball. It is very rapid process which resulted in melting of corneal structure into a cloudy gelatinous mass may be of dead-white or dirty-yellowish color. Loss of vitreous and extrusion of the lens may also happen. In untreated cases, endophthalmitis not infrequently supervenes. Particularly in children, keratomalacia may swiftly develop without the presence of characteristics changes, defined earlier, in the conjunctiva [213].

19.5.7 Corneal Scarring

The consequences of corneal ulceration and keratomalacia results in corneal scarring with the symptoms of phthisis bulbi (shriveled up eyes) or staphylomas (forward bulging) that depends on the magnitude of pomology in the cornea. Mostly signs of VAD are symmetrical and bilateral, so can accelerate blindness [214].

19.5.8 Age Related Macular Degeneration (AMD)

AMD is the major retinal ocular disorders investigated in elder population between 50 to 60 years and is the chief cause of irreversible vision loss. Due to degeneration of RPE, rods and cone cells in macula the central vision is disturbed as a result sharp vision is affected resulting in complete blindness [215]. The major causes of AMD are smoking, oxidative stress and genetics. Inflammation of macula is the major symptom diagnosed in AMD [216]. Results of eye disease study (AREDS) involving

60 to 80 years patients were proved that risk of AMD would be reduced by supplementation of lutein and zeaxanthin. It has been observed that persons taking 27% lutein and 35 to 55% zeaxanthin have a lower chance to develop hefty or extensive intermediate geographic atrophy, drusen and neovascular AMD [217].

It is admitted fact that dietary antioxidants, especially lutein and zeaxanthin, play a crucial role in the prevention of cataract formation and oxidation of lens proteins. Anti-oxidant supplementation with zinc has perceived to slow down the disease progression [218].

Carotenoids 's ability to accept nascent/atomic oxygen and other abilities like removal of free redicals and ROs, per-oxidation inhibition in phospholipids present in membranes and inhibition of formation of lipofuscin are the evidences of their anti-oxidant properties. Lutein and Zeaxanthin are considered as playing main role of protection against oxidative damage caused by light as they are the chief carotenoids present in lens ad macula of eyes. It was confirmed by Khachik et al. that Lutein and zeaxanthin are the main precursor for the anti-oxidative activity in eye. He found the presence of oxidative metabolites of lutein and zeaxanthin in retinal part of eye which was not originated by any dietary sources thus confirming he antioxidant activity of both compounds. [219, 220].

It was proved by evidences that zeaxanthin and lutein are the main dietary carotenoids that are involved in reducing and preventing Age related macular degeneration. It was showed by a study conducted in USA, a multi-center eye disorder case control type study, conducted by five different ophthalmology centers, that risk of AMD is mainly associated with concentration of lutein and zeaxanthin in eye macula and retina [221]. The outer retina is rich in PEF (polyunsaturated fatty acids) of which composition can be changed at a dangerous level by production of free redicals and their oxidative reactions. This damage can be prevented or reduced by some dietary nutrients which block this damage. These anti-oxidants also help in maintaining the choroidal blood vessels integrity. These vessels supply blood to the macular region of retina [222]. Epidemiological studies were performed to evaluate the relationship of dietary intake of lutein and zeaxanthin and their concentration level in blood. The results were not consistent but they indicated a defensive relationship between them. Seddon et al. used the date of EDCCS (Eye Disease Case-Control Study) and stated that the risk of AMD is inversely associated with the high dietary intake of carotenoids as well as it adjusted the other risk factors for AMD [223].

Moreover, current studies have also shown that visual functions in AMD patients were improved by the supplementation of lutein and zeaxanthin. Effects of antioxidants supplementation including lutein on visual performance were studied by Richer et al. Study was double blind placebo controlled performed on 90 patients of atrophic AMD who were studied for 1 year [224]. Lutein, Xanthophylls and zeaxanthin, present in human eye tissues specially in macula and lens have specific pattern of distribution. These xanthophylls, in macula and lens, thought to provide some exclusive function. Currently, association of low risk of AMD with high levels of lutein and zeaxanthin in blood serum and diet has been showed by a large number of epidemiological and clinical studies and trials. Laboratory data has also shown the role of both carotenoids against the protection of photo oxidative damage of neural retina as well as defense action against other common visually disabling diseases. These molecules do so by removing ROs and absorption of blue light. These evidences suggest that lutein and zeaxanthin work against AMD and other pigmentary abilities by delaying the oxidative events occurring in retina and lens [225].

It was found in a recent study that a high dose of antioxidant combination including vitamin C, E, β- carotene and zinc has reduced the risk of AMD but has not any noteworthy effect on progression or development of cataract. It was found in a study that people having a high risk of development of advanced stage AMD including patients with intermediate AMD and advanced AMD in only one eye, has reduced their risk by 25% by taking high dose of above-mentioned combination. It also reduced the risk of vision loss caused by AMD by almost 19% in the same group of patients. In normal subjects participating in the study has shown no effect on dietary intake of this combination. Due to the reason that single nutrient effect was not evaluated, specific effects couldn't be determined. As β- carotene is not present in eye macula or lens, it is understood that the results were positive due to other nutrients present in the combination [226, 227].

19.5.9 Cataract

Cataract is an ocular disorder which is age related commonly diagnosed at the age of 60 years or above. The protein breakage is the cause of opaque and cloudy lens. Because the lens is necessary to focus on close or far away objects, damage to it leads to blurry vision with decreased color and shape sensitivity. Several factors such as exposure to ultraviolet (UV) light, eye injury, smoking, diabetes and family history are recognized factors of cataract formation [235]. Many people don't have access to the surgeons and remain untreated and the ultimate result is complete blindness. This ocular disorder contributes about 51% of world blindness. Surgery is only option to treat cataract in which defected eye lens is replaced by artificial intraocular lens [236]. Recent data have shown that consumption of botanical compounds containing strong antioxidants may protect the degradation of eye proteins thus minimizing the effects of cataract [237].

Cataract is a condition in which lens become opaque in the perceiving of light which hinders the pathway of light. It results in visual impairment often [238]. Cataract is becoming more mundane with incrementing age within two decade and it is the consequential cause of incapacitation among old aged all over the world. More than 1,000,000 cataract surgeries are performed in a single year only in USA [239]. It is thought that pervasiveness of cataract will be increased by almost 50% as the number of elder Americans will be increased [231]. In the meantime, cataract is widely spread among almost 37% Chinese above the age of 50 [240]. It has been hypothesized that obviation of lens proteins oxidation and cataract prevalence is mainly due to the dietary carotenoids including zeaxanthin and lutein.

Epidemiological studies investigating the cognation of dietary as well as blood levels of zeaxanthin and lutein with the risk of cataract were performed which proposed a trend toward a protective cognation (Table 19.3). A long-term supplementation system of lutein was designed by Olmedilla et al. It was used to study the effects of lutein on the function of eyes of the cataract patients [241]. It was shown by the results that levels of lutein and its metabolites elevated in serum in a significant way and visual performance, activity and sensitivity of glare of test patients ameliorated. Moreover, in a double blinded placebo-controlled supplementation study of lutein and α- tocopherol as well as placebo in cataract patients, Snellen visual perception and glare sensitivity was ameliorated but only in lutein supplemented patients after two-year period of supplementation, while there was a trend toward the constant level and gradual decrease in visual perception with α-tocopherol and placebo supplementation, respectively [242].

It should be noted that the roles of lutein and zeaxanthin in the human ocular perceiver are not thoroughly understood, and the results from such studies have not been entirely consistent. Although we know that Xanthophyll binding proteins are responsible in the storage and stabilization of xanthophylls but up till now we did not know exactly about their exact capturing and stabilization mechanism and the identity of possible lutein-binding protein has thought to be vaguer. Further evidence must be accumulated to elucidate for highly affinitive and selective uptake of the carotenoids. [243]

β-carotene also plays a vital role in decrementing the jeopardy of cataract formation. Cataract has been linked to oxidative stress and free redicals damage. A survey study was done involving patients of cataracts versus controlled persons for the investigation of relationship between alimental factors and the antioxidant barricade of lens region of perception [163]. The threat of cataract development was inversely related with serum carotenoids level that is individuals having low serum level of carotenoids were more prone to cataract threat (almost 5.5 times) and vice versa. Although the lens contains little β-carotene, but β-carotene may forfend against cataract formation by decrementing the overall oxidation level in the body and, indirectly, in the lens [244].

19.5.10 Retinitis Pigmentosa (RP)

Another retinal disease is Retinitis pigmentosa (RP) which resulted in affected retinal rod and cone cells, accounts for decreased vision while blindness in severe cases [245]. Genetic predisposition is the main risk factor. It is very rare disease affecting only 1 in 4000 people in the United States [246–248]. As RP is rare disorder, yet it has been considered as significant problem due to lack of current treatments. However, studies in animal models are suggested that high doses of β-carotenoids such as vitamin A palmitate, have ability to slow down the disease [249]. The term "Vitamino therapy" is used for the treatment of retinitis pigmentosa which includes the combination of vitamins A and E. In vitamin therapy usually 15,000 units of vitamin A and 400 units of vitamin E per day are recommended for 5 to 12 years. Antioxidant and tropic effects of both vitamins protect cone and rod cells of eye [82]. Literature also proved that supplementation of vitamin A or E may decelerate the p42rogression of RP and ERG amplitude [246, 250]. Many plant extracts have proved significant effects on the treatment and prevention of both RP and DR. Still, limitations of clinical trials leave many reservations in the promising benefits of such supplementation.

19.5.11 Retinopathy of Prematurity (ROP)

It is a type of disease that causes the blindness in infants with low weights. ROP is associated with gestational age at birth and with weight. Even in the presence of recent treatment, ROP is still a devastating disease [251]. In retinal neoangiogenesis oxygen play a well characterized role. The regulators of retinal angiogenesis such as hypoxia-inducible factor-1(HIF-1) and vascular endothelial growth factors are regulated by low and high level of oxygen [252]. With increase in metabolic demand avascular retina then becomes hypoxic that mediate the pro-angiogenic factors to express that will initiate the aberrant angiogenesis, that will cause the complications in intravitreal neovascularization [70]. Due to imbalance between reactive oxygen species and developing retina in infants cause oxidative damage [253]. Retinal Tissue having long chain of poly-unsaturated fatty acid may also damage by lipid peroxidation [254]. Full term infants have high levels of anti-oxidants than pre-term infants, because endogenous anti-oxidants system is overcome in pre-term infants that create a pre-oxidative state that will lead to oxidative damage to structures of the cells [255].

Oxidative damage in retinal cells can be overcome by anti-oxidants shown by study about animal models of ROP by inhibiting micro-vascular degeneration [256]. Recent study proves that zeaxanthin inhibit the expression of VEGF and accretion of HIF-1 α protein that is due to hypoxia present in primary culture of retinal pigments in epithelium cells of human retina [257]. Lutein is the major carotenoids than Zeaxanthin and MZ during fetal development [258]. At birth lutein also present in umbilical cord that shows that there is placental transfer [259]. Anti-oxidants supplementation in pre-term infants can be used to suppress the progress of ROP. A random trial was done on 150 newborns showed that neonatal supplementation of lutein can decrease the level of hydro-peroxide by increasing the level of anti-oxidant [259]. Further relation between xanthophyll and ROP is demonstrated by four random trials. Due to prevalence of infant's retina, lutein can be used in primary supplementation of xanthophyll. A study work about supplementation of lutein and zeaxanthin through formula- fed infants and breast milk fed was done to inhibit the ROP in pre-term infants [260]. This study could not show statics about supplements; however, it reveals that supplemented infants showed that there is low progression in ROP from early to threshold ROP stages by 50% [261]. To check the effect of weight base dose a third trial was done, but better outcomes about carotenoids doses were given by AMD trial. But weight-base dose trial did not show the prevalence of ROP due to its limitation by small size sample [258]. Then to check the level of carotenoids in pre-term infants given formula-fed with or without lutein, lycopene and beta carotene in compare to full term infants fed with breast milk. The occurrence of ROP was similar in pre-mature formula-fed infants, but there was decrease level of ROP at various stages in supplemented infants. This study also shows the comparison between lutein concentrations with activity of photoreceptors that indicate that lutein level is related to sensitivity of photoreceptors at 50 weeks of age [258]. The authors showed that carotenoids supplementation of lutein and zeaxanthin is responsible for visual development by maintaining retinal health. Further clinical trials reveled that lutein have great effect against ROP progression [258].

19.5.12 Diabetic Retinopathy (DR)

Patients with both type-I and type-II diabetes may develop diabetic retinopathy. 140 million individuals all over the world are suffering from blindness affected by DR [262]. There are abnormal changes in the retinal blood vessels such as swelling or leakage that may be detected on the surface retina. DR pathology is described by four stages: mild non-proliferative retinopathy, moderate non-proliferative retinopathy, severe non-proliferative retinopathy, and proliferative retinopathy. Due to lack of treatment at first three stages patients should control their blood sugar levels, blood pressures, blood cholesterol level and by maintaining healthy life-style to control the DR [263].

During DR oxidative stress is caused by continuous hyperglycemia many different pathways [264]. Therefor experiments on animal's model suggest that Z and L can overcome such oxidative stress and maintain the health of retinal function and inhibit the DR [265]. Recent studies also suggested that lutein have role in neuroprotective activities and prevent the neural damage in diabetic retina [266]. The serum of type-II mellitus patient contains high amount of lutein and zeaxanthin and lycopene as compared to carotene family [267]. A paper published by Lima et al. 2016 showed patient suffering from diabetic had remarkable low level of MPOD than age-matched control. DR patient also has low amount of MPOD than healthy subjects, and those levels are related with glycosylated hemoglobin [268]. Then a paper was published by Hu et al. found that daily intake of Zeaxanthin and lutein have increased the level of MPOD in patients suffering from non-proliferative DR and also improve VA and macular edema in contrast to placebo [269]. Studies showed that MP role against DR is limited but lutein and zeaxanthin play a protecting role against DR by overcoming oxidative stress. Current researches suggest that dietary supplements show a healthy effect against retinal disease. Many plants extracts are now being used for the treatment and prevention against ocular disorders like RP and DR.

19.6 Carotenoids as Translational Medicine

The term translational medicine was coined first time in 2015 by European Society for Translational Medicine (EUSTM) as branch of the biomedical science. Translational medicine ensures the improvement in prevention, diagnosis and therapies to expand the glob of healthcare organization. For visual development different carotenoids are predominantly being used as translational medicine [270].

19.6.1 L and Z Role in Visual Function in Healthy Adults

Macular pigment improves visual function by light-filtering properties whereas lutein improves visual development by increasing neural proficiency in young and old adults. Lutein supplementation with zeaxanthin was found to improve contrast acuity thresholds at high mesopic levels for 12 months and with xanthophyll supplementation improves visual performance during night driving [271]. Macular pigment density, visual performance in glare function test and better contrast sensitivity were observed to be increased with 12 mg of lutein per day for six months [272]. Lutein and zeaxanthin supplementation also protects our eyes from hazardous effects of long term computer usage. Eye sightedness can be overcome with 5 mg lutein along 1 mg zeaxanthin and 200 mg black currant extract in healthy subject of 22 to 45 age groups years [273].

19.6.2 L and Z Role in Visual Function in Age-Related Diseased System

Clinical trials of 3640 patients of 55 to 80-year age groups given orally supplementation of Vitamin E (400 International Units), vitamin C (500 milligram), ZnO (80 milligram), β-carotene (15 milligram) and Cu_2O (2 milligram) for five years proved the 25% drop in the rate of AMD and 19% decrease in peril of restrained visual loss. Clinical experiments have also proved that oral supplementation of 2 mg zeaxanthin with 10 mg lutein; 650 mg eicos-apentaenoic acid and 350 mg docosahexaenoic acid could be used to overcome age-related macular degeneration [274]. In a double-blinded, controlled in a placebo nature study of cataract patient supplementation of 15 mg lutein for 3 times per week improves glare sensitivity and visual performance [275]. Supplementation of lutein (20–40 mg/day for twenty six weeks) improves visual acuity as well as mean visual-field area while in retinal degeneration patients; improvement was observed after 2–4 weeks of intrusion but plateaued at six to fourteen weeks [276].

19.6.3 Role in Early Life

Limited data is available for role of zeaxanthin and lutein in early neural growth [277]. The concentration of lutein varies in infant brains as it was observed from brain tissues of 30 children whose parents were died in the first of their life [278]. It has also been noted that parts of brain capable to store memory, control emotions, vision and hearing had significant accumulation of carotenoids. These carotenoids were lutein, zeaxanthin, cryptoxanthin and β-carotene [279]. However, later on literature proved that infants of 2–11 months had 43% β-carotene as major carotenoid, 28% lycopene, 13% α-carotene and 12% lutein respectively while decedents had 16% β-carotene, 2% lycopene and 59% lutein. The adult brains have twice amount of lutein (59% vs 13%) [4]. It was also reported that breast milk fed infants had great amount of lutein than formula-fed infants indicating that mode of feeding played a significant role in brain lutein concentration.

19.6.4 Role of Other Carotenoids in Visual Cycle

Literature revealed that zeaxanthin and lutein had played a pivotal role in deterrence of age related macular degradation, cataract and other visual disorders in retina and lens [4]. Yet it is well known fact that carotenoids are present in other tissues of eye e.g., ocular tissues and uveal tract of the eye provides the great visual approach (Fig. 19.6) [271, 272]. Recent data (Table 19.4) revealed that Beta carotene, γ-carotene, α-carotene and beta-cryptoxanthin are the basic nutritional precursor of the retinoids, significantly important for visual cycle of rhodopsin lightening and revival [280–285] (Fig. 19.6).

Table 19.4 Different levels of dietary mono-hydroxy carotenoids and hydrocarbon carotenoids in pooled extracts from human ciliary body and RPE/choroid

Sr. no.	Type of dietary carotenoids	Pooled ciliary body (n = 30)	RPE/choroid (n = 20)	References
1	α-Cryptoxanthin	1.36	N. D	[284]
2	β-Cryptoxanthin	0.36	N. D	[285]
5	γ-Carotene	4.48	N. D	[286]
6	Lycopene	7.80	8.64	[287]
7	α-Carotene	1.60	2.97	[288]
8	β-Carotene	2.72	10.80	[4]
	Total	22.82	22.41	[287]

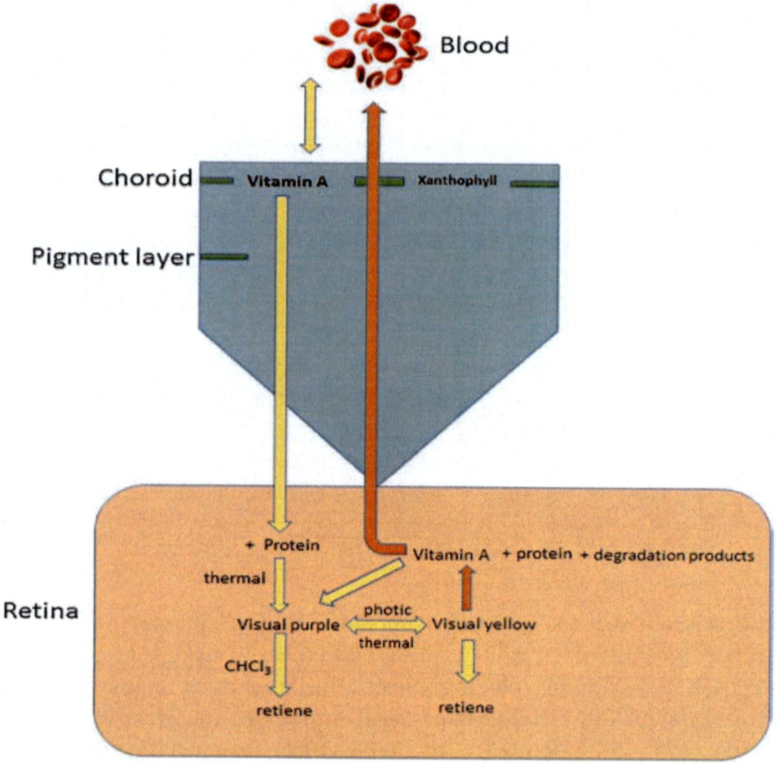

Fig. 19.6 Role of other carotenoids in eye (retina)

19.7 Conclusion

Carotenoids have gigantic variety of compounds ranging from β-carotenes to lutein, xanthine and vitamin A derivatives. For many years, the combination therapy of commercially available synthetic carotenoids with the natural may experience alleviate the treatment efficacy in humans with the precise examinations under experimental conditions. Relations of dietary carotenoids with naturally present in human body including β-carotenes or other carotenoids are always a hot issue for a biologist working on retinal disorders. Several clinical trials and studies indicated, from advanced AMD to cataract as well as Xerophthalmia and other macular diseases have been become consistent in current longitudinal epidemiological studies with respect to carotenoids. This suggests strong evidence that carotenoids protect the macula from oxidative stress, inflammation and exposure to harmful rays of sun light. Now it is utmost need of future research in field of understanding molecular mechanisms of carotenoids in human body and their implications in the treatment of macular degenerative diseases. Carotenoid therapy for macular cataract, ulceration and other retinal diseases should be analyzed at molecular level comprehensively which can be further enhanced by combining with latest equipment to enhance the quality of treatment. One example could be the use of light activated drugs as carotenoids are light pigments in nature to enhance its efficacy, but further research is required for the detailed study of its future prospects and side effects in humans.

References

1. Khachik F, Beecher GR, Goli MB, Lusby WR, Smith JC Jr (1992) Separation and identification of carotenoids and their oxidation products in the extracts of human plasma. Anal Chem 64(18):2111–2122
2. Nolan JM, Meagher K, Kashani S, Beatty S (2013) What is meso-zeaxanthin, and where does it come from? Eye 27(8):899
3. Bernstein PS, Delori FC, Richer S, van Kuijk FJ, Wenzel AJ (2010) The value of measurement of macular carotenoid pigment optical densities and distributions in age-related macular degeneration and other retinal disorders. Vis Res 50(7):716–728
4. Lima VC, Rosen RB, Farah M (2016) Macular pigment in retinal health and disease. Int J Retina Vitreous 2(1):19
5. Saini RK, Nile SH, Park SW (2015) Carotenoids from fruits and vegetables: chemistry, analysis, occurrence, bioavailability and biological activities. Food Res Int 76:735–750. de Oliveira VE, Castro HV, Edwards HG, de Oliveira LFC (2010) Carotenes and carotenoids in natural biological samples: a Raman spectroscopic analysis. J Raman Spectrosc, 41(6):642–650
6. Liu RH (2013) Health-promoting components of fruits and vegetables in the diet. Adv Nutr 4(3):384S–392S
7. Mecocci P, Polidori MC, Troiano L, Cherubini A, Cecchetti R, Pini G et al (2000) Plasma antioxidants and longevity: a study on healthy centenarians. Free Radic Biol Med 28(8):1243–1248
8. van Breemen RB, Pajkovic N (2008) Multitargeted therapy of cancer by lycopene. Cancer Lett 269(2):339–351
9. Rao PS, Kalva S, Yerramilli A, Mamidi S (2011) Free radicals and tissue damage: role of antioxidants. Free Radicals Antioxidants 1(4):2–7
10. Koushan K, Rusovici R, Li W, Ferguson LR, Chalam KV (2013) The role of lutein in eye-related disease. Nutrients 5(5):1823–1839
11. Khandhadia S, Lotery A (2010) Oxidation and age-related macular degeneration: insights from molecular biology. Expert Rev Mol Med 12:1–28
12. Cooper DA, Eldridge AL, Peters JC (1999) Dietary carotenoids and certain cancers, heart disease, and age-related macular degeneration: a review of recent research. Nutr Rev 57(7):201–214
13. Eye Science Labs 2019 493 Village Park Dr Powell, Ohio. 43065, 877-393-8482 or 877-EYE-VITA
14. Buscemi S, Corleo D, Francesco DP, Maria LP, Angela S, Giulio M (2018) The effect of lutein on eye and extra-eye health. Nutrients 10(9):1321
15. Krinsky NI, Johnson EJ (2005) Carotenoid actions and their relation to health and disease. Mol Asp Med 26(6):459–516
16. Priyadarshani AMB, Jansz ER (2014) A critical review on carotenoid research in Sri Lankan context and its outcomes. Crit Rev Food Sci Nutr 54(5):561–571
17. Snell S (1896) On nyctalopia with peculiar appearances on the conjunctiva. Trans Ophthalmol Soc U K 18:55–102
18. Nolan J, Landrum JT (2013) Carotenoids and retinal disease. CRC Press, Taylor and Francis Group, LLC
19. Hubbenet M (1860) Observations sur l'hemeralopie. Ann Ocul (Paris) 44:293

20. Sommer A (1982) Nutritional blindness: xerophthalmia and keratomalacia. Oxford University Press, New York
21. Bhuyan DK, Podos SM, Machlin LT, Bhagavan HN, Chondhury DN, Soja WS, Bhuyan KC (1983) Antioxidant in therapy of cataract II: effect of all-roc-alpha-tocopherol (vitamin E) in sugar-induced cataract in rabbits. Invest Ophthalmol Vis Sci 24:74
22. Sommer A, Tarwotjo I, Djunaedi E, West KP, Loeden AA, Tilden R, Mele L, the Aceh Study Group (1986) Impact of vitamin A supplementation on childhood mortality. A randomised controlled community trial. Lancet 1:1169–1173
23. Jacques PF, Chylack LTJ (1991) Epidemiologie evidence of a role for the antioxidant vitamins and carotenoids in cataract prevention. Am J Clin Nutr 53:352S–355S
24. Snodderly DM (1995) Evidence for protection against age-related macular degeneration by carotenoids and antioxidant vitamins. Am J Clin Nutr 62(Suppl):1448S–1461S
25. Eye Disease Case-Control Study Group (1993) Antioxidant status and neovascular age-related macular degeneration. Arch Ophthalmol 111:104–109
26. Seddon JM, Ajani UA, Sperduto RD, Hiller R, Blair N, Burton TC et al (1994) Dietary carotenoids, vitamins A, C, and E, and advanced age-related macular degeneration. JAMA 272(18):1413–1420
27. Seddon JM, Ajani UA, Sperduto RD, Hiller R, Blair N, Burton TC, Farber MD, Gragoudas ES, Haller J, Miller DT (1994) Dietary carotenoids, vitamins A, C, and E, and advanced age-related macular degeneration. Eye disease case-control study group. JAMA 272:1413–1420
28. Bernstein PS, Khachik F, Carvalho LS, Muir GJ, Zhao D-Y, Katz NB (2001) Identification and quantitation of carotenoids and their metabolites in the tissues of the human eye. Exp Eye Res 72:215–223
29. Sommer A, West KP (1995) Vitamin A deficiency: health, survival and vision. Oxford University Press, New York
30. Mares-Perlman JA, Brady WE, Klein BE, Klein R, Palta M, Bowen P, Stacewicz-Sapuntzakis M (1995a) Serum carotenoids and tocopherols and severity of nuclear and cortical opacities. Invest Ophthalmol Vis Sci 36:276–288
31. Khachik F, Bernstein PS, Garland DL (1997) Identification of lutein and zeaxanthin oxidation products in human and monkey retinas. Invest Ophthalmol Vis Sci 38:1802–1811
32. Hammond BR Jr, Johnson EJ, Russell RM, Krinsky NI, Yeum K-J, Edwards RB, Snodderly DM (1997) Dietary modification of human macular pigment density. Invest Ophthalmol Vis Sci 38:1795–1801
33. Bone RA, Landrum JT, Friedes LM, Gomez CM, Kilburn MD, Menendez E, Vidal I, Wang W (1997) Distribution of lutein and zeaxanthin stereoisomers in the human retina. Exp Eye Res 64:211–218
34. Sommerburg O, Keunen JEE, Bird AC, van Kuijk FJGM (1998) Fruits and vegetables that are sources for lutein and zeaxanthin: the macular pigment in human eyes. Br J Ophthalmol 82:907–910
35. Brown L, Rimm EB, Seddon JM, Giovannucci EL, Chasan-Taber L, Spiegelman D, Willett WC, Hankinson SE (1999) A prospective study of carotenoid intake and risk of cataract extraction in US men. Am J Clin Nutr 70:517–524
36. Yeum KJ, Shang FM, Schalch WM, Russell RM, Taylor A (1999) Fat-soluble nutrient concentrations in different layers of human cataractous lens. Curr Eye Res 19:502–505
37. Age-Related Eye Disease Study Research Group (2001b) A randomized, placebo-controlled, clinical trial of high-dose supplementation with vitamins C and E, beta carotene, and zinc for age-related macular degeneration and vision loss: AREDS report no. 8. Arch Ophthalmol 119:1417–1436
38. Chylack LT Jr, Brown NP, Bron A, Hurst M, Köpcke W, Thien U, Schalch W (2002) The Roche European American Cataract Trial (REACT): a randomized clinical trial to investigate the efficacy of an oral antioxidant micronutrient mixture to slow progression of age-related cataract. Ophthalmic Epidemiol 9:49–80

39. Miyazaki M, Nakamura H, Kubo M, Kiyohara Y, Oshima Y, Ishibashi T, Nose Y (2003) Risk factors for age related maculopathy in a Japanese population: the Hisayama study. Br J Ophthalmol 87:469–472
40. Falsini B, Piccardi M, Iarossi G, Fadda A, Merendino E, Valentini P (2003) Influence of shortterm antioxidant supplementation on macular function in age-related maculopathy: a pilot study including electrophysiologic assessment. Ophthalmology 110:51–60, discussion 61
41. Richer S, Stiles W, Statkute L, Pulido J, Frankowski J, Rudy D, Pei K, Tsipursky M, Nyland J (2004) Double-masked, placebo-controlled, randomized trial of lutein and antioxidant supplementation in the intervention of atrophic ate-related macular degeneration: the Veterans LAST study (Lutein Antioxidant Supplementation Trial). Optometry 75:216–230
42. SanGiovanni JP, Chew EY et al (2007) The relationship of dietary carotenoid and vitamin A, E, and C intake with age-related macular degeneration in a case-control study: AREDS Report No. 22. Arch Ophthalmol 125(9):1225–1232
43. Weikel KA, Garber C, Baburins A, Taylor A (2014) Nutritional modulation of cataract. Nutr Rev 72(1):30–47
44. Wu J, Cho E, Willett WC, Sastry SM, Schaumberg DA (2015) Intakes of lutein, zeaxanthin, and other carotenoids and age-related macular degeneration during 2 decades of prospective follow-up. JAMA Ophthalmol 133(12):1415–1424
45. Manayi A, Abdollahi M, Raman T, Nabavi SF, Habtemariam S, Daglia M, Nabavi SM (2016) Lutein and cataract: from bench to bedside. Crit Rev Biotechnol 36(5):829–839
46. Shetty PS (2010) Nutrition, immunity and infection. CABI, Wallingford
47. Gul K, Tak A, Singh AK, Singh P, Yousuf B, Wani AA (2015) Chemistry, encapsulation, and health benefits of β-carotene-a review. Cogent Food Agric 1(1):1018696
48. O'Connell OF, Ryan L, O'Brien NM (2007) Xanthophyll carotenoids are more bioaccessible from fruits than dark green vegetables. Nutr Res 27(5):258–264
49. Shi J, Le Maguer M, Kakuda Y, Liptay A, Niekamp F (1999) Lycopene degradation and isomerization in tomato dehydration. Food Res Int 32(1):15–21
50. Hayes JE, Stepanyan V, Allen P, O'grady MN, Kerry JP (2010) Effect of lutein, sesamol, ellagic acid and olive leaf extract on the quality and shelf-life stability of packaged raw minced beef patties. Meat Sci 84(4):613–620
51. Žnidarčič D, Ban D, Šircelj H (2011) Carotenoid and chlorophyll composition of commonly consumed leafy vegetables in Mediterranean countries. Food Chem 129(3):1164–1168
52. Pennington JA, Fisher RA (2010) Food component profiles for fruit and vegetable subgroups. J Food Compos Anal 23(5):411–418
53. Saini RK, Shetty NP, Giridhar P (2014) Carotenoid content in vegetative and reproductive parts of commercially grown Moringa oleifera Lam. cultivars from India by LC–APCI–MS. Eur Food Res Technol 238(6):971–978
54. Kubola J, Siriamornpun S (2011) Phytochemicals and antioxidant activity of different fruit fractions (peel, pulp, aril and seed) of Thai gac (Momordica cochinchinensis Spreng). Food Chem 127(3):1138–1145
55. Kirti K, Amita S, Priti S, Mukesh Kumar A, Jyoti S (2014) Colorful world of microbes: carotenoids and their applications. Adv Biol 2014:837891
56. Tanumihardjo SA (ed) (2012) Carotenoids and human health. Springer Science & Business Media; 2012 Nov 28
57. https://www.nutrition-and-you.com/Nutrition-facts-blog.html
58. Hussain A, Larsson H, Kuktaite R, Olsson ME, Johansson E (2015) Carotenoid content in organically produced wheat: relevance for human nutritional health on consumption. Int J Environ Res Public Health 12(11):14068–14083
59. Bai C, Capell T, Berman J, Medina V, Sandmann G, Christou P, Zhu C (2016) Bottlenecks in carotenoid biosynthesis and accumulation in rice endosperm are influenced by the precursor–product balance. Plant Biotechnol J 14(1):195–205

60. Stahl W, Sies H (2012) Photoprotection by dietary carotenoids: concept, mechanisms, evidence and future development. Mol Nutr Food Res 56(2):287–295
61. Abbo S, Molina C, Jungmann R, Grusak MA, Berkovitch Z, Reifen R et al (2005) Quantitative trait loci governing carotenoid concentration and weight in seeds of chickpea (Cicer arietinum L.). Theor Appl Genet 111(2):185–195
62. Larsen MK, Fretté XC, Kristensen T, Eriksen J, Søegaard K, Nielsen JH (2012) Fatty acid, tocopherol and carotenoid content in herbage and milk affected by sward composition and season of grazing. J Sci Food Agric 92(14):2891–2898
63. Hadley CW, Miller EC, Schwartz SJ, Clinton SK (2002) Tomatoes, lycopene, and prostate cancer: progress and promise. Exp Biol Med 227(10):869–880
64. Bauernfeind JC (1981) Carotenoids as colorants and vitamin A precursors; technological and nutritional applications. Academic Press, New York, pp 1–45
65. Ollilainen V, Heinonen M, Linkola E, Varo P, Koivistoinen P (1989) Carotenoids and retinoids in Finnish foods: dairy products and eggs. J Dairy Sci 72(9):2257–2265
66. Handelman GJ, Nightingale ZD, Lichtenstein AH, Schaefer EJ, Blumberg JB (1999) Lutein and zeaxanthin concentrations in plasma after dietary supplementation with egg yolk. Am J Clin Nutr 70(2):247–251
67. Mahmoud BH, Hexsel CL, Hamzavi IH, Lim HW (2008) Effects of visible light on the skin. Photochem Photobiol 84(2):450–462
68. Carvalho AP, Silva SO, Baptista JM, Malcata FX (2011) Light requirements in microalgal photobioreactors: an overview of biophotonic aspects. Appl Microbiol Biotechnol 89(5):1275–1288
69. Eisenreich W, Schwarz M, Cartayrade A, Arigoni D, Zenk MH, Bacher A (1998) The deoxyxylulose phosphate pathway of terpenoid biosynthesis in plants and microorganisms. Chem Biol 5(9):R221–R233
70. Pfander H (1992) [1] Carotenoids: an overview. Meth Enzymol 213(1992): 3–13
71. Villalobos-Castillejos F, Cerezal-Mezquita P, Jesús HD, Lourdes M, Barragán-Huerta BE (2013) Production and stability of water-dispersible astaxanthin oleoresin from Phaffia rhodozyma. Int J Food Sci Technol 48(6):1243–1251
72. Pérez-Gálvez A, Mínguez-Mosquera MI (2005) Esterification of xanthophylls and its effect on chemical behavior and bioavailability of carotenoids in the human. Nutr Res 25(7):631–640
73. Schweiggert U, Kurz C, Schieber A, Carle R (2007) Effects of processing and storage on the stability of free and esterified carotenoids of red peppers (Capsicum annuum L.) and hot chilli peppers (Capsicum frutescens L.). Eur Food Res Technol 225(2):261–270
74. Shahidi F, Zhong Y (2010) Lipid oxidation and improving the oxidative stability. Chem Soc Rev 39(11):4067–4079
75. Khoo HE, Prasad KN, Kong KW, Jiang Y, Ismail A (2011) Carotenoids and their isomers: color pigments in fruits and vegetables. Molecules 16(2):1710–1738
76. Amengual J, Lobo GP, Golczak M, Li HNM, Klimova T, Hoppel CL et al (2011) A mitochondrial enzyme degrades carotenoids and protects against oxidative stress. FASEB J 25(3):948–959
77. Weiter JJ, Delori F, Dorey CK (1988) Central sparing in annular macular degeneration. Am J Ophthalmol 106(15):286–292
78. Bhosale P, Li B, Sharifzadeh M, Gellermann W, Frederick JM, Tsuchida K, Bernstein PS (2009) Purification and partial characterization of a lutein-binding protein from human retina. Biochemistry 48(22):4798–4807
79. Seabra LMAJ, Pedrosa LFC (2010) Astaxanthin: structural and functional aspects. Rev Nutr 23(6):1041–1050
80. Batista AP, Raymundo A, Sousa I, Empis J (2006) Rheological characterization of coloured oil-in-water food emulsions with lutein and phycocyanin added to the oil and aqueous phases. Food Hydrocoll 20(1):44–52
81. Gorusupudi A, Bernstein PS (2016) Macular carotenoids: human health aspects. In: Carotenoids: nutrition, analysis and technology, vol 59

82. Ma L, Lin XM (2010) Effects of lutein and zeaxanthin on aspects of eye health. J Sci Food Agric 90(1):2–12
83. Polívka T, Zigmantas D, Sundström V, Formaggio E, Cinque G, Bassi R (2002) Carotenoid S1 state in a recombinant light-harvesting complex of photosystem II. Biochemistry 41(2):439–450
84. Dall'Osto L, Lico C, Alric J, Giuliano G, Havaux M, Bassi R (2006) Lutein is needed for efficient chlorophyll triplet quenching in the major LHCII antenna complex of higher plants and effective photoprotection in vivo under strong light. BMC Plant Biol 6(1):32
85. Pandhair V, Sekhon BS (2006) Reactive oxygen species and antioxidants in plants: an overview. J Plant Biochem Biotechnol 15(2):71–78
86. Abdel-Aal ESM, Akhtar H, Zaheer K, Ali R (2013) Dietary sources of lutein and zeaxanthin carotenoids and their role in eye health. Nutrients 5(4):1169–1185
87. Redmond TM (2017) RPE65 takes on another role in the vertebrate retina. Proc Natl Acad Sci 114(41):10818–10820
88. Stahl W, Sies H (2005) Bioactivity and protective effects of natural carotenoids. Biochim Biophys Acta (BBA) Mol Basis Dis 1740(2):101–107
89. Gammone MA, Riccioni G, D'Orazio N (2015) Marine carotenoids against oxidative stress: effects on human health. Mar Drugs 13(10):6226–6246
90. Yu BP (1994) Cellular defenses against damage from reactive oxygen species. Physiol Rev 74(1):139–162
91. Johnson EJ (2002) The role of carotenoids in human health. Nutr Clin Care 5:56–65
92. Bone RA, Landrum JT, Fernandez L, Tarsis SL (1988) Analysis of the macular pigment by HPLC: retinal distribution and age study. Invest Ophthalmol Vis Sci 29:843–849
93. Castenmiller JJ, West CE, Linssen JP et al (1999) The food matrix of spinach is a limiting factor in determining the bioavailability of beta-carotene and to a lesser extent of lutein in humans. J Nutr 129:349–355
94. Yeum KJ, Russell RM (2002) Carotenoid bioavailability and bioconversion. Annu Rev Nutr 22:483–504
95. O'Neill ME, Thurnham DI (1998) Intestinal absorption of beta-carotene, lycopene and lutein in men and women following a standard meal: response curves in the triacylglycerol-rich lipoprotein fraction. Br J Nutr 79:149–159
96. Kostic D, White WS, Olson JA (1995) Intestinal absorption, serum clearance, and interactions between lutein and beta-carotene when administered to human adults in separate or combined oral doses. Am J Clin Nutr 62:604–610
97. Albanes D, Virtamo J, Taylor PR et al (1997) Effects of supplemental beta-carotene, cigarette smoking, and alcohol consumption on serum carotenoids in the Alpha-Tocopherol, Beta Carotene Cancer Prevention Study. Am J Clin Nutr 66:336–372
98. Alberg A (2002) The influence of cigarette smoking on circulating concentrations of antioxidant micronutrients. Toxicology 180:121–137
99. Brady WE, Mares-Perlman JA, Bowen P, Stacewicz-Sapuntzakis M (1996) Human serum carotenoid concentrations are related to physiologic and lifestyle factors. J Nutr 126:129–137
100. Williams AW, Boileau TW, Erdman JW Jr (1998) Factors influencing the uptake and absorption of carotenoids. Proc Soc Exp Biol Med 218:106–108
101. Dagnelie G, Zorge IS, McDonald TM (2000) Lutein improves visual function in some patients with retinal degeneration: a pilot study via the internet. Optometry 71:147–164
102. Le Marchand L, Hankin JH, Bach F et al (1995) An ecological study of diet and lung cancer in the South Pacific. Int J Cancer 63:18–23
103. Granado F, Olmedilla B, Gil-Martinez E, Blanco I (1998) Lutein ester in serum after lutein supplementation in human subjects. Br J Nutr 80:445–449
104. Gonzalez de Mejia E, Ramos-Gomez M, Loarca-Pina G (1997) Antimutagenic activity of natural xanthophylls against aflatoxin B1 in Salmonella typhimurium. Environ Mol Mutagen 30:346–353

105. Rosenthal JM, Kim J, de Monasterio F, Thompson DJ, Bone RA, Landrum JT, de Moura FF, Khachik F, Chen H, Schleicher RL, Ferris FL, III, Chew EY. Dose-ranging study of lutein supplementation in persons aged 60 years or older. Invest Ophthalmol Vis Sci. 2006;47(12):5227–5233. van den Berg H (1998) Effect of lutein on beta-carotene absorption and cleavage. Int J Vitam Nutr Res 68, 360–365
106. Hendler SS, Rorvik DR (eds) (2001) PDR for nutritional supplements. Medical Economics Company, Inc., Montvale
107. Tang G, Serfaty-Lacrosniere C, Camilo ME, Russell RM (1996) Gastric acidity influences the blood response to a beta-carotene dose in humans. Am J Clin Nutr 64:622–626
108. Shegokar R, Mitri K (2012) Carotenoid lutein: a promising candidate for pharmaceutical and nutraceutical applications. J Dietary Suppl 9(3):183–210
109. Hammond BR Jr (2008) Possible role for dietary lutein and zeaxanthin in visual development. Nutr Rev 66(12):695–702
110. van der Made SM, Berendschot TT, Kijlstra A, Plat J (2017) One-year daily consumption of buttermilk drink containing lutein-enriched egg-yolks does not affect endothelial function in fasting and postprandial state. Sci Rep 7(1):1353
111. Berendschot TT, Makridaki M, van der Veen RL, Parry NR, Carden D, Murray IJ (2011) The clear (combination (of) lutein effects (on) aging retina) study; lutein supplementation improves visual acuity and night vision in early AMD; a two-centre, placebo-controlled study. Invest Ophthalmol Vis Sci 52(14):3631–3631
112. Evans JR, Lawrenson JG (2017) Antioxidant vitamin and mineral supplements for slowing the progression of age-related macular degeneration. Cochrane Database Syst Rev 7:1–108
113. Chew EY, Clemons TE, SanGiovanni JP, Danis R, Ferris FL, Elman M et al (2013) Lutein+ zeaxanthin and omega-3 fatty acids for age-related macular degeneration: the Age-Related Eye Disease Study 2 (AREDS2) randomized clinical trial. JAMA 309(19):2005–2015
114. Donaldson MJ, Pulido JS (2006) Treatment of nonexudative (dry) age-related macular degeneration. Curr Opin Ophthalmol 17(3):267–274
115. Weigert G, Kaya S, Pemp B, Sacu S, Lasta M, Werkmeister RM et al (2011) Effects of lutein supplementation on macular pigment optical density and visual acuity in patients with age-related macular degeneration. Invest Ophthalmol Vis Sci 52(11):8174–8178
116. Dawczynski J, Jentsch S, Schweitzer D, Hammer M, Lang GE, Strobel J (2013) Long term effects of lutein, zeaxanthin and omega-3-LCPUFAs supplementation on optical density of macular pigment in AMD patients: the LUTEGA study. Graefes Arch Clin Exp Ophthalmol 251(12):2711–2723
117. Bahrami H, Melia M, Dagnelie G (2006) Lutein supplementation in retinitis pigmentosa: PC-based vision assessment in a randomized double-masked placebo-controlled clinical trial [NCT00029289]. BMC Ophthalmol 6(1):23
118. Brinkmann CK, Rothenbuehler SP, Wolf-Schnurrbusch UEK, Wolf S (2009) Enhancement of macular pigment by oral lutein supplementation study (EMPOLS): first results. Invest Ophthalmol Vis Sci 50(13):1708–1708
119. Gong X, Rubin LP (2015) Role of macular xanthophylls in prevention of common neovascular retinopathies: retinopathy of prematurity and diabetic retinopathy. Arch Biochem Biophys 572:40–48
120. Vinson JA (2006) Oxidative stress in cataracts. Pathophysiology 13(3):151–162
121. Bian Q, Gao S, Zhou J, Qin J, Taylor A, Johnson EJ et al (2012) Lutein and zeaxanthin supplementation reduces photooxidative damage and modulates the expression of inflammation-related genes in retinal pigment epithelial cells. Free Radic Biol Med 53(6):1298–1307
122. Landrum JT, Bone RA (2004) Mechanistic evidence for eye diseases and carotenoids. Oxidative Stress Disease 13:445–472
123. Winkler BS, Boulton ME, Gottsch JD, Sternberg P (1999) Oxidative damage and age-related macular degeneration. Mol Vis 5:32
124. Cantrell A, McGarvey DJ, Truscott TG, Rancan F, Böhm F (2003) Singlet oxygen quenching by dietary carotenoids in a model membrane environment. Arch Biochem Biophys 412(1):47–54

125. Minatelli JA, Hill WS, Moerck RE (2013) U.S. Patent Application No. 13/889,392
126. Ahmed SS, Lott MN, Marcus DM (2005) The macular xanthophylls. Surv Ophthalmol 50(2):183–193
127. Trevithick-Sutton CC, Foote CS, Collins M, Trevithick JR (2006) The retinal carotenoids zeaxanthin and lutein scavenge superoxide and hydroxyl radicals: a chemiluminescence and ESR study. Mol Vis 12(12):1127–1135
128. Stahl W, Sies H (2003) Antioxidant activity of carotenoids. Mol Asp Med 24(6):345–351
129. Subczynski WK, Wisniewska A, Widomska J (2010) Location of macular xanthophylls in the most vulnerable regions of photoreceptor outer-segment membranes. Arch Biochem Biophys 504(1):61–66
130. Landrum JT, Bone RA (2001) Lutein, zeaxanthin, and the macular pigment. Arch Biochem Biophys 385(1):28–40
131. Mortensen A (2009) Supplements. In: Carotenoids, Basel, pp 67, Birkhäuser–82
132. DM Pharma, Sco 166, Sector 38-C, Chandigarh, 160014, India
133. The Natural Medicine Company, 1980s. Burgage industrial estate, Blessington, co. Wicklow, Ireland
134. Solgar, Inc 500 Willow Tree Road Leonia, NJ 07605 U.S.A.
135. Throne Research, Inc P.O. BOX 25, Dover, Idaho 83825 U.S.A.
136. Bronson Laboratories, 1960. 3350 south 400 West # 102 Lindon, Utah 84042–1931 untied states of America
137. Boots pharmaceuticals, 1849.United Kingdom
138. Dr. Whitaker wellness, 1979. USA
139. Advance Physician Formulas, 2003. Irvine, California USA
140. Nutrilite, 1939. California
141. Solaray, 1973. Park city
142. Herbs and Nature Corporation, unit 601 Regalia Park Towers 150 P. Tuazon cor. EDSA Cubao, Quezon City
143. Doctor's best Inc, 1990. 18100 Von Karman Ave, Ste 800 Irvine CA 92612 United States
144. Webber natural's pharmaceuticals ltd. Canada
145. Bausch + Lomb, 1853. Legacy Tower, Rochester, New York, United States
146. Natural brand. Florida and Canada
147. Nutriza select. USA
148. Jarrow formulas, 1988. USA, Mexico, Canada
149. Solgar, 1947. Aldburv, Tring, Herts, HP235PT. UK
150. Natural genetics company USA, PO BOX 18996, Fairfield OH 45018513-570-1296
151. Puritan's Pride Philippines, 1977. USA
152. BE WHOLE 5840Hwy 60 East Plant City, FL 33567
153. Sonika Plus, 2011. India
154. Pharma Nord, 1981. B.V. — Editiestraat 9 – 1321 NG Almere – Nederland
155. Li B, Zhao H, Liu J, Liu W, Fan S, Wu G, Zhao R (2015) Application of ultra-high performance supercritical fluid chromatography for the determination of carotenoids in dietary supplements. J Chromatogr A 1425:287–292
156. Skeie G, Braaten T, Hjartåker A, Lentjes M, Amiano P, Jakszyn P et al (2009) Use of dietary supplements in the European prospective investigation into Cancer and nutrition calibration study. Eur J Clin Nutr 63(S4):S226
157. Miller RA, Shelton TB (2016) U.S. Patent Application No. 14/890,262
158. Fang Z, Bhandari B (2012) Spray drying, freeze drying and related processes for food ingredient and nutraceutical encapsulation. In: Encapsulation technologies and delivery systems for food ingredients and nutraceuticals, pp 73–109
159. Heart Protection Study Collaborative Group (2002) MRC/BHF Heart Protection Study of antioxidant vitamin supplementation in 20 536 high-risk individuals: a randomised placebo-controlled trial. Lancet 360(9326):23–33

160. Miller RK, Hendrickx AG, Mills JL, Hummler H, Wiegand UW (1998) Periconceptional vitamin A use: how much is teratogenic? Reprod Toxicol 12(1):75–88
161. Olson JA, Krinsky NI (1995) Introduction: the colorful, fascinating world of the carotenoids: important physiologic modulators. FASEB J 9(15):1547–1550
162. Rao AV, Rao LG (2007) Carotenoids and human health. Pharmacol Res 55(3):207–216
163. Canene-Adams K, Erdman JW (2009) Absorption, transport, distribution in tissues and bioavailability. In: Carotenoids. Birkhäuser, Basel, pp 115–148
164. Vachali PP, Besch BM, Gonzalez-Fernandez F, Bernstein PS (2013) Carotenoids as possible interphotoreceptor retinoid-binding protein (IRBP) ligands: a surface plasmon resonance (SPR) based study. Arch Biochem Biophys 539:181–186
165. Debier C, Larondelle Y (2005) Vitamins A and E: metabolism, roles and transfer to offspring. Br J Nutr 93(2):153–174
166. Erdman JW, Bierer TL, Gugger ET (1993) Absorption and transport of carotenoids. Ann New York Acad Sci *691*(1):76–85
167. Puppel K, Kapusta A, Kuczyńska B (2015) The etiology of oxidative stress in the various species of animals, a review. J Sci Food Agric 95(11):2179–2184
168. Parker RS (1996) Absorption, metabolism, and transport of carotenoids. FASEB J 10(5):542–551
169. During A, Harrison EH (2004) Intestinal absorption and metabolism of carotenoids: insights from cell culture. Arch Biochem Biophys 430(1):77–88
170. Hedren E, Diaz V, Svanberg U (2002) Estimation of carotenoid accessibility from carrots determined by an in vitro digestion method. Eur J Clin Nutr 56(5):425
171. Perera CO, Yen GM (2007) Functional properties of carotenoids in human health. Int J Food Prop 10(2):201–230
172. Kijlstra A, Tian Y, Kelly ER, Berendschot TT (2012) Lutein: more than just a filter for blue light. Prog Retin Eye Res 31(4):303–315
173. Roodenburg AJ, Leenen R, van het Hof KH, Weststrate JA, Tijburg LB (2000) Amount of fat in the diet affects bioavailability of lutein esters but not of α-carotene, β-carotene, and vitamin E in humans. Am J Clin Nutr 71(5):1187–1193
174. van het Hof KH, West CE, Weststrate JA, Hautvast JG (2000) Dietary factors that affect the bioavailability of carotenoids. J Nutr 130(3):503–506
175. Mamatha BS, Baskaran V (2011) Effect of micellar lipids, dietary fiber and β-carotene on lutein bioavailability in aged rats with lutein deficiency. Nutrition 27(9):960–966
176. Reboul E (2013) Absorption of vitamin A and carotenoids by the enterocyte: focus on transport proteins. Nutrients 5(9):3563–3581
177. Von Lintig J (2010) Colors with functions: elucidating the biochemical and molecular basis of carotenoid metabolism. Annu Rev Nutr 30:35–56
178. Borel P (2012) Genetic variations involved in interindividual variability in carotenoid status. Mol Nutr Food Res 56(2):228–240
179. Du Sud C. Mechanisms of provitamin a (Carotenoid) and vitamin a (Retinol) transport into and out of intestinal CACO-2 cells. J Lipid Res 48(10):2283–2294
180. During A, Harrison EH (2007) Mechanisms of provitamin A (carotenoid) and vitamin A (retinol) transport into and out of intestinal Caco-2 cells. J Lipid Res 48(10):2283–2294
181. Harrison EH, Hussain MM (2001) Mechanisms involved in the intestinal digestion and absorption of dietary vitamin A. J Nutr 131(5):1405–1408
182. During A, Harrison EH (2006) Digestion and intestinal absorption of dietary carotenoids and vitamin A. In: Physiology of the gastrointestinal tract, 4th edn, pp 1735–1752
183. Desmarchelier C, Borel P (2017) Overview of carotenoid bioavailability determinants: from dietary factors to host genetic variations. Trends Food Sci Technol 69(2017):270–280
184. Li B, Vachali P, Frederick JM, Bernstein PS (2011) Identification of StARD3 as a lutein-binding protein in the macula of the primate retina. Biochemistry 50(13):2541–2549
185. Li B, Vachali PP, Gorusupudi A, Shen Z, Sharifzadeh H, Besch BM et al (2014) Inactivity of human β, β-carotene-9′, 10′-dioxygenase (BCO2) underlies retinal accumulation of the human macular carotenoid pigment. Proc Natl Acad Sci 111(28):10173–10178

186. Bhosale P, Bernstein PS (2005) Synergistic effects of zeaxanthin and its binding protein in the prevention of lipid membrane oxidation. Biochim Biophys Acta (BBA) – Mol Basis Dis 1740(2):116–121
187. Li B, Vachali P, Bernstein PS (2010) Human ocular carotenoid-binding proteins. Photochem Photobiol Sci 9(11):1418–1425
188. Fernández-García E, Carvajal-Lérida I, Jarén-Galán M, Garrido-Fernández J, Pérez-Gálvez A, Hornero-Méndez D (2012) Carotenoids bioavailability from foods: from plant pigments to efficient biological activities. Food Res Int 46(2):438–450
189. Ötles S, Ozgoz S (2014) Health effects of dietary fiber. ActascientiarumpolonorumTechnologiaalimentaria 13(2):191–202
190. Mudgil D, Barak S (2013) Composition, properties and health benefits of indigestible carbohydrate polymers as dietary fiber: a review. Int J Biol Macromol 61:1–6
191. Goltz SR, Ferruzzi MG (2013) Carotenoid bioavailability: influence of dietary lipid and fiber. In: Carotenoids and human health. Humana Press, Totowa, pp 111–128
192. Hur SJ, Lim BO, Decker EA, McClements DJ (2011) In vitro human digestion models for food applications. Food Chem 125(1):1–12
193. Beatty S, Nolan J, Kavanagh H, O'Donovan O (2004) Macular pigment optical density and its relationship with serum and dietary levels of lutein and zeaxanthin. Arch Biochem Biophys 430(1):70–76
194. Tanaka Y, Sasaki N, Ohmiya A (2008) Biosynthesis of plant pigments: anthocyanins, betalains and carotenoids. Plant J 54(4):733–749
195. Harrison EH (2012) Mechanisms involved in the intestinal absorption of dietary vitamin A and provitamin A carotenoids. Biochimica et BiophysicaActa (BBA)-Molecular and Cell Biology of Lipids 1821(1):70–77
196. Cuttriss AJ, Cazzonelli CI, Wurtzel ET, Pogson BJ (2011) University of Hawai ‘i at Manoa, Honolulu, HI, USA {Department of Biological Sciences, Lehman College, The City University of New York, Bronx, New York, USA {ARC Centre of Excellence in Plant Energy Biology, Research School of Biology, Australian National University
197. Ariviani S, Anggrahini S, Naruki S, Raharjo S (2015) Characterization and chemical stability evaluation of β-carotene microemulsions prepared by spontaneous emulsification method using VCO and palm oil as oil phase. Int Food Res J 22(6):2432–2439
198. Lindqvist A, Andersson S (2002) Biochemical properties of purified recombinant human β-carotene 15, 15′-monooxygenase. J Biol Chem 277(26):23942–23948
199. Al-Babili S, Beyer P (2005) Golden rice–five years on the road–five years to go? Trends Plant Sci 10(12):565–573
200. Poliakov E, Gentleman S, Chander P, Cunningham FX, Grigorenko BL, Nemuhin AV, Redmond TM (2009) Biochemical evidence for the tyrosine involvement in cationic intermediate stabilization in mouse β-carotene 15, 15′-monooxygenase. BMC Biochem 10(1):31
201. Shmarakov IO, Yuen JJ, Blaner WS (2013) Carotenoid metabolism and enzymology. In: Carotenoids and human health. pp 29–56. Humana Press, Totowa, NJ
202. Balagamwala M, Gazdar H (2013) Agriculture and nutrition in Pakistan: pathways and disconnects. IDS Bull 44(3):66–74
203. Lia B (1992) Ontogeny and ontology: Ontophyletics and enactive focal vision. Behavioral and Brain Sciences 15(1):43–44
204. Jeyakumar SM, Yasmeen R, Reichert B, Ziouzenkova O, Ziouzenkova O (2013) Metabolism of vitamin A in white adipose tissue and obesity. In: Carotenoids and vitamin A in translational medicine. CRC Press, Boca Raton, pp 23–51
205. Klepáček I, Malá PZ (2012) “Bochdalek's” skull: morphology report and reconstruction of face. Forensic Sci Med Pathol 8(4):451–459
206. Eghrari AO, Riazuddin SA, Gottsch JD (2015) Overview of the cornea: structure, function, and development. Prog Mol Biol Transl Sci 134:7–23. Academic
207. DelMonte DW, Kim T (2011) Anatomy and physiology of the cornea. J Cataract Refract Surg 37(3):588–598

208. Bonanno JA (2012) Molecular mechanisms underlying the corneal endothelial pump. Exp Eye Res 95(1):2–7
209. Roberts F (2015) The sclera. In: Eye pathology. Springer, Berlin/Heidelberg, pp 155–172
210. Khaw PT, Shah P, Elkington AR (2013) ABC of eyes. 4th ed. John Wiley & Sons. pp. 1–86
211. Pardo-López D, Gallego-Pinazo R, Mateo C, Rohrweck S, Suelves AM, Dolz-Marco R et al (2011) Serous macular detachment associated with dome-shaped macula and tilted disc. Case Rep Ophthalmol 2(1):111–115
212. Singh MS, Issa PC, Butler R, Martin C, Lipinski DM, Sekaran S et al (2013) Reversal of end-stage retinal degeneration and restoration of visual function by photoreceptor transplantation. Proc Natl Acad Sci 110(3):1101–1106
213. Kevany BM, Palczewski K (2010) Phagocytosis of retinal rod and cone photoreceptors. Physiology 25(1):8–15
214. Hubel DH (1988) Eye, brain, and vision, vol 22. Scientific American Library, New York
215. Schwartz SD, Hubschman JP, Heilwell G, Franco-Cardenas V, Pan CK, Ostrick RM et al (2012) Embryonic stem cell trials for macular degeneration: a preliminary report. Lancet 379(9817):713–720
216. Strauss O (2005) The retinal pigment epithelium in visual function. Physiol Rev 85(3):845–881
217. Bok D (1993) The retinal pigment epithelium: a versatile partner in vision. J Cell Sci 1993(Supplement 17):189–195
218. Von Lintig J (2012) Metabolism of carotenoids and retinoids related to vision. J Biol Chem 287(3):1627–1634
219. Quazi F, Molday RS (2013) Differential phospholipid substrates and directional transport by ATP-binding cassette proteins ABCA1, ABCA7, and ABCA4 and disease-causing mutants. J Biol Chem 288(48):34414–34426
220. Kiser PD, Golczak M, Maeda A, Palczewski K (2012) Key enzymes of the retinoid (visual) cycle in vertebrate retina. Biochim Biophys Acta (BBA) Mol Cell Biol Lipids 1821(1):137–151
221. Saari JC (2016) Vitamin A and vision. In: The biochemistry of retinoid signaling II. Springer, Dordrecht, pp 231–259
222. Sommer A, Vyas KS (2012) A global clinical view on vitamin A and carotenoids. Am J Clin Nutr 96(5):1204S–1206S
223. Jaenicke H, Höschle-Zeledon I (2006) Strategic framework for underutilized plant species research and development: with special reference to Asia and the Pacific, and to Sub-Saharan Africa. Bioversity International. Gunarathne Offset Limited, Colombo, Sri Lanka. ISBN: 955-1560-02-7
224. Sherwin JC, Reacher MH, Dean WH, Ngondi J (2012) Epidemiology of vitamin A deficiency and xerophthalmia in at-risk populations. Trans R Soc Trop Med Hyg 106(4):205–214
225. Jastrzebska B, Palczewski K, Golczak M (2011) Role of bulk water in hydrolysis of the rhodopsin chromophore. J Biol Chem 286(21):18930–18937
226. Thorne-Lyman AL, Fawzi WW (2012) Vitamin A and carotenoids during pregnancy and maternal, neonatal and infant health outcomes: a systematic review and meta-analysis. Paediatr Perinat Epidemiol 26(s1):36–54
227. Khachik F, de Moura FF, Zhao DY, Aebischer CP, Bernstein PS (2002) Transformations of selected carotenoids in plasma, liver, and ocular tissues of humans and in nonprimate animal-models. Invest Ophthalmol Vis Sci 43:3383–3392
228. Seddon JM, Ajani UA, Sperduto RD, Hiller R, Blair N, Burton TC, Farber MD, Gragoudas ES, Haller J, Mille DT et al (1994) Dietary carotenoids, vitamin A, C and E, and advanced age-related macular degeneration. Eye Disease Case-Control Study Group. JAMA 272:1413–1420
229. Organisciak DT, Wang HM, Li Z, Li ZY, Tso MOM (1985) The protective effect of ascorbate in retinal light damage of rats. Invest Ophthalmol Vis Sci 26:1580–1588
230. Snellen ELM, Verbeek ALM, van den Hoogen GWP, Cruysberg JRM, Hoyng CB (2002) Neovascular age-related macular degeneration and its relationship to antioxidant intake. Acta Ophthalmol Scand 80:368–371

231. Richer S, Stiles W, Statkute L, Pulido J, Frankowski J, Rudy D et al (2004) Double-masked, placebo-controlled, randomized trial of lutein and antioxidant supplementation in the intervention of atrophic age-related macular degeneration: the Veterans LAST study (Lutein Antioxidant Supplementation Trial). Optometry 75:216–230
232. Age-Related Eye Disease Study Research Group (AREDS) (2001) A randomized, placebo-controlled, clinical trial of high-dose supplementation with vitamins C and E, beta-carotene, and zinc for age-related macular degeneration and vision loss. Arch Ophthalmol 119:1417–1436
233. Yeum KJ, Taylor A, Tang G, Russell RM (1995) Measurement of carotenoid, retinoids, and tocopherols in humans lenses. Invest Ophthalmol Vis Sci 36:2756–2761
234. Huang LL, Coleman HR, Kim J, de Monasterio F, Wong WT, Schleicher RL et al (2008) Oral supplementation of lutein/zeaxanthin and omega-3 long chain polyunsaturated fatty acids in persons aged 60 years or older, with or without AMD. Invest Ophthalmol Vis Sci 49:3864–3869
235. Landrum JT, Bone RA, Joa H, Killburn MD, Moore LL, Sprague KE (1997) A one year study of the macular pigment: the effect of 140 days of a lutein supplement. Exp Eye Res 65:57–62
236. Parisi V, Tedeschi M, Gallinaro G, Varano M, Saviano S, Piermarocchi S (2008) Carotenoids and antioxidants in age-related maculopathy Italian study: multifocal electroretinogram modifications after 1 year. Ophthalmology 115:324–333
237. Trieschmann M, Beatty S, Nolan JM, Hense HW, Heimes B, Austermann U et al (2007) Changes in macular pigment optical density and serum concentrations of its constituent carotenoids following supplemental lutein and zeaxanthin: the LUNA study. Exp Eye Res 84:718–728
238. Cangemi FE (2007) TOZAL study: an open case–control study of an oral antioxidant and omega-3 supplement for dry AMD. BMC Ophthalmol 7:3
239. Rosenthal JM, Kim J, de Monasterio F, Thompson DJ, Bone RA, Landrum JT et al (2006) Dose-ranging study of lutein supplementation in persons aged 60 years or older. Invest Ophthalmol Vis Sci 47:5227–5233
240. Richer S, Stiles W, Statkute L, Pulido J, Frankowski J, Rudy D, Pei K, Tsipursky M, Nyland J (2004) Double-masked, placebo-controlled, randomized trial of lutein and antioxidant supplementation in the intervention of atrophic age-related macular degeneration: the Veterans LAST study (Lutein Antioxidant Supplementation Trial). Optometry 75:216–230
241. Aleman TS, Duncan JL, Bieber ML, de Castro E, Marks DA, Gardner LM, Steinberg JD, Cideciyan AV, Maguire MG, Jacobson SG (2001) Macular pigment and lutein supplementation in retinitis pigmentosa and Usher syndrome. Invest Ophthalmol Vis Sci 42:1873–1881
242. Nelson KD, McSoley JJ (2011) Clinical findings and management of conjunctival intraepithelial neoplasia. Optom J Am Optom Assoc 82(1):15–21
243. Krzyzanowska J, Czubacka A, Oleszek W (2010) Dietary phytochemicals and human health. In: Bio-farms for nutraceuticals. Springer, Boston, pp 74–98
244. Lin D, Barnett M, Grauer L, Robben J, Jewell A, Takemoto L et al (2005) Expression of superoxide dismutase in whole lens prevents cataract formation. Mol Vis 11:853–858
245. Wong TY, Klein BE, Klein R, Tomany SC, Lee KE (2001) Refractive errors and incident cataracts: the Beaver Dam Eye Study. Invest Ophthalmol Vis Sci 42:1449–1454
246. Chiu CJ, Morris MS, Rogers G, Jacques PF, Chylack LTJ, Tung W et al (2005) Carbohydrate intake and glycemic index in relation to the odds of early cortical and nuclear lens opacities. Am J Clin Nutr 81:1411–1416
247. Li T, He T, Tan X, Yang S, Li J, Peng Z et al (2009) Prevalence of age-related cataract in high-selenium areas of China. Biol Trace Elem Res 128:1–7
248. Olmedilla B, Granado F, Blanco I, Vaquero M, Cajigal C (2001) Lutein in patients with cataracts and age-related macular degeneration: a longterm supplementation study. J Sci Food Agric 81:904–909

249. Olmedilla B, Granado F, Blaco I, Vaquero M (2003) Lutein, but not alphatocopherol, supplementation improves visual function in patients with age-related cataracts: a2 year double-blind, placebo controlled pilot study. Nutrition 19:21–24
250. Jacques PF, Hartz SC, Chylack LTJ, McGandy RB, Sadowski JA (1988) Nutritional status in persons with and without senile cataract: blood vitamin and mineral levels. Am J Clin Nutr 48:152–158
251. Gaby SK, Singh VN (1991) Beta carotene. In: Gaby SK, Bendich A, Singh VN, Machlin LJ (eds) Vitamin intake and health: a scientific review. Marcel Dekker, New York, pp 929–957
252. Castenmiller JJM, West CE (1997) Bioavailability of carotenoids. Pure Appl Chem 69(10):2145–2150
253. Britton G, Liaaen-Jensen S, Pfander H (eds) (2008) Carotenoids, vol. 4: Natural functions. Springer Science & Business Media; 2008 Nov 14
254. McLaren DS, Oomen HAPC, Escapini H (1966) Ocular manifestations of vitamin-A deficiency in man. Bull World Health Organ 34(3):357
255. Joint, W. H. O., & World Health Organization. (1976). Vitamin A deficiency and xerophthalmia: report of a Joint WHO/USAID meeting
256. Oláh E (1940) Lines on the bulbus beneath the conjunctiva. Arch Ophthalmol 24(4):792–795
257. Phillips RE (2017) The physical exam: an innovative approach in the age of imaging. Springer International Publishing AG 2018. https://doi.org/10.1007/978-3-319-63847-8
258. Pauleikhoff D (2005) Neovascular age-related macular degeneration: natural history and treatment outcomes. Retina 25(8):1065–1084
259. Ratnapriya R, Chew EY (2013) Age-related macular degeneration—clinical review and genetics update. Clin Genet 84(2):160–166
260. Parekh N, Chappell RJ, Millen AE, Albert DM, Mares JA (2007) Association between vitamin D and age-related macular degeneration in the Third National Health and Nutrition Examination Survey, 1988 through 1994. Arch Ophthalmol 125(5):661–669
261. Krinsky NI, Landrum JT, Bone RA (2003) Biologic mechanisms of the protective role of lutein and zeaxanthin in the eye. Annu Rev Nutr 23(1):171–201
262. Chylack LT, Silverstone B, Lang MA, Rosenthal BP, Faye EE (2000) Age-related cataract. In: The lighthouse handbook on visual impairment and vision rehabilitation. Oxford University Press Inc, New York, pp 33–51
263. Ridley H (1952) Intra-ocular acrylic lenses: a recent development in the surgery of cataract. Br J Ophthalmol 36(3):113
264. Huynh TP, Mann SN, Mandal NA (2013) Botanical compounds: effects on major eye diseases. Evid Based Complement Alternat Med 1–12
265. Hartong DT, Berson EL, Dryja TP (2006) Retinitis pigmentosa. Lancet 368(9549):1795–1809
266. Bunker CH, Berson EL, Bromley WC, Hayes RP, Roderick TH (1984) Prevalence of retinitis pigmentosa in Maine. Am J Ophthalmol 97(3):357–365
267. Grøndahl J (1987) Estimation of prognosis and prevalence of retinitis pigmentosa and Usher syndrome in Norway. Clin Genet 31(4):255–264
268. Novak-Lauš K, Kukulj S, Zoric-Geber M, Bastaic O (2002) Primary tapetoretinal dystrophies as the cause of blindness and impaired vision in the republic of Croatia. Acta Clin Croat 41(1):23–27
269. Bernstein PS, Li B, Vachali PP, Gorusupudi A, Shyam R, Henriksen BS, Nolan JM (2016) Lutein, zeaxanthin, and meso-zeaxanthin: the basic and clinical science underlying carotenoid-based nutritional interventions against ocular disease. Prog Retin Eye Res 50:34–66
270. Musarella MA, MacDonald IM (2011) Current concepts in the treatment of retinitis pigmentosa. J Ophthalmol 2011:1–8
271. Berson EL, Rosner B, Sandberg MA, Hayes KC, Nicholson BW, Weigel-DiFranco C, Willett W (1993) A randomized trial of vitamin A and vitamin E supplementation for retinitis pigmentosa. Arch Ophthalmol 111(6):761–772
272. Chen J, Smith LE (2007) Retinopathy of prematurity. Angiogenesis 10(2):133–140

273. Semenza GL (2001) Hypoxia-inducible factor 1: oxygen homeostasis and disease pathophysiology. Trends Mol Med 7(8):345–350
274. Gariano RF, Gardner TW (2005) Retinal angiogenesis in development and disease. Nature 438(7070):960
275. Weinberger B, Laskin DL, Heck DE, Laskin JD (2002) Oxygen toxicity in premature infants. Toxicol Appl Pharmacol 181(1):60–67
276. Patterson E, Wall R, Fitzgerald GF, Ross RP, Stanton C (2012) Health implications of high dietary omega-6 polyunsaturated fatty acids. J Nutr Metabol 2012:1–12
277. Asikainen TM, White CW (2004) Pulmonary antioxidant defenses in the preterm newborn with respiratory distress and bronchopulmonary dysplasia in evolution: implications for antioxidant therapy. Antioxid Redox Signal 6(1):155–167
278. Kermorvant-Duchemin E, Sennlaub F, Sirinyan M, Brault S, Andelfinger G, Kooli A et al (2005) Trans-arachidonic acids generated during nitrative stress induce a thrombospondin-1–dependent microvascular degeneration. Nat Med 11(12):1339
279. Nagineni CN, Raju R, Nagineni KK, Kommineni VK, Cherukuri A, Kutty RK et al (2014) Resveratrol suppresses expression of VEGF by human retinal pigment epithelial cells: potential nutraceutical for age-related macular degeneration. Aging Dis 5(2):88
280. Scripsema NK, Hu DN, Rosen RB (2015) Lutein, zeaxanthin, and meso-zeaxanthin in the clinical management of eye disease. J Ophthalmol 2015:1–13
281. Perrone S, Tei M, Longini M, Santacroce A, Turrisi G, Proietti F et al (2014) Lipid and protein oxidation in newborn infants after lutein administration. Oxidative Med Cell Longev 2014:1–7
282. STOP-ROP Multicenter Study Group (2000) Supplemental therapeutic oxygen for prethreshold retinopathy of prematurity (STOP-ROP), a randomized, controlled trial. I: primary outcomes. Pediatrics 105(2):295–310
283. Klein R. (2008) The Epidemiology of Diabetic Retinopathy. In: Duh E.J. (eds) Diabetic Retinopathy. Contemporary Diabetes. Humana Press. https://doi.org/10.1007/978-1-59745-563-3_3
284. Chu J, Ali Y (2008) Diabetic retinopathy: a review. Drug Dev Res 69(1):1–14
285. Evans JL, Goldfine ID, Maddux BA, Grodsky GM (2002) Oxidative stress and stress-activated signaling pathways: a unifying hypothesis of type 2 diabetes. Endocr Rev 23(5):599–622
286. Li Q, Dinculescu A, Shan Z, Miller R, Pang J, Lewin AS et al (2008) Downregulation of p22phox in retinal pigment epithelial cells inhibits choroidal neovascularization in mice. Mol Ther 16(10):1688–1694
287. Ozawa Y, Sasaki M, Takahashi N, Kamoshita M, Miyake S, Tsubota K (2012) Neuroprotective effects of lutein in the retina. Curr Pharm Des 18(1):51–56
288. Nathan DM (1993) Long-term complications of diabetes mellitus. N Engl J Med 328(23):1676–1685

Chapter 20
Carotenoids and Cardiovascular Diseases

Sadia Javed, Saqib Mahmood, Muhammad Arshad, Shumaila Kiran, and Hanadi Talal Ahmedah

20.1 Introduction

Atherosclerosis related cardiac diseases is the largest reason of death, globally [1]. The incidence of these diseases is increasing day by day in developing countries [2] in spite of many discoveries in medical and surgical treatments. Despite the fact, cardiac diseases death rate has decreased in the last decades in different countries and it comprises 40% of the total death rate [3, 4]. The cardiovascular diseases including cerebrovascular accidents and acute coronary syndromes (ACS) leads to atherosclerotic plaque generation due to inner predisposing factors (hypertension, hypercholesterolemia and diabetes) as well as foreign agents like diet and lifestyle(physical inactivity, diet rich in fat, smoking, stress) (Fig. 20.1) [5]. All these internal and external factors can contribute to degeneration of vascular endothelium, the early stage of atherosclerotic process [6–9] and some clinical manifestations, such as stroke, acute coronary syndrome and peripheral arterial disease [10].

S. Javed
Department of Biochemistry, Government College University, Faisalabad, Pakistan

S. Mahmood (✉)
Department of Botany, Government College University, Faisalabad, Pakistan

M. Arshad
Department of Biochemistry, University of veterinary and Animal Sciences, Lahore, Pakistan

S. Kiran
Department of Applied Chemistry, Government College University, Faisalabad, Pakistan

H. T. Ahmedah (✉)
Department of Medical Laboratory Technology, Faculty of Applied Medical Sciences, King Abdulaziz University, Jeddah, Saudi Arabia
e-mail: hanadi_mt@hotmail.com

© Springer Nature Switzerland AG 2021

M. Zia-Ul-Haq et al. (eds.), *Carotenoids: Structure and Function in the Human Body*, https://doi.org/10.1007/978-3-030-46459-2_20

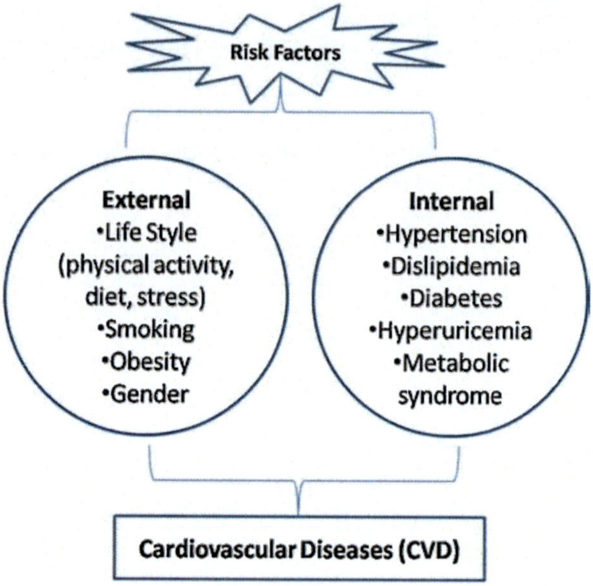

Fig. 20.1 Risk factors that responsible for cardiovascular diseases (CVD)

The leading causative agent for atherosclerotic plaque formation is the several oxidants including ROS and LDL [11]. Oxidized LDL has the ability to produce macrophage colony stimulating factors, macrophages colony-vitalizing factors and monocyte chemotactic protein 1 [12]. Inflammatory reactions in vascular walls and LDL peroxidation are increased by these molecules and LDL carries negative charges. These oxidized LDL is identified by macrophages, capable to capture them to form "foam cells", which are responsible to increase atherosclerotic process [12] Many studies suggested the use of carotenoids in food as a supplement plays role in cardiac diseases (CVD) [13–16]. Carotenoids offer numerous advantages to prevent the cardiovascular diseases by its beneficial effect such as enhanced immune response, prevent inflammation, intervene in cell cycle and neutralize the reactive oxygen species (Fig. 20.2).

Even though human body has its own scavenger system but from time to time it is unable to completely deal with oxidative attack [17]. That's why foreign antioxidants can be inserted to help scavenger system of human body against oxidation, by introducing them in diet such as fruits and vegetables. However, it should be considered that an antioxidant excess could not always regarded as advantageous to our body. Several studies [18] demonstrated that β-carotene supplementation highly elevated the death rates and CVD fear burden. Thus, more studies are important to completely validate such assumptions.

A diet rich in plant food increases health by providing different essential nutrient components and mechanisms. Naturally occurring compounds such as phytochemical are referred to as chemo preventers which possess anti-carcinogenic and

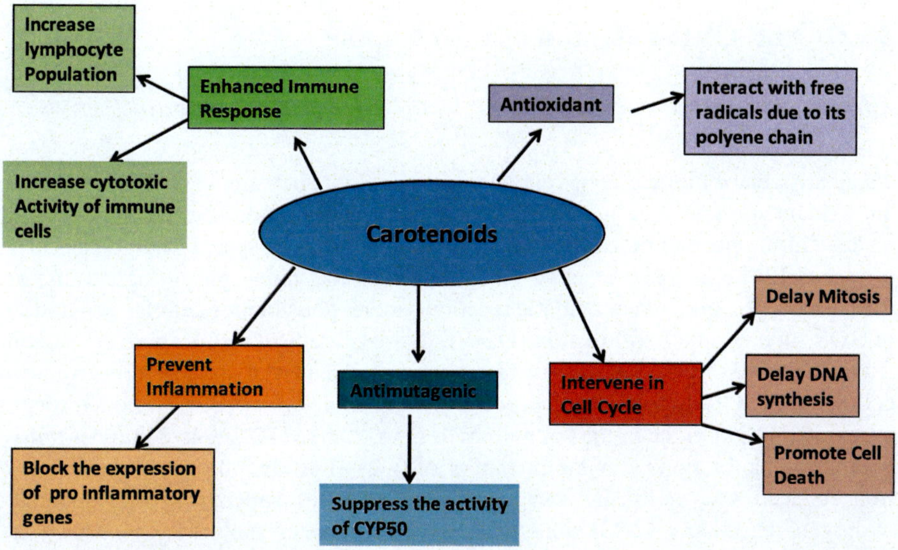

Fig. 20.2 General mechanism of action of carotenoids

other beneficial properties. These phytochemical provide protective action by means of their antioxidant activity and scavenger system. Some plant polyphenols, vitamins and pigments such as flavonoids, chlorophylls, carotenoids and betalains are chemo preventers. Specific antioxidants and other components of fruits and vegetables deserve major attention due to their potential protective roles. Carotenoids have been considered as \useful components, present in diet, are primarily derived from plants such as seeds, shoots, roots, leaves, flowers and fruit. Almost 50 carotenoids are consumed in human diet out of 600 carotenoid compounds that have been characterized [19, 20]. Consumption of high fruits and vegetables dilates blood antioxidant concentration in body tissues and possibly protects tissues and cells against oxidative damage. Generally carotenoids are consumed in diet and they are present in quantifiable concentrations in tissues and blood of humans [20, 21]; the most common areβ-carotene, lycopene, lutein, β-cryptoxanthin, α-carotene and zeaxanthin [21]. Carotenoids exhibit various biological effects, especially their antioxidant activity and other mechanisms [22]. Carotenoid plasma concentrations are recognized as effective biomarkers of complete intake of fruits and vegetables [23]. Studies concerning health effects of carotenoids are very diverse because carotenoids are a complex compounds and it is really hard to assume a brief systematic review or even a meta-analysis related to the health outcome of carotenoids.

20.2 Anti CVD Potential of Carotenoids

20.2.1 Pro-Oxidant and Antioxidant Effect of Carotenoids

There are a lot of biotic effects related to carotenoids. They are well known for their antioxidant and scavenging activity. They can act as regulator of intracellular redox status. During normal metabolism in body, radical oxygen species are consistently produced [24] and these reactive oxygen species has many physiological effects [25]. Enzymatic and non enzymatic reactions are responsible for cellular generation of ROS such as superoxide anion (O2•–), hydroxyl radical (HO•), peroxyl radical (ROO•) and alkoxylradical (RO•). Mitochondria considered as the foremost intercellular area of ROS production, especially superoxide anion (O2•–) and hydrogen peroxide (H2O2) in the organs of mammals (Fig. 20.3). ETC situated in inner mitochondrial membrane is a prime precursor of (O2•–) production when single electron reduced the molecular oxygen. Dismutation of superoxide can generate hydrogen peroxide (H2O2) and ultimately form water molecule, as shown in Table 20.1 [26].

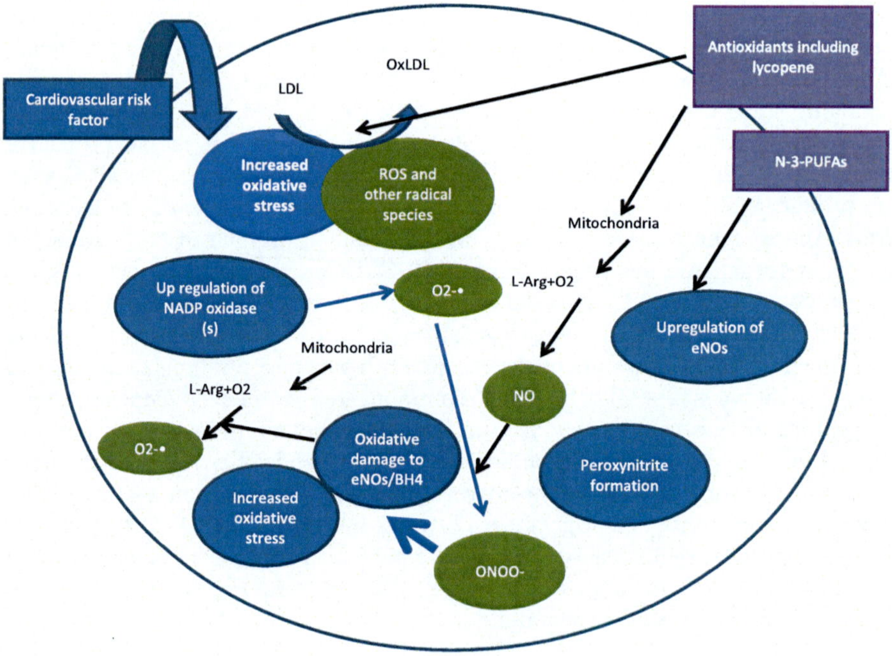

Fig. 20.3 shows the general role of antioxidants, lycopene and 3-nPUFAs (omega-3 polyunsaturated fatty acids) in the disruption of oxidation-sensitive events responsible for up regulation of eNOS (endothelial nitric oxide synthase) and oxidative damage to eNOS and/or BH4 ((6R)-5,6,7,8-tetrahydro-L-biopterin) and NADPH, (nicotinamide adenine dinucleotide phosphate) production of ROS (reactive oxygen species)

Table 20.1 Pro-oxidant and antioxidant activity of carotenoids

Carotenoids	
Pro-oxidant activity	Antioxidant activity
(1) Advanced generation of bioactive atomic species. (2) Modification of the antioxidant shields. (3) An elevation in oxidative damage (bio-molecules oxidation).	(1) Scavengers of ROS under physiological situations. (2) Functional deactivators of electronically elevated sensitizer molecules. (3) Defense of cellular membranes against oxidative damage. (4) Combined action with many antioxidants (*i.e.*, cooperative synergistic effects of vitamins C, E and β-carotene scavenging RNS).

The carotenoid act as energy acceptor and$^1O_2$ acts as an energy donor in quenching system is depend upon an energy transfer [27]. Moreover, no. of conjugated double bonds in carotenoids contributes to scavenging ability of carotenoids [28]. Stahl et al. [21] suggestedthatC-40 dialdehyde has higher scavenging ability than C-20-dialdehyde. GJC is capable to decrease the multiplication of chemical transformed cell [21] as well as it promoting the thought that the unsaturated bonds allows to capture the oxygen radicals. Finally, whole carotenoids contain unique potentials: few are with higher working intensity and few are with lower working intensity [29–31]. Lycopene is one of the most powerful among various substances, the most powerful one is lycopene to combat $^1O_2$, α-carotene isomer in the absence of pro vitamin is present mainly in tomatoes and the products derived from them [29–31]. Carotenoids are powerful scavengers due to huge abundance in double bonds. Therefore, carotenoids have the potential to prevent oxidative damage to DNA, lipid and proteins. Moreover, carotenoids can also serve as Pro oxidant atoms and lessen the augmentation of aggregate radical yield. The carotenoid concentration and oxygen partial pressure (pO_2) is an important element that regulate the switch of carotenoids to pro oxidant from antioxidant. At higher partial pressure of oxygen ,carotenoid radical combines with O_2 and produce a peroxyl radical which can fill in as an Pro oxidant and triggers the oxidation of unsaturated lipids. ROS generated oxidative stress may develop oxidized LDL, which is important in the progression of atherosclerosis. Atherosclerosis generally leads to ischemic stroke and cardiac arrest [32]. Specifically, the endothelial cells dynamically take part in the progression of inflammation. The deployment of WBCs to inflammation sites, secretion of chemical attractants by endothelial cells and amplified expression of sticking molecules that interact with WBCs and surface proteins. Cytokines and arachidonic acid derived from vessel wall cells activate endothelial secretion of many of these molecules.

20.2.2 Anti-inflammatory Activity

Several epidemiological studies exhibited the relation between dietary intake of carotenoids and prevention of CVD. CVD is associated with the relationship between circulating carotenoids and various oxidative stress, endothelial dysfunction and inflammation markers. Atherosclerosis is an inflammatory disease and carotenoids may help to reduce these inflammatory responses. NF-κB inflammatory pathway activation stimulates regulation of the expression of VCAM-1, ICAM-1 and E- selectin partially regulated by reactive oxygen species (ROS), causative agent of CVD. Lycopene derived from natural sources, inhibited LPS-induced nitric oxide and interleukin-6 production with reduced mRNAs of inducible nitric oxide synthase and interleukin-6 but had no impact on TNF-α. Furthermore, lycopene also inhibited lipo-polysacchride induced IκB phosphorylation, IκB degradation and NF-κB translocation. Nevertheless, lycopene obstruct the phosphorylation of extra-cellular flag controlled kinase and p38 MAPK yet not c-Jun amino terminal kinase [33]. Riso et al. [34] exhibited that after dietary admission of tomato-based drink, blood convergences of TNF-α cytokine in the volunteers were diminished. Napolitano et al. [35] reported that macrophage foam cells formation may reduce by lycopene in response to change in low density lipoprotein by reducing lipid biosynthesis in the cells. However, these positive effects are coinciding by a remarkable reduction in the creation of the Interleukin-10 (calming cytokine),which can leads to rise in pro-inflammatory profile of macrophages. Lutein deploys powerful anti-inflammatory and antioxidant effects in aortic tissue that may shield against formation of plaque in guinea pigs [36]. Inflammatory mediators such as interleukins IL-1β, and IL-8 and TNF-α raise the sticking of LDL to endothelial cells and up regulate the expression of WBCs adhesion molecules on endothelial cells while plaque formation. Moreover, dietary lutein and fat has various effects on lipo-polysacchride induced nitric oxide synthase mRNA levels in chicken macrophages. Finally, dose dependent lutein and fat act through the activated receptor PPARγ and RXR pathway to change the expression of inducible nitric oxide synthase (iNOS) [37]. Anti-inflammatory activity in male King ming mice on the impact of lycopene utilizing oil-actuated ear edema was dictated by a noteworthy basic screening strategy [38]. Lycopene amongst all carotenoids was found to be most effective on inflammation, when given as tomato oleoresin. Ohgami et al. [39] established the effects of intravenously administered astaxanthin, including anti-inflammatory [reductions in NO, TNF-α, and PGE2 levels and direct iNOS hampering function. Furthermore, Astaxanthin resists inflammation by inhibiting the Cox-1 and Cox-2 (Cyclooxygenases enzymes) in another pathway. In addition, astaxanthin suppresses nitric oxide, tumor necrosis factor-α, prostaglandins E-2, Cox-2 enzyme, as well as the nuclear factor kappa-B andIL-1β in an experiment done in vitro as well as in vivo [40].

20.2.3 Endothelial Dysfunction

Elevated ROS production is due to inhibition of NO in endothelial cells and can promote cardiovascular diseases (Fig. 20.4). In the vascular endothelium, the free radical NO• is produced by nitric oxide synthase (NOS) from arginine, converted L-arginine to L-citrulline:

$$\text{L-arg} + O_2 + \text{NADPH} \xrightarrow{\text{NOS}} \text{NO}^{\circ} + \text{citrulline}$$

Calmodulin, tetrahydrobiopterin (BH4) and NADPH are required as cofactors. Prevention of TNF-α expression by B-Carotene and lycopene supplementation was related to reduce inflammatory response and nitro oxidative stress in endothelial cells [41]. Meanwhile, lycopene arrests endothelin-1 expression by inducing heme oxygenase-1 expression (HO-1) and inhibiting ROS generation [42] in human endothelial cells during arresting (TNF)-α–induced NF-κB activation, monocyte endothelial adhesion and ICAM-1 expression [15]. Additionally, lycopene repressed endothelial brokenness in STZ-upgraded diabetic rats by bringing down oxidative pressure, which is included for the improvement of medications to anticipate diabetic vascular difficulties [16]. Moreover, astaxanthin arrests NO and ROS production by repressing NF-κB pathway [43]. Therefore, diseases associated with oxidative stress could be treated with carotenoids, such as CVD [46]. Martin et al.

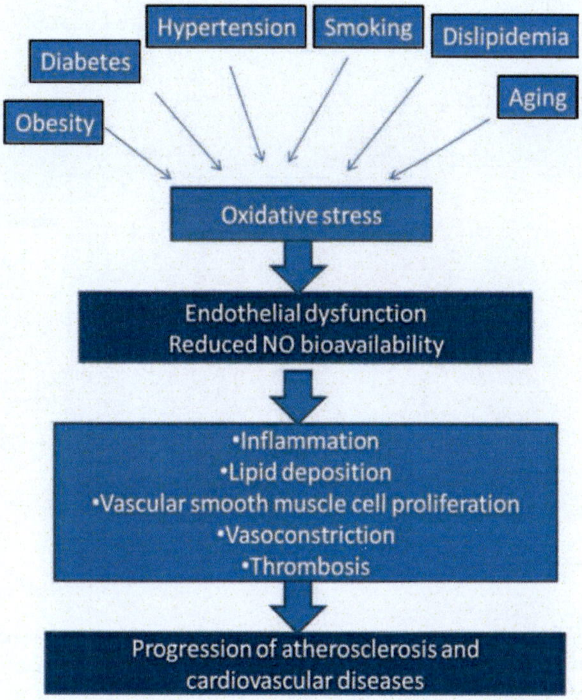

Fig. 20.4 Clinical manifestations of endothelial dysfunction

[43] examined the effect of carotenoid on endothelial cells cultures of human aortic. Pre incubation of lycopene reduced the expression of the chemotactant, a molecule that helps in the aggregation of white blood cells. Development of CVD due to endothelial activation has been associated to enhanced production of tissue factor, resulting in the clot formation in veins. Akt-particular inhibitor turned around the inhibitory impact of carotenoids on tissue factor action, suggesting that carotenoids repressed tissue factor activity in endothelial cells and enhanced phosphorylation of Akt by this mechanism [44].

20.2.4 Hypocholesterolemic Activity

The hypercholesterolemia is established as a chronic risk factor for CVD. Hyperlipidemia refers to elevated cholesterol, elevated TG or both. Hereditary factors may also responsible for this disease, but generally it is due to an acquired reason. In USA, approximately one in every six adults has hyperlipidemia. People with high cholesterol levels have double the risk for heart disease, but many are unaware of their condition due to no symptoms. Atherosclerosis is the development of plaques or clot in the arterial vessels [45]. This plaque finally becomes calcified, make them narrow and rigid.

The pathogenesis of atherosclerosis is partially understood. The progression of a plaque in the arteries exhibit the pathogenesis of atherosclerosis (Fig. 20.5). Blood composition, arterial wall abnormalities are generally considered to be responsible

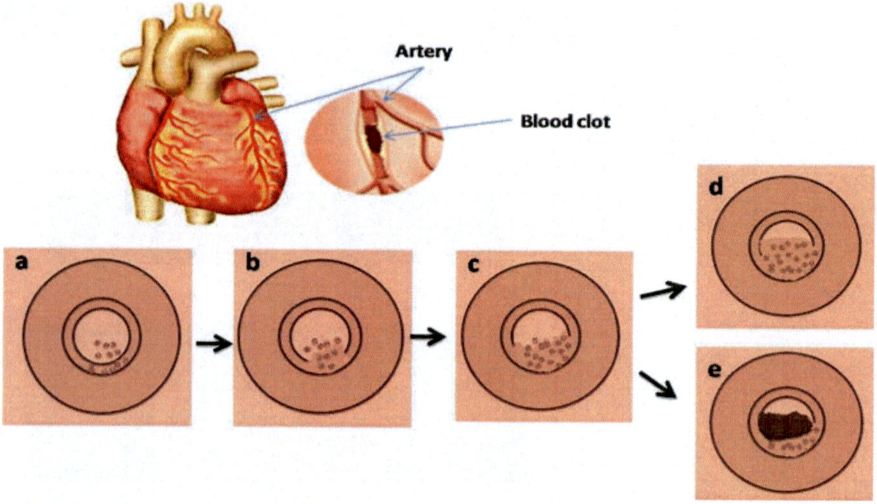

Fig. 20.5 represents plaque formation and growth (**a**) cholesterol accumulation in the inner wall of artery (**b**) development of plaque (**c**) plaque grows (**d**) plaque continue to grow (**e**) formation of blood clot that completely blocks blood flow

for Virchow's triad [46]. There is strong evidence for a premature progression of CVD in younger individuals with familial hypercholesterolemia due to hypercholesterolemia and atherosclerosis [47]. It is evidenced from animal models; the effect of higher amount of low-density lipoprotein-cholesterol [46] eventually starts up the series of events finally leading to atherosclerosis [48]. In hemo dynamically regions, Micro injuries of the endothelium due to shear-stress together associated with micro inflammation of the plaque with local coagulation activation results in plaque rupture [46]. This local clot leads to heart failure. A no. of trials has been confirmed a reduction in mortality due to cardiac arrest and myocardial incidents approximately 35% by cholesterol-lowering medications in post infarction patients [49].

The molecular mechanisms underlying the athero protective effects of carotenoids in cardiovascular disease have been ascribed. A current report has shown that β-carotene manages the outflow of transcriptional instrument [9]. Fuhrman et al. [50] demonstrated that inhibition of biosynthesis of cholesterol and elevated macrophage LDL receptor activity was due to macrophage enriched with lycopene or with β-carotene which improved the LDL clearance from blood plasma, and thus carotenoids might be surrendered as cholesterol bringing down agents. An inverse correlation between matrix metalloproteinase- 9 and plasma levels of provitamin A carotenoids was established, suggesting that the benefits of these nutrients can be imputable to reduced degradation of the extracellular matrix in the arterial wall [51].

Some dietary (spaghetti sauce, tomato juice and tomato oleoresin) intake reduced the oxidation of serum lipid and LDL-C. It is also observed that it did not decrease the serum values of total cholesterol, LDL-C, or HDL-C [52]. Astaxanthin has shown positive effects on overweight and obese adults by increasing the LDL-C, ApoB, and oxidative stress biomarkers in a placebo-controlled study [53]in non obese subjects the astaxanthin consumption increases HDL-C and serum adiponectin and lowers TGs [54]. An inverse relationship rose between serum carotenoids level and inflammation, endothelial dysfunction (soluble P-selectin, soluble intercellular adhesion molecule-1 (sICAM1) and oxidative stress marker [55].

Carotenoids are considered to help stability of plaque and reduce the susceptibility of blood clotting [9]. In 1982, 22,071 male of United States physician's study which had no record of CVD and cancer was under experimental study for over 13 years and they analyzed the relationship between dietary carotenoids and cardiovascular phenomenon that includes acute myocardial infarction. Every person of the experimental group was given the optimum amount of aspirin, beta-carotene or placebo. Lutein, α-and β-carotene, β-Cryptoxanthin, and lycopene blood accumulation in addition to the quantity of retinol andα- andβ-tocopherol. Significant lowering of AMI risk was not viewed in any information [51]. Smokers have lower ability to reduce risk on higher levels of β-carotene [51]. Lutein a protein gave similar result that was encountered not to be safe for heart attack risk [51]. These unpredicted conclusions could be because of biases underlining study sketch. Primarily metabolic tests were taken through blood samples, *i.e.* when the contributors were recruited; a single measurement was made that was not showed accurate levels of carotenoids for a longer period. Secondly the contributors who participated were from the American population that has lower rate of CVD mortality. Furthermore, smokers have mild protective outcome of β –carotene in the whole population

(14.7%) examined. In this subgroup, it is also possible that carotenoids rich diet would be acted if other nutrients are supplied in it [51]. In addition, the global CVD risk profile and normal response of vascular walls to carotenoids would change if patient's pharmacological anamnesis data have added. According to other examinations [51], cerebrovascular disease death would reduce in the presence of α-and β-carotene. On the basis of Dutch men sample, study fulfilled for 15 years that confirmed such assumptions. In fact, when elderly men adapted smoking, BMI, physical activity, dietary factors and vitamin supplements, cardiovascular diseases (CVD) death became inversely proportional to the α-and β-carotene[52]. However, though that population used to take drugs which were produced by elderly men because of their co morbidities.

20.2.5 Effect on Gap Junction Communication

Connexins are the fundamental proteins of gap junction (GJ) channels that allow the exchange of small metabolites <1–2 kDa in size and signaling molecules between adjacent cells and are a group of ≥20 highly conserved proteins with developmental and tissue-specific expression patterns. Gap junction communication plays a principal role in normal growth and physiology with some changes ensnared in various human infections and pathologies. Myocardium is mandatorily activated by electrical signal to supply successful pumping of blood and rely on the sequential transfer of current(coordinated spatial and temporal) from one cell to another in the heart [56]. All cells have the ability to promptly control the amount (transcriptional regulation), arrangement (assembly regulation) and function (posttranslational regulation) of connexins to attain cell-cell communication in response to physiologic demands. Particularly, C-43 is vital for normal cardiac working. There is substantial attraction in the increase in GJC in response to carotenoids, particularly lycopene [57]. In the artery wall, maximum communication between endothelial cells is also prudent; this improvement of gap junction communication by lycopene may retain a perfect arteries. Carotenoids are also known for repair of GJC in mammals and regulators of connexin 43 productions [58]. Various carotenoids have been considered to upregulate gap junction communication and connexin 43 protein levels [59].

Furthermore, recent researches [47–49] explained the carotenoids functions in CVD prevention. Moreover, atherosclerosis enhances connexins, *i.e.* proteins able to connect cells with one another forming channel for communication among them [48]. Particularly, vasodilation capability elevated by connexin 40 while connexins 43 and 37 responsible for increase in high turbulent flow [48]. However, extensive studies have to be needed to understand the expression of these connexins in response to carotenoids and their effect on atherosclerosis progression. There is immense need to determine when carotenoid has the effect whether in the early stages of atherosclerosis or in advanced stages of disease or in both times.

20.2.6 *Effect on Ischemia-Reperfusion Injury*

IR injury is a series of complex events, usually imputed the drastically damage to organ by reflow of oxygenated blood to a previously ischemic area. Various studies revealed that cell death during ischemic reperfusion takes place via apoptotic and necrotic pathways [60] majorly damages the heart (Fig. 20.6). Some proteins like mitogen-activated protein kinases, Protein kinase B and the cyclic guanosine 3′,5′-monophosphate with nitric oxide, all of these play key roles in cardiac I/R injury. Thus, two researcher steam found the effect of carotenoids derived red palm oil on I/R injury. It is evident that two pathways may contribute in the prevention from I/R injury induced by red palm oil [61, 62]. Esterhuyse et al. [61] found that prevention may occur at the time of ischemic event and cyclic guanosine 3′,5′-monophosphate with nitric oxide (NO-cGMP) may be involved in this process. Moreover, Engelbrecht et al. [62] reported a role of the Protein kinase B (PKB/Akt) and the MAPKs process in the cardio-protective effects of red palm oil during reperfusion. Likewise, cardio-protection and myocardial reclamation by Cardax was also studied. After parenteral administration in vivo, phenomenal myocardial reclamation is explained for a novel non esterified carotenoid derivative free astaxanthin [63]. The mechanism(s) of action of astaxanthin has not been well reported in cardio-protection. Although, Aoi and his colleagues revealed that in mice astaxanthin damage the

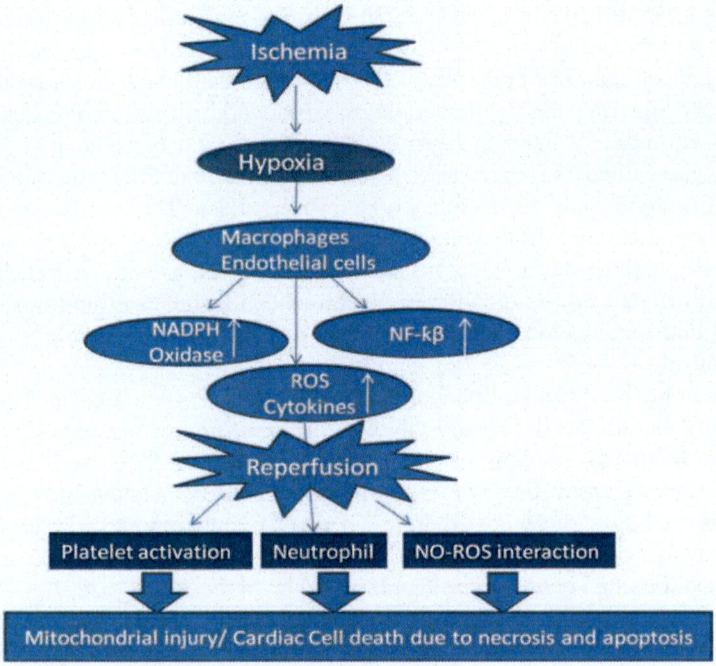

Fig. 20.6 shows cellular events of ischemic reperfusion injury

skeletal and cardiac muscle [64]. It is also reported that free astaxanthin accumulated in myocardium of rodents after oral administration of astaxanthin [65]. In seminal study, a cardio-protective action was viewed in reactive oxygen species interceded strenuous exercise damage display, with calming and antioxidant effects determined in that seminal study. Lutein blocked lipopolysaccharides and hydrogen peroxide induced increase in PI3K activity, NF-κB-inducing kinase (NIK),PTEN inactivation and Akt phosphorylation but did not alter the relationship between MyD88, Toll-like receptor 4 and the activation of MAPKs.

This study revealed that hydrogen peroxide regulates IκB kinase dependent NF-κB activation by initiating the activation of the PI3K/PTEN/Akt and NIK/IKK pathways [66]. Ischemia reperfusion injury may responsible for cardiac cell death by activating apoptotic signals [67]. The ability of lycopene (isolated from rat alveolar macrophages) was acquainted with cigarette smoking concentrate in in increasing PPARγ expression, inhibiting NF-kB/p65 nuclear translocation, IL-8 production, and redox signaling [68].

20.3 CVD Promoters

20.3.1 Smoking

Smoking is the critical risk factors for the progression of atherosclerosis, contributing to premature mortality in the United States. In United States, almost 30% of all dose related coronary heart disease (CHD) deaths are attributable to cigarette smoking per year. Smoking also nearly doubles the risk of coronary heart diseases (CHD). Smokers are more inclined to heart assault when contrasted with non smokers. Smoking generally damages the arteries, forming a fatty material (atheroma) which narrows the arteries and results in a stroke or heart attack. The tobacco smoke contains carbon monoxide (CO) which lowers the oxygen content in the blood. Thus, heart has to work harder to supply the oxygenated blood to body. Adrenaline production stimulated by the nicotine in cigarettes can faster heart and raises blood pressure. In this case, blood tends to clot more often, which raises the risk of stroke or heart attack.

Different life styles and individual characteristics give rise to the smoking statue and obesity that affects differently the result of carotenoids inpatients [73]. Blood carotenoid levels and smoking has inverse relationship [54, 55] as well as the later stages of atherosclerotic disease [51, 69, 70]. On Comparison between smokers and nonsmokers studies, there is variability in results: smoking studies showed that smoking actually influences carotenoids levels [51, 55, 69, 70]; non smoking studies revealed the exact opposite results [20, 71, 72]. In the Heart Protection Study of 20536 British adults with CHD, diabetes or other arterial disease, aged 40–80 years were fed with antioxidant (β-carotene) and vitamin supplementation (vitamin C and vitamin E per day) or placebo [51]. It did not reduce any mortality rate or the occurrence of any type of cancer, vascular disease, or any other major disease, this supplementation only increased the blood vitamin concentrations. In the ATBC Study with 1862 male smokers (aged 50–69) had myocardial infarction but the danger of coronary illness brought up in the gatherings that having either β-carotene or the mix of β-carotene and vitamin E contrasted and the fake treatment [52].

For the estimation of blood lycopene levels and CCA-IMT on the middle-aged men, a study was organized in Finland in 2003 [55], in which they found that there was remarkable indirect relationship with CCA-IMT values in models adaptation on the basis of confounding characteristics including smoking by increasing concentrations of lycopene. Secondly, the significance of this corporation became stronger in smokers and vice versa, if reports were evaluated. Alike, the Rotterdam study that presumes the relation among plasma carotenoids level (α-and β-carotene, Cryptoxanthin, lycopene, lutein) and the progression of atherosclerosis in abdominal aorta, it demonstrated that the lycopene has inverse relationship with atherosclerosis, in contrast to remaining carotenoids (higher plasma level of this molecule belonged to a lower incidence of (AOA). Actually, when changes were made for different sex and age, they obtained the values of different ratio; the ratio was from quite higher to lower quartile (0.55). This correlation (OR = 0.35) was much prominent in smokers then past smokers [69]. MI increased as the level of α-carotene reduced but accurate findings for smoking provided proof that the increased chances of AMI in relation with low levels of carotenoids was restricted to people who smoke [70].

One more experimental study [51], conducted in the United States for low risk population of men which outlined that no corporeal difference among the alteration occur in levels of α-carotene of plasma and the risk of stroke, although the association was common in those who were passive smokers. The chain smokers, which had higher concentrations of α-carotenoid to have a lower risk of AMI (0.55) if they ever smoked earlier, and 0.46 if they become chain smoker by making it a hobby when data was collected; the relationship did not appear in the non-smokers OR = 0.96 [51]. Again Rissanen et al. [20] revealed higher level of CCA-IMT in smokers as compared to non smokers along with the presence of the highest levels of lycopene compared to the lower. The detection of the same association was not possible in relation to others type of carotenoids. EURAMIC is a study [71] that support such results: the humans who suffered AMI have low concentration of lycopene in adipose tissues than subjects in control group. Accuracy for smoking showed that the outcome of lycopene was dominant in nonsmokers (OR = 0.33) [56]. The information leads the scientists to various results and conclusions: (a) in studies of this type, smoking was created confusion in interpretation of data; (b) proper examination is to be needed to understand whether smoking might have an action in regulating the blood antioxidant levels and hence, affecting the CVD risk. Aforesaid point, if smoking is the reason of negative action on antioxidants, related hypotheses is given: (1) smoking is an additional foreign source of free radicals, must be consumed those antioxidants which were added with food resulting in imbalance associated with surplus pro oxidants, that alters the low density lipoprotein initiation and continuous production of foam cells [72]. To maintain equilibrium, high amount of carotenoids are required for smokers. While in those who do not smoke don't require high amount of carotenoids. Politer et al. [73] shows an increase in plasma antioxidant while a fast antioxidant consuming before smoking cessation.

1. Addiction may not help from the inclusion of antioxidants in the diet [20].
2. People who smoking usually take a diet which contains carotenoids in small amount than non smokers [20].

20.3.2 *Obesity*

Adipose tissue is an endocrine organ and also a storage site for fats. Its secretion is the key factor for the atherosclerosis induction. This tissue to produce cytokine, interleuin-6 and TNF-α which results in development of plaque in the vessels or high insulin resistivity or causes the formation of other substances like highly sensitive C-reactive protein (hs-CRP) in liver, homocysteine and lipoprotein-a. These inflammatory biomarkers are present in high levels in obese people as compared to normal weight. Diet and lifestyle increases the chances for cardiovascular disease. In fact, imbalanced diets for longer time period alter lipid metabolism that leads to the deposition of visceral fat, thus resulting in metabolic diseases, such as cardiovascular pathologies hypertension and dyslipidemia. The mechanism is less understood beyond the interaction of amount of adipose tissues and carotenoids. Some assumptions are there; diet containing fewer amounts of antioxidants may give to the obese person. In the body, it may happen due to changing supply of carotenoids. In fact carotenoids (as lipophilic molecule) are generally distributed in plasma and in adipose tissue. In case of high concentration of fat that conquering the large amount of carotenoids. Consequently, carotenoids quantity decreases in plasma as compared to those added in diet. As a result, the high concentration of pro-oxidizing substances among obese persons leads to the excessive depletion of these substances [30]. This low amount of availability of antioxidant would leads to atherogenesis due to accumulation of LDL [74, 75].

Some studies reveled that cardiovascular diseases risk factors such as diabetes, inflammation, triglycerides, LDL concentration obesity, high blood pressure etc. remarkably reduced by fucoxanthin [74, 76]. The main objective of medical research is the identification of substances that can reduce obesity. Uncoupling protein-1 (UCP1) is responsible for adaptive thermogenesis, a physiological defense against obesity [74]. UCP1 dysfunction contributes to the development of obesity which helps body to utilize energy [77].

20.3.3 Gender

The relationship between carotenoids and cardiovascular disease is also affected by gender [78, 79]. According to Finnish study on Women Health Study (WHS) of middle aged and older women demonstrated that there was a decreased risk in cardiovascular disease in women with the increase of lycopene, independently from lifestyle, diet and clinical trials [78]. In the later study by same author selected men for the physician health study (PHS). The results were different from women health study (WHS) concluded that with increased concentrations of lycopene the chances of cardiovascular disease risk with decreases. Even in this study lifestyle and cardiovascular disease risk confounded the results [79]. It was observed from above study, men had lower carotenoid serum concentration of lycopene (9.3 μg/dL) as compared to females (16.5 μg/dL). The fact was that women were younger than men (58.8 vs. 69.7 years on average) since the increase of age was inversely proportional and depends upon the concentration of lycopene[30]. Brady et al. [30] conducted a study to evaluate the serum status of 180 men with the age 50–84. It was observed that physiologic and life style factors influenced on serum carotenoid level. In addition to lycopene lower serum carotenoids were associated with male gender. The other factors that also contributed to the lower serum level of carotenoids were younger age, smoking, cholesterol, high ethanol level, BMI and lower non HDL cholesterol. Concentrations of lycopene with CCA-IMT (Common Carotid Artery Intima Media Thickness) in women and men registered in the Atherosclerosis Supplementation in Antioxidant Prevention Study was compared in Rissanen et al. [80]. They found variations in the relationship between carotenoids and cardiovascular risk profile within both genders, although in cases with higher concentrations of lycopene the values of CCA-IMT were lowered in both genders [80]. These findings may suggest that this randomness may depend upon the behavioral habits or associated with gender differences in body physiology. Moreover, women should be more conscious about diet as compared to the men or women have high quality antioxidant system functioning than men [81].

20.3.4 Life Style

Sedentary lifestyle it is the second major cause of death in the United States [31]. Physical inactivity is responsible for increased risk of morbidity or worsening of many chronic diseases such as cardiovascular disease (CVD), congestive heart failure, obesity, type 2 diabetes, stroke, certain cancers, osteoporosis and hypertension and health conditions(Fig. 20.5) [28]. Sedentary lifestyle e.g. excess television viewing time is adversely associated with metabolic risk factor (independent from overall physical activity levels). The impacts of expanded times of stationary conduct in generally physically dynamic people have started to be outlined, and they give off an impression of being portrayed by metabolic modifications for the most part saw in diabetogenic and atherosclerotic profiles [4, 18, 22, 27]. In 2008, the Physical Activity Guidelines for Americans Advisory Committee recommended that grown-ups ought to collect 75 min of incredible power physical movement or 150 min of direct force physical action, or a mix of both every week [81]. Research has likewise demonstrated that gathering these rules is related with better CVD chance profiles k [82], and additionally decreased danger of mortality [83]. In 2005, the CDC assessed that 37.7% of the United States populace did not take an interest in the prescribed measure of physical action required for medical advantages, while an extra 14.2% did not take an interest in over 10 min of direct or fiery physical action all through the normal week [49]. Healthy lifestyle reduces the risk of coronary events and strokes in developing countries [5]. The healthy lifestyle factors e.g. high physical activity, intake of fruits and vegetables, less intake of alcohol, limited red meat; non smoking and low adiposity etc. were associated with reduced cardiovascular (CVD) events. Globally, significant gains made in tobacco control but there is immense need to do more work on tobacco control. The awareness towards tobacco control has been strong and consistent. It is also considered that trends in smoking declines better than trend for exercise and diet. The message is clear that smoking not only injurious to your health but also harmful for others around you. Recently, several approaches have been evaluated or implemented to improve lifestyle associated risk factors [84]. Consequently, we must remember that what our grandmother might say, "Everything in moderation leads to healthy life."

20.4 Case Studies

20.4.1 Epidemiological Study

Scientists have been interested to study the relationship between diet and health outcomes. Some epidemiological studies provide a yardstick regarding high consumption of fruits and vegetables protect against many chronic diseases including cardio vascular diseases. The risk levels of carotenoid health index by Donaldson are as per the following: high hazard: <1 κM, high hazard: 1–1.5 κM, direct hazard: 1.5–2.5 κM, okay: 2.5–4 κM, and generally safe: >4 κM [84]. He analyzed sixty two plasma samples under prospective cohort study or population based case control studies. On the basis of carotenoid health index, 95% of the American population falls into moderate to high risk [85]. In western cultures, a cardiovascular malady is the essential driver of death and adds to 33% death rate around the world. Mediterranean area has lower rate of deaths from CVD than northern European or other Western countries due to Mediterranean diet rich in plant derived bioactive phytochemicals [86]. Therefore, formulation of correct dietary guidelines depends on recognition of active constituents of the Mediterranean diet. Epidemiological studied based reports shows relationship between CVD and carotenoid plasma, several based on role of carotenoids in the protection from CVD while interventional trials reported contradictory results. It has been already suggested that supplementation of low doses as preventive measure for improving and combating of CVD.

In comparison between 139 Cretan men (aged 79 years) and Zutphen, Cretan men had more good health than Zutphen with lycopene four times higher as well as oxidative stress lower and higher antioxidants in plasma [87]. Later study demonstrated that low concentration of β- carotene concentration may increase the mortality rate due to higher risk of CVD among Finnish men [87]. Moreover, Karppi et al., [88] assessed serum tests from 349 subjects in connection to convergences of conjugated dienes in low thickness lipoprotein (LDL). The lycopene content in plasma inversely correlated with the substance of conjugated dienes. Therefore, dietary carotenoids ended up being altogether lower LDL level in vivo [56, 89]. Karppi et al. [89] also examined that relation between sudden cardiac deaths (SCD) and the serum concentration of carotenoids in middle aged men. They selected a population 1031 Finnish man aged 46–65 years enrolled in the Kuopio Ischemic Heart Disease Risk Factor (KIHD) cohort. According to their findings, low β-carotene concentration increases the risk of SCD. In addition, low serum carotenoids may lead to CVD followed by total mortality. The interrelation between food intake of β-carotene and decreased risk of CVD has been declared through many studies [91, 92]. Tavani et al., [92] revealed the inverse association of acute myocardial infarction in women with diet containing β- carotene. In a population based study, the dietary intake of β-carotene was inversely associated with a lower risk of CVD among current smokers, in the American Health Professional's Study led on 39,910 US guys [93]. Later studies suggested the association between risk factors for development of atherosclerosis and serum carotenoid. In a case control study in women, daily intake of diet containingβ-carotene is inversely proportional to the risk of nonfatal acute myocardial infarction [94]. In the Rotterdam examine focusing on the elderly, a populace based accomplice think about; the dietary admission of β-carotene was contrarily connected with coronary heart disease. In addition, the American professional study conducted on 39,910 US males among smokers and nonsmokers both, one taking carotene; chances of Coronary heart diseases (CHD) are lower [94]. Xu et al., [95] studied forty early atherosclerosis patients without clinical cardiovascular events to search for association between serum carotenoids and hazard factor for improvement of atherosclerosis and comparably healthy controls, resulted in bring down serum centralizations of lutein and zeaxanthin than solid subjects. Serum carotenoids were related with diminished danger of atherosclerosis. Another study on 573 middle aged women and men suggested that elevated plasma oxygenated carotenoids levels and α-carotene may be defensive against early events of atherosclerosis [96]. Moreover, it was demonstrated that diet induces a decrease in endothelial injury and dysfunction, associated with an improvement in the regenerative capacity of the endothelium in elderly subjects [97]. In addition, fruits and vegetables intake has favorable effect on marker of inflammation and oxidative stress is already present by elderly puberty [98]. However, Azzini et al., [99] showed that amelioration of multiple risk factors is significantly associated with the Mediterranean diet pattern including reduced oxidative stress; more cardiovascular risk profile and controlling influence of inflammation but confirmation about health benefits of carotenoids are difficult. A study on effect of serum carotenoids and their interaction on mortality (CVD and cancer) follow-up through 2006 (N = 13,293) was conducted in National Health and Nutrition Examination Survey (NHANES) III, 1988–1994, on US adults aged ≥20 years, revealed that associations of serum α-carotene, lycopene and total carotenoids with different causes of deaths ($P < 0.001$). However, lycopene in low amounts is one of the factors for all cause mortality, followed by very low total carotenoids. α- carotene/lutein+zeaxanthin, lycopene/lutein+zeaxanthin and α-carotene/β-Cryptoxanthin, combinations were significantly related to all-cause mortality ($P < 0.05$), according to a random survival forest analysis. It was interesting that the only carotenoid associated with CVD mortality ($P = 0.002$) was low α-carotene. The risk factors for mortality may be due to very low lycopene concentrations, α-carotene and serum total carotenoid. For confirmation, studies of balanced combinations of carotenoids are required [100]. Furthermore, before several years, the EURAMIC study recommended that lycopene from common food source can serve as the defensive impact of vegetable use on myocardial localized necrosis chance [71]. Further, De Waart et al. [101] prescribed that serum levels of single carotenoids are conversely proportional to all-cause mortality. Bohm [102] in the past few days briefly reviewed data relating lycopene and its related CVD health advantages. In another study, lycopene contents were analyzed in a total of 264 serum samples obtained from Korean women [103]. Additionally, arterial stiffness was determined by brachial-ankle pulse wave velocity (baPWV). Lipid profile, high-sensitivity C – reactive protein and contents of oxLDL (oxidized low density lipoproteins) in serum were also analyzed. An inverse correlation was established between BaPWV and lycopene and lycopene and oxLDL. Hence, lycopene may be accountable for decreased oxidative alteration of LDL. This can be particularly one mechanism through which lycopene reduces the arterial stiffness and the possibility of cardiovascular diseases. Moreover, 299 Korean men were scrutinized to find out the relationship between arterial stiffness, the risk of metabolic syndrome and antioxidant status. It was analyzed, among other parameters, lycopene content, baPWV, oxLDL and lipid profile. baPWV negatively correlated with serum lycopene content. An inverse correlation was also found between lycopene and oxidized low density lipoprotein. Hence, an interrelationship was found between baPWV, metabolic syndrome and circulating lycopene [103]. A questionnaire based study and to give serum samples were conducted on a total of 3061 participants. Lycopene contents in serum of those who died due to CVD were tended to be low than survivors [104]. In a case-control study with 682 controls and 760 cases represented a low risk of MI with high intake of β-cryptoxanthin, α-carotene and β-carotene but no relation with lycopene [105]. The CARDIA (Coronary Artery Risk Development in Young Adults) Study with 4580 people demonstrated that healthy lifestyle people have elevated lycopene contents as compared to less healthy lifestyles. Individual and total carotenoids of serum were inversely associated with markers of oxidative stress, endothelial dysfunction and inflammation [106]. The MHH (Minnesota Heart Survey) Study with 6070 women and 5369 men used a 24 h dietary recall. The researchers succeeded to develop a Heart Disease Prevention Index. Hence, during the last 5 years period, overall the quality of diet has been moderately improved. Moreover,β-cryptoxanthin, zeaxanthin and β-carotene significantly increased but ingestion of lycopene [107] was not significantly increased only. Therefore, controversial epidemiological data exist about the effect of lycopene on CVD.

20.4.1.1 Animal Study

Under high complex physiological situations the animal models are permit to study the effect. The main disadvantage of animal studies is that animals produce carotenoids by metabolism separately from humans. Several studies have been conducted in last 10 years to recognize the productive cardiovascular effect of several animal models *i-e.* mice, rats, hamsters and rabbits *etc.* either carotenoids managed separate or may combine with other molecules of nutritional interest.

Bansal et al. [108] arranged male wistar rats and fed with lycopene dissolved in olive oil for 31 days, this study showed that lycopene reduced effectively oxidative worry in rats when contrasted with different investigations; Lycopene reduced the lipid peroxides levels and increases the level of glutathione and glutathione peroxidase (GSHPx) activity. In another experiment lycopene was administered to female rats for 2 weeks. Several different doses of lycopene that reduce the activity of glutathione reductase, GSH-Px and super oxide dismutase (SOD) and there was no effect on catalase (CAT) [109]. The role of yellow tomato, red tomato or lycopene was studied in a rat display with light oxidative pressure incited by low vitamin E consume less calories. Rats were fed with lyophilized yellow color tomato (16%), lyophilized red color tomato (16%) or lycopene beadlets (0.05%) for 6 weeks could not alter the plasma cholesterol concentration. Triacylglycerol (TGs) levels reduced in response to red tomato compared to controls, yellow tomato and lycopene beadlets while thiobarbituric acid-reactive were also reduced on feeding red and yellow tomatoes, than to controls and beadlets in heart. The result from above study was that to alter oxidative the stress related parameters, tomatoes (due to phytochemical present in tomatoes) had greater potential than lycopene [110].

In another experiment antioxidant mechanism and lipid lowering effect of tomato paste was observe in hamsters. They found significant reduction in total cholesterol and LDL level after 8 weeks of feeding with 9% tomato paste but HDL cholesterol increased by 28.8%. Moreover MDA was reduced in plasma by 89.3%. In addition, antioxidant enzymes (SOD, CAT, GSH, Px) activities significantly increased on feeding 9% tomato paste for 8 weeks [111]. Few authors explained the effect of food having different carotenoids (Astaxanthin, lycopene *etc.*) a large portion of the investigations partner to the impact of separated carotenoids from various plant and animal sources.

In this context, Astaxanthin administered to rats intravenously (5, 10 and 20 mg/kg) and orally (100 and 200 mg/kg). Pharmacokinetic study of astaxanthin revealed that Intravenous administration of was dose dependent while oral administration was dose independent. The absorption of Astaxanthin followed flip flop model after oral administration. It was unstable after up to 4 h incubation of rat gastric juices and 24 h incubation in different buffer solution having pH 1–13. Recently effect of astaxanthin pro drug (CDX-085) was studied using a mouse model of arterial thrombosis. The CDX-085 fed group showed significant increases in basal arterial blood flow and thrombus formation delayed, when compared to control mice. Wistar-Kyoto rats who fed on astaxanthin arranged meaningfully elevated NO levels and decreased levels of peroxynitrite from which platelets and endothelial cells are isolated. It was observed from this study that CDX-085 and its metabolites have the potential to treat or prevent thrombotic cardiovascular complications. Moreover, diet containing astaxanthin administered to hypertensive rats that improves cardiovascular variables and decreases BP. These effects are accompanied by improvements in NO bioavailability and a reduction in oxidative stress [112]. A study arranged on streptozotoin induced diabetes in rats demonstrated the effect of astaxanthin on endothelial dysfunction by inhibiting the ox-LDL-LOX-1-eNOS pathway indicated that this treatment might be useful for diabetic manifestation associated with endothelial dysfunction [113]. Another study conducted on Watanabe heritable hyperlipidemic (WHHL) rabbits to evaluate the effect of astaxanthin and alpha tocopherol on atherosclerotic lesion formation and LDL oxidation. They conclude that LDL oxidation lag time was prolonged by only alpha tocopherol in WHHL rabbits; these two antioxidants did not save atherogenesis.

Some studies revealed that lycopene has a protective effect on cardiovascular diseases on oral administration. Rabbits in New HMG-CoA reductase activity along with cholesterol diet and cholesterol diet for 3 months. The diet containing Lycopene appreciably decreased cholesterol level and HMG-CoA reductase activity along with acyl- CoA-cholesterol acyl transferase activity in serum but enhance HDL cholesterol level. A plaque area in aorta was significantly decreased (64.3%) by highest dose of lycopene. Lee et al., [107] studied the inhibitory effect of lycopene in both cellular and mouse animal models on HMGB1-mediated pro inflammatory responses

In addition, plaque area in the aorta reduced significantly (64.3%) in response to high dose of lycopene. Inhibitory effect of lycopene was also reported by Lee et al. [114] on HMGB1-mediated pro inflammatory responses in both mouse animal models and cellular (HUVECs). The comparison between fluvastatin and lycopene impacts on atherosclerosis delivered by high fat nourishment in 40 male rabbits was reported by HU and his colleagues [115]. The level of TGs and cholesterol Interleukin-1 was increased by high fat diet. Lycopene showed better results about to reduce the changes in these parameters than fluvastatin. Fluvastatin and lycopene also minimize the chances of clot formation in the aorta as contrasted with the rabbits sustained on eat less in high fats [116].

Another study conducted on 65 male watanabe heritable hyper lipidemic rabbits for 16 weeks fed on control consume less calories, a control eating regimen supplemented by plant oil or control diet supplemented with tomato extract but the tomato extract had no effect on triacylglycerol and cholesterol level in plasma, on cholesterol on aortic atherosclerosis and in lipoprotein fraction. It was also noted that the oxidation of plasma lipids had no effect by taking of tomato extract [117]. Lorenz et al. confirmed these results [118]. According to this study, lycopene supplementation for 4 weeks increased plasma level of lycopene and significantly reduced the total and LDL cholesterol serum level as well as cholesteryl ester in the aortae in New Zealand white rabbits but no difference was found in initial lesions to the aorta. A dietary mixture of fish oil, catechin, lycopene, vitamin C, resvertrol and D-α-tocopherol upgrade inflammatory and lipid fear influences CVD in human model of disease [119]. These results help multi target access for complex multifactorial diseases like diabetes 2. Moreover, Zhu et al. [120] reported chronic lycopene treatment in streptozotocin-induced diabetic rats showed endothelial dysfunction by reducing oxidative stress indicating it may be valuable in stifling the diabetic vascular troubles identified with endothelial dysfunction.

It can be inferred from above animal studies that possible beneficial effect of dietary lycopene was evaluated. Different lycopene dosages were used in different experiment. So, these studies cannot be comparable. Lycopene dose 10 mg/kg used in experiments cannot be transferred to human situations. It must be in consideration that experiments performed with hamsters, rabbits and rats cannot directly applied on human organism with same conditions.

20.4.1.2 Human Interventions

Recently, A major publication on effect of cardiovascular fear influence and consequences of few routine dietary components: tea, carotenoids, red wine, grapes, flavonoid-rich cocoa, garlic coffee and omega-3 fatty acids was examined [121]. High intake of few of them can decrease the mortality or occurrence of hypertension, stroke and myocardial infarction. Anyhow, although they prove in vitro studies, study of animals and observational studies viewed relationship between saving of cardiovascular disease and carotenoids and is helpful to the beneficial effects for most of these food supplements discovered from natural products received as part of the food. In fact, several studies have been suggested a relation between the prevalence of CVD and β-carotene, some of experiments has neglected to clarify any decrease in CVD with β-carotene admission. As indicated by the MRC/BHF Heart Protection Study, organization of β-carotene in mix with vitamin C and vitamin E has no preferred standpoint on mortality in people at high-hazard [122]. In another ATBC study designed on 1862 male smokers with the history of myocardial infarction, no significant difference has been obtained in the coronary events between placebo and supplementation group. However, the chances of coronary heart disease was high in the β-carotene and combined β-carotene and α-tocopherol groups than placebo group [123]. For instance, According to the Women's Antioxidant Cardiovascular Study (WACS), there is not any reduction in cardiovascular disease risk was found in women at high risk, whether using β-carotene 50 mg every alternate day, or vitamin E 600 IU every other day or vitamin C 500 mg daily [124]. Anyhow, in doctors' wellbeing study suggested that the use of carotenoids rich vegetables was interconnected with a diminished dread of CHD [125], but there was no significant effect of supplementation of β-carotene 50 mg on CVD, cancer, or overall mortality among primarily non-smokers after 12 years of follow-up [126]. Moreover, no relation between increasing concentrations of plasma lycopene and the fear of CVD was searched [78]. In controlled clinical studies, no definite evidence for CVD prevention with lycopene was found in well-defined subject populations [127]. Effect of zeaxanthin and lutein-rich supplements and foods on serological markers of inflammation, oxidation and endothelial activation and macular pigment levels (MPL) in healthy volunteers was compared. Graydon et al. [128] concluded that this 8-week may positively improve macular pigment level in the highest serum responders and in those who has initially low macular pigment level. In a case control study involving 20 coronary CHD patients, the outcomes of lycopene (from cooked tomatoes) on lipid profile, serum cell reinforcement compounds and lipid peroxidation rate were analyzed. They were feed 200 g cooked tomatoes every day for 60 days. Tomatoes significantly reduced malondialdehyde (MDA) levels, resulting in a bring down rate of lipid peroxidation and high cell reinforcement chemicals (GSH-Px, GSH, SOD)levels, whereas lipid profile was not affected [129]. The above mentioned authors have been also studied the outcomes of lycopene from cooked tomatoes on lipid profile, lipid peroxidation rate and plasma antioxidant enzymes, and resulted that tomato lycopene may not be used as a lipid lowering agent in hypertension but may have appreciable natural therapeutic antioxidant potential [130]; they suggested that a relatively high daily consumption of lycopene supplements (10 mg/day) or tomato-based products (equivalent to 32–50 mg lycopene/d) is ineffective in reducing conventional CVD risk markers in healthy, moderately overweight, middle-aged individuals. The effects of a tomato-based drink on markers of immunomodulation, oxidative stress and inflammation were studied in a randomized, double-blind, placebo-controlled, crossover study. Tomato based drink (ingesting 5.7 mg lycopene per day) was supplemented to 26 healthy men and women for 26 days per period. This drink significantly reduced the tumor necrosis factor-α production in whole blood. Conversely, damage to DNA and urinary 8-iso-PGF2α concentration was not affected by supplementation of tomato drink [55]. Outcomes of lycopene addition on markers of endothelial function and oxidative pressure in 126 healthy men aged 22–57 Y with (6–15 mg) lycopene daily for 8 weeks were recently published in another placebo- controlled, randomized, double-blind study [131]. This lycopene supplementation resulting in increased lycopene contents in serum. SOD activity and DNA damage were reduced but increase in oxidative stress. Endothelial function was significantly improved by lycopene (15 mg per day). This dosage also remarkably reduced the hs-CRP (inflammatory marker) content in serum. Plasma levels of adhesion proteins sVCAM-1 and sICAM- 1 were significantly decreased by lycopene. Moreover, supplementation with 15 mg lycopene for each day more than 2 months was powerful to limit the oxidative pressure and to enhance the endothelial capacity. This investigation depended on Korean men (moderately aged) yet the outcomes can't be limited to ladies. Therefore, the calming and antioxidative effects of lycopene were resulted [132]. Contrastingly, one recent intervention with a group of 31 non-smoking healthy postmenopausal women, 70 g tomato paste per day (46 mg lycopene/day) did not alter the endothelial function. However, dilation of arteries did not alter during the study period as the plasma lycopene concentration was increased [133]. A meta-analysis organized on human subjects explored the outcomes of lycopene on blood pressure and lipid profile [134]. According to this meta-analyses, 25 mg lycopene on daily basis effectively decreased serum LDL and total cholesterol. The clinical difficulties declared that they were low in number (54) to provide any strong evidence concerning the particular lycopene role in the modulation of blood pressure. However, they recommended that in specific hypertensive subjects, lycopene has a blood pressure minimizing effect but more extensive studies are required to verify these conclusions.

Astaxanthin is a nutrient with specific cell membrane actions and diverse clinical benefits with excellent safety and tolerability. Astaxanthin has remarkable antioxidant powers, established primarily on experimental results. This nutrient provides clinically important antioxidant necessary for human, especially in those subjects who endangered to oxidative stress, like overweight, obese and the smokers. In a Korean prospective, unplanned, double-blind stubs and the smokers. In a Korean imminent, impromptu, twofold visually impaired

Examine, y, 23 adults with Body Mass Index (BMI)> 25.0 kg/m^2 enrolled in this study. They were randomly allocated to two dose categories *i.e.* astaxanthin 5 mg and 20 mg once daily for 3 weeks, and sample size 10 as compared to control group with normal body weight (BMI <25.0 kg/m^2) who received no intervention [135]. SOD, isoprostane, malondialdehydeand total antioxidant capacity were measured at baseline and 1, 2 and 3 weeks after astaxanthin administration. Compared with baseline, SOD and total antioxidant capacity levels were remarkably higher in two groups whereas, the malondialdehyde and isoprostane levels were remarkably reduced, after the 3 week intervention. Supplementation of astaxanthin for 3 weeks has improved the oxidative stress biomarkers by stimulating the activity of the antioxidant defense system and suppressing lipid peroxidation. Another double-blind, unplanned effort was organized by the same group [136]. In this study, 39 heavy smokers and 39 non-smokers were included. Astaxanthin at 5-, 20-, or 40 mg/day dose were given to randomly allocated smokers for 3 weeks. Plasma SOD, isoprostane, malondialdehyde, and total antioxidant potential were measured at bas1-α-initiated intercellular and ICAM-1 and VCAM-1eline and after 1–3 weeks of treatment. In comparison with baseline, dismutase levels and total antioxidant capacity increased, whereas the plasma malondialdehyde and isoprostane levels decreased in all astaxanthin intervention groups over the 3-week period. In particular, isoprostane levels showed a significant dose-dependent decrease after astaxanthin intake. The results reveal that astaxanthin supplementation might prevent oxidative damage in smokers by stimulating the activity of the antioxidant system and suppressing lipid peroxidation. In the Park randomized controlled, double-blind trial [137], astaxanthin also remarkably reduced the C-reactive protein and improved certain blood lipids in elevated serum triglycerides [138]. In this study, 61 healthy persons (BMI <28 kg/m^2), aged 20–65 years with fasting triglycerides in the range 120–200 mg/dL, were randomly assigned to take astaxanthin at 6, 12, or18 mg/day, or a placebo for 12 weeks [139]. Astaxanthin remarkably increased the HDL-cholesterol at the doses of 6 mg/day ($p < 0.05$) and 12 mg/day ($p < 0.01$), also remarkably reduced the TGs at doses of 12 mg/day and 18 mg/day ($p < 0.05$ for both) as compared to placebo with no effect on LDL-cholesterol at any dose.

Astaxanthin also significantly increased blood adiponectin (hormone produced by adipose tissue, cardiac and skeletal muscle, and vessel endothelia) levels. Adiponectin level in serum tend to be reduced in obese and/or diabetic subjects, smokers and individuals with metabolic syndrome, patients with coronary heart disease [140]. This study suggests that 12 weeks of supplementation did not affect BMI and a normalization of adiponectin levels. Further, investigation is required under better controlled conditions in order to clarify astaxanthin utility for this condition. Fasset and Coombes [141], the bioavailability safety, and effects of astaxanthin on inflammation and oxidative stress have been estimated in numerous clinical studies. On astaxanthin administration, reduction in inflammation and oxidative stress biomarkers has been reported with no side effects. Prior to induction of ischemic event, administration of astaxanthin either orally or intravenously protects myocardium problem. We do not know at this stage whether administration of astaxanthin after a cardiovascular event is beneficial or not because no clinical cardiovascular studies in humans have been completed and/or reported. Cardiovascular clinical trials are revealed the safety profile and preliminary experimental cardiovascular studies on astaxanthin.

Most intervention studies [131, 133, 134, 142] conducted using lycopene were performed on healthy subjects as widely reviewed by Bohm [102]. Thus, these authors investigated the conceivable essential preventive impact of tomatoes inferred lycopene or their items. Moreover, the duration (between 7 days and 8 weeks)and dose (from 5.7 up to 46.2 mg per day) of intervention trials also varied and the authors used lycopene dosages. Another factor affecting the results could be the matrix of the intervention products, such as, tomato oleoresin capsules, raw tomatoes, tomato purée and tomato-based drink. Thus, no direct comparison among the studies is found.

20.4.2 In Vitro Studies of Carotenoids

Carotenoids rich diet minimizes the risk of CVD. There are several advantages to assess the carotenoids effects on oxidative and inflammatory processes. These investigations include the specific types of cells and specific concentrations of carotenoids under well-defined conditions. These types of analysis are good for hypothesis building and studying mechanical outcomes. Although, these cellular studies gave some restrictions; complexity in long term study because of short life of cell and missing knowledge regarding interaction with other in vivo cells. Multiple in vitro studies depend on cellular models developing a link between carotenoids, inflammation and oxidative stress.

In last 10 years, in vitro studies have seen that carotenoids remarkably suppress tumor necrosis factor-α-induced intercellular and ICAM-1 and VCAM-1(vascular adhesion molecules) circulation in endothelial cells. It has suppressing and block integrity activity on cell migration and adhesion to endothelium by blocking the activation of NF-κB, CD14 and toll like receptor-4 circulation and production of tumor necrosis factor-α [143].

Furthermore, some studies recommended that administration of β-carotene or lycopene remarkably decrease the nitrotyrosine (an index of ONOO-) and reactive oxygen species (ROS) levels, therefore, developing NO bioavailability and cGMP levels. Furthermore, It down- control the monocyte–HUVEC interaction and (NF-kB)-dependent adhesion molecule expression. That's why, results recommended that treatment with β-carotene or lycopene decreases the inflammatory answer in TNF-α-treated HUVECs through the preservation of NO bioavailability and redox balance protection.

Many in vitro studies also consider that maintenance of endothelial NO bioavailability is beneficial to endothelial function and more in general to vascular health. Tumor necrosis factor-α-stimulated endothelium NO rapidly reacts with superoxide anion (O2−) to produce a strong oxidant ONOO- leading to reduced vascular relaxation. It also participates to the up-regulation of nuclear factor kappa particle B cells (NF-kB) dependent cellular response [144]. In the past days, Sung et al. [145] explained that endothelin-1 is a strong vasopressor assumes a key part in the pathophysiology of CVD. Besides, they found that lycopene smother E-1 articulation through the restraint of ROS manufacture and group of hemeoxygenase-1 in HUVECs. Therefore, carotenoids are normally measured to be as potential hostile to oxidant controllers of endothelial reaction to provocative boosts.

Tang et al. [146] examined the antioxidant and apoptotic effect of lycopene in endothelial cells. Lycopene dose dependent treatment significantly reduced the malondialdehyde (MDA) contents in H2O2 -treated cells and responsible for H2O2 induced apoptosis. It also suppresses the up direction of p53 delegate ribonucleic corrosive (mRNA) and caspase3 mRNA. It can be inferred from the above study that beneficial effect of cardiovascular related lycopene protects the endothelial cells from oxidative injury. It supports the hypothesis that carotenoids may offer protective role in response to oxidative pressure. Rossoni-Junior et al. [29] investigated the generation of ROS in neutrophils from diabetic and non diabetic animals and found that diabetic animals produce significantly more reactive oxygen species and nitric oxide than non diabetic and intake of β-carotene and annatto (a carotenoid exerting antioxidative activity) is responsible for the production of these species.

Moreover, Carotenoids may be considered efficient antioxidant that can prevent from atherosclerosis although exact mechanism of action is still unknown. Moreover, Alternate mechanisms have been reported as relevant effects *e.g.* Regulation of lipid metabolism by controlling the cholesterol synthesis [147].

Palozza et al. [144] followed the above mentioned hypothesis that carotenoids may protect against atherosclerosis. They demonstrated that dose dependent lycopene (0.5–2 mM) lowered the total cholesterol in THP- 1 cells. Lycopene has the ability to reduce ROS production and apoptosis by restrictingcaspase-3 activation in THP-1 cells [93]. Furthermore, lycopene inhibited cytokine secretion and expression in the same cellular model joined by hindrance of oxysterol-incited ROS creation, NF-κB activation and mitogen- activated protein kinase (MAPK) phosphorylation. Additionally, the carotenoid expanded peroxisome proliferator-actuated receptor γ levels in THP-1 macrophages. This study suggested that lycopene have anti atherogenic properties for atherosclerosis prevention [148]. Carotenoids also regulate cell proliferation. It has likewise been demonstrated that lycopene ties platelet-determined development factor (PDGF)- BB (homo dimer shape), in rat vascular smooth muscle cells (SMC), it is responsible for development of CVD and prevents intracellular signal transduction [149].

Intravenous thrombosis is another key factor in atherogenesis [150]. Lycopene prevented dose dependent aggregations in human platelets and agonist (collagen and arachidonic acid) based ATP-release reaction. These results indicate that tomato containing foods may useful in the inhibition of clot formation [151]. The above mentioned studies [147, 150–152]utilized high centralizations of lycopene up to 20 mM, which is an undue amount physiologically but when lower concentration (between 0.5 mM and 2 mM) for long treatment up to 24 h was used, unphysiologically [153].

Results on HUVECs treated with high carotenoids concentrations (2.5 mmol/L to 1 mmol/L) indicate high concentration of carotenoids suppressing the cell proliferation and decreasing the cell viability and unphysiological amount of carotenoids may bring non specific effect through a general cytotoxic action [154].

These findings also raise the question that in vivo supplementation of carotenoids at pharmacological level may have adverse effect in case of prooxidant activity specifically in pro vitamin A molecule by over activating the retinoic acid signaling [155, 156] while low doses of carotenoids below 2.5 mmol/L both carotenoids reduced the U937–endothelium interaction and reducing vascular inflammation (Table 20.2).

20.5 Conclusion

It can be inferred from the above study that malnutrition and unhealthy life style are the major contributor for the development of cardiovascular diseases. Dietary carotenoids *i.e.* Beta-carotene, lycopene, lutein and zeaxanthin are natural substances and provide health benefits in decreasing the risk of cardio vascular diseases.

Table 20.2 Various studies on effect of carotenoids on cardiovascular disease

Epidemiological								
Patients/ Model	**Sample specification**	**Region**	**Dosage**	**Duration**	**Assessment**	**Aim**	**Results**	**References**
	Men	Eastern Finnish			Serum Status		Low concentrations of serum β-carotene concentrations increase the risk for CVD mortality	[87]
349					Serum Status		Dietary carotenoids remarkably lower oxidative stress induced LDL in vivo	[88]
1031	Men, Age 46–65	Finnish			Serum Status		Low serum β-carotene may enhance the risk of SCD. Additionally; low serum-carotene concentrations may be associated with the risk of CVD and total mortality.	[89]
	Women				Serum Status		A relationship between a high dietary admission of β-carotene and lessened frequency of CVD	[90, 91]
	Women				Serum Status		The danger of nonfatal intense myocardial localized necrosis (MI) in ladies was contrarily connected with every day admission of β-carotene-containing diet.	[92]

(continued)

Table 20.2 (continued)

					Serum Status		Dietary intake of β-carotene was inversely associated with the risk of MI [50].	[93]
39,910	Men	USA			Serum Status		Carotene admission was related with a lower danger of coronary illness (CHD) among current smokers but not nonsmokers	[94]
40					Serum status		Serum carotenoids were related with diminished danger of atherosclerosis	[95]
573	Men/Women, age 40–50				Plasma Status		Higher levels of plasma oxygenated carotenoids (lutein, zeaxanthin, beta-cryptoxanthin) and α-carotene may be defensive against early atherosclerosis	[96]
		Mediterranean			Dietary Assesment		Mediterranean diet induces a decrease in endothelial injury and dysfunction	[97]
		Mediterranean			Dietary Assessment		Favorable effects of fruit and vegetable intake on markers of inflammation and oxidative stress are already present by early puberty	[98]

		Mediterranean			Dietary Assessment		Mediterranean dietary pattern is associated with significant amelioration of multiple risk factors, including a better cardiovascular risk profile, reduced oxidative stress and modulation of inflammation	[99]
		US			Serum Status		Very low serum total α-carotene, lycopene and carotenoid may cause mortality	[100]
					Serum Status		Serum levels of individual carotenoids are inversely proportional to mortality	[71]
264	Women	Korean			Serum status		Lycopene may help for reduction in LDL, lycopene may reduced arterial stiffness and CVD	[101]
299	Women	Korean			Serum Status		BaPWV inversely associated with serum lycopene contentA negative connection found amongst lycopene and oxLD. Connection set up between baPWV, circling lycopene and metabolic syndrome	[102]

(continued)

Table 20.2 (continued)

3061					Serum Status		Serum lycopene contents were lower CVD victims	[72, 103]
760					Serum status		Intake of α-carotene, β-carotene and β-Cryptoxanthin decreased the MI risk	[104]
Animal								
Patient/ Model	**Sample specification**	**Region**	**Dosage**	**Duration**	**Assessment**	**Aim**	**Results**	**Reference**
Wistar rats	Male		Lycopene dissolved in olive oil	31 days	Plasma status	To study the Effect of lycopene for the reduction of oxidative stress	Lycopene reduced levels of lipid peroxides and elevated glutathione levels GSHPx activity	[108]
Wistar rats	Female		Lycopene	2 weeks	Serum status	To study the Effect of lycopene for the reduction of oxidative stress	The activity of glutathione reductase, GSH-Px and super oxide dismutase was remarkably influenced the various lycopene doses	[109]
rats			Yellow tomato, red tomato or lycopene beadlets	6 weeks	Plasma Status	Potential role of yellow tomato, red tomato or lycopene beadlets in a rat model fed on diet low in vitamin E with mild oxidative stress	16% freeze-dried yellow, red tomato or 0.05% lycopene beadlets has no affect on plasma cholesterol concentration, tomatoes significantly lowers oxidative stress than lycopene, may be due to synergistic effect of all the phytochemicals in tomatoes	[110]

Hamsters			Tomato paste	8 weeks	Serum status	Characterization of the hypolipidemic effects and antioxidant process of tomato paste	Tomato paste significantlydiminished the serum add up to cholesterol and LDL cholesterol contents and the activities of CAT, SOD and GSH- Px remarkably increased	[111]
Rats			Intravenous (5, 10 and 20 mg/kg) and oral (100 and 200 mg/kg) administration of astaxanthin		Dietary assessment		The absorption of astaxanthin after oral administration followed the flip-flop model. The hepatic and gastrointestinal first-pass extraction ratios were approximately 0·490 and 0·901, respectively	[112]
			Astaxanthin prodrug (CDX- 085)			Cardiovascular protective role of a proprietary astaxanthin prodrug (CDX- 085) on thrombus formation	The CDX-085 fed group showed ample increases in basal arterial blood flow and slow down the formation of occlusive thrombus as compared to control mice	[113]

(continued)

Table 20.2 (continued)

Rabbits	Male		Creatures were bolstered on a typical eating routine, an elevated cholesterol (5 g/kg) eat less carbs and an elevated cholesterol eating regimen containing different measures of lycopene	12 weeks		Protective effect of dietary lycopene on cardiovascular disease	Lycopene in high doses remarkably minimize the plaque area in the aorta	[114]
Both cellular and mouse						Inhibitory effects of lycopene	Inhibitory effects of lycopene on HMGB1-mediated pro-inflammatory responses in both cellular (HUVECs) and mouse animal models	[115]
40 rabbits	Male, hyperlipidemic		Lycopene and fluvastatin	8 weeks		The comparison of effect of lycopene and cholesterol lowering drug fluvastatin on atherosclerosis	Lycopene exhibited better results than fluvastatin to prevent deposition of fats. Both reduced the progression of atherosclerotic plaque	[116]

65 rabbits	Male, hyperlipidemic		Extract of lycopene rich tomatoes 0.25 g tomato extract (consist 6% lycopene)/100 g:15mg lycopene/ 100 g diet)	16 weeks	Plasma status	Effect of lycopene on hyperlipidemia	The tomato extract could not affect plasma cholesterol and TGs levels	[117]
Rabbits			Lycopene supplementation	4 weeks	Serum Status	Effect of lycopene on LDL	Lycopene supplementation increased its plasma levels and strongly reduced total and LDL cholesterol serum levels as well as significantly lowering amounts of cholesteryl ester in the aortae in lycopene-treated rabbits, no significant differences in initial lesions to the aorta were detected,	[118]
			Dietary mix of fish oil, resveratrol, lycopene, catechin, d-α-tocopherol, and vitamin C				Improves lipid and inflammatory risk factors for CVD in humanized models of disease	[119]

(continued)

Table 20.2 (continued)

Rats	Diabetic					Chronic lycopene treatment might be useful in preventing the diabetic vascular complications associated with endothelial dysfunction	Chronic lycopene treatment could attenuate endothelial dysfunction by reducing oxidative stress in streptozotocin-induced diabetic rats	[120]
				Human				
Patients	**Sample specification**		**Type of study**	**Assessment**	**Duration**	**Aims**	**Results**	**Reference**
				Dietary assessment		Consumption of carotenoids, flavonoids rich food and CVD risk	Increased consumption of few of them has been related with decrease mortality or other CVD risk	[121]
						Consumption of β-carotene and CVD risk	20 mg daily β-carotene in combination with vitamin E 600 mg and vitamin C 250 mg, has no effect on morbidity or mortality	[122]

	Men, age 50–69		Alpha-Tocopherol, Beta-Carotene Cancer Prevention (ATBC)	Serum Status		Consumption of β-carotene or the combination ofα-tocopherols and β-carotene and CHD risk in patients who had a previous myocardial infarction	β-carotene and combined α-tocopherols and β-carotene groupshas increased the risk of CHD as compared to the placebo group	[34]
	Women		Women's Antioxidant Cardiovascular Study (WACS)			Consumption of β-carotene 50 mg every other day, or vitamin C 500 mg daily or vitamin E 600 IU every other day and CVD risk	Women at high risk not showed any CVD risk reduction	[123]
			Physicians Health Study (PHS)			Estimation of the relationship between CHD risk and vegetable intake	Carotenoid rich vegetables intake was associated with a reduction in CHD	[124]
			Physicians Health Study (PHS)			Estimation of the relationship between CHD risk and vegetable intake	No effect from supplementation of interchange day β-carotene 50 mg on CVD, cancer, or overall mortality among primarily non-smokers	[125]

(continued)

Table 20.2 (continued)

			Physicians Health Study (PHS)			Increased Lycopene concentration and CHD risk	No association between increasing concentrations of plasma lycopene and the risk of CVD was found	[126]
			intervention studies	Serum status	8 weeks	Supplementation with lutein and zeaxanthin and CVD risk	No critical impact on macular color level (MPL) or serological markers of endothelial activation, inflammation and oxidation in healthy volunteers	[128]
26			Case-control study	Serum status		Effects of lycopene (from cooked tomatoes) and CHD risk	Supplementation with tomatoes significantly reduced MDA levels, representing a lower lipid peroxidation and conversely increased levels of antioxidant enzymes	[129]
			Case-control study	Serum status		Consumption of tomato based Lycopene and CVD risk	Tomato derived lycopene act as an antioxidant but may not be act as a lipid lowering agent in hypertension	[130]
26	Men/women		Randomized, placebo-controlled, double-blind, crossover study	Blood Status	26 days	Tomato- based drink (ingesting 5.7 mg lycopene per day and CVD risk	Tomato-based drink effectively lowers blood TNF-α production. No effect on urinary 8-iso-PGF2-α concentration and DNA damage.	[131]

126	Men, age 22–57		Randomized, placebo-controlled, double-blind study	Serum Status	8 weeks		Lycopene contents increased in serum, while DNA damage, SOD activity and Endothelial function were decreased. hs-CRP content in serum and plasma levels of adhesion proteins sICAM- 1 and sVCAM-1 were significantly decreased by lycopene.	[132]
31	Postmenopausal women						70 g tomato purèe per day (46 mg lycopene per day) did not affect endothelial function	[133]
			Human intervention trials			Effect of lycopene on blood pressure and lipids.	Lycopene (25 mg/day) significantly lowered serum LDL and total cholesterol but further extensive studies needed	[134]
29			Prospective, randomized, double-blind study		Plasma status	Effect of Astaxanthin on oxidative stress biomarkers	Supplements of astaxanthin inhibiting the lipid peroxidation and enhancing the antioxidant activity for 3 weeks	[135]

(continued)

Table 20.2 (continued)

39			Double-blind, randomized controlled trial		Plasma status	Effect of astaxanthin on oxidative stress biomarkers in heavy smokers and non-smokers	Supplementation of astaxanthin supplementation might arrest oxidative harm in smokers by hindering lipid peroxidation and rousing the cell reinforcement action in smokers	[136]
			Park double-blind, randomized controlled trial			Effect of astaxanthin on C-reactive protein	Astaxanthin also significantly lowered C-reactive protein	[137, 138]
					Blood Status	Effect of astaxanthin on blood lipids	no effect on LDL-cholesterol at any dose	[139]
				12 weeks		Effect of astaxanthin on adiponectin	Astaxanthin also significantly increased blood adiponectin levels. Serum levels of adiponectin tend to be reduced in obese and/or diabetic subjects, smokers, patients with coronary heart disease, and individuals with metabolic syndrome	[140]

In vitro								
						Carotenoids may exhibit their protective role via reduction of oxidative stress,	Pre-treatment with lycopene lowers the MDA contents in H2O2-treated cells. Lycopene also reduced the apoptosis in response to H2O2 preventing the up regulation of caspase3 mRNA and p53mRNA.	[141]
						Carotenoids may protect against oxidative stress.	β-carotene has the ability to inhibit redox-based NF-αB activation, due to its potential antioxidant activity.	[143]
						Carotenoids may protect against oxidative stress	Lycopene lowers the intracellular contented of whole cholesterol in THP-1 cells. Decrease in expression of HMG-CoA reductase, key enzyme converts HMG-CoA to mevalonate [73]. As hypercholesterolemia is the main reason of atherosclerosis, lycopene diminished formation of foam cell and reduction in CVD.	[144]

(continued)

Table 20.2 (continued)

						Carotenoids may be responsible for hampering of Vascular (SMCs) aggravation by regulating the molecular pathways involved in cell proliferation and apoptosis.	Lycopene obstructed PDGF-BB induced SMC proliferation in rats.	[145]
						Tomato-based foods power stay specifically obliging in the inhibition of platelet clumping and clot formation	Lycopene arrested clot formation in a dose-dependent manner	[146, 147]
						High carotenoid concentration 20 mM, may inhibit cell proliferation Lower concentrations 0.5 and 2 mM, an unphysiologically long treatment for up to 24 h	Carotenoids tested are effective in inhibiting cell proliferation and lowering cell viability, thus demonstrating that unphysiological doses of carotenoids may bring about their in vitro effects not specifically, but through a general cytotoxic action.	[148]

References

1. World Health Organization (2003) The world health report 2003: shaping the future. World Health Organization
2. Thom T, Haase N, Rosamond W, Howard VJ, Rumsfeld J, Manolio T et al (2006) Heart disease and stroke statistics--2006 update: a report from the American Heart Association Statistics Committee and Stroke Statistics Subcommittee. Circulation 113(6):e85
3. Mathers CD, Loncar D (2006) Projections of global mortality and burden of disease from 2002 to 2030. PLoS Med 3:442
4. World Health Organization (2005) Preventing chronic diseases: a vital investment. World Health Organization
5. Rosengren A, Wallentin L, Simoons M, Gitt AK, Behar S, Battler A, Hasdai D (2005) Cardiovascular risk factors and clinical presentation in acute coronary syndromes. Heart 91:1141–1147
6. Ciccone MM, De Pergola G, Porcelli MT, Scicchitano P, Caldarola P, Lacoviello M et al (2010) Increased carotid IMT in overweight and obese women affected by Hashimoto's thyroiditis: an adiposity and autoimmune linkage? BMC Cardiovasc Disord 10:22
7. Ciccone MM, Marzullo A, Mizio D, Angiletta D, Cortese F, Scicchitano P et al (2011) Can carotid plaque histology selectively predict the risk of an acute coronary syndrome? Int Heart J 52:72–77
8. Ciccone MM, Miniello V, Marchioli R, Scicchitano P, Cortese F, Palumbo V, Primitivo SG et al (2011) Morphological and functional vascular changes induced by childhood obesity. Eur J Cardiovasc Prev Rehabil 18:831–835
9. Ciccone MM, Balbarini A, Teresa PM, Santoro D, Cortese F, Scicchitano P et al (2011) Carotid artery intima-media thickness: normal and percentile values in the Italian population (camp study). Eur J Cardiovasc Prev Rehabil 18:650–655
10. Ciccone MM, Niccoli-Asabella A, Scicchitano P, Gesualdo M, Notaristefano A, Chieppa D et al (2011) Cardiovascular risk evaluation and prevalence of silent myocardial ischemia in subjects with asymptomatic carotid artery disease. Vasc Health Risk Manag 7:129
11. Wustmann K, Klaey M, Burow A, Shaw SG, Hess OM, Allemann Y (2012) Additive effect of homocysteine-and cholesterol-lowering therapy on endothelium-dependent vasodilation in patients with cardiovascular disease. Cardiovasc Ther 30:277–286
12. Diaz MN, Frei B, Vita JA, Kainey JJ (1997) Antioxidants and atherosclerotic heart disease. N Engl J Med 337:408–416
13. Voutilainen S, Nurmi T, Mursu J, Rissanen TH (2006) Carotenoids and cardiovascular health. Am J Clin Nutr 83(6):1265–1271
14. Di Mascio P, Kaiser S, Sies H (1989) Lycopene as the most efficient biological carotenoid singlet oxygen quencher. Arch Biochem Biophys 274(2):532–538
15. El-Agamey A, Lowe GM, McGarvey DJ, Mortensen A, Phillip DM, Truscott TG, Young AJ (2004) Carotenoid radical chemistry and antioxidant/pro-oxidant properties. Arch Biochem Biophys 430(1):37–48
16. Riccioni G, Speranza L, Pesce M, Cusenza S, D'Orazio N, Glade MJ (2012) Novel phytonutrient contributors to antioxidant protection against cardiovascular disease. Nutrition 28(6):605–610
17. Bjelakovic G, Nikolova D, Gluud LL, Simonetti RG, Gluud C (2007) Mortality in randomized trials of antioxidant supplements for primary and secondary prevention: systematic review and meta-analysis. JAMA 297(8):842–857
18. Krinsky NI (1993) Actions of carotenoids in biological systems. Annu Rev Nutr 13:561–587
19. Crews H, Alink G, Andersen R, Braesco V, Holst B, Maiani G et al (2001) A critical assessment of some biomarker approaches linked with dietary intake. Br J Nutr 86(S1):S5–S35
20. Gärtner C, Stahl W, Sies H (1997) Lycopene is more bioavailable from tomato paste than from fresh tomatoes. Am J Clin Nutr 66(1):116–122

21. Stahl W, Sies H (1996) Lycopene: a biologically important carotenoid for humans? Arch Biochem Biophys 336:1–9
22. Bendich AD, Olson JA (1989) Biological actions of carotenoids. FASEB J 3(8):1927–1932
23. Campbell DR, Gross MD, Martini MC, Grandits GA, Slavin JL, Potter JD (1994) Plasma carotenoids as biomarkers of vegetable and fruit intake. Cancer Epidemiol Prev Biomarkers 3(6):493–500
24. Gate L, Paul J, Ba GN, Tew KD, Tapiero H (1999) Oxidative stress induced in pathologies: the role of antioxidants. Biomed Pharmacother 53(4):169–180
25. Hensley K, Floyd RA (2002) Reactive oxygen species and protein oxidation in aging: a look back, a look ahead. Arch Biochem Biophys 397(2):377–383
26. Cadenas E, Davies KJ (2000) Mitochondrial free radical generation, oxidative stress, and aging. Free Radic Biol Med 29(3):222–230
27. Beutner S, Bloedorn B, Frixel S, Hernández BI, Hoffmann T, Martin HD et al (2001) Quantitative assessment of antioxidant properties of natural colorants and phytochemicals: carotenoids, flavonoids, phenols and indigoids. The role of β-carotene in antioxidant functions. J Sci Food Agric 81(6):559–568
28. Kidd P (2011) Astaxanthin, cell membrane nutrient with diverse clinical benefits and anti-aging potential. Altern Med Rev 16(4):355–364
29. Rossoni-Júnior JV, Araújo GR, da Cruz Pádua B, Chaves MM, Pedrosa ML, Silva ME, Costa DC (2012) Annato extract and β-carotene modulate the production of reactive oxygen species/nitric oxide in neutrophils from diabetic rats. J Clin Biochem Nutr 50(3):177–183
30. Brady WE, Mares-Perlman JA, Bowen P, Stacewicz-Sapuntzakis M (1996) Human serum carotenoid concentrations are related to physiologic and lifestyle factors. J Nutr 126:129–137
31. Ford ES, Mokdad AH, Giles WH, Brown DW (2003) The metabolic syndrome and antioxidant concentrations: findings from the Third National Health and Nutrition Examination Survey. Diabetes 52:2346–2352
32. Heler FR, Descamps O, Hondekijn JC (1998) LDL oxidation therapeutic perspectives. Atherosclerosis 137(Suppl):S25–S31
33. Feng D, Ling WH, Duan RD (2009) Lycopene suppresses LPS-induced NO and IL-6 production by inhibiting the activation of ERK, p38MAPK, and NF-κB in macrophages. Inflamm Res 59:115–121
34. Riso P, Visioli F, Grande S, Guarnieri S, Gardana C, Simonetti P, Porrini M (2006) Effect of a tomato-based drink on markers of inflammation, immunomodulation and oxidative stress. J Agric Food Chem 54:2563–2566
35. Napolitano M, De Pascale C, Wheeler-Jones C, Botham KM, Bravo E (2007) Effects of lycopene on the induction of foam cell formation by modified LDL. Am J Physiol Endocrinol Metab 293:1820–1827
36. Jung EM, Jose OL, Ryan D, Joan AS, Richard MC, Maria LF (2011) A lutein-enriched diet prevents cholesterol accumulation and decreases oxidized LDL and inflammatory cytokines in the aorta of guinea pigs. J Nutr 141:1458–1463
37. Selvaraj RK, Klasing KC (2006) Lutein and eicosapentaenoic acid interact to modify iNOS mRNA levels through the PPARg/RXR pathway in chickens and HD11 cell lines. J Nutr 136:1610–1616
38. Zhao Y, Yu W, Hu W, Yuan Y (2003) Anti-inflammatory and anticoagulant activities of lycopene in mice. Nutr Res 23:1591–1595
39. Ohgami K, Shiratori K, Kotake S, Nishida T, Mizuki N, Yazawa K, Ohno S (2006) Efects of astaxanthin on lipopolysaccharide-induced inflammation in vitro and in vivo. Invest Ophthalmol Vis Sci 44:2694–2701
40. Lee SJ, Bai SK, Lee KS, Namkoong S, Na HJ, Ha KS, Han JA, Yim SV, Chang K, Kwon YG et al (2003) Astaxanthin inhibits nitric oxide production and inflammatory gene expression by suppressing I(kappa)B kinase-dependent NF-kappaB activation. Mol Cells 16:97–105

41. Devasagayam TPA, Tilak JC, Boloor KK, Ketaki SS, Ghaskadbi SS, Lele RD (2004) Free radicals and antioxidants in human health: current status and future prospects. JAPI 52:794–804
42. Wendy MR, Ineke AA, Bots ML, Kluft C, Princen H, Henk FJ, Lilian BM, Poppel GV, Alwine FM (2004) Serum carotenoids and vitamins in relation to markers of endothelial function and inflammation. Eur J Epidemiol 19:915–921
43. Martin KR, Wu D, Meydani M (2000) The effect of carotenoids on the expression of cell surface adhesion molecules and binding of monocytes to human aortic endothelial cells. Atherosclerosis 105:265–274
44. Lee DK, Grantham RN, Mannion JD, Trachte AL (2006) Carotenoids enhance phosphorylation of Akt and suppress tissue factor activity in human endothelial cells. J Nutr Biochem 17:780–786
45. Zeman FJ (1991) Clinical nutrition and dietetics. Macmillan, New York
46. von Baeyer H, Hopfenmuller W, Riedel E et al (2003) Atherosclerosis: current concepts of pathophysiology and pharmacological intervention based on trial outcomes. Clin Nephrol 60:S31–S48
47. Ross R (1986) The pathogenesis of atherosclerosis – an update. N Engl J Med 314(8):488–500
48. Sacks FM, Pfeffer MA, Moye LA, Rouleau JL, Rutherford JD, Cole TG, Brown L, Warnica JW, Arnold JM, Wun CC, Davis BR (1996) The effect of pravastatin on coronary events after myocardial infarction in patients with average cholesterol levels. N Engl J Med 335(14):1001–1009
49. (2003) Centers for Disease Control and Prevention Prevalence of physical activity, including lifestyle activities among adults--United States, 2000-2001. MMWR Morb Mortal Wkly Rep 52(32):764–769
50. Fuhrman B, Volkova N, Rosenblat M, Aviram M (2000) Lycopene synergistically inhibits LDL oxidation in combination with vitamin E, glabridin, rosmarinic acid, carnosic acid, or garlic. Antioxid Redox Signal 2(3):491–506
51. Hak AE, Stampfer MJ, Campos H, Sesso HD, Gaziano JM, Willett W, Ma J (2003) Plasma carotenoids and tocopherols and risk of myocardial infarction in a low-risk population of US male physicians. Circulation 108:802–807
52. Buijsse B, Feskens EJM, Kwape L, Kok FJ, Kromhout D (2008) Both α and b-carotene, but not tocopherols and vitamin C, are inversely related to 15-year cardiovascular mortality in Dutch elderly men. J Nutr 138:344–350
53. Rissanen T, Voutilainen S, Nyyssönen K, Salonen JT (2002) Lycopene, atherosclerosis, and coronary heart disease. Exp Biol Med 227:900–907
54. Bjelakovic G, Nikolova D, Gluud LL, Simonetti RG, Gluud C (2007) Mortality in randomized trials of antioxidant supplements for primary and secondary prevention: systematic review and metaanalysis. JAMA 297:842–857
55. Rissanen TH, Voutilainen S, Nyyssönen K, Salonen R, Kaplan GA, Salonen JT (2003) Serum lycopene concentrations and carotid atherosclerosis: the Kuopio Ischaemic Heart Disease Risk Factor Study. Am J Clin Nutr 77:133–138
56. Pashkow FJ, Watumull GD, Campbell CL (2008) Astaxanthin: a novel potential treatment for oxidative stress and inflammation in cardiovascular disease. Am J Cardiol 101:58D–68D
57. Stahl W, von Laar J, Martin HD, Emmerich T, Sies H (2000) Stimulation of gap junctional communication: comparison of acyclo-retinoic acid and lycopene. Arch Biochem Biophys 373:271–274
58. Bertram JS (1999) Carotenoids and gene regulation. Nutr Rev 57:182–191
59. Hix LM, Lockwood SF, Bertram JS (2004) Upregulation of connexin 43 protein expression and increased gap junctional communication by water soluble disodium disuccinateastaxanthinderivatives. Cancer Lett 211:25–37
60. Maulik N, Das DK (1999) Apoptosis, heart failure, ischemic heart disease. Heart Fail Rev 4:165–175

61. Esterhuyse AJ, Van RJ, Strijdom H, Bester D, Du Toit EF (2006) Proposed mechanisms for red palm oil induced cardioprotection in a model of hyperlipidaemia in the rat. PLEFA 75:375–380
62. Engelbrecht AM, Esterhuyse AJ, Du Toit EF, Lochner A, Van Rooyen J (2006) p38-MAPK and PKB/Akt, possible role players in red palm oil-induced protection of the isolated perfused rat heart. J Nutr Biochem 17:265–271
63. Grossa GJ, Lockwood SF (2004) Cardioprotection and myocardial salvage by a disodium disuccinateastaxanthin derivative (Cardax). Life Sci 75:215–224
64. Aoi W, Naito Y, Sakuma K, Kuchide M, Tokuda H, Maoka T, Toyokuni S, Oka S, Yasuhara M, Yoshikawa T (2003) Astaxanthin limits exercise-induced skeletal and cardiac muscle damage in mice. Antioxid Redox Signal 5:139–144
65. Guerin M, Huntley ME, Olaizola M (2003) Haematococcusastaxanthin: applications for human health and nutrition. Trends Biotechnol 21:210–216
66. Kima JH, Naa JH, Kima CK, Kima JY, Haa KS, Leeb H, Chungc HT, Kwond HJ, Kwone YG, Kim YM (2008) The non-provitaminA carotenoid, lutein, inhibits NF-κB-dependent gene expression through redox-based regulation of the phosphatidylinositol 3-kinase/PTEN/Akt and NF-κB-inducing kinase pathways: Role of H2O2 in NF-κB activation. Free Radic Biol Med 45:885–896
67. Simone RE, Russo M, Catalano A, Monego G, Froehlich K, Boehm V, Palozza P (2011) Lycopene inhibits NF-κB mediated IL-8 expression and changes redox and PPAR γ signaling in cigarette smoke-stimulated macrophages. PLoS One 6. https://doi.org/10.1371/journal.pone.0019652
68. Das DK, Maulik N (2004) Conversion of death signal into survival signal by redox signaling. Biochemistry 69:10–17
69. Klipstein-Grobusch K, Launer LJ, Geleijnse JM, Boeing H, Hofman A, Wittemen JCM (2000) Serum carotenoids and atherosclerosis: The Rotterdam Study. Atherosclerosis 148:49–56
70. Street DA, Comstock GW, Salkeld RM, Schuep W, Klag MJ (1994) Serum antioxidants and myocardial infarction. Are low levels of carotenoids and alpha-tocopherol risk factors for myocardial infarction? Circulation 90:1154–1161
71. Kohlmeier L, Kark JD, Gomez-Gracia E, Martin BC, Steck SE, Kardinaal AF et al (1997) Lycopene and myocardial infarction risk in the EURAMIC study. Am J Epidemiol 146:618–626
72. Wang L, Gaziano JM, Norkus EP, Buring JE, Sesso HD (2008) Associations of plasma carotenoids with risk factors and biomarkers related to cardiovascular disease in middle-aged and older women. Am J Clin Nutr 88:747–754
73. Polidori MC, Mecocci P, Stahl W, Sies H (2003) Cigarette smoking cessation increases plasma levels of several antioxidant micronutrients and improves resistance towards oxidative challenge. Br J Nutr 90:147–150
74. Gammone MA, Gemello E, Riccioni G, D'Orazio N (2014) Marine bioactive and potential applications in sport. Marine Drugs 12:2357–2382
75. D'Orazio N, Gemello E, Gammone MA, DeGirolamo M, Ficoneri C, Riccioni G (2012) Fucoxantin: a treasure from sea. Marine Drugs 10:604–616
76. Okada T, Mizuno Y, Sibayama S, Hosokawa M, Miyashita K (2011) Antiobesity effects of Undaria lipid capsules prepared with phospholipids. J Food Sci 76:H2–H6
77. Labruna G, Pasanisi F, Fortunato G, Nardelli C, Finelli C, Farinaro E et al (2011) Sequence analysis of the UCP1 gene in a severe obese population from Southern Italy. J Obes 269:043
78. Sesso HD, Buring JE, Norkus EP, Gaziano JM (2004) Plasma lycopene, other carotenoids, and retinol and the risk of cardiovascular disease in women. Am J Clin Nutr 79:47–53
79. Rissanen T, Voutilainen S, Nyyssönen K, Salonen R, Salonen JT (2000) Low plasma lycopene concentration is associated with increased intima-media thickness of the carotid artery wall. Arterioscler Thromb Vasc Biol 20:2677–2681
80. Donaldson MS (2011) A carotenoid health index based on plasma carotenoids and health outcomes. Nutrients 3:1003–1022

81. (2008) U.S. Department of Health and Human Services. Physical Activity Guidelines Advisory Committee Report, 2008. U.S. Department of Health and Human Services; Washington, DC, p A-7
82. Cornelissen VA, Fagard RH (2005) Effects of endurance training on blood pressure, blood pressure-regulating mechanisms, and cardiovascular risk factors. Hypertension 46(4):667–675
83. Leitzmann MF, Park Y, Blair A, Ballard-Barbash R, Mouw T, Hollenbeck AR et al (2007) Physical activity recommendations and decreased risk of mortality. Arch Intern Med 167(22):2453–2460
84. Healy GN, Dunstan DW, Salmon J, Shaw JE, Zimmet PZ, Owen N (2008) Television time and continuous metabolic risk in physically active adults. Med Sci Sports Exerc 40(4):639–645
85. Mordente A, Guantario B, Meucci E, Silvestrini A, Lombardi E, Martorana GE et al (2011) Lycopene and cardiovascular diseases: an update. Curr Med Chem 18:1146–1163
86. Buijsse B, Feskens EJ, Moschandreas J, Jansen EH, Jacobs DR Jr, Kafatos A et al (2007) Oxidative stress, and iron and antioxidant status in elderly men: differences between the Mediterranean south (Crete) and northern Europe (Zutphen). Eur J Cardiovasc Prev Rehab 14:495–500
87. Karppi J, Laukkanen JA, Makikallio TH, Ronkainen K, Kurl S (2012) Low beta-carotene concentrations increase the risk of cardiovascular disease mortality among Finnish men with risk factors. Nutr Metab Cardiovasc Dis 22:921–928
88. Karppi J, Nurmi T, Kurl S, Rissanen TH, Nyyssonen K (2010) Lycopene, lutein and beta-carotene as determinants of LDL conjugated dienes in serum. Atherosclerosis 209:565–572
89. Karppi J, Laukkanen JA, Makikallio TH, Ronkainen K, Kurl S (2013) Serum beta-carotene and the risk of sudden cardiac death in men: a population-based follow- up study. Atherosclerosis 226:172–177
90. Gey KF, Puska P (1989) Plasma vitamins E and A inversely correlated to mortality from ischemic heart disease in cross-cultural epidemiology. Ann N Y Acad Sci 570:268–282
91. Kardinaal AF, Kok FJ, Ringstad J, Gomez-Aracena J, Mazaev VP, Kohlmeier L (1993) Antioxidants in adipose tissue and risk of myocardial infarction: the Euramic Study. Lancet 342:1379–1384
92. Tavani A, Negri E, D'Avanzo B, La Vecchia C (1997) Beta-carotene intake and risk of non-fatal acute myocardial infarction in women. Eur J Epidemiol 13:631–637
93. Klipstein-Grobusch K, Geleijnse JM, den Breeijen JH, Boeing H, Hofman A, Grobbee DE et al (1999) Dietary antioxidants and risk of myocardial infarction in the elderly: the Rotterdam Study. Am J Clin Nutr 69:261–266
94. Rimm EB, Stampfer MJ, Ascherio A, Giovannucci E, Colditz GA, Willett WC (1993) Vitamin E consumption and the risk of coronary heart disease in men. N Engl J Med 328:1450–1456
95. Xu XR, Zou ZY, Huang YM, Xiao X, Ma L, Lin XM (2012) Serum carotenoids in relation to risk factors for development of atherosclerosis. Clin Biochem 45:1357–1361
96. Dwyer JH, Paul-Labrador MJ, Fan J, Shircore AM, Merz CN, Dwyer KM (2004) Progression of carotid intima-media thickness and plasma antioxidants: the Los Angeles Atherosclerosis Study. Arterioscler Thromb Vasc Biol 24:313–319
97. Marin C, Ramirez R, Delgado-Lista J, Yubero-Serrano EM, Perez-Martinez P, Carracedo J (2011) Mediterranean diet reduces endothelial damage and improves the regenerative capacity of endothelium. Am J Clin Nutr 93:267–274
98. Holt EM, Steffen LM, Moran A, Basu S, Steinberger J, Ross JA et al (2009) Fruit and vegetable consumption and its relation to markers of inflammation and oxidative stress in adolescents. J Am Diet Assoc 109:414–421
99. Azzini E, Polito A, Fumagalli A, Intorre F, Venneria E, Durazzo A et al (2011) Mediterranean diet effect: an Italian picture. Nutr J 10:125
100. Shardell M, Alley D, Hicks DE, El-Kamary SS, Miller RR, Semba RD et al (2011) Low-serum carotenoid concentrations and carotenoid interactions predict mortality in US adults: the Third National Health and Nutrition Examination Survey. Nutr Res 31:178–189

101. De Waart FG, Schouten EG, Stalenhoef AF, Kok FJ (2001) Serum carotenoids, alpha-tocopherol and mortality risk in a prospective study among Dutch elderly. Int J Epidemiol 30:136–143
102. Bohm V (2012) Lycopene and heart health. Mol Nutr Food Res 56:296–303
103. Yeo HY, Kim OY, Lim HH, Kim JY, Lee JH (2011) Association of serum lycopene and brachial-ankle pulse wave velocity with metabolic syndrome. Metabolism 60:537–543
104. Ito Y, Kurata M, Suzuki K, Hamajima N, Hishida H, Aoki K (2006) Cardiovascular disease mortality and serum carotenoid levels: a Japanese population- based follow-up study. J Epidemiol 16:154–160
105. Tavani A, Gallus S, Negri E, Parpinel M, La Vecchia C (2006) Dietary intake of carotenoids and retinol and the risk of acute myocardial infarction in Italy. Free Radic Res 40:659–664
106. Hozawa A, Jacobs DR Jr, Steffes MW, Gross MD, Steffen LM, Lee DH (2007) Relationships of circulating carotenoid concentrations with several markers of inflammation, oxidative stress, and endothelial dysfunction: the Coronary Artery Risk Development in Young Adults (CARDIA)/Young Adult Longitudinal Trends in Antioxidants (YALTA) study. Clin Chem 53:447–455
107. Lee S, Harnack L, Jacobs DR Jr, Steffen LM, Luepker RV, Arnett DK (2007) Trends in diet quality for coronary heart disease prevention between 1980–1982 and 2000–2002: The Minnesota Heart Survey. J Am Diet Assoc 107:213–222
108. Bansal P, Gupta SK, Ojha SK, Nandave M, Mittal R, Kumari S, Arya DS et al (2006) Cardioprotective effect of lycopene in the experimental model of myocardial ischemia-reperfusion injury. Mol Cell Biochem 289:1–9
109. Breinholt V, Lauridsen ST, Daneshvar B, Jakobsen J (2000) Dose-response effects of lycopene on selected drug-metabolizing and antioxidant enzymes in the rat. Cancer Lett 154:201–210
110. Gitenay D, Lyan B, Rambeau M, Mazur A, Rock E (2007) Comparison of lycopene and tomato effects on biomarkers of oxidative stress in vitamin E deficient rats. Eur J Nutr 46:468–475
111. Hsu YM, Lai CH, Chang CY, Fan CT, Chen CT, Wu CH (2008) Characterizing the lipid-lowering effects and antioxidant mechanisms of tomato paste. Biosci Biotechnol Biochem 72:677–685
112. Monroy-Ruiz J, Sevilla MA, Carron R, Montero MJ (2011) Astaxanthin- enriched-diet reduces blood pressure and improves cardiovascular parameters in spontaneously hypertensive rats. Pharm Res 63:44–50
113. Zhao ZW, Cai W, Lin YL, Lin QF, Jiang Q, Lin Z et al (2011) Ameliorative effect of astaxanthin on endothelial dysfunction in streptozotocin-induced diabetes in male rats. Arzneimittelforschung 61:239–246
114. Lee W, Ku SK, Bae JW, Bae JS (2012) Inhibitory effects of lycopene on HMGB1-mediated pro-inflammatory responses in both cellular and animal models. Food Chem Toxicol 50:1826–1833
115. Hu MY, Li YL, Jiang CH, Liu ZQ, Qu SL, Huang YM (2008) Comparison of lycopene and fluvastatin effects on atherosclerosis induced by a high-fat diet in rabbits. Nutrition 24:1030–1038
116. Frederiksen H, Rasmussen SE, Schroder M, Bysted A, Jakobsen J, Frandsen H et al (2007) Dietary supplementation with an extract of lycopene-rich tomatoes does not reduce atherosclerosis in Watanabe Heritable Hyperlipidemic rabbits. Br J Nutr 97:6–10
117. Wang XD (2012) Lycopene metabolism and its biological significance. Am J clin nutr 96(5):1214S–22S
118. Lorenz M, Fechner M, Kalkowski J, Frohlich K, Trautmann A, Bohm V et al (2012) Effects of lycopene on the initial state of atherosclerosis in New Zealand White (NZW) rabbits. PLoS One 7:e30808
119. Verschuren L, Wielinga PY, Van Duyvenvoorde W, Tijani S, Toet K, Van Ommen B et al (2011) A dietary mixture containing fish oil, resveratrol, lycopene, catechins, and vitamins E and C reduces atherosclerosis in transgenic mice. J Nutr 141:863–869

120. Zhu J, Wang CG, Xu YG (2011) Lycopene attenuates endothelial dysfunction in streptozotocin-induced diabetic rats by reducing oxidative stress. Pharm Biol 49:1144–1149
121. Yang Y, Chan SW, Hu M, Walden R, Tomlinson B (2011) Effects of some common food constituents on cardiovascular disease. ISRN Cardiol 2011:397136
122. (2002) MRC/BHF Heart Protection Study of antioxidant vitamin supplementation in 20,536 high-risk individuals: a randomised placebo-controlled trial. Lancet 360:23–33
123. Rapola JM, Virtamo J, Ripatti S, Huttunen JK, Albanes D, Taylor PR et al (1997) Randomised trial of alpha-tocopherol and beta-carotene supplements on incidence of major coronary events in men with previous myocardial infarction. Lancet 349:1715–1720
124. Cook NR, Albert CM, Gaziano JM, Zaharris E, MacFadyen J, Danielson E et al (2007) A randomized factorial trial of vitamins C and E and beta carotene in the secondary prevention of cardiovascular events in women: results from the Women's Antioxidant Cardiovascular Study. Arch Intern Med 167:1610–1618
125. Liu S, Lee IM, Ajani U, Cole SR, Buring JE, Manson JE (2001) Intake of vegetables rich in carotenoids and risk of coronary heart disease in men: the Physicians' Health Study. Int J Epidemiol 30:130–135
126. Hennekens CH, Buring JE, Manson JE, Stampfer M, Rosner B, Cook NR et al (1996) Lack of effect of long-term supplementation with beta carotene on the incidence of malignant neoplasms and cardiovascular disease. N Engl J Med 334:1145–1149
127. Riccioni G, Mancini B, DiIlio E, Bucciarelli T, D'Orazio N (2008) Protective effect of lycopene in cardiovascular disease. Eur Rev Med Pharmacol Sci 12:183–190
128. Graydon R, Hogg RE, Chakravarthy U, Young IS, Woodside JV (2012) The effect of lutein- and zeaxanthin-rich foods v. supplements on macular pigment level and serological markers of endothelial activation, inflammation and oxidation: pilot studies in healthy volunteers. Br J Nutr 108:334–342
129. Bose KS, Agrawal BK (2007) Effect of lycopene from cooked tomatoes on serum antioxidant enzymes, lipid peroxidation rate and lipid profile in coronary heart disease. Singap Med J 48:415–420
130. Bose KS, Agrawal BK (2007) Effect of lycopene from tomatoes (cooked) on plasma antioxidant enzymes, lipid peroxidation rate and lipid profile in grade-I hypertension. Ann Nutr Metab 51:477–481
131. Kim JY, Paik JK, Kim OY, Park HW, Lee JH, Jang Y et al (2011) Effects of lycopene supplementation on oxidative stress and markers of endothelial function in healthy men. Atherosclerosis 215:189–195
132. Stangl V, Kuhn C, Hentschel S, Jochmann N, Jacob C, Bohm V et al (2011) Lack of effects of tomato products on endothelial function in human subjects: results of a randomised, placebo-controlled cross-over study. Br J Nutr 105:263–267
133. Ried K, Fakler P (2011) Protective effect of lycopene on serum cholesterol and blood pressure: meta-analyses of intervention trials. Maturitas 68:299–310
134. Choi HD, Kim JH, Chang MJ, Kyu-Youn Y, Shin WG (2011) Effects of astaxanthin on oxidative stress in overweight and obese adults. Phytother Res 25:1813–1818
135. Kim JH, Chang MJ, Choi HD, Youn YK, Kim JT, Oh JM et al (2011) Protective effects of Haematococcusastaxanthin on oxidative stress in healthy smokers. J Med Food 14:1469–1475
136. Park JS, Chyun JH, Kim YK, Line LL, Chew BP (2010) Astaxanthin decreased oxidative stress and inflammation and enhanced immune response in humans. Nutr Metab (Lond) 7:18
137. Pandolfi A (2005) C-reactive protein: a potential new molecular link between inflammation, thrombosis and vascular cell proliferation? Cardiovasc Res 68:3–4
138. Yoshida H, Yanai H, Ito K, Tomono Y, Koikeda T, Tsukahara H et al (2010) Administration of natural astaxanthin increases serum HDL-cholesterol and adiponectin in subjects with mild hyperlipidemia. Atherosclerosis 209:520–523
139. Kajikawa Y, Ikeda M, Takemoto S, Tomoda J, Ohmaru N, Kusachi S (2011) Association of circulating levels of leptin and adiponectin with metabolic syndrome and coronary heart disease in patients with various coronary risk factors. Int Heart J 52:17–22

140. Fassett RG, Coombes JS (2011) Astaxanthin: a potential therapeutic agent in cardiovascular disease. Marine Drugs 9:447–465
141. Riso P, Visioli F, Grande S, Guarnieri S, Gardana C, Simonetti P et al (2006) Effect of a tomato-based drink on markers of inflammation, immunomodulation, and oxidative stress. J Agric Food Chem 54:2563–2566
142. Tang X, Yang X, Peng Y, Lin J (2009) Protective effects of lycopene against H_2O_2-induced oxidative injury and apoptosis in human endothelial cells. Cardiovasc Drugs Ther 23:439–448
143. Bae JW, Bae JS (2011) Barrier protective effects of lycopene in human endothelial cells. Inflamm Res 60:751–758
144. Cai H, Harrison DG (2000) Endothelial dysfunction in cardiovascular diseases: the role of oxidant stress. Circ Res 87:840–844
145. Palozza P, Parrone N, Simone RE, Catalano A (2011) Lycopene in atherosclerosis prevention: an integrated scheme of the potential mechanisms of action from cell culture studies. Arch Biochem Biophys 504:26–33
146. Lo HM, Hung CF, Tseng YL, Chen BH, Jian JS, Wu WB (2007) Lycopene binds PDGF-BB and inhibits PDGF-BB-induced intracellular signaling transduction pathway in rat smooth muscle cells. Biochem Pharmacol 74:54–63
147. Kleinegris MC, Ten Cate-Hoek AJ, Ten Cate H (2012) Coagulation and the vessel wall in thrombosis and atherosclerosis. Pol Arch Med Wewn 122:557–566
148. Hsiao G, Wang Y, Tzu NH, Fong TH, Shen MY, Lin KH et al (2005) Inhibitory effects of lycopene on in vitro platelet activation and in vivo prevention of thrombus formation. J Lab Clin Med 146:216–226
149. Hung CF, Huang TF, Chen BH, Shieh JM, Wu PH, Wu WB (2008) Lycopene inhibits TNF-alpha-induced endothelial ICAM-1 expression and monocyte- endothelial adhesion. Eur J Pharmacol 586:275–282
150. Agarwal M, Parameswari RP, Vasanthi HR, Das DK (2012) Dynamic action of carotenoids in cardioprotection and maintenance of cardiac health. Molecules 17:4755–4769
151. Sung LC, Chao HH, Chen CH, Tsai JC, Liu JC, Hong HJ et al (2015) Lycopene inhibits cyclic strain-induced endothelin-1 expression through the suppression of reactive oxygen species generation and induction of heme oxygenase-1 in human umbilical vein endothelial cells. Clin Exp Pharmacol Physiol 42:632–639
152. Palozza P, Simone R, Catalano A, Monego G, Barini A, Mele MC et al (2010) Lycopene prevention of oxysterol- induced proinflammatory cytokine cascade in human macrophages: inhibition of NF- kappaB nuclear binding and increase in PPARgamma expression. J Nutr Biochem 22:259–268
153. Palozza P, Simone R, Catalano A, Parrone N, Monego G, Ranelletti FO (2011) Lycopene regulation of cholesterol synthesis and efflux in human macrophages. J Nutr Biochem 22:971–978
154. Di Tomo P, Canali R, Ciavardelli D, Di Silvestre S, De Marco A, Giardinelli A et al (2012) beta-Carotene and lycopene affect endothelial response to TNF-alpha reducing nitro-oxidative stress and interaction with monocytes. Mol Nutr Food Res 56:217–227
155. Fortes C, Virgili F (2008) Antioxidant vitamins are not "just antioxidants": not necessarily harmful when targeted to the right population. Biofactors 33:177–180
156. Rao AV, Rao LG (2007) Carotenoids and human health. Pharmacol Res 55:207–216

Chapter 21
Carotenoids and Bone Health

Muhammad Zia-Ul-Haq, Muhammad Riaz, and Alotaibi O. Modhi

21.1 Introduction

Carotenoids are a key class of isoprenoids which are synthesized photosynthetically (plants) and non-photosynthetically (fungi & bacteria). They are heterogenous bioactive compounds C40H56 core chemical structures. The carotenoids can assume different stereo-configurations. Their chemical structure varies in shape, biophysical characteristics, and interaction pattern with other biomolecules. Innately, despite from few species of aphids, carotenoids are not been synthesized in animals, therefore, animals require them through daily diet. Carotenoids are an important class of phytochemical that possesses proven antioxidant potential. Their use in food matrix retards the process of oxidation and helps in improving the shelf stability of the products. In the human body, carotenoids metabolized in myriad of chemical entities with multifarious properties including scavenging free radicals along with provision of vitamin A activity. Finally, yet importantly, they improve the health of individual by improve the defense capabilities of the body along with providing protection many ailments. The functional features of carotenoids in human body are dependent on various intrinsic and extrinsic factors that influence their absorption and metabolism. The external factors associated with their bioavailability include consumption of fats/oils, nutrient-nutrient interactions, chemistry of the compounds, and alcohol addiction and smoking. In contrast, concentration of retinoic

M. Zia-Ul-Haq (✉)
Office of Research, Innovation and Commercialization, Lahore College for Women University, Lahore, Pakistan

M. Riaz
Department of Pharmacy, Shaheed Benazir Bhutto University Sheringal, Sheringal, Pakistan

A. O. Modhi (✉)
Department of Biology, College of Sciences, Princess Nourah bint Abdulrahman University, Riyadh, Saudi Arabia
e-mail: moutaebe@pnu.edu.sa

© Springer Nature Switzerland AG 2021

M. Zia-Ul-Haq et al. (eds.), *Carotenoids: Structure and Function in the Human Body*, https://doi.org/10.1007/978-3-030-46459-2_21

acid and carotenoids are major internal factors responsible for their poor utilization. Moreover, internal metabolisms, body characteristic (hormones, age and gender) and diseased status are intriguing factors.

Fruits and vegetables are cache of carotenoids and are vital constituent of diet owing to their activity as vitamin-A precursor. Other than this, carotenoids play an important role in promoting body's antioxidative status and immune system. Carotenoids perform various functions in animals such as they scavenge free radicals, modulates immune system, regulates apoptosis & cell cycle, and act as precursor for retinol. Higher concentrations of carotenoids prevent neurodegenerative diseases. Carotenoids possess protective role in visual cycle in vertebrates including humans. Carotenoids are also important in meeting the requirements of Vitamin A in different lifecycle groups especially pregnant women. The conversion of carotenoids to retinoic acid is an important determinant due to its role in development of embryo.

Globally, mainly because of versatility in potent health modulating properties of carotenoids an estimated escalation in its demand has been reported from 1.5 billion dollars (2014) to 1.8 billion dollars (2019). Keeping in view the increased demand of carotenoid rich products numerous scientists has devoted their research in understanding the bioavailability, metabolism, and ultimate fate of this phytonutrient in human body. Various chemically synthesized products are being used in formulation of commercial carotenoids, only a few are made from carotenoids extracted from natural sources. Despite their nature of origin, both carotenoids either synthetic or natural are identical in their molecular structure. Synthetic carotenoids are more stable as compare to natural ones as they are designed to minimize oxidation. They are formulated and distributed in market as dispersions, colloids, colloidal suspensions, and emulsions so that their application in food products is much easier. Consumers have shifted their trend towards carotenoids extracted from natural sources. Change in preference of consumer towards carotenoids of natural origin has increased challenges for the manufacturers. As food matrix contains a variety of biomolecules therefore precise knowledge regarding efficient extraction methods is required for efficient recovery of carotenoids. This chapter gives a detailed overview on chemistry, biosynthesis, bioavailability, metabolism and identification & quantification of carotenoids.

Bone maintains size, shape and structural integrity of skeleton and helps in locomotion. Physical activity and diet are two key factor that have substantial effect on bone growth, preservation, and damage during the bone life cycle [1]. A delicate equilibrium between bone development and bone resorption controls the functional mass of bone. When bone construction lags bone resorption, gradual decline in bone mass density occurs, this condition is called osteoporosis. Suitable amount of vitamin D and calcium sustain bone health in best possible way. However, some other foods ingredients, i.e. carotenoids also perform the same function, sometimes even in better way. It's common knowledge that estrogen is mandatory for healthy bones. When estrogen production declines in postmenopausal women, bones become brittle and fracture easily. Carotenoids maintain normal estrogenic level in bone cells.

21.2 Bone Composition

Bone provides strength and hard protective boundary to the body organs and is helps in the synthesis of red and white blood cells, mineral storage, gives specific shape to skeleton and aids in movement. Bone cells are of three basic types

21.2.1 Osteoclasts

The cells involved in the dissolution of bones are termed as osteoclasts. They are responsible for bone fracture. These are bone-absorbing multinuclear cells produced from hematopoietic mononuclear precursor cells, whose over-stimulation leads to the anomalous bone resorption. The procedure of bone resorption by osteoclasts and production of bones by osteoblasts is involved in bone remolding [2].

21.2.2 Osteoblasts

These are single-nucleus cells responsible for bone-formation. These are present externally on osteon seams that produce protein mixture termed as osteoid that is turned into bone after mineralization [2].

21.2.3 Osteocytes

They are distinguished osteoblastic lineage cells, implanted in a matrix of deposited minerals. In most cases, the inactive form of osteoblast is termed as an osteocyte. *When osteoblasts are trapped by a bone matrix that is produced by them, the osteocytes are synthesized [1].* Bone being an endocrine tissue, secretes 2 key hormones, osteocalcin (Ocn) (FGF23) and fibroblast growth factor 23 [3]. Table 21.1 summarizes these facts.

Table 21.1 Functions and location of bone cells

Cell type	Function	Location
Osteoblasts	Bone synthesis	At the surface of bone tissue
Osteocytes	Maintain mineral quantity of serum	Inside the lacunae
Osteoclasts	Bone resorption	In the pits on bone surface

21.3 Carotenoids and Bone Health

Carotenoids maintain bone health by stabilizing the balance of synthesis and breakage of bone. Dietary carotenoids provide protection against osteoporosis. Following are some important carotenoids involved in bone health.

21.3.1 Carotenoids and Osteoporosis

The condition in which an individual has T-score (T-score highlights individual's BMD in relation to mean of BMD in young adults) less than 70% as strategy provided by the Japan Osteoporosis Society on the management of osteoporosis is referred as osteoporosis [4]. Old age and several pathological mechanisms decreases the bone formation and surge in bone resorption, that in turns induces osteoporosis [5]. While osteoporosis and associated fractures are greater in females than in males, men encounter more noteworthy co-morbidity and death after the hip break than females [6]. Antioxidants derived from a diet containing carotenoids, is proved to enhance the formation of bones in test samples [7]. Osteoporosis, is described by change in bone structure, decreased bone density and bone mass, higher bone fragility and more chances of bone fracture. It is a well-recognized public health problem globally. Women, especially postmenopausal women, are especially prone to it [5]. Only in USA, nearly 53 million people have possibility to develop osteoporosis [8]. Approximately more than 200 million women have osteoporosis [9].

21.3.2 Anti-osteoclastogenic Properties

One of treatment mechanism is to develop agents that regulates extra bone resorption. β-carotene exhibits anti-osteoclastogenic properties by attenuating NF-κB pathway thus suggesting it as a potential drug for treatment of osteoporosis [10]. Diet containing β-carotene can maintain bone health in persons suffering from disabilities [11].

21.3.3 Anti-fracture Properties

Osteoporosis is responsible for 8.9 million fractures annually [8, 12]. Of all osteoporosis-related fractures, hip fracture is a chief and most severe issue [13, 14]. In year 2000, nearly 1.6 million hip fractures happened globally [15] and this number may escalate to 6.3 million by the year 2050 due to a surge in number of aging population [16].

One of preventive strategies against osteoporotic fracture is intake of antioxidants from natural sources especially carotenoids. Increased ingestion of carotenoids particularly β-cryptoxanthin, β-carotene and lutein/zeaxanthin are linked with a decreased potential of hip fracture, which suggests that carotenoids can protect from fracture [17]. The cohort studies and randomized controlled trials have specified the connection between fracture and carotenoids intake [18]. Satisfactory intake of carotenoids can decrease chances of osteoporotic fractures in aged men due to their antioxidant properties which counteract the process of osteoporosis [19].

Deregulation of pro-inflammatory cytokines leads to bone loss or osteoporosis. The drugs available against bone loss are expensive and have their own side effects. Functional foods and nutraceuticals can slow down and/or revert osteoporosis by modulating cell signaling pathways. These possess anti-inflammatory properties and are comparatively less expensive and come without side effects [20].

21.4 Individual Carotenoids in Osteoporosis

21.4.1 Fucoxanthin

Fucoxanthin is more cytotoxic against osteoclasts than osteoblasts indicating that fucoxanthin inhibits osteoclast synthesis by preventing osteoclast differentiation and initiating apoptosis in osteoclasts. Hence fucoxanthin is useful for the inhibition of bone diseases such as osteoporosis and rheumatoid arthritis [21].

21.4.2 β-Cryptoxanthin

The main sources of β-cryptoxanthin are tropical orange -flesh fruits tangerines, certain mangoes and papayas. Epidemiological studies propose that β-cryptoxanthin provide protection from inflammatory diseases like arthritis [22] due to its antioxidant potential at physiological conditions [23]. NF-kB kinase (IKK) β can be its potential target in osteoclast differentiation. Since the IKK complex plays a role in both LPS–TLR4–IKKs–NFκB signaling and RANKL–RANK–IKKs–NFkB signaling, β -cry can stop the LPS-induced RANKL expression in osteoblasts and RANKL-induced osteoclast differentiation in osteoclast precursor cells [24] (Fig. 21.1).

β-cryptoxanthin prevents bone loss by increasing osteoblastic bone formation and preventing osteoclastic bone resorption. It increases bone components and suppresses bone resorption @10^{-6} [25]. Similarly its oral ingestion @10^{-7} in rat femoral tissues culture increased substantially alkaline phosphatase and calcium contents, both of which are important for bone mineralization [26]. It also prevents osteoblastic cells by reducing the receptor activator of nuclear factor kB ligand (RANKL),

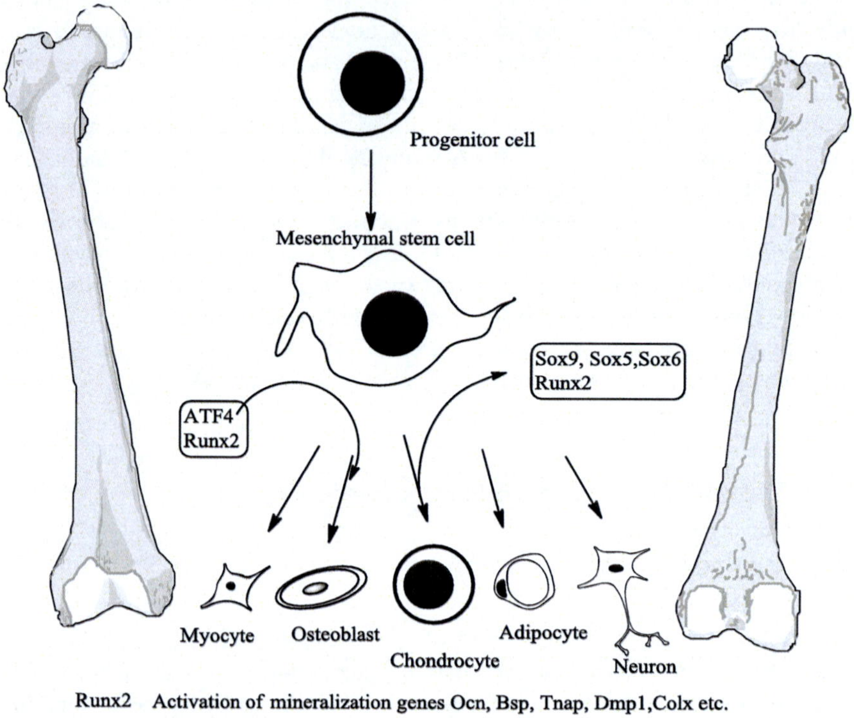

Fig. 21.1 β-cryptoxanthin stimulate cell differentiation and mineralization in osteoblastic cells

suppressing bone resorption factors like prostaglandin E2 and parathyroid hormones connected with osteoclast cell production [26]. β-cryptoxanthin enhances the expression level of runt-related transcriptional factor 2 (Runx 2) collagen type1 alpha 1 (Col1a1), insulin growth factor 1 (IGF-1) stimulating bone formation. β-cryptoxanthin daily oral consumption reduced significantly the osteoclastic activation by limiting the nuclear factor-κB-dependent transcriptional process in OVX-mice [27]. It also prevents the progress of osteoarthritis and suppresses neuropathic pain in mice [28]. It prevents lipopolysaccharides-induced osteoclast differentiation and bone resorption by subduing the NF-κB signaling [24] (Fig. 21.2).

Its daily oral consumption substantially stopped the activation of osteoclastic besides reducing the bone volume in ovariectomized mice. It also halted the differentiation and growth of osteoclasts by suppressing the nuclear factor-kB-dependent transcriptional pathway in *in vitro* studies. Hence β-cryptoxanthin supplementation will be helpful for prophylaxis and for treatment of metabolic bone diseases linked with irregular osteoclast stimulation [29].

The study of cell culture and experiments on rodents suggest that synthesis of osteoclast is enhanced and osteoblast actions are inhibited by the action of β-cryptoxanthin. β-cryptoxanthin is concerned with bone homeostasis [26]. β-cryptoxanthin also stimulates gene expression of proteins concerned with the

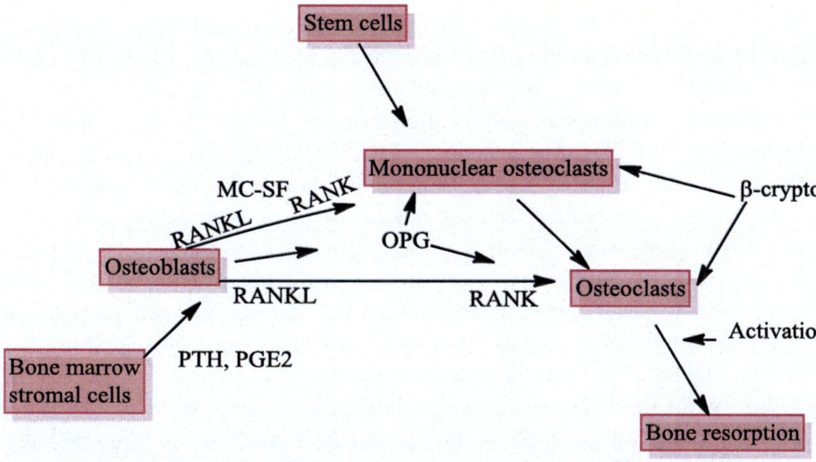

Fig. 21.2 Mechanism by which b-cryptoxanthin inhibit osteoclastic bone resorption

formation of osteoblasts and mineralization of bone, transforming insulin-like growth factor 1, a consequence mediated through protein kinase C or mitogen-activated protein kinase (MAPK) [30]. In vitro, β-cryptoxanthin mitigated the action of parathyroid hormone or prostaglandin E2 that causes bone resorption. It inhibits the formation of osteoclast prostaglandin cell by receptor activator of nuclear factor KB ligand (RANK) [31]. It exhibited an animating impact in the formation of osteoblastic cells in vitro and in vivo [27, 32] by influencing the osteoblastic cell proliferation and differentiation. It also regulates the expression of genes that are specific to osteoblasts like that including Runx2 that is a major regulator to differentiate the osteoblastic cells [27, 32, 33]. Increased level of β-cryptoxanthin in serum is linked with a decreased possibility of osteoporosis or potentially osteopenia. low concentrations of β-cryptoxanthin resulted in increased calcium contents as well as alkaline phosphatase action in the tissues of femoral diaphyseal and femoral-metaphyseal of immature rats in vitro [26]. They show a positive impact on the synthesis of bone and negative impact on bone resorption in a tissue culture [25].

21.4.3 Lycopene

Lycopene prevents the bone resorption *in vitro* [34]. Lycopene (10^{-5} M) restrains basal and parathyroid hormone (PTH)- empowered mineral resorption of osteoclasts as well as arrangement of (TRAP) movement in osteoblast having many nucleus [34]. When lycopene was given to postmenopausal females, they enhanced antioxidant potential, reduction of oxidative stress and the bone resorption marker N-telopeptide. It diminishes bone fracture related markers and decreases the chances of osteoporosis [6]. Lycopene administration diminishes oxidative stress leading to bone wellbeing, Lycopene treatment stifled the ovariectomized (OVX) -instigated increment in bone fracture, as shown by variations in biomarkers of bone digestion like serum osteocalcin (s-OC), serum N-terminal pro-peptide of sort 1 collagen (s-PINP), serum cross linked carboxy terminal telopeptides (s-CTX-1), and urinary deoxy pyridinoline (u-DPD). Many changes in OVX-incited harm of bone mass, bone quality, and micro architectural crumbling was seen in lycopene-treated OVX creatures. Lycopene treatment decreases osteoclast separation while increasing osteoblast growth along with glutathione peroxidase (GPx) catalase (CAT) and superoxide dismutase (SOD) actions. The everyday utilization of lycopene might be critical as it goes about as a cancer prevention agent to diminish bone resorption in postmenopausal females and may in this way be advantageous in diminishing the danger of osteoporosis [35]. The studies indicate that Lycopene is more preferred in the diet in Europe or the U.S. so it has significant contribution to affect the bone health in them [36]. Consumption of lycopene-rich tomato sauce triggered the WNT/β-catenin and ERK1/2 pathways, up-regulated RUNX2, COL1A and alkaline phosphatase and down regulated the RANKL Saos-2, helping in inhibition of bone loss in postmenopausal women [37].

21.4.4 *Lutein and Zeaxanthin*

Egg yolk and maize (corn) are rich sources of lutein and zeaxanthin [38].Lutein (3, 10, and 30 μM) prevents bone resorption by stopping RANKL-dependent osteoclast development, and by subduing the survival of mature osteoclasts. It also incites bone formation hence consumption of lutein-rich sources is useful for the preservation of bone mass [39].In an in vivo study, lutein ingestion for 4 weeks (50 mg/kg) downregulates the inflammation and osteoclast-specific marker (NFATc1) expression via Nrf2 activation protecting the ovariectomized rats against osteoporosis induced by oxidative stress [40]. Likewise, lutein in the form of dietary carotenoid, fortifies bone development (expanding the thickness of, cortical bone) by stifling bone resorption [41]. Lutein improved bone mineralization by stifling osteoclast bone resorption [42]. For premenopausal females, the increased areal bone mineral thickness (BMD) of the lumbar spine was identified with a more prominent dietary administration of lutein and zeaxanthin combined. Administration of lutein and zeaxanthin in adults (average age of 75 years) suggested no substantial connection of ingestion of lutein and zeaxanthin with a BMD (at the femoral neck, trochanter, spine, and radial shaft) [43]. It has been observed that regardless of the nonappearance of an impact at the gauge, an increased intake of lutein and zeaxanthin for male subjects was related with less diminishment in trochanter BMD following 4 years [41]. Lutein and zeaxanthin have been identified with an upper bone thickness in later life , it might be interceded by lessening oxidative stress that advances bone fractures(e.g., by keeping up an appropriate cancer prevention agent/oxidant balance) that is important to keep up a legitimate cancer prevention agent/oxidant adjust essential for bone wellbeing [44]. Lutein and zeaxanthin do advance bone wellbeing by bringing down oxidative stress, at that point dietary intake may not instantly affect bone thickness in youthful, firm people who have in all probability achieved top bone mass and not experienced critical bone misfortune. In any case, steady with the youthful mouse models, dietary lutein and zeaxanthin could impact bone improvement in the young. Persons with more noteworthy lutein and zeaxanthin in their eating regimen not just had decreased bone misfortune contrasted with different subjects, their bone thickness was really higher than their gauge estimation [42].

Lutein upgraded the development of mineralized bone knobs in osteoblasts cultures. Lutein reduced 1α, 25-dihydroxy vitamin D3-actuated bone resorption as estimated by pit arrangement in organ culture of mouse calvaria. In co-cultures of bone marrow cells and osteoblasts, lutein smothered 1α, 25-dihydroxyvitamin D3-incited osteoclast development. In cultures of bone marrow macrophages, lutein smothered solvent RANKL that is the receptor activator of nuclear factor-kappaB (NF-κB) ligand, initiated osteoclast arrangement [41]. Lutein fortified bone arrangement and smothered bone resorption in vitro and improved considerably the femoral bone mass in growing male mice in vivo. In vivo studies indicate that mRNA expression of osteocalcin, that is an ordinary marker of developing osteoblasts, was obviously instigated by bone-actuating factors, however, the expression was not swayed by including lutein [45]. Lutein may follow up on developing osteoblasts to empower the procedure of mineralization, as well as direct the procedure of osteocyte separation. In developing mice, the bone arrangement is significantly more articulated than bone resorption. Along these lines, the rise of femoral bone mass by lutein in developing mice might be for the most part because of the incitement of bone development in vivo. In developing male mice, lutein improved bone mass in vivo. When Lutein is given to developing female mice, it additionally improved bone mass. Lutein may appeared to exhibit useful impacts on bone tissues to improve the pinnacle bone mass in young men and young females and to avert bone misfortune because of maturing as well as menopause [41]. Lutein specifically prompted apoptosis in changed yet not ordinary human mammary epithelial cells and shielded typical cells from apoptosis prompted by the chemotherapy agents. Hence, lutein balances cell development and apoptosis in different cells. Lutein, at 3–10 μM did not show any effect in the cell development of mouse osteoblastic cells [46]. The two most conceivable mechanisms have been observed of lutein action in osteoclast arrangement. Initially, lutein acts on macrophages, osteoclast antecedent cells, and smothered RANKL-subordinate osteoclast arrangement. Besides, lutein follows up on osteoblasts and stifles osteoclast separation by hindering the expression of RANKL in osteoblasts, since lutein smothered the mRNA expression of RANKL instigated by a bone resorbing factor in osteoblast. In ovariectomized (OVX) mice, when lutein is given to animals, it partially repaired the bone loss due to estrogen shortage because of the removed ovary. Further studies to define the conceivable part of lutein in the counteractive action of postmenopausal osteoporosis are needed [41].

21.4.5 β-carotene

Addition of β-carotene enhanced expression of RUNX2, SOX2 and osteonectin confirming the osteo-inductive potential of β-carotene on the differentiation of mesenchyme cells (MSCs) to osteoblasts [47]. Carotenoids ingestion declines the danger of osteoporosis by refining bone calcification and osteogenesis and subduing bone resorption, subsequently increasing bone mass. Oxidative stress leads to osteoclastogenesis and apoptosis of osteoblasts and osteocytes leading to bone resorption [48, 49]. A decrease in hip fracture ratio in elderly and middle-aged people has been observed with increased consumption of b-carotene, b-cryptoxanthin and lutein/zeaxanthin [50] (Fig. 21.3).

Recent studies designate that dietary intake of carotenoids will result in less danger of osteoporosis [51–53]. Oxidative stress is accountable for the activation of osteoclasts synthesis and bone resorption [54]. Pepino (*Solanum muricatum*), a fruit enriched in β-carotene is associated with promoting bone repair in an animal model. β-carotene suppresses osteoclastogenesis by inhibiting NF-κB and MAPK pathways as NF-κB and MAPK pathways have significant contribution to osteoporosis. β-carotene provides protection against osteoporosis and has been used for the medication of osteoporosis. β-carotene reduces osteoclasts formation. β-carotene levels have an inverse relation with the occurrence of osteoporosis [36]. β-carotene @ as low as 0.1 μM decreased viability, osteogenesis, and resorption formation and therefore can be used as a powerful remedy against osteoporosis. β-carotene addition at 48 h after RANKL stimulation, inhibited the osteoclastogenesis. On the other hand, addition of anti-resorptive agents after RANKL stimulation, restricted the anti-osteoclastogenic results [28]. This indicates the effectiveness of β-carotene in suppressing the osteoclast-stimulation by RANKL. β-carotene not only act as a dietary source but also an effective drug to prevent of osteoporosis. NF-κB pathway has the major role in the osteoclastogenesis, it is significantly inhibited by β-carotene.

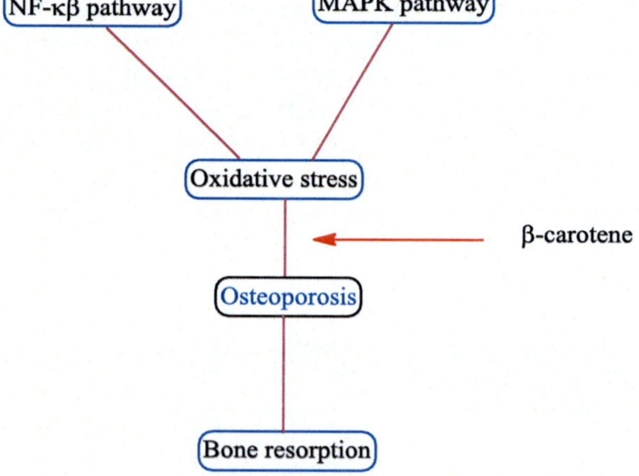

Fig. 21.3 NF-κB and MAPK pathways produce oxidative stress that can leads to osteoporosis but β-carotene inhibits the bone resorption by inhibiting these pathways

NFATc1, c-Fos and CTSK signaling cascade are also crucial elements that regulate osteoclasts formation [55, 56]. The studies reveal that RANKL-activated overexpression of c-Fos and NAFTc has been inhibited by the β-carotene administration, which is also described by the anti-osteoclastogenic efficacy of β-carotene. Further, CTSK is a significant target for the drug in the treatment of osteoporosis [57]. The RANKL-activated NAFTc1 acts with the promoter of CTSK and responsible for CTSK overexpression [58]. CTSK readily degrades type I collagen, the chief element of the bone matrix [59]. β-carotene acts as a CTSK-inhibitor and as an alternative to the CTSK-inhibitor odanacatib, which is presently under clinical observations [60] (Fig. 21.4).

A study in postmenopausal Korean women suggested that β-carotene, vitamin C, zinc and sodium intake are positively linked with BMD. Similarly the incidence of vegetable ingestion is positively linked with femoral neck and total hip T-scores [61].

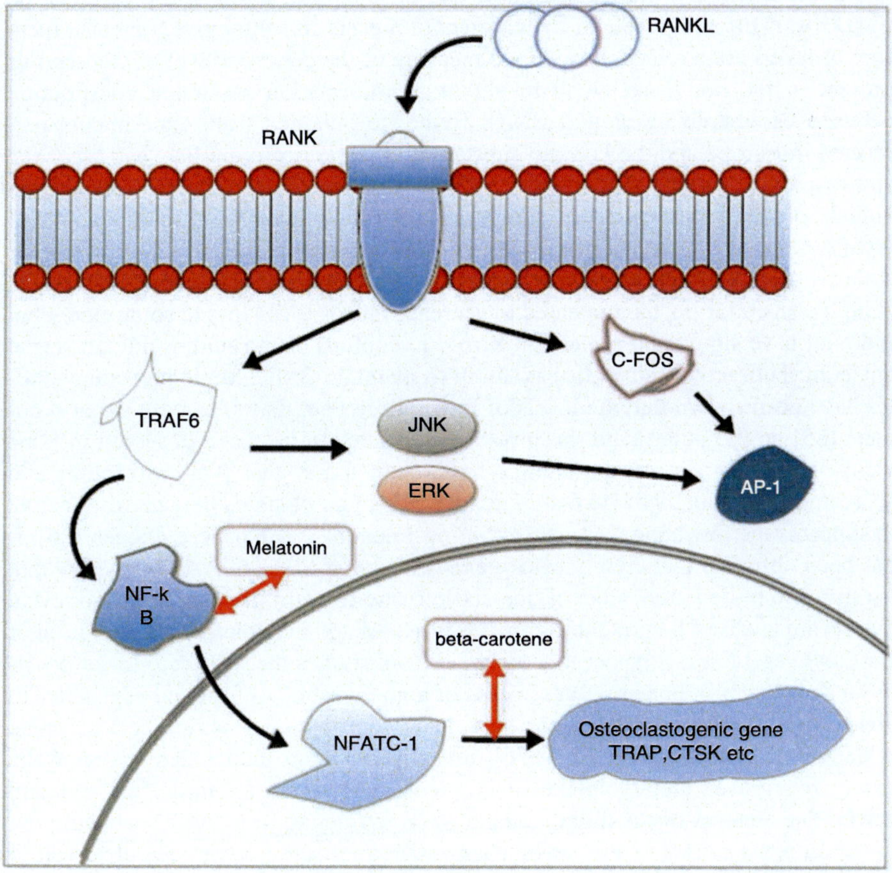

Fig. 21.4 β-carotene suppresses osteoclastogenesis and differentiation of osteoclasts by inhibiting (NFATc1) and osteoclastogenic gene

β-cryptoxanthin and β-carotene present in serum in very low concentrations connected with the outspread density of the total mass of bone in females having undergone menopause [51]. Both of these carotenoids are associated with the avoidance of bone misfortune that may reduce the total mass of bone in Japanese females having undergone menopause [36]. The process that involves the formation of osteoclasts, the bone resorbing cells is called osteoclastogenesis. This process consists of many steps and depends upon various genetic, humoral, and mechanical factors [1, 62, 63]. These factors are terminated in the activation of factors that assist the differentiation of bone-resorbing cells including c-Fos and nuclear factor of activated T cell c1 (NFATc1). A few studies [53, 64] targeting β-cryptoxanthin have been reported in dietary habits of Italian and/or American populations. The Framingham research group found that carotenoid intake over 4 years is linked with changes in BMD and a high consumption of lycopene improves bone health. Most of the studies indicated a longitudinal analysis of the link between serum carotenoid levels and BMD and the stimulatory effect of, β-cryptoxanthin and β-carotene on BMD loss. Different levels of the carotenoid present in serum give a precise measure of levels are a relatively accurate measure of the concrete level of carotenoids present in the body and aided to get more information about the contribution between carotenoid levels and BMD. Evidence obtained from experiments performed in Europe and the U.S, no substantial linkage is present between lycopene and bone health in the individuals under observation. The studies about the individuals of Mikkabi have been shown that they have taken very least amount of lycopene and serum levels of lycopene are also less than the individuals in other countries. In the same way, it has been found that presence of α-carotene in serum at high concentration, has appeared to decrease rate of BMD loss to some extent but it do not have significant impact on BMD. In contrast, some studies did not reveal any contribution of lutein and zeaxanthin to maintain BMD. Some preceding studies have been shown that the levels of lutein more than that of α-carotene or lycopene [65] on the other hand, no involvement of lutein has been observed in bone wellbeing. The most of the studies have been found that β-cryptoxanthin and β-carotene are actually the two significant kinds of carotenoids that are responsible to suppress the low bone mass density in postmenopausal Japanese women [36]. It has been observed that dietary α-carotene conversion to vitamin A is very low that is estimated to lie in the range of one-sixth to one-twelfth). It is important to reveal that serum levels of β-carotene significantly associated with dietary consumption of fruit and vegetables. It has been concluded from studies that there are lower possibilities of danger of bone fracture in case of high ingestion of fruit and vegetables in diet [66]. It has not been found the significant contribution of serum beta-carotene in danger of bone breakage [67] and some observational studies demonstrated the strong relation of dietary intake of carotenes with bone fractures [68, 69] and BMD. The chances of the higher danger of bone fracture in individuals having the lower amounts retinol in diet or biochemical retinol status have been observed in

Table 21.2 Effect of carotenoids in bone diseases conclusions

Sr no.	Carotenoid dosage	Dose	Use	Reference
1	β-cryptoxanthin (23.7 mg/day)		β-cryptoxanthin actually decreases the risk of inflammatory polyarthritis, such as rheumatoid arthritis due to its antioxidant capacity that decreases the danger of chronic inflammation	[77, 78]
2	Lycopene (30–35 mg/day)		Lycopene being a powerful antioxidant, increase the bone health as well as inhibit osteoporosis.	[77]

many studies [70, 71] may associate with a significant part of retinol-binding protein as negative acute phase protein [72]. It has been found in different surveys that individuals belonging to Asian habitually chomp through decrease amounts of vitamin A and most of that is significantly obtained through β-carotene source by conversion [73]. It has been found that the concentrations of circulating retinol have the inverse relation with the bioconversion of β-carotene (to retinol) [74]. Some observational studies have been demonstrated the beneficial impact of high dietary intake of vitamin A [36, 43, 51, 64]. Some observation shows no effect of high levels of vitamin A precursor β-carotene on bone homeostasis [70, 75]. On the other hand, there is no evidence of inauspicious impact of an increased intake of β-carotene on bone health, signifying that retinol produced from β-carotene, or β-carotene itself, appeared to exhibit useful effects on bone homeostasis. Moreover, it has been observed that an elevated proportion of β-carotene to retinol in blood shows significantly useful impact on bones homeostasis so it could be assumed to benefit bone health due to impending damage of extra retinol and the reimbursement of β-carotene [76]. It has been concluded from some observational studies that relatively less concentrations of β-carotene in serum (e.g. 0.1–0.2 μmol/L in serum and 2.76 mg/day in diet) has no impact on bone health [70, 71]. It has been concluded that the second main rather than the top quartile of dietary β-carotene be inclined to show the top hip bone mass density BMD, portentous most complimentary consumptions of β-carotene 5.5 mg/day in women. It has been observed that high β-carotene levels aided to maintain bone homeostasis [76]. Table 21.2 summarizes some of above mentioned details.

There is a range of studies that reveals the association between dietary habits and bone health. Osteoporosis is a silent-killer ailment characterized by a systemic weakening of bone strength and architecture leading to higher possibility of fracture. Food and diet are two major adaptable features that influence the bone health. Carotenoids prevents bone loss by osteoblastic bone synthesis and preventing osteoclastic bone resorption.FundingThis research was funded by the Deanship of Scientific Research at Princess Nourah bint Abdulrahman University through the Fast-track Research Funding Program.

References

1. Yasuda H, Shima N, Nakagawa N, Yamaguchi K, Kinosaki M, Mochizuki S-i et al (1998) Osteoclast differentiation factor is a ligand for osteoprotegerin/osteoclastogenesis-inhibitory factor and is identical to TRANCE/RANKL. Proc Natl Acad Sci 95(7):3597–3602
2. Wheater PR (2006) Wheater's functional histology: a text and colour atlas, 5th edn. Churchill Livingstone, New York
3. Guntur A, Rosen C (2012) Bone as an endocrine organ. Endocr Pract 18(5):758–762
4. Orimo H, Hayashi Y, Fukunaga M, Sone T, Fujiwara S, Shiraki M et al (2001) Diagnostic criteria for primary osteoporosis: year 2000 revision. J Bone Miner Metab 19(6):331
5. Weitzmann MN, Pacifici R (2006) Estrogen deficiency and bone loss: an inflammatory tale. J Clin Invest 116(5):1186–1194
6. Endo Y, Aharonoff GB, Zuckerman JD, Egol KA, Koval KJ (2005) Gender differences in patients with hip fracture: a greater risk of morbidity and mortality in men. J Orthop Trauma 19(1):29–35
7. Kim L, Rao AV, Rao LG (2003) Lycopene II—effect on osteoblasts: the carotenoid lycopene stimulates cell proliferation and alkaline phosphatase activity of SaOS-2 cells. J Med Food 6(2):79–86
8. Looker A, Isfahani NS, Fan B, Shepherd J (2017) Trends in osteoporosis and low bone mass in older US adults, 2005–2006 through 2013–2014. Osteoporos Int 28(6):1979–1988
9. Johnell O, Kanis J (2006) An estimate of the worldwide prevalence and disability associated with osteoporotic fractures. Osteoporos Int 17(12):1726–1733
10. Wang F, Wang N, Gao Y, Zhou Z, Liu W, Pan C et al (2017) β-Carotene suppresses osteoclastogenesis and bone resorption by suppressing NF-κB signaling pathway. Life Sci 174:15–20
11. Matsumoto Y, Tousen Y, Ishimi Y (2018) β-Carotene prevents bone loss in hind limb unloading mice. J Clin Biochem Nutr 63(1):42–49
12. Rachner TD, Khosla S, Hofbauer LC (2011) Osteoporosis: now and the future. Lancet 377(9773):1276–1287
13. Wang J, Wang Y, Liu W-D, Wang F, Yin Z-S (2014) Hip fractures in Hefei, China: the Hefei osteoporosis project. J Bone Miner Metab 32(2):206–214
14. Arceo-Mendoza RM, Camacho P (2015) Prediction of fracture risk in patients with osteoporosis: a brief review. Women Health 11(4):477–484
15. Group Roa WS (1994) Assessment of fracture risk and its application to screening for postmenopausal osteoporosis. World Health Organ Tech Rep Ser 843:1–129
16. Cooper C, Campion G, Melton L (1992) Hip fractures in the elderly: a world-wide projection. Osteoporos Int 2(6):285–289
17. Endo I, Fukumoto S, Ozono K, Namba N, Tanaka H, Inoue D et al (2008) Clinical usefulness of measurement of fibroblast growth factor 23 (FGF23) in hypophosphatemic patients: proposal of diagnostic criteria using FGF23 measurement. Bone 42(6):1235–1239
18. Xu J, Song C, Song X, Zhang X, Li X (2017) Carotenoids and risk of fracture: a meta-analysis of observational studies. Oncotarget 8(2):2391
19. Dai Z, Wang R, Ang LW, Low YL, Yuan JM, Koh WP (2014) Protective effects of dietary carotenoids on risk of hip fracture in men: the Singapore Chinese Health Study. J Bone Miner Res 29(2):408–417
20. Pandey MK, Gupta SC, Karelia D, Gilhooley PJ, Shakibaei M, Aggarwal BB (2018) Dietary nutraceuticals as backbone for bone health. Biotechnol Adv 36(6):1633–1648
21. Das SK, Ren R, Hashimoto T, Kanazawa K (2010) Fucoxanthin induces apoptosis in osteoclast-like cells differentiated from RAW264. 7 cells. J Agric Food Chem 58(10):6090–6095
22. Pattison DJ, Symmons DP, Lunt M, Welch A, Bingham SA, Day NE et al (2005) Dietary β-cryptoxanthin and inflammatory polyarthritis: results from a population-based prospective study. Am J Clin Nutr 82(2):451–455

23. Stahl W, Nicolai S, Briviba K, Hanusch M, Broszeit G, Peters M et al (1997) Biological activities of natural and synthetic carotenoids: induction of gap junctional communication and singlet oxygen quenching. Carcinogenesis 18(1):89–92
24. Hirata N, Ichimaru R, Tominari T, Matsumoto C, Watanabe K, Taniguchi K et al (2019) Beta-cryptoxanthin inhibits lipopolysaccharide-induced osteoclast differentiation and bone resorption via the suppression of inhibitor of NF-κB kinase activity. Nutrients 11(2):368
25. Yamaguchi M, Uchiyama S (2004) β-Criptoxanthin stimulates bone formation and inhibits bone resorption in tissue culture in vitro. Mol Cell Biochem 258(1-2):137–144
26. Yamaguchi M, Uchiyama S (2003) Effect of carotenoid on calcium content and alkaline phosphatase activity in rat femoral tissues in vitro: the unique anabolic effect of β-cryptoxanthin. Biol Pharm Bull 26(8):1188–1191
27. Uchiyama S, Yamaguchi M (2005) β-Cryptoxanthin stimulates cell proliferation and transcriptional activity in osteoblastic MC3T3-E1 cells. Int J Mol Med 15(4):675–681
28. Park G, Horie T, Fukasawa K, Ozaki K, Onishi Y, Kanayama T et al (2017) Amelioration of the development of osteoarthritis by daily intake of β-cryptoxanthin. Biol Pharm Bull 40(7):1116–1120
29. Ozaki K, Okamoto M, Fukasawa K, Iezaki T, Onishi Y, Yoneda Y et al (2015) Daily intake of β-cryptoxanthin prevents bone loss by preferential disturbance of osteoclastic activation in ovariectomized mice. J Pharmacol Sci 129(1):72–77
30. Yamaguchi M, Weitzmann MN (2009) The bone anabolic carotenoid β-cryptoxanthin enhances transforming growth factor-β1-induced SMAD activation in MC3T3 preosteoblasts. Int J Mol Med 24(5):671–675
31. Uchiyama S, Yamaguchi M (2004) Inhibitory effect of β-cryptoxanthin on osteoclast-like cell formation in mouse marrow cultures. Biochem Pharmacol 67(7):1297–1305
32. Uchiyama S, Sumida T, Yamaguchi M (2004) Oral administration of β-cryptoxanthin induces anabolic effects on bone components in the femoral tissues of rats in vivo. Biol Pharm Bull 27(2):232–235
33. Komori T, Yagi H, Nomura S, Yamaguchi A, Sasaki K, Deguchi K et al (1997) Targeted disruption of Cbfa1 results in a complete lack of bone formation owing to maturational arrest of osteoblasts. Cell 89(5):755–764
34. Rao LG, Krishnadev N, Banasikowska K, Rao AV (2003) Lycopene I—effect on osteoclasts: lycopene inhibits basal and parathyroid hormone-stimulated osteoclast formation and mineral resorption mediated by reactive oxygen species in rat bone marrow cultures. J Med Food 6(2):69–78
35. Mackinnon E, Venket Rao A, Rao L (2011) Dietary restriction of lycopene for a period of one month resulted in significantly increased biomarkers of oxidative stress and bone resorption in postmenopausal women. J Nutr Health Aging 15(2):133–138
36. Sugiura M, Nakamura M, Ogawa K, Ikoma Y, Yano M (2012) High serum carotenoids associated with lower risk for bone loss and osteoporosis in post-menopausal Japanese female subjects: prospective cohort study. PLoS One 7(12)
37. Russo C, Ferro Y, Maurotti S, Salvati MA, Mazza E, Pujia R et al (2020) Lycopene and bone: an in vitro investigation and a pilot prospective clinical study. J Transl Med 18(1):43
38. Sommerburg O, Keunen JE, Bird AC, Van Kuijk FJ (1998) Fruits and vegetables that are sources for lutein and zeaxanthin: the macular pigment in human eyes. Br J Ophthalmol 82(8):907–910
39. Tominari T, Matsumoto C, Watanabe K, Hirata M, Grundler FM, Inada M et al (2017) Lutein, a carotenoid, suppresses osteoclastic bone resorption and stimulates bone formation in cultures. Biosci Biotechnol Biochem 81(2):302–306
40. Li H, Huang C, Zhu J, Gao K, Fang J, Li H (2018) Lutein suppresses oxidative stress and inflammation by Nrf2 activation in an osteoporosis rat model. Med Sci Monit 24:5071
41. Takeda H, Tominari T, Hirata M, Watanabe K, Matsumoto C, Grundler FM et al (2017) Lutein enhances bone mass by stimulating bone formation and suppressing bone resorption in growing mice. Biol Pharm Bull 40(5):716–721

42. Bovier ER, Hammond BR (2017) The macular carotenoids lutein and zeaxanthin are related to increased bone density in young healthy adults. Foods 6(9):78
43. Sahni S, Hannan MT, Blumberg J, Cupples LA, Kiel DP, Tucker KL (2009) Inverse association of carotenoid intakes with 4-y change in bone mineral density in elderly men and women: the Framingham Osteoporosis Study. Am J Clin Nutr 89(1):416–424
44. Altindag O, Erel O, Soran N, Celik H, Selek S (2008) Total oxidative/anti-oxidative status and relation to bone mineral density in osteoporosis. Rheumatol Int 28(4):317–321
45. Capulli M, Paone R, Rucci N (2014) Osteoblast and osteocyte: games without frontiers. Arch Biochem Biophys 561:3–12
46. Sumantran VN, Zhang R, Lee DS, Wicha MS (2000) Differential regulation of apoptosis in normal versus transformed mammary epithelium by lutein and retinoic acid. Cancer Epidemiol Prev Biomarkers 9(3):257–263
47. Dabouian A, Bakhshi H, Irani S, Pezeshki-Modaress M (2018) β-Carotene: a natural osteogen to fabricate osteoinductive electrospun scaffolds. RSC Adv 8(18):9941–9945
48. Astley SB, Hughes DA, Wright AJ, Elliott RM, Southon S (2004) DNA damage and susceptibility to oxidative damage in lymphocytes: effects of carotenoids in vitro and in vivo. Br J Nutr 91(1):53–61
49. Almeida M, Han L, Martin-Millan M, O'Brien CA, Manolagas SC (2007) Oxidative stress antagonizes Wnt signaling in osteoblast precursors by diverting β-catenin from T cell factor-to forkhead box O-mediated transcription. J Biol Chem 282(37):27298–27305
50. Cao W-t, Zeng F-f, Li B-l, Lin J-s, Liang Y-y, Chen Y-m (2018) Higher dietary carotenoid intake associated with lower risk of hip fracture in middle-aged and elderly Chinese: a matched case-control study. Bone 111:116–122
51. Sugiura M, Nakamura M, Ogawa K, Ikoma Y, Ando F, Yano M (2008) Bone mineral density in post-menopausal female subjects is associated with serum antioxidant carotenoids. Osteoporos Int 19(2):211–219
52. Hayhoe R, Lentjes M, Luben R, Khaw K, Welch A (2015) Dietary carotenoid intake is positively associated with bone density in individuals of the EPIC-Norfolk cohort. Proc Nutr Soc 74(OCE5)
53. Yang, Zhang, Penniston, Binkley, Tanumihardjo (2008) Serum carotenoid concentrations in postmenopausal women from the United States with and without osteoporosis. Int J Vitam Nutr Res 78(3):105–111
54. Basu S, Michaëlsson K, Olofsson H, Johansson S, Melhus H (2001) Association between oxidative stress and bone mineral density. Biochem Biophys Res Commun 288(1):275–279
55. Yeo H, McDonald JM, Zayzafoon M (2006) NFATc1: a novel anabolic therapeutic target for osteoporosis. Ann N Y Acad Sci 1068(1):564–567
56. Kwak HB, Lee BK, Oh J, Yeon J-T, Choi S-W, Cho HJ et al (2010) Inhibition of osteoclast differentiation and bone resorption by rotenone, through down-regulation of RANKL-induced c-Fos and NFATc1 expression. Bone 46(3):724–731
57. Costa AG, Cusano NE, Silva BC, Cremers S, Bilezikian JP (2011) Cathepsin K: its skeletal actions and role as a therapeutic target in osteoporosis. Nat Rev Rheumatol 7(8):447
58. Rood JA, Van Horn S, Drake FH, Gowen M, Debouck C (1997) Genomic organization and chromosome localization of the human cathepsin K gene (CTSK). Genomics 41(2):169–176
59. Han J, Luo T, Gu Y, Li G, Jia W, Luo M (2009) Cathepsin K regulates adipocyte differentiation: possible involvement of type I collagen degradation. Endocr J 56(1):55–63
60. Bone HG, McClung MR, Roux C, Recker RR, Eisman JA, Verbruggen N et al (2010) Odanacatib, a cathepsin-K inhibitor for osteoporosis: a two-year study in postmenopausal women with low bone density. J Bone Miner Res 25(5):937–947
61. Kim DE, Cho SH, Park HM, Chang YK (2016) Relationship between bone mineral density and dietary intake of β-carotene, vitamin C, zinc and vegetables in postmenopausal Korean women: a cross-sectional study. J Int Med Res 44(5):1103–1114
62. Yavropoulou M, Yovos J (2008) Osteoclastogenesis--current knowledge and future perspectives. J Musculoskelet Neuronal Interact 8(3):204–216

63. Fujikawa Y, Quinn J, Sabokbar A, McGee J, Athanasou N (1996) The human osteoclast precursor circulates in the monocyte fraction. Endocrinology 137(9):4058–4060
64. Maggio D, Polidori MC, Barabani M, Tufi A, Ruggiero C, Cecchetti R et al (2006) Low levels of carotenoids and retinol in involutional osteoporosis. Bone 38(2):244–248
65. Sugiura M, Nakamura M, Ogawa K, Ikoma Y, Matsumoto H, Ando F et al (2009) Synergistic interaction of cigarette smoking and alcohol drinking with serum carotenoid concentrations: findings from a middle-aged Japanese population. Br J Nutr 102(8):1211–1219
66. New SA (2003) Intake of fruit and vegetables: implications for bone health. Proc Nutr Soc 62(4):889–899
67. Michaëlsson K, Lithell H, Vessby B, Melhus H (2003) Serum retinol levels and the risk of fracture. N Engl J Med 348(4):287–294
68. Feskanich D, Singh V, Willett WC, Colditz GA (2002) Vitamin A intake and hip fractures among postmenopausal women. JAMA 287(1):47–54
69. Promislow JH, Goodman-Gruen D, Slymen DJ, Barrett-Connor E (2002) Retinol intake and bone mineral density in the elderly: the Rancho Bernardo Study. J Bone Miner Res 17(8):1349–1358
70. Barker ME, McCloskey E, Saha S, Gossiel F, Charlesworth D, Powers HJ et al (2005) Serum retinoids and β-carotene as predictors of hip and other fractures in elderly women. J Bone Miner Res 20(6):913–920
71. Opotowsky AR, Bilezikian JP (2004) Serum vitamin A concentration and the risk of hip fracture among women 50 to 74 years old in the United States: a prospective analysis of the NHANES I follow-up study. Am J Med 117(3):169–174
72. Stephensen CB, Gildengorin G (2000) Serum retinol, the acute phase response, and the apparent misclassification of vitamin A status in the third National Health and Nutrition Examination Survey. Am J Clin Nutr 72(5):1170–1178
73. Zhang X, Li Y (2002) Relationships of estrogen receptor gene polymorphism with bone metabolism in postmenopausal women. J Tongi Univ (Med Sci)(Chin) 23:1–4
74. Ribaya-Mercado JD, Solon FS, Solon MA, Cabal-Barza MA, Perfecto CS, Tang G et al (2000) Bioconversion of plant carotenoids to vitamin A in Filipino school-aged children varies inversely with vitamin A status. Am J Clin Nutr 72(2):455–465
75. Wolf RL, Cauley JA, Pettinger M, Jackson R, Lacroix A, Leboff MS et al (2005) Lack of a relation between vitamin and mineral antioxidants and bone mineral density: results from the Women's Health Initiative. Am J Clin Nutr 82(3):581–588
76. G-d C, Zhu Y-Y, Cao Y, Liu J, W-q S, Z-m L et al (2015) Association of dietary consumption and serum levels of vitamin A and β-carotene with bone mineral density in Chinese adults. Bone 79:110–115
77. Böhm V, Puspitasari-Nienaber NL, Ferruzzi MG, Schwartz SJ (2002) Trolox equivalent antioxidant capacity of different geometrical isomers of α-carotene, β-carotene, lycopene, and zeaxanthin. J Agric Food Chem 50(1):221–226
78. Sung LC, Chao HH, Chen CH, Tsai JC, Liu JC, Hong HJ et al (2015) Lycopene inhibits cyclic strain-induced endothelin-1 expression through the suppression of reactive oxygen species generation and induction of heme oxygenase-1 in human umbilical vein endothelial cells. Clin Exp Pharmacol Physiol 42(6):632–639

Chapter 22
Carotenoids and Periodontal Diseases

Muhammad Zia-Ul-Haq, Muhammad Riaz, and Hanadi Talal Ahmedah

22.1 Introduction

Carotenoids are phytonutrients found in plants, algae and photosynthetic bacteria. Typically, carotenoids are plant pigments present in many fruits and vegetables and produce red, oranges and yellow colors in plants. They are best absorbed with fat and cooking increases the strength of nutrients. They are unique than other bioactive compounds due to ability of conversion of some of them into Vitamin A. They are important in food, feed, pharmaceutical and cosmetics industry. They usually exert their effects by antioxidant/pro-oxidant mechanism, modulation of signaling pathways (e.g. interaction with nuclear factor- κB (NF-κB) & nuclear factor- erythroid 2–related factor 2 (Nrf2) .They play important role in maintaining proper health as well involved in disease prevention and cure due to their strong antioxidant properties. They have become active area for pharmaceutical investigation by using modern computational technology, high throughput screening systems and bioinformatics applications. Carotenoids, prevent the inflammatory ailments. The reactive oxygen species (ROS) that produce oxidative stress and inflammation are effectively counterbalanced by carotenoids hence inhibiting tissue damages of periodontitis.

Periodontitis is a bacterial infectious ailment which affects the supporting tissues of teeth. When homeostatic mechanism between host and oral micro flora is changed, the periodontium is inflamed leading to tissue damage. This condition is

M. Zia-Ul-Haq (✉)
Office of Research, Innovation and Commercialization, Lahore College for Women University, Lahore, Pakistan

M. Riaz
Department of Pharmacy, Shaheed Benazir Bhutto University Sheringal, Sheringal, Pakistan

H. T. Ahmedah (✉)
Radiological Sciences Department, College of Health and Rehabilitation Sciences, Princess Nourah bint Abdulrahman University, Riyadh, Saudi Arabia
e-mail: htahmedah@pnu.edu.sa

© Springer Nature Switzerland AG 2021

M. Zia-Ul-Haq et al. (eds.), *Carotenoids: Structure and Function in the Human Body*, https://doi.org/10.1007/978-3-030-46459-2_22

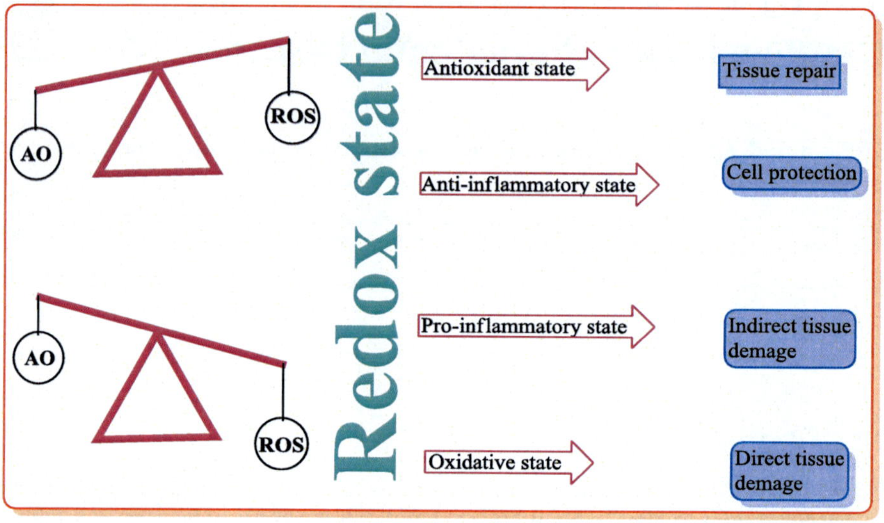

Fig. 22.1 Reactive oxygen species (ROS) and antioxidants (carotenoids; AO) redox state in biological system

called periodontitis. Interaction between host immune response and periodonto-pathogenic bacteria especially *Porphyromonas gingivalis* determine the pathway of disease. Polymorphonuclear (PMN) count and activity are enhanced during periodontitis releasing large number of ROS which increase oxidative stress in periodontal tissues. Reactive oxygen species (ROS) produced due to oxidative stress facilitate tissue destruction [1, 2]. Necessary number of antioxidants like carotenoids are required by periodontal tissues to inhibit this damage. Hence now scientists are focusing on the use of antioxidants in combination with scaling root planning (SRP) on periodontal tissue damage. The consumption of antioxidants like carotenoids is being recommended due to these benefits [3]. Figures 22.1 and 22.2 shows the tentative oxidative stress-mediated inflammatory pathways involved in periodontal tissue damage [4].

22.2 Role of Carotenoids

Substantial quantity of pro-inflammatory cytokines like IL-1, IL-6, prostaglandin E (PGE2), tumor TNF-α, ROS and oxidative stress markers are produced due to periodontitis. A positive association exists between periodontitis and oxidative stress. Carotenoids subside oxidative stress by various mechanisms hence managing periodontal diseases. Decreased profiles of α- and β-carotene, β -cryptoxanthin, and zeaxanthin was noted in the moderate and the generalized severe periodontitis patients indicating the connection between the serum amounts of carotenoids and

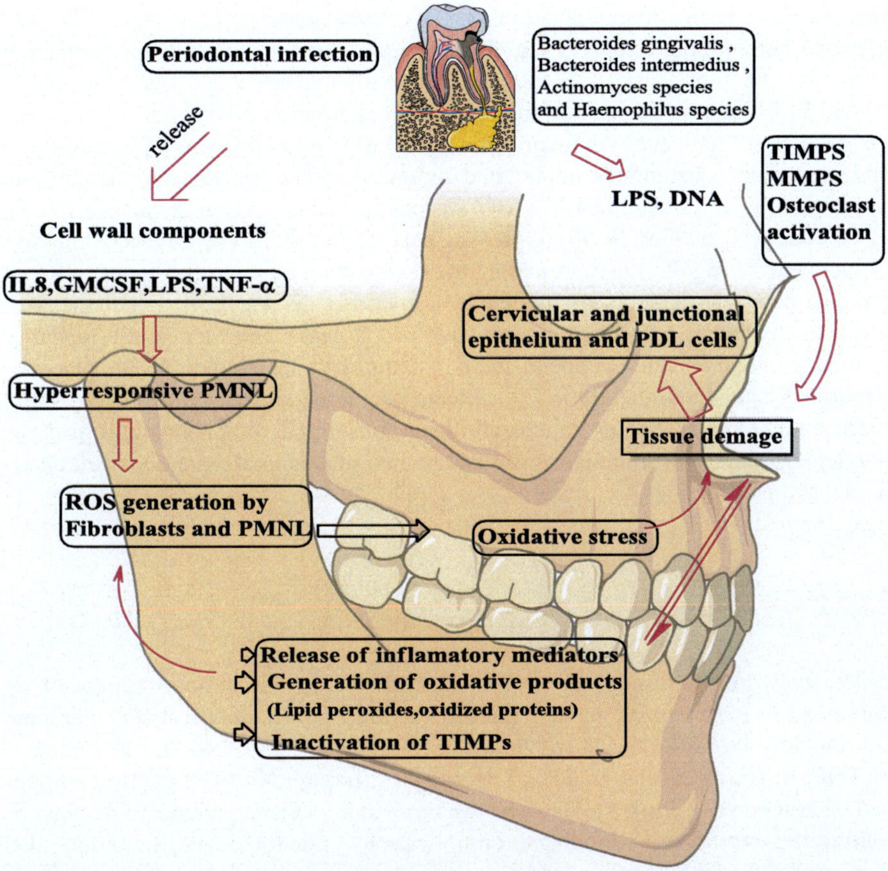

Fig. 22.2 Possible oxidative stress-mediated inflammatory pathways involved in periodontal tissue breakdown. *LPS* lipopolysaccharide, *GM-CSF* granulocyte-macrophage colony-stimulating factor, *IL8* interleukin-8, *TNF-α* tumor necrosis factor-alpha, *PDL* periodontal ligament, *NF-κβ* nuclear factor-kappa B, *ROS* reactive oxygen species, *PMNL* polymorphonuclear leukocyte, *TIMP* tissue inhibitor of metalloproteinases, and *MMP* matrix metalloproteinase

periodontal health [5]. Below is list of individual carotenoids which can be used to treat periodontitis.

22.2.1 Lycopene

Reduced salivary IL-1β and uric acid (UA) profiles were observed in chronic periodontitis patients during investigation of adjunctive usage of systemic lycopene along with nonsurgical treatment [6]. Lycopene administration (8 mg) in chronic periodontitis patients with 40 subjects of type 2 diabetes mellitus undergoing SRP reduced mean serum malondialdehyde (MDA) levels,mean probing depth (PD), and glycated hemoglobin (HbA1c) HbA1c levels at 2 months post therapy indicating protective function of lycopene in decreasing oxidative stress and altered glycemic levels [7].Observational studies suggest that consumption of lycopene (@5–7 mg per day) in normal healthy human is enough to fight oxidative stress [8]. Lycopene and green tea consumption modulated oxidative stress decreased inflammation, improved clinical attachment loss (CAL), plaque index (PI), gingival index (GI), and bleeding on probing (BOP) improving periodontal health [9]. Oral consumption of lycopene (4 mg/day) is a promising therapy as an adjunct to full mouth SRP of the oral cavity in patients of moderate periodontal disease [10]. Lycopene with green tea extract is a promising adjunctive prophylactic and therapeutic modality against gingivitis [11]. Lycopene intake (8 mg/day) in a randomized, placebo-controlled, split-mouth study in 20 subjects decreased significantly gingivitis and bleeding index [12]. Lycopene is useful as a nonsurgical aid against oral diseases like leukoplakia, oral submucous fibrosis, lichen planus, oral squamous cell carcinoma, abd periodontal diseases [13].

22.2.2 *β -carotene*

Chitosan hydrogels containing β-carotene reduced harmful effects produced by common restorative agents in vitro and did not affect bond strength [14]. β -carotene substantially inhibited the *Porphyromonas gingivalis* (Pg)-LPS-induced synthesis of TNF, IL-6 and MCP-1 in THP-1 monocytes through NF-kB signaling without cell destruction of cells [15]. Hence it can be used in diabetic patients to suspend or inhibit the periodontitis-mediated complications. β-carotene consumption also decreases periodontal pocket depth in nonsmokers post periodontal treatment [16].

22.2.3 *β-cryptoxanthin*

Decreased quantity of β-cryptoxanthin is observed in moderate/severe periodontitis patients [17]. β-cryptoxanthin reinstated the alveolar bone loss in an experimental *in vivo* periodontitis models [18]. β-cryptoxanthin prevented the levels of LPS-induced alveolar bone resorption considerably in experimental in vivo periodontitis models [19]. β-cryptoxanthin decreased substantially Pg-induced generation of IL-6 and IL-8 in human periodontal ligament cells [20].

22.2.4 *Fucoxanthin*

Fucoxanthin slightly decreased the TNF, IL-1_ and IL-6 levels using experimental in vivo periodontitis models [21].

22.2.5 *Crocin*

Crocin, an apocarotenoids, supplementation decreased oxidative damage to periodontal tissue due to its strong antioxidant and anti-inflammatory potential [22].

22.3 Conclusions

It is well-known that oxidative stress produced due to abnormal production of ROS is a key element in pathogenesis of different inflammatory diseases including periodontitis. This oxidative stress if not managed properly can harm the cells and tissues. Sufficient amount of antioxidants is needed to cope this oxidative stress to prevent this harm. Carotenoids especially lycopene being strong antioxidants can be used to treat periodontosis locally or systemically. There is need to identify more carotenoids molecules which can be used for this purpose besides lycopene. Further as since combinations of carotenoids with other antioxidants are more effective, hence more fruitful combinations should be sought besides those reported already.

Funding This research was funded by the Deanship of Scientific Research at Princess Nourah bint Abdulrahman University through the Fast-track Research Funding Program.

References

1. Chapple I (1997) Reactive oxygen species and antioxidants in inflammatory diseases. J Clin Periodontol 24(5):287–296
2. Tsai C, Chen H, Chen S, Ho Y, Ho K, Wu Y et al (2005) Lipid peroxidation: a possible role in the induction and progression of chronic periodontitis. J Periodontal Res 40(5):378–384
3. Chapple I, Brock G, Eftimiadi C, Matthews J (2002) Glutathione in gingival crevicular fluid and its relation to local antioxidant capacity in periodontal health and disease. Mol Pathol 55(6):367
4. Gumus P, Huseyinalemdaroglu B, Buduneli N (2016) The role of oxidative stress in the interaction of periodontal disease with systemic diseases or conditions. Oxid Antioxid Med Sci 5(2):18
5. Linden GJ, McClean KM, Woodside JV, Patterson CC, Evans A, Young IS et al (2009) Antioxidants and periodontitis in 60–70-year-old men. J Clin Periodontol 36(10):843–849
6. Reddy PVN, Ambati M, Koduganti R (2015) Systemic lycopene as an adjunct to scaling and root planing in chronic periodontitis patients with type 2 diabetes mellitus. J Int Soc Prev Comm Dent 5(Suppl 1):S25
7. Arora N, Avula H, Kumar Avula J (2013) The adjunctive use of systemic antioxidant therapy (lycopene) in nonsurgical treatment of chronic periodontitis: a short-term evaluation. Quintessence Int 44:6
8. Rao A, Shen H (2002) Effect of low dose lycopene intake on lycopene bioavailability and oxidative stress. Nutr Res 22(10):1125–1131

9. Castro MML, Duarte NN, Nascimento PC, Magno MB, Fagundes NCF, Flores-Mir C et al (2019) Antioxidants as adjuvants in periodontitis treatment: a systematic review and meta-analysis. Oxidative Med Cell Longev 2019:1–24
10. Belludi SA, Verma S, Banthia R, Bhusari P, Parwani S, Kedia S et al (2013) Effect of lycopene in the treatment of periodontal disease: a clinical study. J Contemp Dent Pract 14(6):1054
11. Tripathi P, Blaggana V, Upadhyay P, Jindal M, Gupta S, Nishat S (2019) Antioxidant therapy (lycopene and green tea extract) in periodontal disease: a promising paradigm. J Indian Soc Periodontol 23(1):25
12. Chandra RV, Prabhuji M, Roopa DA, Ravirajan S, Kishore HC (2007) Efficacy of lycopene in the treatment of gingivitis: a randomised, placebo-controlled clinical trial. Oral Health Prev Dent 5:4
13. Gupta S, Jawanda MK, Arora V, Mehta N, Yadav V (2015) Role of lycopene in preventing oral diseases as a nonsurgical aid of treatment. Int J Prev Med 6:70
14. Perchyonoka VT, Zhang S, Oberholzer T (2013) Protective effect of conventional antioxidant (β-carotene, resveratrol and vitamin E) in chitosan-containing hydrogels against oxidative stress and reversal of DNA double stranded breaks induced by common dental composites: in-vitro model. Open Nanosci J 7(1):1–7
15. Kajiura Y, Nishikawa Y, Lew JH, Kido J, Nagata T, Naruishi K (2018) β-carotene suppresses Porphyromonas gingivalis lipopolysaccharide-mediated cytokine production in THP-1 monocytes cultured with high glucose condition. Cell Biol Int 42(1):105–111
16. Dodington DW, Fritz PC, Sullivan PJ, Ward WE (2015) Higher intakes of fruits and vegetables, β-carotene, vitamin C, α-tocopherol, EPA, and DHA are positively associated with periodontal healing after nonsurgical periodontal therapy in nonsmokers but not in smokers. J Nutr 145(11):2512–2519
17. Ebersole JL, Lambert J, Bush H, Huja PE, Basu A (2018) Serum nutrient levels and aging effects on periodontitis. Nutrients 10(12):1986
18. Hirata N, Ichimaru R, Tominari T, Matsumoto C, Watanabe K, Taniguchi K et al (2019) Beta-cryptoxanthin inhibits lipopolysaccharide-induced osteoclast differentiation and bone resorption via the suppression of inhibitor of NF-κB kinase activity. Nutrients 11(2):368
19. Matsumoto C, Ashida N, Yokoyama S, Tominari T, Hirata M, Ogawa K et al (2013) The protective effects of β-cryptoxanthin on inflammatory bone resorption in a mouse experimental model of periodontitis. Biosci Biotechnol Biochem 77(4):860–862
20. Nishigaki M, Yamamoto T, Ichioka H, Honjo K-i, Yamamoto K, Oseko F et al (2013) β-Cryptoxanthin regulates bone resorption related-cytokine production in human periodontal ligament cells. Arch Oral Biol 58(7):880–886
21. Kose O, Arabaci T, Yemenoglu H, Kara A, Ozkanlar S, Kayis S et al (2016) Influences of fucoxanthin on alveolar bone resorption in induced periodontitis in rat molars. Mar Drugs 14(4):70
22. Kocaman G, Altinoz E, Erdemli M, Gul M, Erdemli Z, Gul S et al (2019) Protective effects of crocin on biochemistry and histopathology of experimental periodontitis in rats. Biotech Histochem 94(5):366–373

Chapter 23
Carotenoids and Skin Diseases

Huma Umbreen and Muhammad Zia-Ul-Haq

23.1 Overview of Skin

Skin is the largest organ system, which consists of many layers and different structures. Histological studies divide skin into two major layers called as epidermis and dermis. The epidermis has an outer layer formed of dead cells called stratum corneum that can be peeled off. The peeling is formed by peeling cells which mainly contain keratinocytes, filled with keratin and plays a vital role in skin function as barrier. The dermis is linked to subcutaneous layers of connective tissues and to adipocytes. The structures in subcutaneous layer include hair follicles, sweat glands, blood vessels and fatty deposits [1] (Fig. 23.1).

Skin is not only perceived as sign of health and attractiveness due to its color but is also major organ of immune system. Skin has a judgmental role regarding health, age and allure, owing to color distribution, homogeneity and firmness [2]. Furthermore, it is the outermost physical barrier of our body and protects against strains and hazards of outer environment such as (microbes and radiations) [3]. It is also important in thermoregulation (maintenance of body temperature), water homeostasis and synthesis of vitamin D and other important compounds [4]. Moreover, skin has nerve cells which are sensitive for temperature, pressure, touch and injury [5]. All layers of skin work to function as protective unit against daily challenges of the environment including physical and chemical agents that may be harmful for integrity of the skin (Fig. 23.1) [6].

H. Umbreen (✉)
Institute of Home and Food Sciences, Government College University Faisalabad, Faisalabad, Pakistan

M. Zia-Ul-Haq
Office of Research, Innovation and Commercialization, Lahore College for Women University, Lahore, Pakistan

© Springer Nature Switzerland AG 2021

M. Zia-Ul-Haq et al. (eds.), *Carotenoids: Structure and Function in the Human Body*, https://doi.org/10.1007/978-3-030-46459-2_23

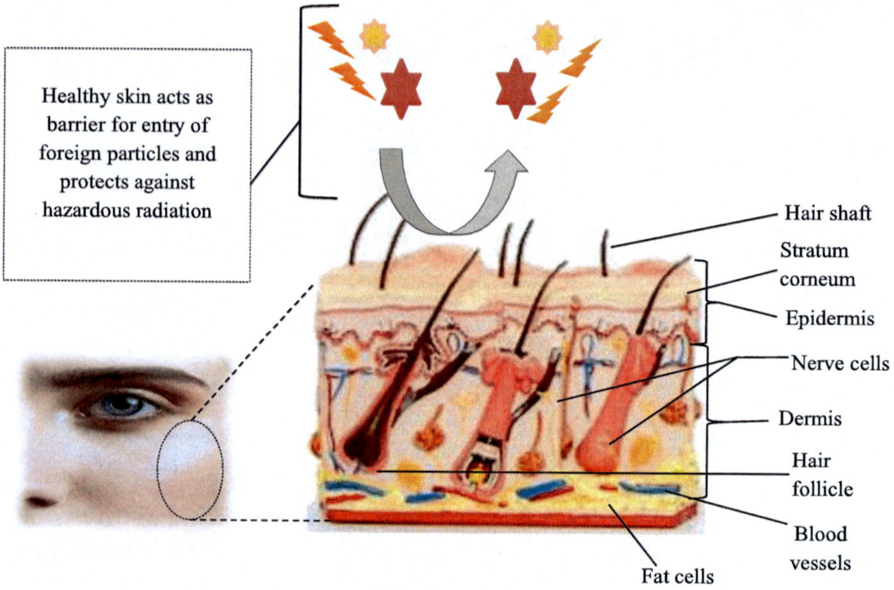

Fig. 23.1 Skin structure and function as barrier

23.2 Carotenoids in Human Skin

Carotenoids have earned their name for being effective antioxidants and these have good role to protect the skin from degenerative skin diseases as sunburn, erythema and photoaging etc. Much is now known about their structure, food sources, absorption and metabolism. Invent of Raman spectroscopy has made it easier to detect their level in skin non-invasively. But still much investigation is needed to know proper role of different carotenoids in prevention of skin diseases as photoaging, photo-immunomodulation and photocarcinogenesis. The famous quote is "Beauty comes from inside" so, the relation between the nutrition and skin health has always been a topic of discussion for scientists (Food Science, Nutrition, medicine etc.) through ancient to present era [7]. Proper structure and function of skin can be maintained through a balanced diet, which not only prevents from skin diseases and early aging but also improves its texture and overall integrity [8, 9]. Therefore, scientists are always in search of interlinks between various diet ingredients and skin and their mechanism of action [10]. Among different dietary ingredients, fruits and vegetables are thought to be as more effective ingredients in increasing health and beauty of skin. This effect is attributed to carotenoids, which are abundantly present in fruits and vegetables and are precursors for important compounds in the body [11]. These carotenoids not only cause change in the color, luminance and health of skin but also act as antioxidants and therefore changes in the skin color may also be indicative of antioxidant status and health of the person [12]. Different layers of skin contain varying amount of carotenoids and can be affected by a number of factors. Furthermore, different techniques are there to determine their levels in skin, which have been discussed below.

23.2.1 Types and Distribution of Carotenoids in Skin

As discussed in previous chapters carotenoids are absorbed via lymphatic system into the systemic circulation and are absorbed into hepatic and extra-hepatic tissues such as skin. Moreover, these can also be derived by skin from hepatic tissues when needed [6]. Concerning the color of skin as described earlier the yellow component of skin is linked with carotenoid levels in skin while red color is associated with iron content [2]. The studies have demonstrated that among the carotenoid; lycopene and β-carotene along with their isomeric forms are abundantly present in skin (about 70%), whereas xynthophyll (zeaxanthin, lutein, α-cryptoxanthin, 2′, 3′-anhydrolutein and β-cryptoxanthin) and its esters with fatty acids (carotenyl mono- and di-fatty acid esters conjugated to linoleate, myristate, stearate, oleate, or palmitate,) present in lower concentrations [4, 13]. The high performance liquid chromatography (HPLC) analysis showed that concentration of lycopene and carotene in epidermis and dermis on weight basis is mostly 0.2–0.3 n mol/g [14].

In nature carotenoids are usually present in all *trans*-form but the case is different in skin, where it usually exits in *cis*-form that may be attributed to uptake as *cis* isomer from systemic circulation or even to absorption from intestine in this form [15]. Although the higher intake of carotenoid may alter its concentration in skin, but overall a gradient is present within skin layers and skin of various parts of body and it is distributed non-homogenously. According to Bayerl et al. [16] within skin layers its concentration is higher in dermis while lower in stratum corneum (lower from inner to outer layer). This effect may be indicative of grander use of carotenoids at epidermis and stratum corneum levels as compared to dermis layer. However, more recently Lademann et al. [17] investigated the concentration of carotenoids within epidermis using Raman spectroscopy and found the highest concentration within 3–4 μm depth, which decreased subsequently until at least 30 μm. It is proposed that subcutaneous tissues act as storage house for carotenoids and encompass skin color that is observed after ingestion of β-carotene, lycopene or canthaxanthin. However, according to Vahlquist [18] compared to subcutaneous, epidermal accumulation is more specific and is of more physiological importance. It is further suggested that carotenoids are laden to epidermal keratinocytes (produced in inner layer and moved to the surface of skin) that transport carotenoids along with them to outer layers [4].

Fig. 23.2 The distribution of carotenoid in different parts and layers of skin (Derived from Bayerl et al. [16])

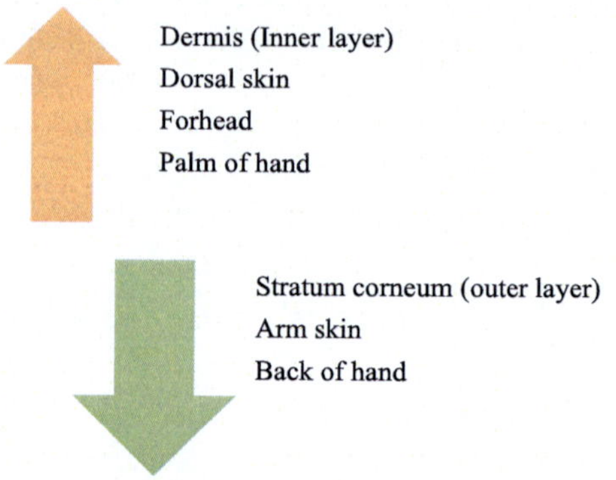

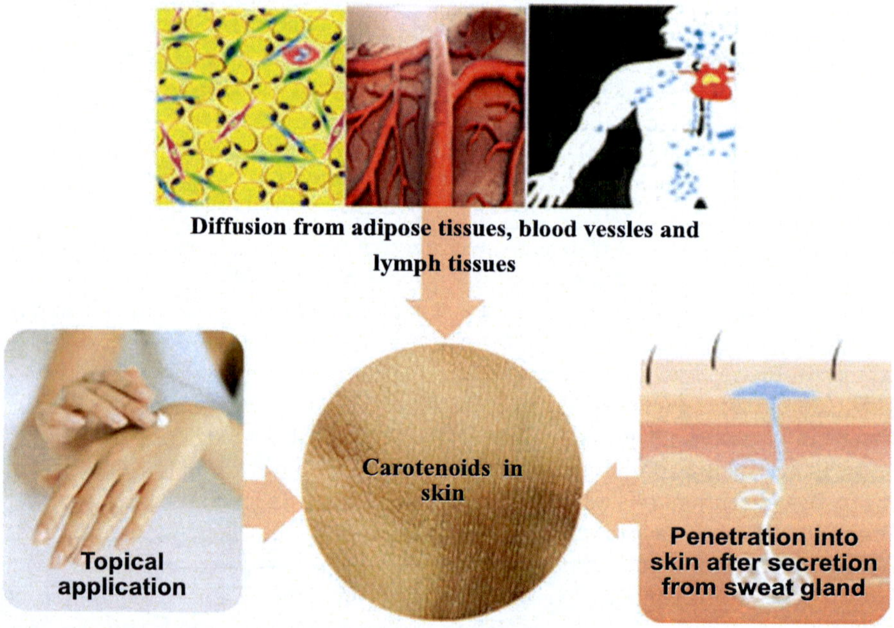

Fig. 23.3 Routes for accumulation of carotenoids in skin

Furthermore, although the exact location of carotenoids still needs further studies and is under discussion, but among various regions of skin, higher concentration is present in forehead, dorsal skin and skin in palm of the hand, whereas lower concentrations are observed in skin regions as arm and back of the hands (Fig. 23.2) [16]. Studies have demonstrated that intake of 30 mg carotenoids (through food or supplements) may result in yellowish skin tone (carotenoderma), which is a reversible condition upon ceasing the supply [6, 19].

23.2.2 *Accumulation of Carotenoids in Skin*

The carotenoids are accumulated in epidermis (where it is more needed) through two pathways from within body. Furthermore it can also be supplied by topical application of carotenoids (Fig. 23.3) as explained below;

(a) Carotenoids from food and food supplements ingested are absorbed through intestine into lymph and are then poured into the blood. From blood these may accumulate into the adipose tissues and diffuse in epidermis. The concentration is the highest on the day following the consumption of carotenoids through food or supplements, which gradually falls down if supply is terminated [20, 21].
(b) The second route from within body comprises of transportation via sweat glands. The sweat along with carotenoids travels through sweat gland onto the skin surface and penetrates into the epidermis as topical application. Therefore, concentration of carotenoids is higher in skin areas where more sweat glands are present [22, 23].
(c) Topical application of carotenoids is of much importance to enhance the skin defense against environmental radiations and hazards. These penetrate into the upper layer of epidermis and enrich the defense against oxidants [17].

23.2.3 *Determination of Skin Carotenoids*

Insight into the level of various carotenoids in skin helps;

- To find the relation between nutritional status and epidemiology of some disease related to skin
- To analyze the effects of stress factors on skin homeostasis by carotenoids
- To study skin pathogenesis and their management
- To understand working of carotenoids
- To find appropriate dose of carotenoids for treatments

Different techniques have been and are being used for determination of carotenoids concentrations in skin including spectrophotometric analysis, HPLC, reflection spectrometry and Raman spectroscopy [24]. For the first time in 1970s and 1980s the skin carotenoids were observed using spectrometric techniques; however no detailed information about composition of carotenoids was found. The studies established a level of 1 n mol/g of wet skin tissue or 10 μg/g of protein in skin [25, 26]. With establishment of HPLC, it became possible to detect the level of carotenoids in different layers of skin with precision and accuracy, but it is an invasive technique and is laborious too. So to find the carotenoids concentration in larger population ethically and practically, non-invasive technique such as reflection spectrometry was developed, which was based on photoacustic profile and remittance analysis. However, these techniques were not found reliable [27–29].

With emergence of latest technologies, highly sensitive and specific technique as Raman spectroscopy paved a way towards accurate determination. Raman spectroscopy is based on vibrational spectroscopy and carotenoids can be excited and thus are detected using laser beam.

This technique may be useful in mass level studies (non-invasively) by applying laser based instrument and getting readings on display. But the problem may be faced in case of skin thickness because this technique can detect carotenoids adequately up to 250 μm, so further investigations are needed by comparing with invasive HPLC technologies [13, 30].

23.2.4 *Factors Affecting Carotenoids of Skin*

Studies show the effect of various factors which may increase or decrease the concentration of carotenoids in skin in one way or the other (Fig. 23.4). These have been discussed as under;

23.2.4.1 Positively Affecting Factors

As described previously, the cutaneous physiology is affected by accumulation of carotenoids in skin. Supplementation with carotenoids may increase their concentration within few days of administration. Diet rich in carotenoids also benefits the levels in skin. Seasons play effective role in carotenoids content in skin. Even without supplementation or special diets, during autumn and summer months the carotenoids content of skin is increased as compared to winter and spring [12]. Darvin et al., [31] showed that average increase in the concentration of carotenoids in

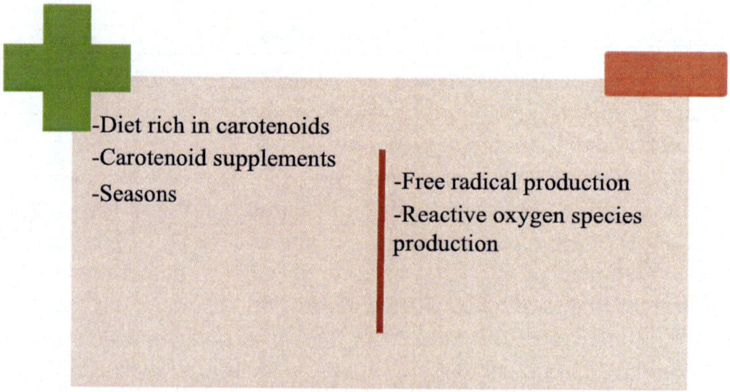

Fig. 23.4 Factors affecting carotenoid concentration in skin

human skin was 1.26 fold with change in season. This may be attributed to content of carotenoids and increased consumption of vegetables and fruits available during the season. Moreover, during summer increased sweating causes the delivery of carotenoids to the skin surface. Topical application of carotenoids increases their content in stratum corneum drastically but their stay is not long term, while combination of topical and systemic application has prolonged protective action [32].

23.2.4.2 Negatively Affecting Factors

Various follow up studies have demonstrated that stress factors (smoking, alcohol consumption, illness, inflammation, sun exposure and environmental hazards) are linked with production of dermal oxidative factors and generation of free radicals (FR) in skin. The effect of ultraviolet (UV) radiation has been extensively studied and it is proposed through all cases that it causes reduction (almost 35 ± 5%) in skin carotenoids (β-carotene and lycopene); however the time for depletion of various carotenoids was found to be different [33].

Darvin et al. [34] used Infrared (IR) radiation during clinical and private applications and found a strongly negative effect β-carotene and lycopene in form of degradation. It was described that the destruction of antioxidants (such as carotenoids) due to IR irradiation was unexpected because the energy of the IR photons is not enough to directly formulate FR [21]. Subsequently, it is proposed that human skin has some structures (inter alia, the mitochondria), which can absorb IR radiation and accumulate energy to ultimately induce the production of free radicals and similar mechanism has also been described for visible radiations (blue-voilet) [35, 36].

Among other factors, it has also been found that consumption of alcohol results in the resilient depletion in concentration of carotenoids. It is caused due to production of FRs related to alcohol consumption. This depletion may recover in 4 days with ingestion of a healthy diet [37]. Furthermore, application of microbial disinfectant such as tissue tolerable plasma on human skin reduces the concentration of carotenoids in the epidermis due to the production of free radicals in form of reactive oxygen species. The study showed that carotenoid concentration reduced in the outer part of stratum corneum to a depth of 10 μm after application of such disinfectant [38].

All the degenerative effects discussed above are caused because; carotenoids cannot continuously neutralize FR produced in the skin. Carotenoids are destructed and lose their potential of antioxidation after neutralization of a small number of reactive oxygen species (ROS) spells. Therefore, amount of destructed skin carotenoids may be indicative of the amount of free radicals produced in the skin [39, 40].

23.3 Functions of Various Carotenoids in Protection of Skin from Diseases

Carotenoids level in skin is indicative of their importance and presents their protective roles against diseases caused by environmental hazards and radiation. The insight into the research studies shows the roles of various carotenoids in different protective functions in skin (Table 23.1).

(a) β-carotene being major precursor of vitamin A plays important function in stratum corneum, where it reduces the inflammation and increases the cell turnover, resulting thin epidermis [41]. Studies show that β-carotene has the highest ability to quench singlet oxygen species ($^1O_2$) and also scavenges (via addition reaction) peroxyl radicals, which are produced by auto-oxidation of cell lipids as a result of UV exposure [42]. β-carotene along with lycopene and lutein helps to prevent the formation of TBARS (Thiobarbituric acid reactive oxygen species) on exposure to UVB [43]. Moreover, it plays significantly important role in healing of skin damaged by sunlight [44].

(b) Lycopene mainly present in tomatoes is red colored non-provitamin A (Chap. 8) carotenoids which is a powerful free radical scavenger and also protects against UV radiations [45, 46]. It is an efficient antioxidant but may behave as pro-oxidative at higher doses (0.15 n mol/mg protein) [47]. The combined effect of β-carotene, lutein and lycopene is more pronounced in photoprotection as compared to lycopene alone [43]. Topical application of lycopene results in penetration in stratum corneum and increases role as antioxidant to tenfolds [48]. Wu et al. [49] showed that lycopene binds with platelet-derived growth factor-BB (PDGF-BB), which otherwise causes cell migration in human cultured skin fibroblast. This suggests that lycopene interferes with stromal cells, tumor cells and their interaction, so helps to prevent tumor [49].

(c) Lutein is member of xanthophyll family and it is proposed to have defensive role for skin through filtration of blue light and as efficient antioxidant [50]. It is also effective in protecting the skin from UV induced erythema (described later) [51]. It has also been established that lutein is present in epidermis and dermis and so contributes to skin color [52]. Lutein also plays important role in skin health and reduces the wrinkling and the risk of skin cancer. The combined effect of topical and dietary administration has more pronounced effect [53] compared to individual application.

(d) Zeaxanthin is a natural carotenoid and also belongs to xanthophyll family. It is rich in green leafy vegetables, peas and egg yolk [50]. Like lutein, it is also involved in protection of oxidation in skin and protection of skin from developing cancer. It is also found to involve in photoaging caused by UVB [53]. Furthermore, in-vitro and in-vivo studies have shown its efficacy in reducing lipid peroxidation in skin within 2 weeks of intake [54, 55].

(e) Canthaxanthin and astaxanthin have a lower rate of free radical quenching but in general act as photoprotectors. Camera et al., [56] described that astaxanthin

Table 23.1 Major source of various carotenoids and their role in disease protection in skin

Sr No.	Carotenoid	Major source	Preventive function in skin	References
1	β-Carotene	Carrots and green leafy vegetables	Upsurges the cellular turnover in stratum corneum	[41, 42, 44]
			Vital for the healing of photo-damaged skin	
			Scavenges peroxyl radicals produced by auto-oxidation of cellular lipids, initiated by UV	
2	Lycopene	Papaya, apricots, guava, pink grapefruit and watermelons	Contributes to UVR protection	[45, 46, 49, 64]
			Quenches singlet oxygen	
			Scavenges oxygen to reduces erythema formation	
			Prevents tumor	
3	Lutein	Green leafy vegetables, peas, broccoli, Brussels sprouts and egg yolk	Plays a role as blue light filters	[50–53]
			Effective antioxidants	
			Protects skin against UV-induced erythema	
			Contributes in skin color	
			Prevents wrinkles and skin cancer	
4	Zeaxanthin	Kale, spinach, brussel sprouts, broccoli, peas and egg yolk	Involved in antioxidation	[50, 53–55]
			Reduces lipid peroxidation significantly in skin	
			Prevents photoaging and developing cancer	
5	Canthaxanthin	Crustaceans, Mushrooms, green algae, paprika and pacific salmon	Enhances the expression of HO-1	[58]
6	Astaxanthin	Red trout, lobster, red seabream, salmon, crabs	Reduces cell culture alteration caused by UVA	[56]
			Shows photoprotective role against UVR	
7	Phytoene and phytofluene	Tomatoes, carrots, red and yellow peppers, apricots, citrus, watermelon, melons, orange, peach,	Absorbs UVA/B directly	[46, 60]
			Contribute in photoprotection	
8	3,3′- Dihydroxy-isorenieratene	*Brevibacterium linens,* used in dairy industry for production of cheeses	Role in the protection against light- induced damage	[61–63]
			Act as a bifunctional antioxidant	
			Reduces MDA formation in skin	

showed an overall photoprotective effect by neutralizing apoptosis, ROS production and altered expression of hemeoxigenase 1 (HO-1) induced by UV radiations. HO-1 is the first rate limiting enzyme in heme degradation and its expression is increased by UVR exposure [57]. Whereas, Trekli et al. [58] proposed that canthaxanthin is powerful enhancer of HO-1 system, which is important deroxifying and antioxidant enzyme system. Fibroblast and melanocytes from human skin cultivated with astaxanthin, remained not only protected from UVA damage but were also able to recover the glutathione content altered due to exposure to radiations [59].

(f) Phytoene and phytofluene are colorless carotenoids present in tomatoes. These act as precursors for production of other carotenoids [3]. These are involved in direct absorption of UVA/B, thus these prevent absorption of UV radiation into the skin so help in photo-protection [46, 60].

(g) 3,3′- Dihydroxy-isorenieratene (DHIR) is a natural carotenoid having an exceptional substructure by carrying two phenolic groups. Mostly, it is present in bacteria and used in food industry for manufacturing products [61]. It is thought to prevent human skin from UV induced DNA damage [62]. It acts as powerful antioxidant even more as compared to other carotenoids and also prevents MDA formation in skin [61, 63].

23.4 Diseases of Skin

There are more than thousand skin diseases which have been identified are becoming a burden on economy in terms of health cost [3]. The carotenoids play important role in protection and management of such diseases and their concentration may change during such disease periods. These diseases and their interrelation with carotenoids have been described under following major categories;

4.1. Diseases Related to Radiation Exposure
4.2. Conditions Related to Skin Color

23.4.1 Diseases Related to Radiation Exposure

As has been previously described, the radiations including Ultraviolet, Infrared and Visible radiations result in production of free radicals but UVR have the most deleterious effect. The UV radiations may be divided into three further categories (UVA, UVB and UVC as given in table below (Table 23.2);

UVA contribute to the most UV exposure and are major causative factor for oxidative damage to human skin. These can penetrate deep into the skin and interact with keratinocytes but are not absorbed by DNA. Whereas UVB can access and damage DNA, so may result in mutagenesis on exposure and induces damage to

Table 23.2 Categories of UV radiation

Sr. No.	Category	Wavelength (nm)	Access to earth (%)	Filtration through glass	Conditions caused
1	UVA	320–400	95	No	Altered gene expression, apoptosis, photo-aging
2	UVB	280–320	5	Yes	Sunburn, mutation, skin cancer, direct damage to DNA
3	UVC	200–280	0	–	–

epidermis causing sunburn [65]. Overall these radiations cause the production of reactive oxygen species which is a general name given to oxygen radical and non-radicals derivatives of oxygen (singlet oxygen $^1O_2$, produced from O_2 by energy rearranging electrons). The ROS are formed in skin and even in the eyes with continuous exposure to light and can damage cellular proteins, lipids and also DNA [66]. However, it is also approved that UVR has positive effect in treatment of some skin disorders such as scleroderma, psoriasis, morphea, vitiligo and atopic dermatitis [67], but it results in decreased level of carotenoids (20–30%) in epidermis [18].

Studies on role of carotenoids in quenching singlet oxygen species in algae, bacteria and plants has made the basis for exploring their role in protection of skin from excited triplet species [68]. The photoprotective effect of carotenoids is attributed to the antioxidant activities, and their efficiency to scavenge the free radicals [69]. Following the intensity and frequency of exposure to UVR, different diseases may be caused with varied levels of severity and can be managed with various carotenoid concentrations, which are described as under;

23.4.1.1 Photosensitivity

A. **Description**

The most common among photosensitivity disorders is erythropoietic protoporphyria (EPP). It is a rare inborn error of blood that results in increased accumulation of endogenous photosensitizer i.e. protoporphyrin in blood, skin, red globules and feces. The diseases is caused by deficiency of ferrochelatase enzyme involved in formation of heme by incorporation of iron into protoporphyrin ring. It can lead to cascade of actions causing photosensitivity when the patient is exposed to sunlight (even visible light) and the patient experiences the symptoms as burning, itching and pain followed by eryrhma, edema and purpura [70]. These annoying and painful effects may even be shown on only a few minutes exposure to the sun. The clinical symptoms are caused by cellular damage resulting because of formation of ROS (excited triplet species) [68]. Other than porphyrins, some endogenously present compounds such as flavins and amino acids also behave as photosensitizing substances [4].

Another photosensitivity disease is “polymorphic light eruption (PLE)”, which is a delayed type of hypersensitivity caused by exposure to sunlight. It is also a genetic disorder like EPP, but its expression is dependent upon gene penetrance. It is usually presented in spring after exposure to sunlight and shows morphological variations such as papulovesicles, papules and plaques [65]. Furthermore, a rare photosensitivity disorder is "hydroa vacciniformia" in childhood and is characterized by repeated formation of vesicles or papulovesicles in area exposed to sunlight [71].

Along with primary photosensitivity disorders, some sensitization is caused by secondary to other diseases. One such condition is, vitiligo (skin disorder) characterized by formation of white macules in which melanocytes disappear, resulting in lack of skin homeostasis, increased photosensitization and altered photoadaptation [72]. Melanocytes are also proposed to be responsible for uptake and storage of carotenoids due to higher affinity of melanin to β-carotene, so absence of melanocytes increases photosensitivity [65]. Similarly, connective tissues disorders also show photosensitivity, as is shown by patients of systemic lupus erythematosus (SLE). Both UVA and UVB may cause sensitivity in 60–70% of patients and form lesions [73, 74] (Fig. 23.5).

B. **Role of carotenoids**

Concerning the studies about the role of carotenoids for EPP, in 1968, a 10 years old girl was treated with carrot oil (equal to 30 mg β-carotene/day) for a month that resulted in improved tolerance to sunlight without any sensitivity sign [75]. Further studies approved the fact that majority of the patients suffering from EPP could bear sunlight exposure after a treatment with β-carotene (higher doses for several months are required). It was observed β-carotene even in larger amount did not show any effect on concentration of porphyrin, so it was concluded that it has no positive

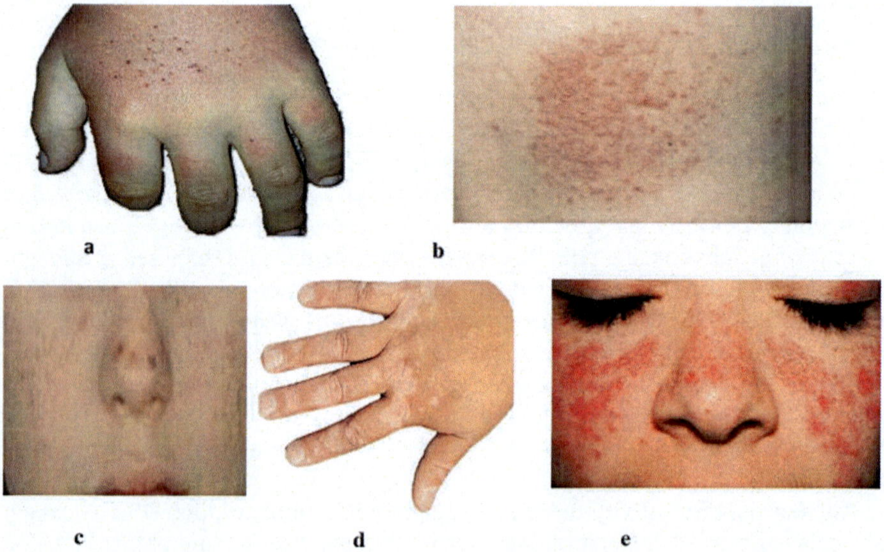

Fig. 23.5 Photosensivity diseases. (**a**) Erythropoietic protoporphyri. (**b**) Polymorphic light eruption. (**c**) Hydroa vacciniformia. (**d**) Vitiligo. (**e**) Systemic lupus erythematosus

impact on biochemical lesions but only improves photosensitivity [76]. Carotenoids have been in use since 1968 for successful treatment of EPP and it is mostly believed that EPP lesions are caused by excited triplet molecules or singlet oxygen species for which β-carotene acts as quencher [77, 78]. In 1975 Food and Drug Administration (FDA) sanctioned the use of β-carotene in treatment of EPP. Moreover, among other carotenoids, canthaxanthin was also found beneficial in alleviating the symptoms of erythropoietic protoporphyria and polymorphic light eruption [79].

Holzle et al. [80] described that β-carotene does not act as effective treatment for PLE. Later it was proposed that β-carotene and canthaxanthin in doses of 10 and 15 mg respectively daily for 4 weeks showed a better tolerance against sun exposure, while different patients behaved differently according to various stages of PLE [81]. However, more recently, Marini et al. [82] evaluated the nutritional supplement based on lycopene, β-carotene, and *L. johnsonii* and observed skin photoprotection and modulated biomarkers for UVR effects not only in normal but also in PLE affected individuals. It was attributed to reduced expression of intercellular adhesion molecule 1 (ICAM-1) mRNA, whose higher expression is involved in pathophysiology of PLE. Similarly, carotenoids did not show any significant effect on hydroa vacciniformia however, 25 mg of β-carotene twice daily, along with topical application of sunscreen has been proposed to stop skin changes [71].

The yellowish tone of skin parted by carotenoids may lower the difference between vitiligo lesion and normal skin; this requires a blood level of β-carotene between 250–500 μg/dL to show the action [83]. Postaire et al. [84] used different levels of α-tocopherols, ascorbic acid, β-carotene and lycopene and observed increase in melanin concentration to a significant level. Likewise, cell culture system (melanocytes) incubated with lycopene from tomato, carotenes, ascorbic acid and α-tocopherols showed increased melanin and also resulted in reduced DNA damage caused by UVR. Active phase of vitiligo also induces the formation of ROS in epidermis due to local and systemic imbalances in enzymatic and non-enzymatic activities [85], which can be treated with a combination of β-carotene (25 mg) and canthaxanthin (35 mg). Moreover this treatment also normalizes the contrast between vitiligo lesion and surrounding skin [86]. Based upon the studies available on the effect of β-carotene on above discussed diseases, it is also proposed that it may also be helpful in skin rashes caused by SLE, but it needs further research studies to prove the effect.

23.4.1.2 Sunburn

A. Description

Sunburn, also called as erythema solare is the first reaction of the skin when it is exposed to UV radiations. It is an inflammatory response of the skin which is dependent upon wavelength of UVR i.e. it decreases with longer wavelength, so UVB are considered as main culprit for the skin condition. The clinical manifestation includes redness on skin and vasodilation, which may take form of blisters or ablations on skin depending on severity of exposure (Fig. 23.6). Histological studies

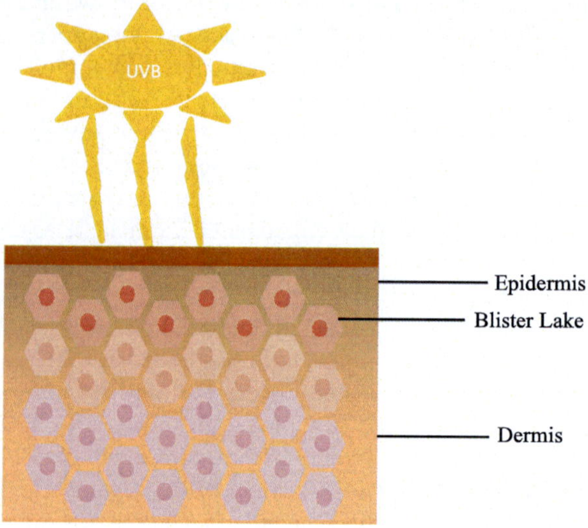

Fig. 23.6 UVB Causing sun burn resulting in darker outer color and blister eruption on severe exposure

demonstrate that keratinocyte exposed to sunburn may pass through programmed cell death within hours of radiation exposure. It is further suggested that inflammatory response is caused by cytokines based action along with vasoactive and neuroactive mediators present in skin. The sunburn signs are shown within 4 h of exposure, peaking up to 8–24 h and may vanish after wards, but persistence is dependent on age and type of the skin of patient [3, 67, 68].

B. **Role of carotenoids**

The studies have shown that carotenoid levels before exposure to UVB reduce as compared to before exposure, presenting that these are used up in quenching the free radical produced by the radiations. The lycopene is used just after exposure while β-carotene level is maintained for 30–60 min proposing lycopene as being more efficient [21]. The studies have recommended doses of β-carotene or mixture with vegetable juices as 24–180 mg/day [87, 88]. The supplements providing different carotenoids including beta-carotene, alpha-carotene, zeaxanthin, cryptoxanthin, lutein in vegetable oil increased the minimal erythema dose (MED) required to cause sunburn by UVB [65]. Furthermore, it was observed that topical application of lycopene (0.03%) reduced the erythema more effectively given along with dietary intervention [89].

23.4.1.3 Photoaging

A. **Description**

The changes in structure and function of skin due to aging are categorized as chronological or intrinsic aging (natural process with passage of time) and extrinsic aging caused by external factors. These external factors include UVR exposure (photo-aging), pollution, smoking, poor nutrition and lack of sleep. Both types of aging can be distinguished histologically and clinically i.e. intrinsic aging results in loss of elasticity and fragility while, aging related to UVR exposure is recognized by irregular hyper-pigmentation, inelasticity, wrinkles, dryness of skin and telean-giectasia (widened capillaries) [7, 68] (Fig. 23.7). The clinical manifestations are shown by destruction of proteins (collagen and elastin) present in extracellular matrix and alterations in melanin production that may result increased vulnerability to skin diseases. In case of photoaging, these signs and diseases may occur

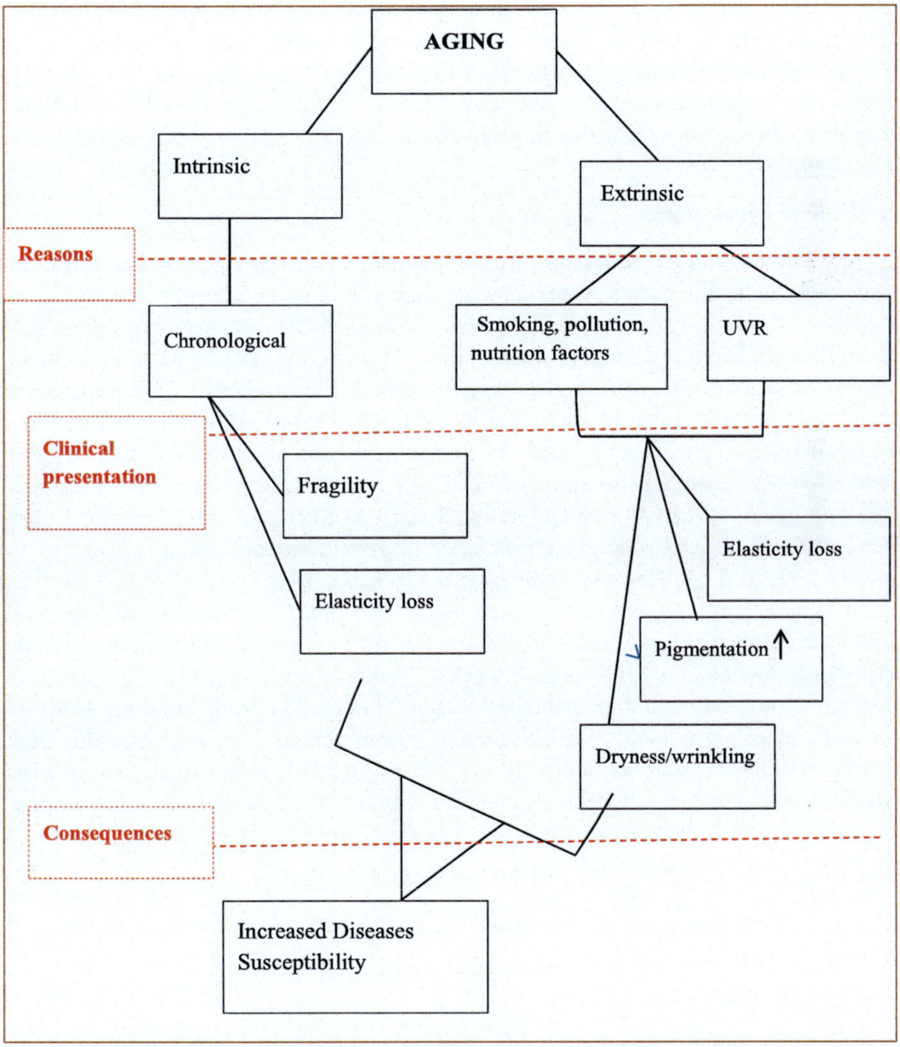

Fig. 23.7 The process of aging and its consequences

premature; causing physiological and psychological distress. Among UVRs, UVA is considered as major contributor towards photoaging. UVA exposure leads to production of ROS resulting in damage of mitochondrial DNA. Mutations in mitochondrial DNA alter the energy metabolism process and start a further chain for ROS production. ROS activates metalloproteases present in skin matrix leading further to protein destruction with pronounced skin sagging, deep wrinkles and even pigmented lesions [68, 90]. Furthermore at molecular level, exposure of UVB causes the activation of transcription factors (activator protein and nuclear factors) for induction of metatlloprotease in skin. Moreover, it is also proposed that lipid hydroperoxides (as cholesterol hydroperoxide) may also activate matrix metalloprotease (MMP). Cholesterol hydroperoxide is added in skin due to photosensitized oxidation by singlet oxygen [90]. Signs of photoaging are more prevalent in body areas more exposed to sun such as face, neck back of hands and forearms [68, 90, 91]. Furthermore it is also suggested that other than UVR, infrared and visible radiations may also cause photo-aging due to production of ROS [92] as is described earlier in this chapter.

B. **Role of carotenoids**

A number of studies have demonstrated that persons having higher concentration of carotenoid in skin have a younger look to age as compared to lower concentration individuals. Similarly, it was observed that the individuals with better lycopene status of the skin had less dermal roughness (sign of photo-aging) [32, 93]. Carotenoids are powerful quenchers of singlet oxygen species (non-radical ROS). β-carotene supplementation in diet increases its level in skin that may inhibit oxygenationation of lipids (caused by UV exposure). Furthermore it was observed that this supplementation also resulted in suppression of MMP and accumulation of cholesterol hydroperoxides [90]. A human study based on SPF-15, usual sunblock and β-carotene (30 mg/day) and placebo showed that occurrence of disease event was as low as 1/3rd in β-carotene group compared to placebo [94].

In another study, a mixture of β-carotene, lycopene, selenium yeast and proanthocyanidins was introduced to volunteers for 10 weeks and then they were exposed to low dose of UVB dose 2 weeks. It was observed that MMP-1 (metalloprotease type) increased in placebo but decreased in treated group, whereas MMP-9 did not change significantly [68]. Similarly, Palombo et al. [55] described that oral ingestion of lutein and zeaxanthin (10 mg and 0.6 mg daily respectively) along with a topical treatment for 12 weeks resulted in improved photoprotection and hydration of skin along with decreased skin lipids peroxidation and better skin elasticity.

23.4.1.4 Photoimmune Modulation

A. **Description**

One of the major functions of skin is to prevent microbial attack from external environment, which is obtained by both innate and adaptive immune function. In this regard, Langerhans cells (LC) present in epidermal cells play key function in both innate and adaptive immunity of skin. Langerhans cells are members of dendritic cells family and antigens have to interact with them. LCs are special cells to sense the environment and these extend their dendrites to stratum corneum [95]. These pick up antigens and carry them to draining lymph node where they also present these antigens to T cells (also act as antigen presenting cells). UVB reach deep into epidermis and act as immunosuppressant by inhibiting or deleting Langerhans cells. Human study explains that UVB radiation impair up-regulation of CD84 expression in Langerhans cells. Furthermore viability of LC is decreased due to apoptic cell death of these cells [96]. Similarly, the UVR (both UVA and UVB) change the microenvironment of epidermal cells in a way that danger signals become ineffective and chemical contact sensitizer (dinitrochlorobenzene) causes immunological tolerance, by modifications in T lymphocytes [97, 98].

B. **Role of carotenoids**

β-carotene (30 mg/day) supplementation has been used as protective factor against delayed type hypersensitivity caused by UVA. Similar dose also has been observed to prevent the loss of Langerhans cells in epidermis due to exposure of UVR [99, 100]. Furthermore, it was observed that lutein and zeaxanthin (0.04% each for 2 weeks) improved UVB induced inflammatory response in female mice [101]. The immunomodulatory effect of carotenoids is also observed in case of psoriasis (chronic inflammatory disease of skin) where these may change the metabolism of retinoids and production of cytokies in psoriatic plaque [102].

23.4.1.5 Photocarcinogenesis

A. **Description**

Photocarcinogenesis consists of number of events originated with solar or artificial light exposure resulting in growth of skin cancer. Skin cancer is the most common type of cancer present all over the world, which may be melanoma (more serious) and non-melanoma (more common). As discussed earlier UVB reaches to DNA resulting in damage to structure e.g. formation of pyrimidone photodimers (between pyrimidine residues) and cyclobutane dimer between two adjacent cytosine and thymine residue). Both lesions are mutagenic and may lead to mutations and ultimately to cancer (Fig. 23.8). UV radiations are powerful carcinogens and they may initiate and then also may promote carcinogenesis. UVA alone has less effect on carcinogenesis but in combination with UVB these become more dangerous. The effect of UVA in DNA damage is considered as indirect i.e. it results in formation of ROS causing damage to DNA. The most common types of cancer present in are squamous cell carcinoma (SCC) and basal cell carcinoma (BCC) (types of non-melanoma cancer). Furthermore, UVR exposure also promotes carcinogenesis by affecting on gene responsible for cell growth [67, 103, 104].

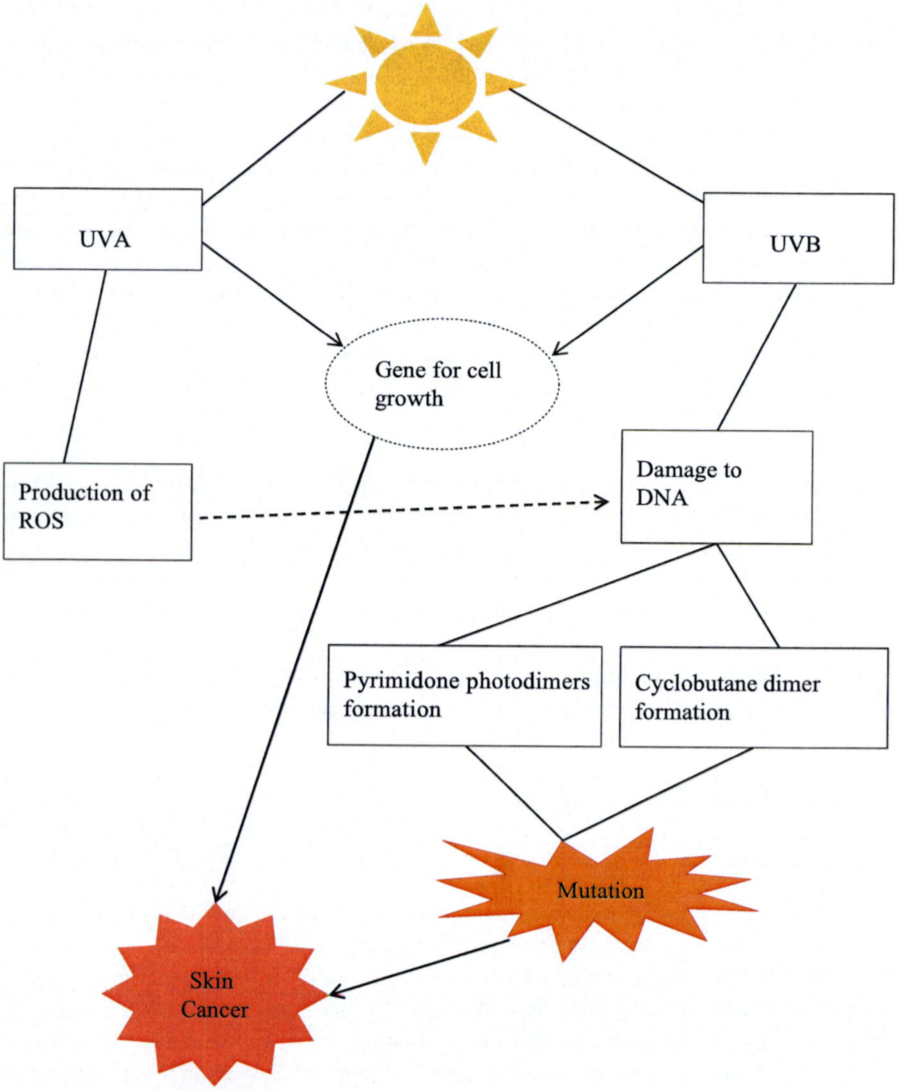

Fig. 23.8 Target of UVA and UVB inside skin cells and possible outcomes

B. **Role of carotenoids**

β-carotene is proposed as most promising and potential candidate for prevention and management of cancer. It has been suggested since1980s that β-carotene may be helpful to prevent cancer and it was observed through animal studies that it can prevent skin cancer caused by UVR and chemical carcinogens [105, 106]. However, no clear prove to set it against human skin cancer has been found and studies show conflicting results for oral intake of β-carotene for prevention from cancer [107].

Some epidemiological studies have depicted no relationship between BCC/SCC and carotenoids in diet [108–110]. Likewise, in a study conducted on human subjects, patients with non-melanoma cancer were tried with 50 mg of β-carotene daily for 5 years but no satisfactory results were obtained [111]. Moreover, similar results were obtained in subjects with incidence of malignant neoplasm treated with β-carotene [112].

23.4.2 Conditions Related to Skin Color

Skin has much importance not only because it protects against environmental hazards and antigens but also it is considered as sign of health status and beauty. The way it appears in color, texture and elasticity is affected by the nutritional status of the person. The major skin pigment is melanin while others as oxy and deoxy-hemoglobin and carotenoids. Carotenoids are not only important as precursor of vitamin A (used in cell differentiation and proliferation in skin), but are also involved in yellow tone of the skin color and excessive of β-carotene and lycopene may result in pigmentation of the skin [3]. Skin tanning is required especially in North Europe and carotenoids have been in use in cosmetics even before their discovery (as Cleopatra used saffron that gives orange color). Different cultures use carotenoids from fruits and vegetables to get different skin tones [113].

Melanin in skin also acts as photo-protector agent, it causes pigmentation on exposure to sunlight and protects from oxidation caused by UVR. Therefore, it is proposed that individuals with darker skin have fewer chances of photocarcinogenesis. But compared to melanin coloration, carotenoids coloration gives a healthier look to the skin [11]. Furthermore, skin rejuvenation is also talk of the town and facial symmetry and proportion are required characteristics. The cosmetic industry is trying to find effective nutrocosmetics for dryness, inelasticity and pigmentation disorders as uneven color and swelling [3]. The loss of color by skin called as vitiligo (described in detail earlier) results in photosensitivity, similarly pigmentation by erythema has also been discussed. Besides these, over use of carotenoids in form of supplements may results in carotenoderma. The condition may also be caused by

(a) ↑ concentration of LDL bound carotenoids in blood
(b) ↓ concentration of 15,15-β-dioxygenase (described in Chap. 8) in gut causing ↑ absorption of carotenoids

Accumulation and elimination of carotenoids are slow processes and may take weeks till normal color of the skin is obtained [18].

23.5 Conclusion

The skin appearance and prevention from different diseases is attributed to health, socioeconomic wellbeing and psychological happiness. Nutrition plays an important role in protecting skin from hazardous effects of environment and also to keep it youthful, elastic, shiny and even toned. The desire for appealing skin has drawn the attention of researcher and cosmeceuticals and nutricosmetical industries. In this regard the natural products from diet such as carotenoids have become important part of intervention studies. It can be interfered on the bases of various studies that carotenoids play marvelous role in protecting the skin from disastrous factors including UVR, inflammatory factors and carcinogenic factors by acting as antioxidants, anti-inflammatory and as anti-mutagenic factors. Further, these may improve skin tone through both systemic and topical application. The skin carotenoids level may be indicative of development of skin diseases. But more insight into their practical application is needed against various diseases. Furthermore, research studies involving human subjects are needed to validate the effect and dose of various carotenoids.

References

1. Proksch E, Brandner JM, Jensen JM (2008) The skin: an indispensable barrier. Exp Dermatol 17:1063–1072
2. Lefevre CE, Ewbank MP, Calder AJ, von dem Hagen E, Perrett DI (2013) It is all in the face: carotenoid skin coloration loses attractiveness outside the face. Biol Lett 9:0633
3. Meléndez-Martínez AJ, Stinco CM, Mapelli-Brahm P (2019) Skin carotenoids in public health and nutricosmetics: the emerging roles and applications of the UV radiation-absorbing colourless carotenoids phytoene and phytofluene. Nutrients 11:1093
4. Stahl W, Sies H (2012) Photoprotection by dietary carotenoids: concept, mechanisms, evidence and future development. Mol Nutr Food Res 56:287–295
5. Fisher GJ, Kang S, Varani J, Bata-Csorgo Z, Wan Y, Datta S et al (2002) Mechanisms of photoaging and chronological skin aging. Arch Dermatol 138(11):1462–1470
6. Thakkar SK, Moodycliffe AM, Richelle MA (2011) Carotenoids and skin. In: Pappas (ed) Nutrition and skin: lessons for anti-aging, beauty and healthy skin. Springer Science, New York
7. Schagen SK, Zampeli VA, Makrantonaki E, Zouboulis CC (2012) Discovering the link between nutrition and skin aging. Dermatoendocrinology 4(3):298–307
8. Schmuth M, Fritsch PO (2011) Cutaneous changes in nutritional diseases. In: Krutmann J, Humbert P (eds) Nutrition for healthy skin. Springer, Berlin/Heidelberg, p 3
9. Krutmann J (2011) Skin aging. In: Krutmann J, Humbert P (eds) Nutrition for healthy skin. Springer, Berlin/Heidelberg, p 15
10. Afaq F (2011) Natural agents: cellular and molecular mechanisms of photoprotection. Arch Biochem Biophys 508:144–151
11. Stephen ID, Coetzee V, Perrett DI (2011) Carotenoid and melanin pigment coloration affect perceived. Evol Hum Behav 32:216–227
12. Whitehead RD, Re D, Xiao D, Ozakinci G, Perrett DI (2012) You are what you eat: within-subject increases in fruit and vegetable consumption confer beneficial skin-color changes. PLoS One 7:e32988

13. Hata TR, Scholz TA, Ermakov IV, McClane RW, Khachik F, Gellermann W et al (2000) Non-invasive Raman spectroscopic detection of caroteinoids in human skin. J Investig Dermatol 115:441–448
14. Walfisch Y, Walfisch S, Agbaria R, Levy J, Sharoni Y (2003) Lycopene in serum, skin and adipose tissues after tomato-oleoresin supplementation in patients undergoing haemorrhoidectomy or peri-anal fistulotomy. Br J Nutr 90(4):759–766
15. Richelle M, Sanchez B, Tavazzi I, Lambele P (2010) Lycopene isomerisation takes place within enterocytes during absorption in human subjects. Br J Nutr 103(12):1800–1807
16. Bayerl C, Schwarz B, Jung EG (2003) A three-year randomized trial in patients with dysplastic naevi treated with oral beta-carotene. Acta Derm Venereol 83(4):277–281
17. Lademann J, Caspers PJ, van der Pol A, Richter H, Patzelt A, Zastrow L (2009) *In vivo* Raman spectroscopy detects increased epidermal antioxidative potential with topically applied carotenoids. Laser Phys Lett 6:76–79
18. Vahlquist A (2007) Carotenoids and the skin: an overview. In: Retinoids and carotenoids in dermatology, Basic and clinical dermatology. Informa Healthcare USA, Inc, New York
19. Bruch-Gerharz D, Stahl W, Gerharz CD, Megahed M, Wingerath T, Sies H et al (2001) Accumulation of the xanthophyll lutein in skin amyloid deposits of systemic amyloidosis (Al type). J Investig Dermatol 116(1):196–197
20. Lademann J, Meinke MC, Sterry W, Darvin ME (2011) Carotenoids in human skin. Exp Dermatol 20:377–382
21. Darvin ME, Haag SF, Meinke MC, Sterry W, Lademann J (2011) Determination of the influence of IR radiation on the antioxidative network of the human skin. J Biophotonics 4(1–2):21–29
22. Hadgraft J (2001) Modulation of the barrier function of the skin. Skin Pharmacol Appl Ski Physiol 14(Suppl 1):72–81
23. Maharshak N, Shapiro J, Trau H (2003) Carotenoderma—a review of the current literature. Int J Dermatol 42:178–181
24. Vahlquist A, Duvic M (2007) Retinoids and carotenoids in dermatology, Basic and clinical dermatology. Informa Healthcare USA, Inc., New York
25. Lee R, Mathews-Roth MM, Pathak M, Parrish J (1975) The detection of carotenoid pigments in human skin. J Investig Dermatol 64:175–177
26. Vahlquist A, Lee JB, Michaelsson G, Rollman O (1982) Vitamin A in human skin: II concentrations of carotene, retinol and dehydroretinol in various components of normal skin. J Investig Dermatol 79(2):94–97
27. Anjo DM, Moore TA (1984) A photoacoustic depth profile of s-carotene in skin. Photochem Photobiol 39(5):635–640
28. Prince MR, Frisoli JK (1993) Beta-carotene accumulation in serum and skin. Am J Clin Nutr 57(2):175–181
29. Stahl W, Heinrich U, Jungmann H, Laar JV, Schietzel M, Sies H et al (1998) Increased dermal carotenoid levels assessed by noninvasive reflection spectrophotometry correlate with serum levels in women ingesting Betatene. J Nutr 128:903–907
30. Hammond BR, Wooten BR (2004) Validity issues with the in vivo measurement of skin carotenoids using Raman spectroscopy. J Investig Dermatol 119:544–546
31. Darvin M, Patzelt A, Gehse S, Schanzer S, Benderoth C, Sterry W (2008) Cutaneous concentration of lycopene correlates significantly with the roughness of the skin. Eur J Pharm Biopharm 69(3):943–947
32. Darvin ME, Lademann J (2013) Health benefits of carotenoids for human skin. In: Yamaguchi M (ed) Carotenoids food sources, production and health benefits. Nova Science Publishers, New york
33. Biesalski HK, Hemmes C, Hopfenmuller W, Schmid C, Gollnick HPM (1996) Effects of controlled exposure of sunlight on plasma and skin levels of β-carotene. Free Radic Res 24:215–224

34. Darvin ME, Zastrov L, Gonchukov SA, Lademann J (2009) Influence of IR radiation on the carotenoid content in human skin. Opt Spectrosc 107:917–920
35. Lee MJ, Popkin BM, Kim S (2002) The unique aspects of the nutrition transition in South Korea: the retention of healthful elements in their traditional diet. Public Health Nutr 5(1A):197–203
36. Schroeder P, Lademann J, Darvin ME, Stege H, Marks C, Bruhnke S et al (2008) Infrared radiation-induced matrix metalloproteinase in human skin: implications for protection. J Investig Dermatol 128(10):2491–2497
37. Darvin ME, Sterry W, Lademann J, Patzelt A (2013) Alcohol consumption decreases the protection efficiency of the antioxidant network and increases the risk of sunburn in human skin. Skin Pharmacol Physiol 26:45–51
38. Fluhr JW, Sassning S, Lademann O, Darvin ME, Schanzer S, Kramer A (2012) *In vivo* skin treatment with tissue tolerable plasma influences skin physiology and antioxidant profile in human stratum corneum. Exp Dermatol 21(2):130–134
39. Ribaya-Mercado JD, Garmyn M, Gilchrest BA, Russell RM (1995) Skin lycopene is destroyed preferentially over beta-carotene during ultraviolet irradiation in humans. J Nutr 125:1854–1859
40. Haag SF, Taskoparan B, Darvin ME, Groth N, Lademann J, Sterry W et al (2011) Determination of the antioxidative capacity of the skin in vivo using resonance Raman and electron paramagnetic resonance spectroscopy. Exp Dermatol 20:483–487
41. Reifen R (2002) Vitamin A as an anti-inflammatory agent. Proc Nutr Soc 61:397–400
42. Cantrell A, McGarvey DJ, Truscott TG, Rancan F, Böhm F (2003) Singlet oxygen quenching by dietary carotenoids in a model membrane environment. Arch Biochem Biophys 412:47–54
43. Eichler O, Sies H, Stahl W (2002) Divergent optimum levels of lycopene, beta-carotene and lutein protecting against UVB irradiation in human fibroblastst. Photochem Photobiol 75:503–506
44. Mukherjee S, Date A, Patravale V, Korting HC, Roeder A, Weindl G (2006) Retinoids in the treatment of skin aging: an overview of clinical efficacy and safety. Clin Interv Aging 1(4):327–348
45. Di Mascio P, Kaiser S, Sies H (1989) Lycopene as the most efficient biological carotenoid singlet oxygen quencher. Arch Biochem Biophys 274:532–538
46. Aust O, Stahl W, Sies H, Tronnier H, Heinrich U (2005) Supplementation with tomato-based products increases lycopene, phytofluene, and phytoene levels in human serum and protects against UV-light induced erythema. Int J Vitam Nutr Res 75:54–60
47. Stahl W, Heinrich U, Aust O, Tronnier H, Sies H (2006) Lycopene-rich products and dietary photoprotection. Photochem Photobiol Sci 5:238–242
48. Lopes LB, VanDeWall H, Li HT, Venugopal V, Li HK, Naydin S et al (2010) Topical delivery of lycopene using microemulsions: enhanced skin penetration and tissue antioxidant activity. J Pharm Sci 99(3):1346–1357
49. Wu WB, Chiang HS, Fang JY, Hung CF (2007) Inhibitory effect of lycopene on PDGF-BB-induced signalling and migration in human dermal fibroblasts: a possible target for cancer. Biochem Soc Trans 35(5):1377–1378
50. Morganti P, Bruno C, Guarneri FM, Cardillo A, Del Ciotto P, Valenzano F (2002) Role of topical and nutritional supplement to modify the oxidative stress. Int J Cosmet Sci 24(6):331–339
51. Stahl W, Sies H (2002) Carotenoids and protection against solar UV radiation. Skin Pharmacol Appl Ski Physiol 15:291–296
52. Alaluf S, Heinrich U, Stahl W, Tronnier H, Wiseman S (2002) Dietary carotenoids contribute to normal human skin color and UV photosensitivity. J Nutr 132(3):399–403
53. Evans JA, Johnson EJ (2010) The role of phytonutrients in skin health. Nutrients 2(8):903–922
54. Rice-Evans CA, Sampson J, Bramley PM, Holloway DE (1997) Why do we expect carotenoids to be antioxidants *in vivo*? Free Radic Res 26:381–398
55. Palombo P, Fabrizi G, Ruocco V, Ruocco E, Fluhr J, Roberts R (2007) Beneficial long-term effects of combined oral/topical antioxidant treatment with the carotenoids lutein and zea-

xanthin on human skin: a double-blind, placebo-controlled study. Skin Pharmacol Physiol 20(4):199–210
56. Camera E, Mastrofrancesco A, Fabbri C, Daubrawa F, Picardo M, Sies H et al (2009) Astaxanthin, canthaxanthin and beta-carotene differently affect UVA-induced oxidative damage and expression of oxidative stress-responsive enzymes. Exp Dermatol 18(3):222–231
57. Obermuller-Jevic UC, Francz PI, Frank J, Flaccus A, Biesalski HK (1999) Enhancement of UVA induction haem- oxygenase expression by beta-carotene in human skin fibroblast. FEBS Lett 460:212
58. Trekli MC, Riss G, Goralczyk R, Tyrrell RM (2003) Beta-carotene suppresses UVA-induced HO-1 gene expression in cultured FEK4. Free Radic Biol Med 34:456–464
59. Lyons NM, O'Brien NM (2002) Modulatory effect of an algal extract containing astaxanthin on UVA irradiated cells in culture. J Dermatol Sci 30(1):73–84
60. Britton G (1995) Carotenoids. In: Britton G, Liaaen-Jensen S, Pfander H (eds) Spectroscopy, vol 1B. Birkhäuser, Basel, p 19
61. Martin H, Kock S, Scherrers R, Luttern K, Wagener T, Hundsdörfer C et al (2009) 3,3′-Dihydroxyisorenieratene, a natural carotenoid with superior antioxidant and photoprotective properties. Angew Chem Int Ed Engl 48:400–403
62. Wagener S, Völker T, De Spirt S, Ernst H, Stahl W (2010) 3,3′-Dihydroxyisorenieratene and isorenieratene prevent UV-induced DNA damage in human skin fibroblasts. Curr Nutr Food Sci 6:36–43
63. Skehan P, Storeng R, Scudiero D, Monks A, McMahon J, Vistica D et al (1990) New colorimetric cytotoxicity assay for anticancer-drug screening. J Natl Cancer Inst 82(13):1107–1112
64. Canene-Adams K, Campbell JK, Zaripheh S, Jeffery EH, Erdman JW (2005) The tomato as a functional food. J Nutr 135:1226–1230
65. Krautheim A, Gollnick HPM (2007) Beta-carotene in the treatment of skin disorders. In: Retinoids and carotenoids in dermatology, Basic and clinical dermatology. Informa Healthcare USA, Inc., New York
66. Halliwell B (2006) Reactive species and antioxidants: redox biology is a fundamental theme of aerobic life. Plant Physiol 141:312–322
67. Pérez-Sánchez A, Barrajón-Catalán E, Herranz-López M, Micol V (2018) Nutraceuticals for skin care: a comprehensive review of human clinical studies. Nutrients 10(4):e403
68. Goralczyk R, Wertz K (2009) Skin photoprotection by carotenoids. In: Britton G, Liaaen-Jensen S, Pfander H (eds) Carotenoids: nutrition and health. Birkhäuser Verlag, Basel
69. Sies H, Stahl W (1995) Vitamins E and C, β-carotene, and other carotenoids as antioxidants. Am J Clin Nutr 62(6):1315S–1321S
70. Kou J, Dou D, Yang L (2017) Porphyrin photosensitizers in photodynamic therapy and its applications. Oncotarget 8:81591–81603
71. Pfau A, Hein R, Karrer S, Landthaler M (1996) Hydroa vacciniforme. Dtsch Med Wochenschr 121(16):517–520
72. Taieb A (2011) Vitiligo as an inflammatory skin disorder: a therapeutic perspective. Pigment Cell Melanoma Res 25(1):9–13
73. Hasan T, Nyberg F, Stephansson E, Puska P, Hakkinen M, Sama S et al (1997) Photosensitivity in lupus erythematosus, UV photoprovocation results compared with history of photosensitivity and clinical findings. Br J Dermatol 136(5):699–705
74. Edlich RF, Winters KL, Lim HW, Cox MJ, Becker DG, Horowitz JH et al (2004) Photoprotection by sunscreens with topical antioxidants and systemic antioxidants to reduce sun exposure. J Long-Term Eff Med Implants 14(4):317–340
75. Mathews-Roth MM, Pathak MA, Fitzpatrick TB, Harber LC, Kass EH (1970) Betacarotene as a photoprotective agent in erythropoietic protoporphyria. Trans Assoc Am Phys 83:176–184
76. Mathews-Roth MM (2004) Therapeutic uses of carotenoids in skin photosensitivity diseases. In: Krinsky N, Mayne S, Sies H (eds) Carotenoids in health and disease. Marcel Dekker, Inc, Basel

77. von Laar J, Stahl W, Bolsen K, Goerz G, Sies H (1996) B-carotene serum levels in patients with erythropoietic protoporphyria on treatment with the synthetic all-trans isomer or a natural isomer mixture of b-carotene. J Photochem Photobiol B Biol 33(2):157–162
78. Badminton MN, Elder GH (2002) Management of acute and cutaneous porphyrias. Int J Clin Pract 56(4):272–278
79. Mathews-Roth MM (2000) Erythropoietic protoporphyria: treatment with antioxidants and potential cure with gene therapy. Methods Enzymol 319:479–484
80. Holzle E, Plewig G, von Kries R, Lehmann P (1987) Polymorphous light eruption. J Investig Dermatol 88(3 suppl):32S–38S
81. Tutrone WD, Spann CT, Scheinfeld N, Deleo VA (2003) Polymorphic light eruption. Dermatol Ther 16(1):28–39
82. Marini A, Jaenicke T, Grether-Beck S, Le Floc'h C, Cheniti A, Piccardi N (2014) Prevention of polymorphic light eruption by oral administration of a nutritional supplement containing lycopene, β-carotene, and *Lactobacillus johnsonii*: results from a randomized, placebo-controlled, double-blinded study. Photodermatol Photoimmunol Photomed 30(4):189–194
83. Arya V, Grzybowski J, Schwartz RA (2003) Carotenemia. Cutis 71(6):441–442. 448
84. Postaire E, Jungmann H, Bejot M, Heinrich U, Tronnier H (1997) Evidence for antioxidant nutrients-induced pigmentation in skin: results of a clinical trial. Biochem Mol Biol Int 42(5):1023–1033
85. Colucci R, Dragoni F, Moretti S (2015) Oxidative stress and immune system in vitiligo and thyroid diseases. Oxidative Med Cell Longev 2015. Article ID 631927:7
86. Raab WP, Tronnier H, Wiskemann A (1985) Photoprotection and skin coloring by oral carotenoids. Dermatológica 171(5):371–373
87. Wolf C, Steiner A, Honigsmann H (1988) Do oral carotenoids protect human skin against ultraviolet erythema, psoralen phototoxicity, and ultraviolet-induced DNA damage? J Investig Dermatol 90(1):55–57
88. Garmyn M, Ribaya-Mercado JD, Russell RM, Bhawan J, Gilchrest BA (1995) Effect of beta-carotene supplementation on the human sunburn reaction. Exp Dermatol 4(2):104–111
89. Andreassi M, Stanghellini E, Ettorre A, Di Stefano A, Andreassi L (2004) Antioxidant activity of topically applied lycopene. J Eur Acad Dermatol Venereol 18(1):52–55
90. Terao J, Minami Y, Bando N (2011) Singlet molecular oxygen –quenching activity of carotenoids: relevance to protection of the skin from photoaging. J Clin Biochem Nutr 48(1):57–62
91. Rittie L, Fisher GJ, Rittié L, Fisher GJ (2015) Natural and sun-induced aging of human skin. Cold Spring Harb Perspect Med 5(1):a015370
92. Akhalaya MY, Maksimov GV, Rubin AB, Lademann J, Darvin ME (2014) Molecular action mechanisms of solar infrared radiation and heat on human skin. Ageing Res Rev 16:1–11
93. Jacobi U, Chen M, Frankowski G, Sinkgraven R, Hund M, Rzany B (2004) In vivo determination of skin surface topography using an optical 3D device. Skin Res Technol 10(4):207–214
94. Battistutta D, Williams DL, Green A (2000) Effectiveness of daily sunscreen application and bcarotene intake for prevention of photaging: a community-based randomised trial. 13th International Congress on Photobiology, Abstr. 58
95. Clayton K, Vallejo AF, Davies J, Sirvent S, Polak ME (2017) Langerhans cells—programmed by the epidermis. Front Immunol 8:1676
96. Rattis FM, Concha M, Dalbiez-Gauthier C, Courtellemont P, Schmitt D, Péguet-Navarro J (1998) Effects of ultraviolet B radiation on human langerhans cells: functional alteration of CD86 upregulation and induction of apoptotic cell death. J Investig Dermatol 111(3):373–379
97. Kreutz M, Karrer S, Hoffmann P, Gottfried E, Szeimies RM, Hahn J et al (2012) Whole-body UVB irradiation during allogeneic hematopoietic cell transplantation is safe and decreases acute graft-versus-host disease. J Investig Dermatol 132(1):179–187
98. Newell L, Polak ME, Perera J, Owen C, Boyd P, Pickard C et al (2013) Sensitization via healthy skin programs Th2 responses in individuals with atopic dermatitis. J Investig Dermatol 133(10):2372–2380

99. Fuller CJ, Faulkner H, Bendich A, Parker RS, Roe DA (1992) Effect of beta-carotene supplementation on photosuppression of delayed-type hypersensitivity in normal young men. Am J Clin Nutr 56(4):684–690
100. Gollnick H, Hopfenmuller W, Hemmes C, Chun S, Schmid C, Sundermeier K, Biesalski HK (1996) Systemic beta carotene plus topical UV-sunscreen are an optimal protection against harmful effects of natural UV-sunlight: results of the Berlin-Eilath study. Eur J Dermatol 6:200–205
101. Gonzalez S, Astner S, An W, Goukassian D, Pathak MA (2003) Dietary lutein/zeaxanthin decreases ultraviolet B-induced epidermal hyperproliferation and acute inflammation in hairless mice. J Investig Dermatol 121(2):399–405
102. Lima XT, Kimball AB (2011) Skin carotenoid levels in adult patients with psoriasis. J Eur Acad Dermatol Venereol 25:945–949
103. Soehnge H, Ouhtit A, Ananthaswamy ON (1997) Mechanisms of induction of skin cancer by UV radiation. Front Biosci 1(2):538–551
104. Ascenso A, Ribeiro H, Marques HC, Oliveira H, Santos C, Simões S (2014) Chemoprevention of photocarcinogenesis by lycopene. Exp Dermatol 23(12):874–878
105. Bollag W (1970) Vitamin A and vitamin A acid in the prophylaxis and therapy of epithelial tumours. Int Z Vitaminforsch 40(3):299–314
106. Krinsky NI (1989) Carotenoids as chemopreventive agents. Prev Med 18(5):592–602
107. Godic A, Poljšak B, Adamic M, Dahmane R (2014) The role of antioxidants in skin cancer prevention and treatment. Oxidative Med Cell Longev 2014. Article ID 860479:6
108. Fung TT, Hunter DJ, Spiegelman D, Colditz GA, Speizer FE, Willett WC (2002) Vitamins and carotenoids intake and the risk of basal cell carcinoma of the skin in women (United States). Cancer Causes Control 13(3):221–230
109. Fung TT, Spiegelman D, Egan KM, Giovannucci E, Hunter DJ, Willett WC (2003) Vitamin and carotenoid intake and risk of squamous cell carcinoma of the skin. Int J Cancer 103(1):110–115
110. McNaughton SA, Marks GC, Green AC (2005) Role of dietary factors in the development of basal cell cancer and squamous cell cancer of the skin. Cancer Epidemiol Biomark Prev 14(7):1596–1607
111. Greenberg ER, Baron JA, Stukel TA, Stevens MM, Mandel JS, Spencer SK et al (1990) A clinical trial of beta carotene to prevent basal-cell and squamous-cell cancers of the skin. N Engl J Med 323(12):789–795
112. Hennekens CH, Buring JE, Manson JE, Stampfer M, Rosner B, Cook NR et al (1996) Lack of effect of long-term supplementation with beta carotene on the incidence of malignant neoplasms and cardiovascular disease. N Engl J Med 334(18):1145–1149
113. Britton G, Liaaen-Jensen S, Pfander H (2017) Carotenoids: a colourful history. CaroteNature GmbH, Bern

Chapter 24
Application of Carotenoids in Cosmetics

May Bin-Jumah, Suaad S. Alwakeel, Marius Moga, Lavinia Buvnariu, Nicu Bigiu, and Muhammad Zia-Ul-Haq

24.1 Introduction

Carotenoids are natural pigmented compounds and abundantly present class of isoprenoids which are synthesized photosynthetically (plants) and non-photosynthetically (fungi & bacteria). Innately, despite from few species of aphids, carotenoids are not been synthesized in animals, therefore, animals require them through daily diet. Carotenoids are an important class of phytochemical that possesses proven antioxidant potential. Their use in food matrix retards the process of oxidation and helps in improving the shelf stability of the products. In the human body, carotenoids metabolized in myriad of chemical entities with multifarious properties including scavenging free radicals along with provision of vitamin A activity. Finally yet importantly, they improve the health of individual by improve the defense capabilities of the body along with providing protection many ailments. The classification of carotenoids is complex and many hypotheses are there in this regard. Based on chain length, cyclic and acyclic carotenoids are two categories, while presence and absence of oxygen, 600 known carotenoids could be broadly classified as non-oxygenated (carotenes) and oxygenated (xanthophylls) carotenoids. Xanthophylls (zeaxanthin & lutein) contain oxygen as a functional group in the form of epoxy, keto & hydroxyl other than their basic hydrocarbon structure,

M. Bin-Jumah (✉) · S. S. Alwakeel
College of Sciences, Department of Biology, Princess Nourah bint Abdulrahman University, Riyadh, Saudi Arabia
e-mail: mnbinjumah@pnu.edu.sa

M. Moga · L. Buvnariu · N. Bigiu
Faculty of Medicine, Transilvania University of Brasov, Brașov, Romania

M. Zia-Ul-Haq
Office of Research, Innovation and Commercialization, Lahore College for Women University, Lahore, Pakistan

© Springer Nature Switzerland AG 2021

M. Zia-Ul-Haq et al. (eds.), *Carotenoids: Structure and Function in the Human Body*, https://doi.org/10.1007/978-3-030-46459-2_24

whereas, carotenes only contain parent hydrocarbon structure for instance lycopene, α-carotene & β-carotene.

On the other hand, deficiency of carotenoids causes corneal ulcer, night blindness, conjunctival xerosis and keratomalacia eventually resulting in blindness. Other than this, owing to their potent UV-rays protection properties they are most commonly being used by cosmetic industry in formulation of various products.

Globally, mainly because of versatility in potent health modulating properties of carotenoids an estimated escalation in its demand has been reported from 1.5 billion dollars in 2014 to 1.8 billion dollars in 2019. Keeping in view the increased demand of carotenoid rich products numerous scientists has devoted their research in understanding the bioavailability, metabolism, and ultimate fate of this phytonutrient in human body. Various chemically synthesized products are being used in formulation of commercial carotenoids, only a few are made from carotenoids extracted from natural sources. Despite their nature of origin, both carotenoids either synthetic or natural are identical in their molecular structure. Synthetic carotenoids are more stable as compare to natural ones as they are designed to minimize oxidation. They are formulated and distributed in market as dispersions, colloids, colloidal suspensions, and emulsions so that their application in food products is much easier. In spite of these benefits, they possess carcinogenic& teratogenic properties and are reported to be toxic therefore increases hesitancy in consumer. Consequently, consumers have shifted their trend towards carotenoids extracted from natural sources. Change in preference of consumer towards carotenoids of natural origin has increased challenges for the manufacturers. As food matrix contains a variety of biomolecules therefore precise knowledge regarding efficient extraction methods is required for efficient recovery of carotenoids. This chapter gives a detailed overview on chemistry, biosynthesis, bioavailability, metabolism and identification & quantification of carotenoids.

The functional features of carotenoids in human body are dependent on various intrinsic and extrinsic factors that influence their absorption and metabolism. The external factors associated with their bioavailability include consumption of fats/oils, nutrient-nutrient interactions, chemistry of the compounds, and alcohol addiction and smoking. In contrast, concentration of retinoic acid and carotenoids are major internal factors responsible for their poor utilization. Moreover, internal metabolisms, body characteristic (hormones, age and gender) and diseased status are intriguing factors. The largest organ of human body is skin and its well-being is basis of a healthy life. Skin color affects attractiveness and is a perception of healthy human. Yellow skin color is a sign of good healthy [1]. Melanin and dietary carotenoids impart yellowness to skin [2]. Fruits and vegetable consumption containing carotenoids provide moisture to the skin and produce novel effects on overall health condition [3]. After absorption, carotenoids are delivered through the circulatory system to tissues being targeted. Carotenoids accumulates in skin layers e.g. epidermis, dermis, and stratum corneum. However, this accumulation depends upon carotenoids intake and their bioavailability from dietary source [4–6]. Carotenoids being the most abundant antioxidants in skin, quench free radicals thus reducing oxidative stress created by photodynamic reactions in the skin.

24.2 Carotenoids and Photo-Protection

Carotenoids offer protect from sunlight damages and erythema [7, 8]. They also defend from premature skin photo-aging comprising of pigmentation, wrinkle formation, inelasticity and skin dryness. Increased skin lycopene level decreases skin roughness substantially [9, 10]. Skin contains more lycopene and beta carotene (hydrocarbon carotenoids) than lutein and zeaxanthin. These hydrocarbon carotenoids substantially protect the skin from the sun [11].

Depending upon skin requirements, various diet supplements have been formulated (including dermo-cosmetic and dermatologic based products) for reduction of wrinkles and skin aging by scavenging free radicals [12]. The nutricosmetics perform antioxidant function as they possesses carotenoids antioxidants as active ingredients. The most significant antioxidants used in nutricosmetics are lutein, zeaxanthin, beta-carotene, lycopene, and astaxanthin [13]. Albert Kligman first presented the idea of cosmeceuticals in 1984 that skin color can be changed by applying products topically. These products cannot be placed in drugs or cosmetics category [14].

24.3 Xanthophyllic Carotenoids

Xanthophylls increases significantly skin (stratum corneum) hydration. Xanthophylls treatment enhances the efficacy of skin moisturization due to lutein and zeaxanthin penetration in the corneocyte membrane and intercellular lipids [15–17]. Since oxycarotenoids are not synthesized in our body, the only way to get them is by taking it from food enriched in lutein [18]. Oxycarotenoid belongs to a group of carotenoids called xanthophylls, which primarily includes AST, β-cryptoxanthin, canthaxanthin, lutein, and zeaxanthin. Topical and oral treatment with lutein and zeaxanthin for various cosmaceutical purposes have shown their remarkable significance in providing anti-oxidants effects, skin elasticity and skin layer hydration [19]. Intake of lutein and zeaxanthin isomers protects skin against oxidative damages and defense to blue-light damages [20].

The xanthophyllic carotenoids (lutein and zeaxanthin) scavenge reactive oxygen species due to their anti-oxidant property and can be applied orally and topically on the skin [19]. Upon exposure to UV light, free radical scavengers capability get reduced [21, 22]. The reduction in antioxidant potential can be managed by administrating antioxidants systematically or orally in the skin [23, 24]. Daily oral ingestion of antioxidants mixture containing lutein and zeaxanthin @6 mg and 0.3 mg respectively increases lipids on the skin surface and properly hydrates the skin [17].

24.4 Lutein in Cosmetics

Lutein is the most researched polar oxygenated xanthophyll due to its antioxidant capacity [25]. It is found on cell membrane lipids where it interrupts lipid peroxidation processes and scavenges free radicals formed during various biochemical reactions. Lutein containing formulations defend skin from photo-aging and skin damage [26, 27]. National Cancer Institute human observational study revealed that high dietary intake of lutein significantly decreases (44%) the melanoma risks [28]. Lutein imparts orange color to marigold flower due to this reason deciduous trees turn yellow in autumn [29–31]. Marigold flowers (*Tagetes erecta*) extract mainly produces a brand of lutein known as FloraGLO® lutein marketed by Kemin industry. FloraGLO® lutein can be taken by oral supplementation and can be applied topically. Dietary supplementation of lutein helps thickening of the skin (epidermal hyperplasia) and skin sunburn (apoptotic cell) formation [32]. Daily consumption of 10 mg of FloraGLO® lutein increases skin elasticity, hydration and skin lipids. Likewise, topical administration of FloraGLO® lutein also has beneficial effects for skin [18].

24.5 Beta Carotene

β-Carotene is the most commonly found carotenoid in palnts, accounting for 25–30% of the total carotenoid content of plants [33] or even more in some of them. Beta- carotene protects from skin sunburn caused upon exposure to sunlight [34]. It is well-known that wrinkle formation and skin sagging is due to increased metalloproteinase-9 contents in skin. β-carotene consumption prohibits metalloproteinase-9 activation and increases 5-α-hydroperoxide synthesis thus inhibiting skin sagging and wrinkle formation. It also protects from photosensitive disease i.e. erythropoietic protoporphyria and sunburn diseases [9, 35]. Topically applied cosmetic sunscreens contain beta carotene due to its property of improving skin health and providing long-term protection against phenomenon of sunburn reaction [36]. Orally consumed beta- carotene protection mechanism is different from topically applied beta-carotene cosmetics. It is a slow process and requires several weeks for the result. Sun protection factor (SPF) of oral supplemented beta-carotene is approximately 4 while SPF for topical sunscreen ranges from 10 to 40 and in some products its range may increase upto 90 [37, 38]. Skin photoprotection effect of oral supplemented beta- carotene is more homogenous than topical cosmetics [37]. Topical cosmetics are applied on selected skin areas where the consumer visualizes UV irradiation effects while orally supplemented protect all parts of skin till the whole day [39].

Oxidative damages is expressed as wrinkles, skin aging and actinic lentigines [17]. Beta-carotene cosmetics enhances skin turn-over the rate and skin regeneration. Skin tone is also improved as it adds glowing pigment upon topical application. Beta Carotene due to its tinting ability, is used in suntan creams and lotions. It heals skin scratches, protects against scarring and reduces skin irritation and itchiness. However topical application of beta carotene does not fulfill the claim of reversing skin aging signs. It is better to take antioxidants orally as compared to the topical application.

24.6 Lycopene

Lycopene is an isomer of beta-carotene and possess great free radical quenching capability. Free radical quenching potential depends on the number of conjugated double bonds a molecule can possess [40, 41]. Lycopene is open chain pigment and comprises of 11 trans-configurated conjugated double bonds [42]. Skin roughness correlates with the concentration of lycopene. The increase in anti-oxidant level significantly decreases skin roughness [10]. Lycopene neutralizes free radicals much efficiently as compared to other carotenoids in skin and therfor makes skin smooth [43]. Lycopene improves skin moisturization, skin texture, skin elasticity, and skin superficial structure [44]. Lycopene protects defensive mechanism against toxic and harmful effects of light and oxygen [45].

Lycopene is used by skin care industries for formulating several cosmetics. The collagen fibers supports dermis layers of skin layer and is replaced constantly although the replacement decreases due to some physiological changes i.e., age, corticoid hormone hyper secretion, vitamins deficiency and menopause which resulted a decrease in skin thickness (skin thinning) and removes dead cells of the skin. Inhibition or reduction of collagenase causes several skin damages. Lycopene cosmetics or lycopene containing compounds inhibits mucous membrane damages like erythema and skin roughness [46].

Lycopene improves the dull skin. Increase in the lycopene concentration significant reduces skin roughness [10]. Skin cancer can be protected by applying several lycopene containing products like jojoba oil which can easily be solubilized with lycopene.

Lycopene can be formulated in the form of the capsule by administrating as tomato oleoresin extracted from several tomatoes varieties (LYC-O-MATO) formulated by LycoRed Natural Products Industries Ltd. The formulated capsules contain lycopene (15 mg), phytoene (1.5 mg), beta-carotene (0.4 mg), phyto-fluene (1.4 mg), tocopherol (5 mg) and 1.5 mg phytoene [47]. Lycopene is the most commonly used sun screening and sunburn protection agent [48]. Lycopene protects against sunburn and cancer induced by progressive effect of sun exposure by minimizing the damaging effects induced by exposure to sunlight. Lycopene is commonly used in cosmetics formulation due to its extraordinary effects in regeneration, renewal and thickness of the skin [49]. Increase of lycopene concentration in skin is beneficial for the improvement of uneven and rough skin [10]. It is used as a face mask especially for dry skin as it provides the skin with softness and resistance, along with-it lycopene reduces the wrinkles. In order to protect skin against UV radiations, free radicals and other skin damaging and aging factors lycopene containing cosmetics and antioxidants were used and proved to have beneficial effects [50].

24.7 Astaxanthin

Aquaculture industry was the first to use Astaxanthin for treating pigmentation, later Astaxanthin got acceptance as food supplements in 1991 [51]. Dermatological effects involve repression of hyper-pigmentation [52] melanin synthesis inhibitions and skin photoaging inhibition [53]. Astaxanthin is helpful for the treatment of skin wrinkles, aging spots and to increase moisture contents of corneocytes. Several studies indicate the useful effects of Asthaxanthin in both male and female skin [54]. Unlike betacarotene and lycopene it does not have pro-oxidative nature [55]. Astaxanthin also inhibits melanin production and signs of skin photo-aging [53].

The property of accumulation in skin after oral intake marks it as best for skin health. Frequent application of astaxanthin cream (@0.7 mg/g) reduced wrinkles [56]. Pre-incubation with astaxanthin inhibits ultraviolet-A (UV-A)-induced declines and changes in cellular glutathione level and SOD activity respectively [57]. Topical or oral use of astaxanthin hinders the effects of UV-A radiation such as wrinkles [57, 58], Astaxanthin improves skin condition in both males and females [54]. It delays skin stiffening and decreases collagen against skin injury induced by UV-A [59, 60]. Astaxanthin dietary consumption improves elasticity and barrier integrity in photoaged facial skin [61].

In 1983 Kuhn and Sorenson first identified astaxanthin. It is an intricate molecule is very difficult to be synthesized and exceptionally expensive product usually cost the US $2000 [62]. Antioxidant property of astaxanthin is much greater than other carotenoids like beta- carotene, lutein, lycopene and alphatocopherol [63]. Astaxanthin taken through drinks provide preventive mechanism against several diseases like arteriosclerosis, ischemic heart disease, and inflammation which in turn leads to chronic heart diseases [64]. The oral supplemented material contains specified concentration of (AstaREAL® Oil), astaxanthin canola oil capsules and *H. Pluvialis* extract while external used products also designed with a specific concentration of, (H pluvialis extract, AstaTROL™- Hp, Astaxanthin by Fuji Chemical Industry Co, Ltd.) [54] Several experimental tests and assumptions were made on astaxanthin, the first test is the topically applied, cream test containing astaxanthin and other effective ingredients [56]. The second test includes a placebo controlled double blind study in which supplements of diet that contains astaxanthin and tocotrienol was used [65]. In third trial, only astaxanthin containing dietary supplement effects was reported in placebo controlled single bind study [66]. In both first and second case collagen fiber recovery undergoes improvement in elasticity. Significant Melanogenesis inhibition was also reported [52, 53]. By repressing the epidermal inflammation and melanocytes polymerization. Treatment with astaxanthin improves rough skin is the first discovery. Increase in moisture contents was observed in the result of improvement, especially in the dry skin. The moisture contents range of dry skin was about 20 μS whereas at first, its range was 12–15μS. In second study moisture contents in cheeks which was >17 μS at the very start was increased with the greater tendency. Treatment involves topical application cause deeper refinement of rough skin as compared to oral supplementation. The dead epidermal cell mainly comprised of corneocytes, treatment with astaxanthin, normalizes the condition of corneocyte and protects the differentiation of keratinocytes and oxidative damages of cornification like epidermal inflammation. Corneocyte can also be normalized by undergoing significant improvements in TWEL (A marker which undergoes barrier functionality among ab layer of corneocyte. Sebum oil was produced more in men as compared to women's excess production of sebum oil cause skin roughness and aging. Astaxanthin protects peroxidation of sebum oil in this way it helps in the reduction of skin roughness and aging. It was concluded that *H. pluvialis*can, derived astaxanthin undergoes improvement in all layers of skin by both topical and oral treatment. Orally supplemented astaxanthin cause skin improvement in both men and women.

Asian people use astaxanthin for skin lightening purpose. It reduces upto 40% melanin synthesis (melanogenesis), also reduces age spots, freckles and dark circles of eyes. Study suggested that AstaREAL astaxanthin (2 mg) administration for @ twice daily for 6 weeks reduced skin lines, wrinkles and enhanced skin moisturization and elasticity. Topical treatment is the single way to reverse the reactive oxygen species effects. Astaxanthin extracted from *Haematococcus* microalgae, protects each skin layer from reactive oxygen species (ROS) damages by penetrating in the skin. In addition to this astaxanthin have an amazing activity in blood flow enhancement i.e. skin circulation became much better which leads to increase in the turnover rate of cell and retention of water as well as enhanced skin elasticity. Astaxanthin protects skin from inflammation, oxidation, and damages because it can scavenge multiple radicals (ROS) at once. In short carotenoids have been proved helpful in cosmetics formulation, due to their anti-oxidants and cell proliferation property.

24.8 Conclusions

Carotenoids based cosmetics possess aesthetic appeal, increased topical absorption, performance and sensory characteristics. Epidemiological reports, health benefits and mechanistic studies have proved health effects of carotenoids as human cosmetics. Extensive *in-vivo* and *in-vitro* experimentation in order to understand the precise mechanism of action followed by well-controlled clinical trials to assess carotenoids as effective cosmetics. Generation of further knowledge involving the mechanisms of carotenoids will help to identify novel carotenoids, establish their clinical relevance and evaluate its potential application in the treatment of skin disorders.

Funding This research was funded by the Deanship of Scientific Research at Princess Nourah bint Abdulrahman University through the Fast-track Research Funding Program.

References

1. Stephen ID, Smith MJL, Stirrat MR, Perrett DI (2009) Facial skin coloration affects perceived health of human faces. Int J Primatol 30(6):845–857
2. Stephen ID, Coetzee V, Perrett DI (2011) Carotenoid and melanin pigment coloration affect perceived human health. Evol Hum Behav 32(3):216–227
3. Whitehead RD, Ozakinci G, Perrett DI (2014) A randomized controlled trial of an appearance-based dietary intervention. Health Psychol 33(1):99
4. Alaluf S, Heinrich U, Stahl W, Tronnier H, Wiseman S (2002) Dietary carotenoids contribute to normal human skin color and UV photosensitivity. J Nutr 132(3):399–403
5. Mayne ST, Cartmel B, Scarmo S, Lin H, Leffell DJ, Welch E et al (2010) Noninvasive assessment of dermal carotenoids as a biomarker of fruit and vegetable intake. Am J Clin Nutr 92(4):794–800
6. Canene-Adams K, Erdman JW (2009) Absorption, transport, distribution in tissues and bioavailability. In: Carotenoids. pp 115–148. Springer
7. Stahl W, Heinrich U, Wiseman S, Eichler O, Sies H, Tronnier H (2001) Dietary tomato paste protects against ultraviolet light–induced erythema in humans. J Nutr 131(5):1449–1451
8. Stahl W, Heinrich U, Jungmann H, Sies H, Tronnier H (2000) Carotenoids and carotenoids plus vitamin E protect against ultraviolet light–induced erythema in humans. Am J Clin Nutr 71(3):795–798
9. Eicker J, Kürten V, Wild S, Riss G, Goralczyk R, Krutmann J et al (2003) Betacarotene supplementation protects from photoaging-associated mitochondrial DNA mutation. Photochem Photobiol Sci 2(6):655–659
10. Darvin M, Patzelt A, Gehse S, Schanzer S, Benderoth C, Sterry W et al (2008) Cutaneous concentration of lycopene correlates significantly with the roughness of the skin. Eur J Pharm Biopharm 69(3):943–947
11. Scarmo S, Cartmel B, Lin H, Leffell DJ, Welch E, Bhosale P et al (2010) Significant correlations of dermal total carotenoids and dermal lycopene with their respective plasma levels in healthy adults. Arch Biochem Biophys 504(1):34–39
12. Rona C, Berardesca E (2008) Aging skin and food supplements: the myth and the truth. Clin Dermatol 26(6):641–647
13. Anunciato TP, da Rocha Filho PA (2012) Carotenoids and polyphenols in nutricosmetics, nutraceuticals, and cosmeceuticals. J Cosmet Dermatol 11(1):51–54
14. Kligman A (2005) The future of cosmeceuticals: an interview with Albert Kligman, MD, PhD. Dermatol Surg 31(7):890–891
15. Traikovich SS (1999) Use of topical ascorbic acid and its effects on photodamaged skin topography. Arch Otolaryngol Head Neck Surg 125(10):1091–1098
16. Fitzpatrick RE, Rostan EF (2002) Double-blind, half-face study comparing topical vitamin C and vehicle for rejuvenation of photodamage. Dermatol Surg 28(3):231–236
17. Morganti P, Bruno C, Guarneri F, Cardillo A, Del Ciotto P, Valenzano F (2002) Role of topical and nutritional supplement to modify the oxidative stress. Int J Cosmet Sci 24(6):331–339
18. Morganti P (2007) Where nutriceuticals meet cosmecuticals. J Appl Cosmetol 25(3):111
19. Palombo P, Fabrizi G, Ruocco V, Ruocco E, Fluhr J, Roberts R et al (2007) Beneficial long-term effects of combined oral/topical antioxidant treatment with the carotenoids lutein and

zeaxanthin on human skin: a double-blind, placebo-controlled study. Skin Pharmacol Physiol 20(4):199–210
20. Zastrow L, Groth N, Klein F, Kockott D, Lademann J, Renneberg R et al (2009) The missing link–light-induced (280–1, 600 nm) free radical formation in human skin. Skin Pharmacol Physiol 22(1):31–44
21. Shindo Y, Witt E, Han D, Epstein W, Packer L (1994) Enzymic and non-enzymic antioxidants in epidermis and dermis of human skin. J Investig Dermatol 102(1):122–124
22. Thiele JJ, Dreher F, Packer L (2002) Antioxidant defense systems in skin. J Toxicol Cutan Ocul Toxicol 21(1–2):119–160
23. Mathews-Roth M, Pathak M, Parrish J, Fitzpatrick T, Kass E, Toda K et al (1972) A clinical trial of the effects of oral beta-carotene on the responses of human skin to solar radiation. J Investig Dermatol 59(4):349–353
24. Aust O, Stahl W, Sies H, Tronnier H, Heinrich U (2005) Supplementation with tomato-based products increases lycopene, phytofluene, and phytoene levels in human serum and protects against UV-light-induced erythema. Int J Vitam Nutr Res 75(1):54–60
25. Lu S, Li L (2008) Carotenoid metabolism: biosynthesis, regulation, and beyond. J Integr Plant Biol 50(7):778–785
26. Morganti P, Bruno C, Fabrizi G, Valenzano F, Del Ciotto P, Morganti G (2004) The antioxidant network of skin and eye. Efficacy of carotenoids. J Appl Cosmetol 22(3):133–142
27. Morganti P (2004) Dry skin and moisturization. SÖFW-J 130(7):29–36
28. Millen AE, Dodd KW, Subar AF (2004) Use of vitamin, mineral, nonvitamin, and nonmineral supplements in the United States: the 1987, 1992, and 2000 National Health Interview Survey results. J Am Diet Assoc 104(6):942–950
29. Greaves JA. Plants as phyto-phactories: connecting the Iowa value chain
30. Daniell H, Streatfield SJ, Wycoff K (2001) Medical molecular farming: production of antibodies, biopharmaceuticals and edible vaccines in plants. Trends Plant Sci 6(5):219–226
31. Hood EE, Witcher DR, Maddock S, Meyer T, Baszczynski C, Bailey M et al (1997) Commercial production of avidin from transgenic maize: characterization of transformant, production, processing, extraction and purification. Mol Breed 3(4):291–306
32. Lee EH, Faulhaber D, Hanson KM, Ding W, Peters S, Kodali S et al (2004) Dietary lutein reduces ultraviolet radiation-induced inflammation and immunosuppression. J Investig Dermatol 122(2):510–517
33. AM M. Carotenoides. Medellín: Universidad de Antioquia (2001). http://farmacia.udea.edu.co/~ff/carotenoides2001. Accessed May 2018
34. Köpcke W, Krutmann J (2008) Protection from sunburn with β-carotene—a meta-analysis. Photochem Photobiol 84(2):284–288
35. Mathews-Roth M (1984) Treatment of erythropoietic protoporphyria with beta-carotene. Photo-Dermatology 1(6):318–321
36. Stahl W, Krutmann J (2006) Systemische photoprotektion durch karotinoide. Hautarzt 57(4):281–285
37. Lim HW (2017) Clinical photomedicine. Routledge
38. Gilchrest BA, Krutmann J (2006) Skin aging. Springer
39. Albanes D, Heinonen OP, Taylor PR, Virtamo J, Edwards BK, Rautalahti M et al (1996) Alpha-tocopherol and beta-carotene supplements and lung cancer incidence in the alpha-tocopherol, beta-carotene cancer prevention study: effects of base-line characteristics and study compliance. J Natl Cancer Inst 88(21):1560–1570. https://doi.org/10.1093/jnci/88.21.1560
40. Foote CS, Denny RW (1968) Chemistry of singlet oxygen. VII. Quenching by. Beta.-carotene. J Am Chem Soc 90(22):6233–6235
41. Foote CS, Chang YC, Denny RW (1970) Chemistry of singlet oxygen. X. Carotenoid quenching parallels biological protection. J Am Chem Soc 92(17):5216–5218
42. Campbell JK, Engelmann NJ, Lila MA, Erdman JW Jr (2007) Phytoene, phytofluene, and lycopene from tomato powder differentially accumulate in tissues of male fisher 344 rats. Nutr Res 27(12):794–801

43. Di Mascio P, Kaiser S, Sies H (1989) Lycopene as the most efficient biological carotenoid singlet oxygen quencher. Arch Biochem Biophys 274(2):532–538
44. Stahl W, Sies H (2012) β-Carotene and other carotenoids in protection from sunlight. Am J Clin Nutr 96(5):1179S–1184S
45. Paiva SA, Russell RM (1999) β-Carotene and other carotenoids as antioxidants. J Am Coll Nutr 18(5):426–433
46. Ribaya-Mercado JD, Garmyn M, Gilchrest BA, Russell RM (1995) Skin lycopene is destroyed preferentially over β-carotene during ultraviolet irradiation in humans. J Nutr 125(7):1854–1859
47. Walfisch Y, Walfisch S, Agbaria R, Levy J, Sharoni Y (2003) Lycopene in serum, skin and adipose tissues after tomato-oleoresin supplementation in patients undergoing haemorrhoidectomy or peri-anal fistulotomy. Br J Nutr 90(4):759–766
48. Singh P, Rani B, Chauhan A, Maheshwari R (2012) Lycopene's antioxidant activity in cosmetics meadow. Int Res J Pharm 3(1):46–47
49. Viljanen K, Sundberg S, Ohshima T, Heinonen M (2002) Carotenoids as antioxidants to prevent photooxidation. Eur J Lipid Sci Technol 104(6):353–359
50. Lowe NJ (1996) Sunscreens: development: evaluation, and regulatory aspects. CRC Press
51. Miki W (1991) Biological functions and activities of animal carotenoids. Pure Appl Chem 63(1):141–146
52. Yamashita E (1995) Suppression of post-UVB hyperpigmentation by topical astaxanthin from krill. Fragrance J 14:180–185
53. Arakane K (2002) Superior skin protection via astaxanthin. Carotenoid Sci 5:21–24
54. Tominaga K, Hongo N, Karato M, Yamashita E (2012) Cosmetic benefits of astaxanthin on humans subjects. Acta Biochim Pol 59(1)
55. Martin H, Ruck C, Schmidt M, Sell S, Beutner S, Mayer B et al (1999) Chemistry of carotenoid oxidation and free radical reactions. Pure Appl Chem 71(12):2253–2262
56. Seki T, Sueki H, Kono H, Suganuma K, Yamashita E (2001) Effects of astaxanthin from Haematococcus pluvialis on human skin-patch test; skin repeated application test; effect on wrinkle reduction. Fragrance J 12:98–103
57. Lyons NM, O'Brien NM (2002) Modulatory effects of an algal extract containing astaxanthin on UVA-irradiated cells in culture. J Dermatol Sci 30(1):73–84
58. Suganuma K, Nakajima H, Ohtsuki M, Imokawa G (2010) Astaxanthin attenuates the UVA-induced up-regulation of matrix-metalloproteinase-1 and skin fibroblast elastase in human dermal fibroblasts. J Dermatol Sci 58(2):136–142
59. Hama S, Takahashi K, Inai Y, Shiota K, Sakamoto R, Yamada A et al (2012) Protective effects of topical application of a poorly soluble antioxidant astaxanthin liposomal formulation on ultraviolet-induced skin damage. J Pharm Sci 101(8):2909–2916
60. Santos SD, Cahú TB, Firmino GO, de Castro CC, Carvalho LB Jr, Bezerra RS et al (2012) Shrimp waste extract and astaxanthin: rat alveolar macrophage, oxidative stress and inflammation. J Food Sci 77(7):H141–H1H6
61. Yoon H-S, Cho HH, Cho S, Lee S-R, Shin M-H, Chung JH (2014) Supplementing with dietary astaxanthin combined with collagen hydrolysate improves facial elasticity and decreases matrix metalloproteinase-1 and-12 expression: a comparative study with placebo. J Med Food 17(7):810–816
62. Olaizola M (2003) Commercial development of microalgal biotechnology: from the test tube to the marketplace. Biomol Eng 20(4–6):459–466
63. Naguib YM (2000) Antioxidant activities of astaxanthin and related carotenoids. J Agric Food Chem 48(4):1150–1154
64. Miki W, Hosoda K, Kondo K, Itakura H (1998) Astaxanthin-containing drink. Japanese patent. 10155459
65. Yamashita E (2002) Cosmetic benefit of the supplement health food combined astaxanthin and tocotrienol on human skin. Food Style 21(6):112
66. Yamashita E (2005) The effects of a dietary supplement containing astaxanthin on skin condition. Food Style 21 9(9):72

Chapter 25
Carotenoids in Women and Infant Health

Suaad S. Alwakeel, May Bin-Jumah, Khansa Imam, Marius Moga, and Nicu Bigiu

25.1 Introduction

Nutritional requirements escalate during pregnancy and lactation and inadequate supply of nutrients during these crucial life stages impacts negatively impact on the mother and developing child. It is well-known that young, pregnant and breastfeeding women are at risk of having shortage of essential nutrients including carotenoids and vitamin A. The inadequate supply of carotenoids and vitamin A is a risk factor for pregnant and lactating women. The intake of both carotenoids and vitamin A should be increased to 1/3rd times during pregnancy and breastfeeding tenure. Nutrition during pregnancy is essential for development of fetus, pregnancy outcomes, and child and mother health before and after birth [1]. Carotenoids are a key player in outcome of pregnancy and in inhibition of various problems during pregnancy caused by increased oxidative stress [2–4]. Carotenoids assist in sustaining good health during childhood as these are involved in the development and maintenance mechanisms of cognition and vision. One group says that enhanced demand can be met from a balanced diet and maternal reserves of a well-nourished mother. Infant and developing fetus totally depend on mother intake of carotenoids for provitamin A requirements. However, there is limited research except suggestions of provitamin A supplementation in pregnancy and lactation. Encouraging consumption of carotenoid rich foods during pregnancy and breastfeeding should continue

S. S. Alwakeel (✉) · M. Bin-Jumah
College of Sciences, Department of Biology, Princess Nourah bint Abdulrahman University, Riyadh, Saudi Arabia
e-mail: ssalwakeel@pnu.edu.sa

K. Imam
Department of Education, University of Education, Jauharabad Campus, Jauharabad, Pakistan

M. Moga · N. Bigiu
Faculty of Medicine, Transilvania University of Brasov, Brasov, Romania

© Springer Nature Switzerland AG 2021

M. Zia-Ul-Haq et al. (eds.), *Carotenoids: Structure and Function in the Human Body*, https://doi.org/10.1007/978-3-030-46459-2_25

simultaneously with further research explaining the significance of them during these critical life periods.

25.2 Prevention of Pregnancy Complication by Carotenoids

Hormonal concentrations and energy metabolism change during pregnancy. The placenta becomes operative leading to an increased level of Reactive Oxygen Species (ROS). Increased ROS generation, inappropriate development of placenta and inadequate antioxidant defense leads to oxidative stress. Increased ROS levels damage cells and tissues structurally and functionally, thereby acting as pro-inflammatory agents. This may cause pregnancy abnormalities and complication. Carotenoids protect humans from oxidative stress and resulting complications [1, 2].

Carotenoids intake in pregnant women varies from country to country. It is well established now that oxidative stress increases the risk of pregnancy-initiated hypertension, impulsive abortion, preeclampsia, gestational diabetes mellitus (GDM), and preterm birth [2, 5]. The shielding effects of carotenoids against preeclampsia have been well-studied [6, 92]. Preeclampsia affected pregnant women suffer from increased stress and inflammation and decreased level of carotenoids [7–9]. Risk of incidence of preeclampsia decreases with increased lutein plasma concentration [10]. Higher blood concentration of β-carotene is linked with decreased risk of pre-eclampsia up to 50% in Zimbabwe pregnant women (n = 359) [11].

Out of 3 clinical studies in Indian women regarding influence of lycopene intake (@2 mg/day) up to 2nd trimester of pregnancy, I study indicated very small decrease risk of development of preeclampsia [12], while remaining 2 studies did not show any link between the two [13, 14].

Oxidative stress is also linked with GDM [15]. Antioxidant potential of plasma in GDM-women is very less as compared to healthy pregnant women [16]. An intervention study suggested that intake of carotenoid supplements did not decrease the hydroperoxide concentration in pregnant women serum [17]. However, a substantial difference in oxidative stress in a 24 women group (12 suffering from GDM) values between newborn to untreated mothers at 2 h of life and newborns to mothers treated with lutein which disappeared after 48 h. The women had taken lutein supplements (10 mg) and zeaxanthin (2 mg), however since no placebo or randomization was used and due to lack of anthropometric and demographic data of participating women, the strength of results are not strong enough for convincing [17]. Mortality and morbidity ate increases due to preterm birth. Increased oxidative stress has been observed in premature births [18]. Premature rupture of membranes (PROM) leads to about 30–40% preterm birth .Further, increased status of oxidative stress biomarkers have been observed in amniotic fluid during the development of such pregnancy [19]. Mothers who give birth to premature newborns, have decreased carotenoid serum concentrations. Increased concentration of lycopene, β-cryptoxanthin and α- and β-carotene decrease risk of preterm births [20]. Preterm birth risk depends upon the dietary pattern of pregnant mothers. Lower intake of α- and β-carotene are associated with increased risk of premature births in USA [21, 22]. Subjects fed lycopene supplements (@2 mg/day) in a randomized controlled trial from the first gestation trimester indicated a weak but substantial continuation of gestation period in subjects [12]. A diet containing carotenoids prevents preterm births in women [23].

Infants who are small for gestational age (SGA) are mostly hyotrophic and few of them are born form pregnancies with problems of intrauterine growth rate (IUGR). Low antioxidant potential and higher oxidative stress biomarkers are characteristics of both IUGAR and SGA [24, 25]. Risk of birth of SGA babies decreases in mothers having increased plasma concentration of α- and β-cryptoxanthin, lutein, zeaxanthin and β-carotene [26]. Decreased levels of carotenoids have been detected in breast milk and serum of mothers giving birth to babies having IUGR [27]. Another study revealed that lycopene intake (@2 mg/day during 16–20 Hbd) lowers the incidence of happening of IUGR [12].Study has revealed that maternal carotenoids level or their concentration in cord blood did not influence the birth parameters of newborn babies [25]. These results were also confirmed in an intervention study (randomized placebo-controlled trial) involving β-carotene intake from first gestation trimester to 3rd month post-birth [28]. However, minor negative impact of β-carotene intake on birth weight needs further investigation [28]. Similarly opposing result of lycopene intake on prevalence of low birth weight was observed in another study although it was a weak study [13]. A positive correlation was found between β-carotene intake in 4th month of pregnancy and head perimeter at time of birth however no association was detected between seasonal differences in nutrients supplementation and birth characteristics [29].

25.2.1 Maternal-Fetal Transfer of Carotenoids

Fetal development in terms of its vitamin A needs greatly dependent on mother intake during pregnancy, as the liver store of the child only lasts for a few days and can be speedily emptied due to abrupt strains for the development. Furthermore, the insufficient intake of vitamin A by the mother during pregnancy can also influence the postpartum supply of the child through breast milk.

Absorption and transportation of carotenoids is analogous to that of fats in human body. Carotenoids when absorbed from small intestine initially circulate through the lymphatic system, and then make a complex with the chylomicron offcuts in the blood system and finally transported to the liver. In liver, this combination of carotenoid and chylomicron remnants is modified, can be stored storage, or may be conveyed back to the blood circulation along with lipoproteins. Xanthophyll transportation is assisted mostly by high-density lipoprotein (HDL) and to a smaller extent by very low-density lipoprotein (VLDL) while carotenes are carried by low-density lipoprotein (LDL). Association of various types of lipoproteins to carotenoids and the quantity and shares between their receptors in several tissues regulate the changes in saturation level of various organs in carotenoids. For this reason, testes, liver and adrenal glands contain the largest quantity of carotene since LDL receptor is mainly located at these 3 places. Since central nervous system (CNS) and nervous tissue of eye retina are hub of HDL, therefore, xanthophylls are preferably transported there [22, 30, 31]. Carotenoids transportation take place by various proteins like lactoglobulin and albumin [32–34].The mode of transfer, metabolism and exploitation of carotenoids in fetus development is still unclear. Increased concentration of maternal lipoproteins during pregnancy helps uptake of carotenoids by placenta. Cord blood contains lipoprotein fractions in order of HDL > LDL > VLDL.

Unlike in adults, the HDL of cord blood is involved in β-carotene transport more than LDL (55% vs. 45%) [32–35] Animal model study suggested that administration of β-carotene controlled transcription, MTP activity and apoB thus increasing β-carotene transfer to placenta [36]. Vitamin A shortage reduces carotenoid levels in placenta which may be due to their reduced translocation to placenta [37].Optimum concentration of carotenoids (276 ng) and total albumin (1.42 mg) was found nearly 20–22 Hbd in tissues of vitreous body of aborted fetuses. Carotenoids and albumin levels were highest during weeks 16–17 and 17 correspondingly of prenatal progress. Carotenoid quantity reduced slowly after that and by 31 week of gestation, was lower than detection threshold [33].

25.2.2 Carotenoid Status in the Newborn

Carotenoid concentration in mother serum and cord blood is the key biomarker of carotenoid status. Majority of studies suggest that carotenoid level in newborn directly relies on their concentration in mother which is mostly many times lower. Carotenoid level in mother blood is many times higher than carotenoid quantity in carotenoid blood [25–27, 38–43]. Minor differences between polar carotenoids level in mother blood and cord blood were noted .The reason is that HDL is the major lipoprotein fraction in initial lifespan of fetus till 1st week post-birth of newborn [41]. Variation in carotenoid levels especially β-carotene and other pro-vitamin A carotenoids between fetus and mother cannot be ascribed to restrictions in placental transport of carotenoids. Inadequate β-carotene storage capacity, transformation of β-carotene to retinol and storage of retinol esters in liver and intensive fetal metabolism may be potential causes of it [44, 45].

It was observed that plasma zeaxanthin of newborn and mother was linked with macular pigment optical density (MPOD) in infants. However, no such correlation existed for the main macular pigment lutein. Transformation of lutein to meso-zeaxanthin detected in macula of retina by an immature enzyme system is reasonable for it thus highlighting importance of zeaxanthin in macular pigment development in infants [27].The newborn nutritional rank relies upon the delivery week of newborn. It may be associated with dynamics and mechanism of fetus growth which is maximum during pregnancy third trimester. Examination of lutein and metabolite 3′-oxolutein quantity in cord blood of full-term and preterm neonates suggested that lutein level peaks at start of third trimester and starts decreasing from 37 Hbd attaining its minimum level at 41–42 Hbd [46]. Lutein quantity in male newborns and newborns delivered from numerous pregnancies was found less [47]. Another element that reduces carotenoids concentration especially β-carotene in cord blood is smoking during pregnancy [38, 48, 49].

It is well-known that oxidative stress increases during child delivery. Child birth through planned cesarean section is associated with elevated oxidative stress although this experiment results are considered poor [50, 51]. No differences were observed in β-carotene level with regard to type of child birth [38] however declined lutein level was observed in cord blood of newborns delivered by cesarean section [46].

Since infant's eye lens is more transparent than adult, therefore their eyes are more vulnerable to blue light damage. Enhanced zeaxanthin and lutein consumption defends human eye by antioxidant and blue light filtration capacity of these carotenoids. β-carotene intake reduced weights of newborns in A study conducted among 450 women from a New Zealand clinic [29]. More than 30 carotenoids in various forms have been reported in breast milk [52], However, 6 carotenoids including lutein, lycopene,α- and β-carotene, Zeaxanthin, and β-cryptoxanthin have been more frequently reported and detected.

Global Exploration of Human Milk study was conducted in China, USA and Mexico (n = 240 samples, n = 60 donors). Milk collected longitudinally in 2,4,13 and 26 weeks postpartum from 20 women. Results indicated that β-carotene, lutein, lycopene and β-cryptoxanthin contents of milk are substantially linked with maternal and neonatal plasma carotenoids level. Provitamin A potential of β-cryptoxanthin, α- and β-carotene and to some extent α-cryptoxanthin play a significant role for mom and neonate. Since vitamin A is needed for immune and visual development for which provitamin A carotenoids are key source in developing tissues. Lack of standardized protocols for determination of carotenoids in milk as well as reference values of carotenoids for each country are 2 major problems [53].

Milk composition is best guide for appraisal of infant carotenoid requirement. Levels of carotenoids and tocopherols were assessed in milk of 509 healthy mothers. Lactation phase, regional variations and socio-economic factors were linked with human milk contents in health Chinese mothers [54].

25.2.3 Carotenoids in Breast Milk

It is well-established fact that breast milk composition varies depending upon phase of time of day, rate of breastfeeding during the year, lactation period and time of single breastfeeding. It also varies from person to person, gestation week, birth week and rate of breast emptying. Dietary practices of lactating women influences nutrient concentration of breast milk including carotenoids [55–59].

Changes occur very frequently in breast milk fat and single breastfeeding phase and lactation phase are mainly responsible for it. Fat contents of mature milk and colostrum are 4.1 g/100 mL and 2.6 g/100 mL respectively [55, 56, 58, 60]. Lutein, α- and β-carotene, β-cryptoxanthin, lycopene and zeaxanthin are major carotenoids of breast milk [53, 60–65]. Average contents of carotenoids, i.e. lycopene, β-cryptoxanthin, β-carotene and lutein in breast milk were 33.7, 33.8, 49.4 and 114.4 nmol/L respectively [53]. Provitamin A carotenoids contents were 62% while other carotenoids accounted for less quantity [61]. The results differences of various studies may be due to different protocols for milk fat extraction and different units of milk fat expression, variations in seasons of breast milk collection and dietary pattern of females [53, 64, 66]. Just like fat contents, foremilk contains less carotenoids (by 25%) than hind milk [64]. Breast milk composition even varies within the same day at various times [63, 64]. Carotenoids contents decreases with duration of breastfeeding. Variations in carotenoids level can be between 82.7–91.3% (for lycopene) and 32.5–52.0% for zeaxanthin [53, 60, 65, 79, 85, 115]. Carotenoid concentration greatly decreases between 2nd and 4th week of breastfeeding but remains same between 4th and 16th week of breastfeeding [53, 67]. Non-provitamin A carotenoids (lutein and lycopene) [53] and less polar carotenes are extra prone to variations [60].

Dietary practices of mother greatly influences carotenoid contents of breast milk. Variation in breast milk composition noted in numerous studies may be due to various dietary practices of populations scrutinized [53, 61, 17, 79]. Milk carotenoid concentration changes easily with dietary intake patterns.

A 3-day intervention study with intake of tomato paste or carrot paste containing 15 mg β-carotene and lycopene respectively by 26 women led to enhanced concentration of both carotenoids in milk even after first day of intervention. Optimum β-carotene and lycopene contents comprising 200% and 130% of initial value was observed in subjects after 2nd and 4th day of intervention respectively [68]. Lutein, β-carotene and zeaxanthin contents increased 2.6X,1.7X and 2.7X after intake of chlorella supplements during 16–20 Hbd [69].The carotenoids content in breast milk is 10 or sometimes even 20 times less than their plasma level and degree of association between two varies depending upon carotenoid or population investigated. Differences also exist in proportion of milk carotenoids to their plasma concentration. A randomized placebo-controlled trial suggested that a 6 week lutein intake (6 or 12 mg/day) enhanced in its serum level by about 170% and 250% and in breast milk by 140% and 250% correspondingly. Lutein serum level amplified by 180% and 330% in case of 6 mg/day and 12 mg/day after lutein consumption [70]. Serum and breast milk carotenoid ratio is variable and amounts to 560–600% for β-cryptoxanthin, 270–300 for α- and β-carotene and 133% for lutein. Intensive turnover and uptake by newborn tissues may be the reason for decreased lutein value [53]. Serum carotenoids contents and their composition in plasma lipoprotein is also observed. Variations in ratio of breast milk carotenoid to serum carotenoid with regard to polarity indicates that a unique mechanism operates for carotenoids transfer to milk which is different from fat transfer mechanism [60]. Breast milk components can be secreted by 5 different mechanism, including 1 para-cellular mechanism with assistance of tight-junction connections and 4 trans-cellular pathways (milk fat transport, intracellular vesicle transport, membrane pathway and transport through

Golgi apparatus and milk secretion via secretive cells through exocytosis) [53, 71]. Carotenoids transfer to breast milk may consist of preferred uptake by lipoproteins and intracellular transport. All kinds of lipoproteins are found in breast tissue but mammary alveolar epithelium cells prefer HDL fraction [53, 65, 72].

25.2.4 Infant Feeding Method and Infant carotenoid Status

Carotenoids cannot be produced de novo in humans. Further carotenoids occur in traces in infant formulas, hence method of infant feeding plays a major role in determination of carotenoid level in infant [27, 67, 73]. Totally breastfed Infants possess superior nutritional status than mixed or artificially fed infants. Carotenoid contents of artificially fed infants after many months are many times less than vales noticed in breastfed infants and sometimes even less than detection threshold [8, 45, 70, 73–76]. It highlights the significance of fortifying infant formulas with carotenoids especially lutein to boost the nutritional status of artificially fed newborn infants. Lutein enriched infant formulas are well-tolerated by neonates and enhance their dietetic level [77]. Experimental studies have established that rhesus macaques fed with a mixture of carotenoids-supplemented foods not only enhances the saturation level of lutein in brain tissue but also elevates zeaxanthin, lycopene and β-carotene from negligible and below detection level to substantial level thus highlighting the importance of lutein for developing brain increases the saturation degree [78]. It is important to increase lutein level 4–5 times than its level in breast milk because of limited bio-accessibility of lutein from infant formulations [73, 77, 79].

25.2.5 Carotenoids and Infant Health and Development

Lot of studies suggest importance of carotenoids in infant health and growth due to their well-accepted health benefits and preferred uptake by breast milk and fetus [31, 80]. These health benefits in infants are due to their antioxidant potential and their role in cognitive and visual development [80, 81].

25.2.6 Visual Development

Macular pigment consist of lutein, zeaxanthin, meso-zeaxanthin, isomers of zeaxanthin and lutein derivatives [82]. Eye carotenoid contents are not equally distributed. Carotenoid optimum quantity is detected in central foveal region and they decline with growing space from fovea being 100 times less in peripheral area. Carotenoid contents are present in different ratios in various retinal regions eg.in central foveal region zeaxanthin dominates while in peripheral region lutein dominates [83–85].

Age also affects macular pigment structure and density. Infants have very less or undetectable mezzo-zeananthin contents and contradictory lutein:zeaxnanthin ratio until 2 years of age. It may be due to immaturity of enzymes responsible for transformation of lutein to mezzo-zeaxnathin [27, 83–85]. Further premature infants have undetectable MPOD and depleted carotenoids due to decreased parental development [86]. Infant MPOD depends upon mom carotenoid level and after birth, baby feeding method is important for maintaining carotenoid concentration [27]. Breastfed neonates have increased MPOD than artificially fed infants. Recent research indicates that early carotenoid exposition can be a major indicator of MPOD in adulthood [87, 88]. Oxidative damage is a potential threat for eye retina due to extreme metabolic activity, extensive vascularity and increase LC PUFA concentration [89]. Newborns suffer more risk of damage due to undeveloped autoregulation of blood flow within the choroid and their extra permeable lens that permits to pass increased quantities of energetic short-wave light [90, 91].

Premature newborns are extra prone to oxidative stress which may cause progress of premature retinopathy [92]. Macular carotenoids shield retina by: (1) absorption of 40–90% of incident blue light which defends retia from photo-injuries [93]; (2) antioxidant potential i.e. a combination of zeaxanthin, lutein and mezo-zeaxanthin may extinguish more singlet oxygen than separate carotenoids [94]; (3) neuroprotective capacity [95] and (4) anti-apoptotic and anti-inflammatory capacity of carotenoids [96]. Zeaxanthin and lutein also help in transfer and managing of visual information by: (1) increasing the gap junctional communication between neurons and glia [97, 98],(2) stabilizing microtubules in cytoskeleton [99]; uplifting visual parameters including light scattering and scoptopic noise [100]; and (4) involvement in oxygen consumption from foveal region [76, 101].Studies also confirm important part of macular carotenoids in appropriate visual and eye development [89]. Animal studies suggest the carotenoids importance for appropriate retina development [102, 103].Proper cognitive development depends upon appropriate visual development of infants [80, 104]. Further lutein status in retina is associated with brain lutein contents in humans [105], primates [106] and to cognitive functioning in children and old age population [107].

25.2.7 Brain and Cognitive Development

Nervous tissues capture carotenoids especially with lutein constituting 59% of all carotenoids in infants while its share in adults is only 31% [78]. In brain, carotenoids are carotenoids are concentrated in frontal and occipital cortex, hippocampus and areas linked with cognitive process. Optimum lutein contents are detected in cellular membranes and axon terminals of neurons, and structure of neurons's cell membranes and amonal axonal projections rely upon brain areas in which they are present [99].Lutein functions in nervous tissue by (1) increasing intracellular communications [97, 108];(2) neuroprotective potential [109]; (3) modifying cell membranes including their stability, fluidity, ion exchange, and oxygen diffusion [108];and (4) contribution in metabolic pathways of brain [110]. However another study did not endorse the effect of infant breastfeeding method on carotenoids contents in brains of full term neonates [111]. The brains of pre-term newborns have decreased carotenoids level than full-term newborns [107, 111]. Studies suggest the importance of carotenoids in cognitive functions and their intake enhances cognitive functioning in adults and elderly people [107].Current study held among 55 exclusively breastfed infants from USA suggested that increased lutein contents in breastmilk of 3 months can enhance cognitive performance in infants at 6 months of life [112].The relationship observed can be due to lutein transport to brain through high density lipo-proteins (HDL) [113].

25.2.8 Preterm Infants

Infants are extra prone to oxidative stress due to increased oxygen intrauterine environment and extreme metabolic activity [81]. Undeveloped antioxidant defense mechanism, allied disorders and extensive medical mechanism make preterm newborns even more susceptible to oxidative damage [81]. Lutein supplementation in infants @ 0.28 mg at 6 and 36 h of life suggested that it can protect against oxidative stress by enhancing biological antioxidant system and diminishing oxidative stress [114]. Lutein intake is well-accepted by preterm neonates and can decrease the danger of occurrence of diseases linked with pre-term birth e.g. necrotizing enterocolitis (NEC),retinopathy of prematurity (ROP) and bronchopulmonary dysplasia (BPA) [115].

25.2.9 Long-Term Studies in Infants and Children

Carotenoids supplementation during pregnancy can have long lasting effects on infants. Study performed among moms of babies suffering from sporadic retinoblastoma and health controls suggested that insufficient intake of fruits and vegetables and zeaxanthin and lutein derived from them can enhance the possibility of sporadic retinoblastoma in children [116]. Carotenoids particularly provitamin A can change the various parameters in progressing immune system including natural killer cell functions or T-cell proliferation [117]. Supplementation of mothers by zeaxanthin and lutein decreased the possibility of respiratory infections in 2 years old kids [118]. A study involving 763 mother-infants dyads in Japan suggested that β-carotene intake during pregnancy reduced possibility of infantile eczema but not wheeze [119]. A cross-sectional study was conducted well-fed, healthy kids aged 5.75 years living in Vancouver, Canada probing lutein supplementation. Results did not endorse lutein importance in cognitive performance evaluated by Peabody

Picture Vocabulary Test (PPVT) and Kaufman Assessment Battery (KABCC-II) [120]. The absence of impact of lutein on cognitive function can be due to inclusion of well-nourished population while principal functional impact of lutein can be critical for those with comparative insufficiency; choice of weak biomarker (plasma contents) for lutein in brain and improper choice of cognitive assessments for computing diet impact on brain growth [121]. Another study on 2044 healthy Dutch children did not confirm the supposition that lutein supplementation in initial life of 1 year has positive effects on later anthropometrics, body measures and cardiometabolic health at age of 6 years [122].

25.2.10 Safety of Carotenoids and the Intake Recommendations for Pregnant Women and Infants

Although carotenoids act as antioxidant however their antioxidant potential declines when oxygen pressure mounts probably because of auto-oxidative processes [123]. This characteristic may be potential reason of reports of adverse health effects of increased carotenoid intake. ABTS clinical trial performed in male smokers of Finland indicated that increased duration (5–8 years) intake of β-carotene (@20 mg/day) led to 18% enhancement of lung cancer incidence and an overall 8% increased mortality [124]. Another study indicated that this intake augmented post-trial of a first nonfatal myocardial infraction [125]. Some studies involving non-smokers however did not suggest any relationship between carotenoid intake (β-carotene @ 50 mg/day) and increased possibility of cardiovascular mortality or morbidity [126]. β-carotene oxidation by cigarette smoke can be responsible for carotenoid adverse health effects which supported formation of toxic β-carotene oxidation products [127]. Carotenodermia having specific yellow discoloration of skin may also occur due to more lutein uptake [128]. Only 1 study suggested adverse health effects of lycopene intake (2 mg/day since 15.7 Hbd). No other study reveals adverse health effects of carotenoid consumption in pregnant women, newborns and infants although these studies were not long-term or used low doses of carotenoids. Since carotenoids are not essential nutrients so there is not any recommended dietary intake for any population group. However, it is being increasingly suggested that dietary intake recommendations should be fixed for zeaxanthin and lutein. Optimal lutein intake for eye health is supposed to be 6 mg/day and no toxic symptoms appeared in clinical trial even when intake was 3 times more than 6 mg [80]. No recommended intake of lutein exists for breastfeeding women or infants, however EFSA confirmed that lutein @ 250 μg/L is safe in infants' formulas [128]. Diet rich in fruits and vegetables being carotenoid source should be suggested to pregnant and lactating women and infants as they have good effects on health and no toxic effects have been reported [30]. Complete breast-feeding upto 6 months should be recommended for infants and overall breastfeeding period should be at least 2 years as breast milk is better carotenoid source than formula and it leads to numerous health results [129].

25.3 Conclusions

Epidemiological, interventional, clinical, animal and human studies strongly suggest relationship between sufficient carotenoid intake, from fruits and vegetables or from supplements and formulas and decreased chances of age-related decline of cognitive performance and chronic disease. However, no strong relationship has yet been confirmed between carotenoid intake and decrease risk of pregnancy and allied disorders. Even studies exist which suggest inverse relationship between beneficial effects of carotenoid intake and pregnancy problems. Since most of RCT studies were small and biased therefore further research is needed in this area. Lutein is one of most-researched carotenoid for its beneficial effects. Its sufficient intake during neonatal period is of vital importance due to its antioxidant and anti-inflammatory activities and its role in development of nervous and vision system. In initial life phases, visual stimuli are necessary factors for brain development cognitive processes of child. As most infant formulas lack carotenoids, therefore carotenoid level of newborns and infants relies upon the nutritional status of mother and her breast-feeding method. Zeaxanthin and lutein role in development of visual and nervous systems has been observed only in animal models. Further studies are required to establish role of carotenoid in newborn, infant, child and pregnant women health.

Funding This research was funded by the Deanship of Scientific Research at Princess Nourah bint Abdurrahman University through the Fast-track Research Funding Program.

References

1. Berti C, Cetin I, Agostoni C, Desoye G, Devlieger R, Emmett P et al (2016) Pregnancy and infants' outcome: nutritional and metabolic implications. Crit Rev Food Sci Nutr 56(1):82–91
2. Agarwal A, Aponte-Mellado A, Premkumar BJ, Shaman A, Gupta S (2012) The effects of oxidative stress on female reproduction: a review. Reprod Biol Endocrinol 10(1):49
3. Al-Gubory KH, Fowler PA, Garrel C (2010) The roles of cellular reactive oxygen species, oxidative stress and antioxidants in pregnancy outcomes. Int J Biochem Cell Biol 42(10):1634–1650
4. Thorne-Lyman AL, Fawzi WW (2012) Vitamin A and carotenoids during pregnancy and maternal, neonatal and infant health outcomes: a systematic review and meta-analysis. Paediatr Perinat Epidemiol 26:36–54
5. Pereira AC, Martel F (2014) Oxidative stress in pregnancy and fertility pathologies. Cell Biol Toxicol 30(5):301–312
6. Mol BW, Roberts CT, Thangaratinam S, Magee LA, De Groot CJ, Hofmeyr GJ (2016) Pre-eclampsia. Lancet 387(10022):999–1011

7. Barden A, Beilin LJ, Ritchie J, Croft KD, Walters BN, Michael CA (1996) Plasma and urinary 8-iso-prostane as an indicator of lipid peroxidation in pre-eclampsia and normal pregnancy. Clin Sci 91(6):711–718
8. Staff AC, Halvorsen B, Ranheim T, Henriksen T (1999) Elevated level of free 8-iso-prostaglandin F2α in the decidua basalis of women with preeclampsia. Am J Obstet Gynecol 181(5):1211–1215
9. Mert I, Sargın Oruc A, Yuksel S, Cakar ES, Buyukkagnıcı U, Karaer A et al (2012) Role of oxidative stress in preeclampsia and intrauterine growth restriction. J Obstet Gynaecol Res 38(4):658–664
10. Cohen JM, Kramer MS, Platt RW, Basso O, Evans RW, Kahn SR (2015) The association between maternal antioxidant levels in midpregnancy and preeclampsia. Am J Obstet Gynecol 213(5):695.e1–695.e13
11. Williams MA, Woelk GB, King IB, Jenkins L, Mahomed K (2003) Plasma carotenoids, retinol, tocopherols, and lipoproteins in preeclamptic and normotensive pregnant Zimbabwean women. Am J Hypertens 16(8):665–672
12. Sharma J, Kumar A, Kumar A, Malhotra M, Arora R, Prasad S et al (2003) Effect of lycopene on pre-eclampsia and intra-uterine growth retardation in primigravidas. Int J Gynecol Obstet 81(3):257–262
13. Banerjee S, Jeyaseelan S, Guleria R (2009) Trial of lycopene to prevent pre-eclampsia in healthy primigravidas: results show some adverse effects. J Obstet Gynaecol Res 35(3):477–482
14. Antartani R, Ashok K (2011) Effect of lycopene in prevention of preeclampsia in high risk pregnant women. J Turkish Ger Gynecol Assoc 12(1):35
15. Lappas M, Hiden U, Desoye G, Froehlich J, SH-d M, Jawerbaum A (2011) The role of oxidative stress in the pathophysiology of gestational diabetes mellitus. Antioxid Redox Signal 15(12):3061–3100
16. Grissa O, Atègbo J-M, Yessoufou A, Tabka Z, Miled A, Jerbi M et al (2007) Antioxidant status and circulating lipids are altered in human gestational diabetes and macrosomia. Transl Res 150(3):164–171
17. Lorenzoni F, Giampietri M, Ferri G, Lunardi S, Madrigali V, Battini L et al (2013) Lutein administration to pregnant women with gestational diabetes mellitus is associated to a decrease of oxidative stress in newborns. Gynecol Endocrinol 29(10):901–903
18. Clerici G, Slavescu C, Fiengo S, Kanninen T, Romanelli M, Biondi R et al (2012) Oxidative stress in pathological pregnancies. J Obstet Gynaecol 32(2):124–127
19. Longini M, Perrone S, Vezzosi P, Marzocchi B, Kenanidis A, Centini G et al (2007) Association between oxidative stress in pregnancy and preterm premature rupture of membranes. Clin Biochem 40(11):793–797
20. Kramer MS, Kahn SR, Platt RW, Genest J, Rozen R, Chen MF et al (2009) Antioxidant vitamins, long-chain fatty acids, and spontaneous preterm birth. Epidemiology 20:707–713
21. Huber JC, Brender JD, Zheng Q, Sharkey JR, Vuong AM, Shinde MU et al (2013) Maternal dietary intake of nitrates, nitrites and nitrosamines and selected birth defects in offspring: a case-control study. Nutr J 12(1):34
22. Kotake-Nara E, Nagao A (2011) Absorption and metabolism of xanthophylls. Mar Drugs 9(6):1024–1037
23. Englund-Ögge L, Brantsæter AL, Sengpiel V, Haugen M, Birgisdottir BE, Myhre R et al (2014) Maternal dietary patterns and preterm delivery: results from large prospective cohort study. BMJ 348:g1446
24. Saker M, Mokhtari NS, Merzouk SA, Merzouk H, Belarbi B, Narce M (2008) Oxidant and antioxidant status in mothers and their newborns according to birthweight. Eur J Obstet Gynecol Reprod Biol 141(2):95–99
25. Weber D, Stuetz W, Bernhard W, Franz A, Raith M, Grune T et al (2014) Oxidative stress markers and micronutrients in maternal and cord blood in relation to neonatal outcome. Eur J Clin Nutr 68(2):215–222

26. Cohen J, Kahn S, Platt R, Basso O, Evans R, Kramer M (2015) Small-for-gestational-age birth and maternal plasma antioxidant levels in mid-gestation: a nested case–control study. BJOG Int J Obstet Gynaecol 122(10):1313–1321
27. Henriksen BS, Chan G, Hoffman RO, Sharifzadeh M, Ermakov IV, Gellermann W et al (2013) Interrelationships between maternal carotenoid status and newborn infant macular pigment optical density and carotenoid status. Invest Ophthalmol Vis Sci 54(8):5568–5578
28. Christian P, Klemm R, Shamim AA, Ali H, Rashid M, Shaikh S et al (2013) Effects of vitamin A and β-carotene supplementation on birth size and length of gestation in rural Bangladesh: a cluster-randomized trial. Am J Clin Nutr 97(1):188–194
29. Watson P, McDonald B (2007) Seasonal variation of nutrient intake in pregnancy: effects on infant measures and possible influence on diseases related to season of birth. Eur J Clin Nutr 61(11):1271–1280
30. Krinsky NI, Johnson EJ (2005) Carotenoid actions and their relation to health and disease. Mol Asp Med 26(6):459–516
31. Fiedor J, Burda K (2014) Potential role of carotenoids as antioxidants in human health and disease. Nutrients 6(2):466–488
32. Herrera E, Amusquivar E, Lopez-Soldado I, Ortega H (2006) Maternal lipid metabolism and placental lipid transfer. Horm Res Paediatr 65.(Suppl. 3:59–64
33. Panova I, Tatikolov A, Sukhikh G (2007) Correlation between the content of albumin and carotenoids in human vitreous body during prenatal development. Bull Exp Biol Med 144(5):681–683
34. Bhattacharya N, Stubblefield PG (2016) Human fetal growth and development: first and second trimesters. Springer, Cham
35. Mango M, Tamai H, Ogihara T, Mino M (1992) Distribution of circulating β-carotene in human plasma lipoproteins. J Nutr Sci Vitaminol 38(5):405–414
36. Costabile BK, Kim Y-K, Iqbal J, Zuccaro MV, Wassef L, Narayanasamy S et al (2016) β-Apo-10′-carotenoids modulate placental microsomal triglyceride transfer protein expression and function to optimize transport of intact β-carotene to the embryo. J Biol Chem 291(35):18525–18535
37. Gomes MM, Saunders C, Ramalho A (2010) Placenta: a possible predictor of vitamin A deficiency. Br J Nutr 103(9):1340–1344
38. Scaife AR, McNeill G, Campbell DM, Martindale S, Devereux G, Seaton A (2006) Maternal intake of antioxidant vitamins in pregnancy in relation to maternal and fetal plasma levels at delivery. Br J Nutr 95(4):771–778
39. Masters ET, Jedrychowski W, Schleicher RL, Tsai W-Y, Tu Y-H, Camann D et al (2007) Relation between prenatal lipid-soluble micronutrient status, environmental pollutant exposure, and birth outcomes. Am J Clin Nutr 86(4):1139–1145
40. Karakilcik AZ, Aksakal M, Baydas G, Sozen R, Ayata A, Simsek M (1996) Plasma -carotene concentrations in pregnancies, new born infants and their mothers. J Pak Med Assoc 46:77–79
41. Yeum K-J, Ferland G, Patry J, Russell RM (1998) Relationship of plasma carotenoids, retinol and tocopherols in mothers and newborn infants. J Am Coll Nutr 17(5):442–447
42. Kiely M, Cogan P, Kearney P, Morrissey P (1999) Concentrations of tocopherols and carotenoids in maternal and cord blood plasma. Eur J Clin Nutr 53(9):711–715
43. Franke A, Lai J, Morrison C, Pagano I, Li X, Halm B et al (2013) Coenzyme Q10, carotenoid, tocopherol, and retinol levels in cord plasma from multiethnic subjects in Hawaii. Free Radic Res 47(9):757–768
44. Goldberg JS (2012) Monitoring maternal beta carotene and retinol consumption may decrease the incidence of neurodevelopmental disorders in offspring. Clin Med Insights Reprod Health 6:CMRH. S8372
45. Spiegler E, Kim Y-K, Wassef L, Shete V, Quadro L (2012) Maternal–fetal transfer and metabolism of vitamin A and its precursor β-carotene in the developing tissues. Biochim Biophys Acta (BBA)-Mol Cell Biol Lipids 1821(1):88–98

46. Picone S, Ritieni A, Fabiano A, Troise AD, Graziani G, Paolillo P et al (2012) Arterial cord blood lutein levels in preterm and term healthy newborns are sex and gestational age dependent. Clin Biochem 45(18):1558–1563
47. Schulz C, Engel U, Kreienberg R, Biesalski HK (2007) Vitamin A and β-carotene supply of women with gemini or short birth intervals. Eur J Nutr 46(1):12
48. Chelchowska M, Ambroszkiewicz J, Gajewska J, Laskowska-Klita T, Leibschang J (2011) The effect of tobacco smoking during pregnancy on plasma oxidant and antioxidant status in mother and newborn. Eur J Obstet Gynecol Reprod Biol 155(2):132–136
49. de Barros Silva SS, de Carvalho Rondó PH, Erzinger GS (2005) β-Carotene concentrations in maternal and cord blood of smokers and non-smokers. Early Hum Dev 81(4):313–317
50. Hung T-H, Chen S-F, Lo L-M, Li M-J, Yeh Y-L (2011) The associations between labor and delivery mode and maternal and placental oxidative stress. Reprod Toxicol 31(2):144–150
51. Noh EJ, Kim YH, Cho MK, Kim JW, Kim JW, Byun YJ et al (2014) Comparison of oxidative stress markers in umbilical cord blood after vaginal and cesarean delivery. Obstet Gynecol Sci 57(2):109–114
52. Khachik F, Spangler CJ, Smith JC, Canfield LM, Steck A, Pfander H (1997) Identification, quantification, and relative concentrations of carotenoids and their metabolites in human milk and serum. Anal Chem 69(10):1873–1881
53. Lipkie TE, Morrow AL, Jouni ZE, McMahon RJ, Ferruzzi MG (2015) Longitudinal survey of carotenoids in human milk from urban cohorts in China, Mexico, and the USA. PLoS One 10(6):e0127729
54. Xue Y, Campos-Giménez E, Redeuil KM, Lévèques A, Actis-Goretta L, Vinyes-Pares G et al (2017) Concentrations of carotenoids and tocopherols in breast milk from urban Chinese mothers and their associations with maternal characteristics: a cross-sectional study. Nutrients 9(11):1229
55. Ballard O, Morrow AL (2013) Human milk composition: nutrients and bioactive factors. Pediatr Clin 60(1):49–74
56. Gidrewicz DA, Fenton TR (2014) A systematic review and meta-analysis of the nutrient content of preterm and term breast milk. BMC Pediatr 14(1):216
57. Kent JC, Mitoulas LR, Cregan MD, Ramsay DT, Doherty DA, Hartmann PE (2006) Volume and frequency of breastfeedings and fat content of breast milk throughout the day. Pediatrics 117(3):e387–ee95
58. Khan S, Hepworth AR, Prime DK, Lai CT, Trengove NJ, Hartmann PE (2013) Variation in fat, lactose, and protein composition in breast milk over 24 hours: associations with infant feeding patterns. J Hum Lact 29(1):81–89
59. Mitoulas LR, Kent JC, Cox DB, Owens RA, Sherriff JL, Hartmann PE (2002) Variation in fat, lactose and protein in human milk over 24h and throughout the first year of lactation. Br J Nutr 88(1):29–37
60. Macias C, Schweigert FJ (2001) Changes in the concentration of carotenoids, vitamin A, alpha-tocopherol and total lipids in human milk throughout early lactation. Ann Nutr Metab 45(2):82–85
61. Canfield LM, Clandinin MT, Davies DP, Fernandez MC, Jackson J, Hawkes J et al (2003) Multinational study of major breast milk carotenoids of healthy mothers. Eur J Nutr 42(3):133–141
62. de Azeredo VB, Trugo NM (2008) Retinol, carotenoids, and tocopherols in the milk of lactating adolescents and relationships with plasma concentrations. Nutrition 24(2):133–139
63. Giuliano AR, Neilson EM, Yap H-H, Baier M, Canfield LM (1994) Quantitation of and inter/intra-individual variability in major carotenoids of mature human milk. J Nutr Biochem 5(11):551–556
64. Jackson JG, Lien EL, White SJ, Bruns NJ, Kuhlman CF (1998) Major carotenoids in mature human milk: longitudinal and diurnal patterns. J Nutr Biochem 9(1):2–7

65. Schweigert FJ, Bathe K, Chen F, Büscher U, Dudenhausen JW (2004) Effect of the stage of lactation in humans on carotenoid levels in milk, blood plasma and plasma lipoprotein fractions. Eur J Nutr 43(1):39–44
66. Jackson JG, Zimmer JP (2007) Lutein and zeaxanthin in human milk independently and significantly differ among women from Japan, Mexico, and the United Kingdom. Nutr Res 27(8):449–453
67. Jewell V, Mayes C, Tubman T, Northrop-Clewes C, Thurnham D (2004) A comparison of lutein and zeaxanthin concentrations in formula and human milk samples from Northern Ireland mothers. Eur J Clin Nutr 58(1):90–97
68. Haftel L, Berkovich Z, Reifen R (2015) Elevated milk β-carotene and lycopene after carrot and tomato paste supplementation. Nutrition 31(3):443–445
69. Nagayama J, Noda K, Uchikawa T, Maruyama I, Shimomura H, Miyahara M (2014) Effect of maternal Chlorella supplementation on carotenoid concentration in breast milk at early lactation. Int J Food Sci Nutr 65(5):573–576
70. Sherry CL, Oliver JS, Renzi LM, Marriage BJ (2014) Lutein supplementation increases breast milk and plasma lutein concentrations in lactating women and infant plasma concentrations but does not affect other carotenoids. J Nutr 144(8):1256–1263
71. Shennan D, Peaker M (2000) Transport of milk constituents by the mammary gland. Physiol Rev 80(3):925–951
72. Schweigert F (1990) Effect of gestation and lactation on lipoprotein pattern and composition in dairy cows 1. J Anim Physiol Anim Nutr 63(1–5):75–83
73. Bettler J, Zimmer JP, Neuringer M, DeRusso PA (2010) Serum lutein concentrations in healthy term infants fed human milk or infant formula with lutein. Eur J Nutr 49(1):45–51
74. Chan GM, Chan MM, Gellermann W, Ermakov I, Ermakova M, Bhosale P et al (2013) Resonance Raman spectroscopy and the preterm infant carotenoid status. J Pediatr Gastroenterol Nutr 56(5):556
75. Sommerburg O, Meissner K, Nelle M, Lenhartz H, Leichsenring M (2000) Carotenoid supply in breast-fed and formula-fed neonates. Eur J Pediatr 159(1–2):86–90
76. Zimmer JP, Hammond BR Jr (2007) Possible influences of lutein and zeaxanthin on the developing retina. Clin Ophthalmol 1(1):25
77. Capeding R, Gepanayao CP, Calimon N, Lebumfacil J, Davis AM, Stouffer N et al (2010) Lutein-fortified infant formula fed to healthy term infants: evaluation of growth effects and safety. Nutr J 9(1):22
78. Jeon S, Neuringer M, Johnson EE, Kuchan MJ, Pereira SL, Johnson EJ et al (2017) Effect of carotenoid supplemented formula on carotenoid bioaccumulation in tissues of infant rhesus macaques: a pilot study focused on lutein. Nutrients 9(1):51
79. Lipkie TE, Banavara D, Shah B, Morrow AL, McMahon RJ, Jouni ZE et al (2014) Caco-2 accumulation of lutein is greater from human milk than from infant formula despite similar bioaccessibility. Mol Nutr Food Res 58(10):2014–2022
80. Johnson EJ (2014) Role of lutein and zeaxanthin in visual and cognitive function throughout the lifespan. Nutr Rev 72(9):605–612
81. Perrone S, Longini M, Marzocchi B, Picardi A, Bellieni CV, Proietti F et al (2010) Effects of lutein on oxidative stress in the term newborn: a pilot study. Neonatology 97(1):36–40
82. Loskutova E, Nolan J, Howard A, Beatty S (2013) Macular pigment and its contribution to vision. Nutrients 5(6):1962–1969
83. Bernstein PS, Khachik F, Carvalho LS, Muir GJ, Zhao D-Y, Katz NB (2001) Identification and quantitation of carotenoids and their metabolites in the tissues of the human eye. Exp Eye Res 72(3):215–223
84. Bone R, Landrum J, Fernandez L, Tarsis S (1988) Analysis of the macular pigment by HPLC: retinal distribution and age study. Invest Ophthalmol Vis Sci 29(6):843–849
85. Bone RA, Landrum JT, Frides LM, Gomes CM, Kilburn MD, Menenes E et al (1997) Distribution of lutein and zeaxanthin stereoisomers in the human retina. Exp Eye Res 64(2):211–218

86. Hendrickson A, Possin D, Vajzovic L, Toth CA (2012) Histologic development of the human fovea from midgestation to maturity. Am J Ophthalmol 154(5):767–78. e2
87. Liu Z, Meyers KJ, Johnson EJ, Snodderly M, Tinker L, Wallace R et al (2015) Exposure to lutein in infancy via breast milk and later life macular pigment optical density. Invest Ophthalmol Vis Sci 56(7):192
88. Neuringer M, Bone R, Jeffrey B, Bettler J, Zimmer J, Wallace P et al (2009) Lutein in breast-milk and infant formula: effects on serum lutein, macular pigment and visual function. Invest Ophthalmol Vis Sci 50(13):1707
89. Lien EL, Hammond BR (2011) Nutritional influences on visual development and function. Prog Retin Eye Res 30(3):188–203
90. Dillon J, Zheng L, Merriam JC, Gaillard ER (2004) Transmission of light to the aging human retina: possible implications for age related macular degeneration. Exp Eye Res 79(6):753–759
91. Hardy P, Dumont I, Bhattacharya M, Hou X, Lachapelle P, Varma DR et al (2000) Oxidants, nitric oxide and prostanoids in the developing ocular vasculature: a basis for ischemic retinopathy. Cardiovasc Res 47(3):489–509
92. Jewell VC, Northrop-Clewes CA, Tubman R, Thurnham DI (2001) Nutritional factors and visual function in premature infants. Proc Nutr Soc 60(2):171–178
93. Landrum JT, Bone RA, Krinsky N, Mayne S, Sies H (2004) Mechanistic evidence for eye diseases and carotenoids. Oxid Stress Dis 13:445–472
94. Li B, Ahmed F, Bernstein PS (2010) Studies on the singlet oxygen scavenging mechanism of human macular pigment. Arch Biochem Biophys 504(1):56–60
95. Ozawa Y, Sasaki M, Takahashi N, Kamoshita M, Miyake S, Tsubota K (2012) Neuroprotective effects of lutein in the retina. Curr Pharm Des 18(1):51–56
96. Li S-Y, Fung FK, Fu ZJ, Wong D, Chan HH, Lo AC (2012) Anti-inflammatory effects of lutein in retinal ischemic/hypoxic injury: in vivo and in vitro studies. Invest Ophthalmol Vis Sci 53(10):5976–5984
97. Bernstein PS, Balashov NA, Tsong ED, Rando RR (1997) Retinal tubulin binds macular carotenoids. Invest Ophthalmol Vis Sci 38(1):167–175
98. Stahl W, Nicolai S, Briviba K, Hanusch M, Broszeit G, Peters M et al (1997) Biological activities of natural and synthetic carotenoids: induction of gap junctional communication and singlet oxygen quenching. Carcinogenesis 18(1):89–92
99. Crabtree DV, Ojima I, Geng X, Adler AJ (2001) Tubulins in the primate retina: evidence that xanthophylls may be endogenous ligands for the paclitaxel-binding site. Bioorg Med Chem 9(8):1967–1976
100. Hammond BR Jr (2008) Possible role for dietary lutein and zeaxanthin in visual development. Nutr Rev 66(12):695–702
101. Liew SM, Gilbert CE, Spector TD, Mellerio J, Van Kuijk FJ, Beatty S et al (2006) Central retinal thickness is positively correlated with macular pigment optical density. Exp Eye Res 82(5):915–920
102. Leung IY-F, Sandstrom MM, Zucker CL, Neuringer M, Snodderly DM (2004) Nutritional manipulation of primate retinas, II: effects of age, n–3 fatty acids, lutein, and zeaxanthin on retinal pigment epithelium. Invest Ophthalmol Vis Sci 45(9):3244–3256
103. Malinow M, Feeney-Burns L, Peterson LH, Klein ML, Neuringer M (1980) Diet-related macular anomalies in monkeys. Invest Ophthalmol Vis Sci 19(8):857–863
104. Knickmeyer RC, Gouttard S, Kang C, Evans D, Wilber K, Smith JK et al (2008) A structural MRI study of human brain development from birth to 2 years. J Neurosci 28(47):12176–12182
105. Vishwanathan R, Schalch W, Johnson EJ (2016) Macular pigment carotenoids in the retina and occipital cortex are related in humans. Nutr Neurosci 19(3):95–101
106. Vishwanathan R, Neuringer M, Snodderly DM, Schalch W, Johnson EJ (2013) Macular lutein and zeaxanthin are related to brain lutein and zeaxanthin in primates. Nutr Neurosci 16(1):21–29

107. Vishwanathan R, Iannaccone A, Scott TM, Kritchevsky SB, Jennings BJ, Carboni G et al (2014) Macular pigment optical density is related to cognitive function in older people. Age Ageing 43(2):271–275
108. Stahl W, Sies H (2001) Effects of carotenoids and retinoids on gap junctional communication. Biofactors 15(2–4):95–98
109. Nataraj J, Manivasagam T, Thenmozhi AJ, Essa MM (2016) Lutein protects dopaminergic neurons against MPTP-induced apoptotic death and motor dysfunction by ameliorating mitochondrial disruption and oxidative stress. Nutr Neurosci 19(6):237–246
110. Lieblein-Boff JC, Johnson EJ, Kennedy AD, Lai C-S, Kuchan MJ (2015) Exploratory metabolomic analyses reveal compounds correlated with lutein concentration in frontal cortex, hippocampus, and occipital cortex of human infant brain. PLoS One 10(8):e0136904
111. Vishwanathan R, Kuchan MJ, Sen S, Johnson EJ (2014) Lutein and preterm infants with decreased concentrations of brain carotenoids. J Pediatr Gastroenterol Nutr 59(5):659–665
112. Cheatham CL, Sheppard KW (2015) Synergistic effects of human milk nutrients in the support of infant recognition memory: an observational study. Nutrients 7(11):9079–9095
113. Blusztajn JK, Slack BE, Mellott TJ (2017) Neuroprotective actions of dietary choline. Nutrients 9(8):815
114. Perrone S, Tei M, Longini M, Santacroce A, Turrisi G, Proietti F et al (2014) Lipid and protein oxidation in newborn infants after lutein administration. Oxidative Med Cell Longev 2014:1–7
115. Romagnoli C, Tirone C, Persichilli S, Gervasoni J, Zuppi C, Barone G et al (2010) Lutein absorption in premature infants. Eur J Clin Nutr 64(7):760–761
116. Orjuela MA, Titievsky L, Liu X, Ramirez-Ortiz M, Ponce-Castaneda V, Lecona E et al (2005) Fruit and vegetable intake during pregnancy and risk for development of sporadic retinoblastoma. Cancer Epidemiol Prev Biomarkers 14(6):1433–1440
117. Rühl R (2007) Effects of dietary retinoids and carotenoids on immune development*: symposium on 'nutrition influences on developmental immunology'. Proc Nutr Soc 66(3):458–469
118. Litonjua AA, Rifas-Shiman SL, Ly NP, Tantisira KG, Rich-Edwards JW, Camargo CA Jr et al (2006) Maternal antioxidant intake in pregnancy and wheezing illnesses in children at 2 y of age. Am J Clin Nutr 84(4):903–911
119. Miyake Y, Sasaki S, Tanaka K, Hirota Y (2010) Consumption of vegetables, fruit, and antioxidants during pregnancy and wheeze and eczema in infants. Allergy 65(6):758–765
120. Mulder KA, Innis SM, Rasmussen BF, Wu BT, Richardson KJ, Hasman D (2014) Plasma lutein concentrations are related to dietary intake, but unrelated to dietary saturated fat or cognition in young children. J Nutr Sci 3:e11
121. Hammond BR (2014) Lutein and cognition in children. J Nutr Sci 3:e53
122. Leermakers ET, Kiefte-de Jong JC, Hofman A, Jaddoe VW, Franco OH (2015) Lutein intake at the age of 1 year and cardiometabolic health at the age of 6 years: the generation R study. Br J Nutr 114(6):970–978
123. Eghbaliferiz S, Iranshahi M (2016) Prooxidant activity of polyphenols, flavonoids, anthocyanins and carotenoids: updated review of mechanisms and catalyzing metals. Phytother Res 30(9):1379–1391
124. Group A-TBCCPS (1994) The effect of vitamin E and beta carotene on the incidence of lung cancer and other cancers in male smokers. N Engl J Med 330(15):1029–1035
125. Törnwall ME, Virtamo J, Korhonen PA, Virtanen MJ, Taylor PR, Albanes D et al (2004) Effect of α-tocopherol and β-carotene supplementation on coronary heart disease during the 6-year post-trial follow-up in the ATBC study. Eur Heart J 25(13):1171–1178
126. Hennekens CH, Buring JE, Manson JE, Stampfer M, Rosner B, Cook NR et al (1996) Lack of effect of long-term supplementation with beta carotene on the incidence of malignant neoplasms and cardiovascular disease. N Engl J Med 334(18):1145–1149
127. Rahman A, Bokhari SSI, Waqar MA (2001) Beta-carotene degradation by cigarette smoke in hexane solution in vitro. Nutr Res 21(6):821–829

128. Authority EFS (2008) Safety, bioavailability and suitability of lutein for the particular nutritional use by infants and young children-scientific opinion of the panel on dietetic products, nutrition and allergies. EFSA J 6(11):823
129. Victora CG, Bahl R, Barros AJ, França GV, Horton S, Krasevec J et al (2016) Breastfeeding in the 21st century: epidemiology, mechanisms, and lifelong effect. Lancet 387(10017):475–490

Chapter 26
Provitamin A Carotenoids

Shagufta Perveen, Sara Zafar, Naeem Iqbal, and Muhammad Riaz

26.1 Introduction

Of several hundred carotenoids present in nature, only 50 are considered to be found in human diet [1]. Among these 6 carotenoids such as α-carotene, β-carotene, β-cryptoxanthin, lutein, lycopene and zeaxanthin are more abundant of which first three carotenoids are considered provitamin A carotenoids due to conversion into vitamin A (retinol) in human body [2–4]. Of these three naturally occurring carotenoids β-carotene is most important constituents in found [5–8]. Furthermore, carotenoids with 3-dehydroretinylidene and β-retinylidene end groups can easily converted into retinoids and therefore called having provitamin A activity. These carotenoids also give orange, red and yellow color to the vegetables and fruits [9, 10].

Of total carotenoids intake plant sources particularly fruits and vegetables contribute about 80–90% [11]. Carotenoids impart characteristic colors to various fruits and vegetables [12]. The colored vegetables and fruits such as papaya and carrots provide vitamin A to human [13, 14]. Beta-carotene is a valuable carotenoid in carrots and sweet potato [15]. Additional significant sources of provitamin A are squashes and pumpkins; they have much importance due to their production, longer shelf life and accessibility in whole world [16–18]. Leafy vegetables have more provitamin A than fruits, but fruits are commonly well putative by adults and youngsters [7]. In developing countries mango and papaya which are known as tropical fruits considered as significant source of provitamins A carotenoids [17, 19]. Crude red palm oil is well-thought-out the major plant source of provitamin A globally. It achieved from mesocarp of the oil palm (*Elaeis guineensis*) tree [20, 21]. Latest studies show that oil of buriti palm has tenfold more provitamin A than red palm tree oil [22]. A list of plant sources of β-carotenoids is given in the Table 26.1.

S. Perveen (✉) · S. Zafar · N. Iqbal
Department of Botany, Government College University, Faisalabad, Pakistan

M. Riaz
Department of Pharmacy, Shaheed Benazir Bhutto University, Sheringal, Pakistan

© Springer Nature Switzerland AG 2021
M. Zia-Ul-Haq et al. (eds.), *Carotenoids: Structure and Function in the Human Body*, https://doi.org/10.1007/978-3-030-46459-2_26

Table 26.1 Common dietary sources of carotenoids

Food	Common name	Scientific name	β-carotene content (μg/g edible portion)	References
Fruit	Indian dates	*Phoenix sp.*	30 ± 3	[19, 23]
	Buriti	*Mauritia vinifera*	360 ± 32	[24]
	Tucuma	*Astrocaryun vulgaris*	107 ± 31	[25]
Oil	Red palm oil Tenera	*Elaeis guineensis*	363	[26]
Non-leafy vegetable	Carrot	*Daucus carota*	34 ± 15	[19, 23]
Wild leaves	Pahadi pig weed	*Amaranthus blitum*	92	[18]
	Mint	*Mentha aquatica*	37	
	Lamb's quarter	*Chenopodium album*	33	
	Prostrate yarba-detegra	*Eclipta prostrate*	36	
	Edible fern	*Dryopteris cochleate*	44	
	Amaranth	*Amaranthus leucocarpus*	33	
Leafy vegetables	Agathi keerai	*Sesbania grandiflora*	66 ± 22	[27]
	Puliara keerai	*Oxalis corniculata*	60 ± 22	
	Kasini keerai	*Raphanus sp.*	56 ± 12	
	Nerringi keerai	*Tribulus terrestris*	74 ± 9	
	Manathakkli keerai	*Solanum nigrum*	70 ± 37	
	Sirukeerai	*Amaranthus polygonoides*	53 ± 14	
	Pacharisi keerai	*Euphorbia hirta*	62 ± 6	
	Muringa keerai	*Moringa oleifera*	52 ± 22	
	Chakravathy keerai	*Amaranthus sp.*	39 ± 3	
	Modakathan keerai	*Cardiospermum helicacabum*	35 ± 7	
	Arai keerai	*Amaranthus tristis*	34 ± 10	
	Pulichai keerai	*Hibiscus cannabinus*	34 ± 20	
	Molai keerai	*Amaranthus sp.*	31 ± 9	
	Poonangani keerai	*Alternanthera sessilis*	52 ± 5	
	Seemai Poonangani	*Alternanthera sp.*	46 ± 8	
	Nadirsan keerai	*Portulaca wightiana*	36 ± 10	
	Knol khol keerai	*Brassica oleracea* var. *caulorapa*	35 ± 7	
	Spinach	*Spinacea oleracea*	53	[28]
	Panna keerai	*Celosia sp.*	34 ± 10	
	Thandu keerai	*Amaranthus gangeticus*	32 ± 11	
	Macaxeira	*Manihot esculenta*	129	[29]
	Vinagreira roxa	*Hibiscus acetosela*	78	
	Taioba	*Colocasia esculenta*	57	
	Mentruz	*Chenopodium ambrosioides*	55	

(continued)

Table 26.1 (continued)

Food	Common name	Scientific name	β-carotene content (μg/g edible portion)	References
	Chinese kale	*Brassica chinensis*	49	
	African spinach	*Amaranthus sp.*	39	
	Quiabo	*Hibiscus esculentus*	116	
	Vinagreira b ranca	*Hibiscus sabdariffa*	86	
	Indian spinach	*Amaranthus sp.*	63	
	Bertalha	*Basellarubra*	55	
	Tomato	*Lycopersicumesculentum*	55	
	Orelha de macaco	*Alternanthera sp.*	46	
	Jambu branco	*Spilanthes acmella*	39	
	Dried cowpea leaves	*Vignaspp*	23–57	
	Dried pumpkin leaves	*Cucurbita moschata*	90–102	
	Kale	*Brassica oleracea var. acephala*	82–146	[30–33]
	Swiss chard	*Beta vulgaris*	21–46	
	Parsley	*Petroselinum hortense*	50 ± 15	
	Dill	*Anethum graveolens*	45	[34]
	Beet greens	*Beta vulgaris*	19–50	
	Motor sak	*Pisum sativus*	89 ± 9	[35]
	Kalmi sak	*Ipomoea reptans*	83 ± 4	
	Kachu sak	*Colocasia antiquoram*	80 ± 10	
	Lau sak	*Lagenaria vulgaris*	66 ± 3	
	Mulasak	*Raphanus sativus*	63 ± 6	
	Puisak	*Basella alba*	56 ± 7	
	Pat sak	*Chorchorus capsularis*	100 ± 11	
	Sharisha sak	*Brassica campestris*	88 ± 7	
	Helencha sak	*Enhydra fluctuans*	82 ± 3	
	Lal sak	*Amaranthus gangeticus*	80 ± 9	
	Palang sak	*Spinaceaoleracea*	64 ± 10	
	Thankuni sak	*Centellaasiatica*	61 ± 1	
	Nunia sak	*Portulacaoleracea*	54 ± 3	
	Garden cress	*Lepidiumsativum*	58	
	Bitter melon leaves	*Momordica charantia*	34	[36]
	Common fennel	*Eryngium foetidum*	44	
	Amaranth	*Amaranthus viridis*	15–39	
	Garlic leaves	*Allium sativum*	50	
	Lead tree leaves	*Leucaena glauca*	31–33	
	Chinese swamp cabbage	*Ipomoea reptans*	12–42	
	Curry leaves	*Murray koenigi*	93	

(continued)

Table 26.1 (continued)

Food	Common name	Scientific name	β-carotene content (μg/g edible portion)	References
	Ranti	*Solanum nigrum*	70	
	Tapioca shoots	*Manihot utilissima*	57	
	Mint leaves	*Mentha arvensis*	48	
	Pegaga gajah	*Hydrocotyle javanica*	38	
	Chinese chives	*Allium odorum*	35	
	Coriander leaves	*Coriandrum sativum*	32	
	Daun mengkudu	*Morinda citrifolia*	31	
	Daun turi	*Sesbania grandiflora*	136	
	Tanki	*Neptunia oleracea*	114	
	Drumstick leaves	*Moringa oleifera*	75	
	Wolfberry leaves	*Lycium chinense*	59	
	Spinach, red	*Amaranthus gangeticus*	51	
	Chinese kale	*Brassica alboglabra*	41	
	Ceylon spinach	*Basella rubra*	35	
	Cemperai	*Champereia griffithii*	32	
	Agathi	*Sesbania grandiflora*	154 ± 69	[23]
	Amaranth	*Amaranthus gangeticus*	6 ± 28	
	Gogu	*Hibiscus sabdariffa*	58 ± 28	
	Onion leaf	*Allium cepa*	49 ± 2	
	Coriander	*Coriandum sativum*	48 ± 2	
	Palak	*Spinacea oleracea*	32 ± 12	
	Drumstick	*Moringa oleifera*	197 ± 55	
	Methi	*Trigonella foenumgraecum*	92 ± 15	
	Curry leaf	*Murraya koenigii*	71 ± 24	
	Chammakura	*Colocasia antiquoram*	55 ± 19	
	Koyyalakura	*Sueda monoica*	48 ± 6	
	Pudina (mint)	*Mentha arvensis*	43 ± 20	
	Ceylon bacchali	*Talinum triangulare*	30 ± 6	
	Macaxeira	*Manihot esculenta*	129	[29]
	Vinagreiraroxa	*Hibiscus acetosela*	78	
	Indian spinach	*Amaranthus sp.*	63	
	Bertalha	*Basella rubra*	55	
	Tomato	*Lycopersicum esculentum*	55	
	Orelha de macaco	*Alternanthera sp.*	46	
	Jambu branco	*Spilanthes acmella*	39	
	Quiabo	*Hibiscus esculentus*	116	
	Vinagreira branca	*Hibiscus sabdariffa*	86	
	Chinese kale	*Brassica chinensis*	49	
	African spinach	*Amaranthus sp.*	39	
	Chennangiaku	*Cassia sp.*	119 ± 22	[19, 23]

(continued)

Table 26.1 (continued)

Food	Common name	Scientific name	β-carotene content (μg/g edible portion)	References
	Mulla thotakura	*Amaranthus spinosus*	109 ± 12	
	Betel leaf	*Piper beetle*	59 ± 10	
	Uttareni	*Achyranthes aspera*	43 ± 7	
	Chitramulan	*Plumbago zeylanica*	71 ± 11	
	Botla benda	*Abutilon indicum*	126 ± 15	
	Yerramolakakaura	*Amaranthus sp.*	119 ± 15	
	Tulasi	*Ocimum sanctum*		
	Ponnagantikura	*Alternanthera sessilis*	57 ± 16	
	Tummikura	*Leucas aspera*	41 ± 9	
Root crops	Carrot	*Daucus carota*	63	[28]
	Sweet potato	*Ipomoea batatas*	64	
Leafy vegetables	Parsley	*Petroselinum hortense*	50 ± 15	[24, 37]
	Cress	*Nastrutium officinale*	42 ± 10	
	Roquette	*Eruca sativa*	35 ± 13	
	Mustard leaves	*Brassica juncea*	60 ± 15	
	Coriander leaves	*Coriandrum sativum*	47 ± 5	
	Kale	*Brassica oleracea* var. *acephala*	35 ± 13	
	Common chicory	*Chicorium intybus*	34 ± 10	
	Komatsuna	*Brassica campestris*	63	[38]
	Parsley	*Petroselineum hortense*	58	
	Ashitaba	*Angelica keiskei*	51	
	Turnip leaves	*Brassica rapa*	46	
	Chinese chive	*Allium odorum*	61	
	Carrot Kintoki	*Daucus carota*	43	
	Shungiku	*Chrysanthemum coronarium*	41	
	Malabar spinach	*Basella alba*	39	
	Hiroshimana	*Brassica campestris var. Pekinensis*	36	
	Water cress	*Nastrutium officinale*	31	
	Hamaboufuu	*Glehnia littoralis*		
	Celery leaves	*Apium graveolens*	61	
	Collard	*Brassica oleracea var. acephala*	54	
	Mitsuba	*Cryptotaenia japonica*	48	
	Hatakena	*Brassica campestris var. Oleifera*	46	
	Spinach	*Spinacea oleracea*	44	
	Sugukina	*Brassica rapa var. neo suguki*	41	

(continued)

Table 26.1 (continued)

Food	Common name	Scientific name	β-carotene content (μg/g edible portion)	References
	Water dropwort	*Oenanthe javanica*	40	
	Vitamin-na	*Brassica campestris var. Narinosa*	37	
	Broad-leaved mustard	*Brassica juncea*	35	
	Leaf mustard	*Brassica juncea*	31	
	Komatsuna leaves	*Brassica campestris*	86	[39]
	Spinach leaves	*Spinacea oleracea*	66	
	Komatsuna, whole	*Brassica campestris*	33	
	Chingentsui leaves	*Brassica campestris var. Chinensis*	91	
	Chennangiaku	*Cassia sp.*	119 ± 22	[19, 23]
	Mullathotakura	*Amaranthus spinosus*	109 ± 12	
	Betel leaf	*Piper beetle*	59 ± 10	
	Uttareni	*Achyranthes aspera*	43 ± 7	
	Chitramulan	*Plumbago zeylanica*	39 ± 11	
	Botlabenda	*Abutilon indicum*	126 ± 15	
	Yerramolakakaura	*Amaranthussp*	119 ± 15	
	Tulasi	*Ocimum sanctum*	82 ± 11	
	Ponnagantikura	*Alternanthera sessilis*	57 ± 16	
	Tummikura	*Leucas aspera*	41 ± 9	

26.1.1 Leafy Vegetables

It has been established that leafy vegetables are main pro-vitamin A sources. These are comparatively easy to crop, accessible all the year and these are low-cost sources of pro vitamin A for the emerging world. Leafy vegetables have low β-cryptoxanthin and β-carotene contents but these contents vary significantly in different leafy vegetables (Table 26.1) [27]. categorized leafy vegetables into three groups depending upon the quantity of β-carotene contents (Table 26.1). During summer, leafy vegetables contain higher β-carotene. [19, 23] analyzed the leaves by HPLC. They observed that 13 vegetables contains 30–197 μg/g of β-carotene. These HPLC results were advanced than the OCC outcomes for the similar vegetables. [17] recently found by HPLC that no significant seasonal changes in β-carotene level observed in edible leaves. They assessed that few leaves have minor β-carotene content in summer while some leaves have lower β-carotene contents in winter. They found 5 out of 22 leaves have average β-carotene contents of nearly 30 μg/g or somewhat higher. Recent investigation conducted by using HPLC found that 2 leaves of bitter melon and fennel have modest quantities of β-carotene (34 and

44 μg/g correspondingly). The β-carotene contents for the similar leafy vegetables were markedly lesser in older study as compared to recent one [35, 36]. examined 13 edible green leaves which have 54–100 μg/g of β-carotene. It has been found that 2 out of 12 leafy vegetables have 30 μg/g or 58 μg/g of β-carotene [18, 39]. examined β-carotene level of 7 from 24 green leafy vegetables exceeded 30 μg/g (33–91 μg/g). It has been investigated by studying 5 fresh green leaves that one out of 5 leaves had higher a-carotene contents as compared to 30 μg/g level [16]. It has been observed that kale, Swiss chard, spinach, and beet greens are good sources of β-carotene [30, 32, 33]. It was examined that some leafy vegetables have β-carotene content among 34 and 60 μg/g [24, 37]. Though, 15 wild edible leaves presented higher β-carotene concentration [40]. Result achieved by OCC declared that two cultivars of kale which were grown in similar field one of them showed statistically higher β-carotene contents due to seasonal variation between summer and winter [41]. A list of vegetables with β-carotene contents is given in the Table 26.1.

26.1.2 Root Crops

Sweet potato and carrot are vital source of carotenoids and these root crops are accessible in the world. Different concentrations of provitamin A have been found in these root crops. Carotenoids has derived their name from carrot which is the most examined root crop in relations to carotenoids by OCC and HPLC techniques [18, 19, 24, 28, 32, 42–46]. HPLC studies showed that in carrot mean β-carotene level is within the range from 5.3 μg/g [34] to 106 μg/g [43] while in another study β-carotene range from 36 μg/g [141] to 182 μg/g [43]. A list of root crop vegetables with β-carotene contents is given in the Table 26.1.

26.1.3 Fruits

Provitamin-A contents are low in fruits as compared to leafy vegetables [15]. Mango and papaya considered as a very important tropical fruits and provitamin-A centers in evolving countries [17, 19, 24, 36, 47–49]. The range of β-carotene in papaya is minor i.e., 0.4–10 μg/g [16–19, 36, 48, 50]. The bioactivity of β-cryptoxanthin is one-half that of β-carotene and is considered the main source of provitamin-A in papaya [48, 50]. HPLC study revealed that about 1.2 ± 0.3 μg/g β-carotene are present in red-fleshed papayas [49]. It has been found that about 3.2 μg/g, 59 μg/g and 107 μg/g β-carotene occur in peach palm (*Bactris gasipaes*), palm fruits *Acrocomia makayayba* and tucuma (*Astrocaryum vulgare*) respectively [25, 51]. A list of β-carotene containing fruits is given in the Table 26.1.

26.1.4 Squashes and Pumpkins

Squashes and pumpkins are considered as have long shelf life and their leaves and flowers are consumed as main sources of provitamin-A carotenoids [16–18]. Data taken from several countries showed pumpkins and squashes have modest to higher levels of pro-vitamin A carotenoids [17–19, 30, 31, 36, 42, 43, 52].

The noticeable differences detected in the pro-vitamin level among samples of cucurbit variety, might be recognized to extended period. Low levels of β-carotene found due to examines of squashes and pumpkins via HPLC in immature stage. It has been reported that samples of pumpkin and squashes contain 24–84 μg/g contents of β-carotene [33, 42, 53] and one pumpkin sample has 55 μg/g β-carotene contents [43]. A list of vegetables containing β-carotene is given the Table 26.1.

26.2 Properties of Provitamin A Carotenoids

Out of 700 carotenoid compounds found in nature, only 50 are known for provitamin A activity. The most imperative vitamin A precursors in humans A are α-carotene, β-carotene and β-cryptoxanthin [55]. Carotenoids are generally C40 tetraterpenoids with eight C5 isoprenoid units in a conjugated double-bond system [56]. The light-absorbing chromophore reflects red, orange, or yellow color in vegetables and fruits. The β-ionone structure (ring structure with single double bond and three methyl groups) are provitamin A carotenoids [57]. Vitamin A can be made available in the diet as preformed forms (retinol, retinal, retinyl ester, 3-dehydroretinol and retinoic acid) from dairy products or as a provitamin A carotenoids, which are eventually transformed into the vitamin A [57]. Retinol can be made into retinoic acid and retinal in the body. Structurally, β-cryptoxanthin is able to generate one molecule while β-carotene two molecules of the vitamin A when cleaved centrally [58] (Fig. 26.1).

26.2.1 Size and Shape

Acyclic carotenoids (e.g., ζ-carotene, lycopene) are linear, long molecules and may contain a six member ring at both ends (e.g., β- carotene, α- carotene) or at one (e.g., γ-carotene, δ-carotene) end. Carotenoids may possess large number of isomers due to double bonds. Every double bond is able to exist in trans/cis configuration, *(E)* or *(Z)* form according to the IUPAC recommendation. Mostly they exist in more stable (all-*E*)-configuration. The shape and size of *Z* and *E* isomers exert influence on the properties and functions, as *E* isomers are rigid and linear and Z isomers are bent, their potential to accumulate in supramolecular structure and interaction with

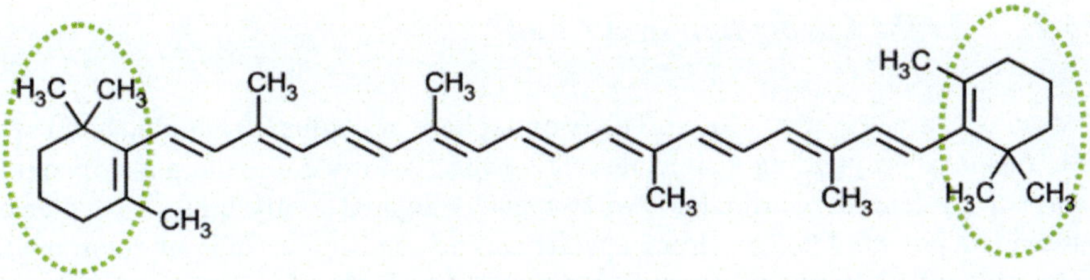

β- Carotene

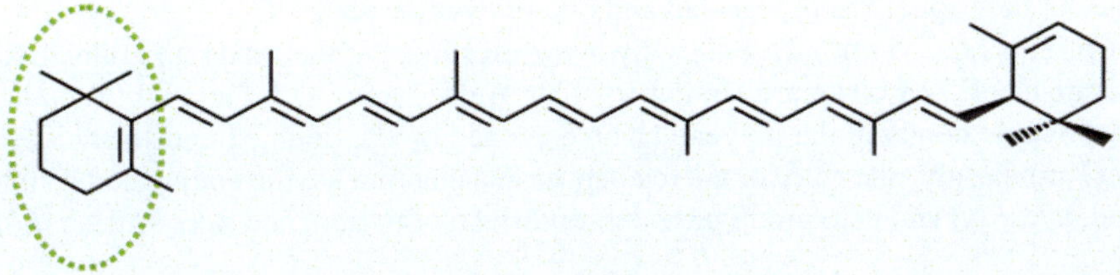

α- Carotene

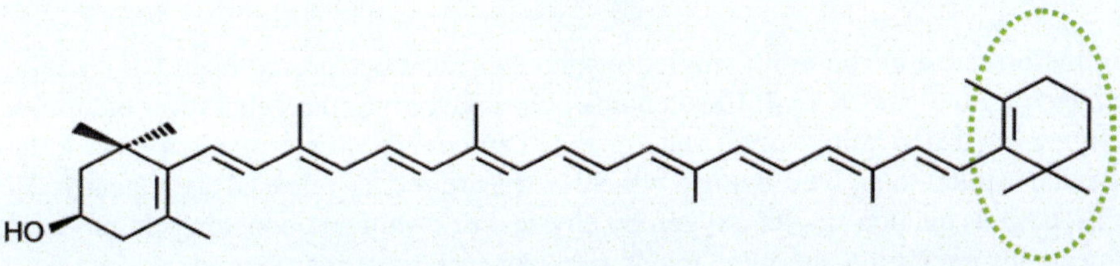

β- Cryptoxanthin

Fig. 26.1 Sources of vitamin A

enzymes differs. Most *Z* isomers aggregate less, possess lower melting point, easily solubilized, absorbed and transported as compared to *E* isomers [59, 60].

26.2.2 Solubility

Carotenoids are generally lipophilic, water insoluble, but have variation of solubility in organic solvents. They are more easily soluble in, hexane, petroleum ether and toluene; xanthophylls are soluble in methanol and ethanol. The solubility of β-carotene and xanthophyll lutein is excellent in tetrahydrofuran. Carotenoids are confined to inner core of membranes (hydrophobic areas) but they can access to aqueous environments in association with protein [61].

26.2.3 Light Absorption and Color

Carotenoids are brightly colored. The chromophore or conjugated double bond system impart color, yielding visible absorption spectra for identification and quantification of carotenoids. Carotenoids with seven conjugated double bonds have greater antioxidant potential and shielding against photochemical alteration of chlorophyll. The carotenoids having three conjugated double bonds i.e. phytoene and five conjugated double bonds as phytofluene, are uncoloured. Acyclic carotenoid such as lycopene is red with eleven (11) conjugated double bonds. The conjugated double bonds in γ- carotene (monocyclic) and β-carotene (bicyclic) are the same as lycopene but they appear as orange-red and orange- yellow [60].

Chromophore is not affected by hydroxy substituents; the mono and dihydroxy derivatives of β-carotene are similar in color. Capsorubin, with the nine (9) conjugated double bonds in the polyene chain extended by the double bonds of two carbonyl groups, give the color of the red pepper. Astaxanthin having conjugated double bond, hydroxyl and keto group provides reddish color to trout and salmon flesh [60].

26.2.4 Antioxidant Properties

In the presence of the light, singlet oxygen ($^1O_2$) can be formed from ($^3O_2$) triplet oxygen by the chlorophyll. Carotenoids help in avoiding the generation of singlet oxygen by deactivating triplet state or they can convert singlet oxygen back to its ground triplet state. The energy which is transferred is released as a heat [60]. Carotenoids quench singlet oxygen by physical or chemical quenching. In physical quenching excitation energy is transferred from $^1O_2$ to the carotenoid; oxygen then proceeds to its ground state and the carotenoid is elevated to its excited triplet state. The energy between the excited carotenoid and the solvent is released by rotation and vibration interactions, resulting in ground state carotenoid and thermal energy. The carotenoid can undergo further cycles of singlet oxygen quenching.

$$^1O_2 + CAR \rightarrow {}^2O3 + {}^3CAR*$$

$$^3CAR* \rightarrow CAR + heat$$

In contrast, the chemical quenching results in the loss of the antioxidant protection and also in the destruction of the carotenoids. Carotenoids can also help in preventing the formation of the $^1O_2$ by quenching the excited triplet state chlorophyll.

$$^3CHL * + CAR \rightarrow CHL + 3CAR*$$

The potential of the carotenoids to quench the singlet oxygen increase, in actual depends on number of the conjugated double bonds, with the utmost protection exhibited by those having nine or more double bonds [62] and can also affected by the cyclic or acyclic end group in structure of the carotenoids. The balance in electron acceptance and electron donation with formation of radical anions and radical cations, respectively, differs in all carotenoids having a major part in the antioxidant networks. The structure of carotenoids gives idea about their reactivity, orientation and location in lipid bilayer, having tendency to self-aggregate under the polar conditions [60]. Lycopene is one of the strong antioxidant and the most effective singlet oxygen quencher twice as effective as compared to the β- carotene [63].

26.3 Classification and Extraction of Provitamin A Carotenoids

Carotenoids classified into two groups such as oxygen-containing carotenoids include xanthophylls: antheraxanthin, bixin, canthaxanthin (give red color), α-cryptoxanthin (give yellow color), beta-cryptoxanthin (give orange color) and zeaxanthin (give yellow orange color), and carotenoids which lack oxygen include α-carotene, β-carotene, delta-carotene, γ-carotenes (provide orange color), lycopene (provide red color) and phytoene (colorless). Some carotenes have linear carbon chain such as zeta carotene and phytoene both are colorless, lycopene, neurosporen have red color [64].

Carotenoids can be extracted by various non-conventional and conventional methods. Non-conventional methods include ultrasound-assisted, microwave-assisted and supercritical carbondioxide extraction, while conventional methods include solvent extraction, centrifugation, soxhlet extraction etc. [65].

26.4 Bioconversion Pathways of Provitamin A Carotenoids

In all plants the pathway shows the same primary steps. The isoprene compounds i.e. DMAPP (dimethylallyl pyrophosphate) and IPP (isopentenyl pyrophosphate) act as a precursor of the carotenoids, tocopherols, monoterpenes, chlorophyll, phylloquinone, gibberellic acid and plastoquinone [66]. For the biosynthesis of carotenoids the MEP (methylerythritol-4 phosphate) pathway provides isoprene units [66]. The production of IPP take place by Mevalonic Pathway (MVA) in the cytosol or methyl erythritol-4-phosphate (MEP) in the plastids [67, 68]. The substrates glyceraldehydes 3-phosphate and pyruvate form the deoxy-D-xylulose 5-phosphate (DXP) is catalysed by the DXP synthase (DXS). The MEP is formed by the intramolecular rearrangement and reduction of the DXP by DXP reductoisomerase (DXR). Both enzymes are rate determining.

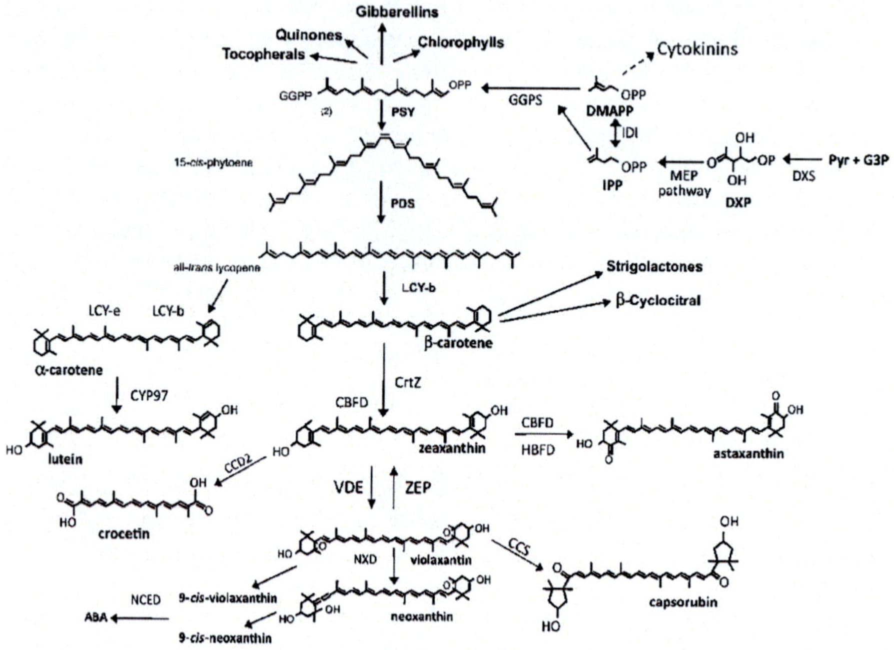

Fig. 26.2 Schematic representation of provitamin A carotenoid pathway in plastids

In Arabidopsis production of carotenoid is increased by the overproduction of DXR and DXS [69]. DMAPP and IPP are converted by isopentenyl diphosphate isomerase (IDI) and condensed into geranyl geranyl diphosphate, which is precursor of the carotenoid biosynthesis, GGPP further by phytoene synthase (PSY) form phytoene (15-cis isomer) (Fig. 26.2).

All the trans isomers of lycopene require specific isomerase enzymes. The activity of an isomerase depends on FAD (flavin adenine dinucleotide binding motif) which can convert cis bonds 7, 9 and 7′, 9′ positions and also converts tetra cis lycopene to all trans lycopene. Condensation of the two molecules of geranylgeranyl diphosphate by phytoene synthase forms phytoene (15-cis isomer) [70].

In plants trans-lycopene generates carotenoid diversity by either the addition of epsilon (ε-type ring) or beta (β-ring), generated by lycopene epsilon-cyclase (LCY-e) and lycopene beta-cyclase (LCY-b) [71–74]. The grouping of the β and ε rings comprises α- carotene and its derivative lutein (3, 3′- dihydroxy- α-carotene) by the action of CYP97 (heme hydroxylase) [75]. Carotenoids having two ε rings are rare in the plants [70, 76].

The β-carotene (orange-colored) in the β, β-branch is hydroxylated by CHY (carotenoid hydroxylases) and carotenoid β-ring 4-dehydrgenase, a non-heme hydroxylase to produce zeaxanthin (yellow). Zeaxanthin is epoxidized by zeaxanthin epoxidase (ZEP) to yield violaxanthin. The final step of the β, β-branch converts yellow-colored violaxanthin into neoxanthin *via* neoxanthin synthase (NXD), [77]. Violaxanthin can be converted back to zeaxanthin by violaxanthin de-epoxidase (VDE). In red pepper and tiger lily, antheraxanthin and violaxanthin are converted by CCS (capsanthin-capsorubin synthase) into capsanthin and capsorubin, generating the characteristic red and orange colors of these species [78, 79]. Hydroxylation/ketolation of β-carotene (beta carotene) by carotenoid b-ring 4-dehydrogenase (CBFD) and 4-hydroxy-b-ring 4-dehydrogenase (HBFD) leads to astaxanthin production in Adonis flowers [73]. Capsanthin-capsorubin synthase has high sequence identity with LCYB and belongs to the lycopene cyclase family [78].

Carotenoids are catabolized enzymatically by a family of carotenoid cleavage dioxygenases (CCDs; carotenoid cleavage oxygenases, CCOs) to produce apocarotenoids, which direct carotenoid turnover contributing to the colors/aromas of flowers and fruits and the production of two main phytohormones, ABA and the strigolactones [80–82].

In various plant species, members of the CCD enzyme family are denominated according to their sequence similarity to the *Arabidopsis* CCD family and are generally divided into two groups: four CCDs (CCD1, 4, 7, and 8) and five 9-*cis*-epoxycarotenoid dioxygenases (NCED 2, 3, 5, 6, and 9) [83]. Different CCDs and NCEDs recognize different carotenoid substrates and cleave at different sites, producing various apocarotenoids. NCEDs specifically cleave 9-*cis*-violaxanthin and 9-*cis*-neoxanthin to yield xanthoxin, which is further modified to ABA [84, 85].

26.5 Mode of Action and Distribution Pattern of Provitamin A Carotenoids

Carotenoids are lipid-soluble pigments that has very wide distribution and ubiquitous in nature mostly present in roots, leaves, flowers, fruits, seeds, vegetables, egg yolk and fatty tissues of animals. Plants, bacteria, algae and fungi are excellent sources of carotenoids [86]. Among microorganisms [87, 88] include algae [89], fungi such as *Penicillium purpurogenum* and *Talaromyces purpurogenum* [90] and *Neurospora intermedia* [91]. Carotenoids with provitamin activity A are present in animals and fish in highest level [92]. In children the sole source of vitamin A is breast-feeding [60].

Provitamin A might take action by means of retinoid pathways during limited metabolism to retinol and additional to retinoic acid [1]. Vitamin A found as retinoic acid form act a significant function in gene transcription. When cell takes retinol it might oxidized to retineldehyde and after that into retinoic acid with the help of retinol dehydrogenases and retinaldehyde dehydrogenases respectively [93]. The procedure of conversion of pro-vitamin A carotenoids to organically active retinol is depend on vitamin A position of the host [55, 94]. Alpha-carotene and β-cryptoxanthin can also convert into vitamin A (retinol) but with half biological activity than that of β-carotene [5]. Presence of at least one β-ionone ring is essential for the exhibition of provitamin A activity. Beta-carotene possesses two β-ionone rings in its structure and generates two molecule of retinol than α-carotene and β-cryptoxanthin that has only one β-ionone ring (Fig. 26.1).

Biological activity is represented by retinol equivalent (RE) that is equal to 6 μg dietary β-carotene or 12 μg other dietary sources. Human bodies can covert these carotenoids with pro vitamin A activity into retinol [95]. B-carotene greater than 1000 μg/100 g found in Chinese vegetables [96]. In addition, chief plant food sources of provitamin include maize with β-cryptoxanthin [97, 98], pumpkin, palm oil and amaranthus [140]. Citrus fruits such as oranges, tangerines, mandarins, mango, papaya, guava, nectarines and apricots are rich source of β-cryptoxanthin [11, 99–102], and deep yellow-fleshed bitter yam (*Dioscorea dumetorum*) [10]. Lycopene is abundant in watermelon, tomato, red grapes and pink grapefruit, apricots and papaya [103]. Asian leafy vegetables namely, water spinach (*Ipomoea aquatica*) and Chinese flowering cabbage (*Brassica parachinesis*) contain β-carotene contents on wet weight basis as 1974–6604 μg/100 g or 329–1101 μg/100 g retinol equivalent (RE) [104].

In recent times, the big struggle to bio-fortify maize during much kind of procedures produced numerous cultivars of maize through improved provitamin A carotenoids which possess eminent β-cryptoxanthin [56, 105]. The continuous endorsement of maize bio-fortified by 5 provitamin A carotenoids might assist improve the occurrence of vitamin A shortage in impecunious areas of Sub-Saharan Africa, Central America, & South-East Asia [106]. Carotenoids are defined as provitamins A, as specific vitamin which is an artifact of the carotenoid metabolism process. The distribution of carotenoids between the diverse plant species displays no clear design [107]. In the leafy vegetables, b-Carotene is very abundant. Zeaxanthin, and antheraxanthin also existing in small extents. In tomato, lycopene is main carotenoid, though fruits comprise changing extents of cryptoxanthin, antheraxanthin and lutein [108].

26.6 Physiological Role of Provitamin A Carotenoids in Plants and Animals

Among all 700 carotenoids that are specified so far only 40–50 are important part of human diet [139]. Some kind of bacteria, fungi and all types of plants can synthesize the carotenoids. Human beings and animals totally depend on their food for its availability.

Carotenoids guard the photosynthetic machinery of the plants from photoxidative damage [109]. They show strong antioxidant potential. Humans metabolized Provitamin A carotenoids into retinal and retinol. These include α-cryptyoxanthin, α & β-carotene [109]. Non-provitamin A carotenoids i.e. lutein, astaxanthin, lycopene and zeaxanthin have no vitamin A activity in human beings [109].

Carotenoids are present in the green tissues and associated with light harvesting complexes as lutein being rich in the photosystem-II and β-carotene rich in the photosystem-1 [110]. Carotenoids are the pigments that absorb light and assist in the photoprotection, dissipation of the excess energy, regulate fluidity of the membrane, regulate communication among the cells and also assemble and stabilize the light harvesting complexes [60].

Vitamin A and retinol promote the cell division, growth, embryonic development, vision, cell division and differentiation. It helps the lymphocytes to fight against the infections very effectively. Deficiency of vitamin A may result in night blindness and its presence promotes the healthy surface lining of the eyes. Large numbers of children in the developing countries are undernourished due to the deficiency of vitamin A. Its inadequacy can results from the limited intake of zinc, calories and protein. It can help in preventing the pneumonia. Its deficiency in the children may increase diarrhea, respiratory infections and slow bone development. Retinoids just behave like vitamin A. Synthetic retinoids are used for the skin disorders [110, 111]. Beta-cryptoxanthin, zeaxanthin and lutein decrease the likelihood of the cancer. It also decreases the risk of rheumatoid arthritis. Lycopene is one of the powerful antioxidant that deactivates the free radicals and reduce the risk of cancer [112].

26.7 Provitamin A Carotenoids Efficacy in Combating Diseases

Carotenoids comprise effective antioxidant properties to defend cell from oxidative strain, free radicals, improvement of resistant role, and repression of many kinds of tumor. Numerous carotenoids have established significant roles in human fitness [113–115]. Carotenoids can convert into biologically active vitamin A called retinol. It is a necessary vitamin for the endorsement of broad development regulation of separation of epithelial tissues, and embryonic progress, maintenance of visual function [116].

It has been reported that about 190 million children and 19 million expecting women world widely were vitamin A lacking in the duration of 1995–2005 [117]. Vitamin A deficiencies leads to blindness. Approximately half million kids become blind due to deficiency of 79 vitamin A which is responsible of xerophtalmia [118]. Beta-carotene is the forerunner of vitamin A which has antioxidant ability and presents a display of fitness benefits like reducing the heart diseases, different kinds of cancer and irreversible blindness. Integration of β-carotene in a range of foodstuff scheme is inadequate by its deprived water solubility [12]. Carotenoid rich diet reduces the chance of unrelieved diseases like cancer [119]. However, high dose intake of carotenoid provitamin A does not reduce the chance of cancer &it may be dangerous for smokers [120].

Inflammation is an important part of immune-surveillance and defense of host. Carotenoids might be inversely connected by inflammatory markers such as TNF-α, IL-6, IL-1β [121]. Controlling of systematic inflammation might be an aim move toward in the avoidance of chronic syndrome. The blood levels carotenoids are inversely linked with possibility of a lot of chronic syndrome, including T2DM, CVDs and numerous types of tumors [122]. Carotenoids proceed by falling complete inflammation [123]. In 2016, the world carotenoid marketplace is expected to be present approximately 1.24 billion US Dollars. It is predictable to enhance to approximately 1.53 billion US Dollars in 2021 [124].

Out of several hundred carotenoids only 50 carotenoids possess pro-vitamin A activity. Among these β-carotene is the main source in diet that is widely distributed in plant kingdom. β-carotene consists of 2 β-ionone rings for vitamin A activity [125, 126]. Lycopene quickly quench singlet oxygen species produced in photosynthesis and convert excess energy of these reactive oxygen species (ROS) to heat thus protect plant tissues [127]. In humans β-carotene in 180 mg/day dose has been used for erythropoietic protoporphyria that is a light sensitive condition [128, 129]. These β-cryptoxanthin with pro-vitamin A activity is found in fruits [130]. Lutine (α-carotene) and zeaxanthin (β-carotene) are used to treat against oxidative damages of blue light [131, 132]. Biofortification of cereal crops such as maize can increase provitamin A carotenoids in endosperms of maize which in return can reduce the rate of vitamin A deficiency (VAD) [97]. Source of carotenoids such as fruits and vegetables have antioxidant properties that combat against diseases like cancer, cardiovascular diseases and muscular degeneration [133]. The prevention method comprises dietary change, protection and supplementation. The viability of put on every preventive plan alongside is someway reliant on lack occurrence and harshness and infrastructure, potential benefits, financial capacity, and safety [134]. It is needed to appreciate that success of every defensive program is consistent to entirely levels, complete of family, district, national and international [135].

26.8 Current and Future Prospects of Provitamin A Carotenoids

Provitamin A is enzymatically converted into vitamin A in small intestine. After absorption β-carotene is converted into vitamin A, however, it depends on the protein nutrition status of the host, fiber content of the food, intestinal health and dietary fat. The important sources of the provitamin A contents are carrot, palm oil, pumpkins, green leafy vegetables, sweet potato, pumpkins, orange or some tropical yellow fruits [60]. Samples of same food can differ due to the cultivar or varietal differences, portion of plant used, climate effects and post-harvest storage. During ripening of vegetables and fruits tropical climate enhances biosynthesis of the carotenoids. Carotenoids are degraded by enzymatic or non-enzymatic oxidation. Thermal decomposition may cause isomerization of the provitamin A from *trans*-to *cis*-state and lowers its content in the foods. Retentions of the provitamins A decreases due to elevated temperatures, longer processing time and cutting of the food [58, 136]. However, cooking with cover on; lessen the time interval between peeling and cooking; minimum storage and processing time improve detention significantly. Thawing long is unfavorable process for long term storage, for preserving the provitamins but refrigerating is favorable. Retention time is increased by protecting the food from traditional sun drying or direct sunlight. During food storage carotenoid destruction is decreased by adding antioxidants and saline treatment. Oxygen-impermeable packaging also protect carotenoids from decomposition [58].

Experiments are underway to generate β-carotene-enriched foods through conventional or bioengineering techniques like golden rice [137], β-carotene enriched yellow maize [138], and β-carotene enriched ground nuts [139].

26.9 Conclusions

Alleviation of vitamin A deficiency needs production, promoting foods and verifying their nutritional composition with provitamin A carotenoids. Carotenoids being fascinating set of pigments are important to the food industries, nutritionists and food scientists with possible roles for a wide range of coloration in nature, basic physiological processes in many living organisms, quenching free radicals, and reducing oxidative stress. Standardized methods of food processing, storage, and preparation, maximize retention and bioavailability of provitamin A carotenoids. The modulation of carotenoid accumulation is yet to be clarified. Plastids govern the regulatory networks of carotenoid metabolism; however, challenges exist in understanding their mechanisms. Transgenic crops with improved carotenoid contents could hold enormous promise to improve human health.

References

1. Khachik F, Nir Z, Ausich RL, Steck A, Pfander H (1997) In: Ohigashi H, Osawa T, Terao J, Watanabe S, Yoshikawa T (eds) Food factors for cancer prevention. Springer, Tokyo, p 204
2. Khachik F, Carvalho L, Bernstein PS, Muir GJ, Zhao DY, Katz NB (2002) Chemistry, distribution, and metabolism of tomato carotenoids and their impact on human health. Exp Biol Med 227:845–851
3. Finley JW (2005) Proposed criteria for assessing the efficacy of cancer reduction by plant foods enriched in carotenoids, glucosinolates, polyphenols and selenocompounds. Ann Bot 95:1075–1096
4. Granado F, Olmedilla B, Blanco I (2003) Nutritional and clinical relevance of lutein in human health. Br J Nutr 90:487–502
5. Institute of Medicine (2001) Food and nutrition board. Dietary reference intakes for vitamin A, vitamin K, arsenic, boron, chromium, copper, iodine, iron, manganese, molybdenum, nickel, silicon, vanadium, and zinc. National Academy Press, Washington, DC
6. de Pee S, West CE (1996) Dietary carotenoids and their role in combating vitamin A deficiency: a review of the literature. Eur J Clin Nutr 50(3):38–53
7. Olson JA (1996) Benefits and liabilities of vitamin A and carotenoids. J Nutr 126:1208S–1212S
8. Pavia SA, Russell RM (1999) Beta-carotene and other carotenoids as antioxidants. J Am Coll Nutr 18:426–433
9. Miki W (1991) Biological functions and activities of animal carotenoids. Pure Appl Chem 63(1):141–146
10. Alamu E, Bussie OMD, Roman FM, Ibironke P, Robert A et al (2016) Characterization and classification of the pro-vitamin a carotenoids of deep yellow- fleshed bitter yam (*Dioscorea dumetorum*) varieties. J Food Nutr Res 4(10):640–645

11. Maiani G, Caston MJ, Catasta G, Toti E, Cambrodon IG et al (2009) Carotenoids: actual knowledge on food sources, intakes, stability and bioavailability and their protective role in humans. Mol Nutr Food Res 53:194–218
12. Gul K, Tak A, Singh AK, Singh P, Yousuf B et al (2015) Chemistry, encapsulation, and health benefits of β-carotene. Cogent Food Agric 1:1018696
13. Fraser P, Bramley M (2004) The biosynthesis and nutritional uses of carotenoids. Prog Lipid Res Elsevier 43:228–265
14. West KP, Darnton-Hill (2008) Vitamin A deficiency. In: Nutrition and health in developing countries. Humana Press, Totowa, pp 377–433
15. Nogueira AC, Georgia AR, Sehn, Rebillato AP, Janclei P et al (2018) Yellow sweet potato flour: use in sweet bread processing to increase β-carotene content and improve quality. Ann Braz Acad Sci 90(1):283–293
16. Pepping F, Vencken CMJ, West CE (1988) Retinol and carotene content of foods consumed in East Africa determined by high performance liquid chromatography. J Sci Food Agric 45:359–371
17. Wasantwisut E, Sungpuag P, Chavasit V, Chittchang U, Jittinandana S, Viriyapanich T (1995) Identifying and recommending vitamin A rich foods in Northeast Thailand. In: Empowering vitamin A foods. Institute of Nutrition, Bangkok, pp 69–90
18. Vaidya Y (1995) Vitamin A food production and use in Nepal. In: Empowering vitamin A foods. Institute of Nutrition, Bangkok, pp 29–44
19. Reddy V, Vijayaraghavan K, Bhaskarachary K, Rani M (1995) Carotene rich foods: the Indian experience. In: Empowering vitamin A foods. Institute of Nutrition, Bangkok, pp 15–28
20. Choo YM (1994) Palm oil carotenoids. Food Nutr Bull 15:130–137
21. Rukmini C (1994) Red palm oil to combat vitamin A deficiency in developing countries. Food Nutr Bull 15:126–129
22. Rains-Mariath JG, Cavalcanti LMC, Santos PLM (1989) Vitamin A activity of buriti (*Mauritia vinifera Mart.*) and its effectiveness in the treatment and prevention of xerophthalmia. Am J Clin Nutr 49:84953
23. Bhaskarachary K, Sankar Rao DS, Deosthale YG, Reddy V (1995) Carotene content of some common and less familiar foods of plant origin. Food Chem 54:189–193
24. Godoy HT, Rodriguez-Amaya DB (1996) Occurrence of cis-isomers of provitamin A in Brazilian vegetables. J Sci Food Agric 46:3081–3086
25. Rodriguez-Amaya DB, Godoy HT, Mercadante AZ, Ramos DMR (1995) Carotenoid composition of underexploited Brazilian fruits. Paper presented at the 9th World Congress of Food Science and Technology, Budapest
26. Trujillo-Quijano JA, Rodriguez-Amaya DB, Esteves W, Plonis GF (1990) Carotenoidcomposition and vitamin A values of oils from four Brazilian palm fruits. Eur J Lipid Sci Technol 92:222–226
27. Begum A, Pereira SM (1977) The beta carotene content of Indian edible green leaves. Trop Geogr Med 29:47–50
28. Abdel-Kader ZM (1991) Determination of carotenoids in foods by high-performance liquid chromatography. Nahrung 35:689–693
29. Penteado MVC, Minazzi RS, Almeida LB (1986) Carotenóides e atividade pró-vitamínica A de folhas de hortaliças consumidas no norte do Brasil. Rev Fac Farm Bioquim Univ Sao Paulo 22:97–102
30. Khachik F, Beecher GR, Whitaker NF (1986) Separation, identification, and quantification of the major carotenoid and chlorophyll constituents in extracts of several green vegetables by liquid chromatography. J Agric Food Chem 34:603–616
31. Bureau JL, Bushway RJ (1986) HPLC determination of carotenoids in fruits and vegetables in the United States. J Food Sci 51:128–130
32. Bushway RJ, Wilson AM (1982) Determination of á- and â-carotene in fruits and vegetables by high-performance liquid chromatography. J Food Sci Technol 15:165–169

33. Quackenbush FW (1987) Reverse phase HPLC separation of cis and trans-carotenoids and its application to â-carotenes in food materials. J Liq Chromatogr 10:643–653
34. Heinonen MI, Ollilainen V, Linkola EK, Varo PT, Koivistoinen PE (1989) Carotenoids in Finnish foods: vegetables, fruits, and berries. J Agric Food Chem 37:655–659
35. Rahman MM, Wahed MA, Mahalanabis D, Sack RB (1995) Preparing and preserving green leafy vegetables for poor communities in Bangladesh. In: Empowering vitamin A foods. Institute of Nutrition, Bangkok, pp 61–68
36. Speek AJ, Speek-Saichua S, Schreurs WHP (1988) Total carotenoid and â-carotene contents of Thai vegetables and the effect of processing. Food Chem 27:45–257
37. Ramos DMR, Rodriguez-Amaya DB (1987) Determination of the vitamin A value of common Brazilian leafy vegetables. J Micronutr Anal 3:147–155
38. Izaki Y, Yoshida K, Hidaka K, Toda K (1986) Chlorophylls, carotenes and tocopherols in green vegetables and their relationships (in Japanese). J Jpn Soc Nutr Food Sci 39:485–493
39. Kon M (1989) The color of green vegetables and fruits and the contents of carotene and chlorophyll in them. Kaseijaku Kenkya 35:99–105
40. Mercadante AZ, Rodriguez-Amaya DB (1990) Carotenoid composition and vitamin A value of some native Brazilian green leafy vegetables. Int J Food Sci Technol 25:213–219
41. Mercadante AZ, Rodriguez-Amaya DB (1991) Carotenoid composition of a leafy vegetable in relation to some agricultural variables. J Agric Food Chem 39:1094–1097
42. Bushway RJ (1986) Determination of á- and â-carotene in some raw fruits and vegetables by high performance liquid chromatography. J Agric Food Chem 34:409–412
43. Khachik F, Beecher GR (1987) Application of a C-â-carotene as an internal standard for the 45 quantification of carotenoids in yellow/orange vegetables by liquid chromatography. J Agric Food Chem 35:732–738
44. Simon PW, Wolff XY (1987) Carotenes in typical and dark orange carrots. J Agric Food Chem 35:1017–1022
45. Heinonen MI (1990) Carotenoids and provitamin A activity of carrot (*Daucus carota* L.) cultivars. J Agric Food Chem 38:609–612
46. Portocarrero L, Serrano JQ, Canfield L, Tarara T, Solomons NW (1992) Carrots and dietary vitamin A adequacy. Food Nutr Bull 14:133–136
47. Godoy HT, Rodriguez-Amaya DB (1989) Carotenoid composition of commercial mangoes from Brazil. Lebensm Wiss Technol 22:100–103
48. Godoy HT, Rodriguez-Amaya DB (1994) Occurrence of cis-isomers of provitamin A in Brazilian fruits. J Agric Food Chem 42:1306–1313
49. Wilberg VC, Rodriguez-Amaya DB (1995) HPLC quantitation of major carotenoids of fresh and processed guava, mango and papaya. Lebensm Wiss Technol 28:474–480
50. Kimura M, Rodriguez-Amaya DB, Yokoyama SM (1991) Cultivar differences and geographic effects on the carotenoid composition and vitamin A value of papaya. Lebensm Wiss Technol 24:415–418
51. Khachik F, Beecher GR, Lusby WR (1989) Separation, identification, and quantification of the major carotenoids in extracts of apricots, peaches, cantaloupe, and pink grapefruit by liquid chromatography. J Agric Food Chem 37:1465–1473
52. Hidaka T, Anno T, Nakatsu S (1987) The composition and vitamin A value of the carotenoids of pumpkins of different colors. J Food Biochem 11:59–68
53. Khachik F, Beecher GR (1988) Separation and identification of carotenoids and carotenol fatty acid esters in some squash products by liquid chromatography. Quantification of carotenoids and related esters by HPLC. J Agric Food Chem 36:929–937
54. Lee CY, Smith NL, Robinson RW (1984) Carotenoids and vitamin A value of fresh and canned winter squashes. Nutr Rep Int 29:129–133
55. Tanumihardjo SA, Palacios N, Pixley KV (2010) Provitamin A carotenoid bioavailability: what really matters? Int J Vitam Nutr Res 80:336–350
56. Davis CR, Jing H, Howe JA, Rocheford TR, Tanumihardjo SA (2008) β-cryptoxanthin from supplements or carotenoid-enhanced maize maintains liver vitamin A in Mongoliangerbils

(*Meriones unguiculatus*) better than or equal to beta-carotene supplements. Br J Nutr 100:786–793
57. Rodriguez-Amaya DB (1997) Carotenoids and food preparation: the retention of provitamin A carotenoids in prepared, processed and stored foods. U.S. Agency for International Development, Arlington
58. De Moura FF, Miloff A, Boy E (2015) Retention of provitamin A carotenoids in staple crops targeted for biofortification in Africa: cassava, maize and sweet potato. Crit Rev Food Sci Nutr 55(9):1246–1269
59. Kiokias S, Proestos C, Varzakas T (2016) A review of the structure, biosynthesis, absorption of carotenoids-analysis and properties of their common natural extracts. Curr Res Nutr Food Sci J 4:25–37
60. Rodriguez-Amaya DB (2015) Provitamin an activity. In: Food carotenoids: chemistry, biology, and technology. Wiley, Chichester. https://doi.org/10.1002/9781118864364.ch10
61. Soares AT, Marques Júnior JG, Lopes RG, Derner RB, Antoniosi Filho NR (2016) Improvement of the extraction process for high commercial value pigments from Desmodesmus sp. microalgae. J Braz Chem Soc 27(6):1083–1093
62. Foote CS, Chang YC, Denny RW (1970) Chemistry of singlet oxygen. X. Carotenoid quenching parallels biological protection. J Am Chem Soc 92:5216–5218
63. Edge R, Truscott TG (2018) Singlet oxygen and free radical reactions of retinoids and carotenoids—a review. Antioxidants 7(1):5
64. Engelmann NJ, Clinton SK, Erdman JW Jr (2011) Nutritional aspects of phytoene and phytofluene, carotenoid precursors to lycopene. Adv Nutr (Bethesda, Md) 2(1):51–61
65. Singh A, Ahmad S, Ahmad A (2015) Green extraction methods and environmental applications of carotenoids-a review. RSC Adv 5:62358–62393
66. Rodríguez-Concepción M (2010) Supply of precursors for carotenoid biosynthesis in plants. Arch Biochem Biophys 504(1):118–122
67. Rodrıguez-Concepción M, Boronat A (2002) Elucidation of the methylerythritol phosphate pathway for isoprenoid biosynthesis in bacteria and plastids. A metabolic milestone achieved through genomics. Plant Physiol 130(3):1079–1089
68. Eisenreich W, Bacher A, Arigoni D, Rohdich F (2004) Biosynthesis of isoprenoids via the non-mevalonate pathway. Cell Mol Life Sci 61(12):1401–1426
69. Estévez JM, Cantero A, Reindl A, Reichler S, León P (2001) 1-Deoxy-D- xylulose-5-phosphate synthase, a limiting enzyme for plastidicisoprenoid biosynthesis in plants. J Biol Chem 276(25):22901–22909
70. Ruiz-Sola MÁ, Rodríguez-Concepción M (2012) Carotenoid biosynthesis in Arabidopsis: a colorful pathway. Arabidopsis Book 10:e0158
71. Cunningham FX, Chamovitz D, Misawa N, Gantt E, Hirschberg J (1993) Cloning and functional expression in Escherichia coli of a cyanobacterial gene for lycopene cyclase, the enzyme that catalyzes the biosynthesis of b-carotene. Fed Eur Biochem Soc Lett 328:130–138
72. Cunningham FX, Pogson B, Sun ZR, McDonald KA, DellaPenna D, Gantt E (1996) Functional analysis of the beta and epsilon lycopene cyclase enzymes of Arabidopsis reveals a mechanism for control of cyclic carotenoid formation. Plant Cell 8:1613–1626
73. Cunningham FX Jr, Gantt E (2011) Elucidation of the pathway to astaxanthin in the flowers of *Adonis aestivalis*. Plant Cell 23:3055–3069
74. Ronen G, Cohen M, Zamir D, Hirschberg J (1999) Regulation of carotenoid biosynthesis during tomato fruit development: expression of the gene for lycopene epsilon-cyclase is down-regulated during ripening and is elevated in the mutant delta. Plant J 17:341–351
75. Goodwin TW (1980) The biochemistry of the carotenoids, 2nd edn. Chapman and Hall, London, pp 1–224
76. Nisar N, Li L, Lu S, Khin NC, Pogson BJ (2015) Carotenoid metabolism in plants. Mol Plant 8(1):68–82

77. Chitchumroonchokchai C, Diretto G, Giuliano G, Failla ML (2016) Bioaccessibility and cellular uptake of carotenoids, carotenoid precursors and a-tocopherolin biofortified golden potatoes (*Solanum tuberosum*). Fed Am Soc Exp Biol J 30(1):914–916
78. Fleshman MK, Lester GE, Riedl KM, Kopec RE, Narayanasamy S, Curley RW et al (2011) Carotene and novel apocarotenoid concentrations in orange-fleshed *cucumismelo* melons: determinations of beta-carotene bioaccessibility and bioavailability. J Agric Food Chem 59:4448–4454
79. Lipkie TE, De Moura FF, Zhao ZY, Albertsen MC, Che P et al (2013) Bioaccessibility of carotenoids from transgenic provitamin A biofortified sorghum. J Agric Food Chem 61:5764–5771
80. Nambara E, Marion-Poll A (2005) Abscisicacid biosynthesis and catabolism. Annu Rev Plant Biol 56:165–185
81. Van Norman JM, Sieburth LE (2007) Dissecting the biosynthetic pathway for the bypass1 root-derived signal. Plant J 49:619–628
82. Xie X, Yoneyama K, Yoneyama K (2010) The strigolactone story. Annu Rev Phytopathol 48:93–117
83. Granado F, Olmedilla B, Blanco I (2003) Nutritional and clinical relevance of lutein in human health. Br J Nutr 90:487–502
84. Kopec RE, Cooperstone JL, Schweiggert RM, Young GS, Harrison EH et al (2014) Avocado consumption enhances human postprandial provitamin A absorption and conversion from a novel high-beta-carotene tomato sauce and from carrots. J Nutr 144:1158–1166
85. Beltran J, Kloss B, Hosler JP, Geng J, Liu A, Modi A et al (2015) Control of carotenoid biosynthesis through a heme-based cis-trans isomerase. Nat Chem Biol 11:598–605
86. Langi P, Kiokias S, Varzakas T, Proestos C (2018) Carotenoids: from plants to food and feed industries. Methods Mol Biol 1852:57–71
87. Cardoso LAC, Kanno KYS, Karp SG (2017) Microbial production of carotenoids—a review. Afr J Biotechnol 16(4):139–146
88. Narsing Rao MP, Xiao M, Li WJ (2017) Fungal and bacterial pigments: secondary metabolites with wide applications. Front Microbiol 8:1113
89. Aluç Y, Kankılıç GB, Tuzun J (2018) Determination of carotenoids in two algae species from the saline water of Kapulukaya reservoir by HPLC. J Liq Chromatogr Relat Technol 41(2):93–100
90. Patil SA, Thakare DB (2017) Impact of fungal pigments (carotenoids) in feed for Cyprinuscarpio fish (KOI KARP). Int J Fish Aquat Stud 5(3):77–82
91. Gmoser R, Ferreira JA, Lundin M, Taherzadeh MJ, Lennartsson PR (2018) Pigment production by the edible filamentous fungus *Neurospora intermedia*. Fermentataion 4(1):11
92. Moore T (1957) Vitamin A. Elsevier, Amsterdam
93. Duester G (2008) Retnoic acid synthesis and signaling during early organogenesis. Cell 134(6):921–931
94. Haskell MJ, Jamil KM, Hassan F, Peerson JM, Hossain MI, Fuchs GJ, Brown KH (2004) Daily consumption of Indian spinach (*Basella alba*) or sweet potatoes has a positive effect on total body vitamin A stores in Bangladeshi men. Am J Clin Nutr 80:705–714
95. NHMRC (2006) National Health and Medical Research Council (AU). Nutrient reference values for Australia and New Zealand including recommended dietary intakes. http://www.nhmrc.gov.au/publications/synopses/n35syn.htm
96. Wills RBH, Wong AWK, Scriven FM, Greenfield H (1984) Nutrient composition of Chinese vegetables. J Agric Food Chem 32(2):413–416
97. Wurtzel ET, Cuttriss A, Vallabhaneni R (2012) Maize provitamin A carotenoids, current resources, and future metabolic engineering challenges. Front Plant Sci 3:29
98. Sowa M, Yu J, Palacios-Rojas N, Goltz SR, Howe JA, Davis CR et al (2017) Retention of carotenoids in biofortified maize flour and β-cryptoxanthin-enhanced eggs after household cooking. ACS Omega 2:7320–7328

99. Irwig MS, El-Sohemy A, Baylin A, Rifai N, Campos H (2002) Frequent intake of tropical fruits that are rich in β-cryptoxanthin is associated with higher plasma β-cryptoxanthin concentrations in Costa Rican adolescents. J Nutr 132:3161–3167
100. Chandrika UG, Jansz ER, Wickramasinghe SMDN, Warnasuriya ND (2003) Carotenoids in yellow- and red-fleshed papaya (*Carica papaya* L). J Sci Food Agric 83:1279–1282
101. Schlatterer J, Breighaupt DE (2005) Cryptoxanthin structural isomers in oranges, orange juice, and other fruits. J Sci Food Agric 53:6355–6361
102. Arscott SA, Howe JA, Davis CR, Tanumihardjo SA (2010) Carotenoid profiles in provitamin A-containing fruits and vegetables affect the bio-efficacy in Mongolian gerbils. Exp Biol Med 235:839–848
103. Zeb A, Mehmood S (2004) Carotenoids contents from various sources and their potential health applications. Pak J Nutr 3(3):199–204
104. Chandra-Hioe MV, Rahman HH, Arcot J (2017) Lutein and β-carotene in selected Asian leafy vegetables. J Food Chem Nanotechnol 3(3):93–97
105. Harjes CE, Rocheford TR, Bai L, Brutnell TP, Kandianis CB et al (2008) Natural genetic variation in lycopene epsilon cyclase tapped for maize biofortification. Science 319:330–333
106. Tanumihardjo SA, Bouis H, Hotz C, Meenakshi JV, McClafferty B (2008) Biofortificationofstaple crops: an emerging strategy to combat hidden hunger. Compr Rev Food Sci Food Saf 7:329–334
107. Coultate TP (1996) Food-the chemistry of its components, 2nd edn. The Royal Society of Chemistry, Cambridge
108. Van den Berg Ç, Faulks R, Granado F, Hirscheberg J, Olmedilla Â, Sandnlanç G, Southon S, Stahl W (2000) The potential for the improvement of carotenoid levels in foods and the likely systematic effects. J Sci Food Agric 80:880–912
109. Arscott SA (2013) Food sources of carotenoids. In: Carotenoids and human health. Humana Press, Totowa, pp 3–19
110. Demmig-Adams B, Adams WW III (1996) The role of xanthophyll cycle carotenoids in the protection of photosynthesis. Trends Plant Sci 1(1):21–26
111. Conaway HH, Henning P, Lerner UH (2013) Vitamin A metabolism, action, and role in skeletal homeostasis. Endocr Rev 34(6):766–797
112. Fiedor J, Burda K (2014) Potential role of carotenoids as antioxidants in human health and disease. Nutrients 6(2):466–488
113. Bendich A (1989) Carotenoids and the immune response. J Nutr 119:112–115
114. Nishino H (1998) Cancer prevention by carotenoids. Mutat Res 402:159–163
115. Rao AV, Rao LG (2007) Carotenoids and human health. Pharmacol Res 55:207–216
116. Underwood BA, Arthur P (1996) The contribution of vitamin A to public health. FASEB J 10:1040–1048
117. World Health Organization (2009) Global tuberculosis control. A short update to the 2009 report. World Health Organization, Geneva
118. Sherwin JC, Reacher MH, Dean WH, Ngondi J (2012) Epidemiology of vitamin A deficiency and xerophthalmia in at-risk populations. Trans R Soc Trop Med Hyg 106:205–214
119. Van PG, Goldbohm G (1995) Epidemologic evidence for beta-carotene and cancer prevention. Am J Clin Nutr 62:1393–1402
120. Omenn GS, Goodman G, Thongquist MD et al (1996) Effects of a combination of beta carotene and vitamin A on lung cancer and cardiovascular disease. N Engl J Med 334:1150–1155
121. Jing L, Xiao M, Dong H, Lin J, Chen G et al (2018) Serum carotenoids are inversely associated with RBP4 and other inflammatory markers in middle-aged and elderly adults. Nutrients 10:260
122. Ruiz RB, Hernandez PS (2016) Cancer chemoprevention by dietary phytochemicals: epidemiological evidence. Maturitas 94:13–19
123. Kaulmann A, Bohn T (2014) Carotenoids, inflammation, and oxidative stress—implications of cellular signaling pathways and relation to chronic disease prevention. Nutr Res 34:907–929

124. Sathasivam RID, Ki J (2018) Marine drugs review a review of the biological activities of microalgal carotenoids and their potential use in healthcare and cosmetic industries. Mar Drugs 16:26
125. Rock CL, Jacob RA, Bowen PE (1996) Update on the biological characteristics of the antioxidant micronutrients: vitamin C, vitamin E and the carotenoids. J Am Diet Assoc 96:693–702
126. Bendich A, Olson JA (1989) Biological action of carotenoids. Fed Am Soc Educ Biol 3:1927–1932
127. Foote CS, Denny RW (1968) Chemistry of singlet oxygen. VII. Quenching by β-carotene. J Am Chem Soc 90:6233–6235
128. Mathews-Roth MM (1986) Systemic photoprotection. Dermatol Clin 4:335–339
129. Mathews-Roth MM, Pathak MA, Fitzpatrick TB, Harber LC, Kass EH (1977) Beta-carotene therapy for erythropoietic protoporphyria and other photosensitive diseases. Arch Dermatol 113:1229–1232
130. Chug-ahuja JK, Holden JM, Forman MR, Mangels AR, Beecher GR, Lanza E (1993) The development and application of a carotenoid database for fruits, vegetables, and selected multi component foods. J Am Diet Assoc 93:318–323
131. Gerster H (1991) Antioxidant protection of the ageing macula. Age Ageing 20:60–69
132. Seddon AM, Ajani UA, Sperduto RD (1994) Dietary carotenoids, vitamin A, C, and E, and advanced age-related macular degeneration. J Am Med Assoc 272:1413–1420
133. Westphal A, Böhm V (2015) Carotenoids, properties, distribution, bioavailability, metabolism and health effects. Ernahrungs-Umschau 62(11):196–207
134. de Benoist B, Martines J, Goodman T (2001) Vitamin A supplementation and the control of vitamin A deficiency: conclusions. Food Nutr Bull 22(3):335–337
135. West KP, Darnton-Hill I (2008) Vitamin A deficiency. In: Semba RD, Bloem MW (eds) Nutrition and health in developing countries, 2nd edn. Humana Press, Totowa, pp 377–433
136. Britton G (1992) Carotenoids. In: Hendry GF (ed) Natural foods colorants. G.F. Blackie, New York, pp 141–148
137. Gayen D, Ghosh S, Paul S, Sarkar SN, Datta SK, Datta K (2016) Metabolic regulation of carotenoid-enriched golden rice line. Front Plant Sci 7:1622
138. Zunjare RU, Hossain F, Muthusamy V, Baveja A, Chauhan HS et al (2018) Development of biofortified maize hybrids through marker-assisted stacking of β-carotene hydroxylase, lycopene-ε-cyclase and opaque2 genes. Front Plant Sci 9:178
139. Britton G, Liaaen-Jensen S, Pfander H (eds) (2009) Nutrition and health. Birkhauser, Basel
140. Nkuba MDE, Joyce K, Wendelin MA, Beatrice E (2017) Provitamin a carotenoids content and bioaccessibility in the modified local diet for children aged 6-23 months in Bukoba, Tanzania. Afr J Food Sci Technol 8(9):171–182
141. Bushway RJ, Yang A, Yamani AM (1986) Comparison of alpha- and beta-carotene content of supermarket versus roadside stand produce. J Food Qual 9:437–443

Chapter 27
Commercialization and Marketing Potential of Carotenoids

Samina Yaqoob, Muhammad Riaz, Aqsa Shabbir, Muhammad Zia-Ul-Haq, Suaad S. Alwakeel, and May Bin-Jumah

27.1 Introduction

Carotenoids are isoprenoid molecules produced de novo in all photosynthetic entities and in some non-photosynthetic bacteria and fungi. The carotenoids share a common a skeleton formed by two isoprenoid unities linked in such a way that the molecule is linear and has inverted symmetry in the center and several c.d.b. in the chain. This basic skeleton of 40 carbon atoms can be modified by hydrogenation, dehydrogenation, cyclization, shortening or chain extension, isomerization, introduction of substituents or by a mixture of these methods. Some 1178 carotenoids have been properly characterized from 700 sources comprising of plants, algae, fungi and bacteria thus exhibiting huge structural diversity and range of physiochemical properties. Macroscopic fungi (e.g. *Blakeslea trispora* for β-carotene & lycopene) or microscopic fungi (e.g. *Rhodotorula* or *Xanthophyllomyces* for torulene & astaxanthin) are also a good source of carotenoids. The algal (e.g.

S. Yaqoob
School of Business and Economics, University of Management & Technology, Lahore, Pakistan

M. Riaz
Department of Pharmacy, Shaheed Benazir Bhutto University, Sheringal, Pakistan

A. Shabbir
Department of Electrical Engineering, Lahore College for Women University, Lahore, Pakistan

M. Zia-Ul-Haq (✉)
Office of Research, Innovation and Commercialization, Lahore College for Women University, Lahore, Pakistan

S. S. Alwakeel (✉) · M. Bin-Jumah (✉)
Department of Biology, College of Sciences, Princess Nourah Bint Abdulrahman University, Riyadh, Saudi Arabia
e-mail: ssalwakeel@pnu.edu.sa; mnbinjumah@pnu.edu.sa

© Springer Nature Switzerland AG 2021

M. Zia-Ul-Haq et al. (eds.), *Carotenoids: Structure and Function in the Human Body*, https://doi.org/10.1007/978-3-030-46459-2_27

Haematococcus pluvialis & *Dunaliella salina* for astaxanthin & β-carotene) and bacterial (e.g. *Flavobacterium* sp. for zeaxanthin) sources are also present. Supplements are usually used to obtain those carotenoids which food is usually deficient like astaxanthin, fucoxanthin and meso-zeaxanthin.

A wide range of pre-clinical and clinical investigations indicate that dietary carotenoids have significant health-promoting effects. Noticeable evidence exists for their ability to reduce the risk of major chronic disorders. It seems that they play roles in both the prevention and treatment of various human diseases and disorders. They are a key symbol of a suitable nutritional condition in birds and fishes suggesting a sign of fitness and consequently increasing sexual pull. In algae and higher plants, they maintain the configuration and task of the photosynthetic complex, quench the chlorophyll triplet states, scavenge ROS, dissipate of excess energy and help in harvesting light. As vital floral pigments, they attract pollinators and seed dispersers due to their striking rich color [1].

The universal carotenoids market volume was valued at USD 1.40 Billion in 2018 and is projected to reach USD 1.85 Billion by the end of 2026 growing at 3.57% compound annual growth rate (CAGR). The volume of world carotenoids consumption in 2007 was equal to 4193 metric tons and has increased within 10 years by about 1500 metric tons, thereby it was equal to 5693.6 metric tons in 2017. Carotenoids added to food totaled 4020.8 metric tons in 2017 [2]. The market of beverages with carotenoids added totaled 1609.8 metric tons (Euromonitor International). Non-food market consumed 63.0 metric tons of carotenoids in 2017 (Euromonitor International). The market prices of carotenoids can vary from 300 to 3000 USD per kg of β-carotene and from 2500 to 10,000 USD per kg of astaxanthin. Ketocarotenoids, such as astaxanthin or canthaxanthin, are amongst the most expensive carotenoids in market. They are abundantly present in algae, whereas they are rarely present in higher plants. They enjoy a characteristic keto functionality on 4 or 4′ position on the β-ionone ring and can also possess OH groups on the 3 and 3′ positions. The key commercial usages of ketocarotenoids are as feed additive in the aquaculture and poultry industry for color and nutritional value.

Various elements like safety, efficacy, price, and packaging volume among others drive the competition in the industry. The global carotenoids market is characterized by intense competitive conditions leading towards merging and acquisition of competitor companies. Subsequently, big pharmaceuticals corporations are taking over smaller companies to strengthen their position. Highest demand of carotenoids is in European and the USA markets. The carotenoids market ecosystem is given in Fig. 27.1. Niche markets exists where customers can pay premium price for health food ingredients like carotenoids.

The food product is the focus of rules in the USA, while the technology utilized to obtain the final food product is concentrated in Europe. The largest market for carotenoids is Europe and it is a home for prominent globally exporting carotenoids manufacturers based in Germany, Netherlands and Denmark. Germany, France, UK and Netherlands are the most significant states within the Carotenoids market in Europe. However, compared to the main regional markets of Japan and the US, the

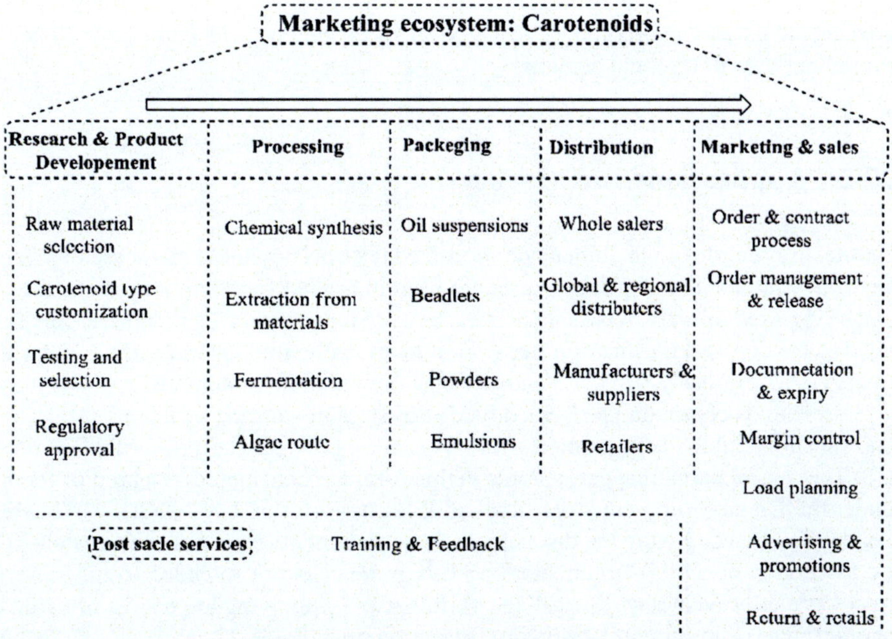

Fig. 27.1 Carotenoids market ecosystem

European markets are less-organized. Various elements limit development of this industry in EU. Food tagging, product devising, food handling, wrapping, advertising, registering and accrediting particulars are rigorously checked in the EU and are recognized as constraining the dimension of the buyer market. Factually the carotenoids markets have developed unevenly in Europe as majority of "Functional Foods" brands have sprung in merely a limited number of countries. Further, multinational food corporations usually offer single products instead of umbrella brands in the functional food (FF) market. North America is the second-largest carotenoids market after Europe. The Asia-Pacific carotenoids market is estimated to accrue commendable proceeds from cosmetics application, growing at a CAGR of 5% over the forthcoming years. Potential markets also exist in oil rich middle-eastern or Gulf States while African markets are not well-structured [2].

The world carotenoids industry is set to achieve over a 4% CAGR up to 2026, supported by rising demand for natural coloring agents, along with growth in end-use applications. Speaking of the application spectrum, the food and beverage application segment of the carotenoids market is estimated to advance at a lucrative pace and show massive growth in the upcoming years. Carotenoids are extensively used in various food and beverages as an additive for maintaining optimal vitamin A levels and boosting immunity, skin health, as well as vision. The Asia-Pacific carotenoids market is estimated to accrue commendable proceeds from cosmetics application, growing at a CAGR of 5% over the forthcoming years. Technological advancements and burgeoning consumption in cosmetics and nutraceuticals are propelling overall regional demand.

27.2 Applications and Market

Carotenoids employed in animal and food-feed uses occupy the lion's share of market size. Carotenoids of plant sources are used in animal food applications as they can be ingested directly bypassing requirements of recovery while microbial sources need extraction and purification steps. Individual carotenoids market size is shown in Fig. 27.2. Similarly, Fig. 27.3 shows usage potential of carotenoids.

Carotenoids containing processed foods and supplements may contains structure and function claims like "promotes cardiovascular, prostate and skin health" in the US. The reason being that carotenoids in this form are considered as a kind of food under the Dietary Supplement Health and Education Act of 1994 (DSHEA). Such goods should accompany the disclaimer "these statements have not been evaluated by the Food and Drug Administration. This product is not intended to diagnose, treat, cure or prevent any disease" [3, 4]. In the EU, following are details of carotenoids under Commission Regulation (EU) 231/2012 (Table 27.1).

The Directive 87/552/EC in 1988, approved the use of astaxanthin @ 100 mg/kg in salmon and trout feed along with canthaxanthin. In 2004 under Regulation 258/97/EC, astaxanthin-rich oleoresin herbal capsules (@ 4 mg/capsule) were

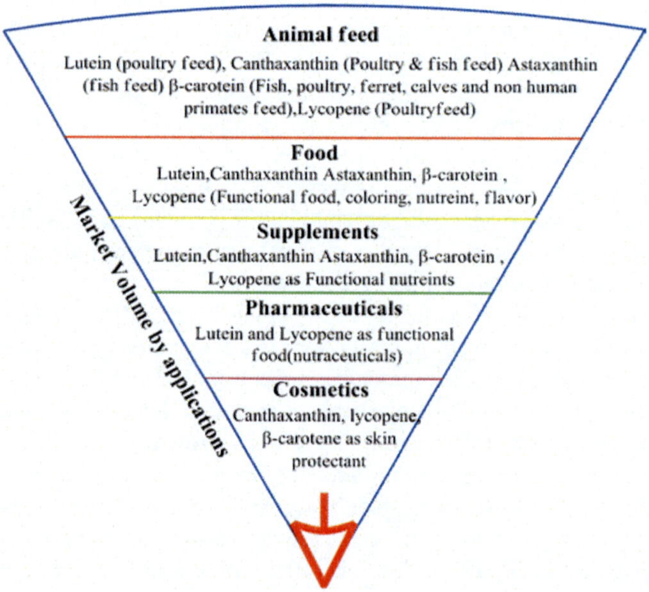

Fig. 27.2 Individual carotenoids market size

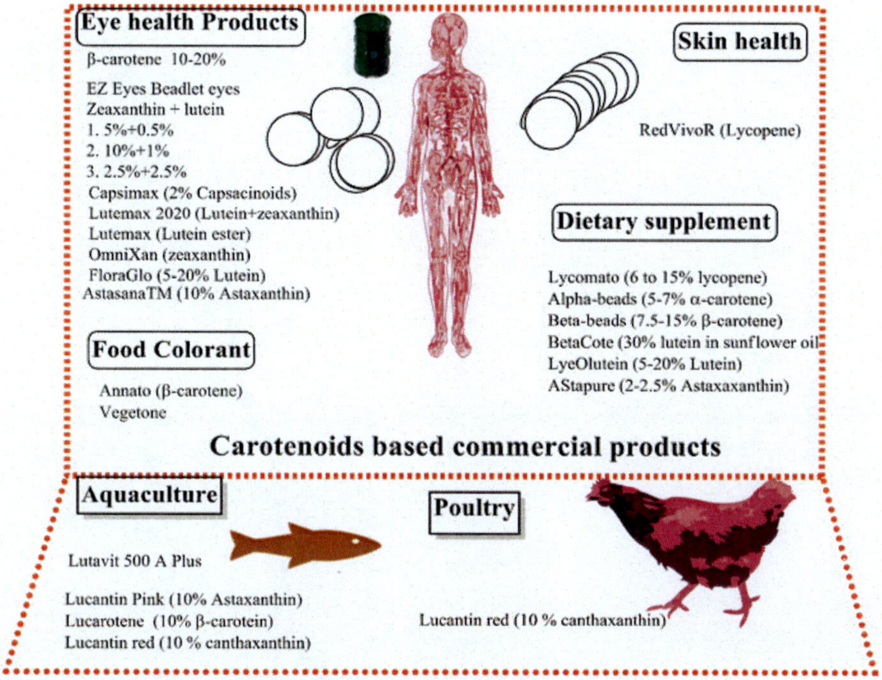

Fig. 27.3 Usage potential of individual carotenoids

Table 27.1 Carotenoids details in Commission Regulation (EU) 231/2012 [3, 4]

Carotenoids as food additive type	Color Index No	Production method
β-carotene (E160a)	75130	Chemically synthesis, extraction from plants, or production by the cultivation of *B. trispora* or *D. Salina*
Lycopene (E160d)	75125	Synthesized chemically, extracted from red tomatoes or obtained from *B. trispora* cultivation
Lutein (E161b)	–	Extraction from edible fruits, grass, Lucerne (*Medicago falcata*) and marigold
Zeaxanthin (E160a)	–	Algal-derived carotenoid preparations

permitted as a novel food ingredient and sold by Herbal Science International (Loughton, UK). Similarly, astaxanthin-rich oleoresin (AstaReal, Sweden) and astaxanthin-rich extracts (AlgaTechnologies, Israel) were permitted under this regulation. According to Regulation 1288/2004/EC, astaxanthin (E161z) obtained from *P. rhodozyma* is used in salmon and trout feed @ 100 mg/kg of the complete feed, while Regulation 393/2008/EC specifies astaxanthin dimethylsuccinate (E161j) for the same purpose. As per Regulation 721/2008/EC, astaxanthin acquired from *Paracoccus carotinifaciens* can be added to feed @ 100 mg/kg of complete feed along with canthaxanthin and adonirubin.

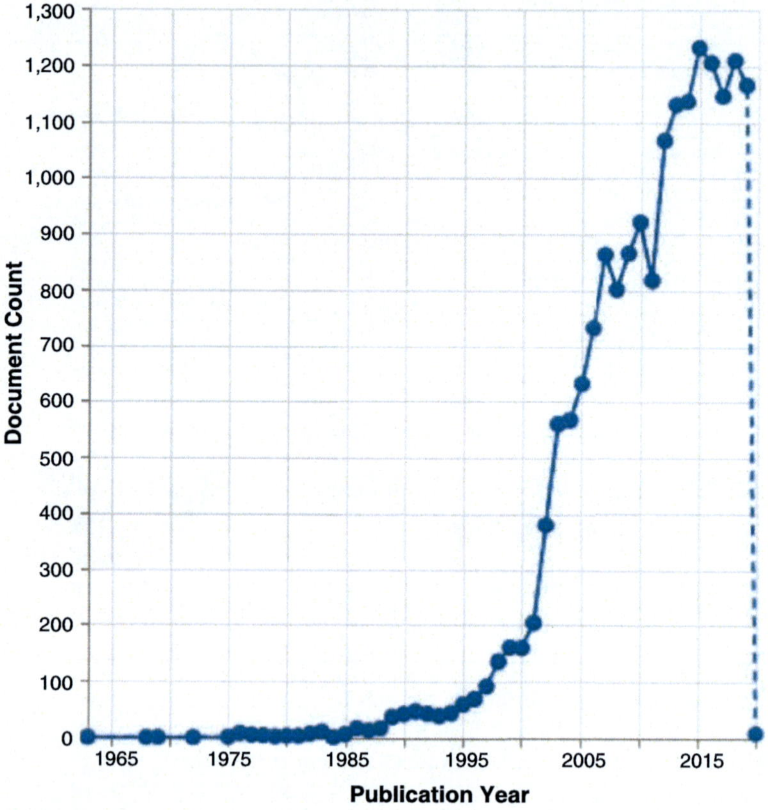

Fig. 27.4 Patents granted to carotenoids inventions with time scale

When "carotenoids" and "cosmetics" are used as keywords, 17,709 patents appear (Fig. 27.4), top owners (Fig. 27.5), top applicants (Fig. 27.6) and top inventors (Fig. 27.7) [5].

When "carotenoids" and "supplements, nutraceuticals" are used as keywords, 5524 patents appear (Fig. 27.8), top inventors (Fig. 27.9), top owners (Fig. 27.10), and top applicants (Fig. 27.11) [5].

When "carotenoids colors" are used as keywords, 31,731 patents appear (Fig. 27.12), top owners (Fig. 27.13), top inventors (Fig. 27.14), and top applicants (Fig. 27.15) [5].

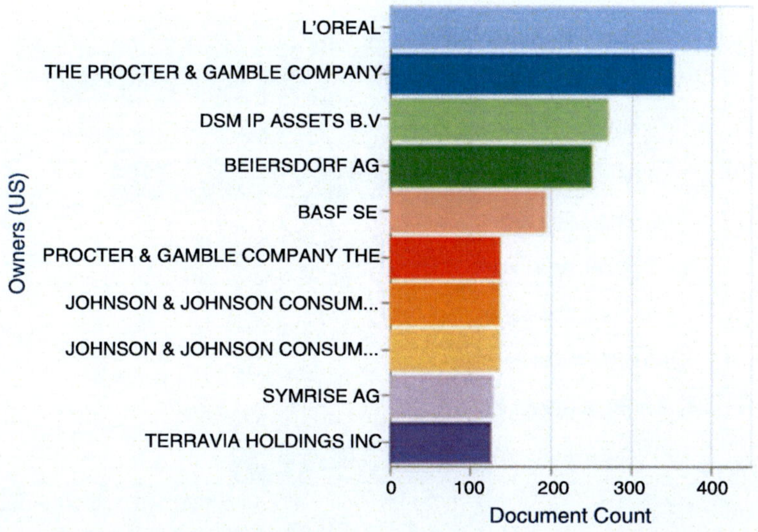

Fig. 27.5 Top owners of patented inventions

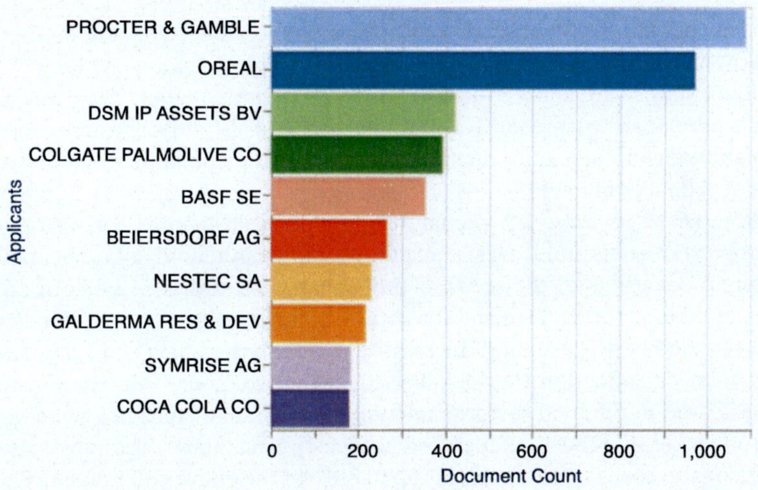

Fig. 27.6 Top applicants

27.3 Coloring Potential of Carotenoids

It is well-accepted saying that we also eat with our eyes, color is thus the first sensory interaction with food products, prior to taste and smell. Carotenoids help in enhancing the overall aesthetic appeal of food products by making them visually appealing. The salmonids characteristic "red, pink or orange" is due to their feed of

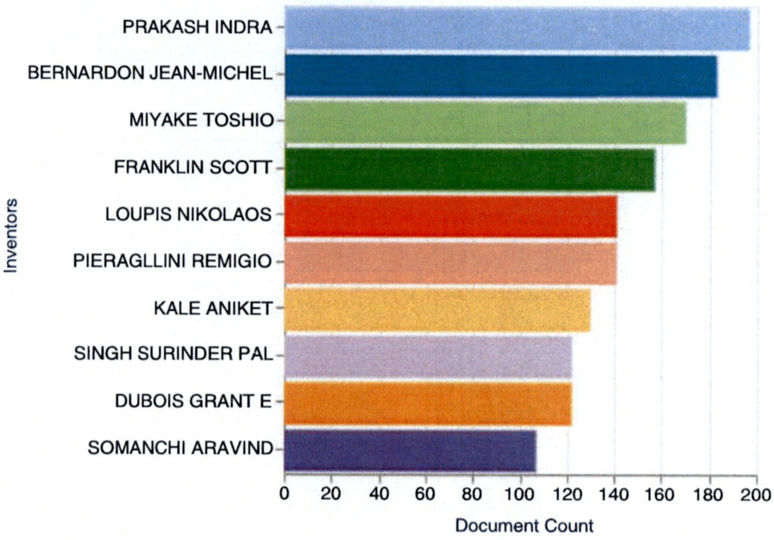

Fig. 27.7 Top Inventors

crustaceans or other fish with small crustaceans in their digestive system. Egg yolks are yellow because they accumulate carotenoids (largely lutein). At least 7 conjugated double bonds are needed for the carotenoid to impart color. Their role as food colorants have been comprehensively studied earlier. Currently, research focus is on microbial/biotechnological production of carotenoids as potential food and as precursors of aroma compound.

Achiote (*Bixa orellana* L.) accumulates several carotenoids derivatives (bixin and norbixin) in seeds and leaves. Annatto, the only carotenoid color obtained from the external coverings of the seeds of the Annatto tree fruits, is being used since decades in various forms. Demand for natural colors to pigment dairy foods with a reddish-orange hue is escalating. Increasing plantations in Asia and Latin America has led to overproduction thus leading to decreased prices for crude products. Annatto is used in the food sector to add yellow or orange color in a wide range of food products (e.g., processed meat, smoked fish, beverages). The most important color of annatto seeds is the cis-bixin (up to 80% of the entire color quantity), while norbixin (cis and trans) represent a insignificant portion of achiote seeds.

The consumption of β-carotene and lycopene as pigments in Food and Beverage division is increasing. β-carotene is extensively used in cheese, spreads, yogurts, soups, sauces, and bakery products, while lycopene is utilized in sauces and soups. Lycopene and canthaxanthin provide red dye for food and beverages are chiefly obtained from tomatoes. Astaxanthin or canthaxanthin are added to animal feed in farmed salmons and trout, giving their meat the characteristics color of the species in nature. In the poultry industry, lutein, bixin, and capsanthin are used to impart typical yellow-red color to chicken skin and egg yolk; for which the natural sources are marigold, achiote, and pepper, respectively. Sales of natural food colors like

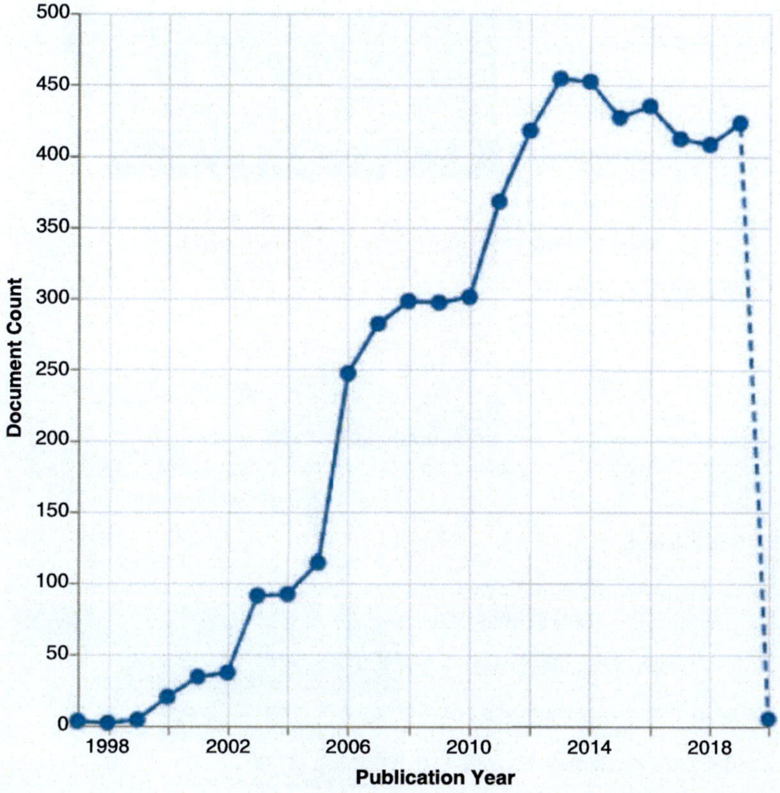

Fig. 27.8 Patented Inventions containing carotenoids over time

carotenoids are growing globally due to growing demand for clean label products. Natural colors are five times expensive than synthetic analogs, particularly for confectionary items, where this difference can reach up to 20 times. Requirement of more raw materials is a key reason for increased cost of natural colors. Cost is also influenced as increased quantity of natural color is required for the production of looked-for color.

27.4 Carotenoid-Based Commercial Products

The carotenoid-based commercial products available as supplements in the market are for (1) provitamin A potential (2) powerful antioxidant and lipid peroxidation preventing action (3) bone and skin health (anti-aging), physical stamina (sports nutrition), vision, and immune system augmentation (4) cancer, cardiovascular, neuronal, and gastrointestinal protecting capacity and (v) animal nutrition (particularly for poultry and fish).

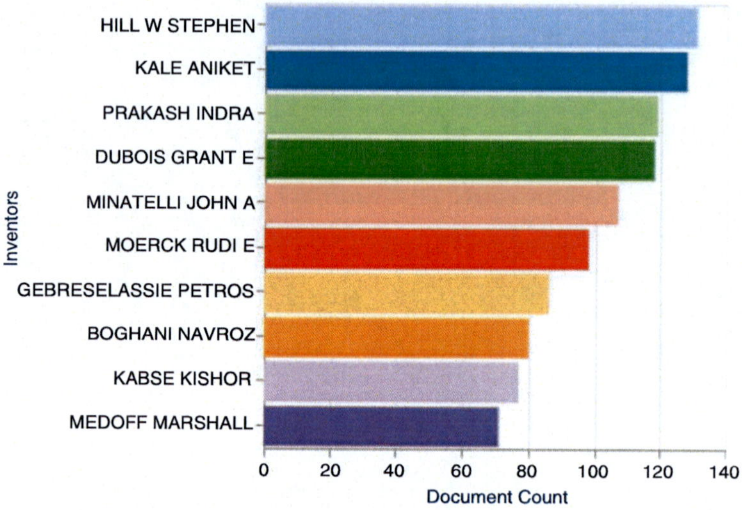

Fig. 27.9 Top inventors

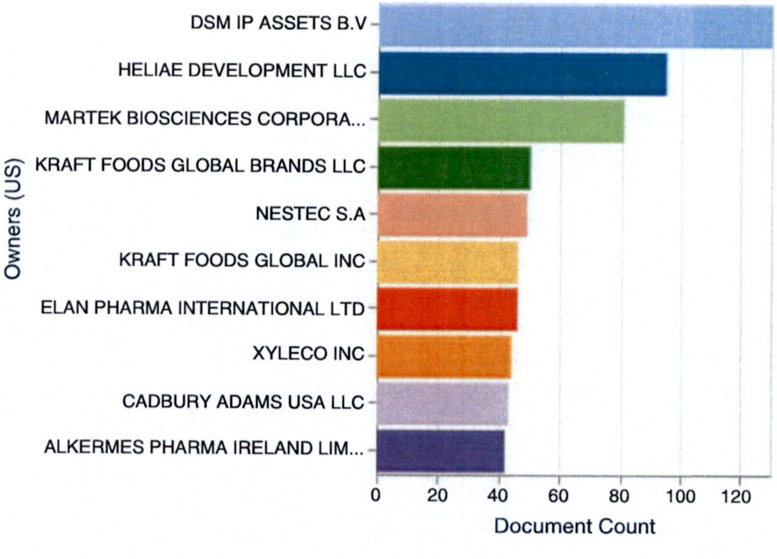

Fig. 27.10 Top owners

Recently, striking red–orange coloration of ornamental aquaculture animals, such as freshwater crustaceans and fishes, is a prevalent advertising element. Astaxanthin and canthaxanthin, are mainly used as nutritional supplements to improve skin pigmentation. Further, carotenoids increase the health and reproduction of ornamental animals, including better embryonic and larval growth, maturation, increased immune response, and photoprotection. Aquaculture feed is the most

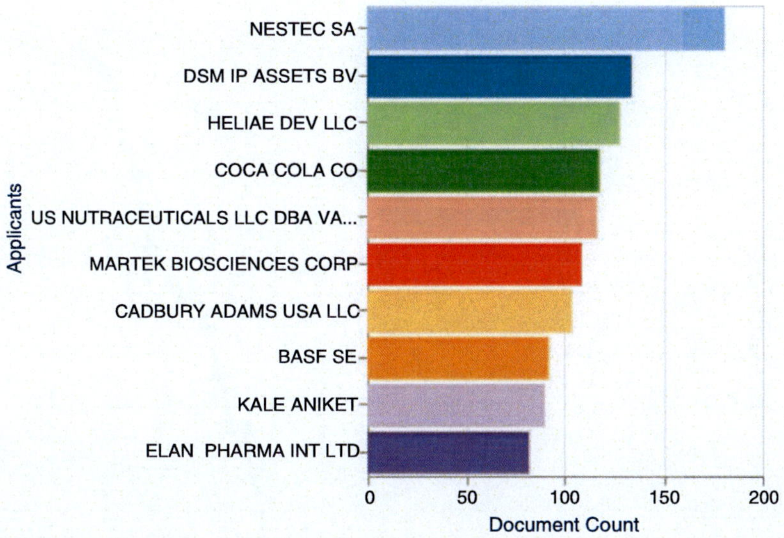

Fig. 27.11 Top applicants

significant segment that is responsible for the substantial revenue earned by astaxanthin and related carotenoids [3, 4].

27.5 Food/Feed Potential

Animal feed and food & beverage (F&B) segments are presently the leading markets for carotenoids worldwide. By 2026 end, F&B business is expected to exceed US$ 2400 million income. The F&B market volume is estimated to grow markedly in near future, due to large-scale usage as food additives worldwide. Animal feed segment, which makeups almost 41% of the total carotenoids market share in terms of size, will grow significantly in coming years. Animals need carotenoids as they cannot make them hence carotenoids are added to animal feed.

Carotenoids are not developed alike in essentially all food groups. Carotenoids products are chiefly demanded in the dairy, confectionery and soft-drinks. The orange-red-brown color of French cheese *vieux-pan* is due the carotenoid (added to increase aesthetic value) produced by *Brevibacterium linens*. In Russia, natural pigments such as lutein are added to infant formulas to increase children's health. In Japan, *Undaria pinnatifida*, fucoxanthin rich seaweed, is marketed as pasta component.

Their use in animal feeds acts as a main precursor for amino acids in animal feeds and improves feed coloration, consequently enhancing their palatability besides increasing the looks of meat, meat products and fish. Feed area is a big segment where carotenoids are required to color fish, broilers and eggs. Suitable fish

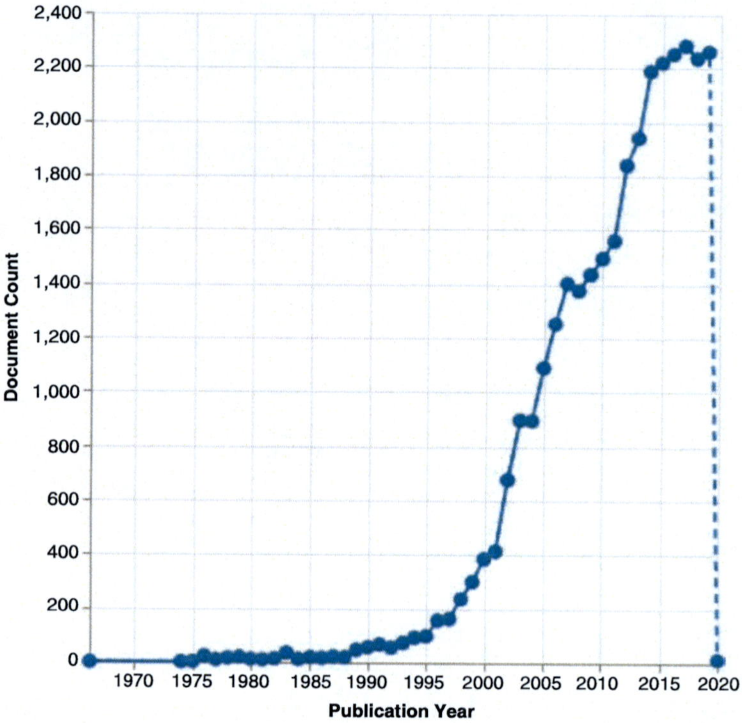

Fig. 27.12 Patented Inventions of carotenoids over time period

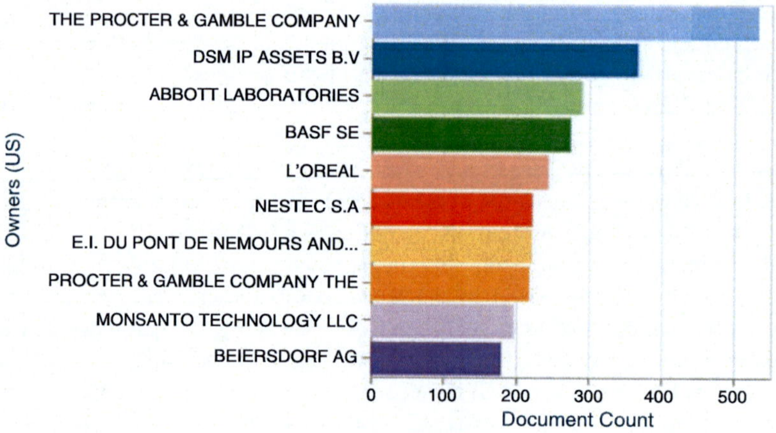

Fig. 27.13 Top owners

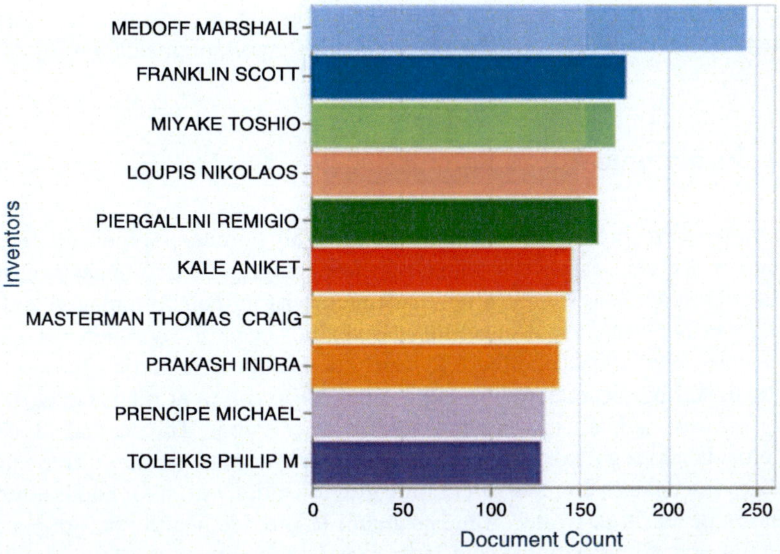

Fig. 27.14 Top inventors

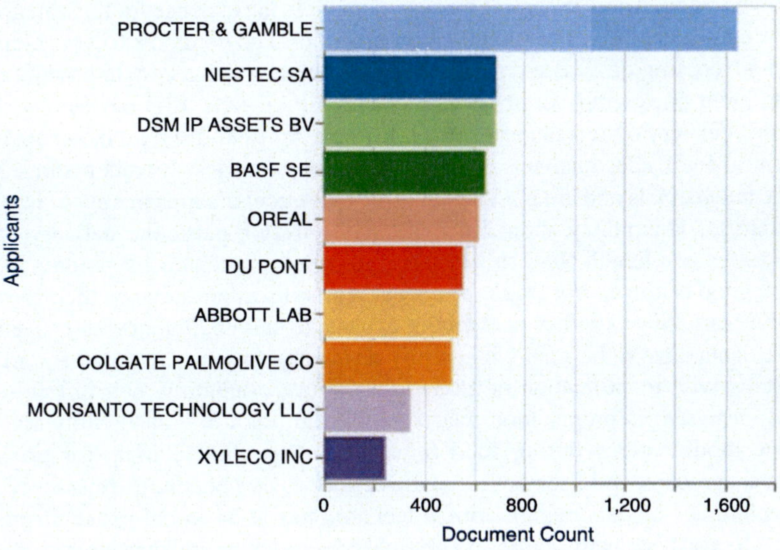

Fig. 27.15 Top applicants

flesh pigmentation and an economically feasible product cannot be obtained without them. Approximately 15–25% of the total feed costs linked with aquaculture production are because of the price of the needed carotenoid feed supplements. There are two chief uses for carotenoids in feed additives. Use of astaxanthin or canthaxanthin to acquire the pink color of the flesh in salmon and trout farming. Carotenoids are used in poultry for the skin and egg yolk pigmentation [6].

27.6 Functional Foods Potential

Their antioxidant potential makes them one of highly purchased product in the over-the-counter sector. Despite the general socio-demographic and behavioral trends favoring carotenoids as FF, peculiar challenges exist in their formulation and marketing. Legally nutraceuticals including carotenoids occur in an in-between domain between pharmaceuticals and food. In nearly all European countries, they are typically regulated by discrete bodies and are subject to different rules, hence a "grey zone" surfaces with an increased degree of uncertainty. The factual processes, requirements and regulatory organizations linked with market entry differ substantially between the two domains. Definition glitches chiefly exist for goods aimed for prevention of nutrition-related ailments and/or to provide health ("health claims"). In the EU and allied national legislations, it is presently not allowed to utilize disease-related traits in customer knowledge or goods flyers for Functional Food. Hence, manufacturers attempt to encompass the nature of claims allowed for Functional Food. Another central success element is the expense for this category of food as compared to "conventional" food products. The common success elements for the advertising of food such as taste, convenience traits, a certain product range and different packaging volumes are valid for FF as well. It is advised to oblige high-size delivery outlets (superstores, wholesale stores & discount merchants) that are central food sales channels in most European countries. Buyers assume FF in such retail channels and are not keen to drive to particular shops merely to buy such FF products. This policy should not eliminate offering particular delivery outlets (pharmacies and health food shops) the equivalent or adapted FF product. As FF warrant good health and/or stop/manage particular disorders suitably, their formulation and marketing is rather compound, expensive and challenging as exceptional requirements have to be met. Noteworthy research efforts are required for all this. This starts with identification of likely carotenoids, evaluating their physiological impact, designing a proper food matrix keeping in mind bio-availability and prospective modifications during food processing, and clinical trials for product's effectiveness to get endorsement for claims of improving health. Businesses require new techniques of recognizing critical technologies at an initial phase of product brand. Merely systematic studies do not make a product popular in emporium. The product should be in a suitable shape so that the customers can receive it easily. Consequently it is essential to sightsee which ailments users are worried about so that the carotenoid product can be fruitful in the market.

Certainly, strong suggestions exist indicating association between carotenoid ingestion and 'good health.' But is it only due to carotenoids? We are still far away from answering this question. In studies using cell and animal models (CAM), carotenoids impact various molecular and cellular mechanisms. It is however difficult to directly associate existing experimental data to human pathophysiology.

First, the findings of many studies require an accurate analysis of the processes by which carotenoids exercise neuro-protective effects. Their antioxidant potential can be one of the reason as increased oxidative stress is one of the characteristic pathologies in neurodegenerative ailments. However, the mechanisms by which carotenoids prevent neuro-inflammation and trigger autophagy have not been well-studied. Second, clinical application studies in human are necessary to observe the causal relationship of the carotenoid effect in human. It is also possible to deduce the connection between carotenoid ingestion and the onset of disease through comparative studies of races taking different diets. Finally, it is possible that merely enhancing the carotenoids intake exercises only limited protective effects to neurons. Hence future studies evaluating other neuro-protective reagents/treatments that exhibit synergistic effects in combination with carotenoids in neurodegenerative ailments will be vital in finding real treatments [7].

27.7 Market Growth Drivers

The carotenoids market growth is primarily driven by

1. The all-natural trend resonating in the global marketplace coupled with increasing demand for naturally obtained colorants like carotenoids in the nutraceutical industry due to their positive impact on various human organs and systems.
2. The alternative treatment methods for diabetes, eye disorders, and other lifestyle diseases are witnessing robust popularity. There is increased expenditure on medical and healthcare insurance facilities coupled with increased life expectancy of aging population. The increasing inclination towards functional foods and dietary supplements is also boosting this industry.
3. Increasing application in personal care and cosmetics especially in face care (anti-aging) and skin care (whitening agents & sun screen) are also a major factor in growth of carotenoid industry.
4. Increasing health awareness thanks to easy access to the information leading to self-care movement among consumers is a key driver in growth of industry. Participants of space science and operations, acknowledgement of FF like carotenoids is increasing due to improving health value and eating routines in long-term flights and tasks.
5. Amendment in government rules and liability are a key driver of growth of carotenoids industry.

27.8 Challenges

1. Deficiency of technical know-how may hinder entry of new players in market.
2. Toxicity symptoms due to the higher carotenoids intake is a key challenge in growth of industry. Currently carotenoids are being considered as a substitute to

antibiotics. Wide use of antibiotics in animal feed has created drug resistance in animals and humans. Hence carotenoids like β-carotene are focus of research to help animals grow by boosting their immunity.
3. Rigid government policies and shortage of awareness in emerging economies is hampering growth of carotenoids industry.
4. Adulteration of natural carotenoids colorants by colors from other sources is a major challenge being faced by food industry.
5. Ingredient compatibility during food manufacturing impacts ingredient interaction. Further, oil-based colors including carotenoids can stain permanently, creating difficulties for F&B processors. Cold storage to retain integrity of natural colors is challenging as opening and closing of storage unit can contaminate and degrade product. The fortified drinks are in high demand. However, fortification can cause unwanted color interactions degrading colors.
6. Procurement of natural colors from different geographical regions where they are grown is an emerging challenge. For example, annatto is chiefly obtained from Latin America and most fruit-based colors originate from Europe. Hostile climatic situations in these areas can upset their supply and hence price.
7. In the last decade, there has been substantial progress in increasing pro-vitamin A carotenoid content in maize. However, maize carotenoid quality during storage depends on temperature and humidity and minute changes in them can degrade carotenoids.

27.9 Production Methods

Carotenoids can be achieved via industrial fermentation using microorganisms, extraction from plants, and chemical synthesis.

Industrial production of carotenoids is done by

(i) biotechnological processes using filamentous fungi, yeasts, bacteria or microalgae
(ii) solid-liquid extraction from plants
(iii) Chemical synthesis

Currently, carotenoids produced by chemical synthesis lead the international commerce [3, 4]. Biotechnological production however is now becoming method of choice due to two strategies (1) isolation after fermentative production in a microorganism and (2) genetic engineering to synthesize the desired carotenoid in plants. There is competition between synthetic and naturally-obtained carotenoids. The biotechnological production is considered superior due to availability of range of microorganisms in nature, flexibility in the usage of substrates and agro-industrial wastes and the option to regulate operational conditions like temperature, pH, dissolved oxygen, and light intensity. These organisms can produce carotenoids from fats and other basic organic metabolic building units. Fatty tissues of animals store carotenoids and entirely carnivores get these molecules from animal fat.

Carotenoids are being made by chemical synthesis since first synthesis in 1950 by Karrer, Eugster, Inhoffean, and Milas. In 1954, synthetic β, β-carotene was made on industrial level. Nearly all key carotenoids including lycopene, canthaxanthin, astaxanthin, β,β-carotene, β-apo-8′-carotenal, β-apo-8′-carotene, and cytranaxanthin have been synthesized either by Wittig reactions or Grignard compounds [8]. Synthetic carotenoids are made from various approaches including reactions of dehydration and elimination, specific condensation of carbonyl compounds and homo-dimerization reaction, and selective coupling reaction of Csp2–Csp2. Chemical synthesis using petrochemical-derived precursors is the preferred production method. Table 27.2 indicates main products of famous companies [2].

The chemical synthesis of over 200 carotenoids has been reported. β-carotene obtained from natural sources is merely 2% of the total global market and remaining from the chemical synthesis. Various rubrics determine strength of both synthetic and natural production.

27.9.1 *Yields*

Very small quantity of carotenoids is obtained from natural sources including plants, animals, and micro-organisms. The maximum producing algal strains yield <10% of carotenoids per DW.

27.9.2 *Cost*

They are quickly created synthetically utilizing low-cost labor and low-priced chemicals and inexpensive harvesting and extraction. The expenses of algae-based carotenoid can range > $7500/kg while its synthetic equivalent can cost half of this amount.

Table 27.2 Main products of famous companies

Company	Country of origin	Main products
BASF SE	German	Lucantin®,Lucarotin®
DSM	Dutch	Carophyll®
Döhler Group	Germany	–
Chr. Hansen	Denmark	NutriPhy®
Kemin Industries	USA	Kem Glo™

27.9.3 Choice of Type

Carotenoid of choice can be manufactured by synthetic way. The algal astaxanthin is >95% esterified while synthetic astaxanthin can be obtained in both esterified and unesterified forms. However due to their chiral nature, natural and synthetic carotenoids cannot be distinguished in a laboratory.

27.9.4 Pros and Cons of Both Methods

The total annual production in nature is estimated at over 1000 million tons. Synthetic carotenoids controlled the source landscape with a market value of 200 million US$ in 2018 and by 2026; they are anticipated to surpass US$ 4700 million income. Synthetic carotenoids account for 90% of the total market while remaining 10% are obtained from natural sources. Synthetic market is dominated by two main companies, the BASF and DSM, which produce 55% of the global market while remaining 45% demand is met by small players.

Synthetic carotenoids have certain alluring traits as compared to natural analogs. Synthetic carotenoids are more stable as they are particularly designed to decrease oxidation or isomerization. They are available as colloidal suspension, emulsification, and dispersion colloids to make carotenoids application in food easier. Despite these advantages, they are less effective with regard to their health-promoting characteristics and are therefore less appreciated and wanted. They manifest high toxicity, carcinogenicity, and teratogenicity characteristics and health-conscious customers are hesitant to use them. The emerging modern standards for a healthy lifestyle and ecofriendly approaches has increased the quest of natural colors as alternatives to synthetic analogs. Hence biomass (vegetables, fruits, yeast and microorganisms) is used as carotenoids source. However naturally sourced carotenoids need complex extraction stages, are more hydrophobic, unstable, experience seasonal fluctuations, and are limited in supply. The yield from natural hosts is very less, commonly a few mg/kg of raw material. Further, plants or microbes commonly yield mixtures of these compounds with same physical and chemical properties e.g. α- and β-carotenes. Thus, their production from natural sources is challenging and quite expensive from process economics and sustainable land-use perspective. Chemical synthesis is also very costly due to structural complexity [9].

27.10 Recent Commercial/Corporate Developments

- ExcelVite announced the collaboration with USP in April 2019 for the publication of new Plant Carotenes Monograph.

- In March 2019, GacLife, solutions by nature beverage brand launched five new daily health beverages, which include the highest amount of antioxidant carotenoids to provide powerful antioxidant protection for the whole body.
- In March 2019, Kemin Industries, Inc. introduced Organic KEM GLO for North American Egg Producers. It is a USDA-certified organic additive that enables organic egg producers to intensify the color of egg yolks. It utilizes the natural characteristics of paprika to evenly distribute color throughout the feed, delivering consistent egg color pigmentation and density.
- In August 2018, BASF Animal Nutrition introduced Lucantin NXT product line in the EU 28 market. Lucantin NXT products offer outstanding stability, longer shelf life, and high homogeneity, while maintaining the efficacy of egg yolk and broiler skin coloring.
- In July 2018, Nutrex Hawaii has received United States Pharmacopeia (USP®) Verified Mark for BioAstin® Hawaiian Astaxanthin® which is the first dietary supplement brand of astaxanthin to receive USP verification. Nutrex Hawaii is the only producer of naturally grown Hawaiian Astaxanthin using open pond technology
- In February 2018, KnipBio established SCP strain for Astaxanthin production. The enterprise declared that it has developed its single-cell protein-based meal containing bio-astaxanthin.
- In January 2018, NextFerm Technologies Ltd., an Israel based biotech start-up company has developed two kinds of products using their fermentation technology—NextFreeze™, a new and improved baker's yeast strain and AstaFerm™ phaffia yeast astaxanthin. NextFerm's AstaFerm™ is a whole astaxanthin extract of *Phaffia rhodozyma*. The Company is aiming to commercialize AstaFerm™ astaxanthin in the market at the last quarter of 2018, and will be available in 10% oil and 5–10% free flowing CWD powder forms.
- DSM & Kemin Industries have partnered to launch natural zeaxanthin as Optisharp® in 2014.

27.11 Overall Scenario

In 2024 the consumption of Natural Carotenoids is estimated to be 2699.8 MT. Western Europe followed by US are likely to be the largest market for natural colors. FDA has approved list of natural colorants including annatto, caramel, β-carotene for use in cosmetics. The capsanthin, astaxanthin, β-carotene, lutein,

Table 27.3 Market share by key carotenoids

Carotenoid	Revenue generated (million) 2017	Expected in 2022 (million)	CAGR (%)	Market share (%)
Capsanthin	USD300	USD385	5.1	20
Astaxanthin	USD288.7	USD426.9	8.1	16

annatto, lycopene, and canthaxanthin together share nearly 90% of the total market value. Astaxanthin and β-carotene are the two most recognized carotenoids and make up almost half of the global carotenoid market. Table 27.3 indicates market share by key carotenoids [3, 4].

27.12 Active Organizations

The positive endorsements by research institutes has greatly impacted the adoption of carotenoid-rich food by customers. International Carotenoid Society, IBERCAROT (the Ibero-American network for the study of carotenoids as FF components), Spanish Carotenoid Network (CaRed), and Eurocaroten are working to increase the carotenoids information and research.

27.13 Key Manufacturers/Players

The market for carotenoids is highly competitive due to existence of numerous players. Moreover, more than half of the market is accounted by two major players DSM and BASF. BASF and DSM occupy the market of β-carotene, astaxanthin, and canthaxanthin, while small companies dominate lutein and lycopene market. These manufacturers are facing stiff competition from the Indian and Chinese manufactures of carotenoids.

The main players in the market comprise FMC Corporation, Chr. Hansen A/S, Kemin Industries, Inc., Allied Biotech Corporation, Cyanotech Corporation, Carotech Berhad, Döhler Group, BASF SE, D.D. Williamson & Co., Inc., Koninklijke DSM N.V., ExcelVite Sdn. Bhd., Brenntag AG, DSM Nutritional Products, Divis Laboratories, Naturex SA, Lycored Ltd. and, Algatechnologies, Ltd.

27.14 Carotenoids Type

Beta-carotene and astaxanthin are dominating in global marketplace while novel and unconventional carotenoids are expected to register higher growth rates. The global market has ample scope of innovative pigments that provide enhanced functionalities – for instance, Phycocyanin from Arthrospira extracts was approved for use in candy, chewing gum, and other types for confection in the U.S. in 2013 and 2014 by FDA.

27.15 Product Type

The worldwide carotenoids market is fractioned on the basis of product kind, application, source, and area (Fig. 27.16).

Based on the product type, β-carotene is one of the hit products between 2019 and 2026. By 2026 end, β-carotene is anticipated to beat US$2000 million income. The astaxanthin, β-carotene, and lutein together occupy almost 60% of total market value. Astaxanthin accounted for a considerable market share in 2019. Synthetic β-carotene chiefly manufactured by BASF or DSM (Heerlen, Netherlands) accounts for 90% of the market while the residual 10% is obtained from natural sources. Aquacarotene Ltd., Cognis Australia Pty Ltd. (a BASF subsidiary) and Nature Beta

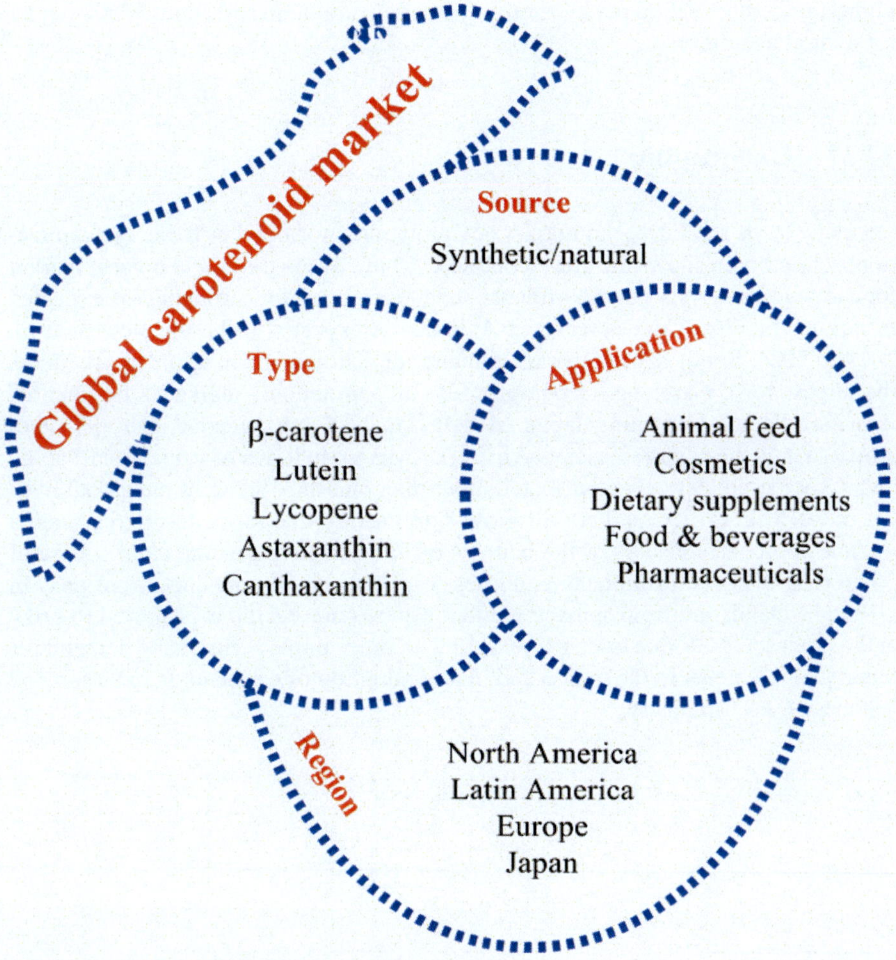

Fig. 27.16 Key segments of carotenoids market

Technologies Eilat use *D. salina* at one of the biggest algal farms in Australia and Israel, respectively to produce β-carotene. Similarly DSM and Vitatene use *B. trispora* while Biotrend of Portugal use Sphingomonas for β-carotene production. The ripe tomatoes and *B. trispora* are the key commercial sources of lycopene although synthetic lycopene (96% pure) is also available in markets [3, 4].

27.16 Price

Prices of all kinds of formulation and origin have dropped substantially owing to the Chinese companies with first-rate products. The egg-pigmentation segment is developing moderately while the spice market is growing quickly. Astaxanthin was the highest priced, @ $2000 per kg for the synthetic compound and about $7000 per kg for natural analog.

27.17 Geographic

Germany leads the European market of Carotenoids for Food & Beverage applications while Spain shows the slowest CAGR. China forms the single biggest market for carotenoids in Asia-Pacific while the Japanese market for carotenoids is expected to nurture rapidly. Their demand in Australia is expected to be the slowest from 2019 to 2024. South Korean market demand for Carotenoids in terms of product is the largest for β-Carotene. Geographically, the carotenoids market is fragmented into Asia Pacific Excluding Japan (APEJ), Japan, North America, Europe, Latin America, and the Middle East and Africa (LAMEA). In terms of value, Asia-Pacific and LAMEA collectively contributed more than one-third share in the global market in 2018. LAMEA market will grow significantly due to increasing per capita income and consciousness of the benefits of FF. Carrots are as the chief source of β-carotene in several European countries. Lutein is chiefly obtained from peas in UK and Ireland, and from spinach in other states. Asia-Pacific is projected to grow at the highest CAGR due to rising per capita income coupled with rising demand for dietary supplements in the region and higher adoption of carotenoids in Indian and Indonesian F&B industry.

27.18 Individual Carotenoids

27.18.1 Beta Carotene

It is an organic, strongly colored red-orange color which is amply present in plants and fruits. β-carotene obtained from *Blakeslea trispora* is used for the pigmentation of butter and margarine, cakes, milk products, and soft drinks. The β -carotene segment, the carotenoid of highest value, is estimated to develop at a compound annual growth rate (CAGR) of 5.1% from 2019 to 2026. The β-carotene market is currently oversupplied. Synthetic type rules the market and expectations that fermentation-derived type will fetch a bigger market stake has not been met, with algal β-carotene almost vanishing from the market [10, 11].

Beta carotene has tremendous health benefits as a provitamin A and is also popular as a colorant for food and beverages applications. Increasing preference towards naturally derived ingredients and cleaner label solutions has instigated the natural beta carotene segment's growth. Beta-carotene market has gained significant momentum over the last few years, due to superior properties of the ingredient such as promoting eye health, healthy skin, and preventing cardiovascular disease. The infusion of this ingredient in diet and supplements has directed product demand across the globe and amplified the carotenoid's industry size. Women, focusing on adopting a healthy and natural diet for the skin rejuvenation, are the foremost target consumers in the natural β-carotene market. As Australia is a geologically important region for the natural beta carotene market, neighboring regions have automatically become target markets. Lower freight costs and storage costs have opened avenues for the natural beta carotene market in Australia's neighboring regions. South Korea makes up nearly 50% of the natural beta carotene demand produced in Australia. India and China have demonstrated immense market potential in fortified beverages market which concurrently aids the natural beta carotene market to grow. The growing demand for natural beta carotene in the fortified food industry is mainly nourished by the research divisions of the natural beta carotene manufacturers.

β-carotene is largely used as colorant for beverage and food applications. In 2017, BASF SE launched a 10% beta carotene powder, Lucarotin 10 CWD (Cold Water Dispersible)/O (Orange) Plus. In the United States, this product replaced azo dyes yellow 5 and 6 in beverages, soups, confections, and nutrition products. In US, β- (95% pure) is allowed as a colorant under 21 CFR 73.95 and 21 CFR 166.110 and a pro-vitamin A nutrition supplement under 21 CFR 182.5245 and 21 CFR 184.1245 [12].

Beverage manufacturers have launched various ready-to-drinks (RTD) containing β-carotene, nature-identical being the most predominant form. For instance, Vitatine which is an antibiotic subsidiary signed an exclusive agreement with B & D Nutritional Ingredient which will bring its natural β-available in the United States for application in food and beverages and dietary supplement industry. Also, beverage brand GacLife has launched the world's first line of gac-based wellness products with functions for skin and vision care. Gac fruit is a fruit found in certain Southeast Asian regions and comprises of high concentration of carotenoids.

27.18.2 Lycopene

Permission obtained for Lycopene from the European Commission for use as a food additive and coloring instrument long-ago but this endorsement could not increase the market size. However, use in traditional supplement sector has augmented. Amounts have declined as numerous Asian dealers sell lycopene obtained from tomatoes with adequate quality. The lycopene product segment is anticipated to witness rapid growth in the future, owing to the high usage of it in cosmetic and pharmaceutical products. Lycopene is efficient in preventing ailments, such as age-related macular degeneration, diabetes, and also protects skin against sunburn. Lycopene industry worth USD 15 million in 2015, will grow significantly in near future due to usage in cosmetics and pharmaceuticals industry.

The hydrophobic character of most carotenoids causes their aggregation and crystallization in aqueous medium, a typical example of lycopene crystals in chromoplasts of tomatoes. A lycopene-based supplement is Lycosome (made by Lycotec, UK) employing lycopene micelles to entrench a whey protein isolate, enhancing its delivery and efficiency. Due to its powerful antioxidant potential and high solubility in alipophilic milieu such as the skin, natural lycopene is also employed in cosmetic goods. Generally recognized as safe (GRAS) certificate has been issued by the FDA to three products of BASF Corporation and Lyc-O-Mato (LycoRed Ltd). Tomato lycopene competes with synthetic analog. Tomato lycopene is employed as a nutraceutical and to red color the food. The Tomat-O-Red (LycoRed) comprises tomato lycopene achieved from oleoresin in crystallized form to provide a fine dispersion in water [13].

Mega manufacture of natural lycopene initiated in Israel in the mid-1990s when LycoRed (a company) initiated its extraction from tomatoes. Today many companies including LycoRed (Israel), Parry Nutraceuticals and Perennial Lifesciences (India), Lycotec (U.K.), Pierre (Italy), and Xi'an Miracle Biotechnology and North China Pharmaceutical (China) are active in this business [13, 14].

27.18.3 Astaxanthin

The current global annual market of astaxanthin is around 250 tons worth $447 million and it is growing rapidly. Due to strong antioxidant potential, it is the third carotenoid in terms of high added value [9]. Its consumption is increasing in aquaculture business for coloring fish and shrimp. Algal astaxanthin is increasingly being used in food supplements. The well-ordered tube systems for creation and usage of microbes that can yield astaxanthin in the dark has multiplied its production. The older adult population face vision damage increasing demand for astaxanthin, as it is useful for eyes. It is also used in the cosmetics sector due to its UV and anti-aging potential.

27.18.4 Canthaxanthin

It provides a red tone in egg yolks, salmonid fishes and shrimp. Canthaxanthin contributed almost 10% of the global carotenoids market share in 2015, and is expected to observe a surge in next few years due to broad usage in cosmetics sector due to its anti-tanning capacity.

Canthaxanthin market trends have shown modest growth lately, primarily driven by increasing consumer preference for foodstuffs such as fruits and vegetables, dairy, baby food, meat, breakfast cereals, snacks, and bakery and confectionery. canthaxanthin also has enormous applications in the cosmetic industry and is used as an artificial tanning agent. However, stringent regulations enforced by the FDA on the use of canthaxanthin in tanning pills might restrict revenue contribution from this segment.

27.18.5 Lutein and Zeaxanthin

The lutein market flourished until 2004. Since then, the market has developed moderately. Lutein is estimated to grow substantially in the near future due to its high demand in pharmaceutical, food, dietary supplements, nutraceuticals, and animal feed sector. Zeaxanthin obtained permission for sale in 2012 is unlocking a considerable market growth.

Table 27.4 Various sources of key carotenoids

Carotenoid	Plant source	Algae source	Bacteria	Fungi
Lutein	Marigold flowers	*Chlorella* Spp.	*Gordonia alkanivorans*	–
β-carotene	Carrot, Palm oil fruit	*D. salina*		*Blakeslea trispora*
Lycopene	Tomato fruit			*Fusarium sporotrichioides*
Capsanthin	Red pepper			
Astaxanthin		*H. pluvialis*		
Bixin	Annatto (*B. orellana*) seeds			

27.18.6 *Microbial Platforms for Carotenoids Production*

Microbial-derived carotenoids metabolites have led to a paradigm shift in carotenoids research leading to new aspects about the impact and role of these metabolites in absorption and biological activity. Plants are the main source of natural carotenoids. Xanthophylls containing acetylene functional groups are unique to algae. Table 27.4 indicates various sources of key carotenoids.

The microbial production of carotenoids has numerous benefits. As from minimum land, water, nutrients and labor, considerably 5–10 times greater growth rate can be achieved. The lutein quantity of dehydrated marigold petals (0.02–2.8 g/100 g) is comparable to microalgae biomass (0.24–0.74 g/kg). Additionally operating dynamics like stress-driven adaptive process influencing the carotenoids contents in microbes can be additionally enhanced for higher yields. The price of the microbial generation can be considerably decreased by using the agro-industrial wastes as cheap substrates. In fact all conditions of production can be regulated and optimized, particularly knowing the metabolic route of each microorganism used [3, 4].

27.19 Algal Production

The market volume of carotenoids obtained from algae has grown five times since the start of the century and its growth has fairly matured now. Algae-based carotenoids are produced on a smaller scale with a bigger market potential, chiefly in Asia, the USA and Australia. While the production is expensive, the quality of carotenoids obtained is superior to those obtained from chemical synthesis or from plants. It is chiefly since the molecules obtained from the algal biomass are more effective for food applications than their synthetic analogs. Earlier market pundits were skeptic about integration of algae-derived carotenoids into market due to great investment needed for open ponds, photobioreactors and allied facilities. However, all these fears were baseless [3, 4].

Natural astaxanthin is economically achievable and competitive with synthetic analog, which has a manufacturing price of USD1000/Kg. In China, astaxanthin can be obtained from *H. pluvialis* @ USD718/Kg, with a cost of algae biomass of merely USD18/Kg. *Xanthophyllomyces dendrorhous* and *Scenedesmus* sp. are used to produce astaxanthin for utilization in the feed and nutraceutical businesses for the production of salmonids (salmon, rainbow trout) and for the development of egg yolk coloring. *Haematococcus pluvialis* accumulates astaxanthin inside their extra-plastidial lipid bodies at a concentration of approximately 4% on a wet weight basis and approximately 300 tons of *Haematococcus* biomass are created yearly as a natural source of astaxanthin.

The most relevant natural source and process to obtain β-carotene is the culture of *Dunaliella salina*, which can accrue up to 12% of β-carotene on a dry weight basis subject to the cultivation settings. Industrial creation of β-carotene from microalgae originated in the 1980s in Israel, Australia, and the United States and expanded later to other countries including India and China.

Key European Algal Production companies are Algalif, Fermentalg, Allmicroalgae Natural Products, BDI-BioLife Science, Qualitas Health, Earthrise, Triton, and A4F. The Cyanotech, established in 1983 dominates the algae-based carotenoid industry and is generating astaxanthin (and cultivating Haematococcus and Spirulina) for years. Sphera (Italy) has combined with AlgaTechnologies Ltd. (Israel) consenting AlgaTechnologies to nurture algae and harvest astaxanthin and fucoxanthin and Sphera for their encapsulation. The Beijing Ginko Group (BGG), famous for its algal manufacturing plant in Yunnan province, has linked with Natural Astaxanthin Association (NAXA) for the manufacturing of the "world's first organic astaxanthin". Moreover, BGG has teamed up with Solix (USA) to found a new company centered in Colorado aiming on the mining of ingredients from algae [2]. The Israeli company Algaennovation has entered into agreement with an Icelandic geothermal plant to obtain eco-friendly electricity, hot and cold water and CO_2 energy for its microalgae farm in Iceland.

Microalgal carotenoids namely astaxanthin and β-carotene are being currently commercialized. The production cost of astaxanthin from *Haematococcus pluvialis* in 718 USD per kg as compared to 1000 USD per kg of synthetic with a market price of over 2000 USD/kg. The *H. pluvialis* can yield astaxanthin at >4% per DW, which is encouraging than the bacterium *Paracoccus carotinifaciens* (2.2% DW), the yeast *Phaffa rhodozyma* (<0.5% DW) of (3R, 3′R)-astaxanthin and shrimp/crab shells (<0.025% DW).

27.20 Conclusions

The hopeful endings stemming from of preclinical studies (CAM) are challenged or at least reduced moving to case-control trials. This limited disappointment is complicated and complicated to address. Different dosages regimes used in animal versus clinical studies, the complex metabolism and biotransformation of carotenoids in the intestine and tissues, the likelihood that dissimilar ingredients of supplemented mixtures can network producing antagonistic, synergistic, or additive influences are some of the opinions. However, another matter contributing confusion is the difference between prevention and therapy. Prevention infers low doses for long time (years), while therapy is linked with upper dosages for smaller time duration. Establishment of new and adequate CAM is essential suggesting more well-organized and prevention-targeted clinical studies.FundingThis research was funded by the Deanship of Scientific Research at Princess Nourah bint Abdulrahman University through the Fast-track Research Funding Program.

References

1. Miyashita K, Airanthi M, Widjaja-Adhi K, Abe M, Hosokawa M (2012) Algal carotenoids as potent antioxidants. In: Handbook of marine macroalgae biotechnology and applied phycology. Wiley, West Sussex
2. Novoveská L, Ross ME, Stanley MS, Pradelles R, Wasiolek V, Sassi J-F (2019) Microalgal carotenoids: a review of production, current markets, regulations, and future direction. Mar Drugs 17(11):640
3. Berman J, Zorrilla-López U, Farré G, Zhu C, Sandmann G, Twyman RM et al (2015) Nutritionally important carotenoids as consumer products. Phytochem Rev 14(5):727–743
4. Saini RK, Keum Y-S (2019) Microbial platforms to produce commercially vital carotenoids at industrial scale: an updated review of critical issues. J Ind Microbiol Biotechnol 46(5):657–674
5. Jefferson R, Connell A, Jefferson O (2020) carotenoids,human health. In: Patent results. https://www.lens.org/lens/search?l=en&st=true&q=carotenoids,human%20health&preview=true&v=analysis&p=0. Accessed Jan 2020
6. Sandmann G (2014) Carotenoids of biotechnological importance. In: Biotechnology of isoprenoids. Springer, New York, pp 449–467
7. Cho KS, Shin M, Kim S, Lee SB (2018) Recent advances in studies on the therapeutic potential of dietary carotenoids in neurodegenerative diseases. Oxidative Med Cell Longev 2018:4120458
8. Bogacz-Radomska L, Harasym J (2018) β-Carotene—properties and production methods. Food Qual Saf 2(2):69–74
9. Zhang C (2018) Biosynthesis of carotenoids and apocarotenoids by microorganisms and their industrial potential. Progress in carotenoid research, pp 85. https://doi.org/10.5772/intechopen.79061
10. Mussagy CU, Winterburn J, Santos-Ebinuma VC, Pereira JFB (2019) Production and extraction of carotenoids produced by microorganisms. Appl Microbiol Biotechnol 103(3):1095–1114
11. Khachatourians GG (2017) Carotenoids: biotechnological improvements for human health and sustainable development. In: Phytonutritional improvement of crops. Wiley Blackwell, Hoboken, p 259
12. Frank J (2015) Beta carotene: a work-horse colorant & pro-vitamin A. Copyright © 2020 Prospector Knowledge Cente, 7930 Santa Fe, 3rd Floor Overland Park, KS 66204 USA. https://knowledge.ulprospector.com/2838/fbn-beta-carotene-work-horse-colorant-pro-vitamin-a/
13. Ciriminna R, Fidalgo A, Meneguzzo F, Ilharco LM, Pagliaro M (2016) Lycopene: emerging production methods and applications of a valued carotenoid. ACS Sustain Chem Eng 4(3):643–650
14. Ilahy R, Siddiqui MW, Tlili I, Montefusco A, Piro G, Hdider C et al (2018) When color really matters: horticultural performance and functional quality of high-lycopene tomatoes. Crit Rev Plant Sci 37(1):15–53

Chapter 28
Past, Present and Future of Carotenoids Research

Muhammad Zia-Ul-Haq

28.1 Introduction

Being the largest groups of lipid-soluble pigments found in plants and particularly fruits, and vegetables, carotenoids are important for human nutrition, health, and wellness. Carotenoids colors sometimes appear in animals consuming them. Key colored-foods like tomato, maize, annatto and capsicum peppers were discovered by Columbus in 1492 during America discovery as all of these food materials and colors are natives of USA. In the start of sixteenth century, these were introduced to Europe and even later to Asia. Annatto, paprika, saffron were earlier extracted colors recognized as carotenoids some 200 years ago. It was immediately followed by recognition of carotene (carrots), lutein (autumn leaves followed by egg yolks) and lycopene (black bryony followed by tomato and water melon) colors. However their structures were elucidated approximately 100 years ago. The spectroscopic techniques and HPLC have made it possible to elucidate structure with as low as 1 mg of isolated carotenoids. Plant and animal sources yield very little amount of individual carotenoids. For example 20 g of carotene is obtained from 600 kg carrots, 2.7 g of lycopene from 125 kg of tomatoes and 4 g of lutein from 6000 egg yolks.

They are composed of 40 carbon atoms designed by the combination of 8 isoprene units linked covalently. They can be totally linear or possess rings at one or both ends. The photo-physical and structural features of carotenoids are due to a conjugated double-bond (c.d.b) system, which is a basic and typical structural element of every polyene. This c.d.b renders them susceptible to isomerization and oxidation. Since they are highly unsaturated, they are colorful (most carotenoids are yellow) and easily crystallized. Their chemistry and their existence as a pH-dependent, metastable system of various chemical forms, is critical in comprehension

M. Zia-Ul-Haq (✉)
Office of Research, Innovation and Commercialization, Lahore College for Women University, Lahore, Pakistan

© Springer Nature Switzerland AG 2021

M. Zia-Ul-Haq et al. (eds.), *Carotenoids: Structure and Function in the Human Body*, https://doi.org/10.1007/978-3-030-46459-2_28

of expressing various shades in plants, association with other molecules in food matrices, and their implication on human health [1, 2].

Their main functions in humans are mediated by antioxidant potential, blue light filtration effect, and pro-vitamin A effect. However the relationship between carotenoids consumption and a healthy life is not linear and is influenced by many environment, genetic and host factors. Still the knowledge of molecular framework of diet and genetics affecting the functioning of carotenoids needs to be refined for science-based recommendations about these bioactives. We can expect another exciting decade of research on the structure and mechanism of carotenoids.

The carotenoids in human diet mostly come from plants which are rich source of carotenoids except egg yolk which is the sole animal source rich in carotenoids. As chlorophyll producing leaves need sunlight and warm temperature, and with disappearing summer, both of these condition fade away decreasing chlorophyll production and starting chlorophyll decomposition. Concomitantly other colors come to the fore affecting perceived coloration. Infact the beautiful fall colors of tree leaves are the result of death of some living colors i.e. chlorophyll (Chl) and the renaissance of second colors i.e. carotenoids. Their vibration ability on absorption of extra energy lead to loss of excessive heat. The release of extra light energy as heat stops creation of ROS and hence protecting leaf from photodamage. Carotenoids are usually found attached to proteins or membranes in the chloroplasts. As chlorophyll can't absorb efficiently in light-blue to green wavelength, carotenoids assist them since they absorb in this range. Red portion of spectrum corresponds to longer wavelengths while blue part encompasses short wavelengths. Carotenoids absorb in green range while chlorophyll absorb blue and red wavelength. As light reflected from leaves to human eye is rich in green wavelength as compared to blue or red, hence leaves appear green to eyes. Carotenoids pass this captured light energy to neighboring chlorophyll molecules as both molecules are positioned close to each other in clusters just like dish antenna. This specific physical organization optimizes capture of photons. The chlorophyll molecules could not harness this energy if they were individually positioned in chloroplasts as they would miss majority of photons. Due to carotenoids helping potential in capturing photons for photosynthesis, they are called accessory pigments.

They also divert excess energy from chlorophyll molecules just opposite to energy capturing property discussed above. The extra energy produced should be dissipated to protect the leaves from photo damage. This dissipation of energy occurs through de-epoxidation and epoxidation mechanism in xanthophyll cycle in light-harvesting antenna where they and chl interact via van der Waals forces. Hence the system works like pressure release valve for energy of the photons. Both carotenoids and Chl are bound non-covalently to proteins as pigment-protein complex. Their composition includes *Chl* a, *Chl* b, lutein, neoxanthin, violaxanthin, zeaxanthin and β-carotene. When sufficient light strikes the leaves, it moves electrons and protons and helps leaves for producing sugars in photosynthesis. However when extra light and consequently extra energy comes in, electron transport chain (ETC) is overburdened. This ETC moves the electrons and comprises of molecules that donate and accept electrons alternatively while moving in the same direction. The energized electrons in overloaded ETC when strike the chlorophyll molecules, can dump extra energy in oxygen molecules, generating superoxide (O^{2-}) radicles. These radicles can react with chloroplast protein and membranes causing severe damage and eventually death of chloroplast membranes and proteins. Similarly the excited chlorophyll molecules in overloaded ETC can produce highly reactive singlet oxygen which can also destroy the leaves. Carotenoids being antioxidants neutralizes these reactions. Leaves can regulate the degree of dissipation balancing between emery capture (conversion of light to chemical energy) and energy utilization (producing sugars in photosynthesis). Plants can't survive outside laboratory in the absence of carotenoids in photosynthetic machinery. In photoprotection they neutralize the triplet state of chl and prevents formation of ROS. Only about 50 types of carotenoids play a role in light-harvesting process of photosynthesis.

They are present in various plant structures (photosynthetic tissues, petals, anthers, stigmas, fruits, seeds, roots), land and water animals (sponges, jellyfish, fish, molluscs, arthropods, reptiles, mammals, birds), macroscopic algae and fungi and numerous microbes including cyanobacteria, one of the earliest residents of earth. They are present in organisms adapted to the most contrasting environmental circumstances, from the bottom of the ocean to glaciers, thermal ponds, hypersaline waters and very dry, oxidizing or radioactive conditions. Nutrition in previous century has seen revolutionary focus on individual components/substances of food. Minerals, fats, carbohydrates, proteins and minerals were focused till 1960s and this focus was shifted to dietary fibre in 1970s. This focus was tilted to omega and *trans* unsaturated fatty acids and phytochemicals like carotenoids. It is well-known now that majority of twentieth century diseases are due to life-style and dietary patterns. It is no exaggeration to say that this information in nutrition world represents the most important advances in field of public health in previous 50 years. This information has been translated into nutrition education as dietary supplements. Consumption of carotenoids is generally larger among people undergoing/recovering from surgeries, elderly people and women. Carotenoids have an auxiliary role in bone formation by enhancing contents of bone-forming proteins. Thus they can be used by elderly population against brittle bones.

Hundreds of books, magazine articles and internet websites have been dedicated to nutrition world still large population of world is indifferent. Even people of north have not paid due attention as can be seen from obesity epidemic in developed countries. The industry spends huge amounts on advertising unhealthy foods to get lion share in market niche simultaneously opposing policies of regulatory bodies of governments especially when they suspect it may lessen their profits. Food industry has been reluctant in lowering salt contents of foods and modification of food guides so that they are based on purely scientific evidences and labeling caloric values on restaurant meals. It is noteworthy that an exhaustive bibliography exists on the biochemical properties of carotenoids suggesting their likely positive effects on human health. Nevertheless, additional studies on their direct and real-time impact on human well-being are essential, in order to substantiate their countless roles [3]. The

Table 28.1 Carotenoids research history

Time scale	Achievements
1900–1970	First golden age of carotenoids dominated by chemistry research
1970–1985	Second golden age of carotenoids research (mainly in biochemistry) indicating inversely proportional relationship between increased carotenoids intake and various disease (especially cancer and age-related macular disease)
1985s–2000	Doubt age decreased interest initially followed by interdisciplinary research leading towards understanding the genetics of the biosynthesis and isolation of many sequenced genes. The decade (1990–2000)was primarily dominated by pharmacological and medical studies like clinical trials
2001–2020	Commercial production, intervention and cohort studies
Post pandemic	Unpredicted yet

in vivo antioxidant and/or pro-oxidant roles of carotenoids is still arguable. Similarly in various diseases, the existing data are vague, however logical clarifications are given for these uncertainties. Scientists are not disappointed because that is the beauty of scientific enterprise and should not be interpreted as "carotenoid fiasco". Most recently intervention and cohort studies are a new research area involving biological effects of carotenoids [4]. The brief history of carotenoids research can be summarized as given in Table 28.1.

They are being investigated since the start of the nineteenth century. Fremy (1860) compiled the carotenoids knowledge. Their nomenclature was abridged by Palmer (1934). Zechmeister (1934, 1962), Lederer (1934), Karrer and Jucker (1948, English translation, 1950), Goodwin (1952, 1976), Cogdell (1978; 1985: for interactions with Chls), Britton and Goodwin (1982), Cogdell and Frank (1987), Mimuro and Katoh (1991), Britton et al. (1995) and Bartley and Scolnik (1995) reviewed research on the carotenoids [5]. Advancements and advent in chromatography (HPLC in 1971) and spectroscopy techniques (mass spectroscopy in 1965) have made identification much easy. By 1902, there were 800 publications on carotenoids.

To date, more than 7558 studies report the effects of carotenoids on human health when "carotenoids" and "human health" are used as keywords (PubMed: 7558; Google Scholar 156,000) (data retrieved on 18 January 2020) [3]. A linear increase has been observed in number of discovered carotenoids since 1948 with an average number 15 per year. The growth curve is not saturated yet, indicating the possibility of exploration of new carotenoids. Paul Karrer and Ernst Jucker in 1948 reported almost 30 fully-characterized and 30–40 partially characterized carotenoids according to Carotenoids Handbook (3). Otto Isler et al. and Otto Straub in their books 'Carotenoids' (4) and 'Key to Carotenoids' (5) respectively, mentioned a total of 273 carotenoids in 1971. Hanspeter Pfander in 1987, complied some 563 carotenoids in the 'Key to Carotenoids' (6). In 1995, 54 new carotenoids were added by Kull and Pfander (7). George Britton, Synnuve Liaaen-Jensen and Hanspeter Pfander, reported 750 carotenoids in the 'Carotenoids Handbook' in 2004 (3). Analysis of scholarly works that were browsed using search term carotenoids, until 18 January 2020 yielded 70,401 works (Fig. 28.1) [6].

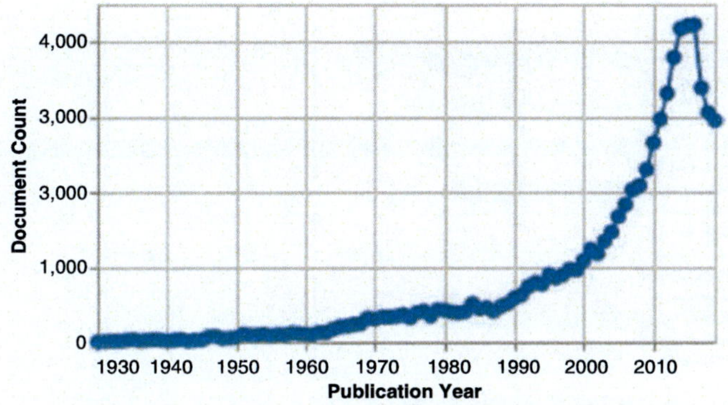

Fig. 28.1 Publication trends over time

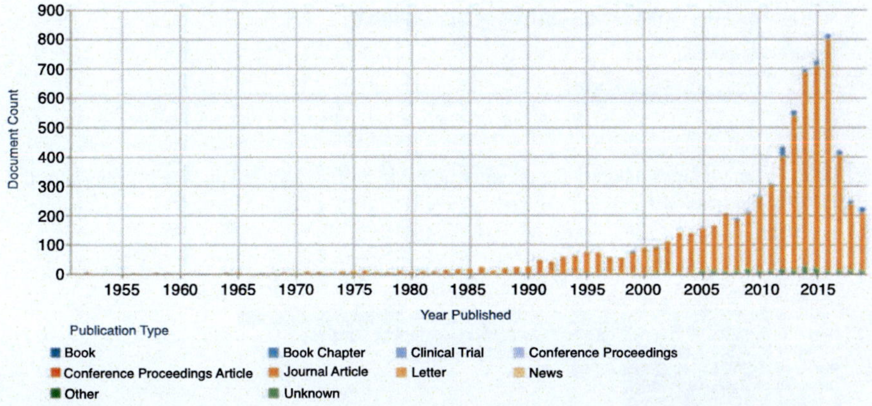

Fig. 28.2 The number of scholarly works over time by their publication type

When "carotenoids" and "human health" are used as keywords, 7007 scholarly works appear (Fig. 28.2), the most active institutions involved with field of study (Fig. 28.3), The most active authors (Fig. 28.4) most cited scholarly works and outliers with unusual ratios of citing patents/scholarly works (Fig. 28.5), patents that cite the scholarly works in the result set and citing patents with large simple family sizes (Fig. 28.6), Top publishers (Fig. 28.7), top journals/source titles for the top publishers by the number of scholarly works in this result set (Fig. 28.8), most active countries/regions (Fig. 28.9) and Top funding agencies trend over time (Figs. 28.10 and 28.11) [6].

When "carotenoids" and "human health" are used as keywords, 17,709 patents appear (Fig. 28.12), top inventors (Fig. 28.13), top patents by scatter plots (Fig. 28.14), type of patents (Fig. 28.15), top applicants (Fig. 28.16), top owners (Fig. 28.17) and the most cited patents (Fig. 28.18) [7].

The books published by famous publishers are given below in Table 28.2.

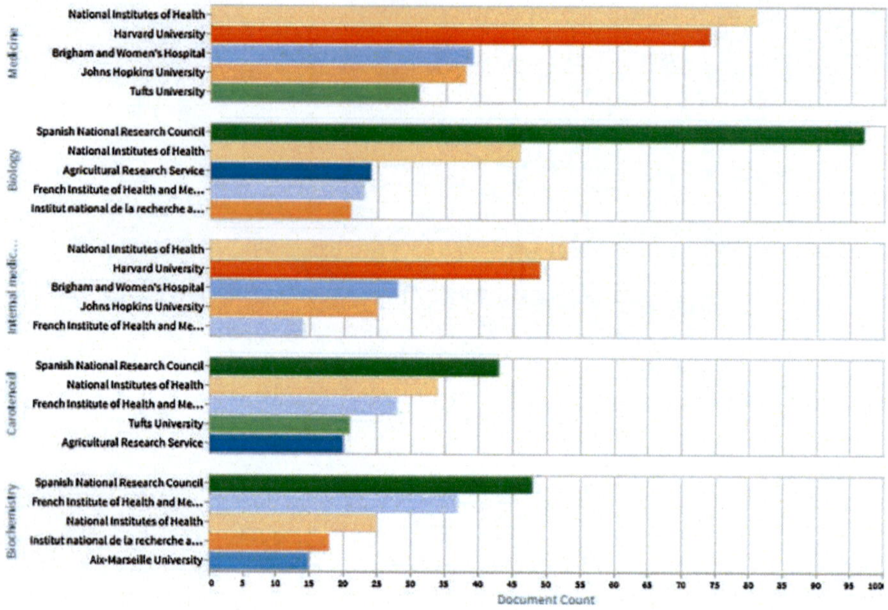

Fig. 28.3 The main fields of study for the most active institutions, based on their number of scholarly works in the result set

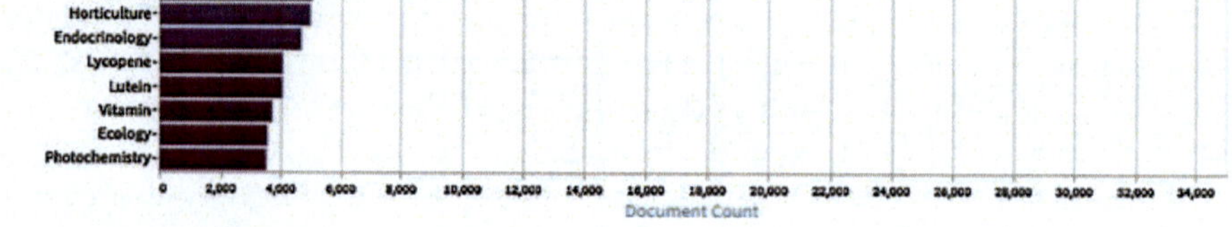

Fig. 28.4 Top fields of studies

28.2 Two Centuries Research Journey

Carotenoids like lutein and β-carotene being abundant in plant leaves are being consumed by humans since dawn of life. Hence β-carotene and lutein perform special "natural functions" of provitamin A and eye and brain health respectively probably due to evolution along with body needs. On the opposite only "pharmacological functions" can be ascribed to other carotenoid molecules which became part of

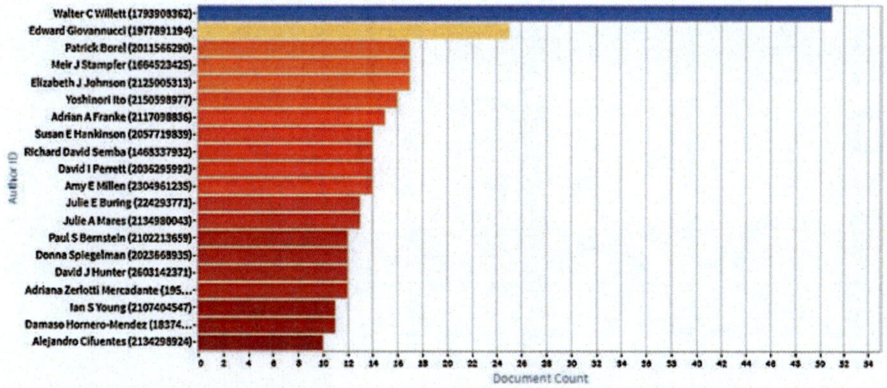

Fig. 28.5 The most active authors by their number of scholarly works in this set of result

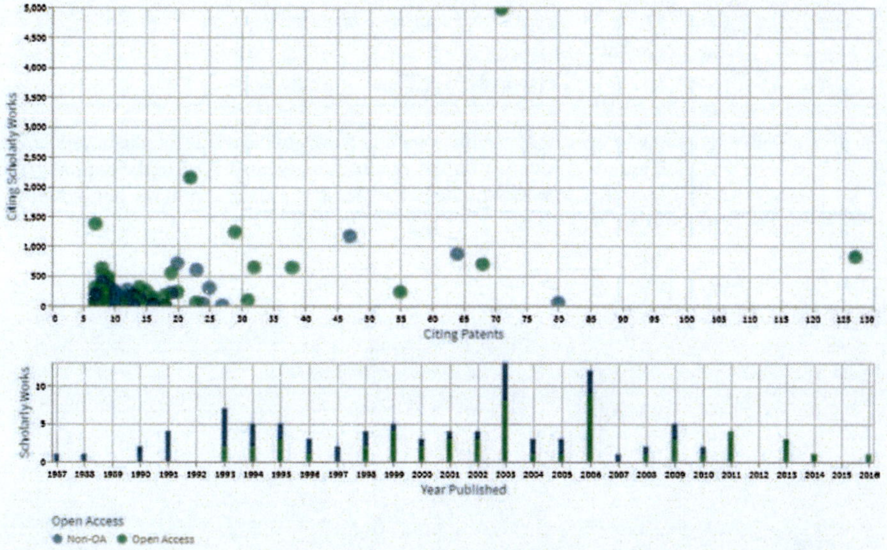

Fig. 28.6 Scatter plot showing most cited scholarly works based on either citing patents or citing scholarly works. Blue circles depict works with toll access and green circles show Open Access works. A timeline for the works' publication year is displayed in the bottom view finder. This graph identifies outliers with unusual ratios of citing patents/scholarly works and in the bottom viewfinder check the variation in open/toll access works per year

human diet much later. Hence they are considered as "foreign elements" as compared to β-carotene and lutein which are treated as "indigenous substances" by human body. Carotenoids research has moved from chemistry to biochemistry to biology since last 200 years. It is essential to have a thorough understanding of their nature and presence in diet as well as metabolism and bioavailability. The carotenoids research can be divided into 3 phases:

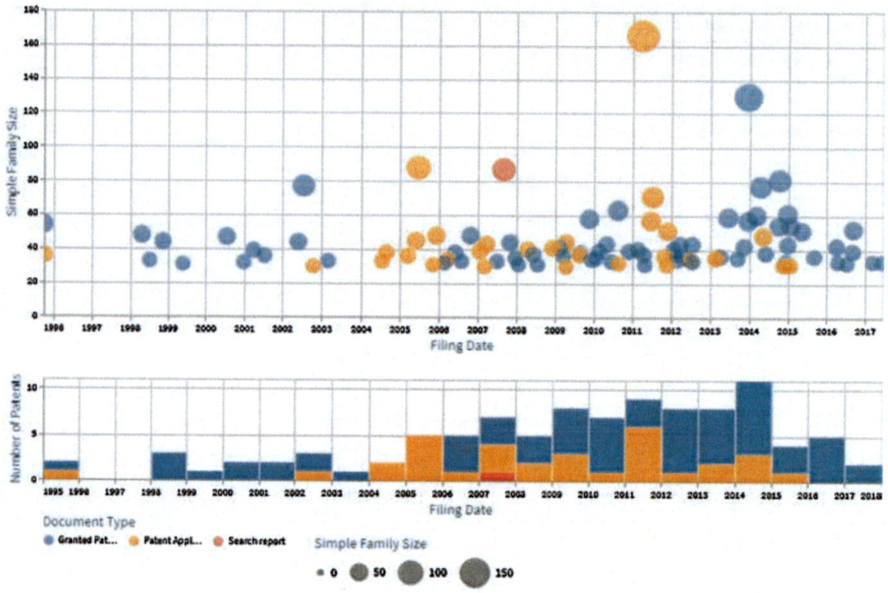

Fig. 28.7 Scatter plot shows patents that cite the scholarly works in the result set, displayed based on the size of their simple family. A timeline for the year of filing is displayed in the bottom view finder. This graph identifies citing patents with large simple family sizes and in the bottom view-finder select the years of filing

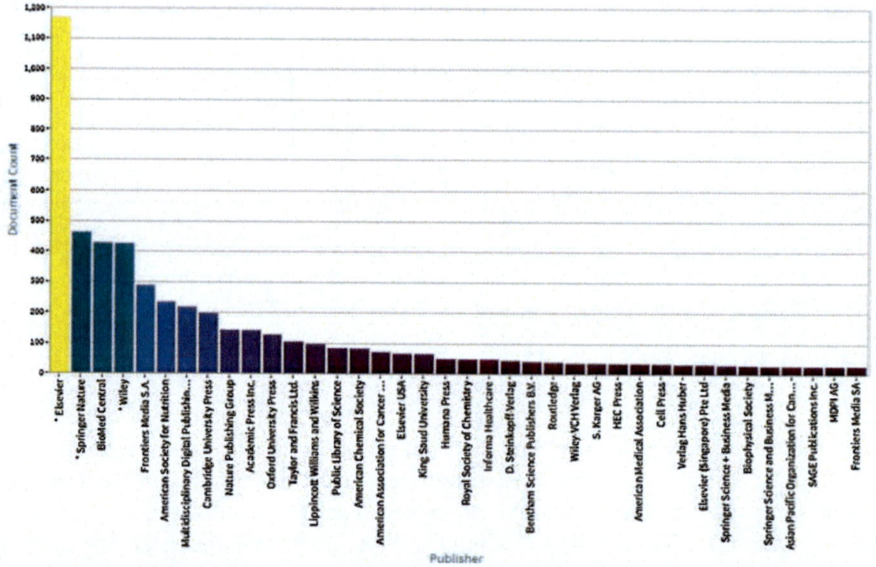

Fig. 28.8 Top publishers

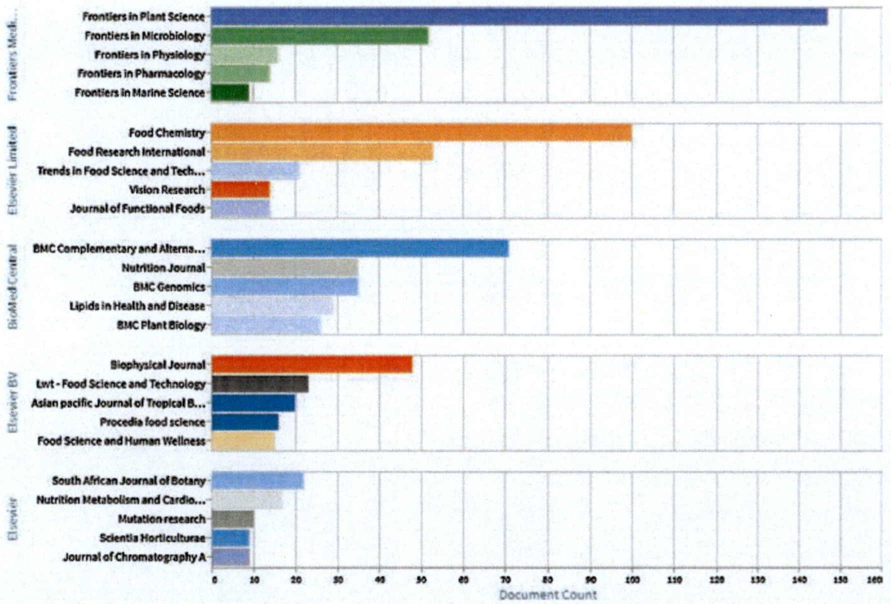

Fig. 28.9 This chart shows the top journals/source titles for the top publishers by the number of scholarly works in this result set

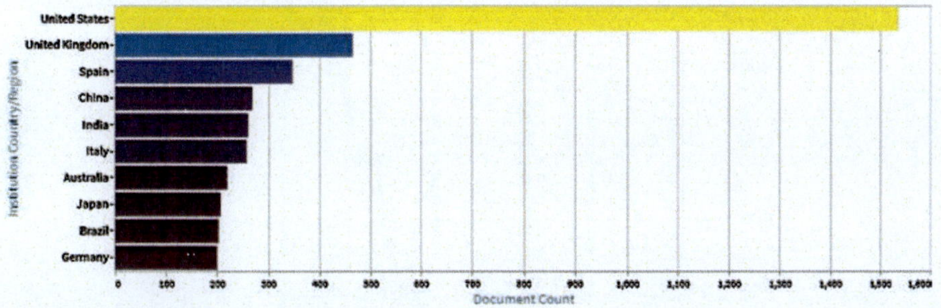

Fig. 28.10 Most active countries/regions

Late nineteenth century only chemists were involved in research. In earlier twentieth century, biochemists and plant scientists started taking interest in understanding how these molecules are synthesized in plants and finally introduced the theories of their biosynthetic pathways. However only limited scholars of clinical biochemistry, pharmacology, toxicology, and cell biology worked with some specific carotenoids. The first ever isolated carotenoid was β-carotene and it was first molecule whose structure was elucidated for the first time in 1930 by Karrer who received Nobel Prize for this work.

By the 60's of twentieth century, some health scientists and cell biologists screened few carotenoids as experimental outfits that effected or interrupted cell functions in particular behaviors. However, it was only in 70's of twentieth century

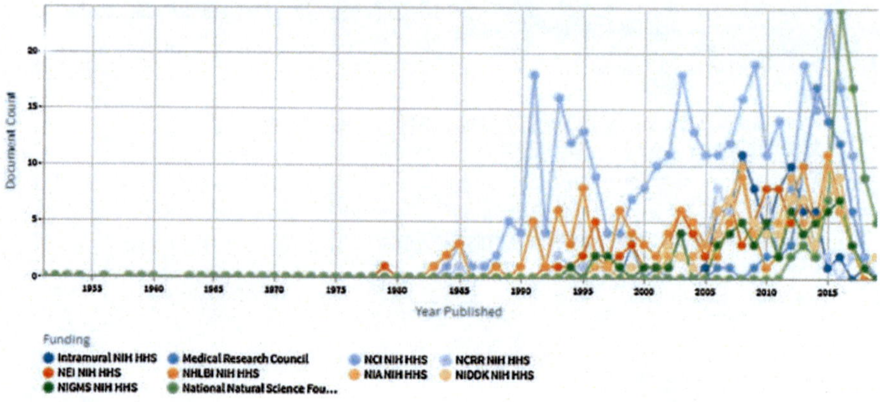

Fig. 28.11 Top funding agencies trend over time

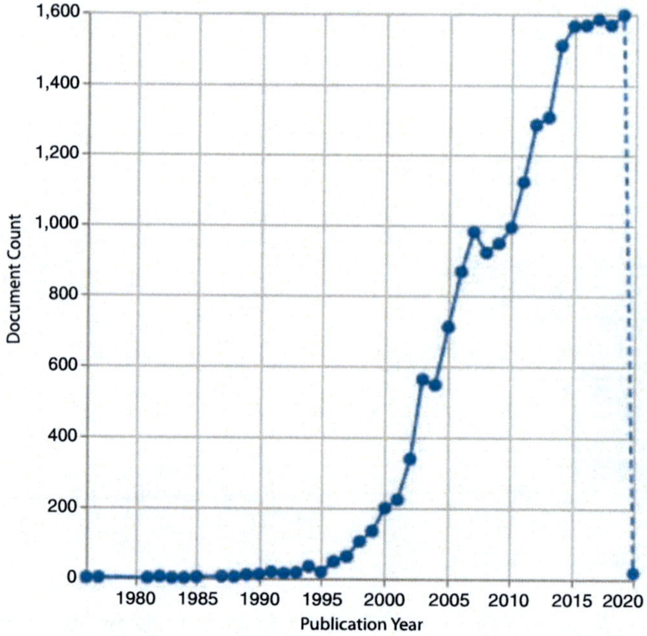

Fig. 28.12 Patents by year

that health effects of carotenoids were discovered and verified by nutritionists and physicians. Provitamin A potential of β-carotene (reported in 1930s) and its usefulness in vision was the sole recognized health benefit of carotenoids till 1970s.From there onward usefulness of other carotenoid molecules and nonprovitamin A carotenoids was realized. Waled in 1945, suggested that yellow color of macula lutea present in human eye was due to “xanthophyll”, however Bone and Landrum in 1985 deciphered this xanthophyll to be made-up of lutein, zeaxanthin and meso-

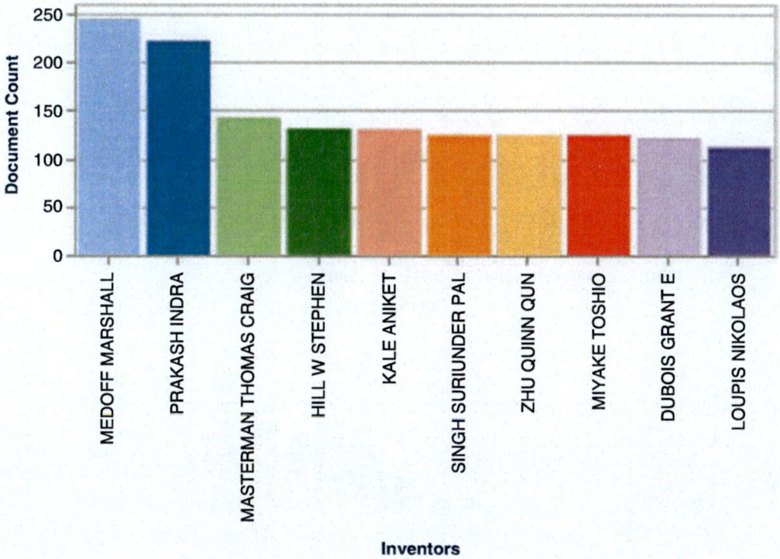

Fig. 28.13 Inventors

zeaxanthin. A clear link between decreased carotenoids contents of macula and age-related macular degeneration was established in 1980s. It was shown that β-carotene can alleviate erythropoietic protoporphyria symptoms in this research decade [12].

Since 2000, biological properties and effects of carotenoids have been divided into functions, associations and actions. Much research has been conducted in understanding mechanism of action, cellular signaling, regulation of metabolism and gene expression and bioavailability and steered the growth of novel concepts in carotenoids research. The chemical, pharmaceutical, and nutraceutical companies manufacturing interest has ensured that carotenoids utility is growing exponentially.

28.3 Lacunas in Research

There are enticing indications of exciting opinions that merit additional search.

I. We notice profligate statements 'augmented by scientific research' daily. We find literature containing carefully chosen materials (sometimes founded on compromised experimental design and/or unconvincing explanation of results) and references which, taken out of context, back many health-promoting claims. This can misguide the inexpert public to take it all at face value but rarely any definite proof. Therefore it is the responsibility of the 'carotenoid world' to strategize this and give informed verdict and direction. Stability, purity, bioavailability, and pharmacokinetics of carotenoids are key complications, which are influenced by food matrix, structural differences of carotenoids, and interactions with other food components.

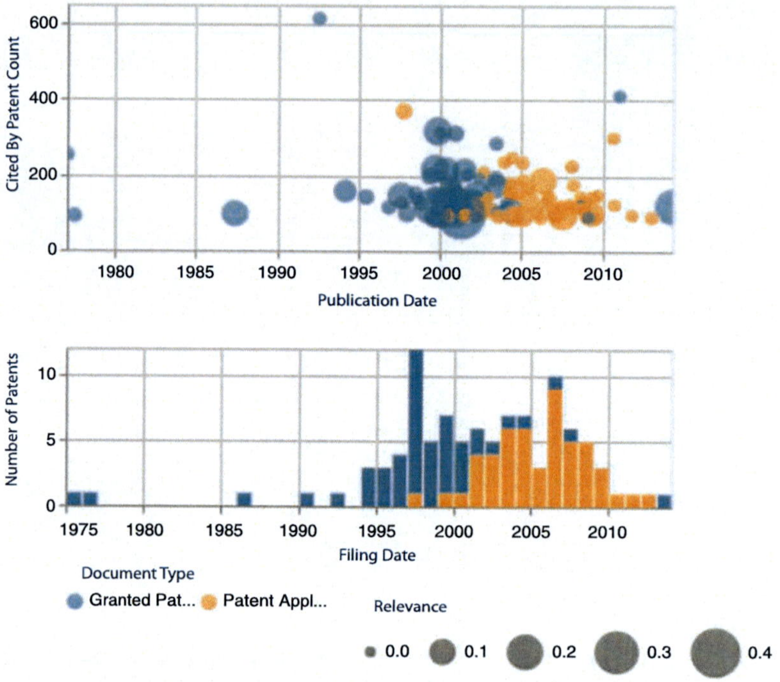

Fig. 28.14 Top patents by scatter plots

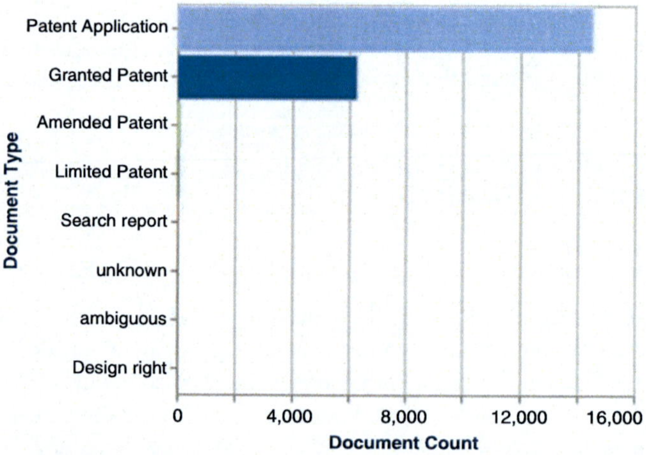

Fig. 28.15 Patent document type

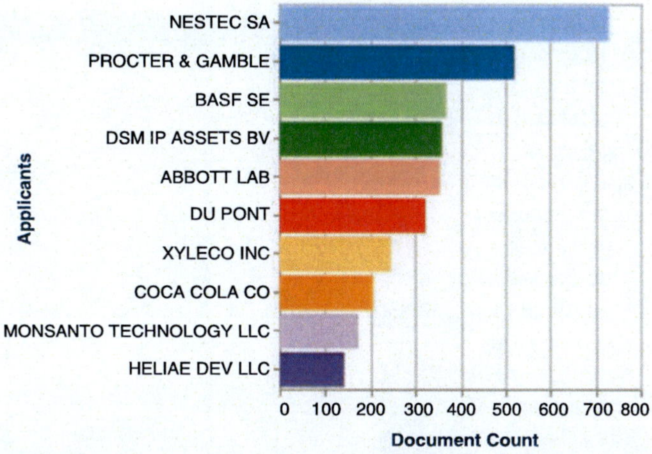

Fig. 28.16 Top applicants

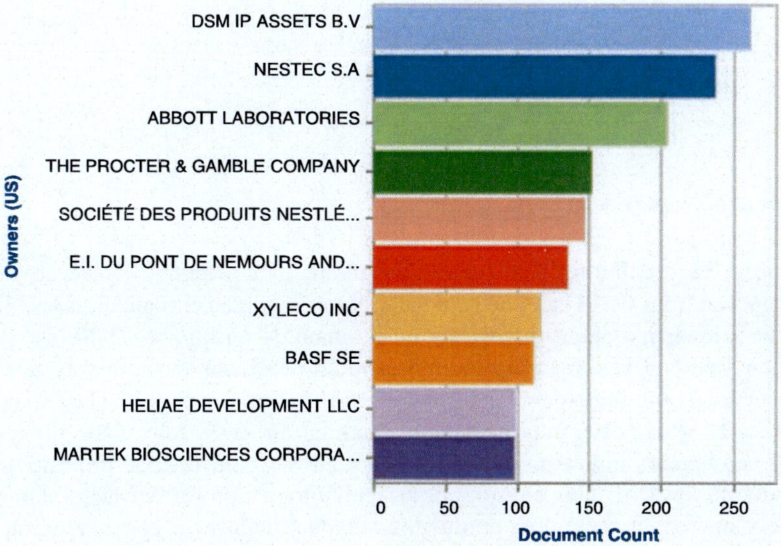

Fig. 28.17 Top owners

II. In some cases, their efficacy is not augmented by placebo-controlled, randomized, double-blind clinical trials in big human populations. Cell and animal models (CAM) are an enticing and just substitute of experiments in humans for preliminary efficacy and mechanistic investigations. Biological activity, bioavailability, pharmacokinetics, and molecular mechanism of these biomolecules can be deduced from CAM which can be helpful in designing human experiments. These in vitro cell and in vivo animal models are practically more suitable being

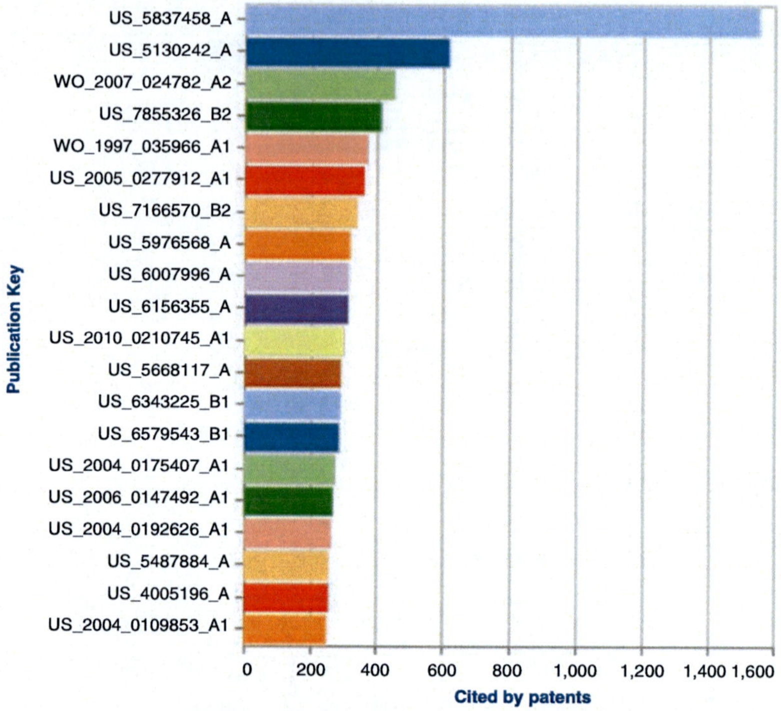

Fig. 28.18 Top patents cited by other patents

cheap, easy culturing, and reasonable throughput potential. However, results achieved from CAM can't be held valid ditto for human chronic diseases [3].

III. The knowledge about the effects of carotenoids on human health should be increased and information presented should be elaborated in a way that it is beneficial both for experienced researcher and common people. The exhaustive research would offer improved understandings about the role of specific classes of carotenoids indicating which of them are best for disease prevention and suitable level of their consumption. Technologies and procedures should be developed to increase their production at industrial level or genetically improve carotenoids-producing-crops so that their demand should be fulfilled.

IV. The figures and numbers indicating carotenoids profile of a specific plant or animal samples indicate amount of carotenoids in that specific food sample. Since same plant or animal samples obtained from different location grown under different agrogeoclimatic conditions may show different numerical values of carotenoids. The numbers however indicate that a specific food is a rich or poor source of carotenoids. Carotenoids consumption varies based on individual, national, regional and even seasonal levels in the same individual. Thudichum nearly one century ago, first time reported β-carotene in human blood and tissues. The carotenoids absorption and deposition ability varies in individuals.

Table 28.2 Book Publications in carotenoids

Date	Author/organization	Description/title of book
1934	Zechmeister	Published the first monograph on the carotenoids
1934	Lederer	Three monographs on carotenoids (plant carotenoids)
1948	Karrer and Jucker	"Carotenoids" mentioned 65 natural carotenoids
1971	Otto Isler	Monograph on Carotenoids
1971	Otto Straub	"List of natural carotenoids" comprise up 273 carotenoids
1973	Butterworth-Heinemann	Carotenoids Other Than Vitamin A–III. Third International Symposium on Carotenoids Other Than Vitamin A held at Cluj, Romania September 1972
1980	Goodwin	Carotenoid biochemistry Vol I, Plants
1981	J.C. Bauernfeind	Carotenoids as colorants and vitamin A precursors: technological and nutritional applications
1982	George Britton and Trevor W. Goodwin (Eds.)	Carotenoid Chemistry and Biochemistry. Proceedings of the 6th International Symposium on Carotenoids, Liverpool, UK, 26–31 July 1981
1984	Goodwin	Carotenoid biochemistry vol II animals
1987	Straub	563 carotenoids were listed in book named *Key to Carotenoids*
1989	Krinsky, Mathews-Roth, Taylor (eds.)	Carotenoids: Chemistry and Biology; provide information about meeting discussion and disseminating scientific research results concerning all aspects of carotenoids
1991	Jeana Gross	Pigments in Vegetables: Chlorophylls and Carotenoids
1992	Lester Packer (Eds.)	Carotenoids Part A: Chemistry, Separation, Quantitation, and Antioxidation; comprehensively describe molecular and cellular methodology needed for pursuing research with carotenoids
1993	Lester Packer (Eds.)	Carotenoids Part B: Metabolism, Genetics, and Biosynthesis
1993	Andrew J. Young and George Britton (eds.)	Carotenoids in Photosynthesis; describes the deeper understanding of the role and function of carotenoids in photosynthesis
1996	Pfander, Liaaen-Jensen and Britton	Carotenoids: Volume 2: Synthesis; devoted entirely to the total synthesis of carotenoids
1999	Frank, Young, Britton and Cogdell	The Photochemistry of Carotenoids
2000	Institute Of Medicine	Dietary Reference Intakes for Vitamin C, Vitamin E, Selenium, and Carotenoids
2002	P. Winterhalter and R. L. Rouseff (Eds.)	Carotenoid-Derived Aroma Compounds
2004	H. A. Frank, A. J. Young, G. Britton, R. J. Cogdell (eds.)	The Photochemistry of Carotenoids; book is correlates between the photochemical behavior of carotenoids in vitro and in vivo
2004	Krinsky, Mayne and Helmut Sies	Carotenoids in Health and Disease (Oxidative Stress and Disease); Describes the role of carotenoids in chronic disease prevention
2004	Britton, Synnove and Pfander	Carotenoids: Handbook; cover almost 700 carotenoids of natural origin isolated till 2001

(continued)

Table 28.2 (continued)

Date	Author/organization	Description/title of book
2007	Vahlquist and Duvic	Retinoids and Carotenoids in Dermatology (Basic and Clinical Dermatology)
2008	G. Britton, S. Liaaen-Jensen and H. Pfander	Carotenoids: Volume 4: Natural Functions; describes the critical discussions of the biochemistry, functions and applications of carotenoid
2009	Pfander, Liaaen-Jensen and Britton	Carotenoids: Vol 5; Nutrition and Health; tells the readers about fundamental chemistry of carotenoids and the basic methods used in carotenoid research
2009	John T. Landrum	Carotenoids: Physical, Chemical, and Biological Functions and Properties
2012	Victor R Preedy	Vitamin A and Carotenoids: Chemistry, Analysis, Function and Effects
2012	José-Luis Barredo (eds.)	Microbial Carotenoids from Bacteria and Microalgae: Methods and Protocols
2013	Winterhalter and Ebeler	Carotenoid Cleavage Products; describe the importance of carotenoid cleavage reactions in plants and animals
2013	O. Sommerburg, W. Siems and K. Kraemer	Carotenoids and vitamin A in translational medicine
2013	Sherry A. Tanumihardjo (eds.)	Carotenoids and Human Health; describes the role of carotenoids in association with diseases
2014	Landrum and Nolan	Carotenoids and retinal disease; outlines the pathological conditions of the central retina in relation with carotenoids
2016	Delia B. Rodriguez-Amaya	Food Carotenoids: Chemistry, Biology and Technology; provide important information about these major compounds which impact both food quality and human health
2016	Baranska and Kaczor	Carotenoids: Nutrition, Analysis and Technology; present information about functional food perspective, including biotechnology and future prospects.
2016	Claudia Stange (eds.)	Carotenoids in Nature: Biosynthesis, Regulation and Function
2018	Jara and Fulgencio	Non-extractable polyphenols and carotenoids importance in human nutrition and health
2019	N. I. Krinsky, H. Sies and S. T. Mayne	Carotenoids in Health and Disease
2019	Charis M. Galanakis	Carotenoids: Properties, Processing and Applications

[5, 8–11]

V. A carotenoid rich diet contains 100–500 fruits and vegetable consumption daily. Processing decrease particle size and release carotenoids from chloroplasts and food matrix for the bioaccessibility. Fruits and vegetables contain small amounts of proteins and fats and higher quantity of water. As fruits and vegetables contain very little amount of lipids hence daily consumption of 3–5 g fat is recommended for maximum absorption of carotenoids. Different consumption ratio across globe is mainly due to heterogeneous dietary patterns, different protocols of collections of this data and variations in carotenoids estimation methods. The differences in results are mainly due to use of different protocols for dissolution of to-be-tested carotenoid molecules, production of oxidation environment, presence of other compounds simultaneously, and use of different detection methodology. Hence it is necessary to harmonize evaluation of food ingestion, evaluation of carotenoids contents in food and validation of results with biomarkers. As carotenoids are consumed along with other dietary antioxidants which definitely leads to interactions of various types at various levels (e.g. at gastrointestinal, biological fluids and cell level) hence these interactions should not be ruled out before any concluding remarks. Similarly long chain fatty acids e.g. oleic acid are more helpful for absorption of carotenes (nonpolar carotenoids) than that of xanthophylls (polar carotenoids) as polar carotenoids are more readily transported from emulsified fat to micelles. Dietary fibre (DF), main component of plant foods, hinders release of carotenoids from food matrix and DF and proteins prevent carotenoids integration in the micelles. Excitingly they are present in less amounts in humans than other antioxidants like vitamin C and tocopherols. Their core hydrocarbon skeleton makes them non-polar and needs fat for absorption in intestine. The lacunas among digestion, metabolism and delivery techniques in biological systems should be carefully filled to get the precise and convincing results. Genetic variations, different techniques of food preparation and interface with other food component during assimilation are responsible for their broad spectrum inter-individual bioavailability diversity.

VI. Three provitamin A carotenoids (β-carotene, α-carotene, and β-cryptoxanthin) and three non-provitamin A carotenoid (lycopene, lutein and zeaxanthin) are usually present in human plasma and tissues. Their constant presence in the diet, fluids, and tissues of humans indicate their effects on human physiology. Complex human physiology, variable dietary pattern, genetic variations, long duration and cost human studies, ethical restraints of human studies make it challenging to indisputably exhibit their health promoting effects in *in vivo*. Similarly dose, formulation and administration regime (results should be of pure compound as their property of being oxidized easily can generate oxidative metabolites and oxidized compounds which can also show results) should be more precisely defined. The colonic transformation of carotenoids is less researched topic and little information exists that how they affect microbiota profile. Nutrigenomics and nutrigenetics will be a valuable tool to discern the relationship between carotenoids intake and decreased chances of chronic diseases.

28.4 From Food to Tissues

Their daily intake varies from 0 to 10 mg, >4 mg being more specific except for lycopene and b-carotenc which are usually more than this range. Their concentration in tissues and plasma is 0–1 nmol/g and 0–2 μmol/L respectively.

28.4.1 Sources, Bioavailability and Conversion

Carotenoids are obtained from food, mainly fruits and vegetables except supplements. Therefore, information about type and quantity of carotenoids present is of utmost importance. Carotenoids bioavailability from green vegetables is usually very small due to the snooping of food matrices. Therefore, the bioavailability of carotenoids from fruits and their forms of consumption is sufficient to improve their level in tissue. Once carotenoids enter the human body, major certainty begins. Bioavailability from supplements is comparatively higher. Since these are not prerequisite for cell viability, their RDI is not established. However they have a key part in shielding cells and tissues from damaging potential of oxidative stress and neuroinflammtion-mediated changes. Their decreased bioavailability is an apparent impediment in achieving the anticipated beneficial effects. The carotenoids with increased bioavailability and stability in stomach should be developed.

28.4.2 Variability Between Individuals

A great and unregulated element about various queries related to bioavailability and conversion of vitamin A is variability between individuals. It is well-documented that there is range of variations in uptake, accumulation and conversion of carotenoids, resulting in excessive inconsistency in carotenoid contents in tissues and blood subsequent to similar intake. Advancement in molecular biology, molecular genetics and omics techniques can be a key to solving these mysteries. Genetic factors that regulate competency of an individual for absorption and storage of carotenoids and conversion into vitamin A is should be identified.

28.5 Carotenoids and Major Diseases: Practical Concerns and General Points

The few connections recognized directly between human health and carotenoid consumption disclose that: (i) being strong antioxidants, they can have antitumor capacity (ii) some of them possess pro-vitamin A potential and (iii) few of them can exhibit regulatory actions at several levels in different tissues, and hence may disclose their protective influence against degenerative diseases, or can inhibit metabolic diseases, like dyslipidemia or type 2 diabetes. Big data and information sorted by statistical tools and algorithm will be of key help in this regard. The big variation among individuals is challenging. The order of biological system complexity is cells < tissues < organs < multicellular organisms. Epidemiological studies suggest a direct relationship between increased carotenoid consumption and decreased risk of various diseases, however there is no proof of direct connection and many queries remain unanswered. Few aspects of experimental processes should be kept in mind while appraising results.

28.5.1 Human Studies

There is factually plethora of studies suggesting their role in human health since 1800s. However big fraction of this research is based on experimental designs that are more suitable for probing drug efficacy and less fit for carotenoids, to which the population is mostly not "naïve". It is notable that merely a small fraction of the plethora of carotenoids has been studied in depth, indicating knowledge gap about their characteristics, activities and concrete applications. Nutritional investigations are typically executed in control subjects ("apparently healthy") but the bioavailability of carotenoids (i.e. response) under subclinical conditions can fluctuate. It is very hard to differentiate whether a valuable outcome is owing to the carotenoid or to the foods containing carotenoids. In intervention studies and clinical trials, the doses administered contain usually much greater amounts of carotenoids than those acquired from the food. Even if the doses administered are analogous to levels in food, since pure forms are comparatively bioavailable to a larger extent thus offering higher quantities. Further, the slow release and uptake of the carotenoids during the digestion process should also be counted [13].

28.5.2 Biomarkers

A strategic building block of this process is the recognition of novel and improved biomarkers of carotenoid effects during disease progression. Biomarkers are clinical, cellular or molecular rubrics suggestive of a specific phase of disease. Better markers are required with extra specificity, obviously defined association to disease and alertness to carotenoid ingestion. Identification of reliable biomarkers for assessment of carotenoid status is of key importance. Carotenoid contents blood or tissue may be a key reliable analytical biomarker, however attempts should be done to devise a non-invasive method i.e. which does not require the blood sample collection or tissue biopsies would be perfect [14]. Non-invasive analytical techniques *e.g.* resonance Raman spectroscopic techniques for fast analysis of carotenoids in the skin can be a potential biomarker for appraisal of carotenoid status. However additional authentication is essential before acceptance of results with sureness and the techniques can be used on a wider scale. Till mow, β-carotene has been used in clinical trials to evaluate the influence of carotenoids on human health almost entirely. However other carotenoids should also be studied to completely comprehend the value of carotenoid supplementation on human health. Their metabolism, mechanism of action, and interaction with other biomolecules *in vivo* should be interpreted. The double-blind, placebo-controlled studies of carotenoids are conducted for fairly short time periods, while it is established that associations between carotenoids and disease states are of a longer term nature. Carotenoids exert small effects that aggregate over years and work synergistically with many other mechanisms of a healthy lifestyle [15].

28.5.3 Animal Models

Much information exists on influence of carotenoids in animals e.g. mice, rats, and guinea pigs, during lab experiments. Since humans and said animals are different from each other in key biological aspects, so the information gathered cannot be accepted with full confidence. This dogma can be resolved by using other primates as models for experimental purposes. However besides the price of trials and concerns of people for using primates in such a fashion should also be born in mind.

28.5.4 High Dose, Low Dose and Balance

Routine daily food of an individual contains about 5 mg carotenoids. However animal models and human trials normally utilize 20–100 mg/day. Such unusual high doses can activate detoxication mechanisms as the body can recognize such dose as foreign substances. Instead recurrent, continued, ingestion of small volumes can be more beneficial comparative to a single huge dose. Appropriate consumption throughout life since early age is healthier than using big amounts as supplements later in life. An equilibrium among different carotenoids as well as balance with other elements, *e.g.* vitamins C and E should be maintained. High-dose supplements interrupt the equilibrium and yield unexpected and entirely changed effects and results.

28.5.5 Safety and Toxicity

Despite great debates, there is not any suggested safe upper limit for carotenoid consumption. Carotenoid supplements should be subjected to food legislation which should not be rigorous as for pharmaceutical products.

28.5.6 *Natural* Versus *Synthetic*

The media and/or public conviction that 'natural is better than synthetic' is based on sentiment and not on facts. Infact natural and synthetic are chemically same if a carotenoid is pure. In similar physical state and formulation both will exhibit similar properties and biological actions.

28.6 How the Effects Might Be Mediated?

28.6.1 Via Antioxidant Action

β-carotene was first molecules whose antioxidant activity was first time reported in 1984 by Burton and Ingold [16]. Although human body is bestowed with an array of defense system (enzymes and endogenous antioxidants) against oxidative stress, it is commonly assumed that nutritional supplementation with antioxidants like carotenoids can be a fraction of a shielding scheme to reduce the oxidative damage. Scientists are yet unable to elucidate the exact mechanisms of this synergism, "1 + 1 = 3". Carotenoids going to be tested must be pure as it may show entirely different results it if contains peroxides. Antioxidant effects observed in non-biological environment are generally accomplished under circumstances that do not match with *in vivo* conditions. Determination of antioxidant potential in a natural system *in vivo* is very challenging; as the human physiological system is very complex. Further circumstances are not even within the system, natural carotenoid contents are less and likely interference with other elements.

28.6.2 Via Metabolites

It is challenging to verify whether effect credited to a carotenoid are due to the integrated carotenoid or its metabolite/breakdown product. As a carotenoid may have many metabolic/intermediate products, existing as complex mixtures. Any such product in isolation or in combination possessing the correct topography to copycat other molecule, *e.g.* vitamin E, can perform as an agonist or antagonist and exhibit same biological property. Similarly the pharmacological properties of breakdown products of carotenoids, especially β-carotene, α-carotene, β-cryptoxanthin, lutein, and Zeaxanthin should be explored. Till now only β-carotene metabolites and their biological actions have been reported [4, 17].

28.6.3 Via the Immune System

They can act as immunity booster. Influence of various carotenoids on several features of the human immune response system have been verified, generally by *in vitro* experiments.

28.7 Final Comments: The Big Questions [18]

(i) ***"Besides being precursors of vitamin A, do they have other roles or tasks beneficial for human health?"***

Many cellular and molecular mechanism have verified actions of carotenoids in cells in culture. However intervention trials indicating a direct connection between administered pure carotenoids ingestion and decreased possibility of diseases have not been very informative.

(ii) ***"Are carotenoids important antioxidants in vivo?"***

Although it is well-known that carotenoids exhibit antioxidant capacity *in vitro*, explicit evidence for their ditto operation *in vivo* is not very convincing. Infact carotenoids contents are less relative to standard antioxidants like vitamins C and E. Hence antioxidant action looks questionable, although certain particular action in a specific sub-cellular environment e.g. membrane or in special tissues, cannot be ignored. Unsatisfactory results of β-carotene diminished the eagerness regarding function of this molecule in the etiology of degenerative diseases, and positive effects of carotenoids in general were quizzed. However, this research enhanced knowledge about carotenoids exponentially leading to the growth of novel notions in carotenoids studies.

(iii) ***"Do they offer protection and decrease risks of serious degenerative diseases?"***

Sure answers and undeniable indications are hard to find, though there is much incidental and suggestive signs about protection potential of carotenoids against degenerative diseases.

(iv) ***"Is it good for health to take carotenoid supplements?"***

The safety of huge consumption, particularly as supplements is not established. Therefore no recommendation can be made. Several developing countries face vitamin A deficiency as a serious public health problem. Although no single solution can be suggested for problem, countries can devise and opt any indigenous cost effective strategy to cope this problem. Some commonly advocated strategies are regular intake of high-dose vitamin A capsules supplements, dietary diversification and food fortification.

28.8 Challenges

1. Since the question "can carotenoids affect cancer rates" first surfaced in 1981, much work has been done to determine if this is indeed the case. The uncertainties about their role in cancer inhibition came to light when two of the three big scale β-carotene intervention trials estimating the chances of getting lung cancer in high-risk populations witnessed a surprising surge in the prevalence of lung cancer. Neither an increase nor a decrease in lung cancer prevalence was reported in 3rd study, in which a relatively healthy population was used. The appraisal of the function of carotenoids in human health was significantly affected by publication of this research. Firstly, it increased information of the function of individual food nutrients

as compared to the influence of whole food. An overwhelmingly positive association was reported between the consumption of foods rich in β-carotene and a reduced risk of lung cancer in published epidemiological research prior to these intervention trials. These observations, coupled with outcomes of β-carotene animal studies and short-term intervention trials in humans, helped formulate hypotheses that one of the active constituents in the food was β-carotene. Although this hypothesis was questioned later, it however disappointed many people. But that is the essence of the scientific enterprise, and cannot be taken as "the β-carotene fiasco" [4, 17].

2. Earlier studies mostly described their general characterizations by using biochemical techniques followed by cell lines or animal models (CAM). Only one-third of the published research has been conducted in humans. There is need to optimize protocols and tools employed for carotenoids examination in human body, as empirical prediction models are used by investigators instead of real-time measurements of carotenoids in human tissues or fluids. Novel protocols are being designed that will help discriminate the antioxidant potential of both the aqueous as well as lipophilic compartments of body fluids and tissues and will help resolve dogma of carotenoid in vivo antioxidant capacity. Newer technologies are increasingly stimulating research progress. One such areas include novel use of non-destructive determination of carotenoids.
3. Although epidemiological studies connecting lutein and zeaxanthin to decreased risk of eye diseases are quite strong, a fundamental association between the onset of eye diseases and comparative deficiency of these two molecules has not been established yet. Same is true for heart and vascular diseases since studies of individual carotenoid like β-carotene in intervention trials, have yielded null results, i.e., neither a positive nor negative outcome. This observation proposes that either β-carotene alone is not the active element in fruits and vegetables, or it acts synergistically with other dietary carotenoids leading to cardioprotection. This dilemma should be resolved. Situation is bit satisfactory in cancer and carotenoids as substantial cell culture work has warranted a persistent attention in the possibility for the utilization of carotenoids or their metabolites in inhibition or treatment of cancer. Due attention should also be paid to discern the association between inflammatory diseases and carotenoids, especially since inflammation yields the oxidizing species that oxidize carotenoids to shorter chain fragments [4, 17].

28.9 Future Directions

I. Till now, carotenoids biosynthesis has been discerned in 594 prokaryotes, although validated pathways have been revealed in only 5% of these microbes. The exploration of genetic and molecular basis of carotenoids accumulations in them can be fruitful for devising appropriate strategies to enhance and efficiently advancement in a domain and same is true for noninvasive estimation of carotenoids. The usage of resonance Raman spectroscopy for eye carotenoids, and Raman and reflectometric techniques for skin carotenoids, are such tools. It is expected that time is not far when individual carotenoids will be identified by noninvasive techniques. The structural diversity, will expand due to advanced analytical methodologies that will help to characterize the minor, although unique, complex carotenoid moieties. This will also stimulate research activities in screening and profiling of natural carotenoids from other plant species that have not been well studied in the years to come [8, 17].

II. The research should focus on synthesis of novel carotenoids augmented with innovative delivery methods. To achieve more environmental-friendly production while reducing the cost of production and downstream processing of carotenoids, as end-user has key attention in Lifestyles of Health and Sustainability (LOHAS) which prioritize their production from sustainable sources. Agro-food by-products are a low-cost source and carotenoids can be obtained from these by-products. In the case of vegetables, leaves, peels (or rinds), and seeds are generally the main studied by-products sources of carotenoids. The main fruit by-products i.e. peels, seeds, fruit pulp, and juice are generally the main studied by-products sources of carotenoids. Organic solvents coupled with lengthy extraction times are necessary in traditional solid–liquid extraction which can increase operational expenses, carotenoid degradation, and environmental apprehensions. Although modern techniques are environmentally sustainable and "green" as they usually need reduced amounts of solvent and decreased processing times, they do require particular equipment and expertise. It can be challenging for food manufacturers, who extract carotenoids from processing byproducts, such as the apple peel during apple juice, to make revenue rather than to incur disposal expenses. Processing by-products are a ready supply and rich source of carotenoids, however the stability of these molecules under specific circumstances and suitable post-harvest handling and processing to reduce their degradation are key challenges. There are many raw materials generated as by-products of food processing that are abundant in carotenoids, and there is potential for their extraction. However considerable challenges exist for the profitable and sustainable industrial extraction of carotenoids from these by-products. Great care should be taken to guarantee suitable post-harvest conditions for stabilization of the waste material as these compounds are susceptible to heat, light, and pH. Extraction techniques can include conventional and assisted technologies that can result in increased yields and reduced solvent usage. Regulatory challenges must be overcome if their full potential as functional food ingredients is to be realized.

III. Previous carotenoids investigations connected to human health was focused on β-carotene, however lutein, zeaxanthin, and lycopene are being studied now. Lycopene due to its beneficial effects against CVD and prostate cancer is under research now a days. Still, there are carotenoids present in the human food that deserve consideration such as phytoene or phytofluene and numerous oxo-carotenoids including capsorubin, capsanthin, or violerythrin. It warrants the formulation of new methods for chemical synthesis and delicate analytical systems to follow the production of these molecules *in vivo*.

IV. Although the estimation of their economical profit results vary, a range from 1.5 to 1.7 US$ billion in 2020 seems to be realistic indicating the industrial relevance of these molecules. The lack of efficient synthetic methods and scarcity of suitable natural sources limits the number of carotenoids offered. Some companies have specialized to market carotenoids for use on a laboratory scale, as powder or crystals (typically 1 mg) for lab trials, or liquefied in a suitable solvent (around 3 μg of carotenoid per 2.5 mL solvent) for HPLC calibration. New uses and applications of carotenoids should be explored, finding of high-value markets (both niche and commodity), and research of rare carotenoids. Myxol a recently discovered carotenoid from freshwater cyanobacteria has more antioxidant potential than β-carotene. Other rare and/or recently defined carotenoids are bacterioruberin, salinixanthin, saproxanthin, siphonaxanthin which should be explored for their activities.

V. Preventive medicine is a buzzword that has encouraged manufacturers and processors to come up with dietary supplements and functional foods that utilize natural ingredients such as carotenoids. Geriatric nutrition is one of the vital sectors for carotenes. DSM markets CaroCare® solutions, which offers a range of natural-sourced beta-carotene formulations.

VI. Numerous signaling pathways and certain crucial cellular processes are responsible for their positive effects and can suggest would-be therapeutic targets and tactics for the treatment of numerous diseases in coming days. Further investigations on the biological activities of carotenoids metabolites are wanted, since carotenoids are conveyed in human serum and urine chiefly as metabolites, and these metabolites can be responsible for the beneficial effects of carotenoids.

VII. Purified carotenoids might exercise diverse biological actions due to their particular chemical structures. The careful and correct characterization of various carotenoids is required to better elucidate the molecular mechanism of their health benefits.

VIII. As only limited work exist on their metabolism in human body, hence substantial studies should be done to better characterize their metabolism. An inclusive understanding of their bioavailability is only recently coming together. Due to their complex chemistry, there are challenges in explicating their destiny in the body as after release from plant material, their chemical environment changes. As they enter the digestion system, their pH changes, leading to generation of diverse chemical moieties in the intestinal tract which are shifted to various body tissues. Scarce and diffuse data is available regarding the colonic metabolism of carotenoids, due to complexity of food composition and inter-individual changes. Bioaccessibility and bioavailability procedures need standardization to merge the discussion of the data obtained in this subject. The outcomes of studies assessing carotenoids bioavailability are mostly challenging since the *in vitro* methods don't include the role played by the individual microbiota present in the human body. Further each person has his/her own microbiota that obviously impedes the bioavailability of carotenoids due to inter-individual variations. Hence, a unified *in vitro* method should be developed to appraise the role of the microbiota in carotenoids bioaccessibility, and future research should address this query by simulating the digestion and absorption processes.

IX. Generally, plasma carotenoids are a decent sign and yardstick for measuring total antioxidant status in human by gold standard technique, HPLC. The carotenoids identification methods can be alienated in two categories: invasive and non-invasive methods. Greatly invasive technique of HPLC is broadly used as a certification tool and biomarker for fruits and vegetable ingestion. Resonance Raman spectroscopy is non-invasive method which is more accurate, sensitive, practical and quick method particularly when handling a large tissue volume. There is need of methods and processes to synthesize carotenoids in lab with special focus on those that are required as analytical standards on a bigger scale for clinical trials. Newer protocols should be developed to follow their role *in vivo*. Carotenoid supplements manufactured and marketed by different companies should be standardized. Their "parent chemical structures" are usually used to produce synthetic analogs having better pharmacological activities, finest bioavailability and pharmacokinetic profiles.

X. Their stability is subject to the nature of the carotenoid (carotene or xanthophyll, *E*- or *Z*-configuration and esterification) and the food matrix (fruit, root, leaf, juice). Existing stabilization efforts are focused on microencapsulation and nanoencapsulation. Encapsulation techniques for carotenoids are extensively reported; spray drying being commonly used in the food industry due to its simplicity. Nowadays targeted delivery in the gastrointestinal (GI) tract is becoming increasingly important area of active research. Current encapsulation research is focused on optimizing wall material preparation and modulating the release of carotenoids; however, new food or health applications and consumer products should be developed to capitalize on the added benefits of encapsulated carotenoids. With better understanding of carotenoids absorption and pharmacokinetics, targeted delivery systems designed specially to be retained in the body, will provide an exciting area for development for the nutraceutical industry. Stability, purity, bioavailability, and information on pharmacokinetics of carotenoids are key challenges, which are influenced by many factors, including the food matrix, carotenoids structures, and interactions with other food components. The combination of molecular, genetic and computational methodologies (genomics, metabolomics and proteomics) can help understand the importance of carotenoids [9].

XI. Selection of a suitable extraction protocol is not a straightforward and simple decision since carotenoids contents from different plant materials vary, they have complex chemistry that is dependent on pH, are sensitive to heat, and can associate with other molecules. It can be challenging to select the most appropriate technique for enhanced yield and stability of the extract. Further, during selecting extraction methods, due attention should also be paid to investment and operation costs, environmental impacts, the chemical composition of the samples, the goal of the extraction, and the intended application of the extract. The best method should use greener technologies or practices by decreasing or avoiding organic solvents, increasing the selectivity of target molecules in extraction, and improving the yield and purity of the carotenoids of interest. Increasing the potential of scale-up and automation of specific extraction processes is necessary for the commercialization of carotenoids. The wide array of structures and poor stability